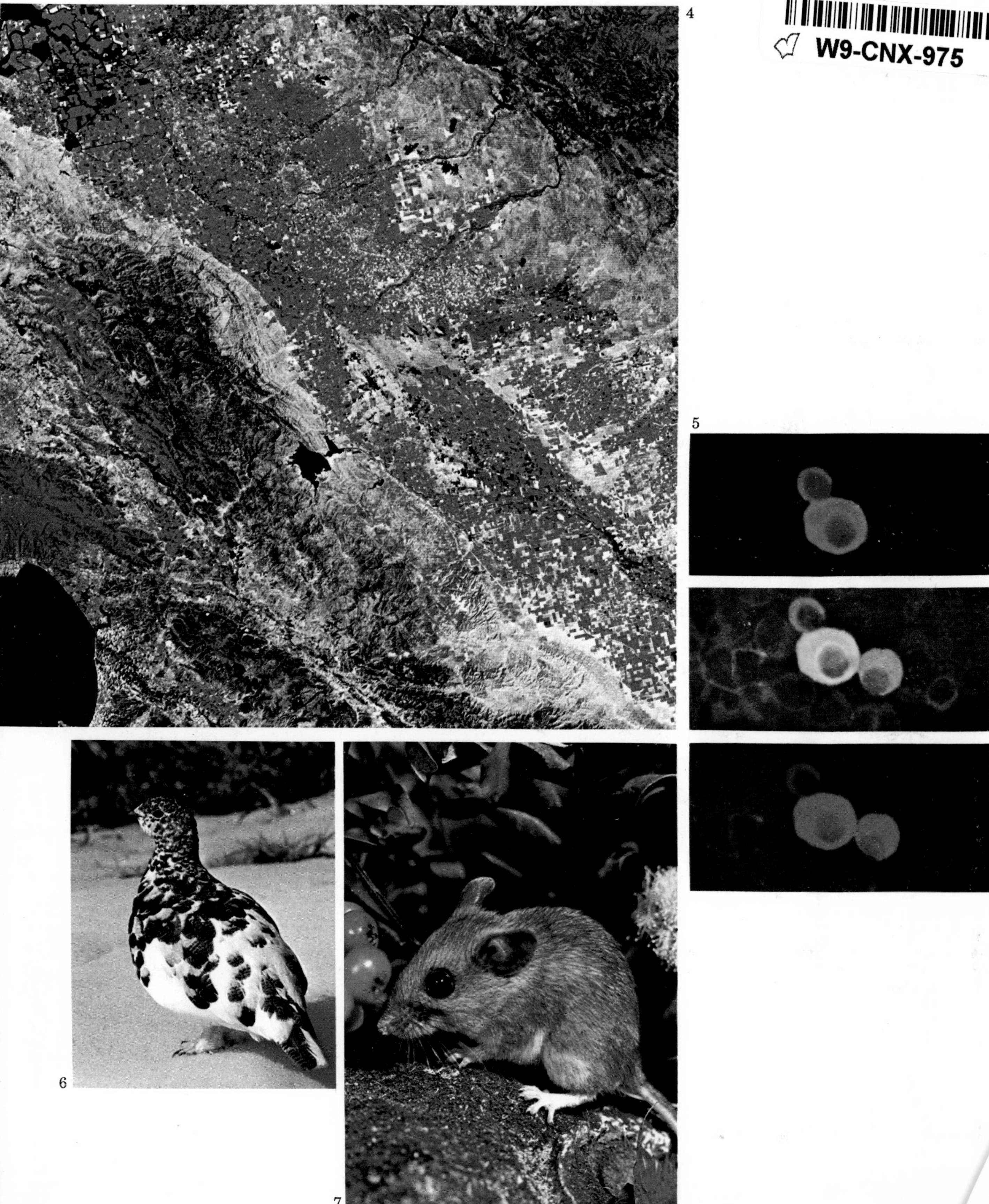
4
5
6
7

BIOLOGY

FOURTH EDITION

JOHN W. KIMBALL

Tufts University

ADDISON-WESLEY PUBLISHING COMPANY

BIOLOGY

FOURTH EDITION

Reading, Massachusetts
Menlo Park, California
London
Amsterdam
Don Mills, Ontario
Sydney

This book is in the
ADDISON-WESLEY SERIES IN LIFE SCIENCES
Harper Follansbee, Consulting Editor

Cover photo: The red-spotted newt (*Noto-opthalmus* sp.) showing the "Red Eft" stage. The generic name of the organism was recently changed from *Diemictylus* to *Notoopthalmus*, which means "black eye." The two newts pictured were observed on the slopes of Grandfather Mountain, North Carolina. (Photo by J. R. Pounds and Marvin Williams.)

Other books by John W. Kimball:

Cell Biology, Second Edition

Man and Nature: Principles of Human and Environmental Biology

Biology: A Laboratory Introduction

Foreign language editions:

Biology, Second Edition
Polish: Published by Pánstwowe Wydawnictwo Naukowe
Spanish: Published by Fondo Educativo Interamericano

Cell Biology
German: Published by Gustav Fischer
Japanese: Published by Maruzen

Second printing, September 1978

 Library of Congress Catalog Card No. 77–074322.

ISBN 0-201-03761-0
DEFGHIJKL-RN-8987654321

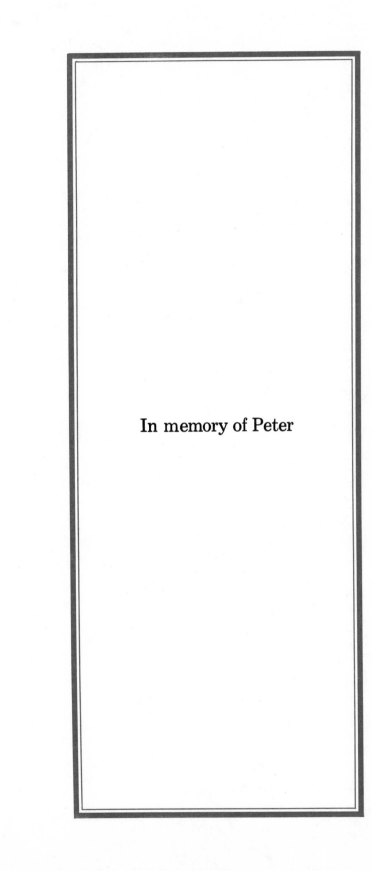

In memory of Peter

PREFACE

This fourth edition of *Biology* was prepared with the same basic goals as the earlier editions. The topics continue to be organized along *functional* lines; that is, the various aspects of biology are examined from the point of view of the features and problems shared by all living things. These range from their molecular and cellular organization to their features of metabolism, reproduction, responsiveness, and evolution. This organization should provide a framework upon which the myriad facts of biology can be organized in a way that makes sense and makes them easier to remember. Perhaps, too, from such an organization will grow a better appreciation for the unity underlying the diversity of life. Such a system also seems to have the advantage of interspersing botanical topics among zoological ones, thus reducing the too-common student antipathy to the former.

No attempt has been made in this edition to single out a few areas of biology as illustrative of the materials and techniques of the field as a whole. Most beginning students seem to want the opportunity to see the whole panorama of the subject. Trying to meet this goal has, however, finally led to what I believe is a significant change from earlier editions. Each of the earlier editions represented what I would have liked to have discussed in *my* year-long course in biology. But there is so much added material in this fourth edition that I would now have to skip some sections, even some chapters. I do hope that each of you will find the topics that you consider important adequately treated. In any case, I have tried to organize the material so that those who wish to use only portions of the book can do so. (One reflection of this is the return to a single section of the previously scattered material on ecology and population biology.)

Introductory biology texts tend to be compilations of conclusions. For many students, this is a congenial arrangement: they prefer to learn the answers and not be concerned with the processes that led to them and the uncertainties that surround them. It is also the only way to keep a book that tries to survey the entire field to a manageable size. But it is not good science. The pursuit of science has some of the characteristics of a detective story: the gathering of evidence and the testing of alternative hypotheses in new situations. Throughout this edition, I have tried to place greater emphasis on the observations and experiments that form the basis of our present understanding and, where the evidence is weak or contradictory, to present alternative hypotheses as well. I think that this will be particularly apparent in the material on photosynthesis, circulation, immunology, and nerve physiology and in the newly written chapter on muscle function.

As I set about the task of preparing this new edition, I was guided by several objectives. One was, of course, to bring the material in rapidly developing areas up to date. Thus you will find a new chapter on the control of gene expression and expanded treatments of cellular differentiation, of evolutionary mechanisms, and of many other topics. A second was to present still more of the observations and experiments upon which our present knowledge is based.

Third, I felt that the time had come to make substantial changes in the treatment of the macromolecules, bioenergetics, and intermediary metabolism. Extraordinary progress has been made in recent years toward understanding cell structure and function in terms of the structure and function of the molecules of which cells are composed. In the hope of achieving a better development of this material, I have completely rewritten the chapters on the chemistry of the elements important to life (Chapter 3) and on the macromolecules made from them (Chapter 4), the material in Chapter 6 on enzyme action, and Chapter 8, on glycolysis and cellular respiration. All of these depend upon an expanded treatment of electronegativity leading to a quantitative treatment of redox potentials, bond energies, and free energy changes. In fact, I hope that the concept of energy—from chemical bond energy to the energy flux through ecosystems—will provide a major unifying theme in this edition.

As any reader of the daily paper knows, biology has long since moved from the quiet of laboratories, museums, and field stations into the mainstream of public debate and political action. In recognition of this, I have added a number of discussions concerning areas where biological knowledge impinges on social problems. These will be found in the new chapters on population dynamics (Chapter 39), the importance of soil and water to human welfare (Chapter 42), the epidemiology of infectious disease (Chapter 44), the use (and abuse) of pesticides (Chapter 45), as well as in sections on such topics as air pollution, mutagenic agents, recombinant DNA techniques, eugenic proposals, and the like.

A major change in this edition is the greatly expanded survey of the diversity of living things. In earlier editions I was guilty of having (mostly unconsciously) allowed the discussions of molecular and cellular biology to crowd out the treatment of the various kinds of living things—their anatomy, their lives, their ecological significance, and their evolutionary relationships. In a book that emphasizes principles, it was deceptively easy to minimize the remarkable diversity of life. But *Homo sapiens* and *E. coli* cannot tell us all there is to know about life. So in this edition you will find an entirely new chapter on the prokaryotes (Chapter 34), as well as expanded treatments of the protists (Chapter 35), plants (Chapter 36), invertebrates (Chapter 37), and vertebrates (Chapter 38).

I am indebted to many people for the help they have provided in the preparation of this edition: To the many scientists who have supplied me with photographs and electron micrographs (93 of them appearing for the first time). Their names appear in the legend accompanying their work. To my colleagues in the Biology Department of Tufts University who assisted with photographs and helpful suggestions. To Sue Simpson, Oxford Illustrators, and Terry Renshaw, Transart, who not only prepared many new drawings and diagrams, but also redrew most of the figures taken over from earlier editions. To the staff at Addison-Wesley for their painstaking efforts in every phase of the book's production.

As each reaches the threshold of retirement from a long and productive career, I want to express my appreciation to two men who are, in large measure, responsible for giving me whatever skills I have brought to Biology and to this book. To Harper Follansbee, Instructor Emeritus in Biology at Phillips Academy for his friendship and guidance in the art of teaching. To Alwin M. Pappenheimer, Jr., Professor of Biology at Harvard University, who taught me how to *do* science.

And, finally, special thanks to my wife, Margaret, for the many ways—both direct and indirect—in which she has made the task of preparing this revision go more smoothly.

Andover, Massachusetts J.W.K.
September, 1977

CONTENTS

PART I

INTRODUCTION

1

THE CHARACTERISTICS OF LIFE

2

THE ORIGIN OF LIFE

PART II

THE ORGANIZATION OF LIFE

3

THE CHEMICAL BASIS OF LIFE: PRINCIPLES

4

THE MOLECULES OF LIFE

PART III

METABOLISM

5 THE CELLULAR BASIS OF LIFE

6 THE METABOLISM OF CELLS

7 HETEROTROPHIC NUTRITION

8

ENERGY PATHWAYS IN THE CELL

PHOTOSYNTHESIS

GAS EXCHANGE IN PLANTS AND ANIMALS

11 ANIMAL CIRCULATORY SYSTEMS

12 THE TRANSPORT OF MATERIALS IN THE VASCULAR PLANTS

PART IV

REPRODUCTION

13 EXCRETION AND HOMEOSTASIS

14 THE REPRODUCTION OF CELLS AND ORGANISMS

15 GENETICS: THE WORK OF MENDEL

16

THE CHROMOSOMES

17

THE CHEMICAL NATURE OF GENES

18

GENE EXPRESSION

THE CONTROL OF GENE EXPRESSION: MODULATION

SEXUAL REPRODUCTION IN PLANTS

21 SEXUAL REPRODUCTION IN ANIMALS

22 DEVELOPMENT: CLEAVAGE, MORPHOGENESIS, AND DIFFERENTIATION

23 DEVELOPMENT: GROWTH, REPAIR, AGING, AND DEATH

PART V

RESPONSIVENESS AND COORDINATION

24 RESPONSIVENESS AND COORDINATION IN PLANTS

25 ANIMAL ENDOCRINOLOGY

26 STIMULUS RECEPTORS

PART VI

EVOLUTION

30 THE IMMUNE RESPONSE

31 EVOLUTION: THE EVIDENCE

32 EVOLUTION: THE MECHANISMS

33 THE CLASSIFICATION OF LIFE

34 THE PROKARYOTES (KINGDOM MONERA)

35 THE PROTISTS AND FUNGI

THE KINGDOM PROTISTA

THE KINGDOM FUNGI

THE PLANT KINGDOM

37

THE INVERTEBRATES

THE VERTEBRATES

PART VII

ECOLOGY: THE BIOLOGY OF POPULATIONS AND THEIR ENVIRONMENT

43 INTERACTIONS BETWEEN SPECIES

44 HUMAN ECOLOGY I: PESTILENCE

45 HUMAN ECOLOGY II: COMPETING FOR FOOD

Our planet as seen from the moon by the crew of Apollo 8. (Courtesy of NASA.)

PART
INTRODUCTION

THE CHARACTERISTICS OF LIFE

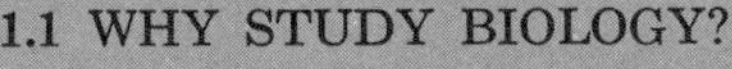

1.1 WHY STUDY BIOLOGY?

One reason stands out above all the others: to learn more about ourselves and the world we live in. We are animals. In many ways we differ but little from other animals. In a few ways we differ so greatly as to occupy a unique position in the world. Although it is a debatable point whether we have any attributes not shown to some degree by other animals, it is perfectly clear that we display some to a far greater degree. One of these is curiosity. *Homo sapiens* is "the man who knows." The desire to know is a hallmark of human existence. Thus we study biology for the same reason we study physics, mathematics, history, literature, and art: to gain knowledge about still another aspect of our lives and our world.

We should also note that productive, rewarding careers can be built on a knowledge of biology. University laboratories are in need of men and women to make the new discoveries that soon will make this book obsolete. Men and women are also needed to apply their knowledge of biology in such practical fields as medicine and agricultural research. Teachers of biology will continue to be needed to pass the knowledge acquired by earlier generations on to the generations to come.

Every citizen will be able to participate more effectively in a democracy if he or she can speak out and vote intelligently on questions that involve both biological principles and human welfare. The use of food additives, drugs, insecticides, radiation, techniques of genetic engineering, and population control measures are just a few of the many ways in which our lives may be altered by biological knowledge. Whether this knowledge is used in ways that increase the value of our lives or ways that rob our lives of value can best be decided by informed citizens. To make effective decisions will require not only a clear understanding of the values worth protecting and developing in our lives but also a knowledge of the physical and biological principles that underlie our lives. Knowledge of the first must come from a study of history, religion, philosophy, literature, and art, in other words, the humanities. Knowledge of the second must come from a study of the sciences. This book is an attempt to aid in the study of one of these: biology, the science of life.

1.2 WHAT IS LIFE?

Life is usually easier to recognize than define. We all recognize that a dog is alive, a stone is not (Fig. 1.1). What, then, are the properties of the dog that distinguish it from the stone?

In 1976 on July 20 at 11:53 Greenwich Mean Time and on September 3 at 10:39 GMT, identical landers of two unmanned spacecraft, Viking 1 and Viking 2, settled gently on the surface of the planet Mars. Contained within each lander were a number of experiments designed to help us learn more about the planet. Five of these experiments sought evidence that would bear on the question, Is there any life on Mars? The experiment taking the most direct approach to the problem was the television camera. If the eye of the camera looked out and saw the eye of a Martian looking in, the question would be answered! However, that did not happen. Nor did the

FIGURE 1.1

camera reveal any signs of the activities of living things. There were plenty of stones but nothing resembling a dog. So what other kinds of evidence of life could we look for?

1. The Complex Organization of Life

Some stones may seem to be rather complex, with various minerals scattered through them. Their organization is simplicity itself, however, when contrasted with that of any living organism. If we examine any part of a dog's body with a microscope, we shall discover that it is made of **cells.** These units, generally too small to be seen with the unaided eye, are organized into **tissues** which, in turn, form **organs** such as the stomach and kidney. Several organs, e.g. the stomach, liver, and intestines, work together as a **system.**

Biologists have studied these levels of the organization of life for many years. More recently they have been able to probe still deeper into the complexity of living things. The electron microscope has revealed a degree of subcellular structure and organization that thirty years ago was hardly suspected (Fig. 1.2). Other tools and techniques are now making it possible to dissect these subcellular structures into the molecules and atoms, that is, the chemicals, of which they are made.

One of the most characteristic features of the chemical composition of living organisms on our planet is the presence of molecules containing carbon atoms. This feature is so characteristic, in fact, that the chemistry of carbon compounds is called organic chemistry. Since life as we know it does not exist without organic molecules, one of the experiments incorporated in the Viking landers was designed to test for the presence of organic molecules in a sample of Martian soil.

Rocks are made of atoms, sometimes even carbon atoms, but unless life processes have put them there, they do not contain organic molecules. Furthermore, the atoms of a rock are more rigid in their organization. The rock we see today contains the same atoms it contained last year. The dog we see today is largely constructed of atoms acquired since we saw him last year. Those he has lost have returned to the environment. This rapid turnover of material in living things is part of their **metabolism.**

2. Metabolism

A dog continually exchanges material with his surroundings. We see him eat and drink. He defecates and urinates. He breathes. With proper equipment, we can show that the air the dog exhales differs in composition from the air inhaled: oxygen has been removed and carbon dioxide added.

All organisms share this property with the dog. Whether rapidly or slowly, they all take in material from and give material back to the environment.

FIG. 1.2 Leaf cells of the sunflower as seen under the electron microscope. Portions of four adjacent cells are shown. Note how the interior of each cell is made up of discrete components (organelles) which in themselves show considerable structural complexity. Structure which can be revealed only by the electron microscope is called ultrastructure. (Courtesy of H. J. Arnott and K. M. Smith.)

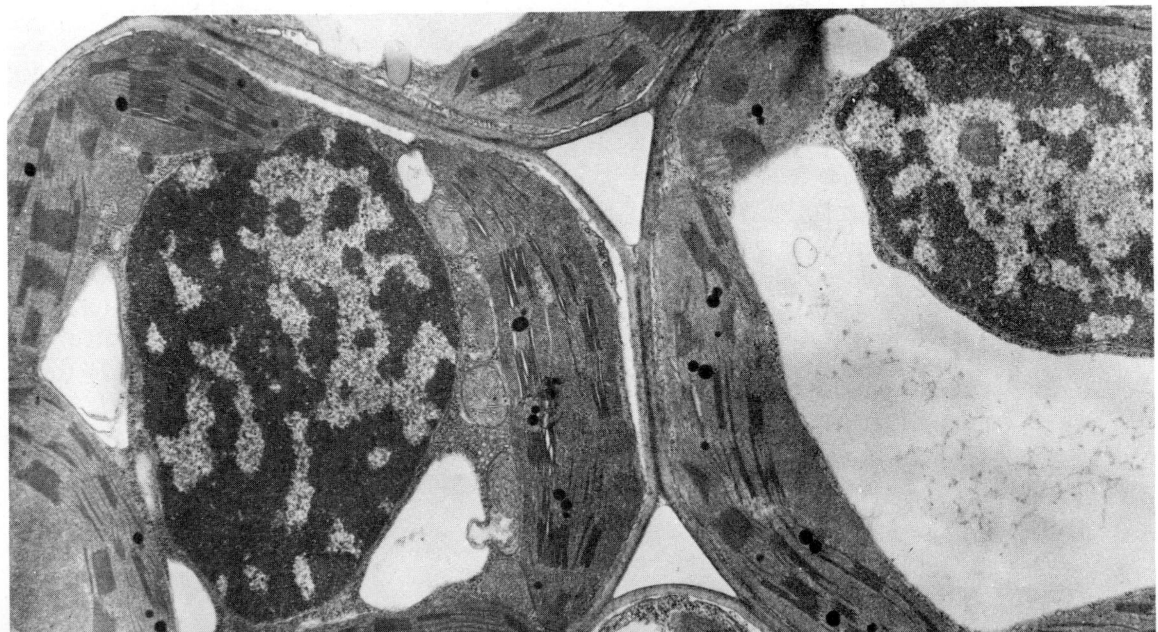

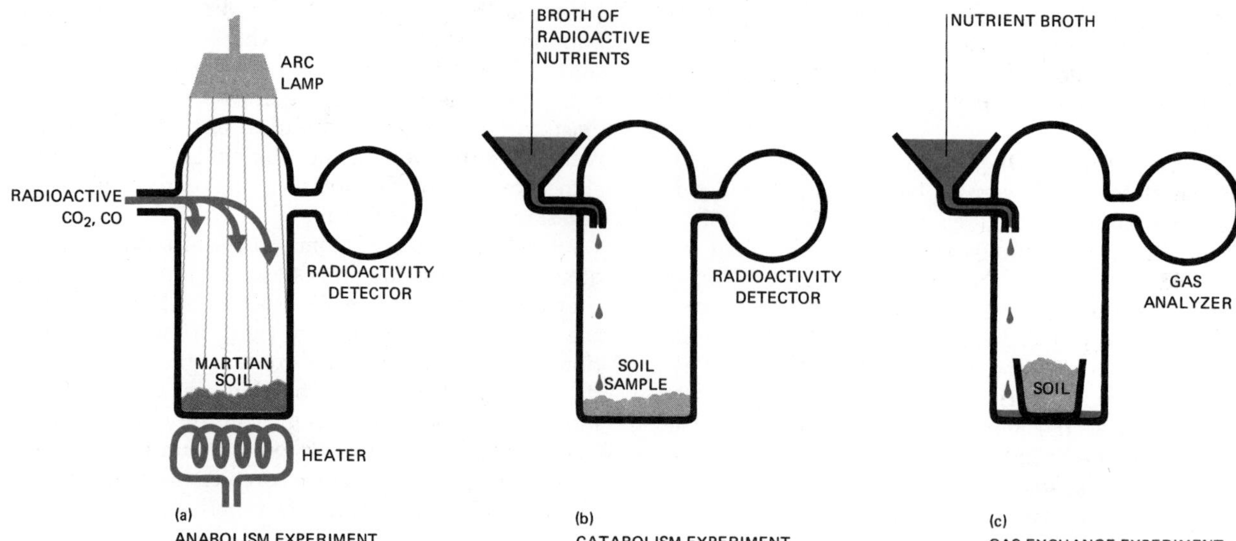

FIG. 1.3 The three life detection experiments performed by the Viking landers. The anabolism experiment (a) looks for evidence of photosynthesis. After the period of illumination is over, any unreacted gases (CO_2, CO) are first flushed through the radioactivity detector. Then any organic molecules that may have been incorporated into living microorganisms by photosynthesis are volatilized by the heater and passed into the radioactivity detector. This experiment is also called the pyrolytic release (PR) experiment.

The catabolism experiment (b) is designed to detect the release of radioactive gases (such as carbon dioxide) from the broth of radioactive nutrients used to wet the soil. This experiment is also called the labeled release (LR) experiment.

In (c), the gas exchange experiment (GEX), a gas analyzer is used to monitor any changes in the composition of the gas mixture in the chamber and thus detect evidence of respiration by living organisms in the soil.

Furthermore, the material given back is not the same as the material taken in. We have seen that exhaled air differs from inhaled air. Many substances found in fecal matter and urine were not present in the food that was eaten. During their relatively brief stay within the dog's body, food materials undergo extensive transformations. As they do so, energy is liberated. Ultimately this energy appears as heat and warms the dog. But before this occurs, the energy may be used for some of the other activities characteristic of life.

Three of the experiments that were included in the Viking landers were designed to look for evidence of metabolism occurring within samples of Martian soil.

One of the experiments looked for evidence that something in the soil could synthesize complex organic molecules from such simple carbon compounds as carbon dioxide (CO_2) and carbon monoxide (CO). In this experiment, a mixture of radioactive CO_2 and CO (in the ratio of 95:5) was introduced into a vessel containing a small sample of Martian soil and Martian atmosphere. The synthesis of organic molecules from simple precursors requires energy, and the chief form of that energy used here on earth is sunlight (used in photosynthesis). Therefore, the incubation mixture was illuminated with a bright arc lamp. After five days, the soil sample was heated to drive off any radioactive organic molecules that might have been synthesized from the radioactive CO_2 (and/or CO) by creatures in the soil. This experiment was thus looking for evidence of metabolism involving the synthesis of complex substances from simple ones. Such metabolism is called **anabolism.**

A second experiment was designed to look for signs of **catabolism,** that is, destructive metabolism. In this experiment, a soil sample was incubated with a dilute soup of radioactive organic molecules (Fig. 1.3). Over the next ten days, the atmosphere above the sample was regularly monitored for the appearance of radioactive gases such as carbon dioxide.

FIG. 1.4 The biology package of the Viking landers. All the equipment to run the three life detection experiments is crammed into a little less than one cubic foot of space. (Courtesy TRW Inc.)

The appearance of radioactive gases would be evidence of the decomposition of the radioactive organic molecules with which the soil samples were soaked.

The third experiment designed to look for evidence of metabolism in Martian soil was the gas-exchange experiment. In this experiment, a known mixture of gases was placed above a soil sample and then periodically analyzed to see if any gases (e.g., CO_2) had disappeared from the mixture or been added to the mixture. If soil alone should fail to bring about a change in gas composition (it did not —large amounts of oxygen were given off at first), then a nutrient soup would be added to the soil. If organisms were there to catabolize the organic molecules in the soup, some of the products of catabolism should be gases and their appearance detected by the equipment.

The underlying assumption in these three experiments, which were run in an apparatus weighing 35 lb (16 kg)—Fig. 1.4, is that a fundamental property of life is metabolism and if there is life on Mars, even of the simplest and most primitive kind, it should reveal its presence there by its capacity to carry on metabolism.

3. Reproduction

The essence of reproduction in a living organism is the self-controlled duplication of the structures characteristic of it. It occurs when the organism takes in more material from its environment than it gives back and organizes these materials into its own structures. We call this kind of reproductive activity **growth.**

To do this, the organism must expend some of the energy released during its metabolism, and it must have a pattern that will guide the building of its structures. If you have studied chemistry, you might argue that the crystals in a rock, once at least, grew in the same way. A crystal placed in a solution of its units does slowly organize these units accord-

FIG. 1.5 Yeast cells growing in culture. The cells reproduce asexually by forming buds (producing a figure-eight shape). The buds soon separate from the parent cell and begin an independent existence.

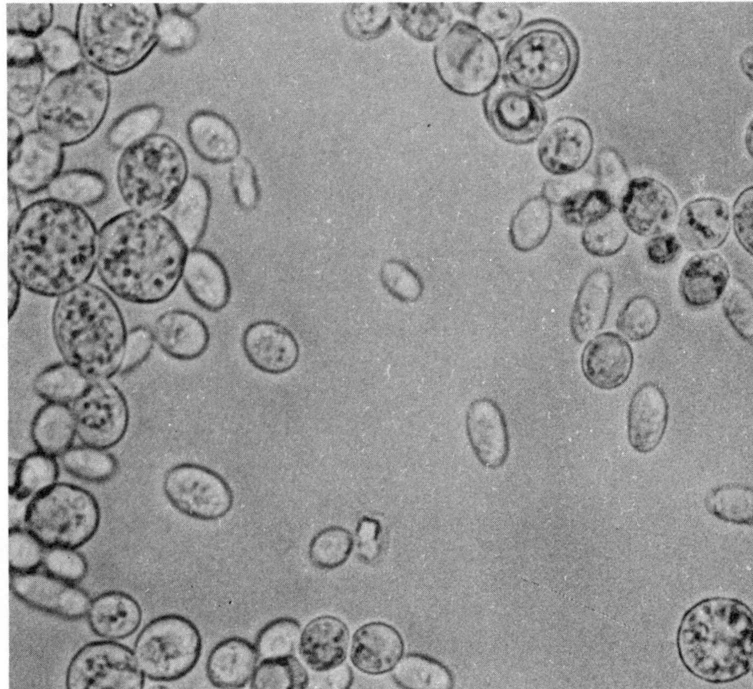

ing to its pattern and thus grows. There is one basic difference, however, between growth of a crystal and growth of a living organism. The crystal grows by accumulating from its surroundings units that are already identical with those in the crystal. The living organism, on the other hand, grows by transforming materials that are not unique to it into those that are. Our dog can thrive on the same diet that we eat. He, however, converts this diet into more "dog," whereas we convert it into more "human."

Reproduction also involves the production, from time to time, of copies of the organism that can live independently of it. All living things must die sometime and, if their kind is to survive, they must make copies of themselves before they die (Fig. 1.5).

Among plants and the less complex animals, this aspect of reproduction may simply be an extension of the growth process. The growing strawberry plant sends out horizontal stems on which develop "daughter" plants. This kind of reproduction is called **asexual** because only one parent is involved. Generally, the daughter plants are identical with the parent.

Almost all organisms (including dogs) engage in another kind of reproduction. **Sexual** reproduction requires that *two* parents contribute to the formation of the new individual. In this way, new combinations of traits can be produced. This is apt to involve greater complexities than asexual reproduction, as the two parents must locate each other and, in many cases, be equipped to care for their offspring until the offspring can care for themselves.

4. Responsiveness

All living things are capable of responding to certain changes **(stimuli)** in their surroundings (Fig. 1.6). Changes in light, heat, gravity, sound, mechanical contact, and chemicals in the surroundings are common stimuli to which living organisms respond. In order to respond to these stimuli, the organism must have a means of detecting them. The eyes, ears, and nose of the dog are effective stimulus detectors, as anyone who has watched these animals can testify.

In order to be effective, responses to changes in the environment must be coordinated. Even the simplest organisms consist of many parts and each of these must do the right thing at the right time for an appropriate action to be carried out. When our dog is called to supper, some of his muscles must contract, others must relax, his digestive glands must begin to function and so on. Each part must work in harmony with the others. A system of **nerves** and a system of chemical regulators called **hormones** coordinate the actions of the dog and many other animals. Plants rely on hormones for their coordination.

The action carried out by an organism in response to a stimulus is accomplished by its **effectors.** Muscles and glands are the most important ones in animals. Like all other parts of the coordinating system, they consume energy. Energy enables the dog's muscles to contract and his glands to synthesize the enzymes with which he will digest his supper.

Organisms respond to changes in their environment by altering their relationship to it. In running

FIG. 1.6 Belle, an adolescent chimpanzee, grooms Bandit's teeth with a small pine twig tool. Shadow (standing) observes. (Courtesy of C. E. G. Tutin and W. C. McGrew.)

to his supper dish, the dog is changing his position in response to your signal. These responses, which often occur in definite patterns, make up the organism's **behavior.** It is an active, not a passive, thing. A dislodged stone rolling downhill is not behaving; it is simply being carried along passively by the force of gravity. Our hungry dog, on the other hand, is *creating* the change in his relationship to his surroundings.

5. Evolution

When organisms reproduce themselves, their pattern is copied with marvelous accuracy. Dogs have puppies, not kittens. Children sometimes share distinctive traits with their mother or father. The little marine animal *Neopilina,* which was dredged up from the bottom of the Pacific in 1952, looks practically identical to its ancestors preserved as fossils over 500 million years ago (Fig. 1.7).

Neopilina is not, however, identical with its ancestors. Nor are any other organisms found on earth today identical with those that lived in ages past. Over long periods of time, changes have occurred. These changes mark the evolution of the organisms. Evolution has often been adaptive; that is, the changes have enabled the organisms to live in their environment—to metabolize, reproduce, and respond—more efficiently than their ancestors could have in the same environment. The many breeds of dogs testify to the ability of organisms to evolve over time, although many of the changes which humans have wrought in dog evolution can hardly be thought of as adaptive.

Evolution involves something else. The number of kinds of organisms that now live on the earth is far greater than the number that were present 500 million years ago. From that time to this, there has been a proliferation of kinds of organisms. This aspect of evolution has occurred as single groups of organisms have given rise to two or more distinct kinds of descendants. All our present dogs, for example, are probably descendants of one ancestral type.

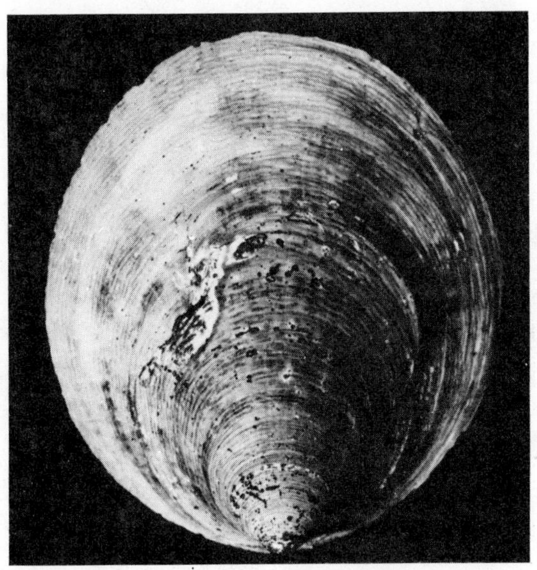

FIG. 1.7 *Neopilina.* This primitive mollusk was discovered in 1952. It differs little from its closest relative, *Pilina,* which has been extinct for 360 million years. (From the *Galathea* Report.)

These then are the major characteristics of living things. Although all share them, how they accomplish or express them differs to some extent from one organism to another. The remainder of this book will be devoted to examining how representative organisms accomplish each of the major activities of life.

EXERCISES AND PROBLEMS

1. Does a candle flame show any of the characteristics of life?
2. In what ways are growth and reproduction similar? In what ways are they different?
3. How is metabolism related to growth?
4. How is responsiveness related to metabolism?
5. What are the differences between sexual and asexual reproduction?
6. Can most animals reproduce asexually? Can most plants?
7. What is a stimulus?
8. What conclusions would you draw if each of the experiments shown in Fig. 1.3 gave positive results but no organic molecules could be detected in Martian soil?

2 THE ORIGIN OF LIFE

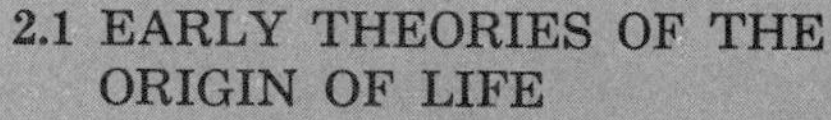

The theory of evolution—a topic we shall examine in Chapters 31 and 32—explains the diversity of life on earth today by the descent of each and every species from common ancestors. Certainly the fossil record tells us that in former times, there were fewer species on the earth and that these were not so complex as most of our modern species. In theory, then, the study of evolution should lead us back to a first form of life from which all others have been descended. What was this first form? We do not know, and the best we can do is to make intelligent guesses about its characteristics. Where did it come from? We do not know the answer to this question either, although every thoughtful person has probably wondered about it at one time or another. Even people who know nothing about or disbelieve in the theory of evolution ask themselves how life first appeared on the earth.

2.1 EARLY THEORIES OF THE ORIGIN OF LIFE

Among the first attempts at answering this question in our civilization were the stories of creation that are found in the Bible. Other cultures, too, have their stories of the creation of life. These stories share two features in common. First, they were created long before we had gained any knowledge of the physical, chemical, and biological principles that are the basis of life. Second, they invoke divine intervention in the creation of life and thus fall outside the scope of scientific inquiry. Scientists assume that the forces governing the world can be known, that these forces act uniformly at all times and in all places, and that their effects can be predicted, at least in a statistical way. While this faith may be ill-founded, it is the only basis upon which scientists can work. If their experiments are subject to unique, supernatural, or capricious intervention, there is no point in performing them. In fact, if some *unrepeatable* result is detected, the scientist assumes that it is due to faulty equipment and/or some error in observation.

The cosmozoa theory. This theory of the origin of life does fall within the scope of scientific inquiry. It explains the presence of life on our earth by assuming it was brought here from elsewhere in the universe, perhaps incorporated in a falling meteorite. Some meteorites do contain organic molecules. While contamination after the meteorite reached the earth can explain the presence of some of these organic molecules, others are not found in living matter here and are thus most likely of extraterrestrial origin. But the arrival of organic molecules from outer space is not the same as the arrival of life. And even if life could withstand the rigors of interplanetary space and the fiery trip through the atmosphere of the earth, the cosmozoa theory really does not answer our basic question. It simply removes it from this planet to some other location.

The theory of spontaneous generation. Until about one hundred years ago, it was commonly believed that life could arise spontaneously from nonliving matter. Van Helmont even gave a recipe—grains of wheat and a dirty shirt in a dark container—for producing mice spontaneously! As more was learned about biology, however, people began to doubt the possibility of spontaneous generation. In 1668 the Italian physician Francesco Redi performed an experiment to show that maggots do not arise spontaneously in decaying meat but are produced from the eggs of flies (Fig. 2.1). Although the spontaneous generation of large forms of life began to be doubted after this, van Leeuwenhoek's discovery of microorganisms reopened the question. Surely these tiny creatures, which appeared so suddenly in rotting food, etc., arose spontaneously! The Italian priest Lazzaro Spallanzani tried to show that even they did not do so. He boiled nutritious broth in glass flasks and then sealed them so that nothing could get into the broth from the outside. The broth remained clear and sterile. Skeptics argued, however, that heating the air within the flask had so

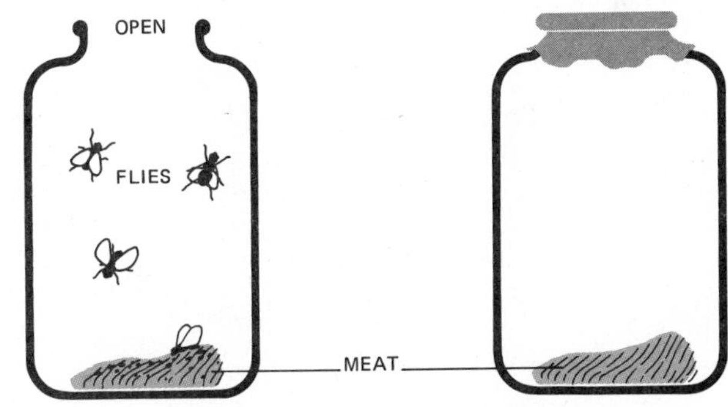

FIG. 2.1 Redi's experiment. Maggots (larval flies) appeared in the meat in the open jar but not in the covered jar.

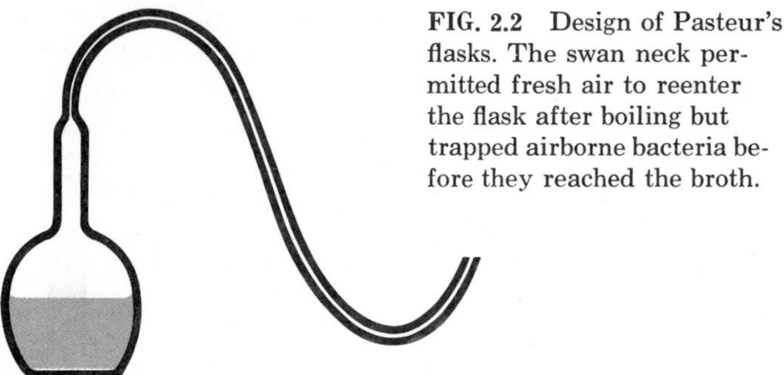

FIG. 2.2 Design of Pasteur's flasks. The swan neck permitted fresh air to reenter the flask after boiling but trapped airborne bacteria before they reached the broth.

altered it that spontaneous generation could not occur.

It was the immortal French biochemist and microbiologist Louis Pasteur who one hundred years ago finally silenced the skeptics. He, too, boiled flasks of broth, but instead of sealing the necks of the flasks, he drew them out into an *S*-shape, leaving the ends open (Fig. 2.2). Now fresh air could reach the interior of the flask but, Pasteur reasoned, any bacteria or other microorganisms floating in it would be trapped in the long neck. Sure enough, the broth usually stayed sterile until he tipped a little into the neck of the flask and then allowed it to run back. Only after doing this, did microorganisms begin to grow in it. Some of the flasks that Pasteur prepared are still in existence, their contents of broth clear and pure (Fig. 2.3).

FIG. 2.3 Flasks used by Pasteur in his experiments on spontaneous generation. (The second flask from the left was prepared by Pasteur but never opened.) (Courtesy Musée Pasteur—Institut Pasteur.)

Two important points about Pasteur's work are often overlooked. First of all, he did not prove that spontaneous generation can never and has never occurred. It is virtually impossible to prove the negative of something. Pasteur, like Redi before him, simply showed that supposed examples of spontaneous generation failed to work if already living organisms were kept away. Actually, we have good reasons for thinking that life is not created spontaneously today but we certainly have no rigid proof of this.

The second point about Pasteur's work is that if it does really mean that the spontaneous generation of life is impossible, then the problem of the origin of life cannot be explained on scientific grounds. Fortunately, a way out of this dilemma was proposed in 1936 by the Russian biochemist A. I. Oparin.

2.2 OPARIN'S THEORY

While conceding that life does not arise spontaneously now, Oparin felt that it might well have arisen spontaneously under the conditions that existed earlier in the history of the earth. According to his theory, the oceans of the early earth contained a rich supply of organic molecules. Over a vast span of time, these molecules became associated with one another in temporary complexes. Ultimately, one such complex developed: (1) some sort of membrane to separate it from the "soup" of organic molecules around it, (2) the ability to take in molecules from this soup and discharge other molecules into it, (3) the ability to incorporate the absorbed molecules into the characteristic pattern of the complex, and (4) the ability to split apart portions of itself that had all these features, too. Such a complex could have been the first living thing. Metabolism, growth, and reproduction, perhaps even responsiveness, would have been present and hence all the attributes of life. From this first form, surely far simpler than even the simplest microorganisms alive today, could have evolved by natural selection the present diversity of life.

The plausibility of Oparin's theory depends greatly on the earth's being somewhat different then than it is now. In order for organic molecules characteristic of life to have accumulated in the oceans, there must have been substantial concentrations of the inorganic ingredients out of which they could be synthesized and a source of energy to accomplish these syntheses. Analysis of the atmosphere of the large, cold planets, Jupiter and Saturn, reveals the presence of methane (CH_4) and ammonia (NH_3). If these two gases, together with water vapor (H_2O), were present in the atmosphere of the early earth, the major elements for the synthesis of the molecules found in living things would have been available. Radioactivity, ultraviolet radiation from the sun, heat from volcanos, and electrical discharge all might have provided energy for the synthesis of various organic molecules from these gases.

Could such a system really work? A student of biochemistry, Stanley Miller, sought to answer this question as part of his work toward a Ph.D. degree. He built the apparatus illustrated in Fig. 2.4 and filled it with water, methane, ammonia, and hydrogen. This mixture was kept circulating by continuously boiling and then condensing the water. The gases passed through a chamber containing two electrodes with a spark passing continuously between them. At the end of a week, chemical analysis of the contents of the flask revealed the presence of several amino acids, the building blocks of proteins, and some other organic molecules. Miller's results encouraged others to vary his procedure, using different mixtures of substances that might have been present on the primitive earth and other sources of energy. Most of the other constituents of living matter have been produced in these later experiments.

None of these experimental re-creations of what might once have been the conditions on the earth has produced anything alive. Undeniably, though, they have produced some of the molecules uniquely associated with life. Could these have eventually associated themselves into a complex with the attributes of life? We may never know. Such an event would be extremely unlikely and a week, or even a million years, might not be long enough for the necessary chance association to occur. Furthermore, there must have been some way in which organic molecules, once formed, could be protected from degradation by the same forces that created them. In Miller's apparatus, the electric spark can break organic molecules back down as easily as it promotes their synthesis. Dissolved in the oceans of the primitive earth or adhering to the surface of minerals, complex associations of organic molecules might have been better protected against breakdown.

In any case, one sure requirement for an event such as the spontaneous creation of life must have been a vast span of time. On uncountable occasions, complexes must have almost reached the point where they would have been alive, only to become degraded again. Was there enough time for chance to finally succeed?

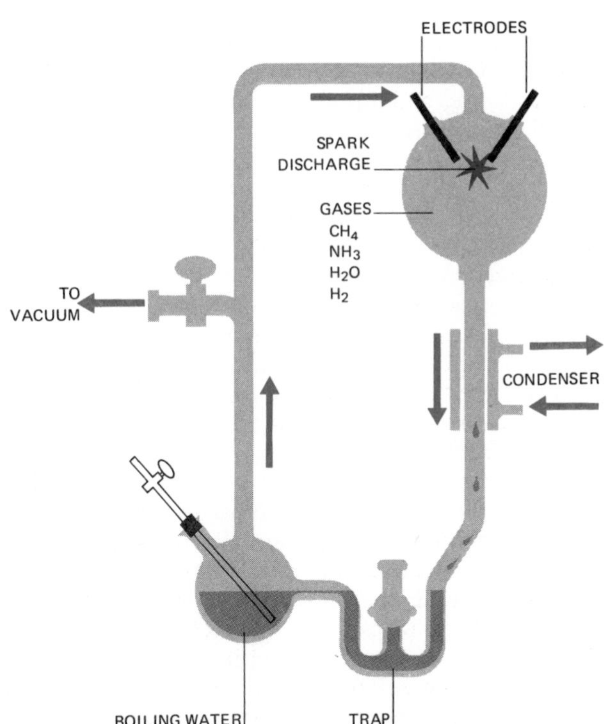

FIG. 2.4 Miller's apparatus. After it had run for one week, a variety of organic molecules had formed in it.

2.3 THE AGE OF THE EARTH

Not until little more than a century ago did people begin to appreciate just how old the earth is. Without a knowledge of the facts of geology and geochemistry, they simply could not conceive of the great length of the history of the earth. One biblical scholar, Archbishop Ussher, calculated from his studies that the earth was created at 9:00 A.M. on October 12 in the year 4004 B.C. However, calculations based on geological studies produced quite a different picture of the antiquity of the earth. One of the first estimates was based on the increasing salinity of the oceans. Dividing the salt concentration of our present oceans by their annual increase in salinity, one arrives at a figure of about 50 million years for their age. Unfortunately, this method requires us to assume that the oceans have been getting saltier at a constant rate through the years, and we have no grounds for believing this. A similar attempt was made to establish the age of the earth by measuring the total thickness of sedimentary rock deposits and dividing this by the estimated increase in the thickness of ocean sediments. Although this method doubled previous estimates, it suffered from the same defect as the first. Perhaps sediments were deposited more or less rapidly at other periods in the history of the earth than they are today. An additional problem was the fact that a continuous record of sedimentary rock is not available, for erosion has erased entire chapters of this portion of the geological history of the earth.

Is there a geological process which we can be sure proceeds at a constant rate? The discovery of the radioactive elements has provided us with one. Radioactive elements contain unstable isotopes which, by emitting subatomic particles such as electrons and alpha particles, become "transmuted" into other isotopes. Atoms of U-238, the most abundant isotope of uranium, undergo a series of transmutations as they "decay" radioactively. Each atom in the series is also unstable and itself decays. Finally, though, an isotope of lead (Pb-206) is formed that is stable, and the process stops. The important thing about these transformations from our point of view is that they take place at a definite, measurable rate. It takes 4.51 billion years for half the atoms in any sample of U-238 to decay. (We call this period the "half-life" of the isotope.) Neither high temperatures nor low, neither gravity, magnetism, electrical currents nor any other physical or chemical force can alter this inexorable rate. Here, then, is a geological clock which keeps perfect time and whose elapsed running time can be computed simply from the ratio of Pb-206 atoms to U-238 atoms in a rock sample (Fig. 2.5). The techniques exist for making these determinations. The oldest rocks to be dated by this method are from Southern Rhodesia. They are 2.7 billion years old. However, they surely were not formed at the time the earth itself was formed. The study of other isotopes in other minerals (as well as in meteorites) indicates that the earth (and the solar system) is some 4.7 billion years old.

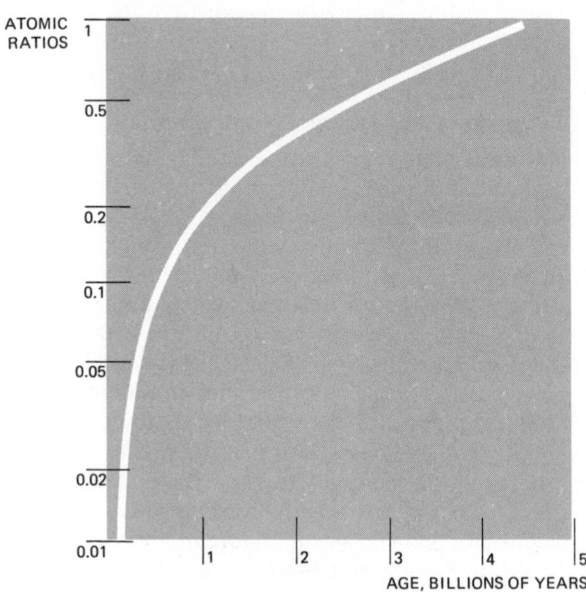

FIG. 2.5 Change in ratio of Pb-206 : U-238 over a period of 4.5 billion years.

2.4 THE DAWN OF LIFE

No one knows exactly when life first appeared on the earth. We would not expect the first forms of life to leave fossil remains for us. However, fossils that resemble bacteria have been found in rocks over three billion years old (Fig. 2.6). This still leaves us with almost two billion years for life to have arisen on the earth. Even the most unlikely event will probably occur sometime. Two billion years provides plenty of time for false starts in the chance production of the first living organism.

Nourishment for the first form of life was no problem. Surrounded by the same soup of organic molecules from which it arose, it had only to use these molecules to supply its needs for energy and the materials for growth and reproduction. This mode of nutrition, which depends upon the intake of preformed organic molecules, is called **heterotrophic.**

But an exclusively heterotrophic way of life could not continue indefinitely. No matter how large the primeval soup of molecules, eventually it would have become exhausted. If this experiment in life was not to cease then, some organisms must have evolved a means of synthesizing new organic molecules from the inorganic substances found in the environment. Such a mode of nutrition is called **autotrophic.** Autotrophic organisms require copious amounts of energy, and their energy probably came from the sun, that is, they made their food by **photosynthesis.** The evolution of autotrophic organisms provided then—as it does today—a constantly renewed supply of organic molecules to meet all the needs of life on this planet.

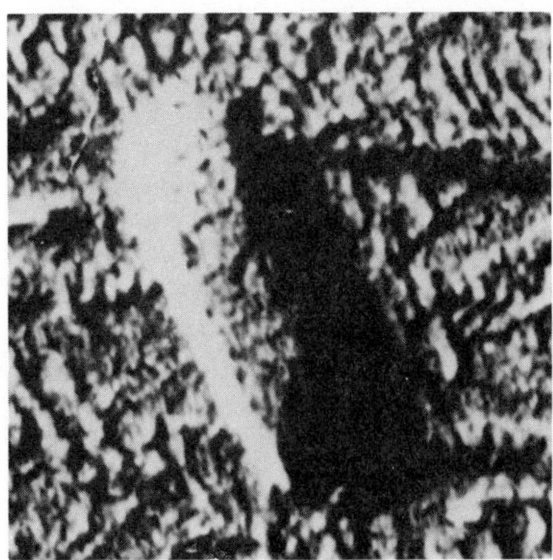

FIG. 2.6 Fossil bacterium. This fossil was found in South Africa in sedimentary rock over three billion years old. It is the oldest evidence of life yet discovered. This prokaryote is about 0.5 μm long and has been named *Eobacterium isolatum.* (Electron micrograph courtesy of E. S. Barghoorn from E. S. Barghoorn and J. W. Schopf, *Science,* **152,** 758–763, May 6, 1966.)

If our theory of the spontaneous generation of life is correct, does it mean that Redi, Spallanzani, and Pasteur were wrong? Not at all. The theory Redi and the others tried to refute was that modern forms of life arise spontaneously. On the other hand, Oparin's theory is that an organism, far more primitive and simple than any that lives today, arose once in the history of the earth. (Our conviction that this event occurred just once is based on the remarkable similarities of biochemistry and subcellular structure in all living things.) We believe that additional spontaneous generations could not occur today. For one thing, there are no longer appreciable quantities of organic molecules in the waters of the earth. Second, any complex of organic molecules that might form spontaneously would quickly be destroyed by the already existing forms of life (for example, bacteria) which would consume our hopeful complex long before it developed to the point where we could consider it to be alive.

"Organic molecules," "amino acids," "isotopes" —is this going to be a book about biology or a book about chemistry? A very good question, and the answer is: both. We simply cannot hope to ever do more than acquire the most superficial understanding of what it means to be alive without some knowledge of the chemical and physical principles that underlie life. This is certainly true of the physiologists, microbiologists, etc., working in their laboratories. But it is also true of the outdoor biologists— the "naturalists" of yesteryear. My outdoor colleagues are today as dependent upon chemical knowledge and techniques as my colleagues at the laboratory bench are. Evolutionary biologists now seek for chemical clues as to the evolutionary relationships between living things. Students of animal behavior have found a multitude of chemical signals by which animals, and plants, interact with one another. And those whose concerns are the broad ones of the functioning of entire ecosystems find that fruitful analysis eludes them until they begin to employ the techniques and tools of chemistry and physics. So why delay? Let us now turn our attention to some of the fundamental chemical and physical principles upon which life is based.

REFERENCES

1. GABRIEL, M. L., and S. FOGEL, *Great Experiments in Biology,* Prentice-Hall, Englewood Cliffs, N.J., 1955. Includes papers by Redi, Spallanzani, and Pasteur on the theory of spontaneous generation.
2. OPARIN, A. I., *The Origin of Life,* Dover, New York, 1953. First published in 1936.
3. WALD, G., "The Origin of Life," *Scientific American,* Offprint No. 47, August, 1954.
4. LAWLESS, J. G., C. E. FOLSOME, and K. A. KVENVOLDEN, "Organic Matter in Meteorites," *Scientific American,* Offprint No. 902, June, 1972.
5. BARGHOORN, E. S., "The Oldest Fossils," *Scientific American,* Offprint No. 895, May, 1971. Tells of the discoveries in Canada, Africa, and Australia of the earliest evidence of life.
6. BERNAL, J. D., and ANN SYNGE, *The Origin of Life,* Oxford Biology Readers, No. 13, Oxford University Press, Oxford, 1972.

Porphyridium, a unicellular microorganism. The electron microscope reveals a wealth of structural details within its single cell. (25,000 ×, courtesy of E. Gantt.)

PART II THE ORGANIZATION OF LIFE

3

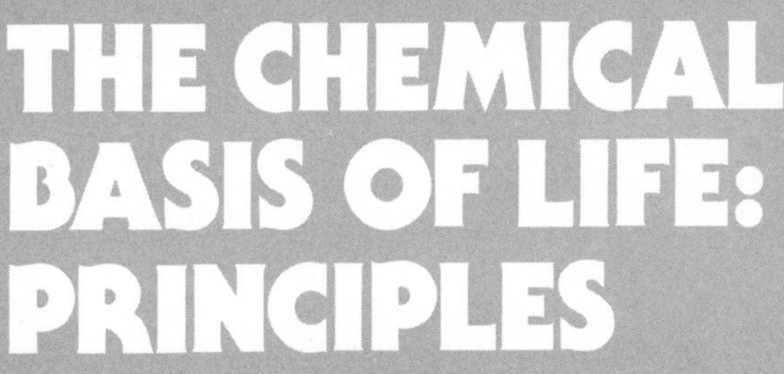

THE CHEMICAL BASIS OF LIFE: PRINCIPLES

You may think it odd that a book on biology should devote space to examining some of the basic principles of chemistry. There was a time when virtually all our knowledge of living things came from the keen observations of biologists working with such tools as preservatives, tissue-slicing machines, a few stains, and microscopes. However, as the analytical techniques of chemists and physicists were turned on biological materials, it became possible to analyze them in terms of structural units far smaller than those visible under the light microscope. These units are atoms and molecules.

As a result of such studies, two things have become clear.

1. Although living things and, to a lesser degree, the cells of which they are composed are quite diverse in their appearance, their basic chemical organization is remarkably similar.
2. Although the chemical organization of living matter is very complex, it is based upon the same materials and principles as those found in the world of nonliving things.

These two discoveries have had far-reaching consequences. They have shifted the attention of biologists from the many ways in which living things differ from one another to the many ways in which they are similar. This shift in approach has gone hand in hand with an enlargement of our way of looking at living things. No longer can biologists be content simply to study **morphology**, that is, the way in which an organism is put together out of cells, tissues, and organs. Now they must ask how these various structural parts of the organism work; that is, they must also examine its **physiology**. To find the answers they must understand the chemical makeup of the living cell and the chemical principles underlying its activity.

3.1 FORMS OF MATTER

As yet the only place in the universe where life, and thus living cells, are known to occur is on our planet. Here in an ever-so-thin layer covering the surface of the earth exists that peculiar organization of matter bearing the attributes of life. Living matter lives surrounded by, and ceaselessly interacts with, its nonliving environment. This includes:

1. the **lithosphere:** the soil and rocks close to the surface of the earth,
2. the **hydrosphere:** the waters—both salt and fresh—of the earth, and
3. the **atmosphere.**

Intersecting with each of these three nonliving "spheres" of the earth is the **biosphere:** that narrow zone where living matter exists and interacts with the lithosphere, hydrosphere, and atmosphere. The biosphere is extremely thin, the vast majority of living things being found in a layer no more than 20 km (12 mi) thick (the earth's diameter is some 12,750 km). Anyone who takes the time to analyze the situation can gain an intellectual appreciation of the dimensions of the biosphere. But it was the space explorations of the sixties and the opportunity they provided to look down on our planet from outer space that dramatized for us the delicate nature of that membrane of air, water, and soil in which we and all other living creatures must coexist (Fig. 3.1).

Of what are the four spheres composed? If we carefully examine a sample of rock or soil, we will be impressed by its heterogeneity, that is, we will find it to be composed of a variety of identifiable ingredients, e.g., grains of sand, flecks of mica. Water appears, and is, more uniform in its composition than soil, but if we evaporate a sample of ocean water we will find a residue that contains several different salts. The atmosphere is even more uniform in its composition than the sea, but here as well, careful analysis reveals a variety of ingredients. If, for example, we place a living mouse in a sealed container of air, the composition of the air will change so that ultimately the mouse can no longer breathe it —the oxygen has gone (and been replaced by carbon dioxide).

All of these kinds of matter—soil, water, air—are examples of **mixtures** (Fig. 3.2). What about living matter? Squeeze the juice from the cells of a living thing, and you will be faced with the most complex mixture of all. One of the major tasks for those who seek to understand the functions of cells is to find ways to separate the components of the cell so that they can be studied in isolation.

Separating the Components of a Mixture

To separate the components of a mixture, we exploit some property of certain of those components that distinguishes them from the rest. If, for example, some of the ingredients of a mixture are larger than others, we can force the mixture through a barrier perforated with openings that permit only ingredients below the size of these openings to get through. Thus by filtering through a sheet of paper, we can easily separate a mixture of sand and water (Fig. 3.3). Or

FIG. 3.1 The earth. (Courtesy NASA.)

FIG. 3.2 The forms of matter: a schematic representation. A mixture is composed of variable proportions of molecules and atoms. The two or more kinds of atoms of which a compound is made are present in fixed proportions. An element is composed entirely of one kind of atom.

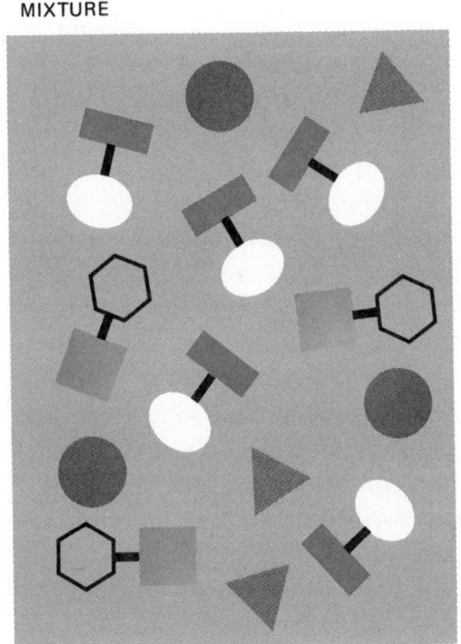

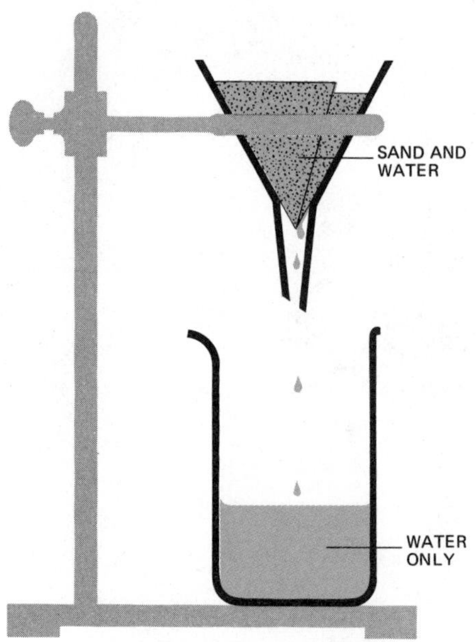

FIG. 3.3 Filtration. Any particles, such as sand, too large to pass through the pores in the paper can be separated in this way from smaller molecules such as those of water.

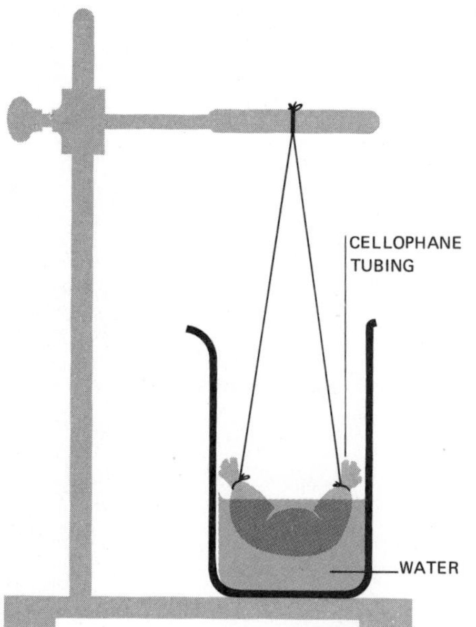

FIG. 3.4 Dialysis. The pores in a piece of cellophane are so small that macromolecules (e.g. starch, egg white) are retained inside while small molecules and ions (e.g. glucose, Na^+, Cl^-) pass out readily into the surrounding water. Dialysis is the separation of macromolecules from small molecules and ions.

perhaps we wish to separate a mixture of sugar and starch. If we place it in a cellophane bag surrounded by water, the small molecules of sugar can escape through the pores in the cellophane, while the larger molecules of starch are retained (Fig. 3.4). Other tactics for separating the components of a mixture exploit differences in the shape, density, solubility in solvents, or electrical charge of the various components.

Figure 3.5 shows a device, called a chromatographic column, used to separate the molecules in a mixture. The column is filled with some solid material which will retard the passage through it of certain molecules more than others. For example, if the column is filled with a negatively charged solid, and a mixture containing both positively and negatively charged molecules is run through it, the positively charged molecules will be retarded in the column. This occurs because of the attraction of opposite electrical charges. By thus exploiting differences in electrical charge, the mixture of molecules can be at least partially purified.

On the other hand, if the column is filled with semisolid beads which will admit small molecules into their interior but not large ones, one can separate a mixture of molecules on the basis of their size. Upon adding such a mixture at the top of the column, the small molecules are able to distribute themselves in a larger volume than is available to the large molecules (Fig. 3.6). Consequently, the large molecules move more rapidly through the column and in this way the mixture can be separated (fractionated) into its components.

Several points should be noted as we apply these separation tactics. First, the techniques are in general gentle ones, that is, they do not involve large inputs of energy. This is because the forces which hold materials together in a mixture are weak. Second, we find that the proportions of the various ingredients in a mixture are variable. Some ores are rich in iron, some poor. The ocean water in some regions is saltier than in others. Even the atmosphere displays slight variations in the amounts of such ingredients as carbon dioxide and water vapor that it contains. As for cells, their water content, to take one example, may range from 60% to 90%.

FIG. 3.5 Column chromatography. A mixture of protein molecules in solution drips onto the top of the chromatographic column on the right. The packing material in the column separates the molecules according to size. As the isolated proteins leave the bottom of the column, they are collected in separate test tubes. (Courtesy of Instrumentation Specialties Company, Inc.)

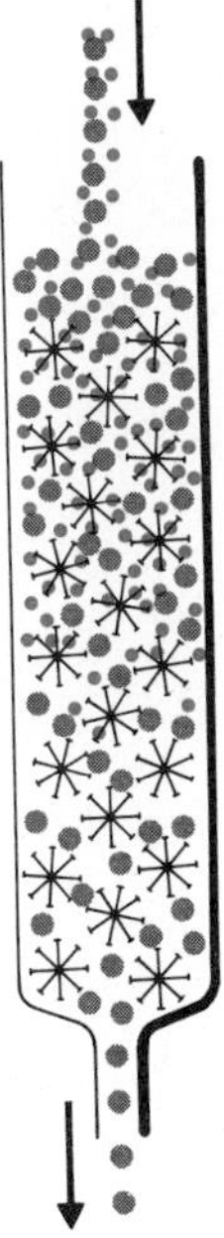

FIG. 3.6 Chromatographic separation by size of molecule. The column is filled with semisolid beads which will admit small molecules (color), but not large molecules (gray) into their interior. Thus when a mixture of the two enters the column (top), the small molecules become distributed in a larger volume of solvent than the large molecules. As the solvent moves down through the column, the large molecules are carried along more quickly than the small ones.

A third important conclusion we draw from our attempts to fractionate mixtures is that the separated ingredients retain the properties that they contributed to the mixture. The ability of air to support combustion is a consequence of the combustion-supporting property of the oxygen that is one ingredient. The chemical properties of table salt (sodium chloride) are the same whether it is part of a mixture (sea water) or isolated as a pure substance. The goal of the separation of the components of a mixture is, then, to isolate the various pure substances of which that mixture was composed.

Pure Substances

Some of the pure substances that we isolate from mixtures cannot be further broken down. Oxygen is an example. The oxygen in the air is one of the fundamental building blocks of matter. It is one of the **elements.** Other pure substances can, however, be further broken down into their constituents. Such substances are called **compounds.** Table salt (sodium chloride) is one example; water is another. If we pass an electrical current through molten table salt, we will produce two new substances: sodium, a shiny metal so reactive that it must be stored out of contact with the air, and chlorine, a yellowish gas that is extremely toxic to life. In this operation, we have decomposed a compound (sodium chloride) into its constituent elements.

What are the differences between this process and that of fractionating a mixture as we did above? First, the process required a substantial input of energy. This is because the forces that hold elements together in a compound are apt to be very strong. Second, the ingredients of a compound are present in fixed proportions. If we pass a current through 58.5 g of sodium chloride, we always produce exactly 23 g of sodium and 35.5 g of chlorine. Similarly, if we pass a current through water, we generate oxygen and hydrogen in an unvarying weight ratio of 16:2. Finally, we must keep in mind that the properties of the components of a compound are *not* the same as those of the compound itself. Both sodium and chlorine are highly hazardous to life; their compound, sodium chloride, is, in contrast, a vital ingredient in animal diets.

3.2 ELEMENTS

By degrading compounds with vigorous chemical methods, we ultimately reduce them to their elements, e.g., the hydrogen and oxygen that make up water. Proceeding in this way with compounds from

COMPOSITION OF LITHOSPHERE	
OXYGEN	47
SILICON	28
ALUMINUM	7.9
IRON	4.5
CALCIUM	3.5
SODIUM	2.5
POTASSIUM	2.5
MAGNESIUM	2.2
TITANIUM	0.46
HYDROGEN	0.22
CARBON	0.19
ALL OTHERS	$<$ 0.1

COMPOSITION OF HUMAN BODY	
HYDROGEN	63
OXYGEN	25.5
CARBON	9.5
NITROGEN	1.4
CALCIUM	0.31
PHOSPHORUS	0.22
CHLORINE	0.03
POTASSIUM	0.06
SULFUR	0.05
SODIUM	0.03
MAGNESIUM	0.01
ALL OTHERS	$<$ 0.01

FIG. 3.7 Elemental composition of the lithosphere and of the human body. Each figure represents the percent of the total *number* of atoms present. For example, 47 out of every 100 atoms found in a representative sample of the lithosphere are atoms of oxygen while there are only 19 atoms of carbon in every 10,000 atoms of lithosphere.

anywhere in the lithosphere, hydrosphere, or atmosphere, we find that no matter how complex the sample may be, it is ultimately reducible to a few representatives from a list of only 90 elements. (Additional elements are known, but these are so unstable that they have not been detected on the earth except as small quantities of them have been deliberately synthesized by the machinery of nuclear physics.) Figure 3.7 lists the relative abundance of the most prevalent elements found near the earth's surface.

The mixtures and compounds of living matter are extraordinarily complex, but ultimately these, too, can be reduced to elements. Their number is not very large—only 25 or so. Figure 3.7 lists some of the more prevalent elements in living matter.

Two points should be made about these lists. First, it is evident that not only does living matter use only a fraction of the elements available to it, but the relative proportions of those that it does acquire from its nonliving surroundings are quite different from their proportions in the environment. To put it another way, one of the properties of life is to take up certain elements that are scarce in the nonliving world and concentrate them within living cells. Hydrogen, carbon, and nitrogen together represent less than one percent of the atoms found in the earth's crust but some 74% of the atoms in living matter. The ability of life to concentrate rare elements within itself is really quite remarkable. Some sea animals accumulate elements like vanadium and iodine within their cells to concentrations a thousand or more times as great as in the surrounding sea water. It has even been proposed that uranium be "mined" from the sea by extracting it from certain algae that can take up uranium from sea water and concentrate it within their cells.

FIG. 3.8 Trace element isolators. Experimental animals raised in these all-plastic chambers are rigorously shielded from environmental contaminants while receiving a carefully prepared diet of precisely determined composition. With this apparatus, it was demonstrated that rats cannot grow normally without small amounts of tin, vanadium, and fluorine in their diet. (Courtesy of Klaus Schwarz, Veterans Administration Hospital, Long Beach, California.)

A second point about the elemental composition of living matter is our uncertainty about the exact number of elements required. Some elements, e.g., selenium and aluminum, are found in vanishingly small concentrations within living things. The question, then, is whether such an element, despite its scarcity, is playing a role essential to life, or is simply an accidental acquisition from the organism's surroundings, for example in its food.

One attempt to answer such questions is shown in Fig. 3.8. Young rats are reared in plastic trace-element isolators. All dust is filtered out of the air the animals receive, and their food is prepared with the utmost care so that contaminating elements are reduced to a minimum. Using this approach, Dr. Klaus Schwarz and his coworkers have been able to demonstrate that rats cannot grow normally unless they receive minute quantities of vanadium, selenium, and fluorine in their diet. With respect to the latter, 0.5 parts of fluorine per million parts of food were able to restore the animals to normal health and growth. There is an irony and an important lesson in this finding. Certain compounds of fluorine are used as rat poisons and, at the concentrations the animal ingests, they do their job well. So here is an element that in large doses is lethal but in small doses is absolutely essential to life.

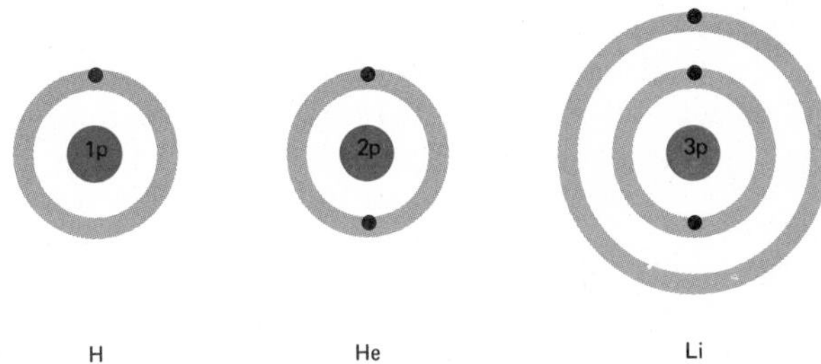

FIG. 3.9 Hydrogen (H), helium (He), and lithium (Li) atoms. In each case, the number of electrons is equal to the number of protons in the nucleus. This number is called the atomic number. (p = proton; neutrons are not shown.)

3.3 ATOMS

Each element is made up of one kind of atom. We can define an atom as the smallest part of an element that can enter into combination with other elements. (More of this in a moment.) For ease in discussing elements and the atoms of which they are composed, we use symbols, one for each element. (The symbols are listed in the periodic table that appears on the next page, Fig. 3.10.)

Structure of the Atom

No one has ever seen an atom, but years of research have provided a model that explains many of the measurable properties of elements. Each atom consists of a small, dense, positively charged **nucleus** surrounded by much lighter, negatively charged particles called **electrons.** The nucleus of the simplest atom, the atom of hydrogen, consists of a single positively charged particle, called the **proton.** Because of its single proton, the atom of hydrogen is assigned an **atomic number** of 1. Associated with this nucleus is a single electron. The charge of the electron is of the same magnitude as that of the proton, so the atom as a whole is electrically neutral. Almost all the weight of the atom is accounted for by the weight of the proton.

The atom of the element helium (He) has two protons in its nucleus. It has, therefore, an atomic number of 2 (Fig. 3.9). The helium nucleus also contains two particles called **neutrons.** Neutrons have the same weight as protons but do not have any electrical charge. Both the protons and neutrons adhere tightly together to form the dense, positively charged nucleus of the atom. Around this nucleus are two electrons so that, once again, the atom as a whole is neutral.

The structure of all the other kinds of atoms follows the same plan. From lithium (At. No. = 3) to hahnium (At. No. = 105) the atoms of each element can be listed in order of increasing atomic number. No gaps appear in this list. Each element has its unique atomic number and its atoms have one more proton and one more electron than the atoms of the element that precedes it in the list.

Isotopes. All the atoms of a given element have the same number of protons in their nuclei. This is the atomic number of that element. The same cannot be said for the neutrons. The nuclei of different atoms of the same element may differ in their number of neutrons. Because (1) neutrons and protons both have substantial (and equal) weight and (2) virtually all the weight of an atom is accounted for by the weight of the nucleus, such atoms differ in their **atomic weight.** Atoms of the same element that differ in their atomic weight are called **isotopes.**

The actual weight of a single atom is infinitesimally small. For greater convenience, atomic weights are expressed in terms of a standard atom: that isotope of carbon that has six protons and six neutrons in its nucleus. This atom is designated ^{12}C. Carbon-12 (^{12}C) is arbitrarily assigned an atomic weight of 12 **daltons** (named after the pioneer in the study of atomic weights—John Dalton). Thus a dalton is 1/12 the weight of an atom of ^{12}C. Both protons and neutrons have weights very close to 1 dalton each. Carbon-12 is the commonest but by no means the only isotope of carbon. Carbon-13 (^{13}C), with six protons and seven neutrons, and carbon-14 (^{14}C), with six protons and eight neutrons, are found in small quantities.

In looking at the table in Fig. 3.10, you may be struck by the fact that most of the atomic weights given there are not integral numbers. This is because natural samples of most of the elements consist of a mixture of isotopes. In natural samples of chlorine, for example, approximately 75% of the atoms are of the isotope ^{35}Cl, while the remainder are ^{37}Cl. A weighted average of these proportions gives a value of 35.5, which approximates the measured value shown in Fig. 3.10.

The existence of isotopes and techniques by which *unnatural* mixtures of them can be prepared provides an invaluable tool for biologists. One can prepare, for example, a carbon compound used by living things that has many of its normal ^{12}C atoms replaced by ^{14}C atoms. Carbon-14 happens to be radioactive. By tracing the fate of the radioactivity within a living organism, one can learn the normal pathway of this carbon compound in that organism. Thus the ^{14}C serves as an isotopic "label" or "tracer." We shall examine several specific examples of this technique in later chapters. The point to bear in mind now is the one made clear by the usefulness of the technique—namely, that the weight of the nucleus of an atom has little or no effect on the

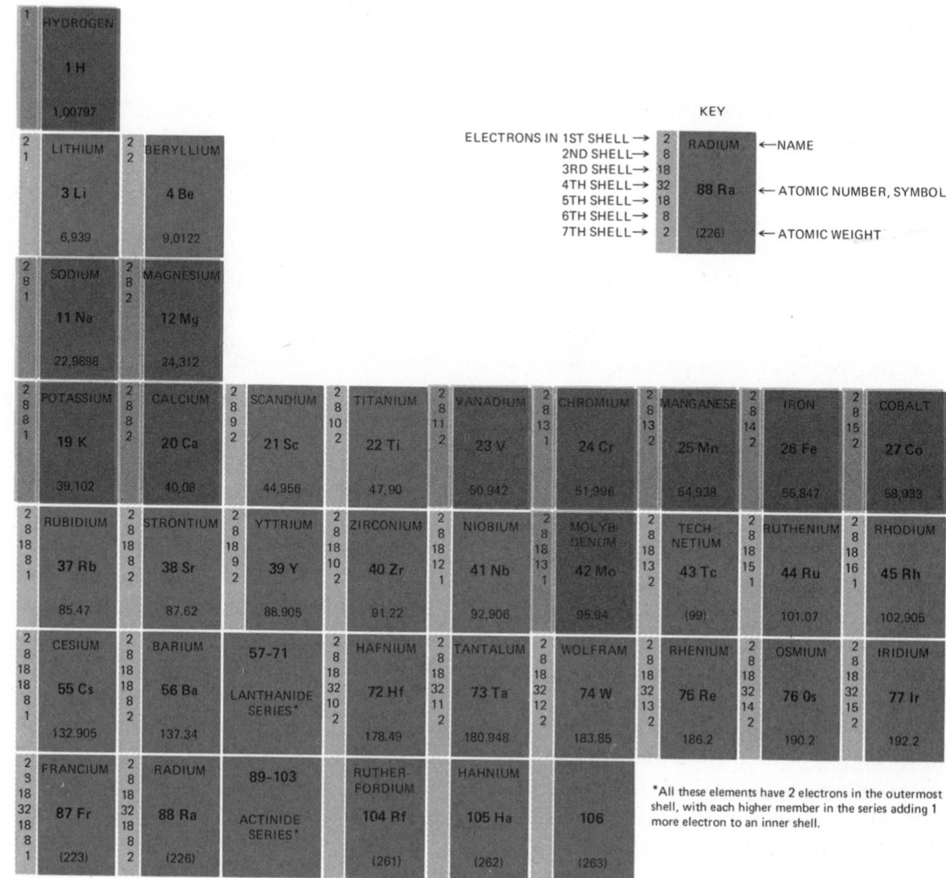

chemical properties of that atom. The chemistry of an element and the atoms of which it is made (whatever their atomic weight) is a function of the atomic *number* of that element. As long as the atom has six protons, it is an atom of carbon irrespective of the number of neutrons. Thus six protons and eight neutrons give us an isotope of carbon, ^{14}C. But seven protons and seven neutrons give us a totally different element, nitrogen-14. Now let us see why this is so.

The Periodic Table: A Closer Look

Figure 3.10 lists most of the elements in order of their atomic number. But the arrangement of the table tells us much more than a simple column would. By arranging the elements as shown, a remarkable property of the list appears. This is the reappearance at *regular intervals* of elements that share many chemical properties in common. In the next to rightmost column, we find a list that includes fluorine, chlorine, bromine, and iodine. These elements are quite similar in their chemical behavior. When dissolved in water, for example, they all produce germicidal solutions. In the second column from the left, we find calcium (At. No. = 20) and strontium (At. No. = 38). One of the worrisome problems created by the atmospheric testing of nuclear weapons was the production of ^{90}Sr, a radioactive isotope that was accumulated in the bones of children because their bodies confused it with calcium (a major component of bone).

What feature is common to all the elements in these vertical columns? The answer is: the number of electrons in their outermost shells. We noted above that the number of protons characteristic of a given element is fixed (and equal to the atomic number of that element). For each proton there is an electron and thus the atomic number of an element is also equal to the number of electrons characteristic of that element. We shall see that it is the num-

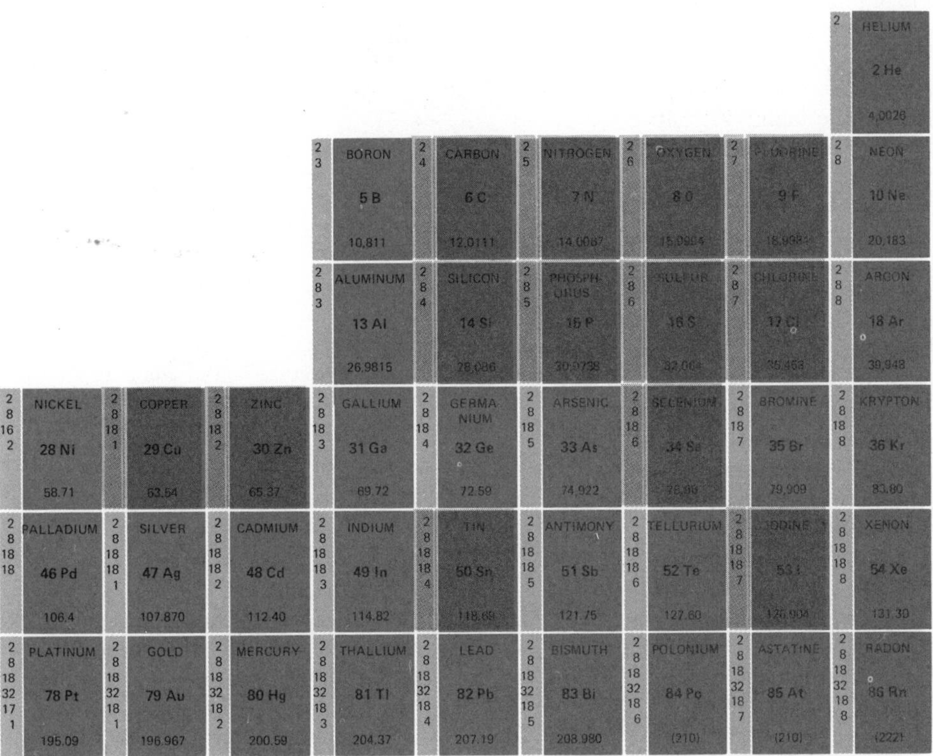

FIG. 3.10 Periodic table of the elements. The elements in each long vertical column have the same number of electrons in their outermost shell and, because of this, share similar chemical properties. Those elements that have been shown to be essential to life are indicated by a darker tone. Elements 57–71 and 89–103 play no role in living matter.

ber and arrangement of the electrons in its atoms that establishes the chemical behavior of an element.

The electrons of an atom surround the nucleus. In one of the early models of atomic structure, it was thought that the electrons moved around the nucleus in orbits—much like the planets moving around the sun. This is an oversimplification, but it is true that electrons are generally confined to relatively discrete regions around the nucleus. The two electrons of helium, for example, are confined to a spherical zone surrounding the nucleus. This is called the *K* shell (or *K* energy level). Lithium (At. No. = 3) has three electrons. Two are found in the *K* shell but the third is located farther out from the nucleus, in the *L* shell. Being farther away from the opposite (+) charges of the nucleus, this third electron is held less stably. Each of the following elements, in order of increasing atomic number, adds one more electron to the *L* shell (Fig. 3.11). This goes on until we reach neon (At. No. = 10), which has eight electrons. Sodium (At. No. = 11) places its eleventh electron in a still higher energy level, the *M* shell. From sodium to argon, this shell is gradually filled with electrons until, once again, a maximum of eight is reached.

Now we are in a position to note two important properties revealed by the periodic table: First, the maximum number of electrons that can be found in the *K* shell is two. The maximum number that any other shell can hold, *when it is the outermost shell,* is eight. (There is one, to us unimportant, exception. Can you find it?)

Second, the elements that are found in each of the long vertical columns not only share similar chemical properties, as we have seen, but also have the same number of electrons in their outermost shell. Those elements toward the left of the table, with one, two, or three electrons in their outermost shells, are the metals. Those elements at the upper right of the table, with four, five, six, and seven electrons in their outermost shells, are nonmetals. And those elements that have eight electrons in their outermost shell (neon, argon, etc.) all share a remarkable property: they are almost totally inert, that is, they have *no* chemical properties in the usual sense. Or, to put it another way, they fail under normal conditions to interact with any other kinds of atoms.

With the exception of these so-called inert gases, all the other elements do interact with one another. And, in examining the ratio in which different kinds of atoms interact, we shall discover another important clue to their chemical properties.

ATOMIC NUMBER	ELEMENT	ENERGY LEVELS *K*	*L*	*M*	*N*
1	HYDROGEN	1			
2	HELIUM	2			
3	LITHIUM	2	1		
4	BERYLLIUM	2	2		
5	BORON	2	3		
6	CARBON	2	4		
7	NITROGEN	2	5		
8	OXYGEN	2	6		
9	FLUORINE	2	7		
10	NEON	2	8		
11	SODIUM	2	8	1	
12	MAGNESIUM	2	8	2	
13	ALUMINUM	2	8	3	
14	SILICON	2	8	4	
15	PHOSPHORUS	2	8	5	
16	SULFUR	2	8	6	
17	CHLORINE	2	8	7	
18	ARGON	2	8	8	
19	POTASSIUM	2	8	8	1
20	CALCIUM	2	8	8	2
21	SCANDIUM	2	8	9	2
22	TITANIUM	2	8	10	2
23	VANADIUM	2	8	11	2
24	CHROMIUM	2	8	13	1
25	MANGANESE	2	8	13	2
26	IRON	2	8	14	2
27	COBALT	2	8	15	2
28	NICKEL	2	8	16	2
29	COPPER	2	8	18	1
30	ZINC	2	8	18	2
31	GALLIUM	2	8	18	3
32	GERMANIUM	2	8	18	4
33	ARSENIC	2	8	18	5
34	SELENIUM	2	8	18	6
35	BROMINE	2	8	18	7
36	KRYPTON	2	8	18	8

FIG. 3.11 Electronic structure of the atoms of elements 1–36. The chemical properties of an element are largely determined by the number of electrons in its outermost energy level (shell).

If, for example, we allow hydrogen atoms to interact with chlorine atoms (they do so vigorously), we find that they unite in a 1:1 ratio. The result of this interaction is a compound, hydrogen chloride. We describe it by the molecular formula HCl. As for lithium, sodium, potassium, and the other atoms in that column, each interacts with chlorine atoms in a 1:1 ratio. Thus are formed the compounds LiCl, NaCl (common table salt), and KCl.

Repeating our procedure with elements in the next column, we find that the ratio now becomes 1:2, that is, the compounds formed are described by the formulas $BeCl_2$, $MgCl_2$, $CaCl_2$, etc. With the elements in the next column, the results are BCl_3, $AlCl_3$, etc. And with carbon, the lightest element in the next column and one of crucial interest to us, the compound formed with chlorine is CCl_4. Proceeding to the next column, we find that atoms of nitrogen and hydrogen combine in a ratio of 1:3 forming the compound NH_3 (ammonia). Continuing on, we find that oxygen and hydrogen form water (H_2O). Sulfur, just below oxygen in the column, follows the same ratio, forming H_2S. Fluorine, like chlorine and all the others of its column, reacts with hydrogen atoms in the 1:1 ratio we have seen. And finally, the elements in the last column will not react at all!

What patterns can be seen? In every case we have examined, the atoms have interacted in such a way that the sum of the outermost electrons contributed by each has been equal to eight. And remembering the great chemical stability of those elements that already have eight electrons in their outermost shells (two in the case of helium), our conclusion appears to be that the atoms of elements interact in ways that will give all participants a configuration of electrons like those of the inert gases.

As we shall see, such configurations can be achieved by acquiring more electrons, losing electrons, or sharing them. The number of electrons that an atom must acquire, or lose, or share, to reach a stable configuration of eight (two for hydrogen) is called its **valence.** Now we must inquire as to which of these three possible ways to express its valence an atom will elect. The answer depends upon the **electronegativity** of that atom.

3.4 CHEMICAL BONDS

The elements differ in the affinity that their atoms have for additional electrons. The greater this affinity for electrons, the greater the electronegativity of the atom. If we distort the periodic table so that the greater an element's electronegativity, the higher it is placed in the table (Fig. 3.12), two generalizations can be made. First, the lower the atomic num-

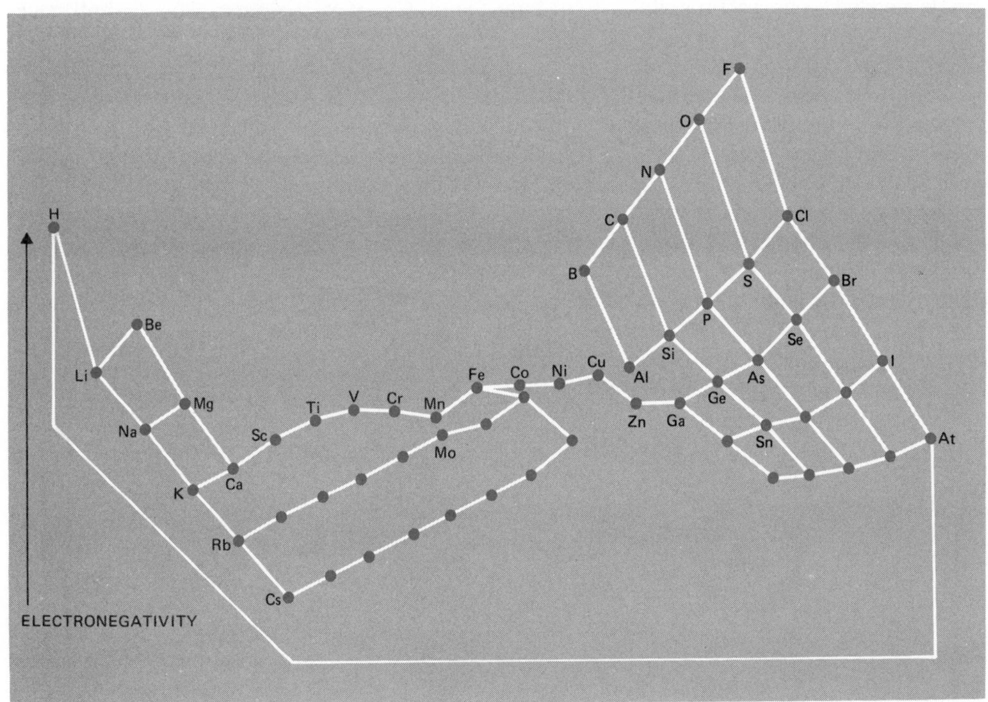

FIG. 3.12 Relative electronegativities. A portion of the periodic table has been drawn to show the relative electronegativities of some of the elements. Although fluorine (F) is the most highly electronegative of the elements, it is the electronegativity of runner-up oxygen that is exploited by living things.

ber of the element in a given column, the greater the electronegativity of its atoms. This makes sense when you remember that each additional shell or energy level is farther away from the attraction of the positively charged protons present in the nucleus. Second, those elements (nonmetals) that need only one or two electrons to fill their outermost shells are more electronegative than those elements (metals) that would need to acquire six or seven electrons to fill a shell. Metals are satisfied, instead, to get rid of the one or two electrons they have in their outermost shell.

The First Case: Ionic Compounds

Fluorine is the most electronegative of all the elements. Sodium is only weakly electronegative. When these two elements are brought together, the electron affinity of the fluorine atom is so great that it completely pulls away the single outer electron of the sodium atom. Sodium, like all the metals, does not have a great affinity for its outer electron so it gives its electron up readily. Having lost an electron, however, it now has a single positive charge (11 protons, but only 10 electrons). Similarly, the fluorine atom now has a single negative charge because of its additional electron (9 protons, 10 electrons). These electrically charged atoms are called **ions.** The mutual attraction of opposite electrical charges holds the ions together by **ionic bonds.** Similarly, the interaction of sodium and chlorine leads to the formation of sodium and chloride ions (Fig. 3.13). In each of these cases, the ratio of the ions is 1:1. However, the ions are not held together in pairs but are stacked in three-dimensional arrays (Fig. 3.14). Each sodium ion is held by six chloride ions (above, below, front, back, left, and right) while each chloride ion is, in turn, held by six sodium ions. The result of this stacking of ions is a cubical crystal of common table salt.

Ionic compounds are widespread in nature. Although we naturally associate them with the nonliving world of rocks and minerals, many ionic compounds are essential to life. Salts of sodium, potassium, calcium, chlorine, and other elements are found dissolved in the water of blood, cell fluid, etc. In this condition, the positive and negative ions become separated or **dissociated** from each other (Fig. 3.15). The essential chemical properties of the substance are not changed in the process, however. Sodium ions (Na^+) and chloride ions (Cl^-) dissolved in water still retain the properties of table salt. They do not regain the poisonous properties of the neutral atoms from which these ions were made.

The Second Case: The Covalent Bond

Carbon and hydrogen atoms have similar electronegativities (Fig. 3.12). When brought together, the only way in which each can achieve a stable electronic configuration is to share pairs of electrons between them. Each of carbon's four *L*-shell electrons associates with the single electron contributed by a hydrogen atom. In this way a molecule of methane is formed (Fig. 3.16). Each shared pair of electrons constitutes a **covalent bond.** We can show this as in Fig. 3.16(a) with the ×'s representing the electrons contributed by the hydrogen atoms and the dots the outermost (*L*) shell electrons of the carbon atom. However, all the electrons are precisely alike, so indicating them all by dots is less misleading (Fig. 3.16b). And if such electron-dot formulas

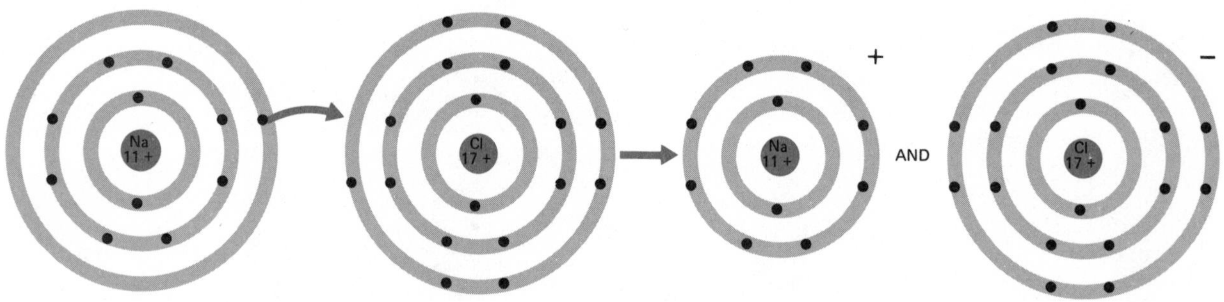

FIG. 3.13 Formation of sodium chloride. In the transfer of an electron from a sodium atom to a chlorine atom, each atom acquires an outer shell of eight electrons and thus attains stability. Each also acquires an electrical charge. Charged atoms are called ions.

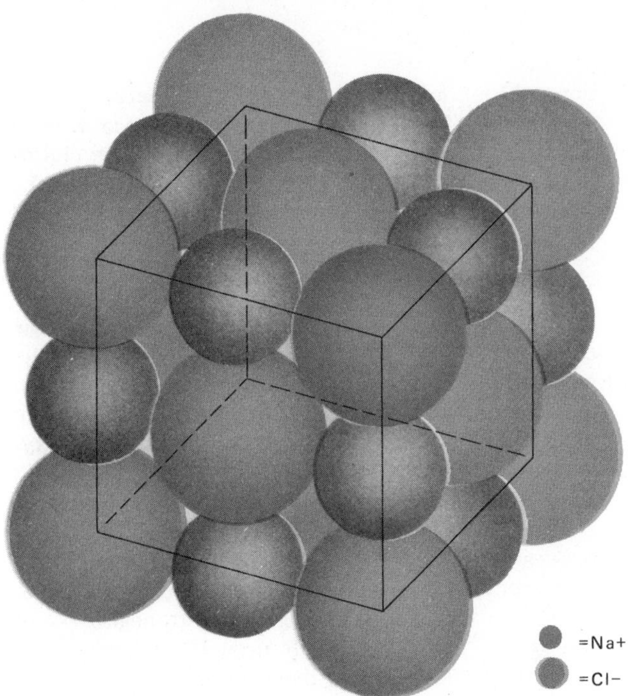

FIG. 3.14 The lattice structure of a crystal of sodium chloride, NaCl. The orderly stacking of Na+ and Cl− ions produces a crystal in the shape of a cube.

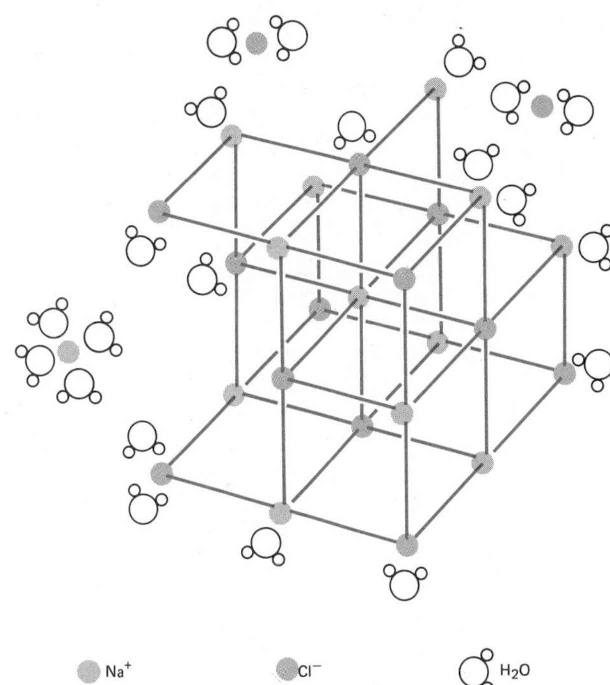

FIG. 3.15 Dissolving the crystal lattice of NaCl. Although the water molecule as a whole is electrically neutral, the uneven distribution of its electrons causes it to be somewhat polar, i.e. negatively charged around the oxygen atom and positively charged around the two hydrogens. Because of this polarity, water is an excellent solvent for ionic compounds. The Cl− ions are attracted to the positively charged region of water molecules, the Na+ ions to the negatively charged region. As a result, the ions become dissociated and the crystal dissolves.

FIG. 3.16 Five ways of depicting the methane molecule. (a) The complete atomic structure. (b) The electron-dot formula. (c) The structural formula. (d) The molecular formula. (e) A three-dimensional representation showing the tetrahedral orientation of the four covalent bonds. The regions shown in color represent the orbitals in which each shared pair of electrons is confined.

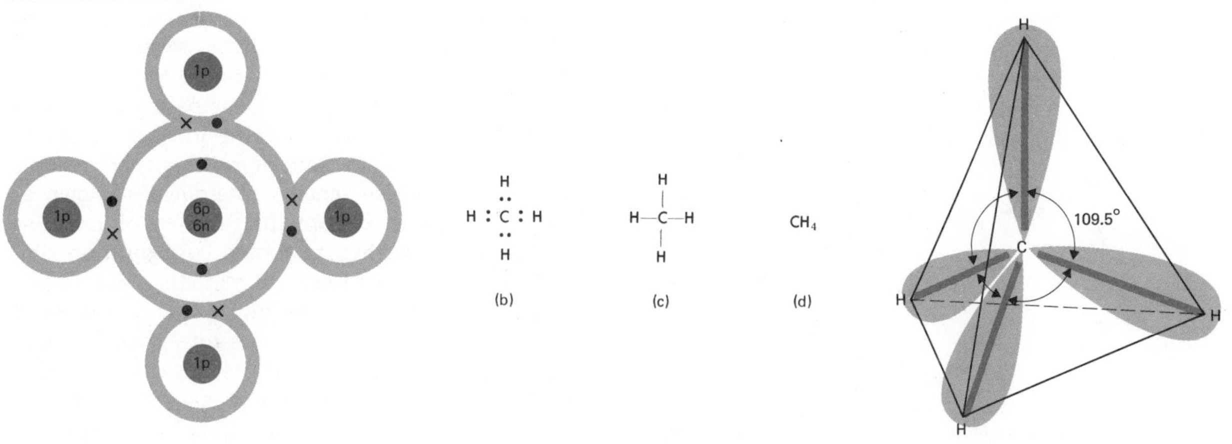

become too cumbersome, we can replace each electron pair by a dash. This gives us the **structural formula** of the molecule (Fig. 3.16c). If that is too much bother or if (as may be the case with more complex molecules) we are not sure just where the covalent bonds are located, we can use a so-called molecular formula, e.g. CH_4. This tells us only the number of each of the different atoms present in the molecule.

Although those of us who spend much time in the two-dimensional world of book pages and blackboards are apt to overlook it, atoms and molecules really exist in three dimensions. So even our electron-dot formula for methane is misleading. Actually the four covalent bonds that are formed between the hydrogens and the carbon atom are directed as far apart from each other as is physically possible. This means that each is directed to one of the four corners of a regular tetrahedron (Fig. 3.16e). Thus each of the four covalent bonds forms an angle of 109.5° with *each* of its three neighboring bonds. The result is a perfectly symmetrical distribution of all the electrons (and nuclei) of which this molecule is composed.

Two additional examples of covalent bond formation between atoms of similar electronegativity are of special interest. Many gaseous elements exist not as individual atoms but as diatomic molecules, i.e., molecules consisting of two atoms. There is no electronegativity difference between identical atoms, so the bonding between them is covalent. Nitrogen and oxygen, which account for over 99% of the atmosphere, are both present as diatomic molecules, with the formulas N_2 and O_2 respectively. Chlorine and hydrogen gas also exist as the diatomic molecules Cl_2 and H_2 (Fig. 3.17).

Carbon atoms provide the backbone of living matter. In the next chapter we shall see several examples of giant molecules without which life as we know it could not exist. All of these giant molecules depend for their size on the ability of carbon atoms to bond together. There being no difference in electronegativity, the bonding is covalent. Thus chains of carbon atoms can be formed. These provide the structural basis for the elaboration of virtually all the molecules unique to living matter.

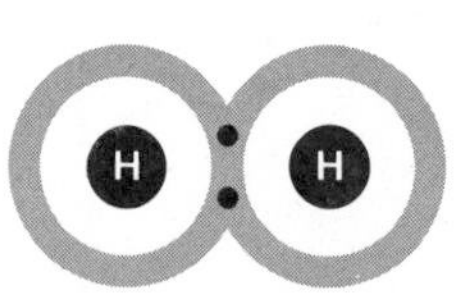

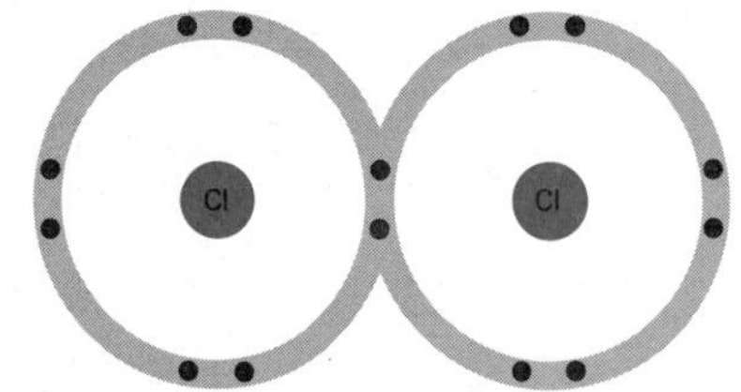

FIG. 3.17 Structure of the molecules of hydrogen and chlorine. The bond between the two atoms of each of these diatomic molecules is covalent.

The Third Case: The Polar Covalent Bond

Oxygen is the second most electronegative of all the elements (Fig. 3.12). Hydrogen is moderately electronegative. It is not as electronegative as oxygen or nitrogen, but is more electronegative than sodium or calcium. What happens when hydrogen and oxygen atoms are brought together? The difference in their electronegativities is not sufficient to yield ions. Consequently, they must share a pair of electrons between them, i.e., a covalent bond is formed. However, some kinds of sharing are more equal than others! In this case, the greater electronegativity of the oxygen atom causes it to draw the electron pairs closer to its nucleus and therefore farther away from the nuclei of the hydrogen atoms. This results in a concentration of negative charges nearer the oxygen atom and thus farther from the positively charged protons that constitute the nuclei of the hydrogen atoms. The bond formed is thus intermediate in nature between a fully ionic bond on the one hand and a purely covalent one on the other. There is a separation of charges, but it is not complete as it is when ions are formed. We indicate the partial charges produced by the symbol δ (Fig. 3.18).

If the orientation of the two hydrogen atoms about the oxygen atom were as shown in Fig. 3.18(a), that is, with the hydrogens opposite and in the same plane, these polar bonds would be of little significance. But, when we examine the situation as it really exists in three dimensions, we find that the distribution of partial charges around the water molecule is not symmetrical. The four pairs of electrons that surround the oxygen atom in a molecule of water are confined to four dumbell-shaped regions in its L shell. These regions (technically known as orbitals) are directed at the four corners of a regular tetrahedron. Thus they take on a configuration similar to that which we found in methane (Fig. 3.16). Two pairs of electrons represent polar covalent bonds to hydrogen nuclei and two are *unshared.* Because of the geometry of the regular tetrahedron, no matter which two corners we elect to place the hydrogen nuclei in, there will result an asymmetric distribution of the partial charges. One side of the molecule (where the hydrogen atoms are) will be partially

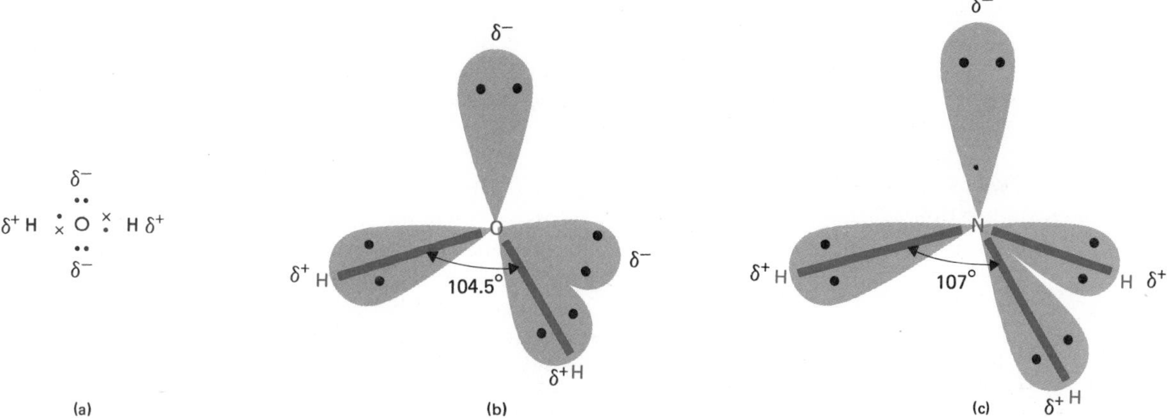

FIG. 3.18 (a) A representation of the water molecule in two dimensions. The symbol δ represents a partial electrical charge. Drawn this way, the distribution of partial charges is symmetrical. (b) A three-dimensional representation of the water molecule. Note that the partial positive charges are located at two corners of a tetrahedron, the partial negative charges at the other two. Thus the molecule as a whole is polar. The areas in color are the orbitals in which both shared and unshared electron pairs are confined. (c) The geometry of the molecule of ammonia. Ammonia, like water, is strongly polar.

positively charged, the other side (with the unshared pairs of electrons) will be partially negatively charged. Thus the molecule *as a whole* is polar.

Actual measurements of the angle between the two polar covalent bonds of water shows that they are 104.5° apart (Fig. 3.18b), slightly less than the 109.5° that we would expect if the two hydrogen nuclei were as far apart as possible. What this probably indicates is that the *unshared* pairs of electrons occupy somewhat more space than the shared pairs, and thus force the two hydrogen atoms somewhat closer together. This would also explain the 107° (instead of the regular tetrahedral angle of 109.5°) found between the three polar covalent bonds in ammonia (Fig. 3.18c). And in this case, too, the overall molecule is polar, with a concentration of negativity at one corner of the tetrahedron and of positivity at the others.

3.5 THE HYDROGEN BOND

The polarity of the water molecule makes possible a mutual attraction between water molecules. Each molecule has a region of δ^+ and a region of δ^-. Because of the attraction of unlike charges, water molecules tend to orient themselves so that the δ^+ regions are brought close to δ^- regions. What this means is that each water molecule attracts four other water molecules to it. Each hydrogen atom is attracted to the oxygen atom of a nearby water molecule. Each oxygen atom can, through its two unshared pairs of electrons, associate with two hydrogen atoms supplied by two other nearby water molecules (Fig. 3.19). While the force of attraction between these polar molecules is not as strong as a covalent or ionic bond, it is nonetheless far from negligible. It is sufficiently important to deserve a name of its own: the **hydrogen bond.** The hydrogen bonding that takes place between water molecules accounts for some of the remarkable and vital properties of water—properties that make the presence of water synonymous with life itself.

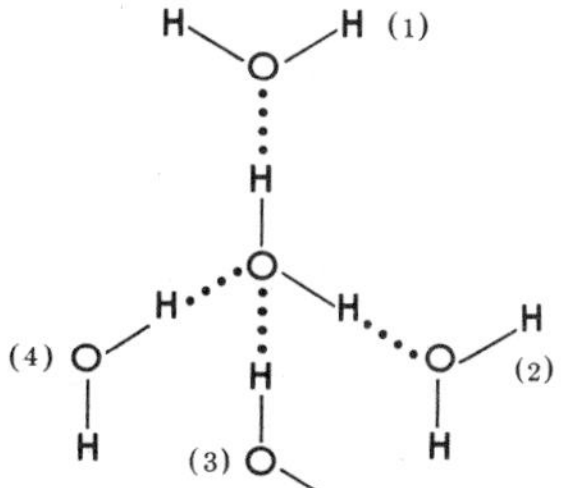

FIG. 3.19 Hydrogen bonding in water. Each water molecule is hydrogen bonded to four others located at the corners of a regular tetrahedron. The extensive hydrogen bonding in water accounts for some of the special properties of this substance.

The ability of water to remain liquid over a large portion of the range of temperatures found on the earth is one such happy outcome. Molecules as small as water that do not form hydrogen bonds (e.g., CO_2) change from liquid to gas at temperatures far below that at which water does so. Hydrogen bonding also accounts for the extraordinarily large amount of energy (heat) needed to convert liquid water—where the molecules are attracted through their hydrogen bonds—to water vapor—where they are not. Thus water is a great stabilizer of temperatures. This accounts for the more moderate climate near large bodies of water like the oceans. This property is also exploited by many living organisms. The production and evaporation of sweat is a vital cooling process for many mammals and depends for its effectiveness on the large amount of heat needed to break the hydrogen bonds between water molecules.

3.6 WATER LOVERS AND WATER HATERS

Nitrogen and hydrogen atoms differ in their electronegativity. Thus these atoms bind by polar covalent bonds. As in the water molecule, the geometry is such that the molecule as a whole is polar, with a δ^+ directed toward three apices of a tetrahedron and a δ^- toward the fourth (Fig. 3.18). And, as in water, there is the opportunity to establish hydrogen bonds between ammonia molecules. But if identical polar molecules can attract one another, why not nonidentical polar molecules? In fact, such attraction does occur. If you bubble ammonia gas through water, the ammonia dissolves readily in the water. The result is household ammonia. In general, any molecule that is attracted to water molecules is said to be **hydrophilic** (water-loving). Such a molecule mixes (i.e., dissolves) readily in water. But if you bubble methane through water, precious little will dissolve. This is because methane is **apolar.** As we noted earlier, there are no polar covalent bonds and thus no δ charges. Thus there is nothing to attract methane molecules to water molecules. Methane is said to be **hydrophobic** (water-hating).

Polar molecules not only attract each other, but they also attract **ions.** This explains the ease with which such ionic compounds as NaCl dissolve in water. Figure 3.15 gives a schematic idea of the dissolution of a crystal of table salt as the Na^+ and Cl^- ions are attracted to the δ^- and δ^+ portions of water molecules respectively.

3.7 ACIDS AND BASES

Consider what happens when you bubble some hydrogen chloride (a gas, HCl) through water. Chlorine is more electronegative than hydrogen, and thus the single bond between them is polar. Polar molecules attract each other, as we have seen, so we would expect HCl to dissolve well in water. And indeed it does. But the story does not end there. Chlorine exerts such control over the pair of electrons "shared" with the hydrogen (i.e., the bonding is so polar) that the single proton is quite deprived. Consequently, it is easily attracted to any *unshared* pair of electrons that might be found on nearby molecules. Water molecules have such unshared pairs, and consequently, the proton leaves the chlorine atom and transfers to one of the unshared pairs on a water molecule (Fig. 3.20a). But note that the hydrogen nucleus (proton) left its electron behind. Thus what started as a single molecule has now become two ions: the hydrogen ion (H^+) and the chloride ion (Cl^-). Thus the polar properties of water have caused the molecules of HCl to *ionize.* The hydrogen ion does not exist alone but occupies one of the unshared pairs of electrons on the water molecule to form the hydronium ion (H_3O^+). The resulting solution (a mixture of H_3O^+ and Cl^- ions) is called hydrochloric acid.

Now let us bubble another polar substance, NH_3, through a hydrochloric acid solution. Ammonia molecules have one pair of unshared electrons and these have a greater affinity for a proton than do the unshared pairs in the water molecule. Consequently, the proton shifts again to form a new ion, the ammonium ion (NH_4^+), and water (Fig. 3.20b).

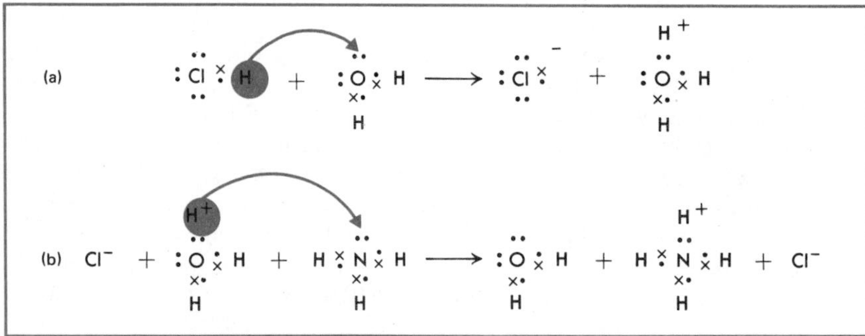

FIG. 3.20 (a) Formation of hydrochloric acid. Any substance that liberates protons is an acid. (b) Reaction between hydrochloric acid and ammonia. The ammonia molecule acts as a base, accepting a proton donated by the hydronium ion. The reaction between an acid and a base is called neutralization.

In general, whenever a substance donates protons to some other substance, we describe the proton donor as an **acid.** Thus in the examples above, the HCl molecule and the hydronium ion (H_3O^+) are acids. Any substance that *accepts* protons is called a **base.** The water molecule in the first example and the ammonia in the second example served as bases.

While HCl is found in living systems (e.g., in the gastric juice created by the stomach), the most common acids found are those that contain the **carboxyl group.**

```
   O
   ‖
 —C—O:H
```

The hydrogen *nucleus* (proton) of the carboxyl group is easily removed, forming the carboxyl ion.

```
   O
   ‖
 —C—O:⁻
```

Acetic acid (CH_3COOH) is a common example of a carboxylic acid. Placed in water, some of the protons on the —COOH group will be attracted to the unshared electron pairs of water molecules. Hydronium ions (H_3O^+) and acetate ions (CH_3COO^-) result. Vinegar is a dilute solution of acetic acid.

Ammonia is also found (in low concentrations) in living matter. But the most common bases are those molecules that contain an **amino group.**

```
   H
   ··
 —N:
   ··
   H
```

The unshared pair of electrons serves as a proton acceptor, as it does in the ammonia molecule. Bicarbonate ions

```
      OH
      |
 O═C—O:⁻
```

also serve as an important base in living tissue.

3.8 MOLECULAR WEIGHT AND THE MOLE

The weight of a molecule is simply the sum of the weights of all the atoms of which that molecule is composed. The unit of weight is, again, the dalton (1/12 the weight of an atom of ^{12}C). Thus the molecular weight (MW) of water is 18 daltons. (We will ignore the tiny error introduced by the presence of traces of ^{17}O, ^{18}O, and ^{2}H atoms among the predominant ^{1}H and ^{16}O atoms.)

Why is it important to know the molecular weight of a compound? For one illustration, let us assume that you wish to study the response of honeybees to solutions of various kinds of sugars. One way to do this would be to make up several different solutions and see which one or ones the bees prefer to harvest. You might offer the bees the choice between, say, a 35% solution of sucrose (common table sugar) and a 35% solution of glucose (a natural component of honey). This would involve, in each case, dissolving 350 parts by weight (e.g. grams) of sugar in 650 parts (g) of water, thus producing 1000 g of each solution. But there is a problem with this approach. The willingness of the honeybee to respond to the presence of sugar dissolved in water is dependent upon the *number* of sugar molecules in a given volume of the solution. And the sucrose molecule (MW = 342) is almost twice as heavy as the glucose molecule (MW = 180). So a 35% solution of glucose would contain almost twice as many molecules as a 35% solution of sucrose. To correct the problem, you should make your solution with the weights of sucrose to glucose in a ratio of 342:180. Then you would have the same concentration of molecules in each, that is, drop for drop each solution would contain the same number of molecules.

If you weigh out exactly 342 grams (g) of sucrose you will have weighed out one **mole** of it. That is, a mole is the quantity of a substance whose weight in grams is numerically equal to the molecular weight of that substance. Thus one mole of glucose weighs 180 g. Furthermore, if you dissolve one mole of a substance in enough water to make one liter (l) of solution, you have made a 1-molar (1-*M*) solution. (A specially graduated flask, Fig. 3.21, is a useful device in which to do this. After the substance is put into the flask, enough water is added to bring the volume up to the etched line on the neck of the flask. The volume of this solution is then exactly one liter.)

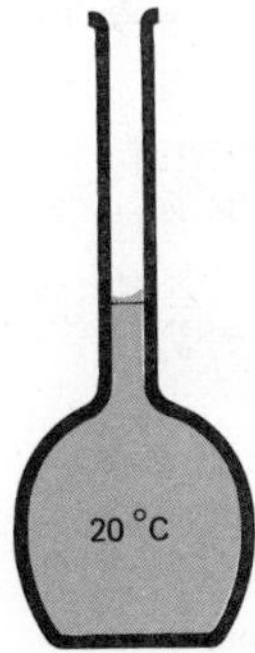

FIG. 3.21 A volumetric flask. When filled to the etched line, it contains exactly one liter. Such a flask is used to prepare solutions of precise molarity.

A 1-*M* solution of these sugars would probably be too strong for your experiment with the bees. It might be better to make up a liter of each solution containing 34.2 g and 18.0 g respectively. Such solutions would be designated one-tenth molar (0.1-*M*) solutions. Drop for drop, these two solutions would still contain exactly the same number of molecules because they are of the same molarity.

How many molecules are there in a mole? The number, as you might expect, is very large: it is approximately 6×10^{23}. (This number is called Avogadro's number after the chemist who first attempted to determine it.) Note that this number applies to a mole of any substance, molecule or ion. Thus we can properly refer to a mole of hydrogen ions (1 g).

3.9 pH

Most cells are extremely sensitive to the degree of acidity of the fluid that surrounds them. Degree of acidity is measured on a scale of pH units. When dissolved in water, hydrogen chloride molecules are almost completely ionized to form hydronium ions (H_3O^+) and chloride ions (Cl^-). Thus a 0.1-*M* solution of HCl contains a 0.1-*M* concentration of hydronium ions. Such a solution is said to have a pH of 1. We define pH as the negative logarithm of the hydrogen ion (really the hydronium ion) concentration measured in moles/liter. Changing 0.1 to scientific notation, we have 1×10^{-1}. The negative of the exponent is 1, so the pH is 1. A solution containing 0.01 moles/liter (0.01 *M*) is 1×10^{-2} *M* and thus has a pH of 2. It is ten times less acidic than the first solution.

To a very slight extent, water molecules themselves dissociate into ions (forming H_3O^+ and OH^-). A liter of pure water contains 10^{-7} moles of hydronium ions and thus has a pH of 7. But pure water is neutral, that is, it is neither acidic nor basic. So any solution of pH 7 is said to be neutral. Basic (alkaline) substances are those (e.g. NH_3, HCO_3^-) that combine with hydronium ions, thus leaving even fewer than are found in pure water. Therefore, they form solutions with pH values greater than 7 (Fig. 3.22).

3.10 CHEMICAL CHANGES

We mentioned earlier that the relative proportions by weight of oxygen and hydrogen in water are 16:2. This is a reflection of the ratio of the numbers of atoms in the molecule: one atom of O (At. Wt. = 16) to two atoms of hydrogen (At. Wt. = 1). If we pass an electric current through water containing enough ions to render it a good electrical conductor, we can decompose the water into constituents: hydrogen and oxygen. This is a chemical change. There has been a rearrangement of atoms. Or, to put it another way, certain bonds are broken and new bonds are formed.

This chemical change can be expressed by an equation. On the left of the equation we write the molecular formulas of all the ingredients or **reactants.** In this case there is only one, H_2O. We then draw an arrow and write in the formulas of all the **products.** Both oxygen and hydrogen when present as elements exist as diatomic molecules. Thus we indicate them by the formulas H_2 and O_2. Now our equation appears $H_2O \rightarrow H_2 + O_2$. But something is wrong. Matter cannot be created or destroyed. Every atom that participates in a chemical reaction must be accounted for in the products and vice versa. So we must use two molecules of water to get one

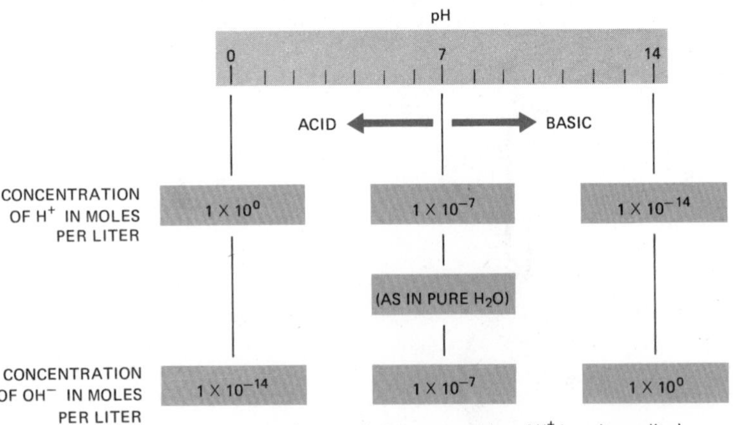

FIG. 3.22 The pH scale. A shift of one pH unit represents a tenfold shift in acidity. A solution of pH 7 is neutral.

molecule of O_2. And the decomposition of two molecules will yield two molecules of H_2. So our final, balanced equation reads:

$$2H_2O \rightarrow 2H_2 + O_2.$$

Note that the prefixes in our equation tell us the relative number of atoms and molecules involved. If we are interested in the proportions by weight, we must multiply by the molecular weights. By doing so, we see that from 36 weight units of water we get 4 parts of hydrogen and 32 parts of oxygen by *weight*. If our unit of weight is the gram, we have decomposed two moles of water and produced two moles of hydrogen and one mole of oxygen. A balanced equation can thus tell us not only the nature of the reactants and products in a chemical reaction but also the proportions in which the reactants are consumed and the products produced. And note, too, that every gram of matter that is used as a reactant reappears in the products.

3.11 BOND ENERGY

To carry out the decomposition of water described above, an electric current must be passed through the water. Electricity is one form of energy. We may define energy as the capacity to do work. The decomposition of water requires the expenditure of energy. The reason for this is that it takes energy to break chemical bonds. But to complete the chemical reaction, new bonds are formed. Energy is released when chemical bonds form. In fact, for any particular bond, say the covalent bond between hydrogen and oxygen, the amount of energy it takes to break that bond is exactly the same as the amount of energy liberated when the bond is formed. This value is called the **bond energy.**

There are many forms of energy: electrical, mechanical, chemical, etc. But all forms of energy are ultimately converted into heat. Therefore it is convenient, at least for biologists, to measure energy in terms of units of heat. The unit we shall use most frequently is the kilocalorie (kcal). A kilocalorie is that amount of heat needed to warm one liter of water one Celsius degree. (A table describing the international system of units—which is used in all scientific work—may be found in an appendix at the end of this book.) The kilocalorie is also the unit used to describe the energy content of foods. It should thus be familiar to anyone who is concerned about diet and achieving or maintaining a particular body weight.

Returning to our example, we find that it takes 118 kcal of energy to decompose two moles of water into its elements. Actually it takes more than 118 kcal to decompose a mole of water into hydrogen and oxygen atoms, but as these atoms immediately bond together to form molecules of hydrogen and oxygen, some of that energy is given back.

The bond energy of the H—O bond is 110 kcal. The bond energy of H—H bonds is 103 kcal and that of O=O bonds 116 kcal. The decomposition of two molecules of water requires the breaking of four H—O bonds and thus the input of 440 kcal. The formation of two moles of hydrogen yields 206 kcal (2×103) and the formation of a mole of oxygen yields 116 kcal. Subtracting the sum of the energy released ($206 + 116 = 322$ kcal) from the energy consumed (440 kcal) gives us the net energy input of 118 kcal that we measured above.

Where has this energy gone? In a real sense, it is now chemical energy stored in the bonds of the hydrogen and oxygen molecules. The energy stored in this reaction is also called free energy because, as we shall see shortly, it is still available to do work. It is useful to have a symbol for free energy, and we shall use the letter F.

What is free energy? Free energy is energy that can be harnessed to do work. The water stored behind a dam has free energy. When allowed to fall through turbines, it can generate electricity (another form of free energy). The tension stored in the mainspring of your watch is free energy. Bit by bit, it is given up to drive the hands. But for biologists, the most interesting form of free energy is the energy stored in chemical bonds. It, too, can be harnessed to do work. When we lift a weight, we are using the free energy stored in the bonds of food molecules to run a machine: our muscles. With this machine, we accomplish the work of lifting the weight. However, the conversion of free energy to work is not 100% efficient. As we contract our muscles, much of the free energy of our food is given off as heat. This energy is no longer free; there is no way we can harness the warmth of our muscles to accomplishing biologically useful work.

If and when you study physical chemistry, you will discover that the notion of free energy involves considerations that are not appropriate or necessary for us to worry about here. Then, you will use several other symbols (e.g. $\Delta G^{\circ\prime}$). But to avoid implying a greater degree of precision than is justified, we will limit ourselves to F. And we shall indicate a change in free energy by the symbol ΔF. In the decomposition of water, the energy is not lost but "stored" in the products. Thus we indicate the free energy change with a plus sign. Therefore the reaction can now be expressed as

$$2H_2O \rightarrow 2H_2 + O_2, \quad \Delta F = +118 \text{ kcal}.$$

If you are fortunate, you may once have seen a chemistry teacher ignite a mixture of hydrogen and oxygen gas for your edification. If not, simply accept my word that the results are dramatic and rather too risky for you to attempt yourself. The result is an awesome explosion. The equation for this chemical reaction is the reverse of the one we have just been studying. It is, then: $2H_2 + O_2 \rightarrow 2H_2O$. And, as the explosion suggests, there is this time a *release* of energy. In fact, the free energy change is once again 118 kcal. This is because it took only 322 kcal to break the H-H and O-O bonds while 440 kcal were liberated by the four moles of H-O bonds that were formed. We express the fact that energy has come out of the reacting system by putting a minus sign before ΔF. Therefore our reaction is now expressed

$$2H_2 + O_2 \rightarrow 2H_2O, \quad \Delta F = -118 \text{ kcal/mole}.$$

What determines the magnitude of the bond energy for a particular kind of bond? There are several contributing factors, but usually the most important from our point of view is the difference in the electronegativity of the two atoms bonding together. Thus the energies of bonds between atoms of substantially differing electronegativities tend to be high, e.g. the 110 kcal of the H—O bond. Another important example is the bonds between oxygen and carbon atoms in carbon dioxide, CO_2. The carbon atom shares two pairs of electrons with each of the oxygen atoms, and each of these **double bonds** has a bond energy of 187 kcal (or about 93 kcal for each shared pair of electrons). The larger the bond energy, the more energy is needed to break the bond. Thus bonds between atoms of differing electronegativities are apt to be very strong and stable.

On the other hand, the energies of bonds between atoms of similar electronegativity tend to be smaller. Of course, where the atoms are the same (e.g. O═O, C—C) there is no electronegativity difference and we are not surprised to find lower bond energy values. Each of the shared electron pairs in the oxygen molecule is worth 58 kcal for a total of 116 in each molecule, as we saw above. The bond energy of the C—C bond is 80 kcal. There is a slight difference in electronegativity between carbon and hydrogen, and a bond energy of 98 kcal. The bonds with smaller bond energies are, by definition, easier to break. Thus these bonds are weaker and less stable.

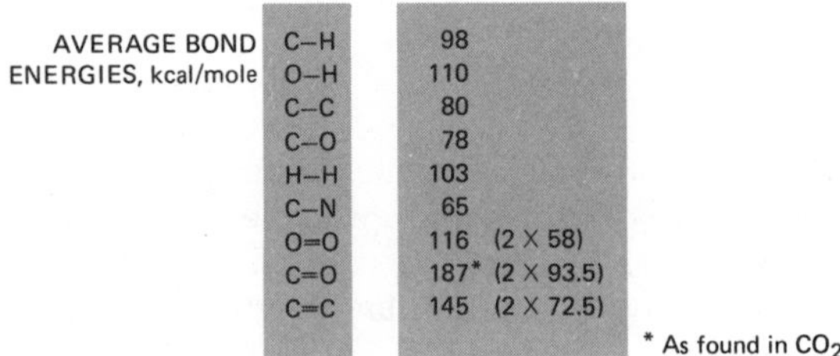

AVERAGE BOND ENERGIES, kcal/mole	
C–H	98
O–H	110
C–C	80
C–O	78
H–H	103
C–N	65
O═O	116 (2 × 58)
C═O	187* (2 × 93.5)
C═C	145 (2 × 72.5)

* As found in CO_2

FIGURE 3.23

We should also note that the energy needed to break a particular bond, e.g. between carbon and oxygen, may also be influenced by the nature of the other atoms attached to the ones of interest. Thus the C═O double bond in carbon dioxide (O═C═O) has a bond energy of 187 kcal, whereas when this bond is found as part of a larger molecule, like

$$H_3C{-}\overset{\overset{\displaystyle O}{\|}}{C}{-}CH_3,$$

the value is closer to 170 kcal. Because of these variations, it is wise to speak in terms of **average bond energies.** Using the table of average bond energies given in Fig. 3.23, we shall be able, as we proceed, to uncover a variety of important concepts with only occasional and minor sacrifices of numerical accuracy.

3.12 OXIDATION-REDUCTION REACTIONS

Most of the chemical reactions that we have been studying involve shifts in the location of electrons. In the reaction of chlorine with sodium, for example, each sodium atom loses an electron to a chlorine atom. In such a case, the atom that loses the electron is said to have been **oxidized.** The atom that gains the electron is said to have been **reduced.** The two processes are inextricably linked. Whenever any substance gives up electrons, it is oxidized. The substance to which it gives its electrons is called the oxidizing agent. Note, however, that any substance acting as an oxidizing agent is itself reduced. Because oxidations always go hand in hand with reductions, we refer to these reactions as **redox** reactions.

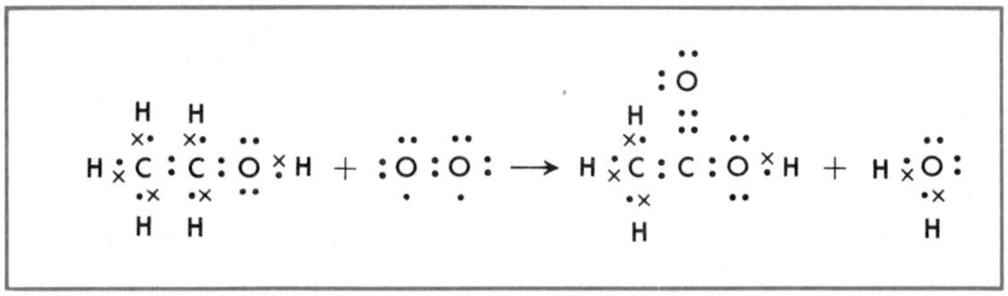

FIG. 3.24 The oxidation of ethanol. Oxidation has occurred both by the addition of an oxygen atom and by the removal of hydrogen atoms. In each case, electrons are moved away from carbon atoms.

You may wonder why the reaction between sodium and chlorine is called an oxidation when oxygen has no part in it. The answer is simply a matter of history. Oxygen has a great affinity for electrons (it is the second most electronegative element) and is an excellent (as well as common) oxidizing agent. Because of its effectiveness in this role, it has supplied the name for all reactions of this type. (Actually, fluorine is even more electronegative than oxygen, and perhaps the process should rightly be called fluoridation. But *that* term has come to be used for quite a different process: the adding of trace amounts of the fluoride ion, F^-, to municipal water supplies.)

Some biological oxidations do indeed involve the addition of oxygen to the molecule being oxidized. When ethanol (the common beverage alcohol) is exposed to air and the action of vinegar bacteria, it is converted into acetic acid and water (vinegar). The equation for this reaction is

$$C_2H_5OH + O_2 \rightarrow CH_3COOH + H_2O.$$

The same equation, using electron-dot formulas, is shown in Fig. 3.24. Note the atom of oxygen that has been added to the ethanol to form acetic acid. Because the difference in electronegativity between carbon and oxygen atoms is not as great as, say, that between oxygen and magnesium, the added oxygen atom does not remove electrons sufficiently far to form ions. It nevertheless does attract two electrons away from the carbon atom and much closer to itself (Fig. 3.24).

The most common kind of biological oxidation is accomplished by the removal of hydrogen atoms from a substance. The oxidation of ethanol also exhibits this kind of oxidation. Two hydrogen atoms, each with its single electron, are removed by the second oxygen atom. These unite to form a molecule of water. Although all oxidations go hand in hand with reductions, we often emphasize the oxidation aspect of the reactions when electrons are moving from a weakly electronegative atom to a strongly electronegative one. The reactions of sodium and chlorine, hydrogen and oxygen, ethanol and oxygen are all examples. In all of these cases, free energy is liberated. This is because weaker bonds have been broken and strong bonds formed. We exploit oxidations characterized by a large $-\Delta F$ to heat our homes, to run our automobiles, and—of special interest to us—to run all the metabolic activities of life.

It is also possible to move electrons back from strongly electronegative atoms to weakly electronegative ones. In this case, ΔF is positive, that is, energy must be put into the reaction. We usually emphasize the reduction aspect of such redox reactions. To take an earlier example, decomposing two moles of water requires the input of 118 kcal of energy. We have oxidized oxygen atoms (they lose electrons) and reduced the hydrogen atoms. We have, in fact, broken strong bonds (H—O) and formed weaker bonds (H—H, O—O). Thus weak bonds can serve as a means of storing energy. This energy is chemical, potential energy. As soon as we permit the electrons which are involved to return to more strongly electronegative atoms, the energy is liberated.

We shall see in Chapters 8 and 9 that life continues to exist on this earth because it is able to exploit redox reactions. It uses them to store the sun's energy and then to liberate the resulting chemical energy as needed to run the activities characteristic of living things. In photosynthesis, the energy of the sun is used to move electrons (associated with protons, thus hydrogen atoms) from strongly electronegative atoms (oxygen) to weakly electronegative ones (carbon). In this process strong bonds (H—O, C—O) are broken and weaker bonds are formed (C—H and especially O—O). The reverse of this process is cellular respiration. In this case, the electrons return to their original partners, and the energy that has been stored is liberated and used to run the organism. Thus the energy differences among chemical bonds make it possible to operate a kind of battery: storing the sun's energy and liberating it as needed to maintain the living process.

3.13 IN CONCLUSION

Perhaps you have become aware, as you studied this chapter, of the pervasive importance of the concept of electronegativity. We have seen the role that electronegativity plays in establishing whether a redox reaction will have a $-\Delta F$ or a $+\Delta F$. Only by coupling the first type to the second type could the orderly, complex molecular systems that *are* life exist on this earth. We have seen the influence of electronegativity on the magnitude of bond energies and thus on the strength or weakness of chemical bonds. We have seen how differences in electronegativity create polar bonds that may lead to the synthesis of polar molecules. The polarity (or absence thereof) of molecules establishes, in turn, how well they dissolve in water, that is, whether they are hydrophilic or hydrophobic. Electronegativity differences and the resulting polarity of molecules or parts of molecules also set the stage for the formation of hydrogen bonds. And, as we shall see in the next chapter, hydrogen bonds play crucial roles in the formation of the giant molecules so characteristic of living matter: the proteins, nucleic acids, and polysaccharides.

EXERCISES AND PROBLEMS

1. Which of the following substances are composed of molecules: (a) oxygen, (b) water, (c) sodium chloride, (d) glucose, (e) steel?
2. Distinguish between an atom and an ion.
3. What ion is produced by all acids?
4. A mole of methane (CH_4) weighs how many grams?
5. How many milligrams (mg) of glucose would you dissolve in a liter of water to make a solution containing one part per million (ppm)?
6. Write electron-dot formulas for the molecules of (a) water, (b) ammonia, (c) methane, (d) ethane (CH_3CH_3).
7. Summarize the differences between mixtures and compounds.
8. Distinguish between organic and inorganic compounds.
9. Which of the following elements would you expect to react with chlorine: (a) hydrogen, (b) neon, (c) sodium, (d) fluorine, (e) calcium, (f) carbon?
10. Chemical analysis shows that the ratio of C atoms to H atoms in a substance is 1 to 3. The molecular weight of the substance is 30. What is the molecular formula?
11. Show by means of electron-dot formulas the combining of a lithium atom (3 electrons) with a fluorine atom (9 electrons).
12. A student needs 200 ml of 0.1 M NaOH solution. What weight of solid is used?
13. What is the molecular weight (MW) of acetic acid (CH_3COOH)?
14. Which of the following would you expect to dissolve readily in water (explain): (a) KCl, (b) CCl_4, (c) N_2, (d) CH_3COOH?
15. What volume of 0.4 M hydrochloric acid will neutralize 200 ml of 0.2 M NaOH solution?
16. Predict whether the pH of the resulting solution will be above or below 7 when each of the following is dissolved in water: (a) HBr, (b) $NaHCO_3$, (c) C_6H_5COOH, (d) CH_3NH_2, (e) NaCl.
17. Convert 98.6°F into °C.
18. What is the pH of a 0.001 M solution of hydrogen ions?
19. Write an electron-dot formula for carbon tetrachloride (CCl_4).
20. When methane (CH_4) burns in oxygen, carbon dioxide and water are the products. Write a balanced equation to represent this chemical change. What is the free energy change (ΔF) of the reaction? Is ΔF positive or negative? What substance has been oxidized? What substance has been reduced?

REFERENCES

1. BAKER, J. J. W., and G. E. ALLEN, *Matter, Energy, and Life: An Introduction for Biology Students,* 3rd ed., Addison-Wesley, Reading, Mass., 1974. A small book that covers topics in chemistry and physics, selected and developed with the needs of biology students in mind.

2. SPEAKMAN, J. C., *Molecules,* McGraw-Hill, New York, 1966. Succinct treatments of atomic structure, valence, chemical bonds, and bond energies. A paperback.

3. ALLEN, T. L., and R. M. KEEFER, *Chemistry: Experiment and Theory,* Harper & Row, New York, 1974. An introductory textbook.

4 THE MOLECULES OF LIFE

MOLECULAR COMPOSITION OF THE HUMAN INFANT	
WATER	66.0%
PROTEINS + NUCLEIC ACIDS	16.0%
FATS AND LIPIDS	12.5%
ASH (MINERALS)	5.0%
CARBOHYDRATES	0.5%
VITAMINS	TRACE

FIG. 4.1 The organic molecules are the proteins, nucleic acids, fats and other lipids, carbohydrates, and vitamins. The others are inorganic.

The one element which, more than any other, is characteristic of living matter is carbon. Carbon is uniquely suited to its role because (1) it can bond covalently to four other atoms, that is, it has a valence of 4, and (2) it can bond covalently to other *carbon* atoms to form chains. These two properties enable carbon to serve as the "backbone" for the formation of an almost limitless variety of molecules. Such variety is essential for such an incredibly complex organization of matter as a living thing. It is no wonder, then, that molecules containing carbon form the very basis of life itself (Fig. 4.1). In fact, the chemistry of carbon compounds is called **organic chemistry** because these compounds are almost exclusively associated with living things.

4.1 HYDROCARBONS

The simplest organic molecules are the hydrocarbons. As the name suggests, these compounds are composed solely of carbon and hydrogen. Methane, CH_4, is the simplest hydrocarbon. But thanks to the ability of carbon atoms to bond to one another, hydrocarbons can exist as chains. Octane, to give a common example (because octane is an ingredient in gasoline) has the structural formula:

```
    H  H  H  H  H  H  H  H
    |  |  |  |  |  |  |  |
H—C—C—C—C—C—C—C—C—H
    |  |  |  |  |  |  |  |
    H  H  H  H  H  H  H  H
```

Hydrocarbons can also exist as rings. Benzene is the commonest example. It has the structural formula:

```
        H
        |
        C
      //  \
  H—C      C—H
     |      ||
  H—C      C—H
      \\  /
        C
        |
        H
```

Note that carbon atoms can join together by double, as well as single, covalent bonds. We shall have a number of occasions to deal with ring compounds. To make their formulas easier to interpret, we shall often show the ring of carbon atoms as a simple geometrical shape, e.g. a hexagon for benzene. Thus, the benzene ring may be depicted as:

4.2 LIPIDS

Lipids are organic compounds that are quite hydrophobic, that is, they dissolve poorly if at all in water. A variety of lipids are found in living cells, but we shall concentrate our attention on three groups: the fats, the phospholipids, and the steroids.

The fat molecule is made up of four parts: a molecule of **glycerol** and three molecules of **fatty acids** (Fig. 4.2). Each fatty acid consists of a hydrocarbon chain with a carboxyl group at one end. The glycerol molecule has three hydroxyl (—OH) groups, and each of these is capable of interacting with the carboxyl group of a fatty acid. A molecule of water is removed in the process, and the fatty acid becomes attached to the glycerol molecule (Fig. 4.2). The three fatty acids in a single fat molecule may be all alike or they may be different. They may contain as few as four carbon atoms or as many as 24. Because fatty acids are synthesized from fragments containing two carbon atoms, the number of carbon atoms in the chains is almost always an even number. In animal fats, 16-carbon (palmitic acid) and 18-carbon (stearic acid) fatty acids are the most common.

Some fatty acids have one or more double bonds in the molecule (Fig. 4.3). Fats formed from these molecules are described as *unsaturated*. This is because the fatty acids could hold more hydrogen atoms than they do. Unsaturated fats have lower melting points than do saturated fats. Because most of them are liquid at room temperature, we call them oils. Cottonseed oil, peanut oil, sesame seed oil, and olive oil are all common kitchen ingredients. As the

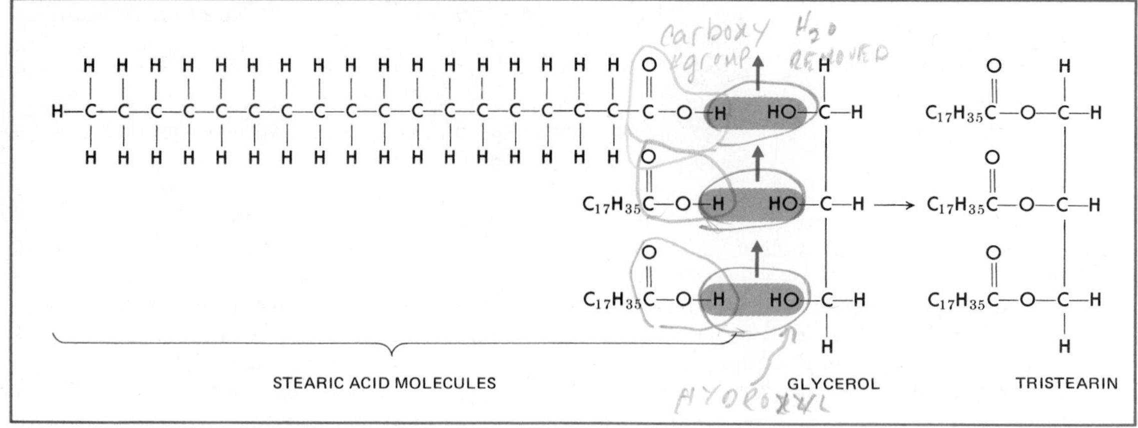

FIG. 4.2 Synthesis of tristearin, a fat.

list suggests, plant fats tend to be unsaturated (therefore "oils") while animal fats tend to be saturated. There is a growing body of evidence suggesting that a diet rich in saturated fat is linked to the development of diseases of the heart and major arteries. Unsaturated fats, on the other hand, seem not to have this effect.

The molecular formula of tristearin, a common animal fat, is $C_{57}H_{110}O_6$. The high ratio of hydrogen atoms to oxygen atoms tells us that this molecule is highly *reduced.* The great majority of the covalent bonds are C—C and C—H bonds, with an average bond energy of about 80 kcal and 98 kcal respectively. These are relatively weak bonds, as we would expect from the very small electronegativity difference between the atoms. This means, then, that fats are rich stores of potential energy. In fact, fats and oils provide the most concentrated energy reserve available to the organism. Figure 4.4 shows a Brazil nut burning like a candle, thanks to its rich stores of oil.

The almost complete absence of strong electronegativity differences also means that fat molecules are not at all polar in nature (we describe them as *apolar*). This is why they are hydrophobic and unable to associate with strongly polar water molecules. Thus a mixture of oil and water quickly separates into a layer of oil floating on the water (in a beef stew, for example).

If we treat a fat with a strong base, such as sodium hydroxide (NaOH), the bonds between the fatty acids and glycerol are broken. The hydroxyl groups (—OH) of the NaOH restore the glycerol molecule, and the sodium ion is attracted to the

FIG. 4.3 Structure of trilinolein, an unsaturated fat. There are two double bonds in each of the fatty acid units. Unsaturated fats liquefy at lower temperatures than do saturated fats and hence are called oils.

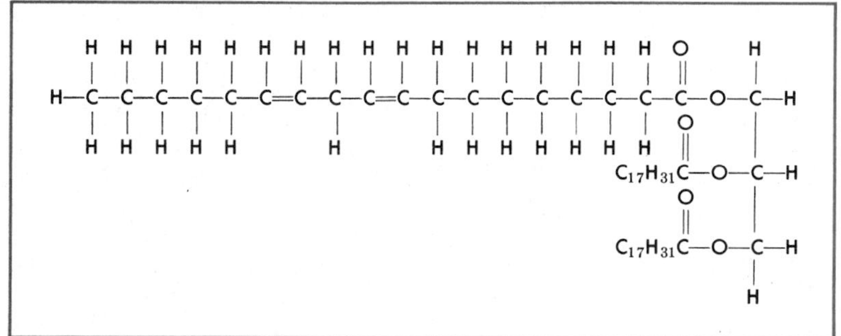

FIG. 4.4 Burning Brazil nut.

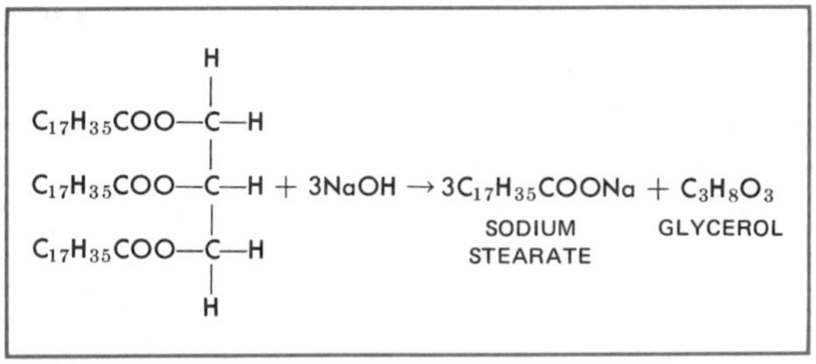

$$
\begin{array}{l}
\quad\quad\quad\quad\quad\ \ H \\
\quad\quad\quad\quad\quad\ \ | \\
C_{17}H_{35}COO{-}C{-}H \\
\quad\quad\quad\quad\quad\ \ | \\
C_{17}H_{35}COO{-}C{-}H + 3NaOH \rightarrow \underset{\text{SODIUM STEARATE}}{3C_{17}H_{35}COONa} + \underset{\text{GLYCEROL}}{C_3H_8O_3} \\
\quad\quad\quad\quad\quad\ \ | \\
C_{17}H_{35}COO{-}C{-}H \\
\quad\quad\quad\quad\quad\ \ | \\
\quad\quad\quad\quad\quad\ \ H
\end{array}
$$

FIG. 4.5 Formation of sodium stearate—a soap—and glycerol from tristearin and sodium hydroxide. The hydrocarbon chain of the soap molecule is hydrophobic; the end with the carboxyl group is hydrophilic.

carboxyl group of each fatty acid. Starting with a molecule of tristearin, for example, we will have synthesized three molecules of sodium stearate as well as one molecule of glycerol (Fig. 4.5). Sodium stearate is soap. And, in fact, preindustrial societies used just the process we have described to make their soap from two accumulating household products: fat and wood ashes (a source of NaOH).

The function of a soap, like that of any detergent, is to make it possible to mix oily materials with water. Looking at the structure of sodium stearate, we can see why it is able to function in this way. One end of the molecule is a hydrocarbon, which means it is hydrophobic and able to mix easily with other hydrophobic materials like oils. The other end is charged and thus polar. It is hydrophilic, interacting with water molecules. The result is an *amphipathic* molecule: hydrophobic at one end, hydrophilic at the other. Mixed with water, it cannot form a true solution but instead forms droplets (Fig. 4.6) with the hydrophilic heads of the soap molecules exposed to the water and the hydrophobic tails in the interior mixed with other hydrophobic materials (like grease). Such droplets constitute an emulsion.

Amphipathic lipids are a major constituent of cells. In the most common of these substances, a polar group is substituted for one of the three fatty acids in a fat molecule. In many cases, this polar substituent includes a phosphate group (PO_4) and thus these compounds are known as **phospholipids.** The most abundant phospholipid in many cells is phosphatidyl ethanolamine or cephalin (Fig. 4.7). We shall see in the next chapter that this phospholipid is a major constituent of the various membranes used in the construction of cells.

The mammalian liver synthesizes and secretes another group of amphipathic lipids, the **bile salts.** These lipids are quite different in structure from the fats and phospholipids. However, their amphipathic nature—hydrophobic at one region of the molecule,

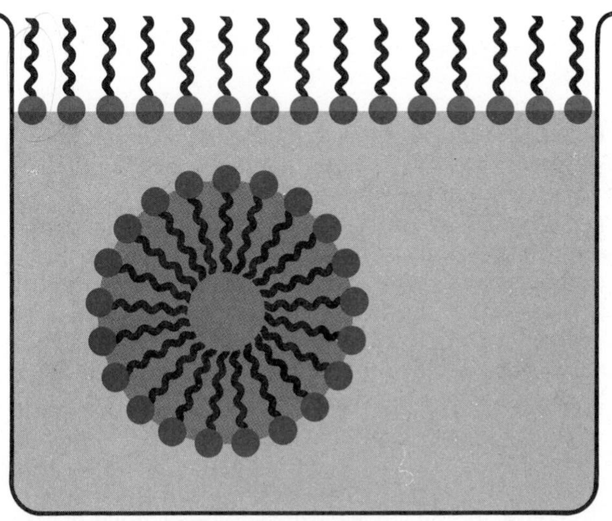

FIG. 4.6 Two ways in which detergent molecules like soap interact with water. At the surface of the water, the soap forms a film with the hydrophobic tails of its molecules oriented away from the water. A droplet of oil beneath the surface is coated with soap molecules oriented with their hydrophilic heads pointed toward the water and their hydrophobic tails extending into the oil droplet. This forms a stable emulsion of oil in water.

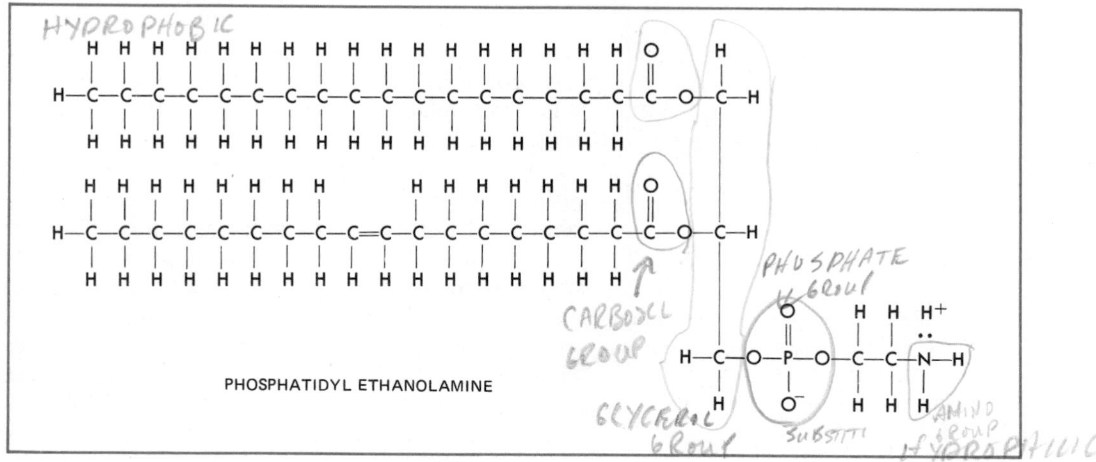

FIG. 4.7 Structure of a common phospholipid, an amphipathic molecule. The hydrocarbon chains are hydrophobic whereas the charges on the phosphate and amino groups make that portion of the molecule strongly hydrophilic. Phosphatidyl ethanolamine is a major constituent of cell membranes.

hydrophilic at another—makes them excellent detergents. They emulsify ingested fats, making far easier the digestion and absorption of fat by the intestine.

Bile salts are members of a large group of closely related lipids called **steroids.** All steroids contain a skeleton of 17 carbon atoms organized into four rings (Fig. 4.8). Each steroid differs from the next in the location and in the nature of attached side groups and, often, in the location of certain double bonds.

FIG. 4.8 Cholesterol, the most abundant steroid in the human body. Like all steroids, it is constructed from a skeleton of 17 carbon atoms organized in four rings (shown in color). Cholesterol serves as the starting material for the synthesis of the various other steroids used by the body.

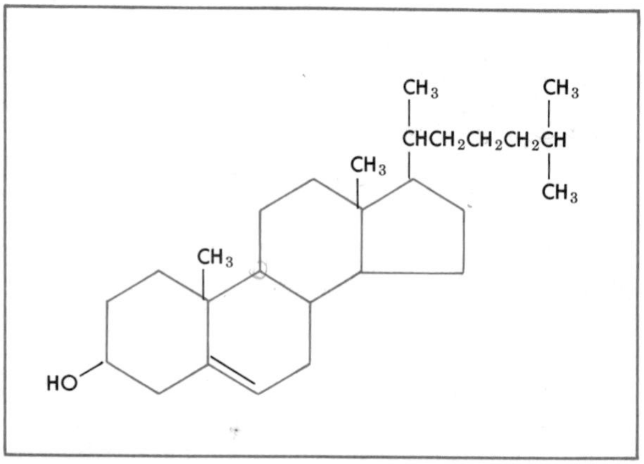

The most abundant steroid in the human is cholesterol (about one-half pound per person). From it are synthesized the bile salts, one of our vitamins (D), and a number of hormones. All the sex hormones, e.g. estrogens, progesterone, and testosterone, are steroids.

4.3 CARBOHYDRATES

Carbohydrates have the general molecular formula CH_2O. Given this expression, you can see why they were thought to represent "hydrated carbon" but, in fact, the arrangement of atoms has little to do with water molecules.

Two common carbohydrates are starch and cellulose. Both of these are enormous molecules with molecular weights in the hundreds of thousands. They are so large that we describe them as **macromolecules.** Each is a polymer. That is to say, both starch and cellulose molecules are constructed from many repeating units much as a chain is constructed from its links. The links in both starch and cellulose are the same: units of a sugar called **glucose.** Glucose molecules are thus the **monomers** out of which each of these polymers is constructed.

Sugars

Glucose has the molecular formula $C_6H_{12}O_6$. Its structural formula is shown in Fig. 4.9. Five of its six carbon atoms and one oxygen atom are bonded

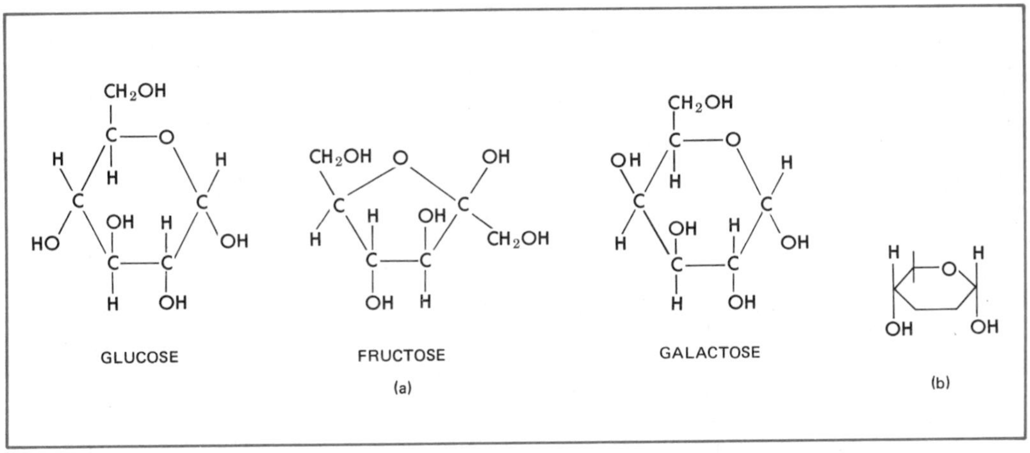

FIG. 4.9 (a) Structural formulas of glucose and two of its isomers. (b) Simplified version of the glucose molecule.

into a ring. The other atoms extend above and below the plane of the ring. Because of its six carbon atoms, glucose is called a hexose. Two other common hexoses are galactose, a sugar found in milk (and yogurt) and fructose, a sugar found in honey. Both galactose and fructose have the molecular formula $C_6H_{12}O_6$, the same as that of glucose. But inspection of Fig. 4.9 will reveal that the actual arrangement of the atoms differs in each case. Galactose differs from glucose in the orientation of a hydrogen atom and hydroxyl group attached to one of the carbons. In fructose, only five atoms participate in the formation of the ring. Substances such as these three, which have identical molecular formulas but different structural formulas, are known as **isomers.**

Two glucose molecules can be linked together with the elimination of a molecule of water between them (see Fig. 4.10). The resulting molecule is **maltose.** It has the molecular formula $C_{12}H_{22}O_{11}$. Because it is constructed from two simple sugar units, maltose is called a **disaccharide.** (Glucose molecules are thus monosaccharides.)

Two other common disaccharides are **sucrose,** the sugar you put in your coffee or on your breakfast cereal, and **lactose,** the major sugar in milk.

Sucrose is constructed from one molecule of glucose and one of fructose (again with the loss of a molecule of water). Lactose contains a molecule of glucose linked to a molecule of galactose.

All sugars are very soluble in water. This is because of the many hydroxyl groups they contain. The polarity established by the highly electronegative oxygen atoms allows for easy attraction to water molecules.

Sugars are an important fuel for many living organisms. They are not as concentrated a fuel as fats because they are not as fully reduced. But their solubility in water enables them to be transported readily in body fluids. Glucose is, in fact, the "blood sugar" of humans. For many cells, it is the most important source of energy.

Starches

Starches are polymers of glucose (Fig. 4.11). These

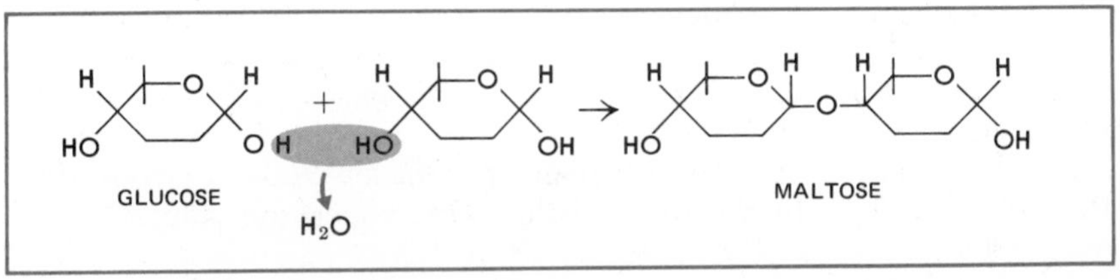

FIG. 4.10 Synthesis of maltose.

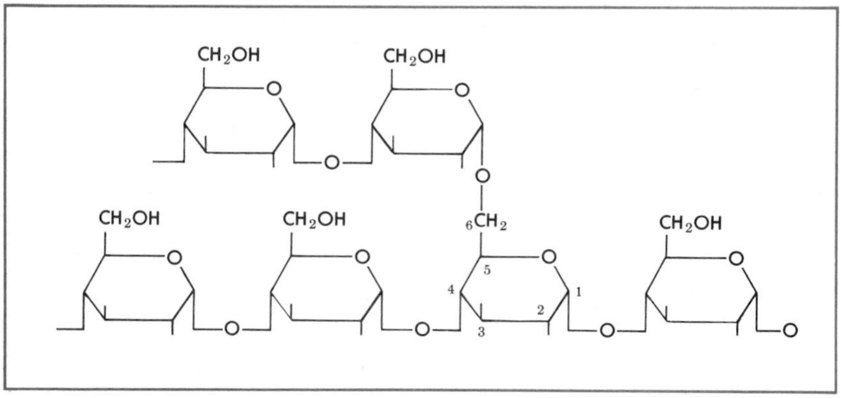

FIG. 4.11 Structure of starch. The glucose monomers are linked through their No. 1 and No. 4 carbon atoms to form straight chains. Short branches are attached to the main chain through carbon atom No. 6.

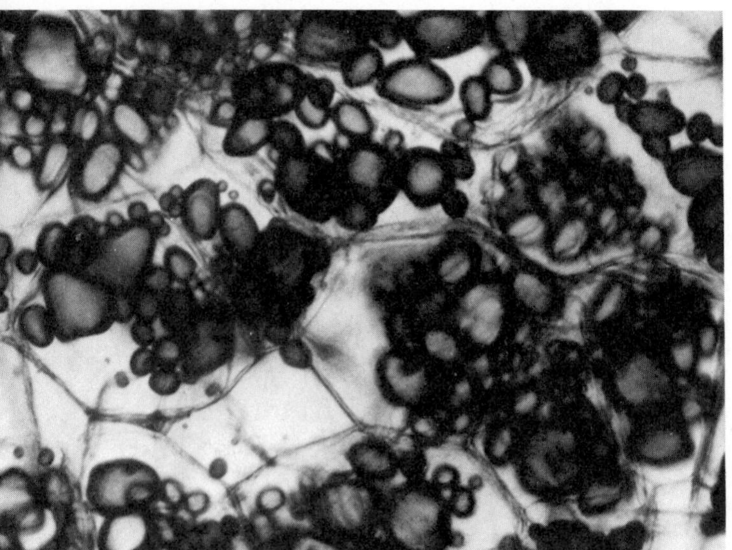

FIG. 4.12 Starch grains in the cells of a white potato. Note the cell walls. The starch grains have been lightly stained with iodine.

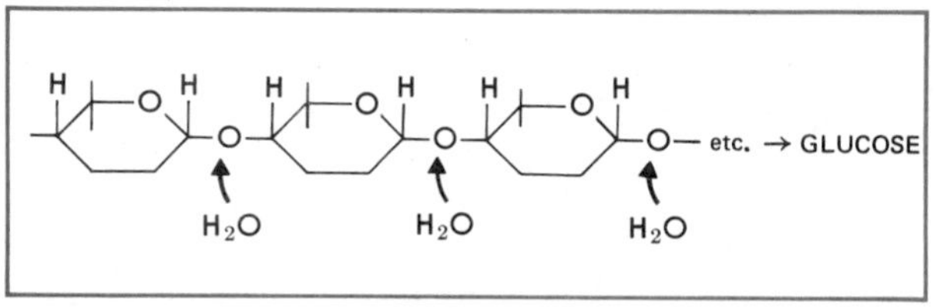

FIG. 4.13 Hydrolysis of starch.

macromolecules are formed both as linear, unbranched chains of several hundred glucose units and as branched chains totaling several thousand glucose units. In the latter case, at about every twelfth glucose residue along the chain of two or three thousand, a short side chain is attached to the No. 6 carbon atom (the carbon atom above the ring). These short chains usually contain about a dozen glucose units each. Because of the many sugar units present, starches are examples of **polysaccharides.**

Starches are insoluble in water and thus can serve as storage depots of glucose. Both plants and animals convert excess glucose into starch for storage. Figure 4.12 shows cells of the white potato, each filled with a number of starch grains. Rice, wheat, and corn are also major sources of starch for human nutrition.

The starch stored in animal cells differs somewhat from plant starch in its properties. For this reason, it is given the special name of **glycogen.** All glycogen molecules are extensively branched; there seem to be no straight chains like those found in plant starch.

Before starches can enter (or leave) cells, they must be digested. This simply means that the long chains are broken back down into their sugar links. This process requires water and starch-digesting enzymes called amylases. With the aid of amylases, water molecules enter at each oxygen "bridge," reforming free sugar molecules (Fig. 4.13). The process of breaking up a molecule with the aid of water is called **hydrolysis.**

Cellulose

Cellulose is probably the single most abundant organic molecule in the biosphere. The reason for its abundance is that it is the major structural material out of which plants are made. Wood is largely cellulose while cotton and paper are almost pure cellulose.

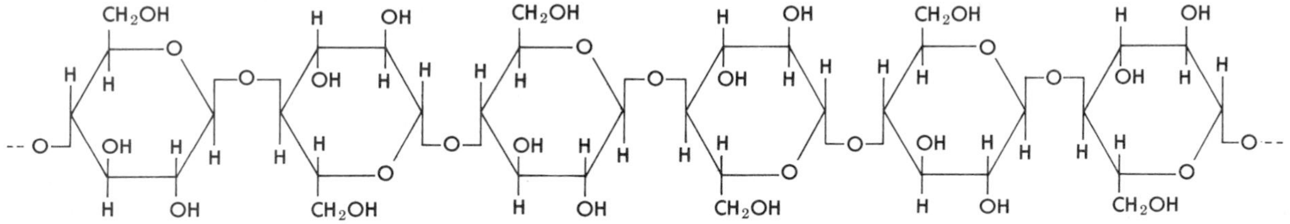

FIG. 4.14 Structure of cellulose. Compare this structure with that of starch (Fig. 4.11).

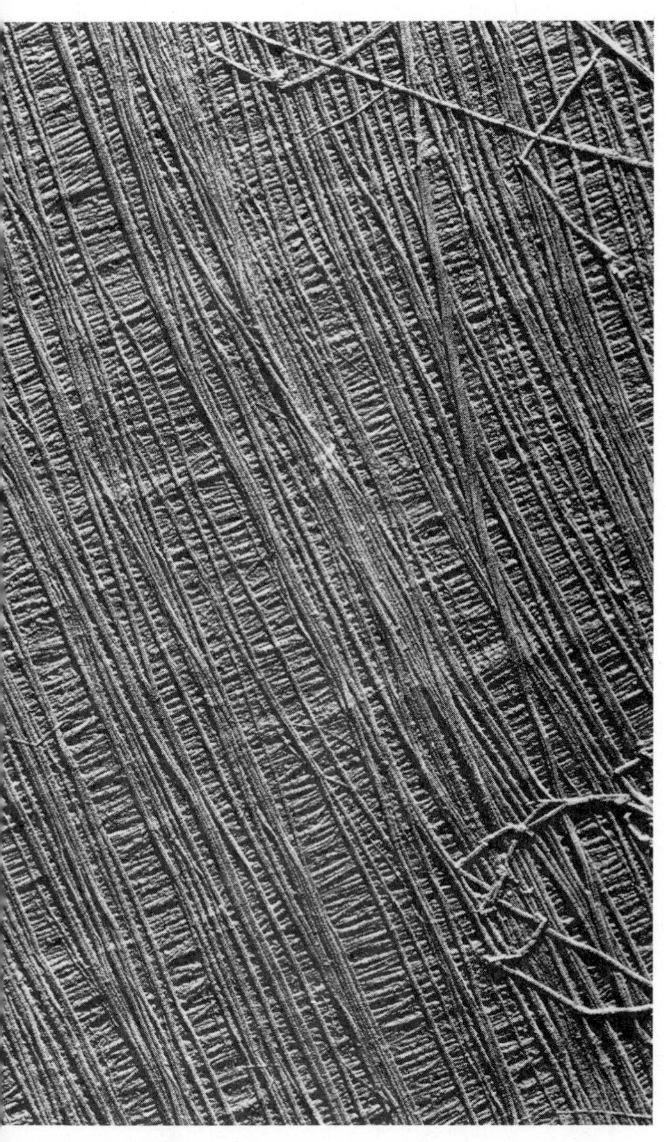

Like starch, cellulose is a polysaccharide with glucose as the monomer (Fig. 4.14). However, the orientation of the bonds linking the No. 1 carbon of one glucose to the No. 4 carbon of the next is different from that found in starch. Whereas in starch, both of the covalent bonds extending from the bridge oxygen atom are directed up to the plane of the rings, in cellulose the bond to the No. 1 carbon is directed down to the plane of the ring. (The bond to the No. 4 carbon is as in starch.) Furthermore, there are no side chains in cellulose as there are in starch. These may seem like small differences, but actually they cause cellulose to differ profoundly from starch in its properties. Because of the orientation of the bonds, the rings of glucose are arranged in a flip-flop manner in the cellulose molecule (Fig. 4.14). This produces a long, rigid molecule. The absence of side chains allows these linear molecules to lie close together. Thanks to the many —OH groups as well as the oxygen atom in the ring, there are many opportunities for hydrogen bonds to form between adjacent chains. The result is a series of stiff, elongated fibrils—the perfect material for building the supporting walls of plant cells. Figure 4.15 is an electron micrograph of cellulose fibrils from the wall of a plant cell. These long, straight, rigid fibrils are a clear reflection of the nature of the cellulose molecules of which they are composed.

4.4 PROTEINS

Approximately 50% of the dry weight of living matter is protein. And protein is not just the simple storage or structural material that the polysaccha-

FIG. 4.15 Fibrils of cellulose in the cell wall of a green alga, 30,000 ×. (Electron micrograph courtesy of R. D. Preston.)

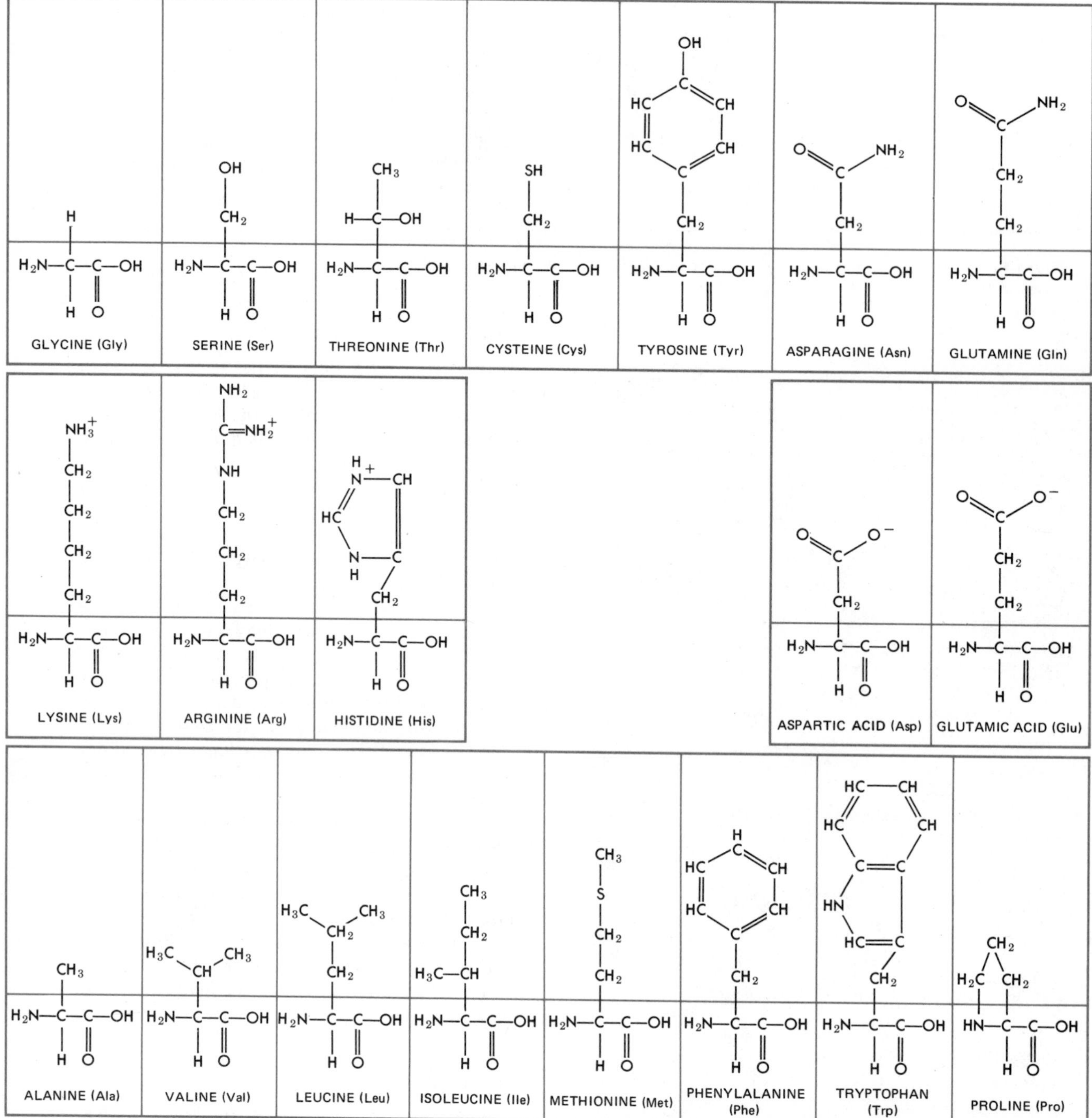

FIG. 4.16 Structures of the 20 amino acids from which proteins are synthesized. Top row: Amino acids with polar R groups. The R groups can hydrogen bond to water molecules, so these amino acids are moderately hydrophilic. Middle row: Basic (left) and acidic (right) amino acids. These are strongly hydrophilic. Bottom row: Amino acids with apolar R groups. These are all hydrophobic.

rides are. The types and functions of proteins are as varied as the functions of life itself. All the thousands of catalysts that make the chemical reactions of living matter possible are proteins. We call these catalysts **enzymes.** Proteins are also the materials responsible for the movement of living organisms. Muscles, for example, are largely composed of precisely ordered protein molecules the interaction of which enables muscular force to be exerted. Proteins are responsible for transporting many materials through the circulatory system. Hemoglobin, which transports O_2 and CO_2 in the blood, is perhaps the most familiar example. The clotting of the blood requires the interaction of a number of different proteins. Antibodies, those extraordinarily varied molecules that can recognize and inactivate virtually any foreign substance that gains access to the body, are proteins. Proteins do also serve as food reserves. Ovalbumin, which is the chief ingredient of egg white, and casein, the major protein in milk, are two common examples. While cellulose is the chief structural material of plants, protein is a major structural material in animals. Collagen and elastin provide the strength and resiliency of connective tissues such as skin and ligaments. Keratin is the major protein in hair and nails.

Proteins are macromolecules. Even the smallest have molecular weights on the order of 6000, and some have molecular weights greater than a million. In addition to the now-familiar carbon, hydrogen, and oxygen, proteins always contain nitrogen atoms and usually sulfur atoms as well. Beta-lactoglobulin, a protein found in milk, has the molecular formula $C_{1642}H_{2652}O_{492}N_{420}S_{18}$. Its molecular weight is 36,684. Faced with a molecule of such a size, have we any hope of ever being able to write a structural formula for it?

Fortunately, all proteins are composed of one or more linear unbranched **polymers.** The monomers of which these polymers are constructed are called **amino acids.** There are 20 different amino acids found in most proteins. These are linked together in chains of, typically, 100–300 amino acids.

Structure of Amino Acids

The 20 amino acids are built on a common plan. Each contains a carbon atom (the "alpha" carbon) to which are attached by covalent bonds:

1. a hydrogen atom,
2. an amino ($-NH_2$) group,
3. a carboxyl group (—COOH),
4. "something else"—this is the "R" group.

It is the nature of the R group that establishes which of the group of 20 a particular amino acid happens to be (Fig. 4.16). The universal presence of an amino group and a carboxyl (acid) group accounts for the name amino acid. Because of its unshared pair of electrons, the amino group is basic: it attracts protons. As we have seen, the carboxyl group is a proton donor and thus an acid.

Figure 4.16 gives the structure of each of the 20 amino acids. In the simplest amino acid, glycine, the R group is simply a hydrogen atom. In alanine, it is a methyl ($-CH_3$) group. Valine, leucine, and isoleucine all have R groups that are simply hydrocarbons. These amino acids are thus quite hydrophobic. Serine and threonine, on the other hand, each have a hydroxyl group as part of their R group. This renders the R group hydrophilic.

Some of the amino acids have charged R groups. Lysine, arginine, and histidine each have an amino group as *part of their R group.* Under normal conditions in the cell, the unshared pair of electrons on the nitrogen atom in these amino groups holds a proton and thus the amino groups carry a positive charge. This charge renders the amino acids hydrophilic—able to interact easily with polar water molecules. Aspartic acid and glutamic acid have carboxyl groups incorporated in their R groups. The carboxyl groups, too, are usually charged (i.e. they have given up their proton) and thus the molecules are rendered extremely hydrophilic.

Cysteine, one of the two sulfur-containing amino acids, deserves special mention. Where two cysteine monomers are in close proximity, they can be oxidized (losing two hydrogen atoms) and bond covalently to each other through the resulting S—S "bridge." In this way, two chains of amino acids (called polypeptides) can be covalently linked (Fig. 4.17). Or, a single chain can be drawn into a loop.

Optical isomers. You remember that the four covalent bonds of a carbon atom are directed toward the apices of a regular tetrahedron. The bonds are thus oriented in three dimensions of space, with an angle of about 109.5° between them. Now let us reexamine the structure of glycine in the light of this fact. We can, of course, make two-dimensional structural formulas of glycine in a variety of ways; for example, we might write any of the following.

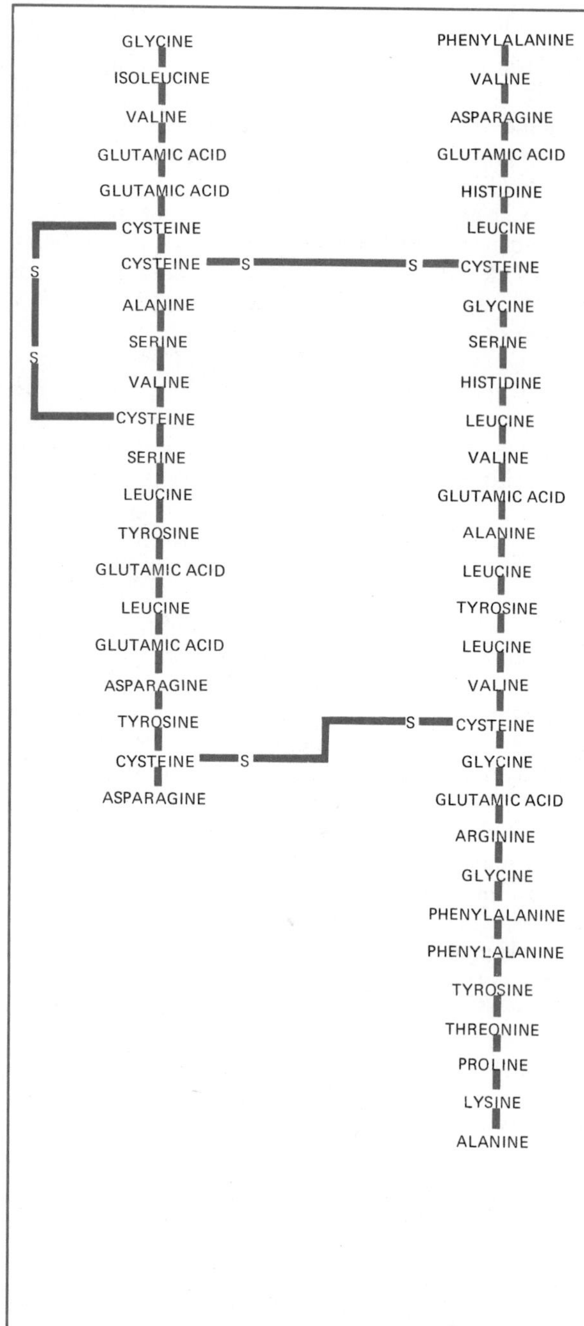

FIG. 4.17 Sequence of amino acids in the insulin molecule. The molecule consists of two polypeptide chains held together by two disulfide bridges.

$$\begin{array}{c} H \\ | \\ H\text{—}C\text{—}COOH, \\ | \\ NH_2 \end{array} \qquad \begin{array}{c} NH_2 \\ | \\ H\text{—}C\text{—}COOH, \\ | \\ H \end{array}$$

$$\begin{array}{c} H \\ | \\ H_2N\text{—}C\text{—}COOH, \text{ etc.} \\ | \\ H \end{array}$$

But these differences have no reality in three dimensions. If we build three-dimensional models of glycine, we find that no matter how many we make and how we hold them in our hands, each can be reoriented to be superimposable on, and thus identical to, the others. Therefore there is only one way to make a molecule of glycine. But not so for alanine! If we start building three-dimensional representations of alanine, we find that we *can* create two distinct forms (Fig. 4.18). If we orient the molecule with the carboxyl group pointing toward us and the amino group away, the R group can be placed either so it projects up to the left and the hydrogen up to the right, or just the reverse. Unlike the situation with glycine, these two forms are not superimposable. They are, instead, mirror images of each other. They have handedness just like the handedness of a pair of gloves. Such molecules are known as enantiomorphs or optical isomers. The second name refers to the property they all have of rotating the plane of polarization of light passing through them. If one form of the molecule should, for example, rotate the light clockwise, its enantiomorph will rotate the light counterclockwise.

Wherever a carbon atom has four *different* things bonded to it, it can form a pair of enantiomorphs. Using an agreed-upon convention, one of these is called the L-form, the other the D. You might well ask whether it is important to life which form is used. The answer is an unequivocal yes. All the amino acids found in proteins, in all the creatures on our planet, are of the L-form. You would starve to death if you tried to fill your amino acid requirements with D-amino acids. To put it another way: the actual orientation of atoms in three dimensions is of crucial importance in many living processes.

Optical isomerism is not restricted to amino acids. If you look back at the structural formulas for the sugars (Fig. 4.9), you will find many carbon atoms that have four different substituents. Thus

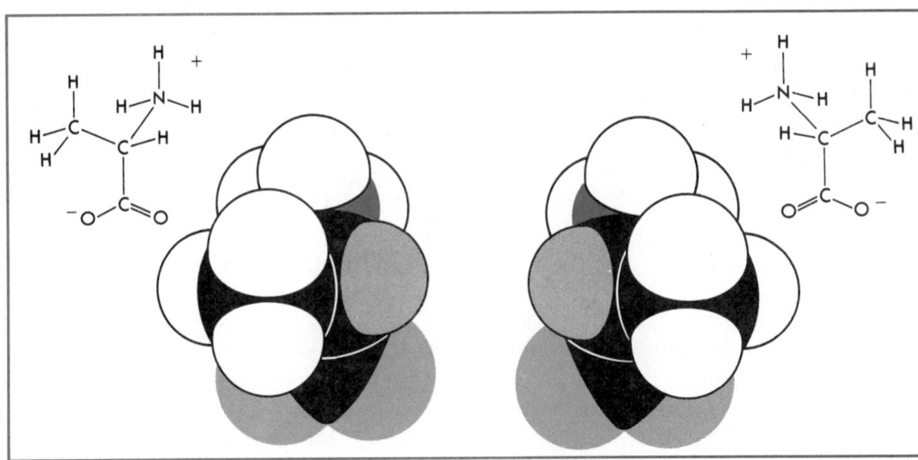

FIG. 4.18 L-alanine and its optical isomer, D-alanine. Only the L isomers of amino acids are found in proteins.

each of these carbons can have its substituents arranged in one of two mirror-image forms. Here again, though, living organisms have definite preferences for one isomer over others.

Structure of Polypeptides

Amino acids are assembled into polypeptides by means of peptide bonds. The carboxyl group of one amino acid interacts with the amino group of the next, with the elimination of a molecule of water from between them (Fig. 4.19). The resulting

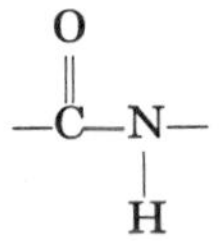

$$
\begin{array}{c} \mathrm{O} \\ \| \\ -\mathrm{C}-\mathrm{N}- \\ \phantom{-\mathrm{C}-}| \\ \phantom{-\mathrm{C}-}\mathrm{H} \end{array}
$$

linkage is called the **peptide bond.** The result is a chain of amino acids or, more precisely, amino acid *residues*. At one end of the chain, there is an exposed amino group. This is called the amino or N-terminal of the polypeptide. The other end, with its exposed carboxyl group, is the carboxyl or C-terminal.

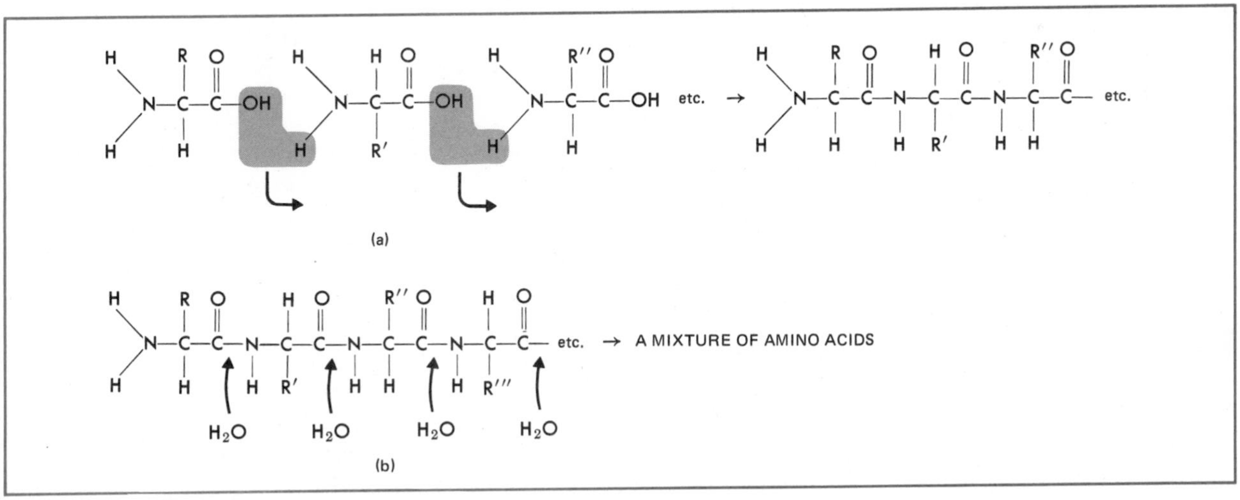

(a)

etc. → A MIXTURE OF AMINO ACIDS

H_2O H_2O H_2O H_2O

(b)

FIG. 4.19 (a) Synthesis of a polypeptide. (b) Hydrolysis of a polypeptide.

Polypeptides can also be disassembled. This occurs by hydrolysis. With the aid of a proteolytic enzyme (protease), a molecule of water is inserted at the peptide bond, breaking the chain at that point (Fig. 4.19). In the human, the digestion (hydrolysis) of ingested protein requires the action of several different kinds of proteolytic enzymes—each capable of cleaving peptide bonds between certain pairs of amino acids but not others.

Primary structure. One of the great achievements of modern chemistry has been the development of techniques for determining the precise sequence in which the amino acids occur in a polypeptide. Only when this is done can a structural formula for the protein be drawn. The first protein to have its sequence of amino acids determined was insulin (Fig. 4.17). This important hormone is small as proteins go. Its molecular formula is $C_{254}H_{377}N_{65}O_{75}S_6$. Nevertheless it took Dr. Frederick Sanger and his colleagues at the University of Cambridge in England ten years (1944–1954) to work out the exact sequence of amino acids and the locations of the crosslinks (S—S bridges between cysteines) in this protein. The protein is constructed of two polypeptide chains held together by disulfide bridges. There are a total of 55 amino acids in the molecule (Fig. 4.17).

Since the days of Sanger's pioneering work, the techniques of protein "sequencing" have developed rapidly. It is now possible, in fact, to have much of the job done automatically by machines. As a result, the sequence of amino acids is now known for a large number of proteins. The sequence of amino acids in a protein, together with the location of any disulfide bridges in that protein, is called the **primary structure** of the protein.

One of the next proteins to have its primary structure worked out was **lysozyme.** Lysozyme is an enzyme found in egg white, tears, and other secretions. It is responsible for breaking down the polysaccharide walls of many kinds of bacteria and thus it provides a measure of protection against infection.

The primary structure of egg white lysozyme is a single polypeptide chain containing 129 amino acids. There are four pairs of cysteines, establishing covalent cross-bridges between positions 6 and 127, 30 and 115, 64 and 80, and 76 and 94 (Fig. 4.20). The existence of these cross-bridges tells us immediately that we cannot accurately represent this polypeptide as a straight rigid chain (like cellulose, for example). The chain must fold on itself to allow these cysteines to be in close proximity.

Secondary structure. To learn more of the actual configuration of the molecule in three-dimensional space, we must turn to another analytical technique, that of X-ray crystallography. A crystal consists of an orderly, stacked array of ions (see Fig. 3.14 in the last chapter) or molecules. If a beam of X-rays is passed through a crystal, some of the X-rays will be reflected by atoms in the crystal. Thanks to the orderly arrangement of the atoms, the reflections will also be orderly. A simple analogy would be the orderly pattern created by a street light at night when viewed through the grid of a window screen. By examining the pattern of reflections produced as the beam is directed at the crystal from a variety of different angles, it is possible (especially with the aid of a computer) to piece together a picture of the arrangement of the atoms in that crystal.

The first proteins to be analyzed in this way were certain structural proteins like the α-keratins of wool, and fibroin, the protein secreted by the silkworm. These proteins were good choices for study because of their very regular, self-repeating structure.

X-ray analysis of α-keratins revealed several interesting facts. The peptide linkage itself turned out to be very rigid, with all its atoms lying in a plane. The only opportunity for flexibility in a polypeptide thus occurs at the bonds to the "alpha" carbon, that is, the carbon that carries the R group. In the α-keratins, this flexibility is exploited by the formation of a helical twist to the polypeptide chain (Fig. 4.21). Several important features of this helix should be noted: (1) The R groups of the amino acids all extend to the outside, (2) the helix makes a complete turn every 3.6 residues, (3) the helix is right-handed; as it recedes, it twists in a clockwise direction, (4) the —C=O group of each peptide bond extends parallel to the axis of the helix and points directly at the —NH group of the peptide bond four amino acids below it in the helix. Both oxygen and nitrogen atoms are quite electronegative, and so a hydrogen bond is formed between these groups: —C=O · · · H—N—. This precise, three-dimensional arrangement of amino acids in the α-keratin polypeptide is called the **alpha helix** (Fig. 4.21). It is an example of the **secondary structure** of a protein.

What makes the alpha helix possible is the large number of hydrogen bonds. A single hydrogen bond has a bond energy on the order of 5 kcal, only a

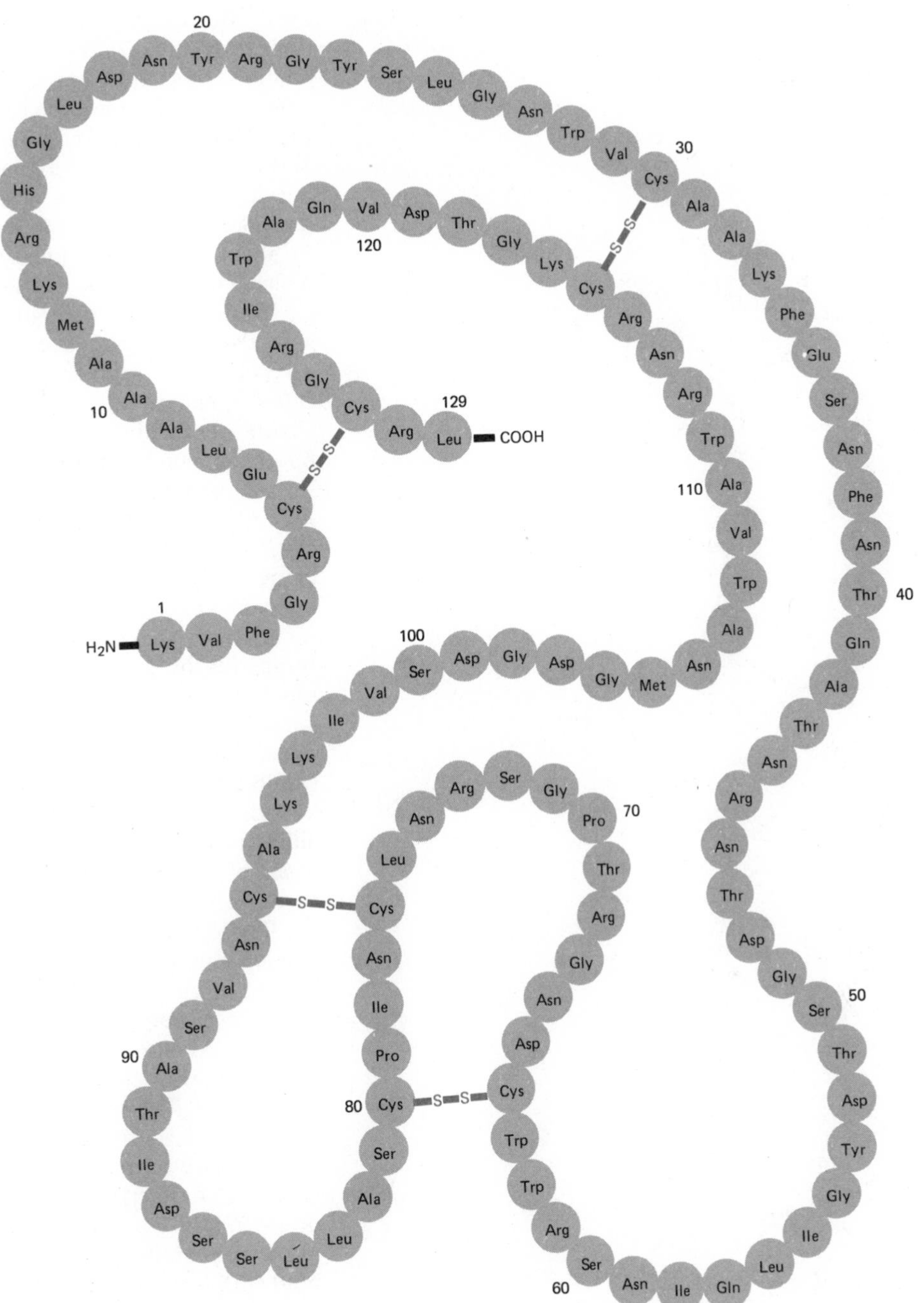

FIG. 4.20 Primary structure of lysozyme. The molecule consists of a single polypeptide chain of 129 amino acid residues. There are four intrachain disulfide bridges.

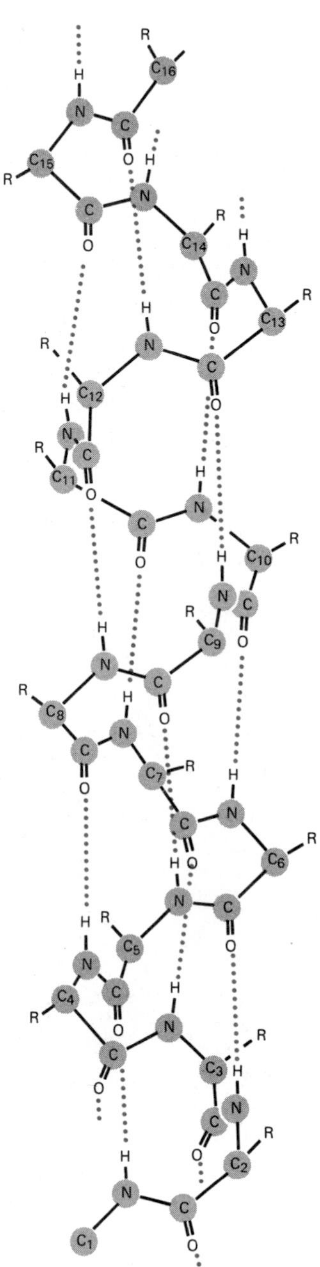

FIG. 4.21 The alpha helix, a common form of secondary structure. This right-handed helix makes a complete turn every 3.6 residues. Note the hydrogen bonds that form between the —C=O groups of each peptide bond and the —N—H group of the fourth peptide bond below it in the helix.

fraction of the bond energy of covalent bonds (~100 kcal). But the presence of a hydrogen bond between each amino acid in the chain and the one that is four amino acids away provides sufficient total bond energy to make the alpha helix a very stable structure. However, it should be possible with a breaking of these intrachain hydrogen bonds to pull the helix out in much the same way that a spring can be extended. And, in fact, wool fibers can be stretched to about twice their normal length, thanks to the molecular properties of the alpha helix.

Fibroin is the protein which silkworms use to spin the threads of their cocoon. It, too, has been subjected to X-ray analysis. In fibroin, the polypeptide chains are extended (Fig. 4.22). A number of chains lie parallel to each other, held together by hydrogen bonds forming between the —C=O and —N—H groups of one chain with the —N—H and —C=O groups of the adjacent chain (Fig. 4.22). In silk, it would be better to describe the chains as being antiparallel, because adjacent chains run in opposite directions, i.e. from N-terminal to C-terminal and vice versa. A series of antiparallel strands lying side by side make up what is called a **beta-pleated sheet.** It is another important and widespread example of the secondary structure of proteins. Because the chains are fully extended, we would not expect to be able to stretch them without breaking covalent bonds. It is for this reason that silk is not extensible. However, the layers of pleated sheet of which the silk fibers are composed make them very supple.

The alpha helix and the beta-pleated sheet are two of the most common examples of secondary structure, but other orderly arrangements of amino acids in a polypeptide chain have also been discovered. When no pattern can be seen along a region of polypeptide chain, the region is said to have a random coil for its secondary structure.

Actually, it is exceedingly doubtful if the three-dimensional configuration of any polypeptide can properly be described as random. All these examples of secondary structure are a direct reflection of the particular amino acids *and their order* that are used in the synthesis of the polypeptide. The chains of fibroin, for example, consist largely of alternating glycine (hydrophilic) and alanine (hydrophobic) residues. This enables the chains to rest close together in the beta-pleated sheet configuration. On the other hand, a chain containing a series of alanine residues will spontaneously assume the alpha helical

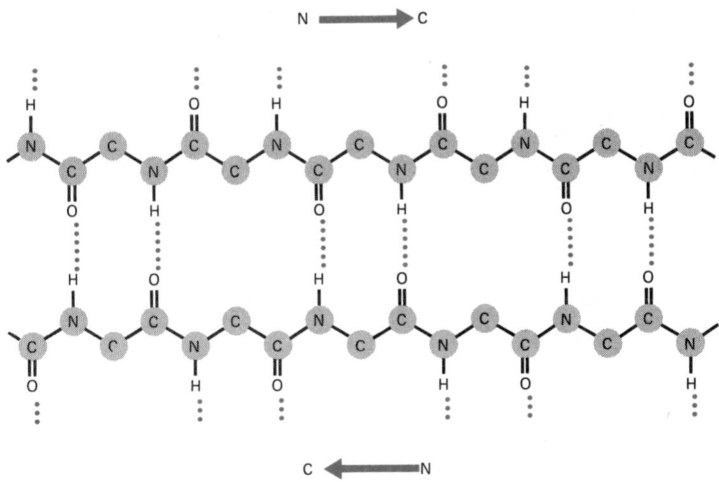

FIG. 4.22 The beta-pleated sheet, another common example of secondary structure. The chains are held parallel to each other by the hydrogen bonds that form between them.

configuration with the $-CH_3$ groups of the alanines projecting to the exterior of the helix.

Tertiary and quaternary structure. The α-keratins and fibroin are examples of fibrous proteins. Fibrous proteins are important in providing the mechanical properties of such structures as skin, horns, hair, etc. Fibrous proteins also play an important role in the contractile machinery of muscles. In all these cases, the proteins are not very soluble in water.

But many proteins (e.g. enzymes, transport proteins) cannot perform their functions unless they are dissolved in water. These proteins tend to be globular, that is, the polypeptide chains of which they are made are folded extensively to form a compact molecule. This makes X-ray analysis difficult but still feasible if the protein can be crystallized. Fortunately, many of them can be. One of the first globular proteins to have its three-dimensional structure determined was myoglobin. Myoglobin is a red, oxygen-binding protein found in muscles. It is responsible for the color of a raw steak.

The myoglobin found in whale muscle consists of a single chain containing 153 amino acids. These are organized into eight regions of alpha helix. At the end of each segment of helix, the chain makes a more or less sharp bend or kink and then another segment of helix begins (Fig. 4.23). The overall result is a globular structure. The *total* conformation of a globular protein is called its **tertiary structure.**

Myoglobin differs in another way from the other proteins that we have discussed up to now.

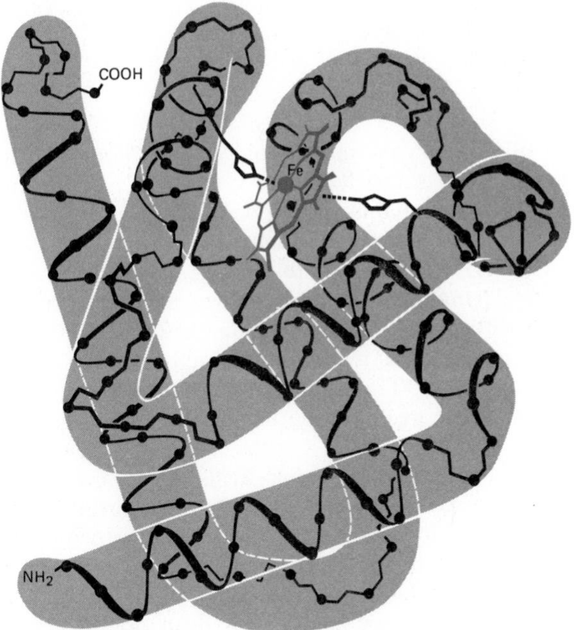

FIG. 4.23 Schematic representation of the tertiary structure of myoglobin. The molecule consists of eight segments of alpha helix bent to form a compact, globular structure. A molecule of oxygen can bind to the prosthetic group, heme (the system of rings at the top center). (Courtesy of R. E. Dickerson.)

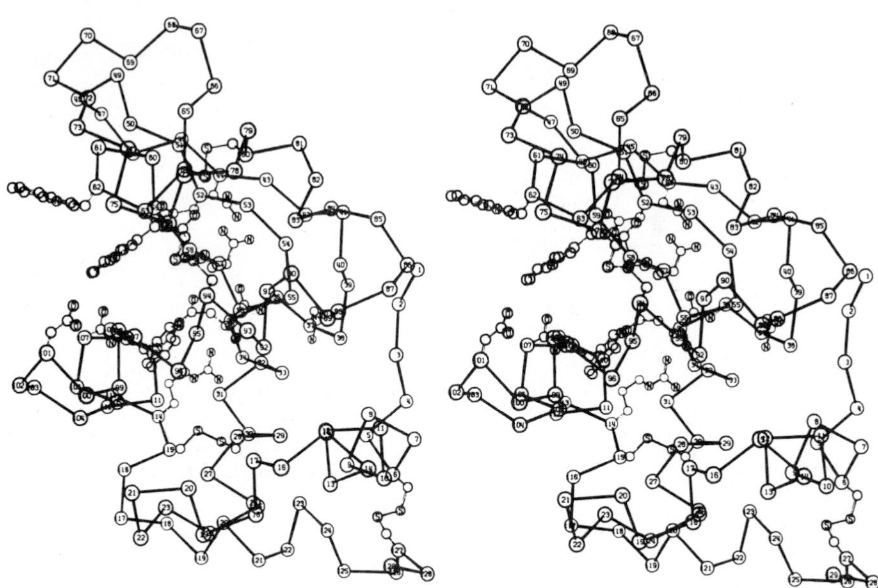

FIG. 4.24 Stereo view of the lysozyme molecule. The easiest way to fuse the stereo images is to use a stereo viewer. But with a little practice, most people can do the job as follows: erect a sheet of stiff paper or cardboard between the two views so that the left eye sees only the left picture and the right eye only the right picture. It helps to have both pictures evenly illuminated and your eyes 18 inches or so away. With continued practice, you may even find that you can fuse the images with no props at all! (Courtesy of Irving Geis.)

Insulin, lysozyme, and fibroin are all "simple" proteins in the sense that they consist of nothing but polypeptide chains. Myoglobin, in contrast, is a **conjugated protein.** Attached to the polypeptide chain is a system of four rings bonded into an overall ring at the center of which is an atom of iron (Fe). This structure (Fig. 4.23) is called the **heme group.** It is the place at which oxygen binds to the molecule. The heme group is one example of a **prosthetic group,** a feature of all conjugated proteins.

The sequence of amino acids in the myoglobin molecule is such that those with hydrophilic R groups are located with their R groups projecting toward the outside of the molecule. It is this that enables myoglobin to exist happily surrounded by water. On the other hand, the hydrophobic R groups project toward the interior of the molecule, and thus water tends to be excluded from the interior of the molecule.

The tertiary structure of **lysozyme** has also been established. The backbone of its 129-amino-acid chain is shown in Fig. 4.24. Actually two slightly differing views are shown. Examining these through a stereo viewer should enable you to fuse the two images into a three-dimensional view of the molecule. In any case, look for the N- and C-terminals of the molecule and for the four disulfide bridges. With the aid of a stereo viewer, you should be able to locate three short stretches of alpha helix. These occur at residues 5–15, 24–34, and 88–96. Lysozyme also has one portion of beta pleated sheet, with residues 41–45 running in one direction and residues 50–54 running in the other (Fig. 4.24).

Two other features of lysozyme should be noted. If you should have the patience to check the actual identity of each of the numbered residues in the stereo view, you would find that (as in myoglobin) those residues located at the surface of the molecule tend to be hydrophilic whereas those tucked in the interior are generally hydrophobic. You should also note the groove or cavity that cuts into this otherwise globular molecule (toward the upper left). As we shall see in Chapter 6, it is here that the polysaccharide upon which lysozyme acts is temporarily bound while the enzymatic action is accomplished.

The stereo views in Fig. 4.24 show only the path

FIG. 4.25 Space-filling model of lysozyme. The cleft at the upper left is where the substrate is bound by this enzyme. (Model by Dr. John A. Rupley; courtesy of Irving Geis.)

taken by the backbone of the polypeptide chain. Only by eliminating almost all of the R groups as well as the atoms of the peptide linkage can we see the overall organization of the molecule. But, in reality, the molecule does not have the empty cage-like structure that Fig. 4.24 implies. If all the atoms are added to the model, the interior of the molecule becomes completely filled. A space-filling model of lysozyme is shown in Fig. 4.25. The tight packing of atoms in the molecule is a general characteristic of globular proteins.

Hemoglobin is a globular, oxygen-binding protein found in the blood. Its red color accounts for the color of blood. It is responsible for the transport of oxygen from lungs (or gills) to all the tissues of the body.

Each hemoglobin molecule consists of four polypeptide chains (Fig. 4.26). Two of these, the alpha chains, are identical. They each contain 141 amino acids. The other two, the beta chains, are also identical. Each beta chain contains 146 amino acids. Each alpha and each beta chain surrounds a hydrophobic interior in which is contained a heme group. The whole molecule thus resembles what we would expect if four myoglobin molecules were brought

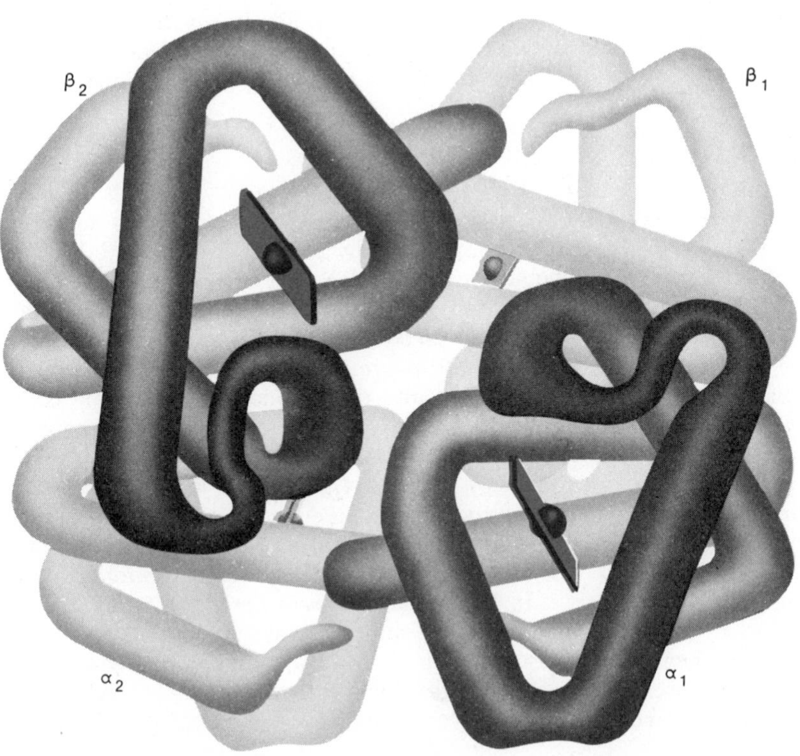

FIG. 4.26 Structure of hemoglobin. The molecule consists of four chains: two alpha chains and two beta chains. Each chain carries a heme to which one molecule of oxygen binds. The assembly of a protein from separate subunits is an example of quaternary structure. (Courtesy of Irving Geis.)

together. And as in myoglobin, the heme groups are the sites at which the oxygen molecules are bound. Thus each hemoglobin molecule can transport four molecules of oxygen.

The four polypeptide chains of the hemoglobin molecule are *not* held together by S—S bridges or any other type of covalent bond. Instead, many weak forces of attraction, including hydrogen bonds, keep the four chains together. The presence in a protein of distinct subunits, like the four chains of hemoglobin, imparts to it a still higher level of organization called its **quaternary stucture.**

Protein Structure and Function: Some Relationships

The function carried out by a protein (except when it is serving as food!) is absolutely dependent upon the three-dimensional structure of that protein. It is possible to expose a protein to one or more substances that will disrupt the secondary, tertiary, and quarternary structure of the protein. For example, a high concentration of ions will break hydrogen bonds. A reducing agent will break the S—S bonds between cysteines. Although these agents do not disrupt or break any of the peptide bonds, and thus do not affect the primary structure of the protein, the treatment nevertheless destroys the functional properties of the protein. The protein has been **denatured.** If we gently denature lysozyme, to take an example, it will lose its ability to attack the polysaccharide upon which it normally acts. Denaturing any enzyme causes it to cease to function. Denaturing antibodies renders them incapable of recognizing and interacting with antigens.

But if function depends upon conformation, upon what does the conformation (i.e. the three-dimensional configuration) of the protein depend? The answer is surely: its primary structure, in other words, the precise sequence of amino acids in the protein. How do we know this? If we take the denatured lysozyme and return it to normal conditions of pH, salt concentration, etc., it will *regain* its characteristic secondary and tertiary structure. And this will be shown by the regaining of its catalytic properties. Similarly, if reversibly denatured antibodies are returned to normal, nondenaturing conditions, they regain their ability to bind to the appropriate antigens.

If the secondary and tertiary structure of a protein is dependent solely on a precise sequence of amino acids, will X-ray crystallography continue to be necessary? If particular sequences of amino acids (e.g. Ala-Gly-Ala) always assume the same orientation in space, could we not predict the total conformation of a protein from a knowledge of its amino acid sequence alone? The orientation that a given amino acid residue will assume is influenced by the residue that precedes it and the one that follows. Theoretically, then, we must be able to specify the angles that will be taken by any of 20 amino acids preceded *and* followed by any of the 20. This requires that we study 8000 possible combinations. In fact, data for well over a thousand of these possible combinations are now available. Attempts have been made to use this information—sometimes with the aid of a computer—to determine secondary and tertiary structure. Where that is already known from X-ray analysis, the two methods have yielded remarkably similar results.

If what a protein does is dependent upon its shape, and if its shape is dependent upon its primary structure, what then determines the sequence of amino acids in its chains? The answer to this question represents one of the greatest scientific triumphs of the twentieth century. The answer is: the precise order of amino acids in a protein is established by the **genetic code.** In Chapters 17 and 18 we shall examine the nature of the genetic code and how it governs the structure of proteins. For now, however, we shall simply say that the genetic code is embodied in and expressed by another group of macromolecules found in all cells: the nucleic acids.

4.5 NUCLEIC ACIDS

Two major kinds of nucleic acids are found in living things: *d*eoxyribo*n*ucleic *a*cid or **DNA** and *r*ibo*n*ucleic *a*cid or **RNA.** Each of these is a linear, unbranched polymer. The molecular weights vary from about 25,000 to as much as 30 million.

Like all polymers, the nucleic acids can be broken down into their monomers. These are called **nucleotides** (Fig. 4.27). Each nucleotide consists of a five-carbon sugar (hence a pentose), to which is attached a phosphate group and a nitrogen-containing ring structure called a base. (They are bases because of unshared pairs of electrons on the nitrogen atoms, which can thus acquire protons.)

In DNA, four different bases are found. Two of these, **adenine** and **guanine,** are double-ring structures called purines. The other two, **thymine** and **cytosine,** are single-ring structures called pyrimidines. For convenience we shall refer to these four

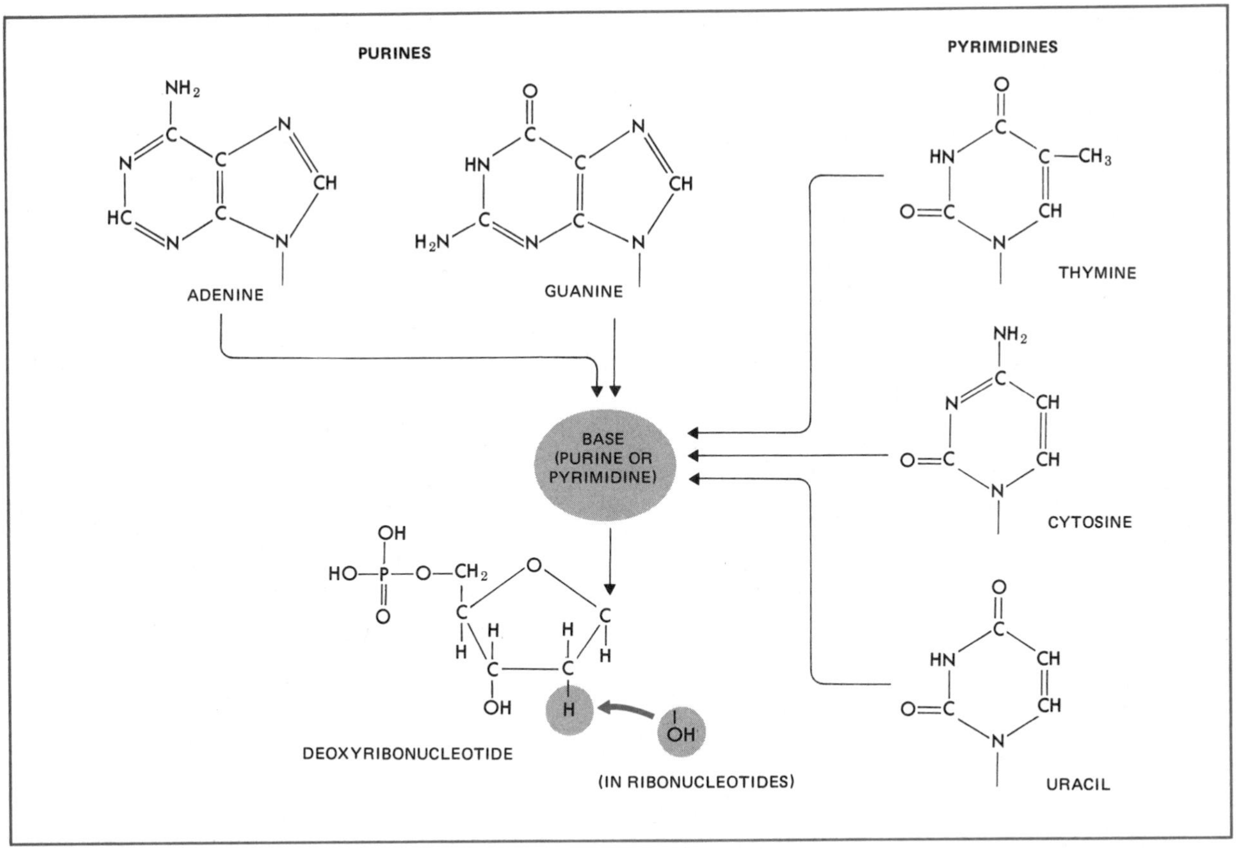

FIG. 4.27 Structures of the nucleotides that serve as the monomers of DNA and RNA.

bases as A, G, T, and C respectively. The sugar in all the nucleotides of DNA is deoxyribose.

Early in the study of DNA, a curious fact became apparent. Although the relative proportions of the four nucleotides varied widely from the DNA of one species to the DNA of the next, whatever the amount of adenine (A) in a given sample of DNA, the amount of thymine (T) was exactly equal to it. Similarly, the amount of cytosine (C) in the DNA of a given species was always equal to the amount of guanine (G). The significance of this finding will become clear when we examine the structure of DNA in Chapter 17.

We should also note that the amount and the base composition of the DNA in any cell (blood, liver, kidney, etc.) of an organism is always the same (except for the gametes—sperm and eggs—which have only half the DNA of other cells). This, too, has far-reaching implications that we shall examine in later chapters.

The pentose sugar in the nucleotides of RNA is ribose. It has one more oxygen atom than deoxyribose (Fig. 4.27). The bases in RNA are chiefly adenine (A), cytosine (C), guanine (G), and the pyrimidine **uracil** (U). There is no thymine in RNA. In contrast to DNA, the amount of any one nucleotide in the molecule is not directly related to the amount of any of the others. RNA molecules are quite heterogeneous in size but, in general, the molecular weights run much smaller than in DNA.

In this chapter and the last, we have examined a hierarchy of structures culminating in the macromolecules so absolutely vital to the structure and function of living things. Now we must attempt to bridge one of the largest gaps in our knowledge of the organization of living matter. This is to determine just how these macromolecules are assembled and organized to build the various structures of the basic unit of life itself: the cell.

EXERCISES AND PROBLEMS

1. Write the structural formula for the neutral fat tripalmitin (palmitic acid = $C_{15}H_{31}COOH$).
2. Write an equation for the conversion of tripalmitin into soap.
3. Distinguish between isomers and isotopes.
4. Distinguish between the primary, secondary, tertiary, and quaternary structure of a protein.
5. How does RNA differ from DNA?
6. Pound for pound, what food yields the most metabolic water when oxidized?
7. Write the structural formula for the tripeptide containing glycine—alanine—phenylalanine.
8. Show by means of structural formulas the hydrolysis of the tripeptide described in Question 7.
9. Which of the following synthetic polypeptides produces the more basic solution: polylysine or polyphenylalanine? Which is more soluble in water?
10. Name an isomer of glucose; of sucrose.

REFERENCES

1. PHELPS, C. F., *Polysaccharides,* Oxford Biology Readers, No. 27, Oxford University Press, Oxford, 1972.
2. PHILLIPS, D. C., and A. C. T. NORTH, *Protein Structure,* Oxford Biology Readers, No. 34, Oxford University Press, Oxford, 1973. Includes a viewer through which many of the illustrations can be seen in three dimensions.
3. DICKERSON, R. E., and I. GEIS, *The Structure and Action of Proteins,* Benjamin, Menlo Park, Calif., 1969. Includes several stereo pairs of models of proteins. A stereo supplement and stereo viewer are also available.
4. THOMPSON, E. O. P., "The Insulin Molecule," *Scientific American,* Offprint No. 42, May, 1955. Describes the techniques used to determine the primary structure of the insulin molecule.
5. KENDREW, J. C., "The Three-Dimensional Structure of a Protein Molecule," *Scientific American,* Offprint No. 121, December, 1961. How X-ray crystallography was used to determine the tertiary structure of the myoglobin molecule.

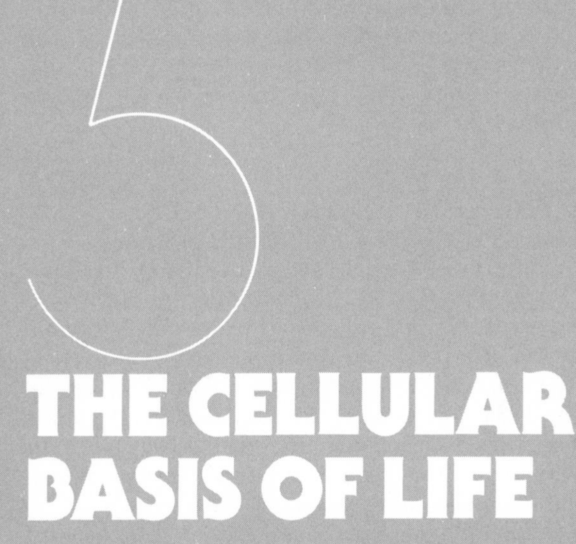

5 THE CELLULAR BASIS OF LIFE

In the last chapter we examined the structure of some of the macromolecules so characteristic of living matter. But cellulose, starch, proteins, even nucleic acids are not alive. Only when these (and other) molecules are organized into the complex interacting systems that make up a cell do we cross the boundary that separates the nonliving from the living. The cell is the minimum unit of life. Our goal in this chapter is to examine some of the structural features of cells. In so doing we shall attempt to narrow what is perhaps one of the broadest gaps in biological knowledge: how the visible structures of the cell are constructed from the macromolecules whose structures and properties we have already examined.

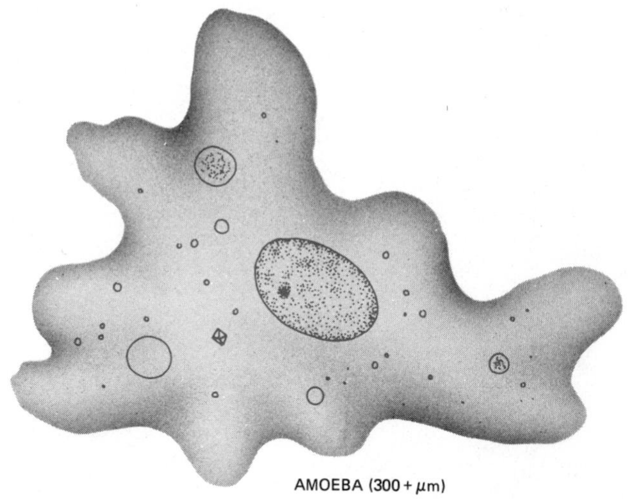

FIG. 5.1 Comparative sizes of various cells. The amoeba is just visible to the unaided eye. Mycoplasmas can be seen clearly only under the electron microscope.

5.1 THE CELL AS THE UNIT OF STRUCTURE OF LIVING THINGS

In many cases, a single cell *is* the living organism. The amoeba (Fig. 5.1), a denizen of fresh water, is such a case. It is a cell about 300 μm across, about the size of a period on this page. Within this package, it houses the machinery with which it carries out all the functions of life. It feeds itself and in other ways exchanges matter and energy with its environment. It responds to stimuli in its environment. It grows and reproduces.

The staphylococcus is a spherical cell about 1 μm in diameter (Fig. 5.1). It is far too small to be seen without the aid of the high-magnification of a microscope. Nevertheless, it, too, contains within its single cell all the machinery of life. Smaller still are the mycoplasmas. These single-celled creatures are so small that they can be seen only with the aid of an electron microscope. But they are still bona fide living organisms, showing all the attributes of life. They are, in fact, the smallest cells and the smallest living things known. Indeed, it would probably be impossible to assemble and organize enough molecular machinery to become alive in a package much smaller than a mycoplasma.

Most of the creatures on this planet are built from many cells. The adult human contains some 6×10^{13} different cells. These occur in a moderately large number (about 100) of different kinds. Two easily studied examples are the epithelial cells that line the inner surface of the cheek and the red cells of the blood (Fig. 5.1). Other examples will be mentioned at the end of the chapter. These cells differ from the cells of singled-celled organisms like the amoeba in one important respect: they are not capable of living an independent existence, i.e., they cannot carry out all the functions of the living organism. Each cell is undeniably alive, but each is specialized to carry out one or possibly a few functions for the organism of which it is a part. Thus each cell is dependent on others to carry out the functions that it cannot perform itself. A nerve cell, for example, rapidly transmits electrical signals in the body but is totally dependent on red blood cells to bring it the oxygen without which it would die.

Despite the diversity of cell types, we find certain architectural and functional features that are common to most (although not necessarily all) cells. Our chief purpose in this chapter is to examine the most prominent of these features.

5.2 THE CELL MEMBRANE

One universal characteristic of cells is an outer limiting membrane. This **cell membrane** serves as the interface between the machinery in the interior of the cell and the watery fluid that bathes all cells. The cell membrane is so thin that it can be visualized only under the high magnification provided by the electron microscope. Figure 5.2 is an electron micrograph showing a slice through the membranes of two adjacent cells. Each membrane is about 100 A thick. Close inspection reveals that the membrane appears to be made of three layers, seen as two dark lines separated by a clear space.

Another way to study the cell membrane is to isolate it from the rest of the cell and examine the molecules of which it is composed. The red blood cell is a convenient source of purified membrane preparations. Chemical analysis reveals that the membrane contains roughly 50% lipid and 50% protein.

The lipids are chiefly phospholipids (like phosphatidyl ethanolamine—look back at Fig. 4.7) and cholesterol. Phospholipids, you remember, are amphipathic molecules, that is, each molecule contains a hydrophilic "head" and a hydrophobic "tail." We have already seen two ways in which these molecules orient themselves when exposed in one direction to water (Fig. 4.6). But the cell membrane is in contact with water on *both* of its sides. The interior of the cell is watery and so is the fluid that bathes the cell. Thus the cell membrane serves as a barrier or interface between two watery phases. How could amphipathic phospholipids accommodate to such a requirement? When a drop of phospholipid is drawn out underwater, the molecules orient themselves in pairs, tail to tail. In this way, their hydrophilic heads face out into the watery medium on each side, while their hydrophobic tails are hidden from contact with water (Fig. 5.3). A **phospholipid bilayer** has formed which does, in fact, possess many of the properties of cell membranes. For example, it permits water molecules to pass through from side to side with relative ease, but is a barrier to the easy passage of ions such as Na^+, K^+, and Cl^-. Such a membrane, which allows certain molecules and ions to pass but not others, is described as differentially permeable. Cell membranes, as well as these artificial membranes, are differentially permeable. The artificial phospholipid bilayer we have constructed is about 45 A thick.

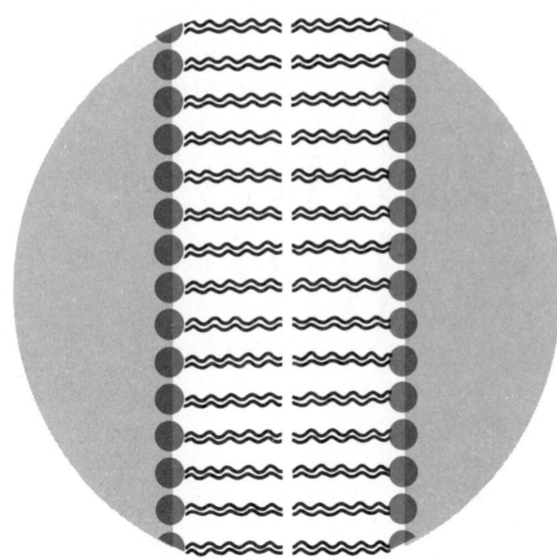

FIG. 5.3 A phospholipid bilayer. This structure establishes a stable barrier between two watery solutions.

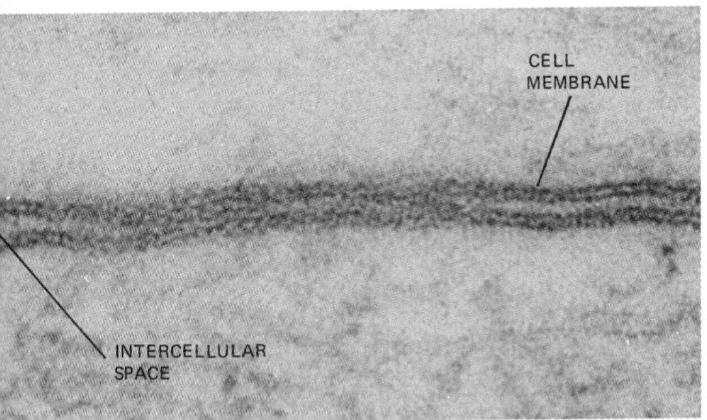

FIG. 5.2 Cell membranes (170,000 ×). The membranes of *two* adjoining cells are shown here. (From Fawcett: *The Cell: Its Organelles and Inclusions,* W. B. Saunders Co., 1966.)

What about the proteins found associated with membranes? Several points now seem well established. First, some membrane proteins are much more easily separated than others from the lipids of the membrane. The proteins that are easily removed from the lipids tend to be quite hydrophilic because of a high content of hydrophilic amino acid residues. Furthermore, some of these proteins contain cova-

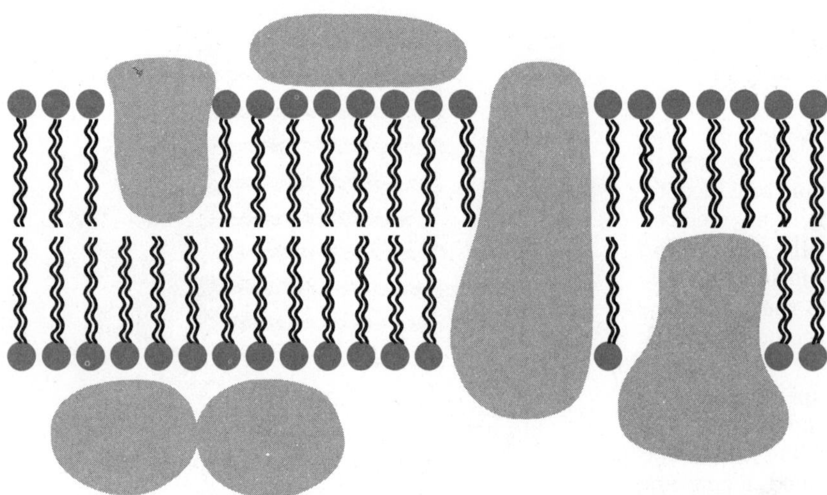

FIG. 5.4 Structure of cell membranes. The globular objects represent proteins adhering to or penetrating into the phospholipid bilayer. The exposed portions of the proteins tend to be hydrophilic while those portions submerged in the phospholipid bilayer tend to be strongly hydrophobic.

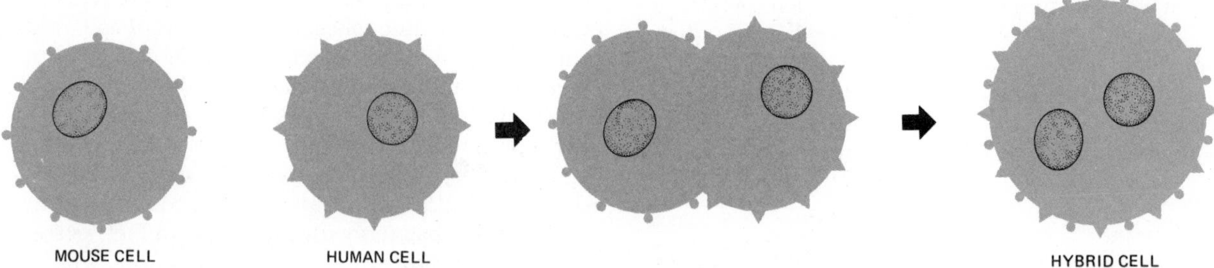

FIG. 5.5 Cell fusion. The speed with which the characteristic cell surface proteins of the mouse and the human become thoroughly intermixed is vivid evidence of the fluidity of the cell membrane.

lently bound sugars and short polysaccharides. The sugars, with their high content of hydroxyl groups, also contribute to the hydrophilic nature of these **glycoproteins.** These easily separated, hydrophilic proteins are described as **extrinsic proteins.** They occur both on the outer and inner surfaces of the cell membrane (Fig. 5.4). (Glycoproteins are especially apt to occur on the outer surface.) The extrinsic proteins probably do not penetrate into the phospholipid bilayer of the membrane.

Other proteins do penetrate into and, in some cases, all the way through the lipid bilayer. These are the **intrinsic proteins** (Fig. 5.4). The portion of the protein molecule that resides within the lipid bilayer is very hydrophobic, thanks to a high content of hydrophobic amino acids. However, those portions of the polypeptide chain(s) that project out from the lipid bilayer tend to have a high percentage of hydrophilic amino acids.

The cell membrane is not symmetrical. Those extrinsic proteins that are associated with the *outer* surface of the membrane are usually quite different from the extrinsic proteins associated with the inner surface.

A lipid bilayer is essentially a film of oil. As such, we might expect that structures immersed in it would be relatively free to float about in it. Such seems to be the case. Laboratory techniques now exist that make it possible to fuse the membranes of two different kinds of cells. When this is done, the proteins characteristic of each cell quickly become intermixed with each other (Fig. 5.5). Another observation that supports the idea that membrane proteins are relatively free to move laterally through the lipid bilayer is the phenomenon of "capping." It is possible to attach "markers"—certain molecules that are easily made visible—to the membrane proteins on certain kinds of cells. The "markers" can be fluorescent or radioactive molecules. At first, the molecules are found evenly distributed across the surface of the cell. Soon, however, they migrate to one portion of the cell. Presumably, they are carried there by the motion of the membrane proteins to which they are attached.

Membranes are also found within the interior of almost all kinds of cells. Under the electron microscope, these internal membranes have the same (or at least a similar) triple-layer appearance as the cell membrane. They, too, contain lipids and proteins. However, the particular lipids are apt to be, and the proteins are sure to be, different from those found in the cell membrane. These differences in composition reflect the special functions that these internal membranes perform. One of these functions is to establish a variety of membrane-bounded spaces or compartments within the cell. The most obvious of these compartments is the one we call the nucleus. In fact, the nucleus is such a striking feature of most cells that everything else in the cell is often simply lumped together under the term "cytoplasm."

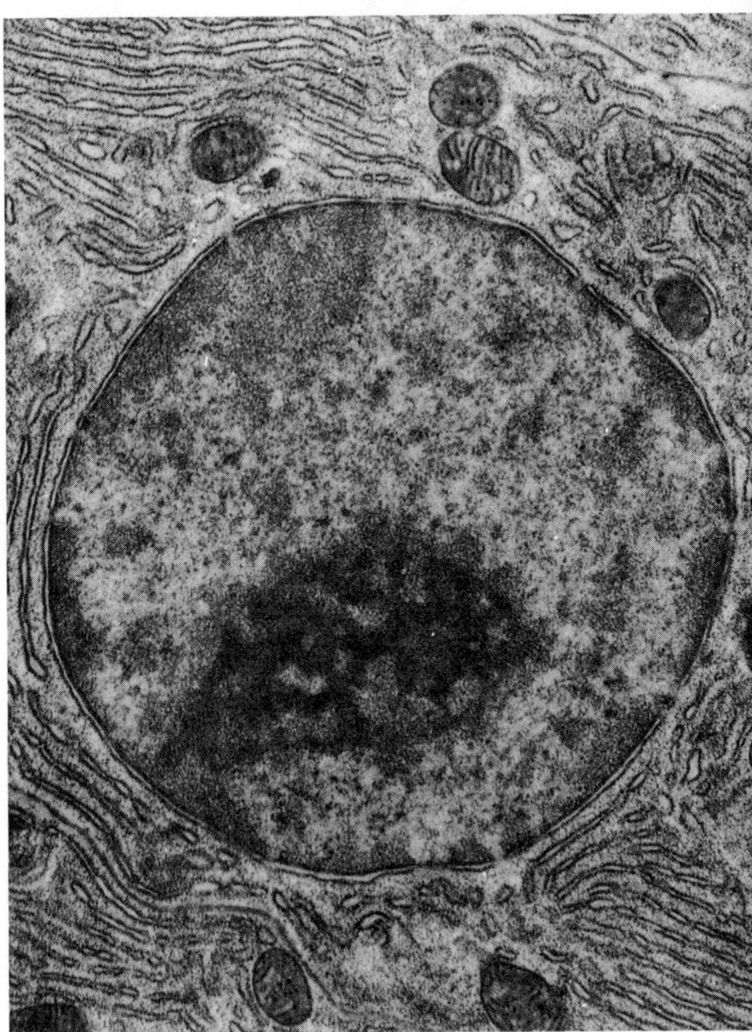

FIG. 5.6 The nucleus and surrounding cytoplasm of a cell from the pancreas of a bat (18,000 ×). Note the double-layer construction of the nuclear membrane and the pores in it. What other cell structures can you identify? (Electron micrograph courtesy of Dr. Don W. Fawcett.)

5.3 THE NUCLEUS

The nucleus is bounded by a pair of membranes. The envelope thus formed is not continuous but, as can be seen in Fig. 5.6, contains pores. These probably permit materials to pass into and out of the nucleus.

Within the nuclear membrane there is a semifluid medium in which are suspended the **chromosomes.** Usually these are present as very elongated structures and cannot be easily observed under the light microscope. The term *chromatin* is used to describe them when they are in this condition.

When a cell is preparing to divide into two cells, the appearance of the chromosomes changes. The long thin strands coil up into thickened, dense bodies, which (with the help of an appropriate stain) are easily visible in the light microscope (Fig. 5.7). During the process of cell division, the chromosomes are distributed in precisely equal numbers to the two daughter cells.

Chemically, the chromosomes are made up of DNA and proteins. It is not yet absolutely certain whether a single chromosome contains only a single DNA molecule or many of them, but the evidence favors the former.

The major proteins associated with the chromosomes are the histones. Histones are basic proteins. This is because they are rich in such amino acids as

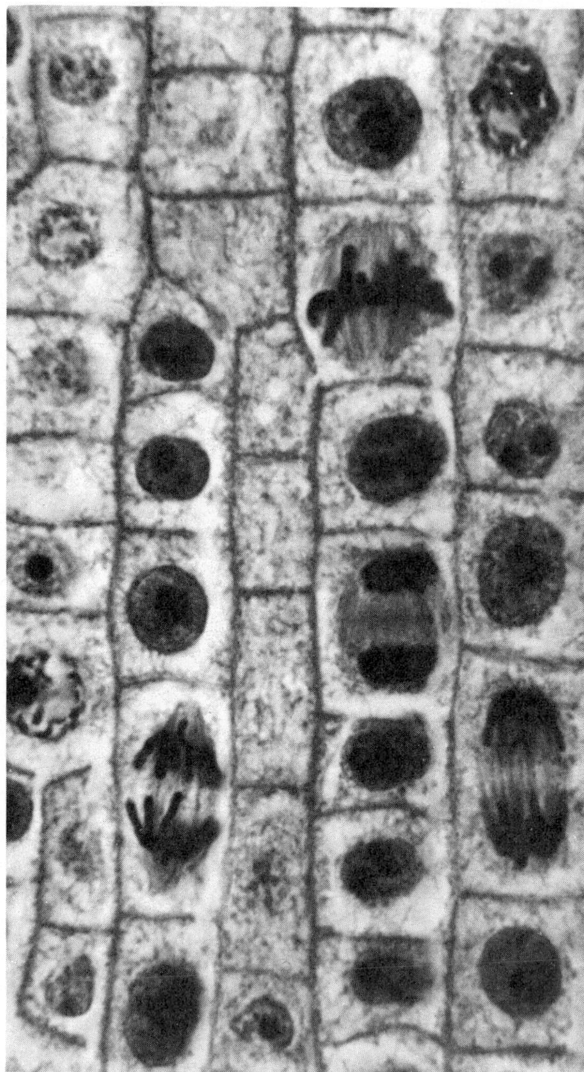

FIG. 5.7 Dividing cells in the root of an onion. The dark, horseshoe-shaped structures are the chromosomes. In most cells, they can be made visible only at the time of cell division, a period called mitosis. (Courtesy CCM: General Biological, Inc., Chicago.)

lysine and arginine. Each of these amino acids has a free amino group, which is able to acquire a proton at its unshared pair of electrons. Thus the histones are positively charged. DNA, because of its many phosphate groups, is negatively charged. Thus it should not be surprising that histones bind very tightly with DNA. There are five kinds of histones found in chromosomes.

When chromatin is prepared in a certain way, the electron microscope reveals long strands containing regularly spaced swellings, giving the appearance of beads on a string. The beads are called nucleosomes (Fig. 5.8). The connecting strand is DNA. Each nucleosome contains both DNA and four of the five kinds of histones.

Other proteins are also found associated with the DNA of the chromatin, although in smaller amounts than the histones. These other proteins are sometimes described as the acidic proteins, but "nonhistone" is probably a more accurate description. The functional interrelations among DNA, histone, and nonhistone proteins will be explored in a later chapter.

During the period between cell divisions, when the chromosomes are in the extended state, one or more of them may have a large, spherical body attached. This body, the **nucleolus,** is easily visible in the light microscope. Here are synthesized several kinds of RNA molecules. Some of this RNA is used in the assembly of **ribosomes.** Ribosomes are essential for protein synthesis in the cell, so it is not surprising that cells that are very active in protein synthesis usually have large nucleoli.

The nucleus is the control center of the cell. If the nucleus in an egg cell is destroyed, the egg cannot go on to develop into a new individual. If the nucleus is removed from an amoeba, the organism continues to live for a few days. However, it can neither feed nor reproduce, and eventually it dies. But if, within a day or two of the operation, a fresh nucleus is injected into the amoeba, normal activity and the ability to reproduce are restored.

Even before the invention of micromanipulators, the importance of the nucleus in determining what the cytoplasm does had been demonstrated by the German biologist Theodor Boveri. By vigorous shaking, Boveri was able to remove the nucleus from the eggs of a species of sea urchin of the genus *Sphaerechinus.* He then allowed these "enucleated" eggs to be fertilized by sperm from sea urchins of the genus *Echinus.* Sperm cells are much smaller than egg cells. They consist of little more than a nucleus and a tail to propel it. In the process of fertilization, the nucleus penetrates the egg. Thus fertilization of enucleated *Sphaerechinus* eggs by *Echinus* sperm resulted in the substitution of one kind of nucleus for another. The stimulus of fertilization caused the egg to undergo cell division and it developed into a sea urchin larva. A glance at Fig. 5.9 shows that this larva possessed the traits of the *Echinus* species

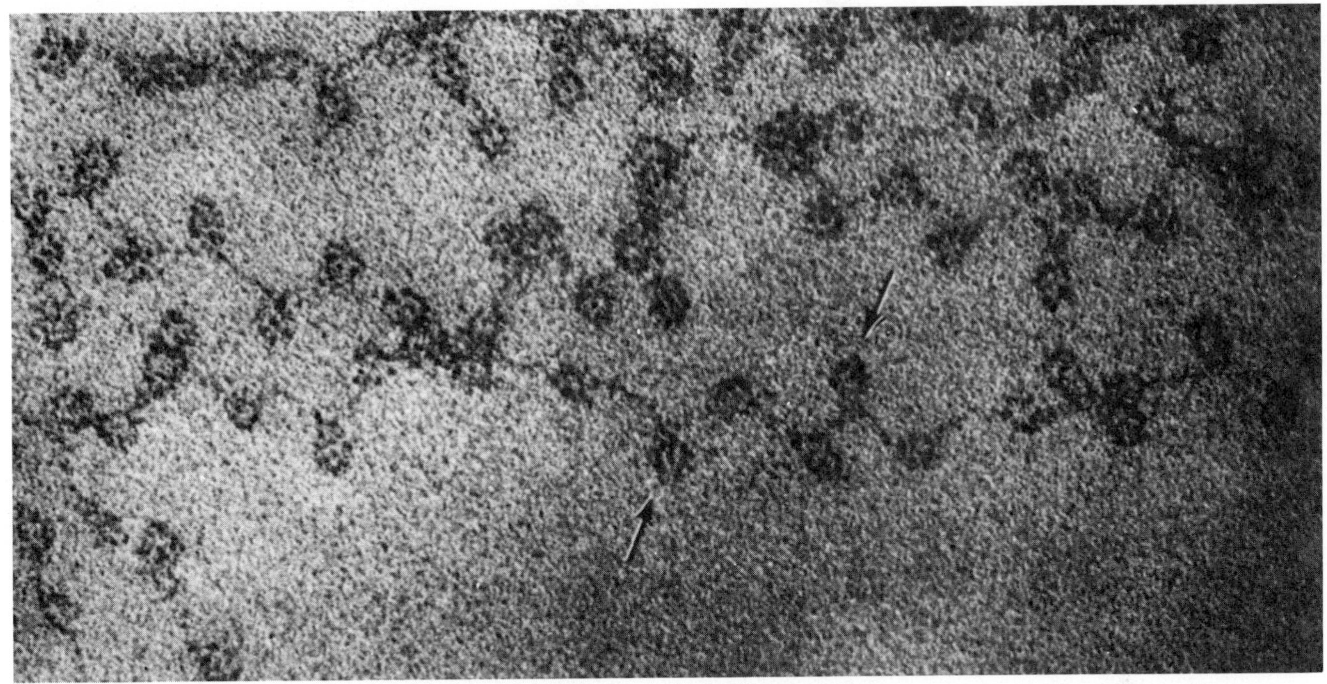

FIG. 5.8 Structure of chromatin from the nucleus of a chicken red blood cell. The globular bodies (arrows) are the nucleosomes. Each consists of a core of histone molecules wrapped in DNA. The nucleosomes are connected by a strand of DNA approximately 140 A long. (Courtesy of Donald E. Olins and Ada L. Olins, University of Tennessee.)

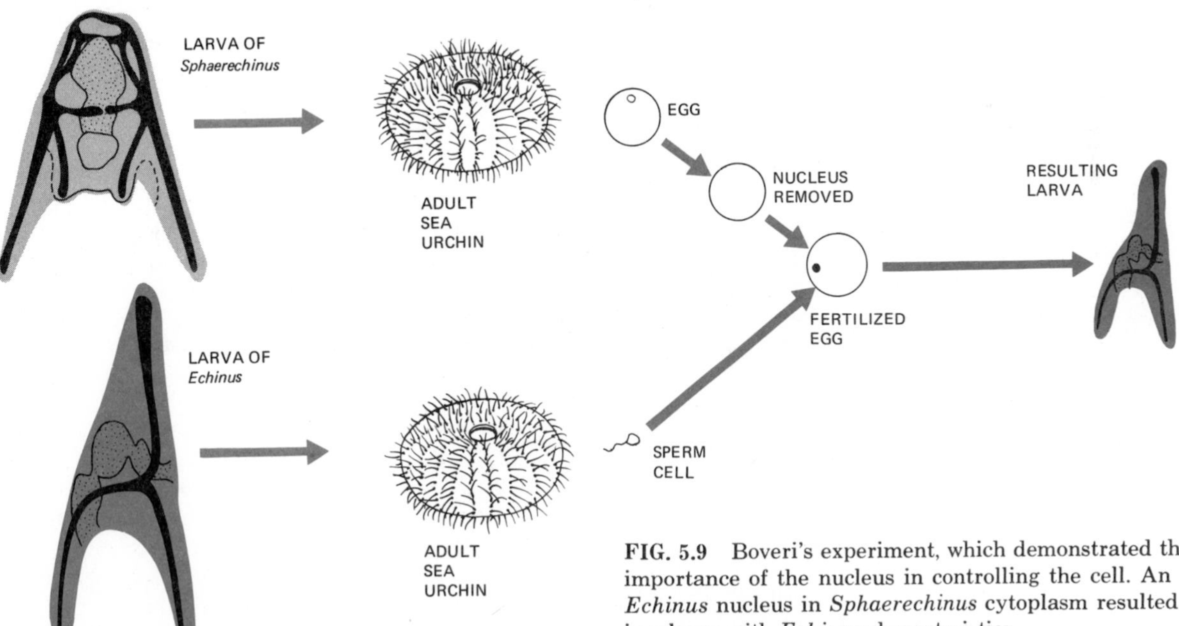

FIG. 5.9 Boveri's experiment, which demonstrated the importance of the nucleus in controlling the cell. An *Echinus* nucleus in *Sphaerechinus* cytoplasm resulted in a larva with *Echinus* characteristics.

rather than of the *Sphaerechinus* species. Although *Echinus* contributed little more than a single tiny nucleus to the system, this nucleus caused the great mass of *Sphaerechinus* cytoplasm to develop according to the *Echinus* blueprint.

5.4 THE CYTOPLASM

The term cytoplasm is traditionally used to describe everything within the cell except the nucleus. In the early days of cytology, when very little was known about the organization of materials outside of the nucleus, the term was a useful one. But with improved methods of studying cells, we have learned of the extraordinary complexity of structures found in the cytoplasmic region. Electron microscopy has revealed elaborate patterns of membranes and membrane-bounded compartments within the cytoplasm (Fig. 5.10). These clearly defined structures are called **organelles.**

Another useful way to study the parts of the cell is to crush a large number of cells, place them in a tube, and rotate the tube in a centrifuge. This imposes a centrifugal force on the organelles in the cell mixture. Large or heavy objects will be thrown to the bottom of the tube more rapidly than small, light objects. Centrifuging a cell mixture for 10 minutes at a force of about 800 times that of gravity will cause the nuclei to be deposited at the bottom of the tube, forming a sediment (Fig. 5.11). Higher forces exerted for longer periods of time will force smaller and lighter organelles to sediment. These organelles can be removed and studied both chemically and under the microscope. After centrifugation for two hours at a force 100,000 times that of gravity, almost all the identifiable organelles of the cell will have been forced to sediment. The fluid above the sediment (the supernatant) represents what is left of the cytoplasm after all the organelles have been removed. This is, then, the material in which the organelles of the cytoplasm are normally suspended. It has been given a variety of names, such as "ground substance," "hyaloplasm," "cytosol," and others. It is largely water in which are dissolved many small molecules and ions as well as substantial amounts of protein. In fact, a number of enzymes crucial to cell metabolism are found here. But most of the functions of the cytoplasm are the functions of the organelles located in it. Let us now examine these.

FIG. 5.10 An idealized view of an animal cell as it might be seen under the electron microscope. Although no single electron micrograph has revealed all the structures shown, the drawing represents a composite view of what many electron micrographs suggest is the organization of the parts of the cell.

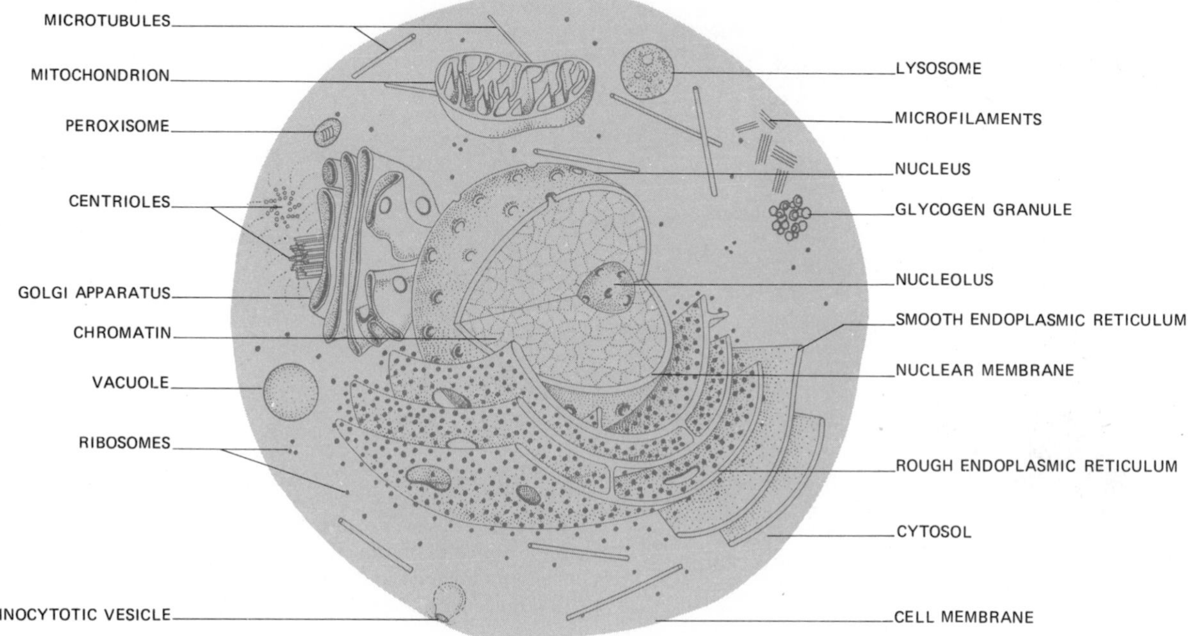

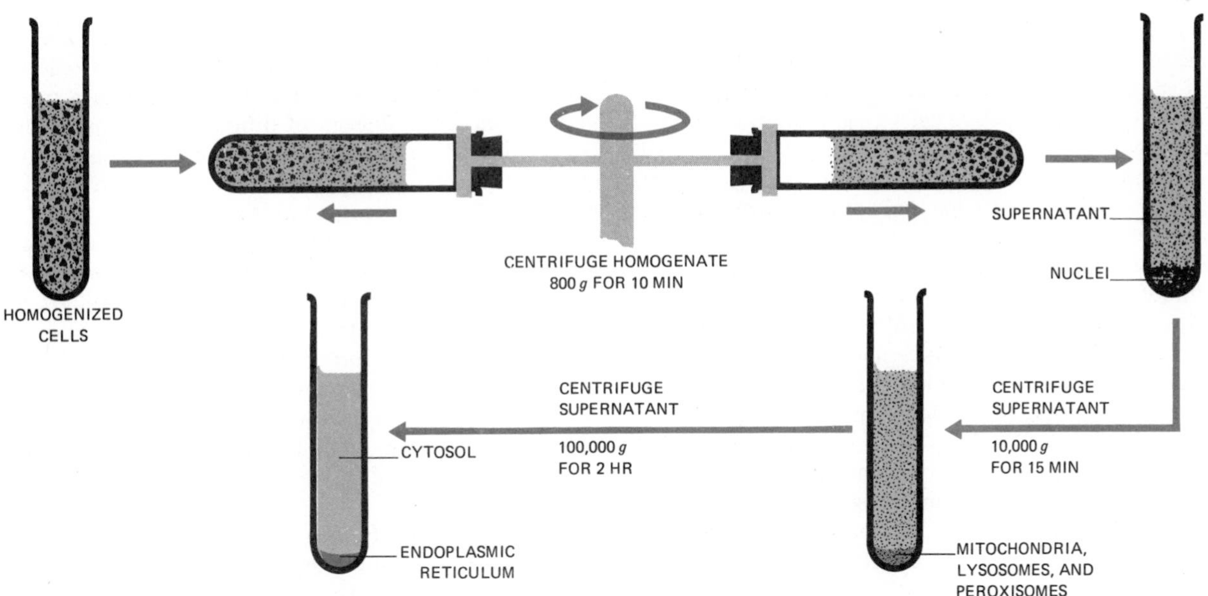

FIG. 5.11 Fractionation of cells by centrifugation. A preparation of cells is first homogenized to rupture the cells and release their contents. Gentle centrifugation of the homogenate forces the most massive cell structures, the nuclei, to form a pellet at the bottom of the tube. The fluid above the pellet, called the supernatant, is then centrifuged more vigorously, forcing smaller organelles, like mitochondria, into a second pellet. After a final centrifugation, the supernatant contains some free ribosomes and the soluble molecules of the cytosol.

FIG. 5.12 A mitochondrion from a bat pancreas cell. Note the double membrane and the way in which the inner membrane is folded into cristae. The dark, membrane-bounded objects above the mitochondrion are lysosomes. (Courtesy Keith R. Porter.)

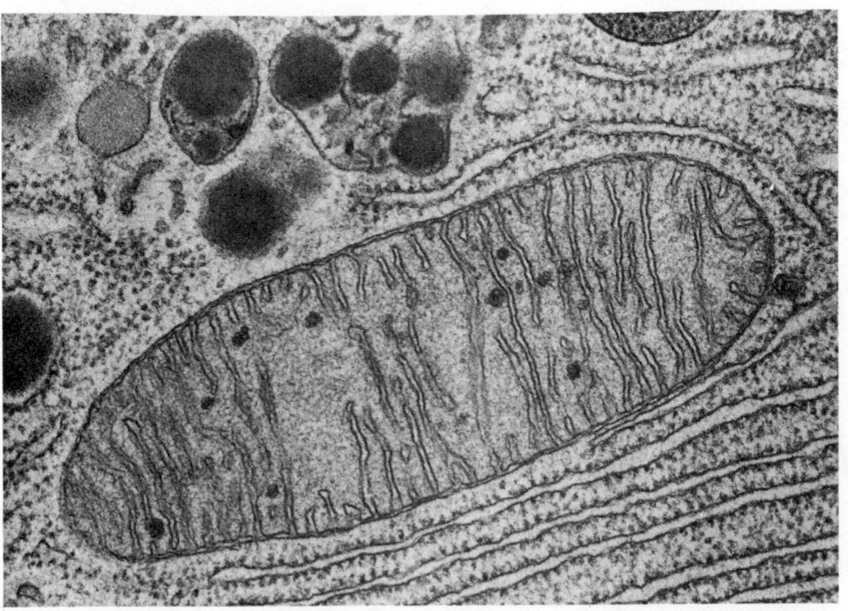

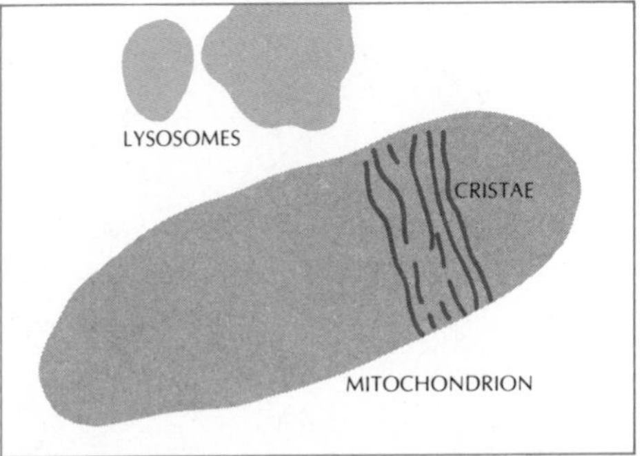

5.5 MITOCHONDRIA

Mitochondria are spherical or rod-shaped bodies that range in size from 0.2 μm to 5 μm. The number in a cell varies but active cells (e.g. liver cells) may have over a thousand of them.

Although the larger mitochondria can be seen under the light microscope, only the electron microscope can reveal their basic structure. Electron micrographs show that each mitochondrion is bounded by a double membrane. The outer membrane provides a smooth, uninterrupted boundary to the mitochondrion. The inner membrane is repeatedly extended into folds that project into the interior space of the mitochondrion (Fig. 5.12). These shelflike inner folds are called **cristae.**

The membranes of mitochondria appear to be similar to the cell membrane. Like it, they contain phospholipid and protein. Some of the protein is extrinsic while much of it is intrinsic, that is, embedded in the lipid bilayer.

The function of the mitochondria is quite clear. They contain the enzymes that carry out the oxidation of food substances. Thus the mitochondria convert the potential energy of different food materials into a form of energy that can be used by the cell to carry out its various activities. In view of this, it is not surprising that mitochondria tend to congregate in the most active regions of a cell. Nerve cells, muscle cells, and secretory cells all contain many mitochondria located in the regions of the cell most actively engaged in transmission of electrical impulses, contraction, and secretion, respectively. The mitochondria have been aptly called the powerhouses of the cell. We shall examine their organization and function in greater detail in Chapter 8.

5.6 CHLOROPLASTS

Chloroplasts are found only in the cells of plants and certain algae. In the cells of plants, the chloro-

FIG. 5.13 Cells from the leaf of a sunflower. Note the nucleus (N), chloroplasts (Chl), mitochondria (M), crystals (Cry), central vacuole, and primary cell wall in these typical plant cells. (Electron micrograph courtesy of H. J. Arnott and Kenneth M. Smith).

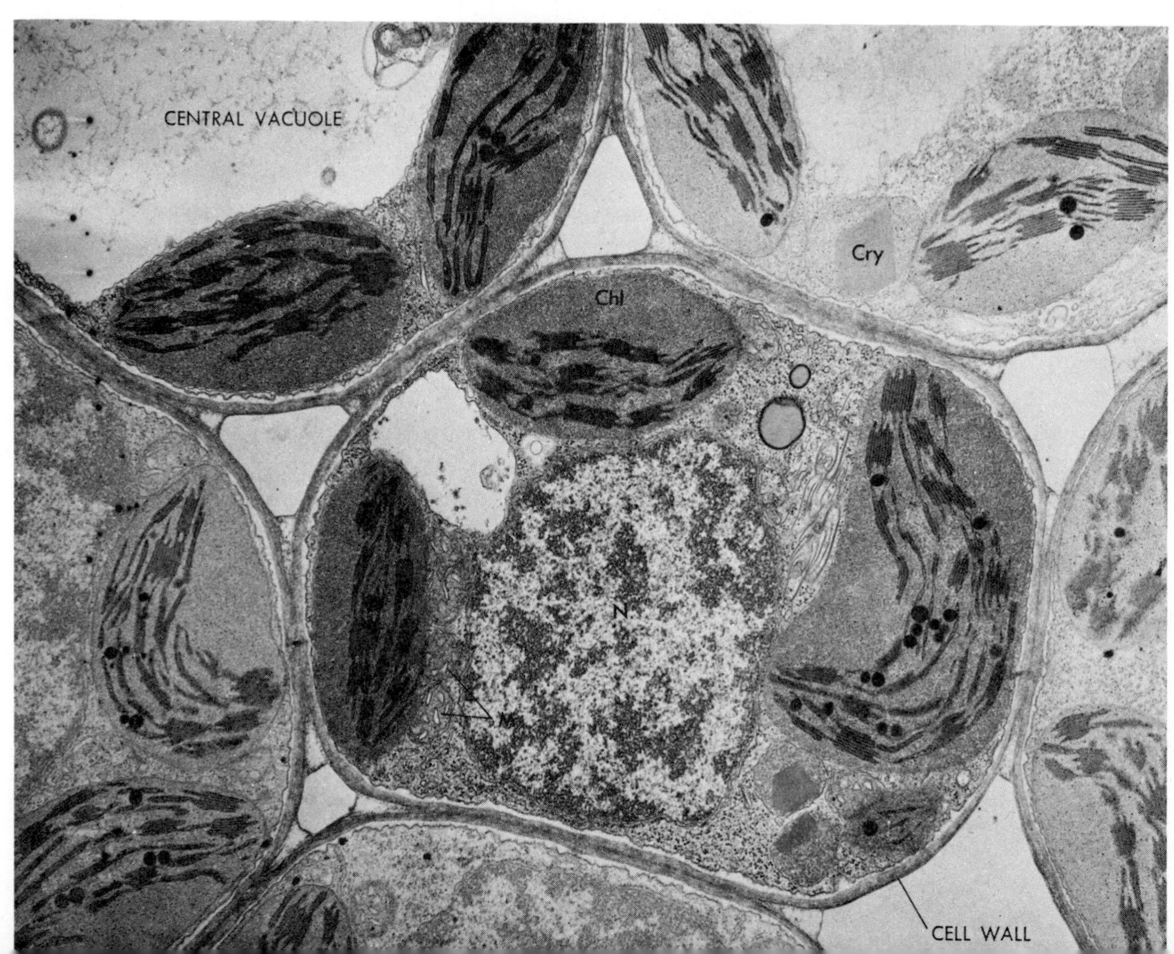

plasts usually occur as disk-shaped structures 5–8 μm in diameter and 2–4 μm thick (Fig 5.13). A single plant cell may contain as many as 50 of them.

The chloroplast, like the mitochondrion, is enveloped by a membrane. Within its outer membrane is an elaborate system of internal membranes embedded in a fluid matrix called the stroma. The internal membranes are rich in phospholipids and in proteins. They also contain pigments, the most prominent of which is chlorophyll. It is the green of the chlorophyll incorporated in their membranes that imparts the green color to chloroplasts and hence to plant cells and tissues exposed to the light.

Chlorophyll traps the energy of sunlight and enables it to be used for the synthesis of food. Thus the chloroplast is the seat of **photosynthesis.** Without these organelles, life as we know it could not exist on the earth. In Chapter 9, we shall examine the process of photosynthesis and, as we do so, we shall also examine the structure and function of chloroplasts in more detail.

5.7 RIBOSOMES

Ribosomes are among the smallest structures suspended in the cytoplasm (Fig. 5.10). These roughly spherical bodies are so small (20 nm) that they can be seen only under the electron microscope. Ribosomes are the site at which protein synthesis takes place. In cells that are especially active in protein synthesis (e.g. liver cells) ribosomes may constitute as much as 25% of the dry weight of the cell.

The ribosomes of certain bacteria have been intensively studied because they are smaller and less complex than the ribosomes of higher organisms. In the common intestinal bacterium *E. coli,* each ribosome contains one copy of each of three different molecules of RNA. In addition, 54 different protein molecules are present, all but one or two of them in single copies. Thus the ribosome can well be considered as a precisely ordered aggregate of macromolecules. Here we can relate the known structure of macromolecules (RNA, protein) to the structure of an organelle in a more precise way than has as yet been possible elsewhere in the cell. Further details of how ribosomes function in protein synthesis will be presented in Chapter 18.

Some of the proteins synthesized within cells are simply added to the fluid of the cytoplasm and perform their function there. The synthesis of such proteins (the hemoglobin in a red blood cell is an example) occurs on ribosomes that are distributed randomly throughout the cytoplasm. In other cases, the newly synthesized proteins are packaged in some type of membrane-bounded organelle. The cells of the liver and pancreas, for example, synthesize large amounts of proteins which are packaged within a membrane and ultimately secreted from the cell. The ribosomes that synthesize these proteins are attached to the membranes of the endoplasmic reticulum.

5.8 THE ENDOPLASMIC RETICULUM

The endoplasmic reticulum is an elaborate system of membranes within the cell. Seen in the sliced cell preparations used for electron microscopy, the membranes appear in pairs, enclosing flattened cavities and tubes (Fig. 5.14). These enclosed spaces are probably interconnected with one another. The membranes have a lipid-protein structure similar to that of other membranes of the cell. Each membrane of the

FIG. 5.14 Rough endoplasmic reticulum. The evenly spaced rows of ribosomes are attached to the surface of the membranes facing the cytosol. The proteins they synthesize are secreted into the interior of the canals formed by the pairs of membranes. (80,000 ×, courtesy of Keith R. Porter.)

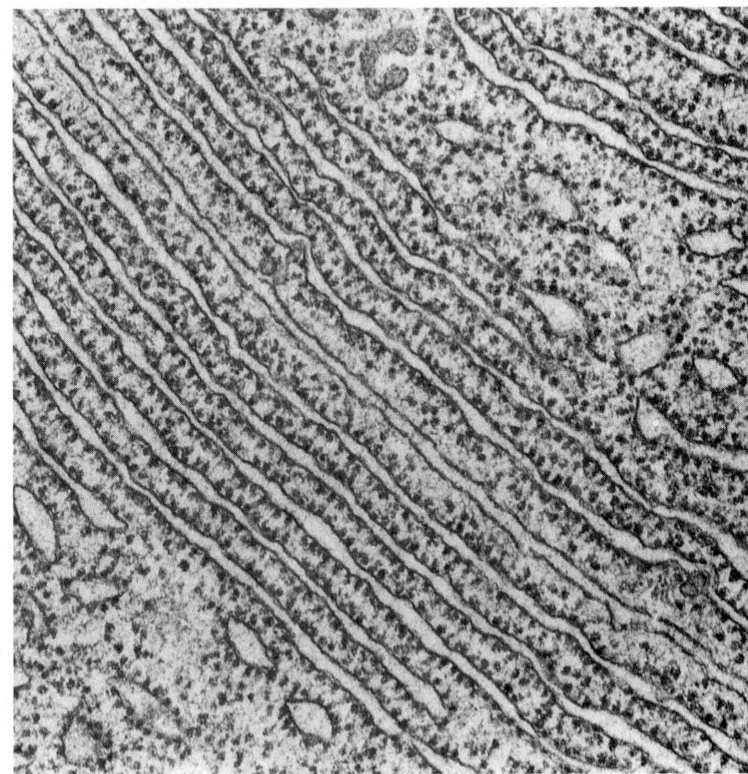

endoplasmic reticulum has one surface that faces the cytosol and one that faces the interior of the cavity.

If you look closely at the electron micrograph of Fig. 5.14 you will see evenly spaced rows of ribosomes adhering to the cytosol-facing surface of the reticulum membranes. Such ribosome-studded endoplasmic reticulum is called *r*ough *e*ndoplasmic *r*eticulum or RER.

Ribosomes are sites of protein synthesis and those attached to the RER appear to deposit their newly synthesized polypeptide chains in the interior space of the reticulum. As this happens, some additional processing of the polypeptide takes place, such as the formation of disulfide bonds between cysteines and the attachment of sugars. The proteins synthesized by the RER are thus packed within a membrane. They may remain to be used within the cell or they may ultimately be secreted from the cell.

Endoplasmic reticulum is also found without adhering ribosomes. This is called *s*mooth *e*ndoplasmic *r*eticulum or SER. The SER is probably involved in the synthesis of other types of molecules such as fats, phospholipids, and steroids. It may also be the source of the membranes that make up the Golgi apparatus.

5.9 THE GOLGI APPARATUS

The Golgi apparatus is found in almost all animal and plant cells. It consists of a stack of flat, membrane-bounded sacs (Fig. 5.15). It is especially prominent in cells that are actively involved in secretion. Proteins synthesized by the RER are transferred to the Golgi apparatus. Here additional carbohydrate may be added to them. In any case, the proteins accumulate within the sacs of the Golgi

FIG. 5.15 Golgi apparatus (arrows). In this cell (from a bat), the Golgi apparatus is used for the final stages in the synthesis of the proteins that are to be secreted from the cell. (Courtesy of Keith R. Porter.)

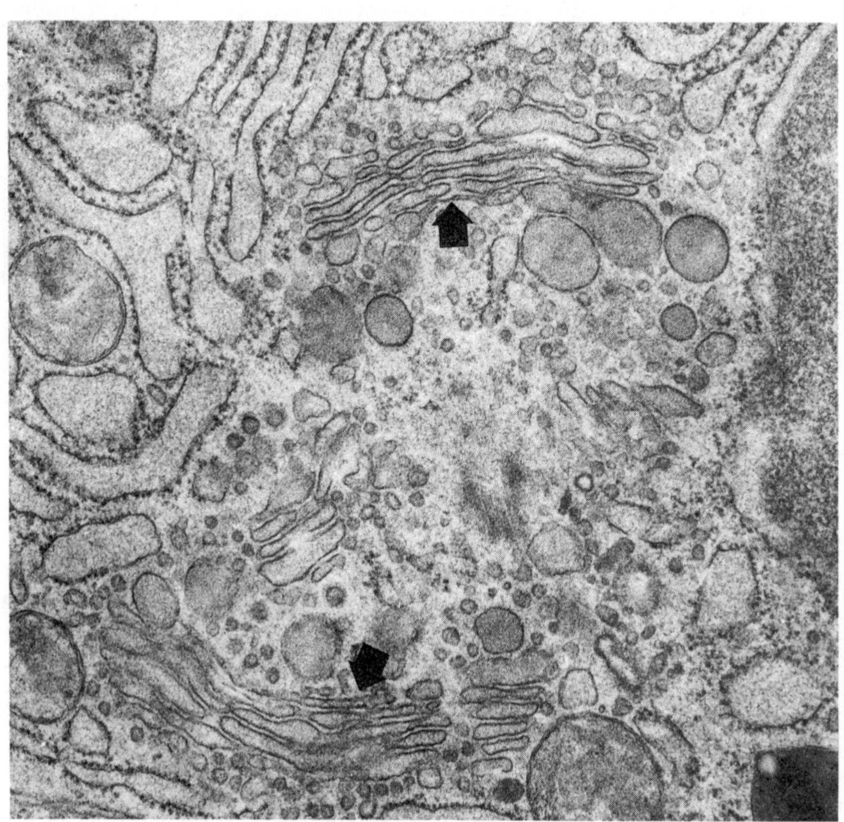

apparatus until finally the sacs are literally stuffed with protein. These protein-filled sacs then may migrate to the surface of the cell and discharge their contents to the outside. Other protein-filled sacs formed in the Golgi apparatus may be retained within the cell as lysosomes.

The Golgi apparatus is also the site where synthesis of polysaccharides, e.g. in mucus, takes place. The cellulose secreted by plant cells to form the cell wall is synthesized within the Golgi apparatus.

5.10 LYSOSOMES

Lysosomes are roughly spherical structures bounded by a single membrane. They are usually about 1.5 μm in diameter, although lysosomes as small as 0.05 μm may occasionally be found. Lysosomes are produced by the Golgi apparatus and are filled with proteins. These consist of some three dozen kinds of hydrolytic enzymes. Enzymes that digest polysaccharides, lipids, phospholipids, nucleic acids, and proteins are all represented. Presumably, by being confined within the lysosome, they are prevented from digesting the other components of the cell.

FIG. 5.16 A neutrophil, one of the white blood cells. The small dark bodies in the cytoplasm are lysosomes. (7500 ×, from Fawcett: *The Cell: Its Organelles and Inclusions,* W. B. Saunders Co., 1966.

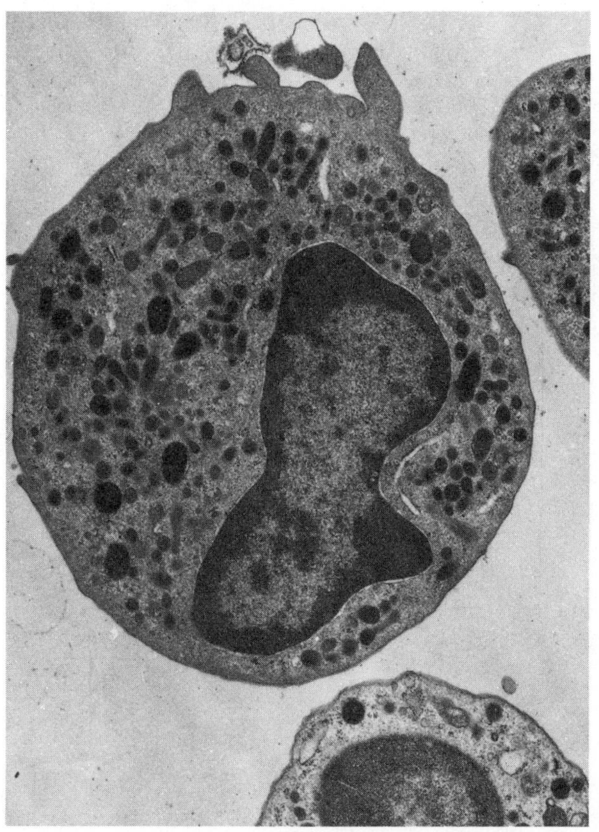

When materials within the cell are to be digested, they are first incorporated into a lysosome. These materials may be other subcellular structures, such as mitochondria, that have ceased to function efficiently. They may be food particles that have been taken into the cell. In the case of the white blood cell shown in Fig. 5.16, they are bacteria or other harmful particles that have been scavenged by the cell.

Lysosomes also play an important role in the death of cells. When a cell is injured or dies, its lysosomes aid in its disintegration. This clears the area so that a healthy cell can replace the damaged one. Cell death is a necessary stage in the life cycle of some organisms. For example, as a tadpole changes into a frog, its tail is gradually absorbed. The tail cells, which are richly supplied with lysosomes, die and the products of their disintegration are used in the growth of new cells in the developing frog.

Lysosomes are found in practically all kinds of animal cells. There is as yet no conclusive evidence that they occur in plant cells.

5.11 PEROXISOMES

Peroxisomes are about the size of lysosomes (0.3–15 μm) and, like them, are bounded by a single membrane. They also resemble lysosomes in that they are filled with enzymes, of which catalase is the most characteristic. Catalase catalyzes the breakdown of hydrogen peroxide (H_2O_2), a potentially dangerous product of cell metabolism. Peroxisomes may also play a role in the conversion of fats into carbohydrates and in the breakdown of purines within the cell.

A number of other metabolic functions have been assigned to peroxisomes in particular tissues or organisms but, except for the activities of catalase, no single function is common to *all* peroxisomes.

In animals, peroxisomes appear to be confined to cells of the liver and kidney. In plants, they may occur in a variety of cell types. The peroxisomes of plant cells often contain material that is crystallized. In both plants and animals, peroxisomes are probably manufactured by the endoplasmic reticulum.

5.12 VACUOLES

Vacuoles are fluid-filled "bubbles" in the cytoplasm. They are bounded by a membrane that is probably identical to the cell membrane. In fact, vacuoles are often formed by an infolding and pinching-off of a portion of the cell membrane (Fig. 5.10). Food materials or wastes may be found inside vacuoles.

A young plant cell contains many small vacuoles, but as the cell matures, these unite to form a large *central vacuole* (Fig. 5.13). Dissolved food molecules, waste materials, and pigments may be found in it.

5.13 THE FUNCTIONS OF INTRACELLULAR MEMBRANES

Most of the cell structures discussed so far have membranes as part of their construction. In general, there seem to be two main functions accomplished by these membranes. One is simply to establish a number of compartments within the cell. The bearers of the hereditary code, the chromosomes, are separated from the rest of the cell by the nuclear membrane. The potentially destructive digestive enzymes in the lysosomes are kept from contact with the cytosol by their bounding membrane. The secretory products of the cell are sequestered in the channels of the endoplasmic reticulum and Golgi apparatus.

The second important role played by the membranes of the cell is to provide for the neat spatial organization of enzymes and pigments. Chlorophyll is incorporated in the internal membranes of the chloroplast. The enzymes which carry out the oxidation of food are neatly arranged on the cristae of the mitochondria. It is quite likely that other enzymes are present as well-ordered systems in the cell membrane and the membranes of the endoplasmic reticulum.

5.14 MICROFILAMENTS

Microfilaments are long, thin fibers, 40–60 A in diameter. The fibers are composed of a protein called actin. Large numbers of microfilaments form clusters or networks at various places in the cell (Fig. 5.17). Their presence is associated with cell movement. When an animal cell divides in two, for example, a band of microfilaments forms and pinches the two daughter cells apart.

In many cells the cytoplasm moves about, a phenomenon known as cytoplasmic streaming. Its motion is dependent on the presence of microfilaments. Microfilaments are also prominent features in cells that migrate from place to place and change their shape. This is true not only for an independent, free-moving cell such as an amoeba but also for most animal cells during the formation of an embryo. The pinching-off of infolded portions of the cell membrane (see Section 5.12) also depends on the contractility provided by microfilaments.

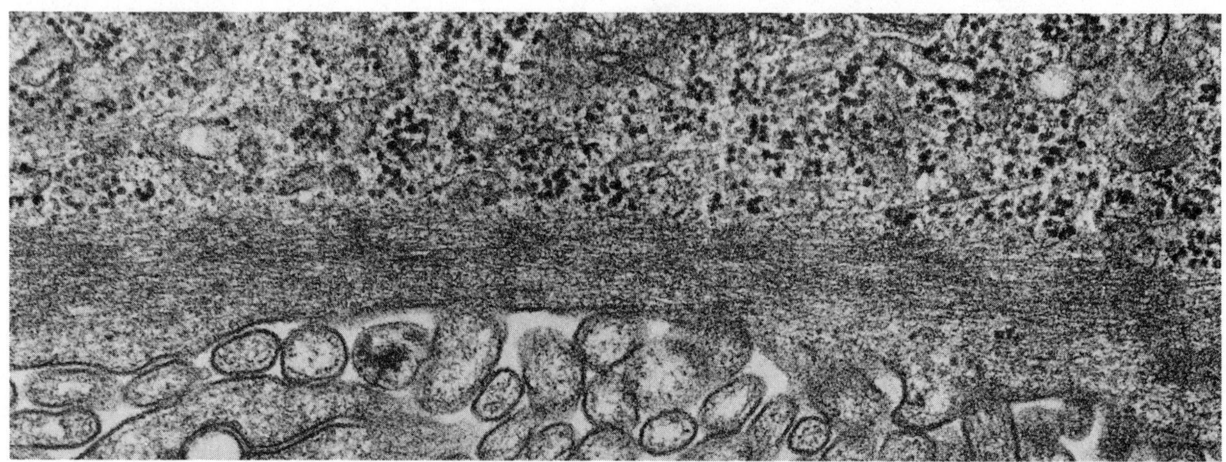

FIG. 5.17 A band of microfilaments (running horizontally) in a cell of the chick oviduct. The dark dots scattered above the microfilaments are ribosomes. (Electron micrograph courtesy of J. T. Wrenn.)

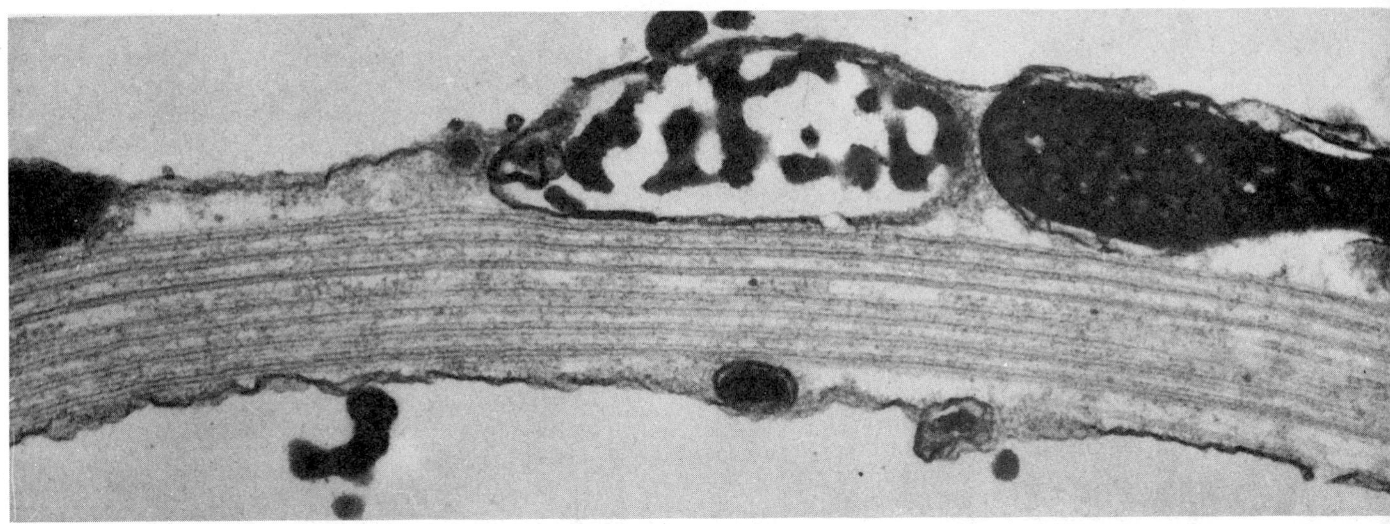

FIG. 5.18 Parallel array of microtubules in an axopod of a freshwater protozoan. Axopods are fine extensions of cytoplasm, and microtubules are essential to maintain their shape. Other views of axopods can be seen in Figs. 35.4 and 35.5. (Courtesy of Lewis G. Tilney, University of Pennsylvania.)

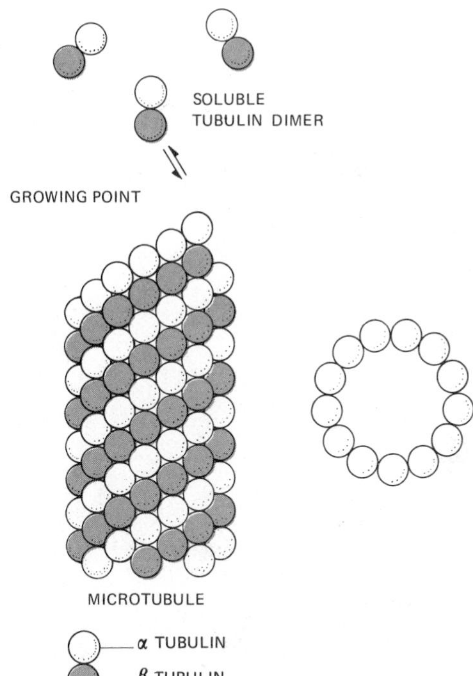

FIG. 5.19 Assembly of a microtubule from the dimers of alpha and beta tubulin. The addition of 13 dimers extends the microtubule by a full turn.

5.15 MICROTUBULES

Microtubules are straight, hollow cylinders of protein which are found in most animal and plant cells (Fig. 5.18). They are about 250 A in diameter. Their length is variable, but it is not uncommon to see microtubules that are over 100 times as long as they are thick (i.e. 2.5 μm or more in length).

The protein out of which microtubules are made is called tubulin. Two types of tubulin (α-tubulin and β-tubulin) are found. The two kinds are of about the same size, each having a molecular weight of approximately 55,000 daltons. Two molecules, one of each type, join together (noncovalently) to form a dimer. This dimer (an example of the quaternary structure of a protein—see Section 4.4) is the building block out of which the microtubule is constructed. The dimers appear to be added one after another to build the wall of the cylinder in a gentle helix (Fig. 5.19). The addition of 13 dimers completes one full turn. Thus, seen in cross section, the microtubule wall appears to be assembled from 13 "protofilaments."

Microtubules appear to be stiff, and it is thought that they provide some rigidity to those parts of the cell in which they are located. They are especially apt to be present in parallel arrays near the margin of the cell and may be acting there as an intracellular skeleton.

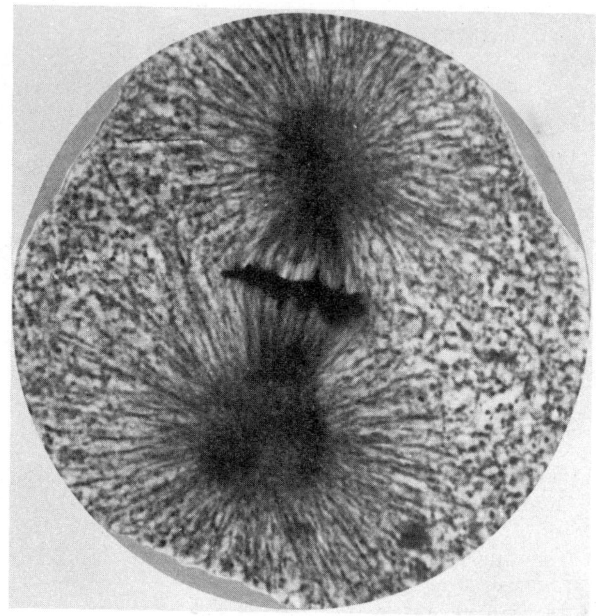

FIG. 5.20 Mitosis of a cell in the embryo of a whitefish. The chromosomes have moved to the center of the cell, where they form a dark, compact mass. The chromosomes are attached to the spindle, the array of fibers extending between the poles of the cell. Around each pole is a system of radiating fibers called the aster. Both the spindle and the asters are built from microtubules. (Courtesy of CCM: General Biological, Inc., Chicago.)

Microtubules also play an essential role in cell division. Successful cell division requires a precise distribution of chromosomes to each daughter cell. Each chromosome moves to its final destination attached to a bundle of microtubules. The entire array of microtubules participating in the process is called the **spindle** (Fig. 5.20). Microtubules are also used in the construction of centrioles, basal bodies, cilia, and flagella.

5.16 CENTRIOLES

Animal cells and the cells of some microorganisms and lower plants contain two centrioles located near the exterior surface of the nucleus. Each centriole consists of a cylindrical array of nine microtubules. However, each of the nine microtubules has two partial (as seen in cross section) tubules attached to it (Fig. 5.21). The two centrioles are usually placed at right angles to each other.

Just before a cell divides, its centrioles duplicate and one pair migrates to the opposite side of the nucleus. The spindle (see above) then forms between them.

In some cells, the centrioles duplicate to produce the basal bodies of cilia and flagella.

5.17 CILIA AND FLAGELLA

Many cells have whiplike extensions, either short ones (cilia) or long ones (flagella). In microorganisms, cilia and flagella are used for locomotion. However, many animals have ciliated cells for which the cilia serve simply to move materials past the cell. The cells lining the inner surface of our trachea ("windpipe") are ciliated.

The origin and structure of cilia and flagella seem to be fundamentally the same. In each case they grow out of basal bodies. These have the same structure as centrioles and are formed by them.

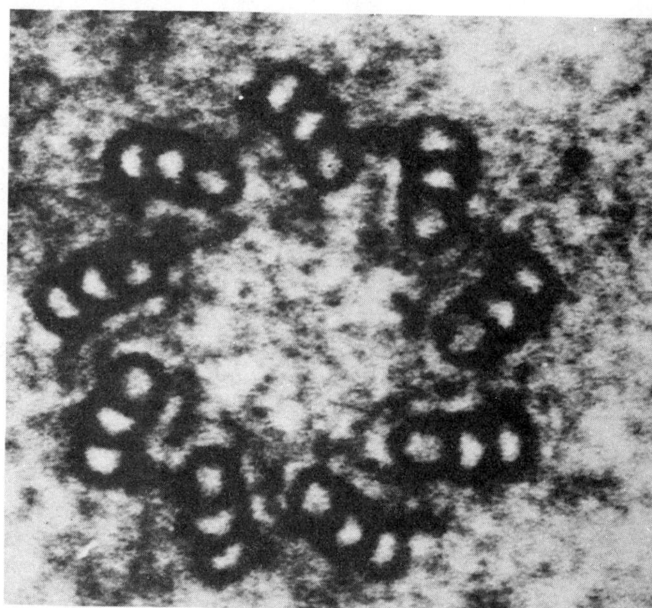

FIG. 5.21 Centriole in a mouse embryo cell as seen in cross section at 305,000 ×. (Courtesy of E. deHarven, Sloan-Kettering Institute for Cancer Research. From *The Nucleus,* Academic Press, New York/London, 1968.)

(Courtesy Peter Satir)

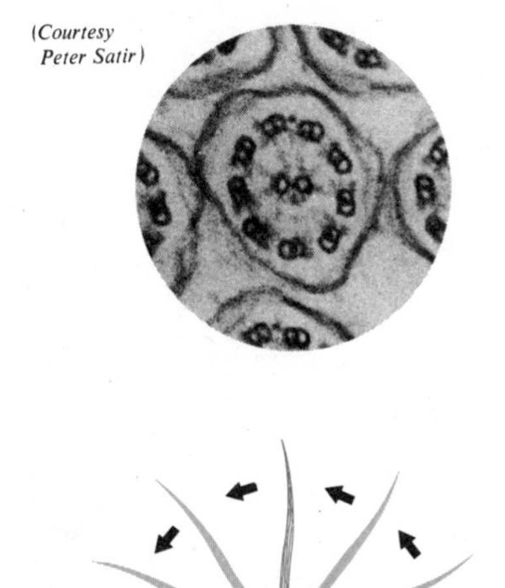

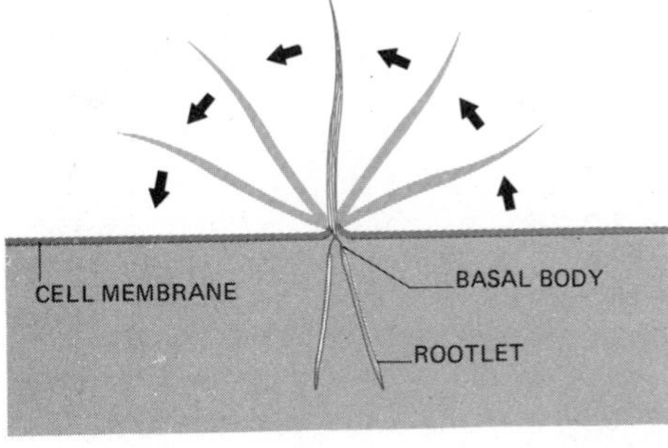

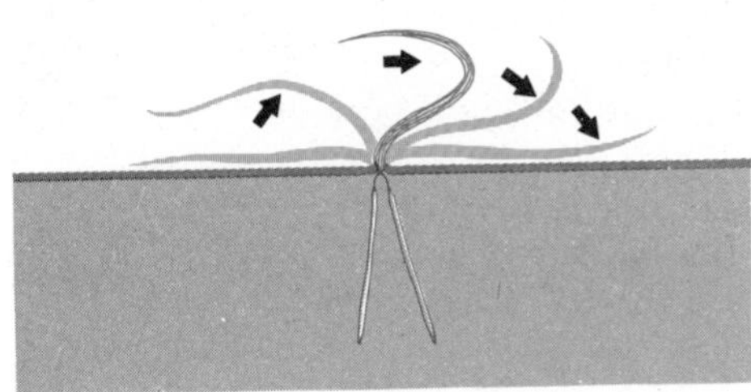

FIG. 5.22 Top: Electron micrograph of a single cilium in cross section. Note the characteristic pattern of microtubules. Bottom: Drawings of a single cilium. The power stroke is shown above, the recovery stroke below.

The cilium or flagellum itself has not only the outer ring of nine microtubules (each now with just *one* accessory tubule attached to it) but also two central fibrils that are identical to microtubules in their construction (Fig. 5.22). In both cilia and flagella, the entire assembly is sheathed in a membrane which is simply an extension of the cell membrane.

The similarities between the structure of cilia, flagella, basal bodies, and centrioles and the structure of microtubules suggest that we are dealing with another one of the fundamental architectural components of cells. The assembly of the tubulin dimers into microtubules appears to require some form of initiation site or template. The logical candidate for this function is the centriole—or the basal bodies derived from centrioles. While many questions of microtubular function still remain, an understanding of the way in which the structure of particular macromolecules (α- and β-tubulin) leads to the structure of the various microtubular organelles of the cell is close at hand.

5.18 CELL COATINGS

Only rarely is the cell membrane the outer surface of the cell. Usually some type of exterior coating is present. In animal cells this appears to be constructed from a protein-polysaccharide complex. It is not rigid but does serve to cement adjacent cells together. In Fig. 5.2, it is represented by the thin line between the two cell membranes. In many cases, however, it is much thicker than shown here. Filaments of the protein collagen (Fig. 5.23) are often embedded in the thicker coatings.

In many algae and in all plants, the exterior coating is made of the polysaccharide cellulose. This forms a rigid, boxlike **cell wall** which is one of the most characteristic features of these cells. In Fig. 5.13, the cell wall that surrounds each of the sunflower cells is clearly visible.

If you look at Fig. 4.15, you will see the orderly manner in which fibrils of cellulose are deposited to form the cell wall. While the exact manner in which cellulose molecules are organized into the fibrils is still uncertain, the linear nature of the molecules and the many opportunities for side-to-side intermolecular hydrogen bonding are just what one would want in order to build long, stiff fibrils.

Rigid walls are also found around the cells of bacteria and fungi. However, polysaccharides other than cellulose are used in their construction.

5.19 PROKARYOTES VERSUS EUKARYOTES

While not *all* cells contain *all* the structures discussed in this chapter, most of these structures are commonly found in the cells of animals, plants, and even in most microorganisms. However, two groups of microorganisms—the bacteria and blue-green algae—have cells that differ from the cells of other organisms in several notable respects. Their cells have no membrane-bounded nucleus. They have no mitochondria, chloroplasts, endoplasmic retic-

FIG. 5.23 Filaments of collagen, a protein, 31,000 ×. (Courtesy of Dr. Jerome Gross.)

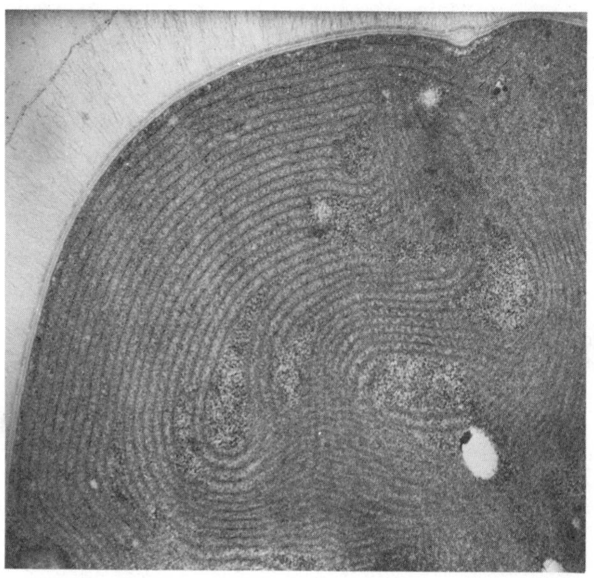

FIG. 5.24 Electron micrograph (25,000 ×) of a blue-green alga. Although internal membranes are found in this prokaryote, these are not organized into the membrane-bounded organelles characteristic of eukaryotes. (Courtesy of Dr. G. Cohen-Bazire.)

ulum, Golgi apparatus, lysosomes, peroxisomes, or vacuoles. In short, they have no *membrane-bounded* organelles, although layers of internal membranes may be present (Fig. 5.24). While some bacteria have flagella, these consist of simply a single strand, like a single microtubule. There is not the multistranded, 9 + 2 arrangement that we find in the cilia and flagella of other organisms.

The term **prokaryotic** ("prenuclear") is often used to distinguish the cells of bacteria and blue-green algae from the nucleus-containing **eukaryotic** ("truly nuclear") cells of all other organisms.

5.20 DIFFERENTIATION

We have seen that living things are made of one or more cells. The microscopic green alga *Chlamydomonas* is a single-celled organism. Within its single cell, it contains all the equipment needed to carry out the various functions of life. From time to time, *Chlamydomonas* divides and forms two individuals where before there was one. Each daughter cell receives a complete set of the nuclear controls present in the parent cell. Prior to actual division of the cell, each chromosome in the nucleus is duplicated. Then, during the process of cell division itself, these duplicated chromosomes become separated. With remarkable precision, a complete set migrates to each of the two daughter cells. **Mitosis** is the term used to describe this important process. It provides a mechanism for the reproduction of single-celled organisms. It also provides a mechanism for growth in multicellular organisms. In both cases, it provides an escape from excessive growth or enlargement of single cells. Small cells can usually operate more efficiently than large ones. One reason for this is that small cells have a larger ratio of surface area to volume than big ones (Fig. 5.25). Thus a more rapid exchange of materials can occur between the cytoplasm and the external environment of the cell.

Among the flagellated green algae are several interesting colonial forms (Fig. 5.26). These species

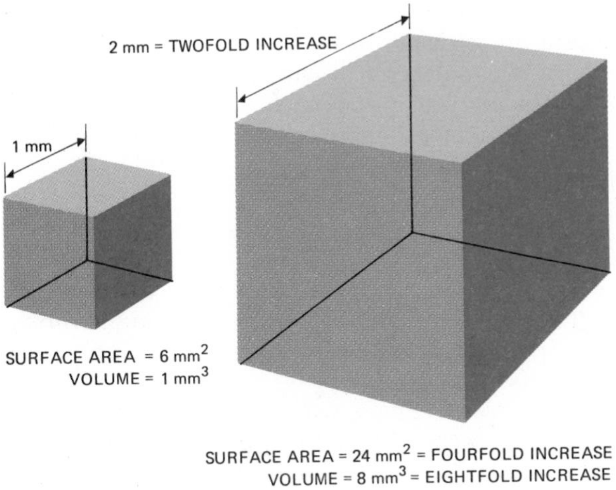

2 mm = TWOFOLD INCREASE
1 mm
SURFACE AREA = 6 mm^2
VOLUME = 1 mm^3
SURFACE AREA = 24 mm^2 = FOURFOLD INCREASE
VOLUME = 8 mm^3 = EIGHTFOLD INCREASE

FIG. 5.25 The ratio of surface area to volume of the cube decreases as its size increases. The same relationship holds for cells and organisms.

FIG. 5.26 A group of flagellated green algae whose constituent cells are strikingly similar and which illustrate the unicellular, colonial, and multicellular levels of organization. The small cells in *Pleodorina* and most of the cells in *Volvox* are incapable of reproduction. The scale of the drawings diminishes from left to right.

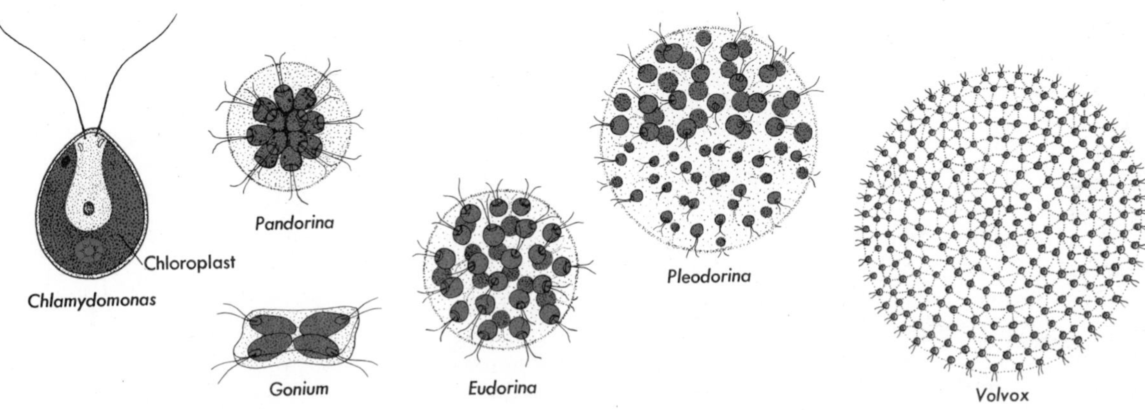

are called colonial because they are simply made up of clusters of independent cells. If a single cell of *Gonium, Pandorina,* or *Eudorina* is isolated from the rest of the colony, it will swim away looking quite like a *Chlamydomonas* cell. Then, as it undergoes mitosis, it will form a new colony with the characteristic number of cells in that colony. The situation in *Pleodorina* and *Volvox* is somewhat different. In these cases, some of the cells of the colony (most, in *Volvox*) are not able to live independently. If a nonreproductive cell is isolated from a *Volvox* colony, it will fail to reproduce itself by mitosis and eventually it will die. What has happened? In some way, as yet unclear, *Volvox* has crossed the line separating simple colonial organisms from truly multicellular ones. Unlike *Gonium, Volvox* cannot be considered simply a colony or cluster of individual cells. It is a *single organism* whose cells have lost their ability to live independently. If a sufficient number of them become damaged for some reason, the entire sphere of cells will die.

What has *Volvox* gained by this arrangement? In giving up their independence, the cells of *Volvox* have become specialists. No longer does every cell carry out all of life's functions (as in colonial forms); instead, certain cells specialize to carry out certain functions while leaving other functions to other specialists. In *Volvox* this process goes no further than having certain cells specialized for reproduction while others, unable to reproduce themselves, fulfill the needs for photosynthesis and locomotion. In more complex multicellular organisms (e.g. ourselves), the degree of specialization is carried to greater lengths. Each cell has one or two precise functions to carry out. It depends on other cells to carry out all the other functions necessary to maintain the life of the organism and thus its own. This process of specialization and division of labor among cells is called **differentiation.** One of the great problems of biology is how differentiation arises among cells, all of which, having arisen by mitosis, share the same nuclear controls. We shall examine some tentative answers to this problem in Chapter 22.

We cannot be certain that *Gonium, Pandorina, Eudorina,* and *Pleodorina* represent stages in the evolution of multicellular *Volvox* from unicellular *Chlamydomonas.* Whether they do or not, these organisms illustrate one way in which colonial forms may have arisen from unicellular ones and multicellular forms from colonial ones. They also illustrate the subtle shift in cell relationships that occurs as one crosses the uncertain boundary between colonies of independent cells and organisms constructed of many interdependent, differentiated cells.

5.21 ANIMAL TISSUES

The cells that make up multicellular organisms become differentiated in a large, but not unlimited, number of ways. In the human animal, for example, there are probably some 100 distinguishable kinds of differentiated cells. One or more types of differentiated cells are organized into tissues. The following kinds of tissues are found in vertebrate animals (Fig. 5.27).

1. Epithelial

Epithelial tissues are made up of closely packed cells arranged in flat sheets. These tissues line the various cavities and tubes of the body. They also form the skin which covers the body.

Epithelial tissues carry out a variety of functions. In every case these functions reflect the fact that epithelia are aways found at the boundary between cell masses and a cavity or space. The epithelium of the skin protects underlying tissues from damage by mechanical abrasion, ultraviolet radiation, dehydration, and bacterial invasion.

Epithelia also function in the transport of materials to and from the tissues and the cavities that they separate. The columnar epithelium of the intestine secretes digestive enzymes into the intestine and also absorbs the end products of food digestion from it. All the digestive glands of the body are lined with epithelium. Epithelium also lines the air tubes and lung cavities. This epithelium secretes mucus to keep itself from drying out and to trap inhaled dust particles. Many of its cells possess cilia on the "free" surface which propel the mucus with its load of foreign matter back up to the throat. The sex cells must be released from the body in order to function in sexual reproduction. Thus we find that they, too, arise from an epithelium, in this case, the germinal epithelium.

FIG. 5.27 Animal tissues. Each of the kinds of cells shown here specializes in carrying out one or a few functions with great efficiency. However, each has relinquished certain functions to the others. Thus all of these specialized cells are interdependent. Their coordinated interactions produce a functioning, multicellular organism. ▶

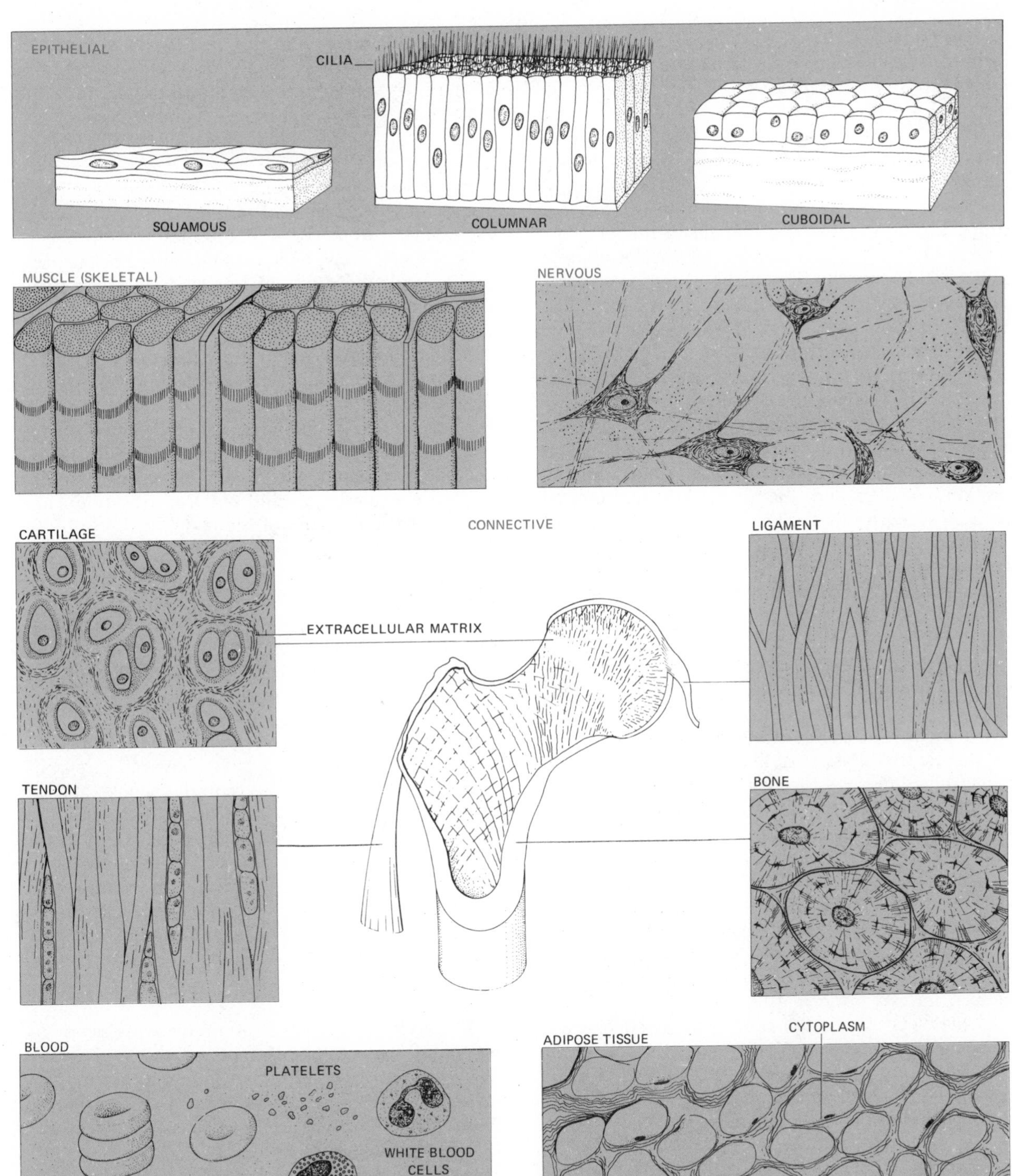
EPITHELIAL
CILIA
SQUAMOUS
COLUMNAR
CUBOIDAL
MUSCLE (SKELETAL)
NERVOUS
CARTILAGE
CONNECTIVE
LIGAMENT
EXTRACELLULAR MATRIX
TENDON
BONE
BLOOD
PLATELETS
WHITE BLOOD CELLS
RED BLOOD CELLS
ADIPOSE TISSUE
CYTOPLASM

2. Connective

Several types of connective tissue are found in the human body. Each of these consists of differentiated cells embedded in a great deal of extracellular material. This **matrix,** as it is called, is secreted by the cells. **Supporting** connective tissue is used to give strength, support, and protection to the soft parts of the body. **Cartilage** and **bone** are the two kinds of supporting connective tissue found in humans. The matrix in cartilage is a protein-polysaccharide mixture, called chondrin. The shape of the outer ear is maintained by cartilage. The matrix of bone contains both fibers of the protein collagen and mineral deposits. The main component of the latter is calcium phosphate, although magnesium, carbonate, and fluoride ions are also present.

Binding connective tissue serves, as its name suggests, to bind body parts together. **Tendons** connect muscle to bone. The matrix is principally the protein collagen and the fibers are all oriented parallel to each other. This gives great strength to the tissue. Tendons are not elastic, however. **Ligaments** attach one bone to another. In addition to collagen fibers, ligaments contain the protein elastin. This protein permits the ligaments to be stretched.

Fibrous connective tissue is found distributed throughout the body. It serves as a packing and binding material for most of our organs. It also provides a pathway for blood vessels and nerves to reach them. Collagen and other proteins are found in the matrix. Fascia is fibrous connective tissue which binds muscles together and binds the skin to the underlying structures. **Adipose tissue** is fibrous connective tissue in which the cells have become almost filled with oil.

Hematopoietic tissue is the source of all the cells that are found in the blood. These include the red blood cells, five kinds of white blood cells, and platelets. The red blood cells transport O_2 and CO_2 in the blood. The white blood cells defend the body against foreign invaders (e.g. bacterial infections). The platelets initiate the process of blood clotting. Bone marrow is hematopoietic tissue in which all the blood cells are produced. Two kinds of white blood cells (lymphocytes and monocytes) are also produced in lymph nodes, the spleen, and the thymus.

3. Muscle

Three kinds of muscle tissue are found in humans. **Skeletal** muscle is made up of long fibers whose contraction provides the force for locomotion and other types of voluntary body movements. **Smooth** muscle lines the walls of the hollow organs of the body, such as the intestine and blood vessels. Its contraction, which is involuntary, reduces the size of the body's hollow organs. **Cardiac** muscle is the muscle out of which the heart is made.

4. Nervous

Nervous tissue is composed chiefly of **neurons.** These are cells that are specialized for the conduction of electrochemical nerve impulses. Each neuron consists of a cell body, which contains the nucleus, and one or more hairlike extensions. It is along these extensions, which in a few cases may extend several feet, that the nerve impulse travels. The tips of the extensions meet either other neurons (Fig. 5.27) or some other kind of tissue, e.g. muscle, that is stimulated by the neuron.

The human brain and spinal cord consist chiefly of nervous tissue. The cell bodies of virtually all our neurons are found here. Their extensions, bundled together in **nerves,** extend out from the brain and spinal cord to all parts of the body.

5.22 PLANT TISSUES

If we examine a mature vascular plant, we can identify several distinct cell types. These are grouped together in tissues. Some tissues consist of only one cell type. Some consist of several.

1. Meristematic. The chief function of meristematic cells is mitosis (Fig. 5.28). The cells are small and thin-walled, with no central vacuole and no specialized features. They are located in tissues (the apical meristems) at the growing points of roots and stems. In some plants, a ring of meristematic tissue, the cambium, is also found within the mature stem. Mitosis in the meristems produces new cells for the growth of the plant. The cells produced by the meristems soon become differentiated into one or another of several types.

2. Protective. The cells of protective tissues are found on the surface of roots, stems, and leaves. They are flattened cells whose top and bottom surfaces are parallel but whose sides may be arranged irregularly (Fig. 5.28). These cells provide protection to the cells lying beneath.

3. Parenchyma. Parenchyma cells are widely distributed throughout plants. They are large, thin-walled cells and usually have a central vacuole. Frequently they become partially separated from one another and gas fills the resulting intercellular

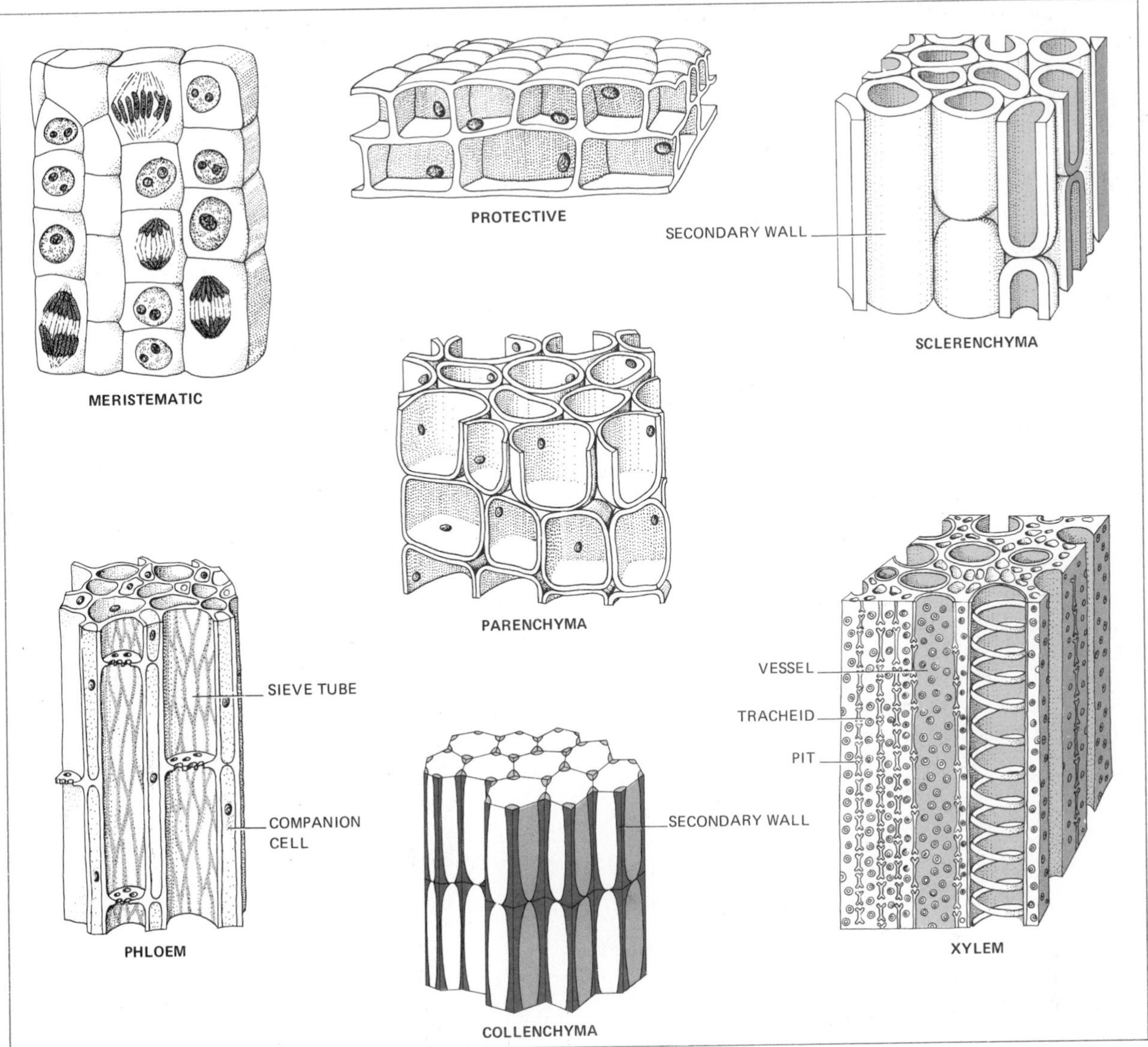

FIG. 5.28 Plant tissues. Each of the differentiated cell types develops from the cells produced in meristems. Several of the meristematic cells are shown in mitosis.

space. They are liberally supplied with plastids. In areas not exposed to light, colorless plastids predominate and food storage is the main function. The cells of the white potato are parenchyma cells. In areas where light is present, e.g. in the leaf, chloroplasts predominate and photosynthesis is the main function.

4. Collenchyma. Collenchyma cells have thick walls that are especially developed at the corners of the cell (Fig. 5.28). These cells provide mechanical support for the plant. They are most frequently found in regions of the plant that are growing rapidly and need to be strengthened. The petiole ("stalk") of leaves is usually reinforced with collenchyma cells.

5. Sclerenchyma. Sclerenchyma cells are a more common type of supporting cell. The walls of these cells are very thick and are built up in a uniform layer around the entire margin of the cell (Fig. 5.28). Sclerenchyma cells are usually found associated with other cell types and give mechanical support to them. In many cases, the protoplasts of sclerenchyma cells die after the cell wall is fully formed. Sclerenchyma cells are found in stems and are also associated with the veins of leaves. They are the exclusive component of the hard outer covering of seeds and nuts.

6. Xylem. The xylem is a "mixed tissue" composed of several cell types. The most characteristic and important of these are the xylem vessels and xylem tracheids. (The xylem of ferns and conifers contains only tracheids.) Xylem vessels have thickened cell walls. The walls are not deposited in a uniform layer but are usually thickened in a pattern of spiral bands (Fig. 5.28). When fully developed, the end walls of xylem vessels dissolve away and the protoplasts die. This produces a long tube. Tracheids differ from vessels in that the cells lack spiral banding and are tapered at the ends. The tapered ends of vertically stacked tracheids overlap and are interconnected by means of many pits. Both tracheids and vessels are used to transport water and minerals from the roots to the leaves. In woody plants, the older xylem ceases to participate in transport and simply serves to give strength to the trunk of the growing plant. Wood is xylem. When one counts the annual rings of a tree, one is counting rings of xylem.

7. Phloem. The phloem is also a mixed tissue. The most important cells in it are the sieve tubes. They are so named because their end walls become perforated (Fig. 5.28). This allows cytoplasmic connections to be established between cells. These presumably aid the cells in carrying out their chief function: the transport of food and hormones throughout the plant. At maturity, sieve tubes do not possess a nucleus. Adjacent to them are nucleated "companion cells," however, and these may take over general control of the sieve tube cells. Frequently sclerenchyma cells also occur in phloem tissue and impart strength to it.

The various plant tissues are themselves organized in definite patterns. The groups of organized tissues make up the organs of the plant. Roots, stems, and leaves are the major organs of higher plants. Their proper functioning depends upon the proper arrangement and coordination of the tissues of which they are composed.

5.23 CELL-TO-CELL JUNCTIONS

In many tissues, e.g. connective tissue, an extracellular coating or matrix separates the cell membrane of one cell from that of the adjacent cells in the tissue. However, in certain tissues, e.g. the epithelium that lines the intestine, the membranes of two adjacent cells are pressed tightly together at several points. These junctions are of three kinds.

1. Tight Junctions

Tight junctions seal adjacent epithelial cells in a narrow band just beneath the exposed surface of the cells. These junctions are probably of considerable physiological importance because they present a barrier to the easy passage of molecules and ions through the space between cells. Where tight junctions are present, materials must actually enter and pass through cells in order to pass through the tissue (Fig. 5.29). This pathway surely provides greater control over what substances are permitted through than would be the case if diffusion *between* cells occurred.

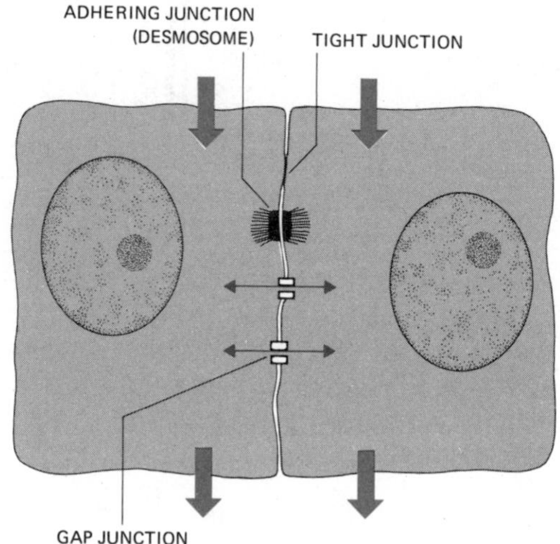

FIG. 5.29 Types of cell junctions characteristic of epithelia. Adhering junctions provide the mechanical strength to hold adjacent cells together. Tight junctions prevent molecules from diffusing across the epithelium by way of intercellular spaces. Gap junctions allow small molecules and ions to pass readily from one cell to another.

2. Adhering Junctions

As the name suggests, adhering junctions provide strong mechanical attachments between adjacent cells. Some adhering junctions are present as narrow bands connecting two cells. Others ("desmosomes") are present as discrete patches holding two cells together.

3. Gap Junctions

Gap junctions serve as pathways of intercellular communication. It has been known for some time that ions (hence electrical currents) and molecules (up to a molecular weight of about 1000 daltons) pass readily from one cell to the next in certain tissues, but do *not* pass easily to and from these cells and the exterior. The structure of gap junctions has recently been studied, both chemically and under the electron microscope. These junctions contain intercellular channels some 15–20 A in diameter through which ions and small molecules could easily pass (in either direction). The widespread occurrence of gap junctions provides further evidence that cells in multicellular organisms are not independent entities but, rather, are the building blocks of an interdependent system: the tissue.

EXERCISES AND PROBLEMS

1. In what ways do typical plant cells differ from animal cells?
2. List all those structures found in a rat liver cell that seem to be constructed from membranes.
3. How do prokaryotic cells differ from eukaryotic cells?
4. Distinguish between organ, organism, and organelle.
5. Assuming them to be spherical in shape, compare the surface areas of a mycoplasma, staphylococcus, and human cheek cell (see Fig. 5.1).
6. How would their *volumes* compare if they were all spherical?
7. What organic molecules are used in the construction of the (a) plant cell wall, (b) chromosome, (c) ribosome, (d) cell membrane, (e) lysosome, (f) oil droplet?

REFERENCES

1. NOVIKOFF, A. B., and E. HOLTZMAN, *Cells and Organelles,* Modern Biology Series, Holt, Rinehart and Winston, Inc., 1970. An excellent survey of cell structure and function.
2. CAPALDI, R. A., "A Dynamic Model of Cell Membranes," *Scientific American,* Offprint No. 1292, March, 1974.
3. COOK, G. M. W., *The Golgi Apparatus,* Oxford Biology Readers, No. 77, Oxford University Press, Oxford, 1975.
4. ALLISON, A. C., *Lysosomes,* Oxford Biology Readers, No. 58, Oxford University Press, Oxford, 1974.
5. JORDAN, E. G., *The Nucleolus,* Oxford Biology Readers, No. 16, Oxford University Press, Oxford, 1971.
6. WESSELLS, N. K., "How Living Cells Change Shape," *Scientific American,* Offprint No. 1233, October, 1971. Discusses the roles played by both microtubules and microfilaments.
7. LOEWENSTEIN, W. R., "Intercellular Communication," *Scientific American,* Offprint No. 1178, May, 1970. Describes experiments showing the physiology of cell-to-cell junctions.

Inner surface of the human small intestine as seen by the scanning electron microscope. The nutrients which fuel our metabolism enter the body here. (Courtesy of Dr. P. G. Toner and Dr. K. E. Carr.)

PART III
METABOLISM

THE METABOLISM OF CELLS

A cell is a chemical machine. It acquires materials and energy from its environment. It transforms these materials within itself by the chemical activities that constitute its **metabolism.** Finally, it returns some of the end products of these metabolic activities to the environment.

FIG. 6.1 Trace element deprivation. The rat at the bottom received the same diet as that at the top except that tin, vanadium, and fluorides were carefully excluded for 20 days. When tin and vanadium were then given to the deprived rat, normal growth still did not occur. But adding 0.5 ppm of potassium fluoride (KF) to its diet restored normal growth and health. (Courtesy of Klaus Schwarz, Veterans Administration Hospital, Long Beach, California.)

6.1 THE CELLULAR ENVIRONMENT

What is the environment of cells? It is always a liquid. This may be perfectly obvious if we consider the amoeba in its pond. It may not be so readily evident when we consider the cells of a terrestrial, multicellular organism like an oak tree or a human being. But in these creatures, too, every living cell is bathed in liquid. The cells of your body, for example, are bathed in an *interstitial fluid* that is derived from the blood. But, you may ask, what of my skin? The skin cells that you expose to the outer world are dead cells. Only underneath this dead, outer husk do we find living skin cells and these, like all others, are continuously bathed in liquid. In fact, where you do have living cells exposed to the external environment, as in the epithelium that lines your air passages and the transparent cornea at the front of your eyes, secretory cells bathe the exposed surfaces in a continuous supply of moisture. The general term for the fluid—whatever its source—that bathes cells is extracellular fluid or **ECF.** Every molecule or ion that a cell requires is secured from its ECF and the products or wastes that a cell manufactures are deposited in its ECF.

6.2 THE COMPOSITION OF THE ECF

The major component of the extracellular fluid is **water.** In this superb solvent are dissolved the molecules and ions upon which cells depend for their functioning. These include:

1. **Gases,** of which oxygen and carbon dioxide are the most important.
2. A variety of **inorganic ions,** of which sodium (Na^+), chloride (Cl^-), potassium (K^+), calcium (Ca^{++}), bicarbonate (HCO_3^-), and phosphate ($PO_4^{\equiv}$) are present in substantial amounts. A number of other inorganic ions are required in only minute quantities. In some cases, e.g., Cu^{++}, Zn^{++}, Mn^{++}, Co^{++}, these "trace elements" are necessary for the activity of certain enzymes. Iodine is incorporated in the hormone thyroxin. Small quantities of fluoride ion (F^-) are important in strengthening the mineralized portions of teeth and bones and essential for normal growth in rats (Fig. 6.1).
3. Organic compounds such as **foods** and **vitamins.** Foods are those substances that serve as a source of energy for the cell as well as a source of materials for growth and repair of the cell. They include lipids, amino acids, and sugars. Vitamins are small organic molecules that the organism cannot manufacture from its food. Vitamins are needed in only very small quantities. They are not used as a source of energy or for the construction of the cell. Instead they per-

form certain special metabolic tasks for cells. Some, for example, serve as the prosthetic group of an enzyme.

In addition to the three types of components just mentioned, the extracellular fluid of multicellular organisms contains **hormones.** These are molecules that are released into the ECF by certain cells and that affect the metabolic activities of other cells. The ECF also serves to carry away the excretory wastes of cells. For animals, the most important of these are the waste products of protein and nucleic acid metabolism. These nitrogen-containing wastes, such as ammonia and urea, are rather toxic and their level in the ECF must not be allowed to exceed certain limits.

The hydrogen ion concentration (pH) of the extracellular fluid and its temperature are also of great importance to the welfare of cells. If the pH of human blood should move out of the range of 7.34–7.44, serious metabolic consequences would follow. Similarly, the temperature of the human ECF is normally maintained within a degree or so of 37.5°C (98.6°F).

Perhaps the great advantage that multicellular organisms with their variety of differentiated, and thus specialized, cells have over single-celled creatures is their ability to regulate the ECF of their cells. The study of the *physiology* of multicellular organisms is, in large measure, the study of the machinery by which the concentration of glucose, O_2, CO_2, Na^+, Ca^{++}, H^+, H_2O, etc., in the extracellular fluid is maintained within narrow limits. The maintenance of relatively unvarying composition and other properties of the extracellular fluid is called **homeostasis.**

Homeostatic control of the ECF reaches its greatest precision in the birds and mammals. In these organisms, the homeostatic machinery that is made possible by the differentiation of cells provides the cells with a relatively unvarying and optimum environment. The birds and mammals owe a great deal of their adaptability to the fact that the properties of their extracellular fluid, their *internal environment,* remain unchanged despite wide fluctuation in their *external* environment. Over a century ago, the French physiologist Claude Bernard discovered several mechanisms by which the mammalian body is able to maintain homeostasis. He was so struck by these findings that he wrote: "The constancy of the internal environment is the condition of a free and independent life."

Let us now examine the ways in which cells exchange matter with their extracellular fluid. There are five known mechanisms by which this occurs: diffusion, osmosis, active transport, endocytosis, and exocytosis.

6.3 DIFFUSION

Divide a chamber into two compartments by means of a sheet of cellophane (Fig. 6.2). Fill one compartment (*A*) with a solution of glucose (or any small molecule or ion whose concentration can be easily measured). Fill the other compartment (*B*) with pure water. Then, every few minutes, remove a sample from each compartment and measure the concentration of glucose in each. You will find that as time goes by, the glucose concentration drops in compartment *A* and, at the same rate, rises in compartment *B*. Finally, a time is reached when the

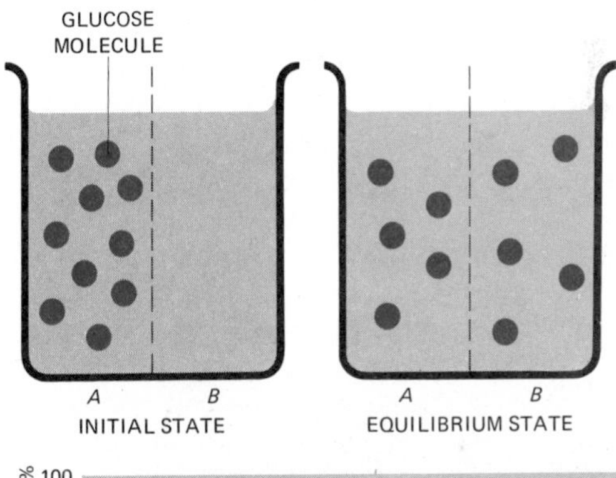

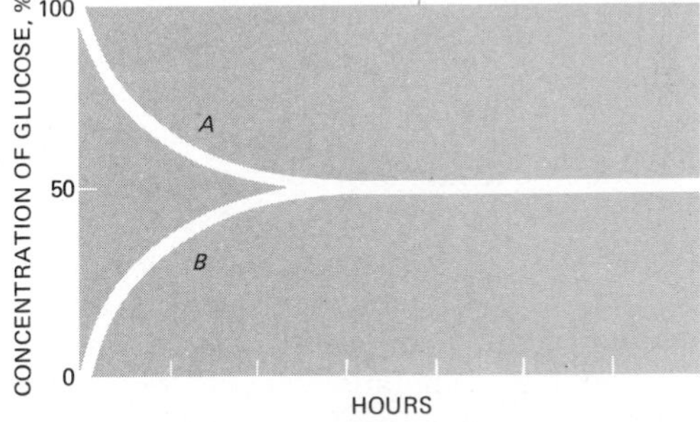

FIG. 6.2 Apparatus with which the diffusion of a small molecule across a cellophane membrane can be measured. At the start of the experiment, all of the glucose molecules are in compartment *A*. With the passage of time, diffusion of the glucose causes its concentration to drop in *A* and, at the same rate, rise in *B*. After three hours, the concentrations in the two chambers are equal and equilibrium has been reached.

concentrations in the two chambers become equal (Fig. 6.2) and remain so.

The process you have been observing is diffusion. It occurs because of the continual random motion characteristic of all molecules that are not held together in a solid. Each molecule travels in a straight line until it collides with another molecule, in this case another molecule of glucose, of water, or of cellulose (in the cellophane). Following each collision, the molecule rebounds in a different direction. The result is random motion of the molecule. When the compartments are first filled, the random motion of the glucose molecules results in many collisions with the cellophane membrane. However, the membrane has pores large enough to allow glucose molecules to pass through, and any molecules rebounding into such a pore will pass through to compartment *B*. As the concentration of glucose begins to rise in *B*, we would expect that glucose molecules would begin to return. Such is the case, but as long as there exists a **concentration gradient** across the membrane —that is, as long as the concentration of glucose molecules is higher in *A* than in *B*—there will be more attempts at penetrating the cellophane from side *A* than from *B* and thus a continued *net* movement of glucose molecules from *A* to *B*. Not until the concentrations become equal will the forward and backward rates of transport between *A* and *B* become equal. At this time, equilibrium has been reached (Fig. 6.2).

The cell membrane is not the simple passive barrier that a sheet of cellophane is. The rate at which materials diffuse through the cell membrane is dependent not only on the concentration gradient across the membrane but on the size, charge, and lipid solubility of the particles in question. In general, lipid-soluble, i.e., hydrophobic, molecules diffuse across the membrane more easily than hydrophilic molecules. Other things being equal, small molecules diffuse across the membrane more rapidly than large ones. In fact, hydrophilic molecules larger than 7–8 A are almost completely blocked by the cell membrane. These observations suggest that the lipid bilayer of the membrane is freely penetrated by hydrophobic molecules but that passage of hydrophilic molecules (including water molecules) can occur only through 7–8-A hydrophilic pores that perforate the membrane. These same pores are probably the route by which ions diffuse through the membrane. In general, cell membranes are far less permeable to ions (e.g., K^+, Cl^-, Na^+) than they are to small molecules. As for large, hydrophilic molecules like proteins, these are unable to pass through the membrane by diffusion and must rely instead on endocytosis (see below).

6.4 OSMOSIS

Osmosis is simply a special case of diffusion. Chemists define osmosis as the diffusion of any solvent through a differentially permeable membrane. Cell membranes, having pores which permit the passage of some molecules but prevent the passage of others, are differentially permeable. The universal solvent in living things, as was mentioned earlier, is water. For our purposes, therefore, osmosis can be defined as the diffusion of water through a differentially permeable membrane from a region of high concentration to a region of low concentration. Note that concentration refers to the concentration of the solvent, water, and not to the concentration of molecules or ions which may be dissolved in the water. The exchange of water between the cell and its environment is such an important factor in cell function that it justifies the special name of osmosis.

An experiment that demonstrates osmosis is shown in Fig. 6.3. The lower opening of the glass tube is covered with a sheet of cellophane. This acts as a differentially permeable membrane, permitting the rapid passage of water molecules but obstructing the passage of larger molecules. The interior of the tube is filled with molasses, a concentrated solution of sugar in water. The whole apparatus is placed in a beaker containing distilled water. In which direction will osmosis occur? The water concentration in the beaker is 100%. The water concentration inside the tube is less than this because a given volume of molasses contains fewer water molecules than the same volume of distilled water. There is, therefore, a net movement of water molecules through the cellophane membrane and into the tube.

As additional water molecules enter the tube, the volume of fluid increases. The molasses is forced up the tube. This force arises from pressure exerted by the diffusion of water molecules into the tube. The pressure is called osmotic pressure. The greater the difference in water concentration on either side of the membrane, the greater the tendency for osmosis to occur and thus the greater the osmotic pressure. In fact, when the column of molasses stops rising, we have a rough measure of the osmotic pressure of the system. The weight of the column of water finally counterbalances the osmotic pressure and osmosis ceases. Note that the water

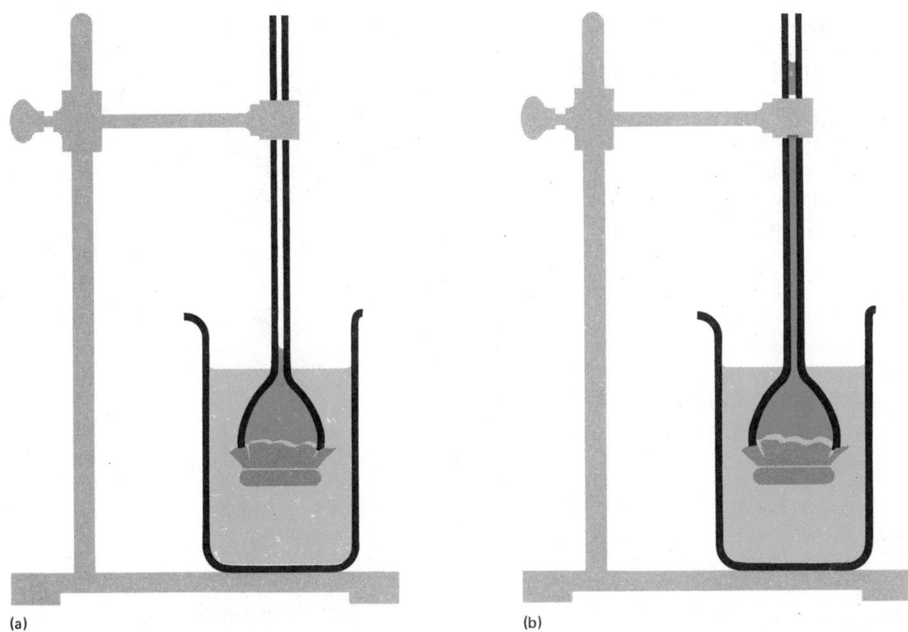

FIG. 6.3 Osmometer. (a) At start. (b) A few hours later. The cellophane is a differentially permeable membrane; water molecules pass through it more readily than do sugar molecules.

concentrations on the two sides of the membrane are still not equal by any means. However, the increase of pressure on the inner surface of the membrane, created by the weight of the column of molasses above, causes water molecules to be forced or filtered back out through the pores. When the rate at which this filtration process occurs becomes equal to the rate at which water molecules are coming in because of the difference in concentration, osmosis ceases.

Another experiment showing the effects of osmosis is depicted in Fig. 6.4. The shell membrane of a hen's egg is a differentially permeable membrane. Its pores are large enough to allow the easy passage of water molecules, but are not large enough to allow

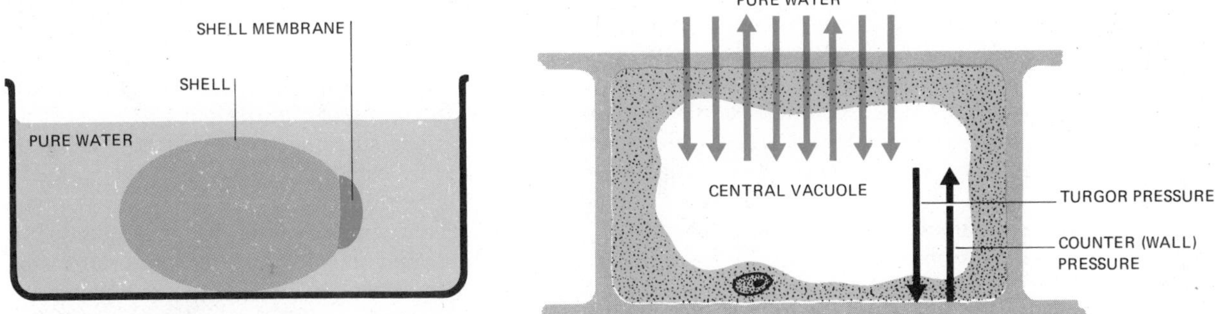

FIG. 6.4 Left: Egg osmometer. The pores of the shell membrane permit water molecules to diffuse into the egg, but are too small to let the macromolecules within the egg diffuse out. Consequently, the volume of materials within the egg increases until increasing pressure ruptures the membrane. Right: Turgor in a plant cell. When the pressure within the cell finally equals the osmotic pressure, the movement of water in and out of the cell will reach equilibrium.

larger fat and protein molecules to get through. If after carefully removing a portion of the waterproof shell (by dissolving it away in dilute acid), one places the egg in pure water, water will diffuse into the egg. This osmosis occurs because there is a greater concentration of water outside the egg (100%) than within. The surrounding water is said to be **hypotonic** to the contents of the egg. As osmosis continues, more and more water accumulates within the egg and this crowding in of additional molecules results in a buildup of pressure. The development of pressure within a cell (and the egg *is* a single cell) as a result of the diffusion of water into it is called **turgor.** Although the remaining portion of the shell can resist the pressure, the unprotected membrane cannot. Ultimately it bursts.

Coping with Hypotonic and Hypertonic Environments

What of organisms that spend their lives in fresh water? Certainly the water concentration of their cytoplasm can never approach that of the pure water surrounding them. In the case of the cells of freshwater plants, water passes into the cell by osmosis and turgor quickly develops. The strong cellulose walls of the cells are, however, capable of withstanding this pressure. Soon the pressure within the cell equals the osmotic pressure even though the two water concentrations are not equal. At this point, osmosis ceases (Fig. 6.4).

Freshwater animals and protozoans lack cellulose cell walls so they must cope differently with life in a hypotonic environment. Water enters their cells continually by osmosis but only slight turgor can safely be developed. The problem is solved by employing energy and some contractile structure to pump the excess water back out into the environment. The single-celled amoeba accomplishes this by means of a contractile vacuole (Fig. 6.5), in which water entering the cell by osmosis is collected. When the vacuole is filled, the amoeba contracts it (with the aid of microtubules) forcing the water out through a pore which forms momentarily in the cell membrane. Note that the water, which had entered the cell as a result of the random molecular activity of osmosis, leaves the cell by flowing out because of a force generated by the cell. The creation of this force requires the expenditure of energy by the cell. In Chapter 13, we shall examine some of the mechanisms by which freshwater animals such as fishes cope with *their* hypotonic environment.

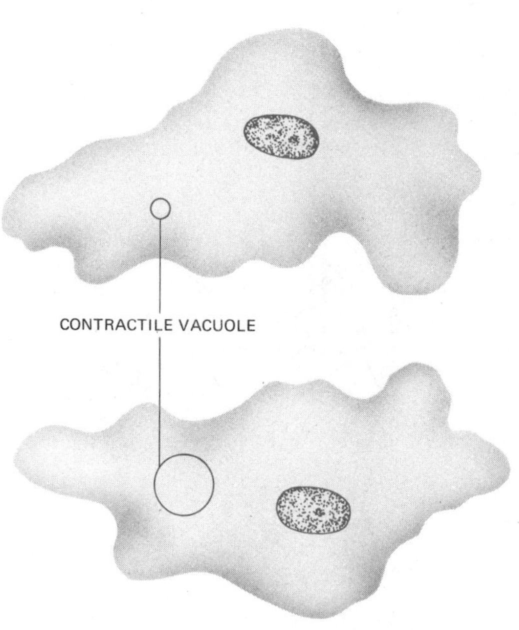

FIG. 6.5 The amoeba bails out the continual influx of water from its hypotonic surroundings by alternately filling and emptying its contractile vacuole. The contents of the vacuole may be discharged at any point on the surface of the cell.

Life in the oceans involves quite different osmotic conditions than life in fresh water. Sea water contains about 3.5% of various ions, especially Na^+ and Cl^-. This results in a water concentration which is approximately the same as that in the cytoplasm of marine plants and the invertebrate animals that live in the sea. Consequently, these organisms are able to exist in a state of equilibrium with respect to the water in their surroundings. They neither gain nor lose water by osmosis. We say that sea water is **isotonic** to their cytoplasm.

That a given volume of sea water is about 3% salt by weight, and thus 97% water, does not mean that the water concentration of the cytoplasm of these marine plants and animals is also 97% by weight. The speed with which diffusion or osmosis takes place is a measure of the difference in the number, not weight, of the molecules or ions involved. When osmosis occurs, the region of higher concentration is simply the region which contains the greater number of water molecules in a given volume

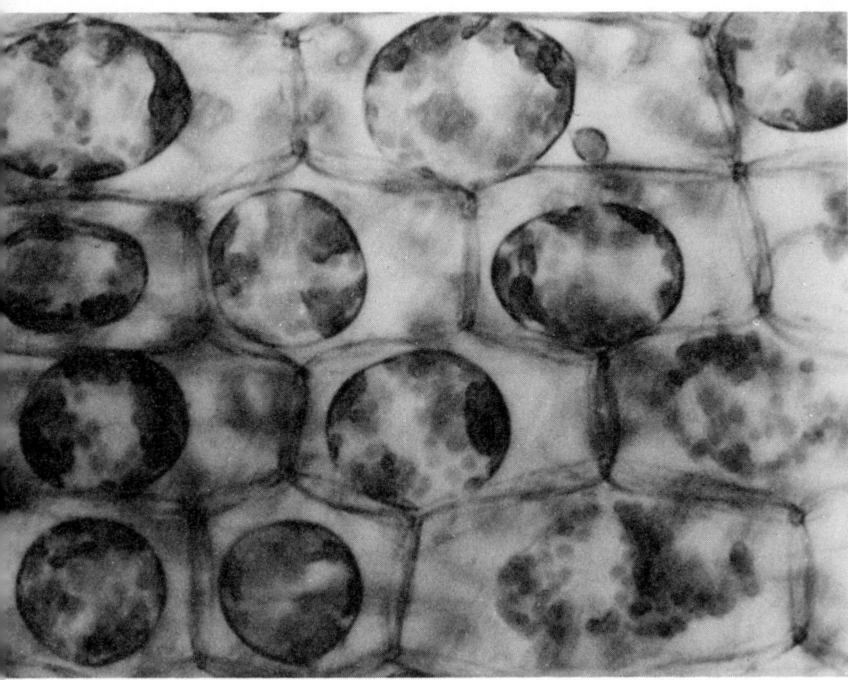

FIG. 6.6 Plasmolyzed cells in the freshwater plant *Elodea,* which has been placed in sea water. The space between the cell membranes and the cell walls has become filled with sea water.

FIG. 6.7 The salt glands of the herring gull. The fluid excreted by the glands is saltier than the blood.

of the mixture. The cytoplasm of these marine organisms may contain as little as 80–90% water by weight. However, much of the remaining material in the cytoplasm consists of protein. These macromolecules make up a substantial fraction of the weight of the cytoplasm but contribute only a minor osmotic effect because of the relatively small number of molecules involved. Similarly, a 0.9% salt solution (99.1% water) is isotonic to human blood plasma although the latter contains only 90% water by weight. The number of water molecules in a given volume of each solution is, however, the same.

When a freshwater (or terrestrial) plant is placed in sea water, its cells quickly lose turgor and the plant wilts. This is because a given volume of sea water contains a smaller number of water molecules than a given volume of the cytoplasm of these plants. The sea water is said to be **hypertonic** to the cytoplasm. As water continues to diffuse from the cytoplasm into the sea water, the cells gradually shrink. This condition is known as **plasmolysis.** Note in Fig 6.6 how the cells have pulled away from their walls, which still retain their original shape.

The ECF of bony fishes has a water concentration which is substantially higher than that of sea water. Thus they live in a medium which is hypertonic. Whereas freshwater organisms have to cope with water passing continually into their body by osmosis, the saltwater bony fish continually loses water by osmosis. Once again, however, survival depends upon the expenditure of energy to combat the force of osmosis. The fish drinks sea water and then uses metabolic energy to desalt it. The salt is transported back into the environment at the gills. Marine birds, which sometimes pass long periods of time away from fresh water, and sea turtles use a similar device. They, too, drink salt water to take care of their water needs and use metabolic energy to desalt it. The salt is extracted by two glands in the head and released (in a very concentrated solution) to the outside through the nostrils (Fig. 6.7). Marine snakes use a similar desalting mechanism.

6.5 ACTIVE TRANSPORT

The cytoplasm of the human red blood cell contains a concentration of potassium ions 30 times greater than that in its extracellular fluid (blood plasma). On the other hand, blood plasma has a concentration of sodium ions almost 6 times as great as that in the red cell. Even if the diffusion of these ions across the cell membrane is slow, how can we account for such steep concentration gradients? If we chill the red cells, or deprive them of glucose, or treat them with some poison that interferes with energy release within the cell, the potassium begins to leak out (by diffusion) and the sodium leaks in until equilibrium is reached. Reversal of any of the inhibitory steps taken will cause these two ions to begin again to move *against* the concentration gradient. This movement of ions and molecules against a concentration gradient is known as **active transport.** It is described as active because—as the experiments above suggest—the cell must expend energy to accomplish the transport against the passive force of diffusion.

The ability of cells to actively transport ions and molecules to and from the extracellular fluid is a widespread one. Marine organisms frequently have certain ions in their cytoplasm in concentrations a thousand or more times greater than in the surrounding sea water. The cells lining the intestine can transport glucose actively from a lower concentration in the intestinal contents to a higher concentration in the blood.

The actual mechanism by which active transport occurs is still the subject of vigorous research. It is clear, however, that this uphill, pumping process requires the expenditure of energy. It also requires the presence of enzymes, presumably located in the cell membrane, which serve to move molecules and ions from the side of low concentration to the side of high concentration. While the active transport of some materials may share a *common* pumping mechanism, other substances, e.g., certain sugars and amino acids, may each require their own specific pumping mechanism.

6.6 ENDOCYTOSIS

Still another mechanism by which the cell transports materials from the ECF into the interior is to engulf them by folding inward a portion of the cell membrane. The pouch that results then breaks loose from the outer portion of the membrane and forms a vacuole within the cytoplasm. In some cells, solid particles like bacteria may be engulfed in this manner. The amoeba derives its nourishment by ingesting smaller microorganisms in this fashion (Fig. 6.8). White blood cells serve the extremely valuable function of engulfing particles, such as bacteria, that may get into the animal's body. This process, which has been observed since late in the last century, has been called **phagocytosis** ("cell eating"). However, in recent years it has been shown that many types of cells are able to engulf droplets of ECF by forming vacuoles, although these are too small to contain solid particles. Whatever the size and contents of the vacuole, though, the mechanism at work (which requires functioning microfilaments) seems to be the

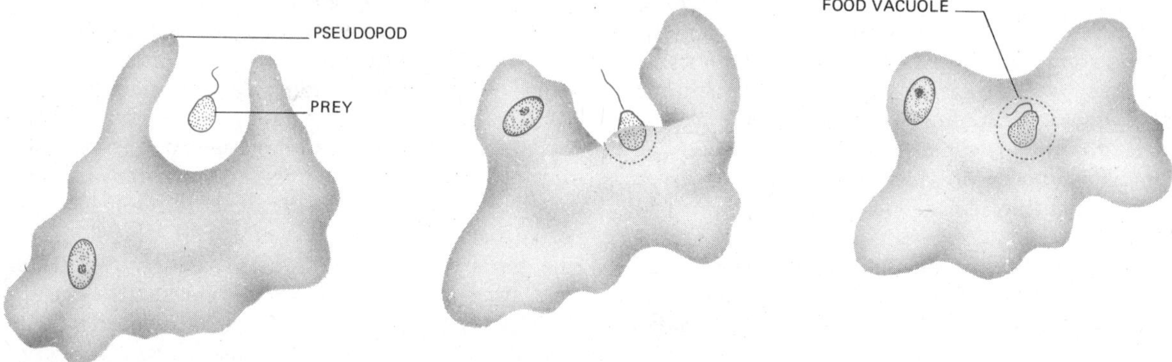

FIG. 6.8 Endocytosis in the amoeba. Even when totally engulfed, the victim is separated from the cytoplasm of the amoeba by a membrane around the food vacuole. Endocytosis of solid particles is also called phagocytosis.

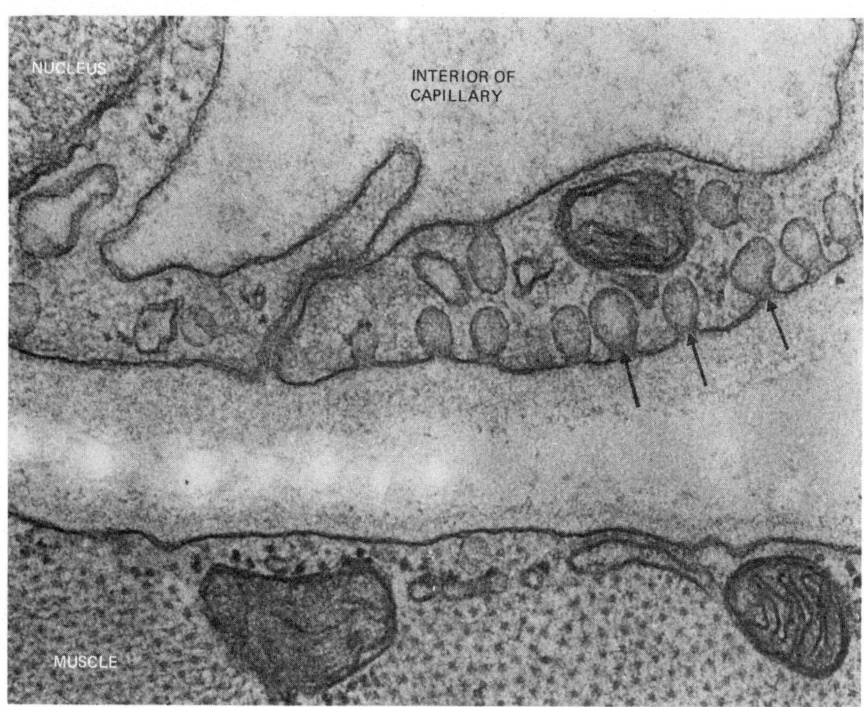

FIG. 6.9 Endocytosis in the thin cell that forms the wall of a capillary. Note that the inpocketings have formed on both surfaces of the cell. (67,500 ×, from Fawcett, *The Cell: Its Organelles and Inclusions,* W. B. Saunders Co., 1966.)

same. Therefore, the all-inclusive term *endocytosis* is now coming into common use for this process.

Endocytosis permits macromolecules like proteins—too large and too hydrophilic for simple diffusion—to gain entry into the cell. Figure 6.9 is an electron micrograph of a section of the wall of a capillary (the smallest type of blood vessel in our bodies). At the top is the interior or bore of the capillary. In the middle is the tissue space separating the capillary wall from a nearby muscle cell (bottom). The small inpocketings of the cell membrane are clearly seen. Most of these are engulfing the ECF of the tissue space, but some can also be seen on the other side of the wall, apparently engulfing fluid from within the capillary.

The mere act of folding in a bit of the cell membrane with its content of materials from the extracellular fluid does not actually get these materials into the cytosol. They are retained within membrane-bounded vacuoles. Quickly, however, one or more lysosomes (see Section 5.10) fuse with the endocytic vacuole and the contents of the organelles merge. The resulting structure is called a *secondary lysosome* (Fig. 6.10). Lysosomes, you remember, are filled with enzymes capable of degrading all types of macromolecules. The soluble products of

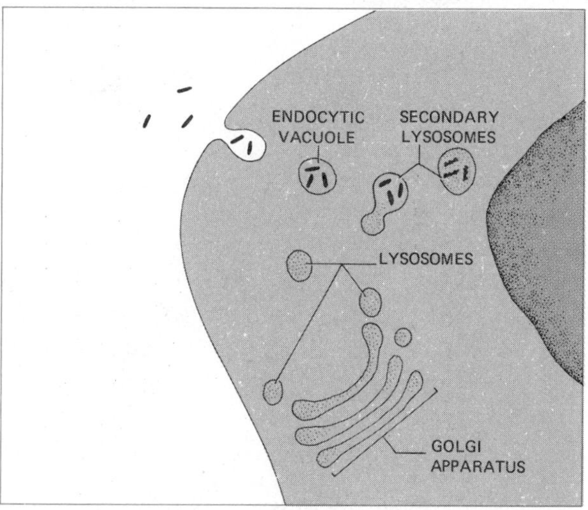

FIG. 6.10 The vacuoles formed by endocytosis fuse with lysosomes. The resulting structures are the secondary lysosomes. Secondary lysosomes can themselves fuse with additional endocytic vacuoles. Lysosomes are manufactured by the Golgi apparatus.

these hydrolyses (e.g., amino acids, sugars, nucleotides) can then pass out of the secondary lysosome and into the cytosol.

This by no means exhausts the capacity of the lysosomal enzymes. Secondary lysosomes can also fuse with newly formed endocytic vacuoles and digest *their* contents.

6.7 EXOCYTOSIS

Exocytosis is the reverse of endocytosis. In cells that secrete large amounts of protein, the protein first accumulates in a membrane-bounded sac within the Golgi apparatus. This moves to the surface of the cell where its membrane fuses with the cell membrane, and it discharges its contents to the outside (Fig. 6.11). The cells lining our intestine synthesize tiny droplets of fat and then discharge these from the cell by exocytosis. It may even be that some of the tiny vacuoles shown in Fig. 6.9 are not taking up material by endocytosis, but are instead discharging material by exocytosis. In other words, the endocytic vacuoles formed at one surface of the cell may, after being detached, move through the cell to the other surface and there discharge their contents. In this way, materials can be moved efficiently through the capillary wall.

6.8 CELL CHEMISTRY

The cell is a veritable chemical factory. Literally hundreds of chemical reactions that occur within cells have been discovered and the list grows longer every year.

In our brief glimpse at chemical reactions in Chapter 3, we learned that some reactions result in a release of free energy ($-\Delta F$) and some require an input of free energy ($+\Delta F$). An example of the former is the burning of any fuel: wood, coal, petroleum, or—of special interest to us—glucose. The equation for the complete combustion of glucose is

$$C_6H_{12}O_6 + 6O_2 \rightarrow 6CO_2 + 6H_2O.$$

ΔF in this case is 686 kcal, that is, the burning of one mole (180g) of glucose liberates 686 kcal of energy. The ΔF value represents the difference be-

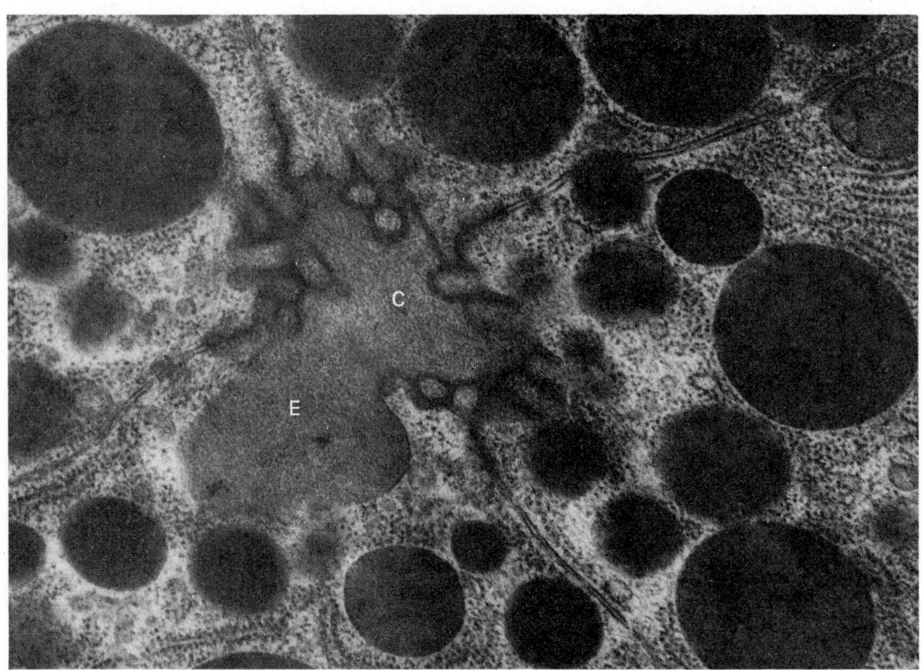

FIG. 6.11 Exocytosis. The large spherical bodies contain digestive enzymes. One is seen in the process of discharging its contents (E) into the intercellular canal (C) where the corners of these four bat pancreas cells meet. (30,000 ×, from Fawcett, *The Cell: Its Organelles and Inclusions,* W. B. Saunders Co., 1966.)

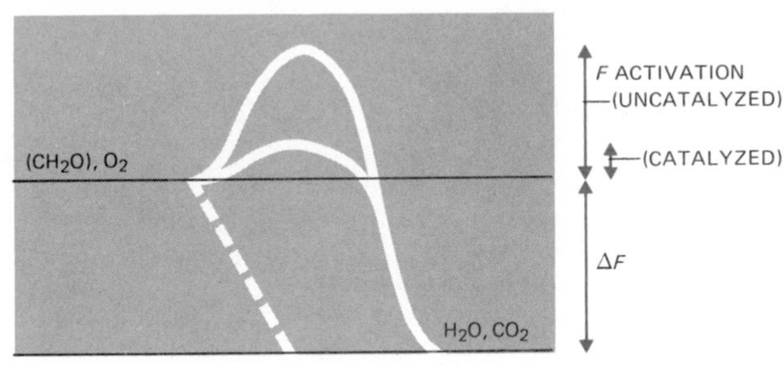

FIG. 6.12 Energy relationships in the burning or respiration of a carbohydrate such as glucose. A catalyst reduces the amount of activation energy needed, thus enabling the reaction to proceed more rapidly.

tween the sum of the bond energies of the reactants and the sum of the bond energies of the products. In Fig. 6.12, this is represented graphically by the two horizontal lines. As the graph suggests, the process is a "downhill" one. It takes less energy to break the bonds of the reactants than you get back by the formation of the bonds of the products. Why, then, does not the process occur spontaneously? Why does not a dish of glucose burst into flame upon exposure to oxygen?

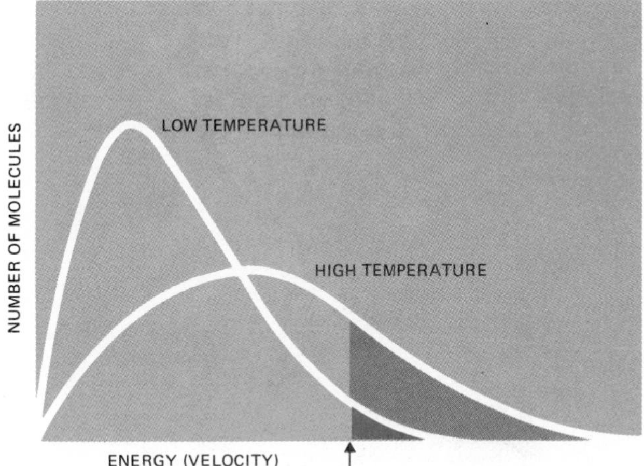

FIG. 6.13 Distribution of energies in two populations of molecules, one maintained at a low temperature, the other at a high temperature. The arrow represents the threshold energy needed for these molecules to react chemically. At high temperatures, a larger proportion of the molecules exceed the threshold energy and the reaction proceeds more rapidly.

The reason is that the process cannot occur without the input of a certain amount of energy called **energy of activation.** One way of looking at this requirement is in terms of the energy input needed to break the bonds of the reactants so that atomic rearrangements can occur. That you will ultimately get back more energy than you put in does not eliminate the need to prime the pump. This can be represented graphically by drawing a hill or barrier which must be overcome before the chemical reaction can occur. Another way of looking at the problem is in terms of the vigor of the collisions which must occur if the reactant molecules are to be able to shift electrons from one orbital to another. At a given temperature, any population of molecules will contain many that are moving at or close to an *average* velocity (and temperature *is a measure of* molecular velocity) with a decreasing number moving at lower and higher velocities (Fig. 6.13).

Let us say that only molecules possessing velocities greater than a value indicated by the arrow in Fig. 6.13 have sufficient energy to collide effectively with each other and accomplish chemical change. Under "low temperature" conditions (Fig. 6.13), then, the reaction might well be expected to go very slowly. This in effect is the situation with a teaspoon of glucose crystals exposed to the air.

How can we speed up the rate at which this otherwise energetically favorable ($-\Delta F$) reaction can take place? One obvious way is to raise the temperature. By so doing, a larger proportion of the molecules will acquire the necessary energy (Fig. 6.13), more and more vigorous collisions will occur, and the reaction will proceed more rapidly. The flame of a bunsen burner applied to the glucose will quickly ignite it, whereupon the glucose will begin to burn merrily without further help from the bunsen burner. The $-\Delta F$ of the ignited molecules will provide the necessary activation energy for the rest, and the combustion will go on spontaneously until the process is completed.

Of course, the temperature of a flame is certainly not compatible with life. How, then, do living cells accomplish their chemical reactions at rates high enough to be useful but at temperatures low enough to be safe? The answer is by using enzymes.

6.9 ENZYMES

Enzymes are catalysts. They are substances (proteins) that temporarily bind to one or more of the reactants. In so doing, they lower the energy bar-

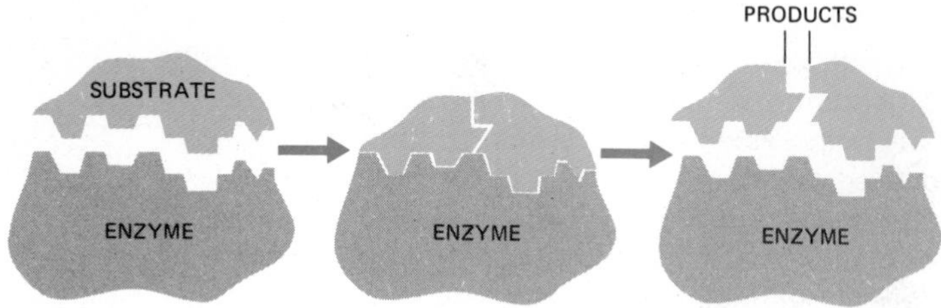

FIG. 6.14 The union of an enzyme with its substrate facilitates the rearrangement of one or more covalent bonds in the substrate by lowering the activation energy needed for breaking these bonds.

rier—the amount of activation energy needed—for the reaction (Fig. 6.12) and thus allow it to take place rapidly. In Chapter 8 we shall examine a whole series of enzyme-catalyzed reactions by which the $-\Delta F$ of the oxidation of glucose is rapidly but safely (i.e., at moderate temperatures) released within the cell.

In order to do its work, an enzyme must unite—even if ever so briefly—with at least one of the reactants. In most cases, the forces that hold the enzyme and its substrate together are not covalent bonds. Instead, hydrogen bonds, ionic bonds, and the mutual attraction of hydrophobic regions of the two molecules may singly or in concert bind the substrate briefly to the enzyme. Most of these interactions are weak and especially so if the atoms involved are not in very close (~1 A) proximity. Consequently, satisfactory binding of enzyme and substrate must require that the two molecules approach each other very closely and over an area large enough to allow the operation of a number of these weak forces of attraction. Thus the substrate molecule must fit into a complementary surface on the enzyme molecule in somewhat the same way as a key fits in a lock (Fig. 6.14).

The requirement for complementarity in the configuration of substrate and enzyme explains the remarkable **specificity** of most enzymes. In general, a given enzyme is able to catalyze only a single chemical reaction or, at best, a few reactions involving substrates sharing the same general structure.

This requirement for a brief, close union between enzyme and substrate also explains the phenomenon of **competitive inhibition.** One of the enzymes needed for the release of energy within cells is succinic dehydrogenase. This enzyme catalyzes the oxidation (by the removal of two hydrogen atoms) of succinic acid (Fig. 6.15). The product of this action is fumaric acid. If one adds *malonic* acid to cells or even to a test tube mixture of succinic acid and the enzyme, the normal action of the enzyme is strongly inhibited. What has happened? The structure of malonic acid is quite similar to that of succinic acid and presumably it binds to the same site on the enzyme. But no oxidation can occur, so there is no speedy release of the products. Thus whenever a

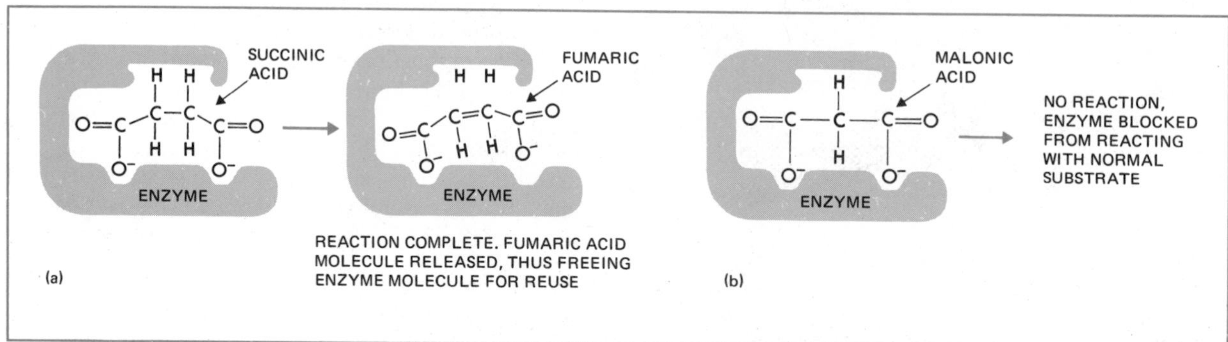

FIG. 6.15 (a) Schematic representation of the action of the enzyme succinic dehydrogenase on succinic acid. (b) The competitive inhibition of the enzyme by malonic acid.

molecule of malonic acid binds to an enzyme molecule, it prevents that enzyme molecule from binding to its normal substrate, succinic acid. The inhibition is called competitive, because if you increase the concentration of succinic acid relative to malonic acid, you will gradually restore the rate of catalysis. In fact, at a 50:1 ratio of succinic acid molecules to malonic acid molecules, the two substances compete on roughly equal terms for the binding (= catalytic) site on the enzyme.

A direct demonstration of the union of substrate and enzyme would be the isolation of such a complex. This, however, is exceedingly difficult to do because of the fleeting nature of the union between the two. For example, the enzyme catalase catalyzes the decomposition of hydrogen peroxide into water and oxygen: $2H_2O_2 \rightarrow 2H_2O + O_2$. It binds to H_2O_2, catalyzes the decomposition, and releases the products so rapidly that a single molecule can break down 5.6 million H_2O_2 molecules in a minute. Carbonic anhydrase, an enzyme found in red blood cells, works even faster. It catalyzes the reaction $CO_2 + H_2O \rightleftharpoons H_2CO_3$ and can process 36 million molecules per minute.

The great speed with which enzymes work is a distinct advantage for the cell, since it enables the cell to get along with only a tiny amount of each enzyme. A single cell might well be capable of carrying out close to 1000 different chemical reactions, most of which require their own specific enzymes. It is the ability to reuse enzymes rapidly that makes it possible for the cell to accommodate such varied enzymatic machinery within a small volume.

Enzyme-Substrate Interaction: An Example

Despite the speed with which enzymes work, it *has* been possible to isolate certain enzyme-substrate complexes. In these cases, covalent bonds unite the two and, with the appropriate tricks, the breaking of these bonds—normally a part of the catalytic action—can be prevented.

In order to demonstrate the complementarity that we believe exists between an enzyme and its substrate, however, it is necessary to know the precise shape of each. Enzymes, being proteins, are very large molecules and determination of their tertiary (three-dimensional) structure is a particularly laborious procedure. Nonetheless, the tertiary structure of a number of enzymes has been established. One of these is lysozyme.

Lysozyme, you may recall, is an enzyme that is found in egg white and such body secretions as tears. It has a strong antibacterial action because its substrate is a polysaccharide that is found in the cell walls of many bacteria. This bacterial polysaccharide consists of long chains of hexose sugars. The sugar units resemble glucose except for the presence of certain side chains, one of which contains nitrogen. The linkage of these so-called amino sugars is like the linkage that occurs in cellulose (Fig. 6.16).

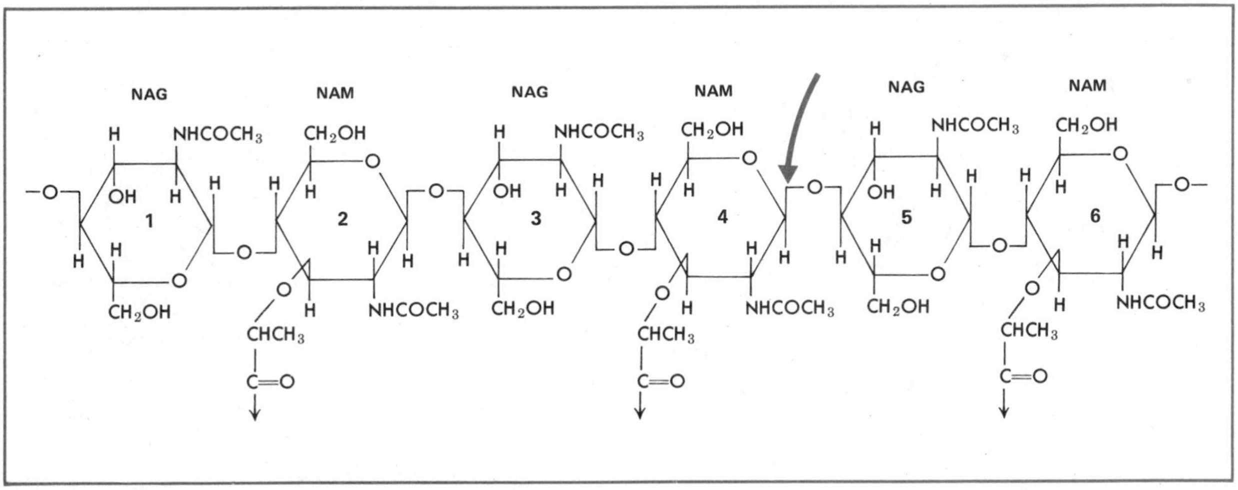

FIG. 6.16 Structure of the polysaccharide cleaved by lysozyme. The alternating monomers are the amino sugars NAG (N-acetylglucosamine) and NAM (N-acetylmuramic acid). The short arrows indicate points of attachment to adjacent chains of the polysaccharide. The large arrow indicates the bond that is broken by lysozyme.

If you look back at the space-filling model of lysozyme in Chapter 4 (Fig. 4.25), you will note that the molecule is roughly globular except for a deep depression or cleft shown at the upper left. It is into this cleft that the substrate fits (Fig. 6.17). Six links of the polysaccharide chain, that is, six amino sugars, just fill the cleft. With the many oxygen atoms in sugars, there is plenty of raw material for hydrogen bonding. Probably 14 hydrogen bonds form between the six amino sugars and certain amino acid side chains, e.g., arginine (#114), asparagine (#37,44), tryptophan (#62,63), aspartic acid (#101), and also with the C=O groups of several peptide linkages. In addition, hydrophobic attractions may also help hold the substrate in position.

X-ray crystallographic analysis reveals two important features of the binding of lysozyme to its

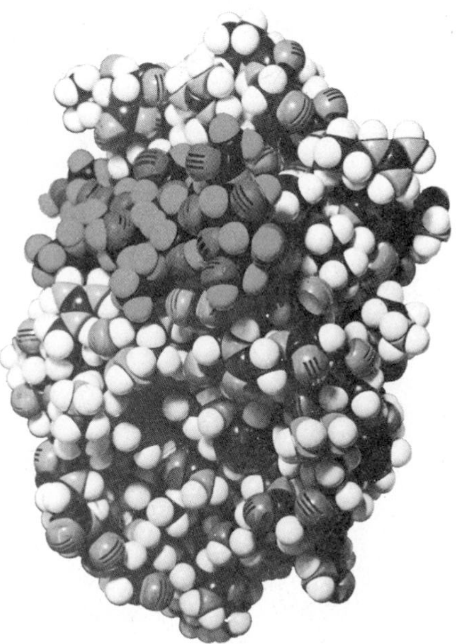

FIG. 6.17 Space-filling model of lysozyme bound to its hexasaccharide substrate (in color). (Model by Prof. John A. Rupley; courtesy of Irving Geis.)

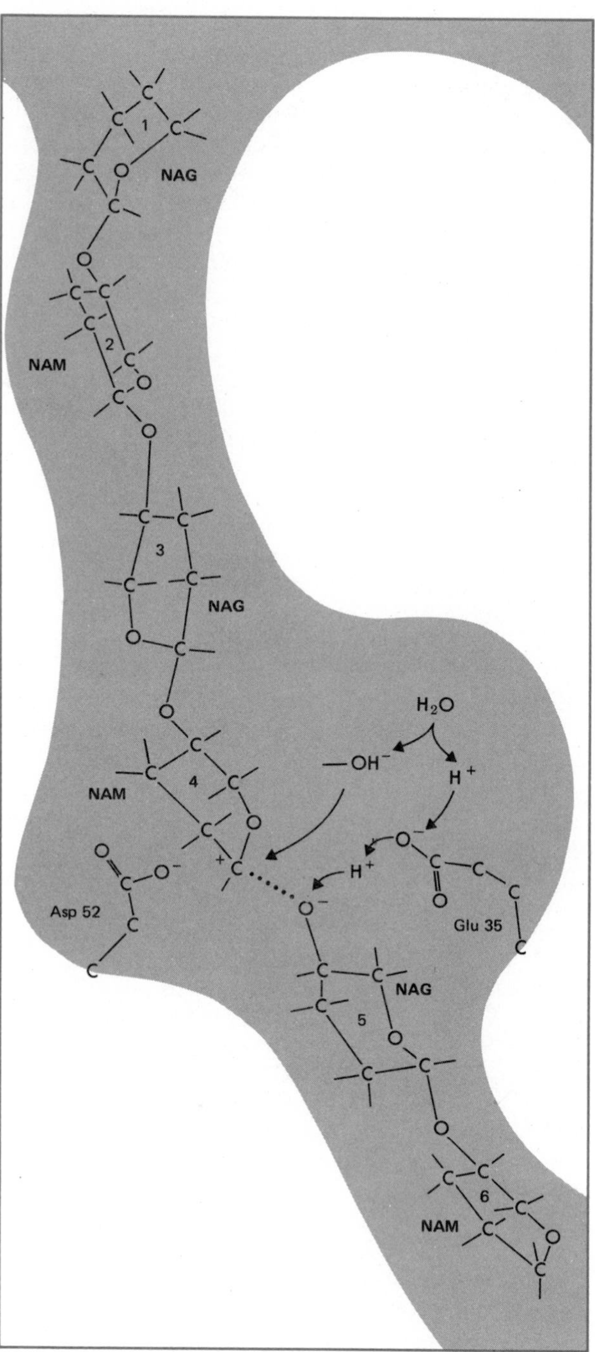

substrate. As the two unite, each is slightly distorted or deformed. The fourth hexose unit (ring 4—see Fig. 6.18) becomes twisted out of its normal configuration. It is thought that this imposes a strain on the C—O bond on the ring 4 side of the oxygen

◀ **FIG. 6.18** Possible mechanism of the catalytic action of lysozyme. Binding of the substrate induces a strain between rings 4 and 5. Residue 35 of the enzyme molecule, glutamic acid, donates a proton to the oxygen atom between the rings. This breaks its bond to the carbon atom of ring 4. The carbon atom, now positively charged, is stabilized (by its attraction to aspartic acid 52) long enough for a hydroxyl group to be attached to it. In this way, the chain is broken by the insertion of a molecule of water, an example of hydrolysis. The products are released from the enzyme and the process can be repeated.

bridge between rings 4 and 5. This is an exciting finding because it is just at this position that the polysaccharide is hydrolyzed. A molecule of H_2O is inserted between these two hexoses with the result that the chain breaks. Here, then, is a physical model of what we mean by saying that enzymes lower activation energy. The energy needed to break this bond is surely lower now that the atoms connected by this bond have been distorted from their normal position.

As for the lysozyme molecule itself, binding of the substrate seems to induce a small (about 0.75 A) movement of certain amino acid residues with the result that the cleft closes slightly over its substrate. Thus we find that our "lock" as well as our "key" changes shape slightly as the two are brought together. It is thought by some biochemists that whenever any substrate binds to its enzyme, it induces a conformational change in the structure of the enzyme.

An examination of the amino acid residues in the vicinity of rings 4 and 5 provides a plausible mechanism for completing the catalytic action. Residue #35, glutamic acid (Glu), is about 3 A from the —O— bridge that is to be broken. The free carboxyl group of glutamic acid is a proton donor and could transfer its proton to the oxygen atom. This would break the already strained bond between the oxygen atom and the carbon atom of ring 4 (Fig. 6.18). Now having lost an electron, the carbon atom acquires a positive charge. Ionized carbon is normally very unstable, but the attraction of the negatively charged carboxyl ion of residue #52 (aspartic acid) could stabilize it long enough for an OH^- ion (from a spontaneously dissociated water molecule) to unite with the carbon. (Remember that even at pH 7.0, water spontaneously dissociates to produce 10^{-7} *M* H^+ and OH^- ions). The proton (H^+) left over can replace the proton lost by Glu-35.

The reaction is now complete: the chain is broken, the two fragments separate from the enzyme, and the enzyme is free to attach to a new location on the bacterial cell wall and to go to work again. In this way, the structure of the bacterial cell wall is broken down.

Requirements for Effective Enzyme Functioning

The activity of enzymes is apt to be affected strongly by changes in temperature and pH. Each enzyme works most effectively at a certain temperature and pH, its activity diminishing at values above or below that point (Fig. 6.19). The protein-digesting enzyme pepsin works most effectively at a pH of 1–2, while another proteolytic enzyme, trypsin, is quite inactive at that pH but functions effectively at a pH of 8. Now that we understand (1) the essential role that tertiary structure, i.e., shape, plays in enzyme function and (2) the role that such weak forces as hydrogen bonds and ionic bonds play in determining tertiary structure, we can understand why enzymes should be so sensitive to temperature and pH. Hydrogen bonds are easily disrupted by increasing temperature. This may in turn disrupt portions of the tertiary structure of the enzyme that are essential for binding the substrate. Changes in pH will alter the state of ionization of charged

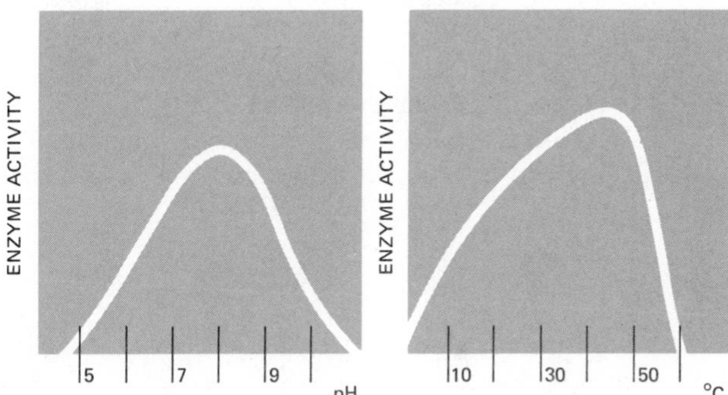

FIG. 6.19 Effects of pH and temperature on the activity of a hypothetical enzyme. The ascending portion of the temperature curve reflects the general effect of increasing temperature on the rate of chemical reactions (see Fig. 6.13). The descending portion of the curve reflects the loss of catalytic activity as the enzyme molecules become denatured at high temperatures.

amino acids (e.g., aspartic acid, lysine) that may play crucial roles in substrate binding and/or the catalytic event itself. Without the un-ionized —COOH group of Glu-35 and the ionized $—COO^-$ group of Asp-52, the catalytic action of lysozyme would cease.

Many enzymes will not work without the presence of an additional, nonprotein substance called a **cofactor.** Cofactors may be a metal ion such as Zn^{++} (a cofactor for carbonic anhydrase), Cu^{++}, Mn^{++}, Mg^{++}, K^+, Fe^{++}, or Na^+. Or the cofactor may be a small organic molecule called a coenzyme. The B vitamins, for example thiamine (B1), riboflavin (B2), and nicotinamide, function as coenzymes. Coenzymes may be tightly (covalently) bound to the protein part of the enzyme as a **prosthetic group.** Others bind more loosely and, in fact, may bind only transiently to the enzyme as it carries out its catalytic function.

6.10 REGULATION OF ENZYMES

Considering the large number of enzymes in the living cell, one might well ask how their activity is regulated. A number of mechanisms play a part in making enzyme action within the cell efficient and well coordinated.

For those enzymes, such as proteinases, which can attack the very substance of the cell itself, we find that their action is inhibited while they are present within the cell. The proteinase pepsin, for example, is manufactured by the cell in an inactive form, pepsinogen. Only when exposed to conditions of low pH outside of the cell is the inhibiting portion of the enzyme molecule removed and the active pepsin produced. Other potentially destructive enzymes are sequestered in the lysosomes (see Section 5.10) and thus isolated from the rest of the cell.

Many of the cell's enzymes cannot move freely within the cell but are, instead, arranged in definite patterns. The enzymes within mitochondria and chloroplasts appear to be organized spatially in such a way that they interact with the greatest efficiency. It is quite likely that spatially organized enzyme molecules are also present on the cell membrane and the membranes of the endoplasmic reticulum.

The activity of enzymes within the cell is also closely regulated by the need for them. If the product of a series of enzymatic reactions (e.g., an amino acid) begins to accumulate within the cell, it specifically inhibits the action of the first enzyme involved in its synthesis (Fig. 6.20). Thus further production of that amino acid is temporarily halted. This phenomenon is called **feedback inhibition.** On the other hand, the accumulation of a substance within the cell may specifically activate an enzyme that sets in motion a sequence of reactions for which that substance is the initial substrate (Fig. 6.20). This action (which is called **precursor activation**) reduces the concentration of the initial substrate to normal levels.

Note that in both these situations, the activity of an enzyme is being regulated by a substance which is *not* its substrate. We might expect that the regulating molecule would have to bind to the enzyme in order to affect its function, and this is indeed the case. Interestingly, the site where the regulatory molecule binds to the enzyme is not the same as the site where the substrate binds. However, when the regulatory molecule binds to its site, it appears to alter the shape of the enzyme in such a way that its activity is altered. This interaction between regulatory and catalytic sites—brought about by a change in the shape of the enzyme—is called an **allosteric** effect. In the case of feedback inhibition, the regulatory molecule exerts its allosteric effect so that the affinity of the enzyme for its substrate is lowered. In the case of precursor activation, the situation is the opposite. The allosteric regulator

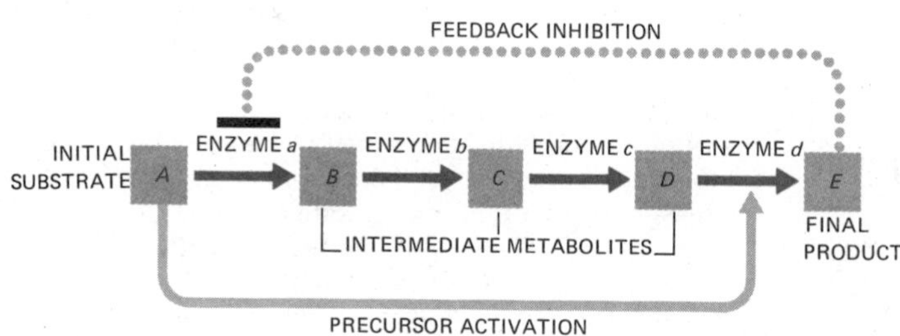

FIG. 6.20 Two distinct homeostatic mechanisms that regulate concentrations of metabolites within the cell. An accumulation of substrate *A* activates enzyme *d*, thus drawing *A* into the metabolic pathway. An accumulation of product *E* suppresses the action of enzyme *a*, thus inhibiting the further synthesis of *E*.

molecule increases the affinity of the final enzyme in the series for its substrate (Fig. 6.21).

The mechanisms mentioned above ensure that the activity of enzymes already present in the cell will be properly regulated. What of enzymes that may not be needed at all or that may be needed but are not present? Here, too, delicate controls are at work. These regulate the rate at which new enzymes are synthesized. If, for example, excess quantities of an amino acid are supplied to a cell from its ECF, the synthesis of all the enzymes by which the cell ordinarily would produce that amino acid for itself will be halted. Thus the cell enhances its efficiency by not producing enzymes that it does not need. Conversely, if a new substrate is made available to the cell, it will stimulate the synthesis of the enzymes needed to cope with it. Yeast cells do not ordinarily ferment the disaccharide lactose and ordinarily no lact*ase* can be detected within their cells. If grown in solutions containing lactose, however, they eventually begin producing lactase and begin to metabolize the sugar.

In these cases where the *synthesis* of enzymes is being regulated, it is clear that this regulation works through the hereditary controls coded in the DNA of the nucleus. The mechanisms by which portions of the hereditary code are thought to be turned on and off in response to the needs of the cell will be examined in later chapters.

Whether acting on the enzymes already present within the cell or on the rate of synthesis of new enzymes, it is clear that these control mechanisms work together to stabilize the levels of substrates and products in the cell. In a sense, then, these mechanisms are homeostatic devices, working within the cell, that regulate enzyme activity with the utmost efficiency and in harmony with the changing needs of the cell.

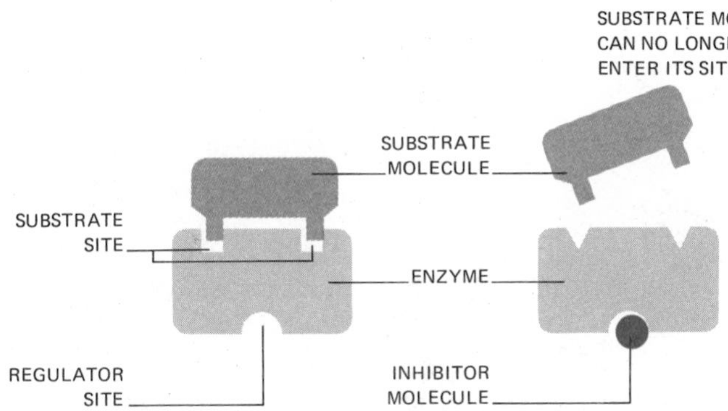

FIG. 6.21 Probable mechanism of feedback inhibition. In the case of precursor activation, the binding of the regulatory molecule would *permit* the substrate to bind to its site. In both cases, the regulator molecule has altered the affinity of the enzyme for its substrate, perhaps through causing a change in the shape of the enzyme. Such an effect is called allosteric.

EXERCISES AND PROBLEMS

1. Describe three different ways in which materials in the human intestine might enter the cells lining it.
2. After the molasses reaches its maximum height (see Fig. 6.3), what will happen? Why?
3. Which will have the higher osmotic pressure, a 1-molar solution of glucose or a 1-molar solution of salt? Why?
4. What would happen if (a) red blood cells, (b) plant cells, (c) an amoeba were placed in distilled water? Explain.
5. What would happen if (a) red blood cells or (b) plant cells were placed in sea water? Explain.
6. What would happen if an amoeba were placed in an isotonic solution?
7. If a mole (28 g) of nitrogen molecules (N_2) and a mole (28 g) of ethylene molecules (CH_2CH_2) were released on opposite sides of a partition that divided a room into equal volumes and then the partition were removed, do you think diffusion of the molecules would occur? Explain.

REFERENCES

1. HOLTER, H., "How Things Get into Cells," *Scientific American,* Offprint No. 96, September, 1961. An excellent review of the forces of passive and active transport across cell membranes.
2. SOLOMON, A. K., "Pores in the Cell Membrane," *Scientific American,* Offprint No. 76, December, 1960. A beautiful illustration of the way physical analysis and experimentation can unlock the secrets of biological function.
3. SATIR, BIRGIT, "The Final Steps in Secretion," *Scientific American,* Offprint No. 1328, October, 1975. On exocytosis.
4. PHILLIPS, D. C., "The Three-dimensional Structure of an Enzyme Molecule," *Scientific American,* Offprint No. 1055, November, 1966. Describes how the shape of the enzyme lysozyme was worked out and how this shape accounts for the antibacterial action of the enzyme.
5. CHANGEUX, J.-P., "The Control of Biochemical Reactions," *Scientific American,* Offprint No. 1008, April, 1965.
6. HOLLAWAY, M. R., *The Mechanism of Enzyme Action,* Oxford Biology Readers, No. 45, Oxford University Press, Oxford, 1976.
7. STROUD, R. M., "A Family of Protein-Cutting Proteins," *Scientific American,* Offprint No. 1301, July, 1974. Describes the mechanism of action of several proteases.
8. KOSHLAND, D. E., JR., "Protein Shape and Biological Control," *Scientific American,* Offprint No. 1280, October, 1973. The importance of allosteric interactions.

7 HETEROTROPHIC NUTRITION

All organisms require a fairly steady supply of materials and energy from the environment in order to stay alive. For many, the chief supply of materials and the only supply of energy come from fairly complex, energy-rich, organic molecules secured directly or indirectly from the environment. (What is the ultimate source of these molecules?) Nutrition that involves dependence upon preformed organic molecules is called **heterotrophic** nutrition, and the organisms using this kind of nutrition are called heterotrophs. The nonchlorophyll-containing microorganisms, the few nongreen plants, and all animals are heterotrophic.

7.1 REQUIREMENTS

The organic molecules that serve as a source of material and energy are the sugars, amino acids, fatty acids and glycerol, and (for purposes of synthesis only) the vitamins. Not all heterotrophs depend upon *all* these organic molecules. Some microorganisms, for example the bacterium *E. coli,* thrive with just sugar as their source of energy. They must, however, take in some inorganic materials such as nitrates in order to synthesize all their other organic constituents.

Humans are quite demanding in their requirements for preformed organic molecules. In addition to a source of carbohydrate, we must take in, ready-made, eight of the twenty amino acids used to synthesize proteins. These are the so-called essential amino acids (Fig. 7.1); from them we can synthesize the other 12. Although we can synthesize saturated fats from sugars, certain unsaturated fats must be included in the diet. Eight or more vitamins complete the list of our dietary needs (Fig. 7.2).

THE ESSENTIAL AMINO ACIDS

ARGININE*
HISTIDINE*
ISOLEUCINE
LEUCINE
LYSINE
METHIONINE
PHENYLALANINE
THREONINE
TRYPTOPHAN
VALINE

*For children only

FIG. 7.1 Of the twenty amino acids used in the synthesis of proteins, these ten must be included in the diet. They cannot be synthesized in the body from other precursors. Many plant proteins do not contain sufficient amounts of lysine and tryptophan to meet human dietary needs.

7.2 INTRACELLULAR DIGESTION

Solid food materials are usually broken down into a solution of relatively small, soluble, organic molecules before they can be used by heterotrophic organisms. This breakdown process is called **digestion.** In some heterotrophic organisms, digestion is intracellular, that is, it occurs after the solid material has actually been engulfed by a cell.

The amoeba engulfs solid particles such as smaller protozoans by endocytosis (see Section 6.6). The prey is incorporated into a **food vacuole** within the cytoplasm of the amoeba. Next, the digestible portions of the food material are digested by enzymes deposited in the vacuole by lysosomes that fuse with it (Fig. 7.3). The soluble food molecules then pass through the vacuolar membrane into the rest of the cell. Indigestible parts are eventually discharged to the outside. Although digestion in the amoeba can properly be described as intracellular, we should keep in mind the definite membrane that persists between the material in its food vacuole and the remainder of the cytoplasm.

It is obvious that endocytosis can occur only if the food materials available to the organism are smaller than its endocytic cells. It is not surprising, therefore, that feeding by endocytosis is restricted to those organisms that are adapted to secure food materials much smaller than themselves.

7.3 EXTRACELLULAR DIGESTION

A second solution to the problem of digesting food is to secrete the digestive enzymes from the cell and digest the food outside the cell, that is, extracellularly. Once the food is digested, the small, soluble molecules produced (e.g. sugars, amino acids) can pass by diffusion or active transport across the cell membrane and into the cell.

Perhaps the simplest approach to extracellular digestion is that employed by **saprophytes.** Saprophytes secure their food from nonliving but organic matter, such as dead bodies of plants and animals, food products, excrement, etc. The nutrition of the common bread mold *Rhizopus stolonifer* is typical of the group. It thrives on a piece of moist bread kept in a dark location (Fig. 7.4). The bread, produced by humans from a once-living wheat plant, supplies all the dietary needs of the mold. The starch

VITAMIN	DEFICIENCY DISEASE	SOURCES	OTHER INFORMATION
A (RETINOL)	NIGHT BLINDNESS	MILK, BUTTER, FISH LIVER OILS, CARROTS, AND OTHER VEGETABLES	PRECURSOR IN THE SYNTHESIS OF THE LIGHT-ABSORBING PIGMENTS OF THE EYE STORED IN THE LIVER TOXIC IN LARGE DOSES
THIAMINE (B_1)	BERIBERI DAMAGE TO NERVES AND HEART	YEAST, MEAT, UNPOLISHED CEREAL GRAINS	COENZYME IN CELLULAR RESPIRATION
RIBOFLAVIN (B_2)	INFLAMMATION OF THE TONGUE DAMAGE TO THE EYES GENERAL WEAKNESS	LIVER, EGGS, CHEESE, MILK	PROSTHETIC GROUP OF FLAVOPROTEIN ENZYMES USED IN CELLULAR RESPIRATION
NICOTINIC ACID (NIACIN)	PELLAGRA (DAMAGE TO SKIN, LINING OF INTESTINE AND PERHAPS NERVES)	MEAT, YEAST, MILK	CONVERTED INTO NICOTINAMIDE, A PRECURSOR OF NAD AND NADP—TWO IMPORTANT COENZYMES FOR REDOX REACTIONS IN THE CELL
FOLIC ACID (FOLACIN)	ANEMIA	GREEN LEAFY VEGETABLES SYNTHESIZED BY INTESTINAL BACTERIA	USED IN SYNTHESIS OF COENZYMES OF NUCLEIC ACID METABOLISM
B_{12}	PERNICIOUS ANEMIA	LIVER	EACH MOLECULE CONTAINS ONE ATOM OF COBALT
ASCORBIC ACID (C)	SCURVY	CITRUS FRUITS, TOMATOES, GREEN PEPPERS	COENZYME IN SYNTHESIS OF COLLAGEN
D	RICKETS (ABNORMAL Ca^{++} AND $PO_4^{\equiv}$ METABOLISM RESULTING IN ABNORMAL BONE AND TOOTH DEVELOPMENT)	FISH LIVER OILS, BUTTER, STEROID-CONTAINING FOODS IRRADIATED WITH ULTRAVIOLET LIGHT	SYNTHESIZED IN THE HUMAN SKIN UPON EXPOSURE TO ULTRAVIOLET LIGHT TOXIC IN LARGE DOSES
E (TOCOPHEROL)	NO DEFICIENCY DISEASE KNOWN IN HUMANS	EGG YOLK, SALAD GREENS, VEGETABLE OILS	MAY ACT AS A REDUCING AGENT IN THE BODY
K	SLOW CLOTTING OF THE BLOOD	SPINACH AND OTHER GREEN LEAFY VEGETABLES SYNTHESIZED BY INTESTINAL BACTERIA	NECESSARY FOR THE SYNTHESIS OF PROTHROMBIN, AN ESSENTIAL AGENT IN THE CLOTTING OF BLOOD

FIG. 7.2 Some of the principal vitamins.

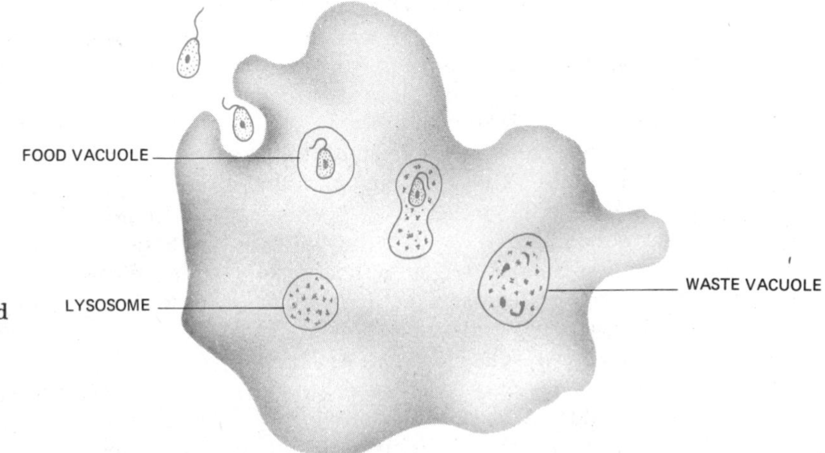

FIG. 7.3 Intracellular digestion in the amoeba. Digestive enzymes stored within lysosomes hydrolyze the food molecules of the prey into smaller molecules that can be absorbed through the vacuolar membrane into the cytoplasm. ►

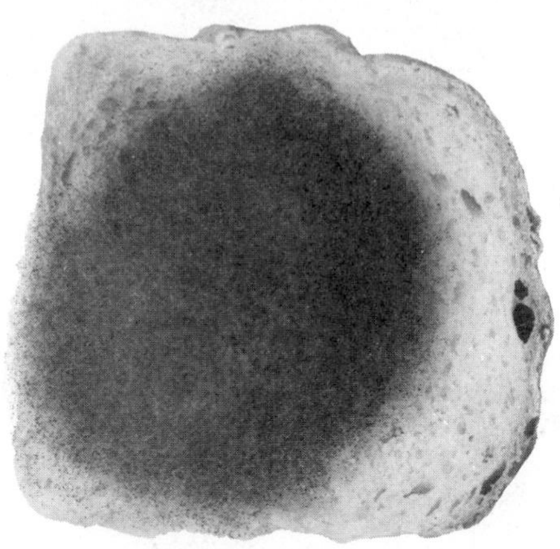

FIG. 7.4 A common mold, *Rhizopus stolonifer*, growing on a piece of bread. This saprophyte secures its nourishment by secreting digestive enzymes on the bread and absorbing the products of digestion.

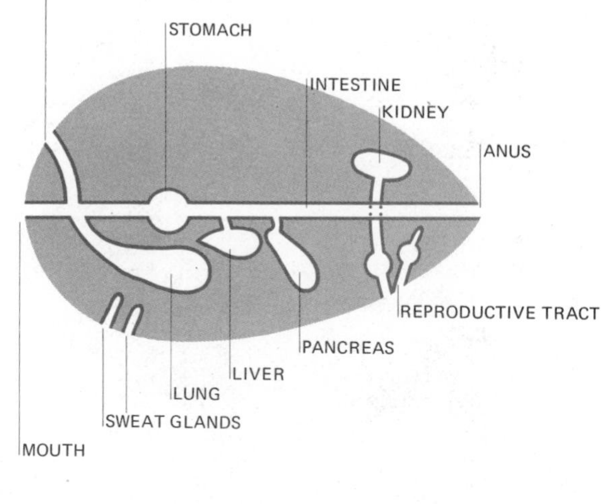

FIG. 7.5 A schematic view of the major topological relationships in a mammal. The ducts of the digestive glands are continuous with the surface of the body.

molecules in the bread are too large, however, to pass directly through the cell membrane. To convert these large, insoluble starch molecules into smaller, soluble molecules that can enter the cytoplasm requires a starch-digesting enzyme, or **amylase.** *Rhizopus* secretes this enzyme onto the bread and thus digestion occurs extracellularly. The sugar molecules produced are then absorbed into the cytoplasm. This pattern of extracellular digestion of foods in the substrate is typical of all fungi and most bacteria.

Most animals, too, digest their food extracellularly. Rarely, though, do they live in a location where they are literally surrounded by a substrate of organic matter. Instead, they secure by one means or another food materials which appear from time to time in their surroundings. They place this food in a pouch or tube within their body, a process called **ingestion.** Then they secrete their enzymes there and digestion takes place. Thus the enzymes are localized where the food is, rather than secreted freely into the surroundings.

Although ingestion is described as the taking of food-containing solids into the organism, these solids are taken in only in a superficial sense. The cavity or tube that receives the solid food materials is really only a portion of the outside of the organism turned inward (Fig. 7.5). A marble swallowed by a baby never enters any cell nor participates in any metabolic activity. A day or two later it emerges at the other end quite unchanged.

7.4 A FILTER FEEDER: THE CLAM

Some freshwater and many marine animals feed by taking in a stream of water and filtering from it any small organisms that may be present. These are then ingested into the alimentary canal for eventual digestion and absorption. The barnacle and the clam are two examples of filter feeders. In the case of the clam, the surfaces of the gills are covered with cilia. The rhythmic beating of these cilia draws fresh water in through an opening, the incurrent siphon (Fig. 7.6). This water then passes through pores in the gills where it gives up oxygen to supply the respiratory needs of the clam. The gills are covered with a sticky mucus, and any solid particles in the water, including diatoms, bacteria, etc., are likely to be trapped in this mucus. Larger particles and indigestible particles drop from the gills and are swept out the excurrent siphon by cilia lining the mantle. The mucus with its load of carefully screened particles is swept forward to the mouth while the now-filtered water passes out through the excurrent siphon. The particle-laden mucus is then ingested. After digestion, any remaining material (the feces) is egested from the anus and passes out with the water flowing out the excurrent siphon.

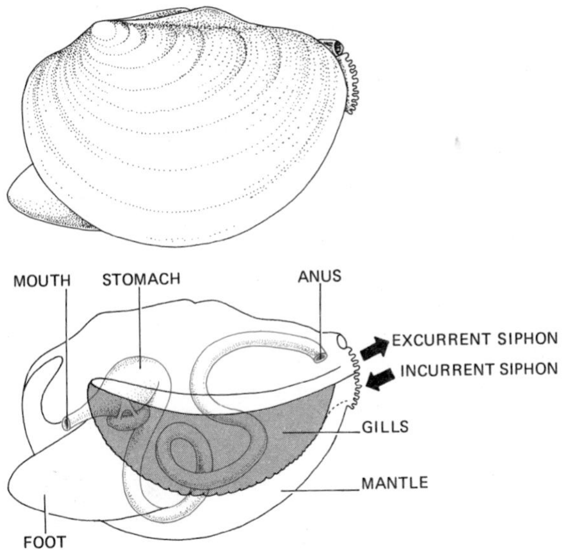

FIG. 7.6 The digestive apparatus of the clam. Food particles are filtered by the gills from the water drawn in through the incurrent siphon.

7.5 ACTIVE FOOD SEEKERS: THE GRASSHOPPER AND THE HONEYBEE

Most insects and most vertebrates are capable of rapid locomotion and thus are able to seek their food actively. The grasshopper moves from plant to plant by walking, hopping, and flying. Specialized appendages around the grasshopper's mouth enable it to consume large quantities of plant leaves quickly and efficiently (Fig. 7.7). The plant material passes from the mouth into the alimentary canal for digestion and absorption. First it passes through the esophagus and into a temporary storage organ, the crop (Fig. 7.7). From here it travels to the muscular gizzard. The gizzard is lined with rigid plates and the muscular action of the gizzard grinds the food material into still smaller particles. The food then passes into the stomach, where **digestion** of it takes place. Digestive enzymes manufactured in six gastric glands (caeca) are secreted into the stomach to carry out this function. Some food may actually enter the caeca during the process. Digested food and a good deal of water are then absorbed in the intestine. The feces are stored temporarily in the rectum before being **egested,** relatively dry, through the anus.

The mouth parts of the honeybee are quite different in appearance and adapted for manipulating a different kind of food. The adult honeybee feeds on liquid food, either the nectar of flowers or the honey stored within the hive. Her maxillae and labium are modified to form a tube through which this food is drawn up into the mouth by the pumping action of the tongue. The whole apparatus is called a **proboscis** (Fig. 7.8).

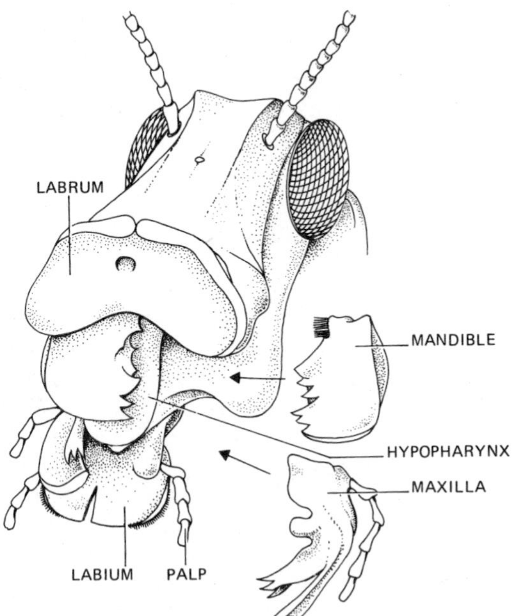

FIG. 7.7 Food-processing organs of the grasshopper. The mouth parts and gizzard accomplish mechanical breakdown of the food. The gastric caeca and the stomach accomplish chemical breakdown, i.e. digestion.

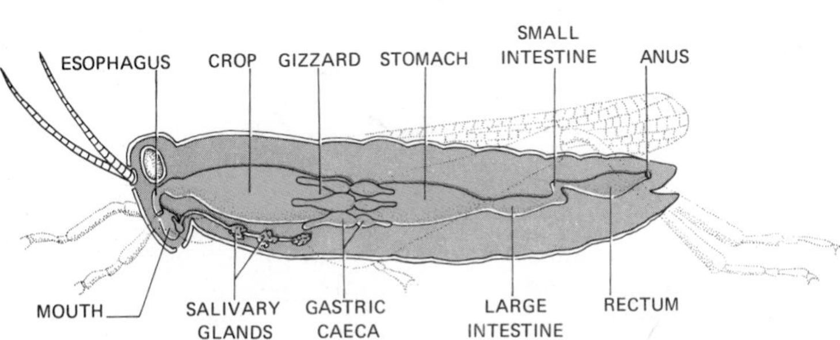

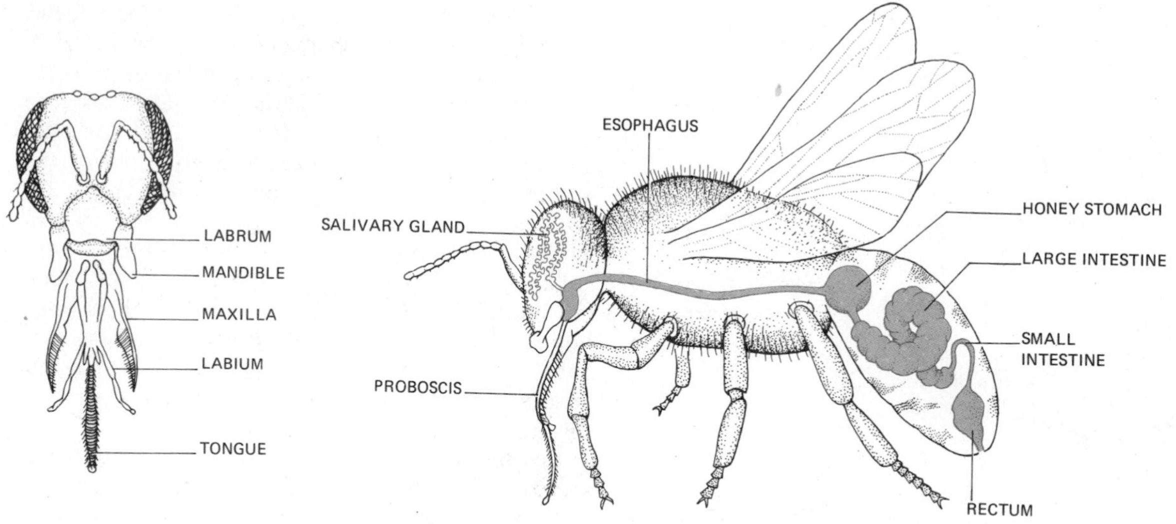

FIG. 7.8 Food-manipulating structures of the honeybee. The maxillae and labium form a proboscis with which the bee sucks up liquid food such as nectar. The honey stomach holds this food while it is being transported back to the hive.

Most of the nectar ingested by a honeybee is destined to be taken back to the hive to be converted into honey. After passing down the long, thin esophagus, it is stored in the honey stomach (Fig. 7.8). The nectar is normally prevented from passing from the honey stomach to the real stomach by a valve located between the two. When the worker gets back to the hive, she regurgitates the nectar, and the process of converting it into honey begins.

The honeybee also uses her mouth to manipulate solids such as the wax out of which the hive is constructed. She uses her mandibles for this process and therefore it is not surprising that these mouth parts are rather like the mandibles of the grasshopper in their structure.

The Human Digestive System

7.6 INGESTION

The strategy by which the human animal digests its food is similar to that of the grasshopper. Once placed within the mouth, the food is ground into small bits by use of our teeth. As in the grasshopper, then, mechanical breakdown of food precedes chemical breakdown (digestion). This is important because it makes the food more easily swallowed and at the same time increases the surface area that will be exposed to the action of digestive enzymes.

While the food is within the mouth, it is moistened by saliva. Saliva is secreted into the mouth from three pairs of glands (Fig. 7.9) that are under the control of the nervous system. The sight, smell, taste, and even the thought of food can all trigger the passage of saliva from the glands through ducts into the mouth. Saliva is a somewhat sticky, "slippery" fluid because it contains carbohydrate-protein molecules called mucins. These enable the saliva to bind the small food particles together into a soft mass which can then be easily swallowed.

The saliva also contains an **amylase,** or starch-digesting enzyme, which catalyzes the hydrolysis of starch into the sugar maltose. This amylase is often called ptyalin, although under the rules for naming enzymes the name salivary amylase is preferred. You can easily demonstrate its action by chewing an unsweetened cracker. After a short time, a distinctly sweetish taste will become noticeable.

Saliva is only the first of a number of secretions that flow into the alimentary canal and aid the process of digestion. In each case, these secretions

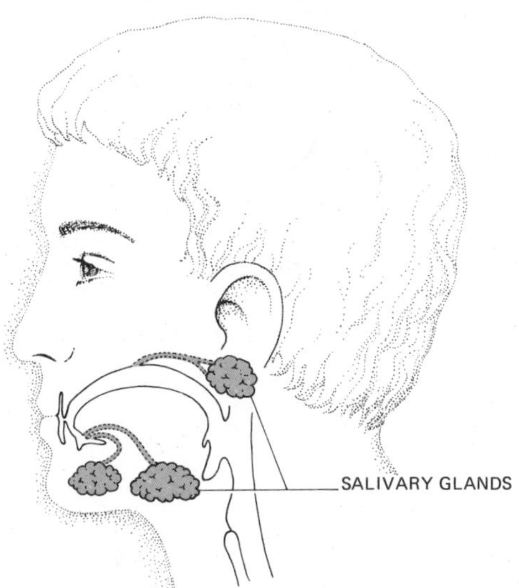

FIG. 7.9 Organization of human salivary glands. As in all exocrine glands, their secretions pass into a system of ducts which lead ultimately to the exterior of the body.

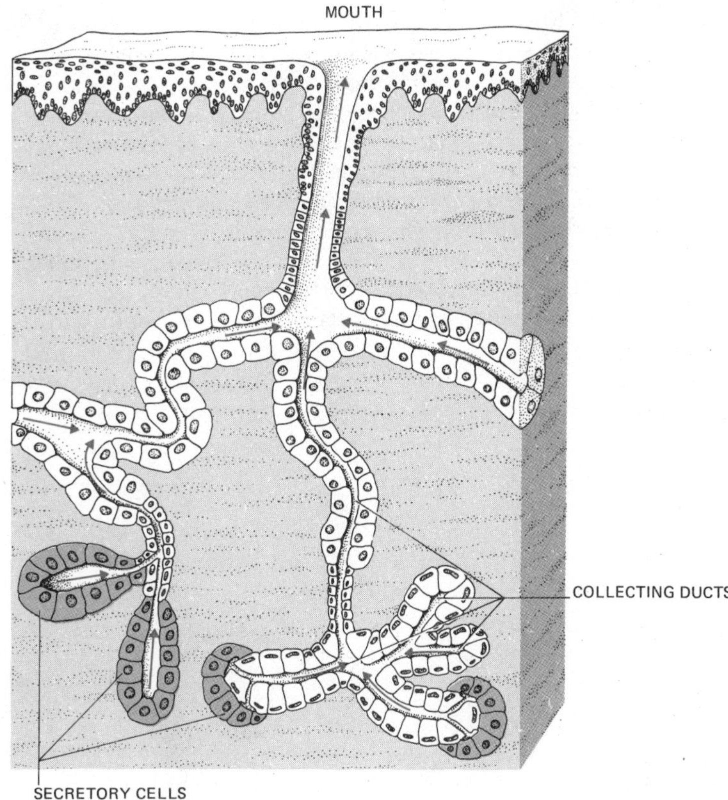

are synthesized within accessory structures called glands. Ducts carry the secretions from the glands to the alimentary canal. The inner surface of each gland is continuous with the inner surface of its ducts (Fig. 7.9) and also with the inner surface of the alimentary canal. In fact, all these digestive glands are formed during embryonic development from outfoldings of the alimentary canal.

When food is swallowed, it passes through the pharynx and into the esophagus. The esophagus is a thick-walled, muscular, straight tube leading from the pharynx to the stomach. It extends through the neck and the chest cavity. Its function can perhaps be best understood as a quick means of conveying food past the large, vital, but nondigestive organs of the chest cavity: the lungs and the heart.

The wall of the esophagus also contains glands that secrete mucin and thus lubricate the passage of food. Once the food mass is well into the esophagus, its movement is controlled by muscles in the wall of the esophagus that are not under our voluntary control. Those just in front of the food mass relax while those just behind it contract. As a result, the food mass is propelled downward. These rhythmic waves of muscular relaxation and contraction are called **peristalsis.**

Below the muscular diaphragm, which separates the chest cavity from the abdominal cavity, the esophagus joins the stomach. A ring of muscle, the cardiac sphincter, surrounds the esophagus at this point. It acts as a valve, relaxing to let into the stomach any food that approaches from above but remaining closed at other times. (The process of vomiting does, of course, also involve the opening of the cardiac sphincter.)

7.7 THE STOMACH

The stomach is a large pouch located high in the abdominal cavity (Fig. 7.10). Its wall is lined with millions of tiny gastric glands, which secrete 400–800 ml of gastric juice at each meal. The incoming food is thoroughly mixed with the gastric juice by vigorous muscular contractions of the stomach.

Three kinds of cells are found in the gastric glands: parietal cells, "chief" cells, and mucus-secreting cells. The parietal cells secrete a solution of hydrochloric acid into the stomach. The concentration of HCl may be as high as 0.15 *M*, giving the gastric juice a pH between 1 and 2. When you consider that the hydrogen ion concentration of the blood nourishing the parietal cells is only about 4×10^{-8} *M*, you can see what a remarkable job these cells do. By active transport, they take in enough hydrogen ions to produce a concentration that is over three million times that value. It is hardly surprising, then, that the parietal cells are profligate energy consumers and literally stuffed with mitochondria.

The hydrochloric acid in the gastric juice has several useful functions: (1) it helps to kill bacteria present in the ingested food; (2) it helps denature proteins, breaking down connective tissue, etc., for easier digestion, and (3) it helps to activate pepsin, the only digestive enzyme secreted by the stomach.

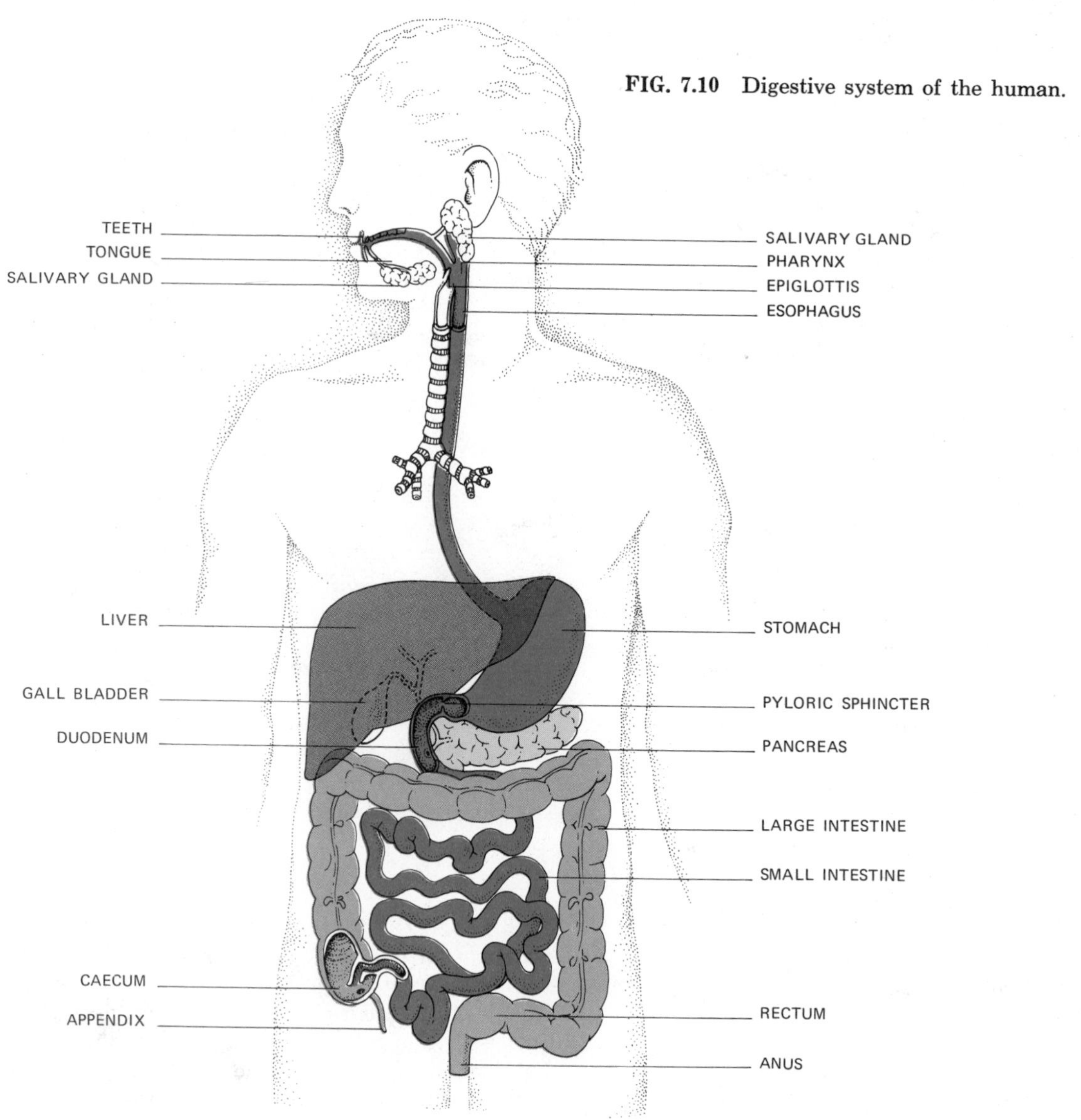

FIG. 7.10 Digestive system of the human.

The "chief" cells of the stomach synthesize and secrete pepsinogen. Pepsinogen is a precursor of the proteolytic enzyme **pepsin.** Once exposed to the acid pH of the gastric juice, a fragment of the molecule is digested away, leaving the active enzyme. Pepsin hydrolyzes not only *ingested* proteins but also additional pepsinogen. This autocatalytic action thus hastens the conversion of pepsinogen to pepsin.

Pepsin exerts its hydrolytic action at certain specific places in polypeptide chains. It is most efficient in cleaving the peptide bonds on the C-terminal side of such amino acids as tyrosine, phenylalanine, and tryptophan (Fig. 7.11). As a result of its action, long polypeptide chains are broken up into shorter lengths.

The stomach wall is largely protein. Why doesn't the stomach digest itself? One factor is a protective coating formed by the mucus-secreting cells of the stomach. In addition, the epidermal cells of the stomach are attached to one another by tight junctions (see Section 5.23), thus preventing access of gastric juice to their interior, unprotected, surfaces. The secretion of pepsin as its inactive precursor pepsinogen blocks proteolytic activity until the enzyme is safely in the cavity (technically known as the lumen) of the stomach. Despite these protective devices, an enormous wastage of epidermal cells occurs, their average life span being only a few days. This necessitates a rapid level of cell replacement. And, of course, these protective devices sometimes fail. The result is gastric ulcer, a localized destruction of the stomach lining.

In the 1890's, the Russian physiologist Ivan Pavlov discovered the mechanisms controlling the release of gastric juice. He found that after he cut nerves leading to the stomach of dogs, the amount of gastric secretion was reduced by about 25%. This showed definite, but incomplete, control of gastric secretion by the nervous system. What about the remaining 75%? Pavlov reasoned that a hormone might be liberated in the blood when food enters the stomach, which would also stimulate the production of gastric juice. He tested this theory by connecting the blood circulatory systems of two dogs. When one dog was fed, it was only a short time before the second dog began to secrete gastric juice also. The second dog, however, secreted only about 75% of the normal amount. We know now that a hormone, **gastrin,** is produced by certain cells in the stomach when food enters the stomach.

These gastrin-secreting cells are especially stimulated by the presence of peptides (and of such drugs as caffeine and ethanol). The hormone is a polypeptide containing 17 amino acids. It enters the bloodstream and is carried through the body. When blood containing gastrin returns to the stomach, it stimulates the parietal cells to secrete hydrochloric acid. In Pavlov's experiment, the hormone also passed into the bloodstream of the second dog, stimulating its parietal cells as well.

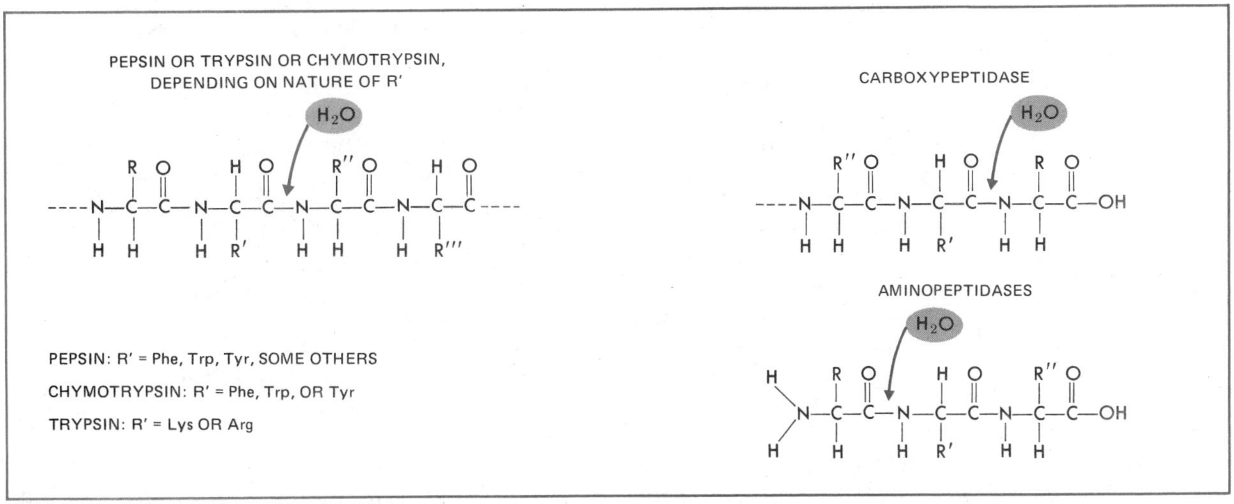

PEPSIN OR TRYPSIN OR CHYMOTRYPSIN, DEPENDING ON NATURE OF R′

CARBOXYPEPTIDASE

AMINOPEPTIDASES

PEPSIN: R′ = Phe, Trp, Tyr, SOME OTHERS

CHYMOTRYPSIN: R′ = Phe, Trp, OR Tyr

TRYPSIN: R′ = Lys OR Arg

FIG. 7.11 Hydrolytic action of various proteases.

Very little absorption of materials occurs in the stomach. Some water, a few ions, and such drugs as aspirin and ethanol are absorbed into the blood (helping to account for the rapid appearance of ethanol in the blood after taking a drink).

As the contents of the stomach become thoroughly liquefied, they pass through a sphincter at the lower end of the stomach, and into the small intestine, the first ten inches of which is called the **duodenum.** An opening in the wall of the duodenum leads to two ducts which, in turn, drain two large digestive glands, the pancreas and the liver.

7.8 THE PANCREAS

The pancreas is an elongated, whitish gland which lies in the loop formed by the duodenum and the under surface of the stomach. Secretory cells in the pancreas produce pancreatic fluid which then passes down the pancreatic duct and into the duodenum. The pancreatic fluid contains the following substances.

a) **Sodium bicarbonate** ($NaHCO_3$). The sodium bicarbonate neutralizes the acidity of the intestinal contents, quickly raising the pH to about 8.

b) An **amylase.** The pancreatic amylase hydrolyzes starch into a mixture of maltose and glucose. Because it has a much longer period in which to act, it plays a more important role in starch digestion than does salivary amylase.

c) A **lipase.** The pancreatic lipase hydrolyzes ingested fats into a mixture of fatty acids and monoglycerides. It does this by catalyzing the splitting off of the fatty acids attached to the No. 1 and No. 3 carbon atoms of the glycerol. The fatty acid attached to the No. 2 carbon remains, yielding a "monoglyceride."

The action of lipase is greatly enhanced by the presence of bile. Bile contains bile salts. These are amphipathic steroids that play an important role in emulsifying ingested fat. The hydrophobic skeleton of the steroid dissolves in the fat while the negatively charged side chain is free to interact with polar water molecules. The mutual repulsion of these negatively charged droplets keep them from coalescing (look back at Fig. 4.6). Thus large globules of fat (liquid at body temperature) are emulsified into tiny droplets (about 1 μm in diameter) which can be more easily digested and absorbed.

d) Two proteinases: **trypsin** and **chymotrypsin.** These enzymes continue the work of protein digestion. Chymotrypsin attacks the same peptide bonds that pepsin does. (The action of pepsin ceases when the sodium bicarbonate raises the pH of the intestinal contents.) Trypsin attacks peptide bonds on the C-terminal side of arginines and lysines (Fig. 7.11). Like pepsin, both trypsin and chymotrypsin are secreted in an inactive form and only when within the duodenum are they converted into their active state.

e) **Carboxypeptidase.** This enzyme removes, one by one, the amino acids located at the C-terminal end of peptide molecules (Fig. 7.11). Thus it aids in the hydrolysis of peptides into amino acids.

f) **Nucleases.** These enzymes hydrolyze ingested nucleic acids (RNA and DNA) into their component nucleotides.

The secretion of pancreatic juice is under hormonal control. When the acidified contents of the stomach enter the duodenum, certain cells in the wall of the duodenum release the hormones **secretin** and **pancreozymin** into the bloodstream. When carried to the pancreas, these stimulate the production and discharge of pancreatic fluid. Secretin chiefly affects the secretion of sodium bicarbonate solution whereas pancreozymin stimulates the production of the digestive enzymes.

7.9 THE SMALL INTESTINE

As the mixture of food, bile, and pancreatic enzymes passes through the small intestine, disaccharides, peptides, fatty acids, and monoglycerides are produced. The final digestion and absorption of these materials is the function of the **villi,** which line the inner surface of the small intestine (Fig. 7.12). The villi increase the surface area of the small intestine to many times what it would be if it were simply a tube with smooth walls. In addition, the exposed surface of the epithelial cells of each villus is covered with projections, the **microvilli** or "brush border" (Fig. 7.13). Thanks largely to these, the total surface area in the intestine is almost 200 square meters, about the size of the singles area of a tennis court and some 100 times the surface area of the exterior of the body.

Incorporated in the surface of the microvilli are a number of enzymes that complete the digestion

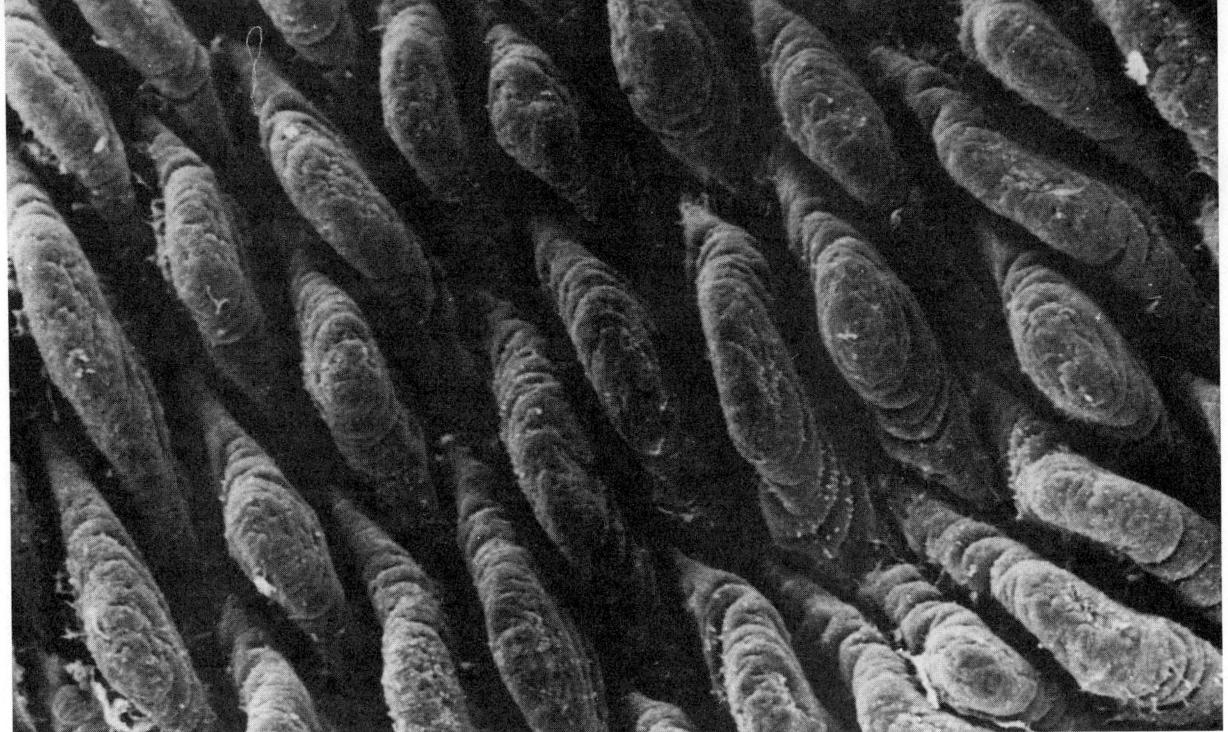

FIG. 7.12 Villi carpeting the inner surface of the small intestine as seen by the scanning electron microscope. (Courtesy of Keith R. Porter.)

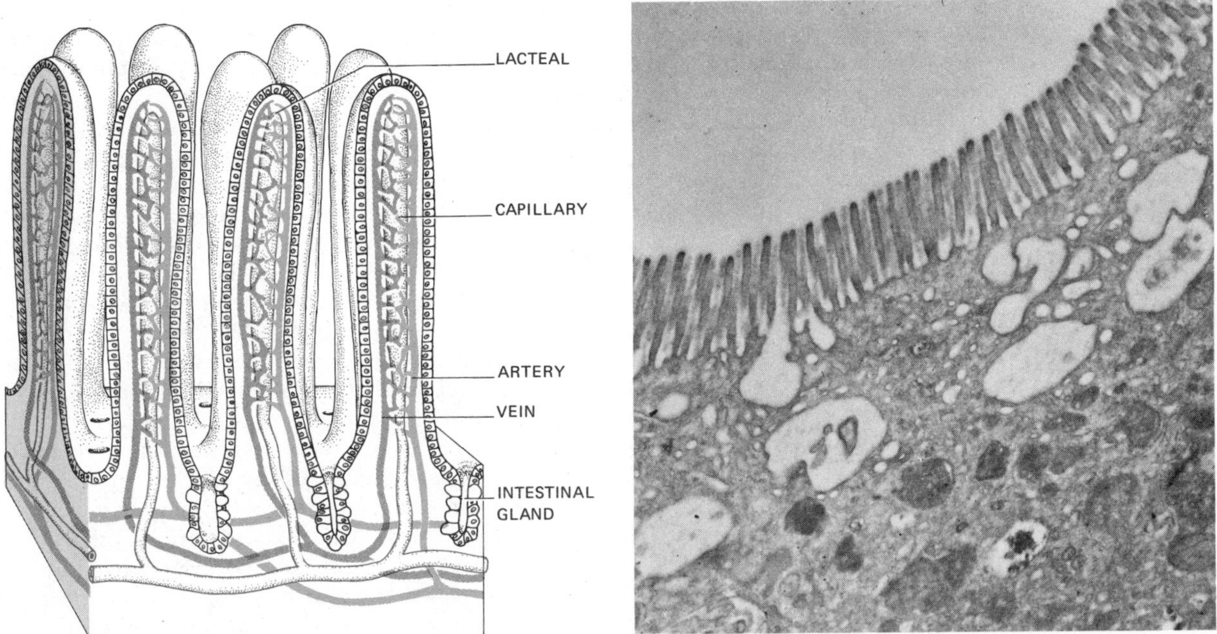

FIG. 7.13 Villi. Left: Absorbed fats enter the lacteals; the other absorbed materials enter the capillaries. Right: Electron micrograph (9000 ×) of the microvilli of a mouse intestine cell. (Courtesy of Dr. Sam L. Clark, Jr. Reprinted by permission from The Rockefeller Institute Press, *Journal of Biophysical and Biochemical Cytology,* 5: 41–50, 1959).

process. Several **aminopeptidases** complete the hydrolysis of peptides into amino acids. The enzymes act in a manner similar to carboxypeptidase but they attack the amino ($-NH_2$-terminal) end of the peptide molecules instead of the C-terminal end (Fig. 7.11).

Three **disaccharidases** localized on or in the microvilli hydrolyze the disaccharides maltose, sucrose, and lactose into their component monosaccharides. Each maltose molecule is split into two glucose molecules and thus glucose is the end-product of the digestion of starch. Sucrose (common table sugar) gives a molecule of glucose and one of its isomers, fructose, while lactose (milk sugar) gives glucose and galactose (Fig. 4.9).

With the action of these enzymes, protein and carbohydrate digestion comes to an end. What were originally macromolecules have now been converted into small molecules (amino acids and monosaccharides) which are ready for passage into the bloodstream.

Each villus is richly supplied with blood which passes through a network of tiny tubes called capillaries. The sugars and amino acids as well as vitamins, salts, and water pass from the intestinal contents into the capillaries of the villus. Although this transport may be accomplished by diffusion in some cases, other mechanisms are involved, too. Glucose, for example, continues to be absorbed even when its concentration in the intestinal fluid becomes less than its concentration in the blood (about 0.1%). This absorption is accomplished by active transport. Some materials may also enter the cells of the villi by endocytosis.

Once within the cells of the villi, the absorbed fatty acids and monoglycerides become resynthesized into fats. This occurs in the smooth endoplasmic reticulum of these cells. The small droplets of lipid are then discharged by exocytosis into the interior of the villus. These droplets do not enter the capillary network. Instead they enter a duct, the lacteal, whose walls are more porous. The lacteals of the villi are part of the lymphatic system. The morphology and physiology of this system will be discussed in Chapter 11. Once inside the lacteals, the fat droplets are carried slowly through the lymphatic system until it joins with the regular blood circulatory system. After a fat-rich meal, the lacteals take on a whitish, milky appearance (*lac* is the Latin word for milk) because of the large numbers of fat droplets contained within them.

7.10 THE LIVER

Although the liver is not strictly an organ of digestion, its secretion, bile, does play an important role in the digestion of fats. Bile is produced continuously by the liver but between meals it accumulates in a storage organ, the gall bladder.

One of the effects of pancreozymin, the enzyme released when food enters the duodenum, is to stimulate contraction of the gall bladder and the discharge of its accumulated bile into the duodenum. (At one time, the hormone that stimulates gall bladder contraction was called cholecystokinin. But it later became apparent that the research workers studying the effects of pancreozymin on the pancreas and those studying the effects of cholecystokinin on the gall bladder were actually studying the same substance: a polypeptide containing 33 amino acids. Like the blind men's divergent descriptions drawn from feeling the opposite ends of an elephant, different research teams with different goals and techniques often end up finding that they have been studying the same thing.)

The action of pancreozymin (or pancreozymin-cholecystokinin as some now call it), like that of gastrin and secretin, illustrates an important fact about hormonal control. In each case, the liberated hormone is carried by the blood to every organ, every tissue, every cell of the body. Only certain organs, however, are capable of responding to the presence of the hormone. These are called the target organs.

Bile contains other materials besides bile salts. Among these are the bile pigments. These are the products of the breakdown of the red blood pigment, hemoglobin, removed by the liver from old red blood cells. The brownish color of the bile pigments imparts the characteristic brown color of the feces.

The formation of bile is just one of several important nutritional functions carried out by the liver. Before the blood that leaves the villi reaches the general circulation, it passes through the liver. Here extraordinary components (e.g. nonnutritive molecules) that are picked up by absorption from the intestine, or excess amounts of ordinary components, are screened out. All monosaccharides other than glucose (e.g. fructose, galactose) are removed by the liver and converted into glucose. And if glucose is present in excess of the normal 0.1% concentration found in the blood, most of this excess is removed and converted into the insoluble polysaccharide glycogen.

Amino acids in excess of the body's anabolic needs are deaminated by the liver. In deamination, the nitrogen-containing amino ($—NH_2$) portion of the molecule is removed. This is then converted into a nitrogenous waste, urea. The mechanism of this process will be discussed in Chapter 13. The non-nitrogen-containing residue of the amino acid can then enter the metabolic pathway of cellular respiration and be oxidized for energy. The details of this mechanism will be considered in the next chapter.

The system of blood vessels that carries all blood received from the intestines through the liver before passing it to the general circulation is called the **hepatic portal** system. In a real sense, this system provides a gatekeeper between the intestines and the general circulation. In this role, the liver screens the blood passing through it so that the composition of the blood when it leaves will be that which is normal for the organism. For example, even after a meal rich in starches and sugars, the glucose level of the blood leaving the liver will not be much higher than the normal 0.1%.

On the other hand, if the concentration of glucose in the blood should start to fall between meals, the liver will hydrolyze enough of its glycogen reserves to maintain the level at 0.1%. In other words, the liver is one of the most important devices in the human body for maintaining the constancy of the ECF and thus for the preservation of homeostasis. In Chapter 25 the liver's role in maintaining homeostasis with respect to the sugar content of the ECF will be explored further.

7.11 THE LARGE INTESTINE

The small intestine leads into the large intestine. A sphincter controls the passage of materials from one to the other. Just beyond this sphincter is a blind pocket, the caecum. Attached to the caecum is a tiny projection, the appendix. It is considered to be the vestigial remains of a structure which may have functioned in cellulose digestion in some very remote, prehuman ancestor. Its chief interest to us is the fact that it may become infected, causing appendicitis. A severe infection may cause the appendix to rupture, thus spreading the infection to the membranes lining the abdominal cavity and supporting the organs (the viscera) within it. This condition is known as peritonitis.

The large intestine receives the liquid residue of material left after digestion and absorption in the small intestine have been completed. This residue contains substantial quantities of water and indigestible substances such as cellulose. Some other food materials are invariably present and these serve to nourish the enormous population of bacteria that live in the large intestine. Usually these bacteria (of which one common species is the much-studied *Escherichia coli*) are perfectly harmless. In the process of their metabolic activities they may produce gases and other odoriferous wastes. They flourish to such an extent that from 10% to 50% of the dry weight of the feces may consist of bacterial cells.

Occasionally, harmful bacteria or protozoans may become established in the large intestine. Typhoid fever, asiatic cholera, and amoebic dysentery are three diseases which are caused by intestinal parasites. These parasites may even invade other organs of the body such as the liver and cause extensive damage.

The chief function of the large intestine is the reabsorption of water. A great deal of water is secreted into the stomach and intestine by the various digestive glands. This must be reclaimed if the individual is not to suffer from dehydration and thirst. Most of the water is reabsorbed in the large intestine during the 12 to 14 hours that the food residues remain there. Sometimes the large intestine becomes irritated and discharges its contents into the rectum and out of the anus before water reabsorption is complete. The resulting condition is called diarrhea. Perhaps you have noticed the thirst which accompanies a case of diarrhea. On the other hand, the large intestine may retain its contents for an extra long period. The fecal material, as it is now called, becomes dried out and compressed into dry, solid masses. This condition is known as constipation.

The feces pass from the large intestine through an S-shaped tube into the **rectum**. Here they remain until the two sphincters guarding the anus relax and vigorous peristaltic waves expel them in the process of defecation.

EXERCISES AND PROBLEMS

1. Outline the digestion of a bacon (protein and fat) sandwich (starch).
2. Draw the structural formula for the "2-monoglyceride" of palmitic acid ($C_{15}H_{31}COOH$).
3. By what mechanisms might amino acids in the small intestine enter the epithelial cells of the villi?
4. Is the digestion of macromolecules a process that occurs only in the digestive tract? Explain.
5. How do cells that secrete extracellular enzymes avoid self-digestion?
6. Do any filter-feeders live on land? Explain.

REFERENCES

1. BEAUMONT, W., *Experiments and Observations on the Gastric Juice and the Physiology of Digestion,* Dover, New York, 1959. Dr. Beaumont's account of his own work.
2. BAYLISS, W. M., and E. H. STARLING, "The Mechanism of Pancreatic Secretion," *Great Experiments in Biology,* ed. by M. L. Gabriel and S. Fogel, Prentice-Hall, Englewood Cliffs, N.J., 1955. Deals with secretin and how it was discovered.
3. DAVENPORT, H. W., "Why the Stomach Does Not Digest Itself," *Scientific American,* Offprint No. 1240, January, 1972.
4. KRETCHMER, N., "Lactose and Lactase," *Scientific American,* Offprint No. 1259, October, 1972. In many human populations, the adults lack lactase and cannot digest milk.
5. MCMINN, R. M. H., *The Human Gut,* Oxford Biology Readers, No. 56, Oxford University Press, Oxford, 1974.
6. YOUNG, V. R., and N. S. SCRIMSHAW, "The Physiology of Starvation," *Scientific American,* Offprint No. 1232, October, 1971.
7. KAPPAS, A., and A. P. ALVARES, "How the Liver Metabolizes Foreign Substances," *Scientific American,* Offprint No. 1322, June, 1975.

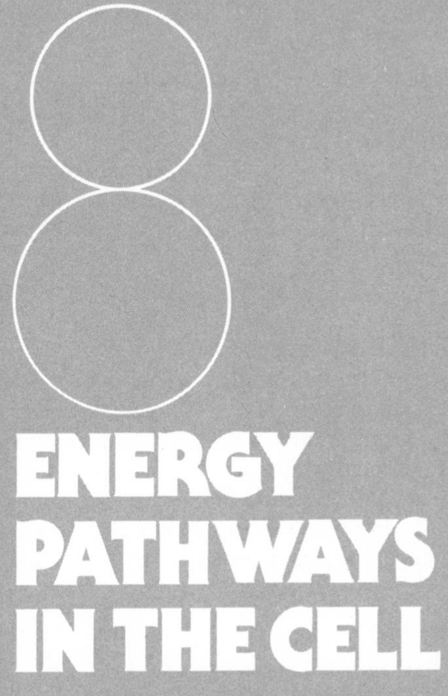

8 ENERGY PATHWAYS IN THE CELL

In the last chapter we examined some of the mechanisms by which heterotrophic organisms *ingest* large organic molecules and then *digest* these into their subunits.

Very little energy is liberated by hydrolysis. The hydrolysis of the disaccharide maltose into two molecules of glucose yields only 4 kcal of free energy:

$$C_{12}H_{22}O_{11} + H_2O \rightarrow 2C_6H_{12}O_6;$$
$$\Delta F = -4 \text{ kcal/mole}.$$

Most of the free energy stored in starches, proteins, and fats is still locked up in the end product of their digestion: sugars, amino acids, fatty acids, and glycerol. The process of digestion does, however, reduce the size of the molecules so they may be absorbed readily from the extracellular fluid into the cytoplasm of the cell.

8.1 ANABOLISM AND CATABOLISM

What is the fate of these small, organic molecules once they enter the cell? Usually, they are further reduced in size, forming still simpler molecules containing, characteristically, two to four carbon atoms. These small molecules then face one of two alternatives (Fig. 8.1). They may proceed back "up" various metabolic pathways and serve as the building blocks of sugars, fatty acids, glycerol, and amino acids. From these can then be assembled the macromolecular components of the cell: polysaccharides, lipids, proteins, and even nucleic acids. This phase of metabolism, in which large, complex molecules are built up from smaller, simpler ones, is called **anabolism.**

The second fate of the intracellular pool of two- to four-carbon molecules is to be still further degraded—ultimately to simple, inorganic molecules like CO_2, H_2O, and NH_3 (Fig. 8.1). The total amount of energy stored in the end products of these decompositions is much less than the amount of energy present in the original molecules. Thus, during the degradative steps, energy is released. This phase of metabolism, in which relatively complex, energy-rich molecules are broken down into simpler, energy-poor molecules, is called **catabolism.**

Anabolic reactions proceed with a gain in free energy. In other words, to convert a two-carbon precursor like acetyl-CoA (Fig. 8.1) to a 16-carbon fatty acid requires the input of energy just as did the electrolysis of water. The ΔF in each case is positive. Put another way, these reactions cannot occur spontaneously. Whether catalyzed or not, all chemical reactions proceed spontaneously only in the direction that yields free energy, that is, when ΔF is negative (look back at Fig. 6.12). How then is anabolism

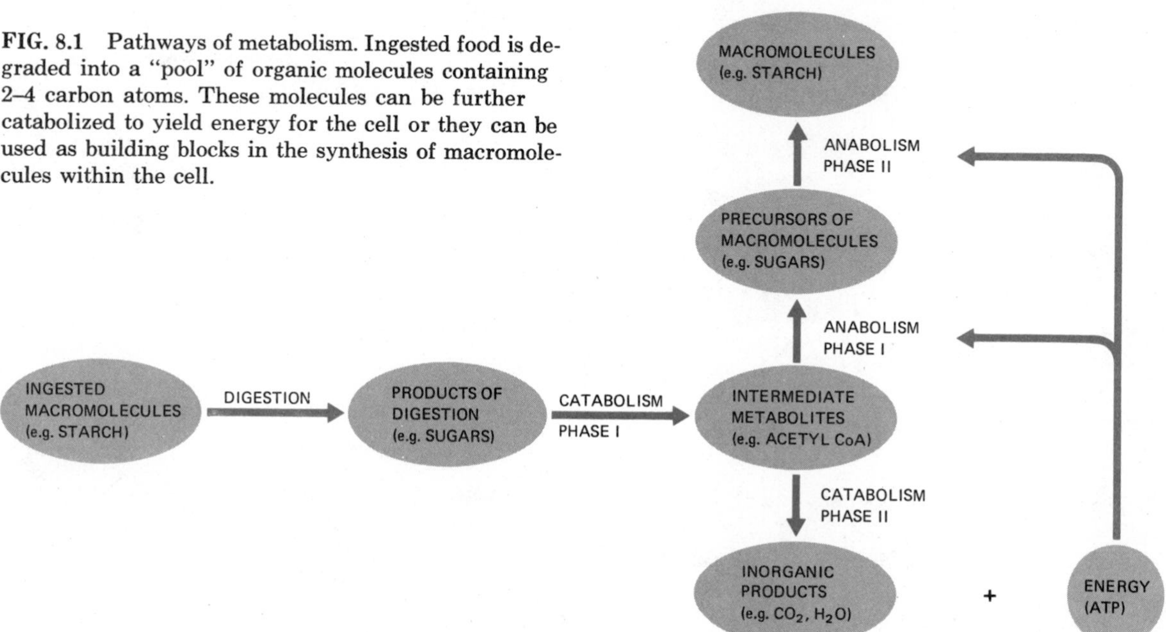

FIG. 8.1 Pathways of metabolism. Ingested food is degraded into a "pool" of organic molecules containing 2–4 carbon atoms. These molecules can be further catabolized to yield energy for the cell or they can be used as building blocks in the synthesis of macromolecules within the cell.

to succeed? The secret is to couple a catabolic reaction, with its negative ΔF, to an anabolic reaction, with its positive ΔF. And because we can never hope to achieve efficiencies approaching 100%, we must choose catabolic reactions that yield more free energy than is needed by our anabolic reaction.

Perhaps the most widely used fuel of living cells is glucose. In Chapter 6, we saw that the complete combustion of glucose, producing CO_2 and H_2O as end products, yields 686 kcal of free energy. But in this reaction, almost all of the free energy is immediately given off as heat. A modest production of heat is fine for keeping cells warm but is of no use for driving anabolic chemical reactions. What the cell must do is to find ways of converting the free energy of glucose to the free energy of other molecules, an amino acid for example, that the cell requires. The strategy of the cell is thus to catabolize the glucose molecule in such a way that not all of the liberated free energy is lost as heat. The key is to proceed by a number of small, discrete steps, steps just large enough that the $-\Delta F$ can be coupled efficiently to the anabolic (and other energy-dependent) needs of the cell. Living cells catabolize glucose so that at least one step, and better several, yields enough free energy to synthesize a molecule of ATP. How this is done is the essential story of this chapter.

FIG. 8.2 Champagne is a product of alcoholic fermentation. Deprived of oxygen, the yeast cells within the bottle have converted sugar to ethanol and carbon dioxide (which gives the product its "sparkle").

8.2 GLYCOLYSIS

The combustion of glucose requires oxygen. But some cells must live where oxygen is not or may not always be available. The yeast cell in a stoppered bottle of champagne has no access to oxygen (Fig. 8.2). There are good reasons for believing that the first cells on this planet lived in an atmosphere lacking oxygen. If so, they, too, faced the problem of extracting energy from fuel without the presence of oxygen. In any case, all cells today retain the enzymatic machinery to catabolize glucose without the aid of oxygen. The anaerobic ("without air," hence without oxygen) breakdown of glucose is called **glycolysis.** The yeast cell in the champagne bottle uses glycolysis to get the energy it needs to live. The products are ethanol (ethyl alcohol) and carbon dioxide. We call the process alcoholic fermentation. An overworked muscle cell also uses glycolysis to meet its energy needs. The end product in this case is lactic acid and we call the process lactic acid fermentation. But most cells *are* able to use oxygen to catabolize glucose. The final products here are carbon dioxide and water, just as they are when glucose is burned. (And the yield in free energy is the same: 686 kcal/mole.) The process is called cellular respiration. But even when a cell is able to respire glucose rather than ferment it, the *initial* steps are still the same: the steps of glycolysis.

The breakdown of glucose by glycolysis proceeds as a series of 11 enzyme-catalyzed sequential reactions (12 in yeast). These enzymes are all found dissolved in the cytosol. In addition to the enzymes, two **coenzymes** are absolutely essential. These are **ATP and NAD.**

8.3 ATP AND NAD

ATP stands for *a*denosine *tri*phosphate. It is a nucleotide. Its pentose sugar is ribose and its nitroge-

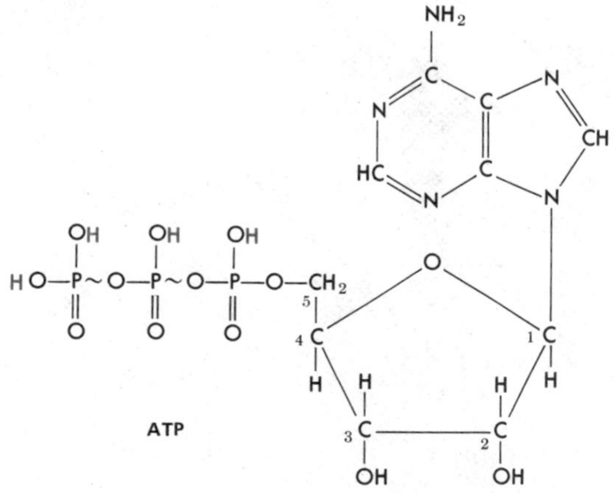

FIG. 8.3 Structural formula of ATP. ATP is very acidic. Under the conditions that exist in the cell (~ pH 7), the four protons shown in color are given up, leaving four negatively charged oxygen atoms (—O: −). These attract positively charged ions in the cell such as Mg^{++}. The bonds attaching the terminal and subterminal phosphate groups yield a large amount of energy when they are hydrolyzed. For this reason, they are often called "high-energy" bonds and are depicted by a wavy line.

nous base is adenine (Fig. 8.3). Attached to the No. 5 carbon atom of the ribose are three phosphate groups (hence triphosphate). The bonds between the second and third and also between the first and second phosphates are known as "high-energy" bonds. They are given this name because of the large amount of free energy that is liberated when they are broken by hydrolysis. Under the conditions existing in cells, the exact amount of energy liberated is difficult to establish. We shall use the value of 7.3 kcal/mole. (This is probably a conservative value.) Thus the hydrolysis is expressed:

$$ATP + H_2O \rightarrow ADP + Pi; \qquad \Delta F = -7.3 \text{ kcal.}$$

(ADP is *a*denosine *di*phosphate. Pi is inorganic phosphate,

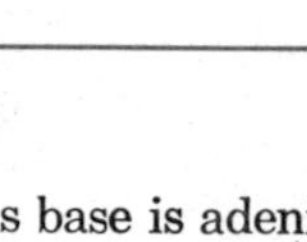

$$HO{-}\overset{\overset{\displaystyle O}{\|}}{\underset{\underset{\displaystyle O^-}{|}}{P}}{-}O^-.$$

H_3PO_4, phosphoric acid, is acidic and is shown here with two of its three protons having been given up.)

It is important to note that in using the expression "high-energy" for these bonds, we are *not* referring to *bond energy*, the concept we discussed at length in Chapter 3. Here the term simply refers to the fact that the products of the hydrolysis of these so-called high-energy bonds have substantially less free energy than the reactants. Really, then, these are weak bonds with low bond energies. However, the concept of the "high-energy" bonds of ATP (and ADP) is so entrenched that we shall retain it. In illustrations, we shall also observe the convention of indicating these bonds by a wavy line (Fig. 8.3).

NAD is *n*icotinamide *a*denine *d*inucleotide. As the name suggests, it is really two nucleotides covalently linked together. One is "adenosine monophosphate," that is, ribose with adenine attached to the No. 1 carbon and phosphate to the No. 5 (Fig. 8.4). The second is a ribonucleotide with the nitrogenous base nicotinamide attached to the No. 1 carbon and phosphate to the No. 5. Nicotinamide is one of the B vitamins and thus an essential component of the human diet. Note in Fig. 8.4 that one of the nitrogen atoms in nicotinamide carries a single positive charge. For this reason, we often represent the entire molecule as NAD^+.

In Chapter 3, we defined oxidation as the removal of electrons from a substance. In cells, most oxidations are accomplished by the removal of hydrogen atoms (each with its electron). NAD^+ plays a crucial role in this. Each NAD^+ molecule can acquire (and thus be reduced by) two electrons. However, only one proton accompanies the transfer, the other being liberated in the surrounding medium. Thus the reduced form of NAD is most accurately represented as NADH. But it *has* gained *two* electrons. The entire reaction is represented as

$$NAD^+ + 2H \rightarrow NADH + H^+.$$

NAD participates in a large number of oxidation-reduction reactions in cells, including those in glycolysis and many of those in cellular respiration. However, some redox reactions in cells exploit the related compound *n*icotinamide *a*denine *d*inucleotide *p*hosphate or NADP. NADP is simply NAD with a

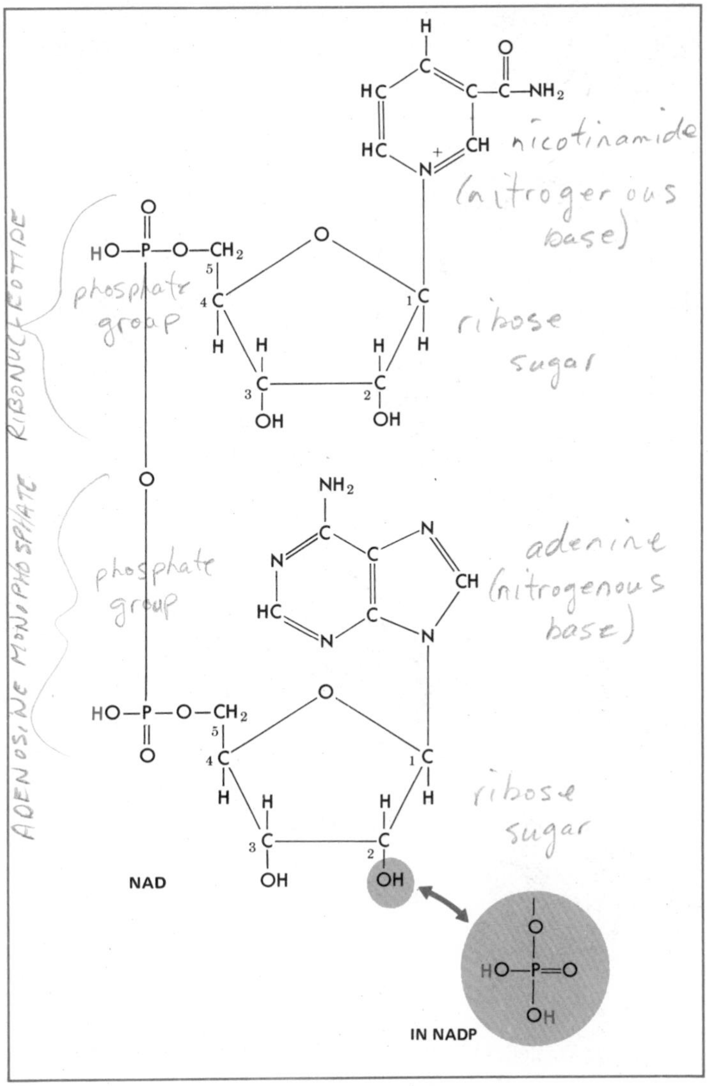

FIG. 8.4 Structural formulas of NAD and NADP. Because of the positive charge in the nicotinamide ring (upper right), the oxidized forms of these important redox agents are often depicted as NAD+ and NADP+ respectively. When reduced, they each acquire two electrons although only one proton. Thus the reduced forms are depicted as NADH and NADPH. Under the conditions existing in the cell, the two hydrogen atoms shown in color are dissociated from these acidic substances.

third phosphate group attached as shown in Fig. 8.4. NADP is reduced in the same manner as NAD. We shall see in the next chapter that NADP is the reducing agent used in photosynthesis. NADP also serves as the reducing agent in other anabolic reactions.

8.4 GLYCOLYSIS: PRIMING THE PUMP

The first step in glycolysis is the transfer of a phosphate group from ATP to the No. 6 carbon atom of glucose (Fig. 8.5). This reaction proceeds with a loss of about 4.0 kcal of free energy. Inasmuch as the breaking of the terminal "high-energy" bond of ATP yields at least 7.3 kcal, we know that the molecule of glucose 6-phosphate that is formed has gained free energy. In a sense, it has been activated.

This step is a good example of the coupling of an energy-yielding reaction to an energy-consuming one. We can express the two reactions as follows:

$$\text{Glucose} \rightarrow \text{glucose 6-phosphate}, \quad \Delta F = +3.3 \text{ kcal}$$
$$\text{ATP} \rightarrow \text{ADP}, \quad \Delta F = -7.3 \text{ kcal}$$
$$\text{Sum: Glucose} + \text{ATP} \rightarrow \text{glucose 6-phosphate} + \text{ADP}, \quad \Delta F = -4.0 \text{ kcal}$$

After an enzyme-catalyzed conversion of the glucose 6-phosphate to its *isomer* fructose 6-phosphate,

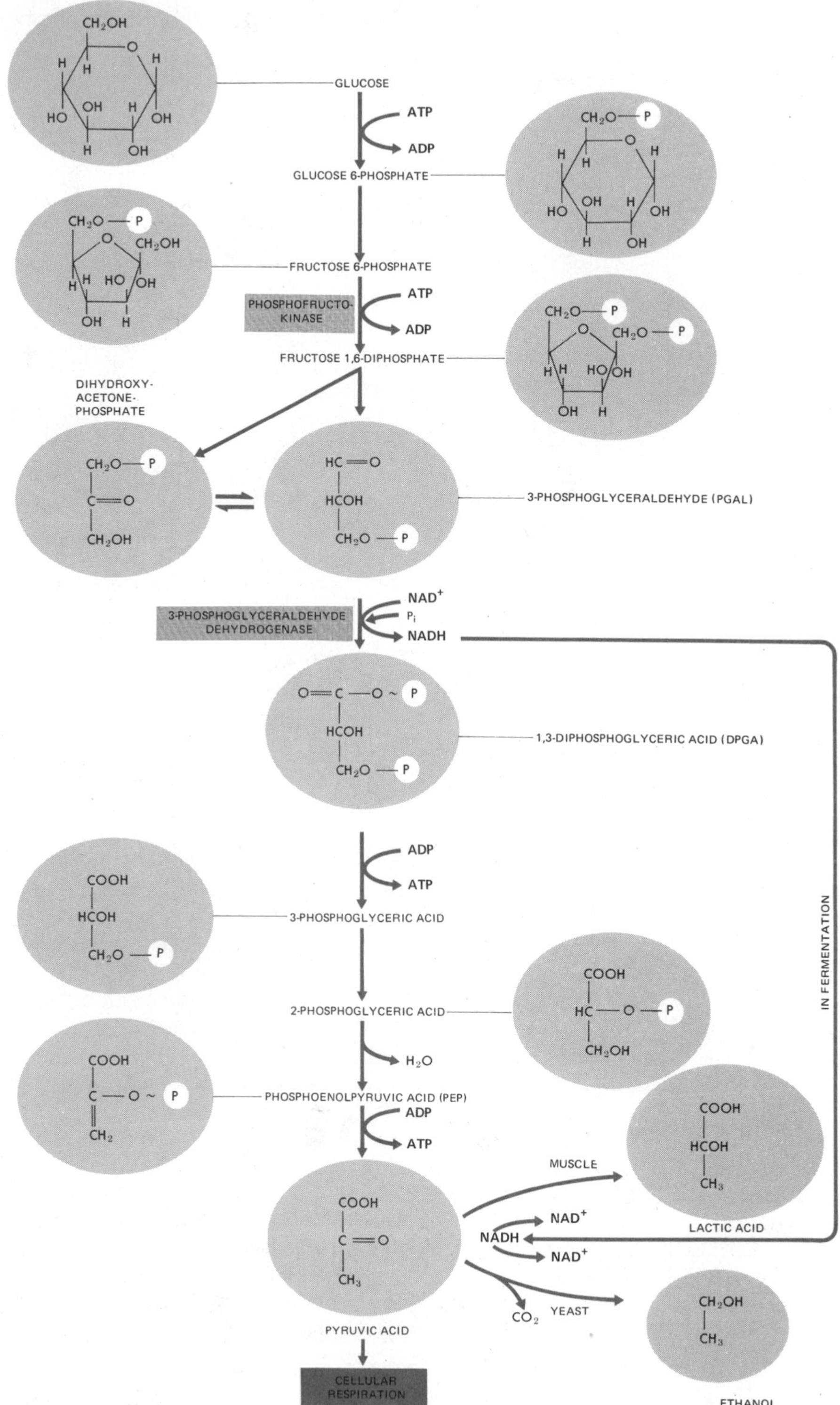

FIG. 8.5 The pathway of glycolysis. All of these reactions take place in the cytosol. In both alcoholic and lactic acid fermentation, the electrons removed from PGAL by NAD+ are used to reduce pyruvic acid. In cellular respiration, however, the electrons are passed into the mitochondria (where each pair gives rise to two molecules of ATP).

another ATP molecule transfers a second phosphate group, this time to the No. 1 carbon. The compound formed, fructose 1,6-diphosphate, again gains a little free energy from the reaction.

The next step in glycolysis is the enzymatic splitting of the fructose 1,6-diphosphate into two fragments. Each of these molecules contains three carbon atoms. One is called dihydroxyacetone phosphate; the other is 3-*p*hospho*g*lycer*al*dehyde or **PGAL.** As inspection of Fig. 8.5 will show, these two molecules are isomers and, in fact, are readily interconnected by yet another enzyme of glycolysis. Note, too, that if we were to remove the attached phosphate groups, we would really be left with two half-glucose molecules ($C_6H_{12}O_6 \rightarrow 2C_3H_6O_3$).

8.5 GLYCOLYSIS: THE FIRST OXIDATION

The next step in glycolysis is of special interest. Not only is it crucial to glycolysis and, ultimately, cellular respiration, but it can serve as a model of the process that occurs repeatedly in the release of energy by cells.

Two electrons (in the form of two hydrogen atoms) are removed from the molecule of 3-phosphoglyceraldehyde (PGAL) and transferred to a molecule of NAD. This is, of course, a redox reaction with the PGAL being oxidized and the NAD reduced. The enzyme that catalyzes this redox reaction is 3-phosphoglyceraldehyde *dehydrogenase*. As the product leaves the enzyme, a second phosphate group is donated to it from inorganic phosphate present in the cell. The resulting molecule is called 1,3-*di*phosphoglyceric *a*cid **(DPGA).**

The oxidation of PGAL to DPGA is an energy-yielding process, with a ΔF of some 64 kcal becoming available to the cell. Of this energy, 52 kcal is stored in the reduced form of NAD. Another 11.8 kcal is trapped in the bond by which the newly arrived phosphate group is attached to the molecule. Thus a "high-energy" phosphate bond is created in this molecule. At the very next step in glycolysis, this phosphate group is transferred to a molecule of ADP, converting it into ATP. The conversion of ADP to ATP requires at least 7.3 kcal, as we have seen. The extra energy stored in the DPGA is thus ample to effect the energy transfer.

After the singly dephosphorylated molecule (3-phosphoglyceric acid) undergoes an isomerization (becoming 2-phosphoglyceric acid), a molecule of water is removed. This process converts the remaining phosphate to "high-energy" form (at the expense of bond energy elsewhere in the molecule). The product, phosphoenolpyruvic acid (PEP), then gives up its "high-energy" phosphate to convert a second molecule of ADP to ATP. The product is pyruvic acid, $C_3H_4O_3$. As you can see, it is equivalent to a half-glucose molecule ($C_3H_6O_3$) that has been oxidized to the extent of losing two electrons (as hydrogen atoms).

What is the fate of pyruvic acid? We shall examine *three* major ways in which cells handle the pyruvic acid that is produced by glycolysis. These are lactic acid fermentation, alcoholic fermentation, and cellular respiration.

8.6 LACTIC ACID FERMENTATION

In animal cells, such as those of an overworked muscle, where oxygen is inadequate for cellular respiration, the pyruvic acid is reduced to lactic acid. The reducing agent is NADH. The energy stored in NADH is transferred back to the carbohydrate fragment.

What has been gained? Remember that we started with a glucose molecule containing a free energy of combustion of 686 kcal. We have produced *two* molecules of lactic acid, each containing a free energy of combustion of approximately 319.5 kcal. Thus there has been a free energy yield of 47 kcal ($686 - 2 \times 319.5 = 47$). Furthermore, although two molecules of ATP were needed to prime the process, four ATPs were produced by glycolysis (one from each of the two DPGAs and one from each of the two PEPs). Thus glycolysis produces a net of two ATP molecules for each glucose molecule. If 14.6 kcal are stored in these ATPs (2×7.3), then about 31% of the available free energy (47 kcal) has been trapped in the form of ATP. Actually, this is probably a conservative estimate of the efficiency.

However, when you consider that the two molecules of lactic acid contain 639 of the 686 kcal originally present in the glucose, the process seems very wasteful. Only 7% of the energy of the fuel has been liberated (with only about one-third of *that* trapped as ATP. Most of the energy in the fuel (glucose) is still trapped in the waste product (lactic acid). The process is potentially dangerous as well. Lactic acid, as its name implies, is an acid. Its production in a fermenting muscle cell can lower the pH to a point where serious disturbance of muscle function results. Wasteful though it is, however, lactic acid fermentation does provide a source of ATP to cells that are desperate for it because they cannot secure all the oxygen needed for cellular respiration (Fig. 8.6).

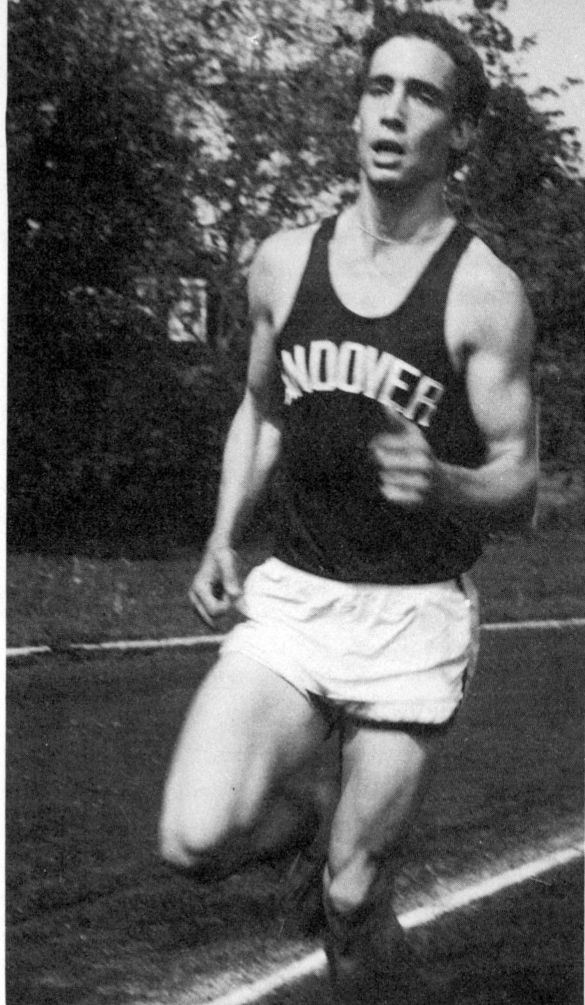

FIG. 8.6 Sprinting toward the finish line, this runner uses the anaerobic process of lactic acid fermentation to supplement the energy yield from cellular respiration. After the race, the "oxygen debt" incurred by fermentation will be repaid.

You might wonder why there is any energy yield in lactic acid fermentation. After all, the molecule of lactic acid is no more oxidized than the glucose molecule is. Looking at its molecular formula, $C_3H_6O_3$, it might seem that all glycolysis has accomplished is a breaking of a glucose molecule into two parts ($C_6H_{12}O_6 \rightarrow 2C_3H_6O_3$). In a sense, that is true. But closer inspection reveals that something more has gone on. One end of the molecule ($-CH_3$) is now more reduced than in a carbohydrate, while the other end ($-COOH$) is more oxidized. Thus there has been an *internal* oxidation-reduction, and it is this process that gives rise to the meager yield of free energy.

8.7 ALCOHOLIC FERMENTATION

Glycolysis in yeast follows the exact pathway we have been discussing except for the final step. In yeast, the pyruvic acid is decarboxylated (a CO_2 is removed) before it is reduced by NADH. The result, then, is a molecule of CO_2 and a molecule of ethanol (actually two of each for each glucose molecule fermented).

$$C_6H_{12}O_6 \rightarrow 2C_2H_5OH + 2CO_2$$

As was true of lactic acid fermentation, the process is very wasteful. Most of the energy stored in glucose is still stored in the ethanol. (This explains why ethanol sees frequent use as a fuel for high-performance engines.) As was true of lactic acid fermentation, the process is also hazardous. Yeasts poison themselves when the concentration of ethanol reaches about 13%. (This explains the maximum alcohol content of fermented beverages like wine. To make beverages with a higher alcohol content requires that the alcohol be concentrated by distillation.) And, as in lactic acid fermentation, note that alcohol fermentation has taken a carbohydrate ($C_3H_6O_3$) and completely oxidized one carbon (to CO_2) while further reducing the others (CH_3CH_2OH).

The third way in which cells use pyruvic acid is as the starting material for cellular respiration. The reduction of pyruvic acid to lactic acid (or ethanol and CO_2) is really an anabolic, energy-consuming process. Pyruvic acid serves as the acceptor of the electrons removed earlier from PGAL. Lactic acid contains more free energy than pyruvic acid, so the process is an uphill, energy-storing one. Most of the benefit gained by oxidizing PGAL has been lost. At best, we are left with an energy yield of only 7% from the fuel we use.

What if we could find a better acceptor for those electrons removed by NAD? By better acceptor, we mean an acceptor that is much more electronegative than NAD. As you know, oxygen is the most electronegative element (except for fluorine). The transfer of two electrons from NADH to O_2 yields about 52 kcal of energy, which could be made available to the cell. Furthermore, if we could find a way of continuing the oxidation of pyruvic acid, the large amount of free energy still locked within it could be made available. Early in evolution, I am happy to report, living things became capable of just these things: (1) the use of O_2 as the final acceptor of electrons removed from carbohydrates and (2) the complete oxidation of pyruvic acid. In eukaryotes, both of these activities take place within mitochondria. We call the process cellular respiration.

8.8 CELLULAR RESPIRATION: THE CITRIC ACID CYCLE

The crucial events of cellular respiration are (1) the complete oxidation of pyruvic acid by the stepwise removal of all the hydrogen atoms, leaving three molecules of CO_2, and (2) the passing on of the electrons removed as part of these hydrogen atoms to molecular oxygen (O_2).

Once within the mitochondrion, the pyruvic acid undergoes another step of oxidation. Like the oxidation of PGAL, this is accomplished by NAD. As before, two hydrogen atoms are removed, and these reduce NAD^+ to NADH. At the same time, a molecule of CO_2 is removed. The resulting fragment covalently bonds to another coenzyme, Coenzyme A, to form the complex called acetyl-CoA (Fig. 8.7). Acetyl-CoA now enters a cyclic series of chemical reactions during which the oxidation process is completed. This series of reactions is called the Krebs cycle or citric acid cycle. The first name honors the biochemist who discovered it. The second name reflects the first step in the cycle: the union of acetyl-CoA with oxaloacetic acid to form citric acid. In this process, a molecule of CoA is regenerated and one molecule of water is used. Oxaloacetic acid is a four-carbon acid with two carboxyl groups. Citric acid thus has six carbon atoms and three carboxyl groups. (Citric acid is the acid found in citrus fruits like lemons and grapefruit.)

After two steps that simply result in forming an isomer of citric acid, another NAD-mediated oxidation takes place. Once again, this is accompanied by the removal of a molecule of CO_2. The resulting compound is α-ketoglutaric acid. It, in turn, undergoes yet another oxidation ($NAD^+ + 2H \rightarrow NADH + H^+$) and decarboxylation ($-CO_2$). These steps are accompanied by the insertion of a molecule of water. The product, then, has one less carbon atom and one less oxygen atom. It is succinic acid.

The conversion of α-ketoglutaric acid to succinic acid is accompanied by a free energy change of about −8 kcal. This is just enough to be coupled to the synthesis of a molecule of ATP ($\Delta F = +7.3$ kcal). The synthesis does, in fact, take place although not directly. First a phosphate group is attached to the nucleotide GDP (guanosine diphosphate) converting it into GTP. GTP then transfers its high-energy terminal phosphate to ADP to form ATP.

The next step in the citric acid cycle is one we examined briefly in Chapter 6. It is the oxidation of succinic acid to fumaric acid. Once again, two hydrogen atoms are removed, but this time the oxidizing agent is a coenzyme called flavin adenine dinucleotide or FAD. FAD is reduced to $FADH_2$.

With the insertion of another molecule of water, fumaric acid is converted into malic acid. Another NAD-mediated oxidation of malic acid produces oxaloacetic acid—the molecule with which this portion of the story began. With the regeneration of oxaloacetic acid, another molecule of acetyl-CoA can enter the cycle and the whole process can be repeated.

You can hardly be blamed for finding this a complex story. But it is an elegant one. Perhaps it will help if you now stand back and see just what a turn of the cycle has accomplished. In essence, we placed a molecule of acetic acid (CH_3COOH, albeit in the form of acetyl-CoA) into the cycle. We added two molecules of water during the turning of the cycle. Two decarboxylations occurred and, on four different occasions, two atoms of hydrogen were removed. Thus the process can be expressed as

$$CH_3COOH + 2H_2O \rightarrow 2CO_2 + 8H.$$

The hydrogens are removed by NAD^+ (FAD for one pair). Their ultimate fate is to be passed on to oxygen (O_2), forming molecules of water. Thus the acetic acid is completely oxidized to carbon dioxide and water. Now let us examine more closely the transfer of the hydrogen atoms to oxygen. This process, too, occurs within the mitochondria.

8.9 THE RESPIRATORY CHAIN

In Section 8.5, I promised that the oxidation of PGAL could serve as a model for many of the steps of cellular respiration. You may recall that DPGA, the product of the oxidation, contained much less free energy than PGAL. Most of the difference (52 kcal of it) was stored in the NAD that had acquired the electrons removed from PGAL.

At each of the other oxidations in cellular respiration, essentially the same situation occurs. From pyruvic acid to acetyl CoA, from isocitric acid to α-ketoglutaric acid, from α-ketoglutaric acid to succinic acid, from succinic acid to fumaric acid, and from malic acid to oxaloacetic acid, the product in each case contains less free energy than the reactant. Most of the difference is trapped in NADH ($FADH_2$ in one case).

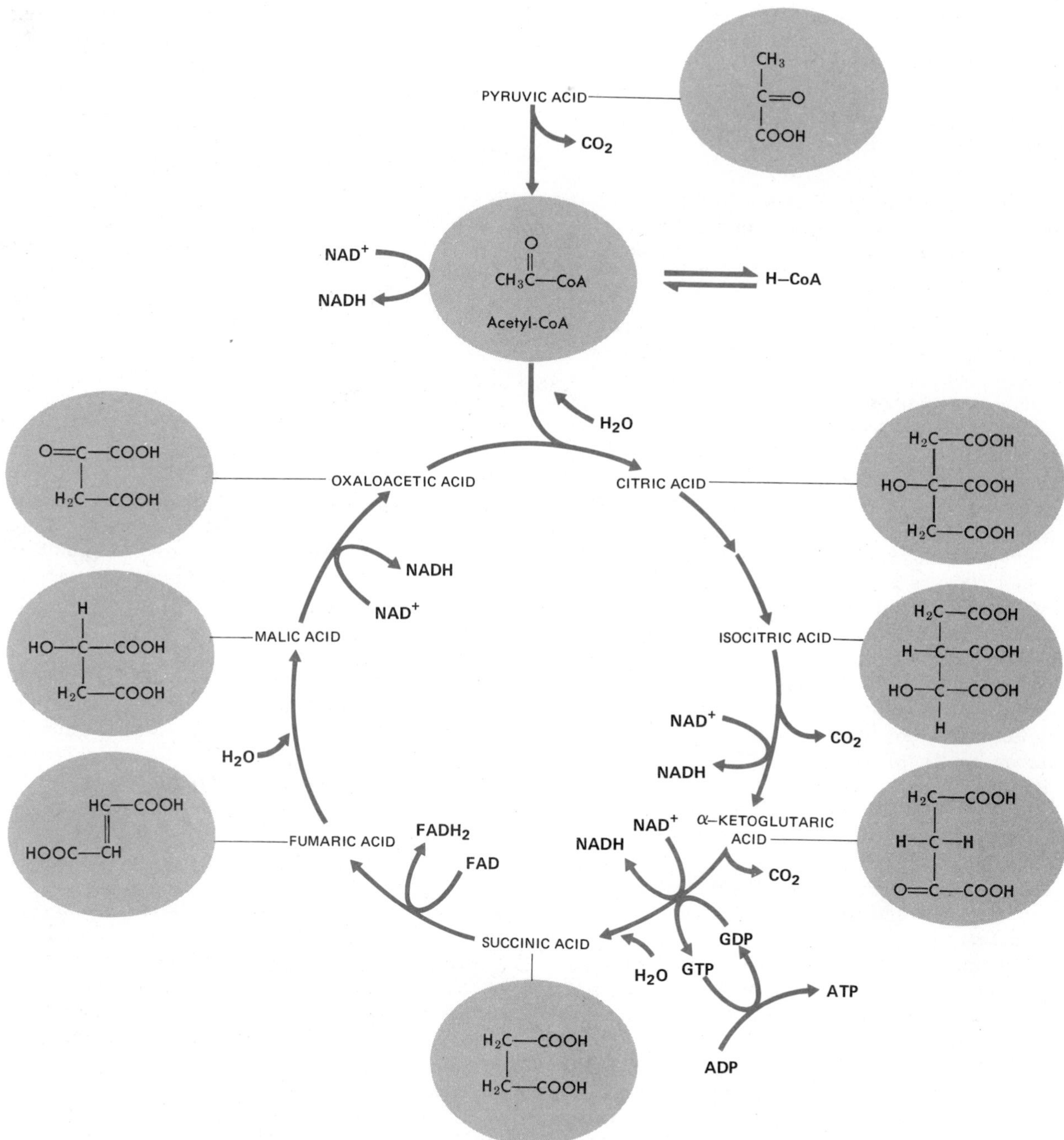

FIG. 8.7 The pathway of cellular respiration. The reactions of the citric acid cycle (also known as the Krebs or tricarboxylic acid cycle) accomplish the complete, stepwise oxidation of the acetyl group of acetyl-CoA. In eukaryotes, these reactions occur within the mitochondria.

The way to get at this energy is to turn around and oxidize the NADH with an oxidizing agent of still greater electronegativity. Because of its great electronegativity and its prevalence, oxygen would be ideal for this. The oxidation of NADH by oxygen will give us 52 kcal. However, this is a value substantially higher than that needed to synthesize a molecule of ATP (something a little larger than 7.3 kcal). So, directly oxidizing NADH with oxygen could result in wasting a substantial amount of free energy as heat. A better approach—and the one used by mitochondria—is to pass the electrons removed from NADH through a linked series of redox substances before finally allowing them to reach oxygen atoms. Each of these redox substances is thus capable of being alternately reduced and oxidized—reduced as it receives a pair of electrons from the substance of lower electronegativity, oxidized as it passes these electrons on to a substance of greater electronegativity. At each successive step, there is a yield of free energy. At *three* of these, the yield is sufficient for the synthesis of an ATP molecule.

The redox substances include: (1) a coenzyme called coenzyme Q, (2) a series of **cytochrome enzymes,** and (3) molecular oxygen (O_2) (Fig. 8.8). Each of the cytochrome enzymes has a molecule of heme as its prosthetic group (Fig. 8.9). Heme is the same iron-containing prosthetic group that we have already seen in the oxygen-carrying pigments hemoglobin and myoglobin (see Fig. 4.23). The whole series of redox enzymes in the mitochondrion is known as the respiratory chain.

The path of the electrons appears to be the following: NADH is oxidized by coenzyme Q. This oxidation yields enough free energy to permit the synthesis of a molecule of ATP from ADP and inorganic phosphate. Coenzyme Q is, in turn, oxidized by cytochrome *b*, which is then oxidized by cytochrome *c*. This step is also coupled to the synthesis of a molecule of ATP (Fig. 8.8). Cytochrome *c* then reduces a complex of two enzymes called cytochrome *a* and a_3. For convenience, we shall refer to the complex simply as cytochrome *a*. Cytochrome *a* is oxidized by an atom of oxygen and the electrons have arrived at the bottom of the respiratory chain. Oxygen is the most electronegative substance in the chain and the final acceptor of the electrons. A molecule of water is produced. In addition, this final oxidation is coupled to yet a third step of ATP synthesis.

Despite extensive research, the precise mechanism by which the energy released by the oxidation of NADH, cytochrome *b*, and cytochrome *a* is cou-

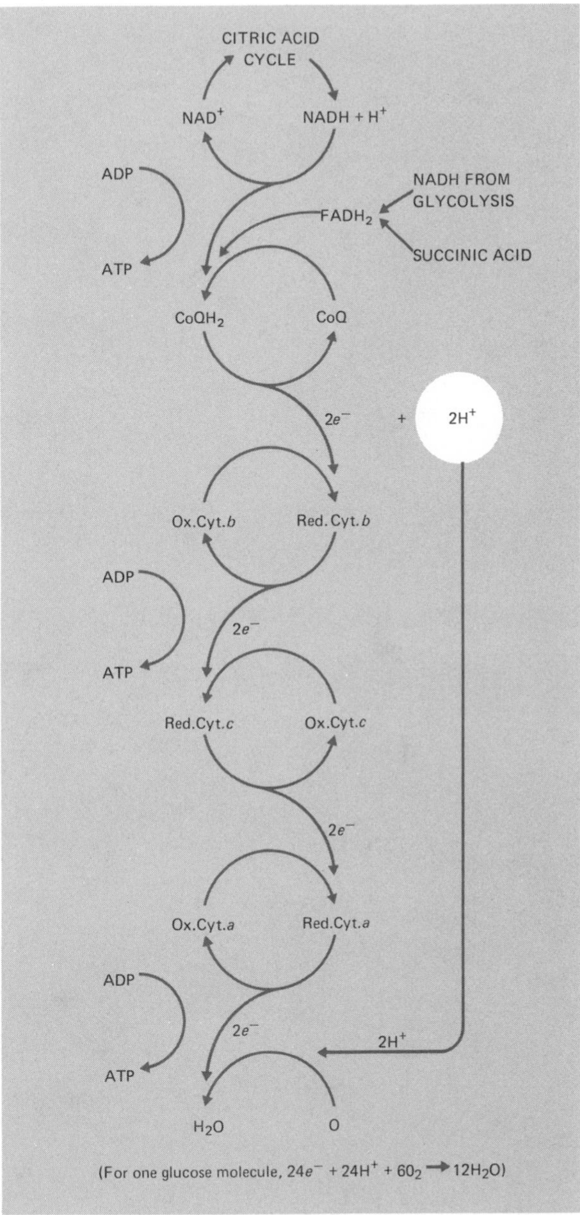

FIG. 8.8 The respiratory chain. As electrons pass down the respiratory chain, they give up energy which is harnessed to the synthesis of ATP.

HEME

FIG. 8.9 Structure of heme. Heme serves as the prosthetic group of the cytochrome enzymes as well as of myoglobin and hemoglobin. Compare its structure with that of chlorophyll (Fig. 9.3).

FIG. 8.10 A single mitochondrion. The way in which the inner membrane projects into the interior to form the cristae shows clearly. Some of the enzymes of the citric acid cycle and all of the enzymes of the respiratory chain are attached to the membranes of the cristae. (80,000 ×, courtesy Keith R. Porter.)

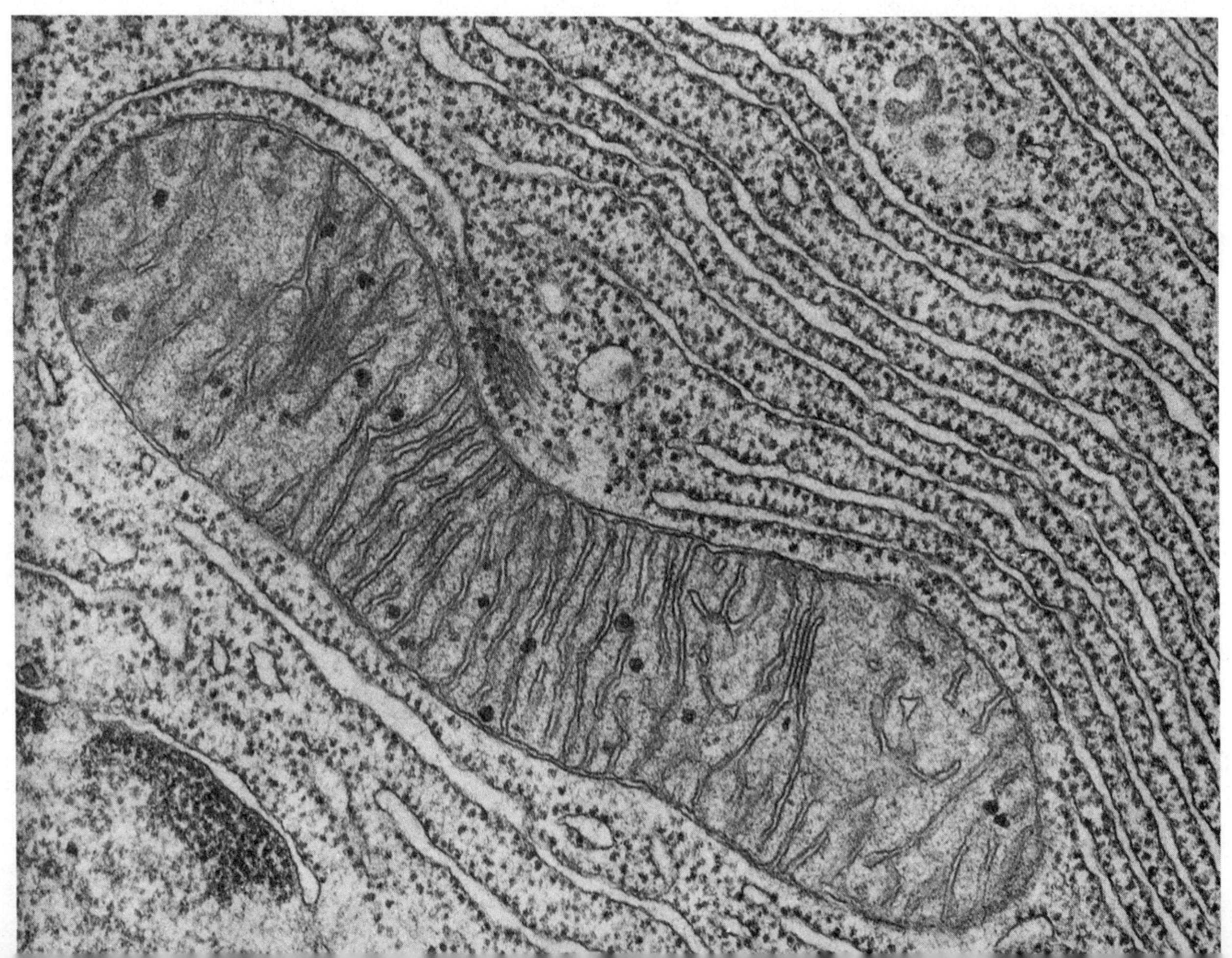

pled to the synthesis of ATP is uncertain. What is clear, however, is that this process traps at least 3×7.3 or 21.9 kcal of the 52 kcal stored in NADH, for an efficiency of at least 42%.

The intricate processes going on in the mitochondria might suggest to you the existence of some structural organization to this organelle. This appears to be the case. Analysis of isolated mitochondria and their fragments indicates that this organelle is not simply a sac filled with enzymes (as is a lysosome). In Chapter 5, we saw that each mitochondrion consists of a smooth outer membrane and a richly-folded inner membrane. The folds of the inner membrane, called cristae, extend into a fluid matrix which fills the interior of the mitochondrion (Fig. 8.10). The enzymes of the respiratory chain are incorporated in the inner mitochondrial membrane. In addition, the enzymes associated with the synthesis of ATP are attached to the inner side of this membrane (facing the matrix). The electron micrograph in Fig. 8.11 shows orderly arrays of stalked knobs on this surface of the inner membrane. There is evidence that each represents one enzymatic package for ATP synthesis. The high degree of order is probably a reflection of what must be an orderly arrangement of all the enzymes participating in electron transport. Some of the enzymes of the citric acid cycle (e.g. succinic acid dehydrogenase) are also incorporated as intrinsic proteins in the inner mitochondrial membrane. Others (e.g. malic acid dehydrogenase) are simply dissolved in the fluid matrix.

FIG. 8.11 Knobs on the surface of the cristae of a mouse liver mitochondrion. Each knob may represent a single enzyme complex for the synthesis of ATP. (200,000 ×, courtesy Dr. Donald F. Parsons, *Science,* **140:** 985, May 31, 1963.)

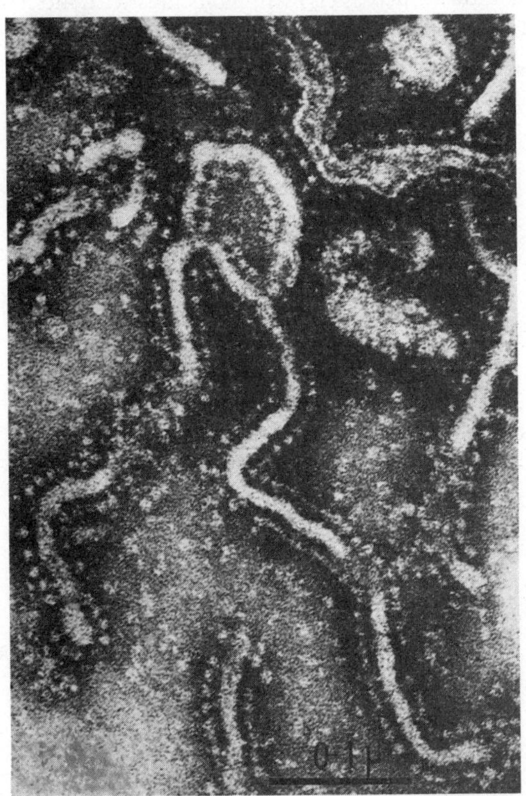

8.10 THE RESPIRATION BALANCE SHEET: MATERIALS

Let us now see if we can account for all the reactants and products of the cellular respiration of one molecule of glucose. A single glucose molecule is cleaved into two three-carbon fragments (PGAL). Each of these undergoes *six* successive oxidations: five dehydrogenations by NAD and one by FAD. Each PGAL thus gives rise to a total of 12 electrons (and H^+); for the *two* PGAL formed from glucose, the total is 24. Thus 24 electrons (and 24 protons) ultimately unite with 12 atoms of oxygen ($6\ O_2$) to yield 12 molecules of H_2O.

Decarboxylations occur at three places in the process (pyruvic acid, citric acid, and α-ketoglutaric acid). Remember that because the cycle turns twice for each molecule of glucose, we get a total yield of $6\ CO_2$. Note, too, that water molecules are inserted at three steps in the process: formation of citric acid, succinic acid, and malic acid. Putting this all together, we come up with an overall equation for cellular respiration:

$$C_6H_{12}O_6 + 6O_2 + 6H_2O \rightarrow 6CO_2 + 12H_2O.$$

What about ATP production? At five different times in the breakdown of one PGAL molecule, NAD removes a pair of electrons and passes them through the respiratory chain. When this occurs *within* the mitochondria (four times), each pair of electrons gives rise to three molecules of ATP. However, the oxidation of PGAL itself, which takes place in the cytosol, yields only two molecules of ATP because the electrons enter the respiratory chain at coenzyme Q (Fig. 8.8). Hence 14 molecules of ATP are produced from each PGAL by

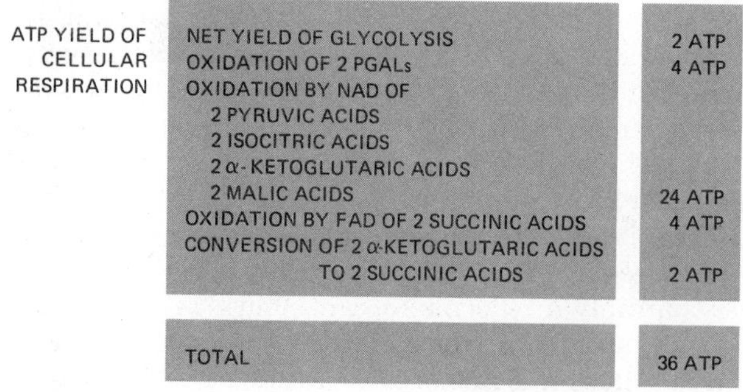

ATP YIELD OF CELLULAR RESPIRATION

NET YIELD OF GLYCOLYSIS	2 ATP
OXIDATION OF 2 PGALs	4 ATP
OXIDATION BY NAD OF 2 PYRUVIC ACIDS 2 ISOCITRIC ACIDS 2 α-KETOGLUTARIC ACIDS 2 MALIC ACIDS	24 ATP
OXIDATION BY FAD OF 2 SUCCINIC ACIDS	4 ATP
CONVERSION OF 2 α-KETOGLUTARIC ACIDS TO 2 SUCCINIC ACIDS	2 ATP
TOTAL	36 ATP

FIGURE 8.12

these oxidations, making a total of 28 ATPs for each glucose molecule. In addition, one pair of electrons is removed by FAD, but these also give rise to only two ATP molecules since they are transferred directly to coenzyme Q and thus bypass the first step of ATP synthesis (Fig. 8.8). That shortage is made up, however, by the single ATP produced in the transformation of α-ketoglutaric acid into succinic acid. Again, as the cycle must turn twice for each glucose molecule oxidized, a total of six ATP molecules are produced by these mechanisms. Adding the *net* yield of two molecules of ATP produced by glycolysis, we end up with a grand total of 36 molecules of ATP produced by the oxidation of a single molecule of glucose (Fig. 8.12).

8.11 THE RESPIRATION BALANCE SHEET: ENERGY

As noted earlier, the combustion of a mole of glucose releases 686 kcal of energy, almost all of it as heat. But the cellular respiration of glucose traps some of this energy by the synthesis of ATP. How much is trapped? Each molecule of ATP stores, as a conservative estimate, 7.3 kcal. Therefore, 36 ATPs store some 263 kcal of free energy. This represents a yield of 38% (263/686) of the free energy of the glucose, a performance far superior to the meager 7% yield from the fermentation of glucose.

In Chapter 3, we examined the basis for the free energy (ΔF) changes in chemical reactions. We found that ΔF represents the difference between the energy liberated as bonds are formed to make the products and the energy needed to break the bonds of the reactants. In the case of cellular respiration, the *reactants* (glucose, O_2) are molecules with atoms of little or no difference in electronegativity. Thus

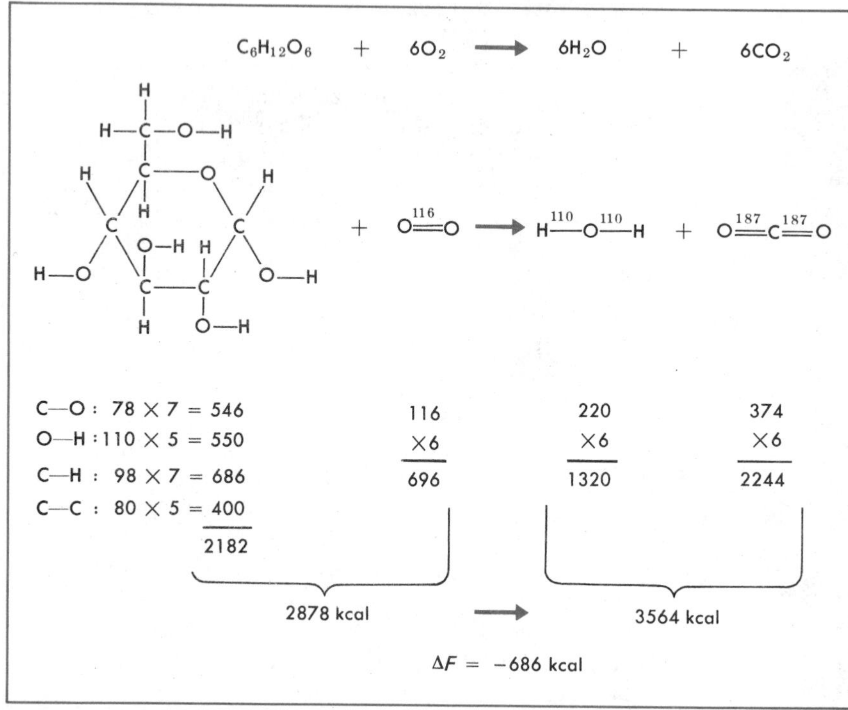

FIG. 8.13 Bond energy relationships in cellular respiration. The difference between the free energy needed to break the bonds of the reactants and the free energy liberated in the formation of the bonds of the products is 686 kcal/mole, the net free energy yield of the reaction.

their bonds are covalent, with average bond energies that tend to be on the low side. This is another way of saying that these bonds are broken with relative ease. The products (CO_2, H_2O), on the other hand, are molecules with atoms of marked differences in electronegativity. They are characterized by polar covalent bonds of high bond energy. These are strong bonds, *broken* with difficulty, and liberating copious amounts of energy when they *form*. Because the sum of the bond energies of the products (= energy released) is greater than that of the reactants (= energy needed) there is a net release of energy. Thus the reaction occurs with a negative ΔF.

Now let us examine the process more quantitatively. Figure 8.13 gives the overall equation of cellular respiration (ignoring the six molecules of water that occur on each side, i.e. as both reactant and product; they "cancel out"). The structural formulas are given as well as average bond energies for each bond involved. As you can see, the 24 covalent bonds of glucose require a total of 2182 kcal to be broken. The six double bonds of oxygen require a total of 696 kcal to be broken. Thus a grand total of 2878 kcal is needed to break all the bonds of the reactants. As for the products, the formation of six molecules of CO_2 involves the formation of 12 double polar covalent bonds, each with a bond energy of 187 kcal. Thus 2244 kcal are liberated in the synthesis of the 6 CO_2. Formation of the two O—H bonds of the water molecule releases 220 (2×110) kcal of energy. For six molecules of water, this is a yield of 1320 kcal. The grand total for the products is 3564 kcal. Subtracting this from the 2878 kcal needed to break the bonds of the reactants, we arrive at −686 kcal, the free energy change of the oxidation of glucose. This value holds true whether we oxidize glucose quickly by burning or in the orderly process of cellular respiration.

8.12 THE STORAGE BATTERY OF LIFE

Throughout this chapter (as well as in Chapter 3), the concept of electronegativity has helped us to understand why electrons flow in the direction they do during redox reactions. It has also given us a rough view of the sign (+ or −) and the magnitude of the free energy (ΔF) changes we can expect during redox reactions. But we can sharpen our analysis still further. The electronegativity of a substance can also be expressed as a redox potential (expressed as E). In this case, there has to be a standard of comparison. The standard we shall use is hydrogen. We shall assign hydrogen a redox potential of 0 ($E = 0$). Any substance (atom, ion, molecule) that is more electronegative than hydrogen is assigned a positive (+) redox potential. Any substance that is less electronegative than hydrogen is assigned a negative (−) redox potential. The scale is quantitative. The greater the *difference* between the redox potentials of two substances (ΔE), the greater the vigor with which electrons will flow spontaneously from the less positive to the more positive (more electronegative) substance. The difference in potential (ΔE) is, in a sense, a measure of pressure between the two. We express ΔE in volts.

If we bring two substances of differing E together so that the opportunity of electron flow between them exists, we have created a battery. Although it may be in a mitochondrion, it is just as much a battery as the lead-acid storage battery in an automobile.

You can imagine that the greater the voltage, ΔE, existing between the two components of our battery, the more energy is available when electron flow occurs. And, in fact, it is possible to determine the amount of free energy available from such a battery. The relationship is:

$$\Delta F = -n(23.062 \text{ kcal})\ \Delta E,$$

where n is the number of electrons transferred and 23.062 kcal is the amount of energy released when one electron passes through a potential drop of 1 volt (called the Faraday constant). Carbon reduced to the extent occurring in carbohydrates like glucose has a redox potential of approximately −0.42 volt. Oxygen, as the most electronegative substance in the system, naturally has the largest E: +0.82 volt. The ΔE is thus 1.24 volts. Allowing 24 electrons to pass through this potential gives us our free energy yield of −686 kcal ($\Delta F = -24(23.062)(1.24)$).

The same kind of analysis may be helpful in understanding the strategy behind the respiratory chain. Figure 8.14 shows the redox potentials of NAD (−0.32 V), coenzyme Q (about −0.05 V), cytochrome *b* (+0.03 V), cytochrome *c* (+0.24 V), cytochrome *a* (+0.29 V), and oxygen (+0.82 V). The free energy released as electrons pass from step to step can easily be calculated using the formula above. At only three places in the respiratory chain does the drop in potential yield enough energy to synthesize a molecule of ATP. These are the transfers from NAD to coenzyme Q ($\Delta E = 0.27$ V and thus $\Delta F = -12.5$ kcal), from cytochrome *b* to cytochrome *c* ($\Delta E = 0.21$ V and thus $\Delta F = -9.7$ kcal),

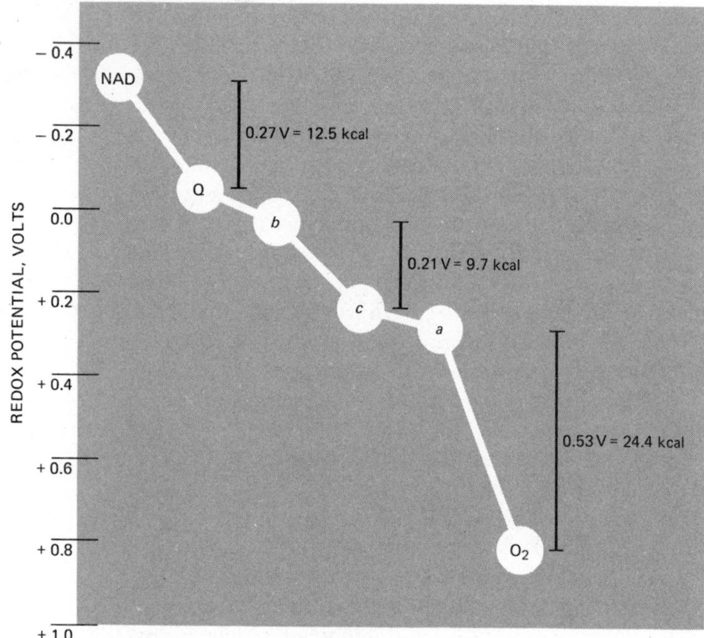

FIG. 8.14 Redox potentials of the components of the respiratory chain. At three steps in the downhill passage of electrons, the drop in potential (ΔE) is sufficiently great to liberate the quantity of free energy required for the synthesis of a molecule of ATP.

and from cytochrome *a* to oxygen ($\Delta E = 0.53$ V and thus $\Delta F = -24.4$ kcal). And these are just the three steps where ATP synthesis takes place.

8.13 WHAT ABOUT OTHER FUELS?

We have examined in some detail an elegant system by which the energy stored in glucose is carefully liberated for use by the cell. But glucose is not the only fuel upon which cells depend. Other carbohydrates, fats, even proteins may in certain cells or at certain times be used as a source of energy. Having seen the elaborate mechanism by which the cell uses glucose, you may fervently hope that similarly constructed systems are not needed for each kind of fuel. And indeed they are not. *One* of the great advantages of the step-by-step oxidation of glucose into CO_2 and H_2O is that several of the intermediate compounds formed in the process link glucose metabolism with the metabolism of other food molecules. For example, when fats are used as fuel, the glycerol portion of the molecule is converted into PGAL and enters the glycolytic pathway at that point. Fatty acids are converted into molecules of acetyl-CoA and enter the respiratory pathway (Fig. 8.15). The amino acids liberated by the hydrolysis of protein can also serve as fuel. First the nitrogen is removed, a process called deamination. The remaining fragments enter the process at several points. For example, the amino acids Gly, Ser, Ala, and Cys are converted into pyruvic acid and enter the mitochondrion to be respired. Acetyl-CoA serves as the entry point for Phe, Leu, Ile, Thr, Lys, Trp, and Tyr. Asp and Asn are degraded to oxaloacetic acid while Glu, Gln, Arg, His, and Pro enter the Krebs cycle at α-ketoglutaric acid (Fig. 8.15). These links thus permit the respiration of excess fats and proteins in the diet. No special mechanism of cellular respiration is needed by those animals that depend largely on ingested fats (e.g. many birds) or proteins (e.g. carnivores) for their energy supply.

8.14 CONTROL OF CELLULAR RESPIRATION

The operation of such an intricate system as glycolysis and cellular respiration must be precisely controlled. For example, the cell must be able to respond effectively to its changing needs for ATP, the changing availability of oxygen, or the changing fuel sources (e.g. glucose → fatty acids) available to it. A number of control mechanisms have been discovered.

Let us examine one of the most important of the control points. This is the step at which fructose 6-phosphate is converted into fructose 1,6-diphosphate. The conversion is catalyzed by the enzyme phosphofructokinase. The catalytic activity of phosphofructokinase on its substrate (fructose 6-phosphate) is strongly influenced by several other substances. An increasing concentration of ADP, for example, accelerates the activity of the enzyme. This makes sense, as it speeds up glycolysis and thus helps to rebuild the cell's available ATP. On the other hand, an increasing concentration of ATP inhibits the activity of the enzyme, thus slowing the production of ATP. This is an example of feedback inhibition (see Section 6.10). An accumulation of citric acid, an intermediate in the Krebs cycle, also inhibits the activity of phosphofructokinase and thus further production of citric acid. Phosphofructokinase is an example, then, of an **allosteric enzyme.** Its activity with respect to its substrate is strongly influenced by the presence of other molecules that presumably bind to it at some site on the molecule

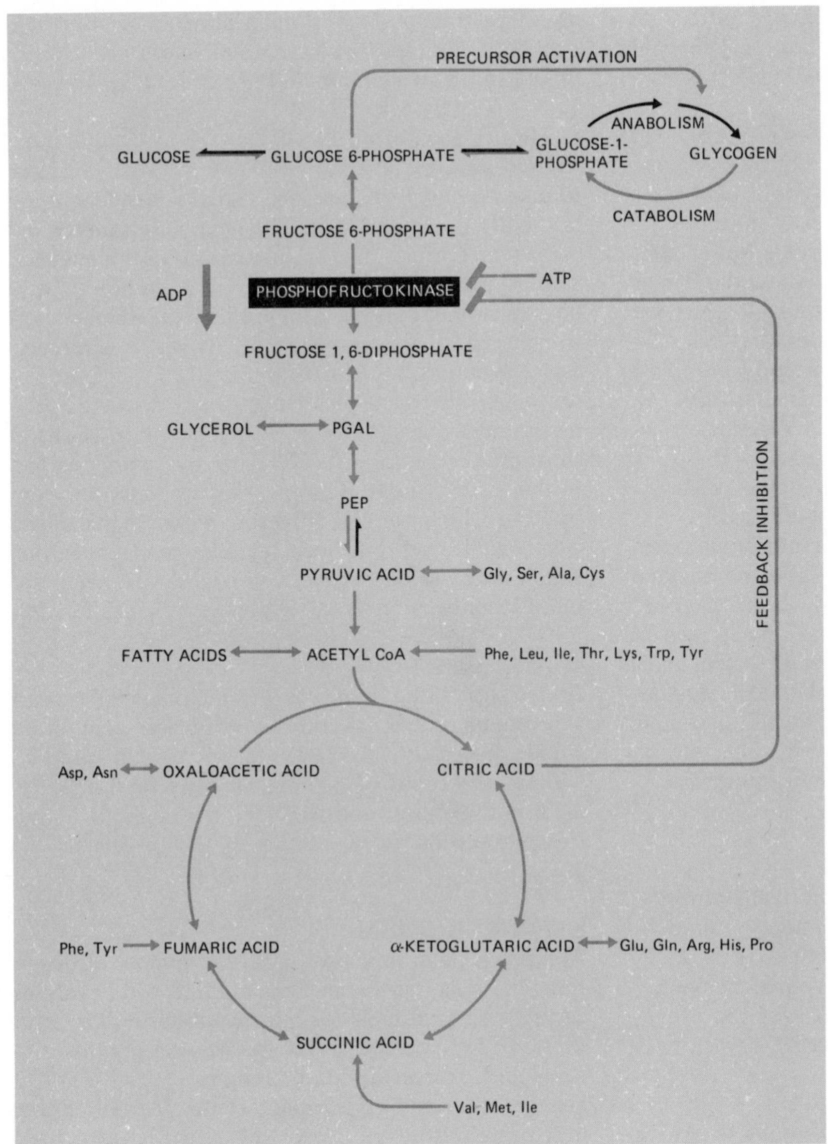

FIG. 8.15 Interconnections between fat, protein, and carbohydrate metabolism. The products of the digestion of fats (fatty acids and glycerol) and proteins (amino acids) enter the pathway of cellular respiration at the points shown. Some of the same points (indicated by double-headed arrows) serve to drain off excess intermediates into anabolic pathways leading to the synthesis of fats and certain of the amino acids. Black arrows indicate those steps where the metabolic pathway can be reversed only by using different enzymes. The figure also shows two major control points in cellular respiration. Enzyme activation is indicated by a heavy arrow (➡). Enzyme inhibition is indicated by a solid bar (⁄——).

other than the catalytic site. In this way, the rate of glycolysis is closely adjusted to the needs of the cell.

8.15 HOW THESE DISCOVERIES WERE MADE

You may be puzzled as to how biochemists have been able to discover all the intermediate steps in glycolysis and cellular respiration. It has required the work of many years and many scientists and the use of a variety of ingenious techniques.

One technique in studying intermediary metabolism is to supply a suspected intermediate substance and see (1) whether it is used up and (2) what new substance begins to accumulate. Sometimes this technique is applied to the intact organism, but usually it is applied to isolated organs, tissues, or even cell extracts. For example, Hans Krebs discovered that citric acid is produced when pyruvic acid and oxaloacetic acid are added to minced pigeon muscle, and that discovery enabled him to work out the reactions in the cycle that now bears his name.

The use of enzyme inhibitors has also provided information on intermediary metabolism. For example, the poisonous effect of malonic acid (see Section 6.9) is due to its inhibition of succinic dehydrogenase, the enzyme that catalyzes the conversion of succinic acid to fumaric acid in the Krebs cycle. When malonic acid is present, the normal substrate, succinic acid, accumulates and may be identified. Then if the substance normally *produced* by that enzyme can be identified, this substance can be added to the poisoned system. If respiration begins again, as it does when fumaric acid (Fig. 8.7) is added to tissues poisoned by malonic acid, another link in the chain has been established.

One of the best techniques for studying intermediary metabolism is to introduce into an enzyme system certain molecules which have been "tagged" with radioactive atoms. The gradual appearance of radioactivity in other chemical substances will indicate the pathway by which the chemical changes have occurred. In the next chapter, we shall see how this technique was used to determine one of the important steps in the process of photosynthesis.

8.16 THE USES OF ENERGY

Glycolysis and cellular respiration transform the free energy of food into the free energy stored in the "high-energy" phosphate bonds of ATP. ATP, in turn, serves as the immediate source of energy for all the energy-requiring activities of the organism. These fall into several categories.

1. Mechanical Work

Locomotion is one of a number of ways in which animals accomplish mechanical work. In each case, the mechanical work is brought about by the contraction of **muscle cells.** The force exerted when muscle cells contract can then be transmitted to structures adapted for locomotion, etc. The energy needed in order for muscle cells to contract is supplied by ATP. The beating of cilia and flagella (see Section 5.17) is also powered by ATP.

2. Active Transport

Cells must expend energy in order to transport ions or molecules against their concentration gradients, i.e. to accomplish active transport. Several examples of active transport were mentioned in Section 6.5. Probably all cases of active transport are closely linked to the production and use of ATP by the cell.

3. Heat Production

Energy is also a source of heat for living things. Mammals and birds are especially dependent upon internally generated heat. They can maintain a fixed body temperature despite fluctuations in the environmental temperature. Generally, the production of heat occurs simply as a byproduct of other energy transformations within the cell. As we have already seen, no energy transformation is 100% efficient. For example, when chemical energy is converted into mechanical energy (as in muscular contraction) a substantial amount (70–80%) of the energy is lost in the form of heat. The involuntary muscular contraction which we call shivering exploits this inefficiency to prevent our body temperature from falling below normal. The immediate source of energy for muscle contraction is, as we have seen, ATP.

4. Anabolism

Anabolism is the synthesis of macromolecules such as proteins, polysaccharides, and nucleic acids from small precursors. Such syntheses cannot proceed without the input of energy. Directly or indirectly, ATP is the energy source for all the anabolic activities of the cells.

8.17 ANABOLISM

The products of anabolism serve a number of essential functions. Glycogen and fats and even proteins serve as reserve fuels for later catabolism. The synthesis of new nucleic acids enables the organism to store and express additional copies of its library of genetic information. Protein, protein-carbohydrate, and protein-lipid molecules are essential structural components of the organism, both intracellular and extracellular. When synthesis of these materials proceeds at a faster rate than their degradation, the organism grows.

All anabolic transformations are catalyzed by enzymes, just as catabolic reactions are. In view of the relative ease with which many enzyme-catalyzed reactions can be reversed, we might expect that the enzymes involved in catabolism could also serve in anabolism by such simple expedients as shifting the relative concentration of substances on each side of the equation. Actually, some anabolic pathways *do* use some of the same enzymes as related catabolic pathways. (A number of examples are indicated in Fig. 8.15 by double-headed arrows.) One of the functions of liver cells is to convert the lactic acid produced by fermentation in muscles back into glucose 6-phosphate and then on to either glucose (for release to the blood) or glycogen (for storage). The liver cells use many of the enzymes of glycolysis for this purpose but not all of them. Probably no catabolic pathway can be completely reversed without the intervention of some enzymes unique to anabolism. In order for glucose 6-phosphate to be synthesized *into* glycogen, three enzymes are required which are not the enzymes that catalyze the conversion of glycogen to glucose 6-phosphate (Fig. 8.15).

Having anabolic and catabolic pathways at least partly separated is good strategy for the cell. Some catabolic reactions are virtually irreversible and an easier route must be found to get back "up" the metabolic pathway. Furthermore, having separate steps in catabolism and anabolism enables the cell to exert separate control mechanisms over each process. For example, a build-up of glucose 6-phosphate might be expected to lead to increased cellular respiration as a simple consequence of increasing the concentration of a reactant in a reversible reaction. But this would be appropriate only if the cell were short of ATP and, in fact, the ATP level does control this catabolic pathway (by means of phosphofructokinase), as we saw in Section 8.14. However, glucose 6-phosphate activates the last enzyme in glycogen synthesis ("precursor activation"—see Section 6.10) and will be drawn off into this anabolic pathway if it is not being used for cellular respiration. It is worth noting that anabolic and catabolic reactions are often physically separated in the cell as well. We have seen the role that mitochondria play in catabolism. Many anabolic reactions require *cytoplasmic* enzymes or enzymes found in the endoplasmic reticulum and elsewhere.

The synthesis of proteins and nucleic acids from their precursors differs in a crucial way from that of polysaccharides and fats. Starch (glycogen) and cellulose are polymers made up of the same monomeric subunit, glucose. A variety of fatty acids may be used in fats, but the structure of the molecules is quite simple. Within limits, these macromolecules are pretty much alike, whatever the organism in which they are found. The situation is quite different for the synthesis of proteins and nucleic acids. These molecules are complex polymers. The absolute sequence in which their monomeric subunits are ordered is crucial to their function. They could not be synthesized by a simple reversal of catabolism because they must be synthesized by a process that assembles their subunits in a specified order. These molecules thus contain *information*—information that is in some cases unique to the organism in which synthesis is taking place. In our bodies we may catabolize beef proteins or bean proteins, but we synthesize only human proteins. The mechanisms by which we synthesize precisely ordered sequences of subunits in our nucleic acids and proteins will be described in Chapter 18.

Just as catabolism releases energy, so anabolism requires it. And, of course, it is the energy released by catabolism—in the form of ATP—that powers anabolism. In order for amino acids to be assembled into a protein, they must first be "activated." The energy for this comes from ATP. The monomers of nucleic acids are nucleotides with high-energy phosphate bonds, one of which is, in fact, ATP. The others (i.e. GTP, CTP, UTP for RNA and their *deoxy*-derivatives for DNA) receive their high-energy phosphate bonds from ATP. In order to assemble glucose molecules into starch or cellulose, they must first be activated. ATP provides the energy for this. Synthesis of fats also requires ATP.

In our discussion of cellular respiration (Section 8.13) we noted the many intermediates which serve to connect carbohydrate catabolism to the catabolism of fats and proteins (Fig. 8.15). These connecting points between different pathways act as two-way valves. They provide points of entry not only for the catabolism (cellular respiration) of fatty acids, glycerol, and amino acids, but for their synthesis as well. Thus the catabolic breakdown of starches can lead (through acetyl-CoA and PGAL) to the synthesis of fat—an all-too-common occurrence! Much of the protein we consume is ultimately converted into glucose to provide fuel for the brain and other tissues. Although all our foods are interconvertible to some degree, they are not completely so. In other words, no one food can supply all our anabolic needs. We can indeed synthesize many fats from glucose, but certain fats (unsaturated) cannot be synthesized and must be taken in directly in our diet. Although we can synthesize twelve amino acids from carbohydrate precursors, we must obtain eight others (the "essential" amino acids) directly.

8.18 SUMMARY

In this chapter we have examined the intricate mechanism by which living cells exploit the free energy of organic molecules like glucose to carry out their various activities. Among the most crucial of these activities are the anabolic reactions by which the complex and ordered macromolecular assemblies of the cell are synthesized. The entire process—and thus life itself—depends on the coupling of catabolic reactions that liberate free energy ($-\Delta F$) to anabolic reactions that store it ($+\Delta F$). But these couplings are nowhere near 100% efficient. Some energy is always lost as heat. Heat can do no work (its energy is no longer "free"). So, if the story we have been examining were the whole story, then life would long since have run out of free energy and ceased to exist.

To choose another metaphor, this chapter has been devoted to examining the way in which the storage battery of life is discharged. But if any battery is to continue to be able to do work, it must from time to time be recharged. The recharging of the storage battery of life occurs by photosynthesis. It is the subject that we shall examine next.

EXERCISES AND PROBLEMS

1. Pound for pound, what food yields the most energy when oxidized?
2. What are the differences between alcoholic and lactic acid fermentation?
3. Using the data in Fig. 3.23, calculate the amount of free energy released ($-\Delta F$) when a mole of palmitic acid ($C_{15}H_{31}COOH$) is burned to yield carbon dioxide and water.
4. What parallels can you draw between the process of bread making and the process of brewing? What differences are there?
5. What would be the net yield of ATPs from respiring a mole of glucose if glycolysis took place within mitochondria rather than in the cytosol?
6. Why is the calorie, a unit of heat energy, also used as a measure of chemical energy?
7. How many moles of oxygen are consumed when two moles of glucose are respired? How many moles of carbon dioxide are produced?
8. What is the ratio of moles of oxygen consumed to moles of CO_2 produced when glucose is respired?
9. What happens to this ratio when fats are used as the fuel in cellular respiration?
10. Why are alcoholic beverages usually brewed in sealed containers?
11. What is the maximum weight of alcohol that yeast cells can produce from 90 g of glucose?
12. What is the net yield of ATP molecules when glycogen rather than glucose is used to produce the glucose phosphate used in cellular respiration?
13. What is the free energy yield when a mole of electron pairs passes from cytochrome *c* to cytochrome *a*? Could this step be used to synthesize ATP? Explain.

REFERENCES

1. CHAPPELL, J. B., and S. C. REES, *Mitochondria,* Oxford Biology Readers, No. 19, Oxford University Press, Oxford, 1972.
2. NICHOLLS, P., *Cytochromes and Biological Oxidation,* Oxford Biology Readers, No. 66, Oxford University Press, Oxford, 1975.

9 PHOTOSYNTHESIS

FIGURE 9.1

FIG. 9.2 Van Helmont's experiment. Over a five-year period the willow gained more than 164 lb, while the weight of the soil was practically unchanged.

The organization and functioning of a living cell depends upon an unceasing supply of energy. The source of this energy is the free energy stored in organic molecules like carbohydrates. Heterotrophic organisms, like the yeast and like ourselves, live and grow by taking organic molecules into their cells. In the last chapter, we examined the mechanisms by which these molecules serve as a source of free energy for the cell and as structural components for building the macromolecules of the cell.

For all practical purposes, the only source of the fuel molecules upon which all life depends is photosynthesis. Green plants and algae are **autotrophs;** that is, they are able to harness the energy of the sun to the synthesis of energy-rich organic molecules from the inorganic precursors H_2O and CO_2. Ultimately, the survival of all life on this earth depends upon photosynthesis. Figure 9.1 reflects a crucial biological truth: heterotrophic organisms depend for their existence upon autotrophic organisms. We may dine on beefsteak, but the steer dined on grass.

9.1 EARLY EXPERIMENTS

Perhaps the first experiment designed to explore the nature of photosynthesis was that reported by the Dutch physician van Helmont in 1648. Some years earlier, van Helmont had placed in a large pot exactly 200 pounds of soil that had been thoroughly dried in an oven. Then he moistened the soil with rain water and planted a five-pound willow shoot in it. He then placed the pot in the ground and covered its rim with a perforated iron plate. The perforations permitted water and air to reach the soil but lessened the chance that dirt would be blown into the pot from the outside (Fig. 9.2). For five years, van Helmont kept his plant watered with rain water or distilled water. At the end of that time, he carefully removed the young tree and found it had gained 164 pounds, 3 ounces. (This figure did not include the weight of the leaves that had been shed during the previous four autumns.) He then redried the soil and found that it weighed only 2 ounces less than the original 200 pounds. Faced with these experimental facts, van Helmont theorized that the increase in the weight of the willow arose from the water alone. He did not consider the possibility that gases in the air might also be involved.

The first evidence that gases participate in photosynthesis was reported by Joseph Priestley in 1772. He knew that if a burning candle is placed in a sealed chamber, the candle soon goes out. If a mouse is then placed in the chamber, it quickly suffocates, because the process of combustion has used up the oxygen in the air—the same gas upon which animal respiration depends. However, Priestley discovered that if a plant is placed in an atmosphere devoid of oxygen, it very soon replenishes the oxygen, and a mouse can survive in the resulting mixture. Priestley thought that it was simply the growth of the plant that accounted for this. It was the Dutch physician Ingen-Housz who discovered in 1778 that the effect observed by Priestley occurred only when the plant was illuminated. A plant kept in a sealed chamber in the dark consumes oxygen just as a mouse (or candle) does.

The growth of plants is accompanied by an increase in their carbon content. A Swiss minister, Jean Senebier, discovered that the source of this carbon is carbon dioxide and that the release of oxygen during photosynthesis accompanies the uptake of carbon dioxide. Senebier concluded (erroneously, as it turned out) that in photosynthesis carbon dioxide is *decomposed,* with the carbon becoming incorporated in the organic matter of the plant and the oxygen being released.

$$\mathrm{CO_2} + H_2O \rightarrow (CH_2O) + \mathrm{O_2}$$

(The parentheses around CH_2O signify that no specific molecule is being indicated but, instead, the ratio of atoms in some carbohydrate, e.g. glucose, $C_6H_{12}O_6$.) The equation also indicates that the ratio of carbon dioxide consumed to oxygen released is 1:1, a finding that was carefully demonstrated in the years following Senebier's work. Using glucose as the carbohydrate product, we can write an equation for photosynthesis as

$$6CO_2 + 6H_2O \rightarrow C_6H_{12}O_6 + 6O_2.$$

9.2 THE PIGMENTS

During his experiments, Ingen-Housz also demonstrated that only the green parts of plants liberate oxygen during photosynthesis. Nongreen plant structures, such as woody stems, roots, flowers, and fruits, actually consume oxygen in the process of respiration. We now know that this is because photosynthesis can go on only in the presence of the green pigment *chlorophyll.*

The structure of the chlorophyll molecule is known. It consists of a porphyrin similar in structure to the porphyrin *heme* that forms the prosthetic group of hemoglobin, myoglobin, and the cytochrome enzymes. The chief differences between chlorophyll and heme are the presence of (1) a magnesium (instead of iron) atom in the center of the porphyrin ring and (2) a long hydrocarbon side chain, the phytol chain (Fig. 9.3).

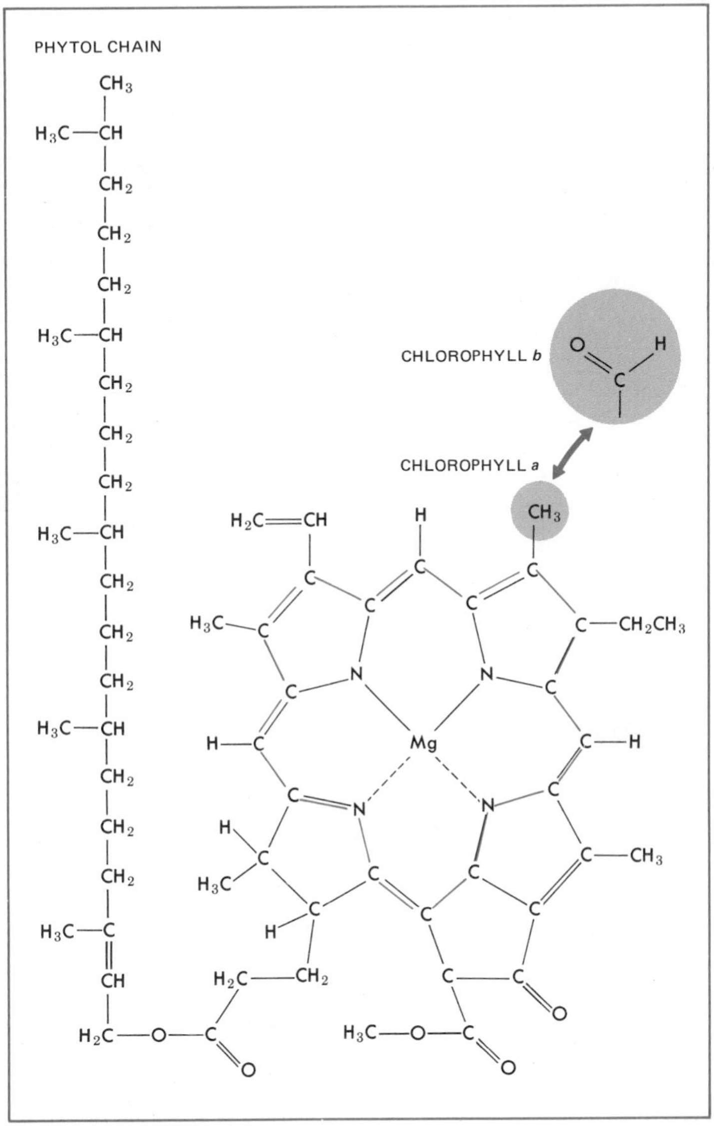

FIG. 9.3 Structure of chlorophyll *a* and *b*. The clustered —C=O groups attached to the porphyrin ring render that region hydrophilic; the phytol chain is strongly hydrophobic. Note the system of alternating single and double bonds that runs around the porphyrin ring (color). The extra electrons implied by the double bonds are not restricted to the positions shown but are, in reality, free to migrate around the system. It is this property that makes the molecule an efficient absorber of light energy.

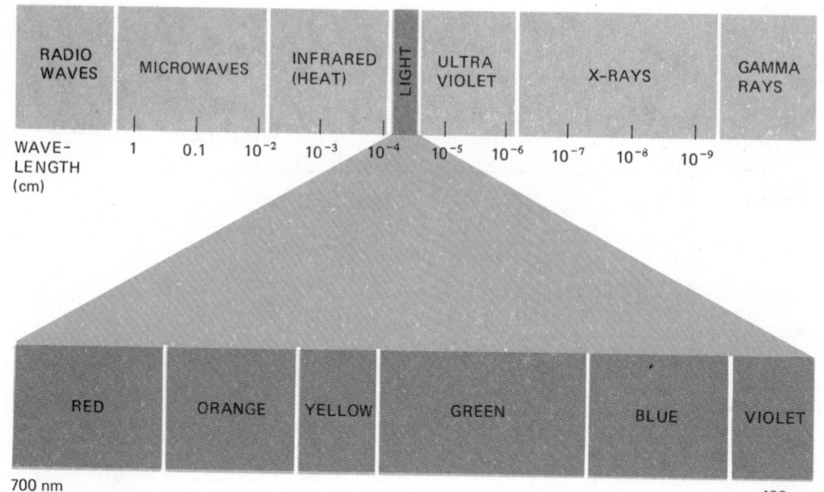

FIG. 9.4 The spectrum of electromagnetic radiation. Approximately one-half of the energy reaching the earth from the sun is light, that is, it is confined to the visible portion of the spectrum. The remainder arrives as heat and a small amount of ultraviolet light.

FIG. 9.5 Absorption spectra of chlorophyll *a* and *b*.

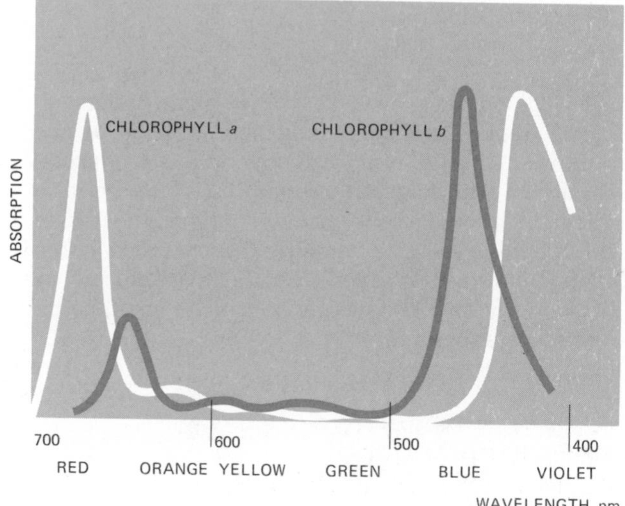

FIG. 9.6 Beta-carotene, one of the most abundant carotenoids. Note again the system of alternating single and double bonds that make this molecule such an efficient absorber of light. Many animals use ingested beta-carotene as a precursor for the synthesis of vitamin A.

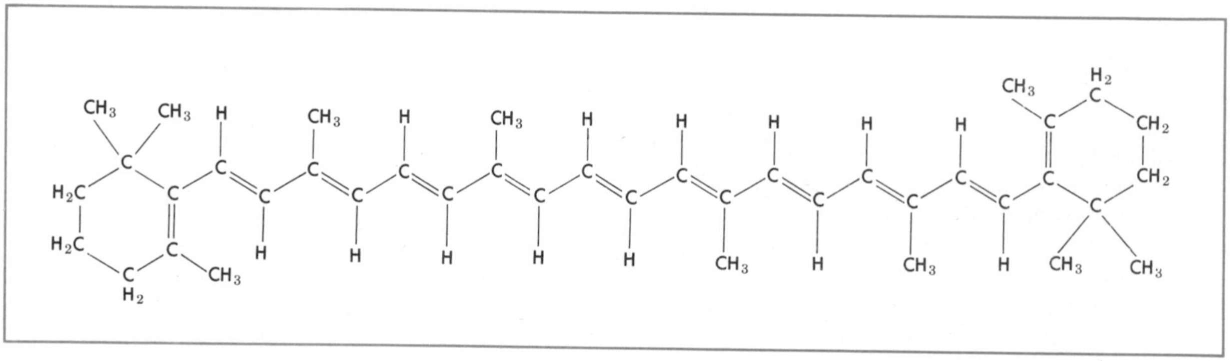

Two types of chlorophyll are found in plants: chlorophyll *a* and chlorophyll *b*. The slight structural difference between these two chlorophylls is shown in Fig. 9.3. In the cell, both chlorophyll *a* and *b* are bound to protein.

The chlorophylls are pigments because they absorb light, that is, electromagnetic radiation in the visible spectrum (Fig. 9.4). White light (such as sunlight) contains all the colors of the visible spectrum from red to violet, but all its constituent wavelengths are not absorbed equally well by the chlorophylls. It is possible to determine how effectively each wavelength (color) is absorbed by illuminating a solution of chlorophyll with monochromatic light (light of a single color) and then measuring with a sensitive light meter the amount of light that passes through the solution. By repeating this process with monochromatic light spanning the entire visible spectrum, it is possible to draw an **absorption spectrum** (Fig. 9.5). Note that both chlorophylls *a* and *b* absorb light most strongly in the red and violet portions of the spectrum. Green light is very poorly absorbed. Hence when white light shines upon chlorophyll-containing structures, such as leaves, green rays are transmitted and reflected, with the result that the structures appear green.

Green plant cells contain, in addition to chlorophylls *a* and *b*, several **carotenoids.** These molecules are also pigments, having colors ranging from red to yellow. They absorb light most strongly in the blue portion of the visible spectrum. Carotenoids are often the dominant pigment in flowers and fruits. The red of the ripe tomato and the orange of the carrot are produced by carotenoids. In leaves, the presence of carotenoids is usually masked by the greater abundance of the chlorophylls. In the autumn, however, as the quantity of chlorophyll in the leaf diminishes, the carotenoids become apparent and produce the brilliant yellows and reds of autumn foliage.

While the structures of the chlorophylls may at first glance seem quite different from those of the carotenoids (Fig. 9.6), close inspection reveals that they share one important attribute. In each case, there is an arrangement of alternating single and double covalent bonds. In the chlorophylls, the system of alternating bonds runs around the porphyrin ring. In the carotenoids, it runs the length of the hydrocarbon chain that connects the terminal ring structures. While we are forced to *draw* the single and double bonds in fixed positions, actually the "extra" electrons responsible for the double bonds are not fixed between any particular pair of carbon atoms but instead are free to migrate throughout the system. It is this property that enables these molecules to absorb visible light so strongly, that is, to act as pigments. And, as we shall see later, it is this property that enables these molecules to absorb the energy of the light that strikes them in a way that can be harnessed to the work of photosynthesis.

While we know which wavelengths of light the chlorophylls and carotenoids absorb best, we have yet to determine which wavelengths are most effective in promoting photosynthesis. In 1881, the German plant physiologist T. W. Engelmann performed

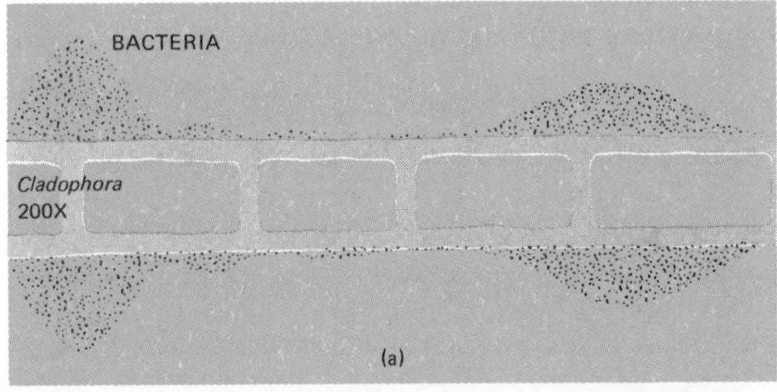

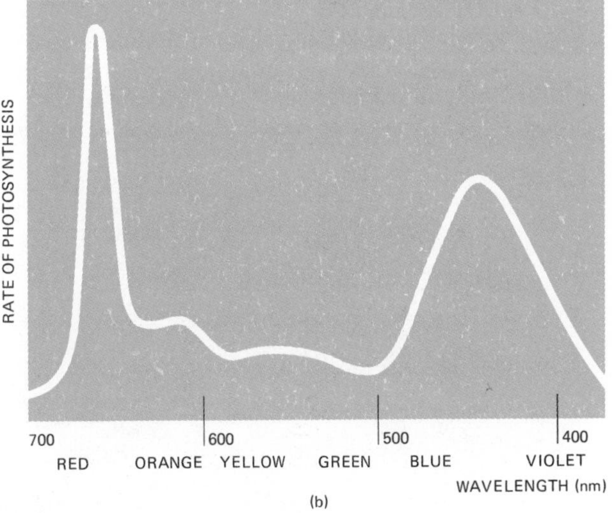

FIG. 9.7 (a) Engelmann's experiment and (b) action spectrum of photosynthesis.

a series of experiments to discover this. He placed a filamentous green alga under the microscope and illuminated the strand with a tiny spectrum. In the medium surrounding the strands were motile, aerobic bacteria. After a few minutes of illumination, the bacteria were found to be congregated and most active around the portions of the filament illuminated by red and blue light (Fig. 9.7). Assuming that the bacteria were congregating in the regions where oxygen was being evolved, Engelmann concluded that the red rays and the blue rays are the most effective colors for photosynthesis. By plotting the effectiveness of each color of light in stimulating photosynthesis, one can draw an **action spectrum** (see Fig. 9.7). The similarity of the action spectrum of photosynthesis and the absorption spectrum of chlorophyll suggests that the chlorophylls are the most important pigments in the process. The spectra are not identical, however, and this suggests that the carotenoids play a role in photosynthesis. They appear to help fill in the absorption gaps so that a larger part of the sun's spectrum can be used. The energy absorbed by the carotenoids is passed to chlorophyll *a*, where it is used in photosynthesis. Chlorophyll *b* probably serves a similar function.

9.3 CHLOROPLASTS

The chloroplasts of plant cells are flattened structures averaging some 7 μm in length and 3–4 μm in width. Each is bounded by a smooth outer membrane. Within the outer membrane is a second, inner membrane and a fluid matrix called the *stroma*. (This organization is quite similar to that of mitochondria—see Section 5.5.) The inner membrane is folded into an intricate pattern (Fig. 9.8). Extensions of the inner membrane are folded together in pairs called *lamellae*. Periodically, the lamellae become enlarged, forming flattened, membrane-enclosed vesicles called *thylakoids*. The thylakoids are arranged in stacks, rather like piles of coins. These stacks of thylakoids are called *grana* (Fig. 9.9).

The membranes, like the other membranes of the cell, are lipid bilayers containing substantial quantities of intrinsic proteins. These include a variety of enzymes, among them cytochrome enzymes and ATP-synthesizing enzymes. The chloroplast membranes also contain all the chlorophyll and some carotenoids as well.

Chlorophyll is an amphipathic molecule, its hydrocarbon phytol chain being strongly hydrophobic while a portion of the porphyrin ring (where the C=O groups are located) is hydrophilic. Presumably, the phytol chain and some of the porphyrin ring is imbedded in the lipid bilayer with the rest of the porphyrin ring projecting above. The carotenoids are strongly hydrophobic and are thought to be totally immersed in the lipid bilayer (Fig. 9.9).

Electron microscopy of the surface of a thylakoid reveals numerous cobblestone-like structures which have been named *quantasomes* (Fig. 9.10). Perhaps each of these represents a precisely organized complex of the enzymes bound to the chloro-

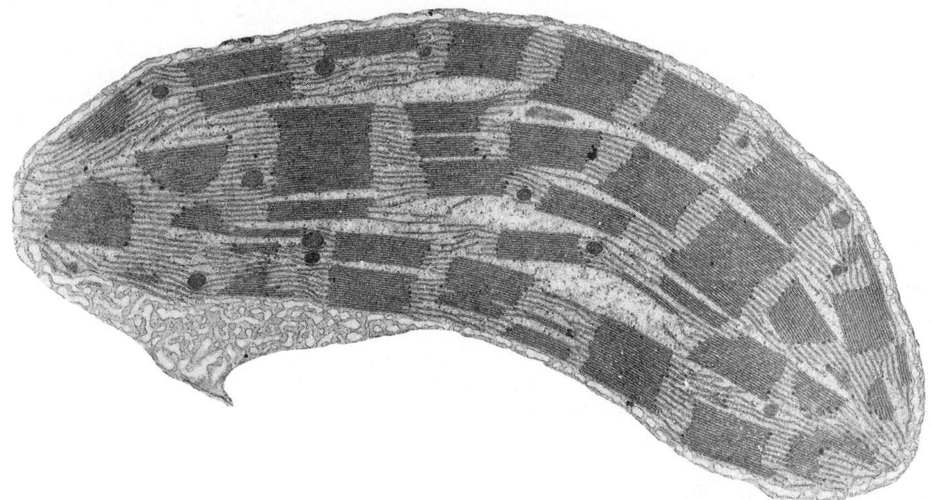

FIG. 9.8 Chloroplast (40,000 ×) from a corn leaf cell. The structures that look like stacks of coins are the chlorophyll-containing grana. (Courtesy of Dr. L. K. Shumway, Genetics and Botany, Washington State University.)

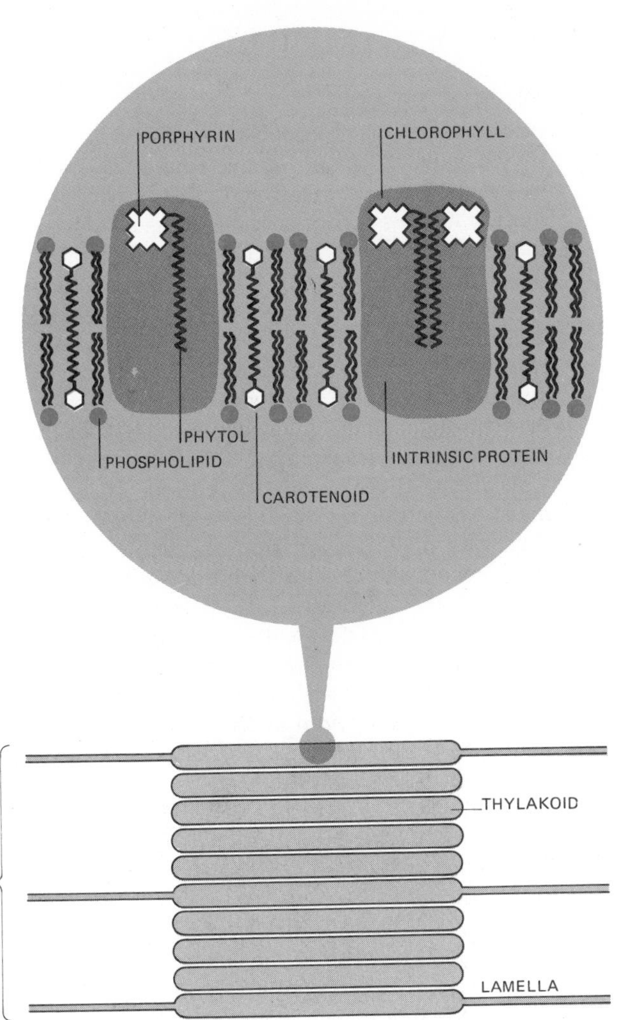

FIG. 9.9 Bottom: structure of a granum. Top: possible organization of the lamellae. The carotenoids and the hydrophobic portions of the chlorophyll molecules are shown buried within the lipid bilayer.

FIG. 9.10 When viewed on its surface, a portion of the thylakoid membrane can be seen to be composed of cobblestone-like subunits, the quantasomes. Each quantasome may represent a precisely ordered array of the pigments and enzymes needed for ATP synthesis by the chloroplast. (Courtesy of Dr. R. B. Park from R. B. Park and J. Biggins, *Science,* 144: 1009–1011, May 22, 1964.

FIG. 9.11 Seedlings of the common garden bean grown in light (left) and in darkness (right). The pale color of the dark-grown plant is caused by its lack of chlorophyll. When the food reserves of its seed are used up, the seedling will die. Each seedling shows three nodes, but the internodes are greatly elongated in the dark-grown seedling, a condition known as etiolation.

plast membrane. There is some evidence that the quantasomes are necessary for ATP synthesis and thus may be the chloroplast equivalent of the knobs on the cristae of mitochondria (look back at Fig. 8.11).

The stroma of the chloroplast is also richly supplied with enzymes. In addition, it contains DNA and numerous ribosomes. These ribosomes are smaller than those in the cytoplasm of the plant cell. The presence of DNA helps to explain the curious autonomy of chloroplasts. Chloroplasts arise either by the division of one chloroplast into two or by the development of tiny, colorless structures called proplastids. Light is necessary for the development of proplastids into chloroplasts. This accounts for the pale color of seedlings grown in the dark (Fig. 9.11). Proplastids are capable of duplicating themselves. In fact, this is the only way in which additional proplastids are formed. The nuclei of green plant cells cannot manufacture them. Thus whenever green plant cells undergo mitosis, it is important that both daughter cells receive proplastids in the cytoplasm as well as the normal chromosome content of the nucleus.

9.4 THE LEAF

In higher plants, chloroplasts are generally confined to the cells of young stems, immature fruits, and leaves. It is the leaves which are the real photosynthetic factories of the plant. A cross section through the blade of a typical leaf reveals several distinct tissue layers (Fig. 9.12). The top surface of the leaf is covered by a single layer of cells which makes up the **upper epidermis.** These cells contain few or no chloroplasts. They are therefore quite transparent and permit most of the light which strikes them to pass through to the underlying cells. They also secrete a waxy, transparent substance called cutin. This material makes up the cuticle. It serves as a moisture barrier on the upper surface of the leaf, thus reducing water loss from the leaf.

Beneath the cells of the upper epidermis are arranged one or more rows of cells which make up the **palisade layer.** These cells are cylindrical in shape and oriented so that their long axis is perpendicular to the plane of the leaf. Each one is filled with chloroplasts (several dozen of them) and, as you might guess, these cells carry on most of the photosynthesis in the leaf. The form of the cells is well suited to this function. They are arranged so that a large number of cells are exposed to the sun's rays. Furthermore, their length increases the chance that the light entering the cell will be absorbed by a chloroplast.

Beneath the palisade layer is the **spongy layer** of the leaf. The cells of the spongy layer are irregular in shape and loosely packed. Although they contain a few chloroplasts, their main function seems to be the temporary storage of the food molecules produced by the cells of the palisade layer. They also aid in the exchange of gases between the leaf and the environment. During daylight hours, these cells give off oxygen and water vapor to the air spaces which surround them. They also pick up carbon dioxide from the air in the air spaces. The air spaces are interconnected with one another and eventually open to the outside of the leaves through special pores called **stomata** (sing., stoma).

In most plants, the stomata are located chiefly in the **lower epidermis** of the leaf. Ingen-Housz himself first demonstrated that photosynthesizing leaves give off oxygen more rapidly from their lower surface than from their upper surface. The presence of as many as 100,000 stomata per square centimeter in the lower epidermis of some oak leaves and none in the upper epidermis easily accounts for this finding.

Most of the cells of the lower epidermis resemble those of the upper epidermis. Around each stoma,

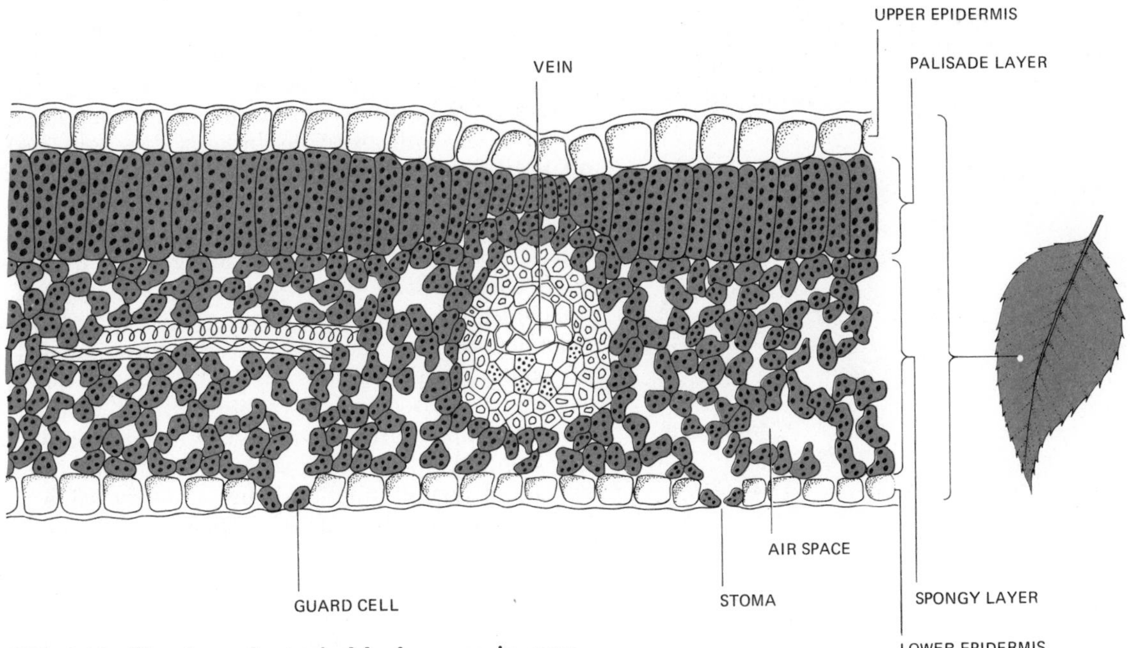

FIG. 9.12 Structure of a typical leaf as seen in cross section. The cells of the palisade layer carry on most of the photosynthesis in the leaf. In some leaves, especially those exposed to bright sun, there may be two or three layers of palisade cells.

however, are located two sausage-shaped cells called **guard cells.** The guard cells differ from the other cells of the lower epidermis not only in their shape but also in their having large numbers of chloroplasts. The guard cells regulate the opening and closing of the stomata. Thus they exert a considerable control over the exchange of gases between the leaf and the surrounding atmosphere.

9.5 FACTORS LIMITING THE RATE OF PHOTOSYNTHESIS

Our equation for photosynthesis

$$6CO_2 + 6H_2O \rightarrow C_6H_{12}O_6 + 6O_2$$

shows the relationship between the substances used in and produced by the process. It tells us nothing about intermediate steps in the process. That photosynthesis does involve at least two quite distinct processes became apparent from the experiments of the British plant physiologist F. F. Blackman. His experimental results can easily be duplicated using the setup shown in Fig. 9.13. The green water plant *Elodea* (available wherever aquarium supplies are

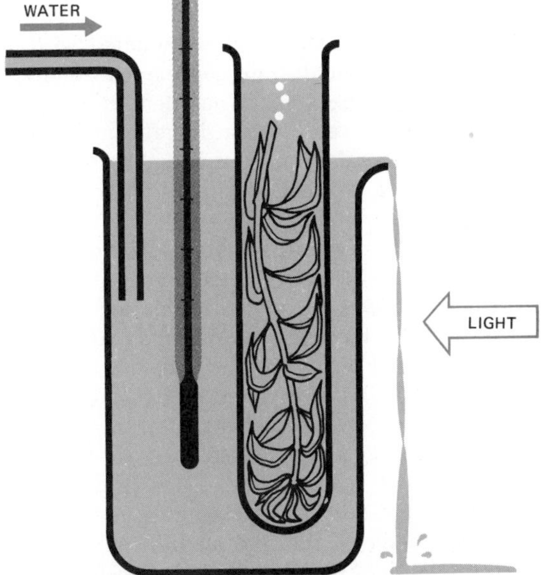

FIG. 9.13 Apparatus for determining the rate of photosynthesis in *Elodea,* an aquatic plant. Measurements are made by counting the rate at which bubbles of oxygen are given off at the stem.

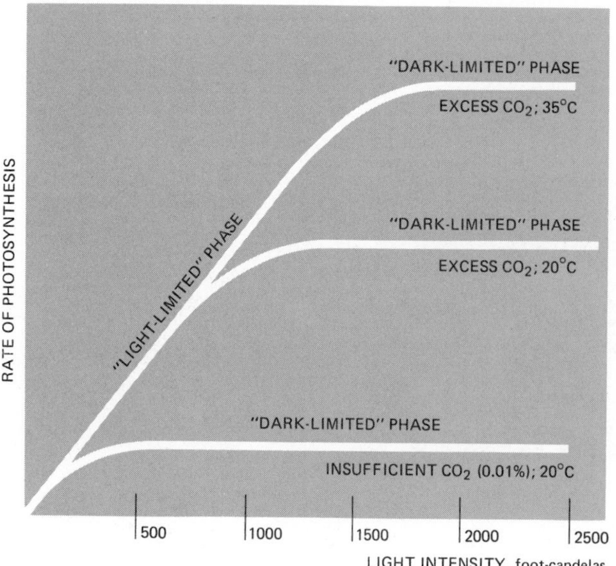

FIG. 9.14 Rate of photosynthesis as a function of light intensity, CO_2 concentration, and temperature. At low light intensities, light is the limiting factor. At higher light intensities, temperature and CO_2 concentration are the limiting factors.

sold) is the test organism. When a sprig is placed upside down in a dilute solution of $NaHCO_3$ (which serves as a source of CO_2), and illuminated with a flood lamp, oxygen bubbles are soon given off from the cut portion of the stem. One then counts the number of bubbles given off in a definite interval of time at each of several light intensities. Plotting this information should produce a graph similar to that in Fig. 9.14.

Since the rate of photosynthesis does not continue to increase indefinitely with increased illumination, Blackman was led to the conclusion that at least two distinct processes are involved; one, a reaction that requires light and the other, a reaction that does not. This latter is called a "dark" reaction although it *can* go on in the light. Blackman theorized that at moderate light intensities, the "light" reaction limits or "paces" the entire process. In other words, at these intensities the dark reaction is capable of handling all the intermediate substances produced by the light reaction. With increasing light intensities, however, a point is eventually reached when the dark reaction is working at maximum capacity. Any further illumination is ineffective and the process reaches a steady rate.

This theory is strengthened by repeating the experiment at a somewhat higher temperature. As we have mentioned, most chemical reactions proceed more rapidly at higher temperatures (up to a point). At 35°C, the rate of photosynthesis does not level off until greater light intensities are present. This suggests that the dark reaction is now working more rapidly. The fact that at low light intensities the rate of photosynthesis is no greater at 35°C than at 20°C also supports the idea that it is a light reaction which is limiting the process in this range. Light reactions depend, not on the temperature, but simply on the intensity of illumination.

The increased rate of photosynthesis with increased temperature does not occur if the supply of CO_2 is limited. As illustrated in Fig. 9.14, the overall rate of photosynthesis reaches a steady value at lower light intensities if the amount of CO_2 available is limited. Thus CO_2 concentration must be added as a third factor regulating the rate at which photosynthesis occurs. As a practical matter, however, the concentration of CO_2 available to terrestrial plants is simply that found in the atmosphere: 0.03%.

9.6 THE DARK REACTIONS

As you might expect, the dark reaction in photosynthesis is actually a series of reactions. These reactions involve the uptake of CO_2 by the plant and the

FIG. 9.15 Apparatus used to follow the fate of $^{14}CO_2$ in the dark reactions. The algal suspension is placed in the "lollypop," supplied with $^{14}CO_2$, and illuminated. The dark reactions are halted by draining the contents of the lollypop into a flask of hot alcohol. (Courtesy of Dr. James A. Bassham.)

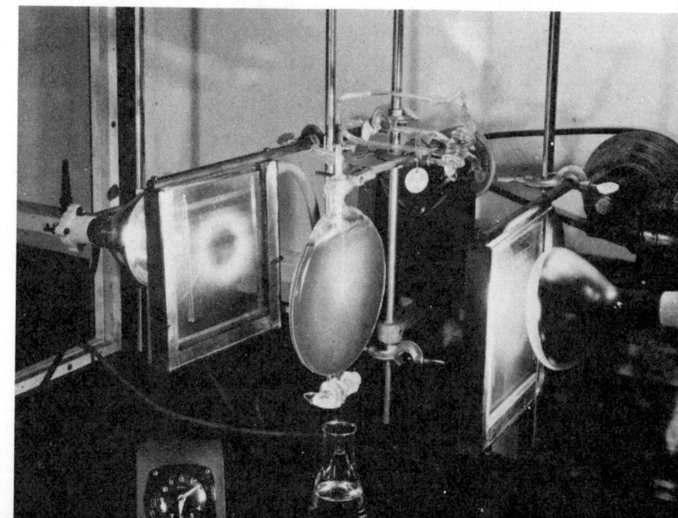

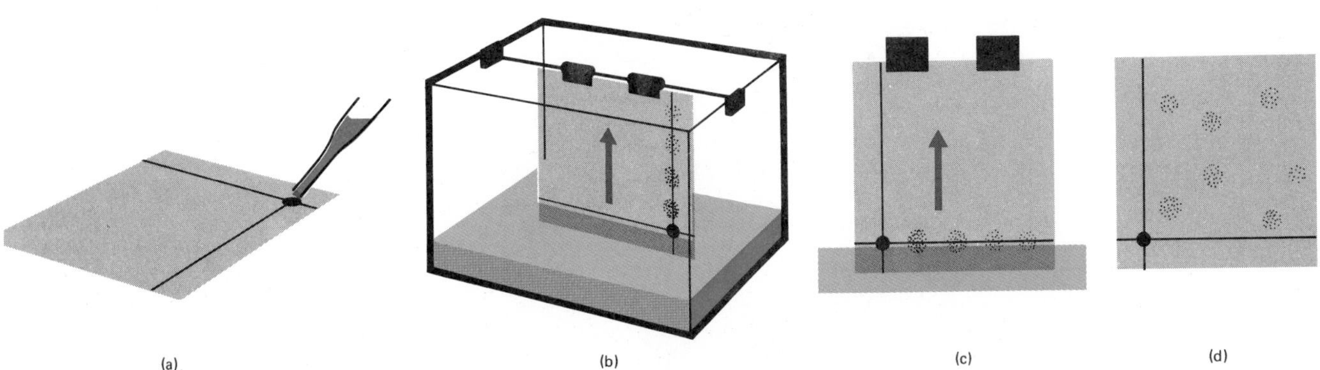

FIG. 9.16 Separation of a mixture by two-dimensional paper chromatography. (a) and (b) Placed in a suitable solvent, the substances in the drop of material at the lower right-hand corner of the paper will be partially separated as they migrate upward. (c) and (d) Further separation can be achieved by turning the paper 90° and using a different solvent.

reduction of CO_2 by hydrogen atoms. Dr. Calvin and his associates at the University of California have devoted years to working out the step-by-step sequence of chemical reactions involved. Their basic experimental procedure has been to expose suspensions of unicellular green algae to light and to radioactive carbon dioxide (Fig. 9.15). The use of radioactive carbon (^{14}C) in the carbon dioxide "tags" the atom and allows its chemical transformations to be studied.

After various intervals of illumination, the algal suspension is inactivated and the contents of the cells extracted. These are then separated by a process called paper chromatography. A drop of the cell extract is placed along one edge of a square of absorbent paper. The paper is then dipped into a solvent. The solvent migrates up the sheet because of capillary attraction (Fig. 9.16). As it does so, the substances in the drop of cell extract are carried along at different rates. Generally, each compound migrates at a unique rate in a given solvent. When the process is completed, the various substances will be separated at distinct places on the sheet of paper, thus forming what is called a *chromatogram*. The identity of each substance may be determined simply by comparing its position with the positions occupied by known substances under the same conditions. Or, the particular portion of the paper can be cut from the sheet and delicate analytical tests run on the tiny amount of substance present.

To determine which, if any, of the substances separated on the chromatogram are radioactive, a sheet of X-ray film is placed next to the chromatogram. If dark spots appear on the film (because of the radiation emitted by the ^{14}C atoms), their position can be correlated with the positions of the chemicals on the chromatogram. Using this technique of *autoradiography*, Calvin found that ^{14}C turned up in glucose molecules within 30 seconds after the start of photosynthesis. When he permitted photosynthesis to proceed for only five seconds, however, he discovered radioactivity in several other, smaller, molecules (Fig. 9.17). Gradually the pathway of carbon fixation was established.

One of the key substances in this process is a five-carbon, phosphorylated sugar called ribulose phosphate. When a second phosphate group is introduced into the molecule by ATP, the resulting compound, ribulose diphosphate, is capable of combining with CO_2. The resulting six-carbon sugar molecule

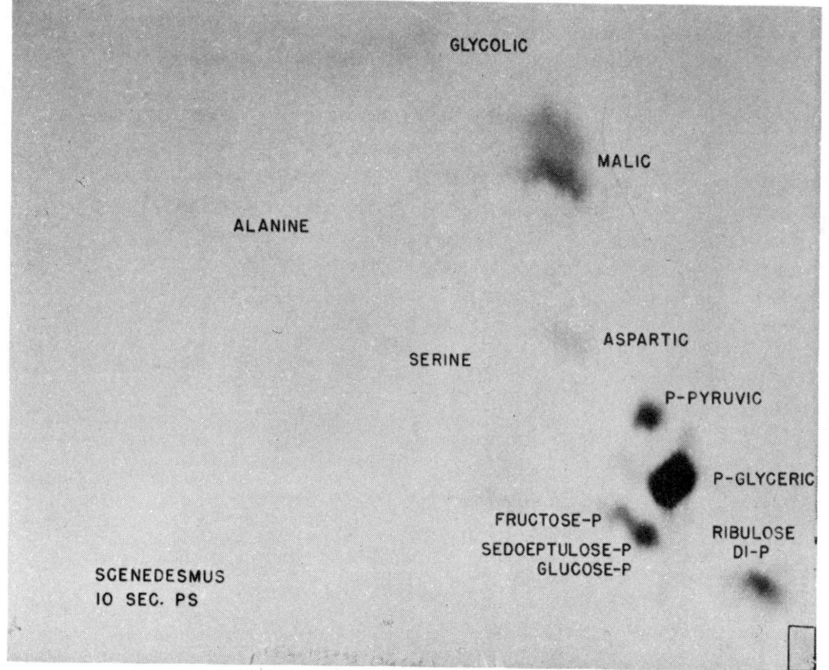

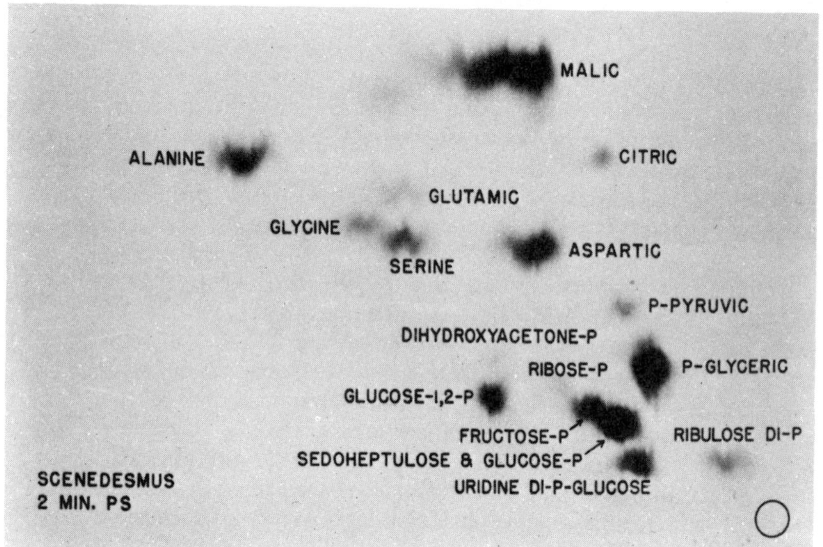

FIG. 9.17 Autoradiographs showing the radioactive substances produced after 10 sec (top) and 2 min (bottom) of photosynthesis using radioactive carbon dioxide. At 10 sec, most of the radioactivity is found in 3-phosphoglyceric acid ("p-glyceric"). At 2 min, radioactive six-carbon sugars (glucose and fructose) have been synthesized as well as a number of amino acids. The small rectangle and circle (lower right-hand corners) mark the spots where the cell extract was applied. (Courtesy of J. A. Bassham.)

then breaks down to form two molecules of 3-phosphoglyceric acid (Fig. 9.18). Each of these receives a second phosphate group (from an ATP molecule), forming two molecules of 1,3-diphosphoglyceric acid (DPGA). These are then reduced to 3-phosphoglyceraldehyde (PGAL). In the process, a phosphate group is given off. The reducing agent is the reduced form of the coenzyme NADP. NADP is just like NAD except for the presence of a third phosphate group (Fig. 8.4). Like NAD, it can be reduced by the acquisition of two electrons. We shall designate the reduced form as NADPH because (again like

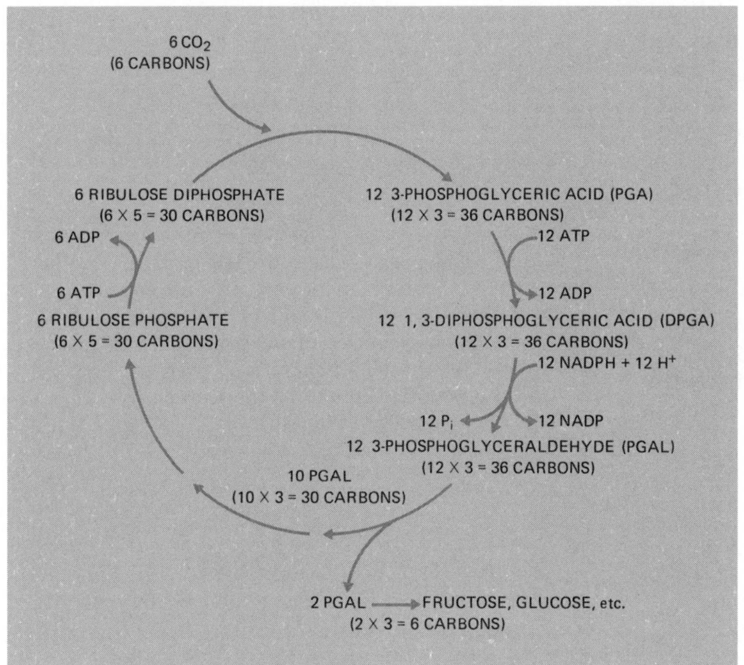

FIG. 9.18 Pathway of carbon dioxide fixation in photosynthesis: the "dark" reactions. This system is often called the Calvin cycle in honor of its discoverer. Compare the steps from 3-phosphoglyceric acid to 3-phosphoglyceraldehyde with the equivalent steps in glycolysis (Fig. 8.5). All the dark reactions occur within the stroma of the chloroplasts.

NAD), only one proton accompanies the reduction. When the oxidized form is to be specified, it will be shown as $NADP^+$.

If you look back at Fig. 8.5 in the previous chapter, you should discover a remarkable fact about the photosynthetic dark reactions described above. From 3-phosphoglyceric acid to PGAL, the steps are the exact reverse of those occurring in glycolysis! In glycolysis, each molecule of PGAL is oxidized to DPGA by NAD^+, a molecule of ATP is synthesized, and 3-phosphoglyceric acid is produced. In photosynthesis, each molecule of DPGA is *reduced* by NADPH, a molecule of ATP is needed, and PGAL is produced (Fig. 9.18). Starting with 6 molecules of ribulose diphosphate, 12 molecules of PGAL are formed. Of these, 10 are ultimately used to re-form the 6 molecules of ribulose phosphate that start the process. The remaining two PGAL molecules can travel up the pathway of glycolysis, finally producing a single molecule of glucose. Thus the carbon atoms from 6 molecules of CO_2 are accounted for in the formation of a molecule of the 6-carbon glucose. Fructose is an intermediate in the pathway, as it is in glycolysis. A molecule of glucose can be combined with one of fructose to form the disaccharide sucrose. This can then be transported to other regions of the plant. The glucose formed by photosynthesis can also be used for the synthesis of starch and cellulose.

What is it that enables green plants but no animals to convert CO_2 into glucose? Many of the crucial steps in the dark reactions are by no means unique to plants. In fact, given enough ATP, some animal cells can take up CO_2 and incorporate it into organic molecules. All that is really needed is an ample supply of reducing activity (NADPH) and ATP. But heterotrophic cells get their reducing activity and their ATP from the catabolism of organic molecules like sugars. At best, the process is only about 40% efficient. So heterotrophic cells can hardly expect to accumulate glucose by this method. No, the secret of autotrophism lies elsewhere, and that is in the light reactions.

9.7 THE LIGHT REACTIONS

It was the American microbiologist Van Niel who first glimpsed the role that light plays in photosynthesis. He arrived at his theory through studying photosynthesis in the purple sulfur bacteria. These microorganisms produce glucose from CO_2 as do the green plants, and they need light to accomplish the synthesis. Water, however, is not used as a starting material. Instead, these bacteria use hydrogen sulfide (H_2S). Furthermore, no oxygen is liberated during this photosynthesis but rather the element sulfur. Van Niel reasoned that the action of light caused a decomposition of H_2S into hydrogen atoms and sulfur. Then, in a series of dark reactions, the hydrogen atoms were used to reduce CO_2 to carbohydrate:

$$CO_2 + 2H_2S \rightarrow (CH_2O) + H_2O + 2S.$$

In these reactions Van Niel envisioned a parallel to the process of photosynthesis in green plants. He reasoned that in green plants the energy of light causes water to break up into hydrogen and oxygen. The hydrogen atoms are then used to reduce CO_2 in a series of dark reactions:

$$CO_2 + 2H_2O \rightarrow (CH_2O) + H_2O + O_2.$$

If this theory is correct, it follows that all of the oxygen produced in photosynthesis is derived from water just as all the sulfur produced in bacterial photosynthesis is derived from the H_2S. This conclusion directly contradicts Senebier's theory (Section 9.1) that the oxygen liberated in photosynthesis is derived from the carbon dioxide. If the equation for photosynthesis in Section 9.1 is correct, then at least some of the oxygen released must come from the CO_2. If, however, Van Niel's theory is correct, the equation for photosynthesis would have to be rewritten:

$$6\ CO_2 + 12H_2O \rightarrow C_6H_{12}O_6 + 6H_2O + 6O_2.$$

Faced with conflicting theories, one tries to devise an experiment that will test them. By deduction, one can make a prediction of how a particular experiment should come out if the theory is sound. In this case, the crucial experiment needed to test the two theories had to await the time when the growth of atomic research made it possible to produce isotopes other than those found naturally or in greater concentrations than are found naturally. In air, water, and other natural materials containing oxygen, 99.76% of the oxygen atoms are ^{16}O and only 0.20% of them are the heavier isotope ^{18}O. In 1941, Samuel Ruben and his coworkers at the University of California were able to prepare specially "labeled" water in which 0.85% of the molecules contained ^{18}O atoms (Fig. 9.19). When this water was supplied to a suspension of photosynthesizing algae, the proportion of ^{18}O in the oxygen gas that was evolved was 0.85%, the same as that of the water supplied, and not simply the 0.20% found in all natural samples of oxygen (and its compounds like CO_2).

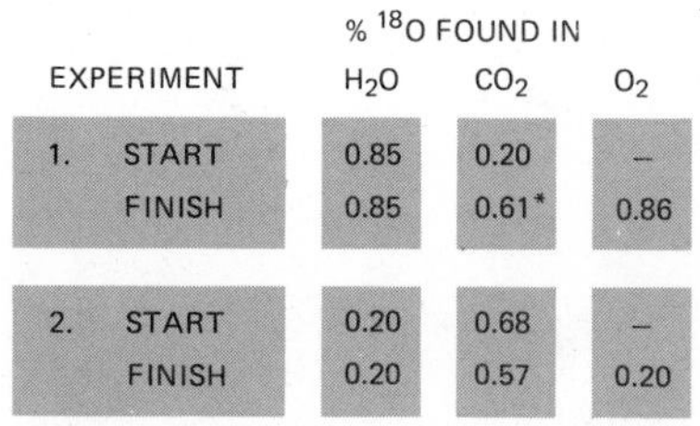

EXPERIMENT	% ^{18}O FOUND IN H_2O	CO_2	O_2
1. START	0.85	0.20	–
FINISH	0.85	0.61*	0.86
2. START	0.20	0.68	–
FINISH	0.20	0.57	0.20

FIG. 9.19 Ruben's results. In each experiment, the percent of ^{18}O found in the oxygen evolved during photosynthesis was that found in the water used. A nonbiochemical exchange of oxygen atoms between the water and the bicarbonate ions used as a source of CO_2 explains the uptake of the isotope by CO_2 in the first experiment (asterisk).

The results clearly demonstrated that Senebier's interpretation was in error. If all the oxygen liberated during photosynthesis comes from the carbon dioxide, we would expect the oxygen evolved in Ruben's experiment to contain simply the 0.20% found naturally. If, on the other hand, *both* the carbon dioxide and the water contribute to the oxygen released, we would expect its isotopic composition to have been some intermediate figure. In fact, the isotopic composition of the evolved oxygen was the same as that of the water used (Fig. 9.19).

Ruben and his colleagues also prepared a source of carbon dioxide that was enriched with ^{18}O atoms. When algae carried out photosynthesis using this material and *natural* water, the oxygen that was given off was *not* enriched in ^{18}O. It contained simply the 0.20% ^{18}O found in the natural water used. The heavy atoms presumably became incorporated in the other two products (carbohydrate and byproduct water).

These experiments lent great support to Van Niel's idea that one function of light in photosynthesis was the separation of the hydrogen and oxygen atoms of water molecules. But there still remained the problem of establishing how the hydrogen atoms were made available to the dark reactions.

We have already seen that the dark reactions of photosynthesis require a copious supply of reduced NADP (NADPH) and ATP. Van Niel's and Ruben's work *suggests* that water serves as the source of electrons for the reduction of $NADP^+$ to NADPH. In 1951, the correctness of their hypothesis was established when several laboratories demonstrated that isolated chloroplasts, *when illuminated,* reduced

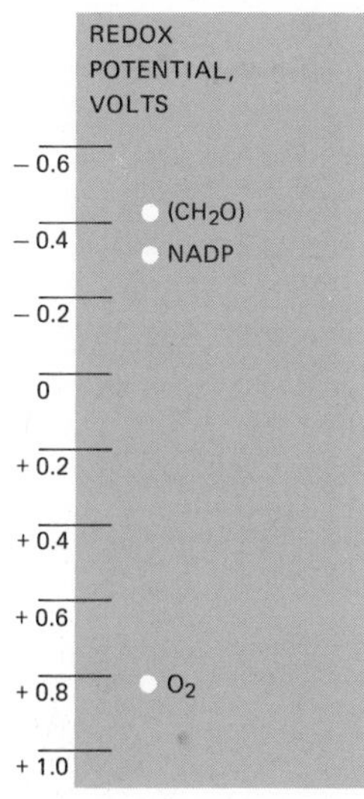

FIG. 9.20 Redox potentials in photosynthesis. To pump electrons from the oxygen atoms of water molecules to NADP requires 1.14 volts, which is equivalent to an input of 52.6 kcal of energy for each pair of electrons.

$NADP^+$ to NADPH with the release of equivalent amounts of molecular oxygen.

$$2H_2O + 2NADP^+ \xrightarrow{\text{light}} 2NADPH + O_2 + 2H^+$$

But what an unfavorable redox reaction this is! Electrons are removed from highly electronegative oxygen atoms and transferred to weakly electronegative molecules of NADP. As we can see in Fig. 9.20, the electrons are moving from a redox potential of about +0.82 volt to one of −0.324 volt. Thus these electrons are being moved against a gradient of 1.14 volts. What oxidizing agent is powerful enough to remove electrons from oxygen atoms so that they can be raised against this substantial gradient?

The answer is: chlorophyll itself. In order to understand how chlorophyll can become a powerful redox agent, we must examine the effect which light has upon chlorophyll.

The Effect of Light on Chlorophyll

When a *solution* of chlorophyll is placed in a beam of white light, it gives off light of a deep *red* hue. This phenomenon is called *fluorescence*. It can be demonstrated easily. A crude chlorophyll extract can be prepared by soaking grass leaves in ethanol. In a beam of white light, the solution fluoresces.

The explanation of the phenomenon of fluorescence is that the energy of the absorbed light is transferred to an electron in the chlorophyll molecule, raising it to a higher energy level. We noted earlier in the chapter the ease with which electrons are free to migrate around the molecule. It is these electrons that are so easily raised to a higher energy level when light is absorbed by the molecule. As long as an electron occupies a higher energy level, the chlorophyll molecule is said to be "excited." In the chlorophyll solution, the excited condition lasts but a fleeting instant. The electron drops back to its former energy level. In so doing, it gives up the energy that had raised it in the first place. Most of this energy is released as red light of a sharply defined wavelength.

When *intact leaves* are illuminated by white light, little or no fluorescence is observed. One possible explanation for this might be that something has removed the high-energy electron in the excited chlorophyll molecule before it could drop back. But if that is so, then the chlorophyll molecule would now be positively charged. And, in fact, there is now good evidence that when chloroplasts are illuminated, *certain* chlorophyll molecules do become oxidized, i.e. acquire a positive charge.

Furthermore, what fluorescence can be seen in the intact system is a red hue characteristic of chlorophyll *a*. Although illuminated solutions of chlorophyll *b* and of carotenoids each display characteristic fluorescence, these wavelengths are not seen when intact leaves or chloroplasts are illuminated. Furthermore, the fluorescence that *is* observed is not even produced by *most* of the chlorophyll *a* molecules. The absorption spectrum of chlorophyll *a* is altered by such factors as the nature of the protein to which it is complexed or, perhaps, its location within the membranes of the chloroplast. In any case, several different forms of chlorophyll *a* are found in chloroplasts—each differing slightly in the wavelength of light it absorbs most efficiently. And only two of these forms of chlorophyll *a* are observed to fluoresce. These forms are designated P_{700} and P_{680} for the wavelengths that each *absorbs* best.

All the other pigments in the chloroplast transfer their energy of excitation to either P_{700} or P_{680}. For this reason, the chlorophyll *b*, the carotenoids, and the bulk of the chlorophyll *a* molecules are sometimes referred to as *antenna pigments*. Each is "tuned" to absorb certain wavelengths of light most efficiently. The energy received is then passed along to P_{700} or P_{680}. Only P_{700} and P_{680} become *oxidized* when light strikes intact chloroplasts. Because of the critical redox role they play, P_{700} and P_{680} are called the *reaction center pigments*.

9.8 PHOTOSYSTEMS I AND II

The discovery of two reaction center pigments is a satisfying one because it helps to explain another characteristic of photosynthesis. When one carefully compares the *action* spectrum of photosynthesis with the *absorption* spectrum of the chlorophylls, a curious discrepancy becomes apparent. Red light longer in wavelength than about 680 nm is inefficient at promoting photosynthesis although chlorophyll *a* still absorbs at that wavelength. In 1956, Robert Emerson and his colleagues demonstrated that the sharp reduction in photosynthetic efficiency at wavelengths beyond 680 nm can be overcome if the chloroplasts are simultaneously illuminated with shorter-wavelength light. With simultaneous illumination from two beams of monochromatic light, e.g. 670 nm and 710 nm, the photosynthetic efficiency was substantially greater than it was with a single wavelength of the same *total* intensity. This observation led to the conclusion that the light reactions must include two distinct processes: one energized by light of longer wavelengths, called *Photosystem I*, and one energized by light of shorter wavelengths, called *Photosystem II*. Not all of the details of the operation of the two photosystems have yet been determined. However, the major steps in each are shown in Fig. 9.21.

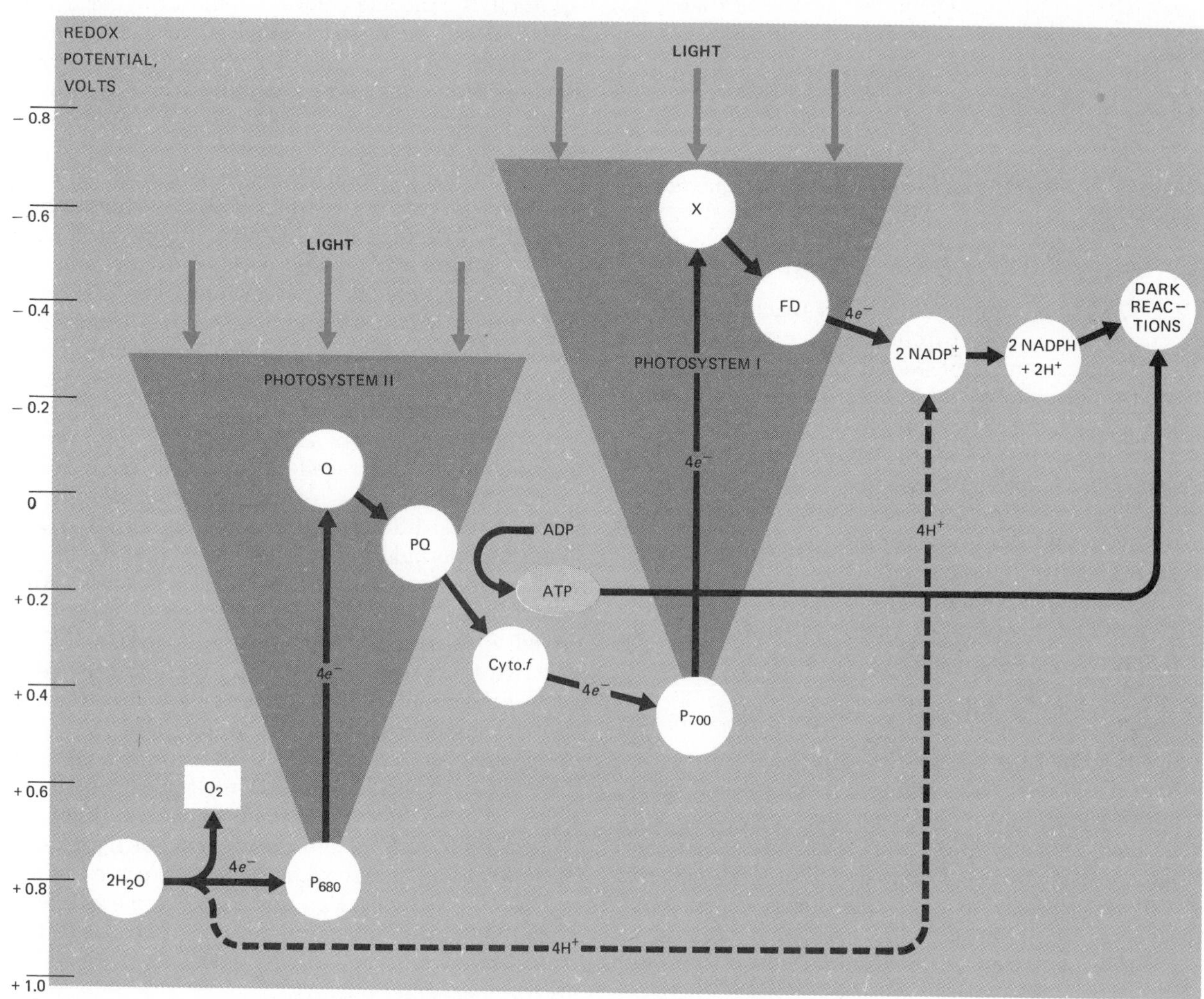

FIG. 9.21 The light reactions of photosynthesis. The relative height of each of the electron carriers reflects its redox potential (y-axis). In cyclic photophosphorylation, electrons cycle ("counterclockwise") from X to PQ and back to X, generating ATP as they go. All the pigments and enzymes of the light reactions are incorporated in the membranes of the grana. FD = ferredoxin.

Photosystem I. The energy harvested by the antenna pigments of Photosystem I is transferred to a molecule of P_{700}. Electrons in P_{700} are raised to such a high energy level (about −0.6 volt) that they can reduce an as yet unknown electron acceptor, shown in Fig. 9.21 as X. Substance X in turn donates its electron to (and thus *reduces*) an iron-containing protein called ferredoxin. Ferredoxin then donates its electrons to $NADP^+$, reducing it and forming the NADPH needed for the dark reactions. In this way, the light absorbed by Photosystem I provides the energy needed to oxidize P_{700} and reduce $NADP^+$ to NADPH.

But the electrons used to reduce $NADP^+$ are then used in the dark reactions for the synthesis of PGAL. If Photosystem I is to continue to operate, it must replace these electrons. It does so by tapping Photosystem II.

Photosystem II. Absorption of light by Photosystem II leads to the oxidation of P_{680} in a manner exactly like that of Photosystem I. But oxidized P_{680} is a more powerful oxidizing agent than P_{700}. With a redox potential of greater than +0.82 volt, it is sufficiently electronegative to acquire electrons from (and thus be reduced by) water molecules. The exact steps involved are still uncertain, but for every four electrons acquired by P_{680}, a molecule of oxygen is liberated.

$$2H_2O \rightarrow 4e^- + 4H^+ + O_2$$

Absorption of light raises these electrons to a sufficiently high energy level that they can reduce the P_{700} in Photosystem I. Thus we have found one of the two essential ingredients of the light reactions, a mechanism that provides the energy enabling electrons to move in an unbroken pathway from water molecules to NADP.

But now we must find the ATP that is the other essential ingredient of the dark reactions. The excitation of P_{680} raises its electrons to a redox potential of about −0.05 volt (Fig. 9.21). This is substantially higher than is needed to reduce P_{700}, whose redox potential is about +0.4 volt. Thus electrons passing from the electron acceptor of P_{680} (called Q) to P_{700} are passing *down* a substantial potential gradient. Just as it takes energy to lift electrons against a voltage gradient, so energy is liberated when they are allowed to pass down a gradient. From *Q* to P_{700}, each electron is passing downhill through a potential of some 0.45 volt. In the last chapter we learned that the free energy available when electrons pass down a potential gradient (ΔE) is given by the expression

$$\Delta F = -n(23.062)\,\Delta E,$$

where n is the number of electrons. With a ΔE of 0.45 volt, passage of each electron yields some 10.4 kcal. This would certainly be enough to couple to the synthesis of a molecule of ATP ($\Delta F = -7.3$ kcal) if only an appropriate mechanism existed.

What could be more appropriate than to enlist the aid of cytochrome enzymes, just as we do in mitochondria? And that is precisely what occurs. In passing from Q to P_{700}, the electrons pass through a chain of redox substances of which at least one is a cytochrome (cytochrome *f*). So the machinery is there. Does it work? In 1954, Daniel Arnon demonstrated that when isolated chloroplasts are supplied with ADP, inorganic phosphate, and then illuminated, they proceed to manufacture ATP. Arnon called this process *photophosphorylation* because it is phosphorylation driven by light. Chloroplasts are thus able to harness the energy of light directly to the synthesis of ATP.

One problem remains. Is the photophosphorylation that occurs as electrons pass from Photosystem II to Photosystem I sufficient to meet the demands of the dark reactions? Let us review what these are. Four electrons are needed for every molecule of CO_2 that is reduced. Looking back at Fig. 9.18, we see why this is so. The uptake of each molecule of CO_2 results in the formation of two molecules of PGA. The reduction of these requires two NADPH, hence the need for four electrons. The process also requires two ATPs. But remember that ATP is also needed to convert ribulose phosphate into ribulose diphosphate (Fig. 9.18). So *three* ATPs are needed for every two NADPH used in the dark reactions. As yet, the only redox step that has been identified as leading to the formation of ATP is that between a substance called plastoquinone (PQ) and cytochrome *f*. The ΔE of this step is 0.265 volt. Four electrons passing through this step would yield about 24.5 kcal of free energy. Needing *at least* 21.9 kcal (3×7.3) to synthesize three ATP molecules (and probably considerably more), we cannot help but wonder whether this step *alone* is adequate for the process. As yet, the answer is uncertain. But if it turns out that only two ATP molecules are synthesized as four electrons pass from Photosystem II to Photosystem I, the chloroplast must have another trick up its sleeve.

One possibility, also discovered by Daniel Arnon, has been named **cyclic photophosphorylation.** In this process the light absorbed by Photosystem I excites P_{700} and expels electrons so that they can be acquired by substance X (Fig. 9.21). But instead of passing its electrons on to ferredoxin, X then reduces plastoquinone (PQ). As these electrons return to P_{700}, some of the free energy they give up is coupled to the synthesis of ATP just as it is in regular photophosphorylation. But this process is truly cyclic because no outside source of electrons is required. Like the photocell in a light meter, Photosystem I is simply using light to create a flow of current. The only difference is that instead of using the current to move the needle on a light meter, the chloroplast uses the current to synthesize ATP. At least the chloroplast does so in isolated test tube preparations.

Whether cyclic photophosphorylation plays a significant role in ATP production in the intact cell is still uncertain. If the answer turns out to be no, other possibilities exist. For instance, it has been suggested that if more ferredoxin is reduced than is needed for the production of NADPH, the excess might be used to reduce the O_2 that is liberated by Photosystem II back to H_2O. At first glance, this might seem a fruitless undoing of all the hard work of photosynthesis. But look again. Although the electrons cycle from water to ferredoxin and back again, part of their pathway is through the ATP-generating step linking Photosystem II with Photosystem I. Here, then, is another way that simply by turning on a light, enough energy is imparted to electrons that they can bring about the synthesis of ATP.

It is a relatively easy task to disrupt isolated chloroplasts and separate the pigment-containing thylakoids from the stroma. When this is done, the isolated thylakoids retain the ability to carry on photophosphorylation, but they cannot convert CO_2 into carbohydrate. That task, however, can be accomplished by the colorless stroma when it is supplied with CO_2, NADPH, and ATP. Thus there is a division of labor within the chloroplast: the light reactions are the responsibility of the grana, while the dark reactions are carried out by the enzymes within the stroma.

C_4 Plants

When the techniques used by Calvin are applied to certain plants (e.g., sugarcane, corn, sorghum), the first products seen are such 4-carbon compounds as oxaloacetic acid, malic acid, and aspartic acid rather than the 3-carbon PGA. Soon, however, radioactive carbon begins to appear in PGA and the Calvin cycle comes into play. It turns out that these plants—called C_4 plants—first fix CO_2 by a 4-carbon pathway present in cells near the surface of the leaf but then release the CO_2 and pass it on to cells in the interior of the leaf where the 3-carbon cycle functions.

The advantage of such a system is at least twofold. The C_4 trapping system works at much lower concentrations of CO_2 (as little as 1–2 ppm) than the C_3 system, which stops functioning at CO_2 concentrations below about 50 ppm. On a very hot day, therefore, a C_4 plant can close its stomata, thus reducing water loss through transpiration, and still acquire fresh CO_2 for photosynthesis. This is because it creates a very steep concentration gradient for the diffusion of CO_2 (from 300 ppm in the air to about 1 ppm in the leaf) and thus can at least partially compensate for its closed stomata. For this reason, C_4 plants are especially well adapted to (and more apt to be found in) habitats with high daytime temperatures, low soil moisture, and intense sunlight. A second advantage is that the C_4 system is able to trap and recycle the CO_2 that is lost in the process of photorespiration (see Section 10.4). This, too, is particularly advantageous in habitats where the daytime temperatures—and hence rates of photorespiration—are high.

9.9 SUMMARY

Although questions still remain unanswered, there can no longer be any doubt that it is the light reactions that are the secret of photosynthesis—the ability to make energy-rich compounds like carbohydrates from energy-poor compounds like carbon dioxide and water. The light reactions make it possible to trap the energy of light and convert it into the free energy of organic molecules. Initially, these organic molecules are carbohydrates. This is because the reducing activity provided by the light reactions is only powerful enough to reduce carbon to the degree found in carbohydrates, that is, to one-half the maximum possible reduction ($CO_2 \rightarrow CH_2O$ vs. $CO_2 \rightarrow CH_4$). Of course, plants do synthesize more fully reduced compounds such as proteins and fats (e.g. $C_{57}H_{98}O_6$, trilinolein), but this is done no differently from the way it is done in heterotrophs. We, too, can synthesize fats from carbohydrates, but we must spend the free energy of ATP to do so. The energy content of fats is much greater than that of carbohydrates and, for plants as for ourselves, the

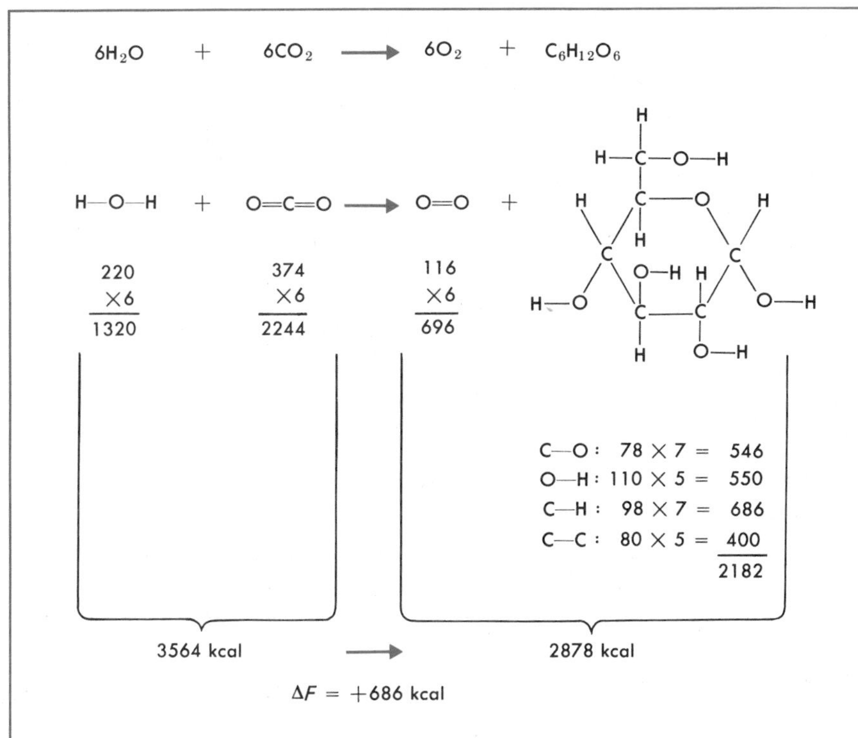

FIG. 9.22 Bond energy relationships in photosynthesis. The difference between the free energy needed to break the bonds of the reactants and the free energy liberated in the formation of the bonds of the products is 686 kcal/mole, the net free energy cost of the reactions. The source of this energy is the light absorbed by the chlorophylls and other antenna pigments.

difference must be paid for by ATP-driven dark reactions.

In the last chapter we examined the overall energy balance of cellular respiration. We did this by finding the sum of all the bond energies of the reactants and subtracting from this value the sum of all the bond energies of the products (look back at Fig. 8.13). Can we do the same for photosynthesis? Indeed we can. To break the bonds of six molecules of water (for our purposes here, we may ignore the six molecules of water that appear on *both* sides of the equation) and six molecules of CO_2 requires the input of 3564 kcal of energy. Formation of the bonds found in a molecule of glucose (2182) and six molecules of oxygen (696) liberates 2878 kcal (Fig. 9.22). The difference, 686 kcal, represents the energy stored. Thus we express the change in free energy as $\Delta F = +686$ kcal. We have, then, simply the reverse of our energy budget in cellular respiration, with a positive rather than a negative ΔF.

Similarly, we can apply an analysis of the changes in redox potential to photosynthesis. If, as we did in Chapter 8, we take the redox potential of carbohydrate as -0.42 volt (a reasonable approximation) and that of oxygen as $+0.82$ volt, photosynthesis has moved electrons through a potential of 1.24 volts (Fig. 9.20). In order to synthesize a molecule of glucose, the light reactions must transport 24 electrons. Using our value of 23.062 kcal for each electron volt (the Faraday constant), we arrive at a value of, once again, 686 kcal.

$$\Delta F = (24)(23.062)(1.24) = 686 \text{ kcal}$$

The only difference here is that ΔF is positive because the electrons are moving *against* the gradient (from positive to negative) instead of with it as they do in cellular respiration. Thus energy must be applied to accomplish the task.

So we see that in many ways photosynthesis and cellular respiration are complementary processes. The transport of electrons lies at the heart of each. In photosynthesis, the energy of the sun is used to remove electrons from oxygen atoms in water molecules. These electrons are transferred to carbon atoms, forming covalent bonds between carbon and hydrogen. Carbon is far less electronegative than oxygen, and consequently, energy is needed to make the transfer. Much of this energy is stored in the covalent bonds established. In this way photosynthesizing organisms form the food molecules upon which all organisms depend.

In cellular respiration, the electrons in food molecules are removed and allowed to return, step by step, to oxygen atoms. As they do so, they give back the energy originally stored by them, a substantial fraction of which is converted into the energy stored in ATP. In the last analysis, then, energy flow in all living things depends on the cyclic transfer of electrons between oxygen atoms and carbon atoms, accompanied by the breakdown (in photosynthesis) and synthesis (in cellular respiration) of water molecules.

In these two chapters, then, we have examined the discharging and the charging of the storage battery of life.

EXERCISES AND PROBLEMS

1. The oxygen released in photosynthesis comes from which of the two raw materials used by the plant?
2. How many kilocalories of light energy are needed to pump one mole of electrons from P_{700} to substance X?
3. Life can be said to be fundamentally a matter of the transport of electrons between energy levels. Defend this view.
4. Distinguish between an autotroph and a heterotroph.
5. What color of light do green leaves absorb least well?
6. Do you think that the rate of photosynthesis would continue to increase indefinitely with increasing temperature? Explain.
7. Do you think that the rate of photosynthesis would continue to increase indefinitely with increasing carbon dioxide concentration? Explain.
8. When students perform the experiment illustrated in Fig. 9.13 starting with minimum light intensity, they often find that the rate of photosynthesis does not level off sharply at higher light intensities. What factor have they neglected to keep constant?
9. Students who have taken an extra-long time to carry out the experiment in Fig. 9.13 often find that the rate of photosynthesis begins to drop at high light intensities. Can you think of an explanation for this?
10. In two columns, contrast photosynthesis and respiration in as many ways as you can.
11. Do green plants need digestive enzymes? Explain.
12. How many electrons are removed from water molecules for each molecule of O_2 produced in photosynthesis?
13. How many electrons are needed for each molecule of CO_2 assimilated in the process of photosynthesis? How many NADPH molecules? How many ATP molecules?

REFERENCES

1. *Great Experiments in Biology,* ed. by M. L. Gabriel and S. Fogel, Prentice-Hall, Englewood Cliffs, N.J., 1955. Contains brief firsthand reports of the experiments of van Helmont, Priestley, Ingen-Housz, Engelmann, and Van Niel as well as the ^{18}O-tracer experiments done at the University of California.
2. CLAYTON, R. K., *Photosynthesis: How Light Is Converted to Chemical Energy,* Addison-Wesley Modules in Biology, No. 13, Addison-Wesley, Reading, Mass., 1974.
3. LEVINE, R. P., "The Mechanism of Photosynthesis," *Scientific American,* Offprint No. 1163, December, 1969.
4. WALD, G., "Life and Light," *Scientific American,* Offprint No. 61, October, 1959. Discusses the properties of light and the chlorophyll molecule that absorbs it.
5. ARNON, D. I., "The Role of Light in Photosynthesis," *Scientific American,* Offprint No. 75, November, 1960. The light reactions.
6. BASSHAM, J. A., "The Path of Carbon in Photosynthesis," *Scientific American,* Offprint No. 122, June, 1962. The dark reactions.
7. LEHNINGER, A. L., "How Cells Transform Energy," *Scientific American,* Offprint No. 91, September, 1961. Emphasizes the reciprocal relationship between photosynthesis and respiration.
8. GOVINDJEE, and RAJNI GOVINDJEE, "The Primary Events of Photosynthesis," *Scientific American,* Offprint No. 1310, December, 1974. The parts played by the antenna pigments and the reaction center pigments of Photosystems I and II.

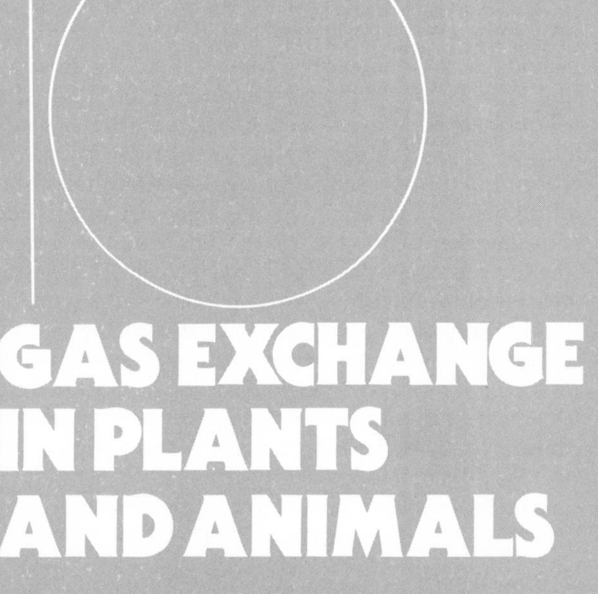

10 GAS EXCHANGE IN PLANTS AND ANIMALS

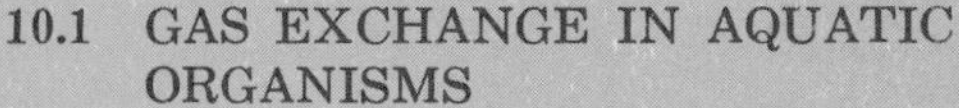

In order to carry on respiration, cells need a continuous supply of oxygen and a means of disposing of carbon dioxide. In order to carry on photosynthesis, green plant cells need a supply of carbon dioxide and a means of disposing of oxygen. Almost all cells are therefore involved in the problem of exchanging gases with the environment.

As we have seen, the environment of cells is the fluid which bathes them, the ECF. For gases to enter and leave cells, they must be dissolved in the ECF.

10.1 GAS EXCHANGE IN AQUATIC ORGANISMS

Among aquatic microorganisms such as the amoeba, the ECF is simply the surrounding water. The ratio of surface area to volume in the amoeba is sufficiently great that simple diffusion of the gases between the cell and the water is adequate to take care of its needs (Fig. 10.1). As the amoeba respires, the consumption of oxygen by its mitochondria lowers the oxygen concentration within the cell. If the concentration of oxygen is greater in the water outside than within the cell, oxygen passes into the cell by diffusion. The oxygen concentration of the water is, in turn, maintained by: (a) diffusion of oxygen from the air into the water and (b) production of oxygen during photosynthesis by aquatic plants (Fig. 10.2).

While the amoeba is consuming oxygen, it is producing carbon dioxide. When the concentration of this gas within the cell becomes greater than its concentration in the surrounding water, it passes out of the cell by diffusion. The carbon dioxide concentration of the water is kept at reasonably low levels by two mechanisms: (1) Aquatic plants consume it in photosynthesis. (2) It diffuses into the air. (Fig. 10.2.)

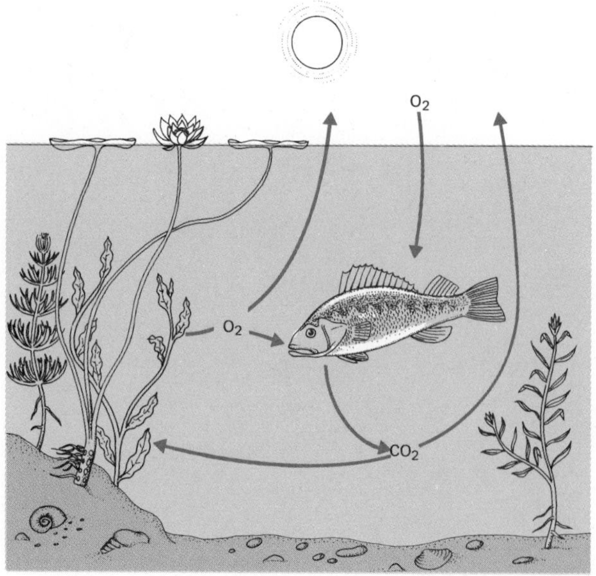

FIG. 10.2 Gas exchange in a freshwater pond. Oxygen is released and carbon dioxide is consumed during photosynthesis. Carbon dioxide is released and oxygen is consumed in cellular respiration.

Even among the smaller plants and animals that live in the water, the surface-to-volume ratio may be sufficiently large so that simple diffusion will take care of the needs of gas exchange. Perhaps the very flat, ribbonlike shape of the planarian (Fig. 10.3) is an adaptation that permits diffusion alone to accomplish the speedy exchange of gases between the interior cells and the environment.

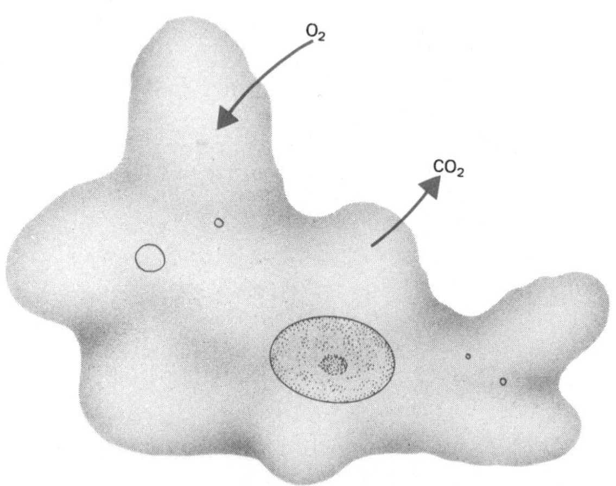

FIG. 10.1 Gas exchange in the amoeba. Oxygen enters and carbon dioxide leaves by diffusion.

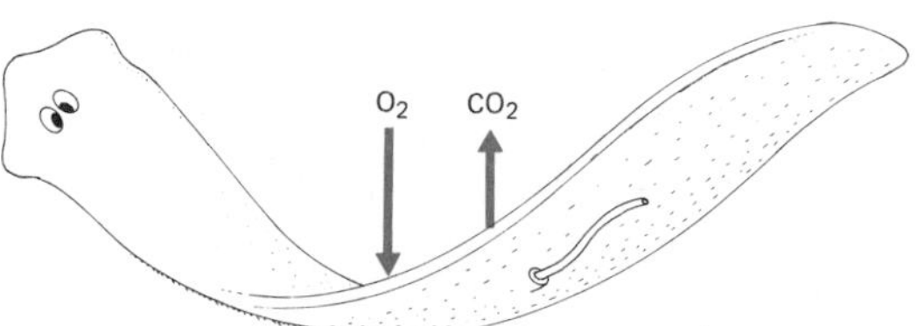

FIG. 10.3 Every cell in the thin planarian body is close enough to the surface for its gas-exchange needs to be satisfied by diffusion.

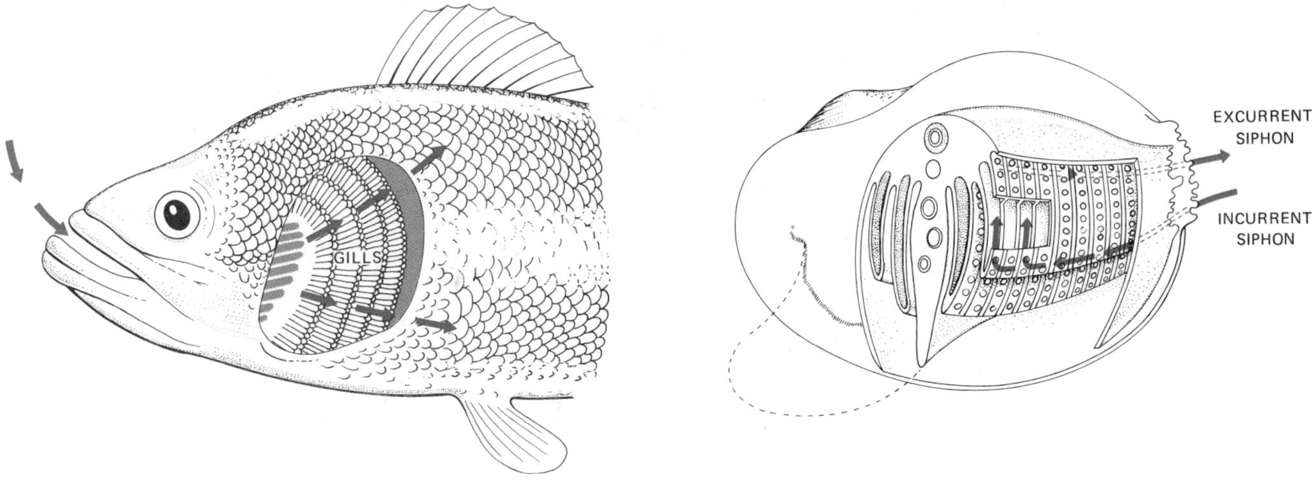

FIG. 10.4 Two gill-breathers: the fish and the clam. Part of the fish has been cut away to reveal the gills. In the clam the left shell and portions of the left gills have been cut away to show the pathway taken by the water.

Among the larger aquatic animals, the ratio of surface to volume is smaller. Many cells of these animals are located too deep within the body to carry on adequate gas exchange with the environment by diffusion alone. An actively respiring interior cell cannot receive oxygen sufficiently rapidly if the oxygen must come to it by diffusion from cell to cell.

The most widespread solution to this problem is to bathe the interior cells in a circulating fluid contained within the organism's body. This fluid bathes each cell and, in so doing, brings oxygen to it and removes carbon dioxide from it. It must also be able to pick up fresh supplies of oxygen from the surrounding water and to discharge its accumulation of carbon dioxide into it. This requires a specialized organ for gas exchange. Among the crustaceans, mollusks, and fishes, the gas exchange organs are **gills.** Gills present a large surface to the water in which they are bathed. Oxygen present in the water diffuses into the cells of the gills. The gills are supplied with blood which transports oxygen from them to the interior tissues of the body. The blood picks up carbon dioxide from the interior tissues and transports it to the gills. From the gills the carbon dioxide passes into the water by diffusion. Thus, there are three distinct and essential parts to the system: (1) **A circulating fluid** which transports the gases. This fluid is kept moving with the help of a muscular pump, the heart. (2) **Gills** which exchange gases between this circulating fluid (the internal environment) and the water (the external environment) (3) A mechanism for drawing a continuous **supply of oxygenated water** over the gills. In the clam, the rhythmic beating of the cilia which cover the gills draws water in through the incurrent siphon, through pores in the gills, and out through the excurrent siphon (Fig. 10.4). In the fish, yawning movements draw water into the mouth. The mouth is then closed. Contraction of the mouth cavity then forces the water through the arches which support the feathery gills. The water is finally expelled through the slits at the rear of the gill chamber (Fig. 10.4).

10.2 WATER VERSUS AIR

In one respect, the problem of gas exchange is simpler for terrestrial organisms than for aquatic ones. Oxygen represents about 21% of a given volume of air. Therefore, a terrestrial organism in the process of passing 100 ml of air over its gas-exchange organ has come in contact with 21 ml of oxygen. On the other hand, even in well-aerated water the oxygen concentration seldom exceeds one ml of gas in every 100 ml of water. Consequently, the aquatic organism must expose its gills to a relatively large flow of water to satisfy its oxygen needs. Perhaps this difference in the availability of oxygen in air and water helps explain why the only warm-blooded animals—animals that can maintain a constant body temperature using heat generated by cellular respiration—are the air-breathing birds and mammals.

On the other hand, a terrestrial organism is faced with the problem of keeping its gas-exchange organ moist even while it is exposed to dry air. As we discussed earlier, diffusion of gas into a cell only occurs from the ECF. The ECF may be only a thin film of moisture, but it must be there. The most general solution to this problem has been to have the gas-exchange organ extend inward into the interior of the organism. In this way, air passing into the gas-exchange organ can be premoistened by accessory membranes and its drying effect lessened.

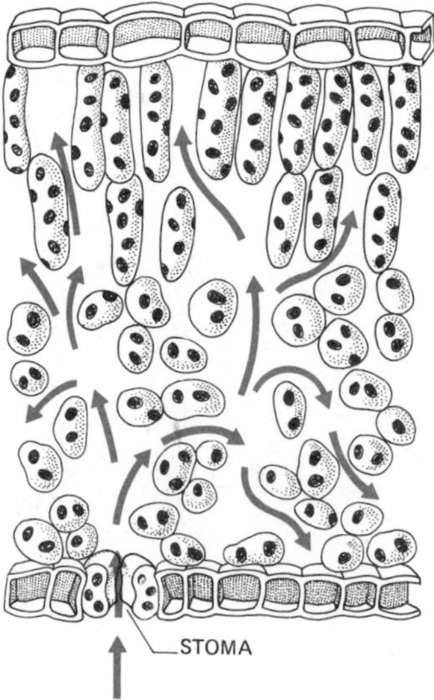

FIG. 10.5 The loose packing of cells in most metabolically active plant tissues (here a leaf) allows for the rapid diffusion of gases.

10.3 GAS EXCHANGE IN ROOTS AND STEMS

Plants have no specialized organs for gas exchange (with the few inevitable exceptions!). There are several reasons why plants are able to get along without them. First, each part of the plant takes care of its own gas-exchange needs. There is very little transport of gases from one part of the plant to another. Although plants have an elaborate liquid transport system (which we shall examine in a later chapter), this system does not participate in gas transport.

Second, plants do not present great demands for gas exchange. Roots, stems, and leaves respire at rates far lower than are characteristic of animals. Only during photosynthesis are large volumes of gases exchanged and, as we shall see, each leaf is well adapted to take care of its own needs during these periods.

Third, the distance that gases must diffuse in even a large, bulky plant are not great. Each cell in the plant is located quite close to the surface of the plant. The plant achieves this in much the same way that the planarian does, that is, by arranging its cells in thin sheets with a resulting high surface-to-volume ratio. "All well and good for leaves," you may argue, "but what of thick, woody stems and roots?" In fact, the situation in these organs is not much different from that in the leaves. The only *living* cells of the stem are organized in thin layers in and just beneath the bark. The cells in the interior of the woody stem are dead and serve only to provide mechanical support.

Finally, most of the cells of a plant have at least part of their surface in contact with air. The loose packing of parenchyma cells in leaves, stems, and roots provides an interconnecting system of air spaces. Gases diffuse through air several thousand times faster than through water. Thus, once oxygen and carbon dioxide gain access to the network of

FIG. 10.6 Lenticels in the bark of a young stem. These openings enable the living tissues beneath the cork to exchange oxygen and carbon dioxide with the air.

intercellular air spaces, they diffuse rapidly through them (Fig. 10.5).

Woody stems and mature roots are sheathed in layers of dead cork cells which are impregnated with a waxy, waterproof (and airproof) substance called suberin. This sheath serves the useful function of reducing water loss from the plant. But cork is as impervious to oxygen and carbon dioxide as it is to water. However, the cork of both mature roots and woody stems is perforated by nonsuberized pores called **lenticels** (Fig. 10.6). These enable oxygen to reach the intercellular spaces of the interior tissues and, similarly, allow carbon dioxide to be released to the atmosphere.

In many annual plants, the stems are green and almost as important in carrying out photosynthesis as the leaves. In these stems, the gas-exchange mechanism is quite similar to that of the leaves.

10.4 GAS EXCHANGE IN THE LEAF

In Chapter 9, we studied how the structure of the leaf adapts it for carrying out its main function, photosynthesis. A leaf photosynthesizing rapidly needs a substantial and steady supply of carbon dioxide. It also must release an equivalent volume of oxygen during the process. The exchange of gases occurs through pores in the leaf surface, the **stomata** (singular, stoma). When the leaf is actively photosynthesizing, the carbon dioxide content of the air in the air spaces of the spongy layer falls below the 0.03% present in the outside air. As a result of the difference in concentration, carbon dioxide then diffuses from the outside air through the stomata into the air spaces of the spongy layer and from there up through the intercellular air spaces between the palisade layer cells. After dissolving in the film of moisture that covers the surfaces of every cell in the interior of the leaf, it diffuses into the cells. The cytoplasm of these cells is richly supplied with carbonic anhydrase which catalyzes the formation of carbonic acid:

$$CO_2 + H_2O \rightleftharpoons H_2CO_3 \rightleftharpoons H^+ + HCO_3^-.$$

This is the immediate source of carbon dioxide for the dark reactions of photosynthesis. Oxygen, produced by the light reactions, diffuses from the cells into the intercellular air spaces and ultimately to the outside through the stomata.

Under most circumstances, the stomata open when light strikes the leaf in the morning and close during the night. A great deal of experimentation has been carried out in an effort to discover how light actually controls the opening of the stomata, but as yet no single scheme explains all the experimental data. The immediate cause seems to be a change of turgor in the **guard cells.** The inner wall of each guard cell is thick and elastic. The outer wall is much thinner. When turgor develops within the two guard cells flanking a stoma, the thin outer walls bulge out and force the inner walls into a crescent shape (Fig. 10.7). This opens the stoma. When the guard cells lose turgor, which can be accomplished experimentally by bathing them in a

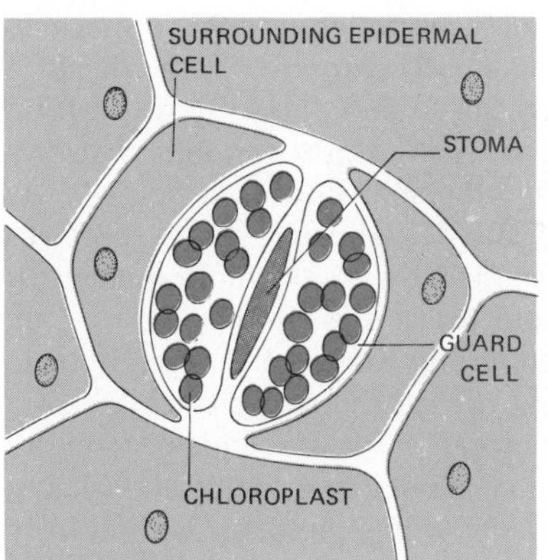

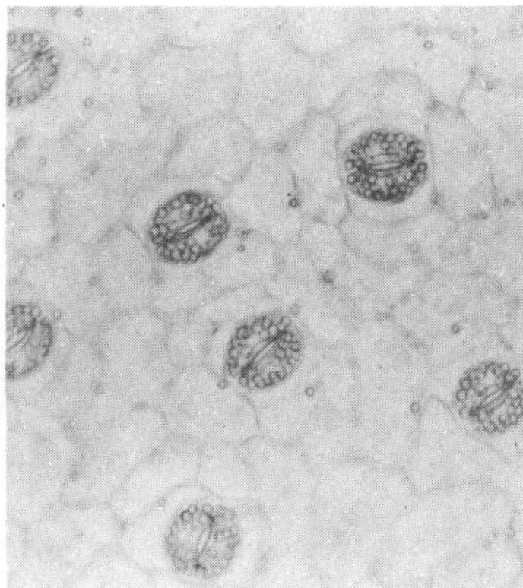

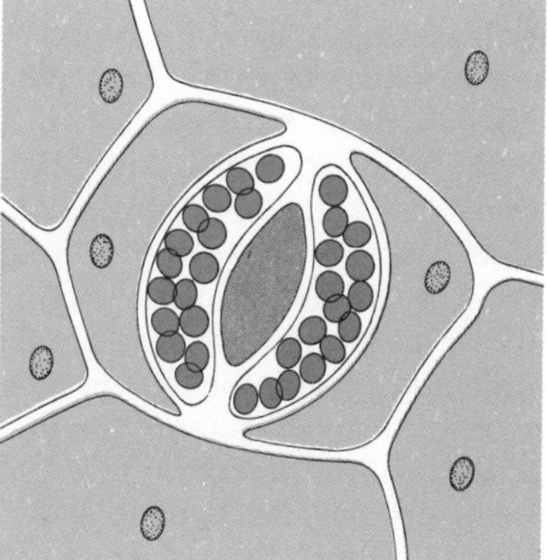

FIG. 10.7 Stomata: closed (left) and open (right). Stomata open when turgor builds up within the guard cells. Light striking the leaf initiates the reactions leading to this effect.

OSMOTIC PRESSURE IN TYPICAL GUARD CELLS	TIME	OSMOTIC PRESSURE, lb/in^2
	7 a.m.	212
	11 a.m.	456
	5 p.m.	272
	12 MIDNIGHT	191

(The osmotic pressure of the other lower epidermal cells remained constant at 150 lbs/in^2 during this experiment.)

FIGURE 10.8

solution isotonic to their cytoplasm, the elastic inner walls restore the original shape and the stoma closes.

The buildup of turgor in a cell depends upon the development of a difference in osmotic pressure between that cell and its surroundings. Figure 10.8 shows the osmotic pressure at different times of day measured within the guard cells of one plant. The osmotic pressure within the other cells of the lower epidermis remained steady at 150 lb/in^2. When the osmotic pressure of the guard cells became substantially greater than that of the surrounding lower epidermal cells, the stomata opened. In the evening, when the osmotic pressure of the guard cells dropped close to that of the surrounding cells, the stomata closed.

In order for the osmotic pressure within the guard cells to increase, their water concentration must decrease with respect to that in the surrounding cells. Put another way, there must be an accumulation of molecules and/or ions in the cytoplasm. This seems to be the case. Upon exposure to light, glucose molecules and potassium ions accumulate quickly within the guard cells.

When leaves are placed in solutions with a pH lower than 6.3, the stomata close. When they are placed in solutions of a higher pH (pH 8 seems to be optimum), the stomata open. Might such pH changes occur naturally? Probably so. A rise in carbon dioxide concentration (which would occur at night) would lower the pH because of the formation of carbonic acid. And the stomata are closed at this time. On the other hand, the carbon dioxide content of the air in an actively photosynthesizing leaf decreases. This, in turn, raises the pH. One bit of evidence to support the idea that it is the amount of carbon dioxide present that controls pH (and thus stomatal opening) is that the stomata of leaves in the dark will open if exposed to air containing less than 0.03% carbon dioxide. Thus it seems that light stimulates the opening of the stomata by (1) stimulating photosynthesis in the leaf, which (2) causes a reduction in the carbon dioxide concentration of the air in the air spaces, which (3) increases the pH of the cytoplasm of the guard cells, which (4) promotes the accumulation of solutes, which (5) causes water to enter the guard cells by osmosis from the surrounding epidermal cells, which (6) causes a build-up of turgor, and the stomata open!

One of the great problems of gas exchange in terrestrial organisms is keeping the cells of the gas-exchange organ or tissue moist in opposition to the drying effect of the air. The solution in the green leaf is typical of the solution in all truly terrestrial organisms. The cells involved in gas exchange are housed in a cavity within the organism. Air is brought to the moist cells by means of openings in the waterproof and otherwise gas-proof exterior. The epidermal cells of the leaf are covered with cutin. This material greatly cuts down water loss from the leaf, but also prevents any appreciable gas exchange from occurring. The stomata, however, allow for the controlled passage of gases to and from the interior of the leaf.

Any air that is not fully saturated with water vapor (100% relative humidity) will tend to dry the surfaces of the cells with which it comes in contact. Therefore, the photosynthesizing leaf, despite its protective modifications, will still lose substantial quantities of water to the air by evaporation. This water vapor then passes out of the leaf (through the stomata) in the process called **transpiration.** The transpired water must be replaced by transport of additional water from the soil to the leaves by way of the roots and stem.

The necessity of keeping the cells of the gas-exchange organ continually supplied with moisture may have some benefits to the plant. Transpiration (like sweating) is a cooling process and thus reduces the rate of photorespiration (see below). As we shall see in Chapter 12, transpiration also drives the transport mechanism by which inorganic nutrients are brought from the soil to the leaves.

When the surrounding air is very dry, the rate of transpiration may exceed the capacity of the root system to supply replacement water. This potentially hazardous situation is usually avoided by the closing of the stomata. Each stoma acts as a humidity sensor and closes when it is exposed to a steep humidity gradient between the interior and exterior of the leaf.

Even when humidity and soil moisture are high, most plants close their stomata during the middle part of hot summer days. The reason is that the rate of cellular respiration rises more rapidly with increasing temperature than does the rate of photosynthesis. Soon a point is reached when the plant is respiring more rapidly than it is photosynthesizing and thus the carbon dioxide concentration begins to rise. This closes the stomata just as it does after sundown when photosynthesis ceases.

The elevated daytime respiration of plants occurs not in the mitochondria but in peroxisomes (see Section 5.11), and results in the oxidation—without accompanying ATP production—of up to 30% of the freshly photosynthesized products. The process is called **photorespiration** because its rate is dependent on the rate of photosynthesis as well as temperature. It is hard to see how it can be anything but a liability to the plant. Interestingly enough, some warm-climate plants (e.g. sugar cane and corn) have biochemical and structural modifications by which they circumvent photorespiration. Not surprisingly, their productivity is very high. It is perhaps also not surprising that crabgrass is another member of this group!

Gas Exchange in Terrestrial Animals

The solution of most terrestrial animals to the problem of gas exchange with the air has been solved in a manner analogous to that used by the green plants. The gas-exchange structures are located within the body cavity. Here they are protected from mechanical damage and, even more important, from the excessive drying effect of the air. Frequently, the gas-exchange organ is liberally supplied with glands which replace the water unavoidably lost by evaporation.

Among the insects and certain other terrestrial invertebrates, gases are exchanged between the tissues and the air by means of an elaborate network of air-filled tubes, the **tracheae.** Among the air-breathing vertebrates, **lungs** are used as the gas-exchange structures.

10.5 TRACHEAL BREATHING

The tracheal system of the grasshopper is fairly typical of that found in all insects (Fig. 10.9). The **tracheae** open to the outside through small holes in the exoskeleton called **spiracles.** The first and third segments in the thorax have two spiracles, one on each side. Another eight pairs of spiracles are found arranged in a line on each side of the abdomen. The spiracles are guarded by bristles which aid in filtering dust and other foreign matter from the air before it enters the tracheae. The spiracles are also guarded by valves. These are controlled by muscles and enable the grasshopper to regulate the opening and closing of the spiracles.

The spiracles lead to major tracheal tubes which lead, in turn, to ever-finer branches. These branches penetrate to every portion of the insect's body. At their extreme ends, they may be only 0.1 μm in diameter. At this end they are filled with liquid. Oxygen, diffusing through the system, dissolves in this fluid and then diffuses into nearby cells. Perhaps every cell in the insect's body is adjacent to at least one of these tiny tubes.

The construction of tracheae is quite interesting. Their walls must be fairly rigid in order not to be forced together by the weight of the surrounding tissues. This rigidity is imparted by spiraling strands of chitin in the walls of the trachea (Fig. 10.9). Nevertheless, there is a limit to the pressures which this kind of construction can withstand without deforming and closing the air passageway. This may

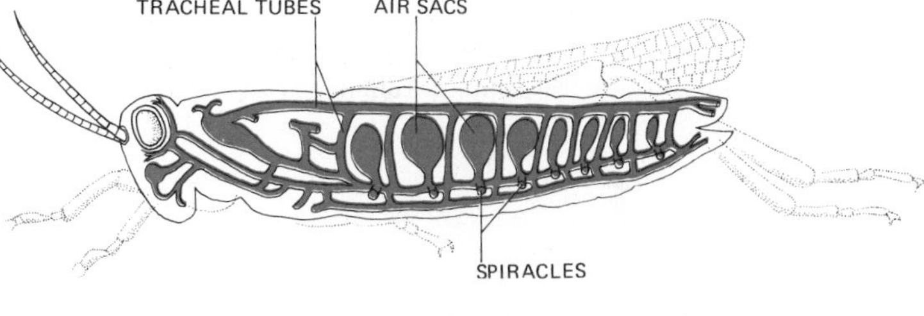

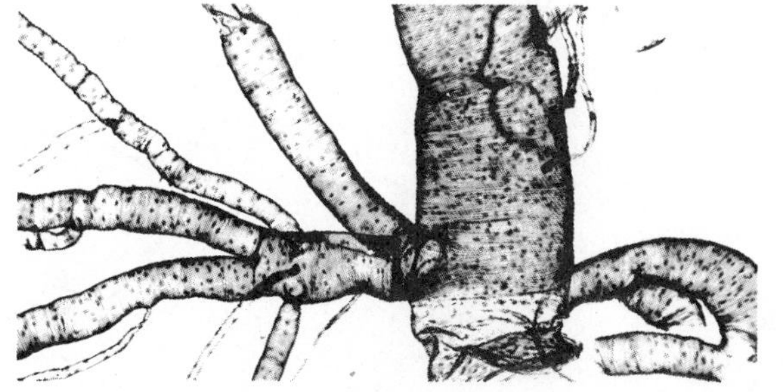

FIG. 10.9 Top: Tracheal system of the grasshopper. Bottom: A photomicrograph of tracheal tubes. (Courtesy CCM: General Biological, Inc., Chicago.)

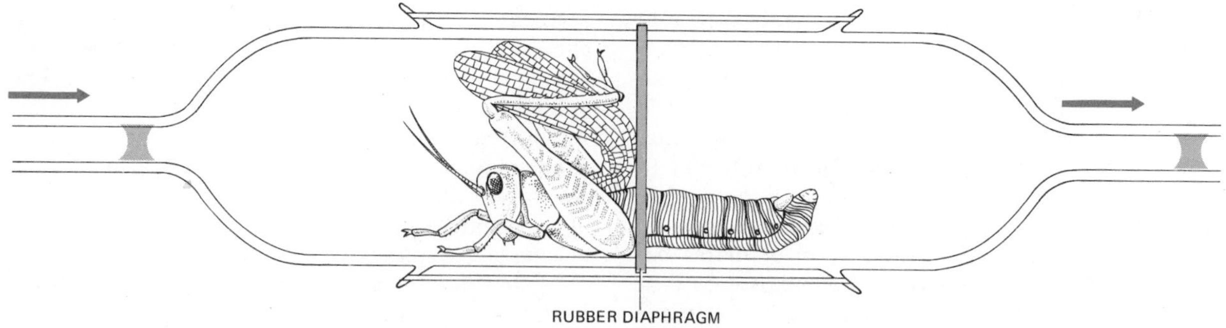

FIG. 10.10 Fraenkel's experiment demonstrating the one-way flow of air through the grasshopper. The liquid seals in the tubing move to the right as air enters the first four pairs of spiracles and is discharged through the last six pairs. The rubber diaphragm prevents air flow except through the grasshopper.

be one reason why insects are all relatively small in size. The increased weight of the tissues of an animal the size of a rabbit, for example, would crush a system of tubes filled only with air.

Among the smaller or less active insects, the passage of oxygen through the tracheal system is by simple diffusion. On the other hand, large or active insects like grasshoppers forcibly ventilate their tracheae. Contraction of muscles in the abdomen compresses the internal organs and forces air out of the tracheae. As the muscles relax, the abdomen springs back to its normal volume and air is drawn into the system. Large air sacs attached to portions of the main tracheal tubes increase the effectiveness of this bellowslike action. By means of an experiment such as the one illustrated in Fig. 10.10 (first performed by the insect physiologist Gottfried Fraenkel), one can show that this action is coordinated so that a one-way flow of air is set up. Air enters the first four pairs of spiracles at the anterior end of the system and is discharged through the last six pairs in the abdomen. This one-way flow increases the efficiency of gas exchange, as oxygen-deficient air can be expelled without mingling with the incoming flow of fresh air.

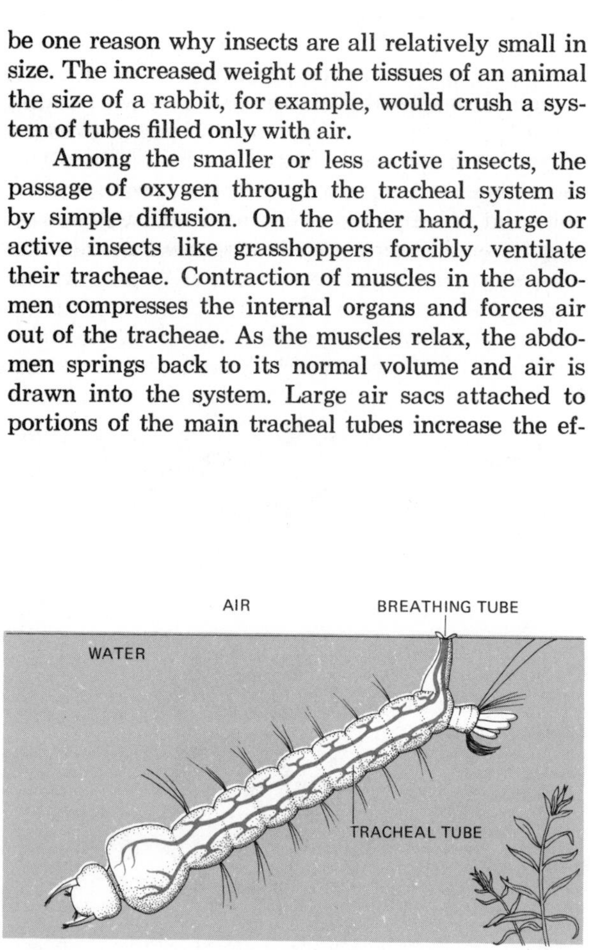

Even the aquatic insects possess a tracheal system. Some of them (e.g. the mosquito "wigglers") secure their air by poking a breathing tube through the surface of the water (Fig. 10.11). This tube is connected with the tracheal system. Other insects can submerge for long periods, but in so doing, they carry with them a bubble of air from which they can breathe oxygen. Still others possess spiracles mounted on the tips of spines. With these they pierce the leaves of underwater plants and obtain oxygen from bubbles formed within the leaves. Even in those aquatic insects that have gills, after the oxygen diffuses from the water into these outgrowths, it then diffuses into a gas-filled tracheal system for transport to the tissues of the body.

FIG. 10.11 The mosquito "wiggler" breathes air even though it lives and feeds under water.

10.6 LUNG BREATHERS

Among the terrestrial vertebrates (amphibians, reptiles, birds, and mammals), the organs of gas exchange are two **lungs.** In the **frog,** the lungs are two thin-walled sacs suspended in the body cavity and connected to the mouth cavity through an opening,

FIG. 10.12 Frog lungs. They are inflated by forcing air down through the glottis. The skin and the lining of the mouth are also used in gas exchange.

the glottis (Fig. 10.12). The surface area of the lungs is increased by a system of inner partitions which are richly supplied with blood vessels. To inflate the lungs, the frog must first fill its mouth with air by opening its two nostrils and lowering the floor of its mouth. Then the frog closes the internal openings to its nostrils, opens its glottis, and raises the floor of its mouth. These actions force air from the mouth into the lungs, inflating them. In fact, the frog may force so much air into the lungs that they expand throughout the entire body cavity.

Oxygen in this air dissolves in the film of moisture on the surface of the epithelial cells of the lungs. The oxygen then diffuses through the epithelial cells and into tiny blood vessels, the **capillaries.** Once in the blood, most of the oxygen enters the myriads of oval-shaped red blood cells. These floating cells contain the red pigment **hemoglobin.** Under the conditions that exist in the lungs, the hemoglobin combines chemically with the oxygen. The oxygen is then distributed throughout the frog's body in the circulating blood. When the blood reaches the various internal organs and tissues, the hemoglobin releases its oxygen. The oxygen is then free to leave the blood and enter the cells of the body to supply the needs of respiration. Carbon dioxide produced by cellular respiration enters the blood at the tissues and is then carried to the lungs and skin for release to the atmosphere.

The **skin** serves as a supplementary organ of gas exchange in most amphibians. It is richly supplied with blood vessels, and permeable to both water and gases. It can, however, function in gas exchange only if it is kept moist. Mucus-secreting cells in the skin assist in this function, but they are not adequate to cope with the drying effects of hot air of low humidity. Thus most amphibians are confined to moist locations such as ponds, swamps, damp soil, etc. This is just one of the reasons why they cannot be considered truly terrestrial animals.

Reptiles possess a scaly, dry skin. It is rather impervious to the passage of water so that little moisture is lost from the animal by this route. Thus reptiles are not confined to damp locations, although many do, in fact, live in such places. Lizards and snakes are abundant in the deserts, one of the driest habitats. While a scaly skin is an adaptation permitting safe exposure to dry air, it is of no use whatsoever as a gas-exchange organ. Reptiles depend upon their lungs for this function. Not only do their lungs have a relatively greater surface area than those of amphibians, but the ventilation of the lungs is considerably more efficient. The lungs are surrounded by a cage of rib bones (Fig. 10.13). These may alternately be spread apart and then drawn together by opposing sets of rib muscles. When they are spread apart, the volume of the rib cage increases. This expansion creates a partial vacuum

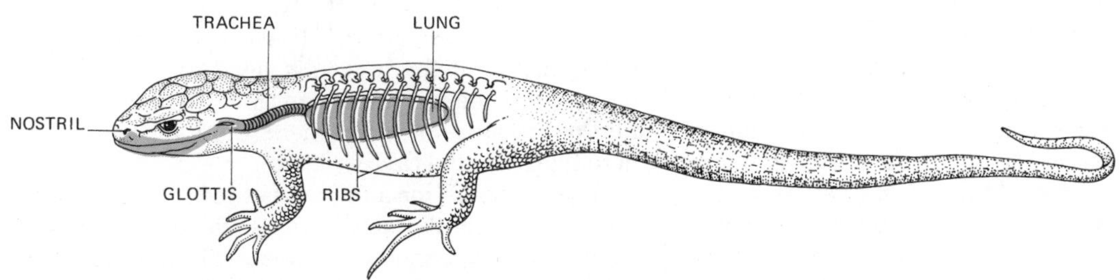

FIG. 10.13 Lizard lungs. They are inflated by expanding the rib cage.

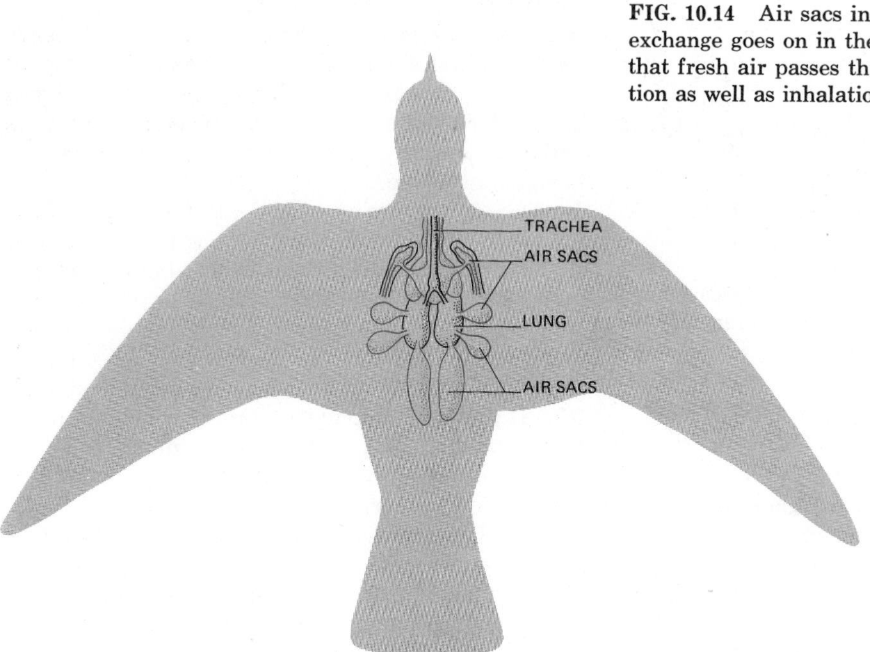

FIG. 10.14 Air sacs in the pigeon. Although no gas exchange goes on in the air sacs, they are arranged so that fresh air passes through the lungs during exhalation as well as inhalation.

within the lungs which is relieved by an inrush of fresh air. The fresh air, of course, brings a fresh supply of oxygen to the moist tissues of the lungs. Contraction of the rib cage then forces air out of the lungs. This exhaled air is depleted of oxygen but enriched with carbon dioxide given off at the lungs.

Although the ventilation of **bird** lungs is similar in principle to that of reptile lungs, the vigor of the action is intensified during flight by the beating of the wings. Unlike reptiles, birds are homeothermic ("warm-blooded"). They maintain a constant body temperature (usually around 40°C) despite wide fluctuations in the temperature of the external environment. Birds maintain their body temperature chiefly through the heat produced by muscular activity. Muscular activity, in turn, depends upon the energy liberated by cellular respiration. It is no wonder, then, that the oxygen demands of a small, active bird are very great. Outgrowths of the lungs, the **air sacs** (Fig. 10.14), help birds satisfy these extraordinary demands. Although no gas exchange actually occurs in them, their anatomical arrangement increases the efficiency of lung ventilation by enabling fresh air to pass through the lungs during *both* inhalation *and* exhalation. The air sacs also aid in reducing the density of the body by substituting air for tissue or fluid in many locations. For example, some of the bird's bones are hollow and penetrated by air sacs.

10.7 MECHANISM OF BREATHING IN HUMANS

Among the mammals, the efficiency of lung ventilation is improved by the presence of a **diaphragm.** This is a dome-shaped, muscular partition that divides the body cavity into two compartments: the abdominal cavity (which contains the viscera—the stomach, intestines, etc.) and the thoracic cavity (which contains the heart and lungs) (Fig. 10.15). The inner surface of the thoracic cavity and the outer surface of the lungs are lined with thin membranes, the **pleural** membranes. With the aid of a film of moisture between them, the lung pleura adhere tightly to the cavity pleura.

The necessity for adhesion between the lungs and the walls of the thoracic cavity is vividly illustrated when air is introduced between the pleural membranes. The force of adhesion is broken and the natural elasticity of the lung causes it to collapse.

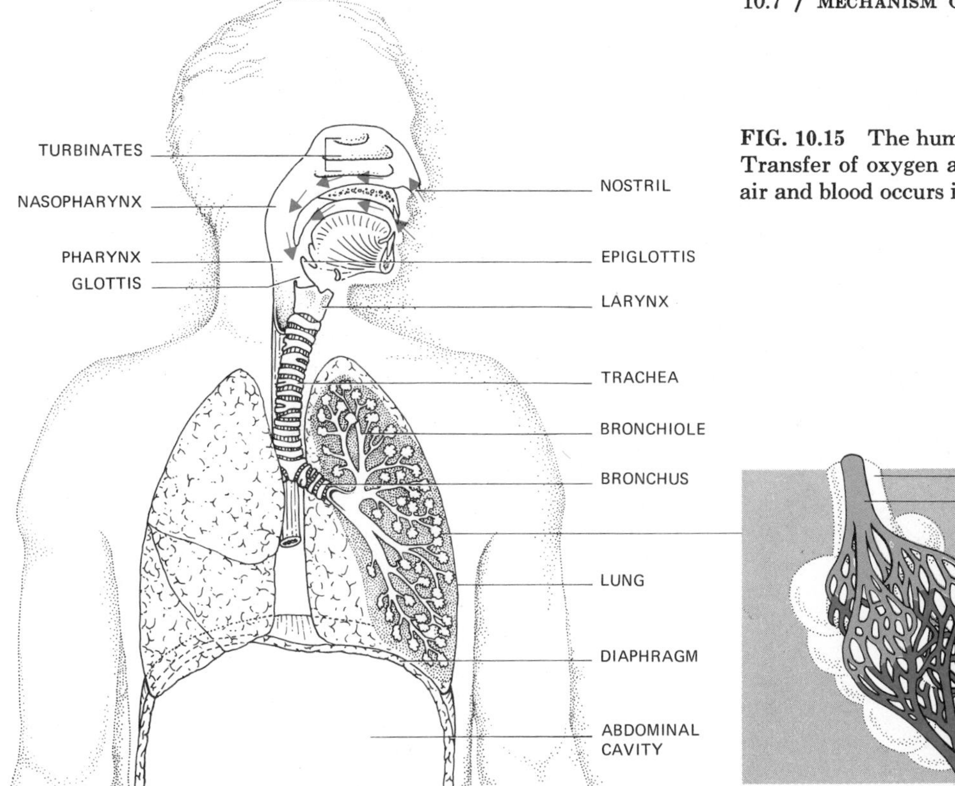

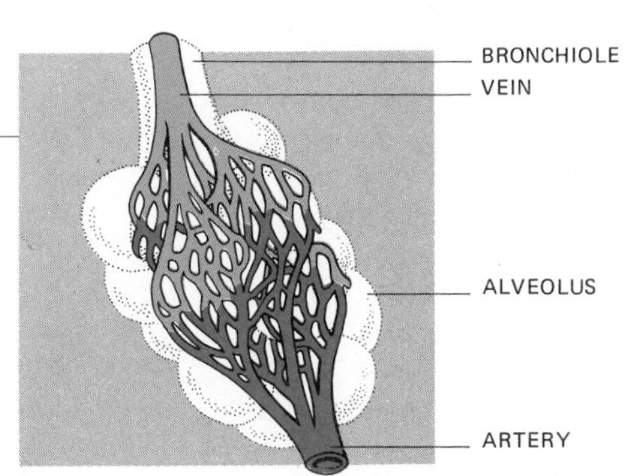

FIG. 10.15 The human gas-exchange system. Transfer of oxygen and carbon dioxide between air and blood occurs in the alveoli.

This procedure is often done deliberately when a lung has been damaged by a disease such as tuberculosis. Collapsing the lung allows it to rest and thus heal more rapidly. Reinflation of the lung is no problem because the air between the pleural membranes is gradually absorbed by the tissues and the lung assumes its normal size once more.

Any action which increases the volume of the thoracic cavity increases the volume of the lungs because of their adhesion to the walls of the cavity. Spreading of the rib cage accomplishes this as it does in the reptiles and birds, but in mammals (including humans) the action is enhanced by contraction of the diaphragm. Contraction lowers the diaphragm, resulting in a corresponding increase in the volume of the thoracic cavity. This, in turn, stretches the lungs, and air rushes into them. When the diaphragm relaxes, it returns to its former position, permitting the lungs to return to their former size. As they do so, air is forced out of them.

The motion of the diaphragm in breathing accompanies, but does not substitute for, the action of the rib cage. The motion of the ribs is controlled by the external and internal intercostal muscles. During inspiration (inhaling), the external intercostal muscles contract, lifting the ribs up and outward (Fig. 10.16). This motion increases the volume of the thoracic cavity and hence draws air into the lungs. During expiration (exhaling), the external intercostal muscles relax and this, aided by the natural elasticity of the lungs, returns the thoracic cavity to its normal dimensions. The cycle of inspiration and expiration is repeated about 15 to 18 times each minute in the resting adult. About 500 ml of air is drawn in and then exhaled during each cycle.

It is, of course, possible to breathe more deeply, either consciously or as the result of exercise. Vigorous inspiration occurs simply as a result of more vigorous contraction of the diaphragm and the external intercostal muscles. Vigorous expiration depends, however, upon more than the elastic recoil of these muscles and the lungs. The *internal* intercostal muscles draw the ribs down and inward, thus reducing the volume of the thoracic cavity still further. At the same time, the muscular wall of the abdomen contracts, forcibly pushing the stomach, liver, etc., upwards against the diaphragm and reducing the volume of the thoracic cavity from that direction. With vigorous inspiration and expiration, an average adult male can flush his lungs with about

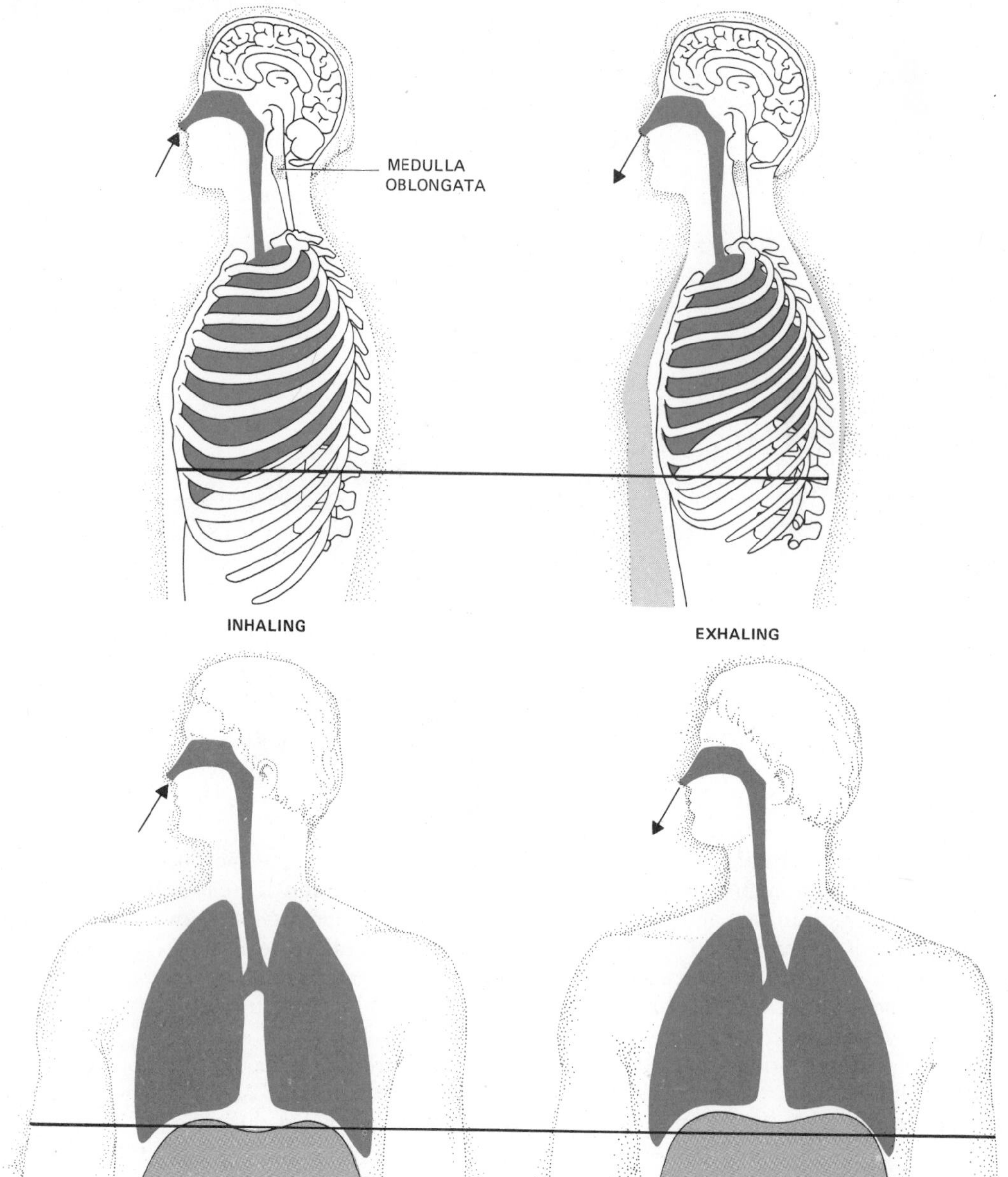

FIG. 10.16 During inspiration (inhaling), the ribs are lifted up and outward; the diaphragm is lowered. The rate and depth of breathing are controlled by nerve impulses originating in the medulla oblongata.

4 liters of air at each breath. This is known as the **vital capacity** of the lungs. Even at the point of maximum expiration, about 1200 ml of air ("residual air") remain in the lungs.

10.8 THE PATHWAY OF AIR

Ventilation of the lungs is, of course, also dependent upon an unobstructed passageway between them and the outside air. During inspiration, air enters the **nostrils,** passes through the **nasal cavities** (one behind each nostril), through the **nasopharynx** and (merging with any air taken in through the mouth) the oral pharynx (Fig. 10.15). From the oral pharynx, the air passes through the **glottis** into the **trachea.** The walls of the trachea are stiffened by horseshoe-shaped bands of cartilage. These bands prevent the trachea from collapsing under the pressure of surrounding tissues. The trachea branches into a right and left **bronchus,** leading to the right

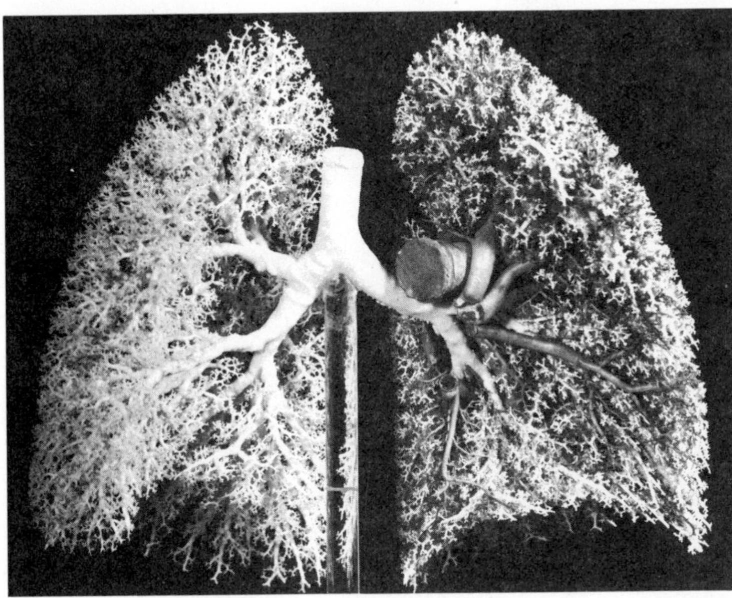

FIG. 10.17 Cast of the human lungs. Note the pattern of dichotomous branching in the bronchial system. (Courtesy Anatomical Institute, Bern.)

and left lungs, respectively. The bronchi, in turn, branch again and again (an average of 22 times) into ever-smaller **bronchioles** (Fig. 10.17). The walls of the bronchi and the major bronchioles are also stiffened with cartilage. Each terminal bronchiole opens into a grapelike cluster of tiny sacs, the **alveoli** (Fig. 10.18). It is only here in the alveoli that gas exchange actually takes place. There are some 300 million alveoli in two adult lungs, and these provide a total surface area for gas exchange of 70–80 m^2. Compare this remarkable value (approximately equal to one-third of a tennis court) with the total surface area of the skin of a 150-lb, 5′10″ human: 2 m^2.

The surface area provided by the alveoli is sufficient to take care of the gas-exchange needs of the body both at rest and during exertion. Under certain conditions, however, the gas-exchange area of the lungs may be seriously reduced. For example, virus or bacterial infections of the alveoli result in **pneumonia.** In pneumonia, lymph and mucus accumulate in the alveoli and bronchioles, reducing the area exposed to air. In critical cases, the patient may turn

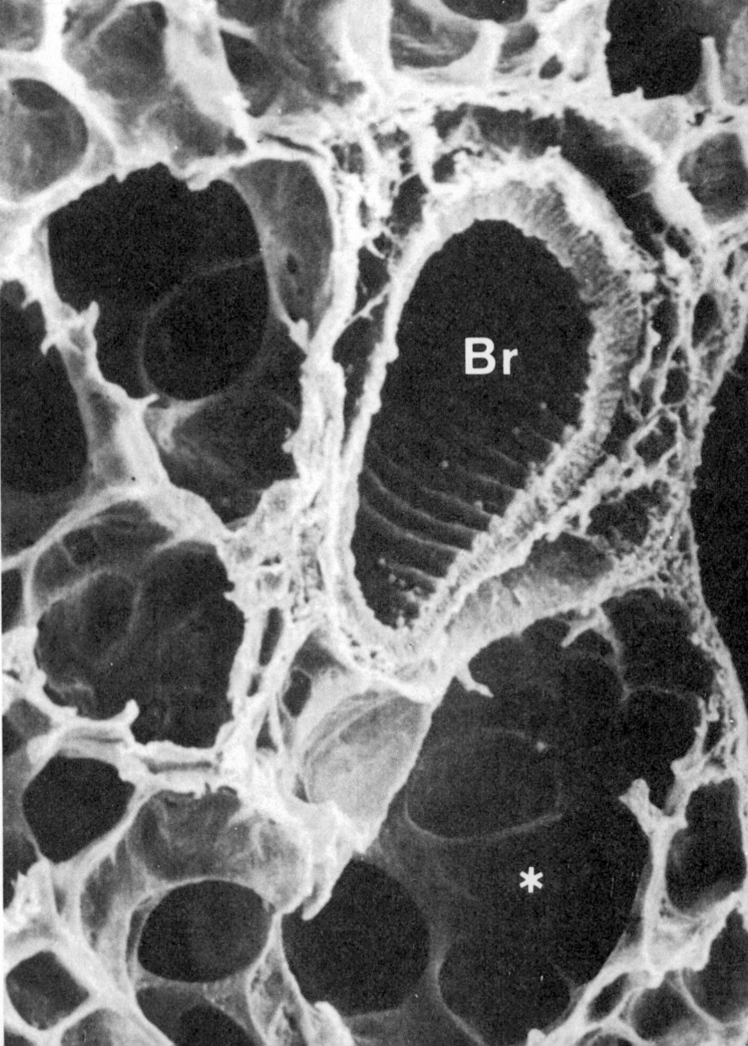

FIG. 10.18 The cut surface of a portion of a mouse lung as seen under the scanning electron microscope. The asterisk indicates a cluster of alveoli. The stiffened wall of a bronchiole can also be seen ("Br"). (Reproduced with permission from Keith R. Porter and Mary A. Bonneville, *An Introduction to the Fine Structure of Cells and Tissues,* 4th ed., Lea & Febiger, Philadelphia, 1973.)

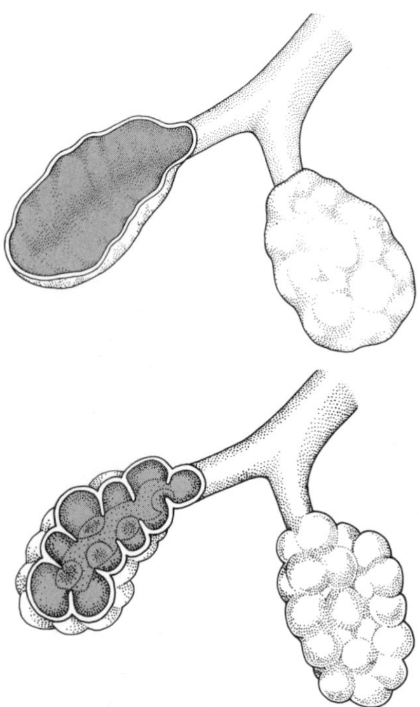

FIG. 10.19 Structure of the alveoli in a normal lung (bottom) and the lung of a victim of emphysema (top). Breakdown of the alveolar walls reduces the surface area for gas exchange.

blue from oxygen starvation. Having the patient breathe pure oxygen under an oxygen tent may save his life. The increased concentration of oxygen inhaled makes possible the most effective use of the gas-exchange area still available.

Another disease of the lungs in which the gas-exchange area is reduced is **emphysema.** In this disease, many of the walls separating the alveoli become irritated and break down. This greatly reduces the surface area available for gas exchange (Fig. 10.19). Unlike pneumonia, this condition develops very slowly and is seldom a direct cause of death. However, the gradual loss of gas-exchange area forces the heart to pump ever larger quantities of blood to the lungs in order to satisfy the gas-exchange needs of the body. The added strain this imposes on the heart often leads to heart failure.

The inner walls of the trachea, bronchi, and bronchioles are all covered with a mucus-secreting ciliated epithelium. Fine dust particles that are not filtered out by the nose are trapped in the mucus. The cilia then sweep this material up the trachea. When it nears the glottis, the forceful exhalation of coughing is initiated, and the material, phlegm, is propelled into the mouth. Particles small enough to reach the alveoli are engulfed by the many phagocytic cells present in the alveolar walls. After a lifetime of city dwelling (and/or cigarette smoking), the accumulation of such material makes the lungs very dark in appearance.

Studies carried out on high-school and college students have shown that the efficiency of lung ventilation is reduced in cigarette smokers. The lungs behave as though the small bronchioles had become partially obstructed. This would occur if there were an excessive production of mucus in the lungs (and cigarette smokers do bring up more phlegm from their lungs—a condition known as chronic bronchitis) or if the rate of mucus removal by the ciliated cells were impaired. The action of the cilia *is* slowed following exposure to tobacco smoke. This has been determined by having subjects inhale an aerosol containing tiny radioactive particles. The rate at which the particles are swept out of the lungs can be followed with a large Geiger counter. It is measurably slowed in subjects who have recently been smoking, although a long-term depression has not been observed.

A glance at the table in Fig. 10.20 shows what happens in a typical case to the composition of the air when it reaches the alveoli. Some of the oxygen dissolves in the fluid that moistens the thin epithe-

COMPOSITION OF ATMOSPHERIC AIR AND EXPIRED AIR IN A TYPICAL SUBJECT

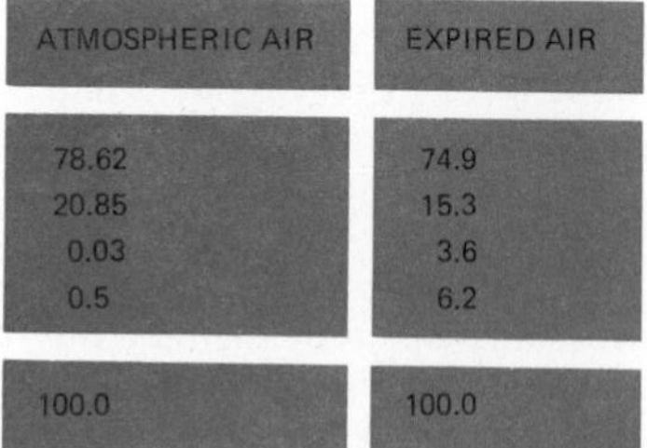

COMPONENT	ATMOSPHERIC AIR	EXPIRED AIR
N_2 (PLUS INERT GASES)	78.62	74.9
O_2	20.85	15.3
CO_2	0.03	3.6
H_2O	0.5	6.2
	100.0	100.0

FIG. 10.20 Note that only a fraction of the oxygen inhaled is taken up by the body.

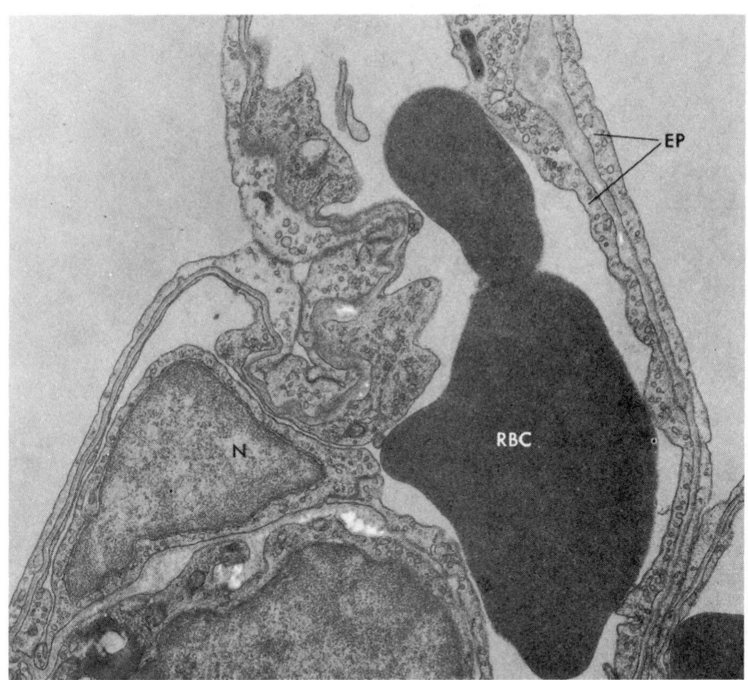

FIG. 10.21 Alveoli and an adjacent capillary from the lung of a laboratory mouse (10,500 ×). Note the thinness of the epithelial cells (Ep) that line the alveolus and capillary (except where the nucleus, N, is located). The dark gray object is a red blood cell. (Reproduced with permission from Keith R. Porter and Mary A. Bonneville, *An Introduction to the Fine Structure of Cells and Tissues,* 4th ed., Lea & Febiger, Philadelphia, 1973.)

lium of the alveoli. From here, the oxygen diffuses into the blood contained within the numerous capillaries present in the walls of the alveoli. As in the frog, most of the oxygen then combines with the hemoglobin contained within the red blood cells. Simultaneously, some of the carbon dioxide in the blood diffuses out into the alveoli, from which it can be exhaled. The circulation of the blood then carries the oxygen to all the cells of the body. In so doing, it picks up carbon dioxide from these cells for transport back to the capillaries of the alveoli. Additional details of the chemical mechanisms by which oxygen and carbon dioxide are transported efficiently by the blood will be considered in the next chapter.

The ease with which oxygen and carbon dioxide pass between the air of the alveoli and the red blood cells in the capillaries can be appreciated by examining Fig. 10.21. Portions of two alveoli are visible at the top, one on the left, the other on the right. Each is lined by a remarkably thin, elongated epithelial cell. The partition between the two alveoli is almost completely occupied by a capillary. Note, too, how thin is the cell that forms the wall of the capillary—except where its nucleus is located. The dark gray object within the capillary is a red blood cell. At the closest point, its surface is only 0.7 μm away from the air in the alveolus, less than one-tenth of the diameter of this tiny cell!

10.9 CONTROL OF BREATHING

The rate at which cellular respiration takes place (and hence oxygen is consumed) varies with the general state of activity of the body. Vigorous exercise may increase by 20 to 25 times the demands of the tissues for oxygen. This increased demand is met by increasing the rate and depth of breathing.

It would be reasonable to assume that lack of sufficient oxygen triggers this response. It is quite easy to show, however, that oxygen deprivation plays a very minor role in regulating the rate at which ventilation of the lungs occurs. Figure 10.22 illustrates an experimental setup by means of which a human subject can inspire various precise gas mixtures. While the subject is inhaling the gas mixture, the rate and depth of breathing can be checked constantly. The subject begins the experiment by breathing pure air (21% oxygen, 0.03% carbon dioxide, and about 79% inert gases by volume), first from the room and then from the tank. This procedure provides a "control." It reveals what, if any, changes in response can be expected just by having the subject breathe from the tank. Nervousness, an unpleasant taste imparted to the air by the equipment, or increased air resistance are just three factors that might alter the subject's responses. These alterations would have to be taken into account in interpreting the results of varying the gas mixtures.

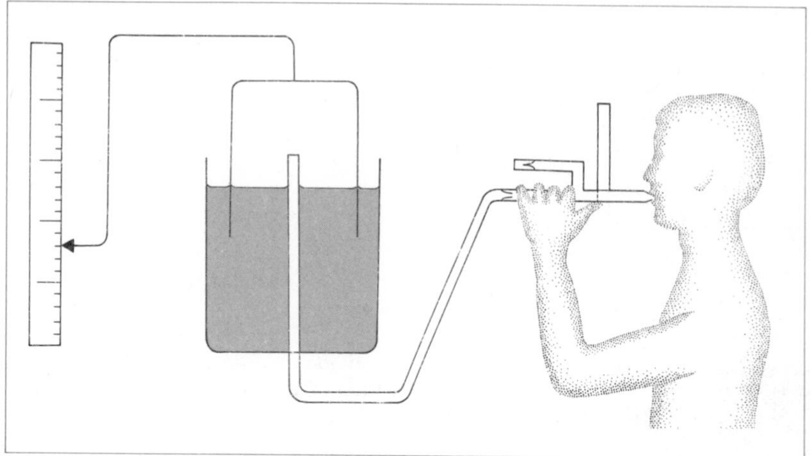

FIG. 10.22 Apparatus for determining the effects of different gas mixtures on the rate and depth of breathing. Gas mixtures are inhaled from the tank and exhaled into the room. The volume of gas inhaled is measured by the gauge on the left.

The two graphs in Fig. 10.23 show that no appreciable change does, in fact, occur in this particular experiment. When 100% oxygen is substituted for air, no marked change in rate or depth of breathing occurs either, although there is a tendency for depth of breathing to decrease slightly. When the subject inhales a gas mixture consisting of 92% oxygen and 8% carbon dioxide, however, a most dramatic increase in rate and depth of inspiration takes place. Note that there is no question here of tissues suffering from a lack of oxygen. The subject is inhaling a gas mixture over four times as rich in oxygen as the air. The experiment suggests that the concentration of carbon dioxide plays a decided part in governing the rate and depth of breathing.

Other physiological experiments have shown that carbon dioxide achieves this effect through its action on a portion of the brain called the **medulla oblongata** (Fig. 10.16). When blood passes through the alveoli, its carbon dioxide content becomes the same as that of the alveolar air. Moments later, the blood reaches the medulla oblongata, which contains cells that are very sensitive to the concentration of carbon dioxide in the blood. If this carbon dioxide content

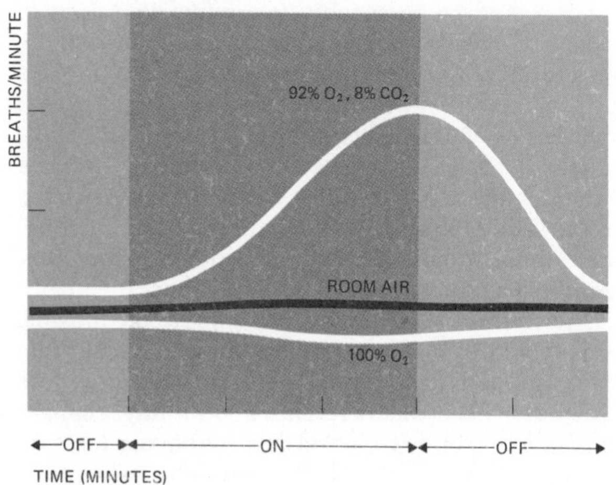

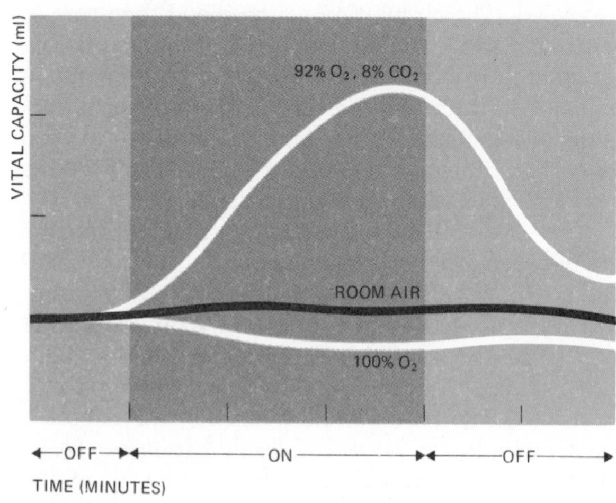

FIG. 10.23 Graphs showing effect of inhaling room air, 100% oxygen, and a mixture containing 92% oxygen and 8% carbon dioxide on breathing rate (left) and depth of breathing (right). Although the oxygen concentration in the 92–8 mixture is over four times as great as in room air, the presence of the carbon dioxide triggers a massive increase in the rate and depth of breathing.

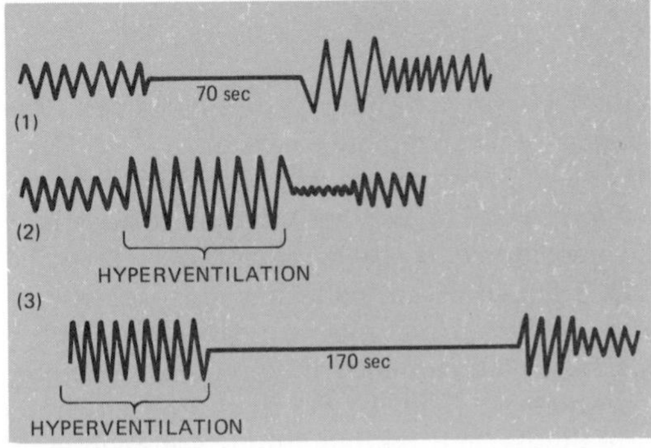

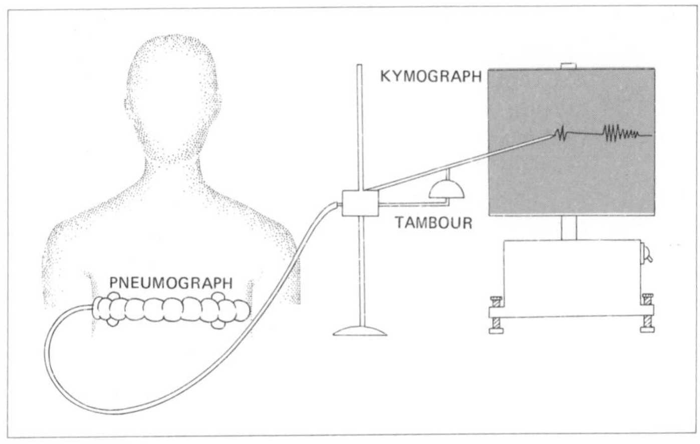

FIG. 10.24 Operation of the pneumograph. The subject's breathing movements are transmitted to the pointer and recorded on the revolving drum (the kymograph). The kymograph records show the effect upon rate and depth of breathing of (1) breath-holding, (2) hyperventilation, and (3) hyperventilation followed by breath-holding. The results in each case reflect the concentration of carbon dioxide in the blood at the conclusion of each activity.

rises above normal levels, the medulla oblongata responds by increasing the number and rate of nerve impulses which control the action of the intercostal muscles and diaphragm. The result is an increased rate of lung ventilation, which quickly brings the carbon dioxide concentration of the alveolar air, and then of the blood, back to normal levels.

Ventilation of the lungs is also controlled locally. The smooth muscle walls of the bronchioles are very sensitive to the concentration of carbon dioxide. A rising level of carbon dioxide causes the bronchioles to dilate. This provides less resistance in the airways and thus makes possible the acquisition of more oxygen at the same time that the increased amounts of carbon dioxide are exhaled.

We all know that rate and depth of breathing are also under conscious control, but this conscious control has definite limits. You can hold your breath for only a limited period of time. Eventually, the carbon dioxide content of the blood reaching the medulla oblongata becomes so high that the medulla overrides conscious control. This is the "breaking point."

Figure 10.24 illustrates an experimental setup with which this type of response can be studied. Movements of the subject's chest are detected by a hollow bellows (the pneumograph) strapped around the chest. Expansion and contraction of the bellows cause decreases and increases in the pressure of the air within. These pressure changes can then be transmitted to a recording stylus, which writes on a slowly revolving, smoked drum (the kymograph). Note that after a period of breath-holding, the rate and depth of inspiration are markedly greater than before the period of breath-holding began. This can be accounted for by the build-up of carbon dioxide during the breath-holding period. The length of time that one can hold his or her breath to the breaking point can be increased substantially by breathing extra fast and extra deeply just prior to the period of breath-holding. Vigorous, forced ventilation reduces the carbon dioxide content of the alveolar air and blood to below its normal value, and thus it takes longer for the carbon dioxide content to build up to the breaking point. If the subject simply breathes naturally after a period of extra-vigorous breathing (hyperventilation), we find that the depth of breathing is markedly less than that which preceded the period of hyperventilation.

The medulla oblongata, then, is an extremely efficient mechanism for controlling the carbon dioxide content of the blood. Any increase in carbon dioxide is quickly reduced by increased ventilation of the lungs. Any reduction in carbon dioxide content is quickly compensated for by a decreased ventilation of the lungs. The medulla is thus one of the homeostatic devices of the body. Through its activity, the carbon dioxide content of the blood is maintained within very narrow limits, and as a result, the constancy of the internal environment with respect to carbon dioxide content is maintained.

It may seem curious that the rate at which one breathes and thus supplies oxygen to the body is controlled by the carbon dioxide content of the blood rather than the oxygen content. Remember, though, that in the breakdown of glucose by cellular respiration, carbon dioxide is produced about as fast as oxygen is consumed. The muscles of the body need increased amounts of oxygen during vigorous exercise, but they produce increased amounts of carbon dioxide at the same time. This triggers increased ventilation of the lungs and thus automatically provides for the greater oxygen need.

There is one situation in which we may find ourselves suffering from a shortage of oxygen without a corresponding increase in the production of carbon dioxide. When transported (e.g. in a nonpressurized airplane) to altitudes above 13,000 ft, we begin to suffer oxygen deprivation. With decreased amounts of oxygen for cellular respiration there is, however, a corresponding decrease in the amount of carbon dioxide. Consequently the medulla is not stimulated to step up the rate of lung ventilation. Fortunately, the body does possess a mechanism to cope with this situation. There are receptors in the aorta and carotid arteries which detect reduced levels of oxygen in the blood. These receptors send impulses to the medulla oblongata and stimulate it to increase the rate of lung ventilation. Although this increase in rate and depth of breathing is not so dramatic as that brought about by carbon dioxide excess, it does bring increased quantities of oxygen to the alveoli.

10.10 AIR POLLUTION AND HEALTH

The alveolar surface provides a large and intimate interface between the internal environment and the air. Over a period of 24 hours, we breathe in some 15,000 liters of air. The purity of this air has become a matter of considerable concern.

A number of ingredients find their way into air, chiefly as a result of human activities, and these may fairly be considered as contaminants or air pollutants. The sources of air pollution are exceedingly varied but can be divided into two categories. One is our industrial and technological activities. Petroleum refineries, smelters (Fig. 10.25), fuel-fired electrical generating plants, and the ubiquitous automobile all release enormous quantities of pollutants into the air. These include:

1. Soot (from unburned fuel) and industrial ashes of other sorts.
2. Sulfur dioxide (SO_2) from the oxidation of fuels (e.g. coal, oil) containing sulfur compounds.
3. Various hydrocarbons (from the incomplete combustion of gasoline). Among these is benzopyrene, a notorious carcinogen (cancer-causing agent).
4. Oxides of nitrogen (e.g. NO_2) produced by the chemical union of O_2 and N_2 (in the cylinders of internal combustion engines).
5. Carbon monoxide (CO) from the incomplete combustion of fuels.

In bright sunlight, nitrogen oxides, hydrocarbons, and oxygen interact chemically to produce powerful oxidants like ozone (O_3) and peroxyacetyl

FIG. 10.25 Copper Basin, Tennessee. The vegetation in this region was killed by deliberate burning and by air pollution from a copper smelter prior to 1907. Over half a century later, continued erosion still hampers attempts at reforestation. Average annual rainfall is less and temperatures are higher than in surrounding regions. (Courtesy of the U.S. Forest Service.)

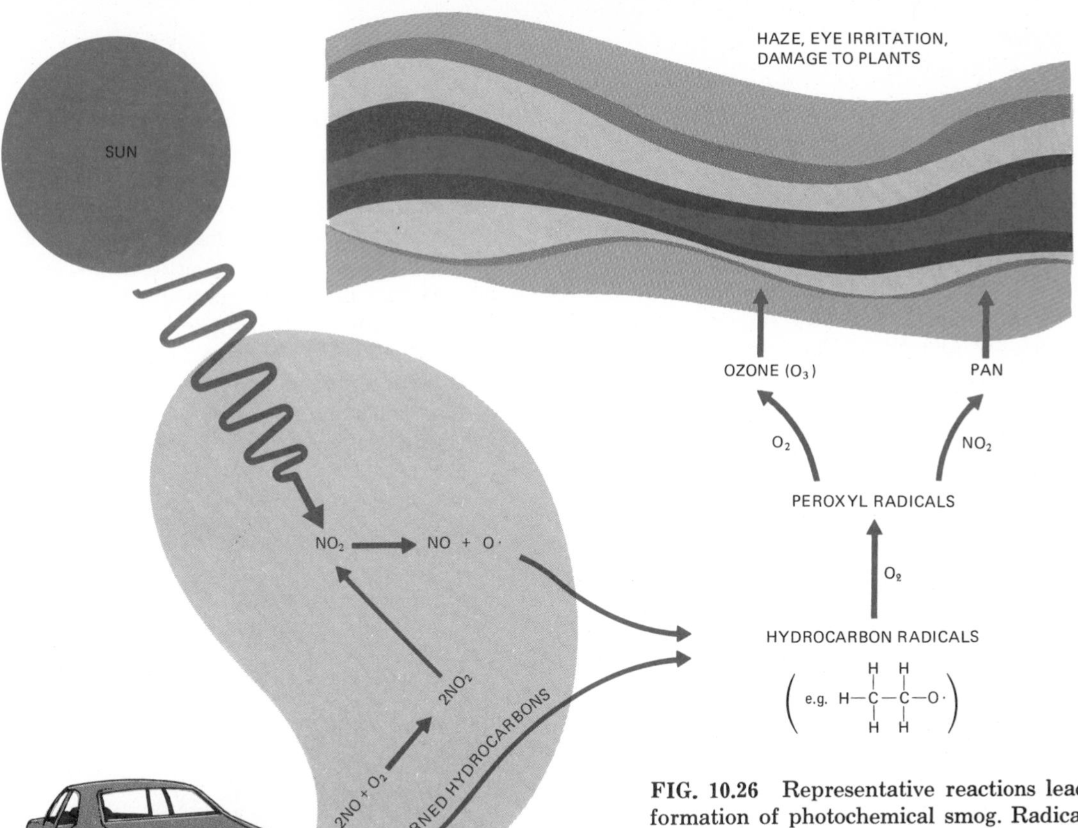

FIG. 10.26 Representative reactions leading to the formation of photochemical smog. Radicals are atoms or molecules with unpaired electrons. They are very reactive chemically. PAN is peroxyacetyl nitrate.

nitrate (PAN) (Fig. 10.26). These *secondary* pollutants are exceedingly damaging to plant life and lead to the formation of smog of the type common in Los Angeles. PAN is primarily responsible for the eye irritation so characteristic of this type of smog.

The other major category of air pollution is cigarette smoke. (Pipes and cigars are usually not inhaled; consequently the concentration of their smoke reaching the lungs is not nearly so high as with cigarettes.) Cigarette smoke contains a number of hydrocarbons, including benzopyrene.

Do the protective mechanisms working in the lungs provide adequate protection from air pollution? The question is not an easy one to answer. The most promising approach to an answer has come from studies in which the rate of human illness has been determined in populations differing in their exposure to air pollutants. Figure 10.27 shows that the rate of lung cancer in Great Britain is substantially higher in urban areas where there is plenty of heavy industry. Unfortunately, data of this sort are a little difficult to interpret because city dwellers are also apt to be heavier cigarette smokers than their rural cousins.

Nevertheless, as more and more data are collected, it is becoming apparent that a variety of human diseases are more common among persons exposed to the air pollution of urban life *or* to the individualized air pollution of cigarette smoking (or, especially, both). These diseases include chronic

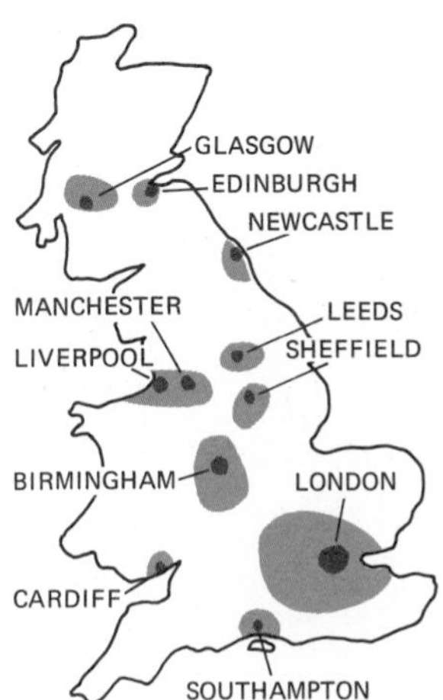

FIG. 10.27 Areas of above-average mortality from cancer of the respiratory system (trachea, bronchi, and lungs) in England, Scotland, and Wales.

bronchitis (and its probable sequel, emphysema), asthma, several circulatory disorders, and lung cancer. In fact, if one analyzes death rates from *all causes* among United States males, the rate is directly proportional to (a) number of cigarettes smoked per day, (b) depth of inhalation, and (c) number of years since smoking began. Fortunately the processes at work—whatever they are—appear to be at least partially reversible. The overall death rate among *former* smokers is inversely proportional to the time elapsed since they gave up the habit.

Chronic irritation of the bronchial epithelium produces a well-defined sequence of cellular changes. These include a thickening of the epithelial layer, a loss of the ciliated cells, and the appearance of cells with bizarre nuclei (Fig. 10.28). The extent of these changes has been clearly shown to be related to the amount of cigarette smoking. Cells with bizarre nuclei, which are probably **precancerous**, are quite rare in nonsmokers and their numbers do not increase with advancing age. In smokers, on the other hand, epithelial cells with bizarre nuclei are quite common and become increasingly so with increasing number of years of smoking. Fortunately, the number of such cells declines slowly once smoking is given up.

FIG. 10.28 Photomicrographs of (a) the normal epithelium of the bronchi (greatly enlarged); (b, c, d) three stages of abnormal changes to the bronchial epithelium commonly found in smokers; and (e) a lung cancer. The drawings show the cellular organization occurring at each stage. (Photomicrographs courtesy of Oscar Auerbach, M.D.)

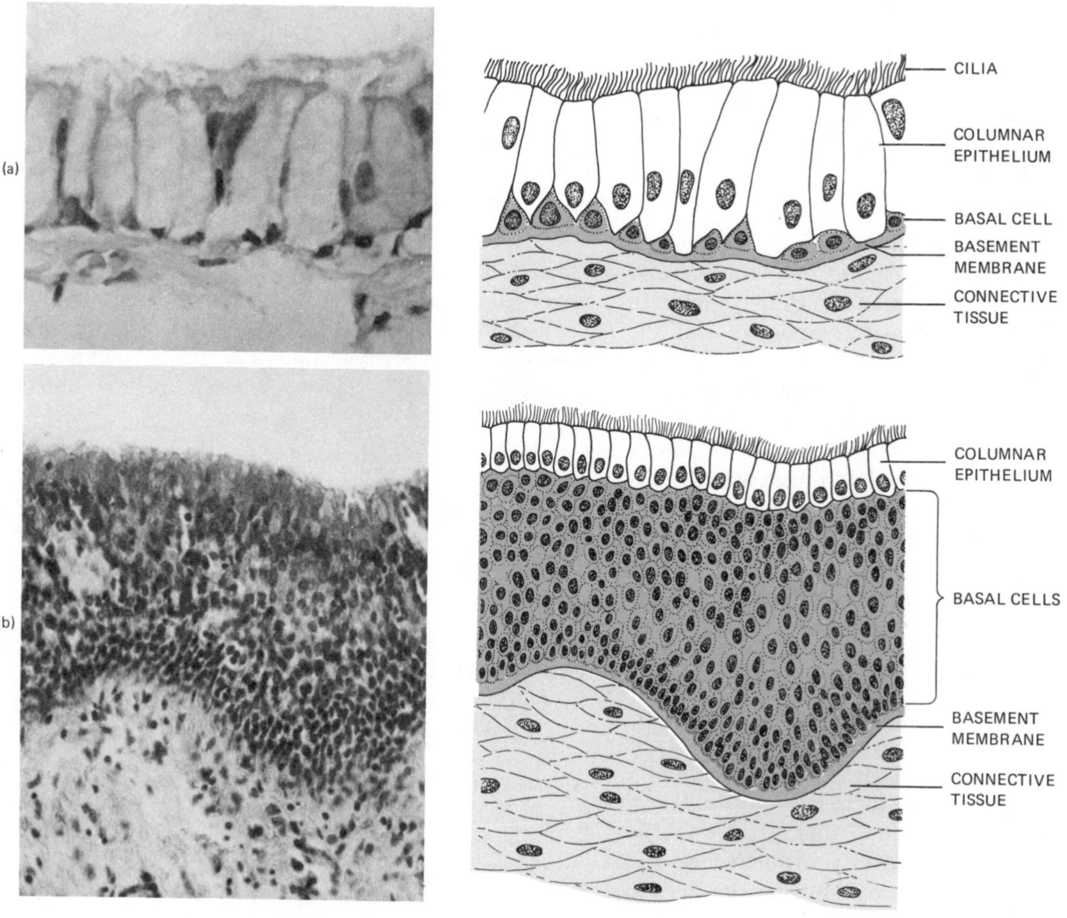

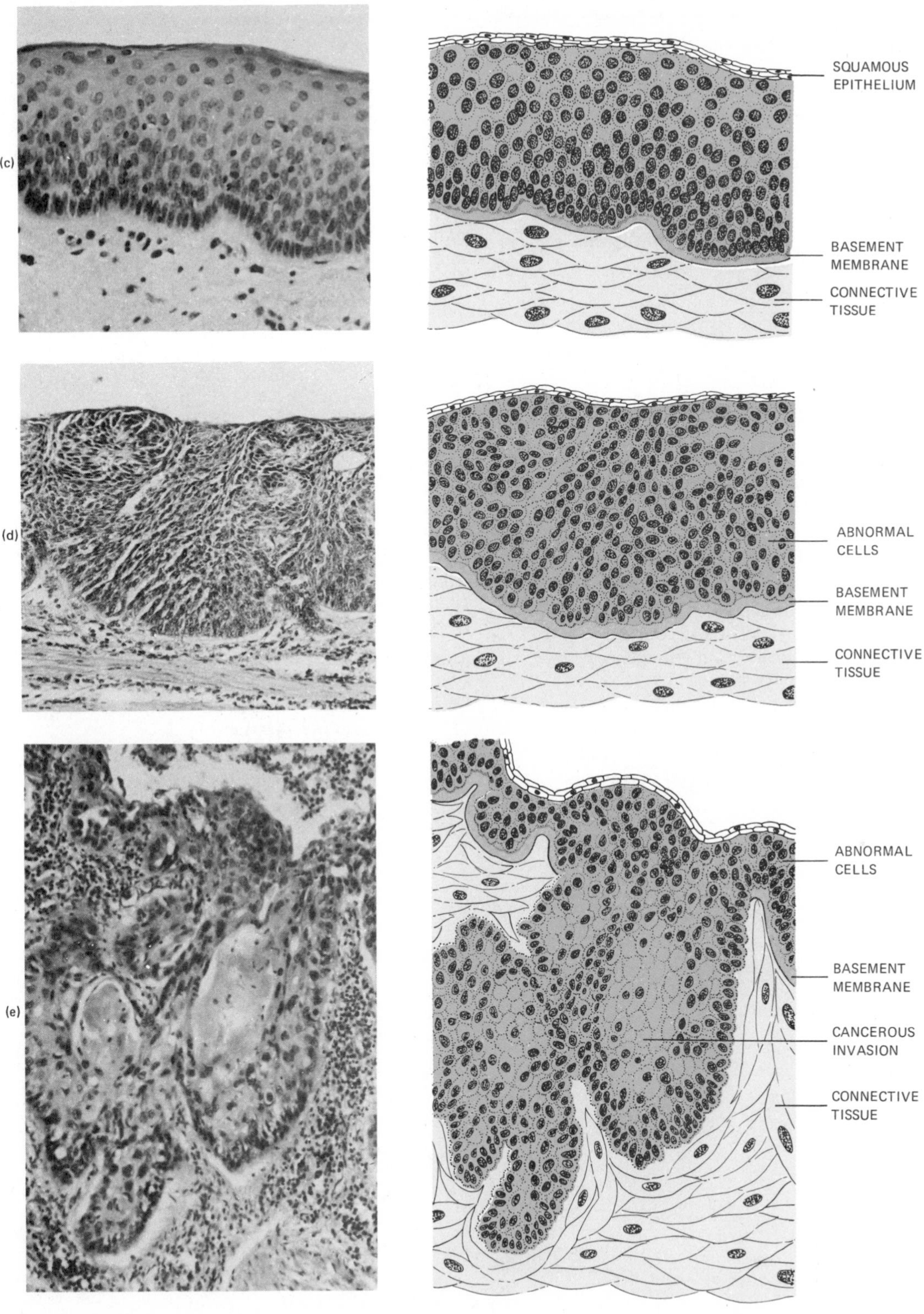
(c)
SQUAMOUS EPITHELIUM
BASEMENT MEMBRANE
CONNECTIVE TISSUE
(d)
ABNORMAL CELLS
BASEMENT MEMBRANE
CONNECTIVE TISSUE
(e)
ABNORMAL CELLS
BASEMENT MEMBRANE
CANCEROUS INVASION
CONNECTIVE TISSUE

EXERCISES AND PROBLEMS

1. Trace the pathway taken by a molecule of oxygen from the time it enters a nostril until it enters a red blood cell.
2. How do reptiles, birds, and mammals improve the efficiency of lung ventilation over that of amphibians?
3. How do amphibians compensate for their low efficiency of lung ventilation?
4. Using a microscope with a 43× objective and a 10× eyepiece, a student sees six stomata on the lower epidermis of a lily leaf. The diameter of the field of view is 0.4 mm. How many stomata are present per square centimeter?
5. Highway construction often interferes with the natural drainage of water, causing water to accumulate in formerly well-drained woods. Soon dead trees are to be seen in these areas. Explain.
6. Why are gills poorly adapted for life on land?

REFERENCES

1. WIGGLESWORTH, SIR VINCENT, *Insect Respiration,* Oxford Biology Readers, No. 48, Oxford University Press, Oxford, 1972.
2. SCHMIDT-NIELSEN, K., "How Birds Breathe," *Scientific American,* Offprint No. 1238, December, 1971.
3. MCDERMOTT, W., "Air Pollution and Public Health," *Scientific American,* Offprint No. 612, October, 1961. Considers the relationship between general air pollution and pulmonary disease.
4. HAMMOND, E. C., "The Effects of Smoking," *Scientific American,* Offprint No. 126, July, 1962. Includes a description of the functional and structural changes of the bronchi and alveoli in smokers.
5. LAVE, L. B., and E. P. SESKIN, "Air Pollution and Human Health," *Science,* Reprint No. 39, 21 August, 1970. A comprehensive review of the various studies that have been made.
6. HUGHES, G. M., *The Vertebrate Lung,* Oxford Biology Readers, No. 59, Oxford University Press, Oxford, 1973.
7. BJÖRKMAN, O., and J. BERRY, "High-Efficiency Photosynthesis," *Scientific American,* Offprint No. 1281, October, 1973. How certain plants circumvent the liability of photorespiration.
8. HEATH, O. V. S., *Stomata,* Oxford Biology Readers, No. 37, Oxford University Press, Oxford, 1975.

11 ANIMAL CIRCULATORY SYSTEMS

11.1 SIMPLE TRANSPORT MECHANISMS

Microorganisms and tiny animals have no need for a special transport system. Diffusion, active transport, and cytoplasmic streaming are sufficient to ensure that every portion of their body is adequately supplied with materials. The food vacuoles formed in the amoeba are swept by streaming cytoplasm throughout the cell as digestion of the food contents occurs. The digested food molecules then pass by diffusion or by active transport into the cytoplasm. Even an animal as complex as the planarian can get along without a true circulatory system. Although there is a small amount of liquid, which bathes the internal organs of the planarian, the only motion imparted to it is that which results from the random body movements of contraction and expansion. We have seen that the shape of the planarian makes it unnecessary to have a special system for the transport of oxygen and carbon dioxide. The necessity for extensive transport of digested food materials is eliminated by a highly branched gastrovascular cavity (see Fig. 37.5). Digestion in the planarian is intracellular, and no portion of the body is far removed from the endocytic cells which line the gastrovascular cavity and accomplish this work.

11.2 A "CLOSED" SYSTEM: THE EARTHWORM

The earthworm is relatively large and complex. It has a circulatory system for the transport of materials, and all of the features essential to an efficient circulatory system are found in it. These are: (1) a fluid in which the materials to be transported are dissolved, (2) a system of conducting vessels or channels in which the fluid flows, (3) a pump for maintaining the flow of the fluid, and (4) specialized organs to carry out exchanges between the fluid and the external environment. These include organs (such as the skin and intestine) that add materials to the fluid, and organs (such as the skin and excretory organs) that remove materials from the fluid and deposit them back into the external environment.

The circulating fluid of the earthworm is blood. It is mostly water in which are dissolved gases, sugars, amino acids, salts, and many other molecules and ions that play a role in the metabolism of the earthworm. The efficiency of the earthworm's blood as a medium for oxygen transport is increased by the presence of the red, oxygen-carrying pigment **hemoglobin.** The earthworm's hemoglobin is not contained within red blood cells as ours is, but is simply dissolved in the blood.

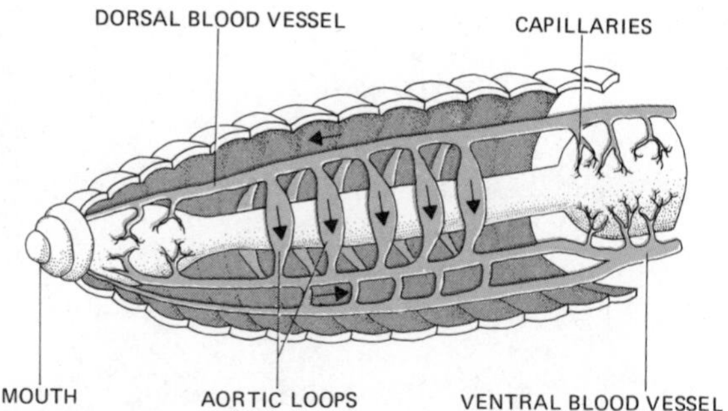

FIG. 11.1 Circulatory system of the earthworm. The blood is confined to the blood vessels at all times. Contraction of the dorsal blood vessel and the five pairs of aortic loops keeps the blood flowing.

The blood of the earthworm is transported by an elaborate system of blood vessels. It can only carry on its function of exchanging materials with individual cells, however, when it passes through the finest of the blood vessels, the **capillaries.** Because the blood is always retained within the system of blood vessels, we say that the earthworm has a "closed" system. The pump that forces the blood to the capillaries consists of five pairs of aortic loops (Fig. 11.1). Muscular contraction of the walls of these aortic loops forces blood into the ventral blood vessel. The ventral blood vessel transports blood toward the rear of the worm and distributes the blood to an elaborate system of lesser blood vessels. All of these terminate in capillaries, and it is here that exchanges between the *exchange organs* and the *blood* and between the *blood* and the *tissues* take place. Once through the capillary beds, the blood is picked up by a second system of vessels leading into the dorsal blood vessel. This vessel contracts rhythmically, forcing the blood back to the aortic loops at the anterior end of the worm.

11.3 AN "OPEN" SYSTEM: THE GRASSHOPPER

The circulatory system found in insects differs from that of the earthworm in one important respect. The blood is confined to vessels during only a portion of its circuit through the body. The remainder of its journey takes place within the body cavity itself. Such a system is known as an "open" circulatory system. The volume of blood required for such a system is kept to a practical level by a marked reduction in the size of the body cavity. This reduced body cavity is called a **hemocoel.** The efficiency of flow and

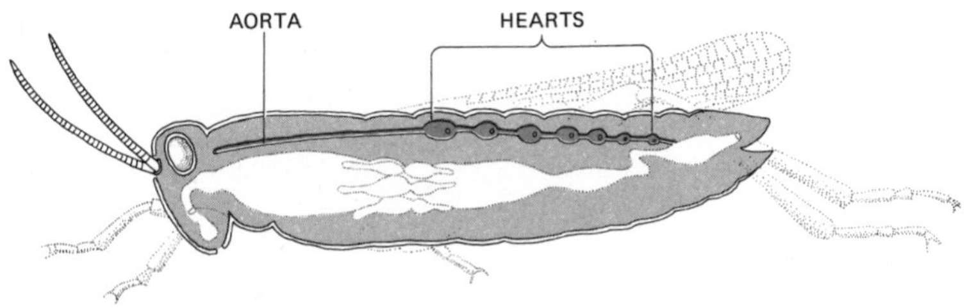

FIG. 11.2 The circulatory system of the grasshopper is "open." The blood is confined to vessels during only a small part of its circuit through the body.

distribution of the blood is maintained by having the hemocoel divided into chambers called sinuses.

In the grasshopper, the closed portion of the system is confined to a single long tubular heart and aorta running along the dorsal side of the insect (Fig. 11.2). The heart pumps blood into the sinuses of the hemocoel, where exchanges of materials take place. Coordinated movements of the body muscles gradually bring the blood back to the sinus surrounding the heart, the dorsal sinus. Between contractions, tiny valves in the wall of the heart open and permit the blood to enter the heart from the dorsal sinus, thus completing the circuit.

The open circulatory system of the grasshopper might seem to be quite inefficient in comparison with the closed system of the earthworm. It should be kept in mind, however, that there is a decided difference in the function accomplished by the two systems. Remember that in the grasshopper the exchange of oxygen and carbon dioxide is accomplished by means of the tracheal system. Blood plays no part in this process. There is not even an oxygen-carrying pigment in the blood of the grasshopper. A good deal of experimentation and study has shown clearly that it is the problem of gas transport that poses the most serious demands upon a circulatory system. In animals such as the grasshopper and other insects, where the gas-exchange system is quite independent of the circulatory system, the demands upon the latter are far less severe than in animals (e.g. the earthworm) where the functions are combined.

11.4 THE SQUID

The closed circulatory system of the squid is especially interesting because there are three separate pumps to maintain circulation (Fig. 11.3). One heart pumps blood to all the internal organs and tissues of the body. The other two hearts simply serve to pump blood from these internal organs and tissues to the gills, where gas exchange takes place. To understand the advantage of such a system, it is important to realize that the pressure produced by the contraction of a heart is almost entirely dissipated when blood enters the capillaries. Although capillaries are tiny, the total cross-sectional area of all the capillaries supplied by a single major blood vessel is considerably greater than the cross-sectional area of the blood vessel itself. The situation can be compared to that which occurs when a rapidly flowing, narrowly confined stream is allowed to spread out over a flat plain. The force and velocity of flow diminish rapidly. The same situation occurs in a network of capillaries. Note, too, that gas exchange must take place in two different locations: the gills and the tissues. In both cases, the exchanges occur only while the blood is passing through capillaries. Thus, in passing through the gills, the blood loses the pressure which could then distribute it quickly to the tissues. On the other hand, in passing through the capillaries of the tissues, the blood loses the pressure which could then force it quickly back to the gills. The squid's system of separate hearts thus copes with the problem very nicely.

It is worth noting here that a closed circulatory system with separate pumps is also found in the birds and mammals, but the two pumps are located together. One half of the bird or mammal heart pumps blood to the lungs; the other half pumps blood to the tissues. The development of separate pumps in these animals was entirely independent of that in the squid.

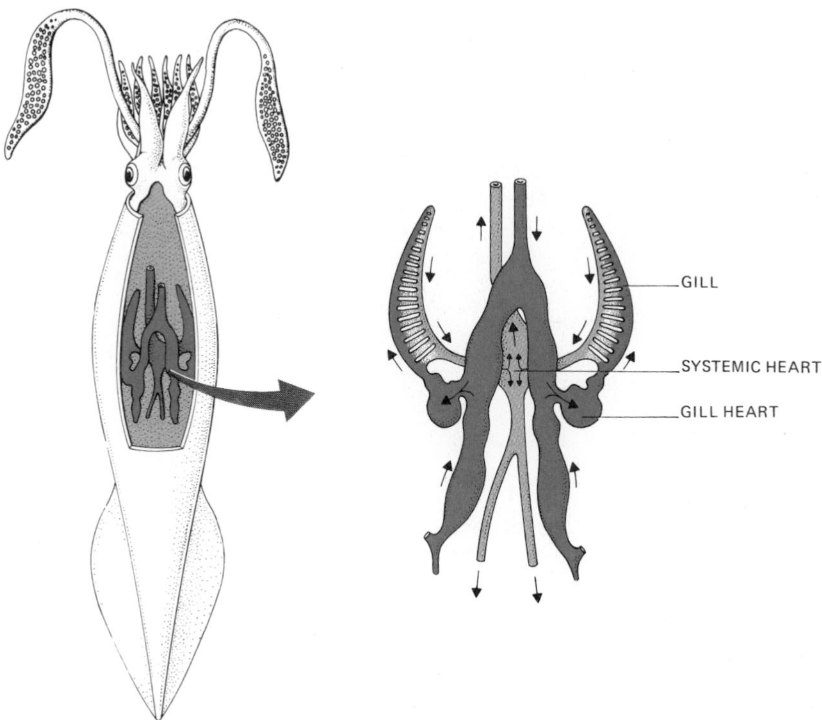

FIG. 11.3 Circulatory system of the squid. The gill hearts pump oxygen-deficient blood (dark color) to the gills where gas exchange takes place. The systemic heart pumps oxygen-rich blood (light color) to all parts of the body.

FIG. 11.4 Circulatory system of a fish. Oxygen-deficient blood (dark color) is pumped forcibly to the gills. Little pressure remains to move the oxygen-rich blood (light color) to the rest of the body.

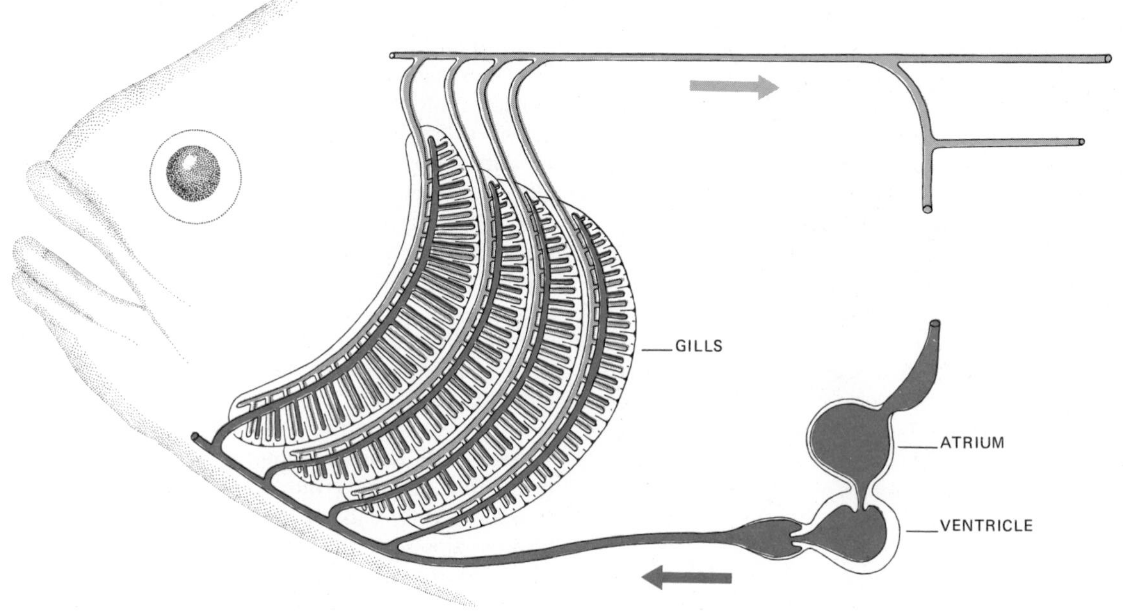

11.5 SINGLE PUMP: THE FISH

Actually, the earliest vertebrates had only a single pump. This arrangement still persists among the modern fishes. Blood collected from throughout the fish's body enters a thin-walled receiving chamber, the **atrium.** During relaxation of the heart, the blood passes through a valve into a thick-walled, muscular **ventricle.** Forcible contraction of the ventricle forces the blood out to the capillary network of the gills. Here gas exchange takes place. From the gills, the blood passes to the many capillary networks in the remainder of the fish's body, where exchanges with the tissues take place. Then the blood returns to the heart. The system is "closed," as the blood is contained within vessels throughout its entire circuit (Fig. 11.4).

The circulatory system of fishes, while obviously adequate to their needs, is not very efficient. As mentioned above, there is a marked pressure drop when blood flows through a capillary network. In the fishes, after the blood passes through the gills, it is no longer transported vigorously by the force of the heartbeat.

11.6 THREE CHAMBERS: THE FROG AND THE LIZARD

The situation in this respect is somewhat improved in the amphibian heart. The frog heart consists of three main chambers, two atria and a ventricle. The right atrium receives oxygen-deficient blood from the blood vessels (veins) that drain the various tissues and organs of the frog's body. Blood from the lungs, rich in oxygen, is carried to a separate atrium on the left. The blood from both atria passes into a single ventricle (Fig. 11.5). Contraction of the ventricle forces the blood into a vessel which immediately divides into a left and a right branch. Each of these immediately branches again into three main arteries. The anterior arteries carry blood to the head and brain of the frog. The middle ones, the systemic arches, carry blood to the internal tissues and organs in the rest of the body. The posterior arteries carry blood to the skin and lungs.

At first glance, it might seem that whatever advantage had been gained by returning oxygen-deficient blood to one part of the heart and oxygen-rich blood to another would be lost by mixing the two

FIG. 11.5 Comparison of an amphibian heart (left) and a reptile heart (right). No mixing of oxygen-deficient blood (dark color) and oxygen-rich blood (light color) occurs in the reptile heart.

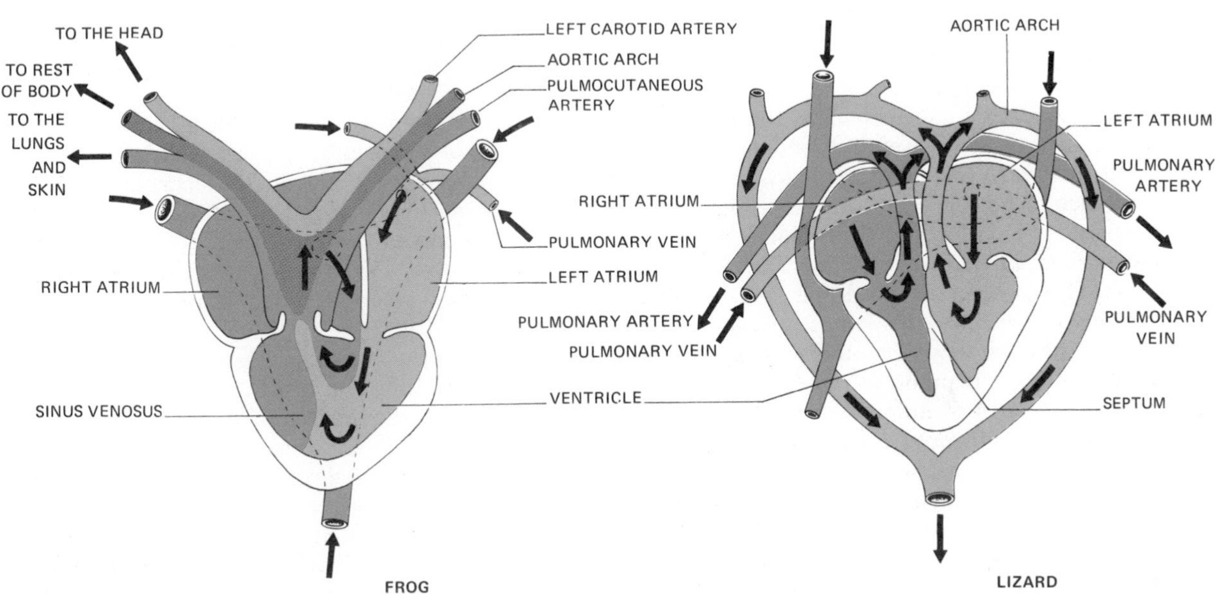

bloods in the single ventricle. To some extent this is true. However, the ventricle is partially divided into narrow chambers which tend to reduce mixing of the two bloods. When the ventricle contracts, most of the oxygen-deficient blood is deflected, with surprisingly little mixing, into the two arteries leading to the skin and lungs. Here a fresh supply of oxygen will be picked up. Oxygen-rich blood from the left atrium is sent, relatively pure, into the arteries leading to the brain. Only the blood passing into the systemic arches leading to the rest of the frog's body has been thoroughly mixed, but even so it contains enough oxygen to supply the needs of the organs to which it is carried.

Note that the problem created by pressure drop in the capillaries has been solved by the arrangement of the frog circulatory system. Both the gas-exchange organs *and* the interior tissues of the body receive blood under full pressure from the contraction of the ventricle.

Reptiles have a further modification of the heart. In the lizard, a muscular **septum** partially divides the ventricle (Fig. 11.5). When the ventricle contracts, the opening in this septum closes and the ventricle is momentarily divided into two entirely separate chambers. This prevents mixing of the two bloods. The left half of the ventricle pumps oxygen-rich blood (received from the left atrium) to the body. The right half of the ventricle pumps the oxygen-deficient blood (received from the right atrium) to the lungs.

11.7 FOUR CHAMBERS: BIRDS AND MAMMALS

In the hearts of birds and mammals, the septum is complete. This provides two entirely separate pumps. The right atrium receives oxygen-deficient (deoxygenated) blood from the body, and the right ventricle pumps this blood forcibly to the lungs, where it gives off carbon dioxide and picks up a fresh supply of oxygen. This oxygenated blood then returns to the left atrium, passes into the left ventricle, and is pumped out forcibly to all the other organs and tissues of the body. It is probably no coincidence that the only two groups of animals to develop warm-bloodedness (homeothermy) are the birds and mammals, with their two separate circulatory systems. One system is responsible for gas exchange with the environment; the other is responsible for gas exchange with the tissues. The efficiency provided by having these separate systems makes possible the high rate of cellular respiration upon which homeothermy depends.

The Pathway of Circulation in Humans

11.8 THE HEART

The heart is located roughly in the center of the chest cavity. It is surrounded by a protective membrane, the **pericardium** (Fig. 11.6). Deoxygenated blood from the body enters the right atrium. From here it flows through the **tricuspid valve** into the right ventricle. The name *tricuspid* refers to the three flaps of tissue that guard the opening between the right atrium and the right ventricle. Contraction of the ventricle then closes the tricuspid valve but forces open the pulmonary valve at the entrance to the **pulmonary artery.** Blood enters the pulmonary artery, which immediately divides into right and left branches leading to the right and left lungs, respectively. These arteries branch and rebranch to form **arterioles.** The arterioles supply blood to the capillary networks of the lungs. Here the blood gives up carbon dioxide and takes on a fresh supply of oxygen. The capillary networks of the lungs are drained by vessels called **venules.** These serve as the tributaries of the **pulmonary veins.** Four pulmonary veins (two from each lung) carry the oxygenated blood to the left atrium of the heart. This completes the portion of the circulatory system that is known as the **pulmonary system.**

From the left atrium the blood passes through the **bicuspid valve** into the left ventricle. Contraction of the ventricle closes the bicuspid valve and opens the aortic valve at the entrance to the **aorta.** The first branches from the aorta occur just beyond the aortic valve. Two openings lead to the right and left **coronary arteries** which supply blood to the heart itself.

Although these coronary arteries arise where the aorta is still within the mass of the heart, they pass directly out to the surface of the heart and extend down across it. They lead to arterioles which in turn supply blood to the network of capillaries that penetrate every portion of the heart. These capillaries are drained by venules which lead to two coronary veins that empty into the right atrium. This portion of the circulatory system is known as the **coronary system.**

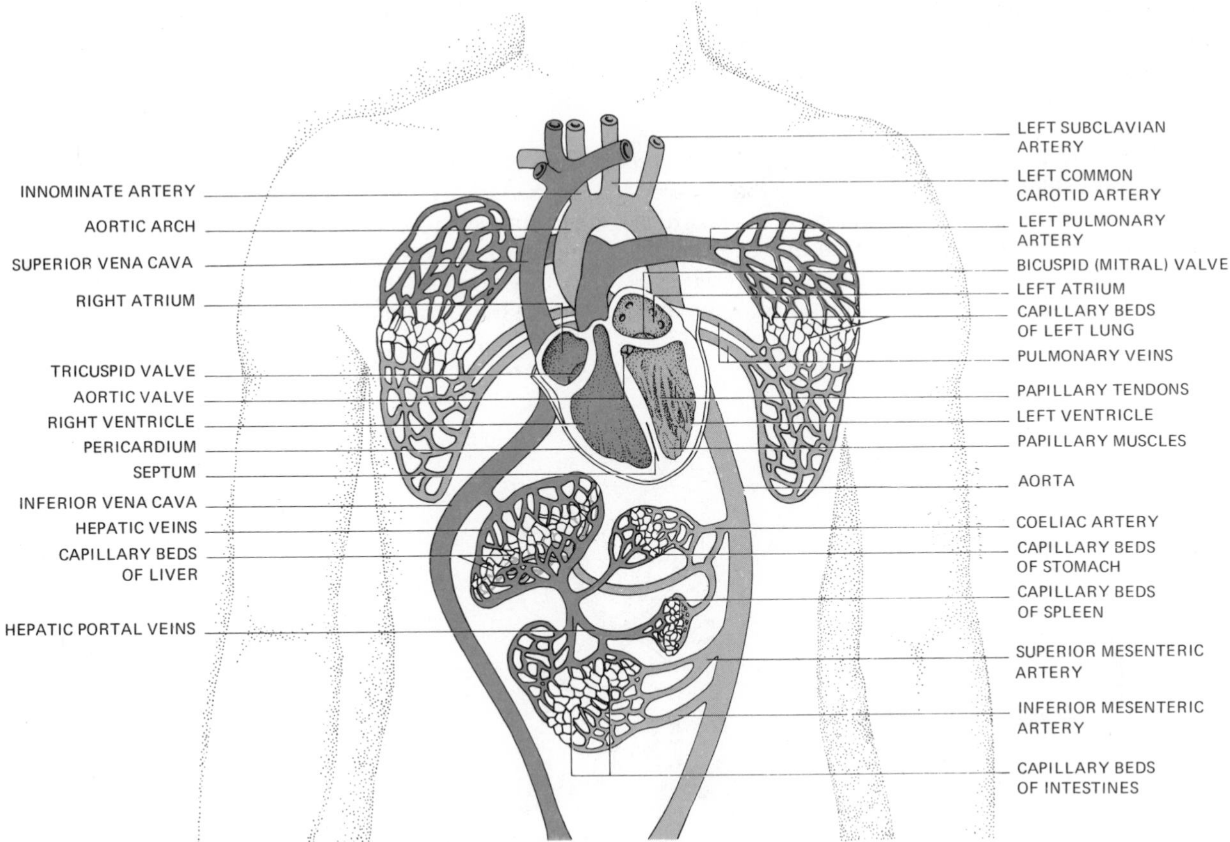

FIG. 11.6 The human heart with a schematic view of the pathway of blood through the lungs and the internal organs. Oxygenated blood is shown in light color; deoxygenated blood in dark color. Note that the blood draining the stomach, spleen, and intestines passes through the liver before it is returned to the heart. Here harmful or surplus materials picked up from those organs are removed.

Its importance can be seen from the fact that the heart muscle must have a continuous supply of oxygen in order to continue beating. Anything which interferes with the proper operation of the coronary system will have drastic, and often fatal, effects upon the organism. Unfortunately, the coronary system is especially liable to malfunctions. The coronary arteries arise at the point of maximum blood pressure in the circulatory system. The continual high pressure to which they are subjected all too frequently causes a premature wearing of these vessels. This may take the form of a loss of elasticity in the arterial walls, which reduces the amount of blood that can surge through the coronary arteries and hence diminishes the oxygen supply to the heart. Frequently, fatty deposits accumulate on the interior wall of the coronary arteries, reducing their bore and thus the amount of blood that they can carry. In either case, the oxygen supply of the heart may be inadequate in times of stress. Painful symptoms of **angina pectoris** result.

Fatty deposits in a coronary artery or arteriole may also lead to the formation of blood clots. These solid masses stop the flow of blood through the vessel and the capillary network it supplies. This is called a **coronary occlusion,** or heart attack. That portion of the heart whose blood supply is cut off quickly dies of oxygen starvation. If the area is not too extensive, the remaining portions of the heart, can, in time, compensate for the damage. All too often, however, the heart fails and death follows. Some 38% of all deaths in the United States in 1975 were caused by diseases of the heart, with males being twice as likely as females to succumb to heart disease.

11.9 THE SYSTEMIC BLOOD VESSELS

The remainder of the human circulatory system is referred to as the systemic or body system. Figure 11.7 shows the path of the major branches of this system. Blood from the aorta passes into various major arteries which lead to all portions of the body. The blood moves because of the force exerted by the contraction of the left ventricle. The surge of blood which occurs at each contraction is transmitted through the muscular, elastic walls of the entire arterial system, where it can be detected as the **pulse.** Even during the moments when the heart is relaxed (diastole) there is a definite pressure in the arterial system. When the heart contracts (systolē), the pressure increases.

The measurement of blood pressure is a common clinical test. This measurement is always expressed as a fraction, e.g. 120/80. The numerator of the fraction represents the pressure of the arterial blood during systole. The unit of measure is the **torr,** in this example the pressure equivalent to that produced by a column of mercury 120 mm high. The

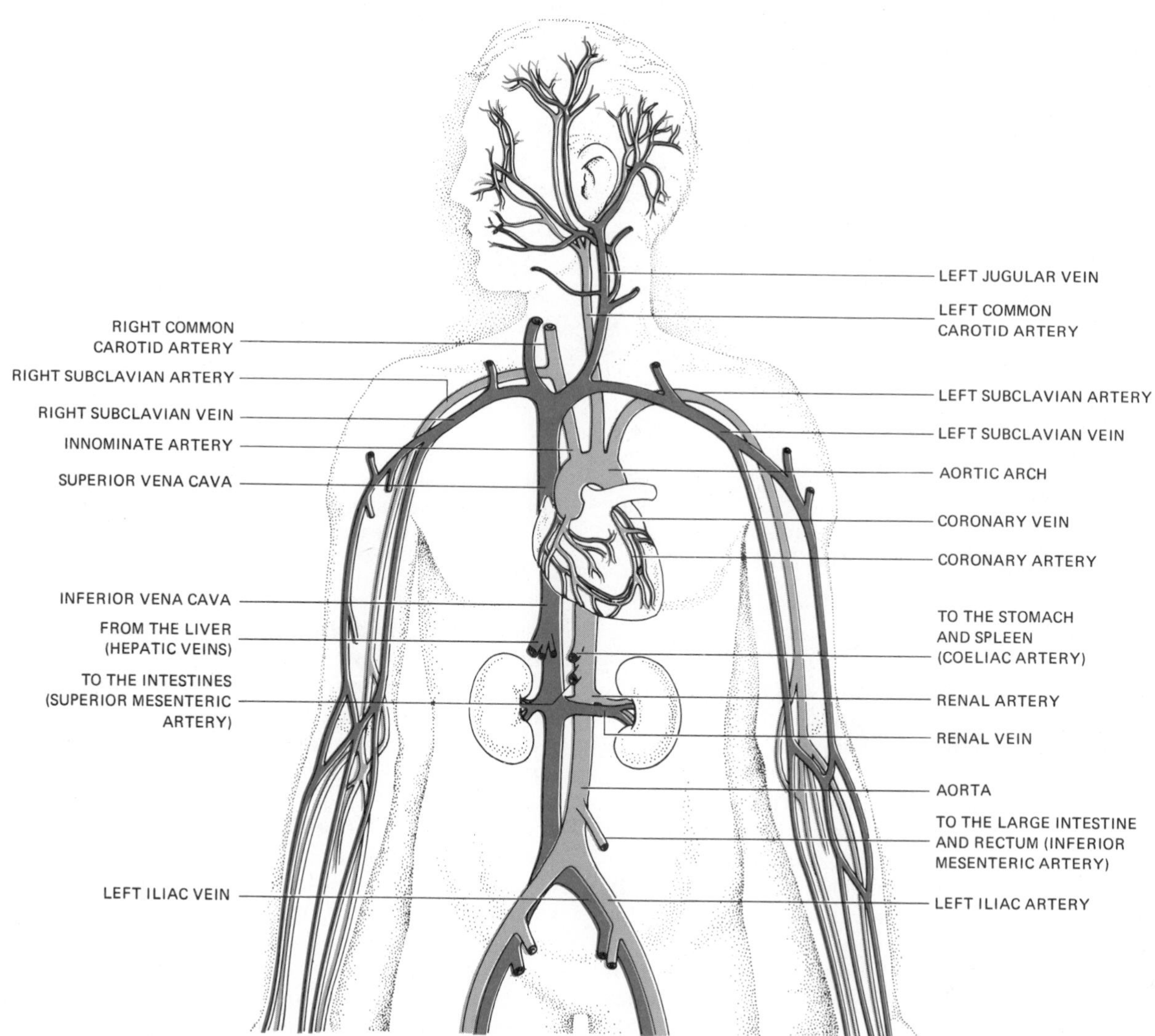

FIG. 11.7 Major blood vessels in the human. Arteries are shown in light color; veins in dark color.

denominator represents the pressure during diastole. Although blood pressure varies considerably in a given individual at different times, continual high pressure may be the symptom or cause of a variety of ailments.

When an artery is severed, the blood flows out in spurts because of the pressure exerted by the left ventricle. Severe bleeding may result. Fortunately, arteries are apt to lie somewhat deeper in the body tissues than veins and hence are not so likely to be damaged. If serious arterial bleeding should occur, however, it is important that pressure be applied on the side of the break *nearest* the heart.

11.10 THE CAPILLARIES

The pressure of the arterial blood is largely dissipated when the blood enters the capillaries. Capillaries are tiny vessels whose diameter is just about that of a single red blood cell (7.5 μm). Hence these must pass through the capillary in single file. Although the diameter of a single capillary is quite small, the number of capillaries which arise from a single arteriole is sufficiently great that the total cross-sectional area available for the flow of blood is increased. It has been estimated that there are 60,000 miles of capillaries in the adult human.

The walls of the capillaries consist of a single layer of epithelial cells. Through these walls all the exchanges of materials between the blood and the tissues take place. Although networks of capillaries are not so conspicuous as the heart and major vessels, it is in these networks that the exchange functions of the circulatory system are carried out. In a real sense, the heart and major blood vessels are just accessory parts of the circulatory system. They simply serve to supply blood to and remove blood from the vast network of capillaries.

It has been estimated that the total surface area available for exchanges in the capillary networks is 800–1000 square meters (an area greater than three doubles tennis courts). And, in addition to impressive length and area, the capillary networks (or "beds" as they are often called) of the human body have considerable volume. In fact, their total volume is roughly the same as the total volume of blood (about 5 liters) in the body. If, then, the heart and major vessels are to be filled with blood at all times, it is apparent that all of the capillaries *cannot* be filled at once. Indeed, this is the case. There is a continual redirecting of blood flow from organ to organ in accordance with the changing needs of the body.

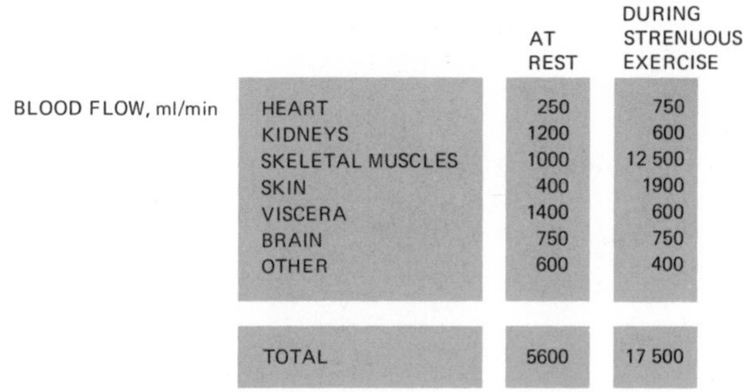

BLOOD FLOW, ml/min	AT REST	DURING STRENUOUS EXERCISE
HEART	250	750
KIDNEYS	1200	600
SKELETAL MUSCLES	1000	12 500
SKIN	400	1900
VISCERA	1400	600
BRAIN	750	750
OTHER	600	400
TOTAL	5600	17 500

FIG. 11.8 Distribution of blood in the human body at rest and during vigorous exercise. Note the increase in blood supply to the working organs (skeletal muscle and heart). The increased blood supply to the skin aids in the dissipation of the heat produced by the muscles. Note also that the blood supply to the brain remains constant. The total blood flow during exercise increases because of a more rapid heartbeat and a greater volume of blood pumped at each beat.

During vigorous exercise, for example, capillary beds in the skeletal muscles open at the expense of those in the viscera (Fig. 11.8). The reverse is true after a heavy meal.

11.11 RETURN OF BLOOD TO THE HEART

When the blood leaves the capillaries and enters the veins, there is little pressure to force it along. Blood in the veins below the heart is helped back up to the heart by the **muscle pump.** This is simply the squeezing effect that active muscles have upon the veins running through them. Of course, this squeezing effect would be useless if there were not some mechanism to insure that the blood travels in one direction. This mechanism is a system of valves that permit venous blood to flow toward the heart but prevent backflow. You can demonstrate these valves for yourself by firmly stroking the veins near the inner surface of your forearm. If you stroke away from the heart, you will force the blood back against the valves. They will close and a bulge will appear at that point (Fig. 11.9). (The walls of veins are quite thin and flabby.)

The importance of the muscle pump is dramatically illustrated when soldiers faint after standing "at attention" for long periods of time. The loss of consciousness results from an insufficient supply of blood (and hence of oxygen) to the brain. This de-

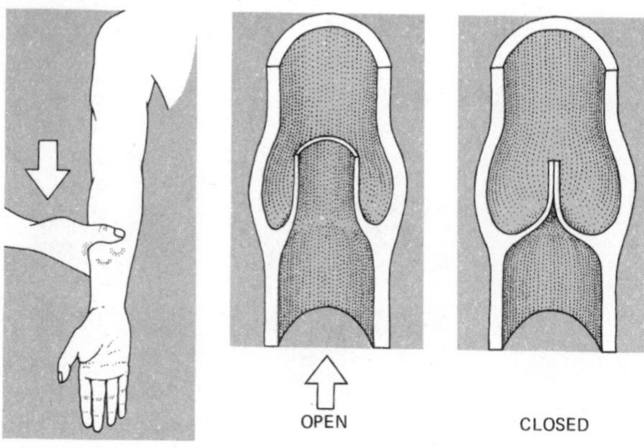

FIG. 11.9 Demonstrating valves in the arm veins. Forcing blood back against a closed valve causes the vein to bulge at that point.

velops because of a gradual accumulation of blood in the legs as a result of inactivity of the leg muscles. Fainting is a very nice self-protective mechanism in these cases. As the body moves from the vertical to the horizontal position, the heart reaches the same level as the legs. The blood in the legs no longer has to return several feet against gravity, and soon proper circulation is reestablished. Fainting is more apt to occur among recruits than among seasoned soldiers. The latter know that unlocking the knees slightly and inconspicuously flexing the calf muscles will prevent fainting. These actions aid in maintaining the flow of venous blood back to the heart.

The development of high-speed aircraft (including space capsules) has posed a number of problems of human physiology for the occupants. One of these is the gravity (or "G") force set up in pulling out of a dive or on leaving and reentering the earth's atmosphere. These G forces can seriously impede the return of blood from the lower portion of the body to the heart and cause fainting or "blackout" at a very critical time. This problem has been solved by the development of anti-G suits. The walls of these suits are inflated with compressed air when the number of G's begins to build up. The pressure exerted on the body (and hence on the veins, which generally lie quite close to the surface) forces the blood back to the heart in opposition to the G forces.

Because of the modest forces involved in getting blood through the veins, the flow is quite even. Blood leaves a cut vein in a smooth, steady stream. For this reason and because the veins lie near the surface, it is relatively easy to check the bleeding. This must be done, however, by applying pressure on the side of the cut away from the heart.

It is a common error to think of arteries as the blood vessels that carry oxygenated (bright red) blood and the veins as the vessels that carry deoxygenated (dark red) blood. While this is usually the case, there is one striking exception in humans. Can you think of what it is? Arteries must therefore be defined as those vessels that carry blood *away from* the heart. Veins are those vessels that carry blood *toward* the heart.

The Components of Blood

The medium of transport in the circulatory system is blood. Not only does the blood transport oxygen and carbon dioxide to and from the tissues and lungs, but it also transports other materials throughout the body. These include food molecules (e.g. glucose, amino acids), metabolic wastes (e.g. urea), ions of various salts (e.g. Na^+, Ca^{++}, Cl^-, HCO_3^-), and the hormones. Blood also serves to distribute heat in the body. In addition to its function as a transport agent, blood plays an active part in combating infective disease agents (e.g. certain bacteria) that gain access to the body.

Blood constitutes about 8% of the weight of the body. For the traditional 150-lb (about 70 kg) man, this represents a volume of about 5.4 liters. Although sophisticated techniques now exist for establishing this value, the modern techniques give about the same result as that determined in 1856 by the enterprising physiologist T. L. W. Bischoff. Bischoff took a most straightforward approach to the problem of determining the quantity of blood in a man: carefully weighing a condemned criminal before and after his execution by the guillotine.

11.12 THE BLOOD CELLS

Blood is a liquid tissue. It consists of cells (and cell fragments) suspended freely in a watery medium, the plasma. The cells and cell fragments constitute the so-called "formed" elements of the blood. They are sufficiently large that they can be observed under the light microscope. There are three types of "formed" elements: the red blood cells, or erythrocytes, the white blood cells, or leucocytes, and the platelets, or thrombocytes.

1. The **red blood cells** (RBCs) are the most numerous of the three types. Normal women possess about 4.5 million of these cells in each cubic millimeter of blood. In normal men, the average runs somewhat higher: about 5 million. However, these values can fluctuate over a considerable range, depending upon such factors as the altitude at which the individual lives (Peruvians living at an altitude of 18,000 feet may have values as high as 8.3 million) and the individual's state of health.

The RBCs are disk-shaped (Fig. 11.10). They have a diameter of 7.5 μm and a thickness at the rim of 2 μm. The center of the disk is thinner (1 μm) than the rim. This interesting "biconcave" shape speeds up the exchange of gases between the cell and the plasma.

In the adult, the RBCs are produced by "stem" cells (Fig. 11.11) located in the marrow of bones, particularly in the ribs, sternum (breast bone), and vertebrae. When first formed, the RBCs have a nucleus and not very much hemoglobin. However, as they mature, the quantity of hemoglobin in the cell increases until some 280 million molecules—representing about 90% of the dry weight of the cell—are

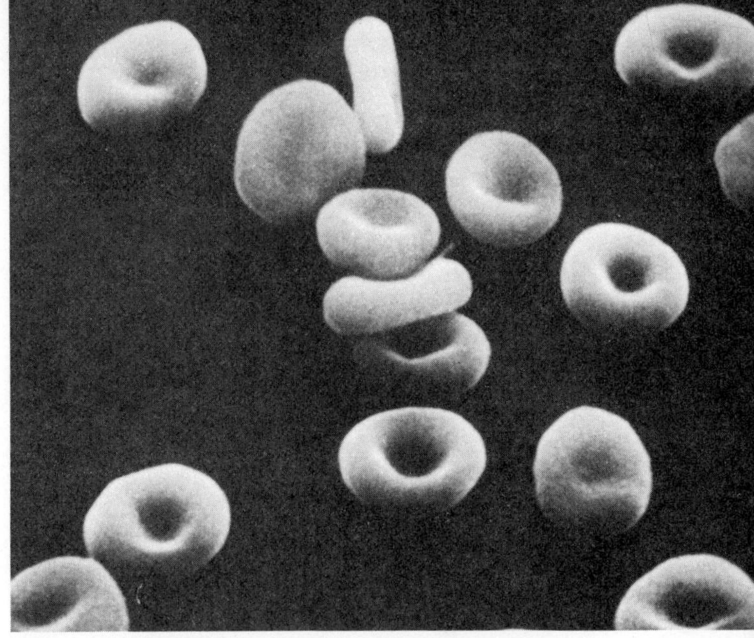

FIG. 11.10 Red blood cells seen under the scanning electron microscope. Note their characteristic biconcave shape. (Courtesy Dr. Marion I. Barnhart, Wayne State University School of Medicine. Originally published in R. M. Nalbandian, ed., *Molecular Aspects of Sickle Cell Hemoglobin, Clinical Applications,* Charles C. Thomas, Springfield, Ill., 1971.)

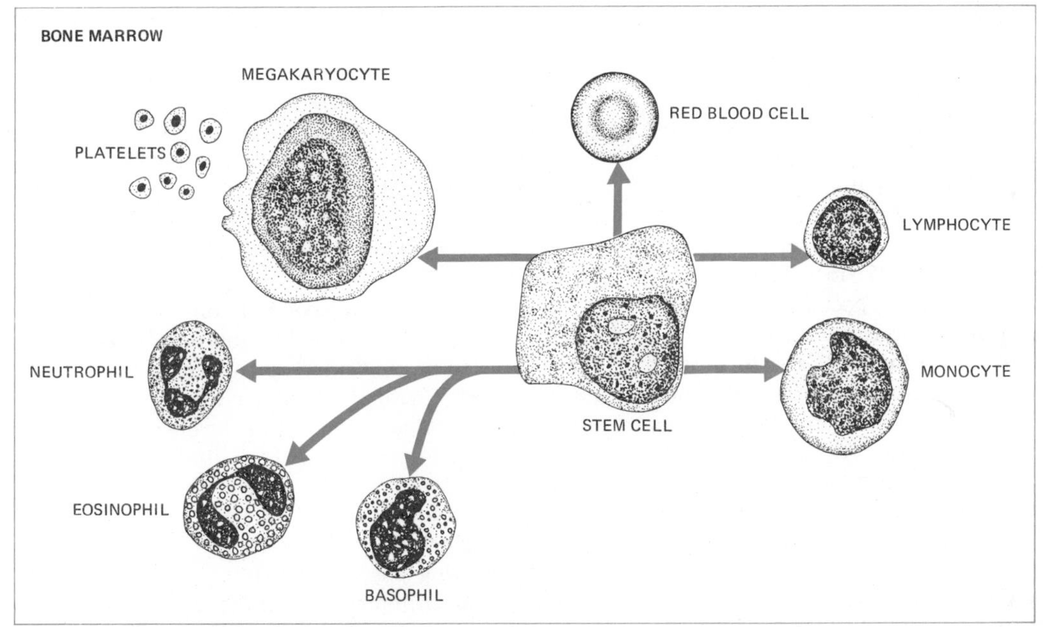

FIG. 11.11 The cells of the blood. Red blood cells transport oxygen and carbon dioxide; platelets participate in blood clotting; and the rest of the cells aid the body in combating infection. All of the blood cells are manufactured in the bone marrow, although lymphocytes and monocytes are produced in other locations (e.g. lymph nodes) as well.

present. Toward the end of this process of hemoglobin synthesis, the nucleus is squeezed out of the cell.

The life span of these cells is about 120 days. The death of a red blood cell occurs when its cell membrane ruptures as the cell is squeezed through a capillary. The fragments of the ruptured cell are then ingested by phagocytic cells present in the liver and in a sac-shaped structure called the spleen. Most of the iron of the hemoglobin is reclaimed for reuse. The remainder of the hemoglobin molecule is broken down. Some of the breakdown products—the bile pigments—are excreted by the liver in the bile. It has been estimated that three million RBCs die and are scavenged by the liver and spleen every second.

The continual loss of RBCs is normally compensated for by the action of the bone marrow. In fact, healthy bone marrow can, when necessary, produce RBCs at four or more times the normal rate of cell destruction. Thus, after severe bleeding (or blood donations), the bone marrow quickly brings the RBC content of the blood back to normal.

Under some circumstances, the rate of RBC loss exceeds the rate of RBC formation. When this occurs, the concentration of RBCs in the blood decreases, a form of **anemia.**

2. The **white blood cells** (WBCs) are much less numerous than the red, the ratio between the two types being approximately 1:700. There are actually five distinct types of WBC found in the circulating blood (Fig. 11.11). All possess nuclei. They range in size from the lymphocytes, which are not much larger (10 μm) than RBCs, to the monocytes, which may be three times as large (25 μm). The shape of the WBCs is quite variable, especially when they are being carried through the capillaries.

The general function of the WBCs is to protect the body from infection. The **neutrophils** and **monocytes** accomplish this by ingesting, by endocytosis, foreign particles (e.g. bacteria) that get into the body. To carry out this function these leucocytes, in response to a chemical attractant, squeeze themselves through the capillary walls in areas where tissue damage is occurring. Once free in the tissues, they begin phagocytosis. The bacteria or other particles are engulfed in vacuoles. These fuse with lysosomes, the enzymes of which may destroy the bacteria. The granules which are so abundant in the cytoplasm of neutrophils are the lysosomes. Usually the battle ends with the death of the white blood cell. Pus is an accumulation of dead white blood cells and the products of tissue destruction.

The life span of neutrophils is relatively short. Even in a healthy person, they die within a few days. You might well ask why these cells die so soon if there is no active infection to combat. The answer is that they are constantly engaged in the control of bacteria that are permanent residents in our mouths, air passages, large intestine, and elsewhere. Generally, this control is sufficiently good that we can tolerate the presence of these bacteria. However, a variety of stresses may temporarily lower our resistance and permit these organisms to invade our tissues. A dramatic illustration of this is the disease *agranulocytosis.* In this disease, the production of new leucocytes ceases. Heavy exposure to radiation is one of several causes of agranulocytosis. With leucocyte production stopped, the bacteria normally held in check in our bodies soon begin to multiply. After two days, ulcers of the mouth and large intestine, or serious lung infection, result. These infections soon spread throughout the body. Unless treatment with antibiotics is started quickly, death follows within a few days. Agranulocytosis is one more peril in our atomic age. It is also a fine illustration of the importance of our leucocytes in maintaining an uneasy truce with the various species of bacteria we harbor within our bodies.

Lymphocytes generally do not carry on endocytosis. They do, however, combat disease by participating in the formation of **antibodies.** Antibodies are proteins that are produced whenever foreign macromolecules enter the body. Such foreign molecules are called **antigens.** Foreign protein, polysaccharide, and nucleic acid molecules can all act as antigens.

The relationship between antigens and antibodies is very specific. Each antigen stimulates the production of antibody molecules that are capable of combining directly with that antigen and, generally, no other. When antigen and antibody do combine, the resulting complex has properties different from those of its components. For example, the complex may settle out of solution and/or be more easily engulfed by endocytic cells.

An invading parasite, such as a bacterium or a virus, has macromolecules on its surface that are foreign to the host and hence antigenic. The ability to manufacture antibodies against these antigens and thus deactivate them is clearly an important weapon against invasion by disease-producing organisms. The result is immunity. The role that lymphocytes play in the immune response will be discussed in Chapter 30.

The percentage of **eosinophils** and **basophils** in the blood is normally very low. Their function is not yet fully understood although there is evidence that each plays a role in combating disease. The number of eosinophils rises markedly in certain diseases, especially those caused by parasitic worms. The number of basophils also increases during infections. Those at the site of infection release histamine, which increases the flow of blood in the area. Although there is a great deal yet to be learned about the functions of the various white blood cells, it seems fairly certain that each in its own way contributes to the body's defense against disease.

All the white blood cells are manufactured in the bone marrow. Rarely, a precursor cell becomes cancerous. As a result, the number of one or more kinds of leucocytes becomes far greater than normal, a condition known as **leukemia.** A variety of harmful symptoms (often including severe anemia) follow. In some forms of the disease, medical care can slow down the development of symptoms and prolong the patient's life for many years. But as yet there is no cure.

3. The **platelets** are cell fragments produced by large cells (megakaryocytes) in the bone marrow. Platelets are disk-shaped and much smaller (2 μm) than the RBCs. Normally, there are 150,000–400,000 platelets in each cubic millimeter of blood. They are very important in the process of blood clotting.

11.13 THE PLASMA

The fluid in which the blood cells are suspended is a straw-colored liquid called plasma. The major component of blood plasma is water (Fig. 11.12). Dissolved in it is a large variety of molecules and ions. These include glucose, which serves as the prime source of energy for our cells, and amino acids. After a fat-rich meal, fat droplets are transported in the plasma. In addition to food molecules, there are the waste products of cell metabolism. Vitamins and hormones are also present in the blood. A number of ions are present of which sodium (Na^+) and chloride (Cl^-) ions are the most abundant. Most of these materials are in transit in the blood, that is, they are being transported from a place where they are added to the blood (a "source") to a place where they will be removed from the blood (a "sink"). Sources include exchange organs, such as the intestine, and depots or reserve supplies within the body. The liver, for example, stores a number of substances such as glucose and two vitamins, for release to the blood as needed (i.e. between meals). Every cell in the body serves as a sink for certain materials. In addition, exchange organs, like the kidney, lungs, and skin remove materials from the blood for deposit back into the external environment.

CHEMICAL COMPOSITION OF BLOOD PLASMA

COMPONENT	PERCENT
H_2O	90
INORGANIC SALTS	LESS THAN 1
MAJOR PROTEINS	7*
OTHER SUBSTANCES (FOODS, WASTES, HORMONES, ETC.)	2

* Serum albumin 4%, serum globulin 2.7%, fibrinogen 0.3%

FIGURE 11.12

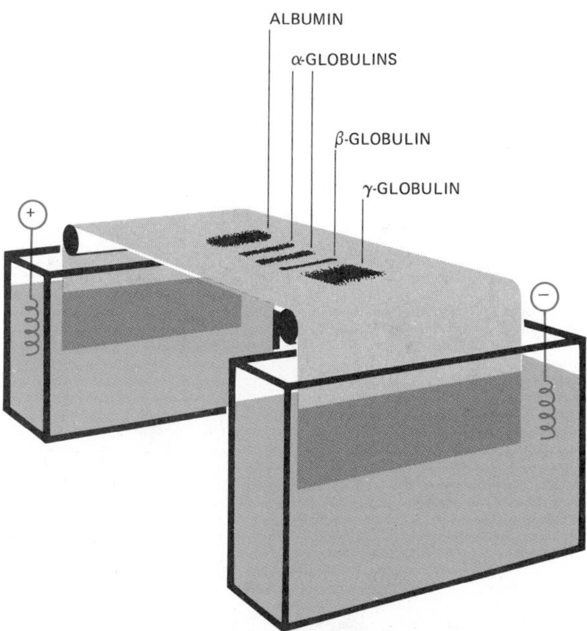

FIG. 11.13 Separating serum proteins by electrophoresis. The albumin molecules move most rapidly while the current is turned on. The slowest-moving proteins are the gamma globulins. Most of the gamma globulins are antibodies.

Some 7% of the plasma consists of protein molecules. These include **fibrinogen,** an essential component of the clotting process. After blood is withdrawn from a vein and allowed to clot, the clot slowly shrinks. As it does so, a clear fluid, called **serum,** is squeezed out of the clot. Serum is basically blood plasma without fibrinogen.

One convenient way to separate the various proteins that still remain in the serum is by a technique known as electrophoresis. A drop of serum is placed on a thin sheet of supporting material (paper is good) which has been soaked in a slightly alkaline, dilute salt solution. Under these conditions all the serum proteins are negatively charged, but some more strongly so than others. A direct current can flow through the paper, thanks to the conductivity of the salt solution with which it is moistened. As the current flows, the serum proteins move toward the positive electrode (Fig. 11.13). The stronger the negative charge on a particular protein, the faster it migrates. After a period of time, the current is turned off, and the proteins are stained to make them visible (most are otherwise colorless).

Inspection of the electrophoretogram reveals a number of distinct bands, each of which represents protein molecules of a particular charge. The most prominent of these bands and also the one that moves closest to the positive electrode is **albumin** (Fig. 11.14). Albumin is manufactured in the liver. Its chief function is to maintain normal blood volume and hence blood pressure. How it does this will be described later in the chapter.

The other protein bands that are found are various **globulins.** Of special interest to us are the least negatively charged globulins. These, the gamma globulins, are particularly abundant following infections or immunizations. This is because antibodies are gamma globulins. Sometimes gamma globulins which have been separated from the blood of many donors are given to persons exposed to certain diseases such as mumps or polio. In this way, the patient acquires temporary protection against the disease because of the antibodies present.

Functions of Blood

The two major functions of the blood are: (1) to transport materials to and from all the tissues of the body and (2) to defend the body against infectious disease. The second function was mentioned in the discussion of the white blood cells and will be considered further in Chapter 30. The first function concerns us here. It is the great solvent power of the water in the plasma that makes blood such an effective transport medium. Glucose, amino acids, short-chain fatty acids, vitamins, hormones, nitrogenous wastes (e.g. urea), and many ions are all

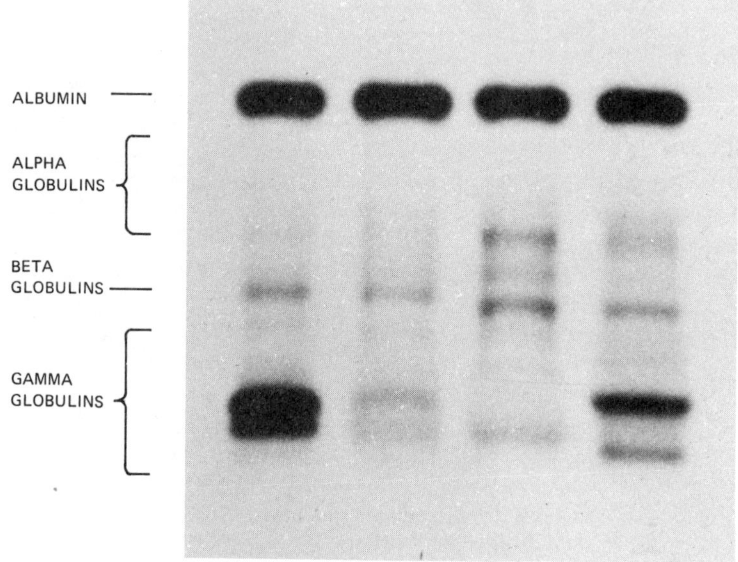

FIG. 11.14 Results of the separation by electrophoresis of the serum proteins from four different rabbits. Each rabbit had been intensively immunized with a vaccine containing killed pneumonia bacteria. This accounts for the large quantities of gamma globulins (especially pronounced in the first and fourth samples) visible at the bottom. The protein bands at the top are albumin. Alpha and beta globulins are also visible.

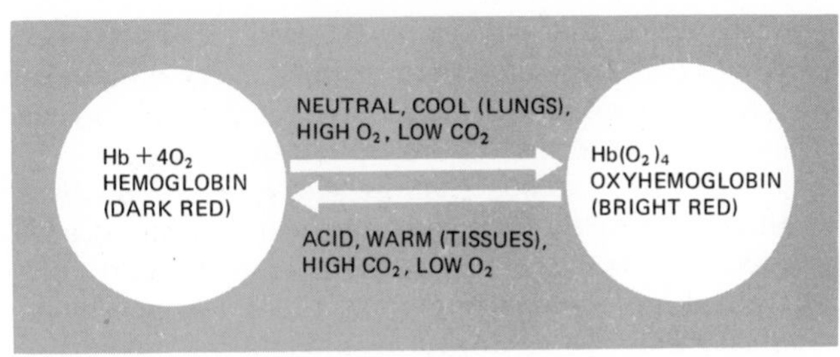

FIG. 11.15 Reversible reaction between oxygen and hemoglobin. Were it not for its *variable* affinity for oxygen, hemoglobin would be useless as an oxygen transport agent. Hemoglobin also transports carbon dioxide from the tissues to the lungs.

transported dissolved in the plasma of the blood. Only in the case of gas transport is the plasma incapable of handling the job alone. It is the RBCs that carry out the bulk of oxygen and carbon dioxide transport.

11.14 OXYGEN TRANSPORT

As much as 90% of the dry weight of a red blood cell consists of the red pigment **hemoglobin.** Hemoglobin (Hb) is a protein consisting of four polypeptide chains to each of which is attached a prosthetic group, *heme* (Fig. 8.9). At the center of each heme is an atom of iron.

Oxygen and hemoglobin combine readily, one molecule of oxygen combining with each heme group. Thus the presence of Hb in the bloodstream greatly increases the amount of oxygen that can be carried by it. The affinity of Hb for oxygen is, however, only part of the story of oxygen transport. Many substances combine vigorously with oxygen. What makes hemoglobin useful as an oxygen transport agent is that it releases the oxygen again to the tissues. In other words, the reaction between hemoglobin and oxygen is freely reversible (Fig. 11.15). Under the conditions of temperature, pH, and increased oxygen pressure existing in the capillaries of the lungs, the reaction proceeds to the right. The purple-red hemoglobin of the venous blood becomes converted to the bright red oxyhemoglobin of the arterial blood. Under the conditions of temperature, pH, and reduced oxygen pressure that exist in the capillaries of the tissues, the reverse reaction is promoted and the oxyhemoglobin gives up its oxygen (Fig. 11.16).

One liter of arterial blood carries about 200 ml of oxygen. (Of this, only 3 ml is carried in solution in the plasma; all the rest is carried in the red blood cells.) In the resting human, only about one-fourth

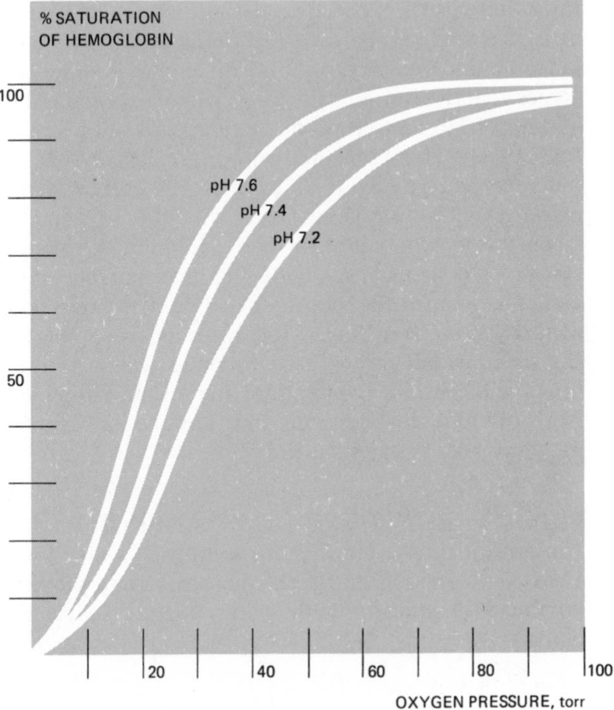

FIG. 11.16 Reversible reaction between hemoglobin and oxygen as a function of oxygen pressure and pH. The pressure of oxygen in the lungs is 90–95 torr; in the interior tissues it is about 40 torr. Therefore, only a portion of the oxygen carried by the red blood cells is normally unloaded in the tissues. However, vigorous activity can lower the oxygen pressure in skeletal muscles below 40 torr, which causes a large increase in the amount of oxygen released. This effect is enhanced by the high concentration of carbon dioxide in the muscles and the resulting lower pH (7.2). The lower carbon dioxide concentration (and hence higher pH) at the lungs promotes the binding of oxygen to hemoglobin and hence aids in the uptake of oxygen.

Temperature changes also influence the binding of oxygen to hemoglobin. In the relative warmth of the interior organs, the curve is shifted to the right (like the curve for pH 7.2), helping to unload oxygen. In the relative coolness of the lungs, the curve is shifted to the left, aiding the uptake of oxygen.

(50 ml) of this oxygen is released to the tissues. About 5 liters of blood are pumped every minute by a resting human, and hence about 250 ml of oxygen are released to the tissues in that time.

During vigorous exercise, however, as much as three-fourths of the oxyhemoglobin in the tissues gives up its oxygen. Furthermore, the heart of a trained athlete may pump five times as much blood during exercise as during rest. Thus the amount of oxygen delivered to the tissues may increase to as much as 3750 ml (= 3.75 liters) per minute, or over 15 times that delivered during rest.

Such a performance would be impossible without the presence of the red blood cells. One liter of plasma alone can carry only 3 ml of oxygen. About 1.8 ml of this dissolved oxygen is released in the tissues. A quick calculation will show that if humans depended on plasma alone for oxygen transport they would have to pump 28 times as much blood every minute, even when at rest, to satisfy the demands of their tissues.

The importance of the loose, reversible association between hemoglobin and oxygen is dramatically illustrated in carbon-monoxide poisoning. Carbon monoxide, like oxygen, combines readily with hemoglobin but, in so doing, prevents the Hb from also combining with oxygen. Carbon monoxide is not, however, released readily by the hemoglobin. Thus in an individual exposed to carbon monoxide (such as in a closed garage with a running automobile engine), more and more of the hemoglobin soon becomes inactivated for the transport of oxygen. The affinity of Hb for carbon monoxide is 230 times that for oxygen. Hence, in air containing just 1/230 as much carbon monoxide as oxygen, the carbon monoxide competes on equal terms with the oxygen. In a short time, one-half of the hemoglobin in the bloodstream will become incapable of transporting oxygen. The skin of the victim of carbon-monoxide poisoning develops a cherry-red color because this is the color which hemoglobin becomes when it combines with carbon monoxide. Only prompt removal from the fouled atmosphere and application of oxygen may save the victim's life.

11.15 CARBON DIOXIDE TRANSPORT

Carbon dioxide is much more soluble in water than is oxygen. One reason for this is that carbon dioxide combines chemically with water, forming carbonic acid,

$$CO_2 + H_2O \rightleftharpoons H_2CO_3.$$

This then dissociates into a hydrogen ion (H^+) and a bicarbonate ion,

$$H_2CO_3 \rightleftharpoons H^+ + HCO_3^-.$$

If all the carbon dioxide transported by our bloodstream were simply carried in the plasma this way, the pH of the blood would be lowered from its normal level of pH 7.41 to about pH 4.5. This would be instantly fatal. However, not more than 5–10% of the carbon dioxide produced by the tissues is actually transported in this fashion. The red blood cells accomplish the transport of the rest. About 25% of the carbon dioxide actually combines with hemoglobin in the RBCs, forming carbaminohemoglobin. The carbon dioxide does not join the hemoglobin at the same place on the molecule that oxygen does, but the release of oxygen from hemoglobin in the tissues does increase somewhat its ability to carry carbon dioxide. Similarly, in the lungs, the pickup of oxygen by hemoglobin promotes the release of carbon dioxide from it.

The RBCs assist in the transport of carbon dioxide in still another way. When carbon dioxide enters the red blood cell, some of it combines with water within the cell, forming carbonic acid. The reaction is greatly speeded by an enzyme, carbonic anhydrase, found within the cell. (This enzyme is not present in the plasma and thus, as we have seen, not much carbonic acid is found there.) The hydrogen ions released by the carbonic acid then combine with the protein portion of hemoglobin. Thus combined, they do not lower the pH. Most of the remaining bicarbonate ions then diffuse back out into the plasma. When the RBCs reach the lungs, these reactions are reversed and carbon dioxide is released to the alveolar air. Two-thirds of the carbon dioxide transported by the blood is carried in this fashion. Thanks to these two important mechanisms provided by the red blood cells, we are able to transport carbon dioxide quickly and safely from the tissues to the lungs.

11.16 EXCHANGES BETWEEN THE BLOOD AND THE CELLS

The medium of transport in the human body is blood. Our circulatory system is a closed one, however, and blood does not come in direct contact with the cells. (An exception, of course, is the epithelial cells that line the walls of the arteries and veins and make up the walls of the capillaries.) The number

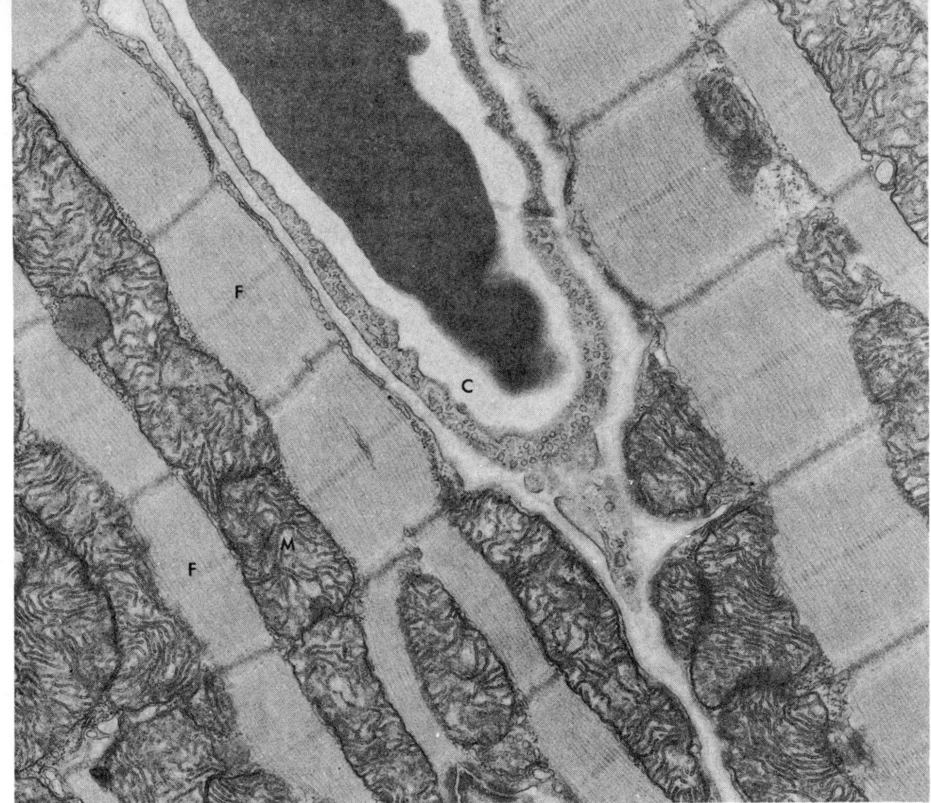

FIG. 11.17 Capillary between two adjacent heart muscle cells (17,000 ×). Rows of mitochondria (M) can be seen between the contractile units (F). Note the thinness of the epithelial cell that forms the wall of the capillary (C). The object within the capillary is a red blood cell. (Reproduced with permission from Keith R. Porter and Mary A. Bonneville, *An Introduction to the Fine Structure of Cells and Tissues,* 4th ed., Lea and Febiger, Philadelphia, 1973.)

and distribution of capillaries is such that probably no cell is ever farther than 50 μm from a capillary (Fig. 11.17). Nevertheless, materials must cross this gap between the blood and the cells. This is accomplished, in large measure, by diffusion. However, some blood components also flow out of the capillaries. When blood enters the arteriole end of a capillary, it is still under the pressure (about 35 torr) produced by the contraction of the ventricle. As a result of this pressure, some filtration of the blood through the walls of the capillaries takes place. Between the epithelial cells that make up the wall of the capillary are tiny pores. It is through these pores that filtration of the blood occurs. The pores are sufficiently small that no formed elements get through (except when migration of WBCs occurs). However, a substantial amount of water and some of the proteins in the blood plasma pass through the capillary wall and into the cell-containing region, the tissue space (Fig. 11.18). This fluid, called **interstitial fluid,** is simply blood plasma minus most of the proteins. In the tissue space, it bathes the cells. Any substances present in this fluid can then pass into the cells by diffusion or active transport. Also, any materials present in excess amounts in the

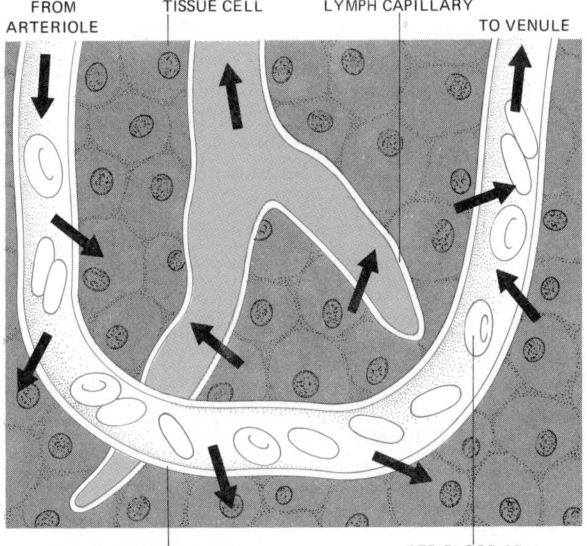

FIG. 11.18 Mechanism of exchange between blood and cells. The cells are bathed in interstitial fluid filtered from the blood. Some returns directly to the blood by diffusion into the venule end of the capillary. The rest enters a lymph capillary and, now called lymph, returns by way of the lymphatic system.

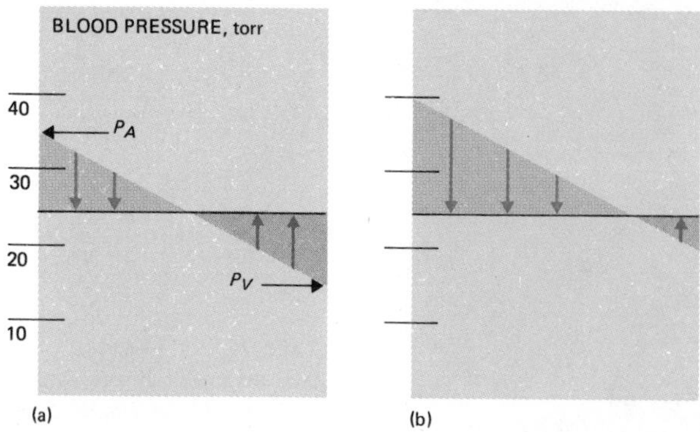

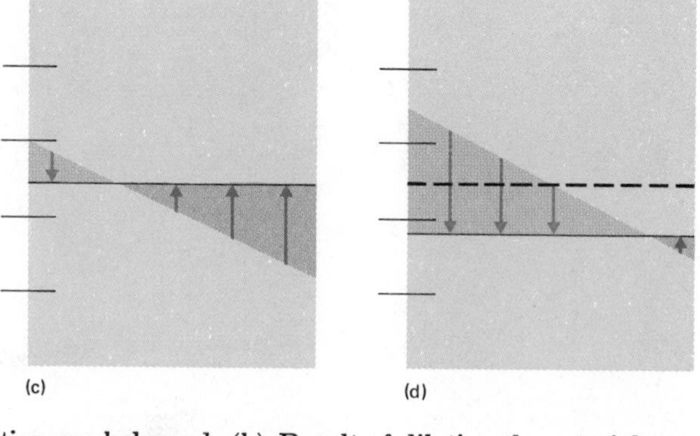

FIG. 11.19 Pressure relations in the capillaries. P_A = blood pressure at the arteriole end of the capillary; P_V = blood pressure at the venule end of the capillary. The horizontal line represents the osmotic pressure of the blood. When the blood pressure is greater than the osmotic pressure, filtration of interstitial fluid occurs (↓). When the blood pressure is less than the osmotic pressure, reabsorption of interstitial fluid occurs (↑). (a) the normal situation. Filtration and reabsorption are balanced. (b) Result of dilating the arterioles. P_A increases and the tissue space becomes engorged with interstitial fluid. (c) Result of constricting the arterioles. P_A decreases and interstitial fluid is withdrawn from the tissue space. (d) Result of a lowered concentration of protein in the blood (such as occurs during prolonged protein malnutrition). Because of the reduced osmotic pressure, fluid accumulates in the tissue space (edema).

cells (e.g. carbon dioxide or nitrogenous wastes) can diffuse out into the interstitial fluid.

Near the venous end of a capillary, the blood pressure is greatly reduced (about 15 torr). Here, another force comes into play. Although the composition of interstitial fluid is similar to that of blood plasma, it does contain a smaller quantity of proteins than plasma and thus a somewhat larger quantity of water. Because of this difference, an osmotic pressure is established. The osmotic pressure is quite small (about 25 torr) but it is greater than the blood pressure at the venous end of the capillary. Consequently, the fluid reenters the capillary here (Fig. 11.19).

It is interesting to note that most capillaries curve around a tissue space in a horseshoe shape. Thus fluid formed at the arteriole end of a capillary flows across the tissue space, bathing the cells, and reenters the same capillary at the venule end. Where this structural arrangement is not found, the venule end of some other capillary is located across the tissue space.

11.17 THE LYMPHATIC SYSTEM

From one-third to two-thirds of the plasma entering a capillary passes into the tissue space. Although most of this returns to the venule end of the capillary, some does not.

The small amount of fluid that remains is apt to be particularly rich in serum proteins which filtered through the capillary walls. These do not pass back into the capillary by diffusion. Fortunately, the bloodstream does not lose this fluid and protein for very long. Excess fluid in the tissue spaces is picked up by tiny vessels called lymph capillaries. The cells forming the walls of these vessels are loosely fitted together, thus making the wall very porous. The cells are also very active in endocytosis and exocytosis. Even large protein molecules pass easily from the tissue space into the interior of the lymph capillary. (The lymph capillaries of the intestinal villi, called lacteals, also pick up fat droplets—see Section 7.9).

The lymph capillaries lead into still larger vessels which make up the lymphatic system. The flow through the lymph vessels is quite slow. It is accomplished in a manner similar to that of venous blood. Muscular activity compresses the lymph vessels and squeezes the fluid, now called **lymph,** along. The lymph can flow in only one direction because of the many valves present.

All the lymph collected from the entire left side of the body, plus the lymph collected from the digestive tract and the right side of the lower part of the body, flows into a single major lymph vessel, the **thoracic duct.** This duct empties about 100 ml of lymph every hour into the left subclavian vein (Fig.

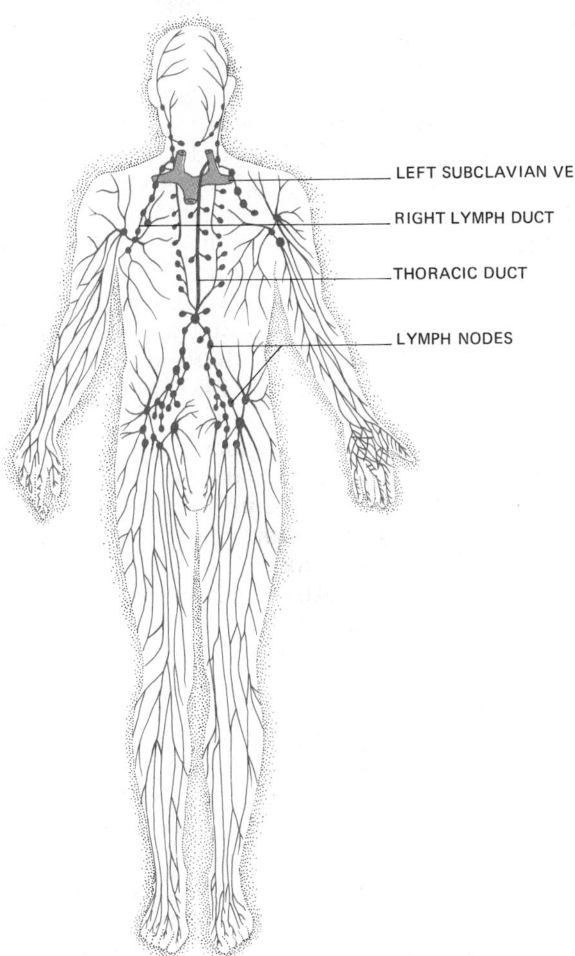

FIG. 11.20 Human lymphatic system. It returns valuable plasma proteins to the blood and aids in combating infection.

FIG. 11.21 Elephantiasis. A heavy infection with filarial worms (nematodes) has so damaged the lymph vessels of one leg that they can no longer drain lymph from it. The resulting chronic edema and thickening of the skin cause increasing deformity. The worms are transmitted from person to person by mosquitoes.

11.20). The lymph produced in the right side of the head, neck, and chest is collected by the right lymph duct and empties into the right subclavian vein. In this way, lymph with its content of protein is returned to the bloodstream.

A number of things may upset the normal production or flow of lymph. An increase in blood pressure in the capillaries or a decrease in plasma proteins (such as may follow prolonged malnutrition) will result in the production of abnormally great quantities of lymph (Fig. 11.19). The lymphatic system may be unable to handle the increased lymph production successfully, and the lymph will begin to accumulate in the tissues and to distend them. This condition is known as **edema.** Another cause of edema is blockage of the lymph vessels. In the tropics this may occur as a result of infection with a parasitic roundworm, the filarial worm. The resulting edema may cause portions of the body, such as the legs or arms, to become grossly enlarged, a condition known as elephantiasis (Fig. 11.21).

Scattered at various places in the lymphatic system (especially in the groin, armpits, abdomen, and neck) are several hundred lymph nodes (Fig. 11.20). These contain cavities, or sinuses, into which the lymph flows. The walls of the sinuses are lined with endocytic cells which engulf any foreign particles that might be present in the lymph. This mechanism protects the bloodstream from invasion by bacteria. It is one of the important body defenses against infectious disease. When combating a heavy infection, the lymph nodes may swell, resulting in "swollen glands." Lymph nodes also manufacture lymphocytes which then enter the bloodstream at the subclavian veins.

The Control of Circulation

Earlier in the chapter we mentioned that the amount of blood pumped by the heart increases with exercise and that the blood supply in the capillary beds varies from time to time over wide ranges. It is not surprising that a system as important as the circulatory system should be flexible in its operation so as to meet the changing needs of the body. This flexibility arises from a well-integrated system of controls.

11.18 THE HEART

The heart itself plays an important role in determining how much blood is pumped in a given period of time. During rest, the heart beats about 70 times a minute in the adult male while pumping about 5 liters of blood. The stimulus that maintains this rhythm is entirely self-contained. Embedded in the wall of the right atrium is a mass of specialized heart tissue called the sino-atrial (S-A) node (Fig. 11.22). The S-A node is often called the **pacemaker** of the heart because it establishes the basic rhythm at which the heart beats. The fibers of heart muscle, like all cells, are charged positively on the outside and negatively on the inside. In the pacemaker this charge breaks down spontaneously about 70 times each minute. This, in turn, initiates a similar breakdown in the nearby muscle fibers of the atrium. A tiny wave of current sweeps over the atria, causing them to contract. When this current reaches the region of insulating connective tissue between the atria and the ventricles, it is picked up by the atrial-ventricular (A-V) node. This leads to a system of branching fibers that carry the current to all parts of the ventricles, which then contract vigorously. The entire operation constitutes **systole.** A period of recovery follows in which the heart muscle and S-A node become recharged. The heart muscle relaxes and the atria fill during this period **(diastole).**

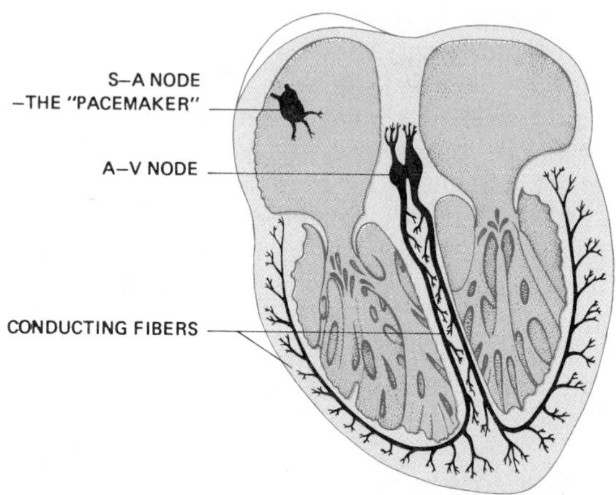

FIG. 11.22 The pacemaker generates electrical impulses that trigger contraction of the heart. Battery-operated pacemakers are now routinely implanted in persons with defects in their own electrical system. (S-A = sino-atrial; A-V = atrial-ventricular.)

The electrical activity of the heart can be detected by electrodes placed at the surface of the body and, after amplification, can be displayed on a cathode ray tube (like a TV tube) or, for a permanent record, on a piece of graph paper. Analysis of such electrocardiograms (ECGs) frequently makes possible accurate diagnosis of the extent of heart damage following a heart attack. This is because death of a portion of the heart muscle prevents electrical transmission through that area and the resulting interruption of the smooth propagation of the impulse alters the appearance of the ECG (Fig. 11.23).

Damage to the pacemaker does not necessarily result in heart failure. Even without the pacemaker, the ventricles can maintain a beat, although it is considerably slower. There is, however, a danger that impulses arising in the ventricles may become disorganized and random. When this happens, the ventricles begin to twitch spasmodically and cease to pump blood. This is known as ventricular fibrillation. Death follows quickly unless immediate measures are taken to start rhythmic beating again. In the past this has meant opening up the chest cavity and massaging the heart by hand. Obviously such heroic measures have usually been confined to cases of ventricular fibrillation occurring while the patient was under surgery.

Ventricular fibrillation is probably the immediate cause of death in 25% of all cases. Although most cases arise spontaneously in damaged hearts, moderately severe electric shocks can also cause fibrillation. Unfortunately, the 60-cycle alternating current in such widespread use today is especially effective at this.

A person who seems to be unduly predisposed to fibrillation, because of a heart attack or for other reasons, can now be given an artificial pacemaker. This is simply a device that generates rhythmic impulses which are transmitted to the heart by fine wires. Thanks to miniaturization techniques and long-lived batteries, modern pacemakers are now implanted in the body, close enough to the skin to be reached through a small incision when maintenance is needed. The more sophisticated pacemakers are even able to adjust the frequency of their output to correspond to the varying demands placed on the heart as the patient alters the level of his or her physical activity.

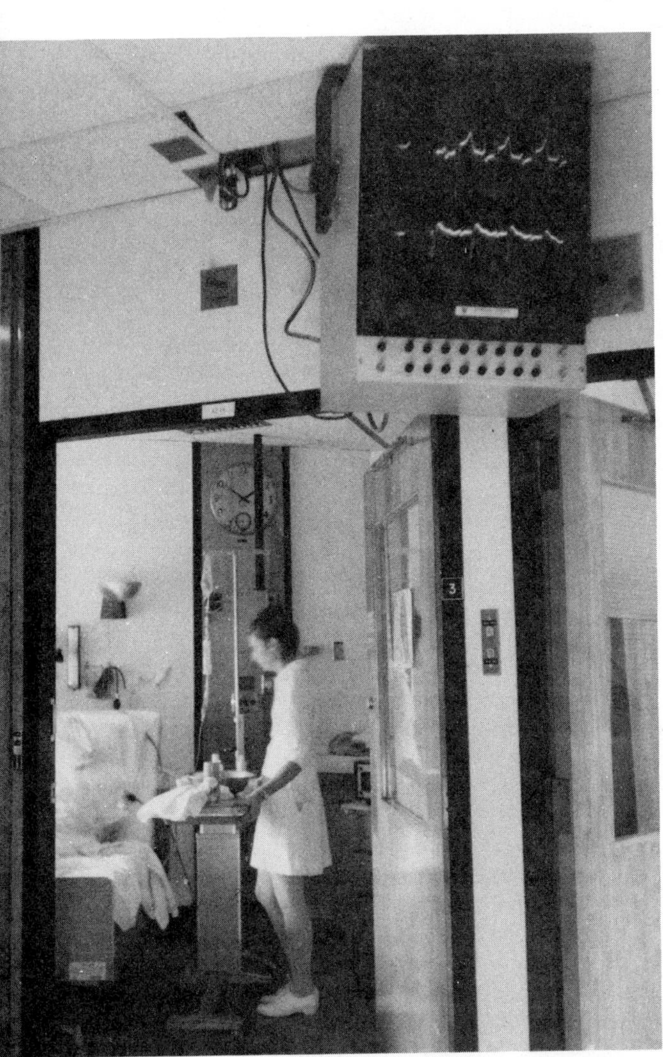

FIG. 11.23 The electrical activity of the hearts of two victims of heart attacks is displayed continuously on the TV-like screen at the top of the photograph. The electrocardiograms (ECG's) of as many as six patients can be displayed simultaneously in this coronary care unit. In addition, individual bedside monitors can be set to trigger an alarm if a patient's ECG changes ominously. The photograph was taken in the S. A. Levine Cardiac Center of the Peter Bent Brigham Hospital in Boston. (Courtesy of John Withee.)

11.19 AUXILIARY CONTROL OF THE HEART

Although the pacemaker sets the basic rhythm of the heart, the rate and strength of beating can be modified by other factors. Located in the medulla of the brain are two auxiliary control centers. These initiate impulses which travel to the heart by way of nerves.

Impulses travelling to the heart by way of the *accelerator* nerves cause an increase in the rate and strength of the heartbeat and hence an increase in the rate of flow of blood through the body. These impulses are most likely to arise when the individual is subject to some stress such as fear or violent physical exertion. The heartbeat may increase to 180 beats per minute. Simultaneously, the strength of contraction increases. The total quantity of blood pumped by the heart may thus increase to as much as 25–30 liters per minute. Have you ever noticed your heart start to "pound" rapidly when you were frightened? The accelerator nerves are responsible for this.

Vigorous exercise accelerates heartbeat in two ways. As cellular respiration increases, so does the carbon dioxide content of the blood. This stimulates special carbon dioxide receptors located in the carotid arteries and aorta (Fig. 11.7), and these transmit impulses to the medulla for relay, by way of the accelerator nerves, to the heart. Furthermore, as muscular activity increases, the muscle pump (Section 11.11) drives more blood back to the right atrium of the heart. The atrium becomes distended with blood, thus stimulating stretch receptors located in its wall. These, too, send impulses to the medulla for relay to the heart. Working together, these two mechanisms thus ensure that the increased demands on the heart during vigorous exercise are met.

Two *inhibitory* nerves also run from the medulla to the heart. Impulses passing down these, the **vagus** nerves, cause a slowing down of the heartbeat. Pressure receptors are present in the aorta and in the carotid arteries. When the blood pressure increases above normal levels, these receptors send impulses back to the medulla. The medulla, in turn, relays these through the vagus nerves to the heart. Heartbeat and blood pressure diminish. The vagus nerves thus act antagonistically to the accelerator nerves and protect the heart from unnecessary or excessive overactivity.

11.20 PERIPHERAL CONTROL OF CIRCULATION

Nerves also affect the flow of blood through the arterioles. The walls of these vessels are muscular. Their constriction reduces the bore of the vessels, thus increasing blood pressure but reducing blood flow. Dilation of the vessels has the opposite effect. For example, in times of danger or other stress, the arterioles supplying blood to the skeletal muscles will be dilated while the bore of the vessels supplying the digestive organs will be decreased. This action is carried out by nervous stimulation and also by a hormone, adrenaline, which is released into the blood itself. On the other hand, after a full meal, the blood supply to the digestive organs will be increased at the expense of that to the skeletal muscles.

Many chemical substances in addition to adrenaline have been discovered in the blood which affect the bore of arterioles and hence blood pressure. The pituitary gland and the kidneys both release substances that constrict arterioles and thus raise blood pressure. On the other hand, cells where infection or other damage is occurring release substances, e.g. histamine, that dilate the arterioles and thus increase blood flow in the area. Basophils and other cells participate in the process. Much research remains to be done on the action and interrelationships of these substances.

Although the walls of the capillaries have no muscles and hence cannot be controlled directly, the overall action of the capillary bed is under precise control. Running between the arterioles and venules is a vessel called the "thoroughfare channel" (Fig. 11.24), from which arise the true capillaries. There is a ring of muscle, the precapillary sphincter, at the point where each capillary branches from the thoroughfare channel. When it is relaxed, blood flows into the capillary. When it is contracted, blood flow through the capillary ceases. Generally, the stimuli which cause constriction of the arterioles in any given portion of the body also cause constriction of the precapillary sphincters. The same is true for stimuli which cause dilation of the arterioles.

There is also a considerable degree of *local* control over the capillary beds. When cells in a tissue space are metabolizing rapidly, they release substances that relax the precapillary sphincters. This is a very nice control mechanism because rapidly metabolizing cells need additional supplies of glucose and oxygen. They also need to have their wastes carried away more quickly. Both these needs are

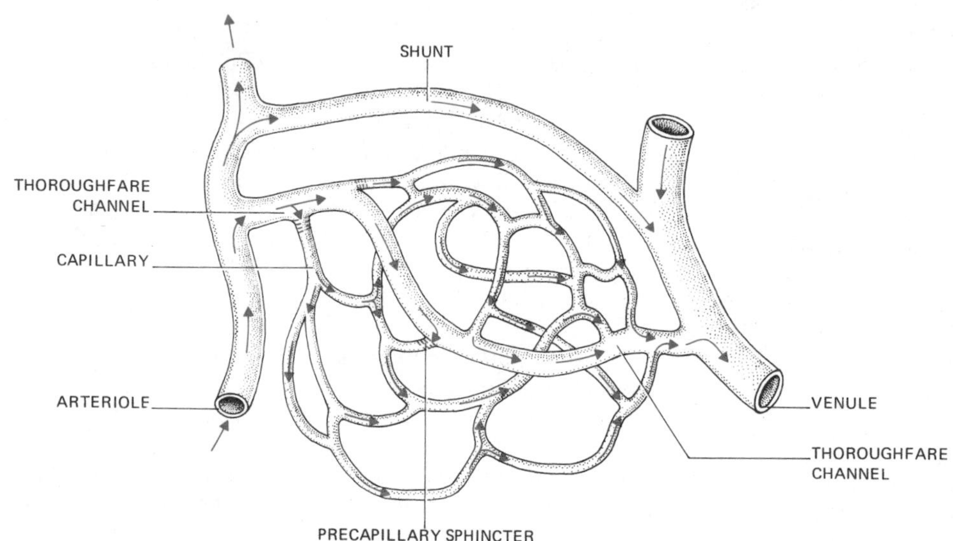

FIG. 11.24 Organization of the capillary bed. Control of blood flow is achieved by changes in the diameter of the arteriole, shunt, thoroughfare channel, and precapillary sphincters. Only *their* walls are muscular.

taken care of by increasing the blood flow through the nearby capillaries.

The total volume of blood in the body could be contained within the capillary beds alone if they were all open at once. In order to maintain adequate blood pressure and blood flow, it is important that as capillary beds open in one portion of the body, they close in another. Under a few circumstances, this may not occur. Some drugs, pain, even emotional upsets may cause a general dilation of the arterioles and opening of capillary beds throughout the body. Although the volume of blood remains normal, the blood pressure decreases markedly and the victim may go into "shock." If the victim is placed in a horizontal position, the brain and heart will continue to receive sufficient quantities of blood and the chances of recovery are excellent. If maintained in a vertical position, however, the victim will eventually die from lack of oxygen in the brain and heart. This may be the mechanism responsible for death by crucifixion.

Shock can also result from severe bleeding. The heart can pump only as much blood as it receives. If insufficient blood gets back to the heart, the output of the heart (and thus blood pressure) drops. The tissues, especially of the brain and heart, fail to receive adequate supplies of oxygen and glucose. To combat the effects of severe bleeding, arterioles become constricted and capillary beds shut down (except in the brain and heart). This reduces the volume of the system and helps maintain normal blood pressure. In the skin, the flow of blood through the capillary beds may be circumvented completely. Short lengths of vessel, normally closed, make a direct connection between arterioles and venules. These shunts can pass arterial blood directly into the venules and thus hasten its passage back to the heart (Fig. 11.24).

Air-breathing vertebrates that spend long periods of time under water (such as seals, penguins, turtles, and alligators) employ a similar mechanism to ensure that the oxygen supply of the heart and brain is not seriously diminished. When the animal dives, the blood supply of the rest of the body is sharply reduced so that what oxygen remains in the blood will be available for those organs that need it most: the brain and the heart. The anatomy of the turtle heart provides an additional mechanism to help with this adjustment. While underwater, a good deal of the blood entering the right side of the ventricle passes through the opening in the septum (see Fig. 11.5) and right back out to the vital parts (e.g. brain) of the body. In so doing, this blood bypasses the lungs, which become less useful anyway as their oxygen content is depleted.

In mammals, the kidney is responsible for monitoring blood pressure and initiating corrective action if the blood pressure begins to drop. It does this by the release of a proteolytic enzyme, **renin,** which has a very specific activity. Its substrate is a globulin, called angiotensinogen, found in the plasma. Renin splits off a peptide (containing 10 amino acids), called angiotensin I, which can then be acted upon by a peptidase already present in the plasma. The final product is a smaller peptide, containing eight amino acids, which is called angiotensin II. Angiotensin II causes the muscular walls of the arterioles to contract. This closes down capillary beds and, by bringing the volume of the functioning blood vessels back into balance with the volume of blood, restores normal blood pressure (Fig. 11.25).

Angiotensin II has a second, indirect effect on blood volume and pressure. Perched atop each kidney is an endocrine gland, called the adrenal gland. When angiotensin II reaches the cells of the outer part of the adrenal gland, it causes them to secrete the hormone aldosterone. Aldosterone acts directly on the kidney tubules, enhancing the reabsorption of Na^+ ions. This in turn enhances the reabsorption of water and thus also aids in restoring normal blood volume and pressure.

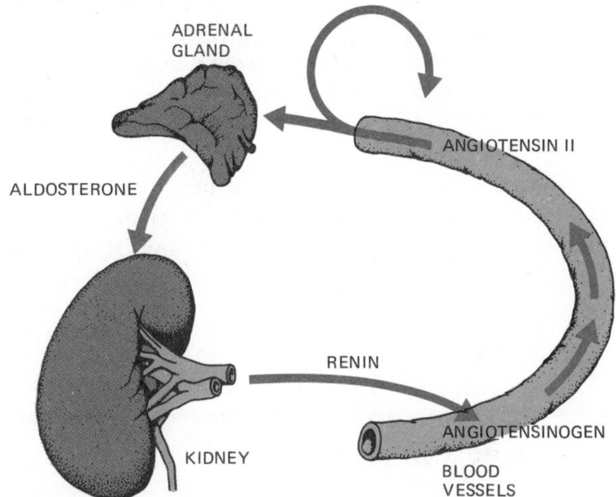

FIG. 11.25 Regulation of blood pressure by the kidney. Angiotensin II causes constriction of the blood vessels, thus raising blood pressure. Aldosterone is a hormone that increases the reabsorption of sodium and water by the kidney tubules. This effect, too, raises blood pressure.

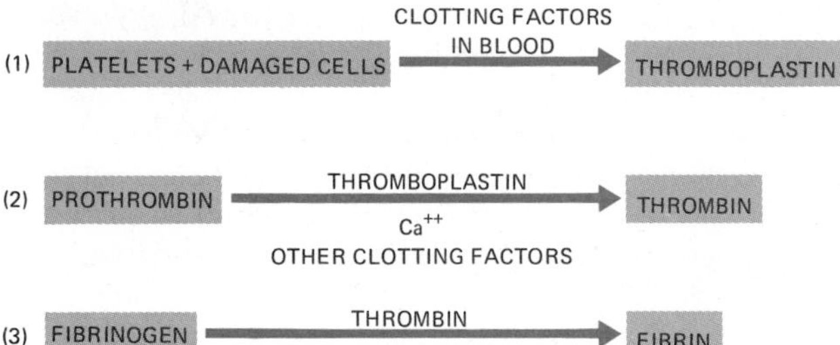

FIG. 11.26 The mechanism of blood clotting.

11.21 CLOTTING OF BLOOD

When blood vessels are cut or ruptured, it is vital that loss of blood from the system be stopped before shock and death follow. Solidification, or clotting, of the blood is able to accomplish this in all but the major vessels. As blood flows out of a damaged vessel, the platelets adhere to the inner surface of the vessel wall. Both they and damaged cells in the area release a fatty material which is activated by certain proteins (clotting "factors") in the blood to form "thromboplastin." In the presence of calcium ions (Ca^{++}) and additional clotting factors in the plasma, thromboplastin catalyzes the conversion of prothrombin (a serum globulin manufactured continuously by the liver) to **thrombin.** Thrombin is an enzyme that catalyzes the conversion of the soluble plasma protein fibrinogen to the insoluble protein **fibrin.** The fibrin gradually forms a mesh in which the blood cells become embedded. Soon a dam (the clot) is constructed which stops loss of blood from the broken vessel.

The various steps in the clotting process are not yet fully understood, and Fig. 11.26 gives no more than an overall picture. A shortage of any component in the process, e.g. platelets, a plasma clotting factor, or prothrombin, may lead to serious external and/or internal bleeding, even with minor injuries such as a slight cut or the rupture of small blood vessels by a bruise.

11.22 SUMMARY

The environment of each of the trillions of cells that make up our bodies is a fluid. This fluid, the interstitial fluid, along with the blood, makes up the ECF of the body. Although interstitial fluid is perhaps a less conspicuous part of our circulatory system than blood, it *is* the internal environment in which our cells live. Interstitial fluid is manufactured from the blood, however, and returns, after bathing the tissues, to the blood. Any change in the physical or chemical properties of the blood immediately results in a corresponding change in the interstitial fluid. In the past few chapters, we have examined several mechanisms that operate to preserve a steady, unvarying chemical composition (and a steady physical state) in the blood. Its temperature, pH, and concentration of glucose, oxygen, and carbon dioxide are all held remarkably constant by means of finely adjusted regulatory mechanisms. This constancy of the blood ensures the homeostasis of the interstitial fluid and hence of the true environment, the internal environment, of our bodies. Despite great changes in the external environment, in diet, and in the general activity of the body, our cells continue to operate under conditions that are relatively unvarying and optimum.

In the chapters ahead, we shall examine still other mechanisms by which homeostasis is maintained in the human body. But, before turning to these, let us examine how plants accomplish the transport of materials throughout their bodies.

EXERCISES AND PROBLEMS

1. Citrate ions ($C_6H_5O_7^{\equiv}$) combine with calcium ions Ca^{++}) to form an insoluble product, $2C_6H_5O_7^{\equiv} + 3Ca^{++} \rightarrow Ca_3(C_6H_5O_7)_2$. Why do you suppose that a small quantity of sodium or potassium citrate is added to blood collected for storage in blood banks?

2. Trace the path taken by a red blood cell from a capillary bed in your right thumb, to your lungs, and then back to the same capillary bed. Include all vessels, valves, and chambers through which it passes.

3. General edema is characteristic of extreme cases of protein deficiency. Can you account for this?
4. Why is it difficult to see platelets in stained blood slides?
5. Which arteries in the human carry deoxygenated blood?
6. How much blood would an adult human need to pump every minute in order to meet his oxygen needs during strenuous exercise if he had no hemoglobin?
7. Summarize the various mechanisms by which our circulatory system adjusts to the demands of strenuous exercise.
8. Why are valves needed in veins but not in arteries?
9. Oxygen-transport pigments are found in crustaceans but not in insects. Explain this difference.
10. Distinguish between blood plasma, serum, interstitial fluid, and lymph.
11. What percent of your hemoglobin has oxygen bound to it when the oxygen pressure is 50 torr and the pH is 7.4 (see Fig. 11.16)? What would be the percent at the same pressure but a pH of 7.2?
12. What oxygen pressure (in torr) would you expect to find in the tissues of a resting human (see Fig. 11.16)? What might be the pressure in the leg muscles during vigorous exercise?
13. On a graph like that in Fig. 11.16, draw your predictions of the relative locations of the hemoglobin saturation curves at 36°, 37.5°, and 39°C.
14. Draw a diagram like that in Fig. 11.19 that shows filtration and absorption in the capillary beds of (a) a person suffering from severe dehydration, (b) a person suffering from congestive heart disease (which causes pooling of the blood in the venae cavae and right atrium).

REFERENCES

1. MUIR, A. R., *The Mammalian Heart,* Oxford Biology Readers, No. 8, Oxford University Press, Oxford, 1971.
2. MAYERSON, H. S., "The Lymphatic System," *Scientific American,* Offprint No. 158, June, 1963.
3. ZWEIFACH, B. W., "The Microcirculation of the Blood," *Scientific American,* Offprint No. 64, January, 1959.
4. SCHOLANDER, P. F., "The Master Switch of Life," *Scientific American,* Offprint No. 172, December, 1963. Describes how air-breathing vertebrates reduce the blood supply to all organs but the heart and brain while underwater.
5. CHAPMAN, C. B., and J. H. MITCHELL, "The Physiology of Exercise," *Scientific American,* Offprint No. 1011, May, 1965. Describes the mechanisms by which the human body meets the demands on the circulatory system that occur during violent exercise.
6. ADOLPH, E. F., "The Heart's Pacemaker," *Scientific American,* Offprint No. 1067, March, 1967.
7. NEIL, E., *The Mammalian Circulation,* Oxford Biology Readers, No. 82, Oxford University Press, Oxford, 1975.

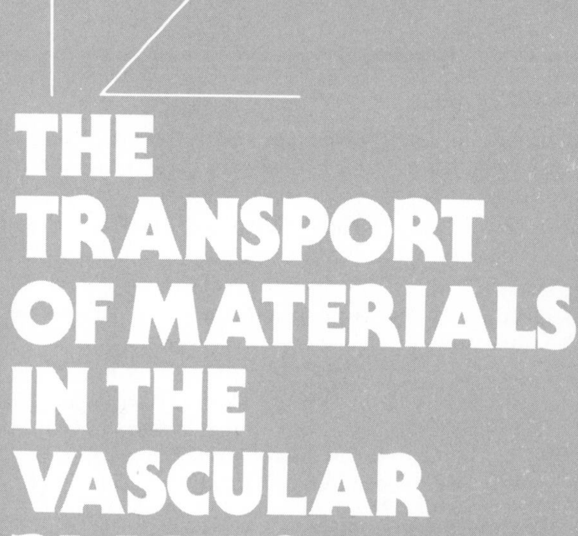

12 THE TRANSPORT OF MATERIALS IN THE VASCULAR PLANTS

12.1 IMPORTANCE

Plants are also faced with the problem of transporting materials throughout their bodies. The problem of supply to and from internal tissues is not so acute in plants as it is in animals. As we shall see, the living tissues of a plant are usually quite close to the surface. The real demand for a transport system comes about because of the curious dilemma in which land plants find themselves. A plant cannot carry on photosynthesis without both water and sunlight. It secures its water by sending a root system into the soil. It secures its light by displaying leaves in the air. The greater its success in displaying its leaves above those of competing plants, the farther it has removed them from their supply of water in the soil. An efficient vascular system is needed, therefore, to bridge the gap and transport water quickly from roots to leaves. The vascular system is also needed to ensure that food is transported efficiently from the leaves to the living cells of the stem and roots in order to satisfy their food requirements for metabolism and growth.

The transport of materials in plants is called **translocation.** It takes place in a special system of conducting vessels. These are located in groups known as **vascular bundles** which extend through all the organs of the plant: root, stem, leaf (in the veins), and flower. All the vascular bundles are interconnected so that transport between organs is carried on rapidly and efficiently. Within the vascular bundles, two distinct types of tissue are found, the xylem and the phloem.

12.2 XYLEM

The most important parts of the xylem tissue of flowering plants are the xylem **vessels** (Fig. 12.1). These are thick-walled tubes which extend vertically through several feet of xylem tissue. They range in diameter from only 20 μm to as much as 700 μm (0.7 mm), depending upon their location and the species in which they are present. The walls of the xylem vessels are thickened with secondary deposits of cellulose and are usually further strengthened by impregnation with a cementing material, lignin. The secondary walls of the xylem are not deposited evenly, but in spirals, rings, etc. The walls of xylem vessels may also be perforated by pits.

The xylem vessels arise from individual, cylindrical cells usually oriented end to end. At maturity, the end walls of these cells dissolve away and the cytoplasmic contents die. The result is the xylem vessel, a continuous nonliving duct. The xylem vessels function in the upward transport of water and minerals.

Also present in the xylem tissue are xylem **tracheids.** These are individual cells about 30 μm in diameter and several millimeters in length. In cross

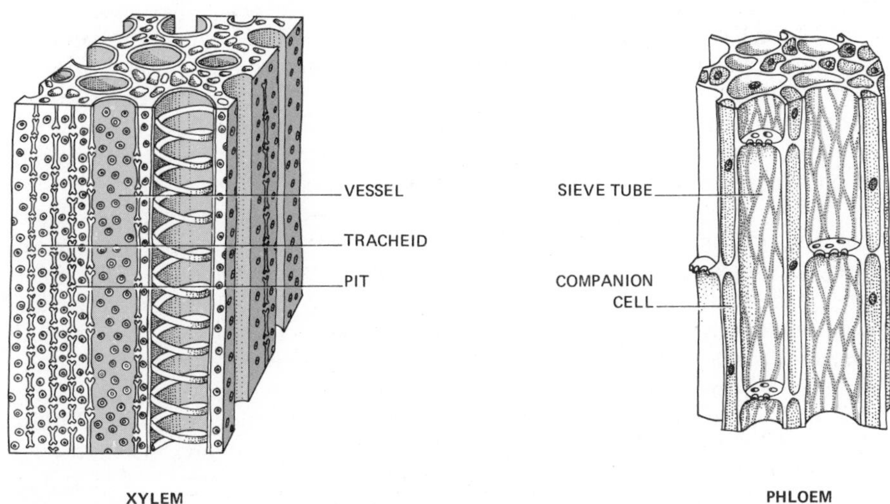

FIG. 12.1 Vascular tissues. Xylem vessels and tracheids (left) contain no cytoplasm at maturity.

section, they can be distinguished from the xylem vessels by their angular walls and, in many cases, smaller size. They taper at each end and the tapered ends of one cell overlap the tapered ends of the adjacent cells. Like xylem vessels, they possess thick, lignified cell walls and, at maturity, no cytoplasm. Their walls are perforated so that water and dissolved minerals can, and do, flow readily from one tracheid to the next. Like the xylem vessels, they transport water and minerals upward through the vascular bundle. In ferns and conifers, the xylem tracheids are the only water transport ducts. Vessels are not present.

12.3 PHLOEM

The major conducting vessels of the phloem are the sieve tubes. These consist of cylindrical cells (about 25 μm in diameter and 100 μm long) oriented end to end. The end walls of mature sieve-tube cells are perforated, permitting strands of cytoplasm to extend between adjacent cells. The appearance of these end walls under the microscope has caused them to be called sieve plates (Fig. 12.1). The side walls of the sieve-tube cells are also perforated. Fine strands of cytoplasm, the **plasmodesmata,** extend through these pores, connecting sieve-tube cells that lie side by side. Like the xylem vessels and tracheids, the sieve tubes form continuous ducts extending from the bottom to the top of the plant.

The sieve tubes differ from the xylem vessels and tracheids in that the constituent cells do not lose their cytoplasm at maturity. They do, however, lose their nucleus. In many plants, the cells of the sieve tubes lie adjacent to companion cells (Fig. 12.1) which do retain their nucleus at maturity and may possibly exert some control over the activities of the cells of the sieve tube. The sieve tubes transport foods and plant hormones both upward and downward in the plant.

The xylem and phloem make up the vascular bundles of the plant. The organization of these vascular bundles is different in the root, stem, and veins of the leaf. Furthermore, the arrangement of vascular bundles within a single organ varies considerably from species to species. In considering the internal organization of the root, stem, and the leaf we shall necessarily restrict ourselves to just a few representative types.

There are two major subdivisions of flowering plants: the dicotyledons (**dicots,** for short) and the monocotyledons (**monocots**). These names arise because of the occurrence of two food storage organs, or cotyledons, in the seeds of the first group and only one in the seeds of the second. Dicots include an enormous variety of plants. Buttercups, cabbages, legumes, cacti, tomatoes, elms, oaks, and maples are just a few of the many kinds of dicots. The monocots include lilies, palms, orchids, tulips, onions, and all the grasses. Grasses, in turn, include corn, wheat, rice, and all the other cereal grains upon which we depend so heavily for food.

Monocots and dicots differ in a number of ways other than in their seed structure. One of these is in the pattern of organization of the vascular tissues. Other differences will be touched on in Chapter 36.

12.4 THE ORGANIZATION OF THE ROOT

Figure 12.2 shows the structure of a typical young dicot root. At the tip of the root is a meristem that produces the cells from which the first (primary) root structures will develop. Mitosis in this meristem increases the length of the root. Because of the frequency with which mitosis occurs in this "embryonic" region, root tips are frequently used to demonstrate and study this type of cell division. The meristem is protected from abrasion and damage in the soil by the root cap.

As soon as cells are manufactured by the meristem, they undergo a period of elongation. The region of the root in which this occurs is known as the region of elongation.

Once the root cells are fully elongated, they begin the process of **differentiation.** Differentiation involves the development of specialized structures. Cells at the surface of the root differentiate to form epidermal cells. Most of these develop long extensions of the wall, the **root hairs.** These greatly increase the surface area of the root and are the main point of entrance for water.

Within the epidermis develops a ring (as viewed in cross section) of parenchyma cells, the **cortex.** The cortex serves as a food storage area. Its inner surface is bounded by a single layer of cells, the **endodermis.** Within the endodermis is the central cylinder containing the vascular bundles. Surrounding the central cylinder is the **pericycle,** from which branches of the root (secondary roots) arise. Within the pericycle of the young root are xylem tissue, phloem tissue, and parenchyma or pith. The xylem tissue is arranged in bundles in a radial or spokelike fashion. The phloem tissue alternates with the xylem tissue (Fig. 12.2).

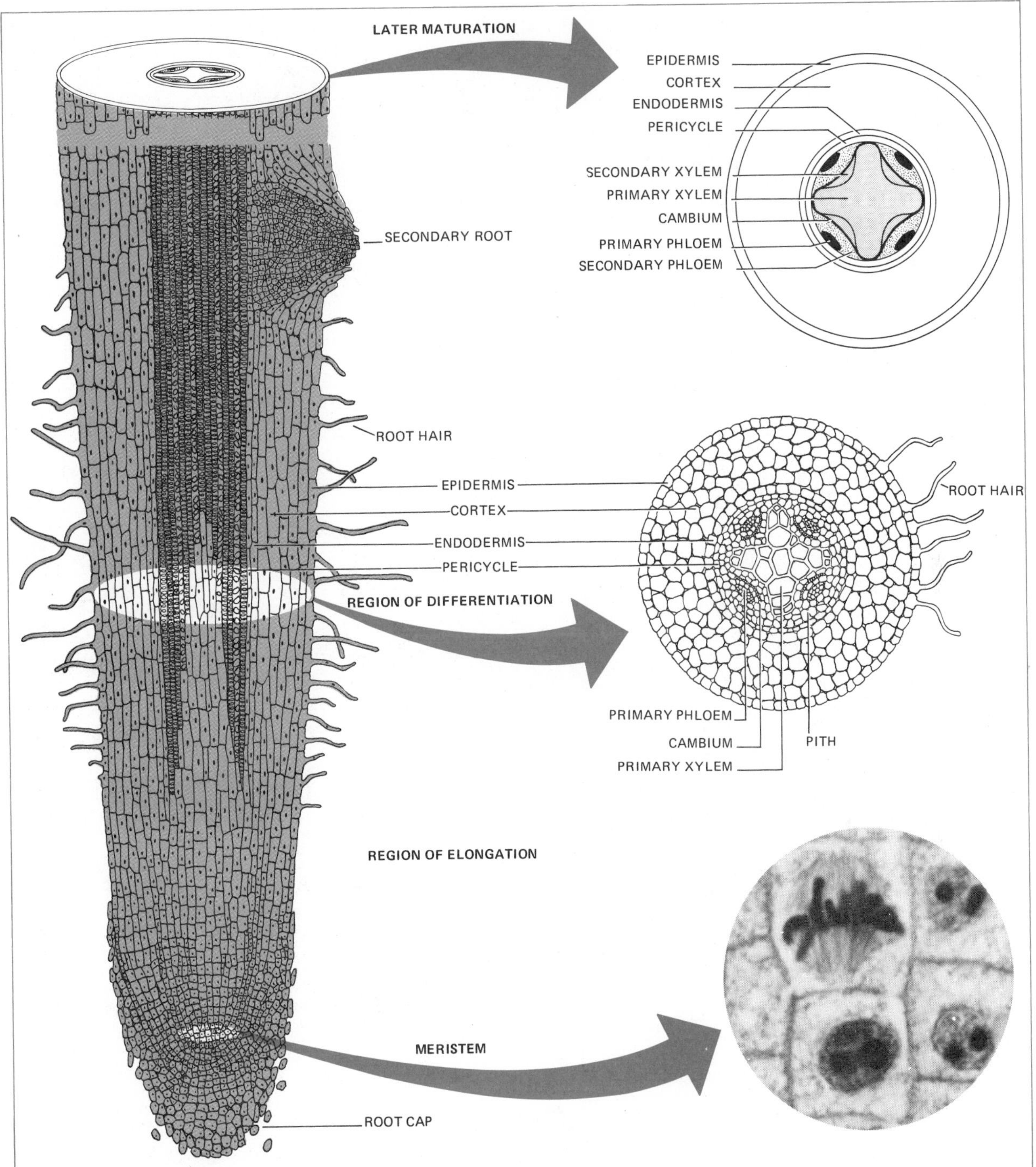

FIG. 12.2 Organization of a young dicot root in longitudinal section (left) and cross sections (right). The longitudinal section has been shortened to enable each stage in root development to be shown. (Photo courtesy CCM: General Biological, Inc., Chicago.)

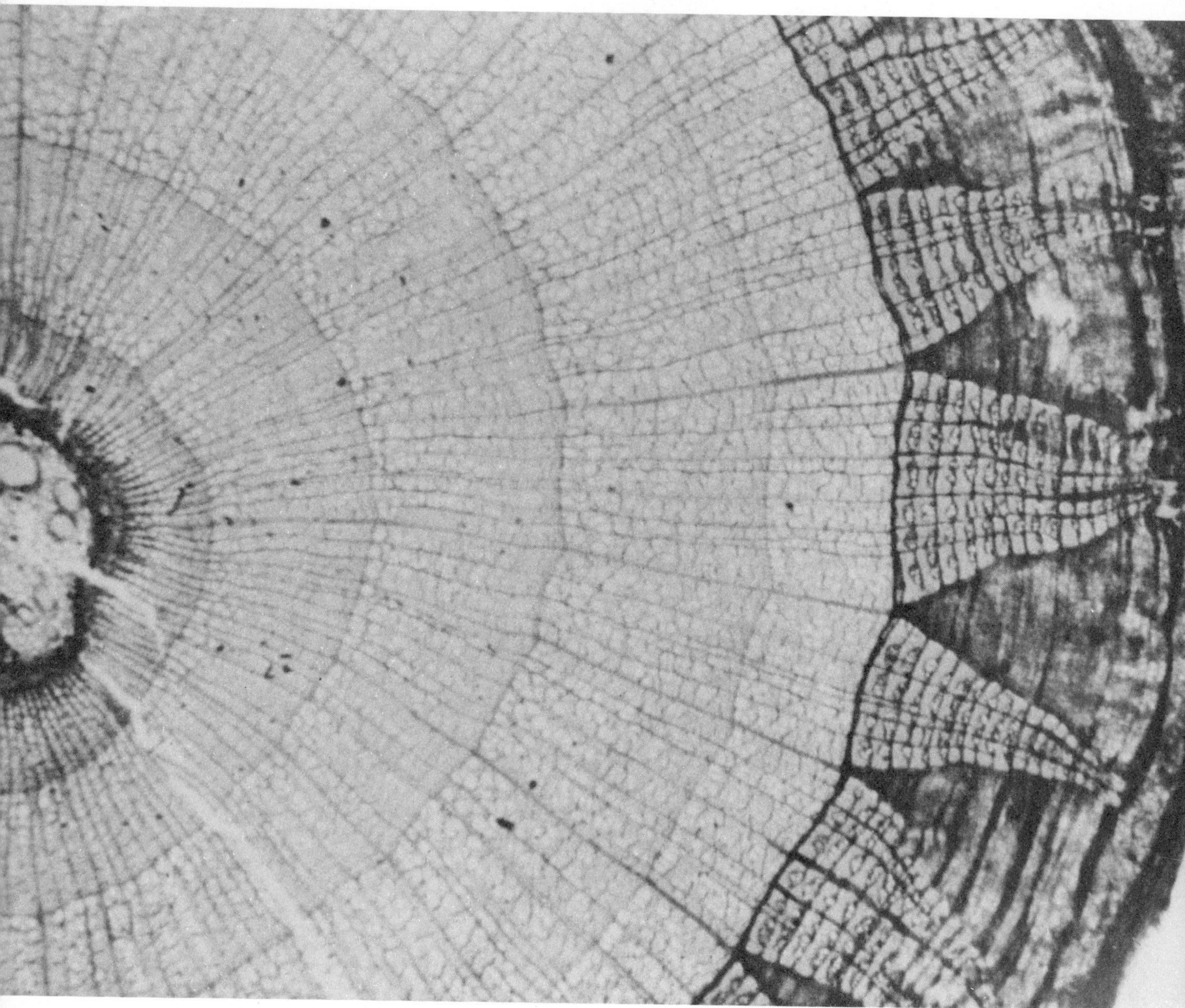

FIG. 12.3 Annual rings in the wood of *Tilia,* the American linden or basswood. How old was this twig when it was cut?

In the older root, another meristem, the cambium, develops between the xylem and phloem. Mitosis in the cambium produces new (secondary) xylem to the inside and new (secondary) phloem to the outside (Fig. 12.2).

The vascular bundles of the stem are simply extensions of those in the roots. They are, however, arranged somewhat differently. Furthermore, the arrangement of bundles in the monocot stem is markedly different from that in the dicot stem. (The organization of tissues in monocot roots and dicot roots is essentially alike.) Even among the dicots, the arrangement of tissues in those that develop woody stems (perennials) is somewhat different from those whose stems are herbaceous (annuals).

12.5 THE WOODY DICOT STEM

A cross section of a young twig of the basswood or American linden (*Tilia*) provides a good example of the organization of tissues in the typical woody dicot (Fig. 12.3). The twig is made up of three distinct regions, the bark, the wood, and the pith. The outer surface of the bark region is protected by layers of dead cork cells impregnated with suberin. Suberin is a waxy material which markedly cuts down water loss from the stem. A great many openings, the lenticels, are present in the cork and it is by means of these openings that ogygen and carbon dioxide can be exchanged between the stem tissues and the air. Beneath the cork are layers of parenchyma cells which make up the cortex. These cells store food in the stem just as they do in the root. In the very young stem (before cork has developed) they may have chloroplasts and carry on photosynthesis. In the older stem a meristem develops between the cortex and the cork. Mitosis in this meristem (the cork cambium) replaces the cork cells that are lost by weathering (Fig. 12.4).

The inner portion of the bark is marked by alternating areas of phloem tissue and parenchyma. The parenchyma makes up the terminal portion of the horizontal rays that run between the pith and the bark. These rays carry on lateral transport of materials between the two regions. The enlarged portion in the bark also serves as a food storage area. The phloem tissue consists of bundles of sieve tubes surrounded by and supported by sclerenchyma cells.

The inner boundary of the bark region is marked by a meristem, the cambium. As a result of its activ-

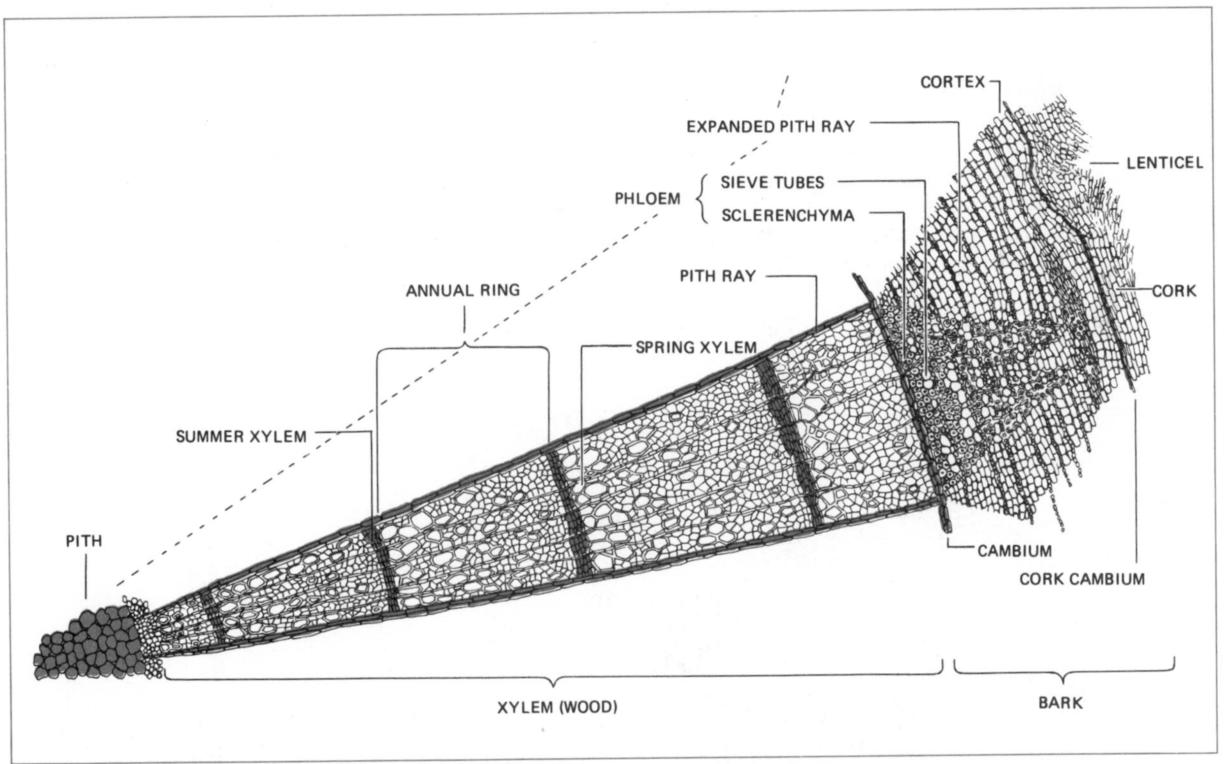

FIG. 12.4 Structure of a typical woody dicot stem. A single vascular bundle of *Tilia* is shown in cross section.

ity, new phloem is continually produced in the bark region during the growing season. The cambium also produces new xylem toward the interior.

Xylem makes up the wood region. The xylem vessels produced during the spring, when water is plentiful, have larger diameters than those produced later in the summer. During the dormant season, no xylem whatsoever is formed. The visual contrast between the summer xylem of one year and the spring xylem of the next year permits the easy counting of annual rings. The entire band of xylem formed in one growing season makes up one annual ring. How old was the basswood twig shown in Fig. 12.3 when it was cut?

The xylem serves a dual function: support and transport. In older stems, only the most recent rings of xylem participate in the transport of materials. These make up the **sapwood.** The inner rings of xylem cease to function in transport, but provide ever-increasing support for the weight of the tree above. This portion of xylem is the **heartwood.**

The innermost portion of the young basswood twig is the pith. It contains parenchyma cells and serves as a food storage area. In the older woody stem, the pith disappears.

12.6 THE HERBACEOUS DICOT STEM

The basic arrangement of the herbaceous dicot stem (Fig. 12.5) is similar to that of the woody dicot, although the surface of the stem is protected only by a layer of epidermis rather than by cork. Beneath the epidermal cells are cortex cells. These often possess chlorophyll and carry on photosynthesis. The interior of the stem is packed with parenchyma cells, the pith. The outer portions of the vascular bundles contain phloem tissue, just as in the woody dicot. On the inner side, the phloem is separated from the xylem by a cambium. Although the cambium does produce additional xylem and phloem during the growing season, no annual rings are formed.

As its name implies, the herbaceous dicot stem lacks the strength and rigidity of the woody dicot stem. This is because there are no rings of woody xylem present. The stem does receive some mechanical support, however, from masses of sclerenchyma located between the phloem and the cortex.

12.7 THE MONOCOT STEM

The organization of the monocot stem is quite different from that of the dicot. A cross section of a corn stem (*Zea mays*) (Fig. 12.6) shows these differences nicely. The corn stem consists of an external rind and an interior filled with pith. The rind provides most of the mechanical support for the stem. Scattered pretty much at random through the pith are the vascular bundles. In cross section, they remind one of a caricature of a human face. Each is surrounded by a layer of sclerenchyma cells that provides mechanical support for the cells within. The phloem is located in the "forehead" region. Careful examination reveals companion cells located between the sieve tubes. Four xylem vessels are present, making up the two "eyes," the "nose," and the "mouth." Much of the bottom xylem vessel is surrounded by air in the air space. Xylem tracheids are present in the "cheek" areas.

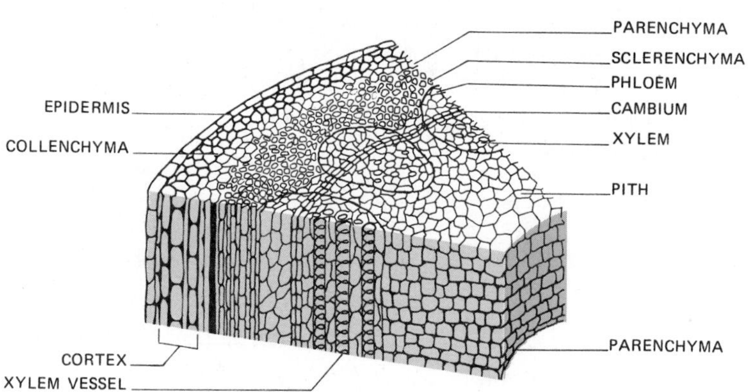

FIG. 12.5 Anatomy of a typical herbaceous dicot stem. Why are there no annual rings?

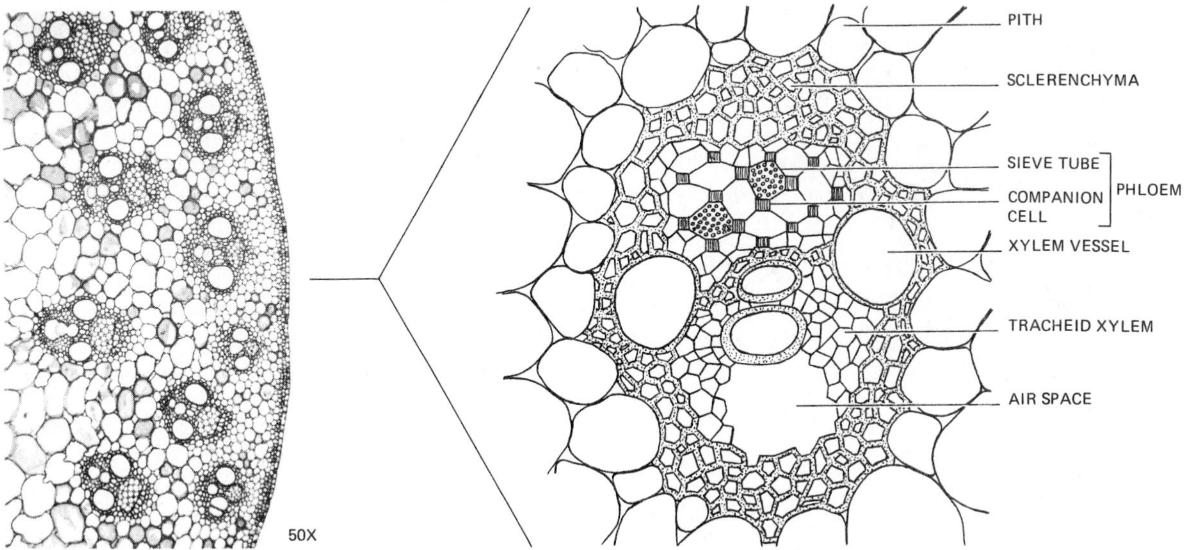

FIG. 12.6 Organization of a typical monocot stem. Left: a cross section of a portion of a corn stem (*Zea mays*). Right: detail of a single vascular bundle. (Photo courtesy CCM: General Biological, Inc., Chicago.)

12.8 THE LEAF VEINS

The vascular bundles of the leaves are direct extensions of the vascular bundles of the stem. They pass from the stem into the leaf stalk and then into the various veins of the leaf. These branch into ever-finer veins (Fig. 12.7). Probably no cell in the spongy layer of the leaf is more than two cells away from the end of a vein. The xylem and phloem of the vein (Fig. 9.12) are often surrounded by layers of sclerenchyma cells which impart strength to the vein. The veins, in turn, provide a rigid framework to support the remaining soft tissues of the leaf blade.

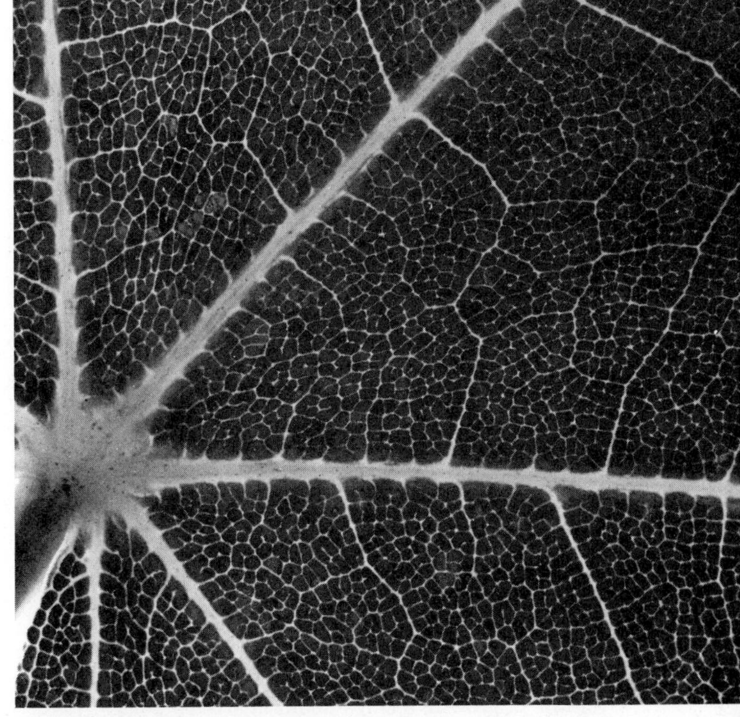

FIG. 12.7 Veins in a maple leaf.

The Transport of Water and Minerals

12.9 THE PATHWAY

Water enters the plant through the root hairs. These extensions of the epidermal cells have sticky walls and adhere tightly to soil particles. This places them in direct contact with the film of water that also adheres to the soil particles of all but "bone-dry" soils

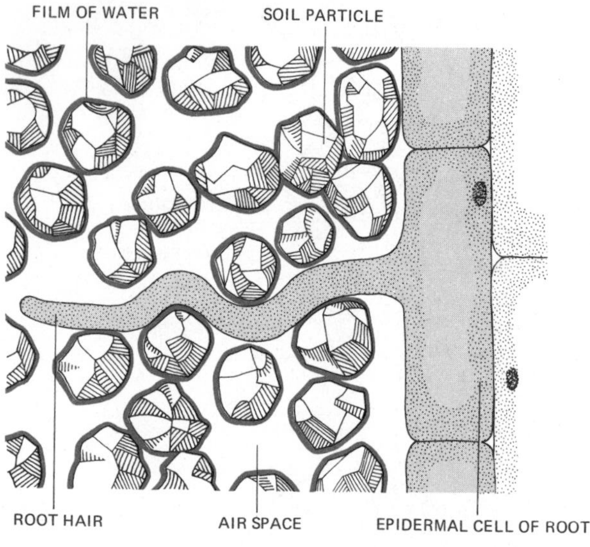

FIG. 12.8 The increased surface area provided by root hairs facilitates the absorption of water from the soil.

(Fig. 12.8). Root hairs are found only at the tips of the roots, but there may be a great many of them. Careful measurements have been made of the root system of the grass, rye. After four months of growth, a single plant was found to have 387 miles of roots. There were approximately 13 million root tips possessing an estimated total of 14 billion root hairs (Fig. 12.9). Once within the epidermal cells, water passes through and between the cells of the cortex. It enters the central cylinder by passing through the endodermis and pericycle. In many roots, the endodermis contains specialized cells, the passage cells, which provide an easy pathway for the movement of water into the central cylinder. In the young root, water then enters the xylem directly. In older roots, it may have to pass first through a band of phloem and cambium. It does this by travelling through horizontally elongated cells, the xylem rays.

One might suppose that minerals enter and move through the plant simply dissolved in water. While water and minerals do share the same final

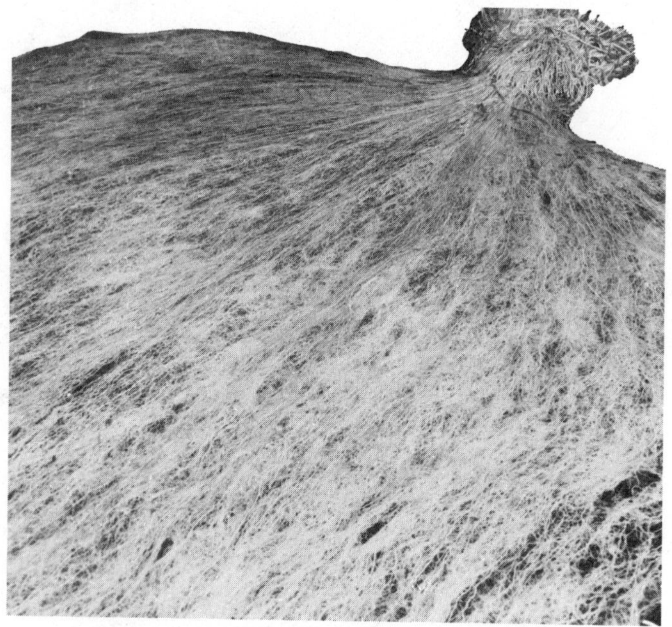

FIG. 12.9 Absorptive area of roots. Left: after two years of growth, a single plant of crested wheat grass had produced 319 miles of roots. (Courtesy of "Science Service.") Right: the surface area for water absorption is increased further by the root hairs at the tip of each root (seen on the primary root of a germinating radish.)

pathway, the uptake of minerals by the root has several special features. Even when no water is being absorbed by the roots, minerals enter freely. Furthermore, they can enter against a concentration gradient, i.e. from a region of lower concentration (the soil) to a region of higher concentration (the cells of the root). Hence, active transport must be at work and, indeed, anything which interferes with active transport processes in general also blocks mineral uptake.

Perhaps it should be emphasized that all the elements needed by the plant are taken up in inorganic form, hence as "minerals." Nitrogen enters as nitrate (NO_3^-) or ammonium ions (NH_4^+), phosphorus as $PO_4^{\equiv}$, potassium as K^+, calcium as Ca^{++}, etc. When one hears the virtues of organic fertilizer being extolled, it should be remembered that such material meets no nutritional needs of the plant until it has been degraded to inorganic form. Organic matter does play an important role in the formation of good soil texture, but only to the extent that it can yield inorganic ions can it meet the nutritional needs of plants.

Once within the epidermal cells of the root, inorganic ions pass inward from cell to cell, probably by way of plasmodesmata. The final step from the cytoplasm of the pericycle cells to the xylem is probably accomplished once again by active transport.

Once in the xylem, water and the minerals that have been deposited in it move upward in the vessels and tracheids. These run up through the root and the stem. At any level the water can leave the xylem and pass laterally to supply the needs of other tissues. At the leaves the xylem passes into the leaf stalk (the petiole) and then into the veins of the leaf. A glance at the scars left on woody stems after the leaves have dropped in the autumn will reveal the severed ends of the vascular bundles that had supplied the leaves.

At the ends of the veins, water leaves the xylem and enters the spongy layer and palisade layer cells of the leaf. Here the water may be used in photosynthesis or may be evaporated from the leaf in the process of transpiration.

Several bits of evidence lead to the conclusion that the upward transport of water takes place in the xylem. For example, if a ring of bark is carefully removed from the trunk of a tree, thus removing all the phloem, the upward passage of water is not interrupted. Such an operation is called girdling. Young fruit trees are all too frequently girdled during the winter by hungry field mice feeding upon the bark. With the arrival of spring, the tree develops leaves normally, suggesting that the upward movement of water and minerals has not been interfered with. Unless a special graft (Fig. 12.10) is made, however, the tree will ultimately die. (Why?) Additional evidence is obtained when the severed top of a plant is placed in water containing a dye. The water is drawn up into the stem for some period of time; microscopic examination of the vascular tissue shows that the dye is localized in the xylem vessels.

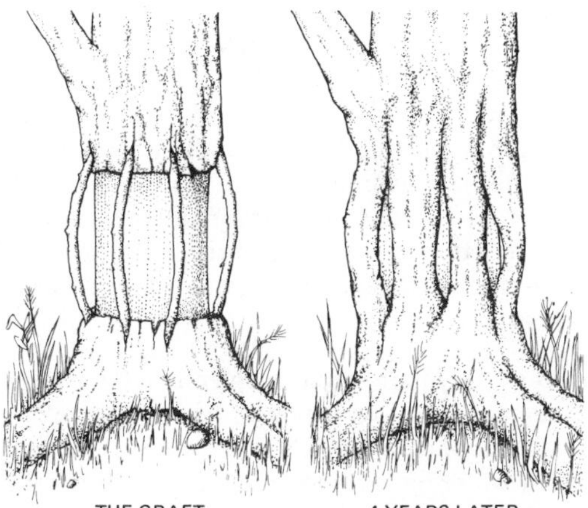

THE GRAFT 4 YEARS LATER

FIG. 12.10 A bridge graft. It is used to save valuable trees that have been girdled. Without the bridge graft, the roots would die for lack of food.

12.10 MAGNITUDE OF FLOW: TRANSPIRATION

Any theory to explain the upward movement of water in the xylem must take into account the large volume of water transported and the speed with which the water travels. Herbaceous plants may absorb a volume of water each day equal to several times the volume of the plant itself. Water that contains a marker (for example a radioactive isotope) has been shown to move up the stem as much as 75 cm each minute.

Only 1–2% of all this water is used in photosynthesis or in other metabolic activities of the leaf cells. The remainder evaporates from the leaf in the process of **transpiration.** In Chapter 10, you learned that water evaporates into the air spaces of the spongy layer. When the stomata open, the water vapor diffuses out of the leaf. If the leaf is to continue to function properly, fresh supplies of water must be

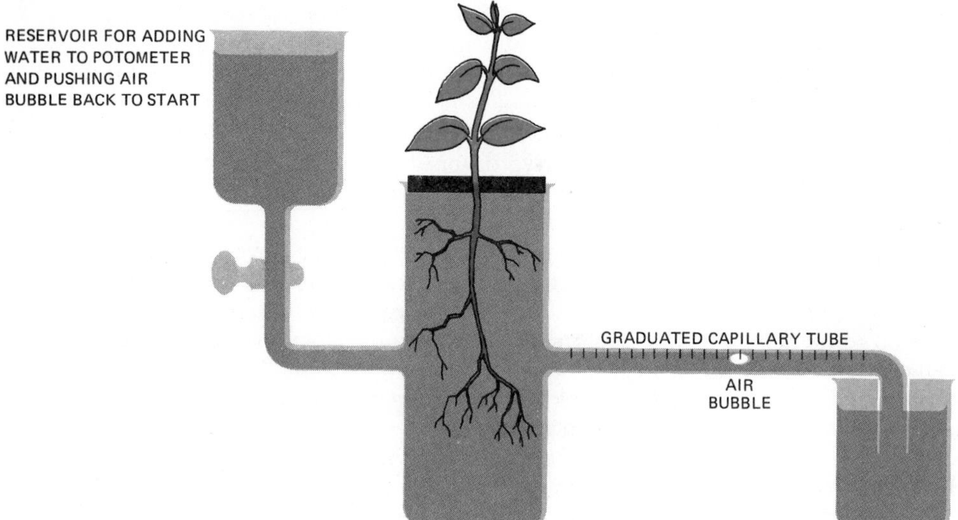

FIG. 12.11 Potometer in use. As water is transpired and otherwise used by the plant, it is replaced from the reservoir on the right. This pushes the air bubble to the left and permits the amount of water used to be measured accurately.

brought to the leaf to replace that lost in transpiration.

Figure 12.11 illustrates a device with which it is possible to approximate the amount of water vapor given off in transpiration. As water is absorbed by the roots of the plant, it is replaced by water drawn through the graduated tube from the reservoir on the right. A small bubble introduced into the system serves to mark the volume consumed. Although this device, a potometer, measures the total water use of the plant, 98–99% of this is accounted for by transpiration.

12.11 FACTORS AFFECTING THE RATE OF TRANSPIRATION

Using a potometer, it is possible to study the effect of various environmental factors upon the rate of transpiration.

1. Light. Plants transpire much more rapidly when exposed to light than in the dark. This is chiefly because light stimulates the opening of the stomata (see Section 10.4) and thus greatly increases the transfer of moisture-laden air from the air spaces of the spongy layer to the outside. Light also speeds up transpiration by warming the leaf.

2. Temperature. Plants transpire more rapidly at higher temperatures than at low. At 30°C a leaf may transpire three times as fast as it does at 20°C. This is because water evaporates more rapidly as its temperature increases and, in this case, thus increases the humidity of the air in the air space with respect to that outside.

3. Humidity. Rate of transpiration is also affected by the relative humidity of the air surrounding the plant. The rate of diffusion of any substance is decreased as the difference in concentration of the substance in the two regions decreases. The reverse is also true. Therefore diffusion of water from the moisture-laden air spaces of the leaf to the outside goes on rather slowly when the surrounding air is quite humid. When the surrounding air is dry, diffusion proceeds much more rapidly.

4. Wind. The presence of gentle air currents also increases the rate of transpiration. When none are present, the air immediately surrounding a transpiring leaf becomes increasingly humid. As it does so, it causes a decrease in the rate of transpiration for the reason just mentioned. When a breeze is present, however, the humid air is carried away and replaced by fresh, drier air.

5. Soil water. A plant cannot continue to transpire rapidly if its moisture loss is not made up by absorption of fresh supplies of water from the soil. When absorption of water by the roots fails to keep up with the rate of transpiration, loss of turgor occurs, and the stomata close. This immediately reduces the rate of transpiration greatly. Often the loss of turgor may extend to other plant parts and wilting occurs.

Under optimum conditions, the loss of water through transpiration is truly impressive. It has been estimated that over the course of the growing season an acre of corn plants may transpire 400,000 gallons of water. In liquid form this would be enough to cover the field with a lake 15 inches deep. The quantity of water transpired per acre by the trees of a mature deciduous forest is probably even greater.

Not only must a theory of water transport in the xylem account for the large quantities of water involved, but it must also account for the great height to which the water may be transported. The sequoias and Douglas firs of the Pacific coast regularly attain heights of 300 feet or more. The weight of a column of water this high exerts a pressure of almost 140 lb/in². Taking into account the speed at which water moves up the xylem and the resistance to this flow created by the walls of the tracheids, it seems likely that a push (or pull) of 300–400 lb/in² is called for.

Any theory to explain the upward transport of water must be based on purely physical principles. Remember that the xylem vessels and tracheids are quite lifeless. This was shown dramatically by a German botanist who sawed down a 70-foot oak tree and placed the base of the trunk in a barrel of picric acid solution. The solution was drawn up the trunk, killing nearby tissues as it went. Three days later water, to which a dye had been added, was still moving successfully to the top of the tree.

Theories of Water Transport

Two theories that have been proposed to explain the upward transport of water in the xylem deserve mention.

12.12 ROOT PRESSURE

When a tomato plant is carefully severed close to the base of the stem, a fluid—sap—oozes from the stump. If a pressure-measuring device is connected to the stump, it is found that the sap is being pushed out under considerable pressure (Fig. 12.12). This is known as root pressure. Although the experiment is not an easy one to perform, careful workers have reported root pressures in the tomato as high as 130 lb/in². While this pressure would not lift sap to the top of a sequoia, it would easily lift it to the top of the tomato plant and even to the top of most trees.

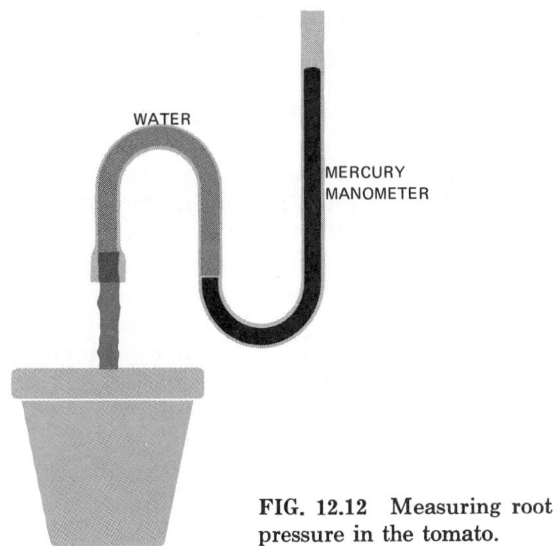

FIG. 12.12 Measuring root pressure in the tomato.

The origin of root pressure is thought to lie in the difference in water concentration of the soil water and the sap in the xylem ducts. Xylem sap may be hypertonic to soil water because of the presence of dissolved sugars, or salts that have been actively transported into the central cylinder. This is particularly apt to be true in woody perennials like the sugar maple when, in very early spring, they hydrolyze the starches stored in their roots into sugar. At such a time, it is theorized, water would pass by osmosis from the root hairs, through the cortex and endodermis, to the xylem ducts. The inflow of water into the sap-filled xylem ducts would create a pressure, which could then be relieved by movement of the sap up the ducts. This is basically the same mechanism by which the molasses solution rose in the osmometer described in Section 6.4.

While root pressure is now a demonstrated fact, it seems unlikely that it can account completely for the upward transport of water in the xylem. Although substantial root pressures have been measured in the tomato, practically no other plants have been found to develop root pressures greater than 30 lb/in². Furthermore, it has been impossible to demonstrate any root pressure at all in many plants.

The volume of fluid transported by root pressure is also not adequate to account for the measured movement of water in the xylem. Furthermore, those plants whose roots produce a reasonably good flow of sap are apt to have the lowest pressure, and vice versa.

There is still another difficulty with the root pressure theory of water transport. Even in those plants that show root pressures, the highest values occur in the spring, when the sap is strongly hypertonic to soil water but the rate of transpiration is low. In summer, when transpiration is proceeding rapidly and hence water is moving through the xylem rapidly, often no root pressure can be detected at all.

Thus, while root pressure may play a significant role in water transport in certain species (recent studies indicate that it may be the most important factor in the coconut palm) and at certain times, we must look for an alternative mechanism.

12.13 THE DIXON-JOLY THEORY

In 1895, the Irish plant physiologists H. H. Dixon and J. Joly proposed another theory to explain the rise of water in plants. They thought that water, instead of being pushed up by root pressure from below, was pulled up by tension (negative pressure) from above. As we have seen, water is continually being used at the leaves. Transpiration accounts for most of this, but photosynthesis and other biochemical activities in the cell account for some consumption of water, too. According to the Dixon-Joly theory, the use of water in the leaf exerts a pull on the water in the xylem ducts and draws more water into the leaf. The pull on the water in the xylem vessels is, in turn, transmitted to the roots and pulls water from the soil into the xylem ducts of the root.

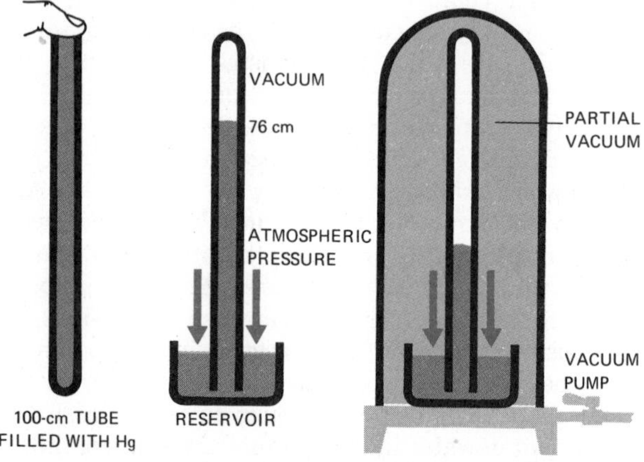

FIG. 12.13 Establishment of Torricellian barometer and effect of placing the barometer in a vacuum chamber.

Anyone who has studied physics knows that even the finest vacuum pump can only pull water up to a height of 32 feet. This is because a column of water that high exerts a pressure (about 15 lb/in^2) just counterbalanced by the pressure of our atmosphere (Fig. 12.13). How, then, can water be drawn up 300 or more feet under tensions of 300–400 lb/in^2?

The answer to this question lies in the property of water molecules to cling to each other. This property is known as **cohesion.** When water is confined to tubes of very small bore, the force of cohesion between the water molecules imparts great strength to the column of water. It has been found experimentally that tensions as great as 3000 lb/in^2 are necessary to break such a column of water. This compares very favorably with the strength of steel wires of the same diameter. According to the Dixon-Joly theory, the great cohesive strength of the tiny filaments of water in the xylem ducts, coupled with the property of water molecules to *adhere* to the walls of the ducts, enables the filament of water to be pulled to the top of the tallest trees without breaking or pulling away from the walls of the ducts. Water use at the leaves, chiefly in transpiration, creates the tension which is then transmitted back down the xylem all the way to the root hairs. (The Dixon-Joly theory has often been called the transpiration pull-cohesion theory.) In a real sense, the tiny filaments of water in the xylem possess the physical properties of solid wires.

12.14 EVIDENCE FOR THE THEORY

What evidence is there to support this theory? Figure 12.14 illustrates an experimental model which shows that it is physically possible to raise water by such a mechanism. Evaporation of water at the porous clay cup creates a tension which is transmitted to the water and mercury in the glass tube. When great care is taken to see that no dissolved air is present in the water, it has been possible to draw the mercury column as high as 226 cm. This is equivalent to raising a column of water almost 100 feet into the air, and is three times greater than the height that atmospheric pressure working against a perfect vacuum could attain.

The model shows only that the phenomenon is possible. It does not prove that it takes place. When a cedar twig is substituted for the porous cup, however, the mercury is again lifted higher than the height (about 76 cm) that could be attained with a perfect vacuum.

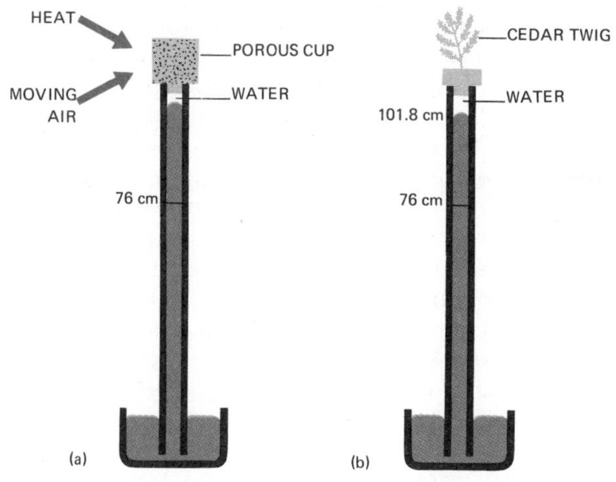

FIG. 12.14 Rise of mercury column in excess of 760 mm as a result of evaporation from (a) a porous cup and (b) a cedar twig.

Accepting the theory that the water in the xylem ducts is indeed under tension rather than pressure, it is possible to deduce certain facts which must follow from this. By checking to see if these predicted facts are indeed so, one is able to test the validity of the theory. If the water in a xylem vessel is under tension, we would expect the column to snap apart if air is introduced into the vessel by puncturing it. This has been found to be so.

Furthermore, if the water in all the xylem ducts is under great tension, there should be a resulting inward pull (because of adhesion) on the walls of the ducts. The inward pull on the walls of all the ducts in the band of sapwood in an actively transpiring tree should, in turn, result in a decrease in the diameter of the tree. This deduction can be *tested* by continuously measuring, with suitable equipment, the diameter of tree trunks. Figure 12.15 shows the results that the American plant physiologist D. T. MacDougal obtained with the trunk of a Monterey pine tree. The diameter of the tree did indeed fluctuate, with its minimum diameter being reached just after midday, the time of greatest transpiration.

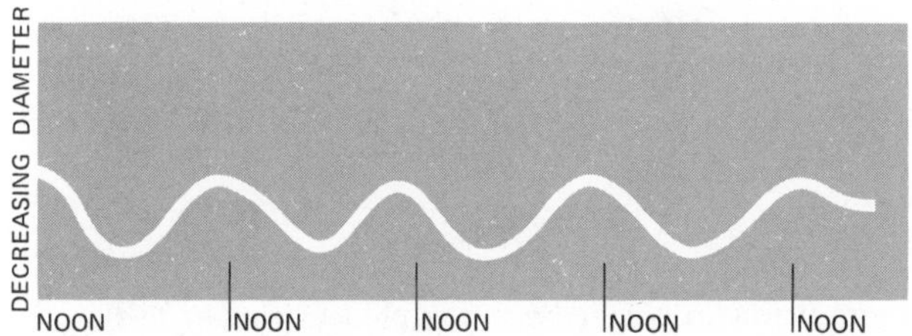

FIG. 12.15 Diurnal (daily) variation in the diameter of the trunk of a Monterey pine. The diameter of the trunk reaches its minimum when the rate of transpiration reaches its maximum.

In 1960 a team of plant physiologists working in the jungles of Northeastern Australia carried out experiments on the rattan vine which lent additional support to the Dixon-Joly theory. The rattan vine climbs upon the trees of the jungle. In order to get its foliage into the sun, it may have to climb as high as 150 feet into the air. When the base of the vine is severed while immersed in a basin of water, water continues to be taken up by the vine. Obviously, root pressure is playing no part in the process, because the vine has been detached from its root system. A rattan vine less than 1 inch in diameter will "drink" water indefinitely at a rate of about 12 ml each minute. If forced to take up water from a sealed container, the vine does so without any decrease in rate even though a high vacuum is quickly produced in the container. In fact, the vacuum becomes so great that the remaining water begins to boil spontaneously. (The boiling temperature of water decreases as the air pressure over the water decreases.)

The Dixon-Joly theory provides an explanation for the ability of certain vascular plants to live in salt water. The roots of the coastal mangrove grow immersed in salt water, a solution markedly hypertonic to the cytoplasm of their cells. On the basis of respective water concentrations, we could expect water to leave the cells of the mangrove with resulting plasmolysis. Remember how the cells of *Elodea* become plasmolyzed in salt water (see Fig. 6.6). However, remarkably high tensions have been measured in the vascular system of coastal mangroves. These tensions (on the order of 500–800 lb/in^2) are so great that they can pull water molecules into the plant against the osmotic gradient. By this mechanism, mangroves literally desalt sea water to satisfy their water requirements.

On the basis of all the evidence, the Dixon-Joly theory seems to provide a satisfactory explanation of the rise of liquids in plants. Though several questions remain to be answered, most plant physiologists feel that the theory has been well tested and best describes the mechanism by which most, if not all, plants transport water and dissolved minerals from the soil to the leaves.

The Transport of Food

12.15 THE PATHWAY

Foods and other substances (e.g. hormones) manufactured by the plant are also transported in the vascular bundles. Sugars (usually sucrose), amino acids, and other organic molecules that have been manufactured in the leaves enter the cells of the phloem. This entry is probably facilitated by the plasmodesmata which connect the photosynthesizing cells of the leaf with the cells of the phloem.

Once within the phloem, the sugar and other organic molecules may be transported either up or down to any region of the plant. In the case of sugar destined to be stored in the roots, it passes in solution through the phloem of the stem and root. Once within the root, the sugar then passes out of the phloem, through the pericycle and endodermis and into the cells of the cortex. Here it can be converted into starch and stored as a food reserve.

That food transport occurs in the phloem is suggested by girdling experiments. In summer, when a tree is girdled so that the phloem is removed but the xylem remains intact, the tree continues to live. There is, however, no further increase in the weight of the roots. Furthermore, chemical examination of the bark just above the girdle shows an extraordinary accumulation of carbohydrates. (This accumulation does not occur after girdling in the dormant season, when no leaves are on the tree.) No accumulation of sugar occurs in the bark below the girdled region.

A more direct demonstration of the pathway taken by organic compounds in the plant is shown in Fig. 12.16. A leaf was supplied with radioactive water (^{3}HOH) and allowed to carry on photosynthesis for 30 min. At the end of this time, the radioactivity in the petiole (through which both xylem and phloem pass from leaf blade to stem) was localized in the sieve tubes and companion cells.

The transport of food materials through the phloem proceeds quite efficiently. Some fruits, such as the pumpkin, receive over half a gram of food each day through the phloem. Other fruits have been found to do even better. Bearing in mind that the food is being transported in a fairly dilute water

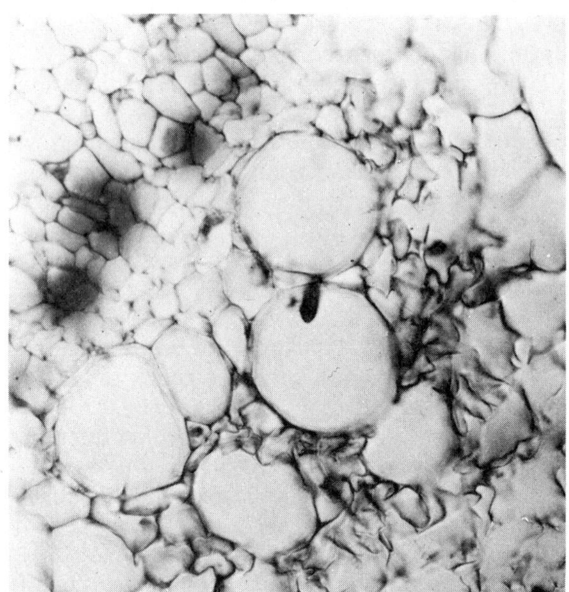

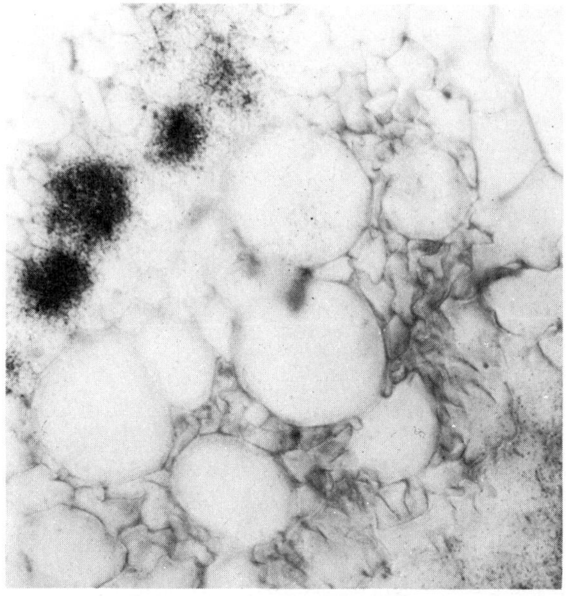

FIG. 12.16 A demonstration by radioautography that the products of photosynthesis are transported in the phloem. A cucumber leaf was supplied with radioactive water (^{3}HOH) and allowed to carry on photosynthesis for 30 min. Then slices were cut from the petiole of the leaf and covered with a photographic emulsion. Radioactive products of photosynthesis darkened the emulsion where it was in contact with the phloem (upper left), but not where it was in contact with the xylem vessels (center). In the photomicrograph on the left, the microscope is focused on the tissue in order to show the cells clearly; on the right, the microscope has been focused on the photographic emulsion. (Courtesy of R. S. Gage and S. Aronoff.)

solution, an appreciable flow must occur. In fact, studies with radioactive tracers have revealed that substances can travel through as much as 100 cm of phloem in an hour.

12.16 MECHANISM OF FOOD TRANSPORT

The mechanism by which sugars and other molecules manufactured by the plant are translocated through the phloem is not yet understood. It seems to be dependent upon the metabolic activity of the phloem cells, because any condition that slows down their metabolism also slows down the rate of translocation. Figure 12.17 illustrates apparatus with which one can show the effect of lowered temperature on the rate of food translocation. Oxygen lack similarly depresses it. Anything that kills the phloem cells puts an end to the process.

Several theories have been proposed to explain the movement of food materials through the phloem.

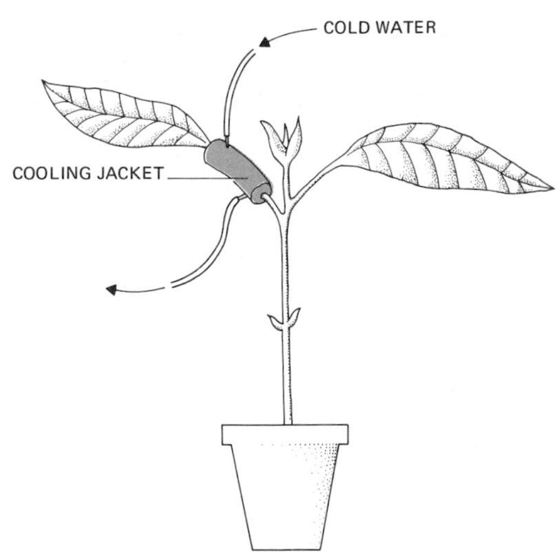

FIG. 12.17 Lowering the temperature of the petiole reduces the rate at which food is translocated out of the leaf.

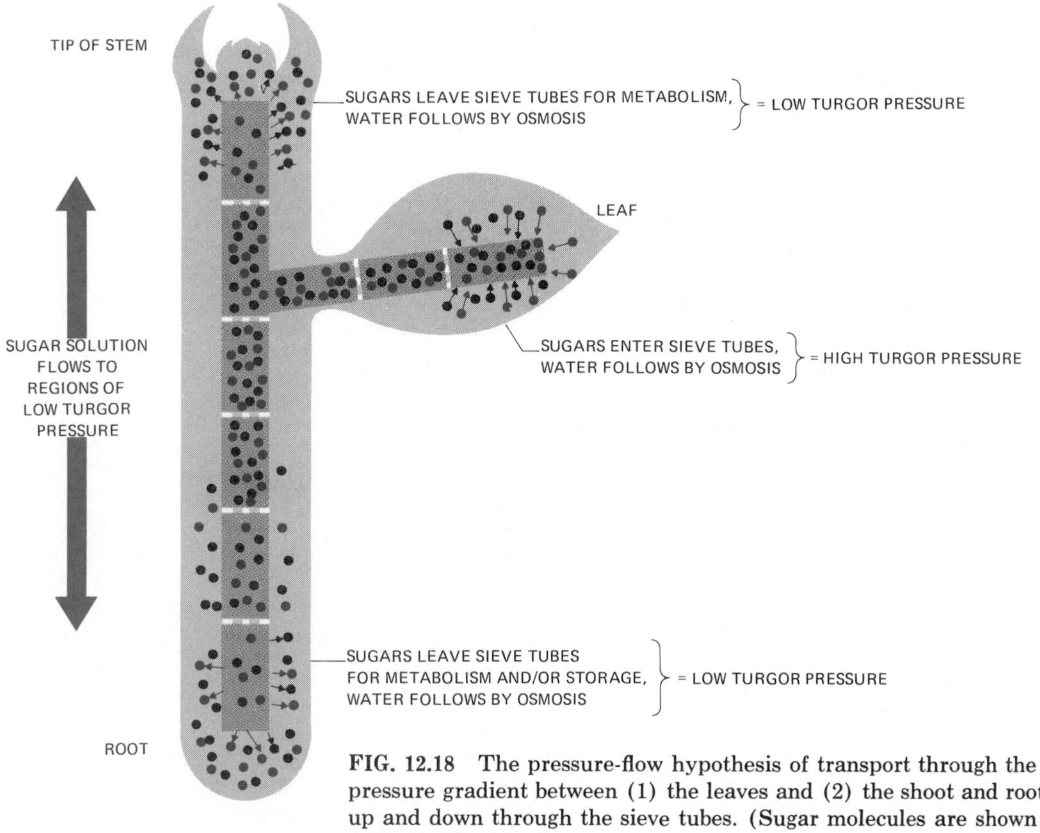

TIP OF STEM
SUGARS LEAVE SIEVE TUBES FOR METABOLISM, WATER FOLLOWS BY OSMOSIS = LOW TURGOR PRESSURE
LEAF
SUGAR SOLUTION FLOWS TO REGIONS OF LOW TURGOR PRESSURE
SUGARS ENTER SIEVE TUBES, WATER FOLLOWS BY OSMOSIS = HIGH TURGOR PRESSURE
SUGARS LEAVE SIEVE TUBES FOR METABOLISM AND/OR STORAGE, WATER FOLLOWS BY OSMOSIS = LOW TURGOR PRESSURE
ROOT

FIG. 12.18 The pressure-flow hypothesis of transport through the phloem. The pressure gradient between (1) the leaves and (2) the shoot and root drives the sap up and down through the sieve tubes. (Sugar molecules are shown in black; water molecules in color.)

According to one of these, water containing the food molecules in solution flows under **pressure** through the phloem. The pressure arises from the difference in water concentration of the solution in the phloem and the relatively pure water in nearby xylem ducts. With the accumulation of sugars and other products of photosynthesis in the phloem, water enters by osmosis (Fig. 12.18). Pressure thus builds up in these cells something like root pressure in reverse. As the sap is pushed down (and up) the phloem, sugars are removed by the cortex of both stem and root and consumed or converted into starch. Starch is insoluble and exerts no osmotic effect. Consequently, the osmotic pressure of the contents of the phloem decreases. Finally, relatively pure water is left in the phloem, and this is thought to leave by osmosis and/or be drawn back into nearby xylem vessels by the suction of transpiration pull.

How well does this theory fit the observed facts? According to it, the contents of the sieve tubes must be under pressure. This has been found to be true. When sieve tubes are punctured by such sap-eating insects as aphids, the sap flows through the insect's mouth parts and into its body without further assistance. In fact, if the body of the aphid is cut away from its mouth parts, sugar-rich sap continues to ooze from the cut end (Fig. 12.19). This theory also requires that the osmotic pressure in the phloem of the leaves be greater than in the phloem of food-receiving organs such as the roots. Generally this seems to be true although apparent exceptions have been found.

On the other hand, one would predict on the basis of this theory that factors affecting the rate of metabolism of the *phloem cells* themselves would have no effect on translocation. The reason is that metabolic energy is needed only at the "ends" of the system, i.e. for depositing sugars in the phloem of the leaves and withdrawing them again from the phloem of the root and the growing shoot (Fig. 12.18). But, as we have seen, transport through the phloem *is* affected by localized inhibition of its metabolism (Fig. 12.17). Furthermore, there is a real question as to whether the measured pressures are sufficient to push sap through the sieve tubes at the velocities that have been measured.

A number of alternative theories have been advanced which avoid the difficulties of the pressure-flow theory (but run into difficulties of their own). Several of these assume some sort of localized pumping mechanism working in relays along the length of the sieve tubes—perhaps at each sieve plate. Such theories *would* explain why chilling a restricted portion of the phloem inhibits translocation. As yet,

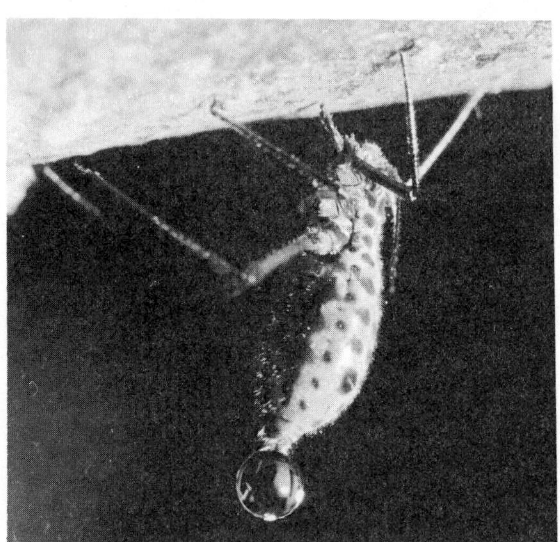

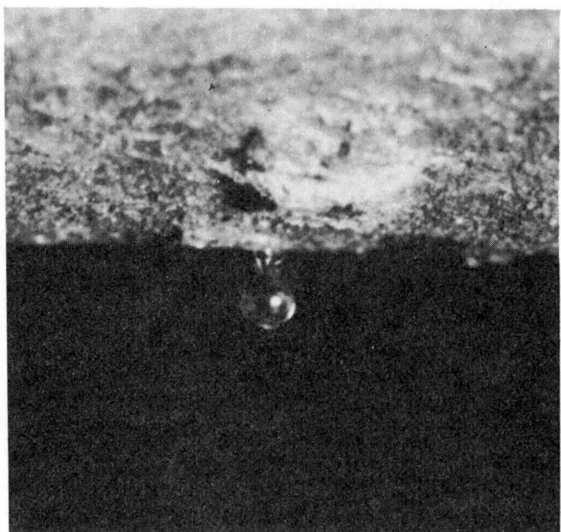

FIG. 12.19 Left: Aphid feeding on the branch of a linden tree. Excess sugar is released as a drop of honeydew that serves as food for ants or bees. The sap in the phloem enters the insect's mouth parts under pressure. Right: It will continue to exude from the mouth parts after the aphid has been cut away from them. (Courtesy of Martin H. Zimmerman, *Science,* **133**, 73–79, Jan. 13, 1961.)

however, no single theory has been proposed that adequately explains all the known facts of phloem transport. Such a theory still awaits fresh observations, better measuring techniques, new experiments and, especially, the creative imagination capable of welding all these into a comprehensive scheme.

12.17 SUMMARY

Xylem and phloem represent one of the most important adaptations of plants to a land environment. In order to carry on photosynthesis, plants must expose a broad chlorophyll-containing surface to the sun's rays. For land plants, this means exposing a broad surface to the drying effect of the air. In order to remain functioning, the land plant must secure water for photosynthesis and replace water lost through transpiration with fresh water from the soil. As we have seen, the structure and functioning of the xylem permits this. On the other hand, the water-absorbing roots need food in order to remain alive. The phloem solves this transport problem. These two tissues have played a crucial role in the successful invasion by the vascular plants of almost all the exposed land on our planet.

EXERCISES AND PROBLEMS

1. Trace the path taken by a hydrogen atom from the time it enters a root hair incorporated in a molecule of water until it is stored in the root cortex incorporated in a molecule of starch.
2. In what ways does the circulatory system of a plant differ from that of a vertebrate?
3. Many low-growing plants secrete drops of water along the edges of their leaves. This phenomenon, called guttation, is particularly apt to occur during damp spring nights. Which of the forces by which water rises in the xylem do you think is responsible for guttation? Why?
4. What functions in the maple tree are carried out by roots?
5. Why is it good practice to prune off some of the branches of a tree or shrub that has just been transplanted?
6. If in counting the annual rings of an ancient elm, several narrow rings are found just within several wide ones, what conclusions might you draw?
7. What is cohesion? What role does it play in the upward transport of water?
8. What is adhesion? What role does it play in the upward transport of water?
9. What is the function of the upward transport of maple sap to the maple tree?

REFERENCES

1. ZIMMERMANN, M. H., "How Sap Moves in Trees," *Scientific American,* Offprint No. 154, March, 1963. A brief, up-to-date review of our knowledge of translocation in both the xylem and the phloem.
2. STEWARD, F. C., *About Plants: Topics in Plant Biology,* Addison-Wesley, Reading, Mass., 1966. Chapter 10 examines the problems of translocation with emphasis on the role played by the active transport of salts and organic molecules.
3. RAY, P. M., *The Living Plant,* Holt, Rinehart and Winston, New York, 1972. Chapters 5 and 7 expand on the topics treated in this chapter.
4. EPSTEIN, E., *"Roots," Scientific American,* Offprint No. 1271, May, 1973. With emphasis on the role roots play in "mining" the minerals of the soil.
5. WOODING, F. B. P., *Phloem,* Oxford Biology Readers, No. 15, Oxford University Press, Oxford, 1971.
6. RUTTER, A. J., *Transpiration,* Oxford Biology Readers, No. 24, Oxford University Press, Oxford, 1972.

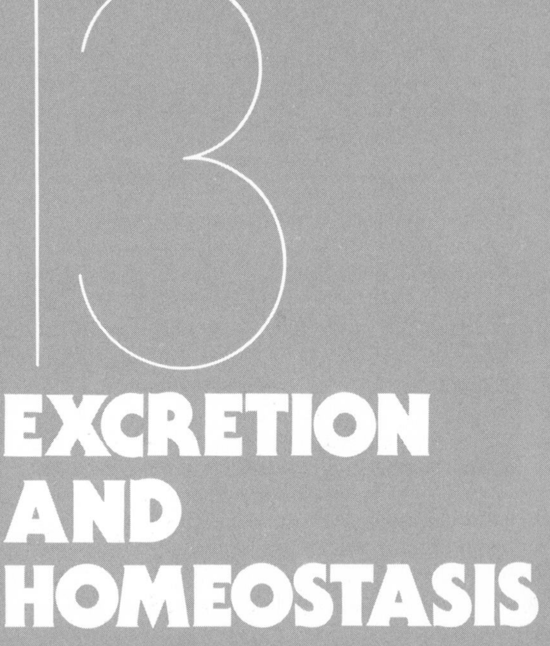

13 EXCRETION AND HOMEOSTASIS

In the preceding chapters of Part III, we have considered the ways in which living things (1) acquire materials (and energy) from their environment, (2) transport materials throughout their bodies, and (3) transform these materials within their cells. These operations are all part of an organism's metabolism. To complete our story of metabolism, we must now examine the ways in which living things return the waste products of their metabolism to the environment. In general, the most abundant products of metabolism are carbon dioxide, water, and ammonia. These substances should not, however, be considered simply as wastes. Each has important roles to play in the regulation of body activities (recall, for example, the effect of CO_2 on the rate of breathing) and in various anabolic cell syntheses. Even ammonia, a very toxic substance, is an essential reactant in a number of biochemical transformations of amino acid metabolism. It is, then, only as these substances accumulate to levels *in excess of the organism's needs* that they must be eliminated. The return of these unneeded products of metabolism to the environment is called **excretion.** As we have already studied (in Chapter 10) the ways in which plants and animals return gaseous wastes to the environment, we shall not stress these in the present discussion.

13.1 EXCRETION IN PLANTS

Excretion in plants does not pose any serious problems. There are several reasons for this. First, the rate of catabolism in plants is generally much lower than in animals of the same weight. Consequently, metabolic wastes accumulate more slowly. Second, green plants use much of the waste products of catabolism in their anabolic processes. Water and carbon dioxide, produced by respiration, are used in photosynthesis. Waste nitrogen compounds, whose excretion makes such demands upon animals, can be used by green plants in the synthesis of new protein. Finally, the metabolism of plants is based mainly on carbohydrates rather than proteins. This reduces their excretory needs, as the end products of carbohydrate metabolism are far less poisonous than the nitrogenous wastes produced by protein metabolism. Of course, plants do produce protein from which a variety of important cell structures and all their enzymes are made. Nevertheless, protein metabolism plays a much smaller role in plants than in animals, whose general body structure is based to such a great extent upon protein.

Among the aquatic plants, metabolic wastes are free to diffuse from the cytoplasm into the surrounding water. No cell is far removed from the water and the concentration of wastes within the cell exceeds the concentration of these substances in the water. The only product of metabolism for which this is not true is water itself. Freshwater plants live surrounded by a water concentration greater than that within their cytoplasm. Not only does this eliminate diffusion (osmosis) as a means of getting rid of waste water, but (and far more important) it also exposes them to a continual inflow of water from the environment. As water enters the cell by osmosis, the pressure within the cell rapidly increases. The rigid cellulose cell wall withstands this pressure and turgor results. When the pressure within the cell becomes equal to the osmotic pressure, equilibrium is achieved between the cell contents (especially the central vacuole) and the environment.

Among the terrestrial plants, the waste products of metabolism, such as salts and organic acids, are simply stored in the plant. These wastes may be stored in solid form in crystals or they may be dissolved in the fluid of the central vacuoles. In herbaceous species, the wastes simply remain in the cells until the tops of the plants die in the autumn. In perennial plants, wastes are deposited in the nonliving heartwood and are also eliminated when the leaves are shed.

13.2 EXCRETION IN THE AMOEBA

The amoeba and many other single-celled organisms live in a watery environment and eliminate their metabolic wastes by diffusion, just as the aquatic plants do. For many, the chief end product of protein metabolism is ammonia. This substance (in the form of its ion, NH_4^+) diffuses readily out of the cell before reaching a dangerous concentration.

However, these organisms cannot handle excess water in this way. Lacking a rigid cell wall, they cannot combat the continual inflow of water by building up turgor either. The problem is solved by the **contractile vacuole.** Energy is used to force water back out of the cell and into the surrounding water. The contractile vacuole probably plays no significant role in the excretion of other substances. Amoebae that live in an isotonic environment (marine and parasitic forms) do not have a contractile vacuole. When placed in a hypotonic environment, however, they may develop one. This evidence suggests that the contractile vacuole serves simply to regulate water balance within the organism.

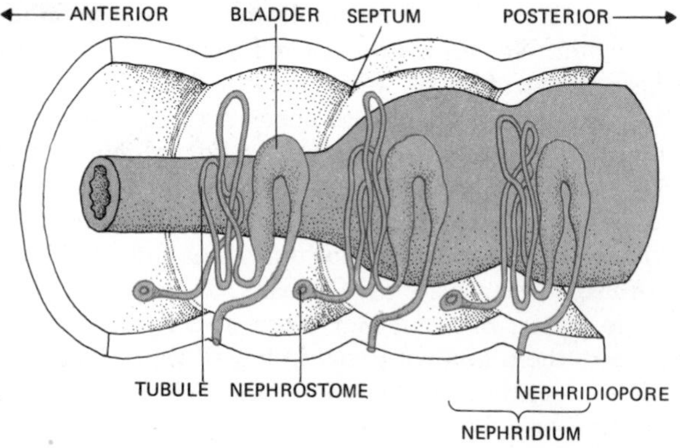

FIG. 13.1 Excretory system of the earthworm. A second nephridium is present in each segment. The nephridia pick up fluid containing both wastes and useful materials, but the useful materials are reclaimed as the fluid moves through the tubule.

13.3 EXCRETION IN THE INVERTEBRATES

The excretory system of the earthworm resembles the excretory apparatus of many other animals, both invertebrate and vertebrate. This does not necessarily mean that all these animals have inherited the structure from a common ancestor. It may be that such an efficient mechanism has arisen more than once in the course of animal evolution.

In the earthworm and many other invertebrates, these excretory structures are called **nephridia** (Fig. 13.1). There is a pair of these in each segment of the earthworm (except the first three and the last). Each nephridium consists of a ciliated funnel, the nephrostome, which is mounted on the septum, the membrane that separates the segments. It leads through the septum into a long tubule which lies coiled in the next posterior segment. The far end (distal end) of the tubule enlarges to form a temporary storage structure, the bladder. The bladder opens to the outside through a pore, the nephridiopore, on the ventral surface of the worm.

The nephrostome lies within the body cavity. This cavity is filled with fluid which is chiefly lymph filtered from the closed circulatory system. This fluid is the ECF of the cells of the earthworm's body. It contains useful materials (e.g. glucose) and useless or harmful materials (e.g. nitrogenous wastes). The fluid enters the nephrostome and passes down the tubule. It is pushed along by cilia and muscular contractions of the tubule. As it moves along, materials useful to the worm are reclaimed by the cells lining the tubule. The tubule is richly supplied with blood vessels so that these useful materials are quickly placed back in circulation. Useless or waste materials are not reclaimed and they ultimately leave the worm by way of the nephridiopores. We shall see that many other animals solve their excretory problems in a similar way. *All* the components of the ECF are sorted over, the useful materials being reclaimed while the rest are not.

In addition to this indirect sort of excretory activity, the nephridia may also function in a more positive fashion. Secretory cells in the tubule probably can transfer wastes from the blood capillaries directly into the lumen of the tubule. This is called tubular secretion.

Among freshwater animals, there are adequate quantities of water to flush away toxic nitrogenous wastes as fast as they accumulate. In fact, the chief problem of these animals seems to be coping with the problem of too much water entering their bodies from the hypotonic environment. Among terrestrial animals, the situation is quite different. Every effort must be made to conserve water. In the terrestrial arthropods, such as the insects, we get our first glimpse of the problems of excretion in a dry environment. The grasshopper solves the problem by means of both structural and biochemical modifications. The nitrogenous waste of the grasshopper (and other insects) is **uric acid.** This biochemical adaptation is extremely important because uric acid combines high nitrogen content with low toxicity. This low toxicity is partly a result of low solubility. Just as soon as the concentration of uric acid begins

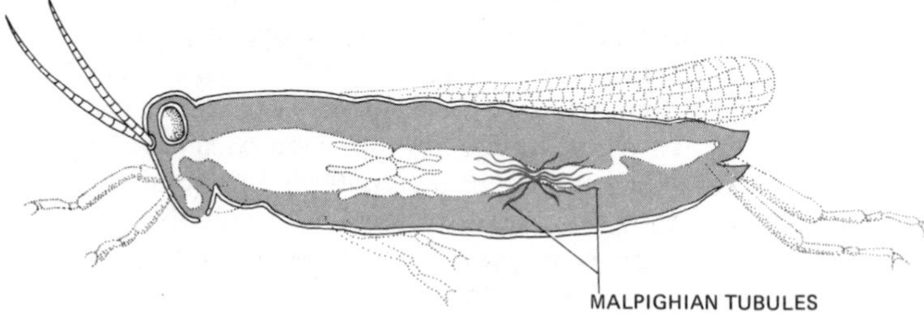

FIG. 13.2 The excretory wastes (chiefly uric acid) of the grasshopper are deposited in the intestine, where almost all moisture is reabsorbed. The grasshopper can thus excrete its metabolic wastes with practically no loss of water.

to increase appreciably, it settles out of solution as a solid precipitate. As a solid, it exerts no biochemical effect and can be removed from the body with only a small amount of water.

Structurally, the insect's excretory system is made up of **Malpighian tubules.** These lie in the hemocoel and are bathed in blood (Fig. 13.2). Although there is no direct opening to the blood, the cells of the tubules extract substances from it and pass them into the lumen of the tubules. Then valuable materials, including water, are reabsorbed. As the water concentration within the tubule decreases, uric acid precipitates. Instead of opening directly to the outside, the Malpighian tubules lead into the intestine where still more water is reabsorbed. The uric acid then passes to the outside with the feces. This excrement is quite dry. Thus the grasshopper is able to dispose safely of its nitrogenous wastes without losing valuable water in the process.

Excretion in Humans

The most important excretory organs of vertebrates are the kidneys. Because more is known about the function of the mammalian kidney than any other type, we shall first examine it, using ours as an example. Later in the chapter we shall examine some of the structural and functional modifications found in the kidneys of other vertebrates.

13.4 STRUCTURE OF THE HUMAN KIDNEY

The human kidneys are two bean-shaped organs, each about the size of a clenched fist. They are located against the dorsal body wall on either side of the backbone.

Although the total weight of the kidneys is only 0.5% of the weight of the body, the kidneys receive an extraordinarily rich supply of blood. Twenty to twenty-five percent of the blood pumped by the heart each minute flows through them. This blood reaches the kidneys by way of a right and left renal artery and leaves by the right and left renal veins (Fig. 13.3).

A transverse section of the kidney shows it to be composed of three distinct regions. The outer region is the **cortex.** Beneath this is the **medulla.** Within the medulla is a hollow chamber, the **pelvis** of the kidney.

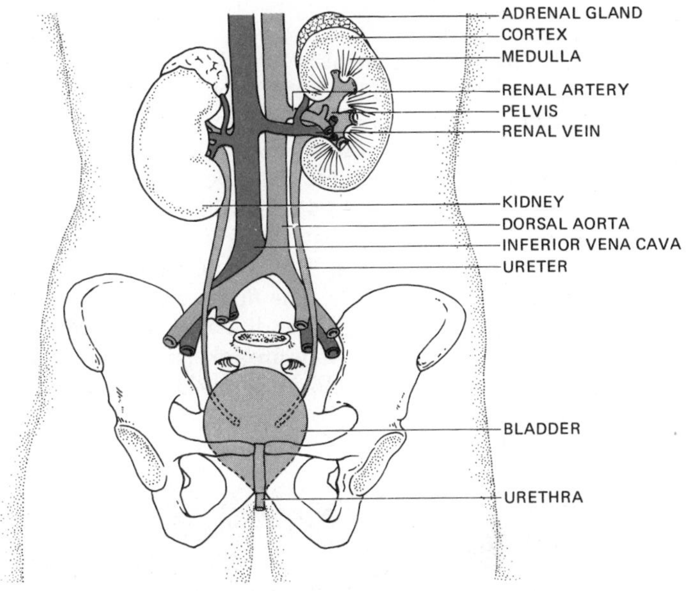

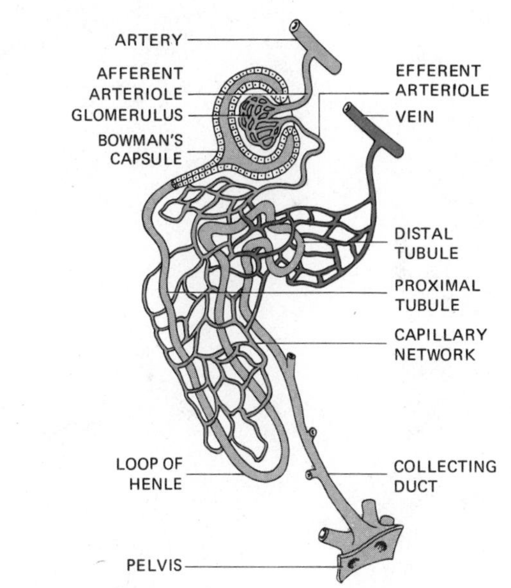

FIG. 13.3 Human excretory system. Above: Gross anatomy. Below: A single nephron (enlarged). There are approximately one million nephrons in each kidney.

The cortex and medulla of each kidney are made up of approximately one million nephrons. The **nephron** is the structural and functional unit of the kidney. To understand the physiology of the kidney as a whole we need only learn the physiology of a single nephron.

The nephron consists of a long (several centimeters) coiled tubule closed at one end and open at the other (Fig. 13.3). At the closed end of the tubule, in the cortex, the wall of the nephron is expanded and folded into a double-walled chamber, the **Bowman's capsule** (Fig. 13.4). Within the infolded portion of the Bowman's capsule is a network of capillaries, the **glomerulus.** The tubule itself consists of three distinct segments. The first segment, the **proximal tubule,** is coiled near the Bowman's capsule. The cells of which its walls are constructed are richly supplied with mitochondria. Many microvilli extend from these cells into the interior (the lumen) of the tubule. The proximal tubule leads to a long, thin-walled segment, the **loop of Henle.** This portion runs down into the medulla, makes a hairpin turn and returns to the region of the Bowman's capsule. Here the tubule expands again to form the **distal** (far) **tubule.** Like the proximal tubule, the distal tubule is highly coiled.

13.5 THE FORMATION OF URINE

The nephron manufactures **urine.** It does this by filtering the blood and then reclaiming useful materials back into the blood. This leaves the useless material to pass out of the nephron in a solution which we call urine.

Each glomerulus receives blood from an afferent arteriole and discharges its blood into an efferent

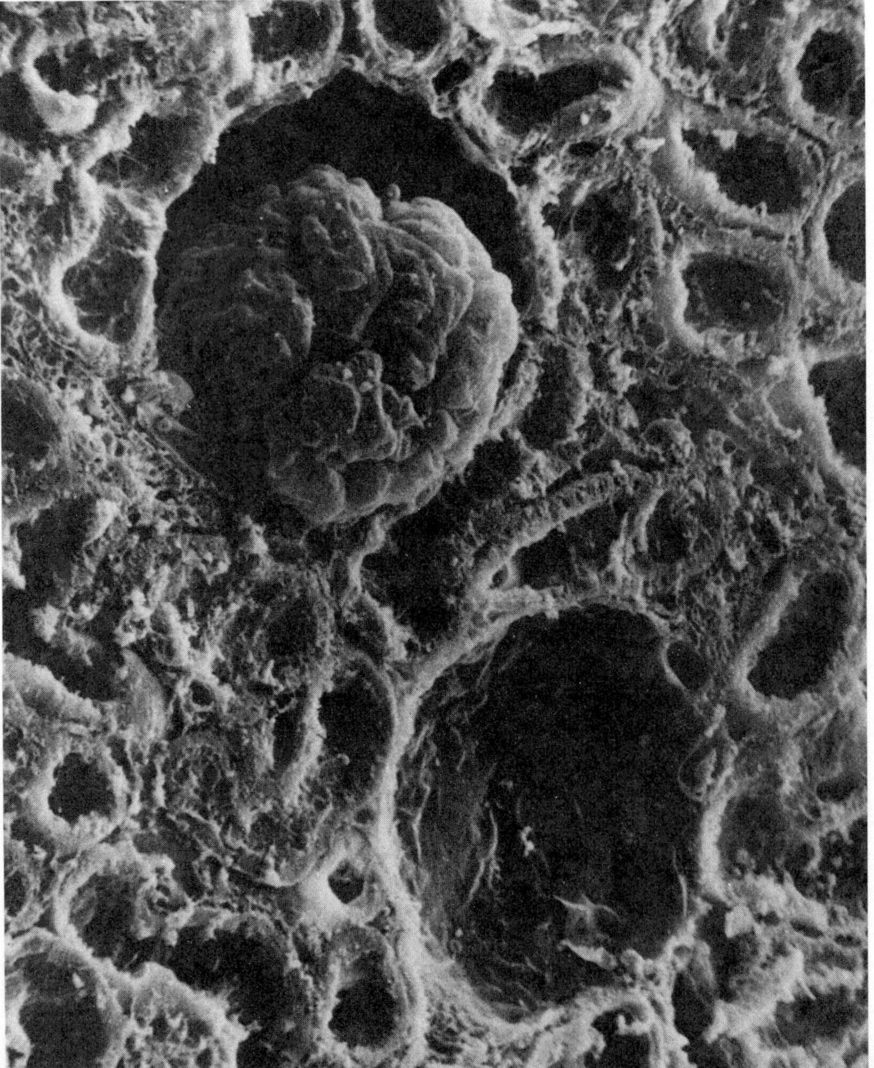

FIG. 13.4 Cut section of the cortex of a mouse kidney as seen under the scanning electron microscope. Near the top, center, can be seen a Bowman's capsule with its glomerulus. Directly beneath is another Bowman's capsule with the glomerulus removed. The remainder of the field reveals the lumens of both proximal and distal tubules as they have been cut at various angles. (Courtesy of Keith R. Porter.)

arteriole. The blood within the glomerulus, like that at the arteriole end of any capillary, is under pressure from the contraction of the left ventricle. This pressure causes water and the small molecules present in the blood (thus excluding proteins) to filter through the capillary walls. The fluid produced is called **nephric filtrate.** A glance at Fig. 13.5 will show that it is simply blood plasma minus almost all of the blood proteins. Essentially, then, it is no different from interstitial fluid (see Section 11.16).

The nephric filtrate collects within the Bowman's capsule and then passes into the proximal tubule. Here the reabsorption of glucose, amino acids, and large amounts of inorganic ions (Na^+, K^+, Ca^{++}, Cl^-, HCO_3^-, $PO_4^{\equiv}$, $SO_4^{=}$) takes place (Fig. 13.6). This reabsorption proceeds by active transport. All the details of how this is accomplished are not yet known, but enzymes and ATP are required. As mentioned above, the cells of the proximal tubule are richly supplied with mitochondria. Their rate of metabolism is equal to that of any cell in the body, including active muscle cells. Furthermore, their microvilli greatly increase the surface area exposed to the nephric filtrate. Thanks largely to the microvilli, the total surface area that participates in urine formation in humans is almost 6 m^2, or three times the surface area of the exterior of the body.

The materials reabsorbed by the proximal tubule are returned to the blood by a capillary bed which surrounds the tubules. In the mammals, this capillary bed receives its entire blood supply from the efferent arteriole of the glomerulus. The bed is drained by venules which lead into the renal vein.

As the various solutes are removed from the nephric filtrate and returned to the blood, a great deal of water follows them. This reabsorption of

COMPOSITION OF PLASMA, NEPHRIC FILTRATE, AND URINE, (g/100 ml OF FLUID)

COMPONENT	PLASMA	NEPHRIC FILTRATE	URINE	CONCENTRATION	% RECLAIMED
UREA	0.03	0.03	1.8	60X	50%
URIC ACID	0.004	0.004	0.05	12X	91%
GLUCOSE	0.10	0.10	NONE	–	100%
AMINO ACIDS	0.05	0.05	NONE	–	100%
TOTAL INORGANIC SALTS	0.9	0.9	<0.9–3.6	<1–4X	99.3%
PROTEINS AND OTHER COLLOIDS	8.0	NONE	NONE	–	–

FIG. 13.5 These are representative values. The values for salts are especially variable, depending on salt and water intake.

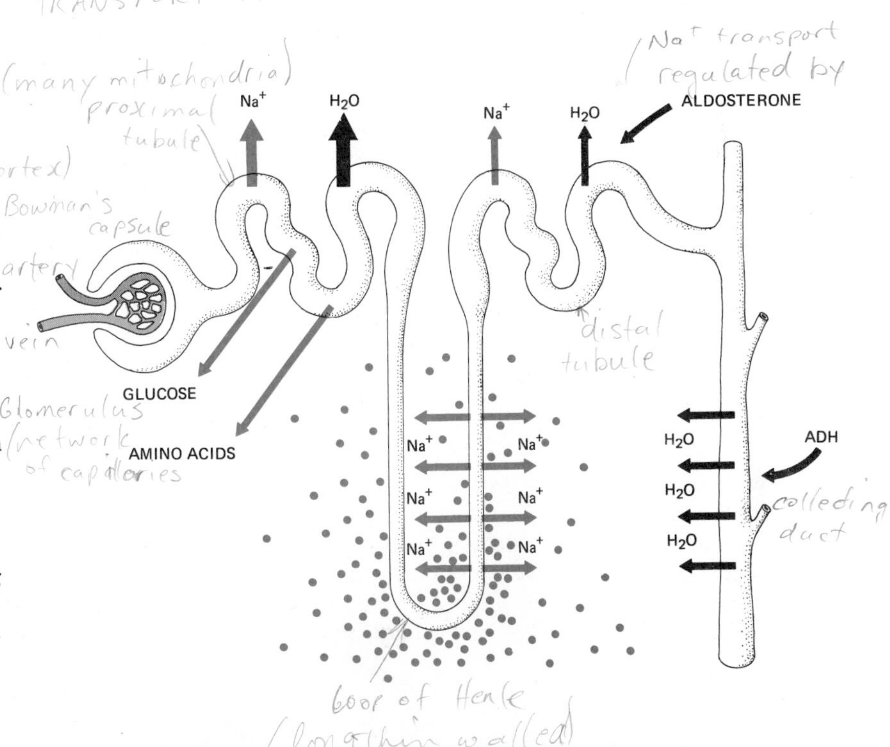

FIG. 13.6 Transport activities in the nephron. Sodium ions, glucose, and amino acids are actively transported out of the proximal tubules and back into the blood. A large volume of water follows these solutes by diffusion. Sodium ions are also actively transported out of the fluid in the loop of Henle and, in amounts regulated by the level of the hormone aldosterone, out of the distal tubules as well. The high concentration of sodium ions in the interstitial fluid around the loop of Henle sets up an osmotic pressure which draws water out of the collecting ducts. The permeability of the walls of the collecting ducts (and distal tubules) and thus the amount of water that diffuses out of them is regulated by ADH.

water is due entirely to the passive process of osmosis. The transfer of solutes from the nephric filtrate to the blood increases the water concentration of the nephric filtrate. Water then passes by osmosis into the blood and restores osmotic equilibrium.

Although the amount of nephric filtrate produced by a single nephron is not great, the quantity produced by two million of them is truly impressive. About 180 liters of nephric filtrate are produced in the kidneys each day. This volume of fluid may represent more than twice the weight of the individual. However, 80–85% of the water in the nephric filtrate is reabsorbed by the proximal tubules. Over a kilogram of NaCl, 400 g of $NaHCO_3$ (approximately equivalent to the contents of a pound box of commercial baking soda), 180 g of glucose, and lesser amounts of other useful materials are absorbed along with the water. It is easy to see that life could not be sustained for very long if this amount of material simply passed out of the body.

Not all of the solutes in the nephric filtrate are reclaimed by the proximal tubule. Some 50% of the nitrogenous waste, urea, remains behind. Substantial quantities of salts also remain dissolved in the fluid within the tubule.

As this fluid passes into the loop of Henle, it is approximately isotonic with the blood. The solutes and water have been absorbed in equivalent proportions by the proximal tubule. This situation is greatly altered as the fluid passes through the loop of Henle and the distal tubule. Here more sodium is actively transported out of the fluid (Fig. 13.6). But the walls of the loop of Henle are quite impermeable to the passage of water, so now water does *not* follow the sodium ions by osmosis. As a result, the interstitial fluid in the medulla becomes very salty. Having lost salt but not water, the fluid that remains within the loop of Henle actually becomes hypotonic to the blood by the time it reaches the distal tubule.

The walls of the distal tubules and of the collecting ducts are *variably* permeable to water. In the distal tubule, additional sodium is pumped out by active transport. If the body needs to reclaim more water to avoid dehydration, the walls are made more permeable to water and the water follows the sodium by osmosis. The collecting duct passes through the very salty, hypertonic medulla. Thus, when the walls of the collecting duct are made more permeable to water, the water passes out by osmosis into the interstitial fluid of the medulla (Fig. 13.6). When the system is working to reclaim the maximum amount of water, as little as 0.5 liter/day of urine may remain of the original 180 liters/day of nephric filtrate. And thanks to the strongly hypertonic interstitial fluid of the medulla, the concentration of solutes in the final urine can be as much as four times what it is in the blood.

Although urine formation in mammals occurs primarily by the filtration-reabsorption mechanism just described, an auxiliary mechanism is also involved. This is **tubular secretion.** The cells of the tubules remove certain substances from the blood in the surrounding capillaries. These substances are then secreted into the fluid within the tubules and thus added to the urine. Two important substances secreted in this fashion are hydrogen ions (H^+) and ammonium ions (NH_4^+). In each case, the secretion of these materials (which occurs in the distal tubules) is accompanied by the absorption—ion for ion—of sodium. This process is very important in helping to preserve the constancy of the pH of the blood. When the pH of the blood starts to drop, the extra hydrogen ions are secreted into the tubules. In the rare instances when the blood becomes too alkaline, the tubules secrete NH_4^+ (a base). In thus helping to maintain the pH of the blood within the normal limits of 7.3–7.4, the kidney can produce a urine with a pH as low as 4.5 or as high as 8.5.

Tubular secretion also plays a role in the elimination of excess potassium ions (they, too, are exchanged for sodium ions), and of certain nitrogenous wastes, e.g. creatinine. But even without tubular secretion, the kidney could probably handle the elimination of these substances easily. Curiously, tubular secretion plays a major role in the elimination of certain abnormal constituents that get into the body such as the dye phenol red. It seems quite remarkable that the cells of our tubules should possess the enzymatic machinery for the active transport of substances to which our species has only recently been exposed.

13.6 CONTROL OF THE KIDNEY

Most of the water filtered at the glomeruli—80–85% of it—is obligatorily reabsorbed in the proximal tubules. Varying amounts of the remainder are reabsorbed in the distal tubules and collecting ducts according to the water needs of the body. This *selective* reabsorption of water is regulated by a hormone which increases the reabsorption of water and thus reduces the volume of urine formed. Because of its action, the hormone is called the *a*nti*d*iuretic *h*ormone (diuresis = increased excretion of urine) or **ADH.** (It is also known as vasopressin.) ADH is a nonapeptide, that is, a polypeptide containing nine amino acids.

ADH is secreted by the pituitary gland, a tiny gland located at the base of the brain. The secretion of ADH is controlled by the water concentration of the blood. Anything that tends to dehydrate the body, such as perspiring heavily, causes a drop in the water concentration of the blood. This is detected by receptors in the brain which stimulate the pituitary to release ADH. ADH is then carried by the blood to the kidney. There it increases the permeability to water of the walls of the distal tubules and collecting ducts. More water leaves by osmosis and returns to the blood. This tends to restore the normal water level in the blood. It also leads to the production of a scanty, concentrated urine. When ADH is being actively secreted, the concentration of solutes in the urine may become as much as four times that of the blood. We share this ability to produce a hypertonic urine with the other mammals and with the birds. No other vertebrates can do it (and no other vertebrates have the loop of Henle that makes it possible).

If the blood should start to become too dilute (as would occur after drinking a large quantity of water), the secretion of ADH is inhibited. The distal tubules and collecting ducts fail to absorb as much water, and a great deal of watery urine is produced. The concentration of solutes in this urine may be as little as one-fourth that in the blood.

Occasionally people lose the ability to secrete ADH. They become victims of the disease **diabetes insipidus.** As you might guess, the most dramatic symptom of this ailment is an enormous production of watery urine. It is accompanied by terrible thirst. Sufferers may urinate as much as 20 liters of urine each day (10 to 20 times the average amount). To avoid fatal dehydration, they must constantly replace this water loss by drinking fresh water.

The reabsorption by the tubules of sodium ions and of calcium ions is also under hormonal control. The details of the action of the two hormones involved will be examined in Chapter 25.

Our study of the physiology of the tubules should make it quite clear that the kidney is far more than just an excretory organ. Certainly it removes nitrogenous and other wastes. However, it also removes foreign substances from the blood and normal plasma constituents that are present in the blood in greater than normal concentration. When excess water, sodium ions, chloride ions, etc. are present, the excess quickly passes out in the urine. On the other hand, the kidney steps up its reclamation of these same substances when they are present in the blood in less than normal concentration. In this way, the kidneys continually regulate the chemical composition of our blood within very narrow limits. Excretory products are eliminated simply because they naturally tend to accumulate to excess levels. The maintenance of a constant blood composition in turn assures a stable interstitial fluid. In short, the kidney is one of the most important of our organs for maintaining constant the composition of the internal environment. It is one of the major homeostatic devices of the body.

13.7 MECHANICS OF ELIMINATION

The process of urine formation continues unceasingly. Each collecting duct drains urine from several nephrons into the pelvis of the kidney (Fig. 13.3). The urine then flows down from the kidney to

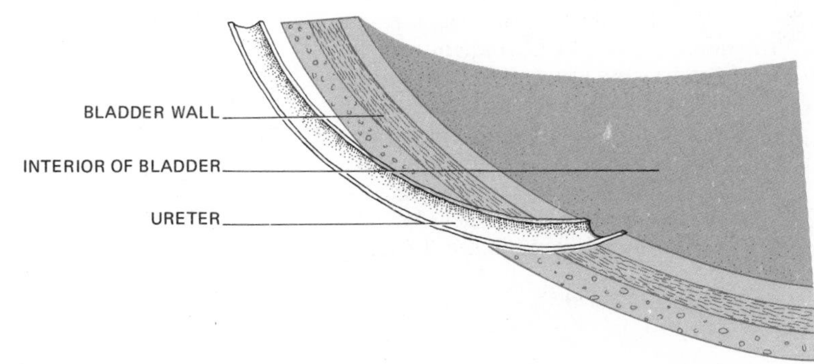

FIG. 13.7 Pressure within the bladder closes the ends of the ureters, causing urine to accumulate in them.

the bladder by means of a duct, the **ureter.** The **bladder** is a hollow, muscular organ which expands as urine flows into it from the two kidneys. When the bladder becomes filled with urine, the muscular sphincter guarding its outlet relaxes and urine flows to the outside through the **urethra.**

The two ureters enter the bladder in such a way that their exits become forced shut when the pressure in the bladder begins to build up (Fig. 13.7). If urination is delayed for some time, urine begins to accumulate in the ureters and even in the pelvises of the kidneys. When the bladder is finally emptied, this accumulation is permitted to flow down. Thus the individual finds that a long-delayed urination is quickly followed by the need for a second.

13.8 KIDNEY DISEASE

As one would expect from the sheer magnitude and crucial importance of the job done by our kidneys, any prolonged interference with their functioning has grave consequences. Unfortunately, there are a number of diseases which result in reduced or even no kidney function. Shock from serious injury and/or excessive blood loss, certain poisons, and the aftermath of certain infectious diseases may bring on a period of acute kidney failure. If the victim can survive the period of shutdown—with its attendant buildup of toxic wastes—the kidneys may begin functioning again as well as before. On the other hand, there are a number of progressive, degenerative diseases of the kidney which lead to a gradual, irreversible loss of kidney function.

One weapon against these life-threatening diseases has been the development of artificial kidneys. These operate on the principle of dialysis (Fig. 13.8). As we saw in Section 6.3, a semipermeable membrane, e.g. a sheet of cellophane, separating two solutions permits small molecules and ions to pass through but blocks the passage of large molecules such as proteins. A kidney machine (Fig. 13.9) simply passes blood (treated with heparin, an anticlotting agent) across one side of a cellophane membrane and a carefully prepared bath fluid across the other. The fluid cannot, of course, simply be water, because this would lead to a rapid loss of all small molecules and ions from the blood. However, by preparing a bath fluid of precise composition, any particular ion or molecule can be removed from (or added to) the blood, as appropriate. In fact, the efficiency of exchange is so great that if no urea is present in the initial bath fluid, the machine will remove urea from the blood more rapidly than normal kidneys can.

Artificial kidneys have proved enormously effective in helping victims of acute kidney malfunction to survive the crisis until their own kidneys resume operation. They have also enabled many

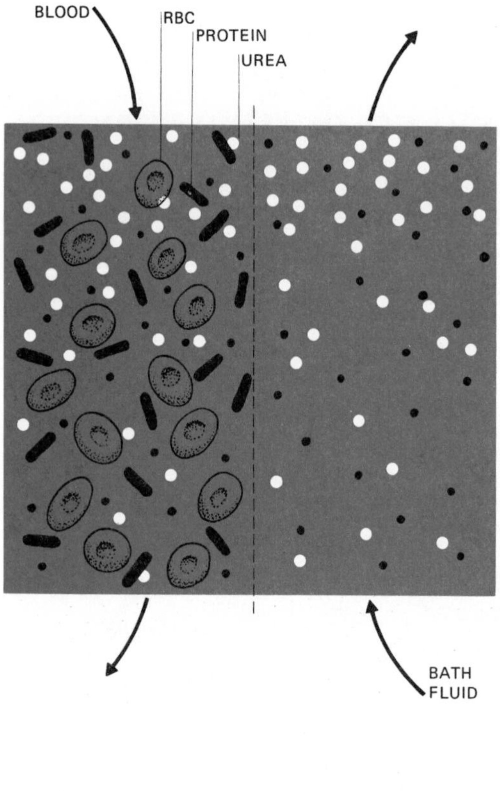

FIG. 13.8 Mechanism of dialysis in the artificial kidney. Small molecules, like urea, and salts are free to diffuse between the blood and the bath fluid whereas large molecules and cells remain confined to the blood. The bath fluid must already have had essential salts added to it to prevent a dangerous loss of these salts from the blood. An anticoagulant is added to the blood so it will not clot while passing through the machine.

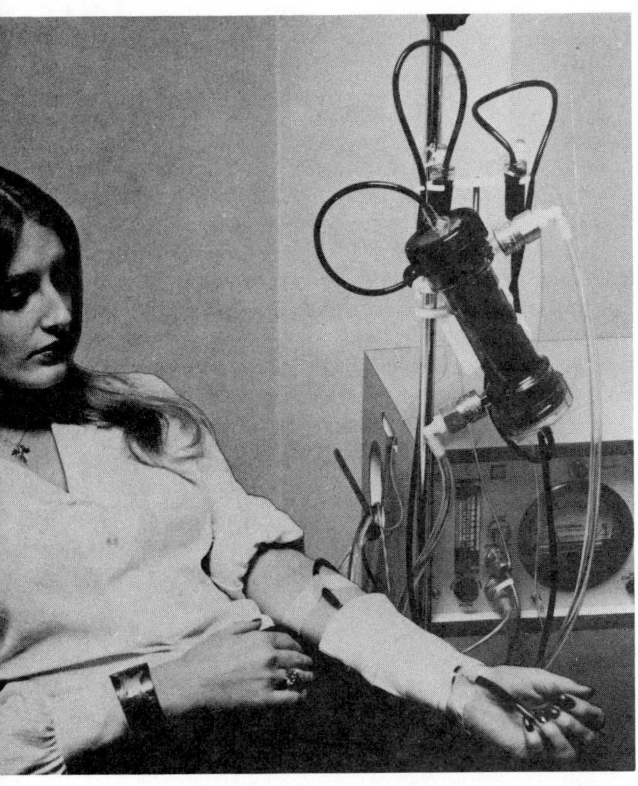

FIG. 13.9 An artificial kidney. Blood is withdrawn from an artery (upper tube), passes through a bubble trap, and enters the artificial kidney at the top. A continuous supply of fresh bath fluid (clear) enters the machine (lower left), passes over the dialysis membranes, and leaves (upper right) to be discarded. Purified blood leaves the artificial kidney (bottom), passes through a second bubble trap, and is returned to an arm vein (lower tube). (Photo courtesy of the Cordis Corporation.)

sufferers from chronic conditions to remain alive, though at an enormous expense of time (often three sessions of six or more hours per week), money, and psychological well-being.

A promising alternative to long-term dialysis for chronic kidney failure is the transplantation of a new kidney. The operation is technically quite simple. The donor kidney is placed low in the abdominal cavity and its artery and vein are connected to an iliac artery and vein respectively. The ureter can easily be connected to the bladder (Fig. 13.10).

The major problem with kidney transplants is the problem of immune rejection. Unless the patient has an identical twin as the donor, his immune sys-

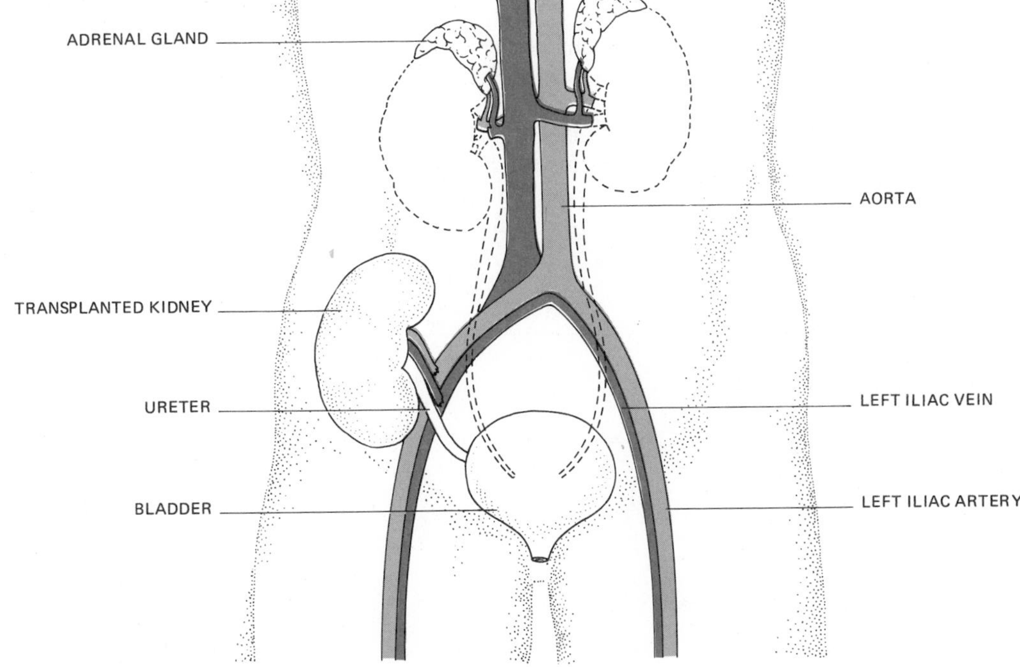

FIG. 13.10 Kidney transplant procedure. The recipient's own, diseased kidneys are usually removed. The renal arteries and veins are tied off (except for the branches supplying the adrenal glands, which are left intact).

tem (to be examined in Chapter 30) will recognize the implanted kidney as "foreign" and proceed to destroy it. However, a variety of drugs have been found to be effective in suppressing the body's immune mechanism. With the careful use of these, many implanted kidneys remain functional for periods of several years. As for transplants between identical twins, there is no need for immunosuppressive drugs, and the new kidney should survive indefinitely (20 years is the record at the time of this writing) unless the same underlying disease process that destroyed the host's own kidneys proceeds to destroy the implant.

13.9 THE NITROGENOUS WASTES OF HUMANS

The chief nitrogenous waste in humans (as in all mammals) is urea. It is manufactured in the liver. Most of the nitrogenous wastes in the body arise from the breakdown of amino acids. This breakdown, **deamination,** also occurs in the liver. Deamination results in the accumulation of ammonia (Fig. 13.11). Ammonia is an extremely poisonous substance and its accumulation in the body would quickly prove fatal. Fortunately, the liver contains a system of carrier molecules and enzymes which quickly converts ammonia (and carbon dioxide) into urea. As each molecule of urea is synthesized, the carrier molecule, ornithine, is regenerated for reuse. Although our bodies cannot tolerate high concentrations of urea, it is much less poisonous than ammonia. As we have seen, the kidney removes it efficiently from the body.

The presence of small quantities of uric acid in the urine is interesting. Uric acid is the main nitrogenous waste of insects, lizards and snakes, and birds. Of all the mammals, however, only humans, the

$$H_2N{-}C(H)(R){-}C(=O){-}OH \xrightarrow[+H_2O]{-2H} O{=}C(R){-}C(=O){-}OH + NH_3$$

(a)

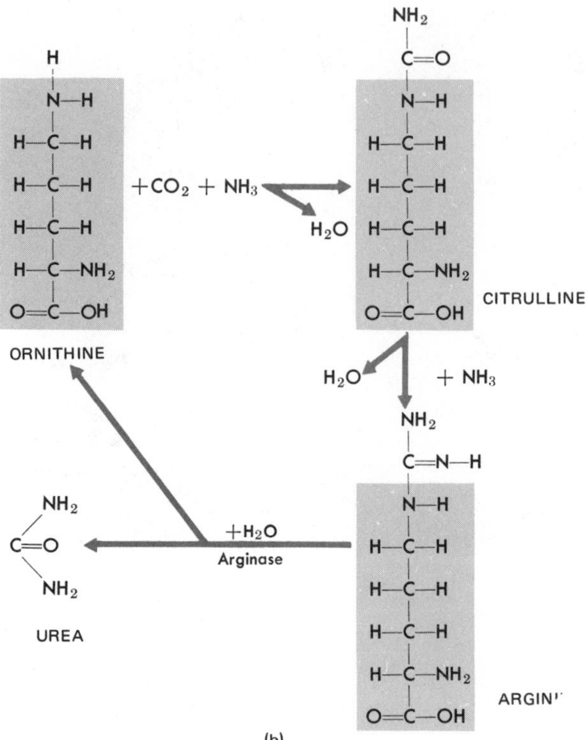

FIG. 13.11 (a) Deamination and (b) the synthesis of urea. The action of arginase on arginine produces a molecule of urea and regenerates ornithine for recycling.

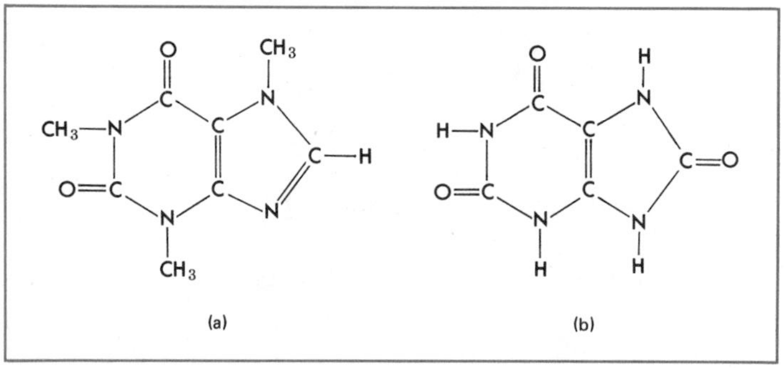

(a) (b)

FIG. 13.12 Structural formulas of (a) caffeine and (b) uric acid.

higher apes, and the Dalmatian dog excrete it. Our uric acid arises from the breakdown of nucleic acids rather than proteins. Other mammals produce uric acid by the breakdown of nucleic acids, but they also have an enzyme which then decomposes the uric acid. As mentioned earlier, uric acid is very insoluble. In some humans, the concentration of uric acid may be sufficiently high that it begins to precipitate out of solution. Needlelike crystals of uric acid accumulate in the joints, producing excruciating pain. The ailment is known as gout.

The uric acid molecule is quite similar in structure to the caffeine molecule (Fig. 13.12). Caffeine is the ingredient in coffee which, for many people, seems to be a mental stimulant. One scientist has suggested that when the loss of the ability to break down uric acid occurred in our early ancestors, their mental activities were heightened by the accumulation of this substance in the blood. Unquestionably, the dominant feature in the evolution of our species has been the development of the brain. Did uric acid play a role in this? We do not know. However, the fact that many of the famous (and brilliant) men of history were known to suffer from gout is used as support for this interesting speculation. Do you see any weaknesses in this notion?

Excretion in the Other Vertebrates

All vertebrates have kidneys. Like the human kidney, they are made up of masses of nephrons. There are, however, some differences in the structure and functioning of the nephrons in the various vertebrate kidneys. These differences can be related to the environment in which the animals live.

13.10 AQUATIC VERTEBRATES

The freshwater fish, like all freshwater animals, suffers a continual inflow of water from its hypotonic environment. The scales of the fish are impervious to water, but the delicate gill membranes provide ready access for water to enter the body. In order to avoid serious dilution of its body fluids, the freshwater fish must excrete this excess water. It does this by substantially the same mechanism as the amoeba: the use of energy to force the water back into the environment. In the fish, ATP provides the energy for the contraction of the heart. The contraction of the heart provides the pressure to force blood out of the glomerulus into the Bowman's capsule. Of course, valuable solutes (e.g. salts, glucose) pass into the nephric filtrate, too. These must then be reclaimed by the tubules. They pass into the blood in the capillaries surrounding the tubules. The blood supply in these capillaries comes from the glomerulus (as in humans) and also from the veins which drain the posterior portion of the fish's body (Fig. 13.13). This part of the blood supply is called the renal portal system. (What other example have you studied of a vein leading to a capillary bed rather than directly back to the heart?)

After salt reabsorption is complete, the urine is little more than just water. Most of the nitrogenous wastes (including large amounts of NH_3) of the fish actually leave the body by diffusion out of the gills. The kidney is really a device for maintaining proper water balance in the animal rather than an organ of excretion.

Marine fishes face just the opposite problem from that of the freshwater fishes. The salt content of sea water is so high that, physiologically, the marine fishes actually live in a dry environment. They are in continual danger of loss of vital body water to the hypertonic environment.

The marine fishes have solved this problem in two ways. The so-called cartilaginous fishes, the sharks, skates, and rays, have developed a tolerance to extraordinarily high levels of urea in their bloodstream. Shark's blood may contain 2.5% urea in contrast to 0.01–0.03% in other vertebrates. This urea level is so high that the shark's blood is isotonic with the sea water. It lives in osmotic balance with its environment and has a kidney which functions similarly to ours (with the exception that far more urea is reabsorbed in the shark's tubules than in ours).

The marine bony fishes have solved the problem differently (Section 6.4). They lose water continuously but replace it by drinking sea water and desalting it. The salt is returned to the sea by active transport at the gills. Living in constant danger of dehydration by the hypertonic sea, the bony fish has no reason to pump out large quantities of nephric filtrate at the glomerulus. The less water placed in the tubules, the less water has to be reabsorbed. It is not surprising that many of these fishes have very small, weakly functioning glomeruli. Some have no glomeruli at all (Fig. 13.13).

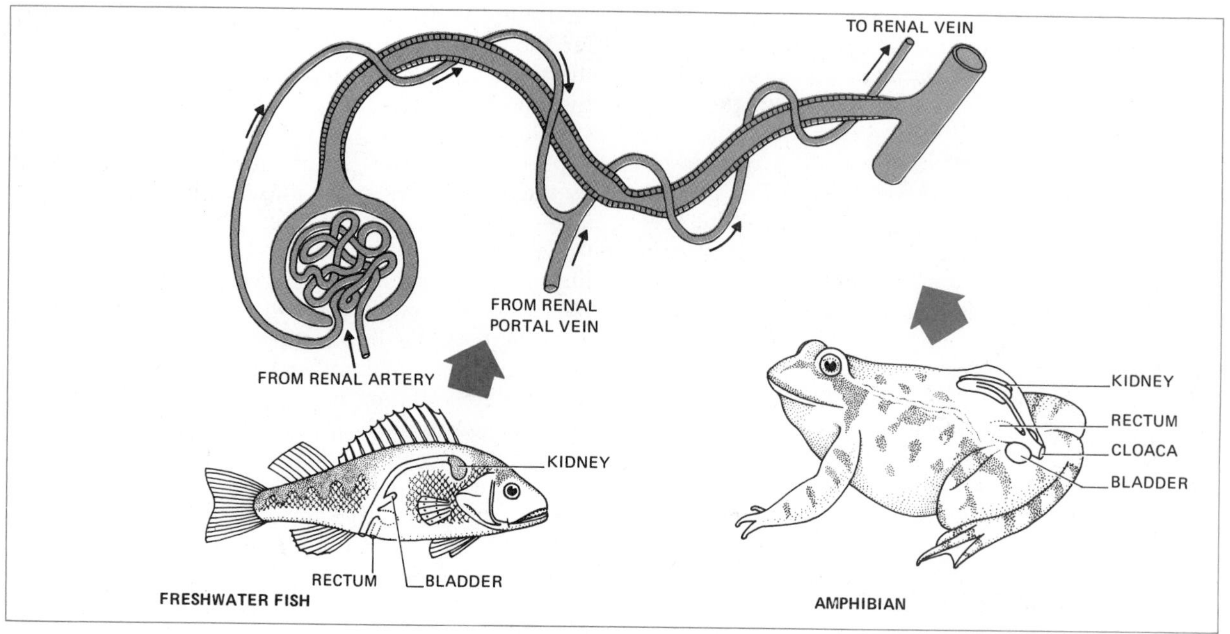
TO RENAL VEIN
FROM RENAL
PORTAL VEIN
FROM RENAL ARTERY
KIDNEY
RECTUM
BLADDER
FRESHWATER FISH
KIDNEY
RECTUM
CLOACA
BLADDER
AMPHIBIAN

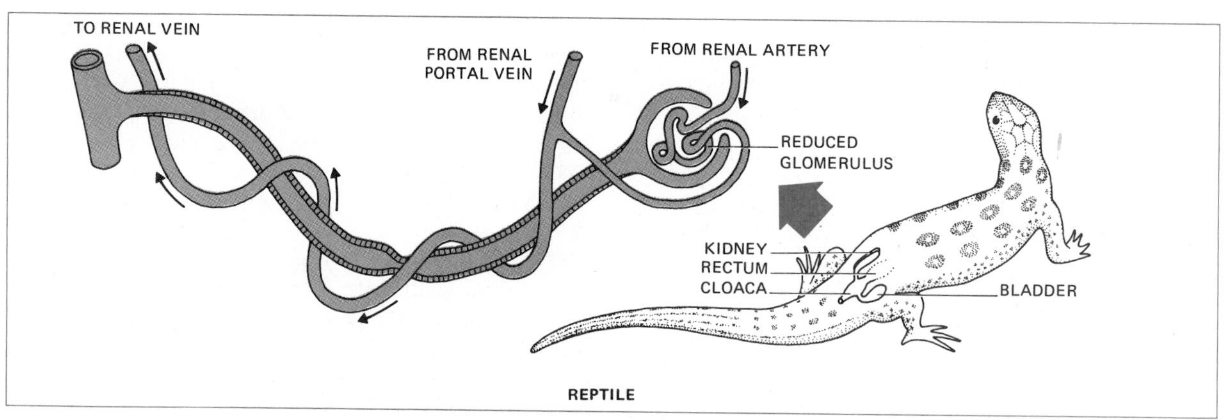
TO RENAL VEIN
FROM RENAL
PORTAL VEIN
FROM RENAL ARTERY
REDUCED
GLOMERULUS
KIDNEY
RECTUM
CLOACA
BLADDER
REPTILE

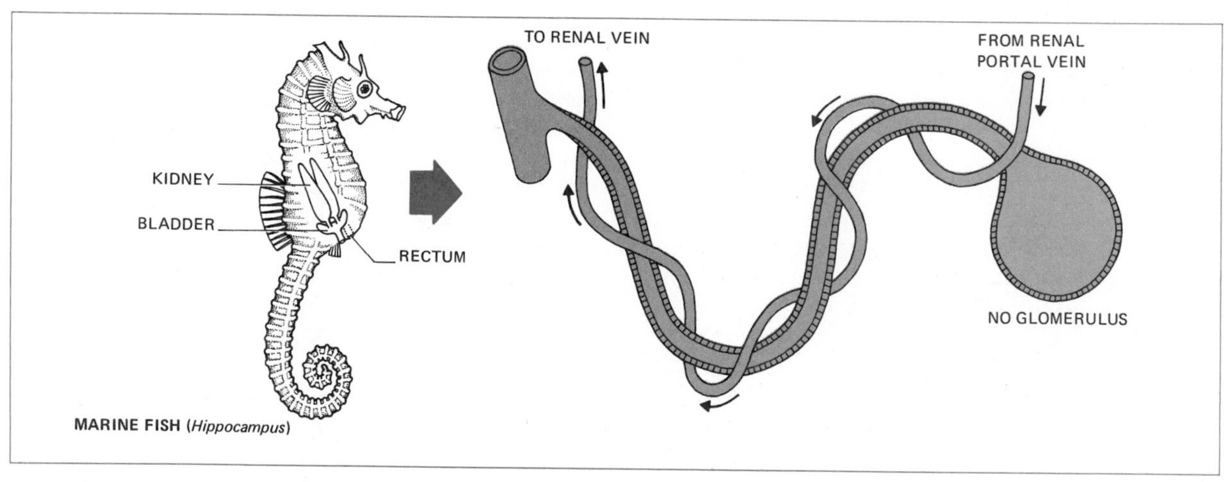
TO RENAL VEIN
FROM RENAL
PORTAL VEIN
KIDNEY
BLADDER
RECTUM
NO GLOMERULUS
MARINE FISH (Hippocampus)

◀ FIG. 13.13 A comparison of vertebrate kidneys. In each case, the structure of the nephron is related to the role it plays in maintaining a suitable water content within the animal.

With a reduction in the filtration-reabsorption mechanism, these fishes rely more on tubular secretion for eliminating undesired solutes. Tubular secretion requires a good blood supply to the tubules. Lacking efficient glomeruli, the renal portal system must carry most of the burden.

13.11 TERRESTRIAL VERTEBRATES

The **amphibian** kidney, like that of the freshwater fish, functions chiefly as a mechanism for excreting excess water. The permeable skin of the frog provides an easy site for the surrounding water to enter by osmosis. As the name amphibian suggests, however, these animals may spend some time on land. The problem then becomes one of conserving water, not eliminating it. The frog adjusts to the varying water content of its environment by adjusting the rate of filtration at the glomerulus. A renal portal system is present to carry away materials reabsorbed by the tubules during those periods when the blood flow through the glomerulus is restricted (Fig. 13.13). Furthermore, the frog is able to use its urinary bladder as a water conservation device. When in the water, the frog's bladder quickly becomes filled with a copious, watery urine. On land this water is reabsorbed into the blood thus replacing water lost by evaporation through the skin. The reabsorption mechanism is under the control of a hormone quite similar to ADH.

Many **reptiles,** e.g. desert rattlesnakes, live in extremely dry environments. Among the many adaptations that enable them to do so is their ability to convert waste nitrogen compounds into uric acid. As we have noted, uric acid is quite insoluble. Therefore, it can be excreted without the use of too much water. Accordingly, the reptile glomeruli are quite small. In fact, some reptiles have no glomeruli at all. Those that do have glomeruli filter just enough fluid to wash the uric acid, which is secreted by the tubules, into the **cloaca.** (The cloaca is a chamber through which the feces and the sex cells, as well as urine, pass on the way to the outside. The name comes from the Latin word for sewer.) Within the cloaca, most of the water is reabsorbed. Periodically, the reptile empties its cloaca. The feces are brownish in color; the uric acid forms a white paste. These water conservation mechanisms enable the reptile to survive even though little or no fresh water is available. The water content of its food coupled with the water produced in cellular respiration is usually sufficient for its needs.

It is important to note that the ability of these reptiles to convert waste nitrogen into uric acid is just as much an adaptation to dry-land living as their waterproof skin and their lungs. One is a physiological adaptation, the others are morphological. But note, too, that the excretion of uric acid is accompanied by structural changes in the kidney. Structure and function in living organisms are completely interrelated. One cannot appreciate the structural organization of cells, tissues, organs, and systems without understanding how they function. On the other hand, the physiological activities of living organisms can only be accomplished by properly organized structures. The student who studies the way in which living things are put together without any concern for the way these structural components function glimpses only half the story of life. Just as restricted is the view of the student who studies physiological activities with no concern for the structures which accomplish these.

Birds solve the problem of dry-land living in a way similar to that of the lizards and snakes. Uric acid is their nitrogenous waste product, too. It is the whitish material that pigeons leave on statues. Although their glomeruli are very small, they do function actively and a substantial amount of nephric filtrate is produced. There are probably two reasons for this. Birds (like mammals) have no renal portal system, so the blood needed for tubular reabsorption and secretion must come from the glomeruli. Furthermore, birds (again like mammals) have high blood pressures and thus cannot avoid substantial filtration rates at the glomeruli. However, the increased production of nephric filtrate is more than compensated for by an increased absorption of water in the tubules. The concentration of uric acid in bird urine may be as high as 21%. This is over 3000 times the concentration of uric acid in the blood. Tubular secretion of uric acid is important in this process of concentration, but the reabsorption of water in the tubules is vital, too.

Although **mammals** are also terrestrial animals, their chief nitrogenous waste is urea. This substance requires a good deal more water for its elimination than does uric acid. As we have seen, mammals produce large quantities of nephric filtrate but are able to reabsorb most of this in the tubules. But even with maximum reabsorption of water in the tubules,

FIG. 13.14 The kangaroo rat (*Dipodomys spectabilis*). This little mammal is so well adapted to life in the desert that it never needs to drink water. (Courtesy of Dr. Knut Schmidt-Nielsen.)

humans must excrete several hundred milliliters each day in order to flush urea out of the body.

Some mammals have developed kidneys more efficient than ours in this respect. Water reabsorption in the tubules of the kangaroo rat (Fig. 13.14) is so effective that it can produce urine 17 times as concentrated as its blood. The most hypertonic urine that we can produce is only about four times as concentrated as our blood. The kangaroo rat also conserves water by remaining in its burrow during the heat of the day, thus reducing water loss through evaporation. This combination of behavioral and physiological adaptations makes it unnecessary for the kangaroo rat to drink water. The small amount of moisture in its food and the water produced by cellular respiration are fully adequate to its needs.

We have been discussing the various kidneys found in vertebrates of today. It is thought that all of these vertebrates have evolved over a vast period of time from a common ancestor. Probably the fishes have evolved the least in this period. The mammals and birds represent a much greater change from the earliest vertebrates. As the various vertebrate classes evolved on earth, new ways of life developed, too. The amphibians took to the land. The birds took to the air. It would be surprising if the evolution of an organ as important as the kidney did not reflect these changes. In this chapter we have seen how the various vertebrate kidneys are adapted to the way of life of their owners. In Chapter 38, we shall reexamine these adaptations as part of the broader picture of the many structural and functional changes that occurred during the evolution of the various classes of vertebrates.

EXERCISES AND PROBLEMS

1. What is the chief nitrogenous waste of the (a) rattlesnake, (b) dog, (c) chicken, (d) shark?
2. Trace the path traversed by a molecule of urea from the time it enters a renal artery until it leaves the body.
3. Compare interstitial fluid and nephric filtrate with respect to (a) composition, (b) method of formation.
4. Synthesis of urea occurs in what organ?
5. Why is the problem of excretion less complex in the maple tree than in the dog?
6. List the most important structures for maintaining water balance in the (a) amoeba, (b) trout, (c) human, (d) *Elodea*.
7. How does the quantity of urine formed by a frog change when it moves from a pond onto land? Would the quantity of urine formed by a beaver change under the same circumstances?
8. Distinguish between excretion and egestion.
9. Carnivorous mammals excrete a higher concentration of urea than herbivorous ones. Explain.

10. During the formation of urine, why does the concentration of salts not increase in proportion to the concentration of urea?

11. The polysaccharide inulin is neither reabsorbed nor secreted by the tubules of the kidney. Inulin is infused into the bloodstream of a human subject at a rate that maintains a concentration in the plasma of 6 mg/liter for 24 hours. All the urine formed during this period is collected and a total of 1.02 grams of inulin is found in it. How many liters of nephric filtrate did the subject form during the 24-hour period? (This value is known as the *g*lomerular *f*iltration *r*ate or GFR.)

12. During the experiment described in the previous question, 25.5 grams of urea was excreted in the urine. What percent of the urea filtered at the glomeruli was reclaimed in the tubules if the plasma concentration of urea was 300 mg/liter throughout the period?

13. During the same 24-hour period described in the previous two questions, the drug p-aminohippuric acid was administered to the subject at a rate that maintained a plasma concentration of 150 mg/liter. A total of 100 grams of the drug was excreted in the urine. Explain the mechanisms by which the p-aminohippuric acid must have been excreted.

REFERENCES

1. SMITH, H. W., "The Kidney," *Scientific American,* Offprint No. 37, January, 1953. This beautifully written article reviews the anatomy and physiology of the mammalian kidney, its evolutionary history, and how the various vertebrate kidneys are adapted to the homeostatic needs of their owners.

2. MOFFAT, D. B., *The Control of Water Balance by the Kidney,* Oxford Biology Readers, No. 14, Oxford University Press, Oxford, 1971.

3. SCHMIDT-NIELSEN, K. and BODIL, "The Desert Rat," *Scientific American,* Offprint No. 1050, July, 1953. Describes the observations and experiments from which it was learned how the kangaroo rat conserves body water.

4. MERRILL, J. P., "The Artificial Kidney," *Scientific American,* July, 1961. How these life-saving devices work in temporarily replacing or supplementing natural kidney function.

5. FERTIG, D. S., and V. W. EDMONDS, "The Physiology of the House Mouse," *Scientific American,* Offprint No. 1159, October, 1969. With the emphasis on its mechanisms for maintaining water balance.

The moment of fertilization of a sea urchin egg as seen with the scanning electron microscope. (Courtesy of Dr. Don W. Fawcett.)

PART IV
REPRODUCTION

14 THE REPRODUCTION OF CELLS AND ORGANISMS

One of the most important aspects of living things is their ability to reproduce their kind. For every organism, there comes a time when its powers of metabolism, growth, and responsiveness are insufficient to maintain its complex organization against other forces. Attack by predators, parasites, starvation, other harmful changes in the environment, or simply those ill-defined processes that we call aging, result ultimately in the death of the organism. However, the species survives for periods far greater than the lifetime of any individual in it. This is accomplished with the production of new individuals by the old before the old die. Many of the major questions in biology concern this ability of living things to produce copies of themselves.

Two quite distinct methods of producing offspring are found among living things. One method is **sexual reproduction.** This is the production of new individuals that combine the hereditary information contributed by two different cells, generally representing two different parents. In most organisms, these cells are the gametes. The other method of reproduction involves only one parent. It is called **asexual reproduction.**

Asexual Reproduction

Asexual reproduction is the formation of offspring without the union of two gametes. Widely used by microorganisms, plants, and lower animals, it may be accomplished by a variety of specific methods.

14.1 ASEXUAL REPRODUCTION IN UNICELLULAR ORGANISMS

The most common method of asexual reproduction among unicellular organisms is **fission.** The organism divides into two roughly equal halves. Each of these then grows to full size and the process may then be repeated. Under ideal conditions, bacteria can divide by fission every 20 to 30 minutes. *Amoeba* and most of the other protozoans also reproduce by this method (Fig. 14.1).

Asexual reproduction in yeast cells is by **budding.** Budding differs from fission in that the two parts produced are not of equal size. In the yeast cell, a bulbous projection, the bud, appears at one portion of the cell wall (Fig. 14.1). The nucleus of the parent cell divides and one of the daughter nuclei passes into the bud. Under favorable conditions, the bud may in turn produce another bud before finally separating from the parent cell.

14.2 ASEXUAL REPRODUCTION IN MULTICELLULAR ORGANISMS

1. Budding

The term budding is also used to describe asexual reproduction in certain multicellular organisms. Undercooked pork may contain living "bladder worms" of the pig tapeworm, *Taenia solium.* These consist of a capsule containing a scolex (Fig. 14.2). When we ingest one of these, our gastric juice dissolves the wall of the capsule. The scolex turns inside out and attaches itself by suckers and hooks to the wall of the intestine. It then proceeds to produce buds, called **proglottids,** at its posterior end. These remain

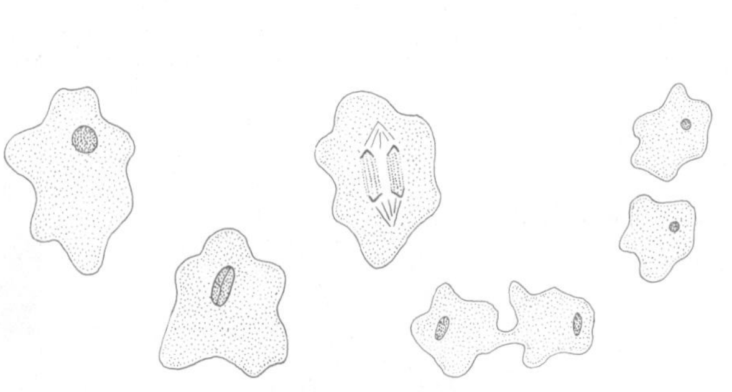

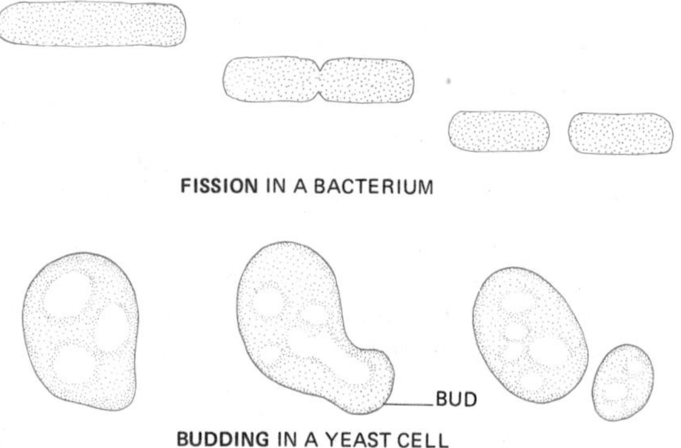

FIG. 14.1 Asexual reproduction in unicellular organisms.

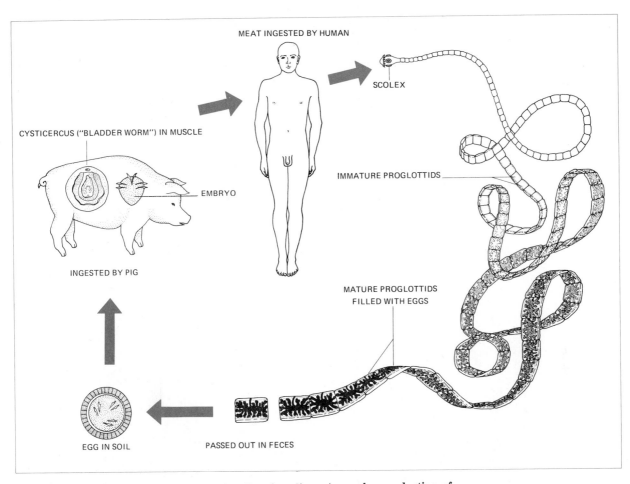

FIG. 14.2 Life cycle of the pig tapeworm *Taenia solium*. Asexual reproduction of proglottids by the scolex may produce a chain 6 meters (20 ft) long containing over 1000 proglottids. Each proglottid can be considered a separate individual. The proglottids reproduce sexually.

attached to one another. As they mature, they develop organs for sexual reproduction. The most mature proglottids eventually break loose and are passed out in the feces. Before this happens, the chain may reach a length of 20 feet and contain over 1000 proglottids. Although there are rudimentary nervous, excretory, and muscular structures shared by the proglottids, each proglottid can be considered a separate individual.

Similar *vegetative* methods of reproduction are found in plants. In some species, horizontal stems form which give rise to new individuals. These stems may extend underground (*rhizomes*) or along the surface of the ground (*stolons*) (Fig. 14.3). The houseplant *Bryophyllum* uses its leaves as organs of

FIG. 14.3 Asexual reproduction in the strawberry. The daughter plants form along the horizontal stems (stolons).

FIG. 14.4 Leaves of the common ornamental plant *Bryophyllum*. Note the miniature plants that have formed by asexual reproduction along their margins.

asexual reproduction. Along the margins, miniature copies of the plant form, complete with roots and stem (Fig. 14.4).

2. Sporulation

Fungi and certain plants carry out asexual reproduction by the formation of spores. These are small bodies containing a nucleus and a small amount of cytoplasm. The spores of terrestrial organisms are usually very light and have a protective wall around them. These two features make sporulation more than just a special reproductory mechanism. The small size and light weight of spores enable them to be carried great distances by air currents. Spores thus serve as agents of dispersal, spreading the organisms to new locations.

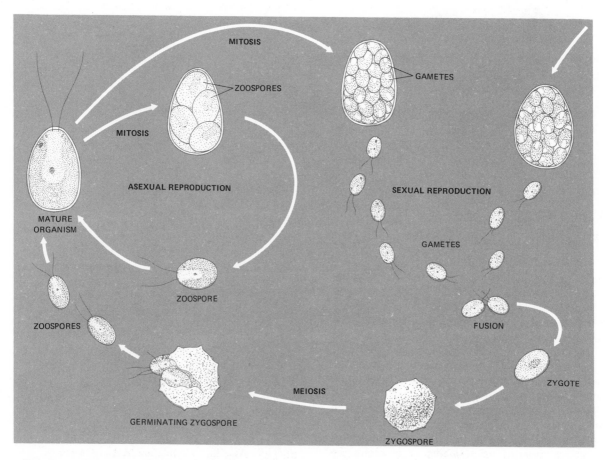

FIG. 14.5 Life cycle of *Chlamydomonas*. Although all its gametes look alike, two gametes from the same parent cell will not fuse with each other.

The resistant coat of the spore often serves another useful function. It enables the species to be maintained in a protected and dormant state over periods of unfavorable conditions which would be fatal to the active, vegetative organism. It is not surprising that this type of spore is produced most rapidly when conditions of temperature, moisture, or food supply become less favorable.

Many bacteria are spore formers. This cannot be considered a reproductory mechanism for them, however, for each cell simply forms a protective case around its cytoplasmic contents. No new individuals are produced. The spore merely provides a means of survival during unfavorable conditions.

In some of the green algae and aquatic fungi, the spores are not resting stages. In *Chlamydomonas,* the contents of a single cell divide one to three times, forming two to eight small **zoospores,** each with its nucleus, cytoplasm, and two flagella (Fig. 14.5). After being released each zoospore grows to the full size of the parent cell. Some nonmotile algae use zoospores not only for reproduction but as a method of dispersal. Swimming by means of their flagella, the spores disperse the species to new locations.

The fungi are prolific spore producers. A single *Lycoperdon* mycelium produces approximately 700,000,000,000 spores in one season in its puffballs (Fig. 14.6). Spores of the wheat rust fungus have been recovered by aircraft flying 14,000 feet above the earth. If one leaves a piece of moistened bread (which does not contain a mold inhibitor) in a warm, dark place exposed to air currents, a luxurious growth of mycelium soon reveals how widely distributed fungus spores are (Fig. 14.7). The mosses, club mosses, and our common ferns also produce enormous numbers of tiny, windblown spores which serve to disperse the species to new locations.

FIG. 14.6 Puffballs of the fungus *Lycoperdon.* Spores are produced asexually within the puffballs. (Courtesy of Jack Dermid.)

3. Fragmentation

Some plants and animals accomplish asexual reproduction by fragmentation. In these species, the body of the organism breaks up into several parts. Each of these can then regenerate all the structures of the mature organism. In some tiny worms, the process occurs spontaneously. After the worm grows to full size, it breaks up into eight or nine pieces. Each of these then develops into a mature worm and repeats the process.

Most cases of fragmentation depend upon outside forces. Green and brown algae at the seashore are often broken into pieces by the action of the waves. Each of these pieces can then grow to full

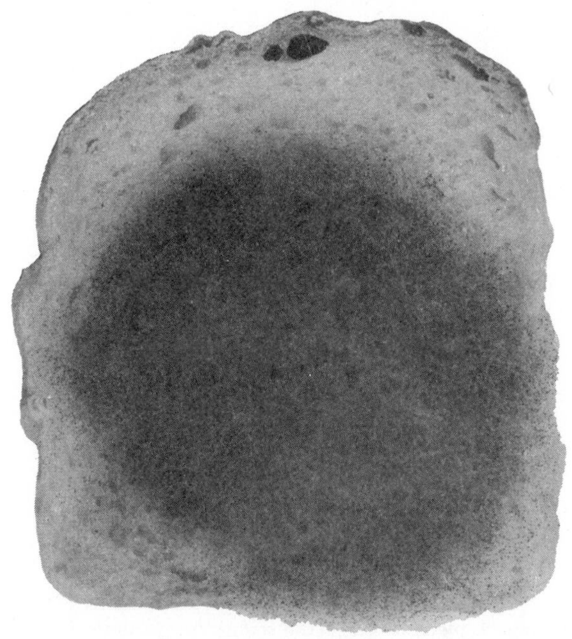

FIG. 14.7 This slice of bread was exposed to the air in a dusty room and then placed in a warm, dark location. The mold is *Rhizopus stolonifer.*

size. Even in fresh water, filaments of algae are frequently broken apart. Fission of the cells remaining in a fragment quickly rebuilds the filament.

Gardeners often fragment plants deliberately in order to reproduce desired varieties asexually. This is often done by making *cuttings*. With proper care, these develop roots and leaves and can take up an independent existence.

14.3 THE NATURE OF ASEXUAL REPRODUCTION

The three kinds of asexual reproduction just considered occur in nature regardless of whether we can exploit them to meet our own needs. **Grafting,** by contrast, is a method for asexual reproduction of plants that has been deliberately devised by humans to produce additional individuals of a desired variety. Only nurserymen deliberately grow apple trees by planting apple seeds. They do not raise these seedlings for the fruit they could produce but simply for a strong root system. After a year's growth, the portion of the plant above ground is removed and a twig (the **scion**) taken from a mature tree of the desired variety is inserted in a notch in the cut stump (the **stock**) (Fig. 14.8). So long as the cambiums of scion and stock are united and precautions are taken to prevent infection and drying out, the scion will grow. It will get all its water and minerals from the root system of the stock. However, the fruit that will be eventually produced will be identical (assuming it is raised under similar environmental conditions) to the fruit of the tree from which the scion was taken.

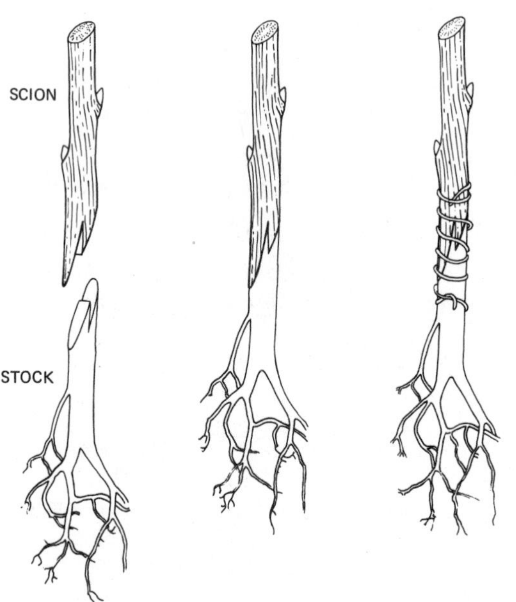

FIG. 14.8 Apple trees are propagated asexually by grafting. A piece of stem (the scion) from the desired variety is inserted into a notch cut in the stump of the stock. The fruit eventually produced will be characteristic of the variety that supplied the scion, not the variety that supplied the stock.

The need to specify similar environments is nicely illustrated in the wine industry. Most French vineyards are planted to grapes propagated vegetatively from California vines. However, the grapes of France (and the wines produced from them) are somewhat different from those of California.

The McIntosh apple is one of many popular varieties of apple grown in the United States and Canada. The first McIntosh tree was found over 150 years ago growing on the farm of John McIntosh in Ontario, Canada. It had grown from a seed. McIntosh's daughter-in-law appreciated the desirable qualities of the fruit. Moreover, she knew that it would be useless to try to grow other trees of the same kind from the seeds of the apples growing on this particular tree. Seeds, as we shall see in Chapter 20, develop as a result of sexual production. Two parents are involved, and while one would be the McIntosh tree, the other parent would probably be some other apple tree growing in the vicinity. The offspring would possess characteristics of both parents. Perhaps it might produce better apples, perhaps worse. But it would not be a McIntosh. The only way in which the McIntosh could be made available to other apple growers was by *asexual* reproduction. Scions removed from the original tree and grafted onto stocks of any apple variety would produce McIntosh apples. Every one of the hundreds of thousands of McIntosh apple trees growing today is descended from a scion of that first tree. Or, to put it another way, all these trees make up a **clone.** They all share an identical genetic heritage because each has been produced by the continued division of the cells of that first tree.

This continuity of traits from one generation of cells to the next is nicely exploited in the brewing industry. The flavor of a beer or ale is dependent on a number of factors, but one of the most important is the particular strain of yeast which is used in the fermentation process. In a typical case, several hundred pounds of yeast cells are added to a vat filled with various ingredients, including carbohydrate as an energy source. After four or five days, the quantity of yeast in the vat will have increased three- or fourfold. A portion of this population is removed from the mixture and carefully saved to be used to start the next batch of ale or beer. At all times, great care is taken to see that the yeast strain does not become contaminated by other microorganisms. Thanks to such precautions, a single strain of yeast may be used for decades to produce a unique ale or beer. Even with the slow growth rate that takes place under the conditions used in brewing, after some 20 years the cells being used are the product of as many as 3000 generations—yet the traits of the original yeast cells have remained unchanged.

These examples of asexual reproduction are useful because they point out the essential feature of this method of reproduction. In all kinds of asexual reproduction, the offspring are identical in every way to the parent so long as they are raised under environmental conditions similar to those of the parents. If a given species is successful in its habitat, any inheritable variation in the offspring may be disadvantageous. Asexual reproduction is a method for producing new individuals which will probably not show any such variations. It tends to preserve the *status quo*.

In all forms of asexual reproduction, new cells are produced by old. As our example of the McIntosh tree has shown, these new cells retain the same hereditary blueprints as their parents. In Chapter 5 we saw how Boveri's experiment showed that these hereditary blueprints reside in the nucleus. What in the nucleus contains them? When a cell divides, how is the information contained in the hereditary blueprints transmitted to its two daughter cells? As we shall see in Chapter 16 the answer to the first question is the *chromosomes*. The answer to the second question is *mitosis*.

14.4 MITOSIS

All the kinds of asexual reproduction discussed earlier are carried out by mitosis. In addition, mitosis is the mechanism responsible for growth, regeneration, and cell replacement in an individual multicellular organism of any kind, whether that organism reproduces itself by sexual or asexual methods.

A convenient tissue in which to study mitosis is the meristem at the growing point of an onion root. Staining with an appropriate dye makes the chromosomes in dividing cells visible (Fig. 14.9).

A newly formed onion root cell contains 16 chromosomes; 8 of these had originally been contributed by the "father" of the onion plant, that is, the plant that supplied the male gamete. These chromosomes are often called *paternal* chromosomes. The other 8 were originally supplied by the onion's "mother," that is, the onion that produced the egg. These are the *maternal* chromosomes. For each of the maternal chromosomes, there is a paternal chromosome which looks just like it. These similar chromosomes make up a **homologous pair.** Each member of a given homologous pair is often referred to as the *homologue* of the other member of the pair.

When a cell is not actually in the process of dividing, the chromosomes (which are housed within the nucleus) cannot be seen under the light microscope. They are too tenuous to absorb very much stain and reveal their true nature. When the chromosomes are in this condition, they are sometimes referred to collectively as the **chromatin** of the nucleus.

In many cells, including the onion, one or more of the chromosomes has a nucleolus attached to it. This *can* be easily observed under the light microscope. The tenuous state of the chromosomes during the period between cell divisions should not suggest that they are inert at this time. Far from it. They are quite active in RNA synthesis and, at some time before the next cell division, DNA synthesis as well. In fact, the DNA content is exactly *doubled* between cell divisions.

The various events which occur during mitosis have been divided into four consecutive phases: prophase, metaphase, anaphase and telophase. The period between divisions is called interphase. It is important to realize that these phases are simply a convenient way of describing mitosis. The actual process involves (with a few exceptions) an unbroken sequence of events which merge smoothly with one another. The photomicrographs in Fig. 14.9 are "snapshots" of these various phases. Motion pictures would impart a better appreciation of the process.

Prophase. The onset of mitosis is marked by several changes. The nucleoli begin to disappear while the chromosomes themselves begin to appear. The previously extended strands of the chromosomes coil up into a helix, that is, like a cylindrical spring (Fig. 14.9). In so doing, they become shorter and thicker and thus more easily visible. At this time, the nuclear membrane begins to disappear.

The significance of the doubling of the DNA content of the cell before mitosis now becomes ap-

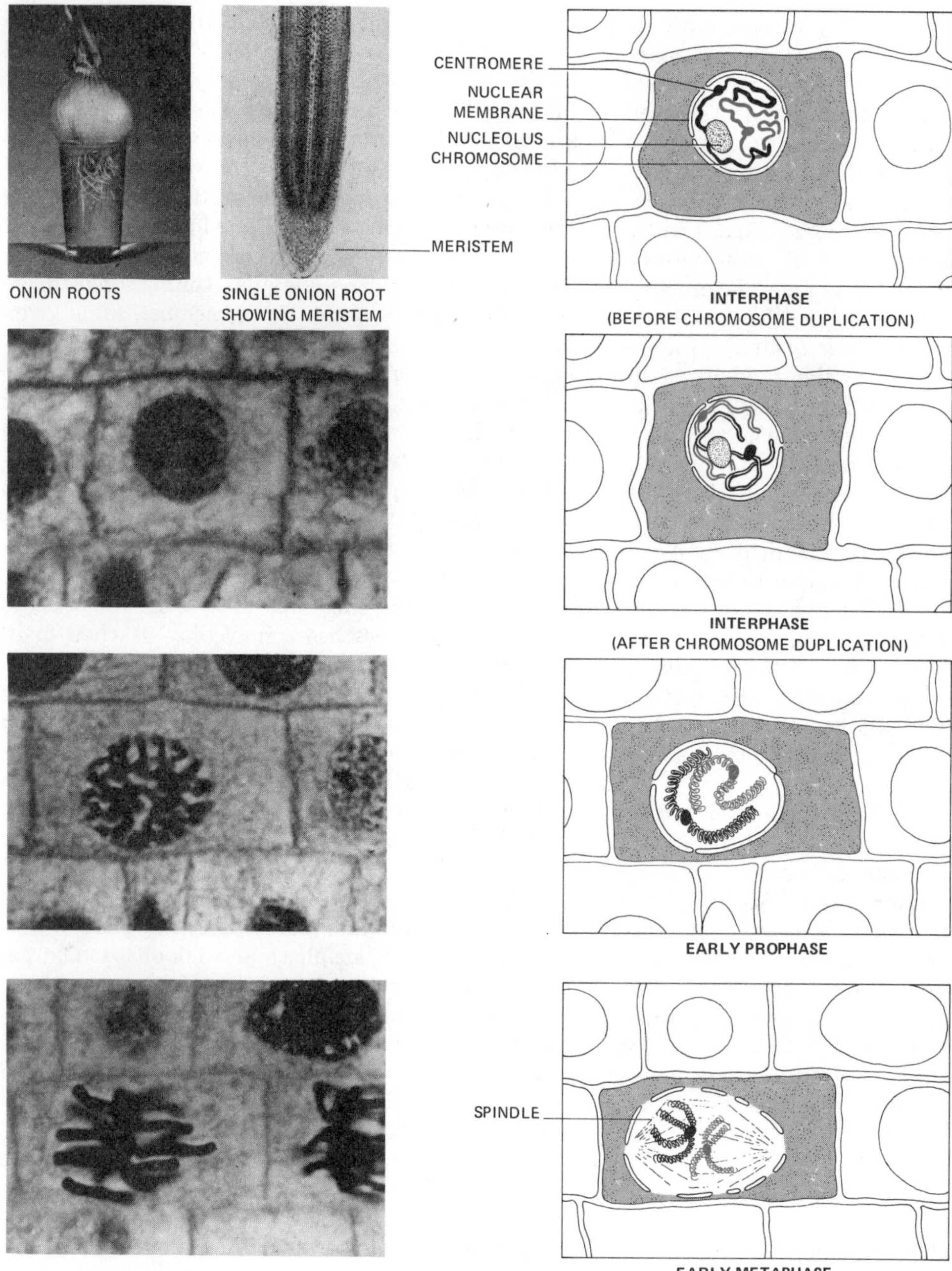

ONION ROOTS

SINGLE ONION ROOT SHOWING MERISTEM

INTERPHASE (BEFORE CHROMOSOME DUPLICATION)

INTERPHASE (AFTER CHROMOSOME DUPLICATION)

EARLY PROPHASE

EARLY METAPHASE

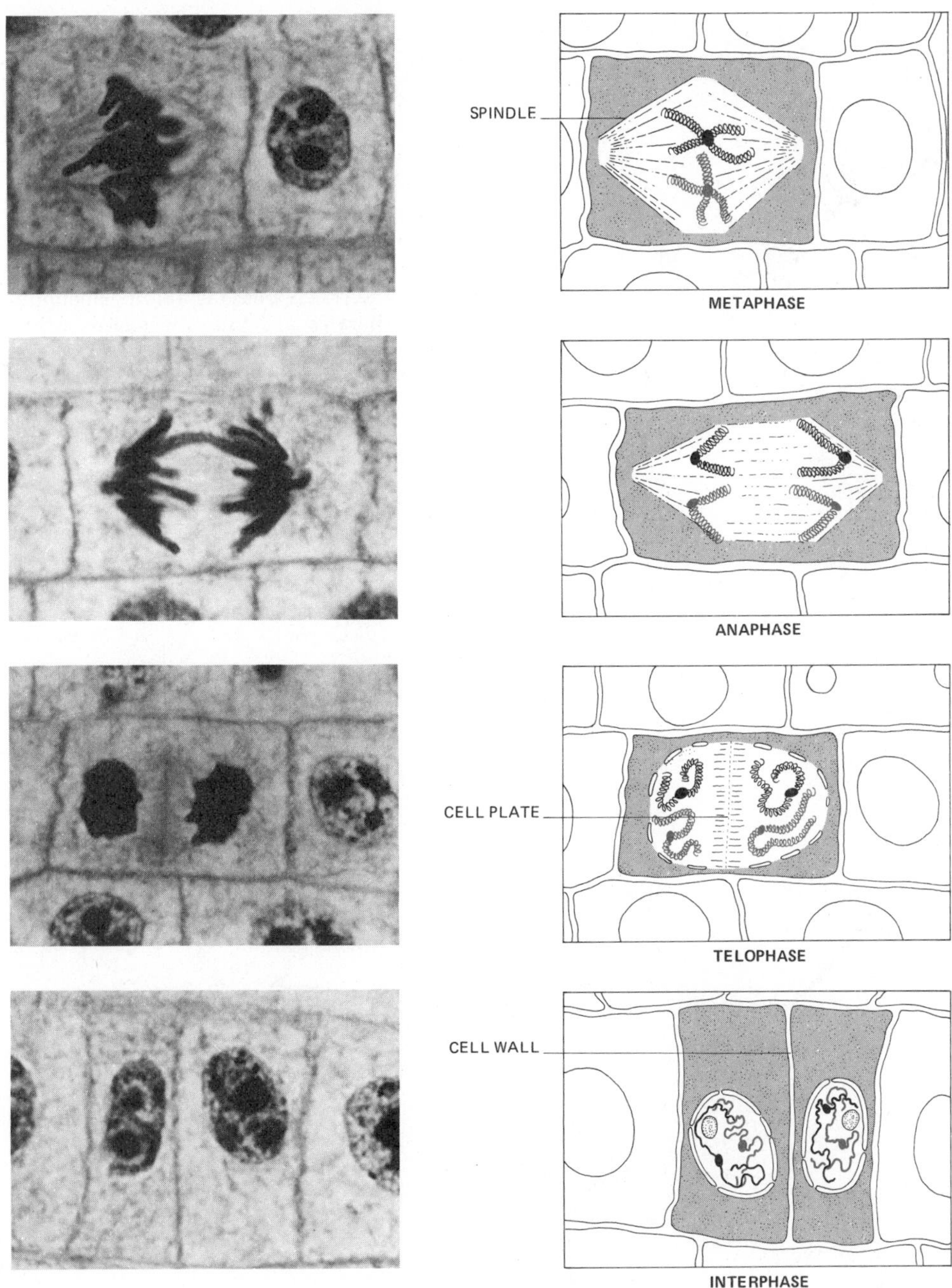

FIG. 14.9 Mitosis in a plant. The photographs show the stages as they occur in the dividing cells of the onion root tip. (Courtesy of the Carolina Biological Supply Co.) The drawings show the stages in a semidiagrammatic fashion. For the sake of clarity, only one pair of homologous chromosomes is shown: one member in black, the other in color.

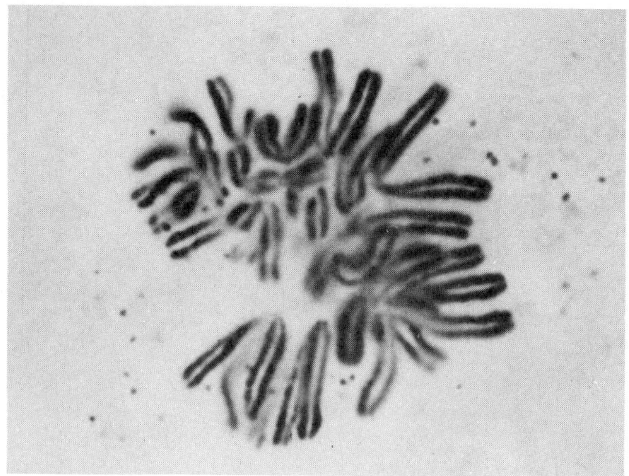

◄ **FIG. 14.10** Chromosomes in a dividing epidermal cell of a salamander. The duplicated chromosomes (sister chromatids) are still attached to each other by centromeres. (Courtesy CCM: General Biological, Inc., Chicago.)

FIG. 14.11 The cell cycle. The timing and location of DNA synthesis can be established by supplying radioactive thymidine (^{3}H-thymidine) followed by autoradiography. Only during *S* phase is a large amount of radioactivity incorporated and this is localized in the nucleus (b_1). If the treated cell is permitted to proceed to mitosis (b_2), the radioactivity can be seen to be localized in the chromosomes. (From D. I. Patt and Gail R. Patt, *An Introduction to Modern Genetics,* Addison-Wesley, Reading, Mass., 1975.) ▼

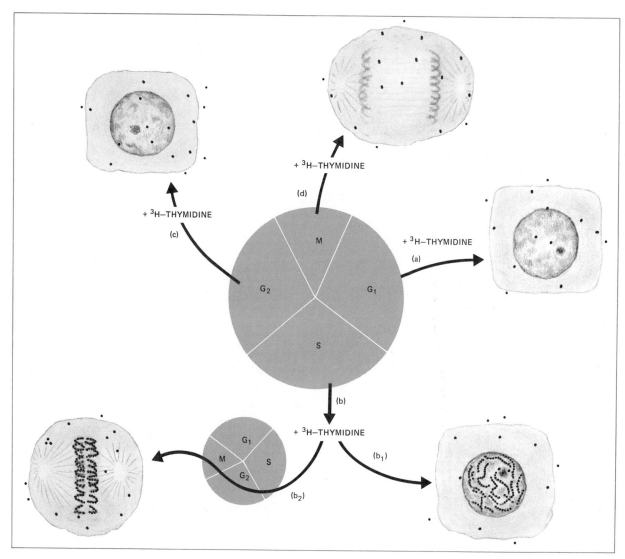

parent. Each of the 16 chromosomes (8 homologous pairs) that was present in the cell when it was first formed now reappears, *doubled* (Fig. 14.9). The duplicates are attached to each other at a constricted region called the **centromere** (also called a kinetochore). Although the term is not a technical one, we can refer to these duplicated chromosomes as making up a *doublet* (Fig. 14.10). The standard practice is to call the entire structure a chromosome and each of its strands a **sister chromatid.** Unfortunately this terminology, while well entrenched, obscures the fact that each member of a doublet, that is, each sister chromatid, is equivalent in every way to the *chromosome* that the cell received when it was formed by a prior mitosis. In this sense, the onion cell in prophase has not 16 but 32 chromosomes consisting of 8 *pairs* of homologous *doublets.*

Metaphase. Metaphase is marked by the appearance of the **spindle.** This structure consists of an array of microtubules (see Section 5.15) which extend between the ends or "poles" of the cell. The centromere of each doublet becomes attached to a microtubule and migrates to a point midway between the poles. The loose ends of the chromosomes may be oriented randomly, but all the centromeres lie exactly in a plane at the "equator" (Fig. 14.9).

Anaphase. Anaphase begins when the duplicated chromosomes (or sister chromatids) of each doublet separate from one another. They now move apart, still on the spindle, and migrate to opposite poles, dragging their free ends behind them. The metaphor seems especially apt because the free ends of the chromosomes now turn toward the equator just as though friction with the surrounding cytoplasm were impeding their motion toward the poles.

Telophase. Telophase is roughly the reverse of prophase. Once the chromosomes reach their poles, they begin to uncoil. The nucleoli reappear. A nuclear membrane begins to form around the chromosomes. Finally, a structure called the **cell plate** appears at the equator. A cell wall is secreted on each side of the cell plate and cell division is thus completed.

Interphase. Cytologists find it convenient to distinguish three periods during interphase. Once mitosis is complete, the cell begins a period of growth (called G_1). This is followed by a period (S) of DNA synthesis during which the chromosomes are duplicated. Then a second growth period (G_2) occurs prior to the next mitosis (M). These four periods in the life of dividing cells make up the "cell cycle" (Fig. 14.11).

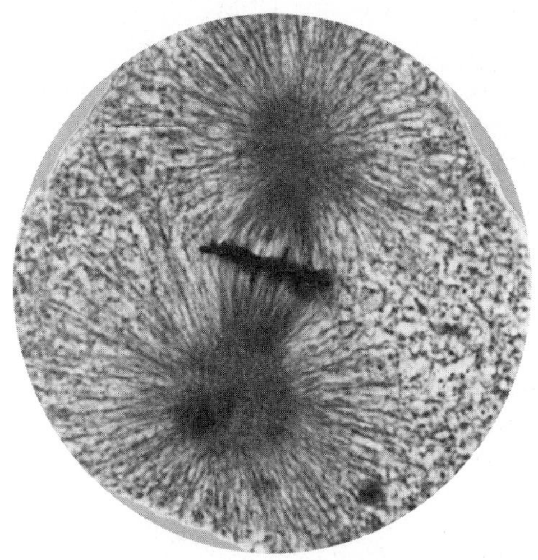

FIG. 14.12 Metaphase in a cell of the embryo of a whitefish. Note the prominent aster at each pole of the cell. (Courtesy CCM: General Biological, Inc., Chicago.)

Mitosis, as described above, is found universally among the plants. Animal cells, too, divide by mitosis. The phases are similar to those in plant cells, and the behavior of the chromosomes is the same. Two striking differences can, however, be observed. One is the appearance of the **asters.** Animal cells contain centrioles (see Section 5.16), and in prophase these migrate to the opposite sides of the nucleus. Here they may help to organize the formation of the spindle. Around each centriole also develops a system of radiating fibers, the aster (Fig. 14.12). The function of the aster is unknown. The second difference is that no cell plate is formed in animal-cell mitosis. Instead, in telophase, a furrow appears in the cell membrane, forming at the equator. The furrow deepens, and the two daughter cells simply become pinched apart. This process depends upon the presence of a belt of microfilaments (see Section 5.14) that extends around the cell and appears to provide the contractile force. The cell plate in plants is probably a consequence of the rigid cell wall which prevents furrowing from taking place.

How long does mitosis take? The entire sequence of phases may be completed in from nine minutes to many hours. The exact length of time varies considerably with cell type, species, and temperature.

What is the significance of mitosis? The most dramatic activity in mitosis is the organized behavior of the chromosomes. In fact, this aspect of the process is so important that we still use the term mitosis even when the chromosomal activity occurs without *cytokinesis,* that is, without actual division of the cell. Nuclear division without cytokinesis occurs under certain conditions in a variety of organisms such as in the production of zoospores by *Chlamydomonas* (Fig. 14.5).

We know that the nucleus contains the information that controls the development and activities of the cell. The evidence of mitosis suggests that the chromosomes play an important part in this. Viewed in this light, mitosis can be considered a device for the orderly duplication (in the *S* phase) and separation (in anaphase) of the chromosomes and thus of the hereditary information. Each daughter cell gets a complete set of chromosomes identical to those of the parent cell. Mitosis thus provides a way of transmitting this information unchanged and undiminished from parent cells to daughter cells.

Sexual Reproduction

14.5 THE NATURE OF SEXUAL REPRODUCTION

In sexual reproduction, new individuals arise from the union of two sets of hereditary information (DNA). Usually, each of these sets is contained within a specialized cell called a **gamete.** In order to combine their hereditary information, the two gametes must first unite, a process known as fertilization. Among some simpler organisms, such as the green alga *Chlamydomonas,* the two gametes are indistinguishable from each other. Such similar-appearing gametes are called **isogametes.**

The nucleus and cytoplasm of a single *Chlamydomonas* cell divide five or six times to form 16 or 32 isogametes. These are simply miniature copies of the adult. The cell wall of the adult breaks open and releases the flagellated gametes into the surrounding water. These then fuse with the gametes produced by other *Chlamydomonas* cells. There are no *visible* differences between the fusing gametes (Fig. 14.5), but gametes from the same parent will not fuse together.

In most organisms, two quite distinct kinds of gametes are produced. **Sperm** are tiny gametes which consist of little more than a nucleus and a flagellum. The flagellum enables the sperm to swim. Sperm are the male gametes. **Eggs,** the female gametes, are larger and nonmotile. They contain substantial quantities of food in addition to the nucleus. Because sperm and eggs are so dissimilar in appearance, they are called **heterogametes.**

The product of fertilization is the **zygote.** The zygote resulting from the union of heterogametes is frequently referred to as a fertilized egg.

One of the most striking features of sexual reproduction is that each of the gametes involved is usually produced by a separate individual. Thus, two individuals, the parents, contribute to the formation of the offspring. If the two parents differ in any way (and often, as you will see, even if they don't), the offspring will possess new combinations of characteristics. Many times these new characteristics may be disadvantageous to the offspring. Many times they will have no effect on the welfare of the possessor. Sometimes the new combination will result in an individual better suited than its parents to the habitat in which it lives. It is this latter category that permits gradual structural and functional improvements in a species, that is, evolution.

In many organisms, both male and female gametes are produced in one individual. Such organisms are said to be *hermaphroditic.* The common earthworm, some fishes and the majority of our flowering plants are examples of hermaphroditic organisms. Even in these cases, two individuals usually contribute to the formation of the zygote. Sperm cells from one individual unite with the eggs of another individual. **Cross-fertilization** is the result, and variability in the offspring is thus promoted. Some hermaphrodites, such as the earthworm, never fertilize their own eggs. Others, such as many of the flowering plants, engage in self-fertilization only when cross-fertilization fails to occur. In Chapter 20 we shall examine some of the mechanisms by which cross-fertilization in plants is promoted and self-fertilization is avoided.

14.6 SEXUAL REPRODUCTION IN BACTERIA

The essence of sexual reproduction, then, is newness, change, variability. This is perhaps best illustrated by the process as it occurs in some bacteria. For many years, bacteria were believed to reproduce by fission only. No evidence of sexual reproduction in them was detected until 1947. In that year, J. Lederberg and E. L. Tatum discovered that the common intestinal bacterium *Escherichia coli* could reproduce sexually. Although the cells were not actually observed in the process for some time after, Lederberg and Tatum did find definite, if indirect, evidence that sexual reproduction was taking place.

The particular strain of *E. coli* that they used is not at all fussy about its food supply. Given glucose as a source of energy and carbon, and some inorganic salts to supply nitrogen, sulfur, and phosphorus atoms for the synthesis of amino acids and nucleic acids, the organism grows well. Within its tiny cell, it contains all the enzymes with which to carry out the biochemical reactions needed to convert glucose and inorganic salts into all the organic molecules necessary to life.

E. coli, like all bacteria, is quite sensitive to exposure to ultraviolet light. A moderately heavy dose is lethal. In fact, this is why ultraviolet lamps are sold as "germicidal" lamps to be used for killing bacteria in hospital operating rooms, etc. Lesser exposures to ultraviolet light may not kill the bacterium outright but may, instead, destroy its ability to manufacture one or more important organic compounds. These compounds must then be added to the bacterium's diet if it is to survive. Furthermore, this loss of synthetic ability is hereditary. All the offspring produced by the fission of such a *nutritionally deficient* bacterium will suffer the same disability.

A strain of *E. coli* has been developed by ultraviolet irradiation that lacks the ability to synthesize one of the B-vitamins, biotin, and one of the amino acids, methionine. These two substances must be added to its diet in order for it to live. Another strain has been developed that can still synthesize biotin and methionine from glucose and salts but is incapable of synthesizing another B-vitamin, thiamine, and two amino acids, threonine and leucine.

If these two strains of bacteria are allowed to mingle freely with each other and then placed on a culture medium containing only glucose and salts, colonies of cells soon appear. These cells thrive despite the absence of the vitamins and amino acids needed by their parents. Furthermore, descendants of these cells continue to be able to live on glucose and salts alone.

What has happened? Evidently, members of each strain have in some way managed to combine their hereditary blueprints. The first strain, with blueprints for making the enzymes needed to synthesize thiamine, threonine, and leucine (as well as all the other organic molecules except biotin and methionine), has combined these blueprints with those of the second strain, which included the instructions for making biotin and methionine. Thus, a bacterium was produced which, like the original strain, is capable of satisfying all its needs from glucose and salts alone. The evidence indicates that a bacterium of one strain actually injects part of its single chromosome into a cell of the other strain when the two cells come in contact.

Note that in this example, sexual reproduction has been reduced to its essence. No gametes are formed; one cell simply attaches temporarily to another and donates some of its hereditary blueprints to it. Nor have *additional* bacteria been created by this process. A *new* bacterium has been created, however, and this bacterium is capable of living under conditions in which neither of the parental types could survive.

A problem remains. Sexual reproduction in bacteria involves the transfer of a piece of chromosome from one cell to another. It is not reciprocal: that is, no exchange is involved. Thus one cell acquires a greater total hereditary blueprint. This must be followed sometime by the elimination of an equivalent amount. Otherwise, over a period of time, there would be a steady accumulation of chromosomal material in the species. It is not yet certain by just what mechanism a bacterial cell reestablishes its normal chromosome content. The same problem faces other sexually reproducing organisms, however, and its solution in them *has* been discovered.

14.7 MEIOSIS

Each species of organism has a characteristic number of chromosomes. The little fly *Drosophila melanogaster* has 8, the onion 16, and humans 46. (Lest you assume from this that the number of chromosomes in a species is necessarily a function of the complexity of the organism, you should note that the crayfish has at least 200 chromosomes in its cells.) All these numbers are even. This reflects the fact that the chromosomes are present in homologous *pairs*. Cells containing a full complement of homologous pairs are said to contain the **diploid** or $2n$ number of chromosomes.

Mitosis results in the formation of cells that have exactly the same number of chromosomes as the parent cell. This would create difficulties if the cells being formed were to serve as gametes. A human sperm with 46 chromosomes uniting with an egg containing 46 chromosomes would result in a zygote with 92 chromosomes—twice the normal number for the species. Development of the zygote by

STAGES IN PROPHASE OF MEIOSIS I

STAGE	FEATURES
1. LEPTOTENE	CHROMOSOMES ELONGATED; APPEAR SINGLE
2. ZYGOTENE	HOMOLOGOUS CHROMOSOMES PAIR, FORMING N BIVALENTS
3. PACHYTENE	BIVALENTS SHORTEN
4. DIPLOTENE	HOMOLOGUES PULL SLIGHTLY APART REVEALING SEPARATE CHROMATIDS AND CHIASMATA
5. DIAKINESIS	HOMOLOGOUS CENTROMERES MOVE APART; CHROMATIDS CONTINUE TO SHORTEN

FIGURE 14.13

mitosis would result in all subsequent cells having the new number. You can readily see that after a few generations of this, there would be no room in the cell for anything but chromosomes.

Actually, this situation rarely arises in living things. At some point between the formation of the zygote and the formation of gametes, a special kind of cell division occurs, called **meiosis.** Meiosis consists of two consecutive cell divisions with but one duplication of the chromosomes. When a cell divides by meiosis, then, four cells are produced, each containing just one-half the diploid number of chromosomes. This half-number is called the **haploid** or n number. The reduction in chromosome number is not a random process. In cells produced by meiosis, only one member of each of the homologous pairs of chromosomes of the diploid cell is present. Thus when two gametes unite, the resulting zygote ($2n$) gets one member of each pair of homologous chromosomes from each gamete and thus from each parent.

Each of the two meiotic divisions can be separated into phases similar to those occurring in mitosis. However, significant differences in the behavior of the chromosomes occur in the first of the divisions.

Prophase I. Prophase of the first meiotic division is a much slower and more complex process than in mitosis. Cytologists subdivide the first meiotic prophase (I) into five stages. The names and main features of these are given in the table in Fig. 14.13. A schematic view of the events in meiosis is given in Fig. 14.14.

When the chromosomes first become visible (leptotene of prophase I), each homologue appears to be a single structure. But most, if not all, of the DNA of the cell has been doubled during the S phase preceding prophase I, so we must conclude that these structures are in reality already doubled. As prophase continues (zygotene and pachytene), each chromosome in the cell pairs up lengthwise with its homologue. This process of **pairing** (also called **synapsis**) is a unique feature of meiosis; it does not occur in mitosis. The paired homologues are called *bivalents.*

Then (diplotene), the two homologues begin to pull apart from each other. At this time, the fact that each consists of a pair of sister chromatids finally becomes visibly evident. Thus, each bivalent contains four strands or chromatids. However, the four strands remain connected by two mechanisms: (1) the sister chromatids of each homologue remain attached to their single shared centromere and (2) at one or (usually) more points, two non-sister chromatids are held together. These points of attachment are called **chiasmata** (sing., chiasma). At each chiasma, the non-sister chromatids have already exchanged segments. This process of exchange is called **crossing over.** It is reciprocal, that is, the portions exchanged by each non-sister chromatid are identical. In Fig. 14.14, for the sake of clarity, only a single pair of homologues is shown and only one crossover between these. However, there often are a number of chiasmata in a single bivalent (Fig. 14.15). With two, three, or more chiasmata, any strand can be involved, i.e., if the strands (sister chromatids) of one homologue are numbered 1 and 2 and of the other, 3 and 4, one can find in a single bivalent one or more of the following combinations: 1–3, 2–3, 1–4, and 2–4. And any combination may appear more than once. The only events that are excluded are: (1) chiasmata between sister chromatids (which would be pointless anyway because they are identical in genetic makeup) and (2) involvement of more than two non-sister chromatids at any given point along the length of the chromosome (i.e., three or four strands cannot exchange parts at the same point).

FIG. 14.14 Meiosis. The chromosomes are shown duplicated in leptotene and zygotene although they almost always appear single-stranded at these stages. However, the fact of their duplication is inferred from the doubling of the DNA content of the cell. For simplicity, the behavior of just a single pair of homologues is shown with one homologue black, the other in color. And only one chiasma is shown although often several are formed. ►

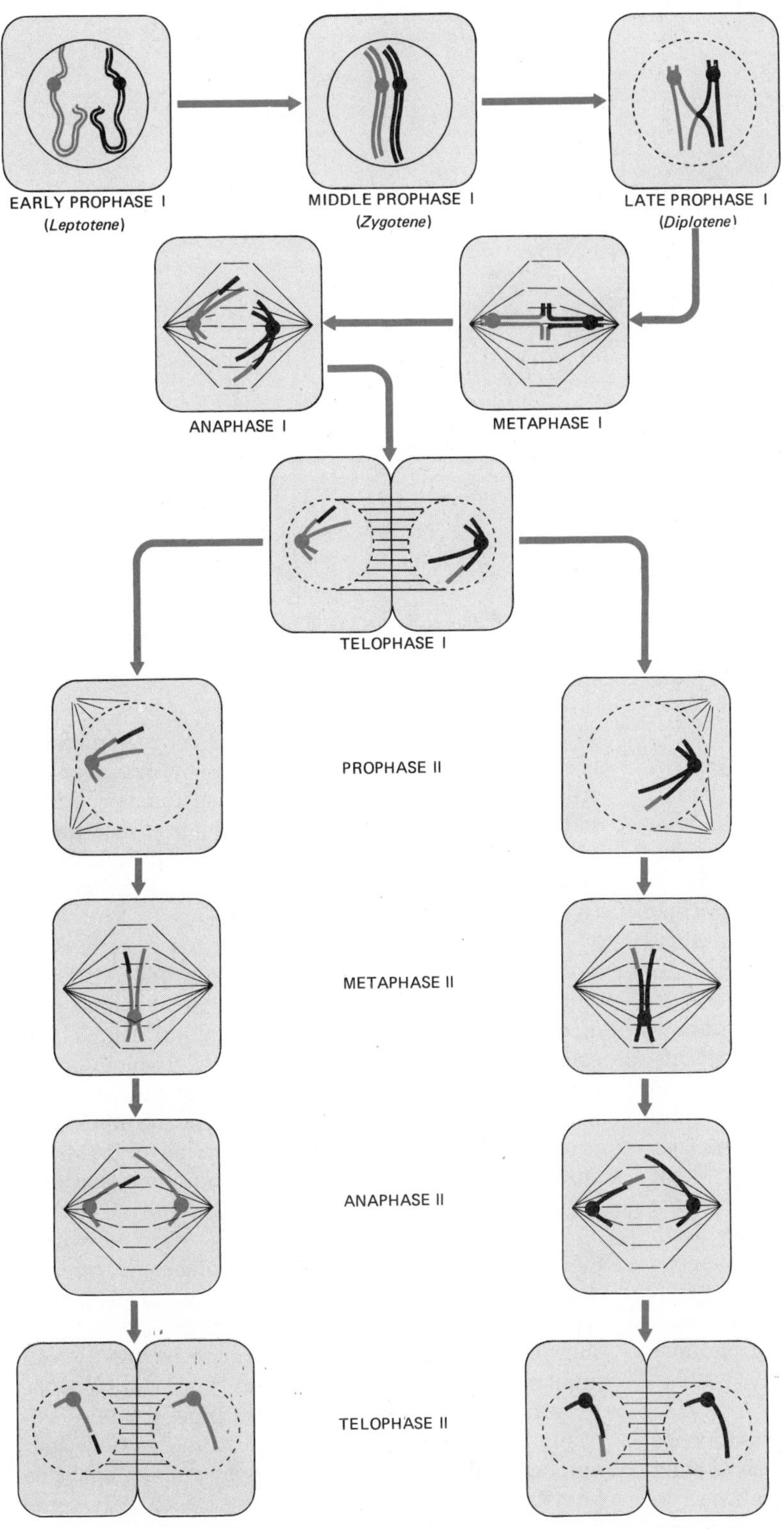
EARLY PROPHASE I
(Leptotene)
MIDDLE PROPHASE I
(Zygotene)
LATE PROPHASE I
(Diplotene)
ANAPHASE I
METAPHASE I
TELOPHASE I
PROPHASE II
METAPHASE II
ANAPHASE II
TELOPHASE II

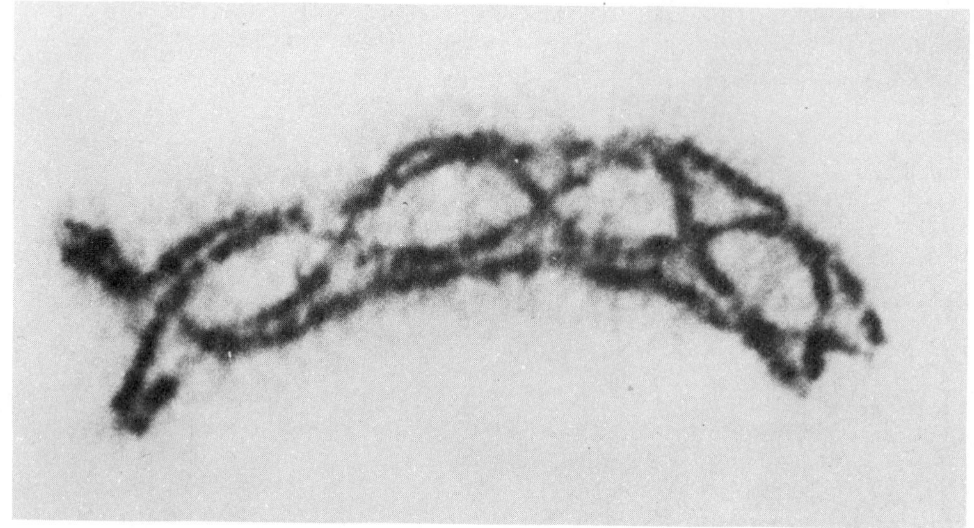

FIG. 14.15 Five chiasmata in a diplotene bivalent of the grasshopper *Chorthippus parallelus*. (Courtesy Professor Bernard John, Australian National University, Canberra.

The location and, to some degree, the number of chiasmata vary from chromosome to chromosome and from cell to cell. (In humans, the *average* number of chiasmata for each bivalent is just over two.)

Metaphase I. Metaphase I of meiosis resembles metaphase of mitosis in the disappearance of the nuclear membrane and the appearance of the spindle. However, it differs in one crucial respect from the metaphase of mitosis. In metaphase I, the centromeres of each pair of homologues become attached to the spindle—one above, the other below the equator.

Anaphase I and telophase I. With the onset of anaphase I, the two centromeres of each bivalent migrate to their respective poles. This separates the bivalents (into "half-bivalents"—Fig. 14.14). Note that there is not the splitting or division of the centromeres that we found in mitotic anaphase. What has happened is that the homologues have become separated. Thus telophase produces two cells, each of which has just one member of each homologous pair of chromosomes that was present in the original cell (although the original homologues have reciprocally exchanged one or more segments of chromatid).

Interkinesis. In some organisms, neither telophase nor interphase intervenes between meiosis I and meiosis II. The cell goes directly from anaphase I to prophase II. However, even in those organisms where an interphase does occur between the two divisions, there is no *S* phase. Thus no additional DNA synthesis takes place.

The second division. The second meiotic division is similar to a mitotic division. The chromosomes are still present as "doublets." The centromeres attach to the spindle and orient at the equator at metaphase II. Division of the centromeres in anaphase II separates the chromatids and each is pulled to its respective pole (Fig. 14.14).

With the completion of the second meiotic division, a total of four cells has been produced. Each contains just one member of each homologous pair of chromosomes present in the original cell. These cells thus contain just one-half (the haploid number) of the chromosomes of the parent cell. Furthermore, in our simplified case, two of the four cells produced have an unchanged maternal or paternal chromosome; the others have a chromosome containing *both* maternal and paternal parts.

Crossing over is not the only way by which meiosis produces cells whose chromosome content differs. In Fig. 14.14, for the sake of simplicity, only one pair of homologous chromosomes ($2n = 2$) is shown. In the vast majority of organisms, the diploid number is larger, often substantially so. But consider the situation in the female *Culex pipens,* one of the commonest mosquitoes. This creature has a diploid number of 6. These six chromosomes make up three homologous pairs. One member of each pair came from the mosquito's mother (the *m*aternal chromo-

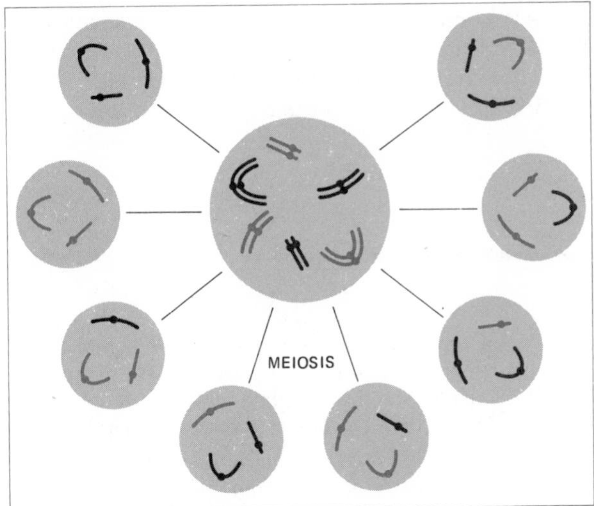

FIG. 14.16 Random assortment of maternal (black) and paternal (color) chromosomes during meiosis in an organism with a diploid number of 6 ($2n = 6$).

somes); the other from its father (the *p*aternal chromosomes). Thus we can indicate the three pairs as A^mA^p, B^mB^p, and C^mC^p. At metaphase I of meiosis, the homologues orient themselves on opposite sides of the equator. If A^m, B^m, and C^m should orient themselves on one side of the equator and thus A^p, B^p, and C^p on the other, the gametes would be the same (neglecting crossing over!) as those the mosquito received from its parents. But six other combinations are equally likely (Fig. 14.16). The fruit fly, *Drosophila melanogaster,* has eight chromosomes, four paternal and four maternal. In distributing one of each homologous pair to the gametes, 16 different combinations of maternal and paternal homologues can be produced. The number of different kinds of gametes produced by this **random assortment** of the chromosomes is equal to 2^n, where $n =$ the haploid number of chromosomes in the organism. Human mothers and fathers, each with a haploid number of 23, can each produce by random assortment alone 2^{23} or 8,388,608 different kinds of gametes. When you consider further that crossing over will have occurred between their own paternal and maternal chromosomes in the first meiotic division, it is not surprising that no two people ever resemble each other exactly. (An exception to this is the special case of identical twins; but it is only an apparent exception since they arise from a *single* fertilized egg.)

Meiosis and Variability

What, then, is the significance of meiosis? First, the number of chromosomes is reduced from the diploid number to the haploid number. This sets the stage for the union of two gametes and thus provides a mechanism by which traits of two different parents can be combined. Second, meiosis provides for variation in the gametes produced by each parent. **Random assortment** of the paternal and maternal chromosomes *plus* the **crossing over** of portions of the maternal and paternal chromosomes make it likely that no two gametes of even one parent will be exactly alike.

How is meiosis used by living things? You have seen the necessity for meiosis to occur sometime between fertilization and gamete formation in sexually reproducing forms. In animals, meiosis leads directly to the production of gametes. Among the fungi and algae, on the other hand, meiosis is apt to occur very soon after the formation of the zygote. The zygote formed by the fusion of two *Chlamydomonas* gametes develops a heavy wall and becomes dormant. When favorable conditions exist, the zygote undergoes meiosis, producing *four* haploid zoospores (Fig. 14.5). These are released and soon grow to full-size *Chlamydomonas* cells. Most of the life cycle of this organism is thus spent in the haploid condition. In plants, too, meiosis is used in the production of spores. Only later are gametes derived from these.

The occurrence of meiosis long before the formation of gametes in many organisms illustrates two additional points about cell division. First, *mitosis* can occur in haploid cells as well as diploid ones. Second, the haploid number of chromosomes is sufficient to control cell functions in *these* organisms (but not in many animals) and reestablishment of diploidy by fertilization simply provides a double set of the hereditary information. This double set does, however, lay the basis for a reorganization of the hereditary information at the next meiosis. Once again, we see sexual reproduction serving one major function: the production of variability in the species.

Earlier in the chapter it was pointed out that asexual reproduction usually results in offspring that are exact copies of their parents. Sexual reproduction, on the other hand, was described as a method of producing offspring with characteristics different from those of the parents and thus of producing variability within the species. Assuming that the information that controls the development of the characteristics of an organism is located in its chromosomes, it is easy to see how mitosis makes the first possible and meiosis, followed by fertilization, makes the second possible.

Throughout this chapter, we have referred often to the hereditary blueprints within the nuclei of cells. One of the most active fields of biological research is that concerned with the nature and operation of these hereditary blueprints. This study is called genetics. It is the topic of the next four chapters.

EXERCISES AND PROBLEMS

1. The haploid number of horse chromosomes plus the haploid number of donkey chromosomes can produce a healthy mule. The mule is sterile. With rare exceptions (see Question 2), it cannot breed with either another mule or a horse or a donkey. Can you explain why in terms of the behavior of the chromosomes during meiosis?
2. Although mules are generally sterile, a few cases are known where a female mule has given birth to a horse (after mating with a horse) or another mule (after mating with a donkey). Using your knowledge of meiosis, can you think of an explanation for these rare events?
3. How does the behavior of the chromosomes in mitosis differ from their behavior in the first meiotic division?
4. Asexual reproduction preserves the *status quo.* Sexual reproduction promotes change. Describe how mitosis makes the first possible and meiosis followed by fertilization makes the second possible.
5. Explain why the minimal medium for growing *E. coli* contains (a) glucose or some other carbohydrate, (b) KH_2PO_4, (c) $MgSO_4$, (d) NH_4NO_3.
6. If a hermaphroditic organism fertilizes its own eggs, must all its offspring be identical? Explain.
7. How many kinds of gametes can the onion ($2n = 16$) produce by random assortment alone?
8. What functions do spores accomplish in the bread mold, *Rhizopus stolonifer?*

REFERENCES

1. JOHN, B., and K. R. LEWIS, *Somatic Cell Division,* Oxford Biology Readers, No. 26, Oxford University Press, Oxford, 1972. Mostly about mitosis.
2. MAZIA, D., "The Cell Cycle," *Scientific American,* Offprint No. 1288, January, 1974.
3. PATT, D. I., and GAIL R. PATT, *An Introduction To Modern Genetics,* Addison-Wesley, Reading, Mass., 1975. Chapter 10 discusses bacterial conjugation and other mechanisms of sexual reproduction in bacteria.
4. JOHN, B., and K. R. LEWIS, *The Meiotic Mechanism,* Oxford Biology Readers, No. 65, Oxford University Press, Oxford, 1973.
5. NOVIKOFF, A. B., and E. HOLTZMAN, *Cells and Organelles,* Holt, Rinehart and Winston, New York, 1970. Mitosis and meiosis are treated in Chapters 4.2 and 4.3 respectively.

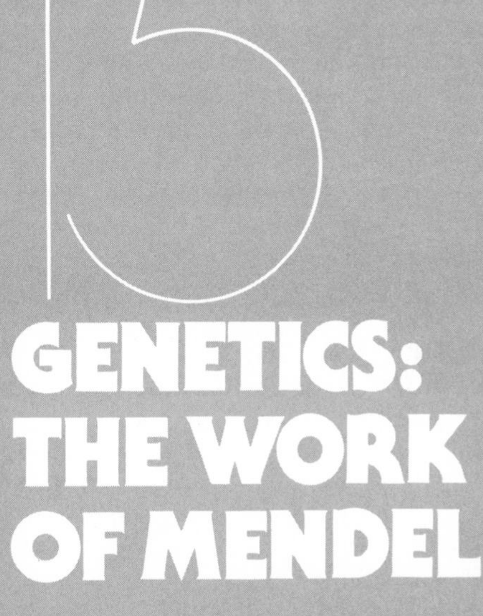

15 GENETICS: THE WORK OF MENDEL

When living things reproduce asexually, their offspring develop into exact copies of their parents so long as they are raised under similar conditions. On the other hand, when living things reproduce sexually, their offspring develop traits different from one another and different from those of either parent. When a collie and a German shepherd mate, their offspring are dogs, not some other species of animal. Their offspring are, however, neither collies nor German shepherds. Long before biologists discovered many of the facts of mitosis and meiosis, they tried to discover rules which would explain how the observable characteristics of offspring are related to those of their parents and even their parents' parents.

Of the several theories that have been formulated to explain how traits are inherited, two deserve special mention. One of these, Mendel's theory, has provided the foundation upon which all later work in genetics has been built. The other, the theory of the inheritance of acquired characteristics, has failed to pass scientific tests but, despite this, has continued to have its defenders.

15.1 THE THEORY OF THE INHERITANCE OF ACQUIRED CHARACTERISTICS

This theory states simply that traits *acquired* by parents during their lifetime can be passed on to their offspring. The theory is usually associated with the name of Lamarck, a French biologist, who used it in an attempt to explain the many striking adaptations to their environment shown by plants and animals. His most famous illustration was the giraffe. He claimed that the long neck of the giraffe has evolved as a result of generations of giraffes stretching their necks to browse on the leaves of trees. Each generation has passed on to its offspring the small increase in neck length caused by continual stretching.

Is there any evidence that such a phenomenon occurs? Despite repeated attempts to prove that changes in the body acquired by an individual can be passed on to its offspring, no evidence for it has yet been discovered. The earliest experiments that attempted to settle the question were those in which some body part, such as the tail of a mouse, was removed surgically. Even after generations of tail removal, young mice were born with tails as long as ever. Actually, the experimenters could have simply looked about them for corroboration of their findings. Sheep raisers have been removing the tails of lambs through uncounted generations, but the process must still be carried out with each new generation. Even when more sophisticated attempts were made to alter heredity by altering the environment, it could not be done.

Why not? In order for changes carried out on the body to be transmitted to the offspring, they would have to become incorporated into the sperm and egg cells, the only link between the bodies of the parents and the bodies of the children. Perhaps such a thing might occur if the specialized cells of the body, upon which some alteration could be made, then produced the gametes. But they do not. For many years it has been known that in animals the gamete-producing cells of the body are set aside very early in embryonic life. In fact, a newborn girl has already set aside and begun the first meiotic division for each and every cell that will someday develop into a mature egg.

The German biologist Weismann expressed these thoughts in the form of his theory of the *continuity of the germplasm*. Multicellular organisms, according to his theory, are made up of gamete-producing cells or **germplasm** and the cells of the rest of the body, the **somaplasm.** Weismann considered the germplasm immortal, an unbroken chain of gametes and embryos going right back to the dawn of life (Fig. 15.1). At each generation, the embryo that develops from the zygote not only sets aside some germplasm for the next generation but also produces the cells that will develop into the body, the somaplasm, of the organism. In this view, the somaplasm simply provides the housing for the germplasm, seeing to it that the germplasm is protected, nourished, and conveyed to the germplasm of the opposite sex in order to create the next generation. The old riddle about which came first, the chicken or the egg, would have been no puzzle to Weismann. In his view, the chicken is simply one egg's device for laying another egg.

The essential truth of Weismann's theory was beautifully demonstrated in 1909 by the Americans W. E. Castle and John C. Phillips. They removed the ovaries from an albino guinea pig and substituted the ovaries from a black guinea pig. Then they mated this albino guinea pig with an albino male. Instead of getting albino babies as might normally be expected, the babies were black. (Matings between albino and black guinea pigs always produce black offspring.) The hereditary blueprints of the eggs had not been altered by their maturation in the body of a different animal.

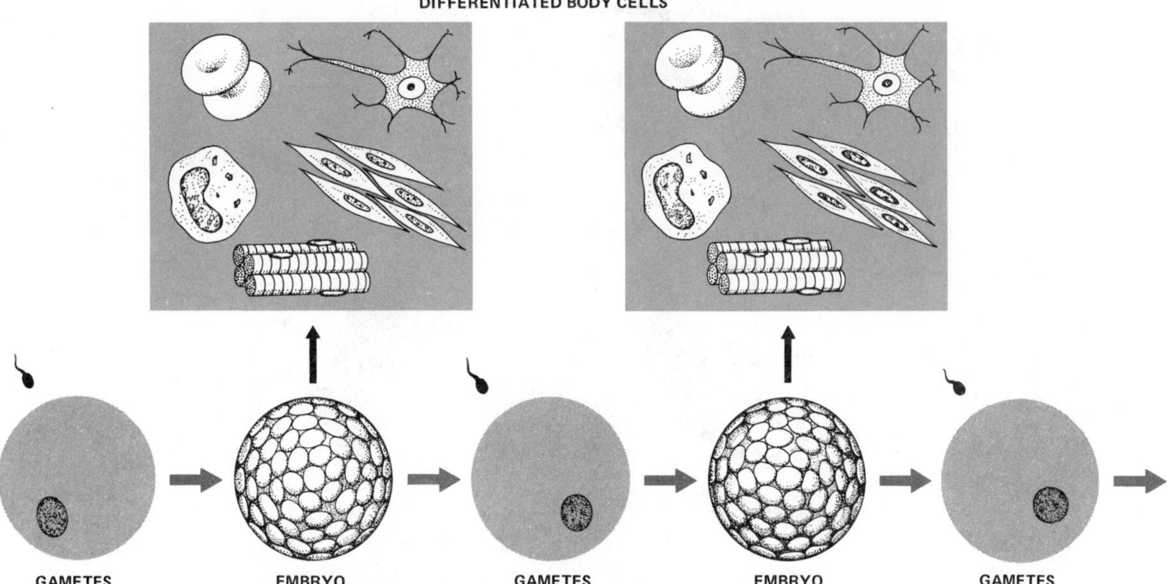

FIG. 15.1 Weismann's hypothesis. While true for most animals, it does not hold true for plants. Plants *can* produce gamete-forming cells from differentiated body cells.

15.2 MENDEL'S THEORY: THE BACKGROUND

Our present theories of inheritance were first worked out by the Austrian monk, Gregor Mendel (Fig. 15.2). From 1858 to 1866, Mendel worked in the garden of his monastery in the town of Brünn (now Brno), breeding garden peas and examining the offspring of these matings.

Mendel's decision to work with the common garden pea was an excellent one. The plant is hardy and fast growing. As in many legumes, the petals of the flower entirely enclose the sex organs. These are the **stamens,** which produce pollen (the bearer of the male gametes) and the **pistil,** which produces the female gamete, the egg. Although insects may occasionally penetrate to the sex organs, self-fertilization is the rule. Mendel found he could open the buds and remove the stamens before they ripened. Then, by dusting the pistil with pollen from another plant, he could effect cross-fertilization between them.

The choice of the garden pea was a happy one also because many varieties existed which differed from one another in clear-cut ways. Some produced (after drying) wrinkled seeds. Others produced smooth, round seeds. Some produced seeds with green cotyledons; others, seeds with yellow cotyledons. Some produced green pods; others, yellow pods. Some produced white flowers; others, reddish flowers. These paired characteristics (and three others) Mendel chose to study because they were so easily distinguishable and because they "bred true" generation after generation. That is, so long as they were maintained by normal self-pollination, these varieties continued to produce offspring identical to the parents in the characteristics being studied.

There were other characteristics in which Mendel's pea varieties differed, such as leaf size and flower size. Mendel wisely ignored these differences in his studies simply because they did not fall into a clear-cut "either-or" classification. Mendel's peas produced either round seeds or wrinkled seeds. There were no intermediate types. On the other hand, the size of leaves and flowers varied over quite a range. There were not just two distinct categories in which they could be placed. Mendel's decision to thus limit the scope of his experiments was certainly an important factor in his success.

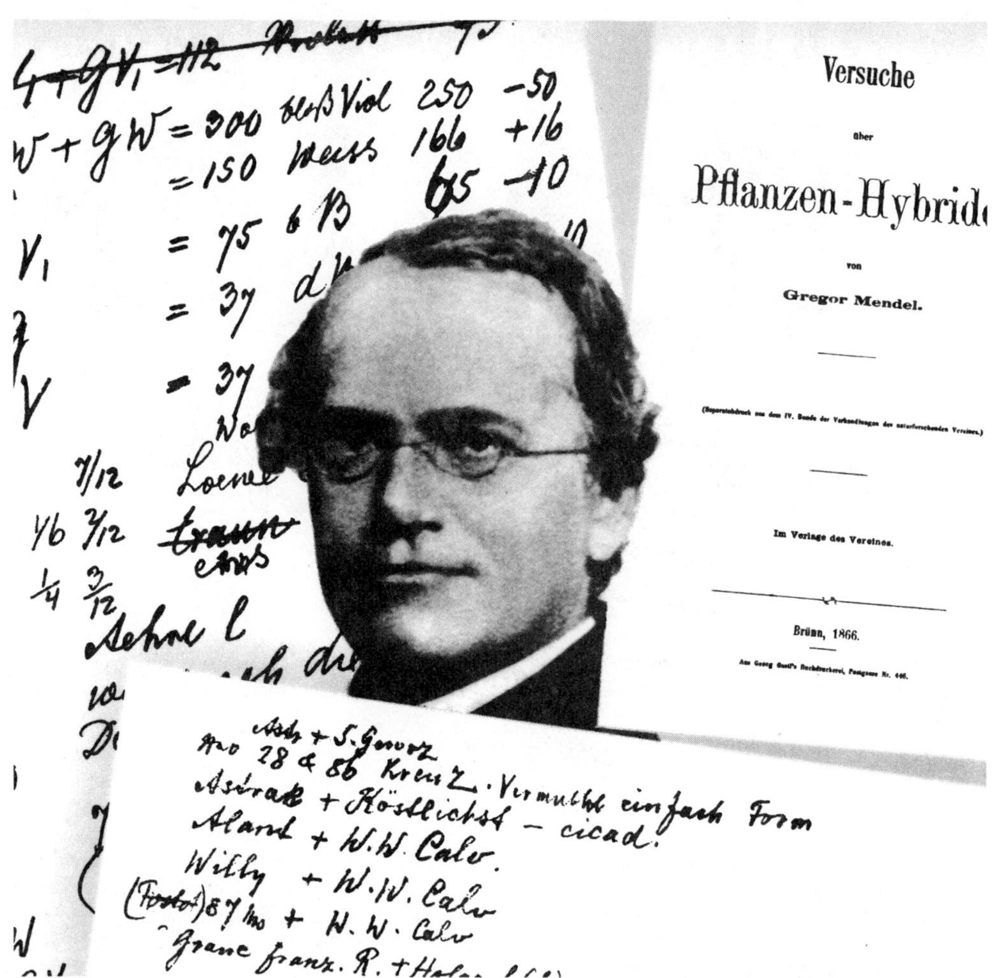

FIG. 15.2 Gregor Mendel (1822–1884), with the title page of the paper in which he published his work, and some of his notes on plant genetics and hybridization. (Photo courtesy of Professor Verne Grant.)

15.3 MENDEL'S EXPERIMENTS

In one of his first experiments, Mendel crossbred a round-seeded variety with a wrinkle-seeded variety. This parental generation we call the P_1 generation. Pollen from the stamens of the round-seeded variety was dusted on the pistils of the wrinkle-seeded variety. The reciprocal cross was carried out: pollen from the stamens of the wrinkle-seeded variety was placed on the pistils of the round-seeded variety. In both cases, every one of the seeds produced by these cross-fertilized flowers was round. There were no seeds intermediate in shape. (Seed shape and cotyledon color were particularly rewarding characteristics to study. Their form could be determined in the same season that the fertilization was carried out. The seeds *were* the next generation. Pod shape, stem length, and flower color in the second generation could not be determined until the following season, when the seeds had germinated and developed into the mature plants.) Mendel called the second generation the **hybrid** generation because it was produced by dissimilar parents. It is also called the F_1 generation.

Mendel planted all his F_1 round seeds; 253 F_1 plants grew to maturity. The F_1 flowers were allowed to self-fertilize themselves in the normal way. In effect, Mendel was breeding the F_1 (or hybrid) generations together. From the pods of these F_1 plants, Mendel recovered 7324 seeds, the F_2 generation. Of these, 5474 were round and 1850 were wrinkled, yielding a ratio of 2.96 round seeds to 1 wrinkled seed.

Mendel then planted some of these two kinds of F_2 seeds. From the wrinkled seeds, he raised plants which produced (by self-fertilization) a new crop (F_3) of seeds. These were exclusively of the wrinkled

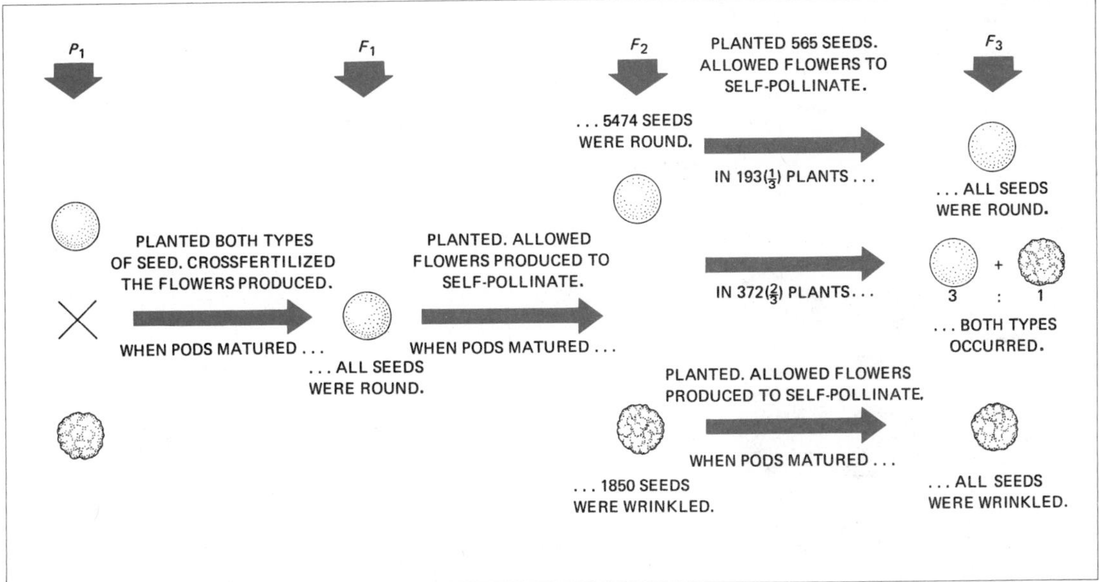

FIG. 15.3 Mendel's results.

type. From the round seeds, he raised 565 plants and allowed these also to produce an F_3 crop by self-fertilization. In this case, 193 of the plants produced round seeds only. The other 372 plants produced both round and wrinkled seeds and in a 3 to 1 ratio (Fig. 15.3).

What is the significance of these facts? Clearly, when round-seeded peas are crossbred with wrinkle-seeded peas, the round-seeded ones pass some controlling factor on to the offspring (F_1). Further, it does not matter whether the round-seeded factor is supplied by the male gamete or by the egg. The results are the same in either case.

The reappearance of wrinkled peas in the F_2 generation can only mean that at least some of the F_1 plants were also carrying a factor for the wrinkle-seeded condition. Its presence was, however, obscured in the F_1 generation. Traits which are transmitted unchanged to the F_1 generation (e.g. round seeds) Mendel called **dominant.** Those which are hidden in the F_1, but reappear in the F_2 (e.g. wrinkled seeds), he called **recessive.**

15.4 MENDEL'S HYPOTHESIS

In an attempt to explain these facts, Mendel made a series of assumptions. These assumptions we call a hypothesis. They were not observations; they were not facts. They were simply statements which, if true, would provide an explanation for the observed results. They were:

1. In each organism there is a pair of factors which control the appearance of a given characteristic. (Today we call these factors **genes.**)
2. The organism gets these factors from its parents, one from each.
3. Each of these factors is transmitted as a discrete, unchanging unit. (The wrinkled seeds in the F_2 generation were no less wrinkled than those in the P_1 generation, although the factors controlling this trait had passed through the round-seeded F_1 generation.)
4. When the reproductive cells (sperm or eggs) are prepared, *the factors separate* and are distributed as units to each gamete. This is often called *Mendel's first law,* the **law of segregation.**
5. If an organism has two unlike factors for a given characteristic, one may be expressed to the total exclusion of the other. Today, we use the term **allele** to describe *alternative forms of a gene* controlling a given characteristic. Thus, in the case discussed, there are two alleles (round-seeded and wrinkle-seeded) of the gene which controls seed shape.

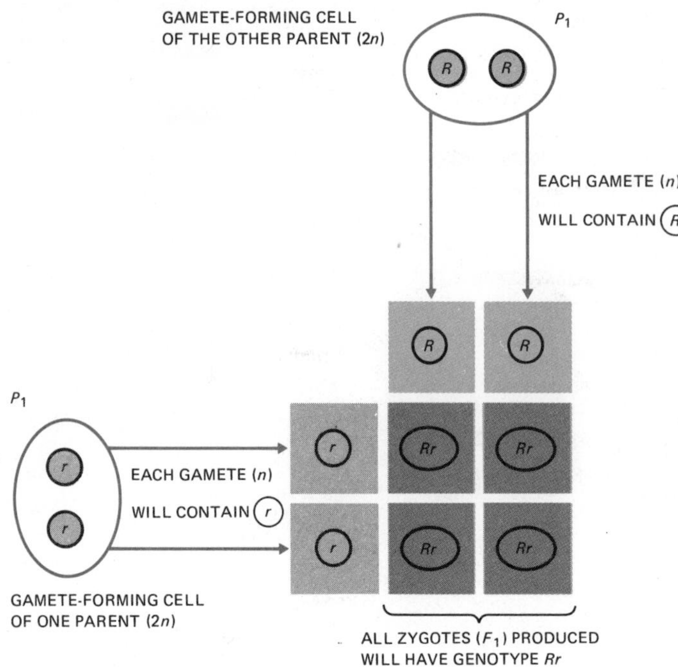

GAMETE-FORMING CELL
OF THE OTHER PARENT (2n)
P1
R R
EACH GAMETE (n)
WILL CONTAIN R
R R
P1
r Rr Rr
r
EACH GAMETE (n)
WILL CONTAIN r
r Rr Rr
r
GAMETE-FORMING CELL
OF ONE PARENT (2n)
ALL ZYGOTES (F1) PRODUCED
WILL HAVE GENOTYPE Rr

FIG. 15.4 How Mendel's hypothesis explains the results of mating a purebreeding wrinkle-seeded pea plant with a purebreeding round-seeded one.

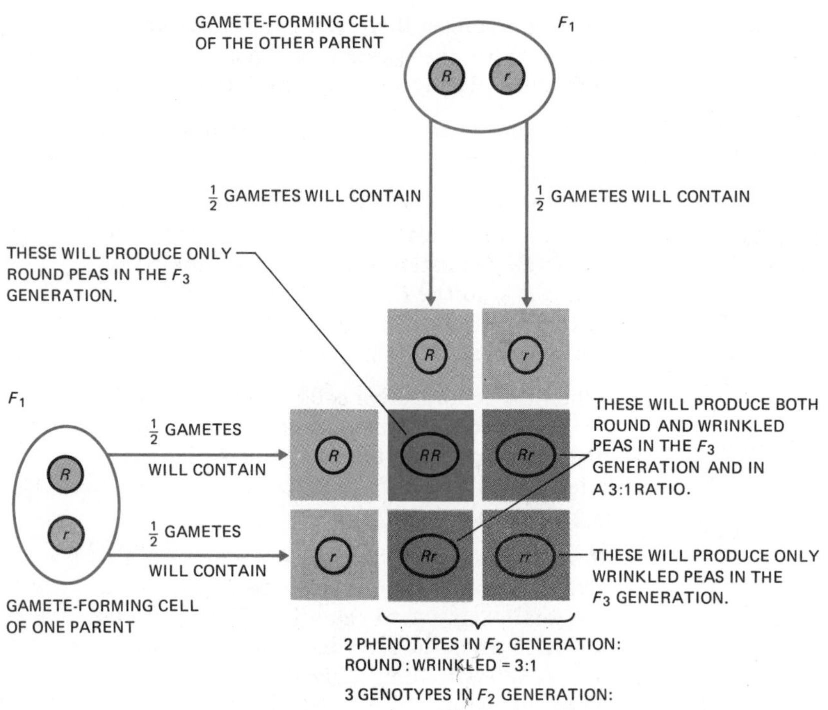

GAMETE-FORMING CELL
OF THE OTHER PARENT
F1
R r
1/2 GAMETES WILL CONTAIN
1/2 GAMETES WILL CONTAIN
THESE WILL PRODUCE ONLY
ROUND PEAS IN THE F3
GENERATION.
R r
F1
1/2 GAMETES
WILL CONTAIN
1/2 GAMETES
WILL CONTAIN
R RR Rr
r Rr rr
THESE WILL PRODUCE BOTH
ROUND AND WRINKLED
PEAS IN THE F3
GENERATION AND IN
A 3:1 RATIO.
THESE WILL PRODUCE ONLY
WRINKLED PEAS IN THE
F3 GENERATION.
GAMETE-FORMING CELL
OF ONE PARENT
2 PHENOTYPES IN F2 GENERATION:
ROUND : WRINKLED = 3:1
3 GENOTYPES IN F2 GENERATION:
RR : Rr : rr
1 : 2 : 1

FIG. 15.5 How Mendel's hypothesis explains the results of mating two F_1 plants.

How well does this hypothesis explain the observed facts? According to Mendel's scheme, the pure-breeding round-seeded plants of the P_1 generation contained two identical genes for round seeds. We can designate these as RR. The pure-breeding wrinkle-seeded plants contained two of the wrinkle-seeded genes, rr. Today we say that each of the P_1 plants is **homozygous** for its respective trait. In the formation of gametes, the genes separate. Since the genes are alike in each plant, however, all the gametes of one plant are the same. Any sperm nucleus or egg from the round-seeded plant will carry the allele R. Likewise, any gamete from the wrinkle-seeded plant will carry the allele r. The only possible zygotes that can be formed when these two varieties are crossbred would contain both, Rr. All the cells of the F_1 plant thus carry one of each of the two alleles. Today we would say that each of the F_1 plants is **heterozygous.** All the F_1 seeds are round according to Mendel because in the heterozygous state, the allele R is expressed to the total exclusion of the allele r. In other words, R is dominant over r. We can describe this cross by means of a so-called Punnett's square (Fig. 15.4).

When the F_1 plants form gametes, the alleles again separate, only one allele being transmitted to a given gamete. This means that half of the total number of gametes formed will contain the allele R; half will contain r. When these gametes are allowed to unite at random, roughly one-half of the zygotes will be heterozygous, one-quarter homozygous for R, and one-quarter homozygous for r (Fig. 15.5). Three different combinations are thus probable (RR, Rr, rr), and the expected ratio would be 1:2:1. Because of the dominance of R over r, however, there is no way to distinguish visibly between seeds containing RR and those containing Rr. Both would have round coats. Today we say that they have the same **phenotype,** that is, *appearance* with respect to some trait. However, they have different **genotypes,** *actual gene content,* for the trait. This explains the interesting results Mendel achieved when he grew his F_3 generation. All the wrinkled seeds bred true. One-third (193) of the round seeds also bred true, indicating that they were homozygous for RR. However, two-thirds (372) of the round seeds produced both round and wrinkled seeds and in a 3:1 ratio just as in the F_2 generation. Thus, they must have been heterozygous.

It is important to note that these ratios are only approximate. Far more pollen is produced than is ever used in fertilization. Many eggs may never be fertilized. One would no more expect that four F_1 fertilizations would always produce $1RR$, $2Rr$, and $1rr$ than one would expect a coin flipped four times to always come up heads twice and tails twice. But as the size of the sample gets increasingly large, chance deviations become minimized and the ratios approach the theoretical prediction more and more closely. Figure 15.6 shows the actual count of round and wrinkled seeds produced by ten of Mendel's 253 F_1 plants. In many cases, the ratio is far from the expected 3:1. By the time the seeds of all the plants have been counted, however, the ratio (3.14:1) is very close to that predicted.

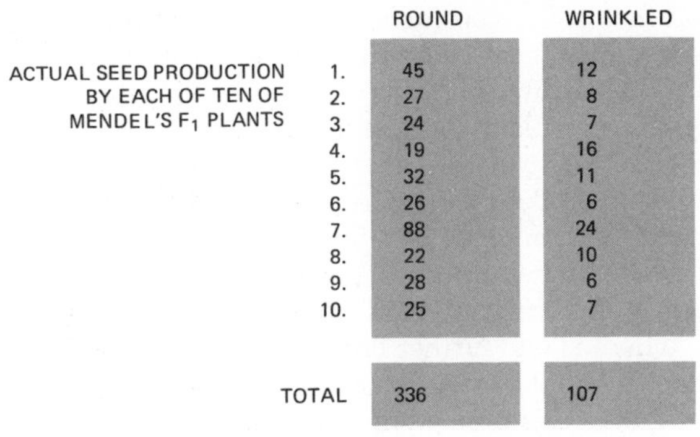

ACTUAL SEED PRODUCTION BY EACH OF TEN OF MENDEL'S F_1 PLANTS	ROUND	WRINKLED
1.	45	12
2.	27	8
3.	24	7
4.	19	16
5.	32	11
6.	26	6
7.	88	24
8.	22	10
9.	28	6
10.	25	7
TOTAL	336	107

FIG. 15.6 Actual seed production by each of ten of Mendel's F_1 plants. Note that individual plants deviated widely from the expected 3:1 ratio but that the group as a whole approached it quite closely.

15.5 HOW HYPOTHESES ARE JUDGED

Mendel's hypothesis provides a reasonable explanation for the results of his breeding experiments. While it is not the only scheme which can be devised to explain these results, it is the simplest scheme that anyone has yet thought up. We cannot ask if Mendel's hypothesis is true. All we can say is that it provides an adequate explanation of all the facts observed and that it is the simplest explanation which does this. "Reasonable," "adequate," "simplest" are all standards which are extraordinarily difficult to measure. Hypotheses, especially when newly created, must frequently be judged by just such subjective criteria as these, however. In the popular view, scientists lead working lives of complete objectivity. When dealing with facts, this is generally true. When dealing with hypotheses, however, they are also guided by subjective, aesthetic, even emotional considerations. The reasoning proceeded in Mendel's case from the concrete details of numbers of different kinds of seeds to the generalizations that explain the facts, that "make sense" out of them. We call this kind of reasoning **inductive.** No simple rules of logic lead in this direction; one needs intuition or insight. The ability to make these inductive generalizations requires more art than science, but science would not exist without them.

There is one further criterion that any good hypothesis should meet. It should be able to predict new facts. This prediction of new facts involves reasoning which is called **deductive.** If the generalizations are valid, then certain specific consequences can be deduced from them.

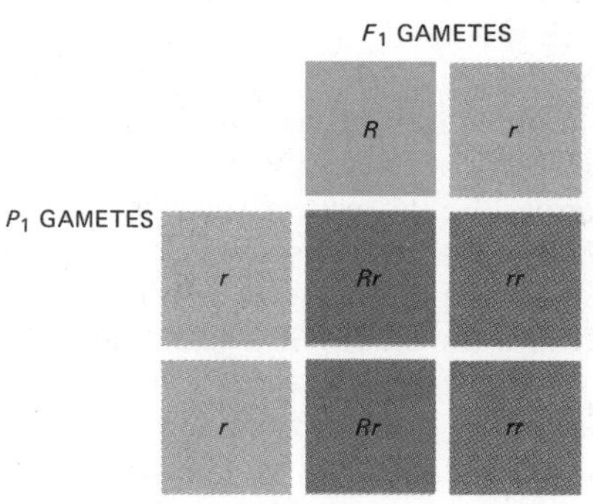

FIG. 15.7 Mendel's testcross. Although it superficially resembled a P_1 mating, Mendel's prediction that two phenotypes would be produced and in equal numbers was verified.

15.6 THE TESTCROSS: A TEST OF MENDEL'S HYPOTHESIS

Mendel appreciated fully the importance of this step. To test his hypothesis, he tried to predict the outcome of a breeding experiment he had never carried out. He crossed his heterozygous round peas (*Rr*) with homozygous wrinkled ones (*rr*). He reasoned that the homozygous recessive parent could only produce gametes with *r*. The heterozygous parent should produce equal numbers of *R* and *r* gametes. Mendel predicted that one-half of the seeds produced from this cross would be round (*Rr*) and one-half would be wrinkled (*rr*) (Fig. 15.7). This kind of mating, using as one parent a known homozygous recessive, is called a testcross. It "tests" the genotype in those cases where two different genotypes (like *RR* and *Rr*) produce the same phenotype. Note that to the casual observer in the monastery garden, the cross appeared no different from the P_1 cross described earlier. Round-seeded peas were being crossed with wrinkle-seeded peas. But Mendel, believing that the round-seeded pea plants in this cross were actually heterozygous, predicted that both round and wrinkled seeds would be produced and in a 50:50 ratio. He performed the matings and harvested 106 round peas and 101 wrinkled peas from his plants.

Mendel's hypothesis had explained all the known facts. It also led to the prediction of other facts as yet undiscovered. When these predicted facts were discovered, his hypothesis was greatly strengthened. A hypothesis that explains all the known facts in the situation and successfully predicts new ones is soon referred to as a **theory.** If a theory continues to serve its explanatory and predictive functions, it may eventually come to be called a **law.** Two of Mendel's assumptions (one of which we have already discussed) are now referred to as Mendel's laws.

15.7 DIHYBRIDS—THE LAW OF INDEPENDENT ASSORTMENT

During the same period of time that Mendel was investigating inheritance in round- and wrinkle-seeded peas, he ran similar experiments with pea varieties differing in six other clear-cut ways. The results from all these other experiments also supported his hypothesis. Next, he crossed pea plants differing in two characteristics. A pure-breeding, round-seeded, yellow-cotyledon pea variety was cross-pollinated with a pure-breeding, wrinkle-seeded, green-cotyledon variety. All the seeds produced from the mating were round and had yellow cotyledons. This bore out Mendel's previous finding that the allele for yellow cotyledons, like the allele for round seed shape, was dominant. (This F_1 generation is said to be *dihybrid* because it is produced by crossing parents differing in two traits.) Then Mendel planted these seeds and allowed the resulting flowers to self-pollinate. Either of two possibilities might be expected. The alleles for round shape and yellow cotyledons, which had been inherited from one par-

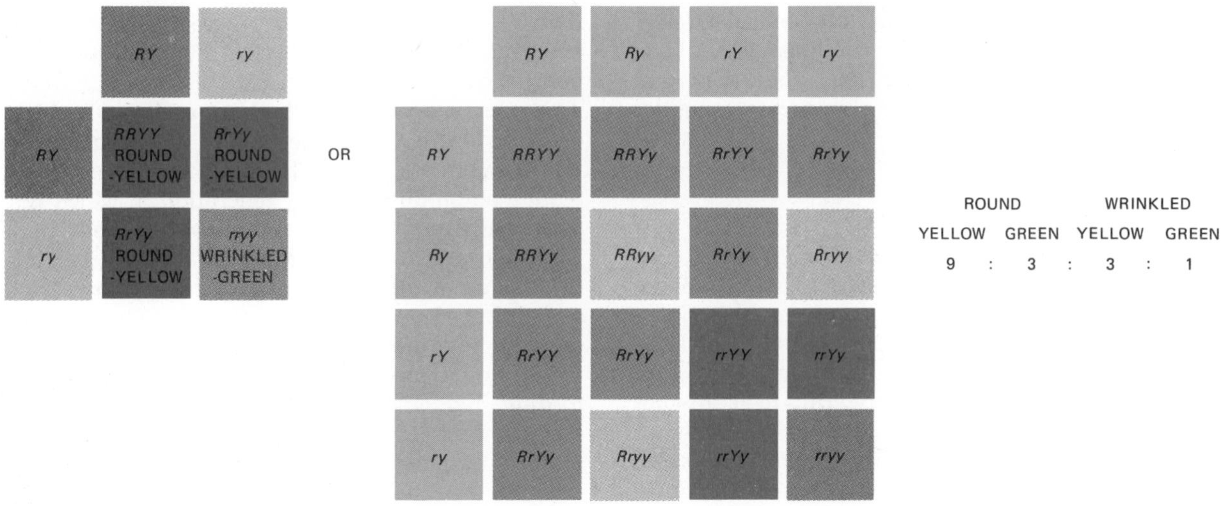

FIG. 15.8 Alternative predictions of the results of mating two dihybrids. In this case, the second prediction (right) turned out to be correct.

ent, might be inseparable and thus passed on as a single unit to the F_2 generation. If the same were also true for the wrinkled-green alleles, one would expect that three-fourths of the F_2 generation would be round-yellow and one-fourth wrinkled-green (Fig. 15.8). If, on the other hand, the genes for seed shape and cotyledon color were distributed to the gametes independently of each other, then one would expect in the F_2 generation to find some peas that were round with green cotyledons and some that were wrinkled with yellow cotyledons as well as those resembling the P_1 types. According to this latter assumption, then, four phenotypes would be produced in the ratio 9:3:3:1 (Fig. 15.8).

Mendel performed this cross and harvested 315 round-yellow peas, 101 wrinkled-yellow peas, 108 round-green peas and 32 wrinkled-green peas. It is characteristic of Mendel's careful work that he then proceeded to plant all of these peas and verify the presence of four separate genotypes among the round-yellow peas and two separate genotypes among each of the peas with the new combination of characteristics. Only the 32 wrinkled-green peas were of a single genotype. These results led Mendel to frame his last hypothesis (Mendel's second law): the distribution of one pair of factors is independent of the distribution of the other pair. This has come to be known as his **law of independent assortment.**

15.8 MENDEL'S THEORY: THE SEQUEL

The experiments described in this chapter were carried on from 1858 to 1866. In 1866 Mendel published the results as well as his analysis of them. Little attention was paid to them by other biologists. No one tried to repeat any of the experiments or carry them on with different traits or different organisms. Mendel himself soon gave up his experiments and became increasingly involved with the administration of his monastery.

Mendel died in 1884. In 1900, thirty-four years after he had published his work and sixteen years after his death, Mendel's work was brought to light once more. Three men, working entirely independently of one another, discovered the same principles we have been describing. Only after their work was done did they discover that they had been anticipated by an obscure monk a third of a century before.

Several reasons have been proposed for the failure of Mendel's work to gain acceptance. Whatever the reasons, it did not. It is indeed ironical that the present development of genetics dates only from the year 1900, not 1866. Mendel's brilliant work failed to become part of the scientific world of his time. When scientists were ready to pursue his findings further, they had already rediscovered them on their own.

What is the status today of Mendel's *laws?* Although important exceptions to them have been discovered in the years since 1900, they still form the foundation upon which the science of genetics rests.

15.9 CONTINUOUS VARIATION: THE MULTIPLE-FACTOR HYPOTHESIS

One of these exceptions arose from a study of the inheritance of traits which vary quantitatively among different individuals. Mendel had restricted his studies to traits which varied in a clear-cut, easily distinguishable, *qualitative* way. But men are not *either* tall *or* short; nor are they *either* heavy *or* light. Many traits differ in a continuous, *quantitative* way throughout a population (Fig. 15.9). Some of the variation can be explained by differences in diet and perhaps other factors in the environment. Environment is not, however, sufficient to explain the full range of heights or weights. No theory of inheritance could be regarded as completely successful until this kind of variation could be accounted for.

By 1908 the information was available which provided a solution to the problem. The Swedish geneticist Nilsson-Ehle studied the inheritance of kernel color in wheat. Using Mendel's methods, he mated pure-breeding red-kernel strains with pure-breeding white-kernel strains. The offspring were all red, but the intensity of the color was much less than in the red parent. It seemed as though the effect of the red allele in the F_1 generation was being modified by the presence of the white allele. This phenomenon, which had not occurred in Mendel's studies, is called **incomplete dominance.**

Furthermore, when Nilsson-Ehle mated two F_1 plants, he produced an F_2 generation in which red-kerneled plants outnumbered white-kerneled plants 15:1. Close examination revealed that the red kernels were not all alike. They could quite easily be sorted into four different categories. One-fifteenth of them were deep red, like the P_1 type. Four-fifteenths were medium dark red, six-fifteenths were medium red (like the F_1 generation) and four-fifteenths were light red. These results could be explained by assuming that kernel color in wheat is controlled by not one, but two pairs of genes, the effects of which add up *without* distinct dominance (Fig. 15.10). Four genes for red produce a deep red kernel. Four genes for white produce a white kernel. Just one red gene out of the four produces a light red kernel. Any two out of the four produce a medium red kernel. Any three out of the four produce a medium dark red kernel. If one plots the numbers of the different colored offspring in the F_2 generation against color intensity, a graph similar to that in Fig. 15.10 results.

In other wheat varieties, Nilsson-Ehle found F_2 generations with a ratio of red kernels to white of 63:1. These could be explained by assuming that three pairs of alleles were involved. In these cases, six different shades of red could be detected. (Try to work out the genetics of this cross for yourself.) The color differences were very slight. Environmental influences also caused alterations in intensity so that in practice the collection of kernels displayed a continuous range of hues all the way from deep red to white.

The occurrence of continuous variation of a trait in a population can thus be explained by assuming it is controlled by several pairs of genes, the effects of which are added together. This hypothesis is called the Multiple Factor Hypothesis. It suggests that (1) when two extreme types are mated (e.g. *AABB* and *aabb*), the offspring are intermediate in type; (2) when two intermediate types are mated, most of their offspring are also intermediate, but some extreme types will be produced; and (3) the results of random matings in a large population will be a wide range of types with the greatest number in the middle range and the fewest at the extremes. These three effects are actually observed in most cases of quantitative variation in living things.

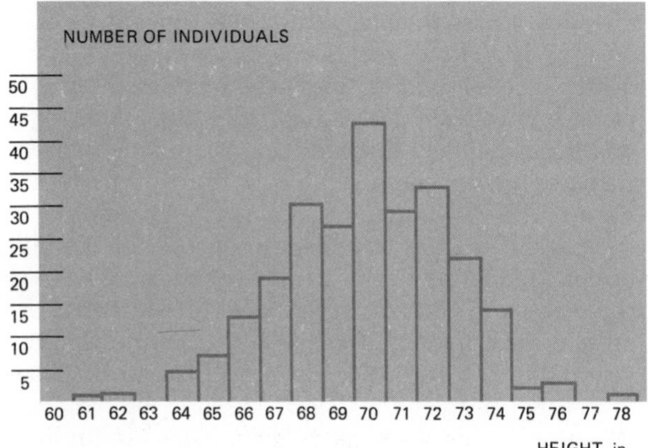

FIG. 15.9 Histogram showing distribution of heights among a group of male secondary-school seniors.

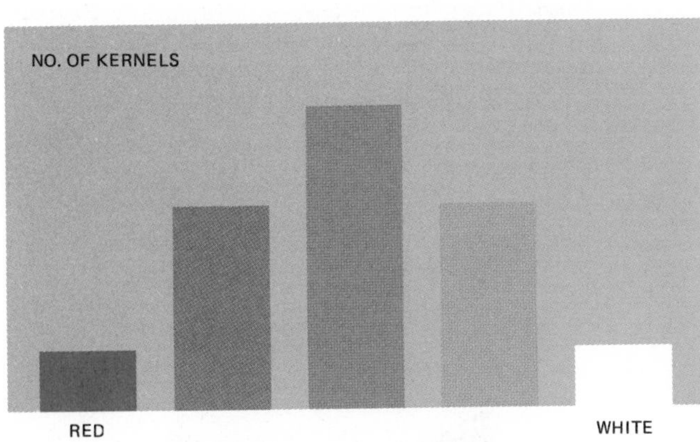

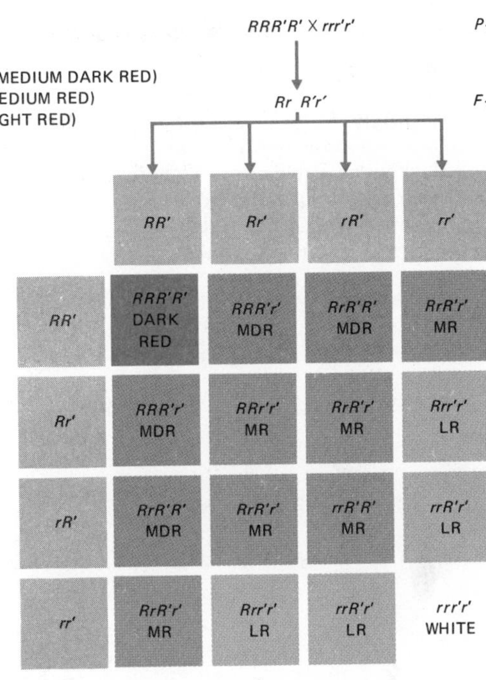

FIG. 15.10 Mutiple-factor inheritance in wheat. The alleles at one locus are indicated with prime marks; at the other, without. They affect the phenotype additively.

When Mendel carried out his experiments on heredity, the nature of the nucleus, the chromosomes, and meiosis was not understood. In formulating the idea of genes, Mendel made no attempt to describe them as physical structures or to say precisely where in the gametes they were located. For Mendel, the genes were simply hypothetical entities which explained observed patterns of inheritance. With the rediscovery of his work, biologists immediately attempted to relate the behavior of his genes to visible features in the cell. Their efforts will be the topic of the next chapter.

EXERCISES AND PROBLEMS

1. What terms did Weismann use to distinguish the body cells of an animal from its reproductive cells?
2. Do these terms apply equally well to bacteria? To plants?
3. Give the gene content of the different kinds of eggs produced by a woman whose genotype is JjKkLl.
4. Is it correct to say that crossing two heterozygous red-flowered peas will produce eight seeds, two of which will be homozygous for the trait? Explain.
5. The gene for tallness in peas is dominant over the gene for dwarfism. A cross between a tall pea and a dwarf pea produced 86 tall plants and 81 dwarf plants. What was the probable genotype of the tall plant? What is this kind of cross called?
6. Distinguish between a fact and a hypothesis. Are the two always easily distinguishable?
7. In *Drosophila* sepia eye is recessive to red eye, and curved wing is recessive to straight wing. If a pure-breeding sepia-eyed, straight-winged fly is mated with a pure-breeding red-eyed, curve-winged fly, what phenotypes will appear in the F_1 generation? If two F_1 flies are allowed to mate, what phenotypes will occur in the F_2 generation and in what ratio?
8. How many different genotypes will occur in the F_2 generation described in Question 7?
9. When plant A (pure-breeding tall) is crossed with plant B (pure-breeding short), the offspring are all intermediate in height. When two of the offspring are crossed, 1/16 of the next generation is as tall as the tall grandparent. How do you account for this?
10. Define a gene.

REFERENCES

1. CASTLE, W. E., and J. C. PHILLIPS, "A Successful Ovarian Transplantation in the Guinea-Pig, and its Bearing on Problems of Genetics," *Great Experiments in Biology,* ed. by M. L. Gabriel and S. Fogel, Prentice-Hall, Englewood Cliffs, N.J., 1955. A brief report of their experiment.
2. MENDEL, G., "Experiments in Plant Hybridization," *Classic Papers in Genetics,* ed. by James A. Peters, Prentice-Hall, Englewood Cliffs, N.J., 1959. A clear presentation of Mendel's experiments and his interpretation of his results.

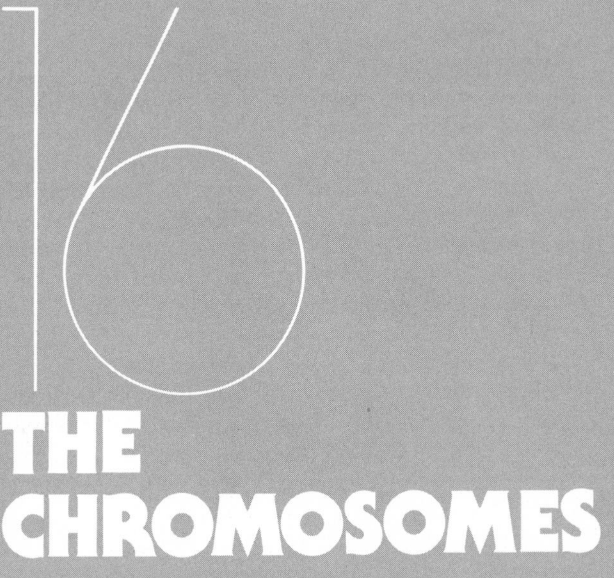

16 THE CHROMOSOMES

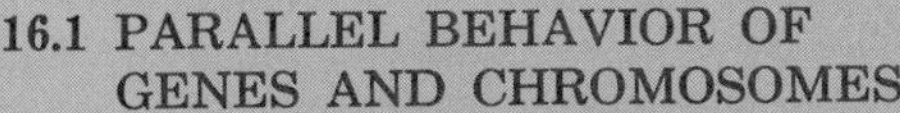

The 35 years during which Mendel's discoveries lay unnoticed saw many important discoveries in cytology—the study of cells. The importance of the nucleus in controlling the traits of the organism was revealed by Boveri's work with sea-urchin eggs, discussed in Chapter 5. The behavior of the chromosomes in mitosis and meiosis, discussed in Chapter 14, was also discovered.

16.1 PARALLEL BEHAVIOR OF GENES AND CHROMOSOMES

Shortly after the rediscovery of Mendel's "laws," several workers, among them H. E. Sutton and Boveri himself, realized that certain of Mendel's rules of gene behavior could be explained if the genes were located on or in the chromosomes. Mendel had said that genes occur in pairs, one member of which is received from each parent. Chromosomes also exist in pairs, one member of which is received from each parent. Mendel had said that when the reproductive cells are produced, the paired factors (genes) separate and are distributed as units to each gamete (*law of segregation*). In meiosis, the homologous pairs of chromosomes separate and only one member of each pair goes to a given sex cell. Mendel had stated further that in a dihybrid cross, the distribution of one pair of factors is independent of the distribution of the other pair (*law of independent assortment*). In meiosis, the distribution of maternal and paternal chromosomes is entirely at random. If we assume that the genes for one trait, say seed shape, are located on one pair of chromosomes and the genes for the other trait (cotyledon color) on another pair, then the independent assortment of the chromosome pairs in meiosis will also result in the independent assortment of the two traits (Fig. 16.1).

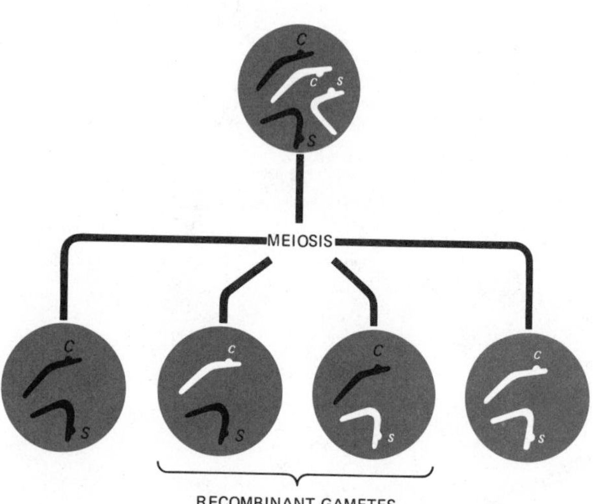

FIG. 16.1 Reshuffling genes as a result of the random assortment of chromosomes during meiosis. The position that a gene occupies in a chromosome is called its *locus*. If, as shown here, the gene loci in a *doubly* heterozygous individual are on different chromosomes, random assortment of these chromosomes produces the four possible gene combinations in equal numbers.

Because there are so many more genes in an organism than there are chromosomes, each chromosome must contain many genes. The site on the chromosome where a given gene is located is called its **locus.** The two alleles controlling a given trait are presumed to be located at corresponding loci on each of two homologous chromosomes.

If we mated dihybrids in which both traits were controlled by genes on the same chromosome, would you expect Mendel's *law of independent assortment* to still hold? Later in the chapter we shall find that the law does *not* hold for dihybrids of this sort. Instead, these genes are said to show "linkage."

For a direct demonstration of the validity of the chromosome theory, it was necessary to be able to associate the presence or absence of a given trait or traits in an organism with the presence or absence of a given chromosome in the body cells of this organism. But the two alleles controlling the expression of a given trait are, according to the chromosome theory, located at corresponding loci on each of two homologous chromosomes. Homologous chromosomes may be visually distinguishable from other chromosomes in the cell but are not distinguishable from each other. So it is not possible to tell by visual inspection of a member of the pair whether it carries a given allele or not.

Tests of the Chromosome Theory

16.2 SEX DETERMINATION

The first solution to this dilemma came as a result of studies on the fruit fly *Drosophila melanogaster* (Fig. 16.2). Thomas Hunt Morgan pioneered in the use of this tiny organism as a subject of research in genetics. It was an admirable choice for several

FIG. 16.2 Stages in the development of the fruit fly *Drosophila melanogaster*.

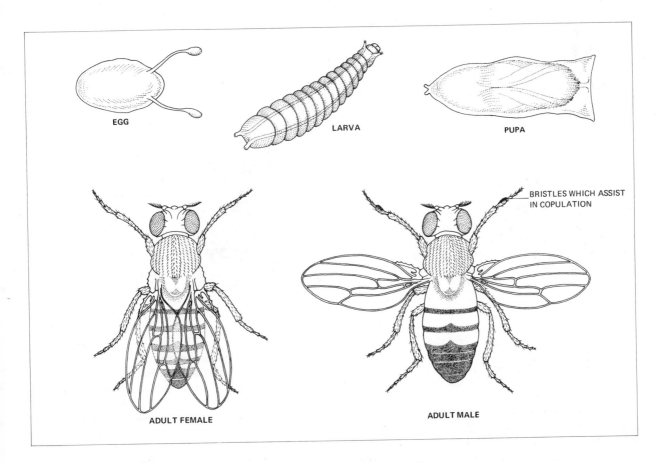

reasons. First, the flies are small and thus sizable populations of them can be reared easily in the laboratory. Second, the life cycle is completed quite rapidly. A new generation of adult flies can be produced every two weeks. Third, the flies are remarkably fecund; that is, a female may lay hundreds of fertilized eggs during her brief life span. The large populations thus produced make statistical analyses easy and reliable. Still a fourth virtue was discovered: the presence of giant chromosomes in the salivary glands of the larvae. These chromosomes (Fig. 16.3) show far more structural detail than do normal body chromosomes. Furthermore, they are present

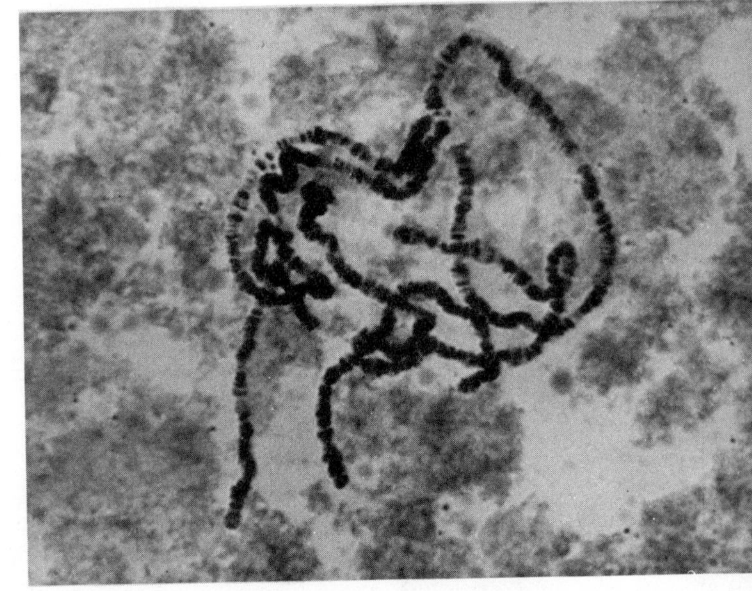

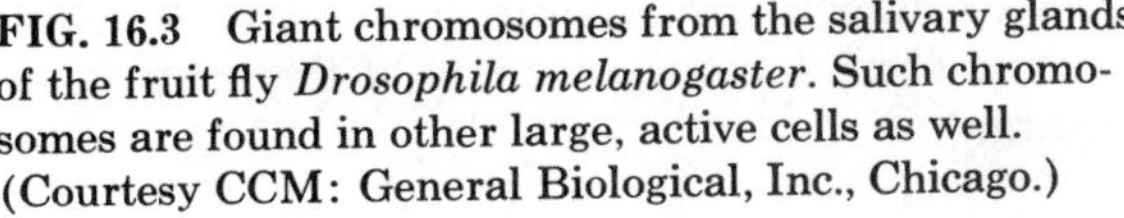

FIG. 16.3 Giant chromosomes from the salivary glands of the fruit fly *Drosophila melanogaster*. Such chromosomes are found in other large, active cells as well. (Courtesy CCM: General Biological, Inc., Chicago.)

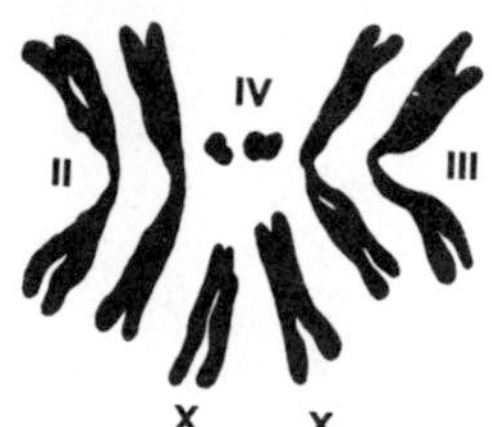

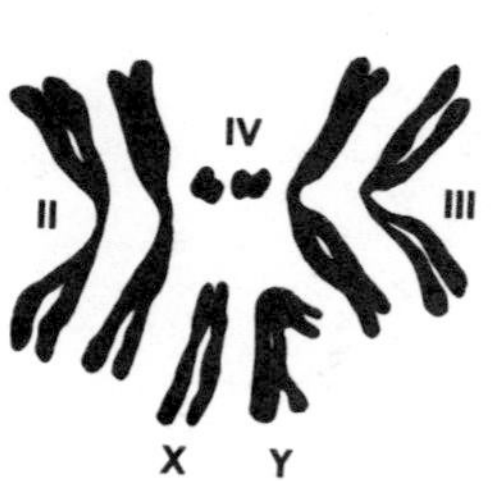

FIG. 16.4 Chromosomes of *Drosophila melanogaster*. Both sexes have three homologous pairs of autosomes. In addition the females have two X chromosomes (top), the males have an X and a Y chromosome (bottom). These are the sex chromosomes.

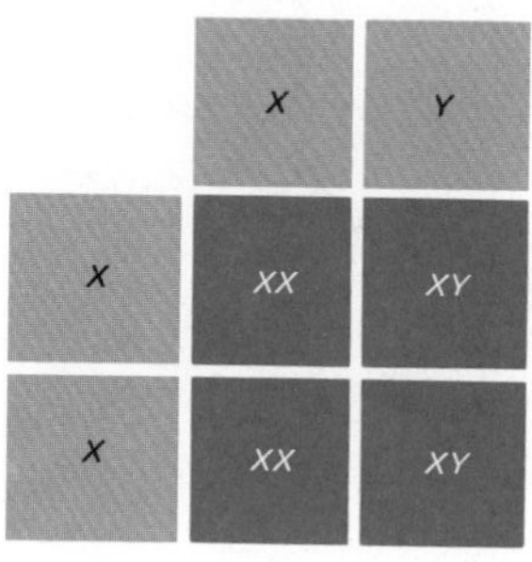

FIG. 16.5 Sex determination. Each gamete contains a single sex chromosome. In mammals, a zygote that receives two X chromosomes develops into a female. A zygote containing one X and one Y develops into a male. Thus it is the sex chromosome in the sperm cell that determines the sex of the child.

during interphase, a time when chromosomes are normally invisible. The origin of the giant chromosomes and their special usefulness in research will be explained in Chapters 19 and 22.

Although a female fruit fly has four pairs of homologous chromosomes (Fig. 16.4), the male fly has only three pairs of homologous chromosomes. The remaining two chromosomes are not homologous. One member of the fourth pair of chromosomes is identical in appearance to the chromosomes of the fourth pair in the females. It is called the X chromosome. The other member is quite different in appearance. It is called the Y chromosome (Fig. 16.4). Collectively, these X and Y chromosomes are called **sex chromosomes** because their presence is always correlated with the sex of the fly. The other chromosomes are called **autosomes.**

As a result of the separation of homologous pairs in meiosis, fruit-fly *eggs* contain one of each of the autosomes plus one X chromosome. The *sperm* cells produced by male fruit flies contain three autosomes and either an X *or* a Y chromosome. We can depict the results of random union of these sperm and eggs by setting up a Punnett's square (Fig. 16.5). (Note that the symbols now refer to whole chromosomes, not genes.) One can quickly see that the offspring should be about equally divided between males and females, and this indeed is the case.

Glancing at Fig. 16.2, we can immediately see that the matter of sex differences is one of inherited traits. The pigmentation on the abdomen of the male, the penis, the bristles on the first tarsal segment of the foreleg are just a few of the visible traits that distinguish the male fruit fly from the female. The mere fact that the presence or absence of these traits is always related to the sex chromosomes present is evidence of the chromosome theory of heredity.

It is tempting to think of the Y chromosome as carrying the genes which determine maleness. In mammals, including humans, this is the case. However, in many insects (including the fruit fly), the Y chromosome seems to be genetically inert. Only a few genes have been associated with it and these are not directly related to sex. Probably the autosomes carry the male determinants and it takes two X chromosomes to mask their effect and produce a female. The fact that the male grasshopper has no Y chromosome at all, just one X chromosome, while the female has two, is one bit of evidence to support this idea. In birds, the XY chromosome constitution is found in the females, the XX in males. Reptiles appear to be like birds in this respect. Among the amphibians and fishes, we find sometimes one system, sometimes the other.

Whatever the specific mechanism involved, sex in the animal kingdom seems to be directly related to the arrangement of specific chromosomes and thus serves to support the general theory that the chromosomes carry the determinants of the traits of an organism.

16.3 *X*-LINKAGE

The discovery of the mechanism of sex determination in fruit flies quickly paved the way for another test of the chromosome theory, this one involving the inheritance of one specific trait. In one of Morgan's *Drosophila* cultures there appeared a male fly with white eyes instead of the brilliant red eyes characteristic of the species. When this fly was mated with normal red-eyed females, the offspring all had red eyes. This suggests that if the white-eyed trait is determined by a specific gene, it must be recessive. According to Mendel's *law of segregation,* all the red-eyed F_1 offspring must have been hybrid for the trait. (Why?) We can then predict that if two red-eyed F_1 flies were mated, white-eyed flies would appear in the F_2 generation in about 25% of the offspring. When Morgan made this cross, he found that this indeed occurred. However, all the white-eyed offspring were *males.* Not a single white-eyed female appeared. Morgan reasoned that this peculiar sex-related type of inheritance could be explained if one assumed that the alleles involved were located on the X chromosomes. The female flies, having two X chromosomes, would have to be homozygous for white eyes to show the trait. The males, on the other hand, having only one X chromosome, would show the trait for whichever allele they had on that chromosome. Even if it were the recessive allele (r), the Y chromosome contains no allele that would be dominant to it. Figure 16.6 shows diagrammatically Morgan's idea of what had occurred in his first two crosses. Again, the symbols X and Y stand for the visible chromosomes and the superscripts R and r for the dominant and recessive alleles, respectively.

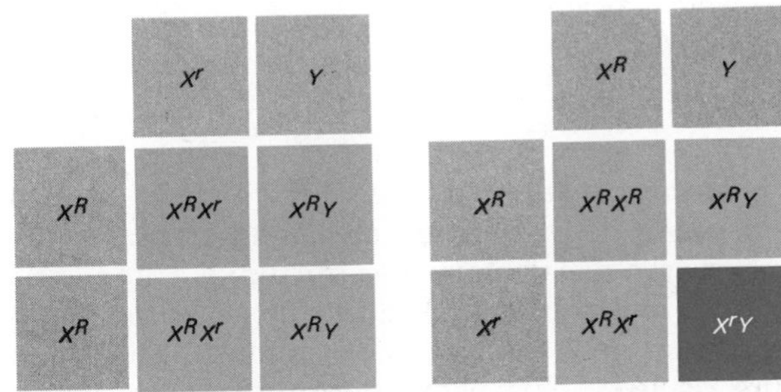

FIG. 16.6 Mechanism of X-linkage. A recessive X-linked allele will be expressed in males because there is no corresponding gene locus on the Y chromosome. R = dominant X-linked allele, producing red eyes in *Drosophila*. r = the recessive X-linked allele, producing white eyes.

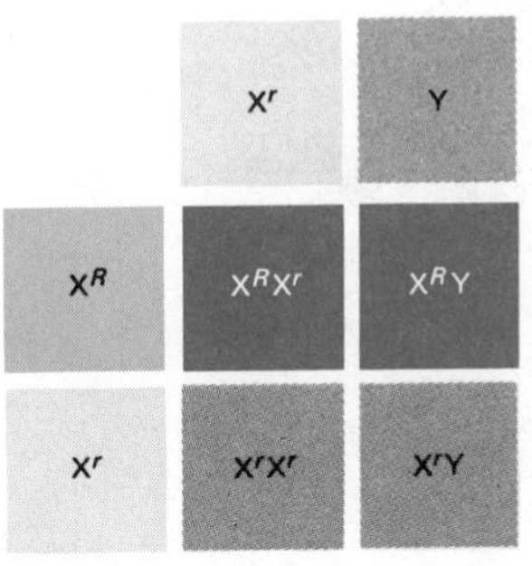

FIG. 16.7 Morgan's testcross. By mating a white-eyed male fruit fly with a "carrier" female, Morgan was able to produce white-eyed females.

If this hypothesis is correct, it should be possible to produce white-eyed females by mating white-eyed males with F_1 females. Morgan made this cross and produced all four kinds of flies in approximately equal numbers (Fig. 16.7). Today we call this type of inheritance *X-linked,* because the genes are located on the X chromosome. In humans, red-green color blindness, the clotting disorder called hemophilia, and over 100 other known traits are produced by X-linked genes. These traits are rare in females because they must inherit the recessive alleles from both their father and their mother in order to show the trait. This would be especially unlikely in the case of hemophilia because until recently the chances that a hemophiliac would ever become a father were quite dim. Figure 16.8 shows a pedigree of the descendants of Queen Victoria. Note that the allele for hemophilia, which has plagued the royal houses of Europe since her time, was nearly always passed on by the mothers who were heterozygous for the trait and thus showed no symptoms. We call them "carriers" because, while free from symptoms, they passed the recessive allele on to approximately one-half their sons, who became bleeders.

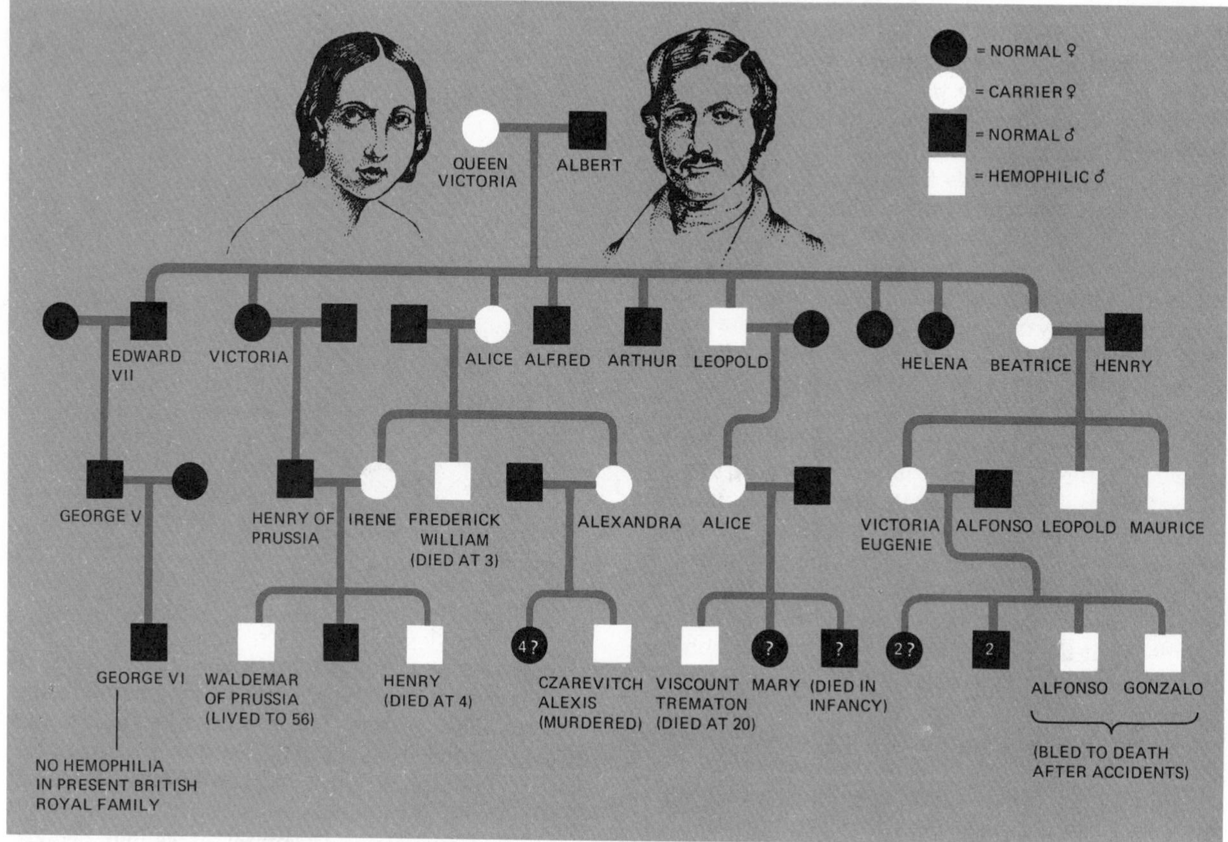

FIG. 16.8 Pedigree showing inheritance of hemophilia, an *X*-linked trait, in the descendants of Queen Victoria. Many of the descendants in the third and fourth generations (third and fourth rows) have been omitted because the mutant gene was not transmitted to them. Offspring whose status is uncertain are indicated with a question mark.

A good deal of disability and grief could be avoided if methods were found by which "carriers" of genetic defects could be detected. Knowledge of their genetic status would provide a basis on which to evaluate the risk of passing a defect on to their children. For many genetic defects (including hemophilia) the presence of one "good" gene permits normal functioning, i.e. the good gene appears to be fully dominant. However, with the development of delicate biochemical tests, it is now often possible to detect a reduced level of gene activity in carriers. In some cases, this is probably because one gene cannot do the job quite as well as two. In the case of *X*-linked genes, however, another mechanism comes into play. This is the inactivation which occurs to *one X* chromosome in each female cell.

During interphase, the chromosomes are too tenuous to be stained and observed by microscopy. However, a dense, stainable structure is found to be present in virtually all the interphase nuclei of *female* mammals. It is not found in males. This structure (called a **Barr body** after its discoverer) turns out to be one of the *X* chromosomes. Its compact appearance is a reflection of its inactivity.

It now appears that early in embryonic development, one or the other of a female's *X* chromosomes becomes inactivated and converted into a Barr body (Fig. 16.9). Which of the two *X* chromosomes becomes inactivated in a given cell is probably a matter of chance (except in marsupials like the kangaroo, where it is the *X* chromosome inherited from the father that is inactivated). However, after inactivation has taken place, all the descendants of that cell will have the same chromosome in the inactive state. We may again use the term *clone* to describe the descendants of such a cell. Although all the cells in one individual probably have the same absolute gene content, X inactivation creates subpopulations (clones) with differing *effective* gene content.

Several X-linked genetic defects are known that involve deficiencies in enzyme synthesis. When cells taken from women who are carriers for such traits are grown in tissue culture, we should expect to find some cells that show the defect and some that do not. In at least four such cases, it has been possible to demonstrate the existence of two such populations. Clones that fail to express enzyme activity are presumably descendants of a cell in which the normal gene, and the X chromosome of which it is a part, became inactivated. An organism whose cells vary in *effective* gene content and hence in the expression of a trait is said to be a genetic *mosaic*.

The discovery and analysis of X-linked traits, and of their peculiar mosaic expression in certain cases, serves to corroborate further the chromosome theory of inheritance. Each example described can now best be explained by assuming that the alleles involved are located on a specific chromosome.

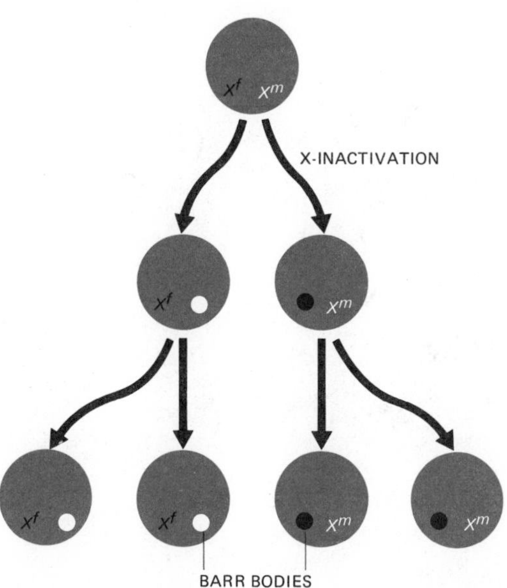

FIG. 16.9 X chromosome inactivation with the formation of a Barr body. Which X chromosome becomes inactivated in a particular cell is a matter of chance. However, all the descendants of that cell will have the same X chromosome inactivated.

16.4 CHROMOSOME ABNORMALITIES

Karyotype analysis (Fig. 16.10) has revealed that in about 3% of human pregnancies, the cells of the child contain abnormal chromosomes. These abnormalities are often associated with severe deformities, although this is not always the case. The chromosomal abnormalities fall into two major groups. These are (1) alterations in chromosome number and (2) alterations in the structure of one or more chromosomes. In humans, most of the disorders of chromosome *number* arise because of **nondisjunction.** You remember that in meiosis the homologues separate, one going to each pole of the cell. On rare occasions, a given pair of homologous chromosomes fails to separate (nondisjunction). As a result, one daughter cell gets both homologues and the other daughter cell gets neither.

If the sperm or egg cell that develops from these aberrant cells is used in fertilization, it will transmit its chromosome excess or deficiency to the zygote. Approximately 0.2% of newborn children suffer from Down's syndrome, which is characterized by a number of defects including severe mental retardation and a high susceptibility to leukemia. Karyotype analysis of these unfortunate individuals reveals 47 chromosomes instead of the usual 46. The extra chromosome is a number 21 (Fig. 16.10). One way in which the zygote can receive three number 21 chromosomes is nondisjunction during egg formation, as a result of which an egg gets both of the mother's number 21 chromosomes. The chance of this occurrence seems to increase with age, older women being somewhat more likely to produce children with this defect than younger ones. The occurrence of children with three number 13 chromosomes or three number 18 chromosomes has also been detected. Each defect is accompanied by severe developmental abnormalities.

Abnormal numbers of sex chromosomes also occur at a low frequency. These chromosome abnormalities do not produce such gross deformities as do abnormal numbers of autosomes. Individuals with XXY and $XXXY$ karyotypes develop as males (because of the presence of the Y) but are sterile. (The condition is known as Klinefelter's syndrome.) XYY males are also known and studies both in England and the United States indicate that this chromo-

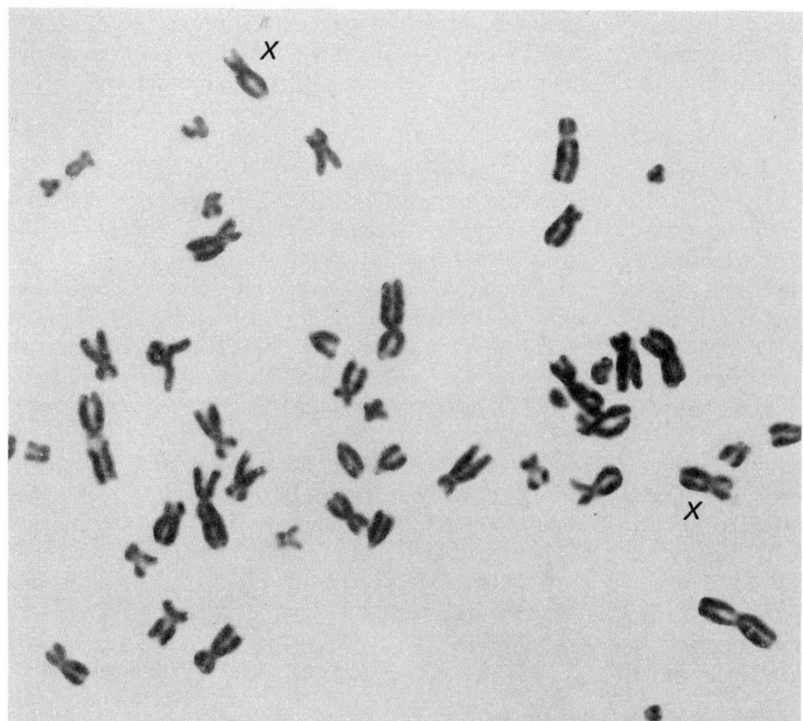

FIG. 16.10 (a) The 46 chromosomes of a human female. The two X chromosomes are marked. Courtesy of Drs. T. T. Puck and J. H. Tjio.) Karotypes can be conveniently analyzed by cutting the photograph into pieces containing single chromosomes and arranging these by homologous pairs.

(b) This is the karyotype of a normal male. The X and Y chromosomes are marked.

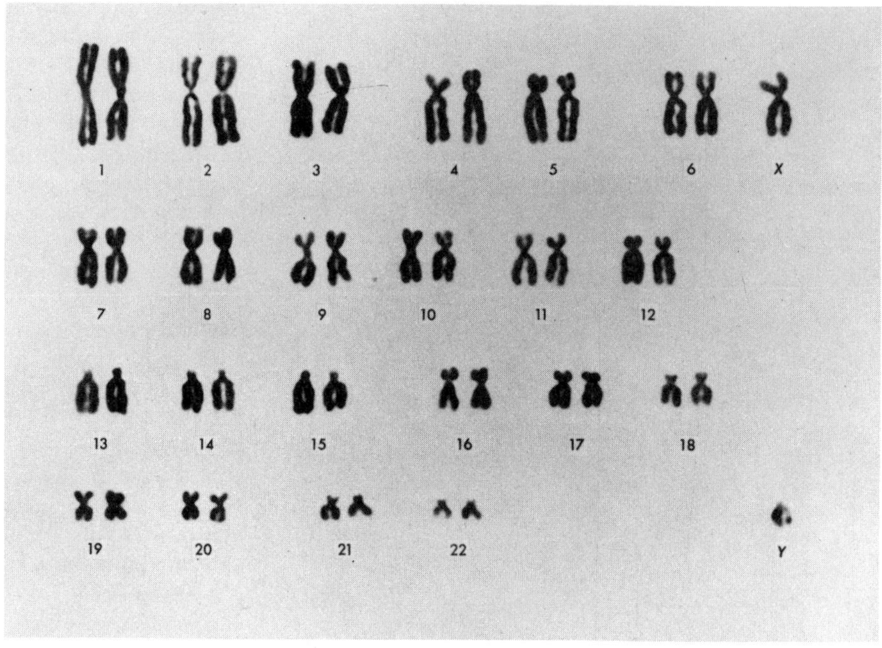

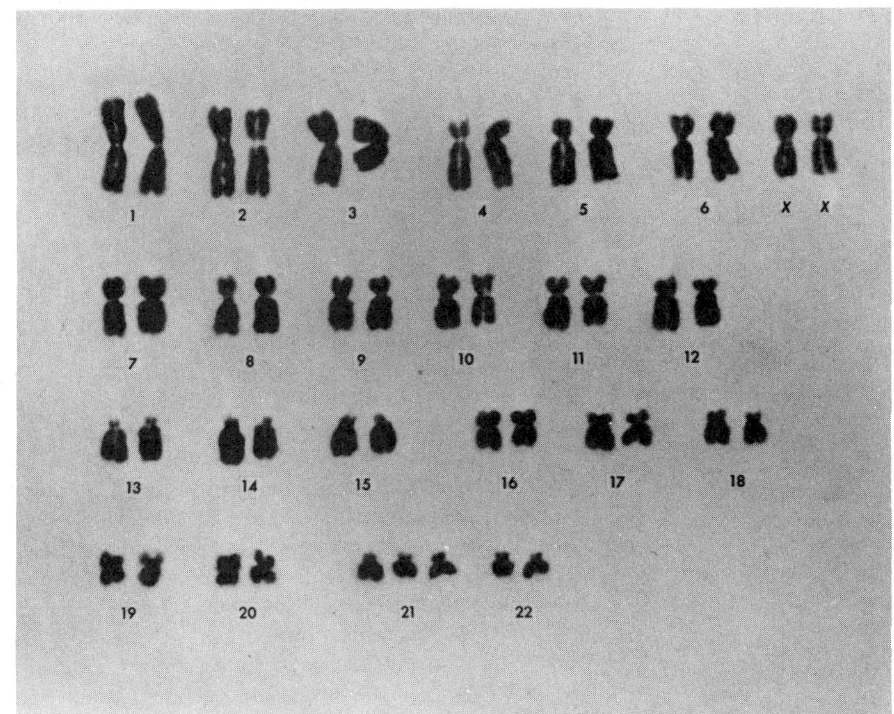

(c) This is the karyotype of a victim of Down's syndrome. Because of the third number 21 chromosome, the condition is also called **trisomy 21.** What was the sex of this individual? (Karyotypes b and c, courtesy of Dr. James L. German, III.)

some constitution is associated with tallness (6 feet or more) and is found more frequently among males committed to prisons and mental institutions than among the general population.

Females with a single X are known. While somewhat abnormal, the fact that they are viable at all lends support to the idea discussed earlier in the chapter, that only one X chromosome is functional in female cells. This would also explain the relative benignity of such chromosome constitutions as XXX, XXXX, and even XXXXX, all of which have been found in females. Whereas normal women have one X chromosome inactivated and hence one Barr body, these women have all their extra X chromosomes inactivated as well. The number of Barr bodies in their cells is equal to the number of X chromosomes minus one (Fig. 16.11).

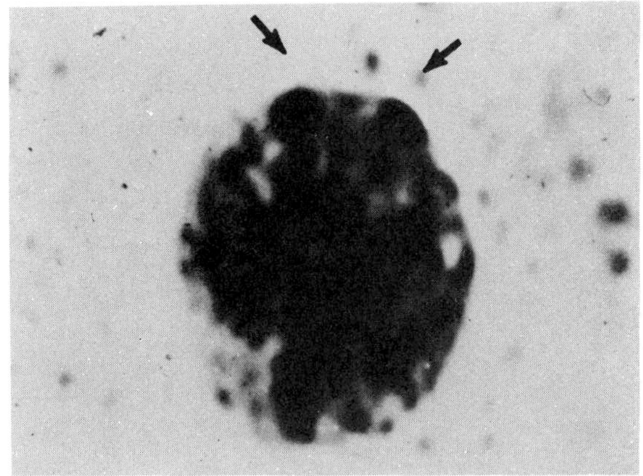

FIG. 16.11 Cheek cell from a woman with three X chromosomes instead of the usual two. Two Barr bodies can be seen (at top). (Courtesy of Dr. Kurt Hirschhorn.)

Among plants and, more rarely, among animals we sometimes find another abnormality of chromosome number, called **polyploidy.** If after the duplication of the chromosomes in a diploid ($2n$) cell, the *cell* itself fails to divide in two, the nuclear membrane will reform around four sets of chromosomes instead of the usual two sets. Such a cell is tetraploid ($4n$) rather than diploid. A cell containing any multiple of the haploid number of chromosomes (above two) is said to be polyploid. Polyploidy is often found in natural plant populations. There is a species of grass in New Zealand that is 38-ploid ($38n$, with $n = 7$).

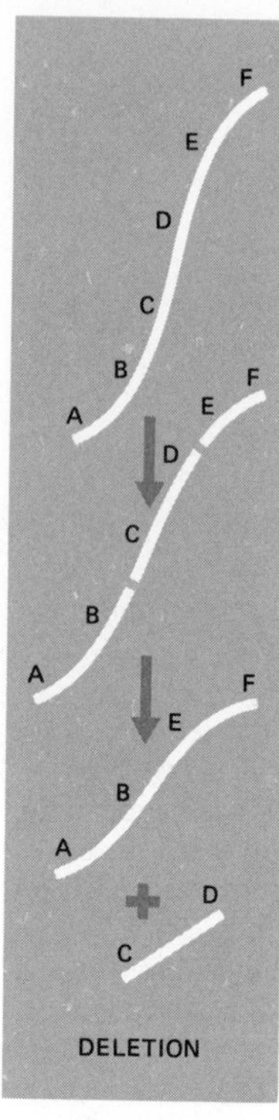

FIG. 16.12 Mechanism of deletion.

Polyploidy can also be induced by treating dividing cells with the substance colchicine. Colchicine disrupts microtubules and thus prevents formation of the spindle. As a result, the duplicated chromosomes fail to separate and move to opposite poles. Onion cells exposed to colchicine for several days have been found with over a thousand chromosomes inside.

Polyploidy in plants is often associated with larger size and greater vigor. For this reason, plant breeders have developed a number of polyploid varieties of commercially important species including corn, watermelons, marigolds, and snapdragons.

Structural alterations in chromosomes seem to occur when a break in a chromosome is followed by a loss or rearrangement of the broken parts. For example, after such a break, a segment of the chromosome may become lost, resulting in a **deletion** (Fig. 16.12). When a chromosome carrying such a deletion undergoes pairing with its homologue, only corresponding portions come together. The result is a loop in the normal chromosome (Fig. 16.13). An organism may reveal such a deletion by a change in its phenotype. For example, a recessive allele on the normal chromosome will be expressed if its dominant

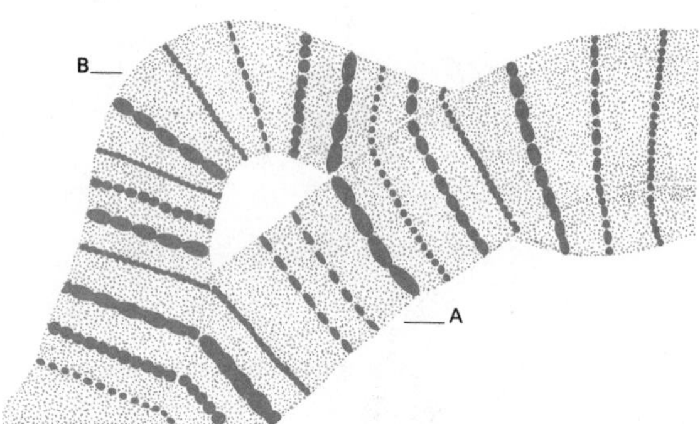

FIG. 16.13 Pairing (synapsis) of a chromosome having a deletion (A) with its normal homologue (B). The buckled portion in B corresponds to the deleted portion of A.

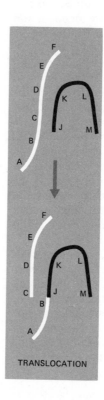

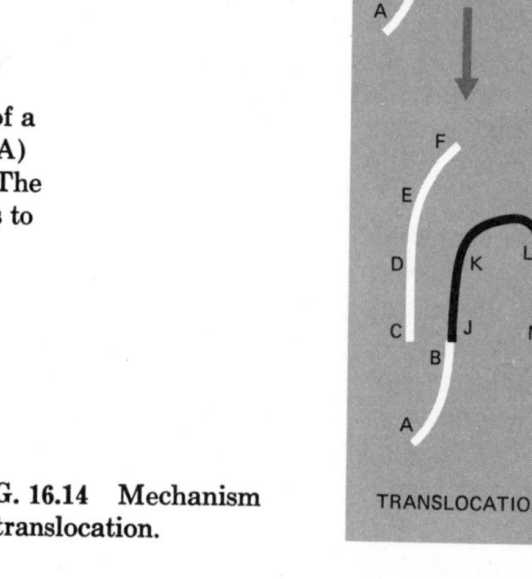

FIG. 16.14 Mechanism of translocation.

allele was located in the missing segment of the homologue with the deletion. One form of leukemia in humans is associated with the presence of a deletion on a number 22 chromosome in the white blood cells. To be able to associate a specific trait with a microscopically observable chromosome provides strong support for the chromosome theory.

Chromosome breaks may also lead to inversions and translocations. In an **inversion,** a segment of a chromosome is turned around *(ABCDE→ADCBE)*. Inversions seldom have drastic effects on the phenotype of their owner because all the genes are still present. Difficulties arise, however, when a chromosome with an inversion attempts pairing and crossing over with its normal homologue during meiosis. Sex cells with defective chromosomes will be produced, leading to partial sterility or serious defects in the offspring. Inversions are easy to detect in giant chromosomes, such as those in the salivary glands of the fruit fly, and were first seen in them. However, recent advances have now made it possible to identify inversions in human chromosomes as well.

Occasionally a segment of one chromosome breaks loose and becomes attached to a chromosome with which it is not homologous. This is called a **translocation** (Fig. 16.14). Organisms with translocations also have fertility problems because half or more of the gametes formed by meiosis will have deletions and may not be able to create viable offspring. Translocations are now commonly found in human chromosomes, usually in people who have had repeated stillbirths or children with serious developmental defects.

A large number of agents have been identified that cause chromosome breaks. These include X-rays and other types of ionizing radiation. Radiation-induced breaks are quite random, that is, they may occur anywhere on a chromosome.

Many chemicals induce chromosome breakage. These include inhibitors of DNA synthesis and certain antibiotics. Most chemically induced breaks occur at specific parts of the chromosome, often near the centromere or other constricted parts of the chromosome. A great deal of publicity has surrounded the discovery that the psychoactive drug lysergic acid diethylamide (LSD) causes nonrandom chromosome breaks in laboratory animals and has been *associated with* such breaks in psychiatric patients receiving the drug. Whether pure LSD in doses apt to be ingested by casual users of the drug represents a hazard *to the chromosomes* is still a matter of controversy.

Chromosome breakage and rearrangements are also associated with virus infections. Measles, chicken pox, and viral meningitis infections have all been implicated.

16.5 LINKAGE

Earlier in the chapter, it was pointed out that the large number of genes and the relatively small number of chromosomes in an organism require that each pair of chromosomes possess many gene loci. In considering the inheritance of two different traits controlled by genes located on the same chromosome, it is obvious that Mendel's *law of independent assortment* can no longer hold. You will remember from the previous chapter that Mendel developed several strains of peas which were dihybrids; that is, they were heterozygous for two different traits, such as seed shape and cotyledon color. When a dihybrid plant *(RrYy)* was mated with the homozygous recessive P_1-type (a testcross), the resulting generation contained not only individuals of the P_1-types, round-yellow and wrinkled-green, but also individuals displaying new combinations of the trait, i.e. round-green and wrinkled-yellow. These latter individuals are called *recombinants.* The four kinds of offspring were produced in approximately equal numbers. Today we know that in several cases this independent assortment of genes occurred because the gene loci for the two traits were located on different pairs of chromosomes. Because of the random arrangement of maternal and paternal chromosomes on either side of the equator in meiosis I, random distribution of the genes resulted. As a result, Mendel framed his second law, the *law of independent assortment:* the distribution of one pair of genetic factors is independent of the distribution of the other pair.

Shortly after the rediscovery of Mendel's work, it became apparent that his second law does not account for inheritance in many dihybrids. In many cases, two alleles inherited from one parent show a strong tendency to segregate together in the formation of gametes and the two alleles inherited from the other parent do the same. This phenomenon is called **linkage.** Let us examine an example of linkage in corn. A strain of corn can be developed which produces kernels that are yellow in color and well filled with a food reserve tissue called endosperm. This strain breeds true, indicating that it is homozygous for these traits. Another strain can be developed which breeds true for colorless kernels that are quite wrinkled in appearance because they contain

GENOTYPE OF ALL GAMETES FORMED BY *ccss* PARENT	GENOTYPES OF GAMETES FORMED BY HETEROZYGOUS (*CcSs*) PARENT			
	CS	*cs*	*Cs*	*cS*
cs	*CcSs*	*ccss*	*Ccss*	*ccSs* ← GENOTYPES OF OFFSPRING
APPEARANCE (PHENOTYPE)	COLORED, SMOOTH	COLORLESS, WRINKLED	COLORED, WRINKLED	COLORLESS, SMOOTH
IF RANDOM ASSORTMENT	25%	25%	25%	25%
ACTUAL RESULTS	48.2%	48.2%	1.8%	1.8%

FIG. 16.15 Demonstrating linkage in corn. Because the gene locus controlling kernel color is located on the *same* chromosome that carries the locus for kernel texture, the genes tend to be inherited together. Only when crossing over separates the loci are *recombinant* gametes formed (see Fig. 16.16).

shrunken endosperm. When the pollen of the first strain (P_1) is dusted on the silks of the second (P_1) (or vice versa), the kernels produced (F_1) are all yellow and full. This tells us that the genes for color (C) and smoothness (S) are dominant over those for colorlessness (c) and shrunken endosperm (s) (Fig. 16.15). The plants grown from the F_1 kernels are then crossed with the homozygous recessive (P_1) strain (*ccss*) to find out what kinds of gametes are produced by the F_1 plants. According to Mendel's second law, the inheritance of the genes determining color should be entirely independent of the inheritance of the genes determining the appearance of the endosperm. The heterozygous F_1 should thus produce the following gametes in approximately equal numbers: (1) *CS*, as inherited from one parent, (2) *cs* as inherited from the other parent, (3) *Cs*, a recombinant, and (4) *cS*, a recombinant. All the gametes produced by the double homozygous recessive in the cross would be *cs*.

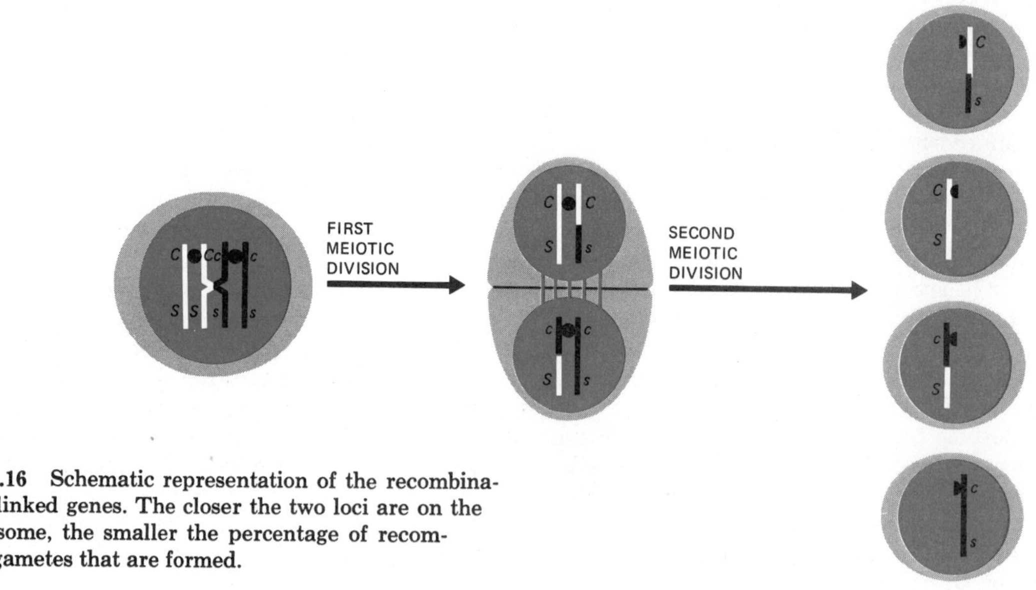

FIG. 16.16 Schematic representation of the recombination of linked genes. The closer the two loci are on the chromosome, the smaller the percentage of recombinant gametes that are formed.

Random union of these gametes would result in four kinds of individuals: (1) colored, smooth, as in one original strain, (2) colorless, wrinkled, as in the other original strain, (3) colored, wrinkled, a recombinant, and (4) colorless, smooth, another recombinant. The genetics of this cross and the actual results achieved are shown in Fig. 16.15. Obviously, Mendel's *law of independent assortment* does not hold. There is a very strong tendency for the gene for colored (C) to remain associated with the gene for smooth texture (S), and similarly, for the gene c to remain associated with the gene s. As we have seen, it can be explained by assuming that the loci of the genes controlling kernel color are on the same chromosome as the loci of the genes controlling kernel texture. In the F_1 generation, when the chromosomes separate in the first meiotic division, the paternal chromosome (CS) goes to one pole and the maternal chromosome (cs) goes to the other (Fig. 16.16).

One question remains. If the gene loci for color and texture are on the same chromosome, how can one explain the recombinant individuals (3.6%) produced by the testcross? The answer is clear when one remembers precisely what happens in the prophase of the first meiotic division. Pairs of duplicated homologous chromosomes unite in synapsis and then one chromatid of each homologue exchanges one or more segments in the process of crossing over. It is this process of crossing over that gives rise to these recombinant gametes. In this case, whenever a crossover occurs between the gene locus for kernel color and the gene locus for kernel texture, the original combination of alleles (CS and cs) will be broken up and a chromosome containing Cs and one containing cS will be produced. Note, however, that the other 2 chromatids do not participate in this crossover. Thus only one-half of the gametes produced after second meiosis will contain the recombinant genotype. Remembering that 3.6% of the total gametes produced by the F_1 generation were recombinants, in what percent of the meiotic divisions did crossing over occur between the C and S loci? (The answer is 7.2%. Do you see why?)

16.6 CHROMOSOME MAPS

Other linked traits have been studied in corn. In some, the percentage of recombinant gametes formed by F_1 individuals is less than the 3.6% found above. In others, the percentage of recombinant gametes is higher. Assuming for a moment that the gene loci are in a linear order from one end of a chromosome to the other, we may deduce that the higher the percentage of recombinant gametes formed for a given pair of traits, the greater the distance separating the two loci. The opposite is also true. Using this knowledge, one can plot a linkage map for as many of the gene loci on a given chromosome as can be discovered. For example, we have seen that about 3% recombinant gametes are produced as a result of crossing over between gene loci C and S. A dihybrid can then be developed for another pair of linked traits, say kernel color (C) and waxy endosperm (wx). This dihybrid produces 33% recombinant gametes, indicating that the loci C and wx are farther apart than the loci C and S. However, one does not yet know whether the locus wx is on the same side of C as S or on the opposite side (Fig. 16.17). The answer can be determined by developing a dihybrid for S and wx. If the percentage of recombinant gametes produced by this individual is less than 33%, then the gene locus wx must be on the same side of locus C as locus S. The opposite is true if the percentage of recombinant gametes turns out to be greater than 33%. Actually the number of recombinant gametes formed is about 30%. Thus we know that the sequence of gene loci on this chromosome is C-S-wx. Furthermore, the fact that the sum of the percents of recombinants between C and S and S and wx is so close to the percent between C and wx lends strong support to the idea that the gene loci are arranged in a line along the length of the chromosome (Fig. 16.17). A straight line is the only geometric arrangement in which this simple numerical relationship can exist.

By pursuing this method with as many linked genes on a given chromosome as can be discovered, it is possible to plot chromosome maps. These maps show the sequence in which the gene loci occur and

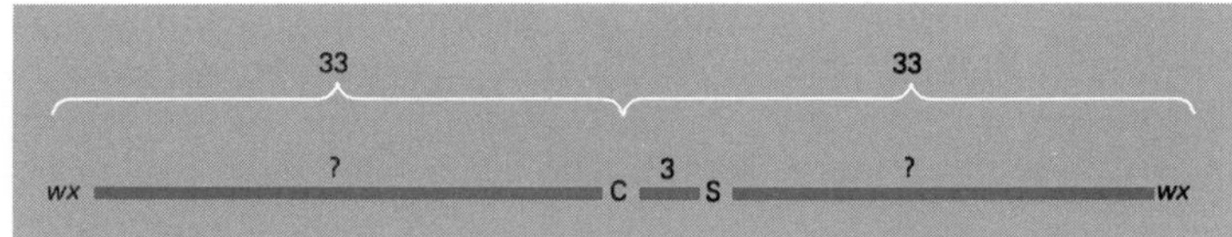

FIG. 16.17 Plotting a linkage map. The production of 30% recombinant gametes as a result of crossing over between the S and wx loci tells us that locus wx is on the same side of locus C as is locus S.

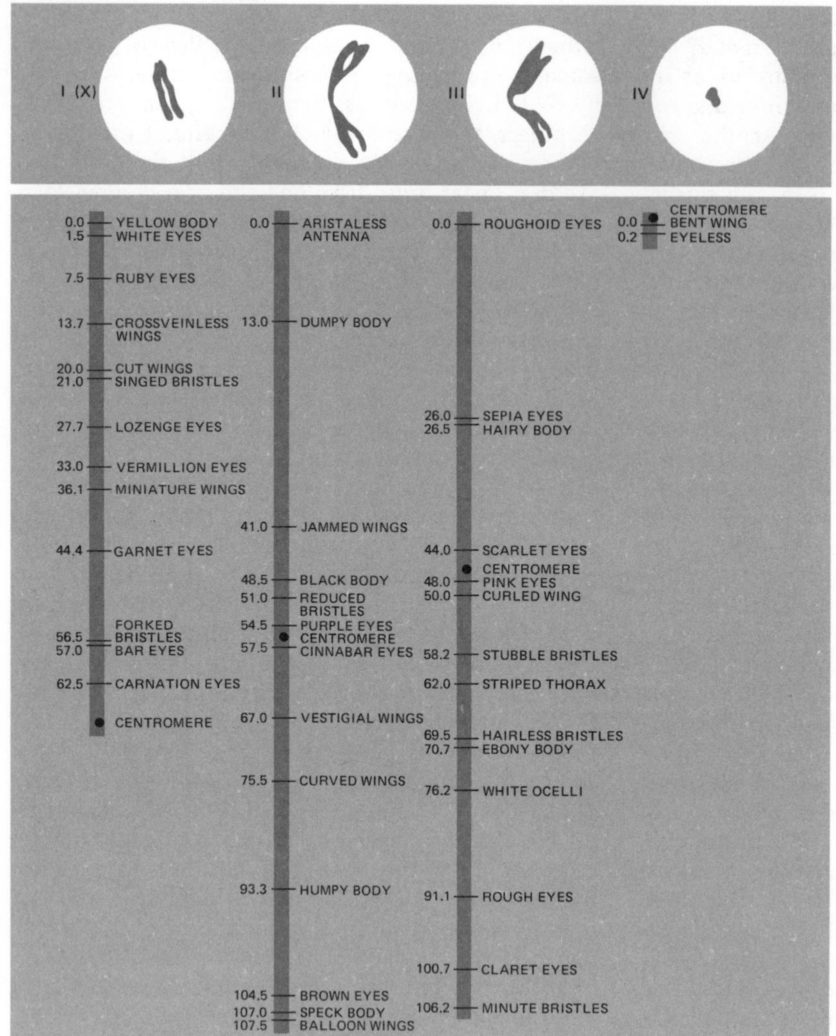

FIG. 16.18 Maps of the chromosomes of *Drosophila melanogaster.* Only a few of the many gene loci known are listed.

the relative spacing between them. Such maps have been produced for the chromosomes of the fruit fly (Fig. 16.18), corn (Fig. 16.19), certain bacteria, the mouse, the silkworm, the tomato, *Chlamydomonas,* and the human *X* chromosome.

As geneticists have established which genes in a particular organism show linkage and which do not, they have been able to assign each gene to a *linkage group.* All the genes in a given group show linkage with respect to each other. They do not show linkage to genes in other groups. The interesting thing about this is that the *number* of linkage groups in an organism is equal to the number of homologous pairs of chromosomes in that organism. Thus *Drosophila melanogaster,* with its four pairs of chromosomes, has four linkage groups. The corn plant has 10 pairs of chromosomes. All of its *genes* fall into one or another of 10 linkage groups. Similarly, in *Neurospora* and the tomato plant, both of which have been exhaustively studied by geneticists, the number of linkage groups is equal to the number of pairs of chromosomes—7 in *Neurospora* and 12 in the tomato. Here, then, is additional evidence that an organism's genes are located in or on its chromosomes.

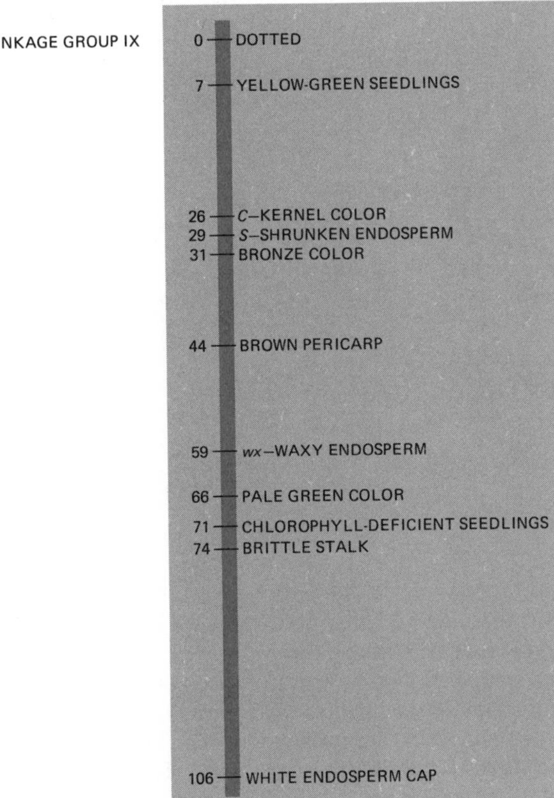

FIG. 16.19 Linkage map of one of the chromosomes found in the corn plant (*Zea mays*).

16.7 THE EVIDENCE OF CREIGHTON AND McCLINTOCK

One of the most elegant demonstrations of the chromosome theory came in 1932 from the work of geneticists Harriet Creighton and Barbara McClintock. During the course of their studies of linkage in corn, they developed a strain of corn that had one chromosome (out of 10 pairs) with two unusual features: a knob at one end of the chromosome and an extra piece of chromosome on the other. This extra piece of chromosome was the result of a translocation that had occurred in an earlier generation. These workers saw immediately the use to which their discovery could be put. Here was an organism with a pair of homologous chromosomes which could easily be distinguished from each other by microscopic examination. Furthermore, this unusual chromosome carried the dominant allele for colored kernels (*C*) and the recessive allele for waxy endosperm (*wx*). Its normal-appearing mate carried the recessive allele for colorless kernels (*c*) and the dominant allele for normal (starchy) endosperm (*Wx*). Thus the plant was a dihybrid for these two linked traits and, in addition, one chromosome of the pair was visibly marked at each end. Creighton and McClintock reasoned that this plant would produce four kinds of gametes: the parental kinds (*Cwx* and *cWx*) and the recombinant kinds (*cwx* and *CWx*) produced by crossing over. Fertilization of these gametes by gametes containing a chromosome of normal appearance and both recessive alleles *cwx* (a typical testcross) should produce four kinds of kernels: (1) colored waxy (*Ccwxwx*) kernels, (2) colorless kernels with normal endosperm (*ccWxwx*) and the two recombinant types, (3) colorless waxy (*ccwxwx*), and (4) colored kernels with normal endosperm (*CcWxwx*). Furthermore, microscopic examination of each of the plants resulting from these four kinds of kernels should reveal the following kinds of chromosomes. In the first case, there should be one normal chromosome and one extra-long chromosome with the knob at the end. In the second case, both chromosomes should be of normal appearance. However, in the third case, where crossing over had occurred, one would hope to find evidence that a physical exchange of parts between the homologous chromosomes of the dihybrid parent had occurred. Either a chromosome of normal length but with a knob at one end should be present or an extra-long chromosome with no knob. Actually Creighton and McClintock found the latter, thus indicating that the gene locus for *wx* was associated with (and thus near) the end of the chromosome with the extra segment. The gene locus for kernel color must then be nearer the end with the knob. Examination of the plants in Class 4 (colored kernels-*C* and normal endosperm-*Wx*) revealed a chromosome of normal length but with a knob at one end (Fig. 16.20).

Thus, behavior of the genes as revealed by study of the phenotypes produced was shown to be directly related to the behavior of the chromosomes as revealed by microscopic examination. The recombination of genes is now shown to occur at the same time that homologous chromosomes exchange parts. This

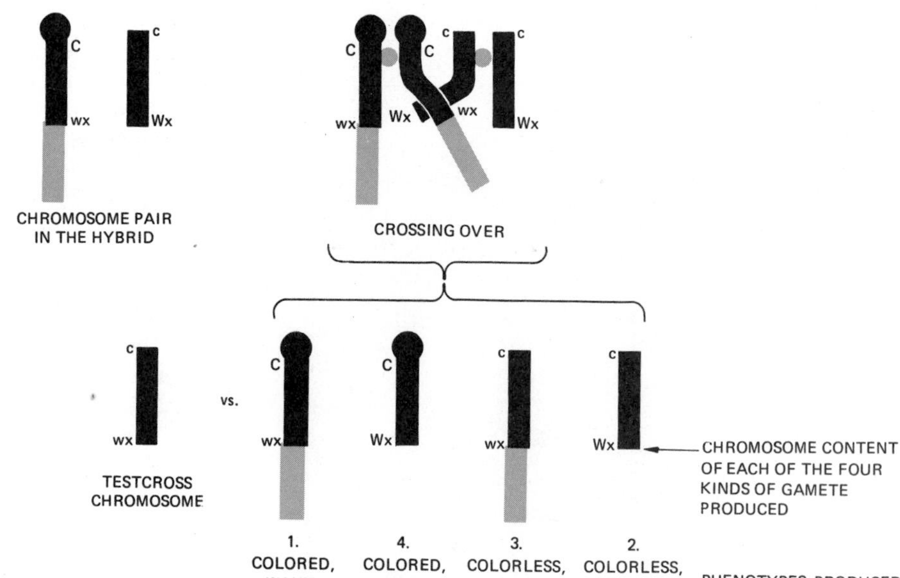

FIG. 16.20 Correlation of cytological crossing over and genetic recombination in corn.

is a very convincing demonstration of the chromosome theory. The determinants of the hereditary traits, genes, whose existence was first hypothesized by Mendel, are shown indeed to be located in these visible structures, the chromosomes.

16.8 ASSIGNING LINKAGE GROUPS TO CHROMOSOMES

If an investigator can establish a linkage group and can assign one gene in it to a particular pair of homologous chromosomes, he or she immediately has a list of traits controlled by those chromosomes. The remarkable materials with which Creighton and McClintock worked enabled them to assign all the known genes of linkage group IX (Fig. 16.19) to one pair of the corn plant's 10 pairs of homologous chromosomes. They were able to do this because they could associate the inheritance of a particular gene with the inheritance of a chromosome that could be visually distinguished from all the others (including its homologue). Their success thus depended upon the discovery of an abnormal chromosome and the ability to breed the plant containing it as they wished. A few associations between the inheritance of a visually distinguishable (and hence abnormal) chromosome and the inheritance of a particular trait have recently been discovered in humans. Other approaches have, however, been more fruitful in enabling us to assign known genes to particular chromosomes. One approach, which is beginning to produce useful information, depends upon the technique of somatic cell hybridization.

Human cells grown in tissue culture can be made to fuse with cells from other animals such as the mouse. At first, such hybrid cells contain a complete set of both human and mouse chromosomes. However, as mitosis continues, many of the human chromosomes become lost (Fig. 16.21). Now, human cells possess a number of biochemical features, e.g. particular enzymes, not shared by mouse cells. If every time a particular human enzyme is lost in a clone of hybrid cells, a particular human *chromosome* is lost as well, we may assume that the gene controlling the synthesis of that enzyme is located on that chromosome. We may also assume that every other gene in the linkage group to which that gene is assigned is also on the chromosome. For example, mouse-human hybrid cells that produce a particular human peptidase always retain a number 1 human chromosome. Conversely, no mouse-human hybrid clones that lack the human number 1 chromosome

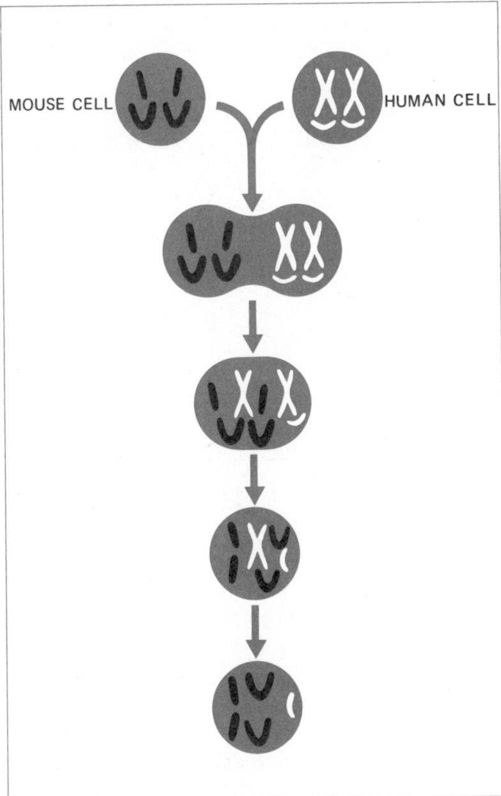

FIG. 16.21 Results of somatic cell hybridization. Following the fusion of a mouse cell with a human cell, there is a gradual loss of the human chromosomes. This makes it possible to correlate the presence or absence of a particular human gene function with the presence or absence of a particular human chromosome.

CHROMOSOME	GENES
1	Rh FACTOR, SALIVARY AND PANCREATIC AMYLASES, 5S rRNA
4	ALPHA AND BETA CHAINS OF HEMOGLOBIN
5	INTERFERON PRODUCTION
6	TRANSPLANTATION ANTIGENS (HLA), PEPSINOGEN
7	HISTONES, SITES OF INTEGRATION OF SIMIAN VIRUS 40 (SV 40)
9	ABO BLOOD GROUP ANTIGENS
12	3-PHOSPHOGLYCERALDEHYDE DEHYDROGENASE, TRIOSEPHOSPHATE ISOMERASE, PEPTIDASE B
X	CLOTTING FACTOR VIII (ABSENCE PRODUCES HEMOPHILIA), OVER 100 OTHERS
15	ISOCITRATE DEHYDROGENASE (MITOCHONDRIAL), PYRUVATE KINASE-3
17	THYMIDINE KINASE, GALACTOKINASE
18	PEPTIDASE A
19	GLUCOSE PHOSPHATE ISOMERASE
22	BETA-GALACTOSIDASE

FIG. 16.22 The chromosome assignments of a few human genes. In fact, at least one gene locus has been assigned to each of the 22 human autosomes. Once a single member of a linkage group can be assigned to a particular chromosome, all the other genes in that linkage group receive the same assignment. With the exception of those that are X-linked, the majority of the genes in this table have been assigned through the analysis of somatic cell hybrids.

produce the human peptidase. Some nine human genes (including those that specify the Rh blood factor) have been shown through family studies to be linked to the peptidase gene. Thus we conclude that all these genes are present on the number 1 chromosomes. A partial list of recent gene-chromosome assignments in humans is given in Fig. 16.22. By the time you read this, the list will probably be much longer.

EXERCISES AND PROBLEMS

1. In humans, normal vision is an X-linked trait and its gene is dominant to the allele for red-green color-blindness. When a color-blind woman marries a man with normal vision, what would be the expected distribution of color vision among (a) their sons and (b) their daughters?
2. How many pairs of homologous chromosomes are found in the human male?
3. Summarize the parallelism between the behavior of genes as described by Mendel and the behavior of chromosomes in meiosis and fertilization.
4. Nondisjunction may lead to zygotes containing XXY. What sex would this produce in a fruit fly? In a human?
5. When tall tomatoes with red fruit are crossed with dwarf tomatoes with yellow fruit, all the offspring are tall-red. When these F_1's are mated, only the two P_1 phenotypes are found in the F_2. What do we conclude about the genetic control of these two traits?
6. Define a gene.

7. For some reason (not yet fully understood) more male babies are conceived than female. By childhood, the number of boys and girls is the same. Can you think of a possible explanation for the higher mortality (both before birth and after) of male babies?

8. By testcrossing the following dihybrids, it is shown that each produces recombinant gametes in the percentage indicated.

AaBb	31%	AaXx	36%
XxYy	22%	AaYy	14%
XxBb	5%		

From these data determine the linear sequence and relative spacing of each gene locus on the chromosome.

9. Predict the sex of humans with the following chromosome abnormalities (all of which have been observed): *XXY, XXX, X, XYY, XXXX, XXXY, XXXXY, XXYY, XXXYY.*

10. How many Barr bodies would you expect to find in the cells of each of the individuals in the previous problem?

REFERENCES

1. HANNAH-ALAVA, ALOHA, "Genetic Mosaics," *Scientific American,* May, 1960. Explains how nondisjunction and crossing over may occur during *mitosis,* providing another mechanism for producing genetic mosaics, that is, individuals with cells of differing genotypes.

2. BEARN, A. G., and J. L. GERMAN, III, "Chromosomes and Disease," *Scientific American,* Offprint No. 150, November, 1961. Describes the mechanisms by which certain chromosome abnormalities arise and the diseases that may result.

3. GABRIEL, M. L., and S. FOGEL, eds., *Great Experiments in Biology,* Prentice-Hall, Englewood Cliffs, N.J., 1955. This paperback contains several original papers on the chromosome theory, including the report by Harriet Creighton and Barbara McClintock of their experiments.

4. MITTWOCH, URSULA, "Sex Differences in Cells," *Scientific American,* Offprint No. 161, July, 1963. Discusses the significance of the sex chromosomes and Barr bodies in both normal and abnormal conditions.

5. MCKUSICK, V. A., "The Mapping of Human Chromosomes," *Scientific American,* Offprint No. 1220, April, 1971.

6. DAVIDSON, RICHARD L., *Somatic Cell Hybridization: Studies on Genetics and Development,* Addison-Wesley Modules in Biology, No. 3, Addison-Wesley, Reading, Mass., 1973.

7. RUDDLE, F. H., and R. S. KUCHERLAPATI, "Hybrid Cells and Human Genes," *Scientific American,* Offprint No. 1300, July, 1974. How somatic cell hybridization enables linkage groups to be assigned to specific chromosomes.

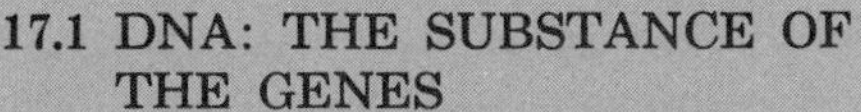

17 THE CHEMICAL NATURE OF GENES

The idea (though not the name) of the gene was developed by Mendel in an attempt to explain certain rules of inheritance. Mendel made no attempt to visualize the gene as a specific structure, in a specific location, with a specific chemical nature and a specific method of action. However, once it was fully established that the genes are located on the chromosomes, it became possible to make this attempt. Chromosomes are visible structures in a specific location. It is possible to isolate them from the cell and to study their chemical composition.

17.1 DNA: THE SUBSTANCE OF THE GENES

The chromosomes of eukaryotes are nucleoproteins; that is, they are made up of both nucleic acids and proteins. The nucleic acid is primarily DNA although small amounts of RNA are found as well. The proteins consist of five kinds of strongly basic proteins called histones and a roughly equivalent amount of a heterogeneous collection of "nonhistone" proteins. If chromosomes contain genes, is genetic information stored in the proteins, the nucleic acids, or both? In 1928, an English bacteriologist, Fred Griffith, made a discovery which, pursued over the next 30 years, ultimately revealed the answer.

The Experiments of Griffith and Avery

Griffith worked with the bacterium that causes bacterial pneumonia, a widespread and dangerous disease before the discovery of antibiotics. One of the most striking features of this organism is the presence around each cell of a gummy capsule made of a polysaccharide. When these bacteria are grown in culture, the presence of the capsule causes the colonies to have a glistening, smooth appearance (Fig. 17.1). Because of this, the cells are referred to as "S" cells. However, after prolonged cultivation outside the living host, some cells lose the ability to make the capsule. The surface of their colonies then appears wrinkled and rough ("R"). With the loss of the ability to make the capsule, the organism also loses its virulence. The R forms cannot cause the disease.

Pneumococci, as these bacteria are known, also occur in a large variety of types: I, II, III, etc. Each type produces its own particular polysaccharide capsule, which can be distinguished from that produced by other types. Unlike the occasional shift of S $\rightarrow$ R, the type of the organism is constant. Mice injected with a few S cells of one type, say Type II, will soon have their bodies teeming with descendant cells of the same type. With respect to type, then, we find the same continuity of traits from generation to generation that we found in yeast.

As mentioned above, injections of R cells are ordinarily harmless. So, too, are injections of S cells *if* they have first been killed (by subjecting them to a high temperature, for example). However, Griffith unexpectedly found that when *living* R cells and *killed* S cells were injected *together* into a mouse, the mouse became ill and *living* S cells could be recovered from its body. In exploring this remarkable phenomenon further, Griffith discovered that the *type* of the S cells recovered from the mouse was

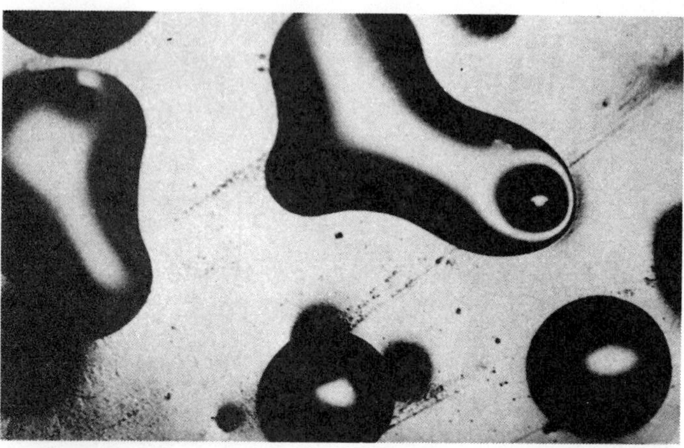

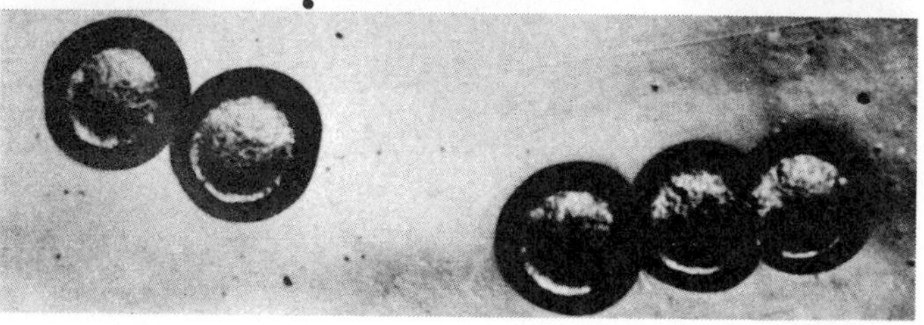

FIG. 17.1 Pneumococci growing as colonies on the surface of a culture medium. Top: the presence of a capsule around the bacterial cells gives the colonies a glistening, smooth (S) appearance. Bottom: pneumococci lacking capsules have produced these rough (R) colonies. (Research photograph of Dr. Harriett Ephrussi-Taylor, courtesy of The Rockefeller University and *Scientific American*.) The appearance of individual encapsulated and nonencapsulated pneumococci can be seen in Fig. 30.4.

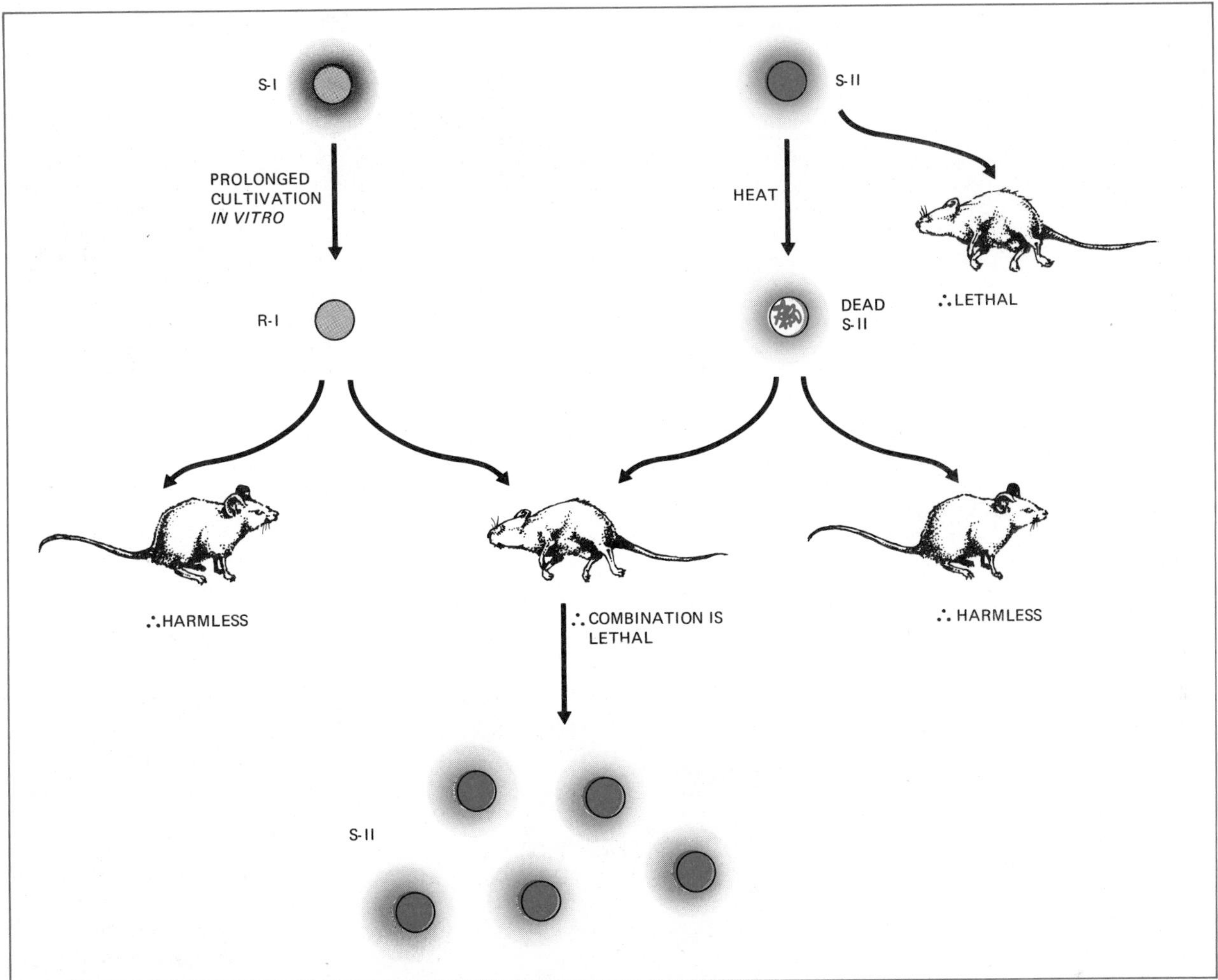

FIG. 17.2 Griffith's experiment. Something in the killed S-II cells converted R-I cells into S-II cells. Such a hereditary modification is called *transformation.* Eventually it was demonstrated that DNA was the substance responsible for transformation.

determined by the type of the dead S cells he had used, not the type of the strain from which the living R cells were derived. In other words, when a mouse was injected with living R cells from Type I pneumococci and dead S cells of *Type II,* the living cells that were recovered after the mouse became sick were S cells of Type II (Fig. 17.2). Furthermore, this change of R-I cells to S-II cells was stable and inheritable. The S-II cells could be cultured indefinitely and remained true to type. Clearly, something in the dead S-II cells had converted the living R-I cells into S-II cells, and that something was passed on from generation to generation.

Within a few years, this **transformation,** as it was called, was accomplished in a test tube. A small percentage of R-I cells, grown in test tubes of broth to which *dead* S-II cells were added, became transformed into S-II cells. Later, an *extract* of S-II cells was successfully used to transform R-I cells into S-II cells in a test tube.

As you would expect, this cell extract had a variety of ingredients in it, including polysaccha-

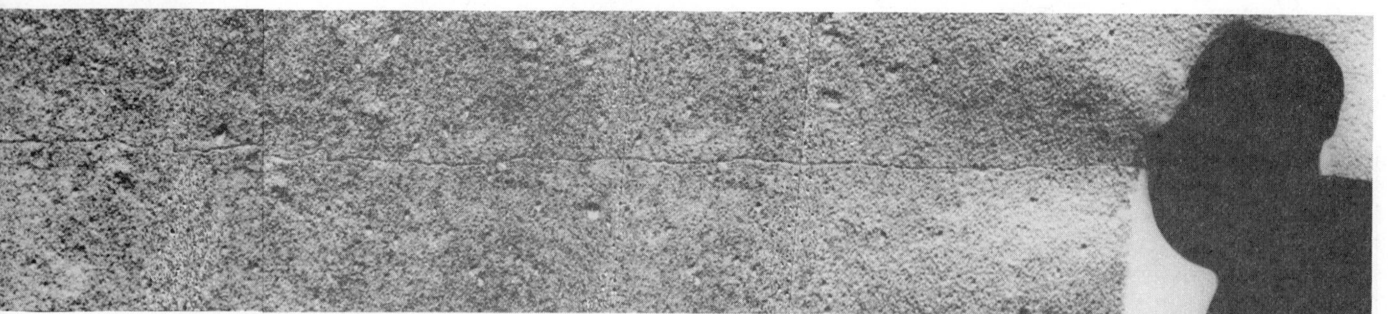

FIG. 17.3 Electron micrograph of a DNA molecule entering a pneumococcus (the large object at the right). This DNA molecule, the long fine line, is approximately 7 micrometers (μm) in length, long enough to include a dozen genes. The process of transformation follows the uptake of such a molecule by the bacterium. (38,000 ×, courtesy of Dr. Alexander Tomasz.)

FIG. 17.4 Cycle of infection of bacterial cells by bacteriophages. After the virus attaches to an uninfected bacterial cell (a), its content of DNA is injected (b). New viral DNA molecules *and* new viral protein coats are produced (c). From these are assembled infective bacteriophages (d), which are released when the bacterial cell bursts (e).

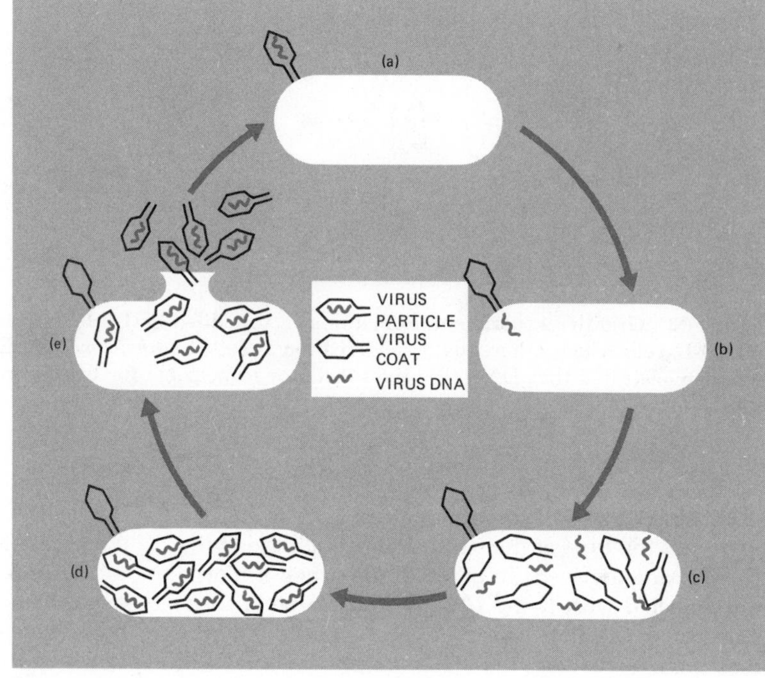

rides, protein, lipids, and the two nucleic acids: DNA and RNA. By selectively destroying one of these ingredients after another and testing the transforming activity of the remaining material, Dr. O. T. Avery and his coworkers were able to show in 1943 that the active ingredient of the transforming principle was DNA (Fig. 17.3). Purified DNA from S-II cells transformed R-I cells into S-II cells. Here, then, was a molecule which changed the function of the recipient cell, making it produce, in this case, a Type II rather than a Type I capsule.

Furthermore, this molecule was self-duplicating. Far more DNA capable of the same transforming activity could be extracted from the offspring of the transformed cells than was used originally to accomplish the transformation.

The Virus Experiments of Hershey and Chase

In 1952, geneticists A. D. Hershey and Martha Chase performed a series of experiments with viruses that gave further proof that DNA is the basic hereditary material. In many ways viruses resemble small chromosomes. They, too, are made of nucleoprotein and when they enter a host cell, the latter goes to work duplicating the virus particles. It is as if the normal genetic control of the cell had been taken over by foreign, invading genes.

Bacteria may be infected with viruses called **bacteriophages.** The process of entry has been observed under the electron microscope. Only an inner portion of the virus particle actually enters the cell (Fig. 17.4). The exterior part, the virus coat, probably serves to inject the infecting portion.

Bacteriophages produced within bacteria growing in radioactive culture medium will themselves be radioactive. If radioactive sulfur atoms are present in the medium, they will be incorporated into the protein coats of the bacteriophages, since two of the twenty amino acids found in proteins contain sulfur atoms. They will not become part of the DNA of the bacteriophage, because there are no sulfur atoms in DNA. On the other hand, radioactive phosphorus atoms will be incorporated into the DNA exclusively, for it alone contains phosphorus atoms. When bacteriophages containing radioactive phosphorus atoms are allowed to infect *non*radioactive bacteria, all the bacterial cells become radioactive. Much of this radioactivity is then passed on to the next generation of bacteriophages. However, when bacteria are infected with viruses containing radioactive sulfur atoms and then the virus coats are removed (by whirling in an electric blender), practically no radioactivity can be detected in the cells. It is retained in the virus coats.

From these experiments we conclude that the DNA component of the bacteriophage is injected into the bacterial cell while the protein component remains outside. The DNA is, however, able to direct the formation of new virus particles *complete* with protein coats. Clearly, DNA is the fundamental hereditary material, the substance of the genes. Thus we must now ask: what are the properties of DNA that enable it (1) to dictate the synthesis of Type II polysaccharide or the coat protein of a virus and (2) to duplicate itself from generation to generation? The answer to the second part of this question was discovered first.

17.2 THE WATSON-CRICK MODEL OF DNA

At the time of Avery's discovery, DNA seemed a most unlikely candidate to serve as the repository of genetic information. Let us review what was then known about the molecule. As we learned in Chapter 4, DNA is a macromolecule and a polymer. Gentle hydrolysis produces its monomeric subunits: the nucleotides. But whereas proteins are built from *twenty* amino acids, only *four* nucleotides are found in DNA. Each nucleotide contains a five-carbon sugar (thus a *pentose*) called deoxyribose. Attached to the No. 5 carbon atom of the pentose is a single phosphate group (Fig. 17.5). Attached to the No. 1 carbon atom is a nitrogenous base. There are four kinds of nitrogenous bases and thus four kinds of nucleotides in DNA. Two of them, the purines, consist of carbon and nitrogen atoms arranged in two interlocking rings. These are adenine (A) and guanine (G) (Fig. 17.5). The other two are single-ring pyrimidines: cytosine (C) and thymine (T).

One other feature of DNA is remarkable. This is the fact that whatever the amount of A found in the hydrolytic digests of an organism's DNA, the amount of T is equal to it (Fig. 17.6). Similarly, whatever the amount of C (and it varies substantially from species to species), the amount of G is the same. When first discovered, this relationship was a little disheartening. It suggested a dull repetitive uniformity to the molecule. Not at all the sort of thing you would want to store information with. Whereas different proteins contain up to twenty amino acids in widely varying mole ratios, DNA has only four com-

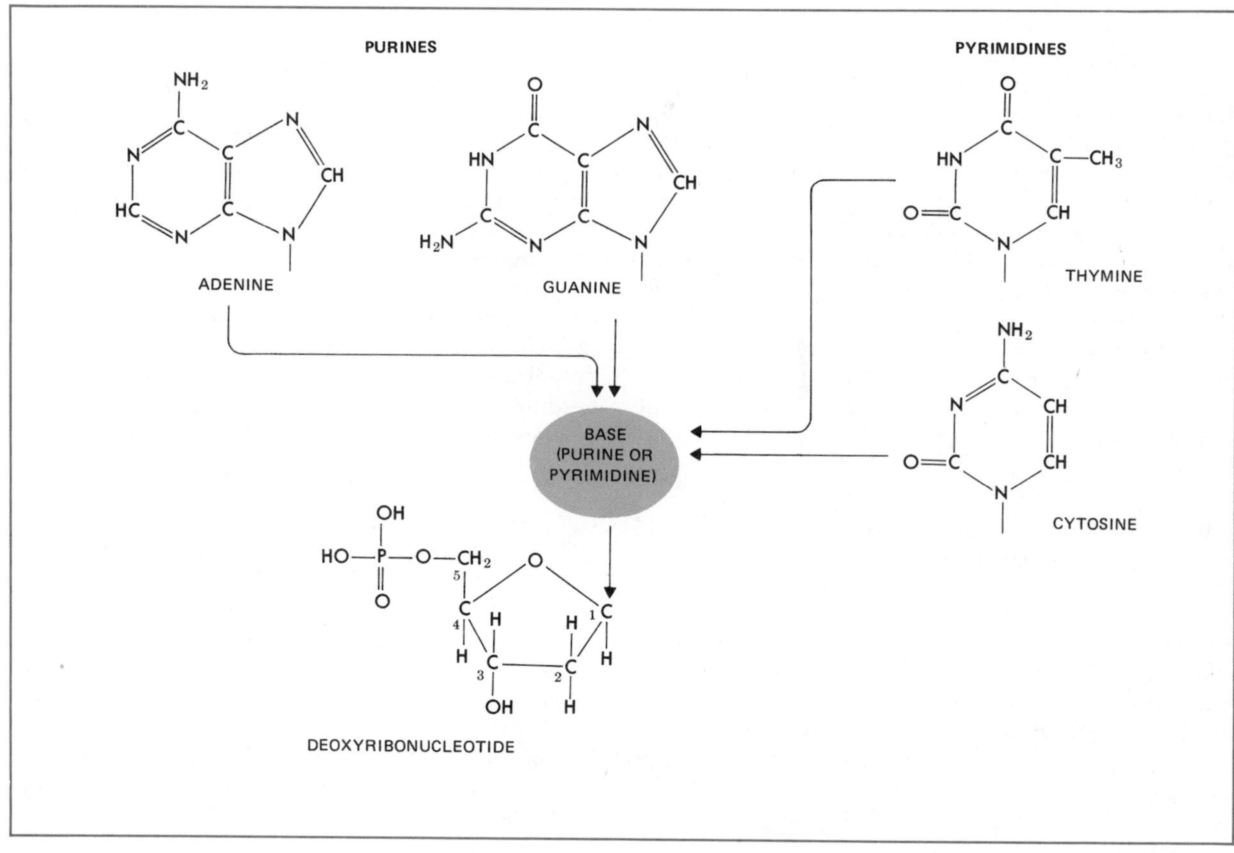

FIG. 17.5 Structure of the nucleotides that serve as the monomers of DNA.

RELATIVE PROPORTIONS (%) OF BASES IN DNA

	A	T	G	C
HUMAN	30.9	29.4	19.9	19.8
CHICKEN	28.8	29.2	20.5	21.5
GRASSHOPPER	29.3	29.3	20.5	20.7
SEA URCHIN	32.8	32.1	17.7	17.3
WHEAT	27.3	27.1	22.7	22.8
YEAST	31.3	32.9	18.7	17.1
E. coli	24.7	23.6	26.0	25.7

FIG. 17.6 The C + G : A + T ratio varies from organism to organism (particularly among the prokaryotes) but, within the limits of experimental error, A = T and C = G.

ponents, and if you know the molar composition of two of them (one purine and one pyrimidine), you automatically know it for the other two. This is not to say that DNA *couldn't* encode genetic information. After all, the International Morse Code has only three units—a dash, a dot, and a pause—and yet, with enough time and patience, the contents of a library of any size can be encoded in it.

To find out how DNA could serve its genetic functions—storing and replicating information—it was absolutely essential that the three-dimensional structure of the molecule be determined. Success at this task came in 1953 to two scientists working in Cambridge, England: James Watson (Fig. 17.7) and

▲
FIG. 17.7 Dr. James D. Watson of Harvard University, codiscoverer of the structure of DNA. (Photo by Charles Harbutt, Magnum Photos.)

FIG. 17.8 The Watson-Crick model of DNA structure. The small spheres represent phosphate groups; the open pentagons represent deoxyribose. The solid planar structures represent the purine and pyrimidine bases. (Photo courtesy of Dr. Donald M. Reynolds.) ▼

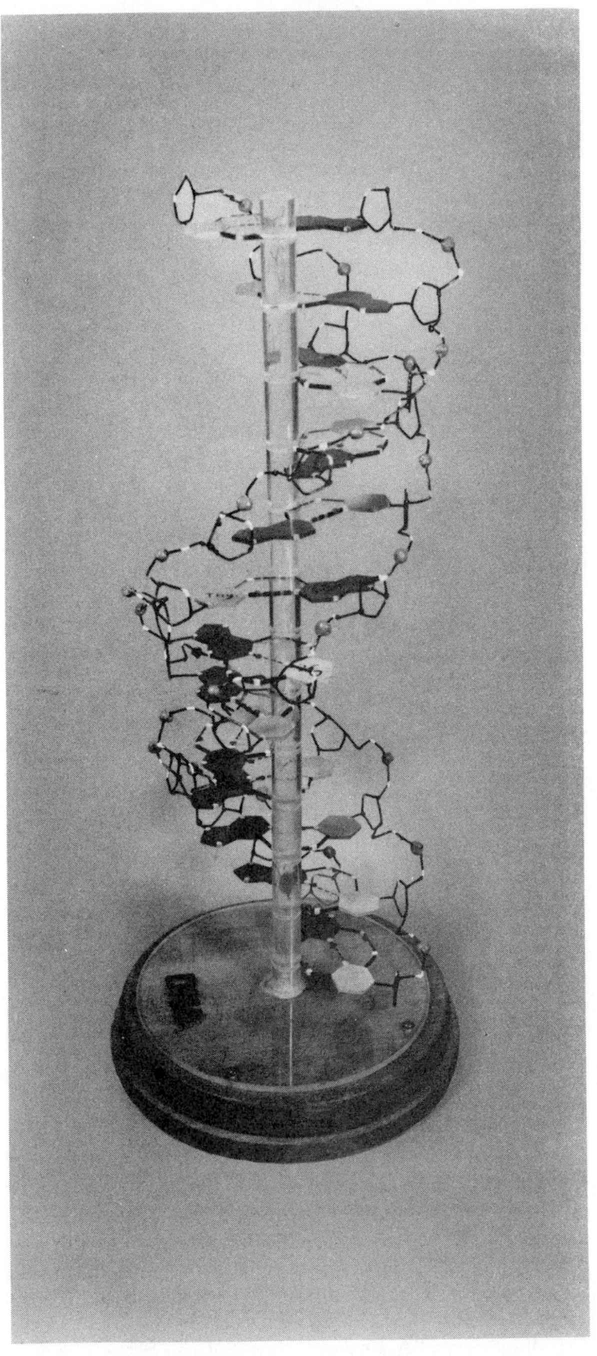

Francis Crick. They had at their disposal the chemical knowledge outlined above, a knowledge of the bond angles and opportunities for hydrogen bonding of the four nucleotides and, perhaps most important of all, patterns produced by X-ray crystallography of DNA preparations. Working with metal models of the nucleotides, they finally succeeded in building a model polymer that instantly explained all the information known about DNA. Not only that but, as they quickly perceived, it also provided a mechanism to explain the self-replicating properties of DNA. This was the Watson-Crick model of DNA (Fig. 17.8). For their epoch-making discovery, they shared a Nobel Prize in 1962.

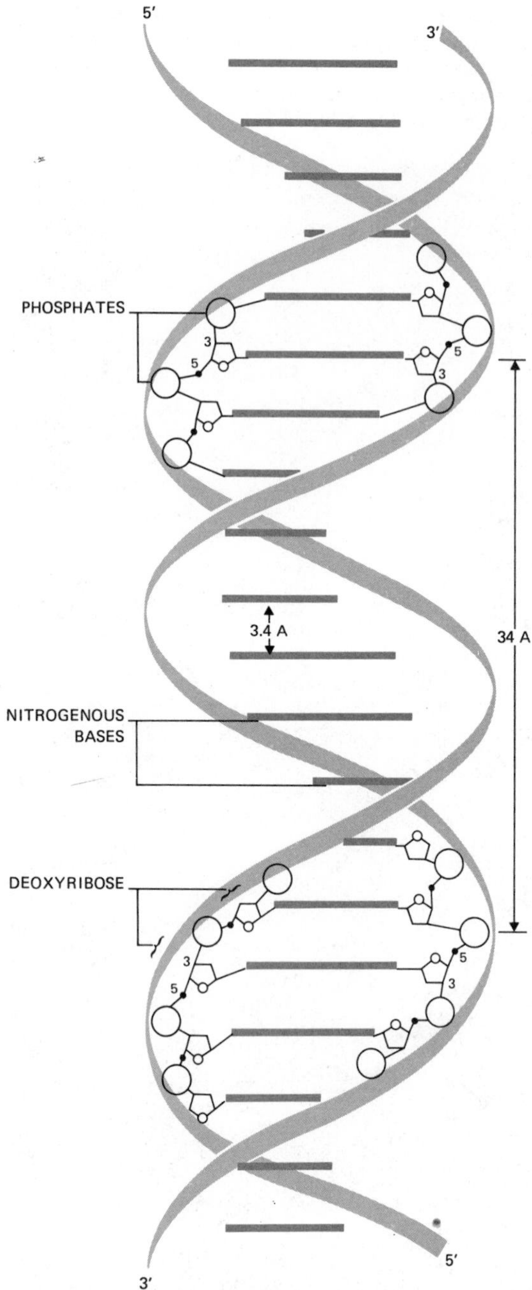

FIG. 17.9 The Watson-Crick model of DNA. Each pair of nitrogenous bases is located in a plane spanning the central axis of the double helix. Note that the two strands of DNA run in opposite directions; that is, one strand runs 3′ to 5′ while the opposite runs 5′ to 3′. Each strand makes a complete turn every 34 A. (Modified from "A Representation of the Structure of DNA" by William Etkin in *BioScience* 23 (November 1973) p. 653, with permission of the author and the American Institute of Biological Sciences.)

Details of the Model

In the Watson-Crick model, the backbone of the polymer consists of alternating sugar and phosphate groups. The phosphate group bonded to the No. 5 carbon atom of one sugar is covalently bonded to the No. 3 carbon of the next. *Two* such strands are twisted around each other something like a double spiral staircase (Fig. 17.9). This is the **double helix.**

The purine or pyrimidine base bonded to each pentose sugar projects in toward the axis of the helix. At each "step" in the staircases, two such bases project toward each other in a plane. The diameter of the double helix is 20 A (Fig. 17.10).

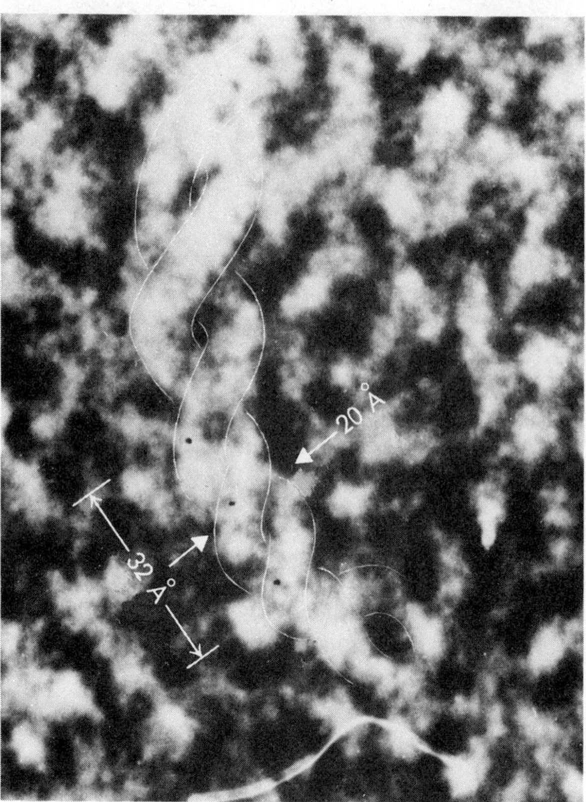

FIG. 17.10 Direct visualization of the helical structure of DNA. Lines have been added to the electron micrograph to delineate the two strands of the molecule more clearly. (Courtesy of Dr. Jack Griffith, Stanford Medical School.)

Now, if two pyrimidines should be present at the same level, they would not span the space between the two backbones of the staircases. Two purines, being of larger size, could not squeeze into a single plane spanning the double helix. However, *one* purine and *one* pyrimidine just fill the available space. We noted earlier that the amount of A (a purine) in DNA was always equal to the amount of T (a pyrimidine), and that of C (a pyrimidine) was always equal to that of G (a purine). This certainly is consistent with the requirement that a purine always match with a pyrimidine to span the helix. But why not A with C and G with T? A close examination of the molecular configuration of each gives the answer. It is a matter of the opportunities present for hydrogen bonding between the purine and the pyrimidine.

Remember that wherever we can bring two strongly electronegative atoms into close proximity (with a hydrogen atom attached to one of them), we can establish a hydrogen bond. The location of the C=O and N—H groups of thymine and the $-NH_2$ group and one of the "ring" nitrogen atoms of adenine are such that nesting the two bases together permits the formation of two hydrogen bonds (Fig. 17.11). If, on the other hand, we try to match guanine with thymine, we cannot establish these bonds. However, when guanine and cytosine are placed in a plane, the resulting proximity of three pairs of strongly electronegative atoms results in the establishment of three hydrogen bonds between them (Fig. 17.11).

Thus the geometry and resulting opportunities for hydrogen bonding explain why the amount of A in DNA is equal to T, and the amount of C equal to G. It is only these **base pairs** that can hold the helix together. A single DNA molecule may contain from hundreds of thousands to millions of base pairs. While hydrogen bonds are individually very weak (less than 5 kcal/mole), the combined effect of two or three hydrogen bonds between each of the base pairs in the molecule provides a structure of great stability.

DNA Duplication

The Watson-Crick model of DNA provides a molecular basis for the two central features of genetic information: a means of storing information and a means of making copies of it. The linear sequence of four bases could be the information storage system—a code made up of four units, or one more than in the Morse code. But note that if that is the case, then a DNA molecule contains two complete sets of the encoded information. This is because if we know the sequence of bases on one strand, the rules of base pairing immediately establish the sequence on the other. Whatever sequence of bases is present on one strand of the DNA molecule, a complementary sequence is present on the other strand. The two sequences have the same relationship to each other as the "positive" and "negative" of a photograph. This redundancy of the DNA structure provides a nice mechanism to explain one of the fundamental properties of genetic information: its ability to duplicate itself.

As proposed by Watson and Crick, DNA duplication begins with an "unzipping" of the "parent" molecule; that is, the hydrogen bonds between the base pairs are broken and the two halves of the mole-

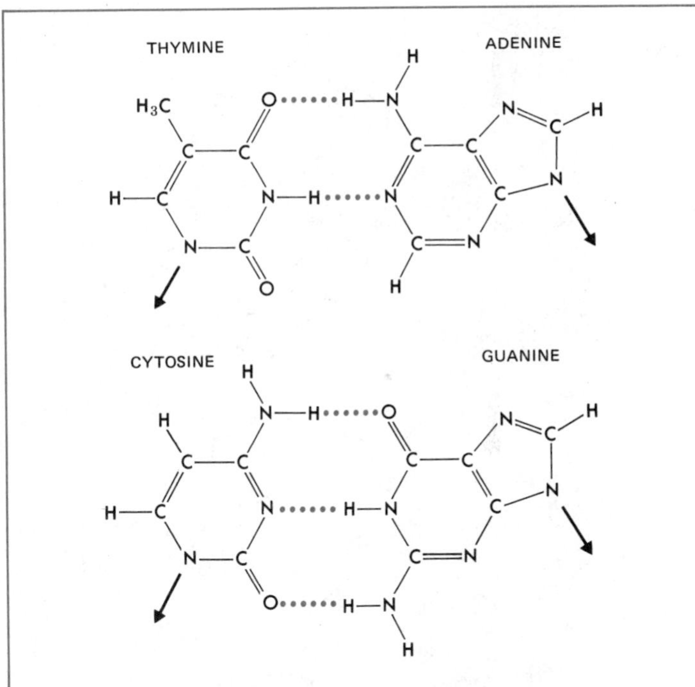

FIG. 17.11 Location of the hydrogen bonds between thymine and adenine and between cytosine and guanine. It is the existence of these multiple hydrogen bonds (~25 per turn) that gives the DNA double helix its great stability. Although their *shapes* are appropriate, the pyrimidine thymine cannot form hydrogen bonds with the purine guanine nor can cytosine form hydrogen bonds with adenine.

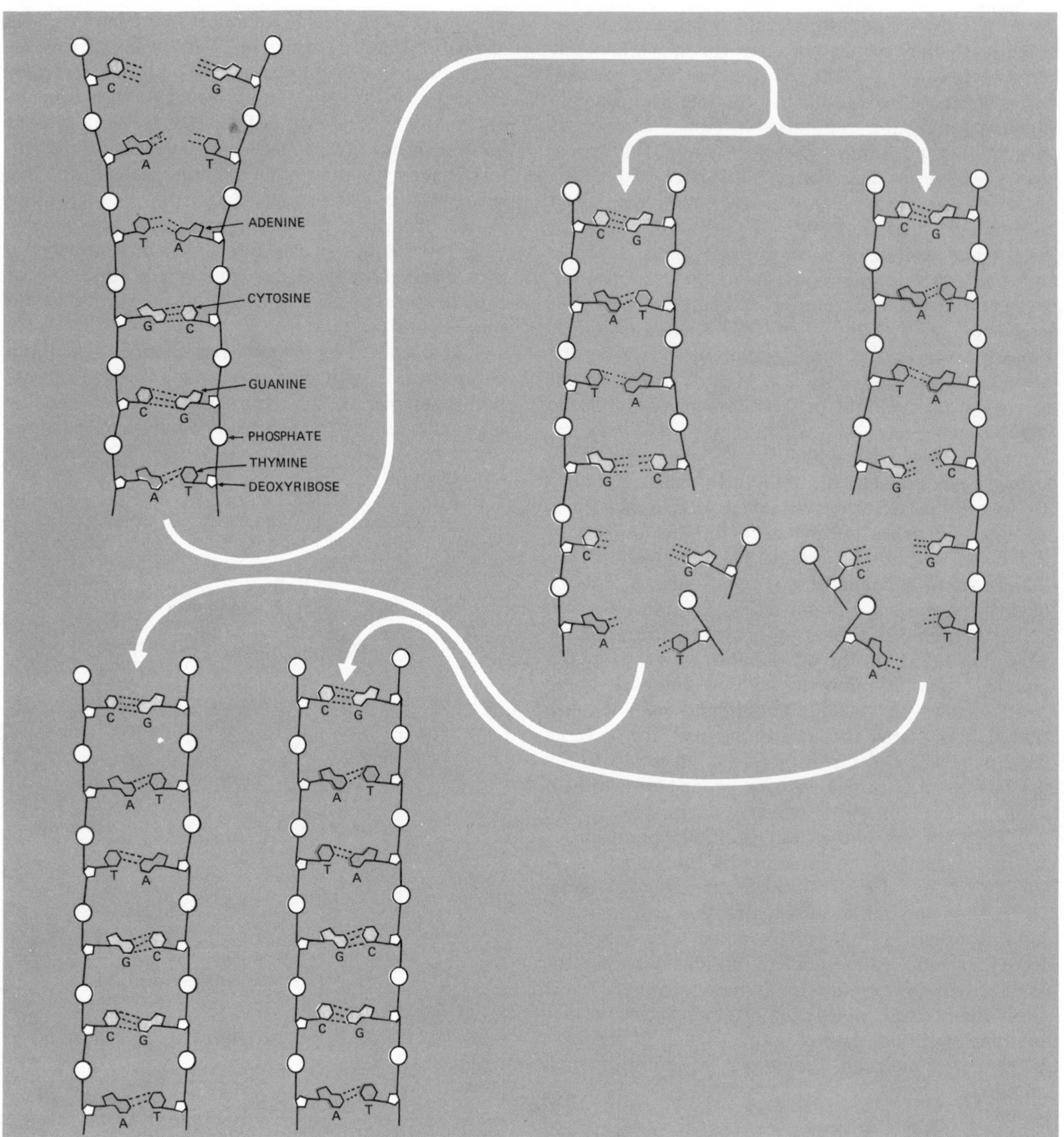

FIG. 17.12 The mechanism of duplication of DNA. After the two strands of the DNA molecule separate, each serves as a template for the assembly of a strand complementary to it. In this way two DNA molecules are synthesized—identical to each other and to the original molecule.

cule unwind. Once exposed, the bases on *each* of the separated strands can pick up the appropriate nucleotides present in the surrounding medium. Each exposed C will pick up a G, each G a C, etc. With the aid of a linking enzyme, called **DNA polymerase,** these newly arrived bases are polymerized into a chain complementary to the chain serving as the template. When the process is completed, two DNA molecules will have been formed, identical to each other and to the parent molecule (Fig. 17.12).

In 1958, this hypothesis received strong experimental support from the work of M. S. Meselson and F. W. Stahl. They used the common intestinal bacterium *Escherichia coli* as their experimental organism. *E. coli* can grow in a culture medium containing simply glucose and inorganic salts. Among the latter must be a source of nitrogen atoms for protein and nucleic acid synthesis. Ammonium ions (NH_4^+) serve nicely. Although the most common isotope of nitrogen is ^{14}N, it is possible to synthesize ammonium ions containing a heavier isotope of nitrogen, ^{15}N. Meselson and Stahl first grew *E. coli* cells for several generations in a medium containing $^{15}NH_4^+$. They found that at the end of this period the DNA of the cells was heavier than normal because of the incorporation of ^{15}N atoms in it. Then they transferred the cells to a medium containing ordinary ammonium ions ($^{14}NH_4^+$) and allowed them to divide just once. The DNA in the new generation of cells was exactly intermediate in weight between the heavier DNA in the previous generation and the normal. This, in itself, is not surprising. It tells us no more than that half the nitrogen atoms in the new DNA are ^{14}N and half are ^{15}N. It tells us nothing about their arrangement in the molecule. However, when the bacteria were allowed to divide *again* in normal ammonium ions ($^{14}NH_4^+$), two distinct weights of DNA were formed: half the DNA was of normal weight and half was of intermediate weight. As shown in Fig. 17.13, this indicates that DNA molecules are not degraded and reformed between cell divisions, but instead, each original strand

FIG. 17.13 The experiment of Meselson and Stahl. The results (right) suggest that during the process of duplication each strand of DNA remains intact and builds a complementary strand from the nucleotides available. This is consistent with the Watson-Crick model of DNA.

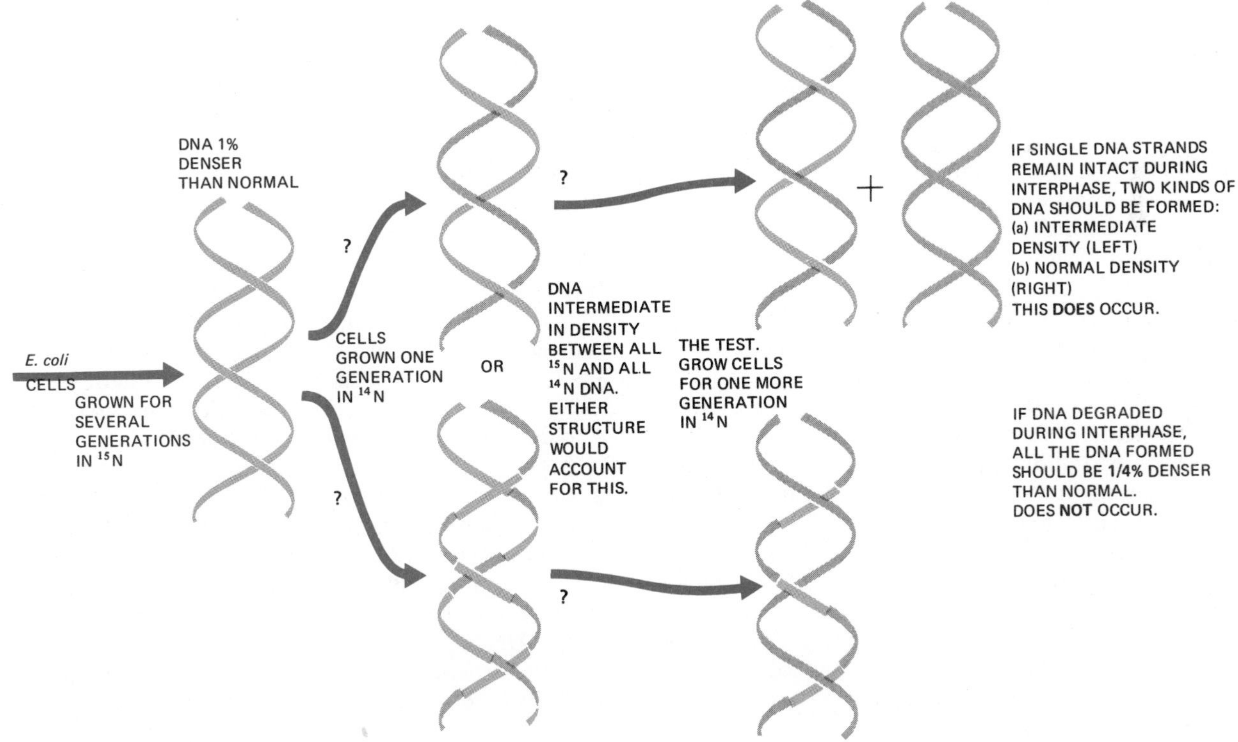

DNA 1%
DENSER
THAN NORMAL
E. coli
CELLS
GROWN FOR
SEVERAL
GENERATIONS
IN ^{15}N
?
CELLS
GROWN ONE
GENERATION
IN ^{14}N
OR
?
?
DNA
INTERMEDIATE
IN DENSITY
BETWEEN ALL
^{15}N AND ALL
^{14}N DNA.
EITHER
STRUCTURE
WOULD
ACCOUNT
FOR THIS.
THE TEST.
GROW CELLS
FOR ONE MORE
GENERATION
IN ^{14}N
?
+
IF SINGLE DNA STRANDS
REMAIN INTACT DURING
INTERPHASE, TWO KINDS OF
DNA SHOULD BE FORMED:
(a) INTERMEDIATE
DENSITY (LEFT)
(b) NORMAL DENSITY
(RIGHT)
THIS DOES OCCUR.
IF DNA DEGRADED
DURING INTERPHASE,
ALL THE DNA FORMED
SHOULD BE 1/4% DENSER
THAN NORMAL.
DOES NOT OCCUR.

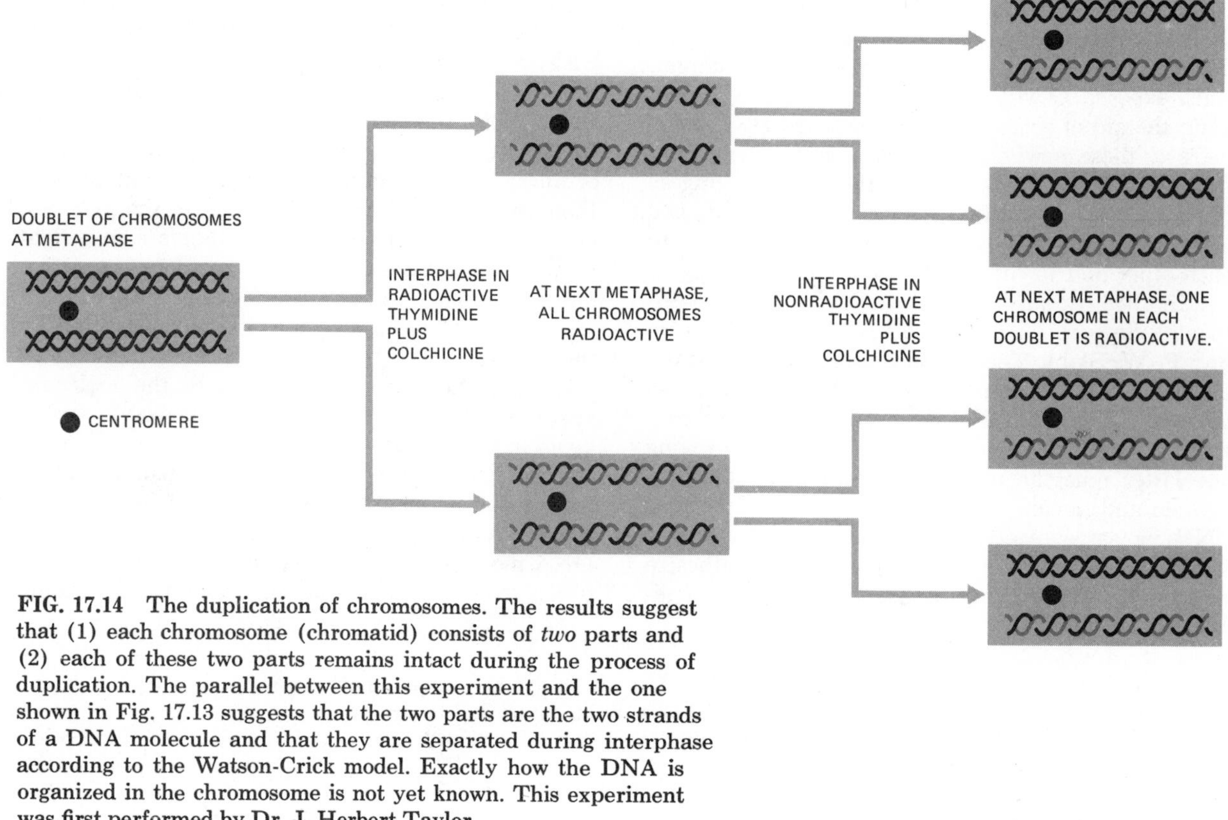

FIG. 17.14 The duplication of chromosomes. The results suggest that (1) each chromosome (chromatid) consists of *two* parts and (2) each of these two parts remains intact during the process of duplication. The parallel between this experiment and the one shown in Fig. 17.13 suggests that the two parts are the two strands of a DNA molecule and that they are separated during interphase according to the Watson-Crick model. Exactly how the DNA is organized in the chromosome is not yet known. This experiment was first performed by Dr. J. Herbert Taylor.

remains intact as it builds a complementary strand from the nucleotides available to it. This method of molecular replication is often described as "semi-conservative" because each daughter DNA molecule is one-half "old" and one-half "new."

Watson and Crick's theory of semiconservative DNA duplication is reflected even at the level of the visible chromosomes. When plant root tips are grown in a solution containing radioactive thymidine (the nucleotide containing thymine, T), the chromosomes in the rapidly dividing cells of the embryonic region become uniformly radioactive. If the root tips are then transferred from the radioactive solution to one containing nonradioactive thymidine and colchicine, chromosome duplication takes place without cell division. After one duplication, the chromosomes are still uniformly radioactive. After a second duplication, however, one chromatid of each doublet is radioactive, the other is not (Fig. 17.14).

DNA, Chromosomes, and Genetic Information

Thorough analysis of the genetics of *E. coli* has made it abundantly clear that all the genes of this organism are incorporated in a single molecule of DNA almost 1 mm long (although only 20 A in diameter!). Such a molecule has actually been isolated from *E. coli* cells. "Sprung loose" might be a better expression, since this one-millimeter *molecule* is folded up in a *cell* that is only about 2 μm long. In *E. coli,* therefore, the chromosome of the organism is simply a DNA molecule. The chromosomes of higher organisms are far larger and more complex in their organization. However, there is evidence (in addition to Taylor's) that a single chromosome in eukaryotes contains only a single unbroken molecule of DNA.

The structure of chromatin (the elongated, interphase chromosomes of eukaryotes) has been sub-

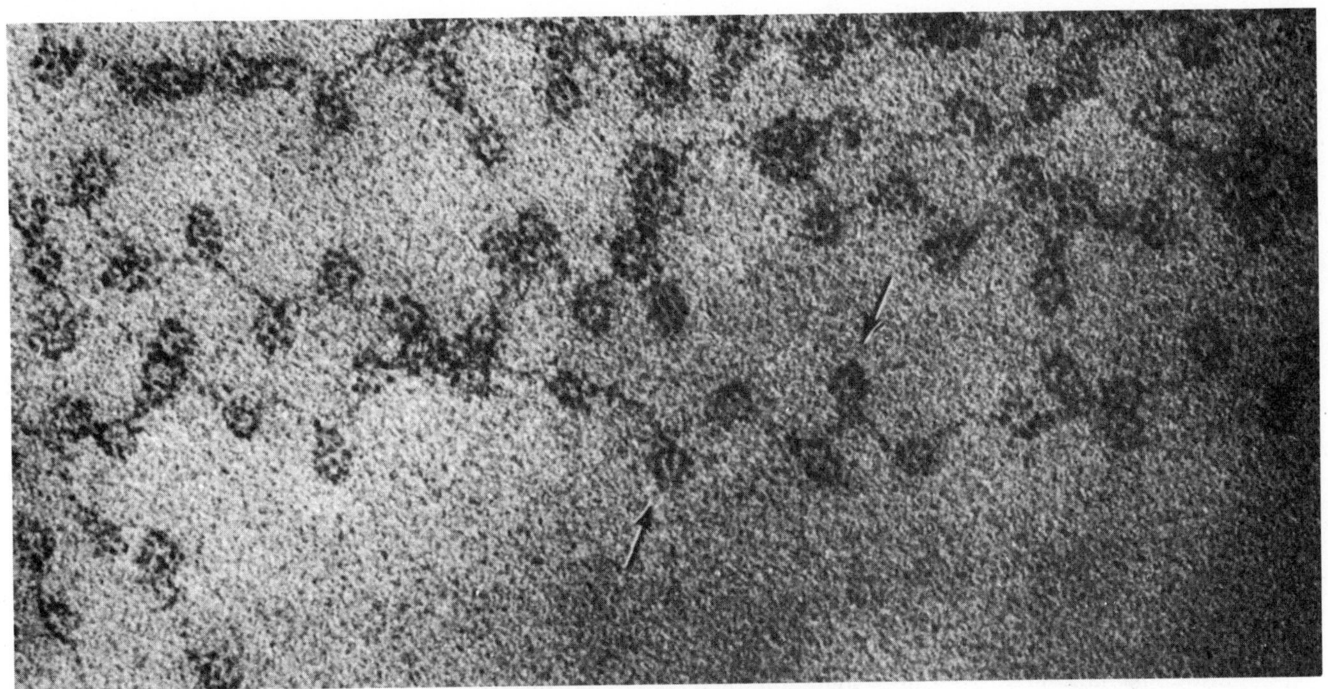

FIG. 17.15 Chromatin from the nucleus of a chicken red blood cell. The arrows point to the nucleosomes. These consist of a core of histone molecules wrapped in DNA. The strand connecting the nucleosomes consists of DNA, probably associated with one type of histone and some nonhistone proteins. (Courtesy of Donald E. Olins and Ada L. Olins, University of Tennessee.)

jected to study by a variety of techniques. Electron microscopy of chromatin from several kinds of cells reveals a linear arrangement of globular structures, called nucleosomes, connected by a fine thread (Fig. 17.15). For obvious reasons, this arrangement has been likened to beads on a string. The beads (nucleosomes) appear to consist of two molecules each of four different kinds of histone around which is wrapped a double helix of DNA approximately 200 base pairs in length. The "string" appears to be a DNA double helix associated with a fifth kind of histone and some other, nonhistone, proteins. The details of how this rather uniform structure expresses genetic information await further research.

The structure of the DNA molecule not only permits accurate self-duplication but also is ideally suited to the long-term preservation of the sequence of bases coded in it. Because the two chains in a DNA molecule are complementary, the information in one is also coded in the other. In a sense, as we have said, each DNA molecule contains two copies of its information. If for some reason the sequence of bases on one chain were disturbed, the correct message could still be determined from the complementary chain. And, in fact, many living cells take advantage of this opportunity to correct errors that occur in their DNA code. By means of enzymes, they remove damaged or incorrect bases from one strand of the molecule. These are then replaced by the correct bases, that is, bases complementary to the bases on the opposite strand. In this way, the genetic code can be preserved intact.

Let us examine one illustration of the great stability of information encoded in DNA. In recent years, studies in South Africa have identified some 8000 persons who are victims of an inherited metabolic disease called porphyria. These persons are unable to metabolize the porphyrins produced by the breakdown of hemoglobin and cytochromes. The disease is inherited as an autosomal dominant gene. Most of the group are inhabitants of South Africa; the others have emigrated from there. Every member of this large assemblage traces his or her ancestry back to one of four people. These are four of the eight children of Ariaantje and Gerrit Jansz, Dutch emigrants who married in South Africa in 1688. Evidently either Ariaantje or Gerrit possessed the defective gene and passed it on to at least four of their children. From them it has spread unchanged through ever-expanding subsequent generations.

17.3 MUTATIONS

Although the organization of the DNA molecule makes it remarkably resistant to changes in its sequence of bases, such changes do occur. These changes are called gene or "point" mutations. The gene for hemophilia that Queen Victoria bequeathed to so many members of the royal houses of Europe arose spontaneously. To call a mutation "spontaneous" is simply to admit that we do not know what caused it. However, we do know of a sizable number of agents that induce mutations.

Mutagenic Agents

In 1927, the geneticist H. J. Muller reported that he had increased by 15 times the rate of mutation in fruit flies (*Drosophila*), by irradiating them with X-rays. Since that time other types of short-wavelength radiation, e.g. gamma rays, have been found to induce mutations. Even ultraviolet light, if it can penetrate to the chromosomes, is mutagenic.

Just how radiation produces mutations is not completely understood. Ultraviolet light can cause breaks in DNA strands and also induce covalent T-T and C-T links, either adjacent in one strand or across the "staircase." Repair of these defects may result in a base shift, e.g. CG $\rightarrow$ TA, and thus a change in the genetic code. X-rays probably cause DNA strand breaks as well. Certainly, they induce visible breaks in chromosomes.

Mutations are also produced by a number of chemicals. The first chemical mutagens to be discovered were the mustard gases used in World War I. These substances, and many others like them, add carbon-containing groups (e.g. CH_3—, CH_3CH_2—) to other organic molecules, a process known as alkylation. Alkylation of the bases in DNA can introduce gaps in the strand. If such gaps are filled by an "incorrect" base, the genetic code is altered. A number of other molecules, which mimic the normal bases, may be incorporated into DNA and lead to changes in the genetic code. The caffeine molecule appears to be one of these. A large number of hydrocarbons such as benzopyrene are also mutagenic. In most cases, their mode of action is unclear.

The Significance of Mutations

How significant are mutations? In order to answer this, we must distinguish between mutations that occur in the gonads, where sperm and eggs are produced, and those (somatic mutations) that occur elsewhere in the body. A mutation in the precursor of a blood cell may kill the cell, may block its ability to divide by mitosis, or perhaps convert it into a cancerous cell. The somatic (general body) response of organisms to radiation—at least at relatively high dose levels—is well known. It includes such effects as an increased probability of falling prey to leukemia, a cancerous proliferation of white blood cells. These data have been derived from careful study of the survivors of Hiroshima and Nagasaki and survivors of a variety of accidental exposures to high levels of atomic radiation.

Several units are used to describe radiation dosage levels. The basic unit of radiation is the roentgen (r) which is named after the discoverer of X-rays. Because different types of radiation differ in their energy content and their effect on human tissue, two related units are frequently used. These are the rad and the rem (*r*oentgen-*e*quivalent-*m*an). For the types of radiation we shall be considering, the three units can be considered to be equivalent.

A dose of 50–100 r distributed over the whole body will make you sick. Half the people exposed to a dose of 400–500 r will die. Total body exposures below 25 r produce no visible effects. Radiation exerts its somatic effects most strongly on tissues where frequent mitosis occurs. Thus the skin, the hair follicles, the lining of the intestine, and the bone marrow (which produces the blood cells) are most apt to suffer damage.

While sublethal doses of radiation may possibly shorten life span or lead to the development of cancers, it is the effect of radiation on the sex cells that is of special concern. Mutations in your body cells die with you, but mutations in the sex cells can be transmitted to future generations, in the sperm or egg. In such a case, the mutant gene will be present in *every* cell of the offspring. While firm data are hard to acquire, it is safe to assume that radiation

SOURCE	DOSE (MREM/YR)
NATURAL BACKGROUND	102
FALLOUT	4
NUCLEAR POWER INDUSTRY	0.003
MEDICAL USES	73
OTHER	3
TOTAL	182

FIG. 17.16 Estimated yearly radiation dosage to the gonads.

doses that have absolutely no visible *somatic* effects are nevertheless capable of producing mutations in the sex cells.

A substantial portion of the medical miseries to which humans are prone are attributable to genetic defects. It is therefore imperative that we learn as much as we can about mutation rates and what effect radiation has on them. Figure 17.16 is a table giving estimates of the yearly radiation dosage to the gonads of a typical U.S. citizen (i.e., not a radiologist). You may find it revealing to study the table for clues as to ways in which human exposure to radiation might most effectively be reduced.

While we should not want to carry any more mutations than we can help, mutations have been an extremely valuable tool for discovering more about genetics. The discovery of X-linkage, discussed in the last chapter, is just one example of how mutant genes make clearer the mechanism of heredity.

Mutant genes have also been of great importance in providing clues to the way in which genes act to produce a given phenotype. Red eyes, polysaccharide capsules, nonclotting blood, waxy endosperm, etc., are not made up of DNA. They are simply the expression or product of gene activity. A discussion of genetics would not be complete without considering the steps by which a given genotype produces a given phenotype. That is the topic of the next chapter.

EXERCISES AND PROBLEMS

1. How many different combinations of bases are possible in a DNA molecule 150 bases long?
2. Which way—clockwise or counterclockwise—does the DNA molecule coil away from you as you look through it from one end? As you look through it from the other end?
3. How many cytosines are present in a DNA molecule 1000 base pairs long if 20% of the molecule consists of adenines?
4. The nucleus of a calf spleen cell contains 6.8 picograms (10^{-12} g) of DNA. If this DNA were present as a single molecule, what would be its molecular weight (in daltons)? (There are approximately 6×10^{23} molecules in a mole.) If each base pair has a molecular weight of 700 daltons, how long would this molecule be? How thick?

REFERENCES

1. CRICK, F. H. C., "Nucleic Acids," *Scientific American,* Offprint No. 54, September, 1957. The author shared a Nobel prize for his work in deducing the organization of nucleic acid molecules.
2. HOTCHKISS, R. D., and ESTHER WEISS, "Transformed Bacteria," *Scientific American,* Offprint No. 18, November, 1956. The alteration of bacterial heredity by treatment with DNA is described in detail.
3. TAYLOR, J. H., "The Duplication of Chromosomes," *Scientific American,* Offprint No. 60, June, 1958. Attempts to relate the structure of DNA to the structure of chromosomes.
4. HANAWALT, P. C., and R. H. HAYNES, "The Repair of DNA," *Scientific American,* Offprint No. 1061, February, 1967. Describes how bacteria are able to exploit the complementarity of the two strands of DNA in the repair of damage to the molecule.
5. MULLER, H. J., "Radiation and Human Mutation," *Scientific American,* Offprint No. 29, November, 1955.

18 GENE EXPRESSION

18.1 THE ONE GENE-ONE ENZYME THEORY

Our knowledge of the method of action of genes grew out of the work of the geneticists G. W. Beadle and E. L. Tatum with the red bread mold *Neurospora*. *Neurospora* is particularly well suited for genetic studies. Its life cycle is shown in Fig. 18.1. Like all fungi, it reproduces by **spores.** Spores are tiny, unicellular bodies that serve to disperse the species to new locations and enable it to survive unfavorable environmental conditions. Each spore can develop into a new individual.

Two kinds of spores are produced by *Neurospora*. The *asexual* spores (conidia) are produced by mitosis of the haploid nuclei of the active, growing fungus. The ascospores, on the other hand, are produced following *sexual* reproduction. If two different mating types ("sexes") of *Neurospora* are allowed to grow together, they will fuse to produce a diploid zygote. Meiosis of the zygote then gives rise to the haploid ascospores.

Note, then, that throughout most of its life cycle the organism is haploid. Thus each gene is present in only a single dose, and the geneticist does not have to worry about the possibility that recessive genes may be masked by dominant ones. Another virtue is that the meiotic divisions occur in a narrow tube, the ascus. The tube is sufficiently narrow that the eight nuclei produced by the first and second meiotic divisions, followed by one mitotic division, are not able to slip past one another. Consequently, if the original diploid nucleus is heterozygous for a pair of alleles (e.g. *A,a*) *and* no crossing over at that locus occurs, the alleles will separate at the first

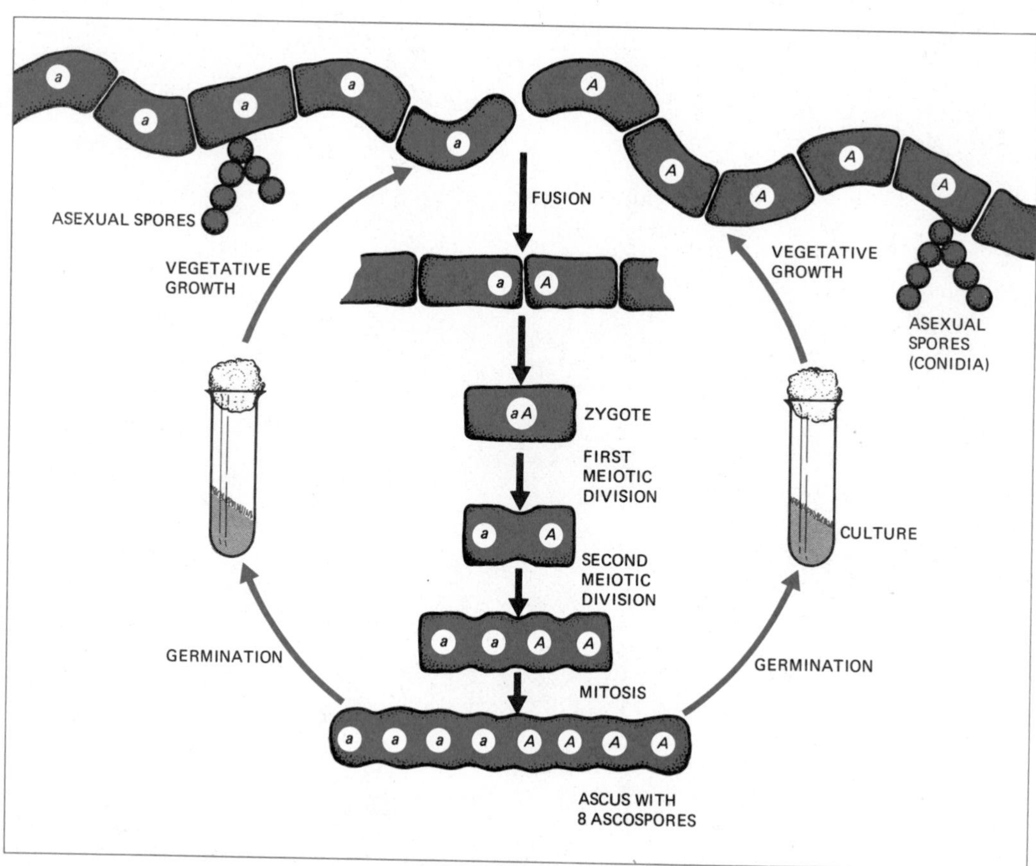

FIG. 18.1 Life cycle of *Neurospora*. The distribution shown here of the *A* and *a* alleles would occur only if no exchange of this locus took place during crossing over in the first meiotic division.

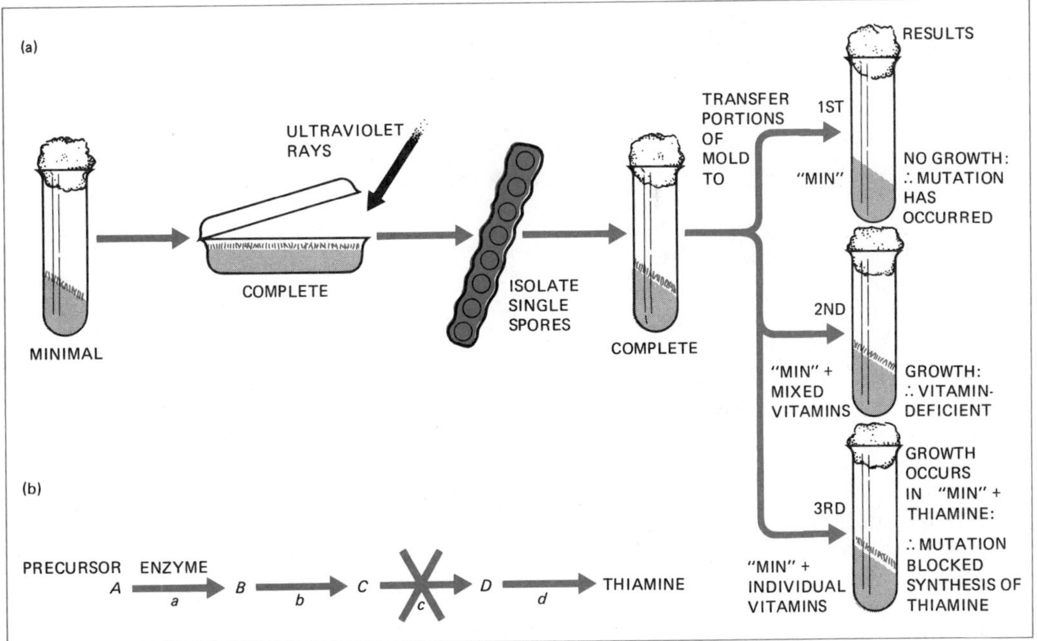

FIG. 18.2 Beadle and Tatum's (a) experiment and (b) hypothesis. Crossing the mutant strain with a normal strain showed that the ability to convert thiamine precursor C into precursor D was controlled by a single gene. Presumably, the mutant gene could not produce the necessary enzyme. Complete medium is enriched with vitamins and amino acids; minimal ("min") medium is not.

meiotic division. After the next two divisions, the ascus will have four spores at one end containing one allele, and four spores at the opposite end containing the other. (What patterns might occur if crossing over of the loci *should* occur in the first meiotic division?)

Neurospora can be grown in test tubes containing a very simple ("minimal") culture medium. Sucrose, a few salts, and one vitamin, biotin, provide all the nutritional requirements for *Neurospora* to live, grow, and reproduce. From these relatively few and simple substances, it is capable of synthesizing all the many complex substances, such as proteins and nucleic acids, necessary for life. Beadle and Tatum exposed some of the asexual spores of one mating type of *Neurospora* to ultraviolet rays, in the hope of inducing mutations. Then individual irradiated spores were allowed to germinate on a "complete" medium, that is, one enriched with various vitamins and amino acids. Once each had developed a vigorous growth of mycelium, it was allowed to mate with the other mating type. The ascospores produced as a result were then dissected out individually and each one placed on an additional quantity of complete medium. After growth had occurred, portions of each culture were placed on a minimal medium (Fig. 18.2). Sometimes growth continued; sometimes it didn't. When it did not, the particular strain was then supplied with various vitamins, amino acids, etc., until growth did occur. Eventually, each deficient strain was found to be capable of growing on a minimal medium to which *one* accessory substance, for example, the vitamin thiamine, had been added. Beadle and Tatum reasoned that ultraviolet irradiation had caused a gene that permits the synthesis of thiamine to mutate to an allele that does not.

The manufacture of thiamine from the simple substances present in the minimal medium does not take place in a single chemical reaction but in a whole series of them. Like all chemical reactions in living things, each one is catalyzed by a specific enzyme. By adding, one at a time, the different precursors of thiamine to the medium in which their mold was growing, Beadle and Tatum were able to locate just which step in the synthesis of thiamine was blocked in their mutant strain (Fig. 18.2). If they added to the minimal medium any precursor further along in the process, growth occurred. Any precursor before the blocked reaction did not permit growth. They reasoned that the change of precursor "C" to precursor "D" was blocked because of the absence of the specific enzyme required. On this basis they created the "one gene–one enzyme" theory of gene action: each gene in an organism controls the production of a specific enzyme. It is these enzymes which then carry out all the metabolic activities of the organism, resulting in the development of a characteristic structure and physiology, the phenotype of the organism.

18.2 INBORN ERRORS OF METABOLISM

The discoveries of Beadle and Tatum have shed new light on a number of human diseases that are known to be hereditary. These diseases have been called "inborn errors of metabolism" because they are inherited and because each is characterized by a distinct metabolic defect. One of these, alcaptonuria, is a rather rare ailment in which the chief symptom is that the urine of the patient turns black upon exposure to air. Biochemical studies have shown that the disease results when the enzyme that catalyzes the conversion of homogentisic acid to maleylacetoacetic acid (Fig. 18.3) is lacking in the individual. These substances are intermediates in the breakdown of the amino acid phenylalanine into compounds that can enter the pathway of cellular respiration. When step 4 is blocked, homogentisic acid accumulates in the blood. The kidney excretes this excess in the urine. Oxidation of the homogentisic acid by the air turns the urine black.

Another inborn error of metabolism, called phenylketonuria **(PKU)**, is caused by a blockage of step 1. In this case, the inability to remove excess phenylalanine from the blood produces—during in-

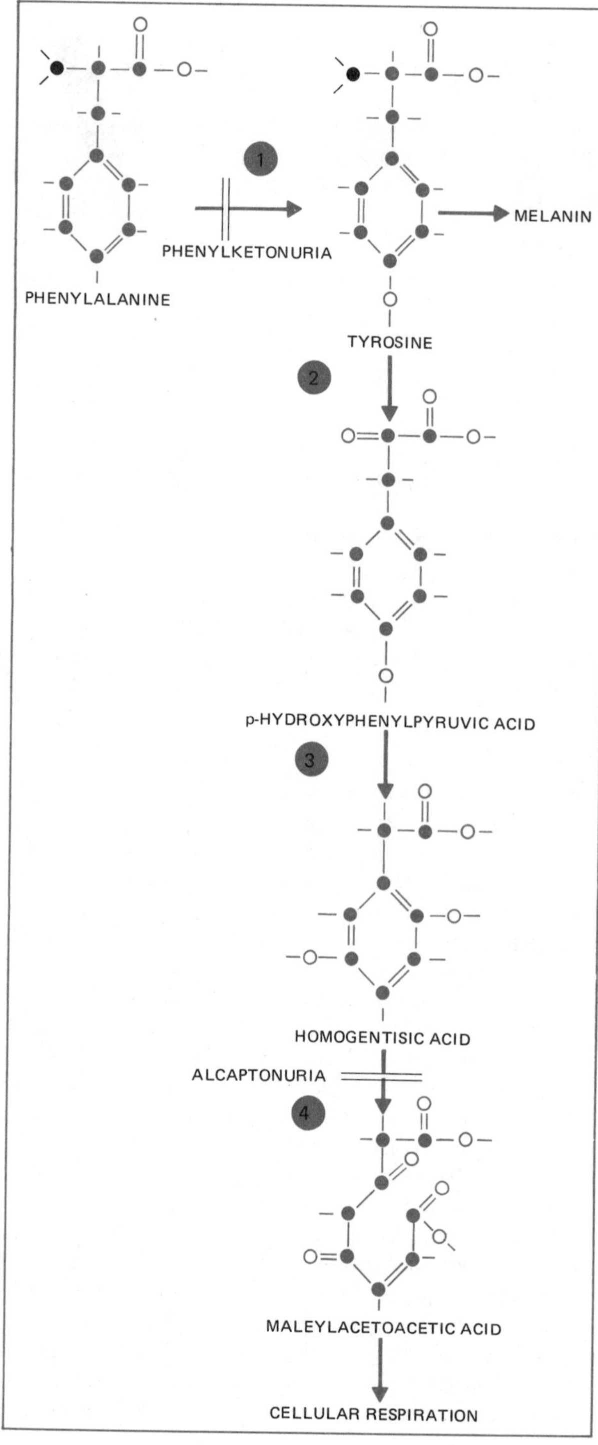

FIG. 18.3 Pathway of phenylalanine metabolism in humans. Phenylalanine that is excess to the body's needs for protein synthesis is broken down as shown here and used in cellular respiration. Victims of PKU lack the enzyme ("1") that converts phenylalanine into the amino acid tyrosine. Victims of alcaptonuria lack enzyme "4." Carbon atoms are shown in color, nitrogen atoms in black, and hydrogen atoms as short dashes.

fancy and early childhood—a serious stunting of intelligence (most sufferers have to be confined to mental institutions), pale skin, and a tendency to epileptic seizures. The pale skin may result from a lack of tyrosine, from which the pigment melanin (responsible for suntan and freckles) is formed. Both alcaptonuria and phenylketonuria are recessive traits. An individual must inherit a defective gene from each parent to show the trait. This is quite consistent with the one gene–one enzyme theory: so long as one nonmutant gene is present, the necessary enzyme is manufactured.

Abnormal Hemoglobins

Perhaps the most thoroughly analyzed hereditary diseases are those in which abnormal hemoglobin molecules are produced. Hemoglobin is the red, oxygen-carrying pigment found in the red blood cells. Each hemoglobin molecule consists of four polypeptide chains, two "alpha" and two "beta" chains, each with an iron-containing heme group to which the transported oxygen molecule is temporarily attached. The alpha chains contain 141 amino acids, the beta chains 146.

The normal hemoglobin of adult humans is known as hemoglobin A (Hb^A). However, over 200 kinds of abnormal hemoglobins have been discovered in humans. One of the most common of these is hemoglobin S (Hb^S). Individuals who manufacture Hb^S exclusively suffer from a disease called **sickle-cell anemia.** This disease is quite common in the malaria-ridden parts of central Africa and is also found among black Americans descended from inhabitants of those regions. The disease receives its name from the fact that the victim's red blood cells become crescent- or sickle-shaped, particularly while passing through the capillaries (Fig. 18.4). The distorted cells are very fragile at such a time and are apt to rupture long before their normal life span (about 120 days) is over. Consequently, victims of this disease suffer from a severe and usually fatal anemia.

Victims of sickle-cell anemia inherit a defective gene from each parent. Individuals who inherit only one defective gene have both Hb^A and Hb^S in their red cells. This is not overly disadvantageous in the

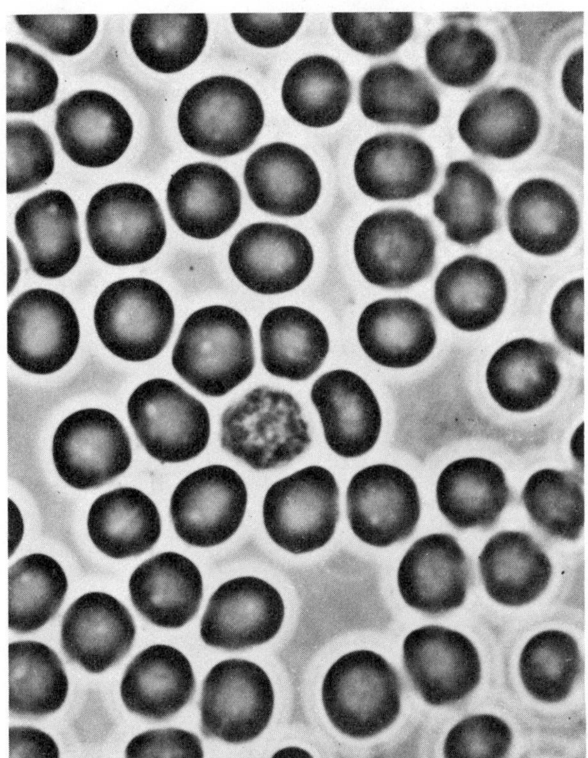

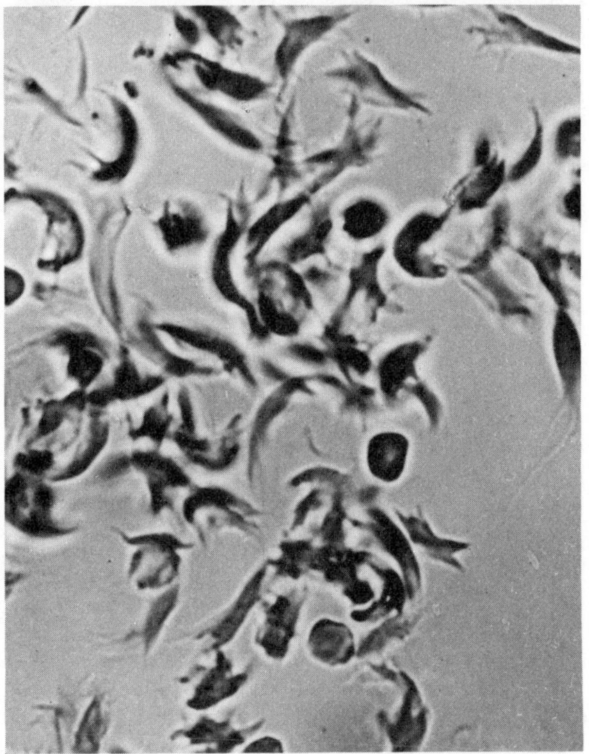

FIG. 18.4 Red blood cells of a victim of sickle-cell anemia. Top: oxygenated. Bottom: deoxygenated. The shape of the cells when deoxygenated causes them to break easily. (Courtesy of Dr. Anthony C. Allison.)

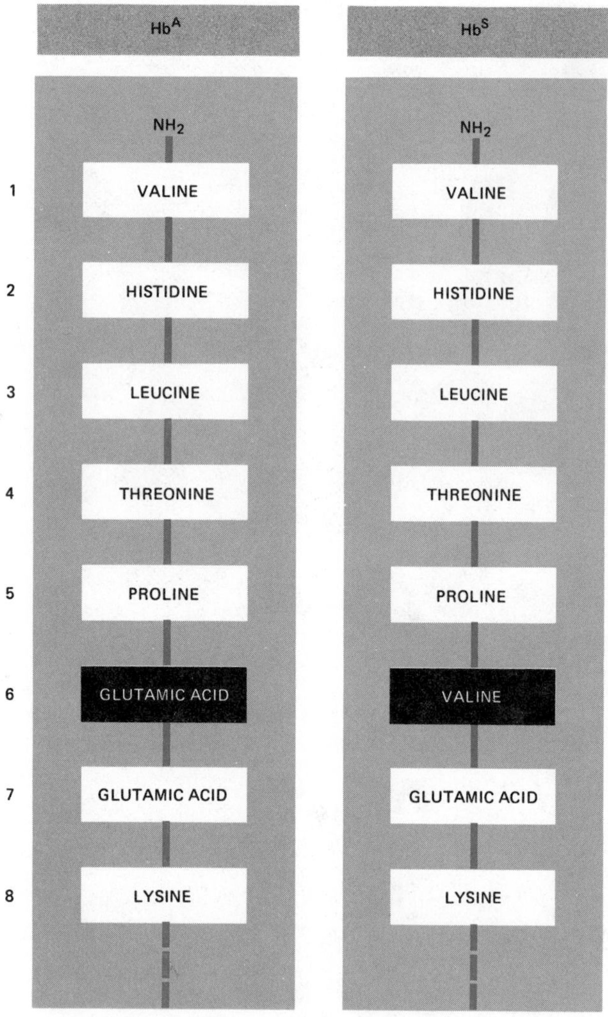

FIG. 18.5 N-terminal portions of normal hemoglobin (Hb^A) and sickle-cell hemoglobin (Hb^S).

malaria-free United States. And in the malarial regions of Africa, the heterozygous condition (Hb^A and Hb^S) is positively advantageous as it seems to confer some resistance to one type of malaria. This probably explains why the mutant gene and the disease are so prevalent in the malarial regions of Africa.

The amino acid sequence of both Hb^A and Hb^S has been determined. It turns out that the alpha chains are exactly the same in each type of hemoglobin. So, for that matter, are the beta chains *except* for the amino acid at position 6 (counting, as always, from the amino terminal). This position is occupied by glutamic acid in normal (Hb^A) beta chains. However, in the beta chains of Hb^S, valine is found instead (Fig. 18.5).

Exactly why this single amino acid change in a chain of 146 amino acids should so drastically alter the properties of deoxygenated hemoglobin is still under intensive investigation. However, it may be worth noting here that in switching from glutamic acid to valine, a strongly hydrophilic molecule has been replaced by a strongly hydrophobic one. Position 6 is located at the surface of the beta chain, where it would normally be exposed to water. Perhaps this switch from a hydrophilic to a hydrophobic region on the surface disturbs the normal solubility of the hemoglobin molecule.

In almost every one of the other cases of abnormal hemoglobins that have been discovered, the defect occurs as a single amino acid substitution at some point on either the alpha or the beta chain. All the abnormal alpha chains are the outcome of different mutations within a single gene. The various beta chains represent different mutations in a separate gene. Thus the one gene–one enzyme theory can now be restated in terms of *one gene–one polypeptide*. This modification also fits our present conviction that not just enzymes but all the proteins manufactured by cells, e.g. structural proteins like collagen, are the products of gene action.

18.3 PROTEIN SYNTHESIS

Beadle and Tatum deduced that the information encoded in a gene finds ultimate expression in the sequence of amino acids in a polypeptide. What, then, is the mechanism by which a particular sequence of nucleotides in DNA leads to a particular sequence of amino acids?

The Ribonucleic Acids of the Cell

Protein synthesis occurs on clusters of tiny (~20 nm) cytoplasmic particles called **ribosomes** (Fig. 18.6). These particles are macromolecular assemblies of **ribosomal RNA (*r*RNA)** and proteins. Each ribosome consists of two subunits: the large and the small. The small subunit contains a single molecule of *r*RNA. In eukaryotes, this molecule is called "18S" RNA because of the rate at which it sediments in a centrifuge tube (Fig. 18.7). In addition, there are two or three dozen polypetides associated with the small subunit.

The large subunit of eukaryotic ribosomes contains one copy of each of three different molecules of *r*RNA (designated 5S, 7S, and 28S—Fig. 18.7).

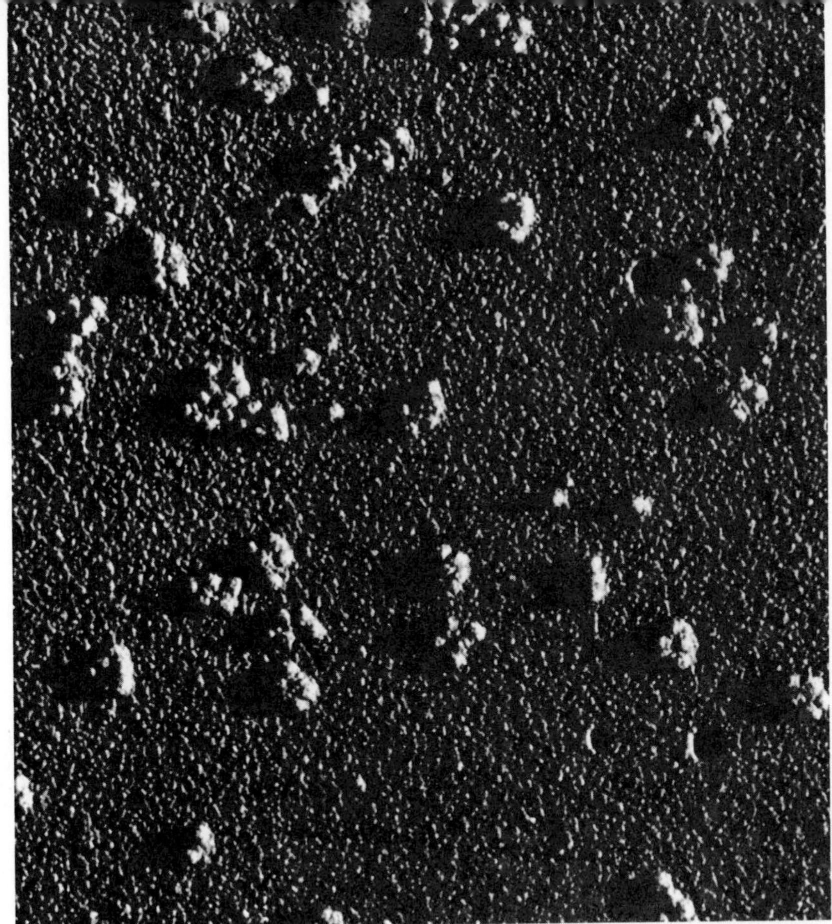

FIG. 18.6 Electron micrograph showing clusters of ribosomes. These clusters, called polysomes, are held together by messenger RNA. (Courtesy of Alexander Rich.)

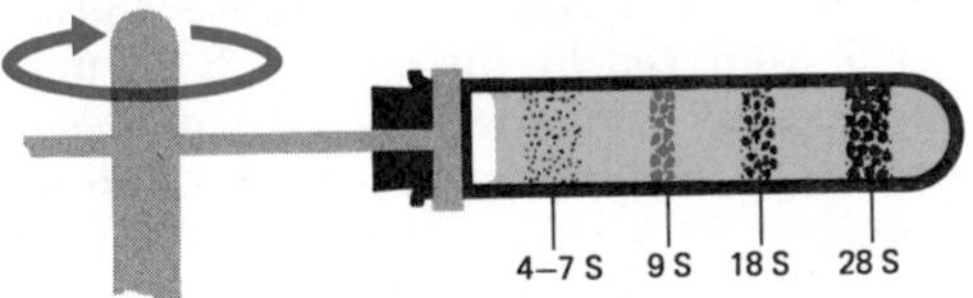

FIG. 18.7 Sedimentation pattern produced by high-speed centrifugation of RNA extracted from the precursors of rabbit red blood cells. The discrete bands represent particular classes of RNA. The transfer RNAs band at about 4S. The ribosomal RNAs of eukaryotes sediment at 5S, 7S, 18S, and 28S. (The larger the sedimentation unit, S, the larger the molecule.) The RNA forming the band at 9S is the messenger RNA for the synthesis of hemoglobin, the major protein produced by these cells. In most types of cells, the messenger RNAs are extremely heterogeneous, with small amounts distributed from 6 to 25S.

Associated with these *r*RNAs are approximately 50 different polypeptides.

In eukaryotes, the ribosomal RNA molecules are synthesized in the nucleolus, using DNA templates (genes). There are many copies of the genes for each kind of *r*RNA (Fig. 18.8). The genes for the lightest *r*RNA (5S) in humans are probably located on chromosome No. 1 (Fig. 16.22).

If we extract all the RNA from a cell, most of it falls into one of several discrete size classes (Fig. 18.7). Ribosomal RNA, as we have seen, occurs (in eukaryotes) as 5S, 7S, 18S, and 28S molecules. Another discrete group is made up of the so-called transfer RNAs (*t*RNAs). These molecules sediment at about 4S (corresponding to molecular weights ranging from 23 to 30 thousand).

In addition to these discrete classes of RNA, there is a small amount of RNA that is extremely heterogeneous in size. It sediments at 6–25S and has molecular weights ranging from 25,000 to one million. Most cells contain only small amounts of this material. However, cells that are actively engaged in synthesizing large amounts of a *single* protein may contain substantial amounts of a particular RNA of a size different from any of the *r*RNAs or

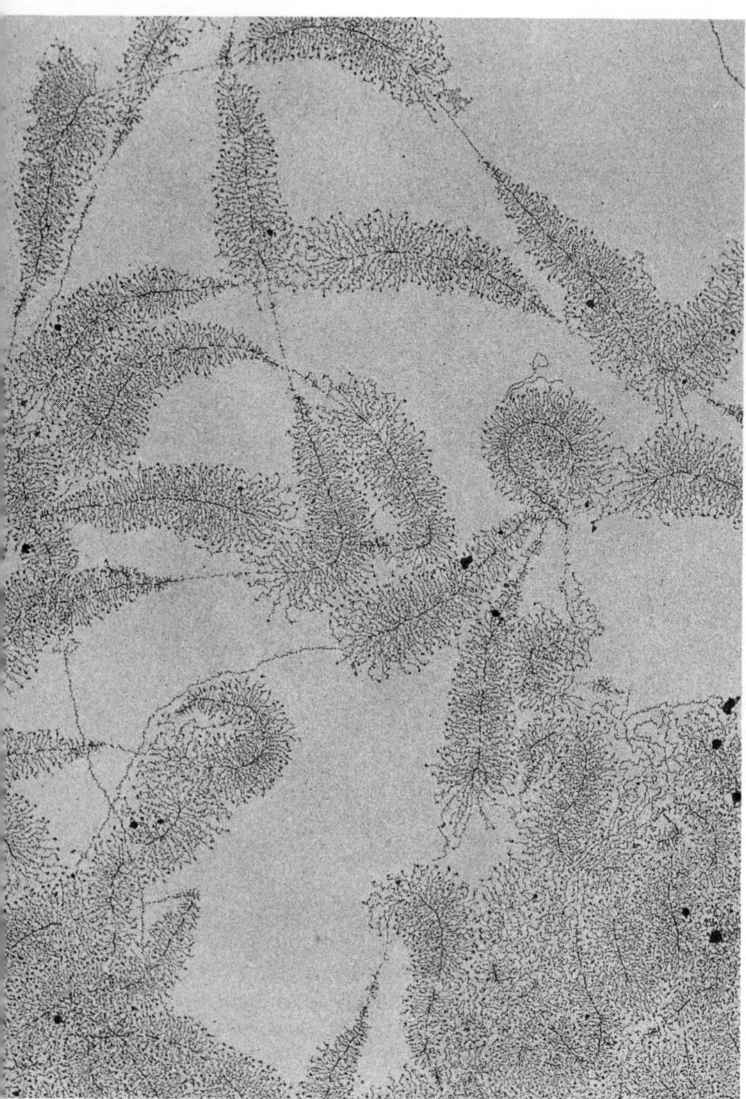

FIG. 18.8 An electron micrograph revealing gene transcription in the developing egg cell of the spotted newt. The long filaments are DNA molecules coated with protein. The fibers extending in clusters from the main axes are molecules of RNA from which the cell's ribosomes will be constructed. Note how transcription begins at one end of each gene, with the RNA molecules getting longer as they proceed towards completion. Note also the large number (up to 100) of RNA molecules that are transcribed simultaneously from each gene. The portions of DNA bare of RNA appear to be genetically inactive. (Courtesy of O. L. Miller, Jr., and Barbara R. Beatty, Biology Division, Oak Ridge National Laboratory.)

*t*RNAs. For example, the hemoglobin-synthesizing red blood cell precursors of the rabbit contain substantial amounts of a 9S RNA (Fig. 18.7). If this RNA is isolated and injected into a frog egg, the frog begins to synthesize rabbit hemoglobin. Only this particular class of rabbit RNA is able to instruct the frog egg to synthesize rabbit hemoglobin. Because of its role, this RNA is called **messenger RNA (*m*RNA).**

Transcription

DNA guides the synthesis of *m*RNA (as well as *t*RNA and *r*RNA—see Fig. 18.8) in the same way that it guides its own duplication (see Section 17.2). A single strand of DNA serves as the template for assembling *ribo*nucleotides (present in the surrounding medium as their triphosphates, e.g. ATP) into a strand of RNA whose sequence of bases complements exactly the sequence of bases in the DNA molecule, that is, they are assembled following the regular rules of base pairing. For every C on the DNA molecule, a G is inserted into the complementary strand of *m*RNA. So, too, every G picks up a C-containing ribonucleotide and every T picks up an A-containing ribonucleotide. The A's on the DNA strand code for the insertion of a uracil-containing ribonucleotide (U). (There is no thymine, T, in RNA).

The synthesis of a molecule of RNA complementary to a strand of DNA is called **transcription.** The process is catalyzed by the enzyme **RNA polymerase.** It proceeds with the liberation of a substantial amount of free energy because the two terminal phosphates of each nucleotide triphosphate precursor are split off as the nucleotide is added to the growing strand. You may remember that both the subterminal and the terminal phosphate bonds of nucleotide triphosphates (like ATP) are so-called "high-energy" bonds.

There is a great deal of experimental evidence for the synthesis of *m*RNA from a DNA template. To cite one example, when a synthetic DNA strand consisting solely of T-containing nucleotides —"poly(T)"—is supplied with all *four* ribonucleotides (ATP, UTP, CTP, and GTP) and RNA polymerase, a molecule of RNA is formed that contains only adenine in the strand—"poly(A)."

Translation

How does the information that has been *transcribed* into a molecule of messenger RNA get **translated** into a sequence of amino acids? This requires the third kind of RNA, called **transfer RNA (*t*RNA).** There are perhaps as many as 40–60 distinct kinds

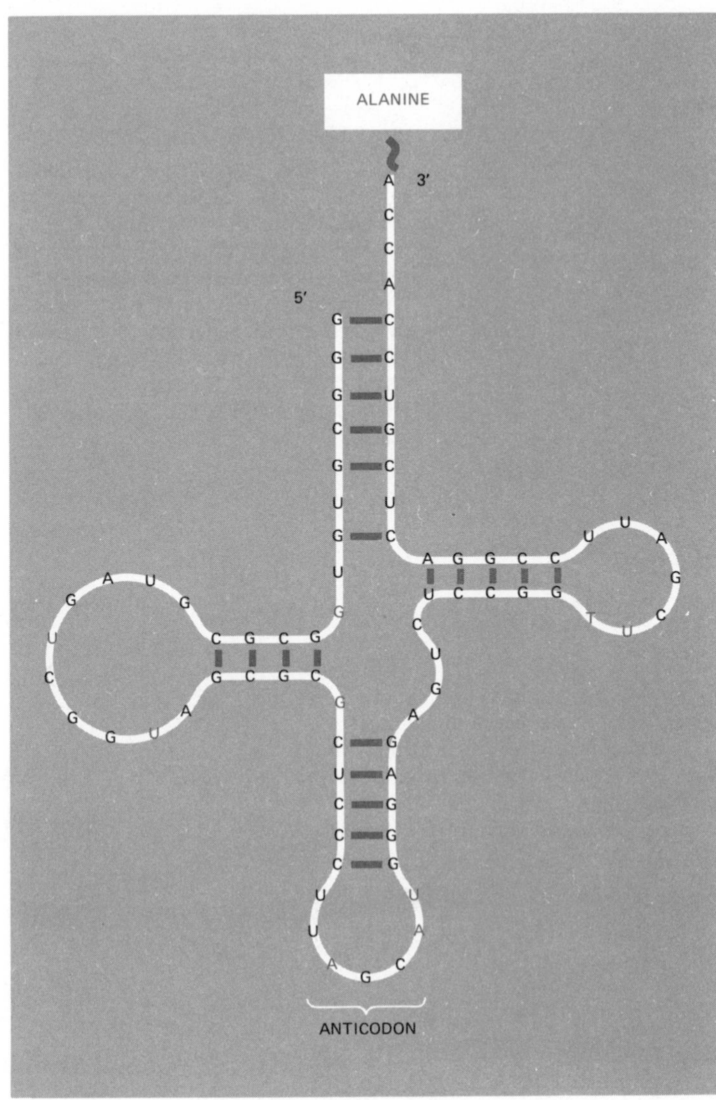

FIG. 18.9 Structure of alanine transfer RNA from yeast. The four helical regions formed by base pairing are shown. The loop on the left is thought to recognize (bind to) the enzyme that catalyzes the attachment (by means of a "high-energy" bond) of alanine to the molecule (top). The anticodon loop carries the three bases that recognize the appropriate codon on the messenger RNA. The loop on the right may be responsible for temporarily binding the *t*RNA molecule to the ribosome. The use of colored letters for certain bases indicates that the actual base is a chemically modified form of the one indicated.

of transfer RNA molecules in the cell. Among these there is at least one kind for each of the 20 amino acids used in protein synthesis. With the aid of a specific enzyme and ATP, each amino acid is activated and attached to a particular transfer RNA molecule.

The structure of several transfer RNA molecules has been determined. Each consists of a chain of 75–90 ribonucleotides (Fig. 18.9). Many of these are the normal RNA nucleotides (A, U, G, and C), and in places these bases link two portions of the chain in a double helix like that of the DNA molecule (Fig. 18.10). Because of this, several loops are formed in the chain. The ends of the chain are exposed, and the amino acid is attached to one of these free ends (Fig. 18.9). *Unpaired* bases at the "anticodon" loop "recognize" complementary exposed bases on the messenger RNA molecule and unite with them according to the usual rules of base pairing. Twenty or more kinds of transfer RNA, each kind carrying one particular amino acid, are able to unite with those portions of the messenger RNA molecule where the bases are complementary to the exposed bases on the anticodon loop of the transfer RNA. In this way, the different amino acids can be assembled in a specific sequence to be joined to form

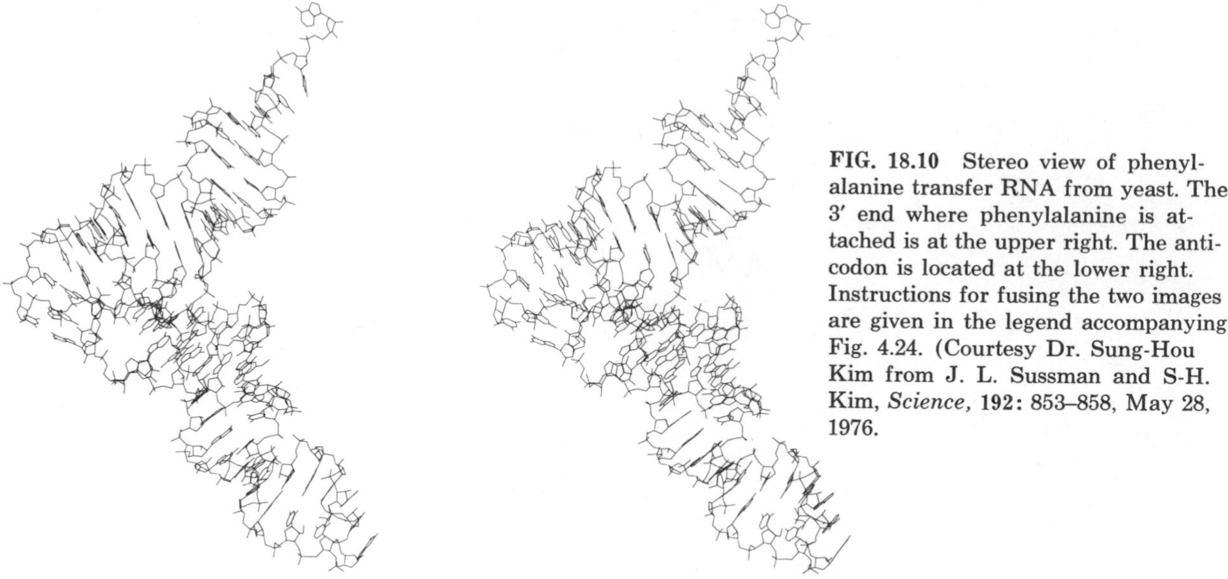

FIG. 18.10 Stereo view of phenylalanine transfer RNA from yeast. The 3′ end where phenylalanine is attached is at the upper right. The anticodon is located at the lower right. Instructions for fusing the two images are given in the legend accompanying Fig. 4.24. (Courtesy Dr. Sung-Hou Kim from J. L. Sussman and S-H. Kim, *Science,* **192:** 853–858, May 28, 1976.

FIG. 18.11 Mechanism of the synthesis of a polypeptide according to the genetic instructions coded in DNA. The use of colored letters for certain bases in the anticodons of the *t*RNAs indicates that the actual base is a chemically modified form of the one indicated, but forms base pairs in the same way. The codon-anticodon interaction of *t*RNA with *m*RNA occurs at the small subunit of the ribosome. Peptide bond formation occurs at the large subunit. The ribosome reads the message from its 5′ to its 3′ end.

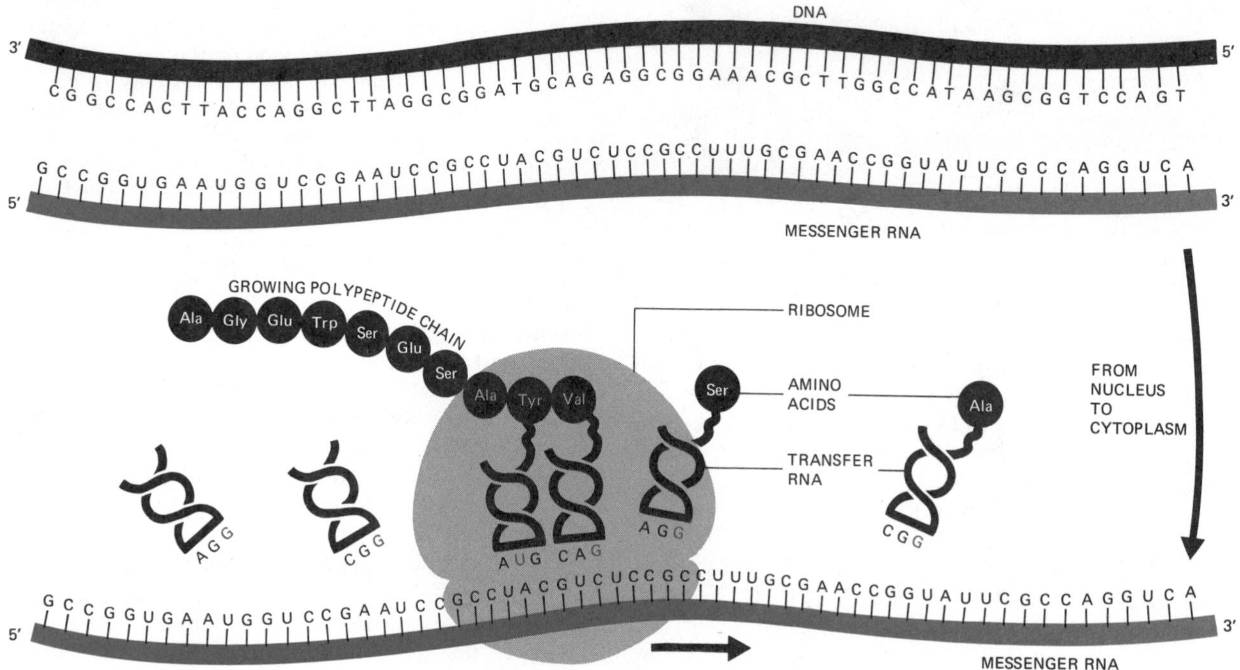

a polypeptide. This sequence of amino acids would be controlled by the sequence of bases in the *m*RNA molecule and thus in the DNA of a gene (Fig. 18.11).

The union of transfer RNA and messenger RNA requires the presence of ribosomes. Electron micrographs and biochemical analysis both show that the ribosomes attach to the end of a strand of messenger RNA and then move along its length, "reading" its sequence of bases. As they do this, they pick up the appropriate transfer RNA molecules, each with its amino acid. One at a time these amino acids are linked together to form the polypeptide chain. When the ribosome reaches the end of the messenger RNA strand, the polypeptide is complete. It and the ribosome are released. In this way, the message in the DNA molecule—its sequence of bases—which had been *transcribed* into a molecule of messenger RNA becomes *translated* into a specific sequence of amino acids in the polypeptide (Fig. 18.11).

In eukaryotes, the transcription of DNA into *m*RNA occurs within the nucleus. However, the translation of the *m*RNA occurs on ribosomes located in the cytoplasm. The RNA molecule synthesized within the nucleus is much larger than the message that will be translated in the cytoplasm. This is because untranslated segments are present at each end of the molecule. Some of this material is removed before the molecule leaves the nucleus. In addition, the messenger RNA molecules have a long (100–200) string of A's—poly(A)—attached at one end. This poly(A) "tail" is retained as the molecule enters the cytoplasm. The longer the tail, the more stable the molecule. Perhaps the length of the poly(A) tail in some way determines how many times that particular messenger will be translated.

Although a single ribosome can manufacture a polypeptide from a messenger RNA molecule, several ribosomes are usually engaged in the process at one time. One messenger RNA molecule with several ribosomes attached to it at various stages in assembling the polypeptide is called a **polysome** (Fig. 18.12).

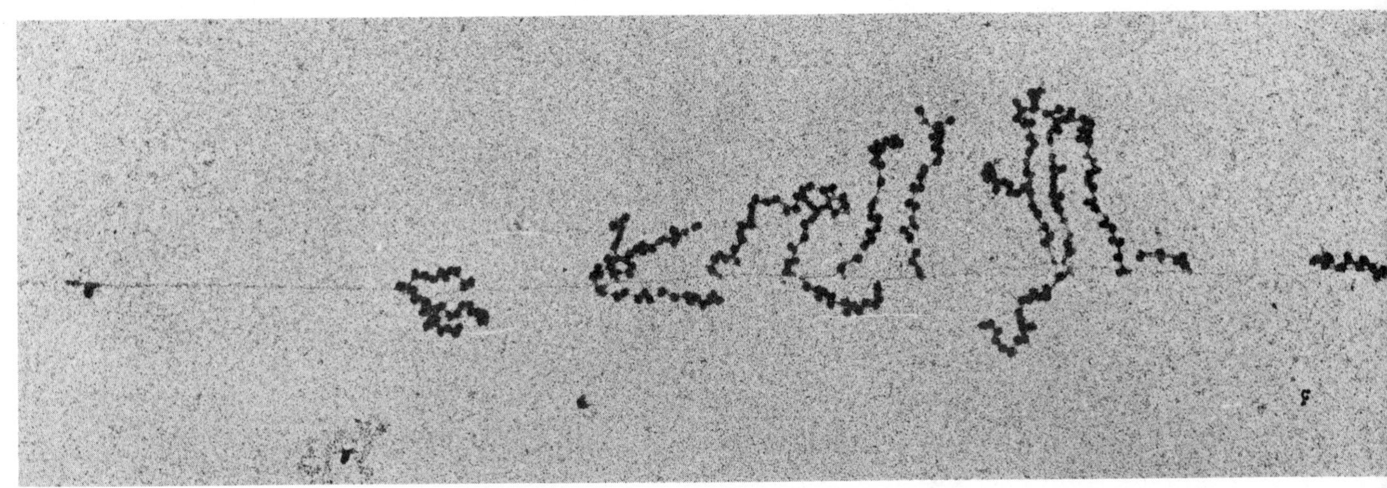

FIG. 18.12 Simultaneous transcription and translation of *E. coli* genes. The long (3 μm) fiber running from left to right is a segment of the *E. coli* chromosome. Extending from it are polysomes the size of which generally increases from left to right. Each polysome consists of a backbone of messenger RNA to which the ribosomes are attached. Each polysome is attached to the DNA fiber by a structure thought to be a molecule of RNA polymerase. Thus it appears that the DNA is *transcribed* by RNA polymerase molecules moving from left to right and the growing messenger RNA strands are *translated* by ribosomes moving along them in a proximal→distal direction. In *E. coli,* then, and perhaps in all prokaryotes, the transcription of DNA into messenger RNA and the translation of messenger RNA into polypeptide chains (not visible here) are closely coordinated. (Electron micrograph, 61,000 ×, courtesy of O. L. Miller, Jr. from O. L. Miller, Jr., B. A. Hamkalo, and C. A. Thomas, Jr., *Science,* **169,** 392–395, July 24, 1970.)

Here, then, is a mechanism by which a unique sequence of bases in the DNA of the chromosomes controls a unique sequence of amino acids in the proteins manufactured by the cell. The DNA of a cell thus constitutes the "master" copy of the hereditary information, while the messenger RNA is the "working" copy. An analogy may help to make this clear. In the recording industry, great care and expense go into the manufacture of a master negative of a record. It is far too valuable to be used in the molds that produce the thousands of wax impressions that will be sold to the public. Instead, working copies of the master are made from it. These copies, which are called "mothers," are then used in the molds to produce the final product.

18.4 THE CODE

A great deal of work has also been done in an effort to determine what sequence of bases in the messenger RNA molecule codes each of the amino acids. There are only four kinds of bases (A, U, C, and G) in the messenger RNA molecule, too few for each one to code a single amino acid. Pairs of bases could be arranged in 16 different ways (AA, UU, UA, etc.), but this is still short of the necessary number. Triplets of bases could be arranged in 64 different ways (Fig. 18.13) and thus would provide ample possibilities for coding 20 amino acids. In fact, we might expect two or three different triplets to code for a single amino acid. Such a triplet code could also account for the presence of more than 20 different transfer RNAs in a cell with, in some cases, two or three different transfer RNAs each bringing the same amino acid to the growing polypeptide.

By manufacturing synthetic messenger RNA molecules and then supplying them with ribosomes, ATP, enzymes and all 20 amino acids, it has been possible to establish the triplets that code for the various amino acids. For example, a synthetic messenger RNA that has only uracil—poly(U)—produces a polypeptide containing the single amino acid phenylalanine. This suggests that the triplet UUU guides the incorporation of phenylalanine into the growing polypeptide chain. Similarly, synthetic poly(A) messenger RNA guides the synthesis of a polypeptide containing only the amino acid lysine. Through the use of other synthetic messenger RNA molecules, it has been possible to correlate one or more triplets with each one of the 20 amino acids (Fig. 18.13). Because of this coding relationship, a triplet of bases is called a **codon.**

Three of the 64 possible codons (UAA, UAG, and UGA) have not been found to code for any amino acid. In *E. coli* (and probably in all organisms) these three codons serve as punctuation marks. When the ribosome reaches them, the growth of the polypeptide chain is halted, and the chain is released, ready to carry out its function in the cell.

The codon AUG serves two related functions. When AUG occurs within the message being translated, it guides the incorporation of the amino acid

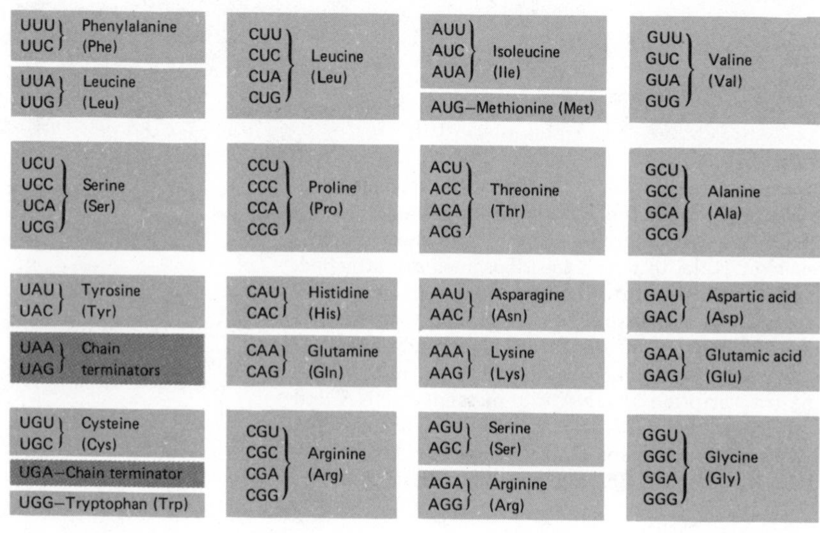

FIG. 18.13 The genetic code. Although most of the codons have been assigned as a result of studies with *E. coli*, there is evidence that the code applies to all organisms.

methionine. However, all "messages" examined so far *begin* with AUG. It appears that AUG thus serves as a starting signal for the translation process. The transfer RNA molecule that binds to AUG at the start of a message is not the same as that which binds in the middle. (It does, however, carry a modified form of methionine which is usually removed from the polypeptide at a later time.) Just how the two AUG-recognizing transfer RNA molecules distinguish between AUG at the start of and AUG within a message is not yet known. Perhaps, as we shall see below, the presence of a short sequence of untranslated nucleotides just before the initiator AUG provides the distinguishing feature needed.

To go back to our one gene–one polypeptide hypothesis, perhaps a mutant gene is simply a section of a DNA molecule in which one base pair has been altered. An alteration of a base pair in the DNA molecule would result in a corresponding alteration in the *messenger* RNA molecule. This, in turn, would provide a codon complementary to a different *transfer* RNA molecule. In some cases this shift might still code for the same amino acid. As you can see in Fig. 18.13, most of the amino acid codons can have the third base of the codon altered without changing the amino acid specified. However, other base changes in the RNA would call for the insertion of a different amino acid into the polypeptide at that point. Such a change is a mutation.

The codons GAA and GAG code for glutamic acid (Fig. 18.13). In our example of Hb^A and Hb^S, the substitution of a single base, a U for an A in the middle position of the codon, would give a GUA or a GUG codon. Both of these have, indeed, been assigned to valine, the amino acid substitution in sickle-cell hemoglobin. Perhaps, then, the primary difference between a sickle-cell individual and a normal one is the presence of adenine (A) instead of thymine (T) at one spot in the part of a DNA molecule that stores the information for the synthesis of the beta chain of hemoglobin. Although this change seems to be a minor one, the resulting substitution of valine for glutamic acid so alters the physical properties of hemoglobin that a fatal anemia is produced in the individual carrying both genes for the trait.

Not all base changes result in an altered polypeptide; nevertheless, you can well appreciate that even in a "small" gene, such as that specifying the beta chain of hemoglobin, the number of possible mutations is very large. And, in fact, as the study of genes and gene action has progressed, it has become quite clear that a given gene may be mutated in a variety of different ways. In just a few years, more than 200 different mutations have been discovered in the two genes controlling alpha and beta chain synthesis in human hemoglobin. Most of these result in a polypeptide with a single amino acid substitution somewhere along its length.

18.5 VERIFICATION OF THE CODE

The brilliant research which led to the codon assignments shown in Fig. 18.13 was carried out in test tubes. Perhaps for this reason and perhaps because the story this research told seemed almost fantastic, a number of biologists questioned whether such a system really operated in living organisms. What was needed to demonstrate this was to be able to determine the exact sequence of nucleotides in a gene (or a messenger RNA molecule) and see whether the sequence of bases corresponded to the sequence of amino acids in the protein coded by that gene (or synthesized by that *m*RNA molecule). Unfortunately, one cell contains, at most, two copies of each gene (with some exceptions). While a given gene may be transcribed into multiple copies of *m*RNA, most cells contain at any one time many kinds of *m*RNA, none of which is present in very large amounts. In order to determine sequences, a chemist needs plenty of purified material with which to work.

A way out of this dilemma was presented by the single-stranded RNA bacteriophages. These viruses have a cycle of infection essentially like that shown in Fig. 17.4. Each virus particle consists of a protein coat inside which is a single molecule of single-stranded RNA rather than the DNA found in so many viruses (and shown in Fig. 17.4). This molecule of RNA has a dual function. It carries all the genes of the virus, of which there are three. One of these RNA genes codes for a subunit of an enzyme, an RNA polymerase, which can then use the entire RNA molecule as a template against which additional copies can be synthesized. This RNA → RNA copying of the genetic material is analogous to the DNA → DNA gene copying process we examined in the last chapter.

A second gene codes for the protein molecule out of which the virus coat is constructed. A third gene codes for the "A" protein which is needed for successful virus assembly.

In providing information which can be translated into these three gene products, the RNA molecule of this bacteriophage is also serving the function of a messenger RNA. As is the case with DNA vi-

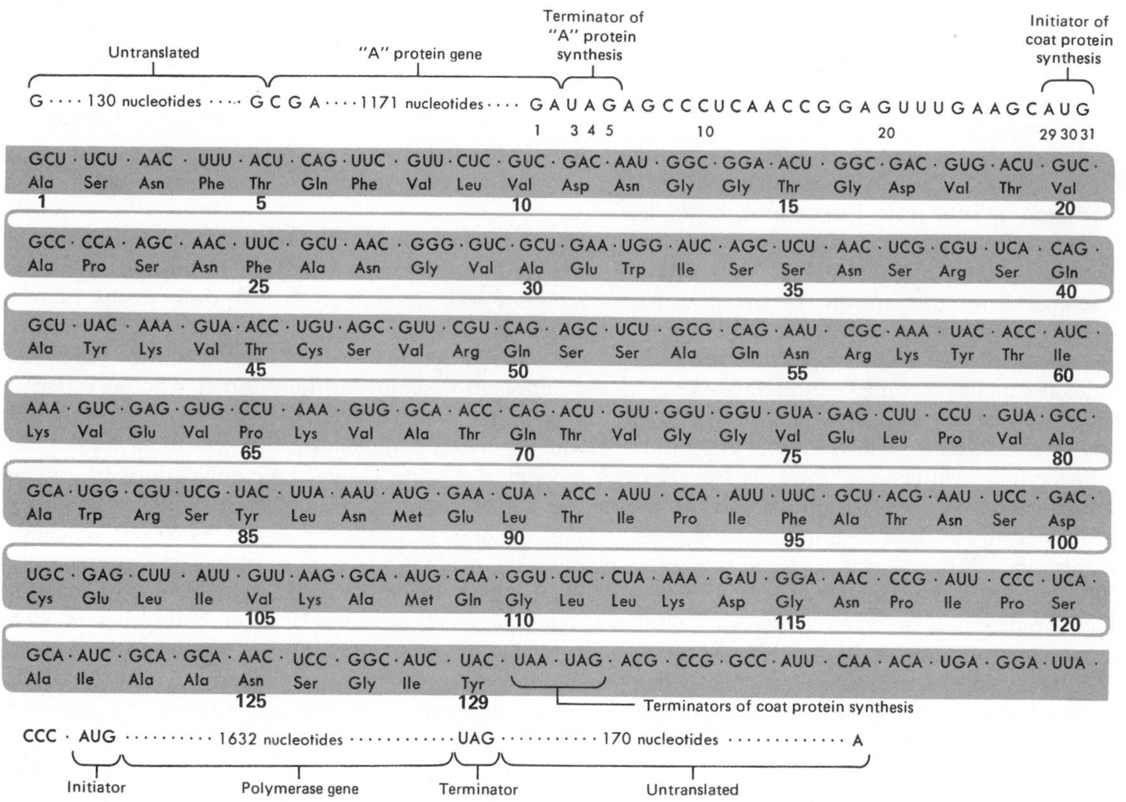

FIG. 18.14 Nucleotide sequence of a portion of the RNA of the MS2 bacteriophage. This sequence includes: (1) the entire gene for the coat protein of the virus (codons Numbers 1–129), (2) its initiation and termination signals, (3) portions of the preceding and succeeding genes, and (4) untranslated sections flanking each of the three genes. Analysis of the proteins synthesized by these genes has shown that the codon assignments given in Fig. 18.13 are, without exception, correct. Of the many synonymous codons, all but 12 are used in the coat protein gene. Analysis of the gene preceding it (only a portion of which is shown here) has revealed the remaining 12; therefore MS2 (like its host *E. coli*) uses every possible codon even though most are redundant. These data were obtained by W. Fiers and his associates.

ruses, the machinery for translation (ribosomes, *t*RNA molecules, amino acids, and several special protein factors) is that of the host cell, in this case, *E. coli.*

The primary structure (i.e. the sequence of amino acids) in the coat protein of several of these RNA bacteriophages has been determined. The structure of the coat protein of the bacteriophage called MS2 is shown in Fig. 18.14. It is a polypeptide containing 129 amino acids. With this knowledge and the ability to grow vast numbers of virus particles from which the RNA can be extracted, it is possible to attempt to verify the code. Over a period of years, W. Fiers and his associates have succeeded in determining the entire sequence of nucleotides (3569 of them!) in the RNA of MS2. A portion of their results is shown in Fig. 18.14.

The molecule begins with a run of 132 nucleotides that is not translated into a protein product. This is followed by the gene, 1176 nucleotides long, for the "A" protein (which contains 392 amino acids). In order to simplify the figure, the next-to-last nucleotide of this gene (number 1307 counting from the start of the molecule) is indicated as No. 1 in Fig. 18.14. Nucleotides Nos. 3–5 are UAG, the terminator for translation of the "A" protein. This is followed by a stretch of 23 nucleotides (Nos. 6–28) that are not translated. The next three nucleotides (Nos. 29–31) are AUG, the start signal for translation of the coat protein. Then, starting with nucleotide No. 32 and reading along in groups of three, we find that the amino acid predicted for each triplet by the codon table given in Fig. 18.13 establishes a sequence that is *exactly* the sequence which independent analysis has determined for the coat protein. Immediately following the 129th triplet appear the triplets UAA and UAG (Fig. 18.14). *Both* of these have been assigned the function of chain termination. (Why two terminators are used is not known—belt *and* suspenders?) Then comes a sequence of 30 untranslated nucleotides. Next is another initiator codon, AUG, followed by 1632 nucleotides, which are translated into the RNA polymerase, and a UAG terminator. The remaining portion of the molecule is an untranslated "tail" of 171 nucleotides.

What Fiers and his associates have done, then, is to reveal the coded information content of three entire genes—complete with start and stop signals. For two of these genes, the "A" protein gene and the coat protein gene, independent analysis of their polypeptide products matches without exception the codon assignments shown in Fig. 18.13. In accomplishing this historic task, these workers have verified in the most direct way possible the validity of the genetic code and thus have placed on an unshakable foundation one of the supreme achievements of the twentieth century.

Is the code universal? Probably so. It appears that the same codons are assigned to the same amino acids and to the same start and stop signals in such diverse organisms as *E. coli,* yeast, the tobacco plant, a species of toad, the mouse, the rabbit, and the guinea pig. As you can see from Fig. 18.13, however, almost all the amino acids are coded by more than one codon. *Which* of the alternative codons for a particular amino acid is most often used may vary from species to species and even from gene to gene.

18.6 MULTIPLE ALLELES

The mutant gene for Hb^S is an allele of the gene for Hb^A. All the other hemoglobin beta chain mutations are also allelic. We use the term **multiple alleles** whenever a variety of mutant forms of a single gene are known. Note that any one individual can have only two alleles (e.g. Hb^A, Hb^S), one on each chromosome. Within a population, however, many alleles may be present. In fruit flies, over a dozen alleles have been found at a single locus that controls eye color.

Using mutagenic agents, some 300 different mutations have been produced in a single gene on a bacteriophage chromosome. In fact, just as crossing over separates linked genes, so it can separate the parts of a gene. As we saw in Section 16.6, the closer two genes are on a chromosome, the less frequently they are separated by crossing over. We might guess from this that the separation of parts of a gene would be rarer still. This is so. It is so rare, in fact, that for any particular gene under study it is usually not detectable in an organism such as the fruit fly, whose offspring number only in the hundreds. On the other hand, this phenomenon *can* be detected among the billions of offspring produced in a short period when bacteriophages infect a culture of their host cells. It has, in fact, been possible to map a linear sequence of intragene mutations for a single bacteriophage gene in the same way that the linear sequence of genes in a single fruit fly or corn chromosome is mapped. Probably any base in the polypeptide can be altered and, if it codes for a different amino acid, a mutation will thus be produced.

If a dominant gene produces one polypeptide, does its recessive allele simply not make any polypeptide at all? Certainly this is not the case with the genes specifying abnormal hemoglobins. A person heterozygous for sickle-cell anemia produces *both* Hb^S and Hb^A in the same red blood cells. Such persons can be distinguished from those homozygous for Hb^A as well as from those homozygous for Hb^S. Many other cases are known where each allele in a hybrid produces its own effect on the phenotype. This situation is called **codominance.**

18.7 REVERSE TRANSCRIPTASE

Among the RNA viruses are a number that infect the cells of mammals and birds, causing these cells to become malignant. That is to say, these so-called RNA tumor viruses cause cancers in a variety of experimental animals (and perhaps in humans). A malignant or cancerous cell is one that divides repeatedly and uncontrollably, giving rise to a tumor. The malignant characteristics of the first cell pass on to all the progeny cells in the same way as any other inherited trait.

The genetic material of cells is DNA, but the genetic material of the RNA tumor viruses is RNA.

How can infection by these viruses lead to the acquisition of a stable inherited trait in the recipient cell? In an attempt to answer this question, Howard Temin and David Baltimore looked for the presence of an enzyme in infected cells that would be able to synthesize DNA complementary to the RNA of the infecting virus. They soon found it. Its discovery demonstrated that genetic information could flow from RNA to DNA as well as in the way which we have been studying, i.e. DNA → RNA → protein. This enzyme was named **reverse transcriptase.**

The discovery of reverse transcriptase also provided a plausible explanation of the well-documented fact that many experimental animals *inherit* the tendency to develop tumors. It is as though among the genes they inherit in the sperm and egg of their parents there is one or more which someday will (1) initiate formation of a cancer and (2) initiate for the first time the production of RNA tumor viruses. Reverse transcriptase provides a mechanism by which the genetic information for both these functions could have been transcribed into DNA, which could then be incorporated with the rest of the genes of the animals' ancestors.

The discovery of reverse transcriptase also provides a tool for making specific genes to order in the test tube. Provided one can secure adequate amounts of purified *m*RNA (for synthesis of hemoglobin A, for example), such a messenger could then serve as a template for the synthesis—catalyzed by reverse transcriptase—of the genes for hemoglobin. In fact, several research groups have reported success in synthesizing DNA complementary to hemoglobin messenger RNA. Could this DNA be incorporated in the cells of a victim of, say, sickle-cell anemia to enable that person to manufacture Hb^A in addition to Hb^S, and thus to acquire the much more desirable phenotype of a heterozygote? Such an achievement would represent but one of a number of possibilities for intervening in human genetics, or, as it is sometimes called, genetic engineering.

18.8 THE PROSPECTS OF GENETIC ENGINEERING

In attempting to introduce new genetic information into a cell, a number of formidable obstacles must be overcome. First, you must identify and isolate the genes you wish to use. Second, a method must be found to get this foreign DNA into the recipient cell. Third, the recipient cell must be able to replicate this newly acquired DNA between each cell division. Fourth, the cell must be able to express its new gene, i.e. it must be able to transcribe and translate the introduced DNA.

In the previous section, we noted how reverse transcriptase can be used to synthesize genetic information. To date, this approach has been most successful when the RNA templates have come from cells that have unusually large amounts of a single kind of RNA. Developing red blood cells, with their almost total commitment to hemoglobin synthesis, are such a source. Synthetic globin genes have been made (by reverse transcriptase) from rabbit *m*RNAs and successfully introduced into *E. coli* (but with no evidence that their new host expressed them).

Most cells have single copies of each of their genes (two copies if they are diploid and homozygous). But there are two important exceptions. Histone genes and the genes for *r*RNA occur in many copies. Ribosomal RNA genes from an amphibian and histone genes from sea urchins have been introduced into *E. coli.* Once inside, these genes have been replicated (and, in the case of the *r*RNA genes, *transcribed*). In 1976 evidence was presented of the successful *translation* of yeast genes that had been introduced into *E. coli.*

Such examples of genetic engineering have been surrounded by controversy. On the one hand, great benefits may come from such work. Imagine, for example, the quantity of human insulin or other such products that could be synthesized by vats of *E. coli* carrying and expressing the appropriate human genes. So far, the limited successes with genetic manipulation have mainly involved inserting genes into prokaryotes like *E. coli* (Fig. 18.15). But what if you could do the reverse, e.g. insert the prokaryotic genes for, say, nitrogen fixation into a eukaryote like corn or wheat or rice? The implications for human nutrition are enormous.

But dangers lurk in such manipulations. What if DNA from tumor-causing viruses became part of the genetic constitution of a normal inhabitant of the human intestine (like *E. coli*)? How will our never-ending battle against infectious disease fare if genes for antibiotic resistance are introduced into human pathogens? Or genes for the production of toxins like diphtheria toxin, cholera toxin, or tetanus toxin? It is the specter of the escape of such organisms that has created much of the controversy surrounding the insertion of foreign DNA into bacteria.

Other approaches to genetic manipulation are also being taken. The technique of somatic cell hybridization, described in Chapter 16, permits some of the genetic information of one species to be expressed in the cells of another. In addition to fusing

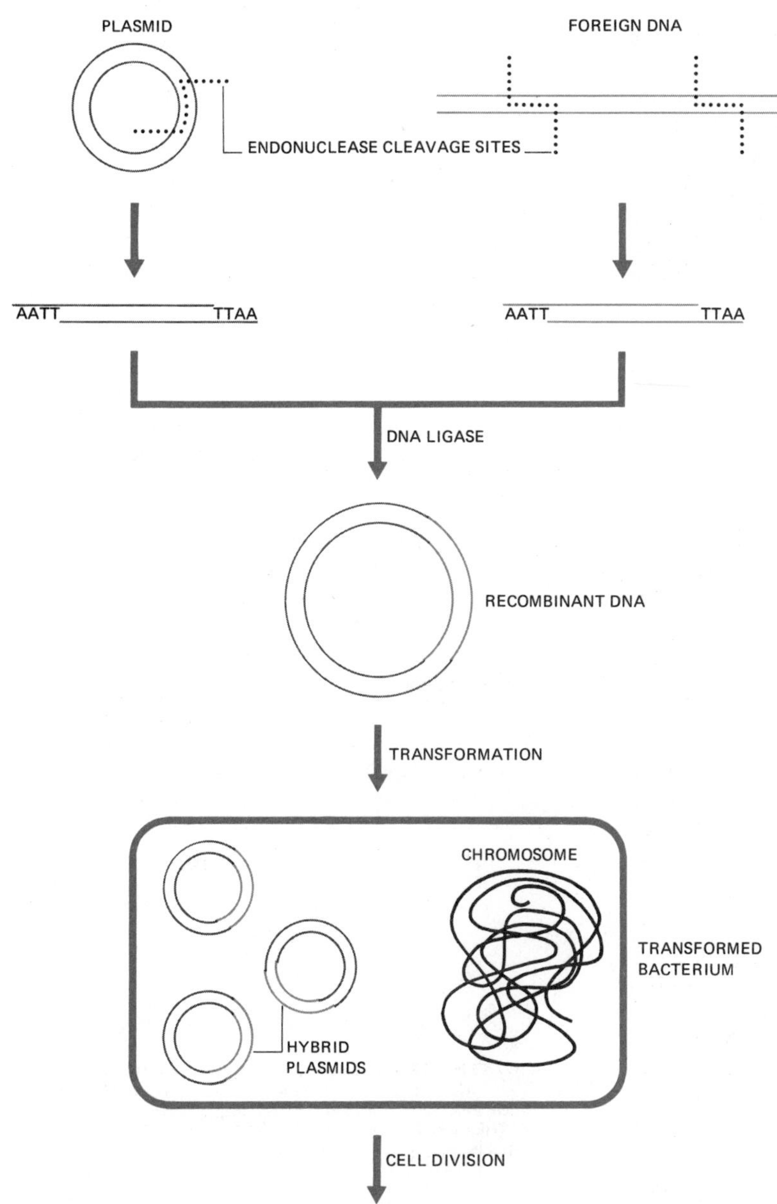

FIG. 18.15 Cloning recombinant DNA using a bacterial plasmid. Bacterial plasmids are small, circular pieces of DNA that are present in the cell in one or more copies and replicate independently of the bacterial chromosome. (They often carry genes for antibiotic resistance.) The plasmids are cleaved by an endonuclease that produces staggered cuts so that single-stranded "sticky" ends are formed at each end. The same enzyme is used to prepare a sample of "foreign" DNA, e.g. from some eukaryote. The two pieces of DNA are then joined with the aid of the complementary bases of their sticky ends and an enzyme called DNA ligase. The hybrid plasmids are then reintroduced into bacterial cells. Individual cells, each with its piece(s) of recombinant DNA, are easily grown into clones containing any desired number of cells. In some cases, the foreign DNA in the plasmid may be transcribed and translated. In all cases, enormous numbers of the gene can be produced and used in biochemical procedures such as sequence analysis. Note, however, that it is not possible to predict, in advance of testing the clone, which particular fragment of the foreign DNA became incorporated in the hybrid plasmid.

whole cells, similar results have been obtained by simply introducing a foreign nucleus into the recipient cell (shades of Boveri!—see Section 5.3). In fact, mouse cells exposed to *isolated* human *chromosomes* have been demonstrated to acquire, transcribe, and express a human gene. There is even some evidence that pure DNA preparations from one species of mammal have been taken up by the cells of another species and successfully replicated, transcribed, and translated. If this is so, this would constitute a transformation entirely analogous to the epoch-making transformation that Avery and his coworkers accomplished with pneumococci in 1943.

While some of the approaches described above may turn out to be impractical (and harmless), the time seems near when we shall be able to manipulate genetic material in ways that a decade or two ago would have been considered impossible. The potentialities that these techniques possess raise serious ethical questions. Technological innovations, the varied uses to which they are put, and the overall impact of them on society have often arrived with little advance notice and opportunity for thorough evaluation. The vistas which recent developments in genetic engineering open up suggest that we would do well to begin now to come to grips with the ethical and legal issues involved rather than wait until the technology is fully developed.

18.9 THE ACTION OF THE TOTAL GENOTYPE

Mendel knew that an understanding of the rules of inheritance could come only from the study of simple, clear-cut traits. Much of the later work on genetics has necessarily followed this tradition. Most of the information discussed in this chapter has come from the study of simplified, and thus more manageable, genetic systems (e.g. the RNA virus MS2). However, clear-cut traits, like hemoglobin type or the presence of a single functioning enzyme, represent only a small part of the story of how genes guide the development and functioning of an organism. The total genetic library of an organism is called its **genome.** As yet, genetics has made only a beginning at relating the total organism to its genome. Nevertheless, three important principles already stand out.

1. Multiple effects of single genes. A single gene may have many effects. Although a single gene produces a single polypeptide (usually), the activity of that polypeptide (or the protein of which it is a part) in different parts of the organism may lead to different effects. Furthermore, effects produced in one part of the organism, as a direct result of gene action, may in turn upset natural balances in other parts of the organism. A whole series of functional and structural changes may thus follow indirectly from some primary gene action. The inheritance of the trait called "frizzling" in chickens illustrates this. Chickens homozygous for the trait have defective feathers which break easily and provide little heat insulation for the bird. Presumably, the mutant alleles produce a protein which does not permit proper feather development. The poor insulating qualities of frizzled feathers lead to increased heat loss from the bird. This, in turn, triggers a whole sequence of homeostatic responses resulting in marked structural and functional changes (Fig. 18.16). The variety of symptoms found in many of the human hereditary

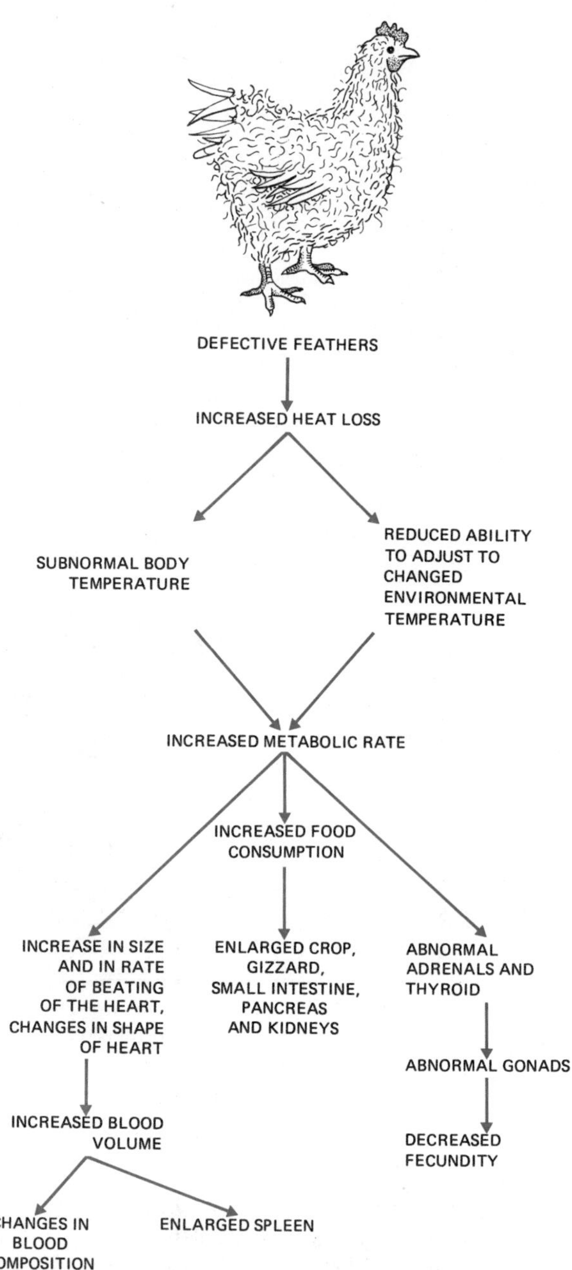

FIG. 18.16 Multiple effects of the gene for "frizzled" feathers in chickens. A whole series of secondary effects stem from the primary effect of the gene. (Adapted, with permission, from W. Landauer's "Temperature and Evolution," *Biological Symposia,* Vol. 6, The Jaques Cattell Press, Inc.)

diseases, for example, the pale pigmentation and seriously lowered I.Q. in phenylketonurics, also illustrates this principle. The creation of multiple effects by a single gene is called **pleiotropy.**

2. The interaction of genes. A second rule is that the expression of any trait is influenced by a large number of genes. One gene may result in a pea flower being red rather than white, but many other genes are necessary to produce the structure and arrangement of the petals. Going still further, no traits for flower color could be expressed without a coordinated activity of many genes guiding the development of functional roots, stems, and leaves.

Even such a simple trait as the ability to metabolize a small molecule like phenylalanine (Fig. 18.3) requires the action of a number of enzymes. Each of these is constructed from one or more polypeptide chains which, in turn, are each synthesized by a specific gene. Studies in a number of microorganisms have shown that in some cases the genes responsible for the synthesis of *related* enzymes are located adjacent to one another on the chromosome. In *E. coli,* for example, the genes that produce the enzymes necessary for the organism to metabolize the disaccharide lactose are located adjacent to each other in one region of the organism's chromosome. At one end of this chain of genes is a special gene, the operator, which turns the others on and off as conditions within the cell dictate. The combination of an operator gene and the genes that it controls is called an **operon.** The close linkage of functionally related genes is surely of value to the organism. By grouping such genes in operons, the production of all the enzymes needed for a given metabolic activity can proceed simultaneously and in accurately coordinated amounts.

How many genes does it take to build an organism? No one knows exactly. One can measure the amount of DNA found in a cell of the organism (and it *generally* increases with the complexity of the organism) and from this determine the number of base pairs present. A typical mammalian cell contains about 5×10^9 base pairs in a *haploid* set of chromosomes. This is equivalent to a total length of about 1.7 meters. This is over 1000 times the amount found in *E. coli.* But do mammals have over 1000 times as many genes as *E. coli?* Probably not.

Let us assume that, on average, each gene codes for a polypeptide of 500 amino acids. This would require 1500 base pairs. Dividing 5×10^9 by 1500 gives us a value of 3.3×10^6 different genes. But such an estimate rests on the additional assumption that all the DNA in the cell is genetically active. It is now quite apparent that such is *not* the case. The cells of organisms other than bacteria contain varying, but always large, amounts of *repetitive* DNA, that is identical—or at least similar—sequences of nucleotides repeated over and over. Some of this repetitive DNA *is* genetically active but represents multiple copies of a single gene (e.g. histone genes). But much of it does not appear to be genetically active. Its function is still a matter of speculation, but it may play important roles in the packaging of genes and in turning them on and off. It appears increasingly likely, therefore, that a very large portion of the genome in eukaryotes is never expressed as a protein product. Perhaps no more than 25% of the nuclear DNA gets transcribed and perhaps only 10% of the transcript formed in the nucleus ever gets translated in the cytoplasm. In our example, then, we are left with a rough estimate of close to 100,000 genes.

With *E. coli,* the situation is clearer. *E. coli* has no repetitive DNA and, except for the genes for *r*RNA, no multiple gene copies. With about 1 mm of DNA, *E. coli* probably contains something of the order of 2000 genes. As for *Drosophila,* the figure appears to be intermediate: some 5000 genes (a figure, interestingly enough, that approximates the total number of bands seen in the giant chromosomes).

3. Environmental influences on gene action. Our *third* rule is that the action of genes is dependent also upon the environment. The ability to manufacture chlorophyll is under gene control, but plants cannot do the job unless they are exposed to light (look back at Fig. 9.11). If victims of the hereditary defect phenylketonuria (PKU) are identified very early, it is possible to prevent the mutant genes from expressing themselves. This is accomplished by reducing the amount of the amino acid phenylalanine in the diet of the patient.

Many farmers today raise hybrid corn varieties. The seeds, produced through extensive breeding programs, are of a very uniform and desirable genotype. The plants one farmer raises from the seeds may, however, vary greatly in phenotype from those raised by another farmer. Soil, temperature, water supply, and fertilizer are just a few of the many environmental influences that affect the ultimate phenotype.

Thus the genes an individual organism inherits determine the potentialities that may be achieved in the course of the entire growth and development of that organism. The environment establishes to what degree these potentialities will actually be achieved.

EXERCISES AND PROBLEMS

1. Distinguish between multiple factors and multiple alleles.
2. Neglecting crossing over, how many different kinds of ascospores can *Neurospora crassa* form by random assortment of its chromosomes? The haploid number is 7.
3. Besides DNA and RNA, what other important biological substances contain the purine adenine?
4. Is it correct to say that in the ascus produced by a heterozygous (*Aa*) *Neurospora* zygote, four spores will contain *A* and four spores will contain *a*? Explain.
5. Why can a definite prediction be made in Question 4 when you could not make a definite prediction in Question 4 at the end of Chapter 15?
6. What sequence of bases in messenger RNA will be coded by the following triplets in the DNA molecule: GCT, GAT, CCA, AAA, AGT?
7. Why are haploid organisms especially useful in genetics studies?
8. Why are most mutations recessive?
9. What inorganic ions would you expect to add to sucrose and biotin to make a minimal medium for growing unmutated *Neurospora*? Why?
10. Minimal medium for *Neurospora* requires the presence of biotin while minimal medium for *E. coli* does not. Does this mean that biotin does not participate in the metabolism of *E. coli* cells? Explain.
11. Define a gene. Is this the same way you defined it at the end of Chapters 15 and 16?
12. Using your knowledge of meiosis, predict what arrangements of *A* spores and *a* spores could occur in the ascus produced by a strain of *Neurospora* heterozygous for these alleles.
13. The first gene of the bacteriophage MS2 ends with a UAG termination signal, shown as nucleotides 3, 4, and 5 in Fig. 18.14. The next gene starts with AUG at positions 29, 30, and 31. Is the number of nucleotides in the intergenic region divisible by 3?
14. Some MS2 phages occasionally misread the UAG terminator (positions 3, 4, 5—see Fig. 18.14) and instead of stopping translation of the first gene, they insert the amino acid serine (Ser) at that point and *continue* with translation. Keeping in mind your answer to the previous question, what is the next possible "stop" signal?
15. Using the table of codon assignments (Fig. 18.13), give the sequence of amino acids that would occur in the aberrant, extra portion of the protein synthesized as described in the previous question. (The existence of such a protein has been verified.)

REFERENCES

1. BEADLE, G. W., "The Genes of Men and Molds," *Scientific American,* Offprint No. 1, September, 1948. The techniques of using *Neurospora* as a tool for studying gene action are described in detail.
2. PETERS, J. A., ed., *Classic Papers in Genetics,* Prentice-Hall, Englewood Cliffs, N.J., 1959. Includes the original papers on the discoveries made by H. J. Muller, Beadle and Tatum, Avery and his coworkers, and Watson and Crick. Other papers on the nature and action of genes are also included.
3. INGRAM, V. M., "How Do Genes Act?" *Scientific American,* Offprint No. 104, January, 1958. Describes the evidence from sickle-cell anemia.
4. ENGELMAN, D. M., and P. B. MOORE, "Neutron-Scattering Studies of the Ribosome," *Scientific American,* October, 1976. Both subunits of the *E. coli* ribosome have been completely dissected into their constituent macromolecules.
5. MILLER, O. L., JR., "The Visualization of Genes in Action," *Scientific American,* Offprint No. 1267, March, 1973.
6. TRAVERS, A. A., *Transcription of DNA,* Oxford Biology Readers, No. 75, Oxford University Press, Oxford, 1974.
7. LANE, C., "Rabbit Hemoglobin from Frog Eggs," *Scientific American,* Offprint No. 1343, August, 1976. How *Xenopus* eggs translate foreign messenger RNA injected into them.
8. HOLLEY, R. W., "The Nucleotide Sequence of a Nucleic Acid," *Scientific American,* Offprint No. 1033, February, 1966. Describes the steps by which the nucleotide sequence of alanine transfer RNA was worked out.
9. NIRENBERG, M. W., "The Genetic Code II," *Scientific American,* Offprint No. 153, March, 1963. Explains

how synthetic messenger RNA is used to determine the "letters" in the genetic code.

10. CRICK, F. H. C., "The Genetic Code: III," *Scientific American,* Offprint No. 1052, October, 1966. The author, one of the codiscoverers of the structure of DNA, summarizes our current knowledge of the nature of the DNA code and how it is translated.

11. GORINI, L., "Antibiotics and the Genetic Code," *Scientific American,* Offprint No. 1041, April, 1966. Summarizes the mechanism of protein synthesis and shows how certain antibiotics interfere with the process by causing errors in the translation of the genetic code.

12. BROWN, D. D., "The Isolation of Genes," *Scientific American,* Offprint No. 1278, August, 1973.

13. COHEN, S. N., "The Manipulation of Genes," *Scientific American,* Offprint No. 1324, July, 1975. One of the pioneers in the cloning of recombinant DNA describes the procedures involved.

14. ADELBERG, E. A., *Bacterial Plasmids,* Addison-Wesley Modules in Biology, No. 8, Addison-Wesley, Reading, Mass., 1973.

15. BRITTEN, R. J., and D. E. KOHNE, "Repeated Segments of DNA," *Scientific American,* Offprint No. 1173, April, 1970.

16. WATSON, J. D., *Molecular Biology of the Gene,* 3rd ed., W. A. Benjamin, Inc., 1976. Already a classic, this superb book was written by the codiscoverer of the structure of DNA. It includes discussions of the molecular genetics of *E. coli* and both DNA and RNA bacteriophages.

17. LEVINE, R. P., *Genetics,* 2nd ed., Holt, Rinehart and Winston, New York, 1968. An excellent introduction to both classical and molecular genetics.

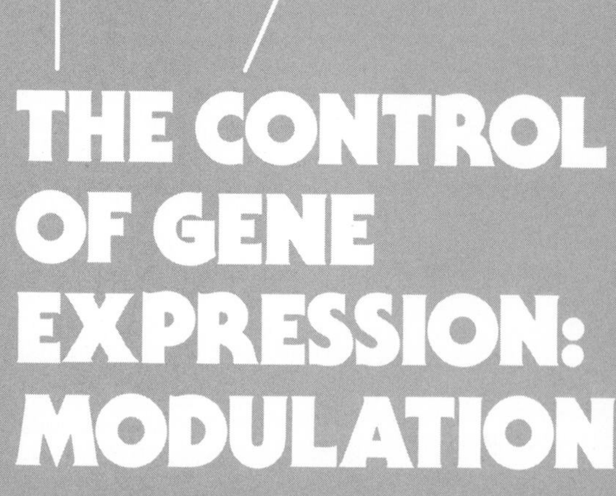

19 THE CONTROL OF GENE EXPRESSION: MODULATION

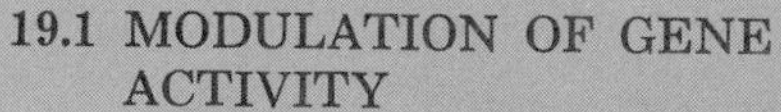

19.1 MODULATION OF GENE ACTIVITY

All cells have the ability to respond to certain signals that reach them from their environment. Consider our now-familiar friend *Escherichia coli.* Within its tiny cell, *E. coli* contains all the genetic information it needs to metabolize, grow, and reproduce. As we saw in Chapter 14, it can synthesize everything it needs from glucose and a number of inorganic ions. We would expect it to need a large number of enzymes to accomplish so many syntheses, and there may well be 600–800 present in the cell under these conditions. Most of these enzymes, such as the enzymes of cellular respiration, are present at all times. Others, however, are produced only when they are needed by the cell. For example, if the amino acid arginine is added to the culture, the bacteria soon stop producing the nine enzymes previously needed to synthesize arginine from intermediates produced in the respiration of glucose. In this case, then, the presence of the *products* of enzyme action *represses* enzyme synthesis.

Conversely, adding a new *substrate* to the culture medium may *induce* the formation of new enzymes capable of metabolizing that substrate. If we take a culture of *E. coli* cells that are feeding actively on glucose and transfer them to a medium containing lactose, a revealing sequence of events occurs. At first, the cells are quiescent: they do not metabolize the lactose, their other metabolic activities diminish in intensity, and cell division ceases. Soon, however, the culture begins growing actively again. The lactose begins to be rapidly consumed. What has happened? During the quiescent interval, the cells have begun to produce three enzymes that they had not been producing before.

One of these enzymes, called a **permease,** transports the lactose across the cell membrane from the medium into the interior of the cell. A second enzyme, called **beta-galactosidase,** hydrolyzes the lactose (a disaccharide) into glucose and galactose. Once induced by the presence of lactose, the quantity of beta-galactosidase in the cell rises rapidly from virtually none to approximately 3% of the weight of the cell. (The significance of the third enzyme in the metabolism of the cell is still uncertain.)

The capacity to respond appropriately to the presence of lactose was always there. The genes for the three induced enzymes are part of the genome of each *E. coli* cell. But until lactose appeared in the culture medium, these genes were not expressed. We shall use the term **modulation** to describe the turning on and off of genetic information within a cell in response to changes in its environment.

Each of these enzymes is encoded by a separate gene. Genetic analysis reveals that these three so-called **structural genes,** which clearly are related in function, are also closely linked on the bacterial chromosome. What causes them to begin the work of producing their enzymes? Or rather, what *keeps* them from producing the enzymes before the need arises?

Another gene, called a **regulator gene,** is responsible for this. The function of the regulator gene is to produce a protein which prevents the structural genes for beta-galactosidase, etc., from expressing themselves. The regulator protein, called the **repressor,** does not inactivate the structural genes directly; rather, it represses a small gene (only 27 base pairs long) immediately adjacent to them, called the **operator** (Fig. 19.1). The combination of the operator and its associated structural genes is called the **operon.** Perhaps the role of the operator, when it is not repressed by the regulator, is to separate the two strands of the stretch of DNA incorporating the three structural genes so that one strand can be transcribed by RNA polymerase into a *single* molecule of messenger RNA. Ribosomes moving down this molecule would *translate* the messages into the polypeptides from which the three enzymes are constructed. (You can see why punctuation codons—UAA, UAG, or UGA—would thus be needed to terminate polypeptide synthesis between the portions of the messenger RNA coding for each of the three enzymes.)

What, then, determines whether the repressor substance produced by the regulator gene is active or not? In the case described, it is the presence of lactose itself that prevents the repressor from acting on the operator. Lactose unites with the repressor, a protein, and as a result, the shape of the protein may be changed enough so that it can no longer combine with and thus inactivate the operator. The synthesis of beta-galactosidase and the other two enzymes may therefore begin.

The mechanism described here was proposed by the French scientists François Jacob and Jacques Monod to explain the genetics of enzyme induction. For this work they shared in a Nobel Prize in 1965.

As we saw above, the action of some genes is repressible. The nine genes involved in the production of the enzymes for arginine synthesis turn out to be clustered in six separate operons. However, a single regulator gene responds to the presence of arginine and represses all six operons. Presumably the regulator gene produces a repressor which blocks all six operators when the repressor has combined with ar-

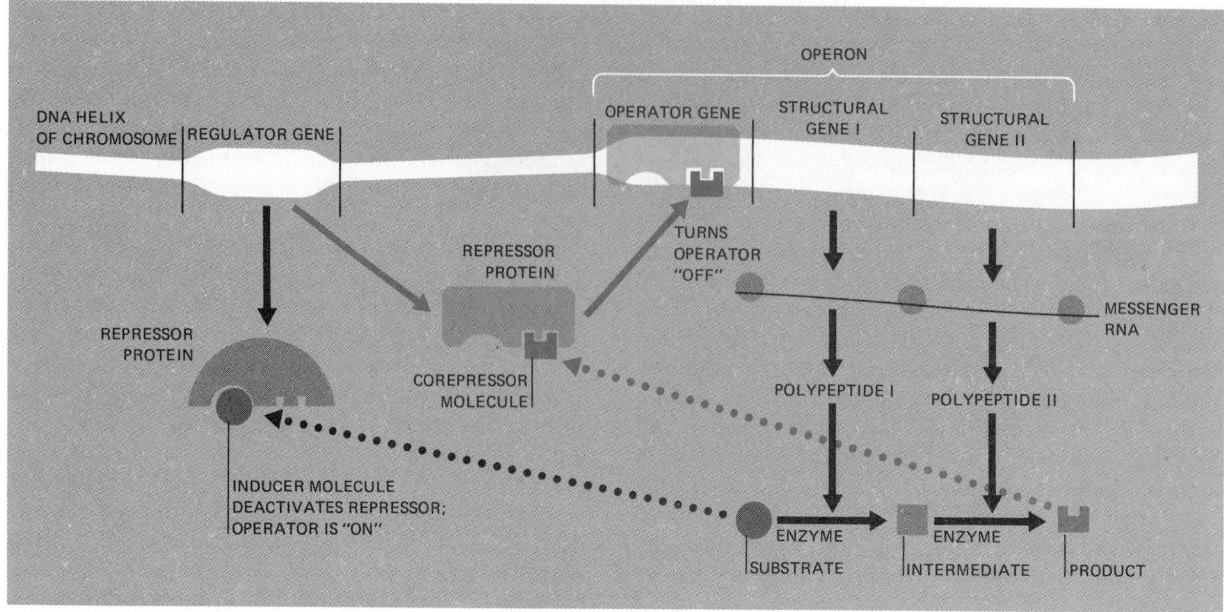

FIG. 19.1 The Jacob-Monod hypothesis of gene control. When an inducer molecule is present, the repressor is unable to attach to the operator gene and polypeptide synthesis takes place. When a corepressor molecule is present, the operon is shut down (color).

ginine, its **corepressor** (Fig. 19.1). The usefulness to the cell of such a mechanism is quite clear. The presence of an essential metabolite turns off the synthesis of the enzymes for its own manufacture and thus stops unnecessary protein synthesis in the cell.

As its name suggests, the repressor is essentially a negative control mechanism. Recently, it has become evident that some gene transcription in bacteria is also under positive control. We have seen that an invading bacteriophage takes over much of the metabolic machinery of the host cell and uses this machinery to make many copies of itself. When *E. coli* begins transcribing and translating the genes of one of its bacteriophages (called T4), one of the first phage proteins synthesized combines with the bacterial RNA polymerase. In so doing, it reduces the efficiency with which the RNA polymerase can transcribe host genes and greatly *increases* the efficiency with which it transcribes other *bacteriophage* genes.

Many bacteria (but not *E. coli*) form resting spores when environmental conditions begin to deteriorate. This involves the expression of genes not hitherto active. Interestingly enough, the molecular structure of the bacterium's RNA polymerase changes at this time, with the result that it switches its transcription from the genes associated with vegetative growth to those responsible for spore formation.

Here, then, are two mechanisms—one negative, the other positive—by which the expression of genetic information in the bacterial cell is modulated. Each works by regulating (repressing or stimulating) the *selective transcription* of genes into messenger RNA molecules.

19.2 MODULATION IN EUKARYOTES

What about eukaryotes? Do they use systems of transcriptional control similar to those in prokaryotes? Certainly eukaryotic cells display modulation. Take, for example, the human oocyte. Several months before she is born, a female fetus sets aside in her ovaries the precursor cells of all the eggs that she will ever form. For a dozen years or so these cells, called oocytes, remain small and quiescent, arrested in the middle of the first meiotic division. Then with the onset of puberty, the oocytes—usually one a month—begin a phase of very rapid growth and development. This includes the completion of the first meiotic division and the start of the second. These dramatic changes are triggered by rising levels of hormones in the blood. Without the stimulus provided by the hormones, the tiny egg precursors would never mature.

Here, then, is an example in a eukaryotic cell of modulation: the new expression of preexisting genetic information in response to changing environmental signals. It tells us that eukaryotic cells, like *E. coli,* contain more genetic information than they use at any one time. By what mechanisms do they achieve such selective gene activity?

Although transcriptional controls undoubtedly exist in eukaryotes, there is little evidence as yet for the existence of operator and regulator genes. Operator-like genes, i.e. genes that affect the transcription of other genes, have to date been identified in only two eukaryotes, both of them fungi.

Perhaps it is foolish to expect to find in eukaryotes the type of transcriptional controls found in prokaryotes. After all, these two sharply contrasting groups of organisms differ significantly in the manner in which they express genetic information. Eukaryotic cells, as the name tells us, have a nucleus and it is here that gene *transcription* takes place. Protein synthesis, on the other hand, i.e. *translation,* occurs out in the cytoplasm. In prokaryotes, transcription and translation occur simultaneously (look back at Fig. 18.12). In eukaryotes, the production of *m*RNA by transcription is followed by the export of that messenger into the cytoplasm before translation can begin.

The messengers of prokaryotes tend to be very unstable. After serving as the templates for a number of rounds of translation, they are degraded. Unless they are replaced by new transcripts, synthesis of that protein will soon halt. Thus transcriptional controls are quickly reflected in the level of translation of a particular protein. On the other hand, the messengers in eukaryotes *may* be quite stable. In the rapidly developing oocyte, for example, maternal messengers are formed that will not be translated for several days after their formation and then only if fertilization has occurred.

The modulation of the developing oocyte occurs because of its exposure to a rising level of certain sex hormones. This example of modulation by hormonal signals is but one of many. Both in animals and in plants, hormones serve as powerful modulators of gene expression. And, it now appears, at least some of the effects of hormones are brought about by their influence on gene transcription.

One of the most thoroughly analyzed systems of hormonal modulation is that involving the female sex hormones, the estrogens and progesterone. Both the estrogens and progesterone are steroids (see Section 4.2). One of the tissues most strongly influenced by estrogens is the inner lining of the uterus, called the *endometrium.* Upon exposure to estrogens, it becomes thicker (in preparation for a possible pregnancy). The cells of the endometrium become very active, first in RNA synthesis, then in protein synthesis. This response, which takes place over a period of days, undoubtedly involves new or at least increased gene activity. The stimulatory effect of estrogens on these cells is inhibited by an antibiotic called **actinomycin D.**

Actinomycin D exerts its effect on cells in a remarkably precise way. It binds to the DNA in the nucleus and prevents the two strands of DNA from separating. Thus the DNA molecule cannot serve as the template for the synthesis of either additional DNA or RNA molecules. Without fresh supplies of RNA, the synthesis of proteins by the cell soon comes to a halt. The inhibition by actinomycin D of the response of endometrial cells to estrogens thus implicates estrogens as modulators of gene transcription.

This conclusion is supported by finding that estrogens accumulate within the nuclei of endometrial cells. In fact, within 15 minutes of giving radioactively labelled estrogens (or progesterone—Fig. 19.2) to a laboratory animal such as a guinea pig or a rat, the radioactivity appears in the nuclei of the endometrial cells. This occurs much sooner than any detectable increase in RNA or protein synthesis by these cells. Furthermore, the accumulation of estrogens does not occur in all types of cells. Liver and blood cells, for example, do not accumulate the hormones in their nuclei. Only certain "target" cells, e.g. endometrial cells, accumulate the hormones and only these cells have their activities altered by the hormones.

What is the mechanism at work? Estrogens, like all steroids, are small hydrophobic molecules. Thus they diffuse freely through cell membranes—all cell membranes. We would expect every cell in the body to be exposed to them as they are transported through the blood. And, in fact, estrogens do diffuse freely into all cells. However, they are just as free to diffuse out again *except* from the cells of the "target" organs. In the target (e.g. endometrial) cells, they become bound tightly to a cytoplasmic protein, called the receptor. The complex of receptor and hormone

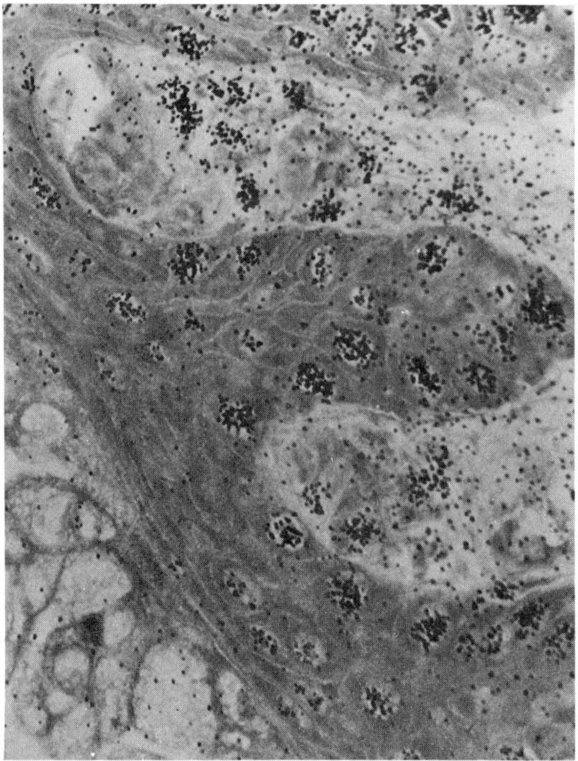

FIG. 19.2 Autoradiograph of endometrial cells of a guinea pig taken from the animal 15 min after an injection of radioactive progesterone. The radioactivity has concentrated within the nuclei of the endometrial cells (as shown by the dark grains superimposed on the images of the nuclei). The same effect is seen when radioactive estrogens are administered. Nontarget cells show no such accumulation of the female sex hormones. (Courtesy Madhabananda Sar and Walter E. Stumpf.)

then migrates into the nucleus (Fig. 19.3). There the receptor-hormone complex binds to the *chromatin* of the nucleus.

Its arrival triggers a burst of RNA synthesis (unless, of course, this effect has been experimentally blocked by actinomycin D). Analysis of the RNA reveals that it includes the messengers for particular proteins that will be synthesized as a result of hormonal stimulation. If, to take another well-studied example, one exposes the cells of the chick oviduct to the steroid **progesterone,** the cells soon begin to synthesize large amounts of the protein ovalbumin (the major constituent of egg white). First, however, the specific *m*RNA molecules coding for ovalbumin make their appearance in the cell. Over a period of a few hours, the number of ovalbumin messengers rises from virtually none to over 10,000 per cell. In this case, a progesterone-receptor complex has turned on the ovalbumin gene, leading to its transcription into *m*RNA.

How does progesterone induce the activity of certain genes (e.g. ovalbumin) and not others? In other words, by what mechanism does it trigger differential gene activity? Only a tentative answer can be given at this time. The progesterone-receptor complex binds to a particular subset of nonhistone proteins found in the target (oviduct) cells.

You may recall that chromatin contains (in addition to DNA) five kinds of histones (Fig. 19.4) and many kinds of nonhistone proteins. Histones are very similar from one tissue to another and even from species to species. In fact, one of the histones found in calf cells differs from a similar histone found in *pea* cells at only two positions (amino acids) out of a total of 102 in the molecule! In the face of such molecular uniformity, it is reassuring that chromatin also contains the far more heterogeneous nonhistone proteins (Fig. 19.4). Furthermore, the content of nonhistone proteins does differ from one cell to another. It is to one or more nonhistone proteins—

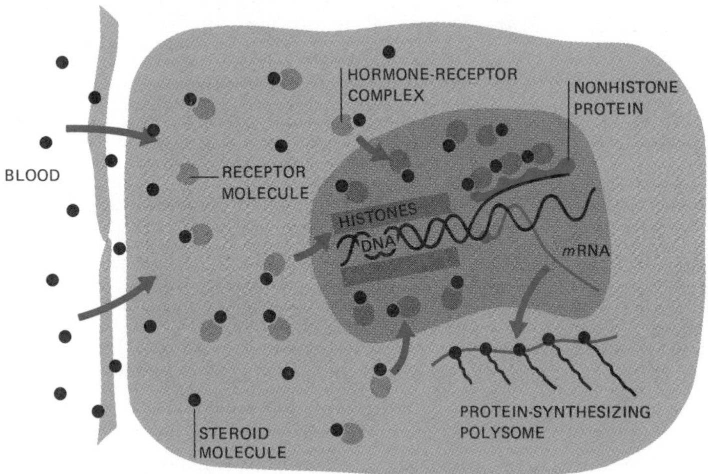

FIG. 19.3 Proposed mechanism of action of steroid hormones. Molecules of the hormone diffuse into the cell where they are bound by specific receptor molecules in the cytoplasm. The hormone-receptor complexes move into the nucleus where they bind to chromatin, probably by means of particular nonhistone proteins. This is followed by transcription, that is, messenger RNA synthesis. Translation of the *m*RNA molecules (which takes place in the cytoplasm) yields proteins that accomplish the function(s) of that particular hormone-activated target cell.

FIG. 19.4 Electrophoretograms of histone and nonhistone proteins isolated from chromatin. There are five kinds of histones (left) and these vary little from one cell type to another or even one species to another. There are many kinds of nonhistone proteins (right) ranging in weight from 10,000 to 150,000 daltons. The kinds of nonhistone proteins vary from one cell type to another and from one species to another. The greater diversity of nonhistone proteins makes them better candidates than histones to carry out specific regulatory functions within cells. (Courtesy Gary S. Stein and Janet Swinehart Stein, University of Florida.)

unique to the target cells of the oviduct—that the progesterone-receptor complex binds. This having been done, sites become available for the attachment of RNA polymerase, and the transcription of the ovalbumin and other appropriate genes begins.

19.3 THE GIANT CHROMOSOMES AND DIFFERENTIAL GENE ACTIVITY

Some of the cells in the larvae of certain insects contain giant chromosomes. In Chapter 16, we noted the presence of giant chromosomes in the cells of the salivary glands of the fruit fly, *Drosophila*. These giant chromosomes are interphase chromosomes and are much more elongated than metaphase chromosomes. The reason they are visible during interphase, when ordinary chromosomes are not, is that they are the product of repeated chromosome duplications without accompanying cell division. The duplicates of both the maternal and paternal homologues lie side by side in perfect register throughout their length. The result is like a multistranded cable. In *Drosophila* salivary glands, each giant chromosome is the product of about nine cycles of replication. Thus there are over 1000 strands in the cable. The giant chromosomes of some insects have as many as 16,000 strands.

The giant chromosomes consist of a linear sequence of alternating dark bands and light interbands (Fig. 19.5). Most of the DNA is in the bands. For several cases, it has been possible to assign a specific gene locus to a specific band. It may turn out, in fact, that each band contains only a single gene locus. There is a total of over 5000 bands in the giant chromosomes of *Drosophila* and these *may* represent all the genes the organism has. In any case, the sequence of genes and their relative map distances (see Section 16.6) approximates quite closely the relative position of those bands that have been identified as the sites of particular gene loci.

The diameter of a giant chromosome varies from place to place. At some locations, called "puffs," the chromosome may appear quite swollen (Fig. 19.6). By supplying the cell with radioactive uridine (U—the nucleotide unique to RNA), it can be shown that the puffs are the sites of intense RNA synthesis (Fig. 19.7). So if the bands represent genes, the puffed bands seem to represent genes that are being transcribed.

The giant chromosomes in the cells of one tissue (e.g. the *Drosophila* salivary gland) have the same general appearance as those in another (e.g. the

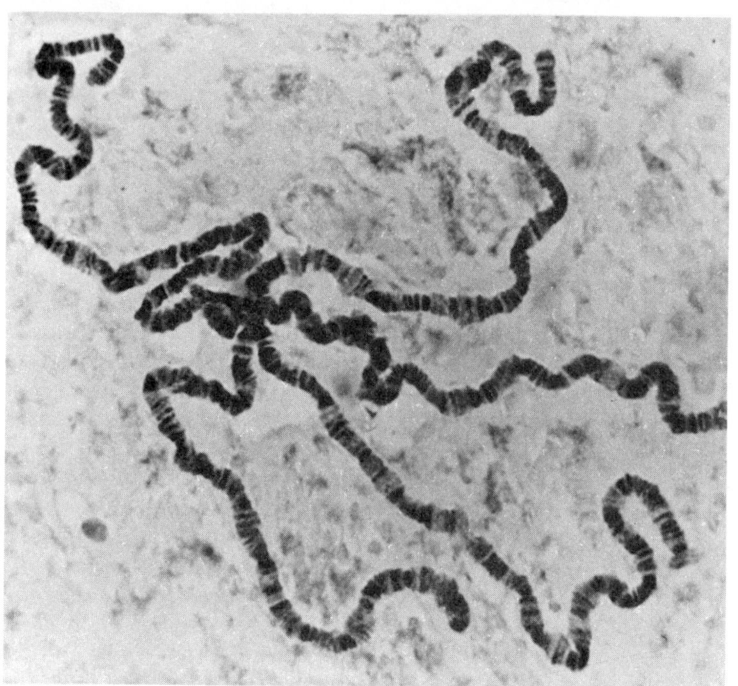

FIG. 19.5 Giant chromosomes from the salivary gland of the fruit fly *Drosophila melanogaster*. The dark bands (there are over 5000 of them) are rich in DNA, and each appears to be the locus of at least one gene. (Courtesy B. P. Kaufmann.)

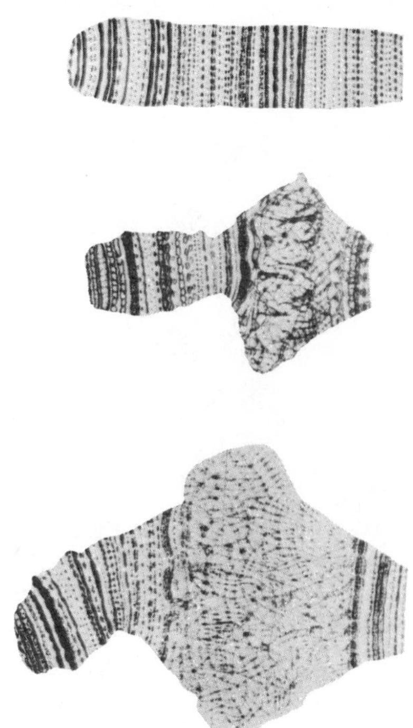

FIG. 19.6 Giant chromosome, unpuffed (top) and with puffs (center and bottom). (Adapted from Breuer and Pavan, *Chromosoma*, **7**: 371–386, 1955.)

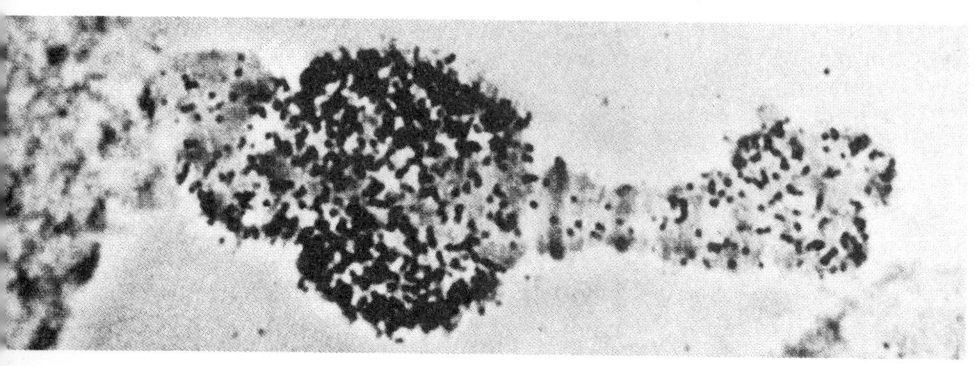

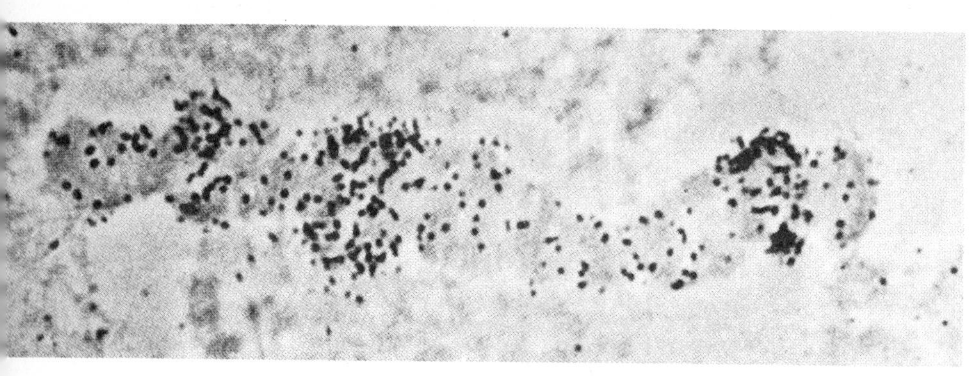

FIG. 19.7 Top: Autoradiograph of a puff in chromosome IV of *Chironomus tentans* after administration of ecdysone and radioactive uridine. The grains clustered over the puff show that it is the site of intense RNA synthesis. Bottom: When actinomycin D is present as well, puffing is inhibited and little RNA synthesis occurs. (Courtesy Claus Pelling, Max-Planck Institute for Biology, Tübingen.)

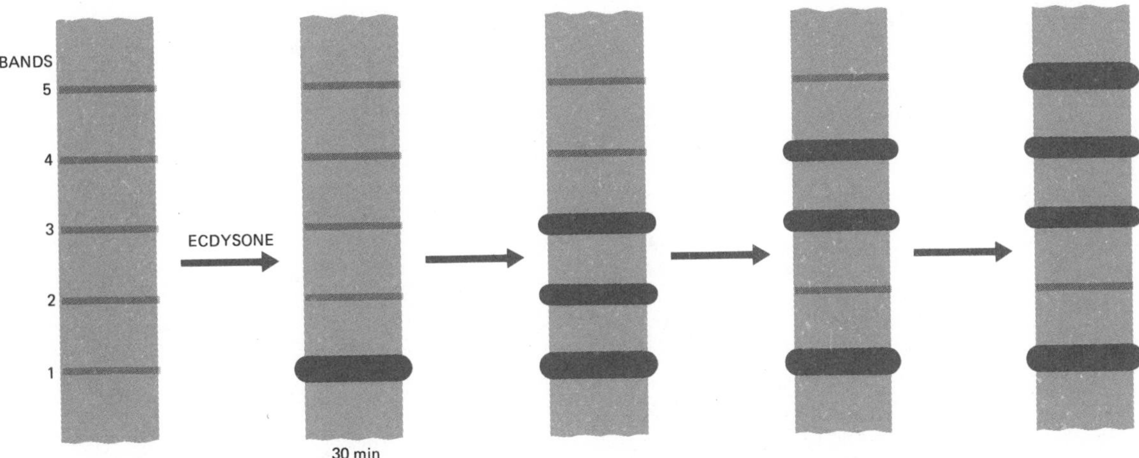

FIG. 19.8 Induction by ecdysone of a specific and sequential pattern of puffing. Perhaps the gene product(s) of certain puffs (e.g. at band 1) act to turn other bands on (and off, e.g. band 2 in the final two stages).

Drosophila Malphigian tubule). But the location of the *puffs* varies from one type of cell to another. This implies that different genes are active in different kinds of cells. And what could be more appropriate?

Furthermore, even in one kind of cell, the pattern of puffs varies from one time to another. For example, each time an insect larva prepares to molt, a definite and predictable sequence of puffing occurs in its chromosomes. Molting in insects is triggered by rising levels of a hormone called **ecdysone.** Ecdysone (like progesterone and the estrogens) is a steroid. If ecdysone is injected into a larva of the tiny midge *Chironomus* shortly *after* it has completed a molt (a time when its own level of ecdysone is very low), the normal premolt sequence of puffing begins again. The first puff appears within 30 minutes and is followed inexorably over the next few hours by the others (Fig. 19.8). It looks as though the presence of ecdysone has turned on a series of genes. And, in fact, an *m*RNA has been extracted from these ecdysone-treated cells that serves as the template for the synthesis of an enzyme not found in the cells of untreated larvae.

The production of puffs following the injection of ecdysone is blocked by the simultaneous administration of actinomycin D. This constitutes additional evidence that the hormone acts by unlocking genetic information stored within the cell.

The evidence would be more direct if the presence of a particular puff could be associated with the presence of a single, specific gene product. Among the species of *Chironomus* are two whose salivary-gland cells differ in a clear-cut way. One (*C. pallidivittatus*) produces certain granules in some of the cells of its salivary glands, whereas the other (*C. tentans*) does not. Furthermore, *C. pallidivittatus* has one extra puff in the chromosomes of these cells that *C. tentans* lacks (Fig. 19.9). Crossbreeding has shown that both these features are inherited as single gene traits and, in fact, chromosome mapping (see Section 16.6) shows that the gene is located at the point on the chromosome where the puff that distinguishes one species from the other is located. Genes code for proteins, and we would expect *C. pallidivittatus* salivary-gland cells to contain a protein not found in *C. tentans*. Fractionation of the proteins produced by these cells reveals six shared by both species and one found in *C. pallidivittatus* only (Fig. 19.9). Here, then, is powerful evidence that a single gene locus produces an RNA-synthesizing puff from which are derived a unique

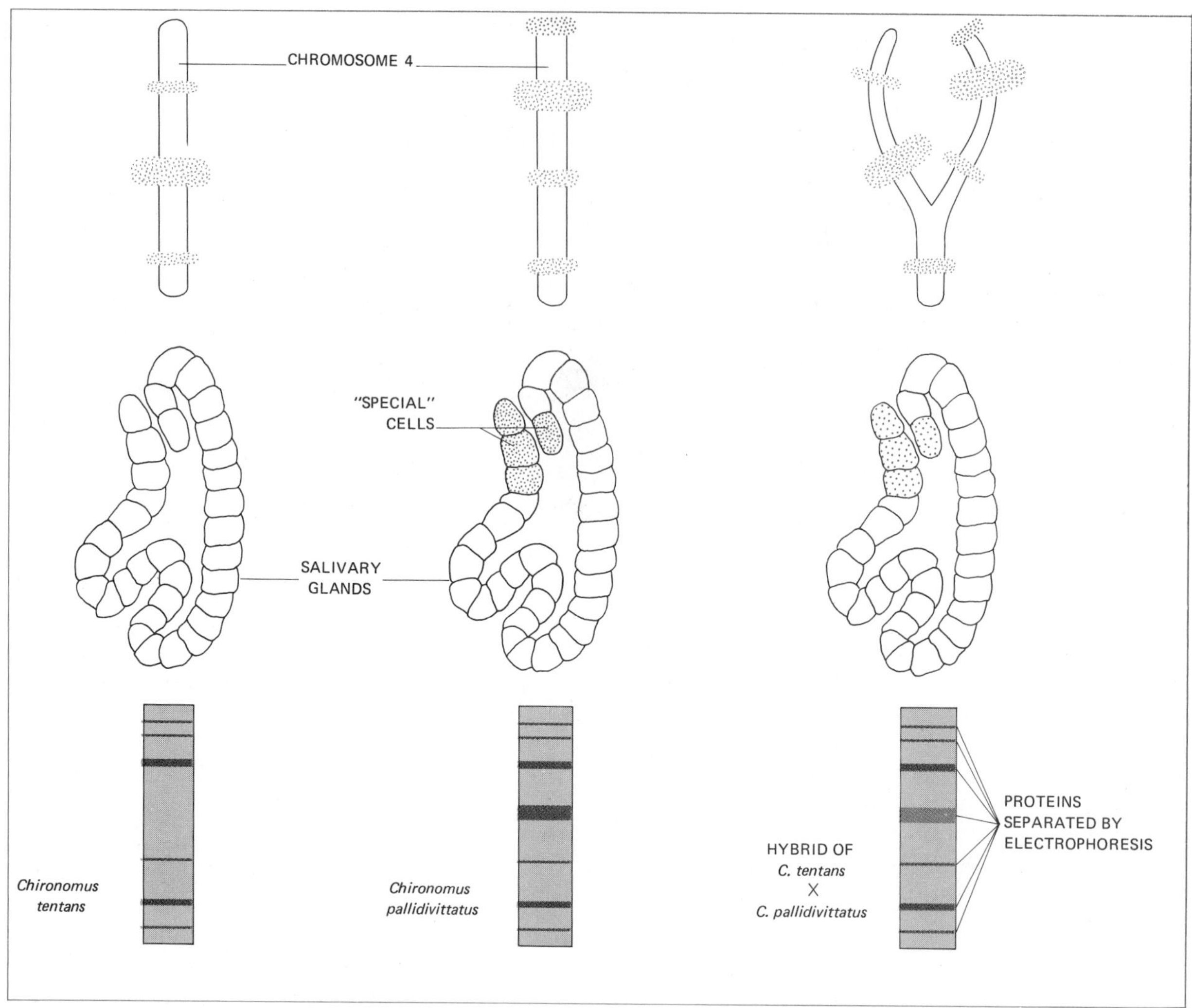

FIG. 19.9 Correlation between the presence of a particular puff (color), granule formation in the "special cells," and the synthesis of a particular protein in *Chironomus*. *C. pallidivittatus* produces a puff, granules, and a protein that *C. tentans* does not. A hybrid of the two produces a small puff and intermediate amounts of granules and the protein. The special puff in *C. pallidivittatus* (and the hybrid) is present only in the four "special cells."

protein product and a unique, visible feature (the granules) in the cell.

19.4 TRANSLATION CONTROLS

The mechanisms of differential gene activity that we have examined up to now have all involved regulation of the rate of gene *transcription*. In prokaryotes, this may be all there is to the story. And in eukaryotes, it may be the most important part of the story. But eukaryotes, in contrast to prokaryotes, are able to produce stable messengers, i.e. messengers that can remain in the cell for long periods. So the opportunity exists to control the rate of gene expression by controlling the rate of messenger *translation*. And, indeed, there is some evidence that modulation in eukaryotes can involve differential messenger translation in addition to differential gene transcription.

A preparation of messengers taken from rabbit red blood cell precursors contains about equal amounts of the messenger for alpha chain synthesis and that for beta chain synthesis (each messenger is transcribed from a different gene, you remember). When incubated in the test tube with all the appropriate materials, this mixture of messengers produces equal amounts of the two kinds of globin chains. If the same preparation is injected into the huge oocyte of a frog, these rabbit messengers are translated by the protein-synthesizing machinery of the frog cell, producing rabbit alpha and beta chains. But now the two kinds of chains are not produced in equal number: the ratio of beta to alpha chain synthesis is about 5:1. It seems that in the frog oocyte, the efficiency of beta *m*RNA translation is greater than that of alpha *m*RNA translation. However, if heme—needed for complete hemoglobin synthesis—is injected along with the messengers, equally balanced production of alpha and beta chains is restored. Thus the presence of heme appears to influence the rate of gene expression by affecting the rate of translation.

A similar situation has been found in the green alga *Chlorella*. The synthesis of one of the enzymes of the dark reactions of photosynthesis is regulated independently of the rate of synthesis of the messenger for that enzyme. Under certain circumstances, the cell contains large quantities of messenger but little or no translation of it occurs. Furthermore, upon adding glucose to cells in which translation is occurring, synthesis of the enzyme (which now has become superfluous) ceases almost immediately. This cessation of enzyme synthesis occurs much more rapidly than would be possible if glucose were simply turning off transcription. The most likely explanation is that glucose has, in a manner yet to be determined, blocked enzyme synthesis at the level of messenger translation.

19.5 SUMMARY

In this chapter we have examined mechanisms by which the expression of genes within a cell is modulated in response to signals reaching the cell from its environment. In multicellular eukaryotes, such an environmental signal may be the arrival at the cell of a hormone. Remember, though, that hormones are transported by blood or sap and thus they quickly reach every cell in the organism's body. But not every cell responds. Only the target cells are capable of responding to the signal. The endometrial cells of the mammalian uterus, for example, respond quickly to the arrival of estrogens, but are not stimulated by the arrival of the kidney hormone erythropoietin. On the other hand, the cells of the bone marrow respond vigorously to erythropoietin (by producing more red blood cells) but are probably not affected by the presence of estrogens in their ECF.

How, then, is it possible that certain genes (e.g. that for ovalbumin synthesis) can be expressed in some cells but not in others? Well, you may rightly say, the oviduct or endometrial cells contain the necessary receptor protein while the cells of the bone marrow do not. True enough. But then, what determines whether or not a given kind of cell possesses the cytoplasmic receptor needed to make that cell responsive to, i.e. a target cell for, estrogens? Surely it is the history of that cell—a history that led to the formation of an endometrial cell as opposed to a bone marrow cell. Thus it is the *pathway of differentiation* that was taken by that cell that ultimately determines which of its genes it will finally be able to express in response to modulating signals. So our search for mechanisms of selective gene activity is by no means over. We have yet to search for the gene control mechanisms which lead to a cell becoming differentiated in a particular way. We shall examine some tentative answers to this crucial question in Chapter 22.

EXERCISES AND PROBLEMS

1. Would you expect mutations in *regulator* genes to be inherited as dominants or recessives? Explain.
2. Would you expect mutations in *operator* genes to be inherited as dominants or recessives? Explain.

REFERENCES

1. PTASHNE, M., and W. GILBERT, "Genetic Repressors," *Scientific American,* Offprint No. 1179, June, 1970.
2. SOBELL, H. M., "How Actinomycin Binds to DNA," *Scientific American,* Offprint No. 1303, August, 1974.
3. O'MALLEY, B. W., and W. T. SCHRADER, "The Receptors of Steroid Hormones," *Scientific American,* Offprint No. 1334, February, 1976.
4. STEIN, G. S., JANET S. STEIN, and L. J. KLEINSMITH, "Chromosomal Proteins and Gene Regulation," *Scientific American,* Offprint No. 1315, February, 1975. Examines the role of histones in keeping genes turned off and of nonhistone proteins in turning them on.
5. MARKERT, C. L., and H. URSPRUNG, *Developmental Genetics,* Foundations of Developmental Biology Series, Prentice-Hall, Englewood Cliffs, N.J., 1971. Chapter 6 includes the *Chironomus tentans-pallidivittatus* story.
6. BEERMANN, W., and U. CLEVER, "Chromosome Puffs," *Scientific American,* Offprint No. 180, April, 1964. Includes a discussion of ecdysone-induced puffs.
7. ILAN, J., *Regulation of Messenger RNA Translation in Development: The Critical Role of Transfer RNA,* Addison-Wesley Modules in Biology, No. 5, Addison-Wesley, Reading, Mass., 1973. Presents evidence of a long-lived messenger RNA whose rate of translation is controlled by the availability of a particular *t*RNA.

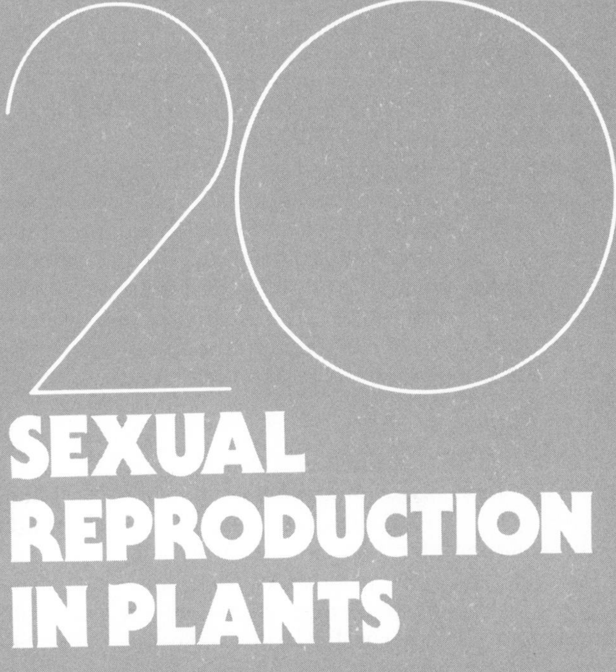

20 SEXUAL REPRODUCTION IN PLANTS

20.1 ALTERNATION OF GENERATIONS

Sexual reproduction involves the two processes of fertilization and meiosis. In fertilization, the nuclei of two gametes fuse, raising the chromosome number from haploid to diploid. In meiosis, the chromosome number is reduced again from the diploid to the haploid condition. Whatever variation there may be in the details from one organism to another, these two activities must occur alternately if sexual reproduction is to continue.

In plants, fertilization and meiosis divide the life of the organism into two distinct phases or "generations." The **gametophyte** generation begins with a spore produced by meiosis. The spore is haploid and all the cells derived from it are also haploid. Among the cells produced by this generation are the gametes. When two gametes fuse, the **sporophyte** generation begins. The sporophyte generation thus starts with a zygote. It contains the diploid number of chromosomes and all cells derived from it by mitosis are also diploid. Eventually though, certain cells will undergo meiosis, forming spores and starting the gametophyte generation once again.

20.2 THE PROBLEMS TO BE SOLVED

Plants probably evolved from aquatic ancestors, the green algae. Although plants can be found today growing in water, they are primarily associated with life on land. The members of each group of plants have a variety of morphological and physiological adaptations which enable them to live away from the aquatic environment. How plants solve the problem of securing water and keeping their cells moist was discussed in Part III. In this chapter, we shall examine the ways in which the plants have solved the problem of reproducing sexually on land.

The problem is really twofold. Gametes are single cells and quite delicate. In cross-fertilization, some mechanism must be present to enable the two gametes to reach each other safely. The problem is compounded in plants by their not having the power of locomotion. The inability of plants to move about also creates a problem of how to disperse the offspring produced by sexual reproduction to locations far enough away from the parent plant that they can receive the sunlight, water, and soil minerals they need. The twin problems of bringing the gametes together in a protected environment and dispersing the species to new locations have been solved in different ways by the various groups of plants.

20.3 MOSSES

A bed of moss consists of masses of leafy shoots which, being haploid, belong to the gametophyte generation (Fig. 20.1). In the common haircap moss, *Polytrichum commune,* the leafy shoots are of three kinds: female, male, and sterile. (The latter do not participate in sexual reproduction and need not concern us here.) The male shoots are easily distinguished from the other two by their flat top. A longitudinal section through the tip of a male plant reveals several of the male reproductory organs, the **antheridia.** Each is filled with sperm. A similar section through the female plant reveals several bottle-shaped **archegonia,** the female reproductory organs. Each of these contains a single egg in a chamber near its base. In early spring, if there is plenty of water present, the sperm are released from the antheridia. The splashing of raindrops distributes the sperm to nearby plants. On female plants, the sperm swim to the archegonium, probably following a concentration gradient of sucrose diffusing from the archegonium. The sperm swim down the canal to the egg, and fertilization takes place. The resulting zygote is the first cell of the sporophyte generation.

Mitotic divisions of the zygote produce the mature sporophyte generation (Fig. 20.1). It consists of:

1. A foot, which absorbs water and minerals from the parent gametophyte. Although the cells of the sporophyte generation contain chlorophyll, some food materials may also be absorbed from the parent gametophyte.
2. A stalk, which grows a few inches up into the air.
3. A **sporangium,** which is formed at the tip of the stalk. Within it are located spore mother cells. The opening to the interior is sealed with a lid, the operculum. The entire sporangium is covered by a calyptra. The calyptra is derived from the wall of the old archegonium. (Thus it is actually a part of the gametophyte generation.) It is the calyptra that is responsible for the common name of this species.

FIG. 20.1 Life cycle of the common haircap moss *Polytrichum commune.* In order for fertilization to take place, there must be surface water in which the sperm can pass from the male plant to the female plant. ►

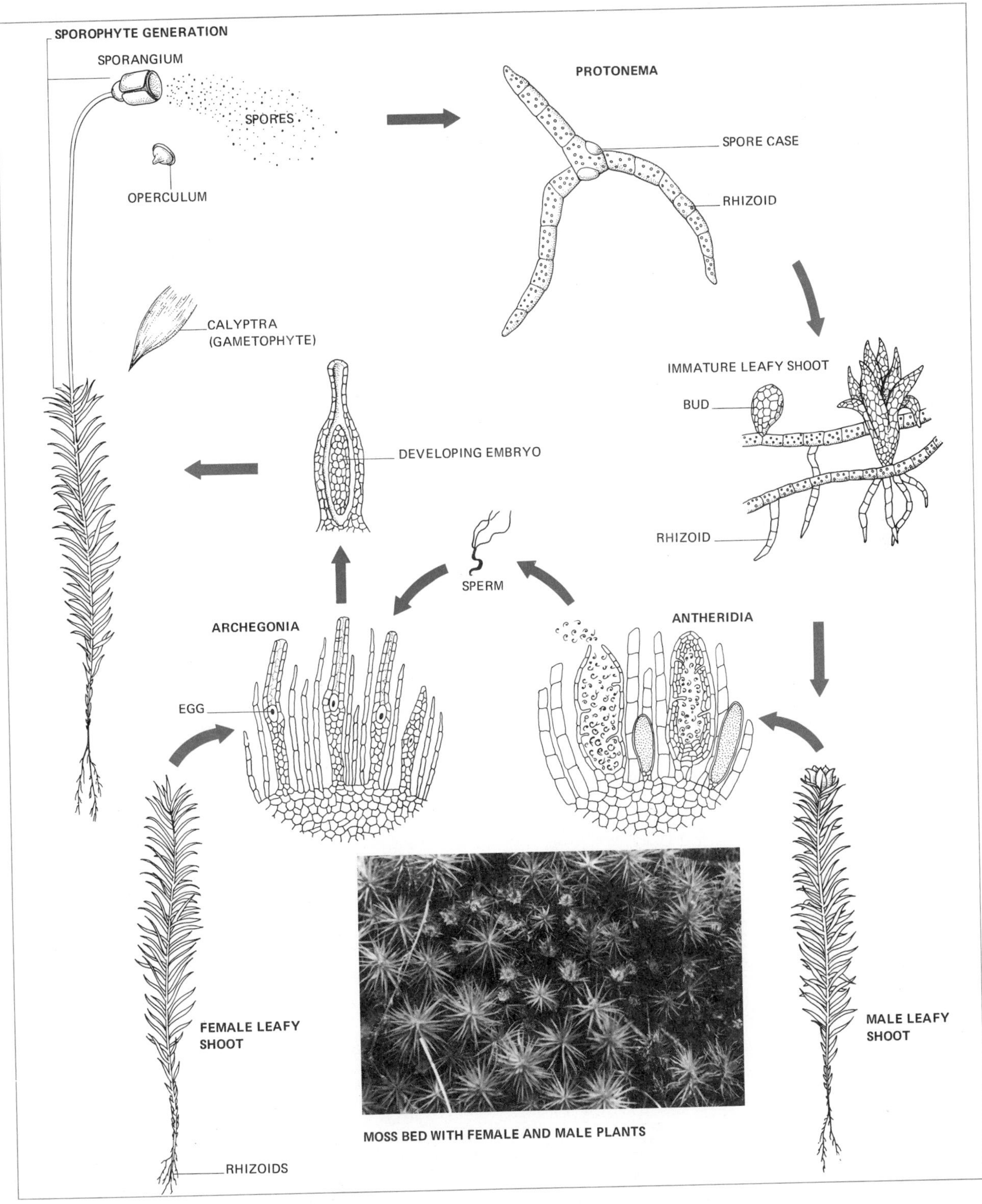

MOSS BED WITH FEMALE AND MALE PLANTS

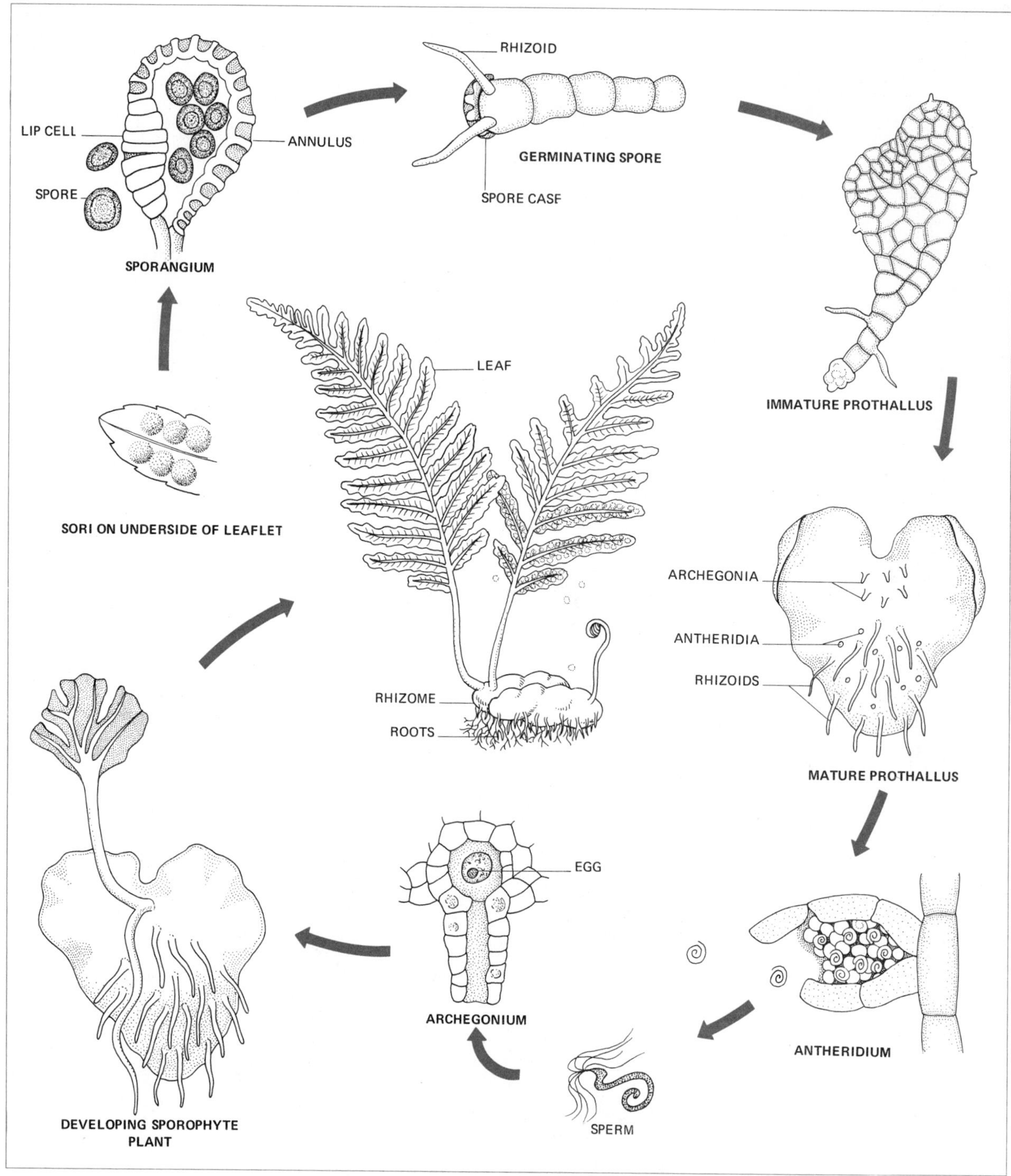

FIG. 20.2 Life cycle of a typical fern. Cross-fertilization is the rule; that is, the sperm swim from one prothallus to another. Surface water must be present for this to occur.

During the summer, each spore mother cell in the sporangium undergoes meiosis, producing four spores, the start of the gametophyte generation. Late in the summer, the calyptra and operculum become detached from the sporangium. Lowered humidity causes the ring of teeth within the opening to the sporangium to bend outward. In so doing the spores are ejected. Their small size enables them to be transported great distances by the wind.

If a spore comes to rest in a suitable location (moist and shady), it will germinate to form a filament of green cells called a **protonema.** A resemblance between the protonema and certain filamentous green algae has suggested to botanists that the mosses evolved from algae. Soon buds appear on the protonema and from these buds develop the leafy shoots—male, female, and sterile—that started our story.

What are the roles played by these two stages in the moss life cycle? The gametophyte generation produces the gametes and is responsible for the carrying out of sexual reproduction. The process requires ample water, however, and in a dry spring sexual reproduction often fails to occur.

Even though it may manufacture some of its own food, the sporophyte generation is completely dependent upon the gametophyte for its water and minerals. Its contribution to the success of the species is the myriads of wind-blown spores which it produces. These disperse the species, enabling it to colonize new habitats.

20.4 FERNS

In ferns, it is the sporophyte generation that we commonly see. The Christmas fern is a typical example (Fig. 20.2). The **leaves** (or fronds as they are often called) are the only part of the plant that is visible above ground. These arise from an underground stem—**rhizome**—from which also extend the **roots.** All these structures make up the mature sporophyte generation. During early summer, brownish spots appear on the underside of the leaflets of the fronds (Fig. 20.3). Each dot is called a **sorus** and consists of many sporangia mounted on stalks (Fig. 20.2). Within each sporangium, the spore mother cells undergo meiosis, producing four spores each. When the humidity drops, the thin-walled lip cells of each sporangium separate, and the annulus slowly straightens out. Then with a quick motion, the annulus snaps forward, expelling spores.

FIG. 20.3 Clusters of sporangia on the underside of fern leaflets.

You can easily demonstrate this action yourself. Soak a few fern leaflets with ripe sori in a small amount of rubbing alcohol for a day or two. Then scrape some of the sori onto a clean microscope slide with the blade of a knife. Under magnification, the details of the sporangia are easily seen. As the heat from the microscope illuminator passes through the moistened material, the alcohol will evaporate and spore ejection will begin.

If the wind-blown fern spores reach a suitable habitat (again moist and shady) they will germinate into filaments of cells. Each grows into a small (about 1 cm), flat, green, heart-shaped structure called a **prothallus** (Fig. 20.2). It grows on the surface of the soil anchored by thin filaments of cells called rhizoids. These also absorb water and minerals from the soil. The cells of the prothallus are haploid and it is the mature gametophyte generation. On its under surface are the sex organs: **antheridia** for sperm production and **archegonia** for egg production. When moisture is plentiful, the sperm are released and swim to an archegonium, usually on another prothallus, because the two kinds of sex organs generally do not mature simultaneously on a single prothallus. This circumstance results in cross-fertilization and thus greater opportunity for variation in the offspring. Fertilization takes place within the archegonium and the new sporophyte generation is initiated. The embryo sporophyte develops by repeated divisions of the zygote. One structure that develops in the embryo but is not found in the mature sporophyte is the foot. This organ penetrates the tissues of the prothallus and derives moisture and nourishment from it until the roots, rhizome, and leaves become self-sufficient. Note that the prothallus, though tiny in comparison with the mature sporophyte, is still an independent, autotrophic plant and even supports the embryo sporophyte during the early stages of its development.

20.5 GYMNOSPERMS

The gymnosperms with which most of us are familiar are the conifers (pines, spruces, etc.). The sporophyte generation (which is the only part the casual observer sees) produces not one, but two, different kinds of spores. There are the small **microspores,** which will germinate to form the male gametophyte generation, and the somewhat larger **megaspores,** which will develop into the female gametophyte generation. Each of these is produced in its own sporangium. In our most common gymnosperms, the conifers, the two kinds of sporangia are produced in the **cones.** Almost everyone is familiar with female cones (Fig. 20.4). The male cones appear in the spring and are much shorter-lived. Within these cones, microspores are produced by meiosis to start the male gametophyte generation. Before being released, mitotic division of the nucleus of each spore takes place with the final production of a four-celled pollen grain. Then the pollen grain is released from the male cone into the air.

In the female cones, the megaspore similarly undergoes a period of development within the cone. The product of this is the mature female gametophyte generation (Fig. 20.4). This small structure

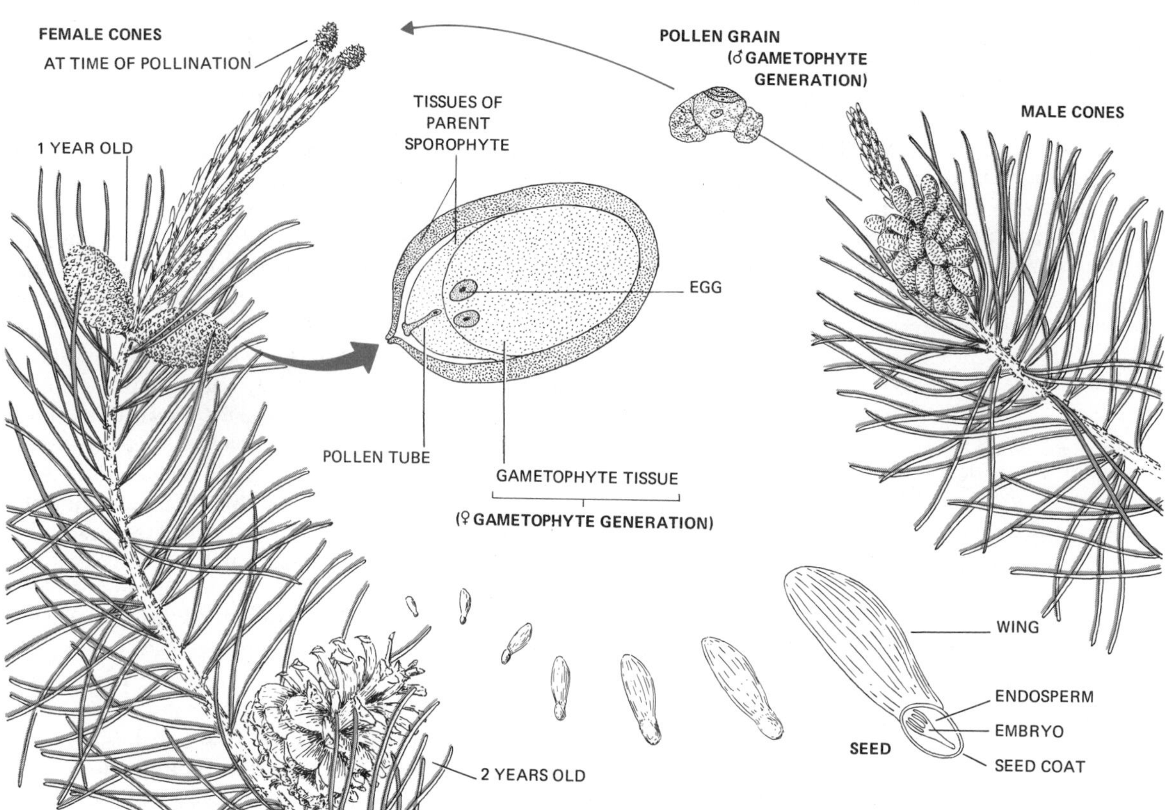

FIG. 20.4 Life cycle of a typical pine. Fertilization of the egg occurs within the tissues of the parent sporophyte.

is not released from the cone but is retained within the tissues of the parent sporophyte. This is seemingly a simple change, but it is one of vast significance to the success of gymnosperms. By retaining the female gametophyte within the tissues of the parent sporophyte, external moisture is no longer necessary for fertilization. The pollen grains are carried by the wind to the female cones. There they germinate. In the earliest gymnosperms, motile sperm were probably released. These could swim to the egg in the archegonium, propelling themselves in fluid supplied by the tissues of the parent sporophyte. This fluid was ultimately derived from a root system extending into the ground. The need for surface water for the sperm was thus eliminated. The ginkgo, commonly planted in city parks, is a gymnosperm that still uses this method of fertilization. For this reason the ginkgo is considered a very primitive type of gymnosperm. In modern gymnosperms, free-swimming sperm are not produced by the male gametophyte. Instead, the pollen grain germinates to form a thin tube, the pollen tube, which grows into the tissues of the female cone until it reaches the vicinity of the egg. (In the pines, this may take a full year.) Then the tube ruptures and a nucleus, the sperm nucleus, fuses with the egg to form the zygote. The basic situation remains unchanged, however. The delicate process of fertilization is carried out deep in the tissues of the parent sporophyte where it is protected from the possible harshness of the external environment.

With the production of two kinds of spores and two kinds of gametophytes, spores of the gymnosperms can no longer serve the function of dispersal as they do in mosses and ferns. In mosses and ferns, a single spore can germinate in the *soil* to form a gametophyte with both kinds of sex organs. Sexual reproduction can follow and soon the organism is established in the new environment. Dispersal of the species has occurred. Gymnosperm spores, in contrast, cannot be agents of dispersal. The only place the wind-blown microspore can germinate is on a female cone of the same plant or, better (why?), another plant of the same species. The microspore is simply transported from one plant to another. This does not accomplish dispersal of the species to a new location.

The function of dispersal is taken over by the **seed.** After fertilization, the zygote develops by repeated mitosis into a tiny embryo sporophyte plant (Fig. 20.4). Around the embryo develops accessory food-containing tissue called **endosperm.** It is derived from the cells of the female gametophyte generation and is haploid. Its food came, however, from the parent sporophyte. Around the embryo and endosperm develops a protective coat of parent sporophyte tissue. This coat usually develops a thin vane or wing on one side of the seed. All these developmental activities require another year for completion in the pines. At the end of this period, the female cone opens, releasing, one by one, its contents of seeds. The "wing" of the seed coat serves as a propeller and increases the distance the seed may be carried by the wind. The seed coat itself also serves to protect the embryo from drying out. If the seed is carried to a suitable location (moderately moist), it will absorb water. The embryo starts metabolizing rapidly and then begins to grow. This resumption of growth is called **germination.** At first, growth occurs at the expense of food stored in the endosperm. When the young seedling grows up into the light, however, chlorophyll develops and the plant begins to manufacture its own food by photosynthesis. The endosperm is totally consumed and the seed coat drops away so that every cell of the maturing plant is descended from the zygote.

Compared with the gametophytes discussed earlier, the gymnosperm gametophyte is little more than a reproductive mechanism. Both the male and female gametophytes are tiny and entirely dependent upon the parent sporophyte for nourishment. The gametes can be brought together only through the use of structures of the parent sporophyte generation. The developing embryo is no longer protected by the gametophyte generation as in the mosses and ferns, but receives its nourishment and protection from the parent sporophyte. (The endosperm is gametophytic, but its food stores come from the sporophyte.) The sporophyte generation continues to accomplish dispersal of the species. This is no longer accomplished by wind-blown spores, however, but by wind-blown seeds.

Angiosperms

The life cycle of the angiosperms, the flowering plants, is similar to that of the gymnosperms. Although minor differences in detail occur in the many species of angiosperms, the major features of the angiosperm life cycle are shared in common.

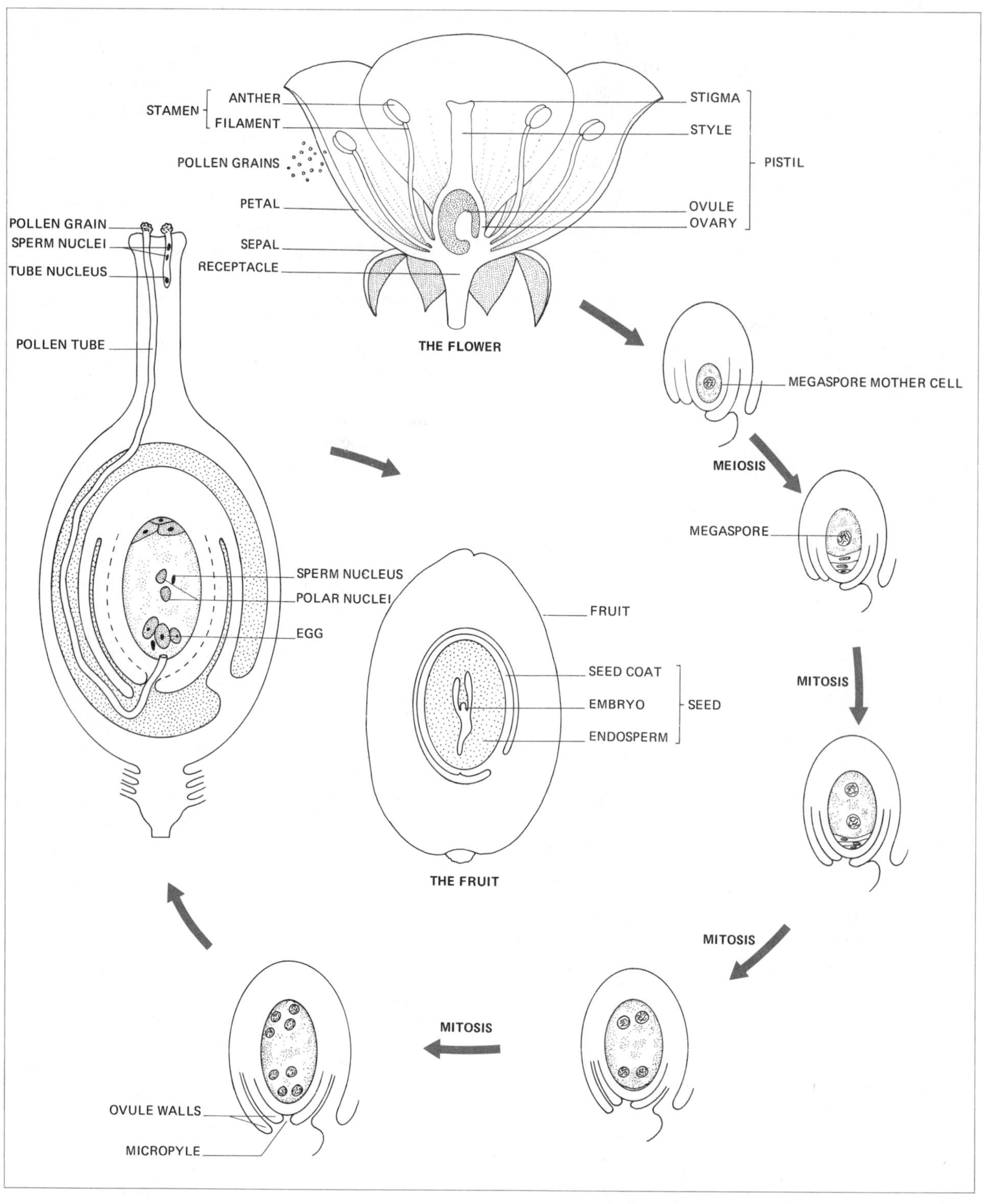
STAMEN
ANTHER
FILAMENT
POLLEN GRAINS
PETAL
SEPAL
RECEPTACLE
STIGMA
STYLE
PISTIL
OVULE
OVARY
THE FLOWER
MEGASPORE MOTHER CELL
MEIOSIS
MEGASPORE
MITOSIS
MITOSIS
MITOSIS
OVULE WALLS
MICROPYLE
POLLEN GRAIN
SPERM NUCLEI
TUBE NUCLEUS
POLLEN TUBE
SPERM NUCLEUS
POLAR NUCLEI
EGG
FRUIT
SEED COAT
EMBRYO
ENDOSPERM
SEED
THE FRUIT

20.6 THE FLOWER AND ITS POLLINATION

In the angiosperms, microspores and megaspores are produced in flowers. In most angiosperms the flowers are *perfect;* that is, each flower has both microsporangia and megasporangia and thus produces both kinds of spores. The microspores are produced in the **stamens.** The megaspores are produced in the **pistil** (Fig. 20.5). The stamen consists of a lobed structure, the anther or microsporangium, supported by a filament. The pollen grains are formed in the anther. Meiosis of microspore mother cells results in the production of four microspores each. These then develop into a two-celled pollen grain with a tough outer wall.

The pistil consists of a stigma, style, and ovary. The ovary contains a chamber within which are located the megasporangia, the **ovules.** The number and arrangement of ovules within the ovary varies greatly from species to species. Meiosis of the megaspore mother cell produces four haploid cells: the large megaspore and three small cells which disintegrate. The nucleus of the megaspore undergoes three successive mitotic divisions. The eight nuclei produced are distributed and partitioned off by cell walls as illustrated in Fig. 20.5. This is the mature female gametophyte generation. The two most important cells are the **egg cell** and the large central cell which contains the two **polar nuclei.** From this latter cell will be derived the endosperm of the seed.

Although wind transfer of pollen grains is employed by certain groups of angiosperms (e.g. the grasses), the angiosperms are particularly noted for the variety of animals which assist in pollination. Associated with the stamen and pistil may be found accessory structures. Most obvious of these are the brightly colored **petals** (collectively called the corolla) supported by a whorl of **sepals** (the calyx). Along with the petals may be found glands which emit fragrant odors. It is these accessory structures which give such aesthetic pleasure to us. In woods and fields, gardens and greenhouses, the extraordinary variety of corolla shapes and colors are a delight. Except as found in certain domesticated species, however, these shapes and colors do not exist for our pleasure. They serve instead to attract certain animals which, as they pass from flower to flower, may incidentally transport pollen grains from the anther of one to the stigma of another.

◄ **FIG. 20.5** Generalized life cycle of an angiosperm. The fruit develops from the wall of the ovary.

Animal-pollinated flowers frequently contain nectaries which secrete a sugary solution (nectar) and reward the animal for its visit. Birds (e.g. the hummingbird) and bats pollinate some flowers, but the great majority are pollinated by insects. Beetles, flies, butterflies, moths, and bees are active pollinators. The relationship between the plant and the animal may be quite loose. Some angiosperms are pollinated by many kinds of insects, and some insects pollinate many kinds of plants. In other cases, the relationship is far closer. There is a tropical orchid that is pollinated by just one species of moth, which itself is limited to foraging that one kind of flower for its nectar. The moth has a 25-cm proboscis; the orchid has a nectary 25 cm (10 in) deep. Other orchids have pistils whose shape and color pattern mimic the abdomens of certain female insects. In attempting to mate with several of these "dummy" females, the male insects effectively transport pollen from one flower to the next.

Generally, insect-pollinated flowers are perfect, both stamens and pistils being found within one flower. Two advantages follow from this. First, there is a greater likelihood of pollination occurring. At each visit by the pollinator, pollen from the last flower is deposited while fresh supplies of pollen are picked up. With **imperfect flowers,** which contain either stamens or pistils but not both, a pollinating insect has to visit staminate flowers and pistillate flowers alternately in order to achieve a comparable efficiency as a pollinator. Second, if pollination between different flowers (cross-pollination) should fail to occur, the flower may still be able to pollinate itself. Seed production will still occur, although the similar heredity of the gametes will diminish the amount of variability in the offspring.

The flowers of some angiosperms are actually modified in such a way that self-pollination is the rule rather than the exception. As we noted in Chapter 15, the corolla of the common pea completely encloses the stamens and pistil so that insects cannot easily reach these structures. Many violets produce, in addition to their conspicuous cross-pollinated flowers, smaller flowers that never open to expose their stamens and pistil to insect pollinators. A glance at Fig. 20.6 will show why only self-pollination occurs in these flowers.

Flower modifications to promote cross-pollination are more common than those to prevent it. In

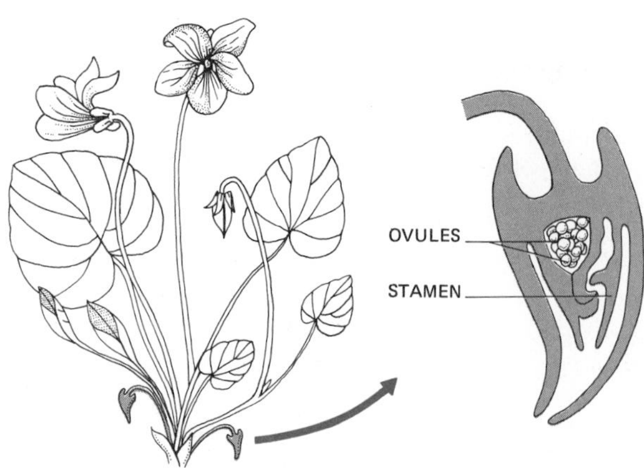

FIG. 20.6 Violet flowers. The petal-bearing flowers are cross-pollinated. The others (enlargement) are self-pollinated.

many species the stamens and pistil of a single flower mature at different times. In some species the floral parts are arranged so that there is little likelihood of a direct transfer of pollen from anther to stigma. The sage (*Salvia*) flower embodies both these mechanisms in addition to an unusual hinged stamen which bends downward to dust pollen on the bee which triggers it (Fig. 20.7). In many angiosperms, e.g. red clover and some apple varieties, pollen will not germinate on a stigma of the same plant. These plants are called self-sterile.

Still another mechanism that ensures cross-pollination is the occurrence of imperfect flowers on separate plants. Species in which this occurs, such as the willows, poplars, and the date palm, are called **dioecious.** (Species with imperfect flowers on one plant are called **monoecious.** Fertilization between these flowers produces no more hereditary variability than self-pollination within a single flower.)

In one locality all the plants of a given species flower at roughly the same time. This usually occurs even if the various individual plants have begun growing at different times. As we shall see in Chapter 24, such synchronous flowering seems to be a response to the changing length of day and night as the season progresses. It, too, is an important factor in encouraging cross-pollination.

Our emphasis on animal-pollinated angiosperms should not obscure the fact that many angiosperms are wind-pollinated. Poplars, oaks, elms, birches, plantains, ragweed, grasses, and many others are wind-pollinated. Their flowers are not apt to have petals, odor, or nectar. They are frequently imperfect (as are the cones of gymnosperms which, you remember, are also wind-pollinated). The stamens of the staminate flowers are exposed to the wind and produce large quantities of light, dry pollen. This wind-blown pollen may cause hay fever in allergic humans. The stigmas of the pistillate flowers are often long and sticky. The ovary usually contains only a few ovules.

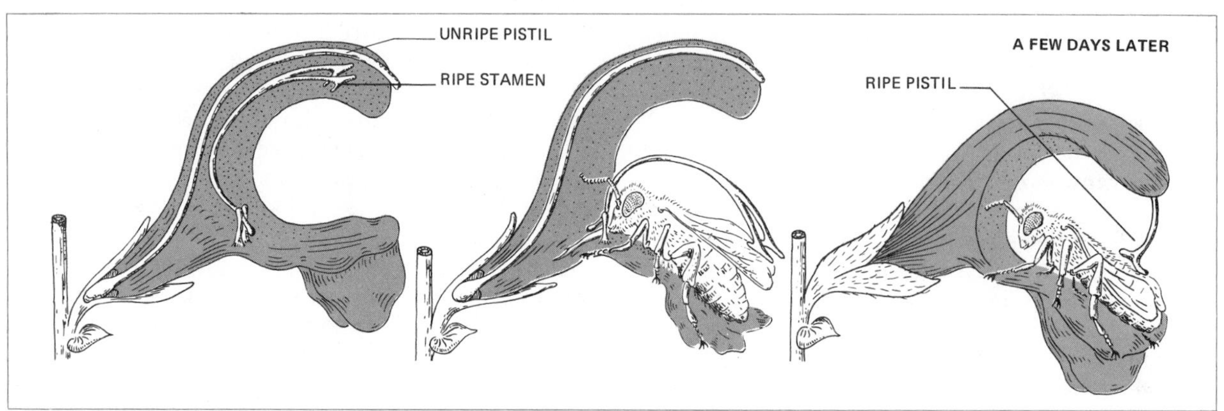

FIG. 20.7 Pollination in *Salvia*. Cross-pollination is assured by (1) ripening of the stamen before the pistil, (2) a trigger mechanism that dusts pollen on the pollinator, (3) later growth of the pistil so that it brushes the bees that continue to forage the flower for nectar.

Although the wind-pollinated flowers do not provide us with much aesthetic pleasure, they are largely responsible for our ability to find time for any aesthetic pleasures at all. All our cereal grains come from wind-pollinated species. As a group, they provide, directly or indirectly, the major portion of the food consumed by mankind.

When, by one means or another, the pollen grain reaches the stigma of a flower of the same species, it germinates into a pollen tube. The pollen tube contains two nuclei: the tube nucleus and the generative nucleus. As the pollen tube starts to grow down the style into the ovule chamber, the generative nucleus divides by mitosis, forming two sperm nuclei. The pollen tube, with its three nuclei, is the mature male gametophyte generation. It enters the ovule through the *micropyle* (Fig. 20.5) and ruptures. One sperm nucleus fuses with the egg, forming the zygote ($2n$). The other sperm nucleus fuses with the two polar nuclei to form an endosperm nucleus containing $3n$ chromosomes. The tube nucleus disintegrates.

20.7 THE SEED

Mitotic divisions of the zygote and endosperm nucleus result in the formation of the seed (Fig. 20.8). It consists of:

1. A **plumule,** made up of two embryonic leaves, which will become the first true leaves of the seedling, and a terminal (apical) bud. The terminal bud is the meristem at which further growth of the stem will occur.
2. The **hypocotyl** and **radicle,** which will grow into the stem and primary root, respectively.
3. The **cotyledons,** which store food that will be used by the germinating seed. This food is derived from the endosperm which, in turn, received it from the parent sporophyte. In many angiosperms (the common bean is one example), the endosperm has been totally consumed and its food stores transferred to the cotyledons by the time development of the seed is complete. In others, the endosperm persists in the mature seed. This is true of some dicots and all monocots. The latter, of course, have only one cotyledon in the seed (Fig. 20.9). The cells of the endosperm are triploid ($3n$) in contrast to the haploid (n) endosperm of the conifers and other gymnosperms (see Section 20.5). While these structures are developing, the walls of the ovule thicken to form the protective seed coats.

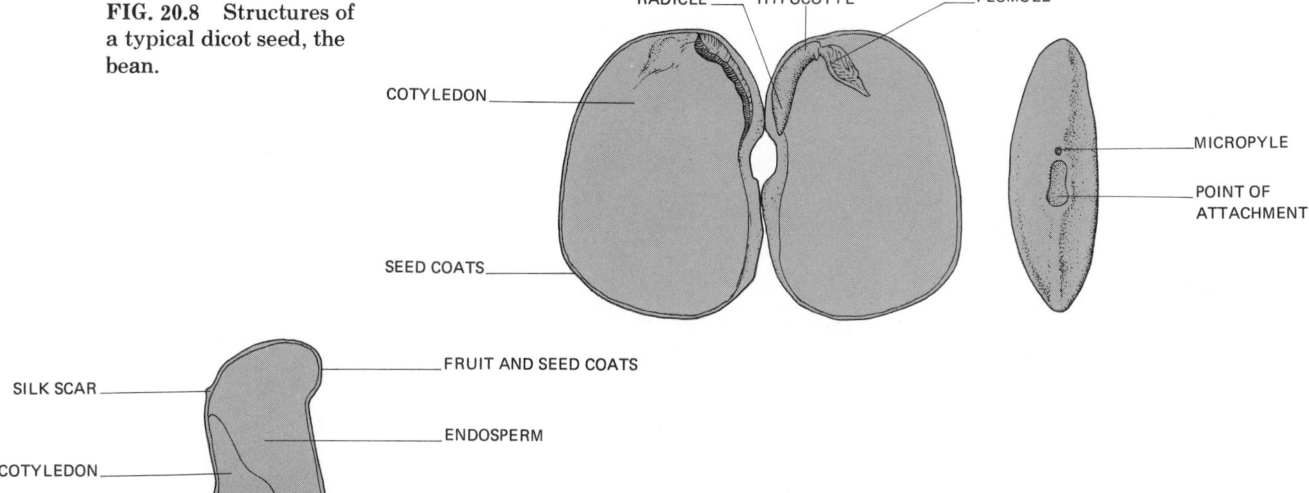

FIG. 20.8 Structures of a typical dicot seed, the bean.

FIG. 20.9 Structures of a corn kernel. Because its outer covering is derived from the ovary wall of the flower, the corn kernel is actually a fruit with a single seed inside.

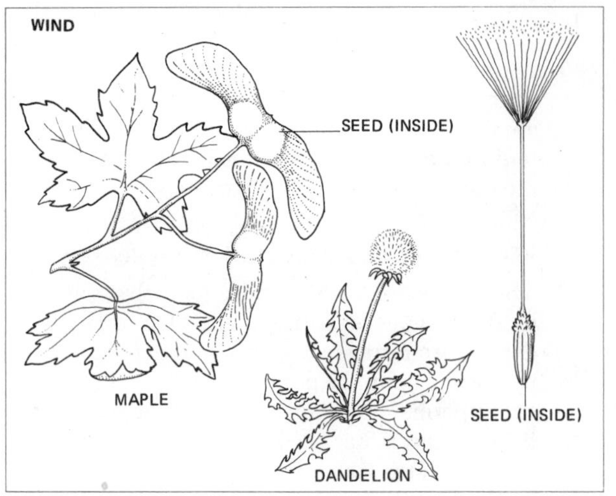

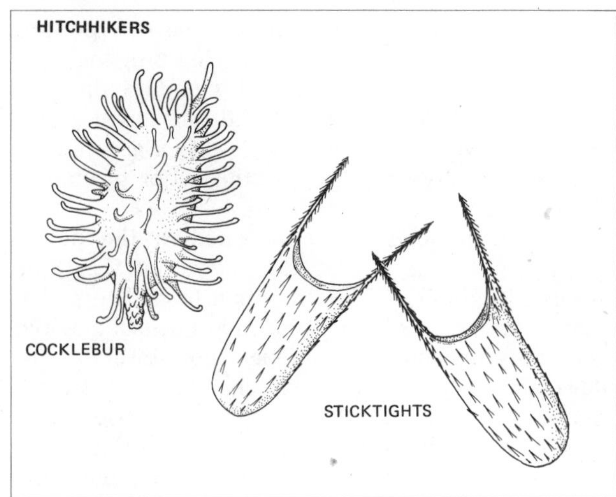

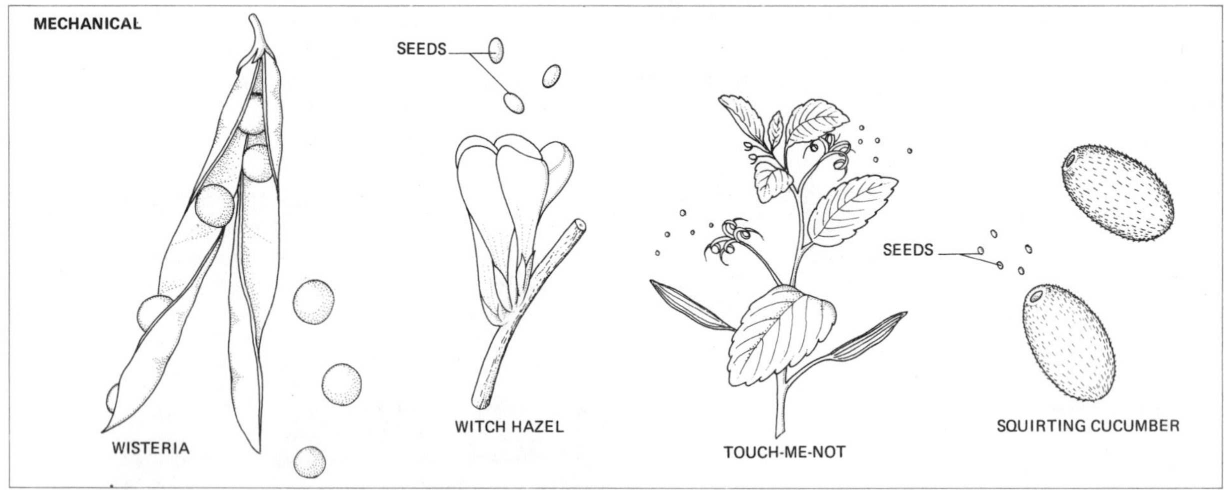

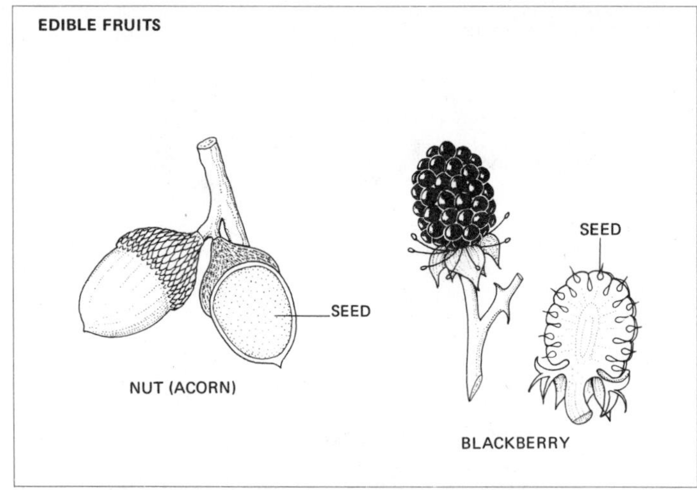

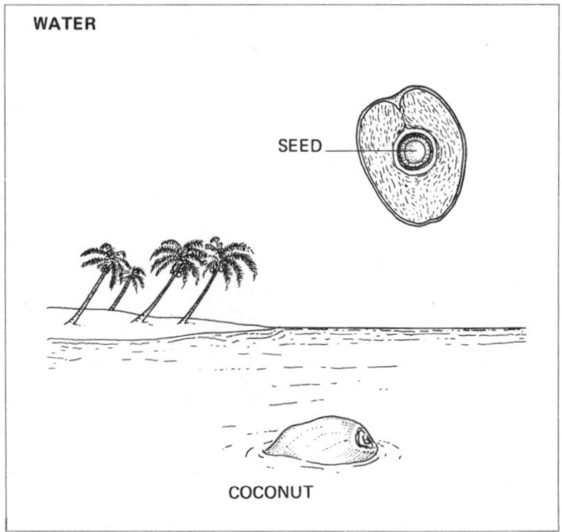

FIG. 20.10 Fruits and their role in seed dispersal.

The seed is thus a dormant embryo sporophyte with stored food and protective coats. Its two functions are: (1) dispersal of the species to new locations and (2) maintenance of the species over periods of unfavorable climatic conditions. "Annuals" (such as the bean, cereal grains, and many weeds) can survive as a species only by the production of resistant seeds. With the onset of freezing weather in the fall, the mature plants die. However, many of their seeds will remain alive, though dormant, throughout the winter. When conditions once more are favorable for the growth of the plant, germination of the seeds occurs and a new generation of mature plants develops.

20.8 SEED DISPERSAL: THE FRUIT

The fruit is a development of the ovary wall and sometimes other flower parts as well. It contains the seeds. To a biologist, the term *fruit* is not restricted simply to those succulent types which we enjoy eating. Figure 20.10 shows some of the many different types of fruits produced by angiosperms.

In every case these fruits aid in the dispersal of their content of seeds. The maple "key" and the dandelion parachute are examples of fruits which facilitate dispersal of the seeds by the wind. The cocklebur and sticktights are fruits which achieve dispersal of their contained seeds by sticking to the coat (or clothing) of a passing animal. The animal's power of locomotion serves to transport the seeds to new locations.

The coconut is a fruit which achieves dispersal of its single seed by *floating* to new locations. Groves of coconut palms ring all the Pacific islands. Trees leaning out over the water may drop their coconuts in. The fruit does not germinate in salt water but it will remain viable (capable of germinating) for long periods as it is carried by ocean currents. When cast up on a beach and washed by rain water, it may germinate and develop into another mature palm.

Some of the most interesting fruit adaptations are those which achieve mechanical dispersal. Legume pods (e.g. wisteria), witch-hazel pods, touch-me-nots, and the squirting cucumber all forcibly eject their seeds as the fruit dries out in the fall. While the distances traveled by these seeds do not achieve rapid dispersal of the species, they do decrease the likelihood that the seedlings will have to compete with the parents for sunlight, moisture, and soil minerals.

Edible fruits are very effective dispersal mechanisms. Squirrels, field mice, and other rodents seldom eat all the nuts and grains they store in the fall. With the coming of spring, the forgotten seeds may germinate and at considerable distances from the parent plant. The various kinds of berries are characterized by tiny seeds which can pass undamaged through the alimentary canal of the bird or other animal that eats them. When the animal expels the seeds in defecation, it is usually in quite a new location. The manure in which the seeds are defecated serves as a rich source of minerals and humus for the developing seedling.

20.9 GERMINATION

Germination is the resumption of growth of the embryo plant inside the seed (Fig. 20.11). Proper temperature, proper amounts of moisture, and an adequate supply of oxygen are all necessary for it to occur. What is proper or adequate for one species

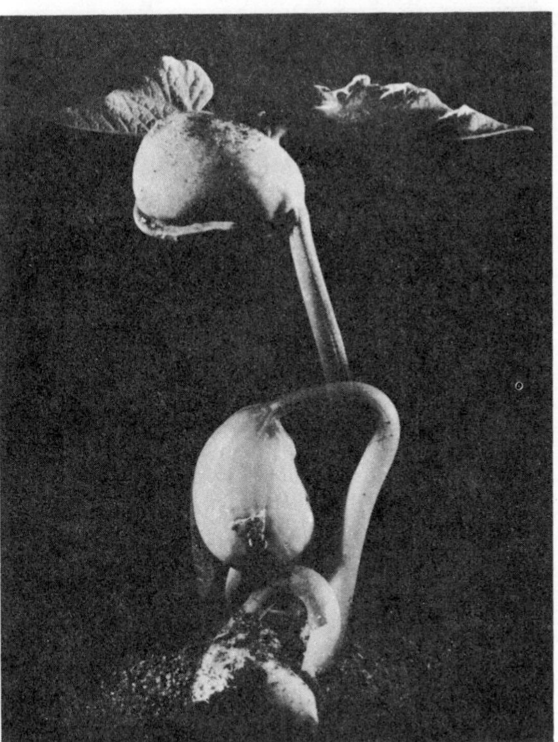

FIG. 20.11 Three stages in the germination of a bean seed, shown by time-lapse photography. The cotyledons protect the plumule from damage as it is lifted through the soil. (Courtesy of the Pittsburgh Plate Glass Co.)

may not be for another, but for each species these three conditions must be met to some degree.

A period of dormancy is also a requirement for the germination of many seeds. For example, the seeds of apples and peaches will germinate only after a prolonged period of cold. There is evidence that a chemical inhibitor is present in the seed when it is first formed. This inhibitor is gradually broken down at low temperatures until finally there is no longer enough to prevent germination when the other conditions become favorable. (What survival value does this mechanism have for the plant?) The seeds of many desert angiosperms possess inhibitors which prevent germination until the substance has been leached away by water in the soil. In this process more water is required than is necessary for germination itself to occur.

Exposure to light of proper duration is also a condition for germination in some cases. The seeds of some plants which grow in marshy locations will germinate only after prolonged exposure to light. On the other hand, the germination of the seeds of some desert plants is inhibited by prolonged exposure to light. What survival value might these responses have?

The angiosperm life cycle is basically similar to that of the gymnosperm. Like that of the gymnosperm, it permits successful existence in a terrestrial environment. The angiosperms, however, are a far more varied and successful group than the gymnosperms. There are approximately 250,000 species of angiosperms today compared with approximately 650 species of gymnosperms. Further, the angiosperms have exploited far more kinds of habitat than have the gymnosperms. Arctic tundra, plains, temperate forests, deserts, and jungles are types of environment where angiosperms grow in far greater variety and numbers than do the gymnosperms. The greater abundance and variety of angiosperms and their wide distribution throughout the world is evidence of the survival value of their two unique structural adaptations: the flower and the fruit.

EXERCISES AND PROBLEMS

1. Where are the sporangia located in the (a) moss, (b) fern, (c) pine, (d) apple tree?
2. In what ways are the gametophyte generations of the moss, fern, and angiosperm alike?
3. In what ways are they different?
4. In what ways are the sporophyte generations of the moss, fern, and angiosperm alike?
5. In what ways are mosses well adapted to life on land? In what ways are they poorly adapted?
6. What three conditions are always necessary for germination of seeds to occur?
7. What additional factors are sometimes necessary for the germination of seeds?
8. Describe the experimental procedure that you would use to test the viability of bean seeds that had been stored for five years.
9. Describe the experimental procedure you would use to determine the optimum temperature for the germination of bean seeds.
10. What plant tissue is normally triploid?
11. Which cells undergo meiosis in the (a) moss and (b) bean?
12. Reproduction in the closed flowers of the violet (see Fig. 20.6) is sexual because it involves the union of two gametes. Does it accomplish what sexual reproduction in the open flowers does? What would be the long-term consequence if the violet reproduced exclusively by means of its closed flowers?

REFERENCES

1. WATSON, E. V., *Mosses,* Oxford Biology Readers, No. 29, Oxford University Press, Oxford, 1972.
2. BOLD, H. C., *The Plant Kingdom,* 2nd ed., Foundations of Modern Biology Series, Prentice-Hall, Englewood Cliffs, N.J., 1964. Chapters 7, 8, and 9 discuss reproduction in the ferns, gymnosperms, and angiosperms, respectively.
3. GRANT, V., "The Fertilization of Flowers," *Scientific American,* Offprint No. 12, June, 1951.
4. BIALE, J. B., "The Ripening of Fruit," *Scientific American,* Offprint No. 118, May, 1954.
5. KOLLER, D., "Germination," *Scientific American,* Offprint No. 117, April, 1959.

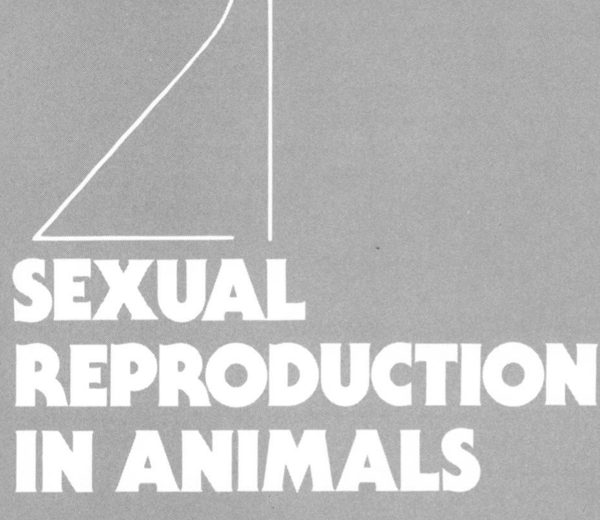

21 SEXUAL REPRODUCTION IN ANIMALS

21.1 THE FORMATION OF GAMETES

Animals, unlike plants, do not exhibit an alternation of diploid and haploid generations. Fertilization is still preceded by meiosis, but the products of meiosis are the gametes themselves. In all animals, **heterogametes** are produced. A moment's reflection will reveal the adaptive value of this modification. To carry out their function most effectively, gametes should be motile (so they can meet and unite) and supplied with food reserves to furnish energy and material for the developing embryo. These two requirements are rather incompatible. The solution is: one gamete, the **sperm,** that is motile and small and one gamete, the **egg,** that is filled with food reserves.

Sperm cells are little more than flagellated nuclei. They are produced in **testes** by specialized cells called spermatogonia. Spermatogonia, which are diploid, may divide by mitosis to produce additional spermatogonia or they may be transformed into *spermatocytes.* Meiosis of each spermatocyte results in the production of four haploid cells, the spermatids (Fig. 21.1). These then become transformed into sperm cells, losing most of their cytoplasm in the process.

A sperm cell consists of (1) a head, which contains the chromosomes in a compact, inactive state, (2) two centrioles, and (3) a "tail." One of the centrioles will help to construct a spindle for the sperm and egg chromosomes following fertilization. The other serves as the basal body of the flagellum, which extends the length of the tail. Mitochondria surround the upper part of the flagellum and supply the energy for its lashing movements (Fig. 21.2).

Eggs are produced in **ovaries.** Diploid oogonial cells divide by mitosis to produce additional oogonial cells (Fig. 21.3). This occurs once a year in most aquatic animals and in the amphibians. Among the reptiles, birds, and mammals, however, the process stops long before birth. In fact, by the time the female human fetus (a developing baby) is 15 weeks old, multiplication of oogonia is almost completed. This certainly justifies Weismann's emphasis on the early isolation of the germplasm from the somaplasm!

The beginning of egg formation occurs when oogonia start growing and become transformed into primary oocytes (Fig. 21.3). These diploid cells enter prophase of the first meiotic division and at that point their development stops. No further development of the primary oocytes takes place until just before the animal is ready to enter a period of reproductive activity. In frogs, this occurs once a year —usually in the spring—after adulthood is reached.

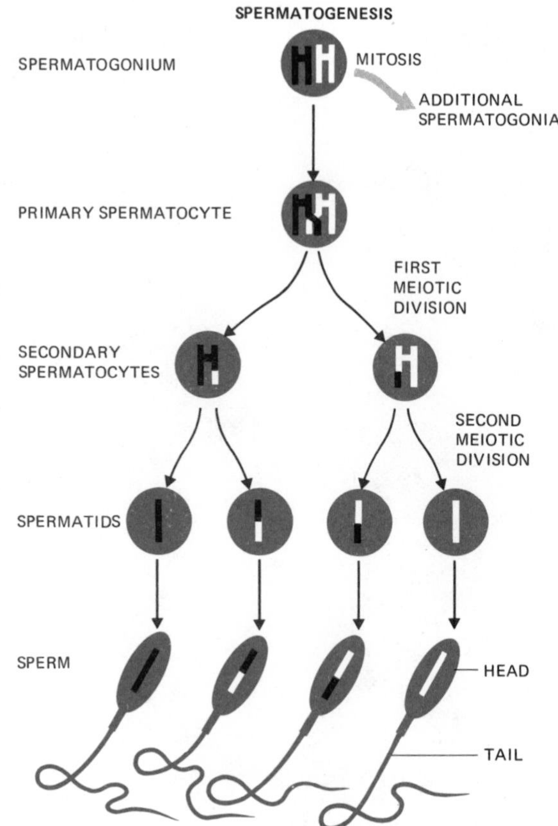

FIG. 21.1 Formation of sperm. For the sake of simplicity, the behavior of just a single pair of homologous chromosomes is shown (cf. Fig. 14.14). With two or more pairs, random assortment of the homologues contributes to the variety of gene combinations in the gametes.

Then thousands of primary oocytes begin a period of marked cell growth. Each is enclosed in a cluster of cells called a **follicle.** Food materials are transferred from the follicle cells to the growing oocyte. The volume of frog eggs increases over a million times during this period.

When this phase of development is completed, the egg cell is a large sphere containing in its cytoplasm large quantities of DNA, RNA, yolk, mitochondria, and oil droplets. These materials are not distributed uniformly throughout the egg of the frog (Fig. 21.4) but in gradients extending from pole to pole. The dark hemisphere of the egg is topped by the so-called **animal pole.** Except for the yolk, most of the egg constituents are concentrated near this pole. The nucleus is also found here. The yolk concen-

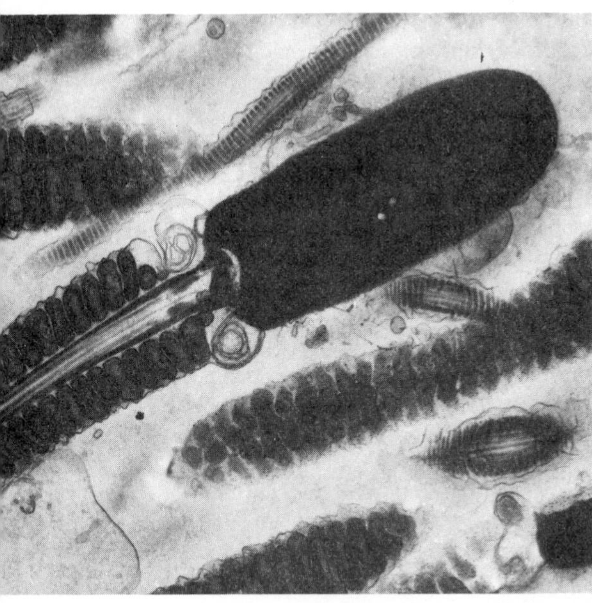

FIG. 21.2 Sperm cell of a bat (12,000 ×). Note the orderly arrangement of the mitochondria in the tail. They supply the energy for its motion. (Courtesy of Don W. Fawcett and Susumu Ito.)

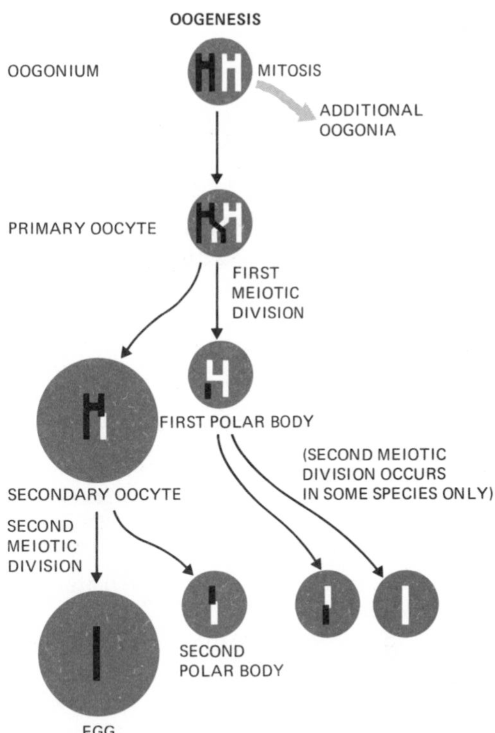

FIG. 21.3 Egg formation. Which chromosomes end up in the egg and which in the polar bodies is entirely a matter of chance.

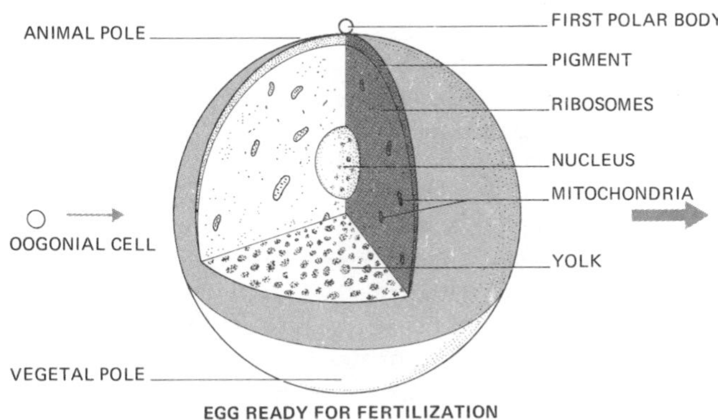

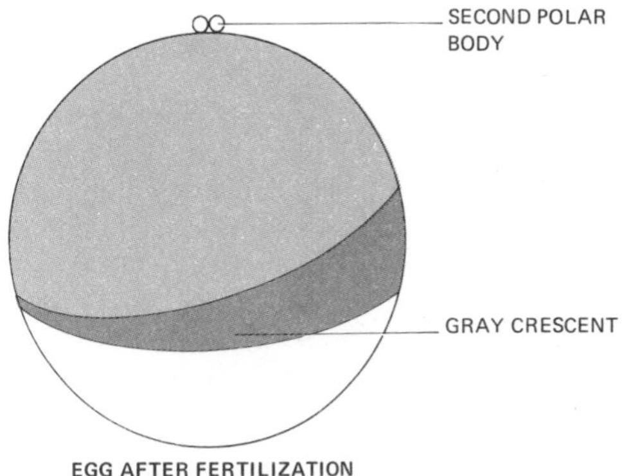

FIG. 21.4 Development of the frog egg. Entrance of a sperm cell is followed by the reorganization of cytoplasmic materials into the gray crescent and then the completion of the second meiotic division.

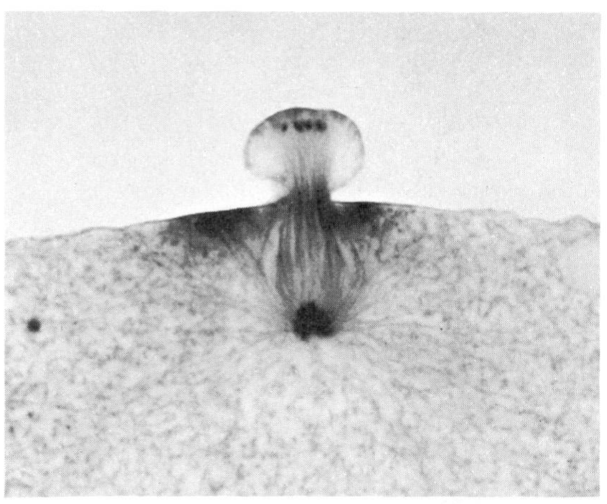

FIG. 21.5 Polar body formation during oogenesis in the whitefish. (Courtesy CCM: General Biological, Inc., Chicago.)

tration increases toward the opposite, light-colored **vegetal pole.**

As growth of the primary oocyte nears completion, the first meiotic division is completed, too. The cytoplasm is not distributed equally to the two daughter cells, but almost all of it passes to just one of them. The other cell is called a **polar body** (Fig. 21.5).

In most vertebrates, the second meiotic division gets only as far as metaphase and then stops. At this time, the egg is ready for release from the follicle, a process known as **ovulation.** The follicle and ovary walls rupture, allowing the egg to pass into the body cavity. It then enters an oviduct where accessory substances, such as gelatinous rings of albumen in the frog, may be added. Further development of the egg awaits fertilization.

The life span of liberated sperm and eggs is usually measured in hours. Furthermore, sperm are definitely limited in the distance they can swim. But, since most animals have the power of locomotion, the sperm themselves do not usually have to travel great distances to reach the egg. Successful locomotion of the sperm and successful fertilization do, however, require moist surroundings.

21.2 BRINGING THE GAMETES TOGETHER

Among aquatic animals, the problem of moist surroundings for fertilization does not exist. Fertilization can occur directly in the water after each parent releases its gametes. To improve the chances of the gametes meeting, they are usually released in fairly close proximity. This requires that the males and females of the species reach full sexual development at about the same time and in about the same place. Synchronous flowering achieved by response to definite photoperiods was discussed in the last chapter. Analogous periodic behavior is observed in many animals. In the California grunion (a fish), the two sexes swim into shallow water to spawn (release their gametes) at the time of full moon and new moon. In other species, a member of one sex (usually the male) carries out certain behavioral activities ("courtship") in the presence of the female. These activities trigger appropriate egg-laying behavior in the female.

The first terrestrial animals were probably scorpions and insects. Life on land created a problem in bringing the gametes together while protecting them from the drying action of the air. It was solved by internal fertilization. A special organ of the male, the penis, is inserted into the female in a process called **copulation** and transfer of the sperm takes place. In insects, the sperm are then stored in sperm receptacles until the female is ready to lay her eggs. As the eggs pass down her genital tract, they are fertilized by the sperm. The critical process of fertilization is thus accomplished entirely within the moist body of the female (Fig. 21.6).

The first terrestrial vertebrates were the amphibians. As the name implies, they are really only semiterrestrial. Most frogs and toads must return to the water to engage in sexual reproduction. Fertilization takes place in the water. The common male bullfrog, for example, clasps the female in an embrace called amplexus. As the female deposits her several thousand eggs into the water, the male deposits sperm over them. No copulation takes place.

The reptiles were the first vertebrates able to live in a truly terrestrial environment. They have several important adaptations, both structural and physiological, which permit them to flourish even in hot, dry climates. Some of these will be discussed in Chapter 38. As for sexual reproduction, the problem of moisture for the gametes is solved by internal fertilization. In most cases, the male copulates with the female, introducing sperm directly into her genital tract. Birds and mammals also solve the problem of bringing the gametes together in moist surroundings by copulation followed by internal fertilization.

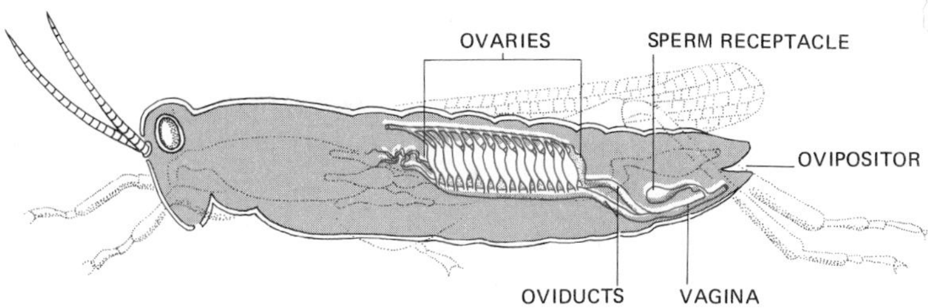

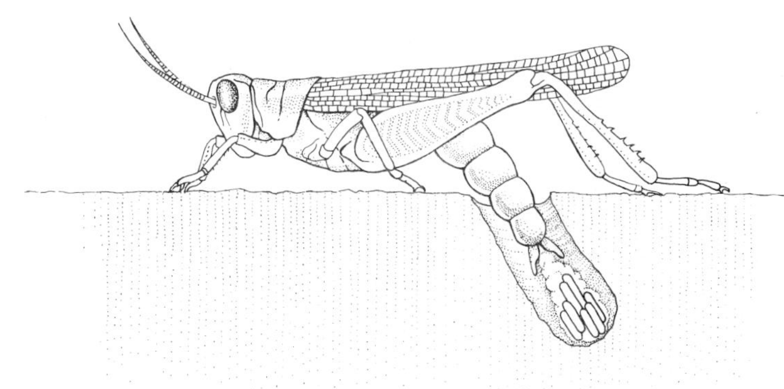

FIG. 21.6 Reproductive organs of the female grasshopper. Sperm are stored in the sperm receptacles following copulation with the male. Fertilization takes place as the eggs pass through the vagina before being deposited in the soil.

21.3 FERTILIZATION

The process of fertilization begins when sperm actually become attached to the egg. In so doing, they may release digestive enzymes that dissolve an opening through the membrane or layer of residual follicle cells that usually cover eggs. Then the nucleus and one centriole of the sperm cell enter the egg. The egg seems to play an important part in this as these structures appear to be drawn inward. Entrance of the sperm head is followed by a rapid and dramatic transformation within the egg itself. Its cytoplasmic constituents quickly become reorganized. In the frog, certain cytoplasmic granules appear at the surface of the egg in a band called the **gray crescent** (Fig. 21.4).

In many eggs, the changes brought about by the entrance of one sperm immediately prevent, in some way, other sperm from entering, too. Just how this is accomplished is not fully understood. In animals with large eggs, such as reptiles, birds, and the duckbill platypus, it doesn't occur at all. Although many sperm gain entrance to these eggs, only one contributes its nucleus to the future zygote. In any case, entrance of a sperm cell also triggers the completion of the second meiotic division, and a second polar body forms.

The final event in fertilization occurs when the sperm centriole initiates the formation of a spindle upon which first the sperm chromosomes and then the egg chromosomes become arranged. At this point, a zygote with the diploid number of chromosomes has been formed and fertilization is complete. Within a short time, the first mitotic division of the cell will take place and embryonic development will begin.

21.4 CARE OF THE YOUNG

An important relationship exists between the number of gametes, especially eggs, produced by different animals and the care given the embryos during development. For a population to continue at a stable size, an average of two offspring should reach maturity for every two parents. Very few aquatic animals take any care of their eggs once they are fertilized, and mortality of the developing young is extremely high. In compensation for this, a mature female American oyster, *Crassostrea virginica,* may release as many as one million eggs per season. The chances of any one egg ever developing into a mature oyster are exceedingly small. Most, even if successfully fertilized, will succumb at some stage of development to predators, infection, or other unfavorable happenings. Some aquatic animals, e.g. many freshwater bivalves and some fish, do provide a certain amount of care for the eggs. The freshwater clam retains fertilized eggs in the gills until they hatch. The male stickleback fish builds a nest and guards and aerates the eggs that the female deposits inside. Characteristically, these species do not produce such enormous numbers of eggs.

The eggs of most aquatic animals are relatively small. In a large number of species, the organism that hatches from the egg is not a miniature replica of the adult but a larval stage quite different in appearance. Usually the larvae are free-swimming and feed on microscopic plant and animal life called plankton. In so doing, they acquire additional food materials for further growth. After a period of growth, the larvae undergo metamorphosis, and the body structure is reorganized on the plan of the adult. Metamorphosis may occur in several stages (e.g. in crustaceans and mollusks) or it may be completed in a single brief, but extensive, transformation (e.g. in the starfish).

An exception to these generalizations on aquatic animals is found in the cartilaginous fishes: the sharks, skates, and rays. Their interesting solution to the problem of dehydration by salt water was discussed in Chapter 13. You remember that the concentration of urea in the bloodstream is maintained at such a high level that the blood is isotonic to the sea water. This physiological adaptation cannot be achieved by their embryos. The cartilaginous fishes have, however, developed two techniques for avoiding fatal dehydration of their embryos. One is to enclose the fertilized egg and a suitable watery medium in a tough, waterproof case. Empty skate egg cases are a familiar sight among the debris washed up on our beaches (Fig. 21.7). The other solution is to retain the developing embryo within the female's body, as some sharks do, until it becomes physiologically capable of maintaining a high urea level. The young then are "born." Either solution requires internal fertilization of the eggs. (The egg case of the skate is as impervious to sperm as it is to water.) The sperm are introduced directly into the body of the female in the process of copulation. Modified fins on the male aid in clasping the female and in transferring the sperm.

Terrestrial animals all need some means of protecting the developing embryo from the drying action of the air. In insects this protection is provided by a waterproof covering deposited around each egg and, in some cases, a capsule or sac which is deposited around a cluster of eggs. The number of fertilized eggs laid each season varies from hundreds to thousands, depending on the particular species and the amount of care taken of the eggs. The smaller

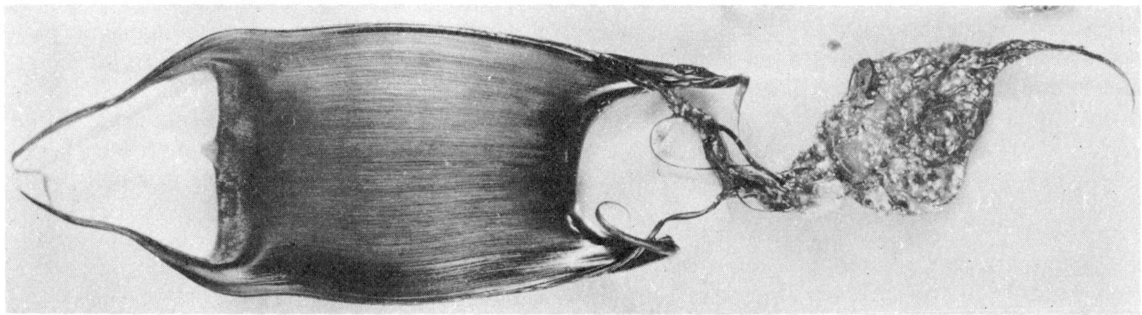

FIG. 21.7 Egg case of a skate (almost life-size). The eggs are sealed inside. (Courtesy of the New York Zoological Society.)

numbers reflect a somewhat greater chance of survival to adulthood for a given embryo than in the aquatic species discussed earlier. Care of the young is developed to an extreme in colonial ants and bees. In these species a very high percentage of the fertilized eggs develops to maturity. Significantly, most of the offspring of these species are sterile workers, which makes the maintenance of a stable population easier.

Most of our common amphibians take no care of their eggs. After fertilization, they are simply left to develop in the water. The gelatinous layers of albumin surrounding the egg absorb water, swell, and provide some physical protection for the egg. They also aid in keeping the egg somewhat warmer than the surrounding water. The clear jelly is transparent to the visible portion of the sun's rays. The dark upper surface of the developing egg absorbs these rays effectively. The light energy is converted into heat which warms the embryo. The jelly is, however, a good heat insulator and less energy radiates from the embryo than was received by it. This physical phenomenon is familiar to anyone who has entered an automobile that has been standing in the sun for some time with the windows closed.

The dark upper surface of the fertilized egg also provides protective coloration. When viewed from above, the egg is hard to distinguish against the dark background of the pond bottom. Similarly, the light underside of the egg blends with the sky when viewed from underneath.

Reptile eggs are fertilized in the genital tract of the female. Then a waterproof shell is deposited around each as it passes down the oviduct. This solves the problem of maintaining the embryo in moist surroundings after the egg is laid. In most species, the eggs are buried in a warm location and abandoned by the mother. There is a plentiful supply of yolk and when the young hatch, they are replicas, although smaller, of their parents. They are also quite self-sufficient.

In some reptiles, e.g. the common garter snake, the eggs are retained in the mother's body until the young hatch. Their nourishment throughout this period is derived chiefly from the yolk of the egg rather than directly from the mother's tissues.

Not long after reptiles first appeared on earth, some species returned to an aquatic environment. Today, one species of lizard (the marine iguana), some snakes and turtles, and the various crocodilians spend most of their lives in water. Nevertheless, those that lay their eggs return to land to do so.

A loggerhead turtle lays about 125 eggs in a season. Although left unprotected by the parents, the mortality rate is sufficiently low so that the population is maintained. Some reptiles are among the most long-lived animals on earth. An old female turtle has had a good many opportunities to contribute to the maintenance of the population.

Birds are also fully adapted to life in dry locations. Like many of the reptiles (from which they are descended), they lay shelled eggs. These are often carefully tended by the parents. The eggs are richly supplied with yolk, and the young hatch with most of the adult structures. The parents continue caring for the newly hatched young until they are ready to care for themselves. Appropriately, birds rarely lay more than 20 eggs in a season and the average is probably closer to four.

The shelled egg is obviously an important adaptation permitting bird and reptile embryos to develop away from water. A close examination of a common bird egg, the hen's egg, will reveal why this is so. The egg itself consists of a large quantity of yolk and a tiny bit of cytoplasm. After fertilization, but while still within the mother's oviducts, the egg is surrounded with thick layers of watery albumin (egg "white") and a shell made of calcium carbonate. The shell is permeable to gases, but practically impermeable to water. As the embryo develops from the zygote, four special membranes are formed (Fig. 21.8). These "extraembryonic" membranes develop from the embryo but remain part of it only during its existence within the egg. These four membranes are the following.

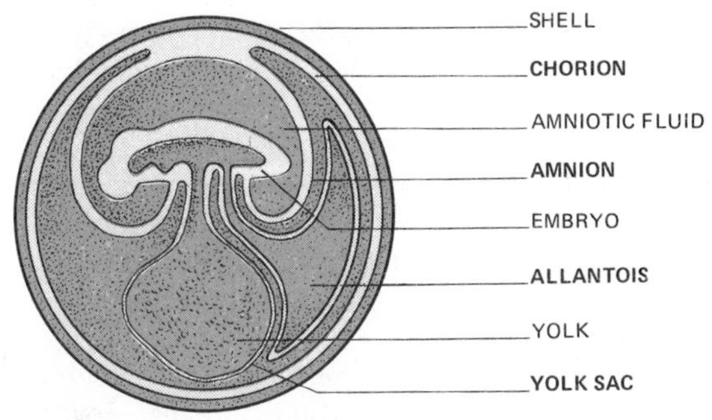

FIG. 21.8 The extraembryonic membranes of the developing chick. Note that the yolk sac and allantois are outgrowths of the embryo digestive tract. The same membranes are produced by the embryos of reptiles and mammals.

FIG. 21.9 Mammals that lay eggs. The duckbill platypus (left) raises its young in a nest; the spiny anteater (right) places them in a pouch on its abdomen. (Courtesy of the New York Zoological Society.)

1. The **yolk sac,** which surrounds the yolk. It connects the embryo with its major source of food. Fats constitute most of the food in the yolk. This is a valuable adaptation, since the oxidation of fats supplies large amounts of both energy and water.
2. The **amnion,** which grows around the embryo, enclosing it in a fluid-filled cavity.
3. The **chorion,** which lines the inner surface of the egg shell. It aids in the exchange of gases (O_2 and CO_2) between the embryo and the outside air.
4. The **allantois,** which serves as a reservoir for the metabolic wastes (chiefly uric acid) excreted by the embryo during its development. As the allantois grows larger, it also participates in gas exchange.

With these four membranes, the developing embryo is able to carry on essential metabolism while sealed within the egg. Surrounded by the fluid in the amniotic cavity, the embryo is kept as moist as the fish embryo in a pond.

These four membranes are produced by the embryos of birds, reptiles, and also mammals. However, in mammals, except for the rare echidnas and duckbill platypus (Fig. 21.9), the membranes are not associated with a shelled egg. Mammalian eggs, which are poorly supplied with yolk, are retained within an oviduct after fertilization. The extraembryonic membranes penetrate the walls of an enlarged portion of the oviduct, the **uterus.** After the small supply of food within the egg is exhausted, additional nourishment is secured through exchange with the mother's circulatory system.

In the marsupials (of which only the opossum is found in this country), the yolk sac serves this function for a time. However, it does not become well established, and as a consequence, the young are born in a very immature state (Fig. 21.10). They are able, though, to crawl into a special pouch on the mother's abdomen. There they attach themselves to her nipples and consume milk from her mammary glands until they are developed enough to fend for themselves.

In all other mammals, the extraembryonic membranes form a **placenta** and **umbilical cord** which connect the embryo to the mother's uterus in a more elaborate and efficient way. The blood supply of the embryo is continuous with that of the placenta. Although the capillaries of the placenta are in close contact with the mother's blood supply to the uterus, actual intermingling of the blood of the mother and the embryo does not normally occur. The placenta

FIG. 21.10 Newborn opossums. Eighteen of them fit easily into a teaspoon. (Courtesy of Dr. Carl G. Hartman.)

extracts not only food but also oxygen from the uterus. Carbon dioxide and other excretory wastes (e.g. urea) are similarly transferred to the mother's circulatory system for ultimate disposal by her excretory organs.

Although development continues within the uterus for a longer time than is characteristic of marsupials, there is a great variation among the placental mammals in the degree of helplessness of the newborn. In all cases, however, a period of feeding on milk supplied by the mother's mammary glands occurs. In some species, care of the young may be extended to include training in behavior patterns characteristic of the species.

Human Reproduction

21.5 THE SEX ORGANS OF THE MALE

The sexual apparatus of the male has two major reproductive functions to accomplish: the production of sperm cells and the delivery of these cells to the reproductive tract of the female.

Sperm production takes place in the **testes** (Fig. 21.11). Each testis is packed with myriads of seminiferous tubules (laid end-to-end, they would extend more than 200 meters). The walls of these tubules consist of diploid spermatogonia, the precursors of the sperm. The process of converting a spermatogonium into sperm includes the two consecutive cell divisions that constitute meiosis. Thus, each spermatogonium gives rise to four sperm cells (Fig. 21.1).

An adult human male manufactures over one hundred million sperm cells each day. These gradually move through the vasa efferentia, which drain the seminiferous tubules, and into the epididymis, where they undergo further maturation and are stored. Despite the rate of sperm production, there is no danger of exhausting the supply of spermatogonia. This is because spermatogonia also divide by *mitosis* and thereby maintain their population.

The testis is a dual-purpose organ. In addition to manufacturing sperm, it is an endocrine gland. Its

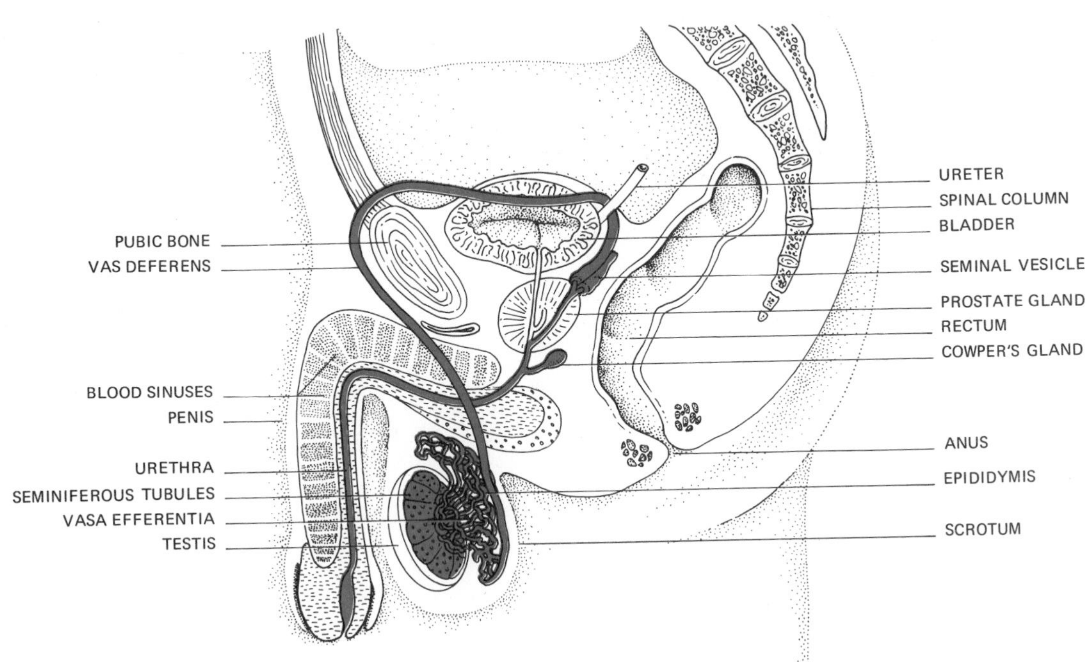

FIG. 21.11 Reproductive organs of the human male.

hormone, **testosterone,** is the chief male sex hormone. It is responsible for the development of the so-called secondary sex characteristics of males, such as the beard, deep voice, and masculine body contours. It is also essential for the production of sperm.

Testosterone is manufactured by cells that lie between the seminiferous tubules. These cells are, in turn, the target cells of a hormone, LH, manufactured by the anterior lobe of the pituitary gland located at the base of the brain. The initials LH stand for *l*uteinizing *h*ormone in recognition of the function which this hormone plays in females and which we shall examine shortly.

A second pituitary gland hormone, FSH (also named for its role in females), acts directly on the spermatogonia to stimulate sperm production. But LH is *indirectly* needed for sperm production because testosterone is also essential for the process, as we have seen.

The activity of the pituitary gland is in turn under certain controls. Special nerve cells in a part of the brain called the hypothalamus synthesize one or more **releasing hormones.** These substances pass directly to the pituitary gland, carried by a system of small veins connecting the two structures. The existence of a direct link between the brain and the hormones involved in sex explains the well-established connection between nervous activity and sexual be-

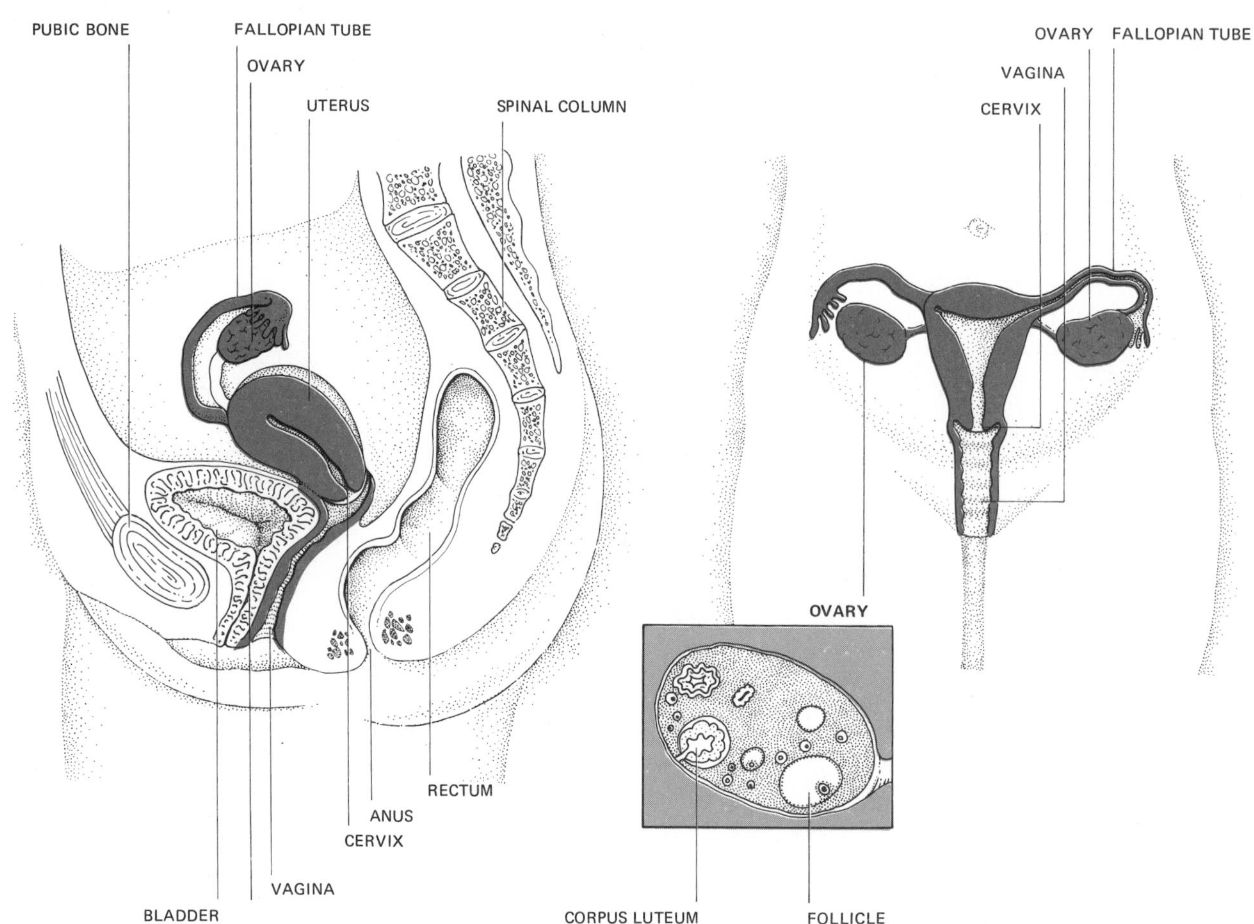

FIG. 21.12 Reproductive organs of the human female.

havior. Sexual behavior in many male vertebrates is controlled or influenced by such external stimuli as length of day, odors released by females of the species, the sight of courtship activity on the part of the female, etc. These stimuli are detected by the nervous system. In addition to leading to the overt events of breeding behavior, they may initiate profound changes in the physiology of the sex organs, testosterone production, etc. In this and other ways, hypothalamic releasing hormones provide links between the coordinating machinery of the nervous system and that of the endocrine system.

21.6 THE SEX ORGANS OF THE FEMALE

The responsibility of the female mammal for successful reproduction is considerably greater than that of the male. It is therefore not surprising to find that her reproductive physiology is considerably more complex than his. She must not only manufacture her sex cells (eggs) but be equipped to: (1) receive sperm from the male, (2) provide the right conditions for fertilization to occur, and (3) be able to nourish the developing baby not only before birth but after.

Egg production takes place in the **ovaries** (Fig. 21.12). In contrast to the situation in males, the initial steps in egg production occur prior to birth. By the time the fetus is 15 weeks old, all the oogonia that she will ever possess have been formed. Hundreds of thousands of these diploid cells enter upon the first steps of meiosis and then stop. No further development takes place until years later, when the girl begins to become sexually mature. Then the eggs begin to complete their development, usually one at a time and once a month. The egg grows much larger and completes its first meiotic division. These events occur within a **follicle,** a fluid-filled envelope of cells surrounding the egg. The ripening follicle also serves as an endocrine gland. Its cells begin to produce a mixture of steroid hormones collectively known as **estrogens.** These are responsible for the development in the girl of the secondary sexual characteristics of a mature woman, such as broadening of the pelvis and development of the breasts. The reproductive organs enlarge and hair develops around the external genitals and in the arm pits. Estrogens also promote the development of fatty tissue and these lead to the more rounded body contours characteristic of adult women.

Estrogens continue to be secreted throughout the reproductive years of women. During this period they play an essential role in the monthly menstrual cycle. About once every 28 days (cycles somewhat longer or shorter or even irregular in length are found in perfectly healthy women) a small amount of blood and other products of tissue disintegration are discharged from the uterus. This process is **menstruation.** It continues for four or five days, during which time a new follicle begins to develop in one of the ovaries. After menstruation ceases, the follicle continues to develop, producing an increasing amount of estrogens as it does so (Fig. 21.13).

The follicle cells are stimulated to produce estrogens by the combined influences of FSH (*f*ollicle *s*timulating *h*ormone) and LH (*l*uteinizing *h*ormone) which are secreted, as in males, by the pituitary gland. And, as we have seen is the case in males, the secretion of FSH and LH is, in turn, regulated by a releasing factor (or factors) secreted by nerve cells in the hypothalamus of the brain. Furthermore, the entire system is under the same sort of negative feedback control that we found for the antidiuretic hormone (ADH—see Section 13.6). The presence of estrogens in the blood (testosterone in males) *suppresses* the releasing factor activity of the hypothalamus. This, then, reduces the secretion of FSH and LH which, in turn, lowers the levels of estrogens (Fig. 21.14). Thus secretion of estrogen in females (and testosterone in males) is partly, at least, controlled by itself—an important homeostatic mechanism.

The rising level of estrogens during the menstrual cycle profoundly influences the structure of the endometrium, the inner lining of the uterus. It becomes much thicker (Fig. 21.13) and more richly supplied with blood. During this period, profound changes occur within the follicle as well. Under the influence of LH, the developing egg completes its meiotic division (having initiated it many years before). In contrast to meiosis in sperm production, the cytoplasm of the cell is not divided equally. One of the two cells produced at each meiotic division is very small (the polar body—Fig. 21.3) and receives little more than one set of chromosomes.

About two weeks after the onset of menstruation, **ovulation** occurs. In response to a sudden surge of LH (Fig. 21.13) the follicle ruptures and discharges the now-mature haploid egg. The egg is then swept into the open end of the Fallopian tube (Fig. 21.12) and begins to move slowly down the length of the tube.

Having discharged its egg, the job of the follicle is by no means complete. Stimulated by LH, it be-

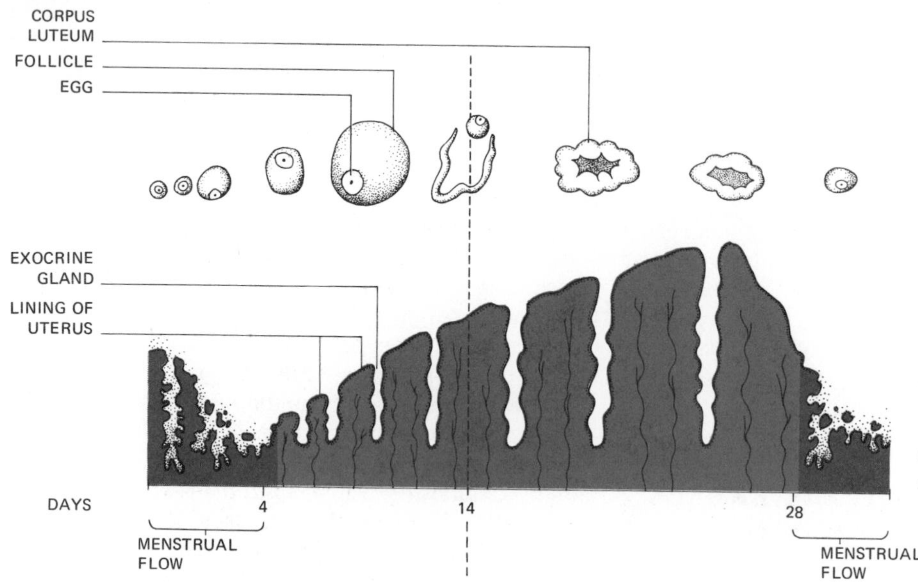

FIG. 21.13 Events occurring in the ovary (top), changes in the lining of the uterus (color), and changes in hormone levels in the blood during the menstrual cycle.

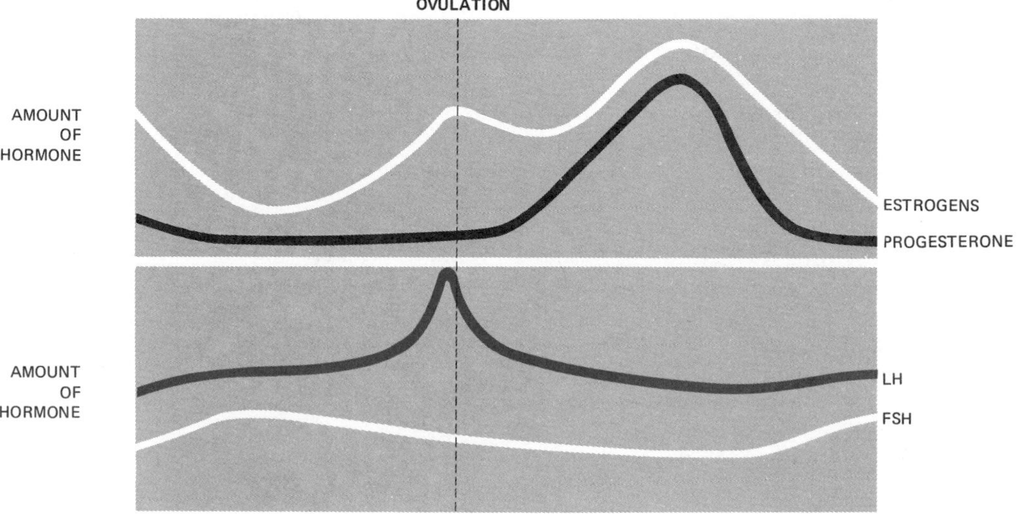

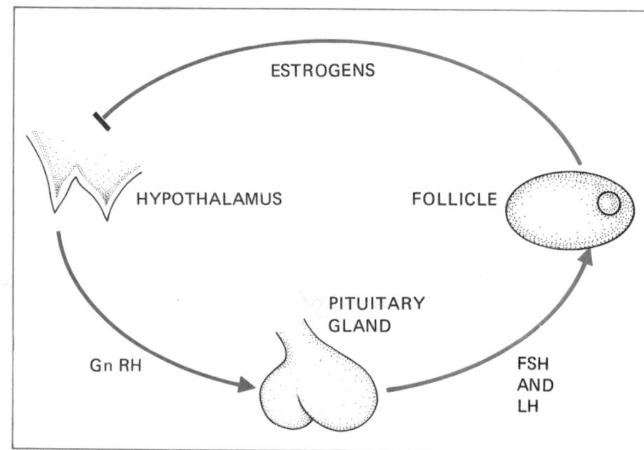

FIG. 21.14 Control mechanisms that regulate the levels of sex hormones in females. A rising level of estrogens *suppresses* the releasing factor activity of the hypothalamus (bar). This leads finally to a drop in the level of estrogens and hence to renewed activity of the hypothalamus. A similar mechanism exists in males. The releasing factor for FSH and LH production is also called gonadotropin releasing hormone (GnRH).

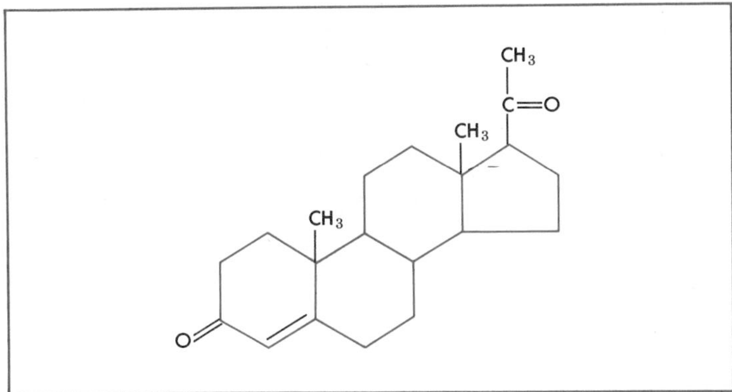

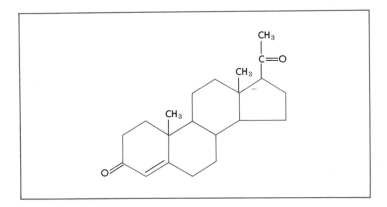

FIG. 21.15 Progesterone, one of the female sex hormones. Like all steroids, it is constructed from a skeleton of 17 carbon atoms (color).

comes transformed into the **corpus luteum** (hence the name *lute*inizing hormone). The corpus luteum is also an endocrine gland, secreting the steroid hormone **progesterone** (Fig. 21.15). For about ten days, the high blood levels of progesterone continue the preparation of the uterus for possible pregnancy, inhibit contraction of the uterus, and inhibit the development of a new follicle.

If fertilization does not occur (which is almost always the case), production of progesterone begins to decline about the 26th day of the cycle. The corpus luteum degenerates, the lining of the uterus begins to break down, and by the 28th day, the menstrual flow begins again. Withdrawal of progesterone lifts the inhibition on uterine contractions (hence the cramps sometimes associated with the first day or two of menstruation) and the inhibition on follicle development. The cycle has come full circle.

The menstrual cycle continues for many years. But eventually, usually between 42 and 52 years of age, the follicles become less responsive to FSH and LH. They begin to secrete less estrogen. Ovulation and menstruation become irregular and finally cease altogether. This cessation of the menstrual cycle is called the **menopause.**

With levels of estrogens now running one-tenth or less of what they had been, the hypothalamus is released from their inhibitory influence (Fig. 21.14). As a result, it now stimulates the pituitary gland to *increased* activity. The concentration of FSH and LH in the blood rises to ten or more times the former values.

A similar shift occurs in aging men with a rise in FSH and LH accompanying a decline in the output of testosterone. However, this transition in males is very gradual. There is not the abrupt shift characteristic of the menopause.

21.7 COPULATION AND FERTILIZATION

For fertilization to occur, sperm must be deposited in the vagina fairly near to the time of ovulation. Sperm transfer is accomplished by copulation. As a result of sexual excitation, the arterioles supplying blood to the penis dilate. Blood begins to accumulate in three cylindrical spongy sinuses that run lengthwise through the penis. The resulting pressure causes the penis to become enlarged and erect and thus able to penetrate the vagina. Movement of the penis back and forth within the vagina causes sexual tension to increase to the point of ejaculation. Contraction of the walls of the ducts propels the sperm to the vas deferens. Fluid is added to the sperm by the **seminal vesicles** and the **prostate gland** (Fig. 21.11). These fluids provide a source of energy (fructose) and perhaps in other ways provide an optimal chemical environment for the sperm. The mixture of sperm and accessory fluids is called **semen.** It passes through the urethra and is expelled into the vagina.

Physiological changes occur in the female as well as the male in response to sexual excitement, although these are not as readily apparent. In contrast to the male, however, such responses are not a prerequisite for copulation and fertilization to occur.

Once deposited within the vagina, the sperm proceed on their journey into and through the uterus and on up into the Fallopian tubes. It is here that fertilization may occur if a ripe egg is present.

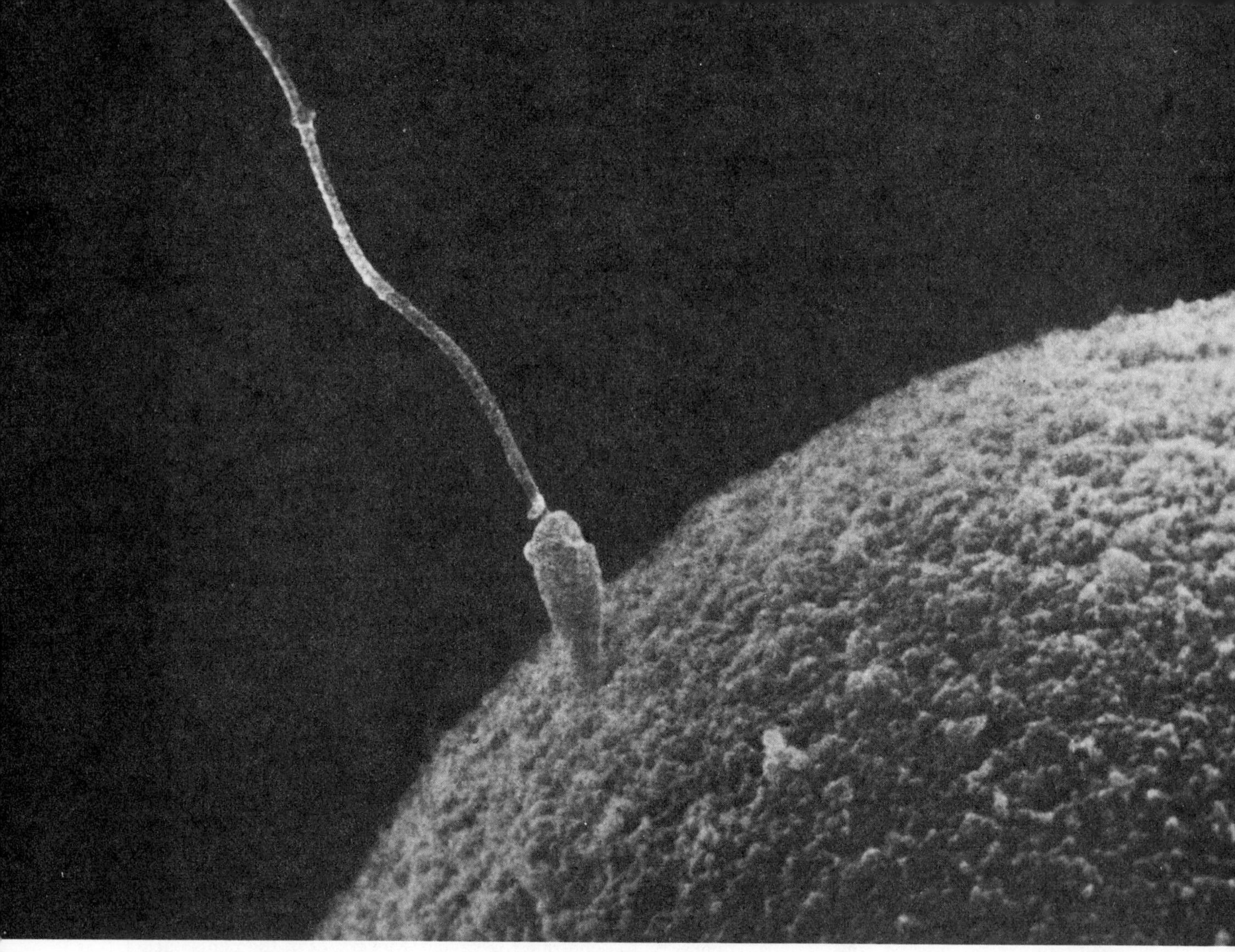

FIG. 21.16 The moment of fertilization of a sea urchin egg as seen with the scanning electron microscope. (Courtesy of Dr. Don W. Fawcett.)

Although sperm can swim several millimeters each second, their trip to the Fallopian tubes is probably assisted by muscular contractions of the walls of the uterus and the tubes. In any case sperm may reach the egg within 15 minutes of ejaculation. The trip is also fraught with heavy mortality. An average ejaculate contains several hundred million sperm but only a few thousand complete the journey. And of these, only one will succeed in entering the egg and thus fertilizing it (Fig. 21.16).

21.8 PREGNANCY AND BIRTH

Embryonic development begins while the fertilized egg is still within the Fallopian tube. The developing embryo continues to travel down the tube, reaching the uterus in two or three days. As a result of repeated mitotic divisions, a hollow ball of cells is formed called the **blastocyst** (Fig. 21.17). Approximately one week after fertilization, the blastocyst imbeds itself in the thickened wall of the uterus, a process called **implantation.** With successful implantation, pregnancy is established.

Development of the blastocyst continues with both rapid cell division and some migration of cells from place to place within the developing embryo. As a result, two major divisions of cells and tissues are produced: (1) the embryo proper, which will ultimately become the baby, and (2) the **extraembryonic membranes,** which will play a number of

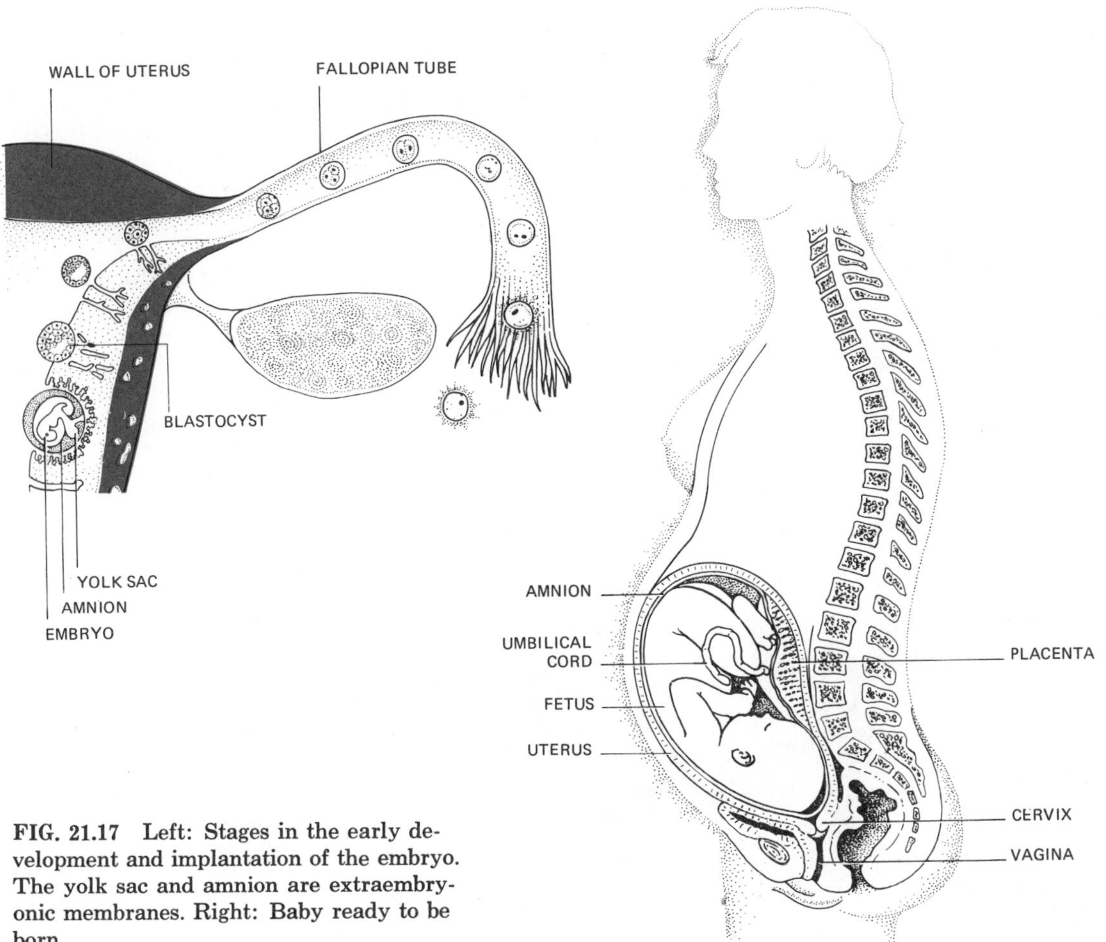

FIG. 21.17 Left: Stages in the early development and implantation of the embryo. The yolk sac and amnion are extraembryonic membranes. Right: Baby ready to be born.

vital roles in the process but which will be discarded at the time of birth (Fig. 21.17). The extraembryonic membranes form the amnion, placenta, and umbilical cord. The **amnion** is a sac that surrounds the embryo and is filled with amniotic fluid. The **placenta** grows in contact with the uterine wall. Its vessels, through which the growing baby's blood flows, are literally bathed by the blood (the mother's) within the uterus. Although there is normally no mixing of mother's and baby's blood, the placenta does make easy the transfer of a variety of materials between the baby and the mother. In a sense, the placenta serves as the baby's intestine (receiving food), lung (receiving oxygen and discharging carbon dioxide), and kidney (discharging urea). Even some proteins, for example the mother's antibodies, can cross the placenta. The **umbilical cord** connects the developing baby to the placenta.

During the first two months of pregnancy the basic structure of the baby is being formed. This involves cell division, cell migration, and the development of cells into the types found in the adult organism (for example, blood cells, nerve cells, etc.). During this period the developing organism, called an **embryo,** is critically sensitive to anything that interferes with the steps involved. Virus infection of the mother, for example by rubella ("German measles") virus, or exposure to certain chemicals may cause serious malformations in the developing embryo. Such agents are called teratogenic ("monster-forming"). The tranquilizer thalidomide, taken by many pregnant European women between 1954 and 1962, turned out to be a teratogen and was responsible for the birth of several thousand grossly deformed children.

After about two months, all the systems of the baby have been established in a rudimentary way. From then on, development of the **fetus**, as it is now called, is primarily a matter of growth and minor structural modifications. The fetus is far less susceptible to the actions of teratogenic agents than is the embryo.

The establishment of pregnancy interrupts the menstrual cycle. Implantation of the embryo prevents the deterioration of the corpus luteum that ordinarily occurs towards the end of the fourth week of the cycle. Consequently, progesterone secretion continues. If, during the first five months or so of pregnancy, progesterone secretion by the corpus luteum should be stopped (e.g., by surgical removal of the ovary), uterine contractions soon begin and the embryo or fetus is born prematurely (aborted). But after this period, the corpus luteum is no longer indispensable. The *placenta* itself secretes progesterone and this is ample to maintain pregnancy for the normal duration.

The placenta has other endocrine functions as well. In addition to progesterone, it secretes estrogens and a hormone somewhat similar to the gonad-stimulating hormones (FSH and LH) of the pituitary gland. (Secretion of FSH and LH is inhibited during pregnancy and thus the development of new follicles is blocked during this period.)

Exactly what brings about the onset of labor is

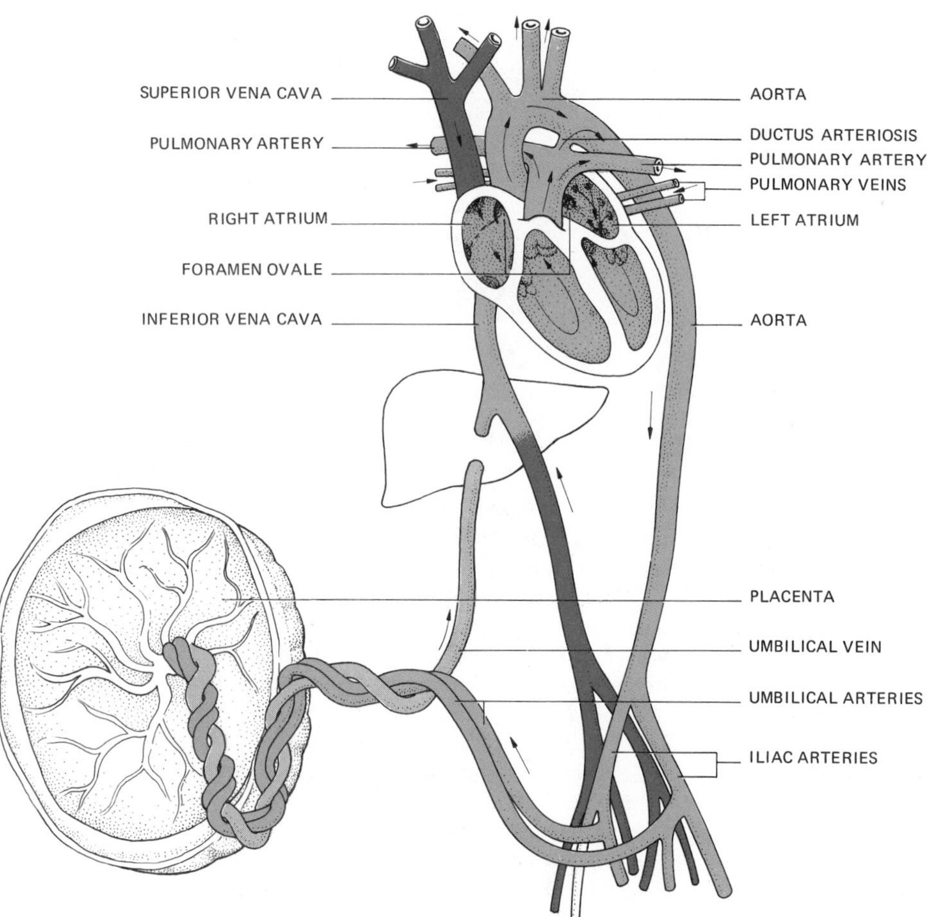

FIG. 21.18 Circulation in the human fetus. Oxygen is picked up at the placenta, not at the lungs. Most of the blood bypasses the lungs by passing through the foramen ovale and the ductus arteriosus.

still not understood. Probably a variety of integrated hormonal controls are at work. Progesterone secretion by the placenta diminishes, and this removes the inhibition that has kept the uterus from contracting during pregnancy. Other hormones, which cause strong contractions of the uterus, make their appearance at this time. One of these is **oxytocin.** Oxytocin (like ADH—see Section 13.6) is manufactured in the hypothalamus of the brain and released to the blood from the posterior lobe of the pituitary gland. There is some evidence that it is the baby itself that is the chief source of oxytocin and, if so, we must give the baby credit for playing a major role in the onset of labor. A number of closely related substances called **prostaglandins** (first identified in semen) also appear in the mother's blood and in the amniotic fluid at the time of birth. Prostaglandins, like oxytocin, cause the uterus to contract vigorously and thus labor begins.

The first result of labor is the opening of the cervix. With continued, powerful contractions, the amnion ruptures and the amniotic fluid flows out through the vagina. Later, the baby is expelled. At this time, the umbilical cord can be severed. The infant's lungs expand, and it begins breathing (sometimes with assistance). While this is a seemingly simple step, it actually entails a major switchover in the circulatory system. Blood flow through the umbilical cord ceases, and the adult pattern of blood flow through the heart, aorta, and pulmonary arteries is established (Fig. 21.18). Shortly after the arrival of the baby, the placenta and the remains of the umbilical cord ("the afterbirth") are expelled.

Within two or three days following birth, the mother's breasts begin to secrete milk. This, too, requires the coordinated interaction of hormonal and nervous stimuli. The stimulation of nursing, working through the hypothalamus, leads to the secretion of several hormones (among them oxytocin) that are essential to milk synthesis and release.

21.9 REPRODUCTIVE ENGINEERING: THE PROSPECTS

The rapid advances that have been made in the understanding of human reproductive physiology raise the prospect of manipulating the process in ways not hitherto possible.

One technological development that is already upon us is the ability to freeze and store human semen for future use. Later (perhaps even after decades) the semen can be thawed and introduced into the female reproductive tract. This procedure was originally developed as a convenient means of enabling prize bulls to inseminate cows anywhere around the world, thus quickly improving cattle breeds. Now the techniques are being used with humans. Several sperm banks have been created in the United States. Although thawed semen is not, as yet, quite as good as fresh semen (an average of 14 injections is needed to make a woman pregnant), the fact that the method works at all means that a man can now father a child after his own death (or after sterilization, which appears to be the more immediate prospect). Furthermore, there is no technical reason why he could not become the father to great numbers of children born to many different mothers.

The transfer of a fertilized egg or early embryo from the female who produced it to a "foster mother" who will carry it until birth is done routinely with mice. In fact, early mouse embryos have been implanted successfully in foster mothers after being frozen and stored for over a week. R. G. Edwards and his colleagues at the University of Cambridge in England have succeeded in removing *human* eggs from their follicles (hence prior to ovulation), fertilizing them *in vitro* (in the "test tube"), and growing them to the blastocyst stage of embryonic development. In 1974, another English scientist reported that three such test tube fertilized embryos had been successfully implanted in their "natural" mothers and had gone on to develop into healthy babies. If this work is substantiated, it would permit mothers with closed Fallopian tubes to bear their own children. Furthermore, if for some reason a woman could not carry a baby to full term, implantation of the embryo in the uterus of a foster mother would still enable couples to have children that were genetically entirely their own.

The ability to maintain early embryos in culture for a short time would also make it possible to determine their sex before implantation. A few cells could be removed and checked for the presence of a *Y* chromosome or a Barr body (see Section 16.3). In this way, only an embryo of the desired sex need be implanted. This technique has already been used successfully with rabbits. If adapted to humans, it would permit, for example, carriers of severe *X*-linked traits, such as hemophilia, to elect not to run the risk of raising an affected son.

Actually, it is already possible to learn a great deal about the genetic makeup of the fetus. During

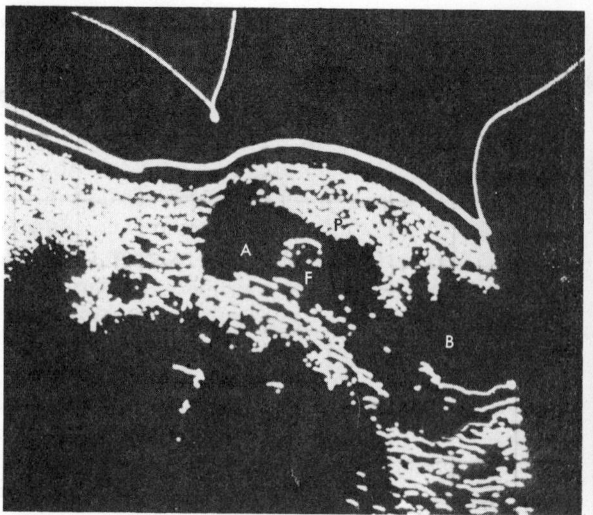

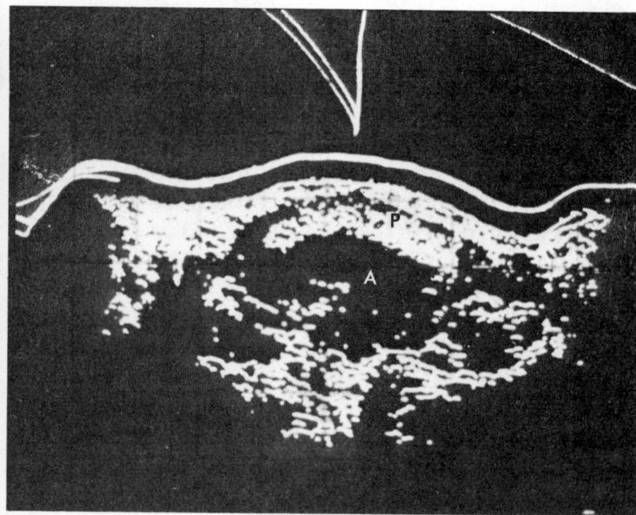

FIG. 21.19 Using ultrasound to locate the position of the placenta prior to amniocentesis. These sonograms are made by recording the echoes received from structures within the abdomen. A, amniotic cavity; B, urinary bladder; F, part of fetus; P, placenta. Both longitudinal (left) and transverse (right) scans are needed for accurate localization of the placenta. (Courtesy of the Downstate Medical Center of the State University of New York.)

its development, the fetus sheds cells into the amniotic fluid. Without too much difficulty, some of the amniotic fluid can be removed, a technique known as **amniocentesis** (Fig. 21.19). Separating the cells and culturing them enables the physician to look for chromosome abnormalities (e.g. the three number 21 chromosomes of Down's syndrome), certain enzymatic defects caused by mutations (e.g. an inability to metabolize galactose, hence milk), and, of course, the sex of the embryo. Approximately 100 genetic disorders can now be diagnosed by amniocentesis, and the pregnancy deliberately terminated if a disorder is present.

Not yet practicable for humans but by no means in the realm of science fiction is the production of children with *four* parents. For a number of years research workers have routinely produced such "tetraparental" mice. Early embryos at the eight-cell stage of development are removed from two different pregnant mice and placed in culture. If two embryos, one from each mouse, are pushed together, they often fuse into a single embryo. After a period

FIG. 21.20 A tetraparental mouse. This mouse had two fathers and two mothers (not including the foster mother that gave birth to it). One pair of parents was black; the other pair was white. Its skin, as well as all its other organs, is made up of a mixture of cells from the black parents and cells from the white parents. Such an animal is called a genetic mosaic. Note that this mouse is not the same as an F_1 hybrid produced by mating a white mouse with a black mouse. In that case, all the cells would be of the same genotype (and the coat would have been a uniform brown). Photo (and mouse) courtesy of Dr. Thomas G. Wegmann.

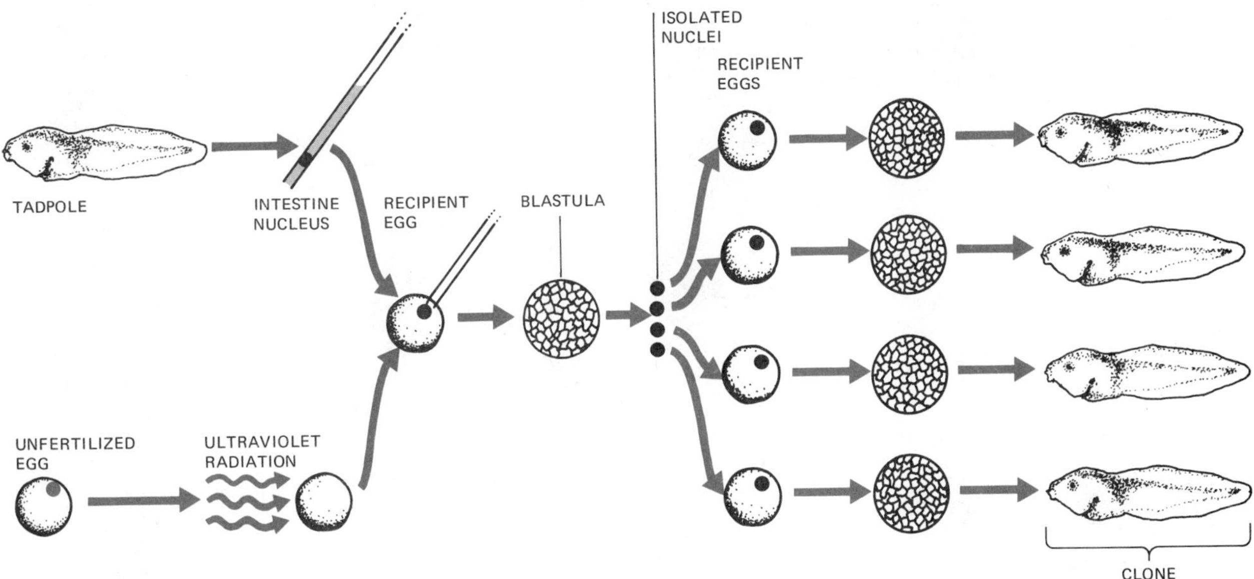

FIG. 21.21 Nuclear transplant procedure. A nucleus withdrawn from a single cell (in this case in the intestine) is implanted into an egg whose own nucleus has been destroyed by radiation. After the egg has developed by mitosis into a mass of genetically identical cells (the blastula), the process can be easily repeated with these as the source of nuclei. In this way a *clone* of animals can be produced.

of further growth, this fused embryo can be implanted in a foster mother and go on to develop normally. The mouse that results is a *mosaic* (see Section 16.3) inasmuch as every organ is made up of some cells derived from one pair of parents and some cells derived from the other pair (Fig. 21.20). Such tetraparental mice have proved to be valuable tools for investigating a number of important biological problems. Whether tetraparental humans would be a blessing or a curse remains to be seen.

Another awesome possibility in the years ahead is the development of a method by which humans could reproduce *asexually*. In accomplishing this, we would be creating **clones** of genetically identical individuals just as we did with McIntosh apples (see Section 14.3). At present, except for a few "lower" forms like certain worms, no isolated *animal* tissues can grow into new individuals in the way that a scion of an apple tree can (Fig. 14.8). Only a fertilized egg has, as yet, the potential to develop into an entire animal. And fertilized eggs, as we have seen, arise by sexual reproduction. Thus they acquire their genes from two parents, each of which has reshuffled its double set of genes during meiosis. Therefore, animals—even though brothers and sisters—are genetically diverse. Only on those rare and unpredictable (usually) occasions when *single* embryos separate into two or more parts, which go on to develop into identical twins, triplets, etc., are small clones of animals created.

But the process can now be accomplished in the laboratory. Many years ago, Boveri showed that an unfertilized sea urchin egg could have its own nucleus removed ("enucleated") and then develop according to the genes inserted into it by another nucleus (look back at Figure 5.9). However, the only way Boveri could insert the new nucleus was by exploiting the natural fertilizing properties of sperm. And these, arising by meiosis, differ genetically even though produced by one individual.

But what if one could collect nuclei that have arisen by *mitosis*—and thus are genetically identical—and place these in enucleated eggs, one in each? This has now been done. J. B. Gurdon, for example,

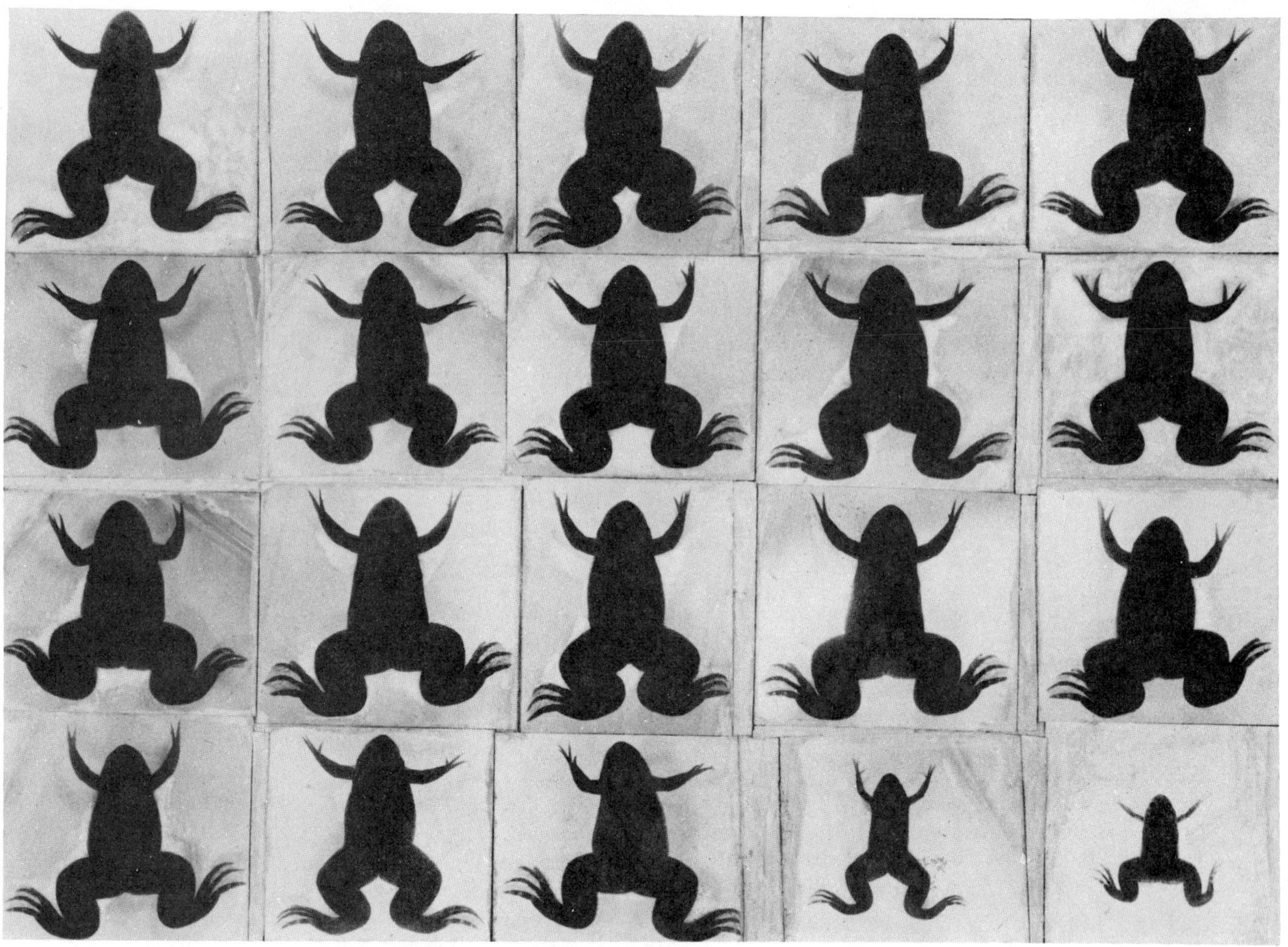

FIG. 21.22 Twenty genetically identical South African clawed frogs. Each animal grew from an enucleated egg that received one of a set of genetically identical nuclei. The smaller frogs were younger than the others but nonetheless possessed precisely the same genes. (Courtesy of J. B. Gurdon, from *J. Hered.* 53 (1), 5, Jan.–Feb., 1962.)

has removed nuclei from epithelial cells of the South African clawed frog and injected these into eggs of that species whose own nucleus has been destroyed (Fig. 21.21). In a small fraction of these nuclear injections, the egg proceeds to develop into a perfectly normal animal.

Furthermore, if one uses cells from an early embryonic stage of development, e.g. the blastula (Fig. 21.21), the success rate is much higher. So high, in fact, that by using nuclei derived from the cells of a single blastula, a substantial number of genetically identical frogs can be created (Fig. 21.22). In this way, a clone can be produced that is entirely analogous to the clones propagated by apple growers.

Amphibian eggs are much larger than human eggs and consequently easier to manipulate. But even if the practical problems of human cloning are, as yet, insurmountable, there appear to be no *theoretical* reasons for believing that what works with the South African clawed frog could not someday work with an outstanding quarterback!

Our discussion of methods of manipulating human reproduction should not ignore the most widespread techniques of all: those contraceptive measures by which couples block the start of pregnancy. These include mechanical barriers to the passage of sperm, the intrauterine device (IUD) which appears to prevent implantation, and the use of estrogen-progesterone preparations (the "pill") which block ovulation. (The function of these hormones will be considered further in Chapter 25.) Finally (and it is at present generally quite final), pregnancy can be avoided by tubal ligation or vasectomy. In tubal ligation, the Fallopian tubes are closed so that no egg can be fertilized (Fig. 21.23). Ovarian function continues, however, including the normal hormonal cycle. In vasectomy, each vas deferens is cut near the top of the scrotum (Fig. 21.11) and tied. This is a minor surgical procedure that is performed in the doctor's office, under local anesthesia, in 30–40 minutes. The hormonal activity of the testes is not interrupted in any way. Nor does the operation stop the production of the various glandular secretions which make up the bulk of the semen. Hence copulation and ejaculation proceed perfectly normally.

Whatever the technological methods employed, reproductive engineering constitutes a special case of genetic engineering (see Section 18.8) and raises the same ethical problems in addition to some special ones of its own. It is to be hoped that the associated ethical and legal questions will be dealt with before the need to do so becomes critical.

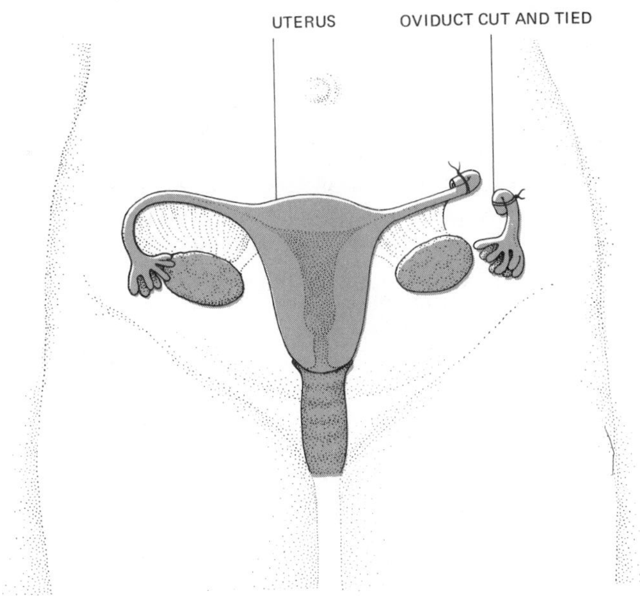

FIG. 21.23 Tubal ligation. Because of the need for an incision in the abdominal wall, this procedure must be carried out under anesthesia. It is a more elaborate operation than vasectomy.

EXERCISES AND PROBLEMS

1. The retention of fertilized eggs within the mother's body occurs commonly among the cartilaginous fishes and the reptiles. Why do you suppose no birds employ this means of protecting their eggs?
2. Distinguish between sperm and semen.
3. In what ways is egg formation in the frog similar to megaspore formation in the bean? In what ways is it different?
4. What is the difference between amplexus and copulation?
5. Where does fertilization occur in (a) the trout, (b) the shark, (c) the queen bee, (d) the frog, and (e) the human?
6. If a family has had seven daughters and no sons, what are the chances that their next child will be a son?
7. Neglecting crossing over, how many kinds of sperm cells can be produced by *Ascaris bivalens* ($n = 2$)? How many kinds of egg cells can be produced?
8. Few animals are hermaphroditic, while most plants are. Can you think of an explanation for this?
9. The number of eggs produced by the members of a species can be related to the amount of care given to the developing young. Is the same true of the number of sperm produced? Explain.
10. What functions do the centrioles of the spermatogonia carry out (a) in sperm formation, (b) in sperm transfer, (c) in fertilization?

REFERENCES

1. MONROY, A., *Fertilization and Its Biochemical Consequences,* Addison-Wesley Modules in Biology, No. 7, Addison-Wesley, Reading, Mass., 1973.
2. EDWARDS, R. G., and RUTH E. FOWLER, "Human Embryos in the Laboratory," *Scientific American,* Offprint No. 1206, December, 1970.
3. FRIEDMANN, T., "Prenatal Diagnosis of Genetic Disease," *Scientific American,* Offprint No. 1234, November, 1971.
4. VANDER, A. J., SHERMAN, J. H., and DOROTHY S. LUCIANO, *Human Physiology,* 2nd ed., McGraw-Hill, New York, 1975. Chapter 14 includes a thorough description of the physiology of the human sex organs, copulation, pregnancy, birth, and lactation.
5. HINTON, H. E., "Insect Eggshells," *Scientific American,* Offprint No. 1187, August, 1970. How their structure permits the free exchange of O_2 and CO_2 while minimizing water loss.
6. PATTON, S., "Milk," *Scientific American,* Offprint No. 1147, July, 1969.
7. SEGAL, S. J., "The Physiology of Human Reproduction," *Scientific American,* September, 1974. Reprinted in *The Human Population: A Scientific American Book.* Freeman, San Francisco, 1974. Available in paperback.

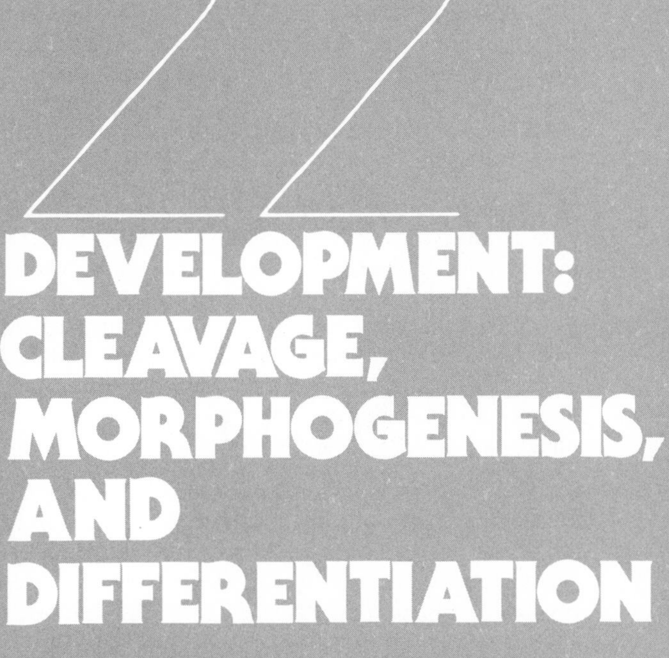

22 DEVELOPMENT: CLEAVAGE, MORPHOGENESIS, AND DIFFERENTIATION

22.1 STAGES IN THE DEVELOPMENT OF THE ADULT

In the past two chapters we have examined the ways in which gametes are manufactured and brought together so that fertilization can occur. Fertilization is not the end of the process of sexual reproduction. It is, in fact, the start of a series of elaborate and well-organized changes which eventually give rise to a new adult of the species. We use the term *development* to describe these changes. Although the exact pattern of development varies from one species to another (especially if they are not closely related) and there is some overlapping, one can usually distinguish the following stages.

1. **Cleavage.** During this stage of development, the zygote nucleus undergoes a series of mitotic divisions. The resulting daughter nuclei usually become partitioned off in separate cells fashioned from the cytoplasm of the zygote. There is little or no growth during this stage.

2. **Morphogenesis.** During this stage, the many cells produced by cleavage continue dividing but also move about and organize themselves into distinct layers and masses. As a result, a definite pattern appears. It is this development of pattern that is called morphogenesis. Although the cells of the embryo during this phase of development are organized into distinct groups, they are all rather similar in structure.

3. **Differentiation.** Soon, however, the cells of the developing embryo begin to take on the specialized structure and functions which they will have in the adult. Nerve cells, muscle cells, etc., are formed. This process is called differentiation. The differentiated cells are organized into tissues, the tissues into organs, and the organs into systems.

4. **Growth.** Even after all the body systems of the organism have been formed, there follows a period of growth. During this period, the organism becomes larger by continued cell division or by cell enlargement, or both. Whatever the exact mechanism, growth depends on the intake of more matter and energy than is needed simply to maintain the normal functions of the organism. There must be an accumulation of proteins, carbohydrates (especially in plants), and the many other kinds of molecules characteristic of living things. While an organism is passing through the first three of these developmental stages, it is called an **embryo.** Let us now examine these stages in more detail.

22.2 CLEAVAGE

As we saw in the last chapter, the entrance of a frog sperm into the frog egg initiates several events. The "gray crescent" appears, meiosis of the egg is completed, and then the sperm nucleus and egg nucleus fuse. Shortly after this vital part of the fertilization process is completed, the first cleavage takes place. The zygote nucleus divides by mitosis, and a furrow appears which runs longitudinally through the "poles" of the egg (Fig. 22.1). This soon divides the egg into two halves.

About an hour after the first cleavage in the frog egg, each of the daughter cells divides again. The resulting cleavage furrow also runs through the poles but at right angles to the first. Each of the four resulting cells then simultaneously divides in a horizontal plane (Fig. 22.1). This plane is located closer to the animal pole than the vegetal pole. As a result, the cells of the animal pole are somewhat smaller than the yolk-filled cells of the vegetal pole. Simultaneous cleavage continues with the production of a 16-cell and then a 32-cell stage.

As cleavage continues, the animal pole cells begin to divide more rapidly than those of the vegetal pole. The former thus become not only smaller but also more numerous. During this period some migration of the cells of the animal pole takes place. These orient themselves so that a fluid-filled cavity, the **blastocoel,** forms inside the mass of cells. This hollow-ball-of-cells stage is called the **blastula.** It marks the end of the cleavage stage of development. Note that during this entire stage there has been no growth of the developing organism. The original mass of the egg has simply been partitioned into smaller and smaller units.

Reptiles and birds produce the largest eggs of all living things. For example, the hen's egg consists of just a tiny patch of cytoplasm resting on the surface of a large ball of yolk. (The "white" of the egg is simply noncellular accessory protein.) When cleavage occurs in the hen's egg, the cleavage furrows do not continue down through the mass of yolk. As a result, each of the cells produced in the early cleavage stages is bounded on the top and on the sides by a cell membrane, but the bottom of the cell is in direct contact with the mass of yolk (Fig. 22.2). This type of partial cleavage is also found in the egg of the zebrafish (Fig. 22.3).

The yolk of insect eggs is concentrated in the center of the egg. Cleavage of the egg does not accompany mitosis of the zygote nucleus. Instead, the daughter nuclei divide repeatedly but remain suspended within the single egg compartment (Fig.

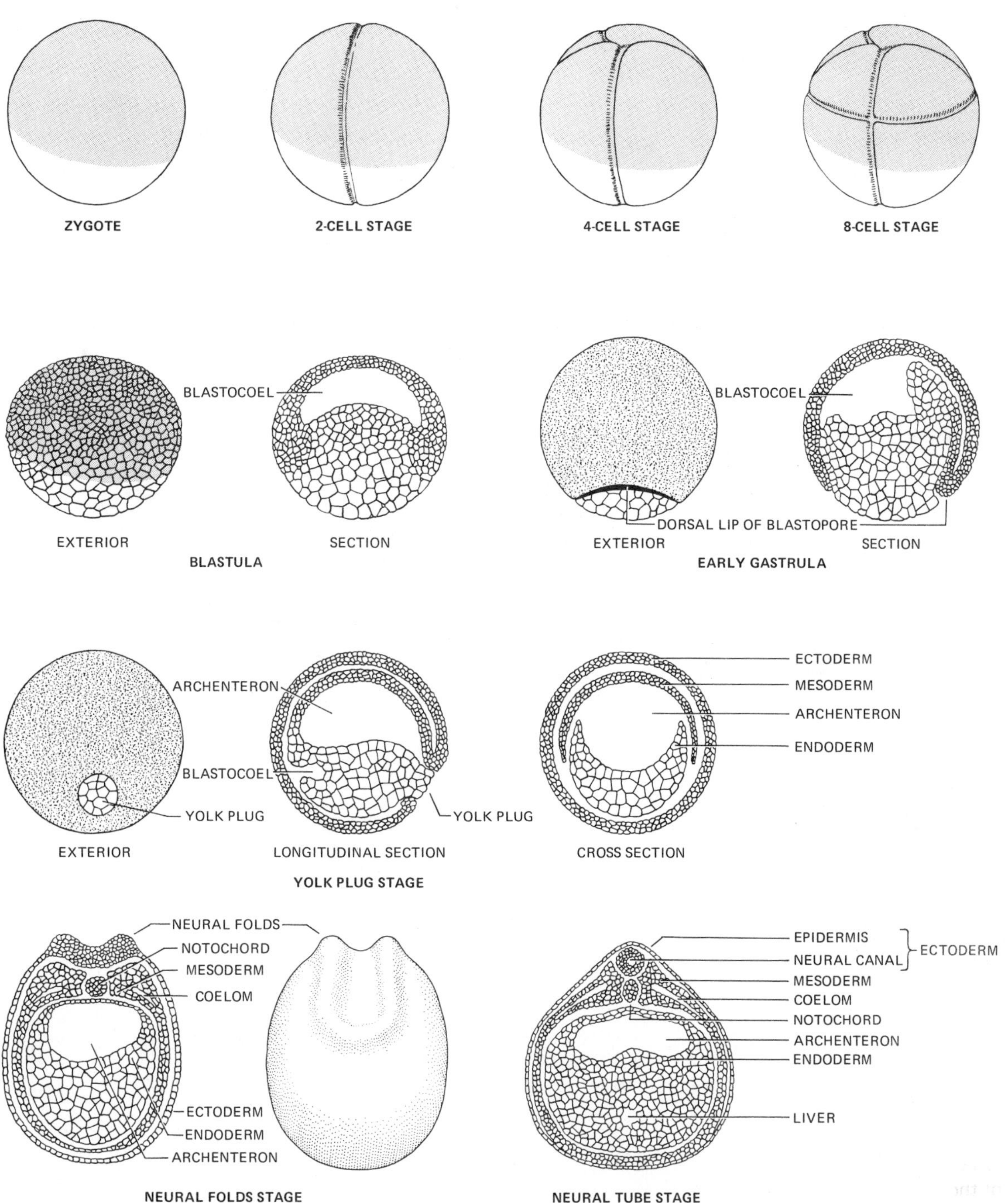

FIG. 22.1 The stages in the development of a frog. (Continued on next page.)

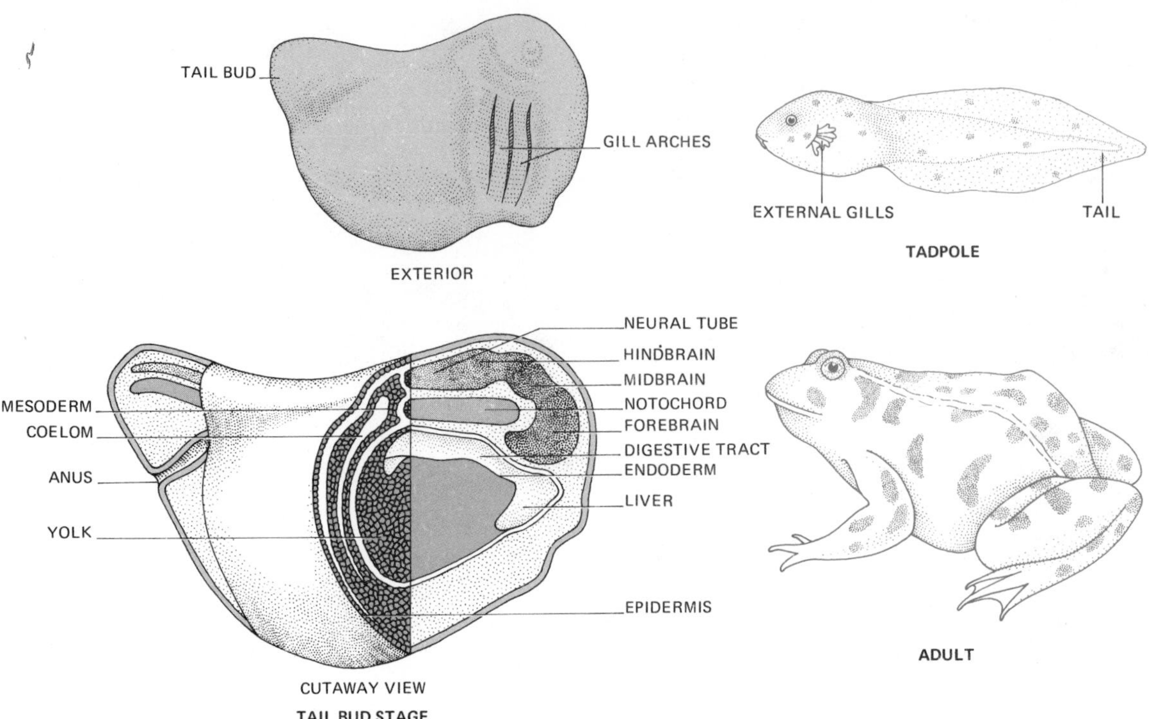

FIGURE 22.1 (Cont.)

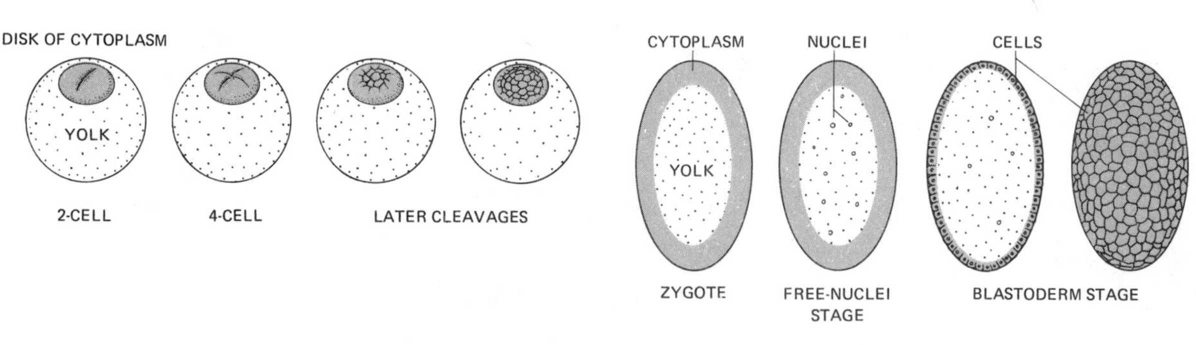

FIG. 22.2 Cleavage in bird and insect eggs.

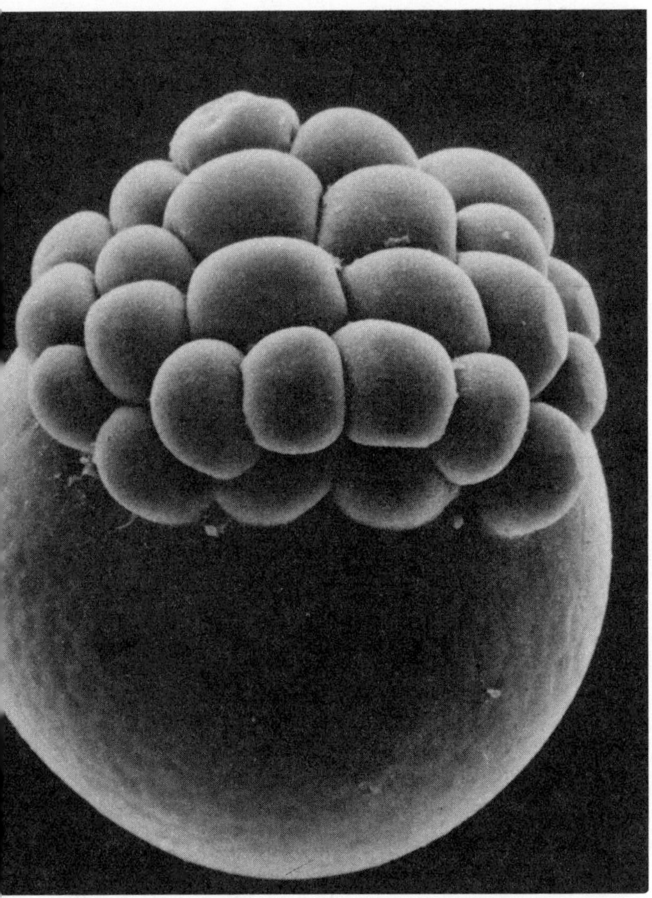

FIG. 22.3 Zebrafish embryo at the 32-cell stage. Note that the cleavage furrows have not continued down through the yolk of the egg. (Courtesy H. W. Beames and Richard G. Kessel.)

22.2). After a large number of nuclei have been produced, they migrate to the cytoplasm-rich margin of the egg. Only then do cell membranes form around each one.

Development of the zygote in plants proceeds somewhat differently than in animals. Mitosis of the zygote is usually accompanied by cell enlargement. In the gymnosperms and angiosperms, some of the daughter cells elongate very rapidly. Their growth pushes the remaining cells up into the food-rich endosperm. The endosperm provides the materials for further development of the embryo (Fig. 22.4).

What does cleavage accomplish in the development of the organism? First, it provides a stockpile of cells out of which the embryo can be constructed. We have seen that cells are the structural unit of almost all living things. Cleavage establishes a supply of them. Second, cleavage establishes a normal relationship between the nucleus and the cytoplasm it regulates. Even small eggs are enormous when compared with other kinds of cells. The volume of the frog egg is about 1.6 million times larger than that of a normal frog cell. But it, too, contains only one nucleus. During the process of cleavage, thousands of new nuclei are produced by mitosis. Each of these eventually becomes established in a cell of normal dimensions. Remember that the frog blastula with its thousands of cells is no larger than the original zygote.

22.3 MORPHOGENESIS

Cell division does not cease with the blastula stage. In the frog the smaller, darker cells of the animal pole continue to undergo rapid mitosis. As the num-

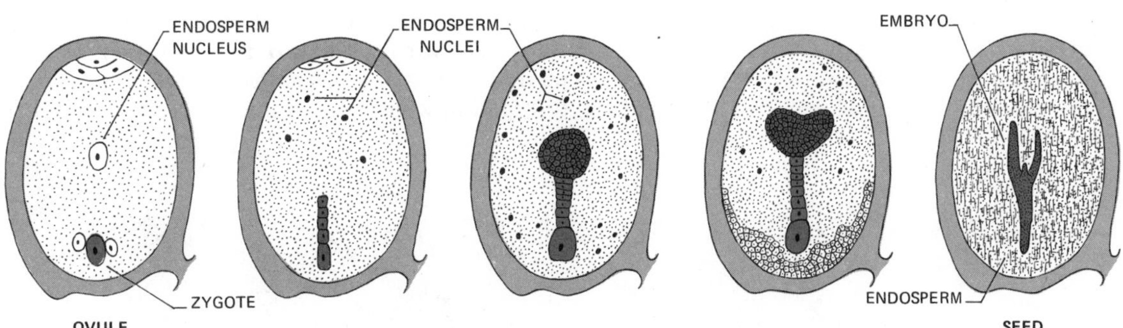

FIG. 22.4 Cleavage and morphogenesis in the angiosperm. Nourishment for the developing embryo is secured from the endosperm. The scale of the drawings decreases from left to right.

ber of these cells increases, they begin to migrate down over the yolk-filled cells of the vegetal pole. Some of the vegetal pole cells also begin to migrate. In their case, they push into the interior of the cell mass, producing a small crescent-shaped depression known as the blastopore (Fig. 22.1). This pushing-in of cells marks the start of **gastrulation.** As the animal-pole cells migrate down over the surface of the cell mass, a large number of them also push into the interior at the region of the blastopore.

The inpocketed vegetal pole cells develop into a sheet of cells which grows upward and eventually encloses a cavity, the archenteron (Fig. 22.1). These cells are known as **endodermal** cells. From the archenteron will develop the alimentary canal, the liver and other digestive glands and, after metamorphosis, the lungs. At this stage, the only opening to the archenteron is at the blastopore. Ultimately, the anus will form here. The inner lining of all these organs will be made of endodermal cells directly descended from the first vegetal-pole cells.

The animal-pole cells growing down over the embryo also migrate inward. This occurs first at the upper (or dorsal) portion of the blastopore. The cells move in directly beneath the overlying animal-pole cells. They are now called **mesodermal** cells. From them develops a rodlike structure, the **notochord,** which runs lengthwise along the dorsal side of the developing embryo.

As gastrulation continues, animal-pole cells crowd in around the entire margin of the blastopore. These late arrivals also become mesoderm. Those nearest the notochord become organized into separate masses called **somites,** from which will develop muscles and the bones of the vertebral column. Nearer the ventral surface of the embryo, the mesodermal cells develop into the lining of the **coelom,** the main body cavity. The cavity itself arises by a splitting of this mesoderm (Fig. 22.1). This stage of development is often known as the yolk-plug stage because only a small disk of yolk cells is still visible at the exterior. The animal-pole cells that do not migrate into the interior are called **ectodermal cells.** From them will arise all the structures of the nervous system as well as the skin of the animal.

Shortly after the formation of the notochord, the ectodermal cells immediately above it undergo a period of rapid cell division. This results in a deep, flat layer of cells called the neural plate (Fig. 22.1). Then the edges of the neural plate grow up while the central region just above the notochord becomes depressed. This so-called neural-folds stage is followed by the formation of a neural tube. The two elevated folds grow over and fuse together along their entire length. From this structure develops the brain and spinal cord. In Fig. 22.1 you can see that three enlargements of the neural tube form at the anterior end of the embryo. These will develop into the brain. The neural tube then drops down into the embryo and is covered over by another layer of ectodermal cells which develops into skin.

Although the details of gastrulation vary considerably from species to species, the variations are superficial rather than fundamental. In every case, the process results in the formation of three primary cell layers (often called "germ" layers), the **ectoderm, mesoderm,** and **endoderm.** From these three germ layers arise the various organs and systems of the animal body. Figure 22.5 lists the germ-layer origin of the various organs and systems in humans. This list could serve as well for any other vertebrates.

The process of gastrulation establishes the basic pattern of the embryo. Cells are stockpiled in specific locations in preparation for the construction of the various tissues, organs, and systems of the organism. Their construction involves more than simply the migration and stockpiling of cells in the three germ layers, however. The unspecialized cells of ectoderm, mesoderm, and endoderm must all take on specialized shapes and internal organization to permit them to carry out the functions of the tissue or organ in which they are located. This conversion

ECTODERM	MESODERM	ENDODERM
SKIN, HAIR, NAILS	MUSCLES	LINING OF ALIMENTARY CANAL
ENTIRE NERVOUS SYSTEM, INCLUDING RECEPTOR CELLS	BLOOD AND BLOOD VESSELS	LINING OF TRACHEA, BRONCHI, AND LUNGS
ADRENAL MEDULLA	CONNECTIVE TISSUE, INCLUDING BONE	LINING OF URETHRA AND BLADDER
	KIDNEYS, URETERS	LIVER
	TESTES, OVARIES, OVIDUCTS, UTERUS	PANCREAS
	MESENTERIES	
	LYMPHATIC SYSTEM	

FIG. 22.5 Germ-layer origin of various body tissues.

of unspecialized cells into specialized ones is called differentiation.

22.4 DIFFERENTIATION

Although the later stages of morphogenesis and the process of cell differentiation overlap to a considerable extent, it is well to keep in mind the distinction between the two processes. Morphogenesis involves the organization of cells into the various layers and groups that will form the structures of the body. It involves cell division and the actual movement of cells from place to place within the embryo. Although the various layers of cells in the frog gastrula have definite and different fates ahead of them, they do not at first reveal these fates by any specialization of structure or function. In fact, it is possible to transplant cells from one location to another in the early gastrula; these transplanted cells readily adapt to their new location and go on to participate in the building of an organ appropriate to it.

As embryonic development proceeds, however, the cells of the developing embryo reach a "point of no return." They become committed to the formation of a specific kind of cell. Gradually they take on the appearance and function of the various kinds of cells discussed in Chapter 5. As they do this, each begins to synthesize a small number of proteins characteristic of that cell type and no others. Differentiating heart muscle cells start synthesizing a special contractile protein called heart myosin. Differentiating red blood cells begin synthesizing hemoglobin. As connective tissue cells differentiate, they become committed to the synthesis of collagen and other extracellular proteins. And so on. This commitment to a specific fate seems to occur first in mesodermal and endodermal cells, then in the ectodermal cells.

Although the processes of cleavage, morphogenesis, and cellular differentiation can be distinguished from one another, the three operations share one fundamental similarity. In each case, unlike things are being made from like things. Beginning with the eight-cell stage of frog cleavage, the cells of the animal pole differ in appearance and chemistry from those of the vegetal pole. From the hollow sphere of the blastula arise three distinct germ layers. From the unspecialized cells of the three germ layers are formed all the differentiated cells of the mature organism.

In many cases, differentiation occurs once and early in development. For example, probably no further differentiation of nerve cells can occur after infancy. In other cases, differentiation goes on throughout life. The several kinds of blood cells are continuously replaced throughout life. There is good evidence that all of them (red cells, lymphocytes, neutrophils, etc.) arise from a single precursor cell called a stem cell. Mitotic division of these stem cells (which are found in the bone marrow) produces daughter cells faced with two options. Some will remain stem cells, thus insuring that the pool of stem cells does not become used up. Others enter on one path of differentiation or another. As they proceed down these paths, they gradually acquire the structure, function, and proteins characteristic of the fully differentiated cells they finally become (Fig. 11.11).

The process of differentiation again raises the question of differential gene action. Certainly the genes needed to build all the differentiated cells of the adult body must be present in the fertilized egg. How then do certain genes get turned on in certain cells (e.g. the genes for the alpha and beta chains of hemoglobin) while different genes are turned on in other cells? How is it that as development proceeds, unlike cells are produced by cells that initially are alike?

One reasonable answer might be that at each mitosis, certain genes are parcelled out to one daughter cell, others to the other. In this way, much of the genome present in the fertilized egg gradually gets broken up and distributed to the different cells of the body. Thus heart muscle cells would receive only those genes needed for their functioning, including the genes coding for the synthesis of heart myosin. The cells destined to become red blood cells would get, among others, the genes for the alpha and beta chains of hemoglobin. While this is a perfectly reasonable hypothesis to explain differentiation and the differential gene activity that must accompany it, a large body of information suggests that it is *not* correct. On the contrary, the evidence leads to quite a different conclusion: namely, that each of the differentiated cells of the body retains the full genome that was present in the fertilized egg. Let us examine this evidence.

22.5 EVIDENCE THAT DIFFERENTIATING CELLS RETAIN THE ENTIRE GENOME

1. The DNA Content of Differentiated Cells

Accurate methods exist for determining the DNA content of individual cells. When this technique is applied to a variety of differentiated cells, a clear

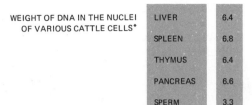

WEIGHT OF DNA IN THE NUCLEI OF VARIOUS CATTLE CELLS*

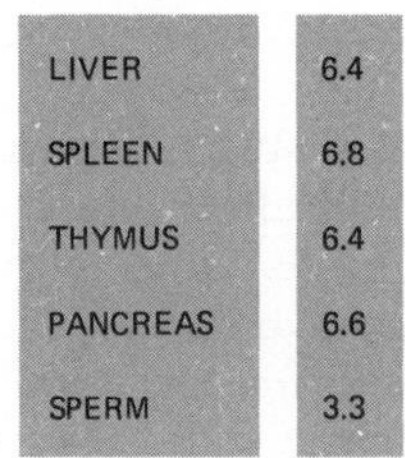

LIVER	6.4
SPLEEN	6.8
THYMUS	6.4
PANCREAS	6.6
SPERM	3.3

FIGURE 22.6

*In picograms (10^{-12}g) per nucleus

pattern emerges (Fig. 22.6). With a few minor exceptions and one major one, the different cells of the body contain the *same* amount of DNA. The minor exceptions are the occasional cells (in the liver, for example) that are tetraploid, having twice the normal number of chromosomes. They arise as a result of chromosome duplication and division without cell division. The major exception is the gametes. These contain one-half the DNA found in the other cells (Fig. 22.6). Since they have arisen by *meiosis,* this is a perfectly reasonable finding.

2. The Chromosome Content of Differentiated Cells

With the development of techniques for karyotype analysis, it soon became evident that (again with a few exceptions) the various differentiated cells of an organism share the same number of chromosomes. T. T. Puck and J. H. Tjio (who supplied one of the photomicrographs in Fig. 16.10) examined the karyotypes of some 1825 individual cells taken from eight different human tissues. In almost every case, the chromosome count was the same: 46. One cell in the group appeared to have 45, another 47 chromosomes. In addition, a few cells (e.g. some in the liver) were found to be tetraploid ($4n$). (Their occurrence was explained in the previous paragraph.) Karyotype analysis of many other species confirms the general rule: normally chromosomes are *not* lost during development and differentiation.

3. The Polytene Chromosomes

In Chapter 19, we studied in some detail the polytene ("giant") chromosomes found in many of the cells of fly larvae. We examined the evidence that each band in these chromosomes represents a gene locus. What is important for the present argument is that the *general* appearance of the giant chromosomes—including the sequence of some 5000 bands—is the same no matter what the cell from which those chromosomes are taken (Fig. 22.7). Certain

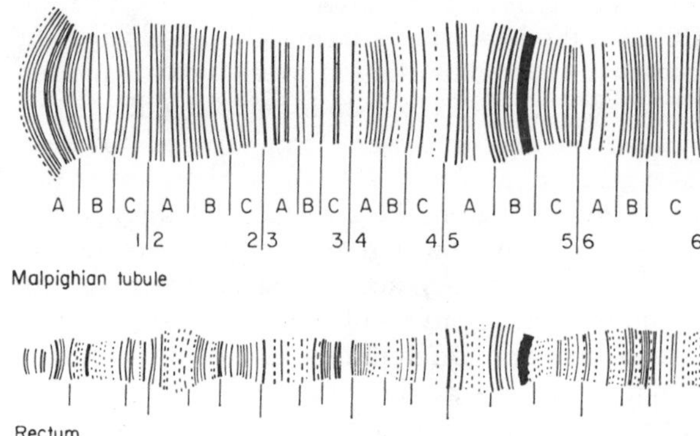

Rectum

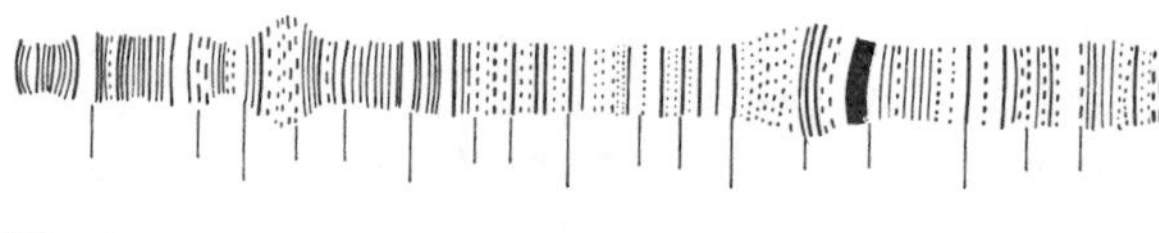

Mid gut

FIG. 22.7 Equivalent portions of chromosome number 3 of *Chironomus tentans* as they appear in four different kinds of cells in the animal. The differences in the appearance of the bands presumably reflect differences in gene activity in the different cells. (From the work of Beermann, courtesy of Springer-Verlag.)

chromosome regions may be puffed in one kind of cell, other regions puffed in other cells. And, indeed, we feel confident that the particular puffs in a given kind of differentiated cell reflect intense gene activity characteristic of that cell. But, even if not puffed, all the bands—and thus all the genes—are present in each and every cell.

4. The Reversibility of Differentiation in Plant Cells

A more direct answer to our question is given by Fig. 22.8. A fully differentiated carrot root cell when grown in a suitable culture medium begins to divide repeatedly, losing its differentiated structure as it does so. Then its descendants begin to differentiate, and they finally form *all* the organs of a mature carrot plant. Certainly the differentiated root cell had not lost any of the genome of the species, al-

FIG. 22.8 The carrot plant in the flask grew from fully differentiated carrot root cells that had been isolated and induced to undergo mitosis. (Courtesy of Roy De Carava and *Scientific American*.)

though prior to the experiment it had only needed to use that portion of it that was appropriate to root cell structure and function. Probably any differentiated plant cell, if placed in surroundings resembling those found in endosperm (Fig. 22.4), can revert to an embryonic type and have all the genetic capabilities of its species once more unlocked.

5. Testing the Capacity of Nuclei to Program Development

The situation is not quite so clear with respect to animals. One of the earliest approaches to the question involved the physical separation of the cells produced in the early stages of cleavage. It is known that each of the two cells produced when the frog zygote undergoes its first cleavage will normally give rise to one-half the embryo. One might expect that in some way each of these two cells is able to use only one-half of the genes of the organism. Physical separation of the cells shows, however, that such is not the case. Reared separately, each goes on to produce a complete, although smaller, embryo.

This discovery provides an explanation for the occurrence of identical twins, those twins that show such a striking resemblance to each other. Identical twins are always of the same sex and even in their body chemistry seem to be identical in every way. We explain these rare individuals by assuming that they came from a single zygote which, after one or two cleavages, was separated into two parts. Identical triplets, quadruplets, and quintuplets undoubtedly arise in a similar fashion. The remarkable similarity of these individuals is understandable when we realize that they were formed from a single zygote and thus a single set of genes.

When the cell-separation experiment is repeated at the four-cell stage of cleavage, the results are quite different. None of the resulting cells, when isolated, can produce a complete embryo. What has happened? Have the daughter nuclei produced during the second cleavage lost some of the genes necessary to construct the total organism? The answer to this question was discovered by the German experimental embryologist, Hans Spemann. Using fine strands of baby hair, he tied tight loops around fertilized newt eggs (which are similar to frog eggs) so that they were constricted longitudinally into two halves (Fig. 22.9). The zygote nucleus was in one half. The other half had no nucleus but was in contact with its nucleus-containing partner by a narrow bridge of cytoplasm. Spemann found that only the half-egg with the nucleus in it would cleave. However, at some point during cleavage, a nucleus usually did get across the narrow connecting bridge of cytoplasm. As soon as this happened, the second half began to cleave, too. Even though its nucleus was the product of as many as five prior cleavages, it went on to form a complete embryo. Clearly, then, the nucleus, at the 32-cell stage at least, has not lost any of the genes for making the organism.

More recently, attempts have been made to answer the question by means of nuclear transplant experiments. Using micromanipulators equipped to operate on single cells, Robert Briggs and Thomas King succeeded in removing nuclei from the cells of frog embryos in various stages of development. To test the potentialities of these nuclei, they transplanted them into unfertilized frog eggs whose own nucleus had been removed. In this way, it could be determined whether a nucleus from the later stages of embryonic development was fully equivalent to

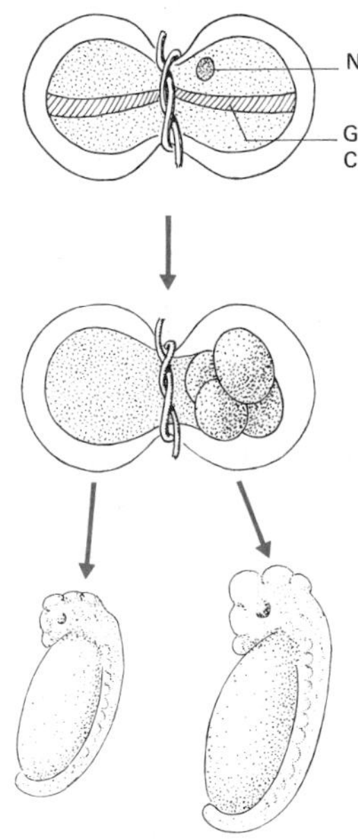

FIG. 22.9 Spemann's experiment. The cytoplasm to the left of the knot did not receive a nucleus until the original zygote nucleus had gone through several mitoses. Nevertheless, when a nucleus finally slipped through, that half-egg went on to develop normally. This showed that no genes had been lost or irreversibly inactivated during the earlier mitotic divisions.

the nucleus of the zygote or early cleavage cells. It was soon found that nuclei from frog blastulas still retain all the potentialities of the zygote nucleus. The nucleus from any one of the thousands of cells of the frog blastula initiates perfectly normal development when transplanted into a frog egg lacking its own nucleus.

This is understandable because there is good evidence now that the chromosomes of cells during cleavage have not begun to carry out any function except their own duplication during the interphases of the repeated mitotic divisions. All the control of the activities of the cleaving cells appears to be provided by the many messenger RNA molecules and ribosomes that were deposited in the unfertilized egg by the mother during oogenesis. In other words, the genetic code supplied by the father in the sperm cell seems to play no role during cleavage. It is only at the start of gastrulation that nucleoli form, new ribosomes are manufactured, and vigorous RNA synthesis—using maternal *and* paternal codes—begins.

When Briggs and King repeated their experiments with nuclei from later embryonic stages, quite different results were obtained. These nuclei were also capable of initiating embryonic development, but with widely varying degrees of eventual success. Many of the resulting embryos ceased developing at one stage or another. Furthermore, those that did develop were often abnormal. This evidence *suggests* that during gastrulation cell nuclei do become altered. Perhaps some of their genes have become permanently repressed. However, the lack of success with these nuclei may simply reflect a failure of experimental method rather than an altered property of the nuclei. Working with the South African clawed frog, *Xenopus laevis,* Gurdon was able—in a modest percentage of cases—to grow a fertile adult animal from an egg whose own nucleus had been replaced by a nucleus taken from a fully differentiated intestinal cell of a tadpole (Figs. 22.10 and 22.11). Perhaps, then, the nuclei from fully differentiated nondividing cells are simply more easily damaged during the manipulations, and are not irrevocably repressed.

The chromosomes of the chick red blood cell appear to be permanently repressed: little or no RNA synthesis can be detected. However, when chick red cell nuclei are introduced into the cytoplasm of an actively metabolizing mammalian cell by the technique of somatic cell hybridization (see Section 16.8), the chromosomes are derepressed and become active once again in RNA synthesis.

Taken together, all the evidence suggests that during embryonic development, genes are *not* parcelled out to the cells entering different pathways of differentiation. Instead, every differentiated cell (with very few exceptions) retains the entire genome that was present in the fertilized egg. So, what alternative mechanism can we invoke to explain the phenomenon of cell differentiation? Once again, we are faced with the problem of differential gene activity. In Chapter 19, we discovered certain factors (hormones) that turned genes on and off in differentiated cells. We called the process modulation. Can we now find factors that selectively turn genes on and off as the cells in which they reside enter one or another pathway of differentiation?

22.6 CYTOPLASMIC FACTORS AFFECTING GENE EXPRESSION DURING DIFFERENTIATION

One of the earliest discoveries of a factor influencing differentiation came from the cell-tying experiments of Spemann, discussed in the previous section (see Fig. 22.9). When Spemann repeated his experiment with the fertilized egg constricted so that all of the gray crescent lay in one half, the final results were quite different (Fig. 22.12). The half lacking the

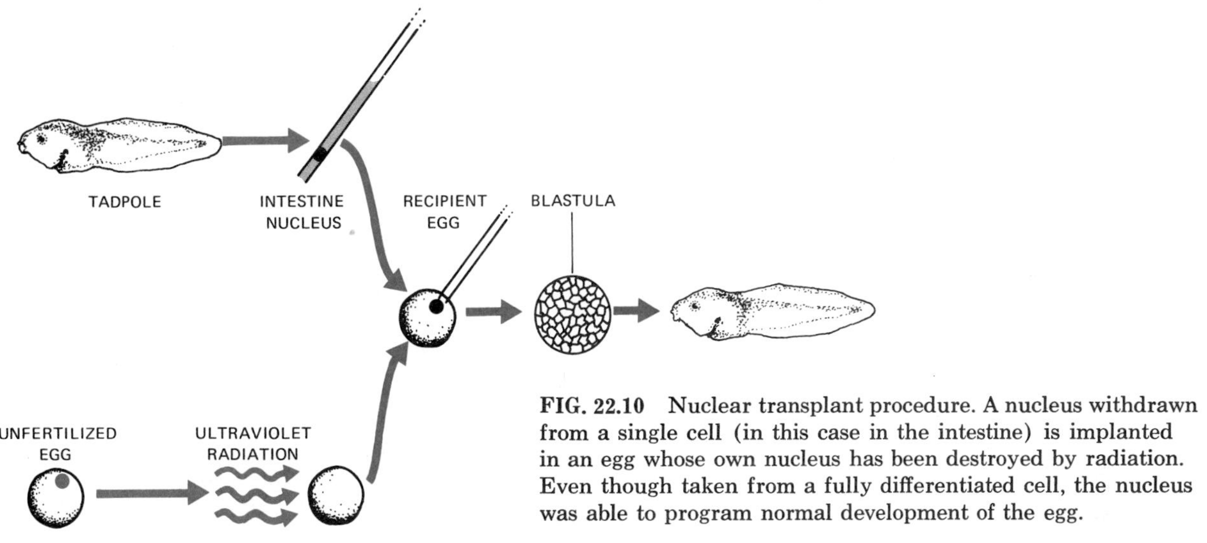

FIG. 22.10 Nuclear transplant procedure. A nucleus withdrawn from a single cell (in this case in the intestine) is implanted in an egg whose own nucleus has been destroyed by radiation. Even though taken from a fully differentiated cell, the nucleus was able to program normal development of the egg.

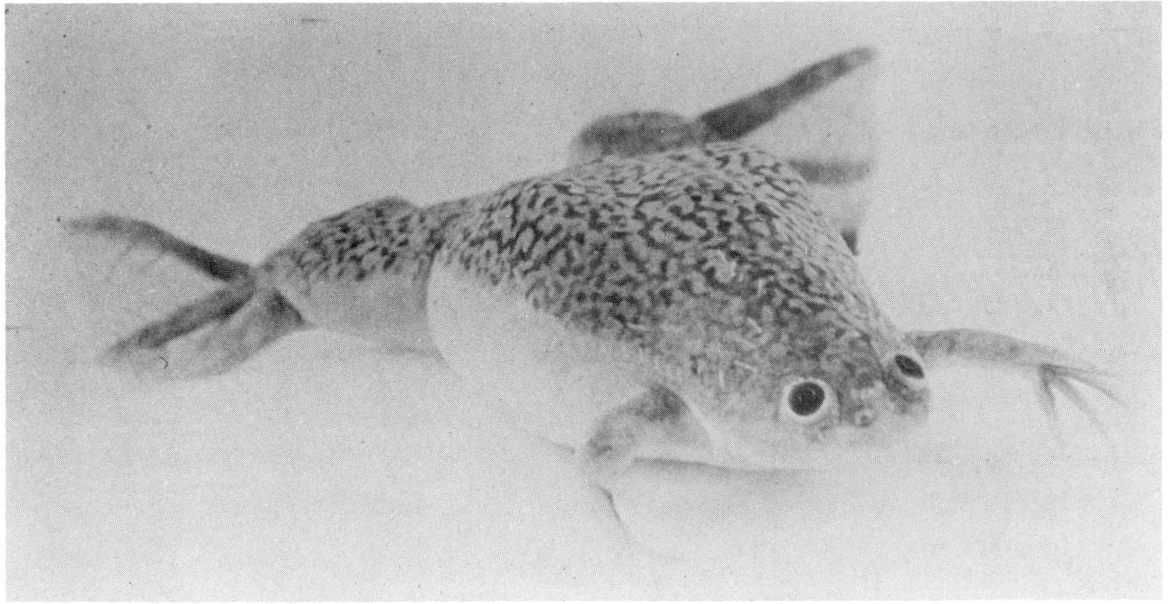

FIG. 22.11 South African clawed frog that developed from an egg whose own nucleus had been replaced with a nucleus taken out of a fully differentiated epithelial cell from the intestine of a tadpole. This animal was fertile and normal in every other way, thus demonstrating that no part of the genetic library of the species was lost during differentiation. (Courtesy of J. B. Gurdon and *Scientific American.*)

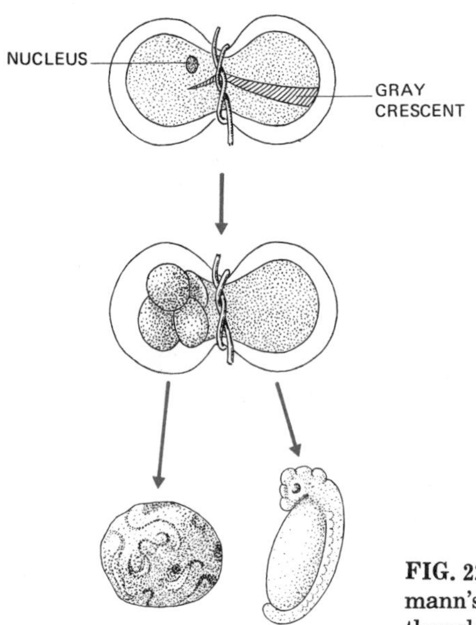

UNORGANIZED DEVELOPMENT NORMAL DEVELOPMENT

FIG. 22.12 Another of Spemann's experiments. Even though it began development first, the embryo lacking the gray crescent never developed beyond an unorganized mass of belly tissues.

gray crescent but containing the nucleus began cleavage right away, yet it never developed beyond an unorganized mass of intestinal, liver, and other abdominal cells. The other half, even though it did not get a nucleus until the fourth mitotic division on the cleaving side, went on to form a perfectly normal embryo. This again was proof that the nuclei, here at the 16-cell stage, had not lost any genes. But why, then, did a normal embryo fail to develop on the side with the original zygote nucleus?

We have already seen that the distribution of cytoplasm, mitochondria, RNA, ribosomes, and yolk in the amphibian egg is not uniform. Shortly after fertilization, some of the cytoplasmic constituents migrate and form the gray crescent. Perhaps the potentialities a nucleus can achieve are regulated by the cytoplasmic environment in which it finds itself. Thus, in Spemann's first experiment, each half of the egg contained all the normal egg constituents, because the longitudinal constriction was perpendicular to the gray crescent. However, in his second experiment, the hemisphere lacking the gray crescent may have lacked some essential cytoplasmic materials.

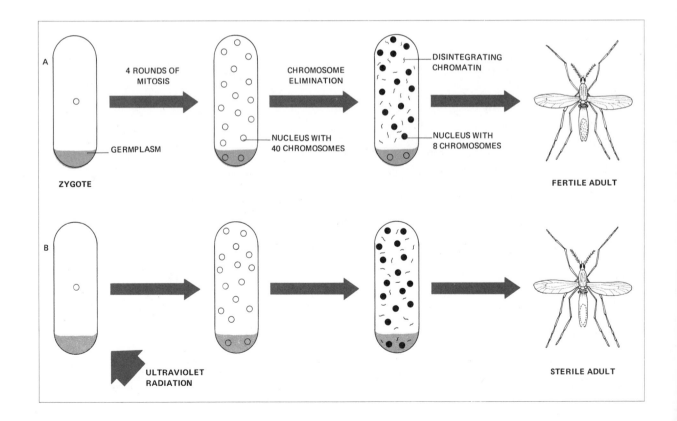

A
4 ROUNDS OF MITOSIS
CHROMOSOME ELIMINATION
DISINTEGRATING CHROMATIN
GERMPLASM
NUCLEUS WITH 40 CHROMOSOMES
NUCLEUS WITH 8 CHROMOSOMES
ZYGOTE
FERTILE ADULT

B
ULTRAVIOLET RADIATION
STERILE ADULT

The importance of the cytoplasmic environment may explain the results of the cell-separation experiments described earlier. When the first cleavage in the frog egg occurs perpendicular to the gray crescent (as it often does), each of the two cells produced contains all of the normal egg constituents. The second cleavage, however, isolates the gray crescent in two of the four cells produced. Although all the cells contain nuclei with the complete hereditary blueprint, we may assume that the action of the genes is now limited by the cytoplasmic surroundings in which they find themselves. Though the nuclei are all alike, the nucleus-cytoplasm combinations are not, and we have our first clue as to the way in which dissimilar cells can arise from a single zygote.

The absence of substantial RNA synthesis prior to gastrulation in the frog egg represents another example of the influence of the cytoplasm on the genetic activity of the nucleus. When nuclei from the late stages of embryonic development, or even from young tadpoles, are transplanted into eggs, they quickly (within 40 minutes) lose their nucleoli and stop synthesizing RNA. Not until they have guided the embryo as far as gastrulation do they resume RNA synthesis. Attempts to isolate the cytoplasmic inhibitor of RNA synthesis indicate that it is a relatively small molecule.

One of the clearest examples of the influence of the cytoplasm on what a nucleus can or cannot do is found in an insect called the gall midge. You remember (Section 22.2) that "cleavage" in insect eggs involves the production of nuclei that remain suspended within the unpartitioned contents of the egg (Fig. 22.2). At the fourth mitotic division in the gall midge egg, two of the 16 nuclei that are produced become pinched off in a small amount of cytoplasm at one end of the egg (Fig. 22.13). At the fifth mitosis, these two nuclei divide normally, producing daughter cells with the full complement of chromosomes ($2n = 40$) of the species. But not so for the other nuclei. When each of *these* reaches anaphase, only 8 of their 40 chromosomes (doublets) separate and move to opposite ends of the spindle. The remaining 32 chromosomes stay at the equator and eventually disintegrate.

◀ **FIG. 22.13** Top: Normal development in the gall midge. The gametes are descended from two nuclei, each with the full diploid set of 40 chromosomes, that become partitioned off in a mass of special cytoplasm (the germplasm). The remaining nuclei lose 32 chromosomes before going on to form the rest of the insect body (the somaplasm). Bottom: Destruction of the germplasm causes the nuclei that move there to undergo chromosome elimination also. The animal that develops is sterile but otherwise normal. This experiment shows that the potentialities of a nucleus are influenced by the cytoplasmic environment in which it finds itself.

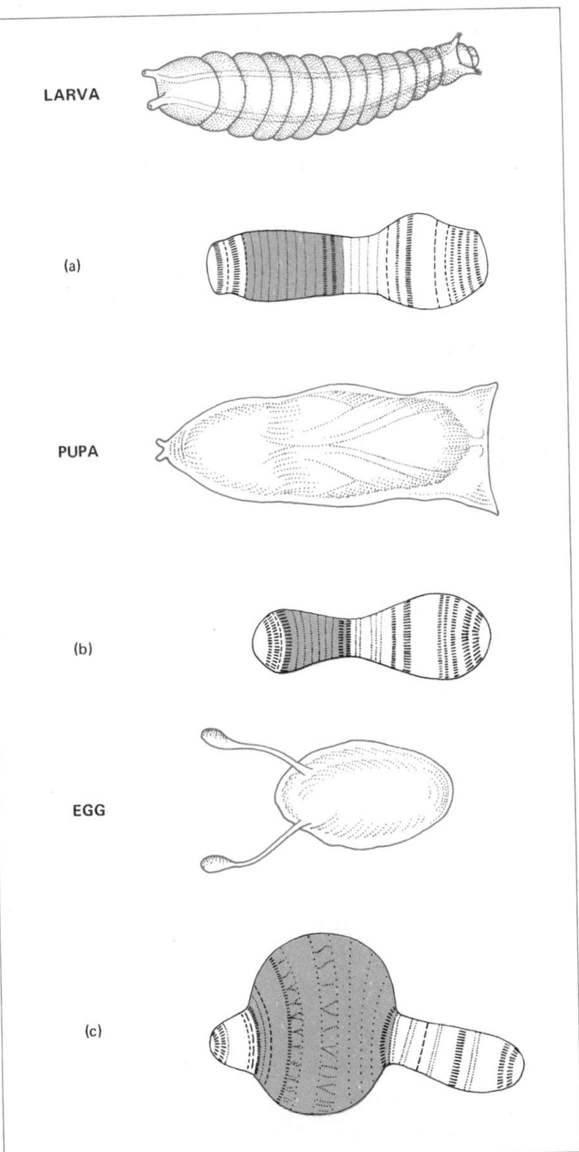

FIG. 22.14 Changes in equivalent portions of a giant chromosome of a fruit fly (a) in the larva, (b) in the pupa, and (c) when transplanted into an egg. (Eggs do not normally have giant chromosomes.) Enlargements ("puffs") in the chromosomes are associated with increased activity.

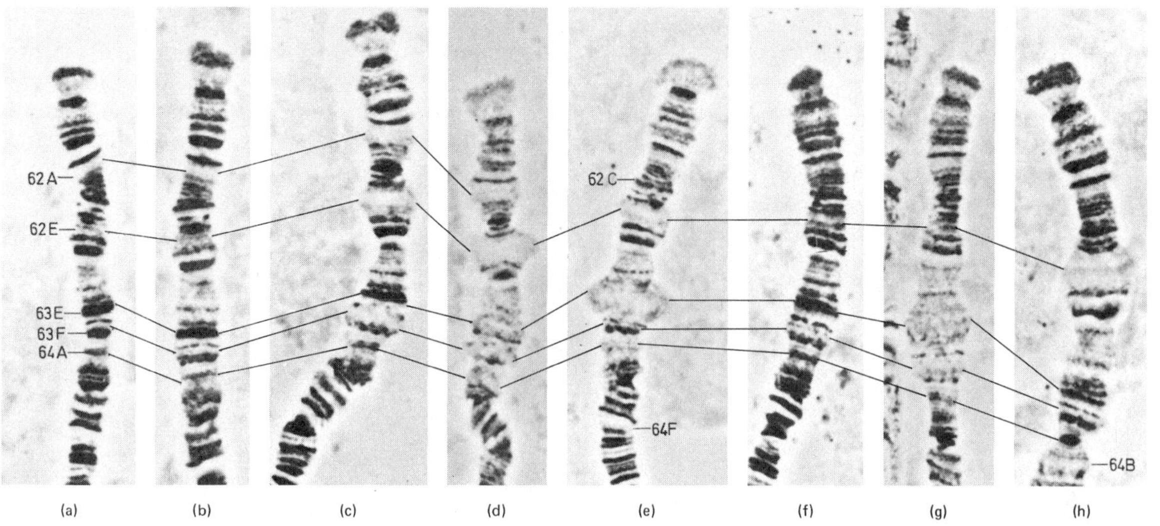

FIG. 22.15 Changes in the puffing pattern in equivalent segments of the number 3 chromosome of *Drosophila melanogaster* over the course of some 20 hours of normal development. Note that during this period, in which the larvae were preparing to pupate, certain puffs formed, regressed, and formed again. However, the order in which they did so often varied. For example, in the larva, 62E becomes active before 63E (c, d, and e), but when pupation begins, the reverse is true (g, h). (Courtesy of Dr. Michael Ashburner, University of Cambridge.)

The descendants of the two "normal" nuclei ultimately differentiate to form the sperm or eggs, i.e. the germplasm. The descendants of the rest of the nuclei, those with the sharply reduced chromosome numbers, go on to form all the other tissues of the insect body, i.e. the somaplasm. (Wouldn't Weismann have enjoyed knowing this story! See Section 15.1.) In fact, if you damage the special region of cytoplasm at the end of the egg (e.g. by irradiating with ultraviolet light), the nuclei that find themselves there will divide abnormally just like their sisters. When the animal reaches maturity, it will be sterile although normal in every other way (Fig. 22.13).

If genes are *selectively* activated by a particular cytoplasmic environment, we might expect the puffing patterns of giant chromosomes to reflect this. We have already seen that at any one time the pattern of puffs in one kind of cell (e.g. salivary gland) may differ somewhat from that in another (e.g. Malpighian tubule). Presumably this is a reflection of their different cytoplasmic surroundings. If, for example, one transplants a salivary gland nucleus into an egg (a cell that does not normally have giant chromosomes), a new pattern of puffs quickly appears (Fig. 22.14).

The same kind of changing puffing activity is seen over the course of the differentiation of a *single* kind of cell. The eight photomicrographs in Fig. 22.15 show how the pattern of banding in the same portion of a giant chromosome changes during the course of the differentiation of a salivary gland cell. Hour by hour, the pattern changes. The bands puff, regress, and may puff again. Here, then, is visible evidence of a sequential pattern of gene activity during the course of the differentiation of a single kind of cell.

22.7 EXTRACELLULAR FACTORS AFFECTING GENE EXPRESSION DURING DIFFERENTIATION

Spemann's egg-tying experiments show that an unequal distribution of substances within the fertilized egg gives rise to cells of diverse fates despite their identical nuclear controls. But is the embryo fully patterned in the frog egg? It is difficult to imagine that the relatively simple gradients in the egg account for all the complex migration and differentiation of cells during embryonic development. Are there other interactions than simply the interactions between a nucleus and the particular cytoplasmic environment in which it is first partitioned off?

It was Spemann, again, who demonstrated that the pattern of development of cells is also influenced by the activities of other cells. You remember he had found that half a newt zygote would develop normally so long as it (1) contained the nucleus and (2) contained some of the region of the cytoplasm known as the gray crescent. Spemann knew that it is the cells that develop in the region of the gray crescent which migrate through the blastopore during gastrulation and form the notochord. With great surgical skill, he succeeded in cutting out the cells of the dorsal lip of the blastopore before their inward migration (Fig. 22.16). Then he transplanted this tiny bit of tissue into the ventral side of a normal newt gastrula. To make it easier for him to follow the fate of the transplant, he used the early gastrula of a pale variety of newt as the donor and the early gastrula of a dark variety as the recipient or host. He discovered that the transplanted piece of dorsal lip developed into a second notochord. Above the site of the transplant a neural groove appeared. This neural groove was made up almost exclusively of host (dark) cells. It continued its development, forming a neural tube and then a central nervous system. Ultimately a two-headed monster was produced.

The fact that most of the tissues of the second head were derived not from the transplanted cells, but from the host cells, suggested to Spemann that the transplanted cells had altered the normal course of development of the host cells around them. Instead of producing the belly of the newt, the host cells above the transplant were stimulated to produce a second central nervous system. We use the term **induction** for this process.

When Spemann transplanted portions of gastrulas other than the cells of the dorsal lip of the blastopore, no induction was observed. The transplanted tissue simply developed according to its new location. Because only the dorsal lip of the blastopore could induce altered development in the host, Spemann called it the *organizer*. He visualized the organizer as initiating the process of morphogenesis and differentiation by inducing the formation of the neural tube. The neural tube then might induce the formation of still other embryonic structures. For example, we know that as the brain develops, two masses of nervous tissue, the optic cups, grow forward from it. As these near the anterior surface of the embryo, skin cells just in front of them differentiate to form lenses (Fig. 22.17). The optic cups become the retinas of the finished products, the eyes.

The theory of induction explains very nicely this transformation of skin cells into lens cells. In fact, it is not hard to visualize the development of the entire embryo by this mechanism. As each structure is induced, it can then induce still other structures. Successive waves of induction could thus account for the complete, organized embryonic development of the animal.

Not long after Spemann's discovery, it was found that the inducing properties of the organizer were retained even if the cells were killed. This immediately suggested that induction is accomplished by the passage of some chemical substance from the organizer to the affected cells. It was thought that the inducing substance instructed the affected cells to organize and differentiate in a specific way. This idea thus provided another plausible answer to our

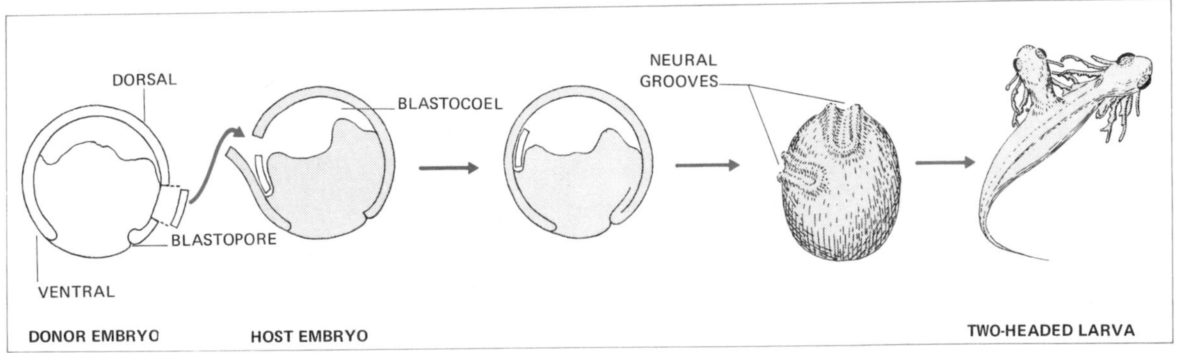

FIG. 22.16 Still another of Spemann's experiments. Donor tissue (white) from the region of the gray crescent developed into a notochord and induced the formation of a second head on its host (gray).

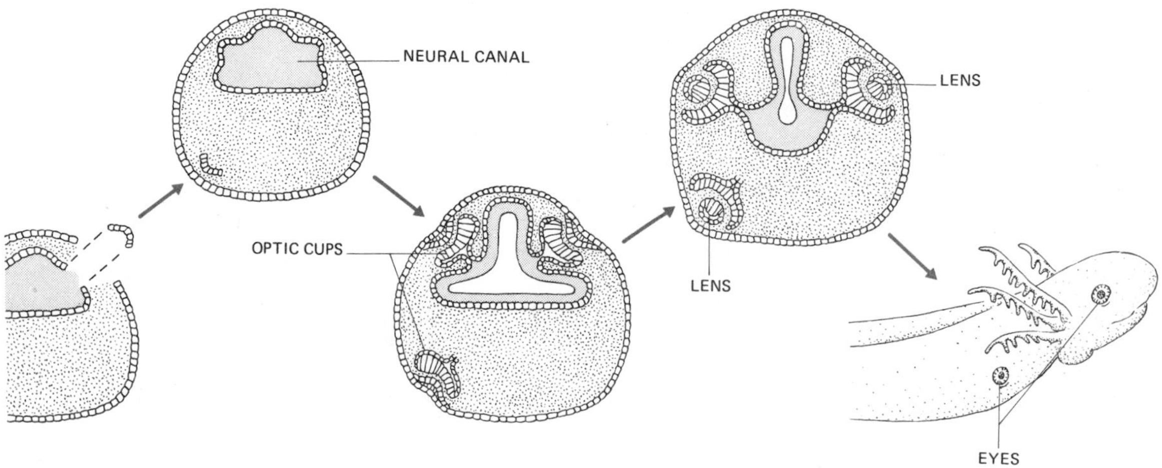

FIG. 22.17 Embryonic development of the salamander eye. A transplanted optic cup induces lens formation in the tissues of its host and an extra eye develops.

question of how diverse cells arise from a single zygote. The peculiar cytoplasmic composition of the gray crescent region might stimulate the blastula cells produced in that region to liberate special chemical regulators. These cells would thus have a feature unique in the embryo and could become the organizer. The diffusion of chemical substances from them might then instruct adjacent cells to develop in a special way. As the latter developed, they in turn could release inducing substances which would program the differentiation of still other cells. In this way the entire embryo could be built.

With such an attractive theory, it was natural that every effort be made to isolate and identify inducing substances, not only in amphibian embryos but in other organisms as well. In a number of cases it was found that induction could occur across a gap filled with agar-agar, but not across a sheet of cellophane. This suggested that a fairly large molecule was involved and, indeed, induction has in some cases been accomplished *in vitro* by treating the tissues with a relatively pure protein preparation. Some workers have also presented evidence that ribonucleoprotein, i.e. a complex of RNA and protein, may have inductive properties.

Although most inducers appear to be macromolecules, there are some exceptions. Vitamin A, a small molecule, can induce altered development in the epidermal cells of a chick embryo. Other small molecules such as certain steroids and even the dye methylene blue (which is not a normal constituent of cells) can induce neural folds formation in the amphibian embryo. The inductive changes produced by these small molecules are not as elaborate as those produced by proteins. Methylene blue simply shifts the development of epidermis to neural tube, both of which are ectodermal structures. Induction by macromolecules can, on the other hand, convert cells that would have developed into ectodermal structures into mesodermal tissues, like muscle, and even into endodermal tissues.

The differentiation of cells seems also to be influenced by *inhibitory* substances that reach them from adjacent cells. When developing frog embryos are placed in cultures containing pieces of adult frog heart, the embryos fail to produce a normal heart. Similarly, embryos cultured with pieces of adult brain fail to produce a normal brain. This would suggest that differentiated cells can produce substances that inhibit adjacent cells from differentiating the same way. Certainly in adult organisms this is the case. Differentiated tissues secrete substances (proteins and/or polypeptides) called **chalones** which inhibit mitosis of the cells of that tissue. Because of their demonstrated role in tissue repair, they will be discussed more fully in the next chapter. Whether these same substances serve as inhibitors in embryonic development remains to be seen. In

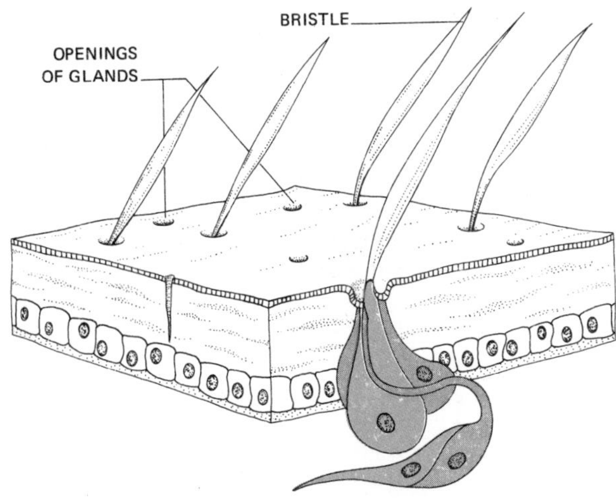

FIG. 22.18 Both the mechanoreceptors and the glands of the *Rhodnius* exoskeleton develop at evenly spaced intervals from one another.

any case, such a phenomenon would account nicely for the fact that as embryonic development proceeds, developing organs become sharply separated from one another. It could also account for the uniform distribution of repeating units in many organisms. For example, the body surface of the larval "kissing bug" (*Rhodnius*) is covered with evenly spaced mechanoreceptors and glands (Fig. 22.18). If a group of these in one area is destroyed (for example, by applying a hot needle to the body surface), replacement cells eventually become differentiated in the same location. These reconstruct the receptors and glands with their normal spacing. No two receptors or glands develop right next to each other. Perhaps as each begins to develop, it gives off an inhibitory substance which prevents nearby cells from developing in the same way.

It is becoming increasingly clear that the course of embryonic development depends not just on the induction or inhibition of some cells by other cells but on *interactions* occurring between both groups of cells. For example, in order for the mesoderm within the developing limbs of an embryo to become differentiated, it must have functioning ectoderm above it. On the other hand, continued differentiation of that ectoderm depends upon the continued activity of the underlying mesoderm. Perhaps, then, normal embryonic development depends not so much on the passage of a specific stimulating or inhibiting substance from one tissue to another but on the proper exchange of materials between the two.

The development of mouse kidney tubules requires two kinds of differentiated cells. One kind makes up the collecting portion of the tubule; the other, the secretory portion. It is possible to grow, in tissue culture, the embryonic cells that normally produce each of these differentiated types. When grown alone, they fail to do so, even when supplied with extracts of the other cell type. But when the two are allowed to grow together, each proceeds to differentiate into its adult form. Not only do they differentiate successfully, but the two kinds of cells then proceed to organize themselves into recognizable kidney tubules.

When, however, cells from two *unrelated* adult tissues are cultured together, the two kinds sort themselves out and unite, like with like, to form recognizable tissues. Some of these have even been grafted successfully into adult bodies and have functioned properly.

The ability of two like cells to recognize each other depends upon the presence of specific substances on their surfaces. As we have seen, embryonic cells migrate extensively, especially during morphogenesis. When they finally reach their destination, they adhere to other such cells, presumably by means of the mutual stickiness of these surface substances. Once together, gap junctions may form between their respective cell membranes so that an easy path is formed for the rapid passage between them of ions, electrical charges, and moderately large molecules. Thus each cell has a means of knowing what is happening in adjacent cells. Interestingly enough, cancer cells lack the sticky cell-surface substances and do not recognize other cells nor stop their migration and cell division. Thus they escape from the controls that regulate the normal cells of the body.

Although the inducing substances that pass from one cell to another may be quite specific or quite general in their instructions, it seems clear that what a differentiating cell does is programmed by its own genes, not those of adjacent tissues. An experiment by one of Spemann's students, Oscar Schotté, vividly demonstrated the truth of this important point.

The larval salamander has bony teeth, and a balancing organ on each side of the head. The tadpole of the frog has horny teeth, and external gills at the sides of the head. Schotté carefully grafted a piece of tissue, destined to become belly skin, from a frog embryo to the part of a salamander embryo that would normally develop into mouth and balancers. The grafted frog tissue underwent development appropriate to its *new* location, but the mouth that was produced had the horny jaws characteristic of tadpoles. Instead of balancers, external gills were formed.

An inducing substance produced by the salamander must have been responsible for changing the fate of what would have been frog belly tissue. However, the instructions provided by the inducing substance must have been quite general. Instead of programming the development of salamander parts, it seems only to have instructed the transplanted tissue to use its own genes, although in a way appropriate to the new location. Thus we must look for the ultimate control of differentiation not in any *organizer* tissue but in the reacting tissue itself. When one considers the large number of substances, even such simple molecules as methylene blue, that can induce neural folds formation in the frog embryo, it begins to look as though the reacting tissue contains all the information required to specialize and simply needs some fairly unspecific influence to enable it to do so. We are thus again left with the central problem of this chapter: what makes a cell in a developing organism express certain of its genes (e.g. those participating in building horny jaws or external gills) and not others (in our example, those for belly skin)?

22.8 THE REVERSIBILITY OF DIFFERENTIATION

As embryonic development proceeds, the ability of bits of embryonic tissue to have their future altered by transplantation to new locations becomes more limited. Their cells become progressively less flexible in their potentialities. Once gastrulation begins, for example, the cells of one tissue layer generally cannot be converted into those of another. The conversion of potential belly skin into mouth parts or the lens of an eye (Fig. 22.17) simply involves a shift from one kind of ectoderm to another. Later, the number of alternatives open to differentiating cells decreases still more. Finally, the cells become committed to a definite fate. If transplanted after this time, they continue to develop as they would have in the old location, not in a way appropriate to the new one. For example, if the limb buds are removed from one salamander embryo and grafted onto a second, they form extra legs on the host. This occurs even though the cells of the limb bud are not visibly differentiated.

Are these progressive losses in the potentialities of differentiating cells utterly irreversible? The question really has two parts. The work of Briggs and King and of Gurdon, which we examined earlier in the chapter, indicates that nuclei may never lose their full developmental potentialities. But what of a differentiated *cell*—a nucleus housed within a particular specialized cytoplasm? As Fig. 22.8 shows, a single differentiated plant cell can be coaxed into **dedifferentiating,** and its descendants will enter upon fresh paths of differentiation culminating in a complete plant.

The situation in animals is not as clear-cut. A wounded animal soon produces a mass of undifferentiated cells at the site of the wound. These eventually enter the various pathways of differentiation to heal the wound—to produce new differentiated structures (e.g. dermis, capillaries) that fit harmoniously with the already differentiated structures at the margin of the wound. The evidence is quite strong that the undifferentiated cells that begin the process of healing are descendants of once fully differentiated cells that become dedifferentiated. What is not clear is to what extent dedifferentiated cells can enter upon any pathways of differentiation other than the one that they or their predecessors had reversed. In other words, a dedifferentiated smooth muscle cell may be limited to reentering that pathway of differentiation leading to another smooth muscle cell.

22.9 SUMMARY

The oocytes in a young woman's body are one example of the perhaps one hundred different kinds of cells of which she is constructed. Some of the others (e.g. muscle cells, red blood cells) were described briefly in Chapter 5. Each type of cell is structurally distinct from the others. Red blood cells are tiny disks that contain no nucleus. The cells of smooth muscle are spindle-shaped. And so on. Each type of cell has its special functions to carry out. Red blood cells transport oxygen and carbon dioxide. Smooth muscle cells contract and, in so doing, reduce the diameter of such hollow organs as the bladder and blood vessels. And finally, many of these cell types are characterized by a particular protein or proteins. The red blood cell is literally stuffed with hemoglobin. The smooth muscle cell contains the two contractile proteins actin and myosin.

Where did these cells come from? The ultimate

answer is the same for all of them. They are descendants of the fertilized egg with which we all began. The development of a fertilized egg into a child requires an average of 41 rounds of mitosis ($2^{41} = 2.2 \times 10^{12}$). But it requires more than that. The daughter cells produced by these mitotic divisions must begin to enter separate pathways of development: some to become red blood cells, some to become oocytes, some to become smooth muscle cells, etc. This is the process of differentiation. It involves the formation of *dissimilar* descendant cells from *similar* precursor cells.

We have established that all of the differentiated cells of the body retain the full genome of the species. Therefore, we must conclude that during the process of differentiation each cell becomes committed to the expression of just a portion of the genome contained within its nucleus. How can we account for the differential gene activity that emerges as cells diverge into separate paths of differentiation?

The problem is really to establish a way in which the two cells produced by a mitotic division can develop different patterns of gene activity while sharing identical genetic information. The most likely solution to the problem is the existence of *different* cytoplasmic environments around *identical* nuclei. We have examined compelling evidence that the ability of a nucleus to express genetic information is controlled by the particular cytoplasm in which that nucleus finds itself. During the earliest stages of cleavage, identical nuclei—produced by mitosis—find themselves partitioned off in different cytoplasms. In most cases, these different cytoplasms are a result of the *nonuniform* distribution of cytoplasmic materials in the cleaving egg. Once these differences are established, our problem becomes one of simply working out the ensuing details. With one combination of nucleus and cytoplasm expressing certain genes while another combination expresses other genes, the stage is set for all the intricate steps to follow.

One of the consequences of differential gene expression during development may be the sending and receiving of chemical signals (e.g. inducing substances, chalones) between one cell type and another. Such signals may further direct the responding cells as to which portions of their genetic repertory they may come to use as they acquire their final differentiated form and function. While many details remain to be worked out (for example, our ignorance of the chemical nature of inductive signals is immense), at least we can now see in theory how a single cell, the fertilized egg, can ultimately give rise to that most intricate of machines, the adult body.

EXERCISES AND PROBLEMS

1. From what embryonic germ layer is the optic nerve derived?
2. Distinguish between identical and fraternal twins.
3. A tadpole ready to hatch is slightly larger than the fertilized egg was. What might account for this?
4. Female armadillos always give birth to four offspring of the same sex. Can you account for this?

REFERENCES

1. WOLPERT, L., *The Development of Pattern and Form in Animals,* Oxford Biology Readers, No. 51, Oxford University Press, Oxford, 1974.
2. EBERT, J. D., "The First Heartbeats," *Scientific American,* Offprint No. 56, March, 1959. The differentiation of cardiac muscle can be detected by biochemical changes even before morphological changes become evident.
3. STEWARD, F. C., "The Control of Growth in Plant Cells," *Scientific American,* Offprint No. 167, October, 1963.
4. SPEMANN, H., "The Development of Lateral and Dorso-ventral Embryo Halves with Delayed Nuclear Supply," *Great Experiments in Biology,* ed. by M. L. Gabriel and S. Fogel, Prentice-Hall, Englewood Cliffs, N.J., 1955. A description of two of the

author's famous experiments with constricted newt eggs.

5. GURDON, J. B., "Transplanted Nuclei and Cell Differentiation," *Scientific American,* Offprint No. 1128, December, 1968.
6. FISCHBERG, M. and A. W. BLACKLER, "How Cells Specialize," *Scientific American,* Offprint No. 94, September, 1961. Shows how the sequence of changes occurring during embryonic development can be traced to specialization within the egg itself.
7. GURDON, J. B., *Gene Expression During Cell Differentiation,* Oxford Biology Readers, No. 25, Oxford University Press, Oxford, 1973.
8. MARKERT, C. L., and H. URSPRUNG, *Developmental Genetics,* Foundations of Developmental Biology Series, Prentice-Hall, Englewood Cliffs, N.J., 1971. Chapter 6 deals with the polytene chromosomes.
9. BEERMANN, W. and U. CLEVER, "Chromosome Puffs," *Scientific American,* Offprint No. 180, April, 1964.
10. WESSELLS, N. K., *Tissue Interactions in Development,* Addison-Wesley Modules in Biology, No. 9, Addison-Wesley, Reading, Mass., 1973.
11. BARTH, LUCENA, J., *Development, Selected Topics,* Addison-Wesley, Reading, Mass., 1964. Examines many of the experiments that bear on the question of the role of nucleo-cytoplasmic interactions in embryonic development.
12. EBERT, J. D., and I. M. SUSSEX, *Interacting Systems in Development,* Holt, Rinehart and Winston, New York, 1970. An excellent treatment of the subject; available in paperback.

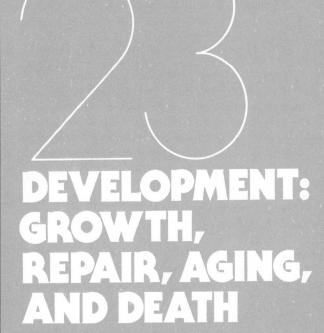

DEVELOPMENT: GROWTH, REPAIR, AGING, AND DEATH

23.1 GROWTH

Plant and animal development is not limited to morphogenesis and differentiation. Development also includes an increase in the size of the organism, that is, growth. Among animals, most growth takes place after the completion of morphogenesis and differentiation. These processes are essentially completed in the human embryo by the time it reaches a weight of 1.5 grams. This weight may, however, be increased some 45,000 times before the adult size is reached. You remember also that the newly hatched frog tadpole is little larger than the original frog egg.

In plants, there exists considerable overlap in these vital developmental activities. Apical growth (in the region of elongation of roots and stems) precedes morphogenesis and differentiation. However, growth of stem thickness occurs by cell enlargement after morphogenesis and differentiation take place.

For growth to occur, the rate of synthesis of the complex molecules of the organism, proteins for example, must exceed the rate of their breakdown. This means that additional supplies of organic molecules (e.g. amino acids, fatty acids, glycerol, and glucose) must be taken in by the organism from its environment. Some of these materials will provide the building blocks for the anabolic reactions. The others will supply the extra energy needed to run them. In photosynthetic organisms, light supplies the energy for anabolism and inorganic molecules are the raw materials.

Growth thus involves gaining more material from the environment than is given back to it in the form of metabolic wastes. This is not just a simple process of accumulation. The amino acids which you assimilate after a meal of beefsteak are not reassembled in your cells into *beef* protein. Instead, you grow by converting relatively unspecific organic molecules taken in from your environment into the specific cell materials that are characteristic of *you.*

The autotrophic organisms are even more remarkable in this respect. In growing, they convert inorganic substances from their environment (e.g. H_2O, CO_2, and nitrates) into the specific structures that form more maple tree, corn plant, etc. This ability of all living things to build more of their own specific, complex organization from relatively simple, disorganized materials in their environment is the ability to grow. It is one of the most characteristic features of life.

Growth in organisms may occur simply by an increase in the number of cells that make up the organism. The adult human is made up of some 60 trillion (6×10^{13}) cells, while the newborn baby contains only about two trillion (2×10^{12}). Growth may also occur as a result of an increase in the size of the cells that make up the organism. Monocots, lacking a cambium, grow in thickness this way. In general, though, organisms grow by both an increase in the number of cells *and* an increase in their size. Growth in woody dicots involves both activities. Tiny new cells are produced by the cambium and apical meristems. These cells then enlarge and differentiate.

Several periods can usually be distinguished in

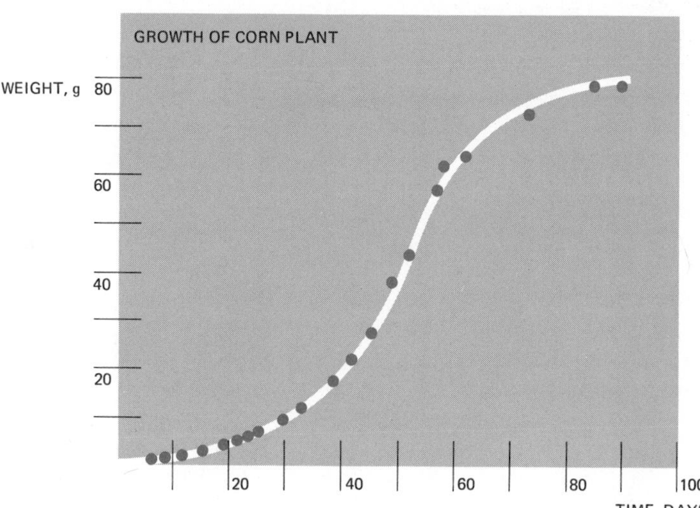

FIG. 23.1 Plotting the growth of a corn plant produces an S-shaped (or sigmoid) curve.

the growth of an organism (Fig. 23.1). The first period, the **lag period,** is characterized by little or no actual growth. During this period, however, the organism is getting prepared for growth. A seed imbibing water, preparatory to germination, and an *E. coli* cell synthesizing the enzymes with which to metabolize its substrate are both in the lag period of growth.

The lag period is followed by the logarithmic or **exponential period** of growth. In this period growth begins, slowly at first, but then more and more rapidly. This gradually accelerating rate of growth is understandable when you remember that in most cases, the product of growth, living material, is itself capable of further growth. Thus the organism enlarges according to a geometric progression, doubling and redoubling in size. Such progressions are expressed in algebra by exponents (logarithms); hence this phase of growth is called the exponential or logarithmic phase. Different organisms vary greatly in the time that it takes them to double their size. The newly hatched housefly larva doubles its weight in 13 hours. The newborn baby usually requires 5 to 6 months to do the same. Whatever the amount of time required, all organisms pass through a constantly accelerating period of growth.

The exponential phase of growth does not continue indefinitely, and it is a good thing that it does not. The housefly larva, doubling its weight every 13 hours, would be an awesome creature indeed after a few days. Fortunately, the fly larva (as well as all other growing organisms) soon enters a period of **decelerating growth.** Growth now proceeds more slowly and finally ceases altogether. At this time the larva prepares for metamorphosis. In many organisms the rate of growth slows down but never ceases entirely. Many fishes and reptiles continue to grow year after year, although more and more slowly, until finally they die. A similar situation occurs in trees, which continue to grow until disease or accident strikes them down.

Mammals, including humans, reach a certain size and then cease growing entirely. At this point, the rate at which more living substance is synthesized by anabolism is exactly counterbalanced by the rate at which it is broken down catabolically. In later years, humans may even enter a period of negative growth, that is, they shrink. Body constituents are then broken down more rapidly than they are synthesized.

If one plots a graph showing the growth of a single organism during the accelerating and decelerating phases, an S-shaped curve results (Fig. 23.1). It is interesting to note that this same curve expresses the growth of populations. If a few bacteria or a pair of rabbits are placed in a suitable and uncrowded environment, they will produce offspring. These offspring will, in turn, produce their own offspring. This will go on in an ever-expanding way until external conditions force a slowdown in the rate of growth of the population. Biologists have a good idea of what forces are involved in the stabilizing of population growth. We shall consider these in Chapter 39. But biologists are not at all sure what forces cause a slowdown in the growth of individuals. Perhaps the controlling factors differ from one group of organisms to another. In any case, until these factors are identified, a large gap will remain in our knowledge of development.

Growth in plants occurs chiefly at the meristems. Here rapid mitosis provides the necessary additional cells. In Section 12.4, we examined the way in which the apical meristem in the embryonic region of the root provides new cells which then elongate and differentiate. A similar pattern of vertical growth occurs at the tips of stems. Mitosis in the apical meristem (Fig. 23.2) of the shoot apex or terminal bud produces a stockpile of new cells. Peri-

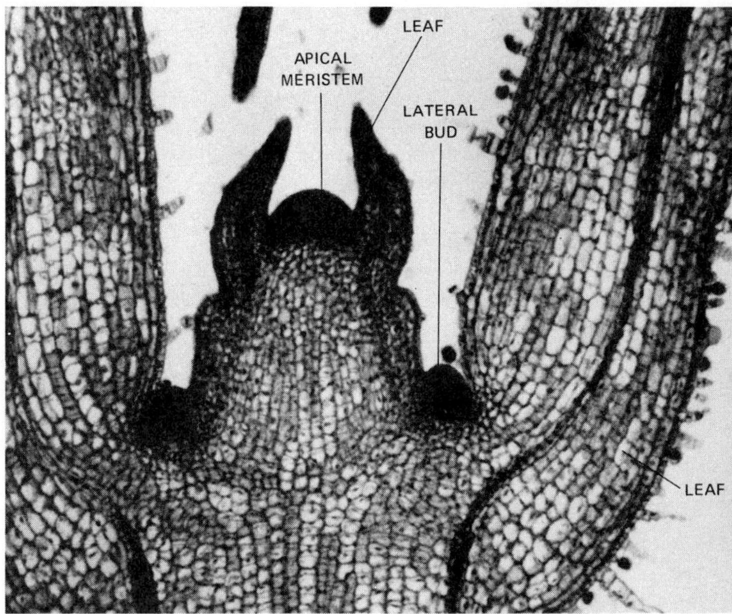

FIG. 23.2 Terminal bud of a coleus stem, cut lengthwise. New cells are produced by mitosis in the apical meristem. Later, branching may occur at lateral buds.

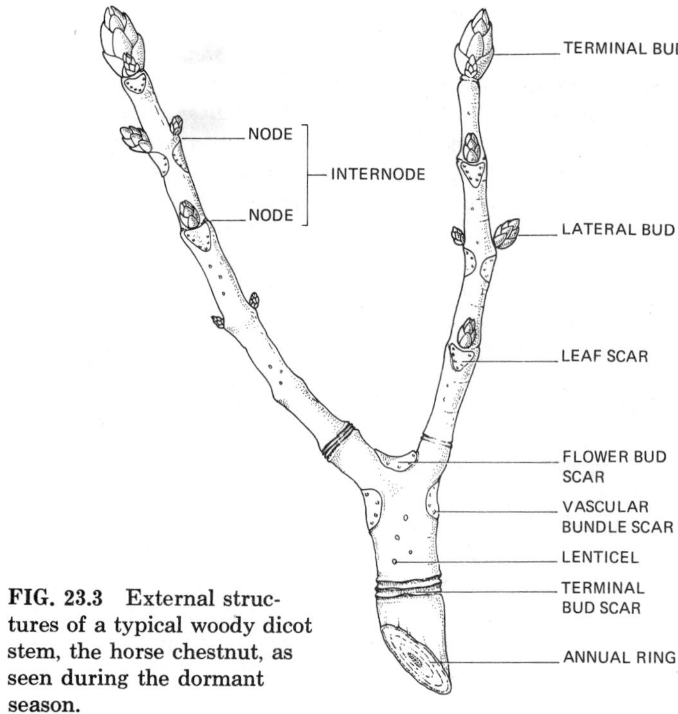

FIG. 23.3 External structures of a typical woody dicot stem, the horse chestnut, as seen during the dormant season.

odically these give rise to leaves. The point on the stem where the leaves develop is called a **node** (Fig. 23.3). The distance between the nodes (the internode) in the terminal bud is very short, and the leaves develop rapidly. As a result, the leaves grow above the apical meristem that produced them and thus protect it.

New meristems, the lateral buds, develop at the nodes, each just above the point where a leaf is attached. When the lateral buds develop, they produce new stem tissue and thus branches are formed.

Under special circumstances (which we shall study in the next chapter), the apical meristem is converted into a flower bud. This develops into a flower and thus enables the plant to carry on sexual reproduction. The formation of a flower bud "uses up" the apical meristem so that no further growth of the stem can occur at that point. Branching may, however, occur from lateral buds directly behind the flower (Fig. 23.3).

In dicots, growth in diameter is accomplished by a band of cambium which separates the phloem from the xylem in both the root and the stem (Figs. 12.2 and 12.4). Except for a few palms and lilies, monocots lack a cambium. As we noted earlier, increase in their stem diameter can occur only by cell enlargement.

Growth in animals does not occur in such localized areas as it does in plants. All of the tissues and organs of the animal body participate in growth. They may not all grow at the same rate, however. Figure 23.4 shows how much faster the trunk and limbs of the human grow from infancy to adulthood than the head. The result is a marked change in body proportions.

Fundamental to the growth of the body is growth of its supporting skeleton. Bones are able to grow and lengthen only so long as they have a cartilaginous, nonbony region where further cell division and elongation can occur. In the long bones, this occurs at the epiphyseal lines just back of the hard ends of the bones.

In humans, the growing regions of the bones become fully "ossified" during the late teens and early twenties. The cartilaginous matrix of these regions becomes replaced with a bony matrix and further skeletal growth ceases. Cessation of growth also occurs in the other mammals but usually at a younger age. On the other hand many of the fishes and reptiles continue to exhibit skeletal growth throughout their lives.

The problem of development is particularly interesting in those animals that undergo metamorphosis. Most aquatic invertebrates (e.g. starfish and barnacles), many aquatic vertebrates (e.g. frogs), and most insects develop from the zygote into a **larva.** Later the larva undergoes the drastic changes of metamorphosis and the adult body is produced.

The significance of this two-stage method of development is not always clear. For those animals, such as barnacles, whose adult bodies are sessile (anchored), free-swimming larvae permit dispersal. In cases such as the lobster and the frog, the larva

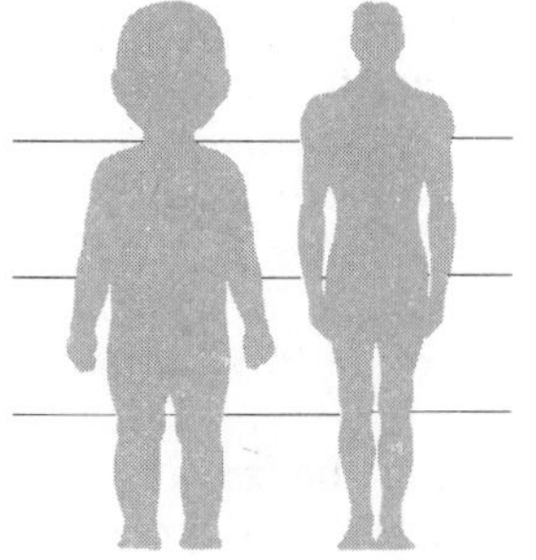

FIG. 23.4 Differential growth in the human. The head of a baby is roughly one-fourth the length of the body, while in the adult the fraction is closer to one-eighth.

serves as an intermediate stage in development during which food (usually of microscopic dimensions, e.g. unicellular algae) can be taken in from the environment. This permits more growth than is provided for by the food reserves in the egg itself and enables the animal eventually to reach the size necessary for it to have the characteristic diet of the adult.

While larvae are a stage in the development of many animals, it is wrong to think of them as incompletely developed or embryonic organisms. For example, the garden caterpillar is a highly specialized creature. It is the product of an elaborate morphogenesis and complete cell differentiation, complete, that is, except for a few masses of embryonic cells, the *imaginal disks,* within its body. At the time of metamorphosis, most of the specialized cells of the caterpillar die. These dying cells supply the nourishment necessary for the cells of the imaginal disks to undergo an entirely new pattern of morphogenesis and differentiation. The result is the adult organism, a moth or butterfly. All growth in insects occurs during the larval stages. In fact, some adult insects do not even feed. They simply mate and die.

Another reason why larvae should not be considered incompletely developed forms is that cases are known in which certain species have abandoned metamorphosis entirely. The axolotl, an amphibian found in Mexico and the American Southwest, spends its entire life as a gill-breathing animal (Fig. 25.6). It reaches sexual maturity and mates without ever undergoing metamorphosis to an air-breathing form. While there is no evidence that the axolotl is going to lead to the evolution of a whole new group of animals, there is evidence that in the past, larval, rather than adult, forms have sometimes evolved into new species. We shall examine some of this evidence in later chapters.

23.2 REPAIR

During the lifetime of an organism, some of its parts may become damaged or lost. Most organisms have, to some degree, the ability to replace defective or missing parts. This process of replacement is called **regeneration** (Fig. 23.5).

FIG. 23.5 Starfish regenerating arm. (Courtesy of Dr. Charles Walcott.)

Plants generally have great powers of regeneration. They can have many, sometimes all, of their branches and foliage removed by pruning. If the root system is healthy, however, buds soon appear on the stem or trunk. The buds develop into new branches, leaves, and flowers. In fact, many hardwoods (maples, birches, etc.) can be cut off right at ground level and sprouting at the margins of the stump will soon produce new stems and leaves.

The ability of animals to regenerate missing parts varies greatly from species to species. Sponges can regenerate the entire organism from just a conglomeration of their cells. This is also true of the hydra. A planarian can regenerate the entire organism from a middle section. Even the starfish can regenerate an entire organism from just one arm and the central disk. At one time oyster fishermen used to dredge up starfish from their oyster beds, chop them up in the hopes of killing them, and then dump the parts back overboard. They soon discovered to their sorrow the remarkable powers of regeneration in this group of animals.

Earthworms, crustaceans, fish, salamanders, and lizards do not possess such powers of regeneration that they can regenerate the whole organism from just one part. They can, however, regenerate fairly substantial parts. The earthworm can regenerate the first four or five segments of its head and even longer sections of its tail. Lobsters and salamanders can regenerate a missing leg. Many lizards will part with their tail if caught by it. They then regenerate a new one at their leisure. Birds and mammals cannot regenerate entire organs. They can, however, regenerate tissues and thus repair damaged or missing parts. The healing of skin wounds and bone breaks is an example of regeneration in humans. The replacement of blood after blood loss is another example. The digestive glands, especially the liver and pancreas, are capable of extensive regeneration after damage.

What happens in the process of regeneration? If a foreleg is removed from a salamander, the first repair process is the healing of the wound by means of skin growing over it. Then a bud of undifferentiated cells appears. This bud has the same appearance as a limb bud in the developing embryo. The rapidly dividing, unspecialized "embryonic" cells of the limb bud probably arise from the **dedifferentiation** of such specialized cells as muscle and cartilage cells. Dedifferentiation means that these cells lose their differentiated structure prior to taking on the task of regeneration. As time goes on, the cells of the regenerating limb become organized and differentiated once again into the muscle, bone, and other tissues which make up the functional leg.

Studies of regeneration have revealed that mature, differentiated cells of a given tissue, e.g. epidermis, synthesize and secrete a substance which actively inhibits mitosis of the young cells of the same tissue. This substance is called a **chalone.** In the early stages of regeneration, there are no mature cells and thus there is no inhibition of cell division. As the tissues of the regenerated structure become differentiated, chalone production begins and, presumably, this gradually stops the growth of the regenerated structure.

Actually, the production of chalones by differentiated cells seems to be a general phenomenon. Virtually every tissue that has been examined (including liver, kidney, lung, uterus, hair follicle, and white blood cells) produces a chalone which inhibits mitosis in the immature cells of that tissue but of *no other* tissue. The chalones do not harm the other activities of these cells in any way. Although chalones have not as yet been well characterized chemically, they behave like proteins or polypeptides. Although their action is exceedingly specific with respect to tissue, it is not so with respect to species. An epidermal chalone from one mammal (even from a fish) will inhibit mitosis in the epidermis of another mammal.

In many ways, the process of regeneration is similar to the process of embryonic development. From rapidly dividing, unspecialized cells arises a complex organization of specialized cells. This involves morphogenesis and differentiation just as embryonic development does. There is, however, at least one way in which the process of regeneration differs from the process of embryonic development. Can you think of what it is?

The similarities between regeneration and embryonic development have caused some embryologists to study regeneration in the hope of gaining an understanding of how embryonic development "works." The discovery of **polarity** has shown that definite organizing forces, probably chemical, are at work in regeneration. A midsection of a planarian will regenerate a new head at the same cut edge at which a head was originally present. A tail will be regenerated at the other edge. It has been found that the cells at the "forward" cut surface have a higher metabolic rate than those at the "rear" edge. It seems to be this difference in metabolic rates in the cells of the regenerating piece that determines polarity. If the rear section of a worm is removed and the head is then treated with a solution containing a metabolic inhibitor, the polarity of the worm can be reversed and a second head will form at the posterior edge (Fig. 23.6).

In comparing the regenerative abilities of different animals, there seems to be some relationship between the complexity of the organism and its ability to regenerate. The powers of regeneration in the sponge are practically complete. In humans, regeneration is limited to repair of certain organs and tissues. We wish we knew just why it is that regenerative powers diminish with increasing structural and physiological complexity.

Within a single organism, at least among the vertebrates, there also seems to be a progressive loss of regenerative ability with increasing age. When legs first appear on a frog tadpole, they can be regenerated easily if lost. After metamorphosis, however, a frog is normally unable to regenerate a missing leg. Everyone knows how much more quickly broken bones and skin wounds heal in a child than in an elderly person. What causes the loss of regenerative power as age increases? In the frog, it may be that the area of amputation fails to receive sufficient amounts of some substance released by the nerves. When extra nerves are moved surgically to the stump of a frog foreleg, regeneration does occur. It is known that the ratio of nerve tissue to other tissues decreases as the tadpole grows into a frog. It is also possible, though, to bring about leg regeneration in the frog simply by irritating the stump. This

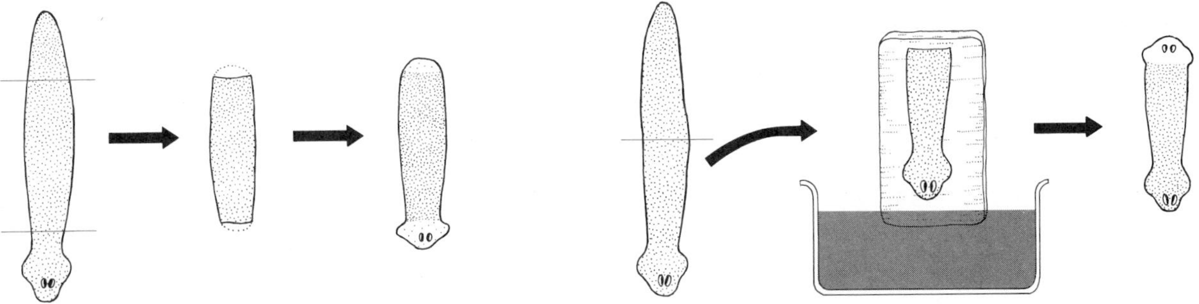

FIG. 23.6 Regeneration in planarians. Left: Each cut surface of the planarian "remembers" its polarity. Right: When the head of a cut planarian is placed in agar and dipped in water containing a metabolic inhibitor, a second head develops at the cut surface.

suggests an alternative explanation. Perhaps aging tissue simply loses the ability to respond to the regeneration-promoting substances released by nerves. If this is the case, irritation of the tissue may restore some of its sensitivity.

Cancer. The gradual development of pattern in both regeneration and embryonic development tells us that tissues normally grow to a certain size and then stop. Further cell division in the tissue awaits, and is regulated by, the need for cell replacement (as in the replacing of old red blood cells and sloughed-off epidermal cells, or the regeneration of damaged parts). Sometimes the process fails. A cell escapes from the normal control mechanism (e.g. chalones) of the body and begins to divide repeatedly and uncontrollably. As it and its progeny continue to divide, they often lose to some degree the characteristic structure and functioning of normally differentiated cells of that type. To some extent, they begin to resemble embryonic cells. Such cells constitute a cancer.

Every cell type known can become cancerous. What converts a normal cell into a cancerous cell is not yet known with certainty. Radiation, many chemicals, and a number of viruses can bring about this cancerous change, but the exact mechanism by which they do so remains to be discovered.

23.3 AGING—THE FACTS

Aging can be defined as the progressive deterioration, with the passage of time, of the structures and functions of a mature organism. This ultimately leads to the death of the organism. Either the progressive loss of function makes the organism less able to withstand infectious disease or, often, the failure of some vital organ precipitates the death of all the rest.

The study of aging has long been neglected. But now that so many infectious diseases of humans have been brought under better control by vaccines, sulfa drugs, and antibiotics, more and more people are living to an age when the disabilities of aging become a problem. Because of this, there recently has been a great surge of interest in the problem of aging. Doctors, social workers, physiologists, biochemists, embryologists, histologists, endocrinologists, and many others are turning their attention to this problem. Not only is aging in humans under direct study, but many other organisms are being examined for clues as to the nature of aging. As a result of these studies, one fact stands out already. The effects of increased age vary widely from one living thing to another.

Among the bacteria, the process of aging does not even occur. When a single cell is fully grown, it divides into two cells, which then repeat the process. Although a wide variety of changes in the environment may kill a bacterium, no cell has ever been known to die because of aging. For this reason, some biologists have described the bacteria as potentially immortal. They have also included all other kinds of single-celled organisms in this category. While the idea that the bacteria enjoy a kind of potential immortality may have some merit, it is probably not true that all single-celled forms can be included.

At one time it was thought that removing cells, or even tissues, from animals and culturing them *in vitro* conferred an "immortality" on them similar to that enjoyed by cultured bacteria. However, the situation is not as simple as that. It is true that certain mammalian cells may be propagated indefinitely, but in every case these cells are *not* normal. They have a number of abnormalities among which are an abnormal karyotype (see Section 16.4) and, generally, the potentiality of growing into a cancerous tumor if reimplanted in the species from which they have been taken. Many successful long-term cell cultures are, in fact, derived from cancers. "HeLa" cells, for example, which are now cultured in laboratories around the world, were isolated many years ago from a human cervical cancer.

Normal mammalian cells *can* be cultured *in vitro,* but generally they pass through no more than about 50 cell doublings before the cells decline in vigor and die. Interestingly enough, cells taken from young mammals are able to pass through more generations than cells taken from old individuals. This would suggest that mammalian cells have an inherent limitation on the number of cell divisions they can accomplish, whether *in vivo* or *in vitro*.

Everyone agrees that multicellular organisms are mortal. Presumably the loss of independence which occurs when cells specialize in a multicellular organism leads to eventual death. In specializing to carry out a certain function, cells abandon other vital functions to neighboring cells. If disease or other damage should strike down any one group of specialists, all the rest will fall, too. Thus heart failure results in the speedy death of all our cells no matter how vigorous and healthy they might otherwise be.

Although all multicellular organisms eventually die, it may not be accurate to say that they all age. Some seem to avoid aging and, in the process, may achieve remarkable records for length of life. Some sea anemones are known to have lived for as long as 78 years. They may achieve their longevity and avoid aging by periodically replacing all old body parts. Woody perennials fall into a similar category. Throughout their lives, they produce new vascular tissues, leaves, and flowers each year. They do not show marked signs of aging, although their rate of growth does decline with advancing years. Finally, disease or inability to support their ever-increasing size against wind or snow load lead to their death. This may not occur for a long period. Tree ring analysis shows that some of the bristlecene pines of eastern California are over 4000 years old (Fig. 23.7). At the moment, this seems to be a record for tree longevity. Note, however, that no *living* cells in these trees are more than a few years old.

Many fish and reptiles also seem to avoid, or at least inhibit, aging by continuing to grow throughout their lives. We do not know just what brings on death, but it may simply be the same factors that cause death in younger members of the species: disease and predation. These challenges to life, acting in a completely random way, will eventually strike down all the members of a given generation (Fig. 23.8). It is probably also true that fish and reptiles become less well adapted to their environment when they exceed a certain size. In any case, the ability to grow steadily, even if slowly, does seem to protect them from the harmful effects of aging. Some marine turtles are estimated to live more than 150 years.

The situation is quite different among the annual plants (such as the grasses and many "weeds") and the mammals. Most annual plants grow vigorously for a period and then form flowers followed by fruits. Fruiting is accompanied by a marked slowing down of growth. As the growth rate slows to a halt, other changes take place, too. A great deal of chlorophyll may disappear from the leaves. The rate of respiration (and thus, catabolism) rises markedly. Many other morphological changes follow. All of these changes constitute aging and end with the death of the plant. It is important to realize that these changes are not necessarily related to adverse changes in the physical environment. Aging in annual plants may occur in the midst of plentiful moisture, soil minerals, sunlight, and warm temperatures.

Mammals, as we have seen, also grow to a certain size and then stop. Some time after the cessation of growth, aging begins. The actual time span involved varies widely from species to species. A three-year-old laboratory rat is very old. In humans, although the deterioration associated with aging can be detected by the age of 30 years, fatal loss of function may not occur until much later.

What are the symptoms of aging? Decreased muscular strength, decreased lung capacity, decreased pumping of blood from the heart, decreased urine formation in the kidney, and decreased metabolic rate are just a few of the many body changes which occur with aging. Figure 23.9 shows still other examples of the structural and functional deterioration that occurs between ages 30 and 75.

Why do the various body organs gradually lose their ability to function well? One answer is that they lose the cells of which they are composed. Many organs of the body lose weight with age (Fig. 23.9), and even those that do not, do lose the specialized cells which enable them to carry out their function. These cells may then be replaced by con-

FIG. 23.7 Bristlecone pines (*Pinus aristata*) growing in the White Mountains of eastern California. Tree-ring analysis reveals that many of the trees are over 4000 years old. (Courtesy of Walter Gierasch.)

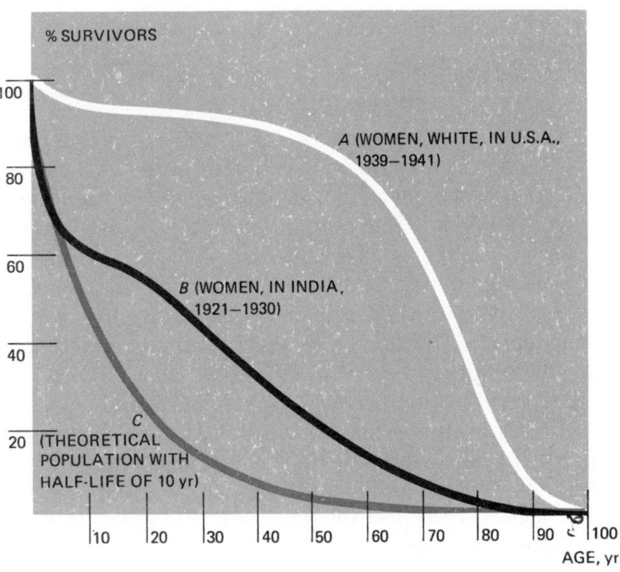

FIG. 23.8 Representative survival curves. Curve A is typical of organisms that age but suffer little random mortality before then. Populations of annual plants have such a curve. Curve B is typical of populations in which such environmental factors as starvation, predation, and disease obscure the effects of aging. The high rate of infant mortality is particularly characteristic of organisms that provide little or no care for their young. Curve C is a theoretical curve for organisms for whom the chance of death is equal at all ages. This might be true for organisms that do not age or those (e.g. many songbirds in the wild) that suffer crushing random mortality before the time of aging.

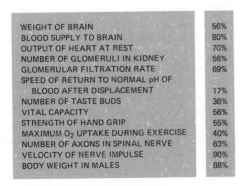

WEIGHT OF BRAIN	56%
BLOOD SUPPLY TO BRAIN	80%
OUTPUT OF HEART AT REST	70%
NUMBER OF GLOMERULI IN KIDNEY	56%
GLOMERULAR FILTRATION RATE	69%
SPEED OF RETURN TO NORMAL pH OF BLOOD AFTER DISPLACEMENT	17%
NUMBER OF TASTE BUDS	36%
VITAL CAPACITY	56%
STRENGTH OF HAND GRIP	55%
MAXIMUM O_2 UPTAKE DURING EXERCISE	40%
NUMBER OF AXONS IN SPINAL NERVE	63%
VELOCITY OF NERVE IMPULSE	90%
BODY WEIGHT IN MALES	88%

FIG. 23.9 Loss of structure and function in aging. Figures represent percentage of a given function remaining in an average 75-year-old man compared with that found in an average 30-year-old man, the latter value taken as 100%.

nective tissue or fatty tissue cells, however, and the overall weight of the organ remains constant. It is interesting to note that the organs which lose so much of their ability to function (and thus show so much aging) are those organs whose cells are no longer dividing actively. These include the heart, brain, kidneys, and muscles. On the other hand, the organs in which cell division is actively maintained (e.g. the bone marrow, the liver, and the pancreas) show far less loss of function with age. The situation reminds us of the perpetual youthfulness of rapidly dividing bacterial cells.

23.4 AGING—THE THEORIES

What causes the degenerative changes of aging? This is one of the many unanswered questions in biology. A great many theories of aging have been proposed, but the known facts of aging are still too few either to substantiate or discredit any of them.

One popular theory explains aging as a consequence of living. That is, the more rapidly the organism lives, the sooner it begins to age. Rapid living is defined as rapid differentiation and growth. It also involves a high rate of metabolism. It is certainly true that mammals, such as rats, which reach maturity weeks after birth, age far more rapidly than mammals, such as humans, that take years to mature.

One way to test this theory is to find some way to slow down an organism's rate of development. It has been found that feeding young rats just enough to keep them alive does just this. When these rats are finally given adequate diets, they usually catch up to their normally raised cousins in almost all respects. In fact, they are apt to be living comfortably long after their well-fed cousins are dead!

Another theory that relates aging to living has been called the "clinker" theory. According to this theory, each cell unavoidably accumulates poisonous wastes during its lifetime. The accumulation gradually reduces the ability of the cell to function, and the cell thus ages. This theory is supported by the finding of large quantities of pigment in the cells of aged people, especially nondividing cells such as those of muscles and nerves. Rapidly dividing cells, such as are found in the liver, contain much less pigment. (You remember that it is the organs in which rapid cell division does not continue throughout life that age the most.) Although the exact chemical nature of this pigment is not yet known, there is evidence that it represents the remnants of old, worn-out cell structures (e.g. mitochondria) that have been incorporated into lysosomes.

Why is the cell unable to rid itself of these harmful substances? It may be that older cells are less able to rid themselves of any substance. Collagen, the chief structural protein of the body, is found in the extracellular coatings or matrix of most of the cells of the body. When collagen is first synthesized, it is quite flexible and easily dissolved. As time goes on, the polypeptide chains of which collagen is made become increasingly linked together. This reduces the solubility and flexibility of the material. It is certainly true that organs and tissues (e.g. skin) in aging animals become less flexible with advancing age. It may also be true that aging collagen provides a barrier to the easy passage of materials in and out of capillaries and, perhaps, in and out of cells. Reduced permeability has been detected in the lungs, kidneys, and skin of aging animals. For example, when histamine is injected under the skin, the capillaries become more permeable, and an increased quantity of protein diffuses into the tissue spaces. However, the magnitude of this response is greatly reduced in old animals. Reduced permeability would also explain why so many of the body's homeostatic mechanisms work more slowly with advancing years.

If the aging of collagen is the clock that times aging for the entire organism, we can see why animals do not age so long as they continue to grow. Growth involves the continual synthesis of new, young collagen and thus slows the ticking of the clock.

But why should the collagen of an old rat that stopped growing two or so years before, be as old as that of a 70-year-old human who stopped growing 50 years before? Perhaps the rate of collagen aging is not just a simple function of time but also of metabolic rate. If so, this would provide a molecular basis for the theory, discussed above, that aging is a consequence of rate of living. Small mammals do, indeed, have higher metabolic rates than large ones.

A few biologists feel that aging is largely a result of adverse changes in the environment. They are particularly concerned with the effects of radiation (e.g. cosmic rays, X-rays) on the genes in the body cells or somaplasm. Mutations in the somaplasm are not as easily detected as mutations in the germplasm. (Virtually the whole science of genetics has been based on the latter.) There is no reason to think, though, that the DNA in body cells is appreciably more resistant to mutations than the DNA in the germ cells. Remembering the one gene–one polypeptide theory (see Section 18.1), we would expect body cells with acquired mutations to produce defective polypeptides and proteins. This would naturally reduce their efficiency and might even cause their death.

In support of this theory, a number of workers have demonstrated that as human cells (growing in tissue culture) begin to age, they do produce defective enzymes. Aging vinegar "eels" (actually tiny worms) and even aging *Neurospora* cultures do the same.

It has been shown that sublethal exposure to radiation lowers the life expectancy of many animals (Fig. 23.10). However, radiation exerts its most drastic effects upon rapidly dividing cells, and it is the tissues with rapidly dividing cells (e.g. liver, bone marrow) that seem the freest from the effects of aging. Cell damage and death in such tissues can be compensated for. The defective cells can be destroyed by lysosomes and phagocytes, and healthy replacements produced by mitosis. In organs such as the brain and muscles, where mitosis does not occur, the damaged cells become a permanent liability.

Many biologists feel that the tendency to age is intrinsic. They claim that even under the best environmental conditions, organisms will age at a rate determined by the nature of their genes. There is a good deal of evidence to support this view. We have seen that annual plants age despite the maintenance of suitable physical conditions around them. The widely different life expectancies from one species of mammal to another are difficult to explain without including hereditary factors. Even within a single species, some family lines show consistently greater longevities than others. The best way to assure yourself a long life is to have long-lived parents! The fact that, as a group, human females have an average life expectancy greater than males also suggests that hereditary factors are at work. (It does not prove this, however. Why not?) Not only do human females tend to outlive the males but female rats, mice, fruit flies (*Drosophila*), spiders, and many fish do, too.

Most biologists accept the idea that aging results from an interaction of both hereditary and environmental factors. Consequently, most of the many current theories on aging assume an interaction of these two forces. But if any aspect of aging is under the influence of the genes, it means that aging is as much a part of the overall process of development as is, say, morphogenesis.

Perhaps the most elegant attempt to weave the various data about aging into a coherent picture has

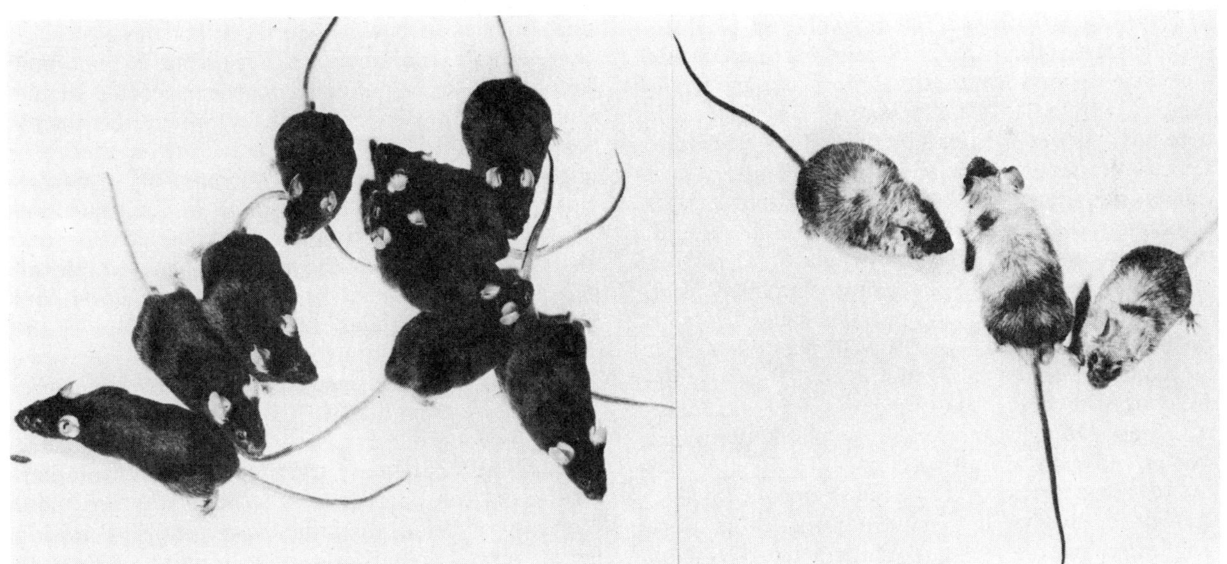

FIG. 23.10 Effect of radiation on aging. These mice are all 14 months old. As young adults, 9 mice were given sublethal doses of radiation and 9 others were left as untreated controls. The control mice (left) are still sleek and vigorous at 14 months, while 6 of the irradiated mice have died and the remaining 3 show signs of extreme aging (right). (Research photograph of Dr. Howard J. Curtis.)

been made by the Nobel Prize winning microbiologist and immunologist Sir Macfarlane Burnet. In his theory, the clock that paces the aging process is the rate at which errors in the process of DNA repair occur. In Chapter 17, we examined some of the agents that cause mutations. We also took note of the enzymes—DNA polymerases and the like—that repair damaged sections of DNA. It is only to the extent that these enzymes fail in their job that mutations occur. In other words, even if a section of DNA is damaged by X-rays or ultraviolet light, no mutation will have occurred if the damage is repaired correctly. According to Burnet, it is somatic mutations that bring about aging. As somatic mutations accumulate within a cell, an increasing number of structural genes begin producing defective proteins. These may reduce the efficiency of the cell in many ways. But if any of these defective proteins should happen to be DNA repair enzymes, the process of somatic mutation will accelerate. As a result, the cell may become cancerous or die. Perhaps the cell loss that is so characteristic of tissues composed of nondividing cells, e.g. the brain, is a consequence of self-destruction by this mechanism.

This theory can accommodate many of the features of the aging process that we have noted. But how can it explain the rapid aging of the rat, with its three-year life span, and the slow aging of humans, with their 70-year life span? According to Burnet, each species—as a result of its evolution—has acquired genes coding for DNA repair enzymes and other key enzymes whose degree of accuracy or efficiency is related to the life span of the species. Is there any evidence of this? In 1974, R. W. Hart and R. B. Setlow reported some intriguing findings that bear on this question. They took cells from various mammalian species and placed them in tissue culture. They then irradiated the various cultures with a uniform dose of DNA-damaging ultraviolet light. Following this, they measured the efficiency with which the enzymes of the cells went about repairing the damage. As the table in Fig. 23.11 shows, the efficiency with which the different species could repair their DNA was directly correlated with the average life span of that species.

For some species, e.g. the house mouse, the best strategy for survival is to mature quickly, reproduce prolifically, and die young. For others, like ourselves, the best strategy is to mature slowly (learning lots of things as we do so), reproduce in moderation, and die when we can no longer contribute to the welfare of our species. According to Burnet, the house mouse

SPECIES	AVERAGE LIFESPAN, yr	RELATIVE EFFECTIVENESS OF DNA REPAIR
HUMAN	70	50
ELEPHANT	60	47
COW	30	43
HAMSTER	4	26
RAT	3	13
MOUSE	2	9
SHREW	1	8

FIG. 23.11 Correlation between lifespan and the relative effectiveness of DNA repair in the cells of certain mammals. In each case, cells growing in tissue culture were irradiated with ultraviolet light and then the efficiency with which they repaired their DNA was determined. (From the work of R. W. Hart and R. B. Setlow, 1974.)

and we have each inherited genes that enable us to carry out our respective strategies: for the mouse, genes coding for enzymes that do their work with many errors; for humans, genes coding for enzymes that are far less prone to making errors in vital cell processes like DNA replication and repair.

You may feel dismayed that there are so many different theories to explain such a vital process as aging. Actually, this is a perfectly normal stage in the development of any science. The fewer the facts that are known, the more theories are plausible. It does not matter now which, if any, of these theories best explains the facts of aging. What is important is that each theory of aging that is concocted should suggest definite experiments by which the theory can be tested. Whether or not a given theory is successful in predicting the outcome of an experiment, it will at least have resulted in the acquisition of new knowledge. The more we know about aging, the fewer will be the theories that can explain all the facts. Only thus, by a process of closer and closer approximations, can we hope to discover the true nature of the aging process.

If aging is even partly controlled by our genes, is there any hope that we can increase our life span through evolution? It is not likely. As we shall see in Chapter 32, the only traits upon which evolutionary forces can operate in a positive way are those that appear before the organism has finished raising its offspring. Unusual slowness in aging or unusual longevity will not be discovered until the chance to pass these traits on to a large number of offspring is gone. Unless the inherited traits that promote longevity also promote vigor during the years of reproduction, there is little likelihood of these traits becoming generally established in the population. In this connection, it is interesting to note how many species of living things begin aging as soon as their reproductive activities are completed.

23.5 DEATH

The outcome of aging is death. As the organs of the body become less efficient, the body becomes less able to cope with the stresses of life. Infections are controlled less easily. Shifts in body chemistry are corrected less easily. Finally, a given organ fails to carry out its vital functions upon which all the other organs depend. Death results.

Although other organs may fail first, we generally mark death in humans at the time the heart stops beating. When you consider the role it plays in supplying food and oxygen to the cells, this should not be surprising. Without food and oxygen the constituent cells of the body die. They do not, however, all die at once. Even after the human body is dead as a functioning organism, various cells survive for short periods of time. Nerve cells are among the first cells to succumb to lack of oxygen. Some of the skin cells are among the last. Eventually, though, every cell of the body dies.

With the development of machines that can sustain circulation and breathing for indefinite periods in patients whose own ability to maintain these functions has failed, it has become necessary to reexamine our definition of death. What, then, can we use as a criterion of death? As we shall see in Part V, it is the central nervous system that coordinates the actions of the body. Without a functioning brain, we are little more than an assemblage of cells. Therefore, irrevocable loss of brain activity is death, of the *organism* even if not of all its constituent cells.

The death of cells is accompanied by a rapid dissolving of the cell constituents. This process of cell degradation illustrates beautifully the need for continual supplies of energy to maintain the complex organization of matter that we call life. Once energy ceases to be made available to the cells, the precise organization of their parts becomes quickly and totally destroyed.

It is hard to visualize how the death of cells can also be considered a normal stage in development. Nevertheless, there are examples where cell death plays an important role in the life of the organism and presumably occurs as a result of definite genetic controls. One of the best examples is found in those insects that undergo complete metamorphosis. In the change from larva to pupa, many of the specialized cells of the larva die. These dying cells then supply the materials which enable the cells of the imaginal disks to develop into the pupa and then the adult. This widespread death of the cells of the larva is clearly a vital stage in the overall development of these insects.

Genetically programmed death may also occur in certain fishes. The sea lamprey and the salmon are well-known examples. When they reach sexual maturity, these animals migrate from the ocean into freshwater rivers and streams. Here the females lay their eggs, and the males fertilize them. Once the task is done, the adults die. The phenomenon is really quite similar to that of the annual plants mentioned earlier in the chapter. The removal of the old generation seems as much a part of development as the embryology that now occurs in the fertilized eggs that are the new generation.

23.6 SUMMARY

In the past ten chapters we have examined many aspects of the process of reproduction. Central to all of these has been the nature and expression of the hereditary controls within cells. These controls, the genes, guide the development of the individual into an adult and guide their own passage to the next generation through the mechanics of reproduction. We have looked for clues as to how the genes are regulated, over time and from cell to cell, so that the organism is maintained and the species preserved. It is not enough that the selective transcription of genes be able to produce a functioning adult organism. The genes must *continue* to act selectively so that the organism can respond to changes that it will meet until death finally occurs. Each differentiated cell in the organism's body must be able to modify the expression of its genetic potentialities in response to changes occurring both outside and within the organism. In other words, the ability of organisms to cope with changes in their environment and to coordinate the multitude of activities that occur in the diverse cells of which they are composed is, in the last analysis, a matter of modulation—the regulation of gene expression. How multicellular organisms integrate the activities of their cells as they respond to changes in their environment is the topic to which we now turn.

EXERCISES AND PROBLEMS

1. Summarize the mechanisms by which the beef protein you eat is converted into proteins characteristic of you.
2. What is the average number of times every cell in the newborn baby has to divide in order to produce an adult?
3. Which best describes an actively growing organism, an arithmetic progression or a geometric progression? Why?
4. A certain restaurant buys 1000 amber water tumblers. An average of one tumbler is broken every day and each broken tumbler is replaced by a new one made of clear glass. What is the shape of the survival curve of amber tumblers?
5. Which of the following organisms would you expect to have a survival curve like that in Question 4: (a) contemporary humans in the U.S., (b) humans living in prehistoric times, (c) parrots in cages, (d) sparrows in the wild, (e) alligators in the wild?
6. What food does a tadpole eat? What food does an adult frog eat? Of what significance is this difference in diet?
7. List all the factors you can think of that affect growth in humans.
8. Which human organs regenerate damaged portions most easily? Which human organs show the fewest degenerative changes in old age?
9. How do the two lists you prepared for Question 8 compare? What explanation can you give for this?
10. In what ways are the processes of regeneration and embryonic development similar? In what way are they different?

REFERENCES

1. ROSE, S. M., *Regeneration,* Addison-Wesley Modules in Biology, No. 12, Addison-Wesley, Reading, Mass., 1974.
2. WOOLHOUSE, H. W., *Ageing Processes in Higher Plants,* Oxford Biology Readers, No. 30, Oxford University Press, Oxford, 1972.
3. SHOCK, N. W., "The Physiology of Aging," *Scientific American,* January, 1962.
4. VERZÁR, F., "The Aging of Collagen," *Scientific American,* Offprint No. 155, April, 1963. Explains how the chemical and physical properties of this important structural protein change with its age.
5. COMFORT, A., "The Life Span of Animals," *Scientific American,* August, 1961.
6. COMFORT, A., *The Process of Aging,* The New American Library, New York, 1964. A paperback.
7. KOHN, R. R., *Principles of Mammalian Aging,* Foundations of Developmental Biology Series, Prentice-Hall, Englewood Cliffs, N.J., 1971.
8. SINGER, MARCUS, *Limb Regeneration in the Vertebrates,* Addison-Wesley Modules in Biology, No. 6, Addison-Wesley, Reading, Mass., 1973.

Teacher and pupils. Who is responding to whom? (Photo by Tom McAvoy, courtesy of LIFE Magazine, © 1955, Time, Inc.)

PART V
RESPONSIVENESS AND COORDINATION

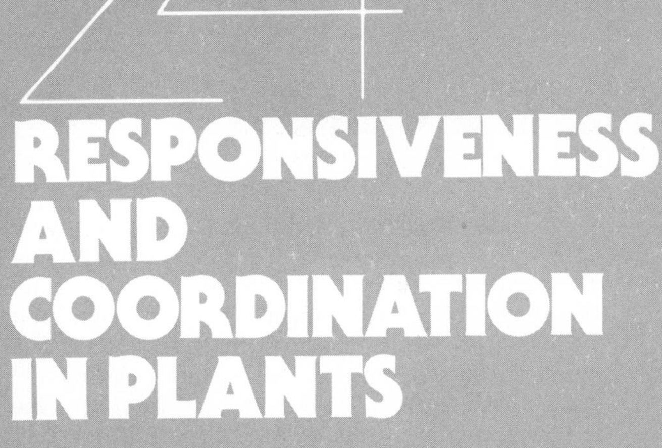

24 RESPONSIVENESS AND COORDINATION IN PLANTS

FIG. 24.1 The sensitive plant, *Mimosa pudica*. Darkness, touch, heat, and certain chemicals cause the leaflets to fold together along the midrib (right).

One of the chief distinguishing features of living things, as opposed to nonliving things, is that living things are capable of actively responding to certain changes in their environment. These environmental changes serve as **stimuli** which trigger a definite response on the part of the organism. When the "sensitive plant" *Mimosa pudica* is darkened, the leaflets along each side of the midrib fold together (Fig. 24.1). When the plant is illuminated once again, the leaflets spread out into the former position. The stimulus is the presence or absence of illumination; the response is the movement of the leaflets.

24.1 IMPORTANCE OF INTERNAL COMMUNICATION

Responsiveness in multicellular organisms requires a proper coordination of parts. Few responses could be accomplished successfully if every cell of the organism responded in the same way to the stimulus. For you to be able to run, you must contract certain muscle fibers while others relax. Your liver must provide additional fuel for your muscles. Your lungs must supply an extra amount of oxygen and take away a correspondingly increased amount of carbon dioxide. In other words, for your entire body to respond as a whole to a given stimulus, its constituent systems, organs, tissues, and cells must respond in a variety of special ways. Furthermore, the response of each of the various parts of your body must be carefully coordinated with the response of all the other parts. This ability of individual cells, tissues, organs, and systems to respond harmoniously to the activities of other parts requires a form of communication among them.

Two different, but related, systems of internal communication are found in animals. One of these, the **nervous system,** is a very fast-acting system. Specialized cells, **neurons,** conduct electrochemical impulses from one part of the body to another. These impulses accomplish quick, localized communication between parts. The impulses are of brief duration and quite separate from one another. Consequently, nervous responses are usually brief, intermittent, and do not continue for long periods unless the stimulus remains.

The second communication system, the **endocrine system,** is usually somewhat slower in its action. Specialized glands, the endocrine glands, release **hormones** to the blood stream or other circulating fluid. Hormones are chemical substances which are carried by the circulatory system to every cell of the body. Sometimes just a few, sometimes many, cells respond to the presence of these hormones. Usually the response occurs as a change in the metabolic activities of the "target" cell, tissue, or organ. These changes may persist for long periods of time.

Plants differ from animals in having no nervous system. Therefore, quick, localized responses are practically nonexistent in the plant kingdom. A few plants, however, do exhibit rapid movement. When *Mimosa* leaflets are pinched, they fold up. This change occurs as a result of a sudden loss of turgor in a special mass of parenchyma cells at the base of each leaflet. If one pinches just the tip leaflets of a *Mimosa* leaf, the leaflets fold up in pairs working from the tip back to the base. It certainly looks as though some stimulus is passing up the leaf. This may be a chemical moving through the vascular bundles. Although nerves are not involved, there is also evidence that a definite electrical impulse does pass up the leaf.

For the most part, plants achieve their responsiveness and coordination through a system of chemical coordinators, the plant hormones. Our knowledge of the nature and interplay of the various chemical regulators in plants is still quite fragmentary. In this chapter we shall examine some of what has been learned from experimentation and some of the current theories on the coordinating mechanisms in plants.

24.2 GROWTH MOVEMENTS

With the exception of turgor movements, plants respond to changes in their environment by growth. Naturally this sort of response takes longer to occur than a turgor response or a nervous-system response in an animal. The growth response may consist in having one part of the plant grow faster than another. Such a response produces definite, if relatively slow, movement. Two kinds of growth movements in response to outside stimuli are recognized in plants.

1. Nastic movements. A nastic movement (or *nasty* as it is sometimes called) is a response that is independent of the direction from which an external stimulus strikes the organism. The opening of certain flowers after sunrise is an example of a nastic movement. Illumination from any direction whatever will trigger the response, and the response is not oriented with respect to the direction of the stimulus. Although most nastic movements involve differential growth, that is, more rapid growth in certain parts than in others, some, such as the response of *Mimosa* leaflets, are turgor movements.

2. Tropisms. A tropism is a growth movement whose direction is determined by the direction from which the stimulus strikes the plant. If the plant part grows in the direction from which the stimulus originates, the tropism is referred to as positive. Growth in the opposite direction constitutes a negative tropism.

It has been known for years that plants respond to both the stimulus of light and the stimulus of gravity. The former response is a **phototropism,** the latter a **geotropism.** Stems exhibit positive phototropism while roots are negatively phototropic. On the other hand, stems are negatively geotropic while roots are positively geotropic (Fig. 24.2). The value to the plant of these responses seems quite clear. Roots growing downwards and/or away from light are more likely to find soil, water, and minerals. Stems growing upward or toward light will be able to expose their leaves so that photosynthesis can occur.

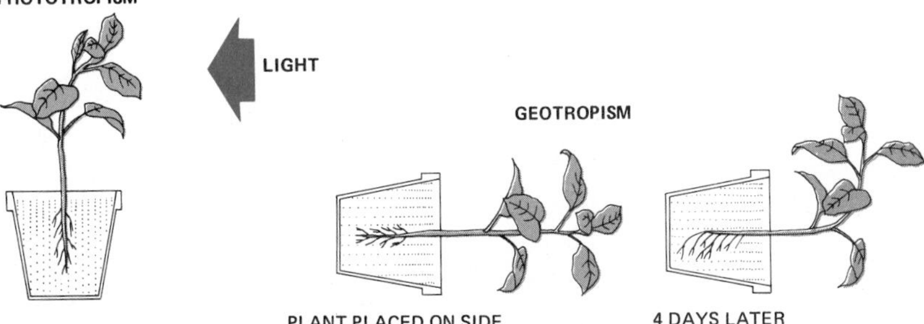

FIG. 24.2 Tropisms. The direction of plant growth is affected by light (phototropism) and gravity (geotropism).

In addition to growth movements, we know that several important *developmental* changes in plants occur in response to environmental stimuli. The germination of seeds, the resumption of growth of perennials in the spring, and the development of flowers all come about as a result of environmental triggers. The problem of understanding how plants respond to changes in their environment is twofold. First, we must discover how the plant detects the specific stimulus. Second, we must discover how the various tissues of the plant are coordinated in carrying out the response.

24.3 THE MECHANISM OF PHOTOTROPISM

The first clues to the coordinating mechanisms in plants came through a study of phototropism reported by Charles Darwin and his son Francis in 1880. They discovered that when the tip is removed from a growing shoot, the shoot ceases to exhibit phototropism. This was particularly surprising in view of their additional discovery: the bending action of the plant stem occurs in the region *behind* its tip. They found that if they placed an opaque covering over the tip of the plant, phototropism failed to occur even though the rest of the shoot was illuminated from one side. On the other hand, when they buried a plant in fine black sand so that only the tip of the shoot was exposed, there was no interference with phototropism. When the tip was illuminated from the side, the buried stem promptly bent in the direction of the light (Fig. 24.3). From these experiments, it seemed quite clear that the stimulus (light) was detected at one location (the tip) and the response (bending) was carried out at another (the region of elongation). This implied that the tip was, in some way, communicating with the cells of the region of elongation.

The Darwins confirmed these results with several different plant species. They found that grass seedlings were particularly easy subjects for this kind of experimentation. When grass seeds germinate, the primary leaf of the plant pierces the seed coverings and soil layers. It is protected from damage as it does so by a hollow, cylindrical structure, the **coleoptile,** which surrounds it (Fig. 24.4). Once the seedling has grown up above the surface of the ground, growth of the coleoptile ceases and the primary leaf pierces it.

The Darwins found that the tip of the coleoptile was necessary for phototropism in grass seedlings just as the shoot tip is necessary in older seedlings or in dicots such as the bean. Also, they found that the actual bending of the coleoptile occurs in the region below the tip. Thus a system of communication seems to be involved here, too.

In 1913, the Danish plant physiologist Boysen-Jensen showed that this communication must be accomplished by means of a chemical substance pass-

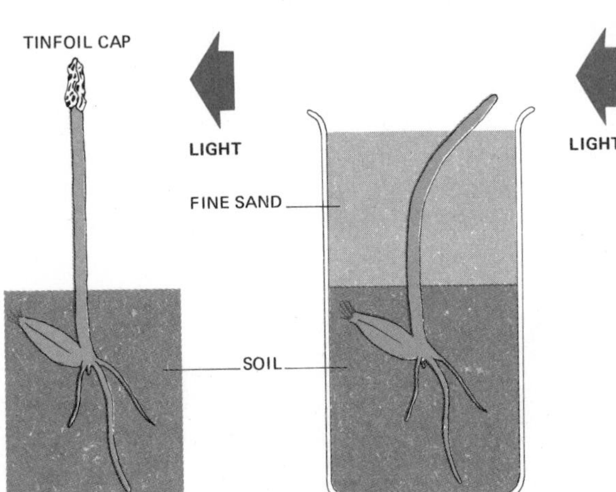

FIG. 24.3 The Darwins' experiments which showed that the response of phototropism is dependent upon light striking the tip of the plant.

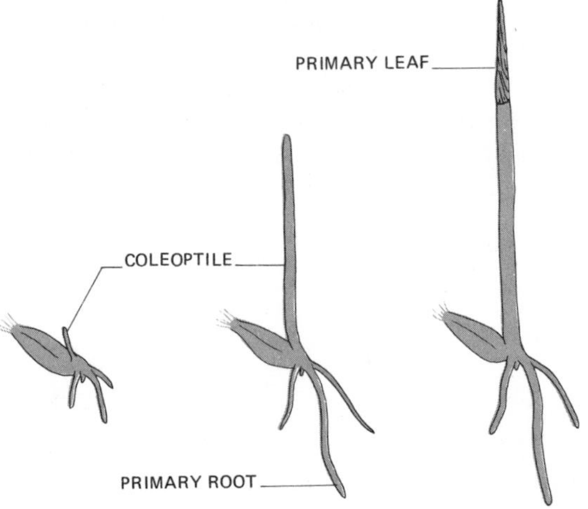

FIG. 24.4 Stages in the germination of an oat (*Avena*) seed.

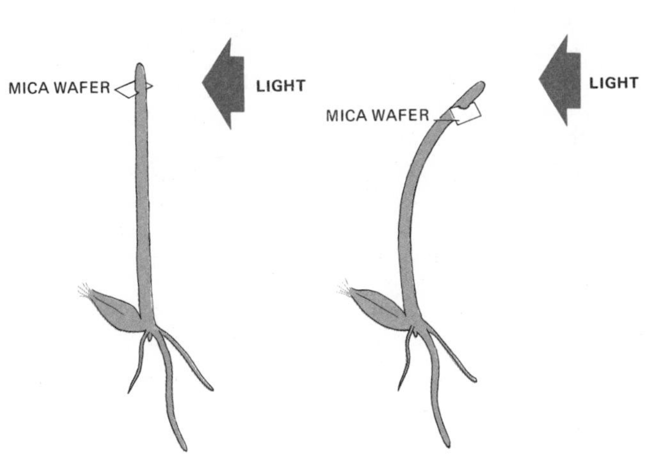

FIG. 24.5 Boysen-Jensen's experiments. The wafer of mica interfered with the phototropic response only when inserted on the shady side.

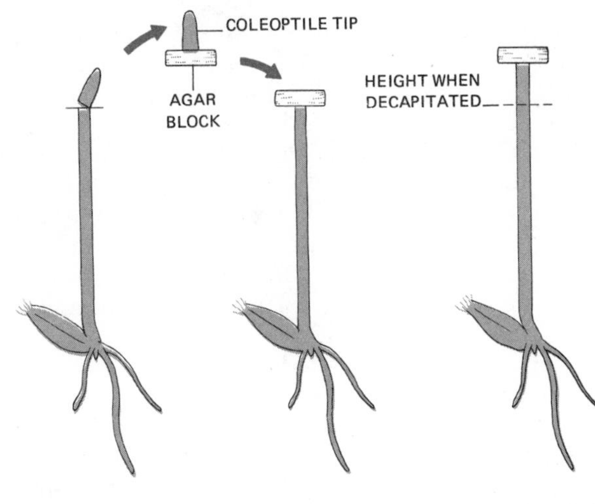

FIG. 24.6 Went's experiment showed that a chemical growth stimulator—auxin—can be extracted from coleoptile tips and still retain its effectiveness. Several tips should be placed on the agar block for a good response.

ing downward from the tip of the coleoptile. He cut off the tip of the coleoptile, covered the stump with a layer of gelatin, and then replaced the tip. Phototropism was still accomplished successfully. However, a piece of impervious material, such as mica, placed between the tip and the stump prevented the phototropic response. Interestingly enough, this interference occurred only when the sheet of mica separated the tip and stump on the shady side of the plant. When a horizontal incision was made in the illuminated side and the mica inserted, no interference with phototropism occurred. This suggested that the chemical coordinator was passing down the dark side of the seedling only. It also suggested that the chemical coordinator was a growth stimulator, as the phototropic response involves faster cell elongation on the shady side than on the sunny side (Fig. 24.5).

24.4 THE DISCOVERY AND ROLE OF AUXIN

F. W. Went first extracted this growth stimulator. He removed the tips from several coleoptiles of the oat plant (a grass), *Avena sativa*. He placed these tips on a block of agar and let them remain there for several hours. At the end of this time, the agar block itself was able to initiate resumption of growth in a decapitated coleoptile (Fig. 24.6). The growth was vertical because the agar block was placed completely across the stump of the coleoptile, and no light reached the plant from the side. This experiment showed that a chemical growth stimulator had diffused from the tips into the agar block. The name **auxin** was given to this material.

Unfortunately the quantity of active material in the coleoptile tips was far too small to be purified and analyzed chemically. Therefore, a search was made for other sources of auxin. In this search, research workers were greatly aided by a technique, developed by Went, for determining the relative amount of auxin activity in a given preparation. The

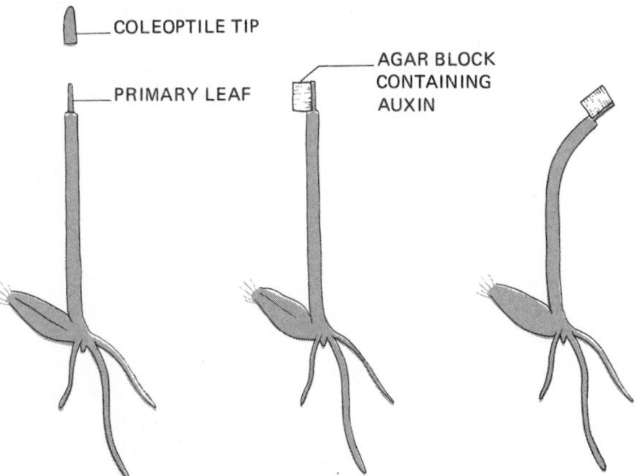

FIG. 24.7 The *Avena* test. The degree of bending is proportional to the amount of auxin activity in the agar block. The use of a living organism to determine the amount of activity of a substance is called a bioassay.

material in question is incorporated into an agar block and the block is placed on one side of a decapitated coleoptile (Fig. 24.7). As the auxin diffuses into that side of the coleoptile, it stimulates cell elongation and the coleoptile bends away from the block. The degree of curvature, measured at the end of 1½ hours, is proportional to the amount of auxin activity (e.g. number of coleoptile tips used) in the agar block. The use of a living organism to determine the amount of a given substance present is called a **bioassay** (Fig. 24.8).

Using Went's technique, it was soon found that auxins occur widely in nature. One of the most potent auxins was actually first isolated from human urine. It is indoleacetic acid or IAA (Fig. 24.9). Although it is only one of many growth-promoting substances which have been discovered, there is good evidence that it is the most important auxin produced by plants.

To Went also goes the credit for discovering that an unequal distribution of auxin is responsible for the bending in phototropism. When a coleoptile tip that has previously been illuminated from one side is placed on two separated agar blocks (Fig. 24.10), the agar block on the side that had been shaded accumulates almost twice as much auxin as the block on the previously lighted side. This explains why cell elongation occurs more rapidly on the shady side of the plant. Further investigation has demonstrated that light causes auxin to be translocated to the shady side.

Unequal distribution of auxin also explains the phenomenon of geotropism. When an oat coleoptile tip is placed on two separated agar blocks, as shown

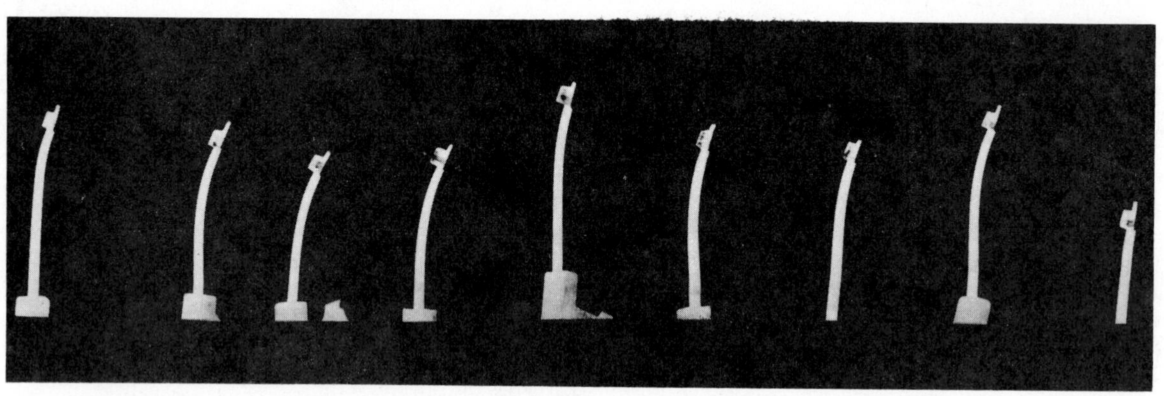

FIG. 24.8 Photographic record of an *Avena* test. (Courtesy of Dr. Kenneth V. Thimann.)

AUXIN (INDOLEACETIC ACID)

2, 4-D

*A CHLORINE ATOM HERE IN PLACE OF THE HYDROGEN ATOM = 2, 4, 5-T

FIG. 24.9 Molecular structure of a natural plant growth hormone (auxin) and the weedkiller 2,4-D. A mixture of 2,4-D and 2,4,5-T was "agent orange" used by the U.S. military forces to defoliate the forests in parts of South Vietnam. The solid circles represent carbon atoms, the short dashes hydrogen atoms.

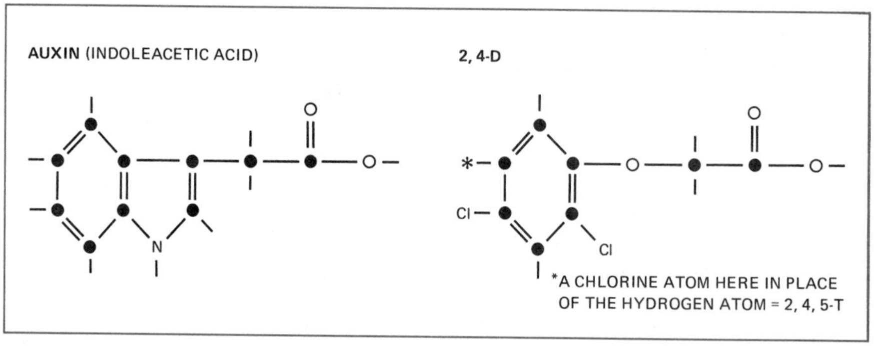

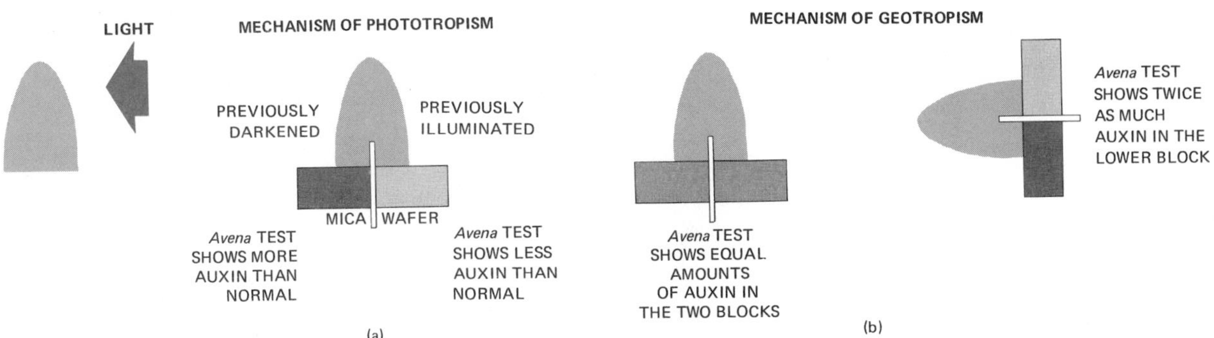

FIG. 24.10 Unequal distribution of auxin in phototropism and geotropism. In the intact plant, the side of the coleoptile receiving the greater concentration of auxin would grow more rapidly, resulting in bending toward light and away from the force of gravity.

on the left in Fig. 24.10(b), there is no difference in the auxin activity picked up by the two blocks. When the two agar blocks are applied to a coleoptile tip placed on its side, however, the lower block accumulates twice as much auxin activity as the upper block. Under natural circumstances, this would result in greater cell elongation on the underside of the coleoptile. Thus the plant would bend upward (Fig. 24.2).

How can one explain the negative phototropism and positive geotropism of *roots?* It seems as though exactly the same mechanism is at work in roots as in stems. The difference lies in the sensitivity of the cells of the root as compared with those of the stem (Fig. 24.11). Concentrations of auxin that stimulate growth in stems actually inhibit growth in roots. Concentrations of auxin which are too low to have any effect upon stems give some stimulation of cell elongation in the root. Therefore it is the illuminated side and the upper side of the root that grow more rapidly. As a result the root bends away from the light and downwards.

Although we have discovered a mechanism that coordinates these plant responses, we do not yet know how the environmental stimuli are actually received. In the case of phototropism, we assume that some light-absorbing *pigment* must be involved. IAA itself cannot fill this role as it does not absorb visible light. Because the phototropic response is most sensitive to blue light, the pigment is probably yellow in color, most likely a carotenoid (see Fig. 9.6).

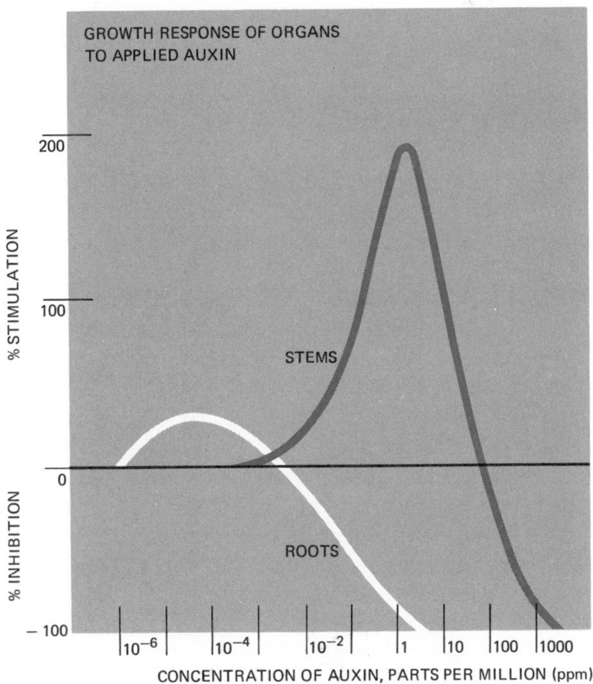

FIG. 24.11 The effect of auxin concentration on root and stem growth. Concentrations of auxin that stimulate stem growth inhibit the growth of roots. (Based on Fig. 2, K. V. Thimann, *Amer. Journal of Botany* **24**, 411, 1937.)

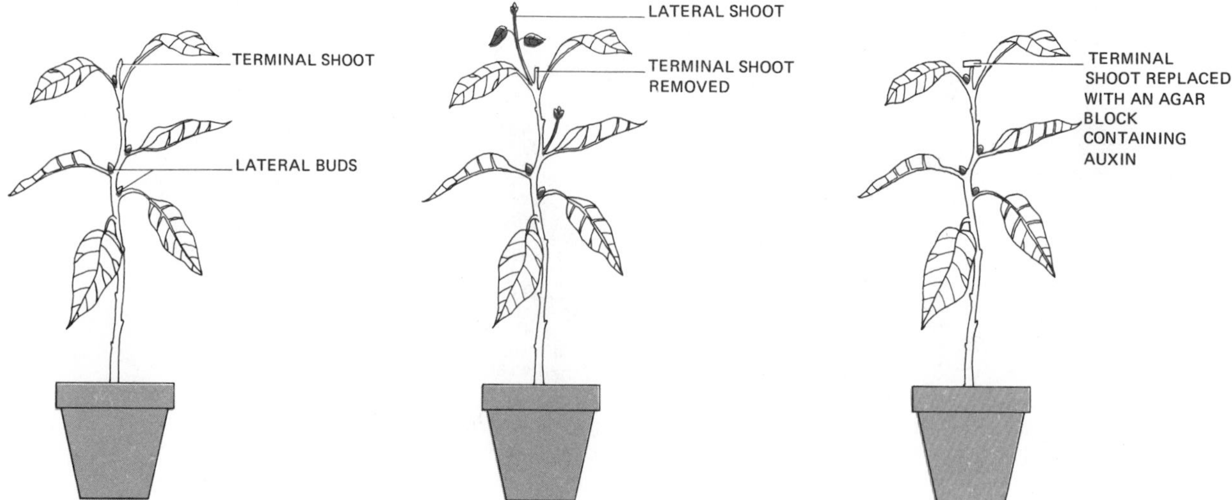

FIG. 24.12 Apical dominance. Translocation of auxin from the terminal shoot (left) or an auxin-containing agar block (right) inhibits development of the lateral buds. Absence of a source of auxin enables the lateral buds to develop (center).

24.5 OTHER AUXIN ACTIVITIES

The discovery of auxins opened up a very active field of research. Over a brief period of time, auxins were found to participate in the coordination of a variety of other plant activities.

1. Fruit development. Pollination of the flowers of angiosperms initiates the formation of seeds. Each seed contains the embryo of a new plant. As the seeds mature, surrounding flower parts form a covering, the fruit, around the seeds. It is now known that as the seeds develop, they release auxin into the surrounding flower parts, thus stimulating the growth of the fruit. In fact, it is possible to stimulate fruit development in nonpollinated flowers simply by the application of auxin to the flower. This is not just a laboratory curiosity. Many tomato and holly growers deliberately initiate fruit development in this way. Not only does this ensure that all flowers will "set" fruit, but it also increases the likelihood that all the fruits will be ready for harvesting at the same time.

2. Apical dominance. As a general rule, growth of the shoot apex (terminal shoot) of a plant inhibits the development of the lateral buds on the stem beneath. This inhibition is called apical dominance. In trees which form a single, straight trunk (e.g. most pines), apical dominance is very pronounced. In low-growing, many-branched, shrubby plants, it is less noticeable. If the terminal shoot is removed, the inhibition is removed with it (Fig. 24.12). The lateral buds then commence growth. Gardeners regularly exploit this principle by pruning the terminal shoot of ornamental shrubs. The release of apical dominance enables lateral branches to develop and the shrub becomes fuller and less spindly. The process must be repeated periodically, though, because one or two of the "laterals" will eventually outstrip the others and reimpose apical dominance.

Apical dominance seems to result from the downward transport of auxin produced in the apical meristem. In fact, if the apical meristem is removed and a block of auxin-containing agar placed on the stump, inhibition of the lateral buds is maintained (Fig. 24.12). A plain block of agar has no such effect.

White pines grown in warm, sunny locations frequently are parasitized by the white pine weevil. This insect lays eggs in the terminal shoot of the tree. After the young hatch, they consume the tissues

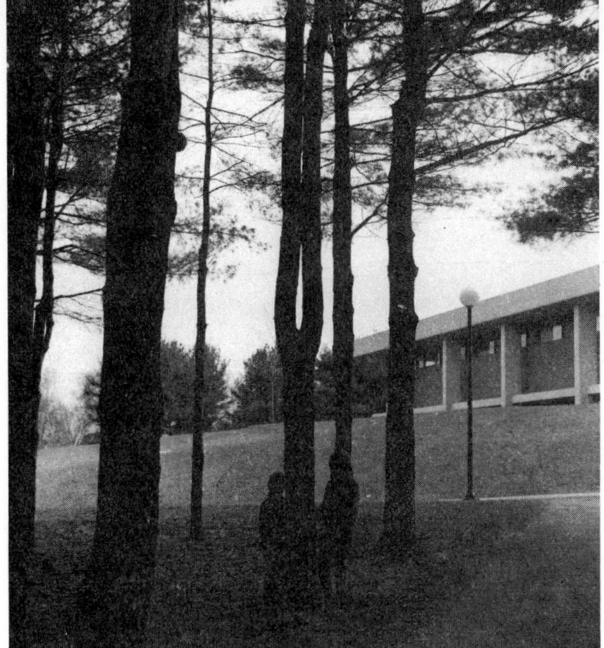

FIG. 24.13 When the white pines in this grove were young, their terminal shoots ("leaders") were killed by the white pine weevil. With the removal of apical dominance, two lateral branches succeeded in replacing the leader on the tree the boys are standing near. (Courtesy of Owen MacNutt.)

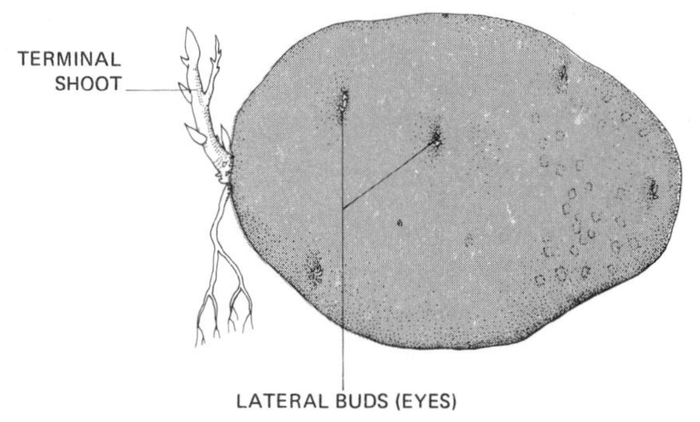

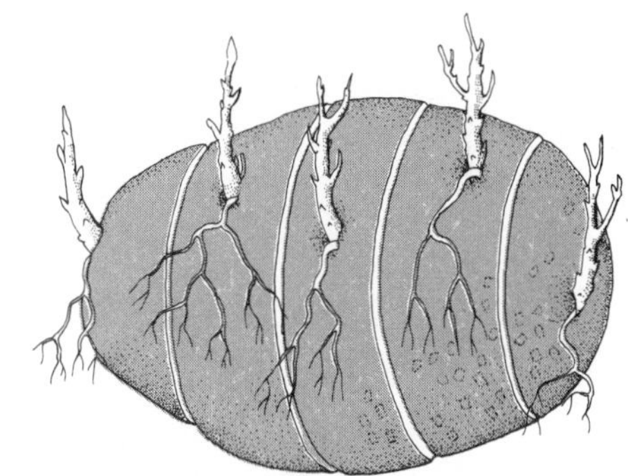

FIG. 24.14 Apical dominance in the white potato. In what ways is this demonstration similar to that shown in Fig. 24.12?

of the shoot and kill it. Death of the shoot removes apical dominance and the lateral branches beneath begin to grow rapidly upward. If left alone, one or two laterals usually succeed in reestablishing apical dominance. Figure 24.13 shows the crotch formed in a white pine by the replacement of one terminal shoot by two.

The principle of apical dominance can easily be demonstrated in the laboratory. The common white potato is really a portion of the underground stem of the potato plant. It has a terminal (apical) bud or "eye" and several lateral buds. After a long period of storage, the terminal bud usually sprouts but the other buds do not. However, if the potato is sliced into sections, one bud to a section, the lateral buds develop just as quickly as the terminal bud (Fig. 24.14).

3. Abscission. It has been found that auxins play an important role in the shedding of leaves and fruit. Young leaves and fruits produce auxin and as long as they do, they remain firmly attached to the stem. When auxin production diminishes, however, a special layer of cells forms at the base of the petiole or fruit stalk. This layer of cells is called the abscission layer (Fig. 24.15). Soon the petiole or fruit stalk breaks free at this point and the leaf or fruit falls to the ground.

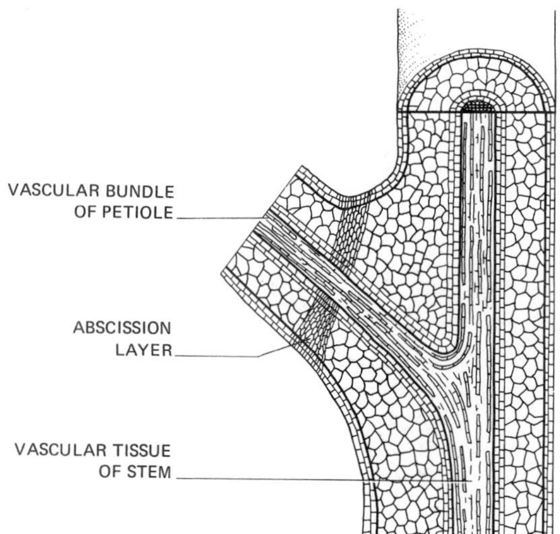

FIG. 24.15 The abscission layer. When leaves and fruits fall from a plant, the separation occurs at the abscission layer.

One can demonstrate this nicely in the laboratory. If the blade of a coleus leaf is removed, as shown in Fig. 24.16, the petiole remains attached to the stem for just a few more days. The removal of the blade seems to be the trigger here, as an undamaged leaf at the same node of the stem remains on the plant much longer, in fact, the normal length of time. If, however, auxin is applied to the cut end of the petiole (Fig. 24.16), abscission of the petiole is greatly delayed.

The ability of auxin to delay abscission has been exploited by apple and orange growers. The fruits of these species often drop off before they are ready for picking. This may result in a severe financial loss to the grower. Carefully applied auxin sprays greatly cut down loss by premature dropping.

Just what causes the lowered production of auxin in mature leaves and fruits is not completely clear. Leaves which are entirely shaded by others above them soon cease to produce auxin and drop off. This is a valuable adaptation of the plant because a shaded leaf needs nourishment but cannot manufacture it by photosynthesis. Thus it becomes a detriment to the plant.

In the fall, all the leaves of deciduous trees drop off. This, too, is a valuable adaptation because it reduces the surface area of the plant, thus diminishing water loss and snow load. As far as the plant's metabolism is concerned, freezing weather is equivalent to a drought. Leaf drop thus serves to conserve water and to minimize snow damage during the winter. (Most nondeciduous, cold-climate species have "needles" for leaves. These are very narrow and have a heavy, waterproof cuticle. The shape aids in the shedding of snow, and the cuticle cuts down on water loss.)

4. Root initiation. Auxins also stimulate the formation of **adventitious roots** in many species. Adventitious roots arise from stems or leaves rather than the regular root system of the plant. Gardeners may propagate desirable plants by cutting pieces of

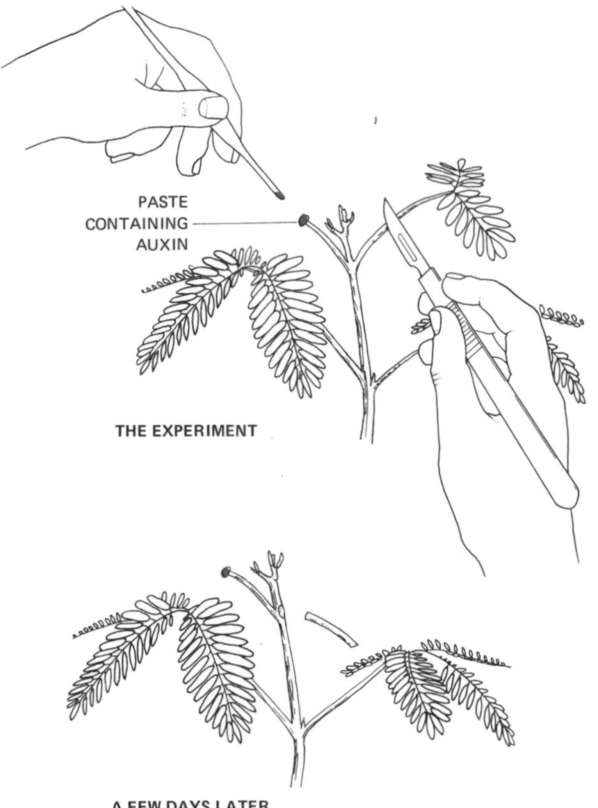

FIG. 24.16 The role of auxin in abscission. The petiole of a leaf whose blade has been removed soon drops from the stem unless auxin is applied to the cut surface.

FIG. 24.17 Top: American holly cuttings which have been treated with a synthetic auxin (beta-indole-butyric acid). Bottom: Untreated controls. (Courtesy of Paul C. Marth, U.S.D.A., Beltsville.)

stem and placing them base down in moist sand. Eventually, adventitious roots grow out at the base of the cutting. Their development seems to depend, among other things, upon the presence of auxin which has been translocated down from the upper portions of the cutting. Many gardeners now hasten the process by pretreating their cuttings with solutions or powders containing synthetic auxins (Fig. 24.17).

24.6 HOW DO AUXINS WORK?

As we have seen, auxins affect a variety of plant tissues in a variety of different ways. Some tissues (e.g. coleoptile, ovary wall of the pistil) are stimulated by auxin. Some (e.g. lateral buds) are inhibited. How can a single small molecule of this sort produce such a variety of effects?

The answer is not yet known with certainty. However, at least some of the effects of auxin on plant cells can be traced to its influence on the genes themselves. The stimulating effect of auxin on coleoptile and stem growth has been shown to be strongly inhibited by the antibiotic **actinomycin D.** As you learned in Chapter 19, actinomycin D binds to DNA and prevents its transcription by RNA polymerase. The inhibition of auxin action by actinomycin D thus suggests that at least one effect of auxin is on gene transcription. This has been further substantiated by the discovery that when isolated plant nuclei are treated with IAA, there soon follows a marked increase in the production of RNA. But when actinomycin D is added to the experiment, this increase is prevented.

Treatment of plant cells with auxin produces an increase not only in RNA synthesis but in protein synthesis as well. This is to be expected in view of the roles that messenger, ribosomal, and transfer RNA molecules play in protein synthesis.

It is not yet known exactly how auxin interacts with the genetic code to unlock genes. There is evidence, however, that auxin is active within the cell only when joined to a special protein molecule. Furthermore, the nature of this auxin-binding protein varies from one type of cell to another. The auxin-protein complex binds to chromatin in the nucleus and when this occurs, RNA synthesis is stimulated. Thus the situation may well be similar to that we found with the chromatin-binding steroid-receptor complex in endometrial and chick oviduct cells (see Section 19.2).

Some of the varied activities of auxin occur too rapidly to be a result of gene activation. Thus auxin must influence cell activities in other, more rapid, ways as well. Exactly how these rapid effects are achieved remains to be established.

24.7 THE GIBBERELLINS

Other growth-promoting substances have been discovered in addition to the auxins. During the 1930's, Japanese scientists isolated such a substance from cultures of a fungus which parasitizes rice plants. They called this substance gibberellin. After the delay caused by World War II, plant physiologists in other countries took up the trail and succeeded in isolating more than 30 closely related compounds. These are now known collectively as gibberellins.

Perhaps the most dramatic effect of the gibberellins is on stem growth. When they are applied in low concentrations to a bush or "dwarf" bean, the stem begins to grow very rapidly. The length of the internodes becomes sufficiently great that the plant becomes indistinguishable from climbing or "pole" beans. Gibberellin seems to overcome the hereditary limitations in many "dwarf" plant types.

Although gibberellins somewhat resemble auxins in their stimulation of stem growth, they are not classified as auxins. Not only is their molecular structure quite different from that of the auxins, but they do not give a response in the oat coleoptile test.

The gibberellins (like the auxins) were first isolated from sources other than the higher plants themselves. This, of course, raises the question of whether they are normal constituents of the higher plants and, if so, what role they play in the normal coordinating mechanisms of the plant. In recent years gibberellins have indeed been found to occur naturally in a variety of plant tissues. In addition to being implicated in the growth of stems, they seem to be the chief stimulants of root growth. The application of gibberellins to certain plants (e.g. spinach, cabbage) promotes the development of flowers. That this reflects a natural activity is suggested by the presence of increased quantities of gibberellins in the tissues of these plants when they are normally ready to flower.

Gibberellins also appear to play a role in the sprouting of buds. When white potatoes are first harvested, neither the terminal nor the lateral buds will sprout (Fig. 24.14). However, the application of gibberellin to the terminal bud will cause it to sprout immediately. In such woody species as the peach, birch, and sycamore, the synthesis of gibberellins in the spring seems to be the trigger for the sprouting of buds, which have been dormant during the winter.

The germination of seeds is another stage of plant life in which gibberellins may play a major role. You remember that most of the food reserves of cereal grains such as barley, wheat, rice, and corn are stored in the endosperm (Fig. 20.9). One of the first steps in the germination process in these monocots is production of gibberellin by the embryo. The gibberellin acts on the cells surrounding the endosperm, causing them to produce a number of specific hydrolytic enzymes (e.g. amylase) that digest the starch and protein of the endosperm and thus make sugars and amino acids available to the growing embryo. These enzymes also break down the seed coats and thus make it easier for the radicle and coleoptile to break through them (Fig. 24.4). This effect of gibberellin can be blocked by actinomycin D, suggesting that the gibberellin is activating genes within the cells surrounding the endosperm. In fact the application of synthetic gibberellin to these cells first produces a burst of RNA synthesis. This is followed by the synthesis of the various hydrolytic enzymes.

24.8 THE CYTOKININS

The cytokinins comprise still another group of plant hormones. Chemically, they each contain the purine adenine as part of their molecular structure. In addition to the naturally occurring cytokinins, a number of synthetic adenine derivatives exhibit strong cytokinin activity.

Cytokinins, when acting together with auxin, strongly stimulate mitosis in meristematic tissues. The food reserves of some seeds contain cytokinins. Presumably these provide the chemical stimulus for rapid mitosis in the developing seedling. Cytokinins also promote the differentiation of the cells produced in the meristems.

In addition to their growth-promoting effects, cytokinins have been shown to slow down the aging of plant parts, such as leaves, and to increase the resistance of plant parts to such harmful influences as low temperatures, virus infection, weed killers, and radiation.

As in the case with auxins and gibberellins, at least some of the effects of cytokinins appear to be brought about by selective gene activation. Two enzymes have been shown to be specifically induced by cytokinins and this induction is blocked by actinomycin D. Furthermore, a marked burst of RNA (probably messenger RNA) synthesis occurs when plant cells or isolated nuclei are treated with cytokinins.

24.9 ABSCISIC ACID

Inhibitors also play a role in the coordination of plant activities. In the fall, the mature leaves of such trees as the birch and sycamore produce a substance which stops growth in the apical meristems of the stems and converts them into dormant buds. The newly developing leaves growing above the meristem (look back at Fig. 23.2) become converted into stiff bud scales that wrap the meristem closely and will protect it from mechanical damage and drying out during the winter months. The substance responsible for the conversion of the apical meristems into dormant buds has been identified and named abscisic acid. Presumably it passes from its place of manufacture, the mature leaves of the plant, to the apical meristems by way of the phloem.

Once a bud has become dormant, it usually cannot be reactivated simply by the return of warm temperatures and abundant moisture. In some cases, gibberellin synthesis is necessary (probably triggered by the lengthening days of spring). In some cases, a period of exposure to cold temperatures is necessary before bud dormancy can be broken. Apple trees and lilacs, for example, cannot be grown in tropical areas because the winters never get cold enough (a few days at 0–10°C) to break bud dormancy.

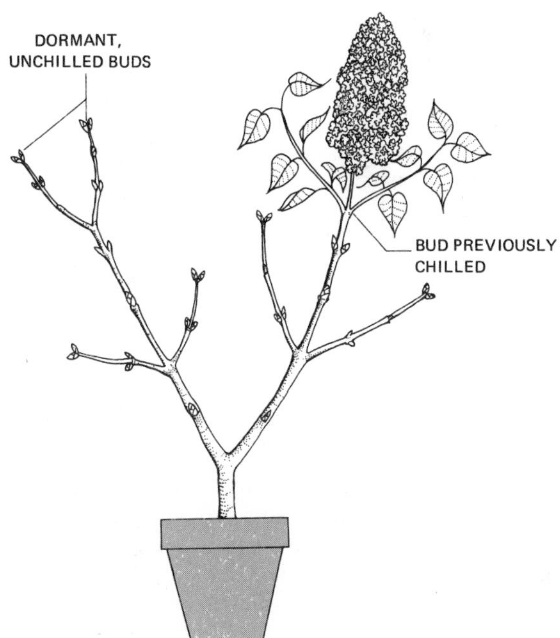

FIG. 24.18 Breaking bud dormancy in the lilac by localized cold treatment. Even under the best growing conditions, lilac buds will not develop without prior chilling.

The value of enforced dormancy to the plant is quite clear. If inhibitors such as abscisic acid were not present in the late fall and early winter, an unexpected warm spell might stimulate bud sprouting. The return of freezing weather would quickly kill this delicate growth. Enforced dormancy, which can be ended either after a period of prolonged cold or with the return of longer days, prevents the plant from being endangered in this way.

The dormancy of buds seems to be a very localized thing. By prior chilling, one bud on a lilac stem can be stimulated to grow while the other, nonchilled buds on the stem remain quite dormant (Fig. 24.18).

Many newly formed seeds must also undergo a period of enforced dormancy before they can germinate. Abscisic acid may be involved in some cases (e.g. in peach seeds) but other inhibitors have been discovered, too. Exposure to cold temperatures or, in some cases, exposure to sufficient amounts of water to leach the inhibitor out of the seed is necessary before germination can occur. The value to the seed of a cold requirement is the same as for buds. As for a water requirement—beyond that needed for the process of germination itself—you remember (see Section 20.9) that this ensures that the seeds of desert plants will not germinate if there has not been sufficient rainfall to enable them to complete their entire life cycle.

Abscisic acid also has been found to speed up the abscission of aging plant parts such as leaves and fruits (Fig. 24.15). It was this effect, which was discovered independently of its effect on dormancy, that gave rise to the name.

24.10 ETHYLENE

When first harvested, lemons are often too green to be acceptable in the market. In order to hasten the development of a uniform yellow color, lemon growers used to store the newly harvested lemons in sheds which were kept warm and humid. The warmth was supplied by kerosene stoves. When one grower tried a more modern heating system, he found that his lemons no longer changed color properly. Following this clue, it was soon found that the important factor in the ripening process was not the heat, but the small amounts of ethylene gas (CH_2CH_2) given off by the burning kerosene.

Since that time it has been found that most fruits produce their own ethylene, and it is this which triggers the ripening process. Among the many changes that ethylene brings about is a change in the permeability of the membranes of the cell. One consequence of this is to allow a chlorophyll-destroying enzyme into the chloroplasts. With the breakdown of chlorophyll, the red and/or yellow pigments in the cells of the fruit are unmasked and the fruit assumes its ripened color.

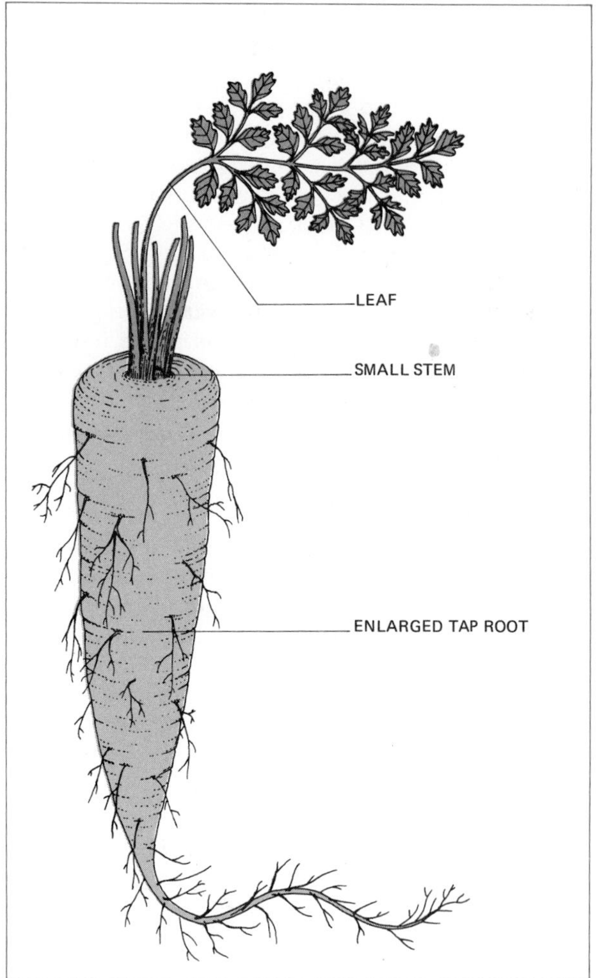

FIG. 24.19 The carrot, a biennial. Flowering occurs during the second growing season.

The Flowering Process

One of the most important developmental activities of angiosperms is flowering. When a plant is in active growth, mitosis in the apical meristem produces cells which go on to form more stem tissues and cells which will form leaf buds (Fig. 23.2). The leaf buds grow into mature leaves. Eventually, though, a time comes (usually as active growth ceases) when the meristems form flower primordia. These are clusters of cells which develop into flower buds. The flowers, into which these buds mature, contain the sex organs of the angiosperms, without which sexual reproduction could not occur.

24.11 FACTORS THAT INITIATE FLOWERING

The importance of the flowering process has led many plant physiologists to try to find out what initiates it. The question is simply, "What causes the plant to stop producing leaf buds and start producing flower buds?"

In some cases, the stimulus seems to be purely internal. Certain tomato varieties automatically produce flower primordia after 13 nodes have been produced by the growing stem. Adequate food reserves must be present in the plant, however. If the plant has not been photosynthesizing actively, it will lack the energy reserves necessary for the flowering process.

In most cases, the stimulus which initiates the flowering process seems to be external. Temperature often serves as the critical stimulus. This is particularly true of biennial species, that is, plants that require two growing seasons in order to complete their life cycle. Beets, carrots, and cabbages are three common biennial plants. In the first growing season, they develop roots, a short stem, and a cluster of leaves (Fig. 24.19). During this season, food is stored in the root system. With the arrival of cold weather, the tops die back. The next season, flowers are produced on the new shoot growth. After the reproductive process is completed, the entire plant dies. Flowering will not occur in the second season, however, unless the plant has been exposed to cold weather during the winter.

Experiments have shown that the entire plant need not be chilled for successful flowering to follow. The terminal bud is the temperature receptor. Its exposure to temperatures in the range of 1–10°C sets the stage for subsequent flowering. Gibberellins may be involved in the process. When gibberellin is applied to a terminal bud that has not been exposed to cold, the plant proceeds to flower normally. Furthermore, when a cold-treated biennial is grafted to a non-cold-treated one, the latter also flowers vigorously. This suggests that the chilling of the terminal bud is followed by the production and translocation of a flower-promoting substance.

Still another factor that triggers the flowering process in many plant species is a change in the daily intervals of illumination to which the plant is exposed. This discovery was made in 1920 by two plant physiologists, W. W. Garner and H. A. Allard, employed by the United States Department of Agriculture. They were perplexed to find that a new variety of tobacco, Maryland Mammoth, did not flower during the summer as other tobacco varieties do. Protected in a greenhouse, however, this variety did flower successfully near Christmastime. After considerable experimentation with artificial lighting in the winter and artificial darkening in the summer, they concluded that the flowering response was governed by the length of the day. They called this behavior **photoperiodism.** Because Maryland Mammoth would flower only when exposed to short periods of light, they called it a short-day plant. Other species such as chrysanthemums, poinsettias, and the cocklebur are also short-day plants. Some plant species, e.g. spinach, sugar beets, and the radish, flower only after exposure to long days and hence are called long-day plants. Still other species, such as the tomato, are day-neutral, that is, their flowering is not closely regulated by the length of their exposure to light.

The discovery of photoperiodism explained several puzzling facts about plant development. First, it provided a mechanism to explain, partly at least, the regular succession of blooms which occurs during the growing season. From the first flowers of early spring to the last flowers in the fall, one plant species after another flowers in a sequence as regular as clockwork. Now one could understand why spinach "bolts" and flowers during the summer months while chrysanthemums produce their blooms in the autumn. Second, photoperiodism explains why the plants of a given species usually bloom simultaneously. Though some individuals may have commenced growth earlier than others in the spring, in a given area they all bloom at the same time. Third, photoperiodism explains why many plant species can be grown successfully only in a rather narrow range of latitude. Spinach, a long-day plant, cannot flower successfully in the tropics because the days never get long enough (14 hours). Ragweed, a short-day plant, fails to thrive in northern Maine because by the time the days become short enough (August) to initiate flowering, a killing frost is apt to occur before the reproductive process is completed and the seeds matured.

24.12 THE MECHANISM OF PHOTOPERIODISM

The discovery of photoperiodism stimulated a great many plant physiologists to explore the process further in an attempt to determine the mechanism of action. They soon found that the terms *short-day* and *long-day* were misnomers. Interrupting the daylight period with intervals of darkness had absolutely no effect on the flowering process. Interrupting the night period with artificial illumination was, however, quite a different story. When short-day plants like the cocklebur were illuminated even briefly at night, they failed to flower. Thus it became clear that it was not the length of day that was important to the flowering process, but the length of night. Short-day plants were really *long-night* plants. For example, a cocklebur, growing at the latitude of Michigan, will flower only after it has been kept in the dark for 8½ hours. If it is illuminated by a flash of light at any time during this period, it fails to flower.

Although the development of flower primordia occurs at the meristems, the photoperiodic stimulus is not detected there. The leaves are the receptors of the stimulus. If only one leaf, in fact just a small portion of one leaf, of the cocklebur is subjected to an 8½-hour period of darkness, the entire plant will flower even though all the rest of its leaves fail to receive a sufficiently long exposure to darkness (Fig. 24.20). This suggests that some stimulus, presumed to be a hormone, is passing from the leaf to the meristems.

Grafting experiments lend solid support to this idea. If a cocklebur exposed to a favorable photoperiod is grafted to a cocklebur exposed to an unfavorable photoperiod, both plants will flower (Fig. 24.20). By grafting a series of plants together, it is even possible to estimate how rapidly the stimulus, which has been given the name **florigen,** is transported. Florigen is manufactured in the leaves and moves through

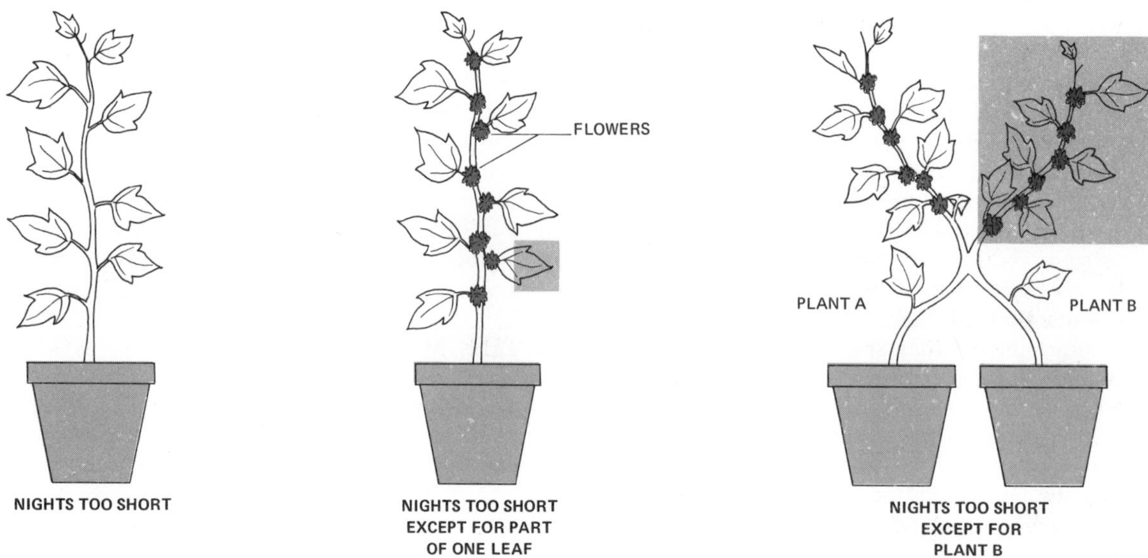

FIG. 24.20 Experimental evidence that the flowering stimulus is initiated in the leaf (center) and is transported by the vascular system to the flower buds (right).

the phloem to the meristems. Its arrival is marked by a number of biochemical changes including a burst of RNA synthesis. These are followed later by the morphological changes associated with the conversion of meristems into flower primordia. Although the molecular mechanisms at work are not yet known, the effect of the arrival of florigen seems to be the derepression of the genes controlling flowering.

In the cocklebur, at least, production of florigen is a one-time affair. Given one night of at least 8½ hours' duration, the plant will begin the flowering process even though it is subsequently placed back under unfavorable, short-night conditions.

The "long-day" plants are also misnamed. Spinach and other members of this group bloom successfully on a short-day schedule as long as the night periods are interrupted by a brief exposure to light. Thus, "long-day" plants are really *short-night* plants. They can bloom only if the nights are not too long.

The mechanism of action in short-night plants seems somewhat more complicated than in long-night plants. When the short-night plant henbane is placed under long-night conditions, it fails to flower. If, however, some of its leaves are removed, or the plant is chilled, or it is placed in an anaerobic atmosphere during the long night, the inhibiting effect of the long night can be overcome and the plant does flower. This suggests that some inhibiting substance (it may be abscisic acid) is produced by the metabolism of the leaves during the hours of darkness. If too much of this substance accumulates, it inhibits flowering. Any interference, however, with the normal metabolism of the plant slows down the accumulation of this inhibitor.

On the other hand, there is evidence that a substance which *stimulates* flowering is released from leaves to meristems when the plant is under short-night conditions. In fact, this substance appears to be identical to the florigen produced by long-night plants. If a short-night plant is grafted to a long-night plant, both plants will flower under a short-night schedule, although the long-night plant would not normally do so.

24.13 THE DISCOVERY OF PHYTOCHROME

Photoperiodism must involve a very sensitive light-detecting mechanism. We have seen that the cocklebur fails to flower on a long-night schedule if the night is interrupted by even a brief flash of light. The light flash can be very dim, exposure to light not much brighter than the light of the full moon being effective.

The most effective light rays for inhibiting flowering in the cocklebur are orange-red rays with a wavelength of 660 nm. The same wavelength is most effective in promoting flowering of spinach when its nights are otherwise too long. These findings suggest that the plants contain a pigment which absorbs orange-red light strongly.

It has also been found that the inhibiting effect of orange-red (660 nm) light on the cocklebur can be completely overcome by exposing the leaves of the plant to far-red light. This is light which is just beyond the visible spectrum. A wavelength of 730 nm is most effective in reversing the inhibitory action of orange-red light.

The system is completely reversible. The 8½ hour night of the cocklebur can be interrupted any number of times by orange-red light alternating with far-red light. Whether flowering is inhibited or not depends upon the wavelength of the last flash. If the last flash is orange-red light, flowering will not occur. If it is far-red light, flowering will occur normally even though the plant has been exposed to substantial quantities of orange-red light prior to the final exposure (Fig. 24.21).

This reversible effect suggests the presence of a reversible pigment. In one condition, the pigment absorbs orange-red light. In so doing, it becomes converted to a form which absorbs far-red light. Exposure of the far-red absorbing form to far-red light reconverts the pigment to the orange-red absorbing form. The pigment has been given the name **phytochrome.** The orange-red-absorbing form is designated P_R. The far-red-absorbing form is P_{FR}. The two forms are interconvertible according to this scheme:

$$P_R \underset{\uparrow\uparrow\uparrow\uparrow \text{ 730 nm light}}{\overset{\text{660 nm light } \downarrow\downarrow\downarrow\downarrow}{\rightleftharpoons}} P_{FR}.$$

Although present in exceedingly small amounts, phytochrome has been isolated from plant tissue. It is a protein to which is attached a prosthetic group which gives it its light-absorbing property. As one could guess from the fact that phytochrome absorbs most strongly in the red end of the light spectrum, its color is blue.

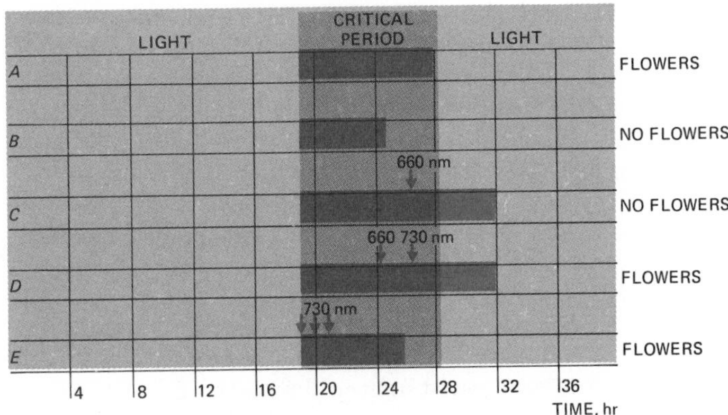

FIG. 24.21 Photoperiodism in the cocklebur. An uninterrupted night at least 8½ hours long is needed for flowering to occur (*A* and *B*). Interruption of an otherwise long night by orange-red (660 nm) light prevents flowering (*C*) unless it is followed by irradiation with far-red (730 nm) light (*D*). An intense exposure to far-red light at the start of the night reduces the dark requirement by two hours (*E*).

The action of phytochrome in controlling photoperiodism seems to hinge on two factors: (1) Sunlight is richer in orange-red light (660 nm) than in far-red (730 nm) light. This means that at nightfall, all the phytochrome in the leaves of a plant is in the form of P_{FR}. (2) P_{FR} is unstable while P_R is stable. In the darkness, P_{FR} gradually becomes reconverted into P_R. This conversion could serve as the clock by means of which the plant measures the length of the night. In fact, it is possible to induce the cocklebur to flower after a night as short as 6½ hours if it is first exposed to a large dose of far-red light (Fig. 24.21). Presumably the far-red light accomplishes immediately the P_{FR} to P_R transition, which would otherwise require two or more hours of darkness. That a large dose of far-red light does not do away entirely with the need for darkness indicates that other chemical reactions must also be carried out in the dark if flowering is to take place. What these are is not yet known.

The behavior of phytochrome provides a reasonable mechanism to explain the long-night requirement of the cocklebur. P_{FR} inhibits the chemical reactions necessary for the release of florigen from the leaves. P_R does not. Therefore the cocklebur needs 8½ hours of darkness in which (1) to convert all the P_{FR} present at sundown into P_R and (2) to carry out the unknown supplementary reactions leading to the release of florigen. If this process is interrupted by a flash of light (containing 660-nm rays), the P_R is immediately reconverted to inhibitory P_{FR} and the night's work is undone. A subsequent exposure to far-red light will convert the pigment back to P_R and steps leading to the release of florigen can be completed. Exposure to far-red light at the start of the night sets the clock ahead two hours or so by eliminating the need for the spontaneous conversion of P_{FR} to P_R.

In the case of short-night plants like spinach, P_{FR} is needed to stimulate the flowering process. A flash of orange-red light in the night will reverse the spontaneous conversion of P_{FR} to P_R which has been going on since sundown. This will permit flowering to occur even though the plant has been maintained under long-night conditions. In fact, spinach will bloom under continuous illumination (no night at all) when presumably all its phytochrome is in the P_{FR} form.

24.14 OTHER PHYTOCHROME ACTIVITIES

Phytochrome has been found to be involved in many other plant activities. The seeds of the Grand Rapids variety of lettuce will not germinate unless exposed to light. If shortly after exposure to visible light they are exposed to far-red light, they fail to germinate. Action spectra (see Section 9.2) for this stimulation and inhibition show that phytochrome is the light-absorbing pigment. P_{FR} stimulates germination while P_R inhibits it. The same mechanism has been found to operate in the seeds of many other angiosperms, both herbaceous and woody, and in some gymnosperm seeds, e.g. those of the Scotch pine. The situation is just reversed in the seeds of the California poppy. P_{FR} inhibits germination and P_R stimulates it. Under natural conditions, these are seeds which will not germinate when exposed to light. Phytochrome is probably responsible for the light-regulated germination of the species mentioned in Section 20.9.

Stem growth is also regulated by phytochrome. It has long been known that plants raised in the dark elongate very rapidly. (Look back at Fig. 9.11). This phenomenon is known as etiolation. It is a mechanism that increases the probability of the plant's reaching the light. Once light shines upon it, the plant produces internodes of normal size and its leaves grow to full size. That this is not brought about by satisfaction of the plant's needs for photosynthesis is shown by the fact that exposure to light too dim to be useful in photosynthesis nonetheless halts etiolation. Orange-red light is especially effective in producing this response. This, plus the fact that exposure to far-red light causes a resumption of etiolation, implicates phytochrome as the receptor in this response, too. The development of a red color in the skins of tomatoes and apples, the breaking of dormancy in some plant buds, and the straightening up of the hypocotyl arch when a dicot seedling grows above the surface of the soil (Fig. 20.11) also are triggered by light acting on phytochrome.

In the last chapter, we discussed the spontaneous aging and death that is so characteristic of annual plants once they have completed flowering and the production of seeds and fruits. In some cases, this process, too, may be a photoperiodic response involving phytochrome. Cocklebur plants that are given long nights age and die even if all their buds are removed before they can develop into flowers. If, however, the debudded plants have their long night interrupted by light, they live several weeks longer than they normally would.

24.15 SUMMARY

Although plants have no nervous system, they do have a variety of mechanisms by which they respond to changes in the environment. They respond to the direction of light, its wavelength, and the duration of exposure to it. They respond to gravity and temperature. All of these responses require a means (e.g. phytochrome, carotenoids) of detecting the stimulus in the environment. The detector mechanism may be located in the terminal bud or leaf or elsewhere. Once the stimulus is detected, these plants then require a communication system to enable all parts of the plant to respond in an appropriate, coordinated way. This communication system is made up of chemical messengers (e.g. auxins, florigen) transported in the contents of the phloem.

Among animals, chemical messengers also play an important role in coordinating the various organs of the body. These substances are the hormones. Their chemical nature and action is the topic of the next chapter.

EXERCISES AND PROBLEMS

1. Why do gardeners not have to worry about planting seeds upside down?
2. List six ways in which auxins affect plant function both in nature and under the control of humans.
3. What term would you use to describe the closing of a crocus flower on a cold spring day?
4. For each of the following responses of the cocklebur, tell (1) where and how the stimulus is detected, (2) where and how the response is carried out, (3) how the information is transmitted from the first region to the second. Cite experimental evidence to support your conclusions.
   a) Phototropism of the shoot.
   b) Production of flowers when nights become at least 8½ hours long.
   c) Geotropism of the root.

REFERENCES

The following three papers have been republished in *Great Experiments in Biology,* ed. by M. L. GABRIEL and S. FOGEL, Prentice-Hall, Englewood Cliffs, N.J., 1955:

1a. "Sensitiveness of Plants to Light: Its Transmitted Effects." CHARLES DARWIN and his son FRANCIS show that the phototropic response originates in the coleoptile (they call it a cotyledon) tip.

1b. "Transmission of the Phototropic Stimulus in the Coleoptile of the Oat Seedling." P. BOYSEN-JENSEN shows that a layer of gelatin separating a coleoptile tip from its base does not interfere with the phototropic response.

1c. "On Growth-accelerating Substances in the Coleoptile of *Avena sativa.*" F. W. WENT proves that a growth-promoting material can be extracted from coleoptile tips.

2. GALSTON, A. W., and PETER J. DAVIES, *Control Mechanisms in Plant Development,* Foundations of Developmental Biology Series, Prentice-Hall, Englewood Cliffs, N.J., 1970. Separate chapters devoted to: (a) phytochrome and flowering, (b) ethylene, (c) auxins and tropisms, (d) gibberellins, (e) cytokinins, and (f) abscisic acid, dormancy, and germination. Available in paperback.

3. BLACK, M., *Control Processes in Germination and Dormancy,* Oxford Biology Readers, No. 20, Oxford University Press, Oxford, 1972.

4. RAY, P. M., *The Living Plant,* 2nd ed., Holt, Rinehart and Winston, New York, 1972. Chapter 9 discusses plant hormones, and Chapter 10 describes the mechanism of photoperiodism.

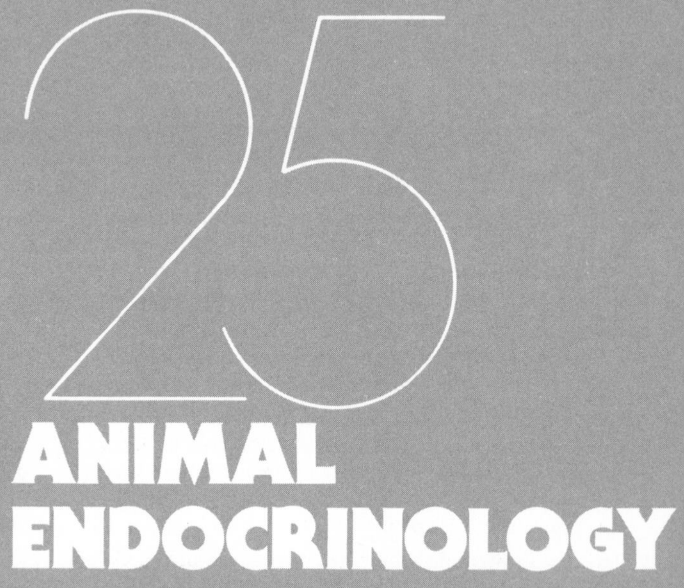

25 ANIMAL ENDOCRINOLOGY

25.1 INTRODUCTION

Multicellular animals, like multicellular plants, must solve the problem of coordinating the activities of all the different kinds of cells of which they are made. Animals, too, need some mechanism by which the various cells, tissues, and organs of the body can communicate. Only in this way can all these structures function in an efficient, well-coordinated way.

Two communication systems exist in most animals. One of these is the **nervous** system. It consists of specialized cells, neurons, which transmit electrical impulses from one part of the body to another. The other is the **endocrine** system. It achieves control of body functions through chemical substances, **hormones,** which are transported throughout the body in the blood. These two systems are not independent of one another. As we shall see in this chapter and the ones to follow, a close connection exists between their activities.

Chemical coordination in animals, like chemical coordination in plants, involves: (1) the release of chemicals from cells into the ECF, (2) the transport, by one means or another, of these substances, and (3) the alteration of the activities of other cells by them. Probably every cell in a multicellular organism participates in chemical coordination of this sort. In fact, we have already studied an example of this in human physiology. Every cell of our body produces carbon dioxide as a result of cellular respiration. This carbon dioxide is released into the ECF and is then carried throughout the body by the bloodstream. When carbon-dioxide-rich blood reaches the medulla oblongata, it triggers the release of nerve impulses to the diaphragm and intercostal muscles. You remember (see Section 10.9) that the rate at which these impulses are generated governs the rate and depth of breathing. This, in turn, maintains a constant level of carbon dioxide in the blood.

There is reason to believe that all multicellular organisms carry on this kind of chemical coordination, that is, coordination achieved by release of chemical substances that are the *byproducts* of other cell activities. Evidence is, however, very scanty on this point. For the present, therefore, we shall narrow our view to those special clusters of cells whose sole function seems to be the release of chemical coordinators in the body.

These clusters of cells are the endocrine glands. They are often referred to as ductless glands because their secretions, the hormones, pass directly into the blood that drains the gland (rather than into a duct as is the case with the *exocrine* glands discussed in Section 7.6). These hormones are then carried to all the other cells of the body. In some cases, they may influence the activities of all these cells. More often, the hormones exert their effect only on certain body structures, the "target" organs. So far, specific endocrine glands have been discovered only in the insects, crustaceans, certain mollusks, and all vertebrates.

25.2 INSECT HORMONES

In insects, the control of growth and metamorphosis has been more carefully studied than any other endocrine activity. Because of their rigid exoskeleton, insects can grow only by periodically shedding the exoskeleton in the process of **molting.** This process occurs repeatedly during the period of larval development. At the final molt, the organism which emerges is the adult. In several insect orders, it bears no resemblance whatsoever to the earlier larval stages. The marked transformation in body structure which occurs in these insects is called **metamorphosis.** It is accomplished during a dormant stage called the **pupa.** Figure 25.1 illustrates the larval, pupal, and adult stages in the development of the domestic silkworm moth, *Bombyx mori.* Metamorphosis occurs within a silken cocoon spun by the mature larva.

If the brain of a mature Cecropia caterpillar, one of several wild silkworms, is surgically removed before the caterpillar spins its cocoon, formation of the pupal stage does not occur. This cannot be simply a result of the shock of surgery because if the brain tissue is reintroduced into some other part of the body, pupation will proceed normally. In fact, just a tiny portion of the brain—about two dozen special cells—will do the trick. This experiment suggests that these special brain cells produce a hormone necessary for pupation to occur. This hormone does not trigger pupation directly, but instead acts upon a pair of glands in the thorax, the prothoracic glands. Because of its action, the brain hormone is called the prothoracicotropic hormone or **PTTH.** When stimulated by PTTH, the prothoracic glands secrete a second hormone, a steroid called **ecdysone.** It is ecdysone that directly initiates molting and the formation of the pupa.

These two hormones, acting together, promote not only the molt from larva to pupa, but also the earlier larva-to-larva molts. What, then, accounts for the sudden change in action which occurs at the time of metamorphosis?

It has been found that if a tiny pair of glands located behind the brain, the **corpora allata** (Fig.

NORMAL PATTERN OF DEVELOPMENT

MOLT

MOLT

MOLT

LARVAL STAGES

MOLT

MOLT

MOLT

PUPA

MOLT

ADULT

EXPERIMENTAL ALTERATIONS OF DEVELOPMENT

CORPORA ALLATA REMOVED

ACTIVE CORPORA ALLATA IMPLANTED

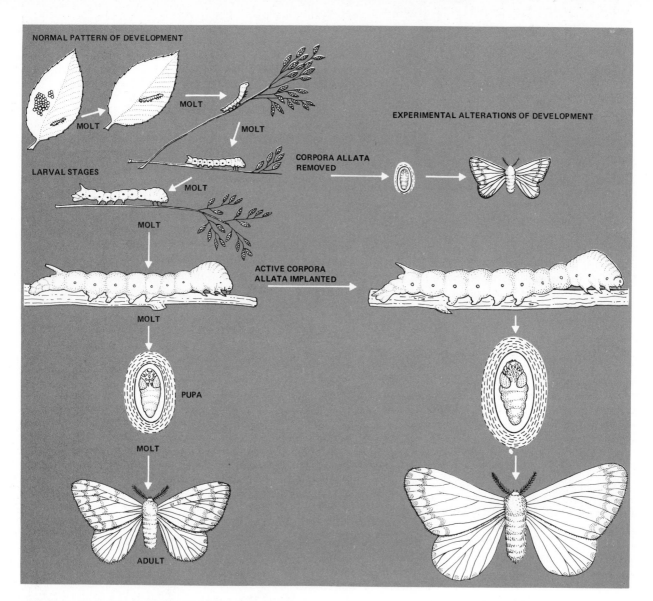

FIG. 25.1 Left: Normal pattern of development in the silkworm, *Bombyx mori.* This sequence can be shortened by removing the corpora allata from a young caterpillar and lengthened by introducing active corpora allata into a mature caterpillar. Pupation does not occur so long as the corpora allata secrete substantial amounts of juvenile hormone.

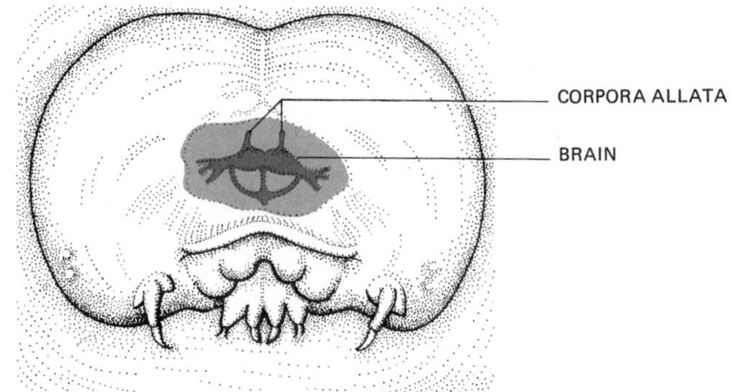

FIG. 25.2 The corpora allata of the silkworm secrete juvenile hormone. The brain itself secretes a hormone that initiates growth and metamorphosis.

25.2), are removed from an immature silkworm, it spins a cocoon and undergoes pupation at its very next molt (Fig. 25.1). This is because the corpora allata produce a *third* hormone which acts as a brake on metamorphosis. As long as substantial quantities of this hormone, which has been named the **juvenile hormone** (JH), are present, ecdysone promotes larval growth. When the amount of juvenile hormone is reduced, ecdysone promotes development of the pupa. Complete absence of the juvenile hormone leads to development of the adult.

This mechanism can be tested nicely. If the corpora allata of a young silkworm are introduced into the body of a fully mature larva, metamorphosis does not occur. Molting simply results in the production of an extra-large caterpillar (Fig. 25.1). Corpora allata from a *mature* caterpillar have no such effect. Thus normal metamorphosis seems to occur when the output of JH by the corpora allata diminishes spontaneously in the mature caterpillar. It is as though JH represses the genes for making adult structures.

The role of JH is beautifully illustrated by an experiment first performed by the English insect physiologist V. B. Wigglesworth. Adult insects do not normally molt, but if extra-normal amounts of PTTH are administered to an adult "kissing bug" (*Rhodnius*), it can be forced to do so. If JH is first applied to the insect's exoskeleton, the regions affected by it revert to larval type at this molt (Fig. 25.3). This shows that the genes for the manufacture of larval structures are present in adult cells although normally their action is repressed.

The development of the characteristic structures

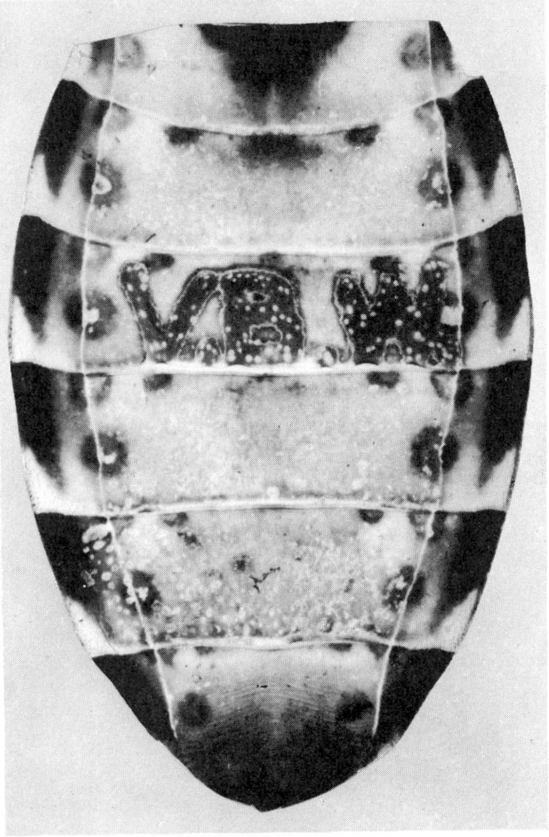

FIG. 25.3 Wigglesworth's experiment. Left: Application of a band of juvenile hormone to the cuticle of an adult *Rhodnius* results in the formation of larval cuticle when the insect is forced to undergo an extra molt. Right: The experimenter has printed his initials with juvenile hormone. (Courtesy of Dr. Wigglesworth.)

of the larva, then the pupa, and finally the adult must require the sequential participation of different sets of genes within the cells of the insect. This raises once again the question of how genes can be selectively activated or repressed. In Chapter 19, we examined some of the evidence that ecdysone modulates gene expression by its differential effects on the *transcription* of DNA into mRNA. As for JH, there is evidence that it modulates gene expression not only by an effect on chromosome puffing (see Section 19.3) but also by controlling the rate of *translation* of particular messenger RNA molecules.

In temperate climates, the pupal stage of many insects is formed in the fall (in response to the shorter days) and remains unchanged throughout the winter. Neither PTTH nor ecdysone is produced during this period of **diapause.** When spring returns, so does the production of PTTH and ecdysone. Metamorphosis is completed, and the adult emerges from its pupal case. In some species of insects, the resumption of PTTH secretion does not occur unless the pupa has first been exposed to a period of cold temperatures and then to warmer temperatures and increasing length of day. Cells in the brain of the pupa of one species detect the light through a transparent area of the exoskeleton just above the brain. When the days become 16 hours long, diapause is ended and development and hatching of the adult are then completed. The phenomena of enforced dormancy, a cold requirement, and photoperiodism that we examined in the last chapter are not, then, restricted to plants.

Insect Hormones and Pest Control

At certain times in the life cycle of an insect, the presence of juvenile hormone leads to abnormal development and death. For example, when solutions containing JH are sprayed on mature caterpillars, or on the foliage upon which they are feeding, normal pupation is prevented. The insect does not molt successfully into a giant larva either. Presumably this method of applying the hormone results in such an uneven distribution within the caterpillar's body that all the tissues do not respond alike. In any case, the animal soon dies. If insect eggs come in contact with even tiny traces of JH, their normal embryonic development is upset. These phenomena have led to speculation that JH, if it could be synthesized cheaply and in large quantities, would make an effective insecticide. There would seem to be little chance that insects could develop resistance to a substance which is a normal constituent of their

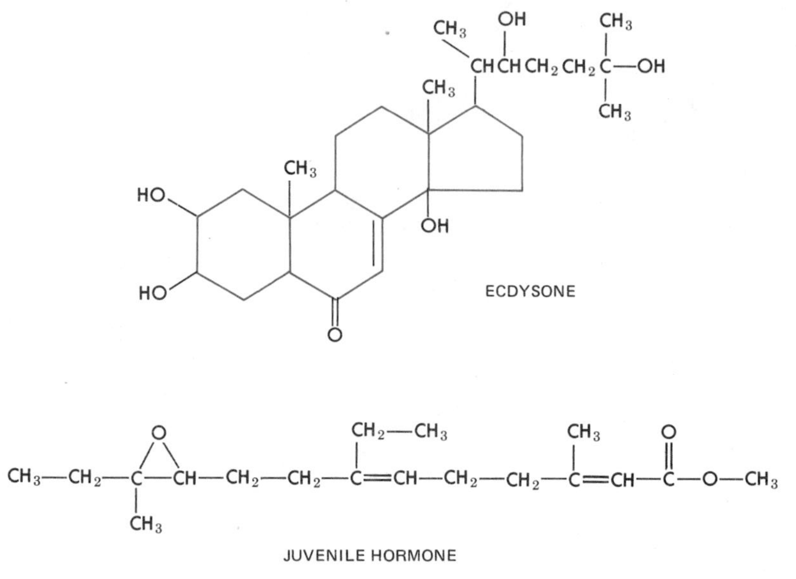

FIG. 25.4 Molecular structures of ecdysone and a juvenile hormone. Ecdysone is a steroid. Compare its structure with that of cholesterol (Fig. 4.8) and progesterone (Fig. 21.15).

ECDYSONE

JUVENILE HORMONE

bodies. Nor does JH seem to have a toxic effect on other organisms (unlike such conventional insecticides as lead arsenate and DDT).

The corpora allata of insects are too small to serve as a source from which JH can be extracted. Fortunately, many insects begin to secrete JH once again after they are fully adult. In the females, the hormone is necessary for successful development of the ovaries. For this reason, adult females who have had their corpora allata removed are sterile. Relatively large amounts of JH accumulate in the abdomen of adult male silk moths, enough to supply biochemists with the quantity needed to determine the chemical structure of the hormone (Fig. 25.4). Actually, three very similar molecules are found, each with powerful JH activity. The structure of these molecules is sufficiently simple that they can be (and have been) synthesized in the laboratory.

As it turned out, when JH was used under field conditions, it proved to be too unstable to serve as a practical insecticide. But once the structures of natural juvenile hormones were known, it did not take organic chemists long to synthesize a variety of structurally related compounds. Some of these JH "mimics" are far more active than JH itself, as well as being more stable. One is now available commercially (for use against mosquitoes and flies) and probably others will follow.

Many insect pests are harmful as larvae, not as adults. To derange the metamorphosis of a tomato hornworm by spraying it with synthetic JH is not of much use if it has already consumed your tomato plant. But one of the often unappreciated benefits of "pure" scientific research is that the more we can learn about the operation of a living system, the better position we are in to intervene in the system in useful ways. The knowledge of the vital roles played by ecdysone and JH suggests that we might be able to control insects by interfering with the action of these hormones. If, for example, an "anti-ecdysone" could be developed, it should stop insect molting and prevent completion of the insect's life cycle. As for an "anti-JH," two of these have already been discovered (in a plant noted for its resistance to insect attack!). Although the exact method of action of these substances is as yet unknown, they show promise of controlling certain insects at stages preceding those affected by JH sprays. Applied to the early larval stages of certain bugs (Hemiptera), they induce premature or precocious metamorphosis (like that induced by surgical removal of the corpora allata—see Fig. 25.1). Because of this effect, these substances have been called *precocenes*.

Not only does precocious metamorphosis cut short the destructive larval phase of the insect, but the adults are abnormal (besides being small). The females, for example, are sterile. This is consistent with the demonstrated need for JH for normal development of the ovaries. And it raises the hope that spraying precocenes on an insect-infested crop would not only put an end to feeding by the larvae (as they underwent precocious metamorphosis) but would prevent the formation of a new generation of the insect, or at least limit its numbers.

25.3 RESEARCH TECHNIQUES IN ENDOCRINOLOGY

The experiments which have led to our knowledge of hormonal control in insect metamorphosis provide an excellent illustration of the methods used in the study of endocrinology. The basic technique is simply this. First, an organ suspected of having an endocrine function is surgically removed from the animal's body. Second, a close observation is made of any changes or symptoms which may then occur. Third, the suspected gland is reintroduced into the animal's body to see if its presence will reverse these symptoms. If it does, the next step is to attempt to prepare an active extract (a mixture) that will duplicate the action of the missing gland. Although the extract is usually made from the gland itself, a few hormones have been extracted from such sources as urine. (The blood seldom contains enough hormone to serve as a source of supply.) Finally, an attempt is made to purify the extract and determine what single chemical substance within it produces the reversal of symptoms. This substance is the hormone.

These techniques have also been used successfully in the study of human endocrinology. Fortunately, many of the hormones found in humans are also found in the other vertebrates, so that the more drastic experimental studies can be performed on some other animal—frequently a dog or laboratory rat—with high hopes that the findings will be applicable to humans. That vertebrates share many of the same hormones is also of great importance in the treatment of human endocrine disorders. Many hormones are too complex to be manufactured synthetically. They can, however, be extracted from the glands of slaughtered cattle, pigs, etc., and used in the treatment of human ailments.

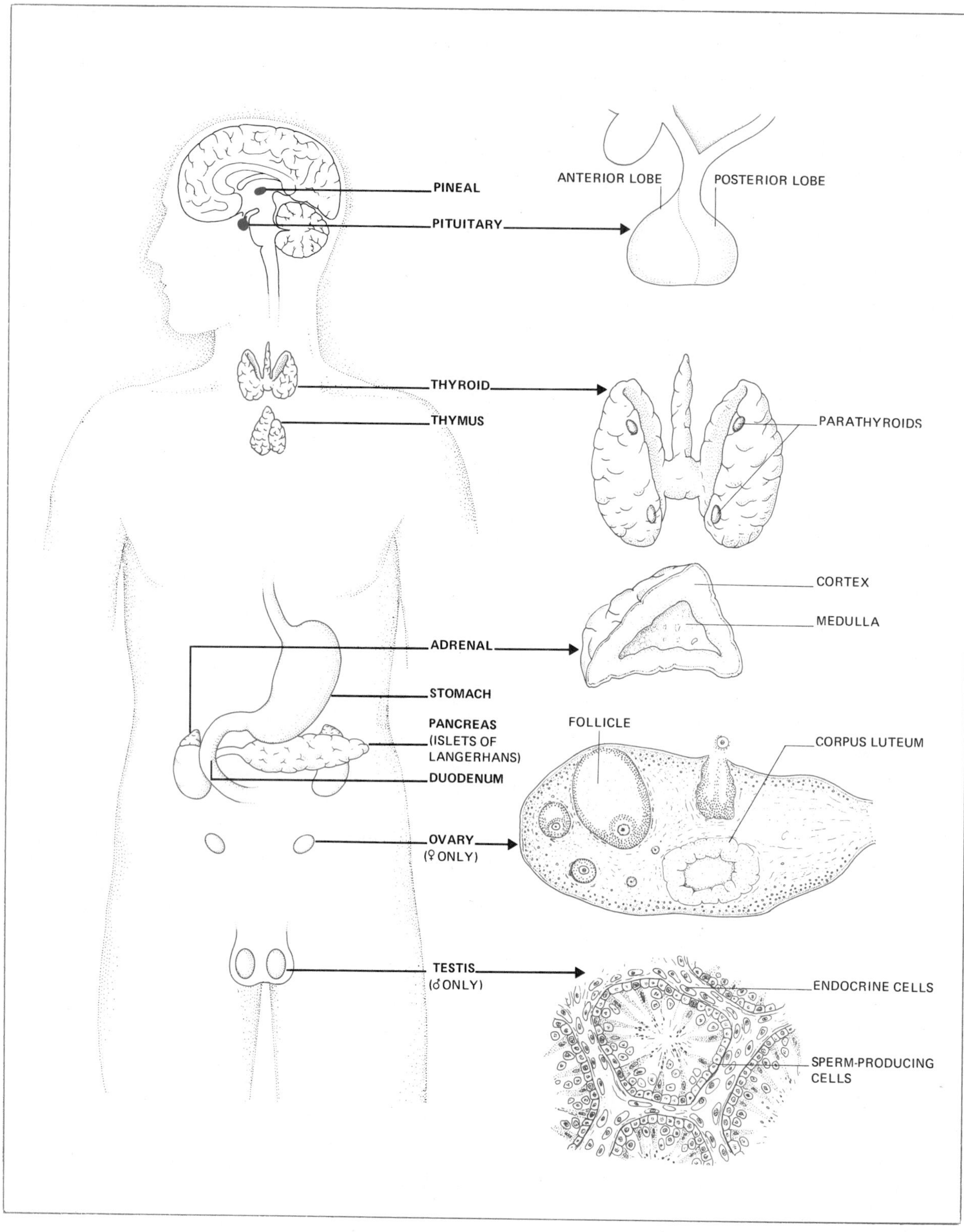
PINEAL
PITUITARY
ANTERIOR LOBE
POSTERIOR LOBE
THYROID
THYMUS
PARATHYROIDS
CORTEX
MEDULLA
ADRENAL
STOMACH
PANCREAS
(ISLETS OF
LANGERHANS)
DUODENUM
FOLLICLE
CORPUS LUTEUM
OVARY
(♀ ONLY)
TESTIS
(♂ ONLY)
ENDOCRINE CELLS
SPERM-PRODUCING
CELLS

The fact that vertebrates have many hormones in common does not mean that in every case they use them in the same way. As we take up the study of the hormones found in humans, we shall find some which have different functions in other vertebrates.

Human Endocrinology

25.4 THE THYROID GLAND

The thyroid gland is a double-lobed structure located in the neck (Fig. 25.5). For its size, it has an extraordinarily rich blood supply. The most important hormone that it releases into the blood is the iodine-containing amino acid *thyroxin*. This substance has been isolated from thyroid tissue and is easily synthesized commercially. Injections of it counteract the symptoms produced when the thyroid gland is removed from experimental animals.

In humans, the most obvious function of the thyroid gland is the control of the rate of body metabolism. When thyroxin is administered, the amount of heat produced by the body is increased. Because energy production is a function of cellular respiration, we might expect the administration of thyroxin to increase the rate of oxygen consumption. Such is the case. In fact, measurements of oxygen consumption are used to diagnose malfunctioning of the thyroid gland.

A special function of the thyroid gland is found in amphibians. Their metamorphosis from the larval to the adult stage is triggered by thyroxin. If thyroxin (or even iodine) is administered to a small tadpole, it will undergo premature metamorphosis into a midget frog. On the other hand, surgical removal of the thyroid gland of a tadpole prevents metamorphosis.

Some races of the common tiger salamander never undergo metamorphosis from the gill-breathing larval stage to the air-breathing terrestrial form. (They reach sexual maturity and reproduce as larvae.) However, administration of thyroxin to these animals causes them to carry out metamorphosis into typical tiger salamanders. It is perhaps no accident that these so-called *axolotl* forms are found in mountain lakes in the western part of our country and in northern Mexico, where very little iodine is found in the soil and water (Fig. 25.6).

Several human diseases are associated with the improper functioning of the thyroid gland. In considering these ailments, it is important to distinguish between those associated with excessive production of the hormone (*hyper*thyroidism) and those asso-

◄ FIG. 25.5 The endocrine glands of the human. They are the same in both sexes except for the gonads, which in the female are ovaries and in the male are testes.

FIG. 25.6 The axolotl, a salamander that reaches sexual maturity without undergoing metamorphosis into an air-breathing form. Note the external gills. (Courtesy of the New York Zoological Society.) ►

ciated with insufficient hormone production (*hypothyroidism*).

Hypothyroidism before maturity results in the disease cretinism. The victim fails to attain either normal physical or mental development. The cause of the disease is not clear. Although it is most prevalent in areas where insufficient iodine is present in the diet, other factors are probably involved, too. In any case, all symptoms can be prevented by the early and regular administration of thyroxin. If treatment is delayed until after the symptoms have become severe, only minor improvement can be secured.

Hypothyroidism in adults causes myxedema. The symptoms of this disease are a low metabolic rate, overweight, and a coarsening of the features. Like cretinism, it is most prevalent in iodine-deficient areas, but other causative factors are probably involved, too. It is cured by administration of thyroxin.

Iodine-deficient areas of the world (e.g. around the Great Lakes and the Pacific Northwest in the United States) are often called **goiter** belts. This expression is used because of the large number of people with simple goiter found in these regions. A goiter is a swelling of the neck caused by a swelling of the thyroid gland. Although there is still disagreement as to the cause of simple goiter, most of the facts point to inadequate amounts of iodine in the diet. Now that iodine-rich foods (e.g. marine fish) are shipped throughout the country and iodized table salt (NaCl with traces of KI added) is used so widely, simple goiter is no longer a serious health problem.

It might seem surprising that inadequate amounts of iodine would lead to an enlargement (and hence extra activity) of the thyroid gland. It is probably a matter of compensation. The rate of activity of the thyroid gland is itself governed by another hormone, TSH, which is liberated by the anterior lobe of the pituitary gland (Fig. 25.5). An increase in the production of TSH causes an increase in the amount of thyroxin. However, an increased amount of thyroxin in the bloodstream depresses the output of TSH. This nice homeostatic device thus ensures a steady supply of thyroxin. When there is insufficient iodine in the diet for the thyroid to synthesize thyroxin, however, this control mechanism breaks down. The pituitary gland is not inhibited and thus produces ever-larger quantities of TSH. This, in turn, stimulates the thyroid gland to work harder even though it has little or no iodine to work with. Thus it becomes enlarged, and a goiter results.

25.5 THE PARATHYROID GLANDS

The parathyroid glands are four tiny structures imbedded in the rear surface of the thyroid glands (Fig. 25.5). This inconspicuous location accidentally led to the discovery of their importance. Early attempts at treating goiters by surgical removal of the thyroid gland sometimes led to unpleasant symptoms of muscle spasms and twitching in the patient. Eventually these symptoms were traced to a loss of the parathyroid glands and a resulting drop in the level of calcium ions (Ca^{++}) circulating in the blood. As we have learned, the calcium ion is one of the most important mineral constituents in the body. In addition to its role in bone formation, an adequate level of Ca^{++} in the ECF is necessary for proper functioning of the nervous and muscular systems. Too low a level of Ca^{++} in the ECF leads to the symptoms associated with removal of the parathyroids.

In 1960 a chemically pure hormone, the parathyroid hormone or PTH, was finally extracted from the parathyroid glands of cattle. It is a small protein (M.W. = 8500) containing 83 amino acid units. It produces the following effects: (1) It promotes the release of Ca^{++} from the bones. (2) It promotes the absorption of Ca^{++} from food in the intestine. (3) It promotes the reabsorption of Ca^{++} in the tubules of the kidneys. All of these actions result in an increase in the level of Ca^{++} circulating in the blood. (The hormone also inhibits the reabsorption of $PO_4^{\equiv}$ in the kidney tubules and thus helps rid the body of the excess $PO_4^{\equiv}$ produced when bone—calcium phosphate—is decomposed to provide Ca^{++}.)

The maintenance of a constant level of Ca^{++} in the ECF (homeostasis) requires that the activity of the parathyroids be under precise control. This control is quite direct. The amount of PTH secreted by the glands is regulated by the level of Ca^{++} in the ECF. When the level of Ca^{++} drops, the glands are stimulated to release the hormone and thus restore the normal level of Ca^{++}. If the level should rise above normal, the hormone output of the gland is depressed. Here, then, is another nicely adjusted mechanism by which constancy of the internal environment is maintained.

In 1961 it was discovered that the response of the parathyroids to excess Ca^{++} (the response being a subsequent reduction in Ca^{++}) is much faster than could occur simply by shutting off the production of PTH. This has led to the discovery of another hormone (named calcitonin) that acts antagonistically to PTH. This second hormone (released by the *thyroid*) provides an additional control mechanism over the Ca^{++} level in the ECF. If, for example, the Ca^{++} level of the ECF drops below normal limits, the parathyroids will be stimulated to release PTH. This is a rather slow response, however, and there is a possibility that the system might overshoot and the Ca^{++} concentration reach too high a level. Of course, depression of the glands would eventually correct this, but it might be some time before fluctuations in the Ca^{++} level were "damped out." However, release of fast-acting calcitonin prevents this overshoot and restores a steady Ca^{++} level more quickly.

Humans rarely suffer from hypoparathyroidism. Most cases in the past have resulted from accidental or unavoidable removal of the parathyroids during thyroid surgery. The unpleasant (and occasionally fatal) symptoms of parathyroid deficiency can be successfully avoided by careful addition of Ca^{++} to the patient's diet. Vitamin D, which duplicates many of the functions of the parathyroid hormone, is also used successfully in treating this disease.

Occasionally one or more of the parathyroid glands becomes enlarged and extra-active. The *hyper*parathyroidism that results produces severe symtoms. The bones become brittle, weakened, and deformed. They break under the slightest stress. The great excess of Ca^{++} in the blood causes some to pass over into the urine where it may precipitate with phosphate ions and form kidney stones. These are dangerous because they occasionally block the urinary ducts. Surgical removal of the diseased glands usually brings about marked improvement.

25.6 THE SKIN

When ultraviolet radiation strikes the skin, it triggers the conversion of dehydrocholesterol (a cholesterol derivative—see Fig. 4.8) into calciferol. Calciferol is also called vitamin D because it is present in a number of foods and such dietary sources can supplement or replace the skin as a source of supply. In its chemical structure and mode of action, however, calciferol meets all the criteria of a hormone. After synthesis in the skin, it is released to the blood, which carries it to the liver, where it undergoes one chemical modification. From the liver it travels to the kidney where it undergoes a second modification (a reaction enhanced by the presence of PTH). The resulting compound (1, 25-dihydroxyvitamin D) enhances the absorption of Ca^{++} from the intestinal contents. Thus this hormone teams up with PTH and calcitonin in the regulation of calcium metabolism. Inadequate amounts of calciferol prevent the normal deposition of calcium in bone. If this occurs during childhood, deformed bones characteristic of rickets are the result. In adulthood, inadequate amounts of calciferol lead to a weakening of the bones, a condition known as osteomalacia.

25.7 THE STOMACH AND DUODENUM

There are cells in the walls of both the stomach and duodenum that secrete hormones. As we saw in Chapter 7, the hormones produced in these cells stimulate a number of digestive processes. **Gastrin,** a polypeptide containing 17 amino acids, secreted into the blood by cells in the stomach wall, stimulates the production of hydrochloric acid by the parietal cells of the stomach. **Secretin** and **pancreozymin** are secreted into the blood by cells in the duodenum. When these hormones reach the pancreas, they stimulate the secretion of the various components of the pancreatic digestive fluid. When they reach the liver and gall bladder, they stimulate the secretion and release of bile.

25.8 THE ISLETS OF LANGERHANS

The islets of Langerhans are small clusters of cells scattered throughout the pancreas in most vertebrates. There are well over a million of these clusters in the human pancreas. The cells of the islets of Langerhans are not connected to the ducts which drain pancreatic juice into the duodenum. They are, however, richly supplied with blood vessels.

As has so often been the case in science, the discovery that the islets of Langerhans are endocrine glands grew out of a chance observation. In 1889, the German physicians von Mering and Minkowski tried to learn more about the digestive functions of the pancreas by observing what digestive upsets occurred after surgically removing the organ from dogs. During the course of their studies, a laboratory assistant noticed many flies collected near the dog's urine. The urine was found to contain a large amount of the sugar glucose. The urine of normal dogs does not.

For many years, all attempts to extract a glucose-regulating hormone from the pancreas failed. However, in 1922 Dr. Frederick Banting finally succeeded in this venture. The hormone, named **insulin,** was found to be a protein. Can you see now why so many workers had failed to extract it from crude preparations of the entire pancreas? Banting succeeded where others had failed by first tying off the pancreatic ducts of his dogs. This operation caused the exocrine portion of the gland to degenerate rather quickly while the endocrine portion remained active. After the exocrine portion was destroyed, extracts could be made free from the digestive action of trypsin.

Insulin acts to lower the level of glucose in the bloodstream. Normally, 100 ml of blood contains about 0.1 g of glucose. After a carbohydrate-rich meal, this level tends to rise. As it does so, it triggers the release of insulin from the islets of Langerhans. The insulin passes immediately to the liver (through the hepatic portal veins), where it speeds up the conversion of glucose into glycogen and fats. As a result, the blood-sugar level is quickly brought back to normal.

One of the effects of insulin is to make the cells of the body more permeable to the entrance of glucose. Once within cells, the glucose can then be metabolized. This effect is not inhibited by actinomycin D. However, insulin also stimulates the synthesis of proteins, including enzymes that participate in carbohydrate metabolism. This action is inhibited by actinomycin D, indicating that at least one step in the inductive process is the transcription of genetic information.

Insufficient production of insulin results in the disease **diabetes mellitus.** Victims of this disease are unable to cope with excess glucose in the blood by converting it to glycogen or fats. Worse still, the reverse reactions are promoted. Glycogen and body fats are converted into glucose, which raises the blood-sugar level further. The kidney tubules fail to reclaim much of this excess glucose and so it passes out in the urine. Even body proteins are converted into glucose and then excreted. The high glucose level in the nephric filtrate exerts a powerful osmotic effect, sharply reducing the transport of water back into the blood. Consequently, victims of the disease urinate frequently and copiously. Unless corrective measures are taken, they waste away, their bodies gradually being converted into a flood of sugary urine. Coma and death eventually follow.

Fortunately, insulin is now available in large quantities from the glands of slaughtered livestock. The molecular structure of these animal insulins is so close to that of human insulin that they may be used to treat diabetes. Unfortunately, insulin must be given by injection rather than by mouth. (Why?) Despite this inconvenience, administration of insulin to a diabetic quickly restores normal sugar metabolism in the body. Of course, insulin injections are no cure. They simply provide temporary relief from the symptoms. Nevertheless, careful attention to diet and periodic injections of insulin have enabled thousands of diabetics to lead active, useful lives. In recent years at least two nationally ranked U. S. tennis players have been diabetics.

The use of insulin is not without some danger. Insulin injected after a period of exercise or long after a meal may drive the blood-sugar level down to abnormally low levels. This can result in an "insulin reaction." The victim becomes irritable, fatigued, and may lose consciousness. It is quite important that those associated with diabetics learn to distinguish between the symptoms of insulin reaction

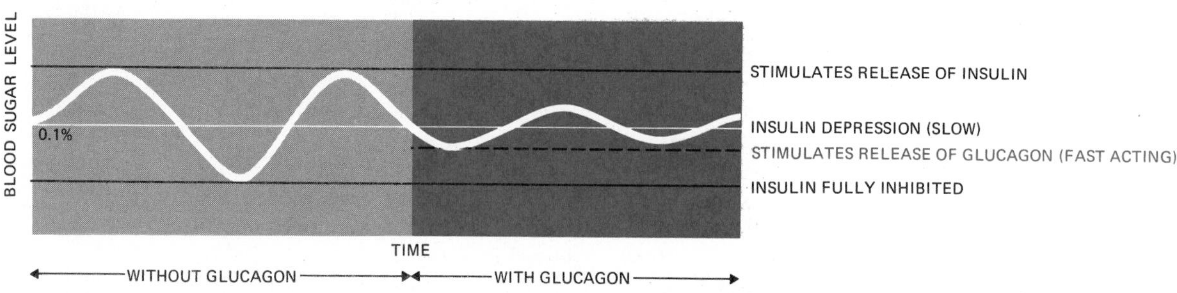

FIG. 25.7 Effect of glucagon on the homeostatic control of blood sugar. Its quick release and fast action suppress the large fluctuations in blood sugar level that would occur if the slow-responding insulin system had to maintain homeostasis by itself. Insulin and glucagon are secreted by different cells in the islets of Langerhans, the β-cells and α-cells respectively.

and diabetic coma so that they can take proper measures if either crisis should occur.

Injection of commercial insulin preparations usually causes a brief rise in blood sugar before the longer-lasting fall. This has been found to be caused by a second pancreatic hormone, **glucagon,** which *stimulates* the conversion of liver glycogen into glucose. In the normal organism, glucagon may act to prevent insulin from lowering the blood-sugar level excessively. Its quick release and quick action dampens the tendency of the slower-responding insulin to overshoot in *its* action (Fig. 25.7). It thus plays an important role in establishing a constant level of glucose in the blood. This homeostatic role resembles that of calcitonin.

Most victims of diabetes mellitus have high levels of glucagon in their blood. In some cases, in fact, it is the elevated level of glucagon rather than a depressed level of insulin that appears to be the major cause of the disease.

25.9 THE PITUITARY GLAND

The pituitary gland is a pea-sized structure located at the base of the brain (Fig. 25.5). In most vertebrates, it consists of three lobes: anterior, intermediate, and posterior. An intermediate lobe is present in the pituitary gland of human infants, but only a vestige remains in adulthood.

Although small in size, the pituitary plays a vital role in the chemical coordination of the body. It is often called the "master" gland because many of its secretions control the activity of other endocrine glands. TSH is one example that has already been mentioned.

The Anterior Lobe

An extraordinary amount of research has been conducted on the pituitary gland. At least seven chemically distinct hormones have been isolated from the anterior lobe alone. These are:

1. Human growth hormone (HGH). HGH is a polypeptide chain containing 190 amino acids. As its name implies, it promotes growth of the skeleton and the body as a whole. It does not appear to do this directly, but rather it stimulates the liver (and perhaps the kidney) to release a polypeptide (called somatomedin) that promotes the growth of muscle, cartilage, bone, and other connective tissues.

It is normally active in this respect only during the years of childhood and adolescence. A hyposecretion of HGH in the child results in stunted growth or dwarfism. A hypersecretion during the same period results in gigantism (Fig. 25.8).

One cause of dwarfism may well be psychological. Orphans and other children raised under conditions where they receive little love and affection are apt not to grow normally even though their diet and other conditions of their nurturing are adequate. Although more data are needed, such children have been found to secrete abnormally small amounts of HGH. When placed in a situation where their emotional needs are met, HGH secretion and skeletal growth are restored to normal rates.

FIG. 25.8 A giant and a dwarf with a man of average height. The amount of growth hormone produced during childhood and adolescence accounts for these differences.

HGH is also secreted throughout adult life, especially during periods of exercise or other stress. What function it serves at these times has been the subject of intensive research. As it did during childhood, HGH continues to promote a wide variety of *anabolic* reactions within the body. In order to do this effectively, thyroxin must also be present. HGH also works synergistically with some other hormones; that is, its presence enhances their activity.

HGH has been shown to have a further function in adults: the stimulation of milk production in women after childbirth. In this it resembles the action of the next hormone to be discussed and, in humans, is very similar or even identical to it.

2. Prolactin. Prolactin is a protein hormone secreted by females during and after pregnancy. It stimulates the development of the mammary glands during pregnancy and, following childbirth, the production of milk by them.

A prolactin-like hormone has even been found in nonmammalian vertebrates. It does not, of course, stimulate milk production in these animals, but instead some kind of maternal activity appropriate to the particular species. In some birds, for example, prolactin stimulates broodiness, that is, an inclination to sit on the nest. In one species of newt, it stimulates the animals to return to the water to lay and fertilize their eggs.

3. Thyroid-stimulating hormone (TSH). This hormone stimulates the thyroid gland to secrete thyroxin. The secretion of TSH is, in turn, depressed by thyroxin; thus there exists a homeostatic control mechanism over the level of thyroxin in the blood. Even the "master" gland has its controls! As we saw in Section 25.4, this mechanism can break down, producing a hypersecretion of TSH that results in goiter.

4. Adrenocorticotropic hormone (ACTH). ACTH is a polypeptide containing 39 amino acids. Its main function is to stimulate the cortex of the adrenal gland to release some of *its* hormones into the bloodstream. The vital role that these adrenal hormones play in human physiology will be considered in the next section.

5. Follicle-stimulating hormone (FSH). FSH acts upon the gonads or sex organs. In females, FSH promotes the development of follicles within the ovary (see Section 21.6). In conjunction with another pituitary hormone, LH, it stimulates the secretion of **estrogens** by the follicle and the ripening of the egg within it.

FSH is produced in human males, too. In them it stimulates the development of the seminiferous tubules and the production of sperm (see Section 21.5).

6. Luteinizing hormone (LH). As a human egg matures, it completes its first meiotic division and reaches metaphase of the second division (see Section 21.1). Then it breaks out of the follicle (ovulation) and is ready for fertilization by a sperm cell. The remaining cells of the follicle become transformed into the *corpus luteum* (see Section 21.6). Every one of these activities is triggered by LH. In some mammals (including humans), LH then stimulates the corpus luteum to secrete *its* hormone, progesterone.

LH is also found in human males. It acts upon the endocrine cells in the testes (the interstitial cells —see Section 21.5) causing them to release male sex hormones (androgens) into the bloodstream.

7. Melanocyte-stimulating hormone (MSH). MSH is a small polypeptide whose sequence of amino acids shows it to be closely related to ACTH. Its "target" cells are the melanocytes, cells which contain the black pigment melanin. Large numbers of them are present in the skin, where they are responsible for moles, freckles, and suntan. Although MSH does not seem to play an important role in the normal behavior of human melanocytes, under certain conditions, such as pregnancy, an increase in its secretion does cause some darkening of the body.

In most vertebrates MSH is produced in an intermediate lobe of the pituitary gland. Its secretion causes a dramatic darkening of the skin of many fishes, amphibians, and reptiles. The darkening occurs because of the spread of granules of melanin through the branches of specialized melanocytes in the epidermis. Figure 25.9 shows these cells (called melanophores) in the skin of a frog. When the melanin is concentrated in the center of the melanophores, the skin has a light appearance. When the melanin is dispersed throughout the branches, the skin becomes much darker (Fig. 25.10). This mechanism is undoubtedly of value in enabling the animal to blend in with its surroundings.

When biologists wish to study the development of frog eggs "out of season," they initiate production of the eggs by injecting female frogs with pituitary

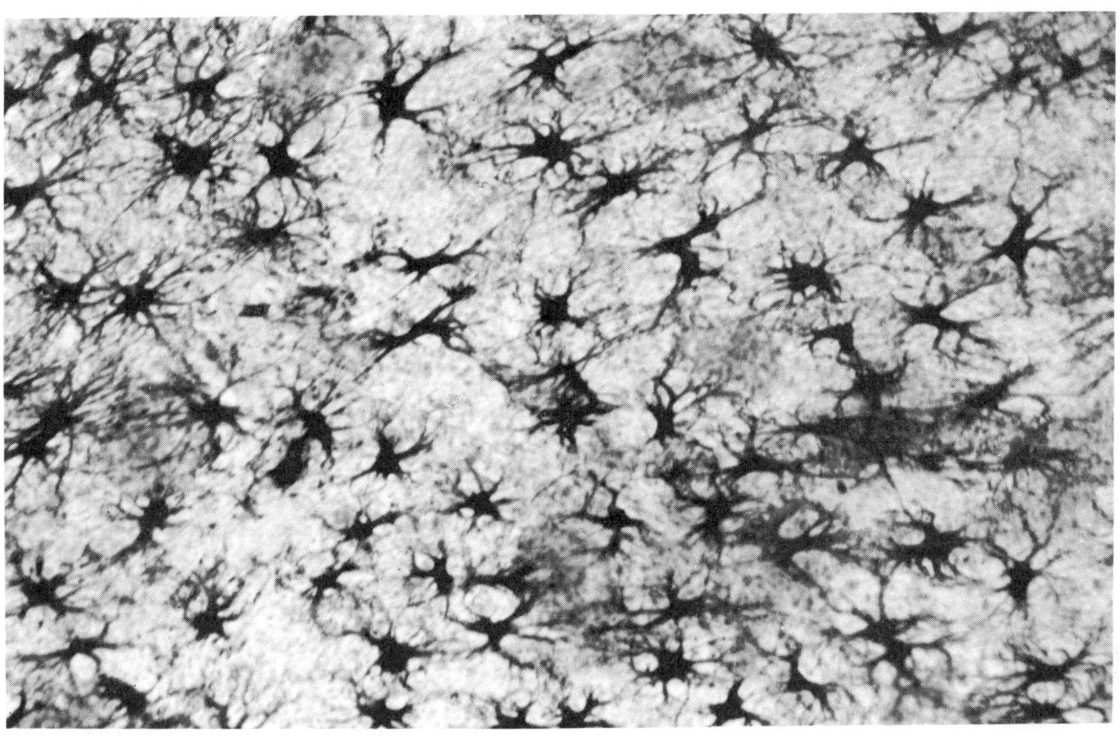

FIG. 25.9 Melanophores in the skin of a frog. The pigment melanin is dispersed throughout the branches of the cells, thus darkening the skin. The general term for any pigment-containing cell is *chromatophore*.

FIG. 25.10 MSH, acting on the melanophores, causes darkening of the frog's skin.

extract. (Why?) Within minutes after the extract is injected into a frog, marked darkening of the skin occurs. This is caused by the MSH which is also present in the extract.

The Posterior Lobe

The posterior lobe of the pituitary is probably not a true endocrine gland at all. Instead of manufacturing hormones of its own, it seems simply to store those produced by nerve cells in the hypothalamus of the brain.

Two hormones have been isolated from the posterior lobe of the pituitary. **Oxytocin** is a polypeptide which stimulates contraction of smooth muscle, particularly the smooth muscle lining the uterus. Its secretion, probably by the baby as well as the mother, plays an important role in childbirth. Injections of this hormone are sometimes given at this time to hasten delivery of the baby and to hasten the return of the uterus to its normal size. The new mother also secretes her own oxytocin, especially if she nurses her baby.

A second polypeptide released by the posterior lobe of the pituitary is called the *a*nti*d*iuretic *h*ormone **(ADH)** or, by some workers, vasopressin. ADH has two functions in humans. It causes the muscular walls of the arterioles to contract. This constricts the bore of these vessels and causes an increase in blood pressure. ADH also stimulates the reabsorption of water from the tubules of the kidneys. As we mentioned when discussing kidney function (see Section 13.6), insufficient production of ADH causes an enormous loss of water through the kidneys. This ailment is called diabetes insipidus. The name is derived from an early diagnostic test for the disease. The copious urine produced by hyposecretion of ADH is very watery and has no marked taste (insipidus). The urine produced by victims of insulin deficiency, however, contains a large amount of glucose and thus has a sweet (mellitus) taste.

25.10 THE HYPOTHALAMUS

Both the location and the function of the pituitary gland suggest that it serves as a vital connecting link between the nervous system and the endocrine system. The pituitary gland is located at the base of a region of the brain called the hypothalamus. The posterior lobe of the pituitary gland has a direct nerve connection to the hypothalamus. The anterior lobe receives blood from the hypothalamus by way of a portal system of veins analogous to the hepatic portal system (see Section 7.10).

Three short polypeptides have been identified which, after synthesis in the hypothalamus, travel to the anterior lobe of the pituitary through these portal veins. One, called *t*hyrotropin-*r*eleasing *h*ormone **(TRH)**, stimulates the anterior lobe to secrete TSH. A second, called gonadotropin-releasing hormone **(GnRH)**, stimulates the anterior lobe to secrete LH (thus accounting for another commonly used name: luteinizing hormone-releasing hormone, LH-RH) and FSH. A third is called *s*omatotropin-*r*elease *i*nhibiting *f*actor **(SRIF)** or somatostatin. As its name implies, it inhibits the secretion of growth hormone by the pituitary. It also suppresses the secretion of TSH, prolactin, insulin, and—with great effectiveness—glucagon. Its dramatic suppressive effect on glucagon release gives promise that SRIF—or a derivative of it—may be of future value in the treatment of diabetes mellitus.

Pituitary secretion can often be traced to the influence of nervous stimulation. Release of ADH from the posterior lobe occurs when special cells in the hypothalamus detect a lowering of the water content of the blood. Release of ACTH from the anterior lobe has been shown to be partly influenced by nervous activity in the hypothalamus and is often associated with emotional states such as anger and fear. In birds, visual stimuli such as increasing length of daylight or frequent sight of a member of the opposite sex have been shown to result in an outpouring of the gonad-stimulating hormones into the bloodstream.

The direct nervous and hormonal connections between the hypothalamus of the brain and the "master" gland of the endocrine system, the pituitary, provide the machinery to link these two major coordinating systems.

25.11 THE ADRENAL GLANDS

The adrenal glands are two small structures situated one atop each kidney (Fig. 25.5). They are richly supplied with blood. Both in anatomy and in function, they consist of two distinct regions. The exterior portion is the adrenal **cortex.** The interior portion of the gland is the adrenal **medulla.**

The Adrenal Medulla

Although a true endocrine gland, the adrenal medulla is also considered to be a part of the nervous system. Its secretory cells seem to be modified nerve cells. It releases two hormones which pass into the bloodstream. The better known of these is adrenaline. It is not yet clear what role, if any, adrenaline plays in the normal metabolism of the body. However, large quantities of this hormone are released into the bloodstream when the organism is suddenly subjected to stress such as anger, fright, or injury. As adrenaline spreads throughout the body, it promotes a wide variety of responses. The rate and strength of the heartbeat is increased, thus increasing blood pressure. A large part of the blood supply of the skin and viscera is shunted to the skeletal muscles, coronary arteries, liver, and brain. The level of blood sugar rises and the metabolic rate increases. The bronchi dilate, permitting easier passage of air to and from the lungs. The pupils of the eye dilate and there is a tendency for the body hair to stand erect. (This is particularly obvious in angered cats and dogs. Humans, being comparatively hairless, display "gooseflesh" instead.) The clotting time of the blood is reduced, and the anterior lobe of the pituitary is stimulated to produce ACTH.

The second hormone of the adrenal medulla, noradrenaline, also causes an increase in blood pressure. It accomplishes this by stimulating the contraction of the arterioles.

Almost all of the body's responses to these two hormones can be seen to prepare the body for violent physical action. We have all heard accounts of heroic deeds accomplished in times of danger or other emergency. The secretion of adrenaline and noradrenaline by the adrenal medulla is an important mechanism for making such action possible.

The Adrenal Cortex

Several different hormones have been extracted from this gland. All of them are steroids (see Section 4.2). They are quite similar in molecular structure and seem to be freely converted from one to another by enzyme action. They fall into two categories.

1. The glucocorticoids. In humans, the most important members of this group are **cortisol** and the closely related corticosterone. These hormones promote the conversion of fat and protein into intermediary metabolites that are ultimately converted into glucose. Thus they cause the level of blood sugar to rise. One of the chief target organs in this response is the liver. If cortisol is administered to an animal whose own adrenal glands have been removed, it induces in the liver the synthesis of a number of specific enzymes involved in protein and carbohydrate metabolism. This induction is blocked if actinomycin D is administered before the cortisol. This suggests that at least some of the effects of cortisol are brought about by the selective transcription of genes.

The glucocorticoids also act to suppress inflammation in the body. Synthetic glucocorticoids have found widespread medical use in the treatment of inflammatory diseases ranging from arthritis to poison-ivy poisoning.

The glucocorticoid hormones are needed to maintain the body during periods of stress after the first brief, adrenaline-triggered response has worn off. This second, longer-lasting response seems to be quite independent of the nature of the stress. Exposure to extremes of temperature, poisoning, severe bodily injury, infection, even emotional turmoil all trigger a definite sequence of responses. The endocrinologist Hans Selye calls these responses the "general adaptation" syndrome. After the initial response to adrenaline, the body goes into a stage of "shock." The levels of glucose and salt in the blood drop sharply. Blood pressure is also reduced. Then, the adrenal cortex begins to pour glucocorticoids into the bloodstream in response to the increased secretion of ACTH by the pituitary gland. The release of ACTH is itself triggered by (1) the emotional state of the subject, (2) the lowered level of steroids in the blood, and (3) the adrenaline released by the adrenal medulla. During this phase of "countershock," the symptoms of the "shock" phase are reversed. The various body functions return to normal or even above-normal activity. The organism then enters a stage of resistance. It has become adapted to the stress.

2. The mineralocorticoids. The chief function of these hormones, of which **aldosterone** is the most important in humans, is to promote the reabsorption of Na^+ and Cl^- ions in the tubules of the kidneys. Not only are these ions valuable for their own sake, but their retention in the blood keeps its osmotic pressure high. This, in turn, assures normal blood volume and pressure.

The rate of aldosterone secretion is controlled by a number of factors. The most important of these is the level of angiotensin II in the blood. In Chapter 11, we examined the mechanism by which a drop in blood pressure causes the kidney to secrete renin, which, in turn, leads to the production of angiotensin II (see Section 11.20). The stimulatory effect of angiotensin II on aldosterone secretion thus provides a negative feedback device by which the concentration of Na^+ in the ECF is maintained. The rate of aldosterone secretion is also enhanced somewhat by the presence of ACTH and a high level of K^+ in the ECF.

Mineralocorticoid hormones are found in other vertebrates, too. In every case, they also function in the control of salt and water balance in the animal. The exact action varies, however, according to the environment of the organism. In the freshwater fishes, they act to retain salts but release water. This action is just reversed in the marine fishes, with the hormones promoting the excretion of salt by the gills and the reabsorption of water by the kidneys. The active excretion of salt by the nasal glands of marine birds and reptiles is also stimulated by them. (See Section 6.4.)

The active transport of Na^+ ions has been intensively studied in the bladder cells of the toad. It has been demonstrated that the phenomenon depends upon the induction, by aldosterone, of enzyme synthesis in the cytoplasm of these cells. However, when aldosterone is administered to the cells, it moves into the nucleus. Only after about an hour is the transport of Na^+ ions increased. The response to aldosterone is inhibited by actinomycin D, and this supports the idea that an early effect of the hormone is on the genetic code.

The glucocorticoids and mineralocorticoids are essential to life. Under laboratory conditions, some animals can survive for about two weeks when their adrenal glands have been surgically removed, but they must be protected from stress. The slightest stress of any kind causes a sudden drop in the sugar and salt levels of the blood, and death follows quickly. Humans cannot survive even under the best conditions without their adrenal cortices. If absolutely no cortical steroids are present, death comes in a few days. A hyposecretion of these hormones results in Addison's disease. The symptoms are lowered blood pressure, loss of appetite, muscular weakness, and general apathy. Unless corrective measures are taken, death follows after about two years. Fortunately, administration of synthetic cortical steroids and/or ACTH can today restore the patient to a relatively normal life.

25.12 THE GONADS

Both the male and female gonads (Fig. 25.5) possess endocrine activity in addition to their prime function of producing the sex cells.

The Testes

The testes of males contain endocrine tissue, the interstitial cells. When stimulated by the anterior pituitary hormone LH, these cells release androgens (e.g. **testosterone**) into the bloodstream. This action commences at the start of adolescence. Testosterone triggers the development of the so-called secondary sexual characteristics found in adult males (see Section 21.5) and is essential for the production of sperm.

Removal of the testes (castration) before adolescence prevents the development of the secondary sexual characteristics. The response of sexually mature males to castration varies, but generally there is only a moderate loss of masculine traits. Castration is frequently used in animal husbandry to provide animals with better meat quality and/or easier temperament. Castrated cattle, horses, and fowl are called steers, geldings, and capons respectively.

A number of synthetic androgens are marketed for certain therapeutic uses. Some evidence has been presented that among the effects of these drugs is an increase in body weight and muscular strength. Despite the fact that many investigators do not agree that these effects occur, use of these drugs is now frequently found among certain athletes, especially weight-lifters, shot-putters, and professional football players. There is a genuine question not only as to whether the drugs do have the anabolic effects that some ascribe to them, but also as to their safety.

The Ovaries

The ripening **follicle** in the ovary not only contains a ripening egg but also acts as an endocrine gland. The accessory cells of the follicle liberate several steroid hormones called **estrogens.** They are stimulated to do so by the combined influence of FSH and LH from the anterior lobe of the pituitary gland.

The estrogens have two major functions in the female body. First, they promote the development early in adolescence of the secondary sexual characteristics (see Section 21.6). Second, they participate in the monthly preparations of the body for possible pregnancy. This includes the preparation of the endometrium (the inner lining of the uterus) to receive an embryo. The endometrium becomes thicker and more richly supplied with blood. Some of the molecular activities involved in these changes were examined in Section 19.2.

The corpus luteum (which is produced after ovulation—see Section 21.6) is also an endocrine gland. Stimulated by LH, it secretes **progesterone** into the bloodstream. This hormone continues the preparation of the uterus for pregnancy and inhibits the development of a new follicle. If pregnancy should occur, the corpus luteum continues to secrete progesterone. However, as the time of birth approches, its secretion of progesterone declines and is replaced by a vigorous outpouring of the hormone **relaxin.** This hormone causes the ligaments between the pelvic bones to loosen, which provides a more flexible passageway for the baby during birth.

The secretion of estrogens is stimulated by FSH and LH, while the secretion of progesterone is stimulated by LH alone. Both estrogens and progesterone act back on the hypothalamus, inhibiting its secretion of gonadotropin releasing hormone (GnRH). Thus production of FSH and LH by the anterior pituitary is inhibited. These homeostatic interactions (similar to that between TSH and thyroxin) serve to coordinate estrogen and progesterone production to the needs of the monthly sex cycle. They also provide the physiological basis for the effectiveness of "the pill" as a contraceptive agent. The pill contains both a synthetic estrogen and a synthetic progesterone. Like the natural hormones, these substances act on the hypothalamus, inhibiting the secretion of GnRH and hence of FSH and LH production by the pituitary (see Fig. 21.14). Without FSH and LH, neither normal follicle development nor ovulation can occur. The pill also inhibits the steps which would otherwise enable an egg to move down the Fallopian tube and, if fertilization had occurred, become implanted in the endometrium.

The pill is generally taken for about three weeks (sometimes estrogens alone are used during the first two weeks of this period). Then, stopping the pill for a week allows normal menstruation to occur.

At the end of a woman's reproductive years (the menopause), the production of eggs and estrogens ceases. Disappearance of the estrogens removes their brake on the pituitary, and large quantities of FSH are liberated into the bloodstream. The control mechanism has broken down. The elevated level of FSH causes a variety of unpleasant physical and emotional symptoms. Many doctors administer estrogens at this time, in order to reimpose a brake on FSH secretion by the pituitary and thus relieve the symptoms.

25.13 THE PLACENTA

Nourishment of the human embryo while within its mother's uterus is accomplished through the umbilical cord and placenta which connect the embryo directly to the mother's circulatory system (Fig. 21.17). After pregnancy is established, the placenta takes on the secondary function of serving as an endocrine gland. It secretes estrogens, progesterone, and a hormone called *h*uman *c*horionic *g*onadotropin (HCG) that is quite similar to the gonad-stimulating hormones of the anterior lobe of the pituitary. These hormones supplement those produced by the corpus luteum and the pituitary gland.

25.14 THE PINEAL GLAND

The pineal gland is a small, pea-sized structure attached to the brain just above the cerebellum (Fig. 25.5). It produces a hormone called **melatonin.** When this hormone is injected into frogs, it causes a marked lightening of the skin. Its action on the melanophores (Fig. 25.9) is thus just the reverse of that of MSH. In the laboratory rat and the hamster, melatonin inhibits the gonads both as gamete producers and as endocrine organs. The secretion of melatonin in these animals is markedly increased when they are placed in the dark and decreased when they are exposed to light. Many mammals have inactive gonads during the winter. With the lengthening days of spring, their gonads become active once again and mating follows. Perhaps the pineal gland, through its secretion of melatonin, serves as a connecting link between the eyes and the gonads in this **photoperiodic** response. As for humans, its normal role is still uncertain.

25.15 THE THYMUS GLAND

The thymus gland consists of two lobes of tissue similar to that found in lymph nodes. It is located high in the chest cavity just under the breastbone (Fig. 25.5). It is large during childhood but then shrinks after the start of adolescence. Many techniques, including surgical removal, have been employed in attempts to find an endocrine function of the thymus, but none has been convincingly demonstrated in adult mammals. However, there is good evidence that the thymus in the infant mammal plays a major role in setting up the lymphocyte-producing machinery of the lymph nodes, thus providing the basis for the development of antibodies. This seems to involve, among other things, the production of one or more hormones (called thymosins). Once the job is completed, the thymus is ordinarily no longer needed.

25.16 THE KIDNEY

The kidney has at least three endocrine functions. In Sections 11.20 and 25.11, we examined the role that its secretion of **renin** plays in maintaining blood pressure. In Section 25.6, we noted that the kidney completes the conversion of calciferol into its most effective form. The kidney also secretes **erythropoietin** into the bloodstream, especially in response to anemia. Erythropoietin acts on the bone marrow to increase its production of red blood cells.

25.17 HORMONES AND HOMEOSTASIS

Our study of the human endocrine system reveals what a variety of important roles are accomplished by chemical coordination. Most (but not all) of these chemical controls act in a rather slow, generalized way. Growth, sexual development, and metabolism are three body processes under hormonal controls that act gradually over a span of time.

The endocrine system also plays a major role in the maintenance of a constant internal environment. The concentration of sugar, water, and various salt ions in the ECF is maintained within narrow limits by hormonal action. There are at least three controlling mechanisms by which hormones maintain homeostasis in the body.

1. The secretion of some hormones is directly controlled by the need for that hormone. A high level of Ca^{++} in the blood suppresses the output of PTH. A low level of Ca^{++} stimulates it. The level of sugar in the blood acts directly on the islets of Langerhans, promoting the appropriate response from them. The osmotic pressure of the blood triggers (with the aid of the nervous system) the production of ADH and thus its own readjustment.

2. In some cases, the response of a gland to the level of the substance it regulates is apt to be slow. The resulting lag in response may cause undesirable fluctuations above and below the desired level. This situation may be improved by a second hormone acting antagonistically to the first. The antagonistic action of such pairs of hormones as insulin-glucagon and PTH-calcitonin provides a system of checks and balances for reestablishing homeostatic equilibrium quickly after any displacement.

3. A third system for achieving self-regulation of hormone production is illustrated by the relationship between TSH and thyroxin. Whenever one hormone stimulates the production of a second, we find that the second hormone acts, in turn, to suppress the production of the first. Another example of this is the way in which high levels of estrogens keep the production of FSH in check. Here again is a self-contained control system (analogous to the action of a governor on an engine) for maintaining homeostasis.

25.18 THE MECHANISM OF ACTION OF HORMONES

All the human hormones we have studied in this chapter are transported by the bloodstream. They are thus carried from the endocrine gland which produces them to every cell of the body. In some cases (e.g. thyroxin), every cell of the body responds to the presence of the hormone. More often, only certain cells, those in the "target organ," respond. This arrangement can be compared with that in radio broadcasting. We live surrounded by the electromagnetic radiation from a large number of broadcasting transmitters. However, only a correctly tuned receiver can respond to this energy. In a similar way, only certain of our cells may be competent to respond to a given hormone circulating in the blood.

The matter of cell competence cannot be overly stressed. Hormones, in themselves, cannot accomplish the various functions we have ascribed to them. Hormones simply release within the target cells their potentialities for dealing with the situation. Insulin added to blood in a test tube has no effect on the glucose content. In the living organism, however, it enables the cells to reduce the blood-sugar level.

Throughout the chapter we have noted that many hormones include among their effects a stimulation of protein synthesis. Usually this effect is inhibited by actinomycin D. In view of the role that actinomycin D plays in preventing DNA transcription (i.e. RNA synthesis), we conclude that these hormonal effects involve the derepressing of gene action.

In trying to discover the mechanisms of hormone action, it has become apparent that our hormones fall into two major categories. The steroid hormones (e.g. estrogens, cortisol, aldosterone, calciferol) *enter* their target cells. In several cases, it has been shown that, once within the cell, the hormone quickly moves into the nucleus bound to a cytoplasmic protein. Once within the nucleus, the hormone-protein complex binds to chromatin (at least in the cases of estrogen, cortisol, and progesterone (Fig. 19.3). These events induce gene activity. Trying to find out

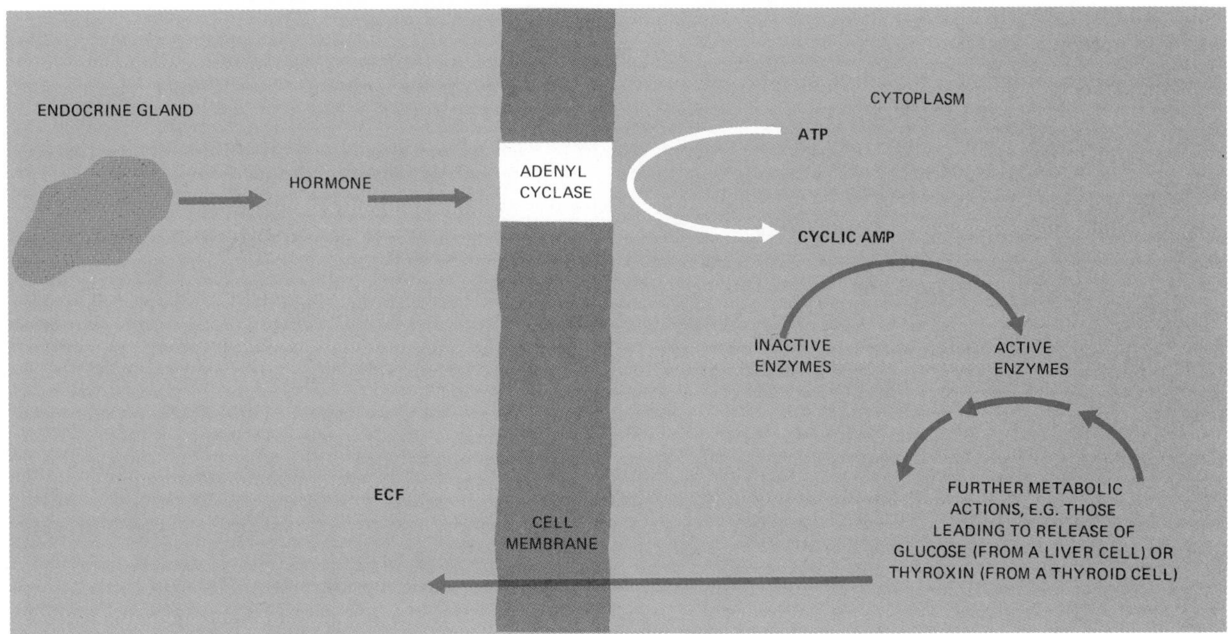

FIG. 25.11 Probable mechanism of action of some hormones. Binding of the hormone (the "first messenger") to the cell membrane stimulates the production within the cell of cyclic AMP, the "second messenger."

just what role these complexes play in the inductive process is an extremely active field of research.

The protein and polypeptide hormones (and also amino acid-derived hormones like adrenaline) make up the second major category. These hormones (which include insulin, MSH, PTH, ACTH, glucagon, TSH, ADH, and LH) may also induce gene activity, but the mechanism at work is quite different. They do not enter their target cells but, instead, bind to specific receptors on the cell surface. In so doing, they activate an enzyme attached to the membrane called adenyl cyclase (Fig. 25.11). Adenyl cyclase catalyzes the conversion of ATP in the cytoplasm into cyclic AMP (Fig. 25.12). An extraordinary variety of cell functions are affected (either stimulated or inhibited) by a rising level of cyclic AMP within the cell. The particular functions affected in any given cell depend on the type of cell involved, that is, on its past history of differentiation. Thus cyclic AMP stimulates the β-cells of the islets of Langerhans to release insulin, kidney cells to secrete renin, liver cells to convert glycogen into glucose, thyroid cells to release thyroxin, and so on. Cyclic AMP and its role as a "second messenger" in hormone action

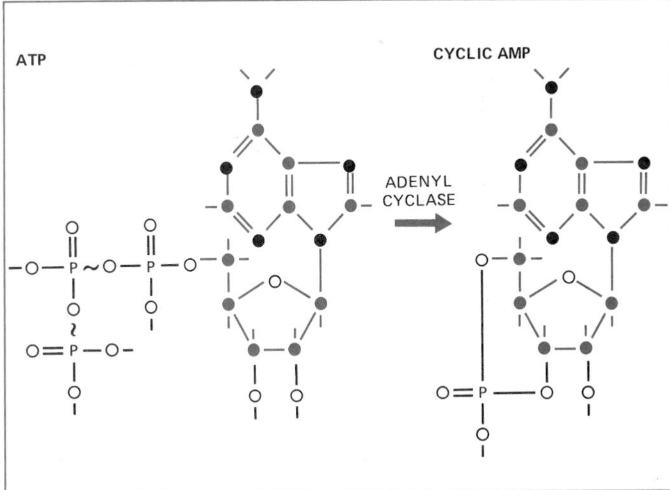

FIG. 25.12 Conversion of ATP to cyclic AMP. (Carbon atoms are shown in color, nitrogen atoms in black, and hydrogen atoms as short dashes.)

were discovered by Earl Sutherland. In 1971 he was awarded the Nobel Prize for these discoveries.

While many (not all) of the actions of hormones can be traced to the activation of genes, the mechanisms at work are surely more complex than those discovered in *E. coli*. Even if some hormones do interact quite directly with DNA, the fact that they can do so only in certain "target" cells (e.g. estrogen in the cells of the uterus) and not in others tells us that something more is involved. We have established that every cell in the organism has a complete library of genetic information. Why, then, can certain genes be expressed in some cells but not in others? The answer is unknown, but surely it is an outcome of the history of the cell, that is, the process of cellular differentiation itself. When we finally understand the gene control mechanisms by which a cell becomes differentiated, we shall be much closer to an understanding of the diverse effects that hormones have on them.

25.19 THE PHEROMONES

Hormones are chemical substances that are released into the *internal* environment (ECF) by *endocrine* glands. Carried throughout the body, they coordinate many of the activities of its various parts. This coordination provides, among other things, for a close regulation of the chemical properties of the internal environment and the activities of our internal organs.

In recent years, an entirely different category of chemical coordinators has also been receiving careful study. These are the **pheromones.** Pheromones are chemical substances which are released into the *external* environment by *exocrine* glands. They provide a means of communication with other members of the same species.

In some cases, the communication is quite subtle. The pheromones released by some members of the species simply initiate physiological changes in the others. These changes usually do not lead to any externally directed response for some time. Adult male grasshoppers of the species that migrates in locust plagues release a pheromone which, when detected by immature members of the species, hastens their growth. This, in turn, speeds the time when a migratory swarm will be formed and a locust plague will get under way. The queen bee in a honeybee colony secretes a substance which prevents the worker bees from developing ovaries and laying eggs. How-

FIG. 25.13 A stick treated with the trail pheromone of an ant (left) can be used to make an artificial trail which is followed closely by other ants emerging from their nest (right). The trail will not be maintained by the other ants unless food is placed at its end. (Courtesy of Sol Mednick and *Scientific American*.)

ever, if the queen is killed or removed from the hive, the disappearance of this pheromone permits some of the workers to take over her egg-laying function.

Other pheromones have been discovered which trigger immediate action when detected. When an ant is disturbed, it secretes from glands in its head a volatile chemical that quickly diffuses in all directions. This chemical can be detected by other ants several centimeters from the spot. They are attracted by low concentrations of the substance and begin to move toward the area of increasing concentration. As they thus get nearer their disturbed nest mate, their response changes to one of alarm. The higher concentration of the pheromone causes them to run about actively as they remedy the disturbance. Unless additional quantities of this pheromone are released, it soon dissipates. This is important, too, so that once the emergency is over, the ants can return quietly to their former occupations.

Several other ant pheromones have been discovered, including one that is laid in a trail by a worker returning to the nest with food. This trail attracts and guides other ants to the food source (Fig. 25.13). It is continually renewed as long as the food holds out. When the supply begins to dwindle, however, trail-making ceases. The trail pheromone evaporates quickly so other ants stop coming to the site and will not be confused by old trails when food is located elsewhere.

Most of the pheromones discovered so far have been in insect species, although a few have been found in other animals. Furthermore, the most elaborate pheromone communication systems have been found in the social insects—the ants, termites, and bees. It would appear that just as hormones help to coordinate the action of the tissues, organs, and systems within the organism, so pheromones help to coordinate the action of the individuals that make up a society.

A response to a pheromone is a response to one of many kinds of stimuli present in an animal's external environment. In order to detect the stimulus and to carry out an immediate, coordinated response, a nervous system must be present. We now turn our attention to the general topic of nervous coordination.

EXERCISES AND PROBLEMS

1. Summarize the mechanisms which participate in the control of the blood-sugar level in the body.
2. Glucocorticoid therapy for children is undesirable because it may result in severe stunting of growth. What mechanism might be responsible for this?
3. Why must insulin be given by injection rather than by mouth?
4. How will each of the following affect the quantity and composition of urine formed in humans: (a) hypersecretion of ADH, (b) hyposecretion of insulin, (c) drinking copius amounts of water?
5. Which endocrine glands are controlled by the secretion of other endocrine glands?
6. What parallels can you find between the relationship of organs in an organism and the relationship of individuals in a society?
7. How is communication among the parts of an organism accomplished?
8. How is communication among the parts of a society accomplished?
9. What is the significance of diapause in insects?
10. Compare the mechanisms by which (a) the body regulates the concentration of water in the blood and (b) a steady temperature is maintained in your home.
11. Distinguish between the molecular activities that take place when each of the following interacts with one of its target cells: (a) a polypeptide or protein hormone and (b) a steroid hormone.

REFERENCES

1. WIGGLESWORTH, SIR VINCENT, *Insect Hormones,* Oxford Biology Readers, No. 70, Oxford University Press, Oxford, 1974.
2. TATA, J. R., *Metamorphosis,* Oxford Biology Readers, No. 46, Oxford University Press, Oxford, 1973. Considers both amphibians and insects.
3. FRIEDEN, E., and H. LIPNER, *Biochemical Endocrinology of the Vertebrates,* Prentice-Hall, Englewood Cliffs, N.J., 1971. An excellent survey. Paperback.
4. RASMUSSEN, H., and PECHET, M. M., "Calcitonin," *Scientific American,* Offprint No. 1200, October, 1970.
5. LOOMIS, W. F., "Rickets," *Scientific American,* Offprint No. 1207, December, 1970. Explains why calciferol should be considered a hormone rather than a vitamin.
6. BANTING, F. G., and C. H. BEST, "The Internal Secretion of the Pancreas," *Great Experiments in Biology,* ed. by M. L. Gabriel and S. Fogel, Prentice-Hall, Englewood Cliffs, N.J., 1955. A description of the method by which insulin was finally isolated. This work, for which Banting shared a Nobel Prize, led ultimately to the development of insulin therapy for diabetics.
7. GARDNER, L. I., "Deprivation Dwarfism," *Scientific American,* Offprint No. 1253, July, 1972. Traces a connection between dwarfism, impaired secretion of HGH, and emotional deprivation in children.
8. GUILLEMIN, R., and R. BURGUS, "The Hormones of the Hypothalamus," *Scientific American,* Offprint No. 1260, November, 1972. Describes the evidence for the existence of several hypothalamic hormones that regulate secretion of anterior-lobe hormones.
9. WURTMAN, R. J., and J. AXELROD, "The Pineal Gland," *Scientific American,* Offprint No. 1015, July, 1965.
10. PASTAN, I., "Cyclic AMP," *Scientific American,* Offprint No. 1256, August, 1972.
11. RANDLE, P. J., and R. M. DENTON, *Hormones and Cell Metabolism,* Oxford Biology Readers, No. 79, Oxford University Press, Oxford, 1974. With emphasis on the activities of cyclic AMP.
12. HÖLLDOBLER, B., "Communication between Ants and their Guests," *Scientific American,* Offprint No. 1218, March, 1971. Including the role played by pheromones.

STIMULUS RECEPTORS

Nervous coordination differs from endocrine coordination in being faster and generally more localized in its action. It enables the organism to respond quickly to changes in the external as well as the internal environment. This is in contrast to the endocrine system which, as we know, is primarily concerned with internal changes. Although both plants and animals carry on chemical coordination by means of transported hormones, nervous coordination is characteristic of animals only.

26.1 THE THREE COMPONENTS OF NERVOUS COORDINATION

The ability of an organism to respond to changes in its environment requires the presence of three different components. *First,* there must be a stimulus **receptor.** This is a structure capable of detecting a certain kind of *change* in the environment and initiating a signal, the nerve impulse, in the nerve cell to which it is attached. Our sense organs are stimulus receptors. They enable us to detect such stimuli as light (the eye), sound (the ear), and chemicals (the odor receptors in the nose).

The *second* component in nervous responsiveness and coordination consists of the impulse **conductors,** the nerves themselves. Nerves are made up of bundles of conducting fibers in much the same way that a telephone cable is made up of bundles of wires. These fibers are greatly elongated extensions of special cells, the **neurons.** Two kinds of neurons make up most nerves. **Sensory** neurons transmit impulses from the stimulus receptor to the central nervous system, the brain and spinal cord. **Motor** neurons transmit impulses from the central nervous system to the part of the body that will take action.

In a few cases, sensory neurons may transmit their impulses directly to motor neurons, the junction occurring in the central nervous system. Most of the time, though, impulses from sensory neurons pass along one or many **interneurons** before finally reaching a motor neuron. The central nervous system is composed of millions of these interneurons. Their complex arrangement provides a virtually unlimited number of routes by which impulses can travel through the central nervous system. This, in turn, enables a great variety of complex actions to be carried out in a well-coordinated way. If the nerves can be compared to telephone cables, the central nervous system should be compared to the switchboards of the telephone exchange.

The *third* component in nervous coordination consists of the **effectors.** These are the structures that carry out some action in response to impulses reaching them by way of motor neurons. The most important effectors in humans are the muscles and the glands (both exocrine and endocrine).

In this chapter we shall examine the various kinds of stimulus receptors. Some of these detect stimuli arising within the body. Often, but not always, the nerve impulses generated by them are not made conscious in the brain. Other receptors detect stimuli in the external environment. Generally, these stimuli give rise to conscious sensations. It is important to realize that sensations exist only in the brain and not in the organ that detects the stimulus. Under general anesthesia, there are no sensations. Stimuli are still detected and nerve impulses are still sent back to the central nervous system. But the brain is unable to assign any interpretation to these impulses and hence no sensation exists. On the other hand, a person whose leg has been amputated may still feel pain ("phantom" pain) in the missing leg. In such a case, remnants of sensory neurons in the stump continue to send impulses back to the brain. It is the brain alone which interprets these impulses as signifying pain in the now-missing structure.

Most animals have receptors of radiant energy **(photoreceptors),** mechanical forces **(mechanoreceptors),** and chemicals **(chemoreceptors).**

Photoreceptors

Radiant energy exists in a range of wavelengths that extends from radio waves that may be thousands of meters in length to gamma rays with wavelengths as short as a million millionth (10^{-12}) of a meter (Fig. 26.1). Over this long span, the only wavelengths that generally serve as stimuli for living things are light rays (about 400–700 nm) and the somewhat longer infrared or heat rays.

Light receptors are common in the animal kingdom. They range in complexity from light-sensitive cells which simply detect the presence of light (as in the earthworm) to eyes that are capable of forming images. The latter are found in certain mollusks (especially the squids and octopuses), most arthropods (insects, crustaceans, and arachnids) and the vertebrates.

26.2 THE COMPOUND EYE

The structure and functioning of the arthropod eye is quite different from that of the mollusk and verte-

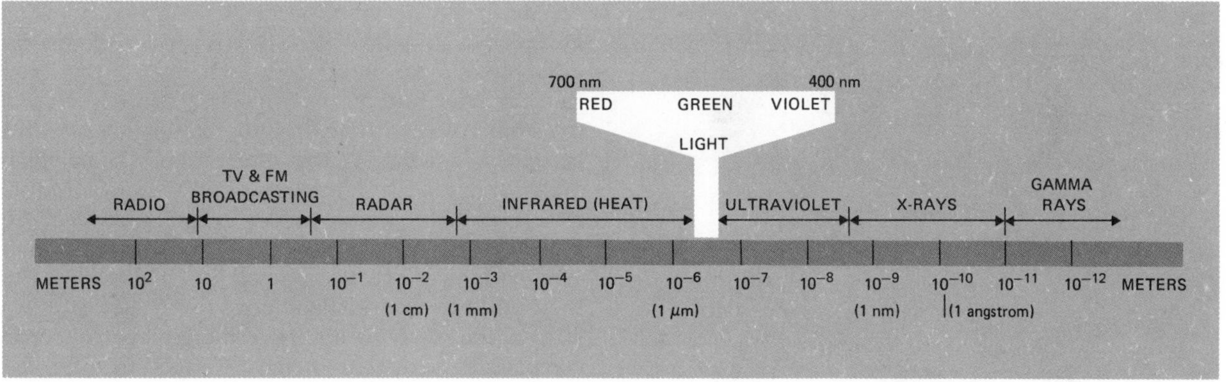

FIG. 26.1 The spectrum of electromagnetic radiation.

brate eyes. It is called a compound eye because it is made up of repeating units, the **ommatidia,** each of which functions as a separate visual receptor. Figure 26.2 shows the arrangement of ommatidia in a compound eye. Each ommatidium consists of (1) a lens (the front surface of which makes up a single facet of the compound eye), (2) a transparent crystalline cone, (3) light-sensitive, visual cells arranged in a radial pattern like the sections of an orange, and (4) pigment-containing cells which separate the ommatidium from its neighbors. The location of the pigment-containing cells is such that only light entering the ommatidium parallel (or almost so) to its long axis reaches the visual cells and triggers nerve impulses. Light entering at an angle is absorbed by the screening pigments. Thus each ommatidium is pointed at just a single area in space. If that area is giving off light of sufficient intensity, the ommatidium (and adjacent ommatidia whose field of "view" overlap its own) will respond to it. If not, there will

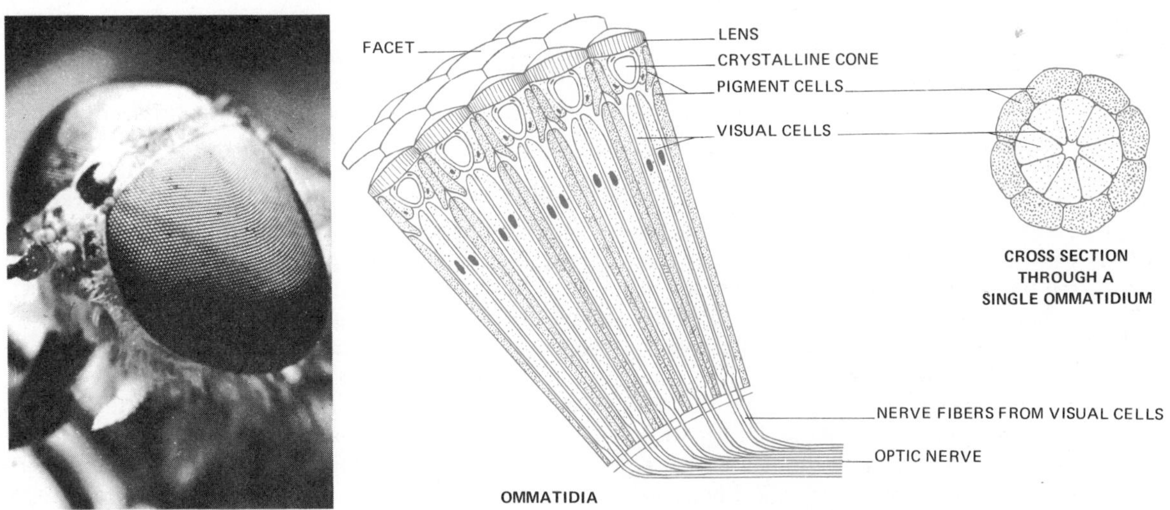

FIG. 26.2 Left: Compound eyes of a tabanid fly. They are made up of thousands of ommatidia (center), each of which serves as a separate light receptor. In some insects, the visual cells of a single ommatidium (right) respond selectively to different colors of light, thus providing the basis for color vision. (Photo courtesy of William H. Amos.)

FIG. 26.3 Portrait of George Washington, as photographed through a portion of the cornea of the compound eye of a water beetle. (Photograph by Prof. Walter E. Flowers, from W. Davis, *Science Picture Parade,* Duell, Sloan and Pearce, 1940.)

be no response. There may be thousands of ommatidia in a compound eye with their facets arranged over most of the surface of a hemisphere.

A favorite trick of nature photographers is to take a picture of some object as it would be seen through the cornea (all the facets, peeled off together) of a compound eye. Unfortunately, the results (Fig. 26.3) are not at all comparable to what the insect sees. In the photographer's setup, each facet acts as a lens, gathering light rays from all parts of the viewed object and passing them back to the film. As we have seen, however, the presence of screening pigments in the intact compound eye permits light from only one area on the viewed object to reach the visual cells. Thus each ommatidium contributes information about only one area on the object. The other ommatidia contribute information about the other areas. The composite of all the responses of all the ommatidia is a mosaic image—a pattern of light and dark dots which make up the total view.

The halftone illustrations used in newspapers (and in this book) are created in a similar fashion. If you look carefully at such an illustration (a magnifying lens will help) you will see a regular array of dots of black ink. The differing sizes of the ink dots provide for intermediate shades of gray and thus a rather faithful, but colorless, reproduction of the original scene. The finer the pattern of dots, the better the quality of the illustration. Grasshopper eyes, with relatively few ommatidia, must produce an exceedingly coarse, grainy image. The bee and the dragonfly possess many more ommatidia in their compound eyes and a corresponding improvement in the ability to discriminate ("resolve") detail. Even at that, the resolving ability of the bee eye is poor in comparison with that of most vertebrate eyes and only 1/60 as good as that of the human eye.

Because the compound eye does not permit good discrimination of points located close together in space, arthropods are quite nearsighted. Two objects which we could distinguish between at 60 feet could be distinguished by the bee only at a distance of one foot. On the other hand, the compound eye is well-adapted to detect motion. As an object moves across the visual field, ommatidia are progressively turned on and off. Because of the "flicker effect" produced, insects respond far better to moving objects than to stationary ones. Honeybees, for example, will visit windblown flowers more readily than still ones. The ability of the honeybee to perceive motion compares quite favorably with our own.

The quality of the compound-eye image gets even worse in dim light. Those arthropods which are apt to be active in dim light (e.g. crayfish, praying mantis) are able to concentrate the screening pigments of the ommatidia into the ends of the pigment cells. This shift then enables light entering a single ommatidium at an angle to pass into adjacent ommatidia and stimulate them, too. With many ommatidia responding to a single area in the visual field, the ability to form an image should deteriorate markedly. The praying mantis is probably able to do little more than distinguish light and dark in the evening. The shift in pigments does, however, make it more sensitive to light than it is during the daytime as more ommatidia can detect a given area of light.

Studies of insect behavior show conclusively that some insects, at least, are able to distinguish colors from varying shades of gray. They have true color vision. In order for this to be possible, two or more light-absorbing pigments must be present in the eye, each of which absorbs best at a different wavelength. A single pigment can only provide information about the amount of light given off by an object, that is, its brightness. In such a case, the individual is totally colorblind, with all objects appearing in varying

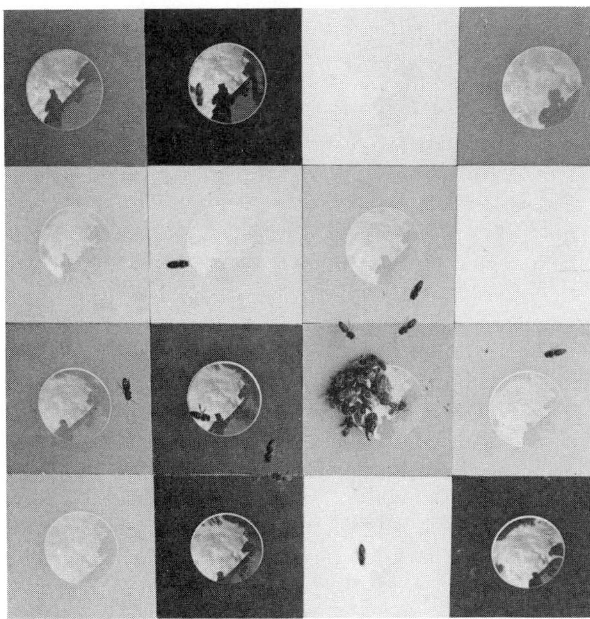

FIG. 26.4 A demonstration of color vision in honeybees. After a period of feeding from a dish placed on blue cardboard, the bees return to an empty dish on a clean blue card. They are able to distinguish the blue card from others of varying shades of gray. (Courtesy of Dr. M. Renner.)

shades of gray. With two or more pigments, however, it becomes possible to distinguish colors because an object giving off rays of predominantly one wavelength will selectively stimulate the receptors containing the pigment which absorbs that wavelength best. Although the actual pigments that permit color vision in certain insects have not been conclusively identified, there is strong evidence for their existence. It has been found, for example, that four of the visual cells in the honeybee ommatidium respond best to yellow-green light (530 nm), two respond maximally to blue light (430 nm), and the remaining two respond best to ultraviolet light (340 nm). This arrangement should enable the honeybee to distinguish colors (except red) and behavioral studies show that this is the case (Fig. 26.4).

Of what value is ultraviolet vision? Television camera tubes are also sensitive to ultraviolet light, as well as visible light, but the lens with which they are usually equipped is opaque to ultraviolet. (This is why you can't get tanned—or synthesize calciferol—from the sunlight passing through window glass.) However, using a special ultraviolet-transmitting lens, Eisner and his coworkers at Cornell have demonstrated that many insect-pollinated flowers appear to the honeybee quite different from the way they appear to us (Fig. 26.5). The revelation of sharp

FIG. 26.5 Blackeyed susan photographed in visible light (left) and ultraviolet light (right). (Courtesy of Dr. Thomas Eisner.)

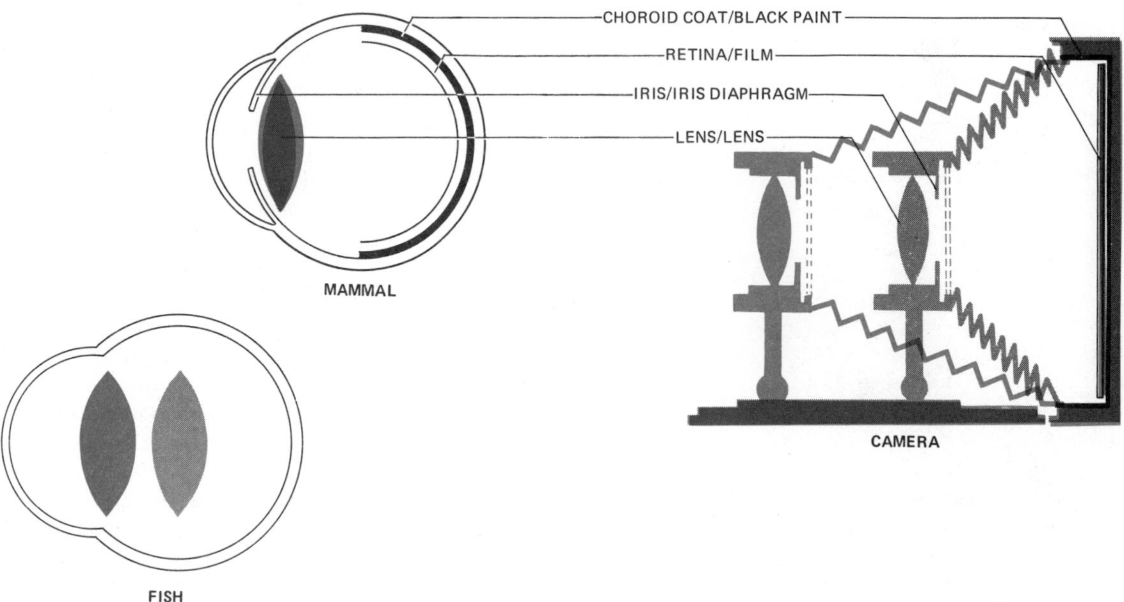

FIG. 26.6 A comparison of two vertebrate eyes and a camera. Each is shown focused on a distant object (gray) and a near object (color). The eyes of amphibians, snakes, and most mollusks work like that of the fish; that is, focus is changed as it is in a camera. Birds and mammals accommodate by changing the curvature of the lens.

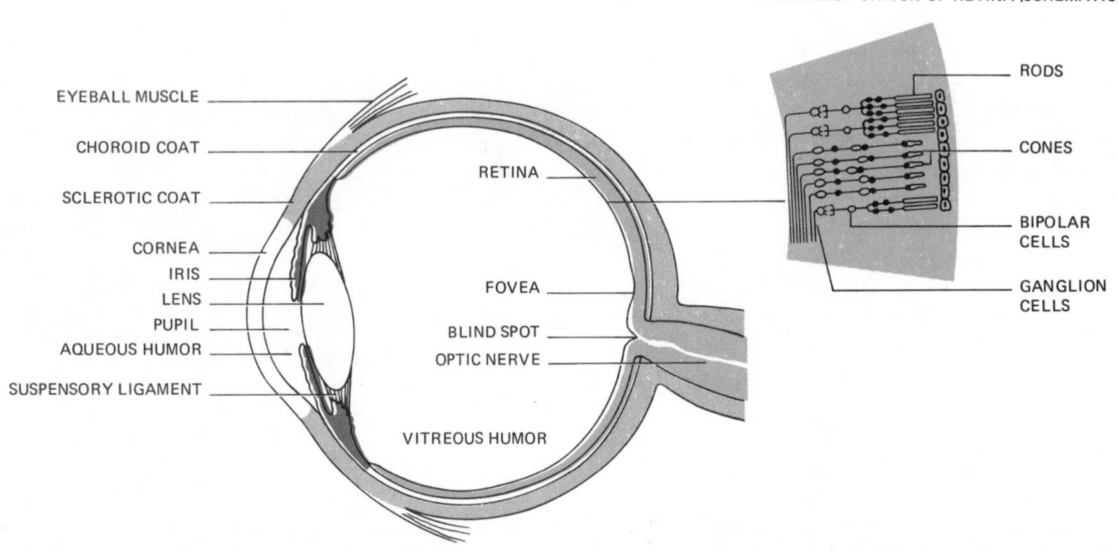

FIG. 26.7 The human eye. The actual light receptors are the rods and cones.

contrasts between flowers that appear similar to us partly explains the efficiency with which honeybees secure nectar from only one species of flower at a time even when other species are also in bloom. It is also interesting to note that flowers that are pollinated by birds and bats (neither of which have ultraviolet vision) do not have special ultraviolet patterns.

26.3 STRUCTURE OF THE HUMAN EYE

The eyes of both mollusks and vertebrates operate on the same basic principle as a camera. A single lens focuses light from all parts of the visual field onto a sheet of light-sensitive cells (Fig. 26.6). Despite the close similarity in structure and function between the eyes of mollusks and vertebrates, all the evidence indicates that they arose and evolved quite separately in the two groups.

The human eye is roughly spherical in shape. It is bounded by three distinct layers of tissue (Fig. 26.7). The outer layer, the **sclerotic coat,** is extremely tough. It is white in color (the "white" of the eye) except in the front. Here it forms the transparent **cornea** which admits light into the interior of the eye and bends the light rays so that they can be brought to a focus. The surface of the cornea is kept moist and dust-free by the secretion from the tear glands.

The middle coat of the eye, the **choroid coat,** is deeply pigmented with melanin and well-supplied with blood vessels. It serves the very useful function of stopping the reflection of stray light rays within the eye. This is the same function accomplished by the dull black paint within a camera.

In the front of the eye, the choroid coat forms the **iris.** This, too, may be pigmented and is responsible for the "color" of the eye. An opening, the pupil, is present in the center of the iris. The size of this opening is variable and under automatic control. In dim light (or times of danger—see Section 25.11) the pupil enlarges, letting more light into the eye. In bright light, the pupil closes down. This not only protects the interior of the eye from excessive illumination, but improves its image-forming ability and depth of field. Photographic enthusiasts, too, make a practice of "stopping down" the iris diaphragm of their cameras to the minimum permitted by the amount of light available in order to get the sharpest possible pictures.

The inner coat of the eye is the **retina.** It contains the actual light receptors, the rods and cones, and thus functions in the same way as the film of a camera.

The **lens** of the eye is located just behind the iris. It is held in position by suspensory ligaments (Fig. 26.7). Ordinarily, these are kept under tension and the lens is correspondingly flattened. However, contraction of muscles attached to these ligaments relaxes them and permits the lens to take on a more nearly spherical shape. These changes in lens shape enable the eye to shift its focus (accommodate) from far objects to near objects and vice versa.

Some people have a difficult time bringing light rays to focus on the retina. If the eyeball is too short, or the lens too flat or too inflexible, the light rays entering the eye will not be brought to a focus by the time they strike the retina (Fig. 26.8). This condition is known as farsightedness because nearby objects are particularly difficult to bring into focus. The use of eyeglasses fitted with convex lenses corrects the condition by helping the eye's own lens to converge light rays more rapidly. Aging people are particularly apt to become farsighted as their lenses become less flexible.

Too long an eyeball or too spherical a lens causes nearsightedness. The image of distant objects is brought to a focus in front of the retina and is out of focus again before the light actually strikes the retina (Fig. 26.8). Nearby objects can be seen more

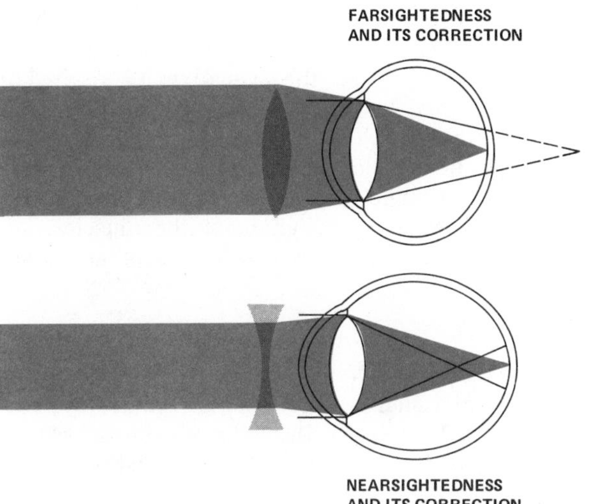

FIG. 26.8 Common eye defects. The path of light rays without glasses is shown in black; with glasses, in color. The lenses of modern eyeglasses are not so simple in shape as those shown but function in the same way.

easily. Eyeglasses with concave lenses correct this condition by diverging the light rays somewhat before they enter the eye.

The method of changing focus by changing the shape of the lens has no parallel in photography. Focus is changed in cameras by moving the position of the entire lens with respect to the film. This method is also used in the eyes of some fishes, amphibians, snakes, and some mollusks.

If either the lens or the cornea has any irregularities in its curvature, all the light rays entering the eye will not be brought to a focus together. This defect is known as **astigmatism.** It is corrected by specially ground eyeglasses which compensate for the irregularities.

For reasons which are still obscure, some people develop cloudy areas in one or both lenses. These are called **cataracts** and they cause partial or complete blindness. Today eye surgeons can usually restore vision by removing the defective lenses surgically and giving the patient extra-"strong" eyeglasses to compensate for their loss.

The iris and the lens divide the interior of the eyeball into two main chambers. The anterior one is filled with a watery fluid, the **aqueous humor.** The posterior chamber is filled with a jellylike material of marvelous clarity, the **vitreous humor.**

Movement of the eyeball is accomplished by three pairs of muscles, the members of each pair working antagonistically. The coordinated action of these muscles enables the eyeball to be rotated in any direction. Thus we are able to train both eyes in a single direction. This produces two slightly differing views of the same scene which our brain is able to fuse into a single, three-dimensional (stereoscopic) image. Improper coordination of the muscles controlling the eye produces such defects as "crosseyes."

26.4 DETECTION OF LIGHT

The actual visual receptors of the eye are the rods and cones. These are cells which are stacked closely together just beneath the surface of the retina.

1. The rods. There are approximately 100 million rods in each eye. They are used chiefly for vision in dim light and are extremely sensitive to light. The image produced by the rods is, however, not a sharp one. Careful microscopic study of the structure of the retina provides an explanation for this. The rods function in groups. In other words, a number of rods share a single nerve circuit to the brain. A single rod can initiate an impulse in that circuit but there is no way for the brain to determine which rod in the cluster was involved (Fig. 26.7).

For light to be absorbed, there must be a light-absorbing substance, a pigment. The pigment in the rods is **rhodopsin.** It is incorporated in membranes that are neatly stacked in the outer portion of the rod (Fig. 26.9). This arrangement is similar to that which we found in the thylakoids of chloroplast grana (see Section 9.3), another light-absorbing device.

Rhodopsin consists of a protein molecule, opsin, to which is attached a molecule of the carotenoid **retinal.** Retinal itself is light orange, but when attached to opsin, the deep reddish-purple color of rhodopsin is produced. When light is absorbed by rhodopsin, retinal is split from the protein, and the rod becomes partially bleached. At the same time, a nerve impulse is initiated. It has been shown that several steps are involved in the breakdown of rhodopsin. The first is a "light" reaction. The later steps are "dark" reactions, that is, they do not require light. A similar situation occurs in photography. Opening of the camera shutter exposes the film, but no visible image is produced. Only the subsequent chemical changes of the developing process can convert this "latent" image into the visible image of the negative.

Although the rods provide us with a relatively coarse, colorless image, they are extremely sensitive to the presence of light. They are capable of detecting light a billion times less intense than that which our eyes receive on a bright, sunny day. The more light which strikes the rods, however, the more rhodopsin is bleached and the less sensitive the rods become. Fortunately, the process is reversible. Some rhodopsin may be resynthesized directly from its breakdown products, retinal and opsin. There is also evidence that fresh retinal is continually being manufactured in the eye by the oxidation of vitamin A. The body reserves of vitamin A thus provide a large reservoir for the synthesis of retinal. It is no wonder, then, that vitamin A deficiency is usually associated with nightblindness, the inability to see in the dark.

In order for the rods to become as sensitive as possible ("dark-adapted") it is necessary that the rate of rhodopsin synthesis exceed the rate of its breakdown. This means that bright light must be excluded from the eye. We all are aware of how difficult it is to see in a dimly lighted room immediately after entering from a brightly lighted one.

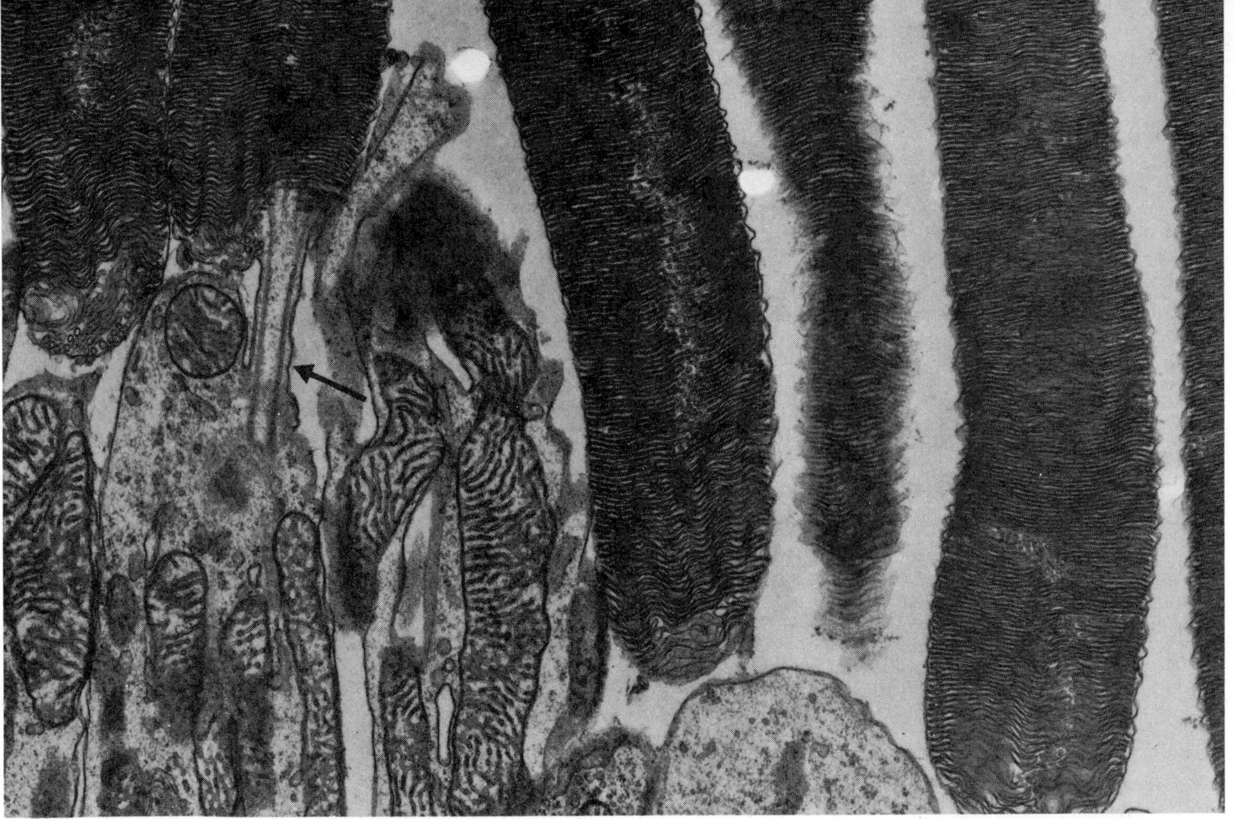

FIG. 26.9 Rod cells of a kangaroo rat (25,000 ×). The outer segments of the rods contain orderly stacks of membranes in which the visual pigment is incorporated. The inner portions contain many mitochondria. The two parts of the rod are connected by a stalk (arrow) that has the same structure as a cilium. (From Porter and Bonneville: *An Introduction to the Fine Structure of Cells and Tissues,* 4th ed., Lea & Febiger, 1973.)

It takes some thirty minutes in the dark for our eyes to become fully adapted to the dark.

For night-flying aviators during World War II, the necessity of remaining in the dark for a period before the start of a dangerous mission was nerve-racking indeed. The realization that the rods are insensitive to red light produced a nice solution, however. Red goggles permitted the aviator to carry on normal activity while still enabling his rods to become dark-adapted. The goggles permitted only red light to reach his retina; this stimulated his cones but not his rods.

2. The cones. We know less about how the cones work than about how the rods work. The cones are especially abundant (about 150,000 in each square millimeter) in a single region of the retina, the **fovea,** a region just opposite the lens. Unlike the rods, the cones operate only in bright light. Furthermore, they enable us to see colors. As we noted in Section 26.2, at least two kinds of cones must be present in order to detect any colors at all. Each must contain a pigment that absorbs a certain wavelength best. Actually, with three kinds of cones, it would be possible to have full color vision if each kind contained a pigment that best absorbed one of the three primary colors: red, green, and blue. Theoretically, the brain could mix three primary color sensations to produce any of the more than 17,000 different hues that the well-trained eye can distinguish. And, in fact, the presence in the fovea of a red-absorbing pigment (with its maximum absorption at 575 nm), a green-absorbing pigment (535 nm), and a blue-absorbing pigment (445 nm) has been demonstrated. Retinal is the prosthetic group for each of these. It is differences in the protein, opsin, to which the retinal is attached that account for the differences in absorption. A single cone contains only one of the three pigments. Working together, the red-, green-, and blue-absorbing cones in the fovea provide the basis for color vision.

These discoveries relate nicely to our knowledge

of color blindness. As you learned in Section 16.3, red-green color blindness is an *X*-linked, recessive trait. Actually, there are two kinds of color-blind people who confuse reds and greens: those lacking the red-absorbing pigment and those lacking the green-absorbing pigment. Since this is an *X*-linked trait, most of the victims in each group are men. Perhaps the recessive gene on their *X* chromosome is coding for a defective opsin and hence a functional red- or green-absorbing pigment is not produced. There is good evidence that the gene for each of these pigments is located at a separate locus on the *X* chromosome.

Far rarer is the case of color blindness involving an absence of the blue-absorbing pigment. The few cases that have been found involve women almost as often as men, so the defective gene is probably located on an autosome. Furthermore, this gene appears to be inherited as a dominant.

In addition to providing the basis for color vision, cones provide us with our most acute vision. The number of cones sharing a circuit to the brain is far fewer than is the case for the rods. Furthermore, the cones are very densely packed in the fovea. Other tissues, such as blood vessels, are absent from this portion of the retina and thus do not interfere with the reception of a distinct image. The image is, however, distinct (and colorful) over just a small area of view. Our ability to direct our eyes quickly to anything in view that interests us tends to make us forget just how poor our peripheral vision is.

All the nerve impulses generated by the rods and cones travel back to the brain by way of neurons in the **optic nerve.** At the point on the retina where the approximately one million neurons converge on the optic nerve, there are no rods or cones at all (Fig. 26.7). This spot, the blind spot, is thus insensitive to light. With the marks in Fig. 26.10 you can demonstrate the presence of the blind spot for yourself. The blind spots of our two eyes do not receive the same portions of the visual image, so that each eye compensates for the blind spot of the other.

26.5 HEAT RECEPTORS

Heat is electromagnetic radiation of wavelengths longer than those of light (Fig. 26.1). The human skin is very sensitive to gain or loss of heat. Distributed discretely through the skin are receptors that, when warmed, give rise to the sensation of warmth. Other receptors give rise to the sensation of cold when they are appropriately stimulated. The location of these two classes of receptors can be mapped by using blunt metal probes dipped in hot water and a salt-ice-water mixture respectively. The skin is richly supplied with sensory neurons, many of which are connected to specialized receptor structures (Fig. 26.11). However, it has not yet been possible to show with certainty whether any of these are exclusively responsible for detecting changes in temperature.

Some snakes contain remarkably sensitive heat receptors located in two pits on the face. These so-called pit vipers include the rattlesnakes (Fig. 26.12), cottonmouth, and copperhead found in North America. The receptors aid the snakes in detecting warm-blooded prey in the dark. Rattlesnakes can strike accurately at a mouse in complete darkness.

The human body also has receptors that detect internal temperature changes. Two sensitive "thermostats" are located in the hypothalamus of the brain. The receptor cells in one of these respond to small (0.01°C) increases in the temperature of the blood. As a result of their response, all the activities by which the body cools itself (dilation of the blood vessels in the skin, sweating, etc.) are brought into play when the temperature of the blood begins to rise. It is this temperature-detecting center which enables us to maintain a constant body temperature (homeothermy) during periods of exertion and/or high environmental temperatures. A second receptor in the hypothalamus maintains homeothermy when the body is exposed to chilling. The action of the skin temperature receptors supplements, but cannot replace, the action of the blood temperature receptors in the hypothalamus.

FIG. 26.10 A demonstration of the blind spot Cover your left eye with your hand and, holding the book at arm's length, stare at the cross with your right eye. What happens to the circle as you slowly move the book toward you?

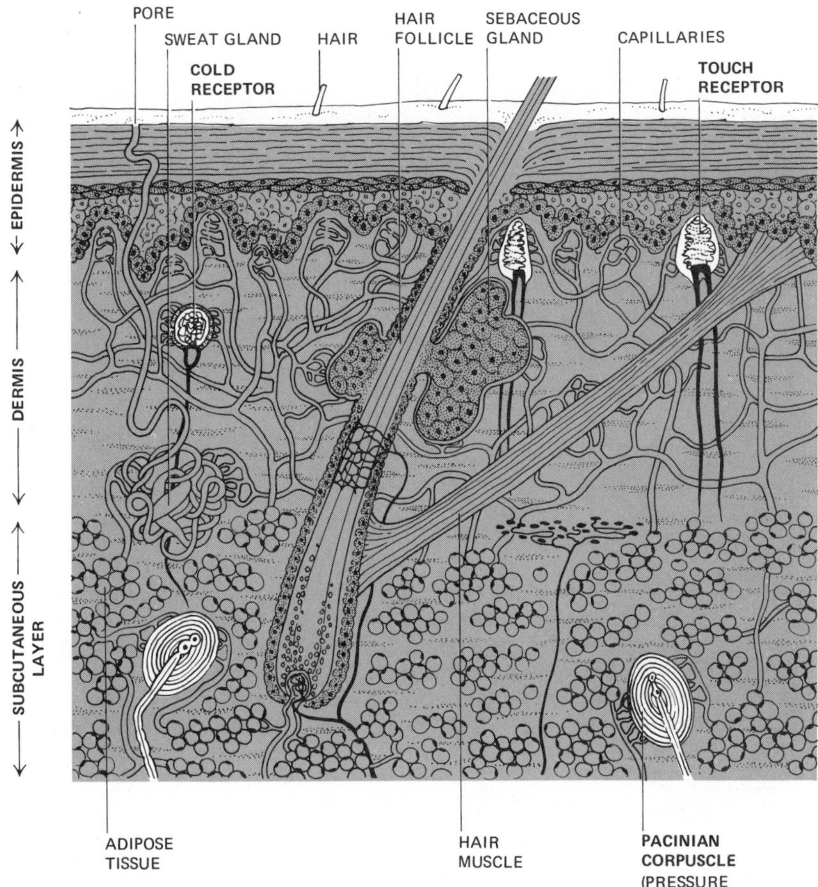

FIG. 26.11 Human skin and its sense receptors.

FIG. 26.12 The western diamondback rattlesnake. Note the sense receptors concentrated in the head. The eyes respond to visible radiation. The pits, which are located below the nostrils, detect infrared (heat) radiation. The tongue samples the air for molecules of odorous substances. (Courtesy of New York Zoological Society.)

Mechanoreceptors

A wide variety of receptors of mechanical stimuli are found in animals. Each initiates nerve impulses when it is physically deformed by an outside force.

26.6 TOUCH AND PRESSURE

Touch. In humans, light **touch** is detected by receptors located close to the surface of the skin. These are often found next to a hair follicle (Fig. 26.11). Even if the skin is not touched directly, movement of the hair is detected by the receptor. Touch receptors are not evenly distributed over the surface of the body. The skin of the fingertips may contain as many as 100 per square centimeter and the tip of the tongue is similarly well supplied. The concentration of touch receptors in other locations is apt to be much lower. The back of the hand, for example, has fewer than 10 per square centimeter. The exact location of touch receptors can be determined by gently touching the skin with a stiff bristle and marking those spots where the subject detects a distinct touch. An interesting variation on this technique can be carried out with a pair of dividers such as are used in mechanical drawing. With a blindfolded subject, determine the minimum separation of the points which will give rise to two separate touch sensations. You will find that the subject's ability to discriminate the two points is far better on the fingertips than on, say, the small of the back.

Pressure. Because of its accessibility and relatively large size, one of the easiest receptors to study is the **Pacinian corpuscle.** These receptors are located in the skin (Fig. 26.11) and also in various internal organs. Like other receptors, each is connected to a sensory neuron. Isolating a single Pacinian corpuscle with its attached neuron provides a convenient means by which to study its properties (Fig. 26.13).

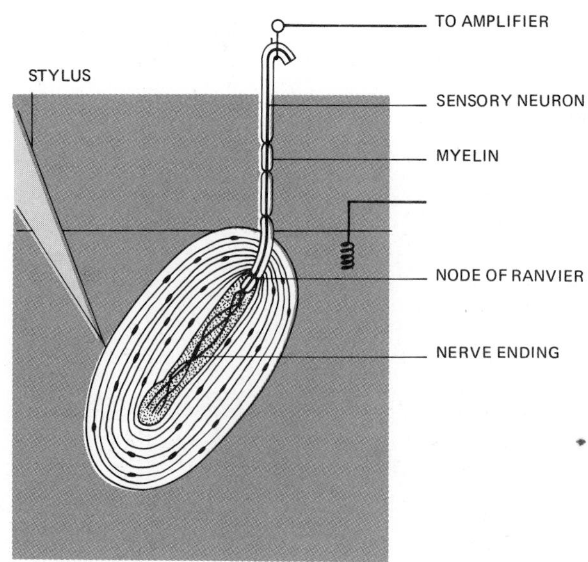

FIG. 26.13 Measuring the electrical response of an isolated Pacinian corpuscle. Mechanical pressure of varying strength and frequency can be applied to the corpuscle by the stylus. The electrical activity is detected by electrodes attached to the preparation. At threshold, a nerve impulse is generated at the first node of Ranvier.

The Pacinian corpuscle is a **pressure receptor.** Applying pressure to the corpuscle deforms it. This creates a tiny current of electricity in the sensory neuron originating within it. The greater the deformation of the corpuscle, the greater this **"generator potential."** If the generator potential becomes sufficiently large, it initiates an impulse in the sensory neuron. We say that the neuron's **threshold** has been reached.

While the generator potential is a graded response, the response of the sensory neuron is not. If its threshold is reached, it fires. If not, it doesn't. Thus a continuously graded input (an *analogue* function, in the terminology of computer science) is converted into an on/off (digital) output. How, then, can the Pacinian corpuscle inform the brain of the size of the stimulus? Two mechanisms are at work. The more massive or rapid the deformation of a single corpuscle, the higher the frequency of impulses generated in the attached neuron. Furthermore, in the *intact* organism a massive stimulus stimulates a number of nearby Pacinian corpuscles, thus increasing the number of activated sensory pathways leading back to the brain.

When pressure is first applied to the Pacinian corpuscle, it initiates a volley of impulses in its sensory neuron. However, with continuous pressure, the rate of impulse propagation decreases quickly and soon stops. The receptor has become *adapted* to the stimulus. Rapid adaptation is a property of most of our sense receptors. It is, for example, readily apparent in the speed with which we cease to detect an odor to which we are exposed. Sensory adaptation is a useful function because it prevents our nervous system from being continuously bombarded with information about such insignificant matters as the touch and pressure of our clothing. Remember that we defined a stimulus as a *change* in the environment, and it is change that our sense receptors detect. In fact, quickly removing the pressure from an adapted Pacinian corpuscle triggers a fresh volley of impulses.

Even if we repeatedly apply and then remove the pressure exerted on a Pacinian corpuscle (e.g. vibrate it 500 times per second for 15 seconds or so), the corpuscle will finally cease to respond. It has become **fatigued.** However, if the corpuscle is rested for a time, it will regain full sensitivity.

All our other sense receptors share most of the properties we have found characteristic of the Pacinian corpuscle. Each type of receptor is constructed in such a way that it normally responds to (has a lower threshold for) one particular category of stimulus (e.g. light, pressure) and not others. Each is connected to a sensory neuron in which it generates nerve impulses, the frequency of which is a measure of the magnitude of the sensory input.

Proprioceptors. In most cases, as we have said, our receptors adapt quickly, i.e. cease to respond, to a constant level of input. However, certain sense receptors do not behave in this manner. Among these are the **proprioceptors.** These are sense receptors which are distributed throughout skeletal muscle and tendons. Stretching or contraction of muscles triggers these receptors to initiate nerve impulses. These, in turn, enable the brain to determine the state of contraction of the muscle. If one starts to lose his balance, the brain is informed by the proprioceptors of the legs and corrective action is taken at once. Complex muscular movements, such as are involved in typing, catching a ball, playing the violin, etc., would be impossible without the proprioceptors. The properly timed, coordinated action of a variety of muscles requires that the brain be continually informed of the performance of each. If you ever have had one or both legs "go to sleep," you have some idea of how difficult locomotion would be without proprioceptors.

Pain. Massive mechanical stimulation of the skin produces the sensation of pain. Excessive heat, cold, and certain chemicals do this also. The sensation may be produced by the stimulation of a network of nerve fibers in the skin that are not attached to specialized stimulus detectors and thus do not respond unless the stimulus is very strong. On the other hand, pain may be felt as a result of a change in the frequency and pattern of the signals passed to the central nervous system by the specialized skin receptors of touch, pressure, heat, and cold. Perhaps both mechanisms participate in the process.

26.7 HEARING

The ability to hear is the ability to detect mechanical vibrations which we call sound. Under most circumstances, these vibrations reach us through the air. The external ear (Fig. 26.14) may aid somewhat in concentrating sound waves. These then pass down the auditory canal and strike the eardrum or **tympanic membrane,** causing it to vibrate. The vibrations of the tympanic membrane are transmitted across the middle ear by three tiny, linked bones, the **ossicles,** which also serve to concentrate the vibrations.

The middle ear is filled with air and is connected to the outside air by means of a **Eustachian tube** opening into the nasopharynx. This opening permits the air pressure on both sides of the tympanic membrane to be kept equal. The "popping" of our ears that we feel when we rapidly change altitude in an unpressurized airplane or an elevator results from the sudden equalizing of pressure when the Eustachian tubes open during swallowing or yawning. Victims of head colds may have inflamed Eustachian tubes which are temporarily prevented from opening. Change in altitude may be very painful at such times because of unequal pressure against the tympanic membranes.

Mechanical vibration of the innermost ossicle, the stirrup, is transmitted through a flexible membrane (the oval window) to the **cochlea** of the inner ear. The cochlea is a tube, about 3 cm in length, which is coiled like a snail shell (Fig. 26.14) and filled with lymph. Running through the cochlea for almost its entire length is a plate of bone and an inner tube which is also filled with lymph. These structures divide the outer tube of the cochlea into two separate chambers. Vibrations of the oval window are transmitted to the fluid in these outer chambers. Because liquids are practically incompressible, it is necessary to have some way of relieving the pressures created when the oval window is pushed in and out. The flexible *round* window accomplishes this by moving the opposite way (Fig. 26.14).

Lying within the inner, or middle, chamber of the cochlea is the **organ of Corti.** It contains thousands of sensitive "hair" cells which are the actual vibration receptors. They are located between the basilar and tectorial membranes (Fig. 26.14). Vibrations in the cochlear fluid cause vibrations in the

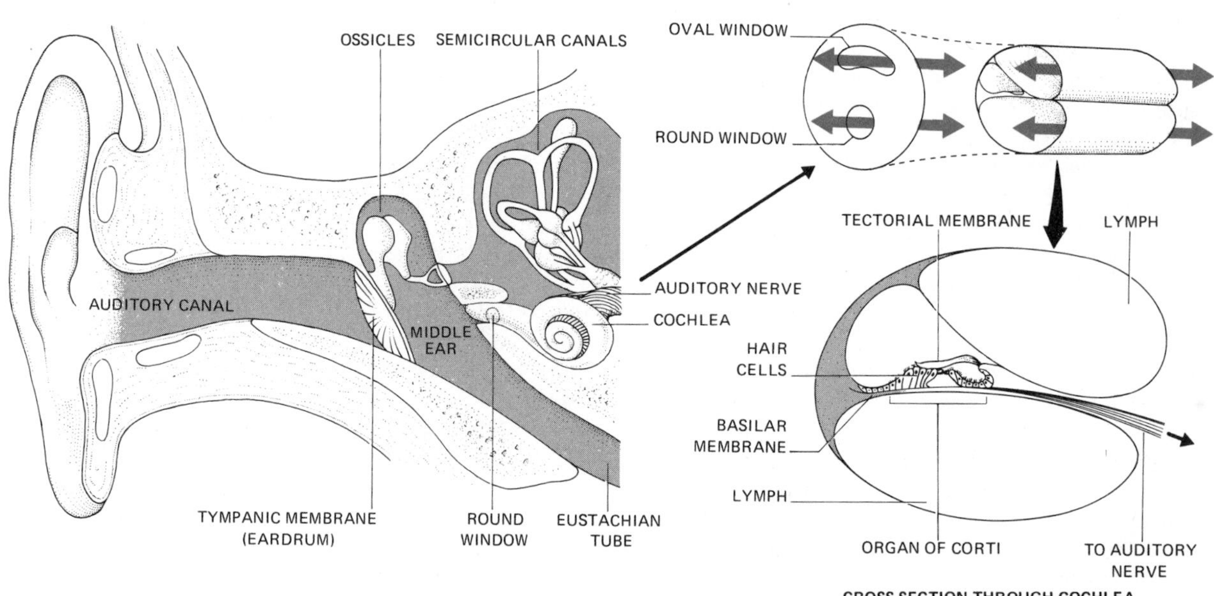

FIG. 26.14 The human ear.

basilar membrane. This moves the sensitive hair cells against the tectorial membrane, thus stimulating them. Electrical impulses arising in these cells then initiate nerve impulses which travel back along the **auditory nerve** to the brain.

The ear is a remarkably precise and versatile sense receptor. Many people, especially when young, can hear sounds with frequencies (pitches) from as low as 16 to as high as 20,000 hertz (cycles per second). Furthermore, the ear can detect sounds over a wide range of intensities. The loudest sound we can hear comfortably is over one trillion times as loud as the faintest we can detect. The faintest sound is so faint that if our ears were any more sensitive, we would probably detect the sound of random molecular collisions (Brownian movement) within the ear. The power of discrimination of frequencies is also great. A trained musician can distinguish about 15,000 pitches.

The way in which the organ of Corti discriminates among different pitches is now fairly well understood. At first glance, it might seem appropriate for the hair cells to send impulses back to the brain at the same frequency as the sound. Something of this sort may actually occur at very low frequencies. It could not occur at frequencies greater than about 1000 hertz because, as we shall see in the next chapter, sensory neurons cannot conduct impulses any faster than that. Actually, even before this limit is reached, the basilar membrane and hair cells begin to respond selectively to the sound frequencies. Low frequencies stimulate the area of the organ of Corti nearest its tip. High frequencies are detected near its base. The intermediate frequencies are detected in an orderly, progressive fashion from one end of the organ of Corti to the other. Evidence to support this view has been obtained by exposing laboratory animals to very intense, pure tones. Eventually the animals become deaf to these frequencies although their ability to hear other pitches is not impaired. In every case, examination of the organ of Corti reveals destroyed hair cells in a single area whose location can easily be correlated with the pitch of the destructive sound.

The hearing ability of bats is extraordinary. The research of zoologist Donald Griffin has shown that bats can hear frequencies as high as 150,000 hertz. Sound at such ultrasonic frequencies travels in fairly straight lines. Bats, flying in complete darkness, are able to locate obstacles (Fig. 26.15) and even insect prey by emitting pulses of this ultrasonic sound and then adjusting their course of flight to the echo which returns to their ears. Such a sysem of echolocation works on the same principle as the underwater sonar devices developed for submarine detection during World War II.

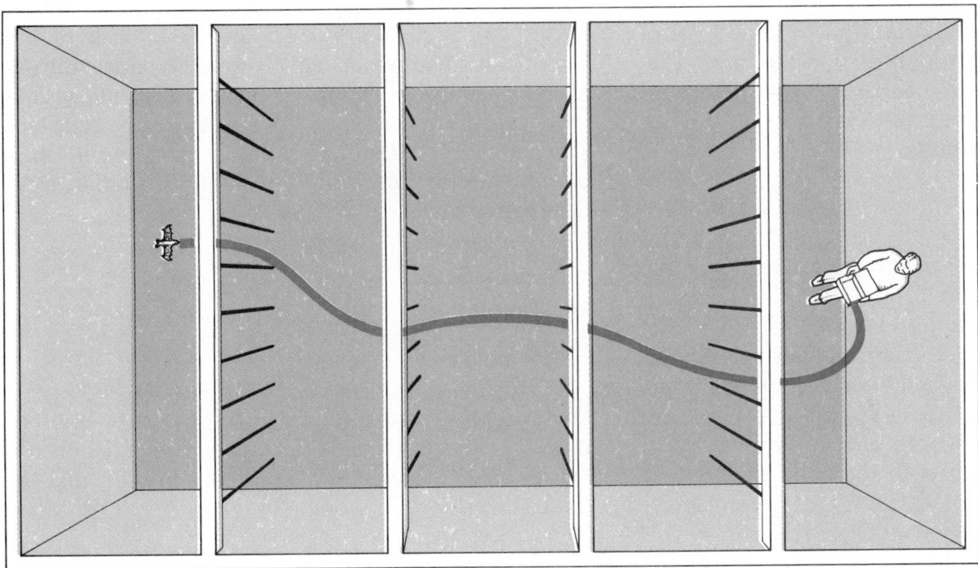

FIG. 26.15 Echo location in the bat. A blindfolded bat can fly between the wires, touching them only rarely. A bat whose ears are plugged collides repeatedly with the wires.

FIG. 26.16 Hunter and hunted. In the top photograph, a moth (bright streak) takes successful evasive action upon detecting the approach of a bat (broad streak across the photograph). (The diffuse image is a tree in the background.) In the bottom photograph, the two streaks intersect, indicating that this time the moth was unable to escape capture by the bat. (Photographs by Frederic A. Webster, courtesy of Professor Kenneth D. Roeder.)

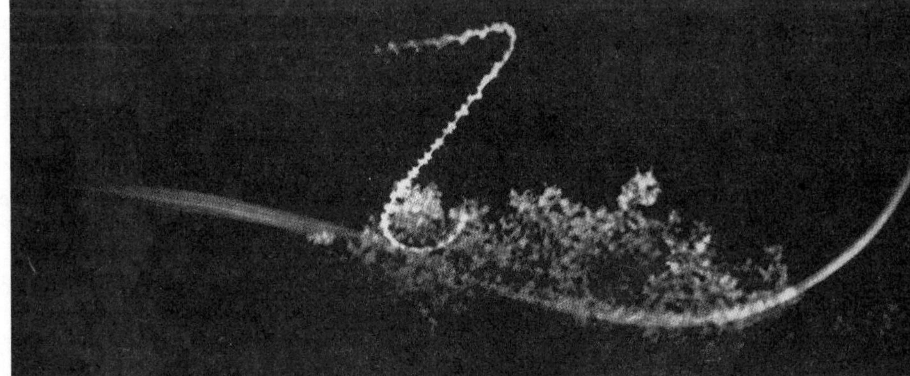

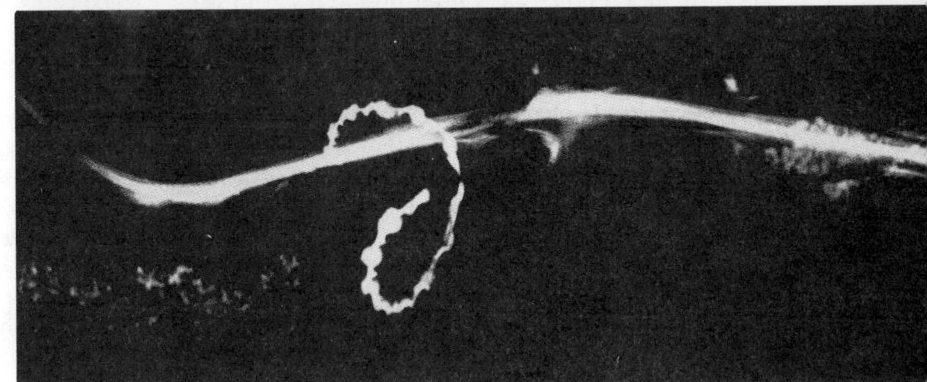

Hearing plays an important part in the lives of other animals, too. Avoidance of predators (Fig. 26.16), the location and courtship of a mate, and the staking out of territorial "claims" may all involve the detection of sounds.

26.8 EQUILIBRIUM

The inner ear also detects: (1) the position of the body with respect to gravity and (2) the motion of the body. These two functions are quite distinct from its function in sound detection and rather distinct from each other. Just above the cochlea are two interconnecting, lymph-filled sacs (Fig. 26.14). These are lined with hair cells which, in turn, are connected to sensory neurons. Attached to the hairs of the hair cells are tiny spheres of calcium carbonate. These are acted upon by gravity and deflect the hairs downward. As the body (or at least the head) is oriented in different directions, the "ear stones" shift their position. The nerve impulses that are initiated by the hair cells are sent back to the brain and inform it of the change.

Analogous (but not homologous) structures, called **statocysts,** are found in many aquatic invertebrates. In the crayfish, these hair-lined sacs contain tiny grains of sand instead of ear stones. When the crayfish sheds its exoskeleton, it also sheds its statocysts. As soon as the new exoskeleton is hardened, however, the crayfish seeks out grains of sand to place within its new statocysts. If iron filings are offered to the crayfish instead of sand, it will use these. When the job is complete, a few minutes of experimentation with a strong magnet will vividly demonstrate the action of the statocysts in maintaining balance. Placing the magnet just above the crayfish draws the iron filings upwards. This, in turn, triggers receptors which would normally be affected only if the animal were upside down. Consequently, the crayfish does, in fact, turn itself upside down in response to the erroneous information received by its central nervous system.

Motion of the human body is detected in the three semicircular canals at the top of each inner ear (Fig. 26.14). These are three fluid-filled tubes, each one of which is oriented in one of the three planes of space. At one end of each canal is a small chamber containing sensory hair cells. Whenever the head is moved, the semicircular canals move, too. The fluid within lags in its motion, however, and consequently there is relative motion between the walls of the canals and the fluid. This motion stimulates the hair cells to send impulses back to the brain. The maintenance of proper balance during athletic activity would be practically impossible without this mechanism. When the hair cells are stimulated in unfamiliar ways, as may occur in a boat or aircraft during rough weather, motion sickness can occur.

Chemoreceptors

Our receptors of chemicals in the external environment are the taste buds, located principally on the tongue, and the olfactory epithelia, located high in the nasal cavity.

26.9 TASTE

In order for a substance to be tasted, it must be soluble in the moisture of the mouth. Only when in solution can it then stimulate the **taste buds.** Four types of these can be distinguished morphologically. Most of them are located on the surface of the tongue although a few are found on the soft palate, high in the back of the mouth.

Most experimenters agree that there are only four primary taste sensations: sweet, sour, salty, and bitter. By using dilute solutions of sucrose, hydrochloric acid, sodium chloride, and quinine sulfate, respectively, one finds that each of the four primary tastes is localized in a special area of the tongue (Fig. 26.17). However, mapping experiments of this sort also show considerable overlapping of taste areas and considerable variation from one person to the next.

The existence of four kinds of taste buds and four primary tastes suggests that each type of bud is responsible for one specific taste. With the possible exception of the bitter taste, however, there seems to be no correlation between bud type and the taste detected.

You may well argue that you can detect more than just four tastes. So you can, but this involves other factors. First of all, combinations of the four primary tastes produce new tastes. More important is the role which smell, temperature, and touch receptors play in the tasting process. As one chews food, vapors probably escape through the oral pharynx into the nasal cavity and are detected by our odor receptors. The marked loss of taste which we experience when our nasal cavities are plugged with mucus during a cold supports this point. You can demonstrate it vividly under more pleasant circumstances by applying water, in which two or three cloves have been boiled, to the tongue of a blindfolded subject. He will taste the mixture readily when his nasal passages are open. When holding his nose, however, he will have a very difficult time distinguishing the clove solution from plain water. The temperature and texture of food also play an important part in our sensation of taste.

Many insects have a well-developed sense of taste. The red admiral butterfly can taste a 0.000078-molar solution of sucrose, which is far too dilute for us to taste. Its taste receptors are located on its legs. Other insects have taste receptors on their antennae and mouth parts.

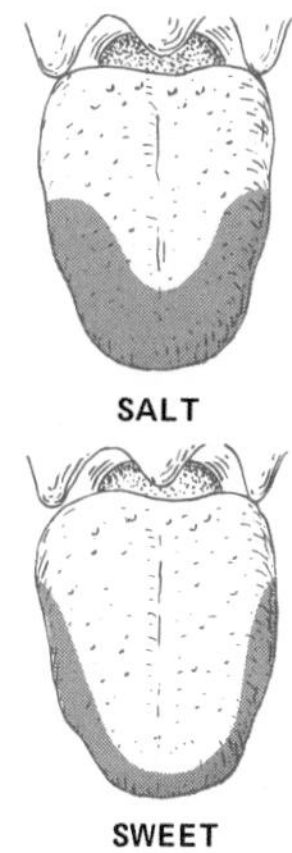

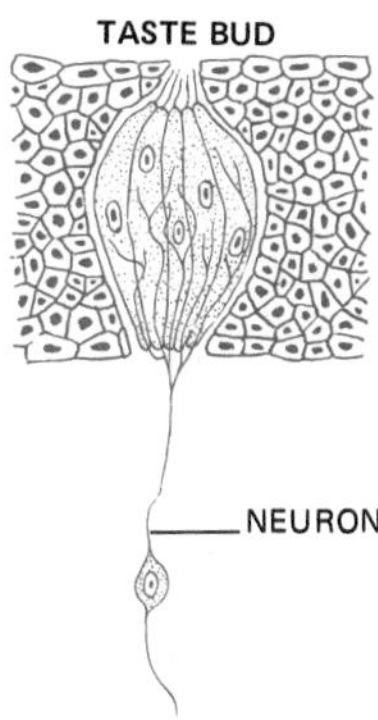

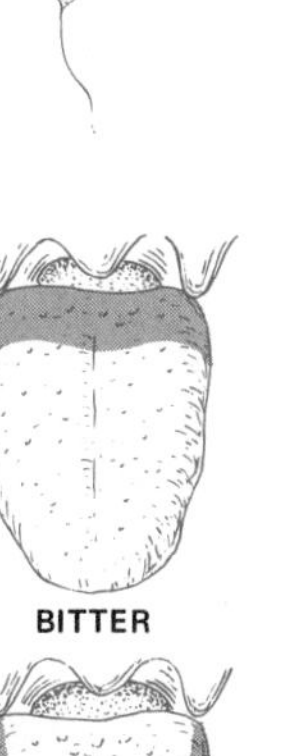

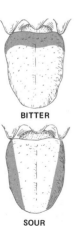

FIG. 26.17 The sense of taste is initiated in the taste buds of the tongue. When stimulated by dissolved chemicals, each initiates one of the four primary taste sensations. While the distribution of taste areas shown is typical, there is considerable variation from person to person.

26.10 SMELL

Humans detect odors by means of receptor cells located in the two **olfactory epithelia** high in the nasal cavity (Fig. 26.18). Each of these areas is about the size of a postage stamp, 250 mm². Air drawn in through the nostrils passes over them. Water- and fat-soluble molecules present in the air dissolve in the mucus layer covering the epithelia and give rise to sensations of smell. Vigorous sniffing improves the exposure of the olfactory epithelia to airborne substances.

It is customary to consider our sense of smell one of our poorest senses. It is true that the sensitivity and the power of discrimination (the ability to distinguish between similar odors) of such animals as the dog and deer are somewhat better than that of humans. Nevertheless, we are able to detect a virtually unlimited variety of odors (but just one at a time!) and in many cases at very low thresholds. We can, for example, detect as little as 0.0000000002 g of vanillin (the active ingredient in vanilla flavoring) vaporized in 1000 liters of air.

The mechanism by which we are able to detect such a great variety of different odors has puzzled scientists for a long time. Only two kinds of receptor cells can be distinguished in the olfactory epithelium by their *structure*. It seems likely, though, that several (perhaps seven) can be distinguished by their *function*. According to one theory, each of these seven kinds of receptors responds to molecules of a particular class. In most cases, the shape of the molecule determines what class it is in and thus to which receptor it will become temporarily attached.

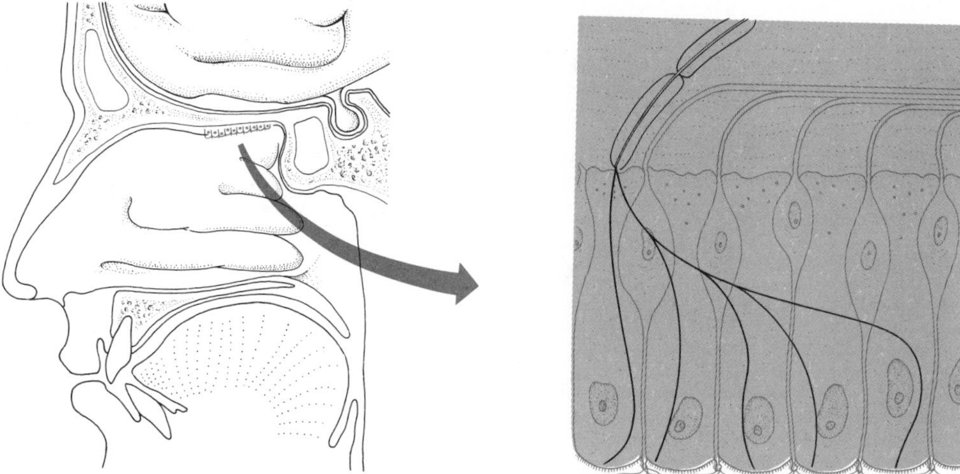

FIG. 26.18 Left: The sense of smell originates in the olfactory epithelium high in the nasal cavity. Right: Although only two kinds of receptor cells can be distinguished, we are capable of discriminating between a wide variety of odorous substances.

Each class of molecules produces a *primary* odor such as musky, pepperminty, pungent, etc. Complex odors arise when the molecules have a shape that permits them to attach to more than one receptor. They can also arise when a variety of molecules are being given off by the odorous substance. Many odors do, in fact, represent the combined effect of a complex array of chemicals. For example, over 100 substances participate in the production of the odor of geraniums.

Part of the explanation for the low regard we have for our sense of smell is that it does not play a very important part in our lives. Other animals, however, depend greatly on smell to enable them to locate mates, locate food, and escape predators. The male silkworm moth can smell the pheromone released by a female moth two or three miles upwind. Its odor receptors, like those of most insects, are located on the antennae.

The remarkable ability of Pacific Coast salmon to return, after a span of four or five years in the sea, to the same freshwater streams in which they were born probably involves the sense of smell (or taste—there really is not much of a distinction between the two for an animal that lives surrounded by water). Odor receptors in the salmon as well as in most bony fishes are located in two small chambers just in front of the eyes. Water enters and leaves each of these chambers through separate openings, the nostrils. It is interesting to realize that nostrils first served an odor-detecting function in our vertebrate ancestors. Only later, when vertebrates became air-breathing, did the nostrils come to be used to take air to and from the lungs.

Snakes and lizards have a well-developed odor (or taste?) receptor organ, Jacobson's organ, located in the roof of the mouth. They alternately stick their tongue out into the air and then into their Jacobson's organ. Thus they "taste" the air and detect the presence of odors (look back at Fig. 26.12).

26.11 INTERNAL CHEMICAL RECEPTORS

Humans also have receptors that detect chemical changes in the *internal* environment. In the carotid arteries (in addition to the pressure receptors mentioned earlier) are cells which are sensitive to increased concentrations of carbon dioxide and other cells that detect decreased concentrations of oxygen. When stimulated, both types initiate nerve impulses that eventually increase the rate of breathing and the rate of heartbeat (see Sections 10.9 and 11.19).

Sensitive carbon-dioxide receptors are also found in the medulla oblongata. These initiate nerve impulses controlling rate and depth of breathing. They provide our most precise control over this function. (See Section 10.9).

Our sensation of **thirst** arises as a result of stimulation of special cells in the hypothalamus of the brain. These cells are remarkably sensitive to changes in the osmotic pressure of the blood. If this increases (through loss of water or extra intake of salts), we quickly become conscious of thirst. In addition, ADH is released from the posterior lobe of the pituitary gland and acts on the tubules of the kidney to produce maximal reabsorption of water. You remember (see Section 25.9) that it is the hypothalamus itself which manufactures ADH for storage in the posterior lobe of the pituitary gland.

Experiments with several laboratory animals, most notably the white rat, have revealed the existence of two pairs of regions in the hypothalamus that regulate feeding. Activation of one of the pairs —let us call them the "hunger" centers—stimulates feeding. Damage to these areas causes the animal to cease feeding even though it is starving. Activation of the second pair—the "satiety" centers—causes a suppression of the "hunger" centers. Thus damage to the satiety centers causes uncontrolled feeding. So uncontrolled, in fact, that the animal becomes grossly obese. Rats have become as heavy as 1 kg after destruction of their satiety centers.

Under normal conditions, activity of the satiety centers is controlled by the concentration of glucose in the blood. Thus the satiety centers are internal chemical receptors ("glucostats"). Elevated levels of glucose (and insulin) activate the satiety centers, which thus suppress the hunger centers and, consequently, inhibit feeding. In this way, the intake of calories is reduced. However, with a decline in blood sugar, the satiety centers are inhibited and feeding is once again stimulated.

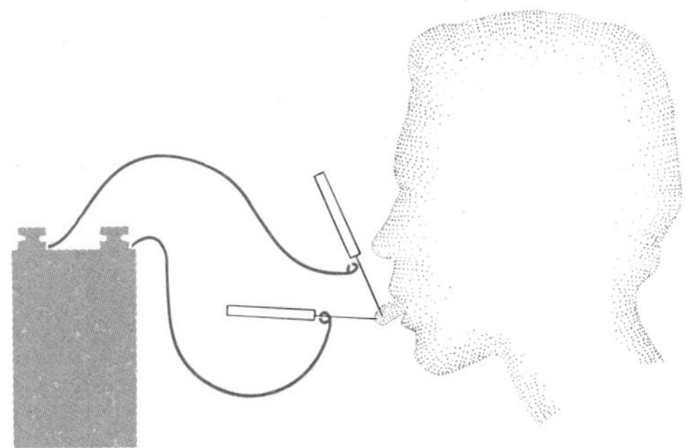

FIG. 26.19 A weak current is detected as a strong taste. When no current is flowing, the dissecting needles are tasteless. What explanation might there be for this?

26.12 SUMMARY

It should now be clear that humans possess more than just the traditional five senses (touch, taste, smell, vision, and hearing). All of our senses depend upon the presence of specific stimulus receptors. These provide a constant flow of information as to the state of both the external and internal environment. In order to accomplish their functions, all stimulus receptors must possess three general features: (1) they must be so constructed as to have a lower threshold for one type of stimulus than for any other; (2) they must be connected to a sensory neuron; (3) they must be capable of initiating nerve impulses in that neuron.

Most of our stimulus receptors have a fourth characteristic which is their property of *adapting* quickly to the stimulus. When the stimulus is first applied, the receptor initiates a volley of impulses in the sensory neuron to which it is attached. With constant exposure to the stimulus, however, the rate of impulse propagation decreases and may eventually cease entirely. We examined one example of such sensory adaptation in our earlier discussion of the Pacinian corpuscle. You might well question whether adaptation occurs in vision but, in fact, it does. However, through constant slight, involuntary movements of our eyes, we shift the position of the image on our retinas and thus continue to observe the scene even though no change may have occurred in it. Only stretch receptors such as are found in our muscles and tendons and in the walls of the aorta and carotid arteries seem to adapt very little to continuous stimulation.

It should be stressed again that when conscious sensation occurs, it occurs solely in the brain. All types of stimulus receptors send the same message to the brain: electrochemical impulses in the sensory neurons. It is the brain that assigns meaning to these impulses. A hard blow to the head may exceed the threshold of neurons in the retina and thus trigger nerve impulses in the optic nerve. Although a mechanical force rather than light has given rise to them, the brain still interprets these in the usual way. Consequently, we "see stars."

This principle can be demonstrated more pleasantly with two clean dissecting needles, two lengths of insulated wire, and a dry cell. When the two needles are applied gently to the tongue, a sensation of touch only is felt. When each needle is wired to a terminal of the dry cell, however, a strong taste is detected (Fig. 26.19). The most likely explanation is that electrical stimulation of the taste buds has given rise to these sensations.

The detection of changes in the external environment is just the first step by which nervous responsiveness and coordination are achieved. The next step involves the conduction of nerve impulses from one part of the body to another. The anatomical and physiological mechanisms by which nerve impulses are generated and circuited throughout the body is the topic of the next chapter.

EXERCISES AND PROBLEMS

1. When trying to see a faint star at night, it is helpful to look slightly away from the spot where the star is. Can you think of an explanation for this?
2. Which of our sense receptors detect events occurring outside of the body? Which detect events occurring within the body?
3. What is a stimulus?
4. Why are we not constantly aware of the touch and pressure of our clothing?
5. What function is common to all sense receptors?
6. In what ways are the compound eye and the human eye similar in function? In what ways are they different?

REFERENCES

1. MILLER, W. H., F. RATLIFF, and H. K. HARTLINE, "How Cells Receive Stimuli," *Scientific American,* Offprint No. 99, September, 1961. With special emphasis on the ommatidia of *Limulus.*
2. WEALE, R. A., *The Vertebrate Eye,* Oxford Biology Readers, No. 71, Oxford University Press, Oxford, 1974.
3. RUSHTON, W. A. H., "Visual Pigments and Color Blindness, *Scientific American,* Offprint No. 1317, March, 1975. Each type of color blindness results from the loss of or an abnormality in one of the three cone pigments.
4. GAMOW, R. I., and J. F. HARRIS, "The Infrared Receptors of Snakes," *Scientific American,* Offprint No. 1272, May, 1973.
5. LOEWENSTEIN, W. R., "Biological Transducers," *Scientific American,* Offprint No. 70, August, 1960. Shows how sense receptors convert environmental stimuli into nerve impulses, with special emphasis on the Pacinian corpuscles.
6. CASE, J., *Sensory Mechanisms,* Current Concepts in Biology Series, Macmillan Company, New York, 1966. A paperback devoted to the sense receptors of both vertebrates and invertebrates.
7. FRIEDMANN, I., *The Mammalian Ear,* Oxford Biology Readers, No. 73, Oxford University Press, Oxford, 1976.
8. ROEDER, K. D., "Moths and Ultrasound," *Scientific American,* Offprint No. 1009, April, 1965. Describes how certain moths are able to detect and respond to the sonar signals of the bats that prey on them.
9. AMOORE, J. E., J. W. JOHNSTON, JR., and M. RUBIN, "The Stereochemical Theory of Odor," *Scientific American,* Offprint No. 297, February, 1964. Presents evidence to support the idea that there are seven primary odors, each triggered by molecules of a particular shape or electrical charge.
10. SCHNEIDER, D., "The Sex-Attractant Receptor of Moths," *Scientific American,* Offprint No. 1299, July, 1974. One molecule of attractant is sufficient to trigger a nerve impulse in the neuron attached to the receptor.

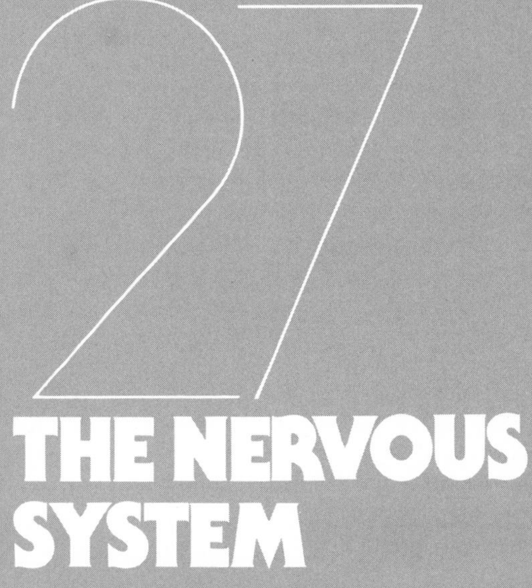

27 THE NERVOUS SYSTEM

The ability of animals to respond quickly and in a well-coordinated way to changes in the environment requires more than the presence of stimulus receptors alone. There must also be a system of conductors to transmit information from the receptors to the structures, chiefly muscles and glands, that will take the appropriate action. This system of conductors constitutes the nervous system.

All stimulus receptors give rise to nerve impulses. As we shall see, these are basically all alike despite the wide variety of stimuli that initiate them. They are then conducted along an elaborate system of neurons. The route taken by these impulses determines what action is produced. The overall response of the organism, the coordination of the separate actions participating in this response, and the conscious perception of the stimulus itself all depend upon the circuits traveled by the nerve impulses.

Probably all living cells are capable of detecting stimuli. When an amoeba is pricked with a pin, the entire cell responds to the stimulus. In fact, electrical activity has been detected in the amoeba which resembles that of the nerve impulse in animals.

Animals, with the possible exception of the sponges, have cells that are specialized for conducting electrochemical impulses. Their structure, function, and organization are surprisingly similar in the various animals. These cells are called neurons.

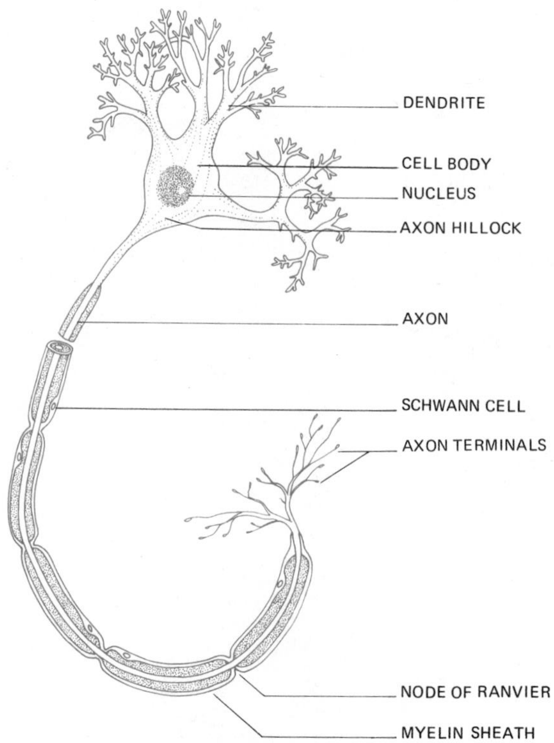

FIG. 27.1 Structure of a motor neuron. Most of the axon has been omitted.

27.1 THE NEURON

A neuron is simply a cell that is specialized to conduct electrochemical impulses over a substantial distance. This function is accomplished by means of hairlike cytoplasmic extensions, the nerve fibers (axons). In a large animal such as a horse, these may be as much as 1 to 2 meters in length although only a few micrometers in diameter. They grow out from the **cell body,** which houses the nucleus of the cell. Destruction of the cell body always results in the eventual death of these fibers.

The length of some nerve fibers is so great that it is hard to see how the cell body could exert any kind of metabolic control over them. Nevertheless, there is a steady transport of materials from the cell body along the entire length of the axon. This flow is probably facilitated by the many microtubules (see Section 5.15) present in the cytoplasm within the axon. There is also evidence that the axon receives materials from accessory cells, called Schwann cells. These cells are spaced regularly along the length of the nerve fibers and practically (but not completely) surround them (Fig. 27.1).

In many neurons the nerve impulses are generated in short branched fibers, the **dendrites,** and also in the cell body. The impulses are then *conducted* along a single long fiber, the **axon.** The axon usually branches several times close to its end (Fig. 27.1).

Many axons are covered with a glistening fatty sheath, the **myelin sheath,** which consists of the greatly expanded cell membrane of a surrounding Schwann cell. The membrane is folded around and around the axon to form the sheath. Where the sheath of one Schwann cell meets that of the next, the axon is unprotected. This region, called the node of Ranvier, plays an important part in the propagation of the nerve impulse, as we shall see shortly.

Structurally, as well as functionally, neurons can be placed in three distinct groups.

1. Sensory neurons. Sensory neurons run from the various types of stimulus receptors (discussed in the last chapter) to the central nervous system (brain and spinal cord). The receptors initiate nerve impulses which then travel the length of the sensory neuron. The cell bodies of the sensory neurons are located in clusters **(ganglia)** next to the spinal cord. The axons usually terminate at interneurons.

2. Interneurons. Interneurons are found exclusively within the spinal cord and brain. They are stimulated by impulses that reach them from sensory neurons or from other interneurons. Interneurons are also called association neurons.

Interneurons form the intermediate link in the pathway of almost all nervous coordination. The vast number of interneurons in our central nervous system (several billion at least) and the unimaginably large number of cross-connections between them provide us with a virtually limitless number of possible circuits for nerve impulses to follow.

3. Motor neurons. Motor neurons transmit impulses from the central nervous system to the muscles and glands which will bring about the response of the body. They are usually stimulated by interneurons although, in some cases, impulses pass directly from a sensory neuron to a motor neuron.

27.2 THE NERVE IMPULSE

With the use of microelectrodes and sensitive recorders (Fig. 27.2), it is possible to study the electrical properties of neurons. A microelectrode is a hollow needle filled with a salt solution which enables it to conduct electricity. If we pierce a neuron with a microelectrode, we discover that the interior of the neuron is negatively charged with respect to the exterior. The size of this charge (sometimes called the **resting potential**) is about seventy millivolts (70 mV). It is maintained only so long as the neuron carries on a slow, but unceasing, oxidation of glucose to produce ATP. The ATP is used to actively transport sodium ions (Na^+) from the interior of the neuron to the extracellular fluid (ECF) and potassium ions (K^+) from the ECF to the interior. The result is a concentration of Na^+ in the ECF ten times as large as that in the cytoplasm and a concentration of K^+ in the cytoplasm thirty times that in the ECF.

These differences in concentration create a strong tendency for the potassium ions to diffuse out of the neuron and the sodium ions to diffuse into it. The membrane of the resting neuron is virtually impermeable to the passage of sodium ions. However, potassium ions do diffuse out and as they do so, the interior of the neuron becomes negatively charged with respect to the exterior. When the charge across the cell membrane reaches about 70 mV, the process reaches equilibrium. The tendency of the K^+ to diffuse out because of the concentration gradient is balanced by the electrical attraction between these positively charged ions and the negatively charged interior.

There are a variety of stimuli such as a negatively charged electrode, heat, mechanical deformation, and certain chemicals that will increase the membrane's permeability to sodium ions and permit them to diffuse back into the neuron. This, in turn,

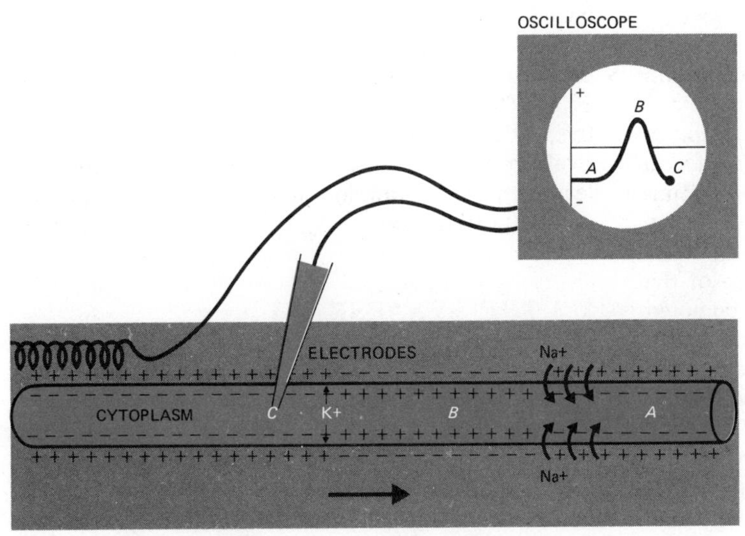

FIG. 27.2 The nerve impulse. In the resting neuron the interior of the axon membrane is negatively charged with respect to the exterior (*A*). As the nerve impulse passes (*B*), the polarity is reversed. Then the outflow of K^+ ions quickly restores normal polarity (*C*). At the instant pictured in this diagram, the moving spot, which has traced these changes on the screen of the oscilloscope as the impulse swept past the intracellular electrode, is at position C.

reduces the voltage between the ECF and the cytoplasm. This change in voltage is called the **generator potential.** In the previous chapter, we examined a good example of the phenomenon in our discussion of the Pacinian corpuscle. When pressure is applied to a corpuscle, a generator potential is created in the tip of the sensory neuron to which it is attached (see Section 26.6). If the stimulus is a weak one, the influx of sodium ions is slight. The generator potential dies out quickly and the normal −70-mV polarity is reestablished. In such a case, the stimulus was **subthreshold.**

However, if the stimulus is strong enough, the depolarization proceeds to a point where the voltage is reduced to about −50 mV. At this value, called the **threshold,** the permeability of the membrane to the inflow of sodium ions increases sharply. Sodium ions flow in with a rush, eliminating the voltage entirely. In fact, they create a momentary overshoot, with the interior of the membrane now becoming positively charged (Fig. 27.2).

The sudden influx of sodium ions at the stimulated point of the membrane has the interesting property of increasing the permeability of adjacent portions of the membrane to sodium ions. Consequently, the process is repeated continuously along the length of the neuron, each portion of the neuron triggering the depolarization of the portion adjacent to it (Fig. 27.2). The resulting wave of depolarization which sweeps down the neuron is the **nerve impulse.** It is also called the **action potential (AP).**

In myelinated neurons, depolarization occurs only at the nodes of Ranvier. However, the depolarization of one node creates at the next node an almost instantaneous generator potential which leads to *its* depolarization. The nerve impulse thus jumps rapidly from node to node. For this reason, myelinated axons conduct nerve impulses more rapidly than nonmyelinated ones.

The strength of an action potential is simply a property of the neuron itself. It has nothing to do with the strength of the stimulus. As long as a stimulus just exceeds the threshold of the neuron, the neuron will "fire." Stimuli of greater strength can do no more than this. We say that the response of the neuron is "all-or-none." However, it should be noted that strong stimuli may give rise to a greater *number* of impulses in a given period of time than weak ones.

Generator potentials and action potentials differ in several respects. Unlike action potentials (which are all-or-none), generator potentials are *graded,* that is, the size of the generator potential is proportional to the size of the stimulus. Generator potentials also differ from action potentials in that they are not self-propagating. The strength of the generator potential declines rapidly at increasing distances along the membrane from the point where the generator potential was created.

The nerve impulse should not be compared to the flow of current in a wire. The processes are not at all similar. In the latter case, an electrical impulse is conducted along the wire at the speed of light. In the neuron, an electro*chemical* reaction is simply moving down the length of the neuron. Even in our fastest neurons, this wave of depolarization proceeds at less than 300 km/hour. In one sense this is very rapid, but of course it does not begin to compare with the speed of light (300,000 km/sec).

A better analogy for the nerve impulse is the fuse to a string of firecrackers. When a burning match is applied to the end of the fuse, its threshold is reached and a spark begins to pass along the fuse. The energy for this movement comes from the energy of the chemicals stored in the fuse itself, not from the energy of the burning match. There is no weakening of the spark with distance traveled. If the fuse branches, the spark travels with undiminished vigor along each branch. (Neither of these conditions occurs in electrical circuits.)

A second stimulus applied to a neuron less than 0.001 sec after the first will not trigger any impulse. The membrane is depolarized and the neuron is said to be in its **refractory period.** In some of our neurons, the refractory period lasts for only 0.001–0.002 seconds. This means that the neuron can transmit 500–1000 impulses every second. Other neurons, especially those of cold-blooded animals, repolarize more slowly. When the −70-mV polarity is reestablished, the neuron is ready to "fire" again. Repolarization is established by the rapid diffusion of potassium ions from within the cell to the ECF (Fig. 27.2). Only when the neuron is finally rested are the sodium ions that came in at each impulse actively transported out of the cell. For each one that leaves, a potassium ion returns, thus maintaining normal polarization.

27.3 THE SYNAPSE

The points at which the axon terminals of one neuron are in contact with other neurons are called **synapses.** Each axon terminal is swollen to form a synaptic knob (Fig. 27.3). The neuron ending in the synaptic knob is called the **presynaptic neuron.** The neuron upon which the knob is placed is called the **postsynaptic neuron.** The synaptic knob contains a

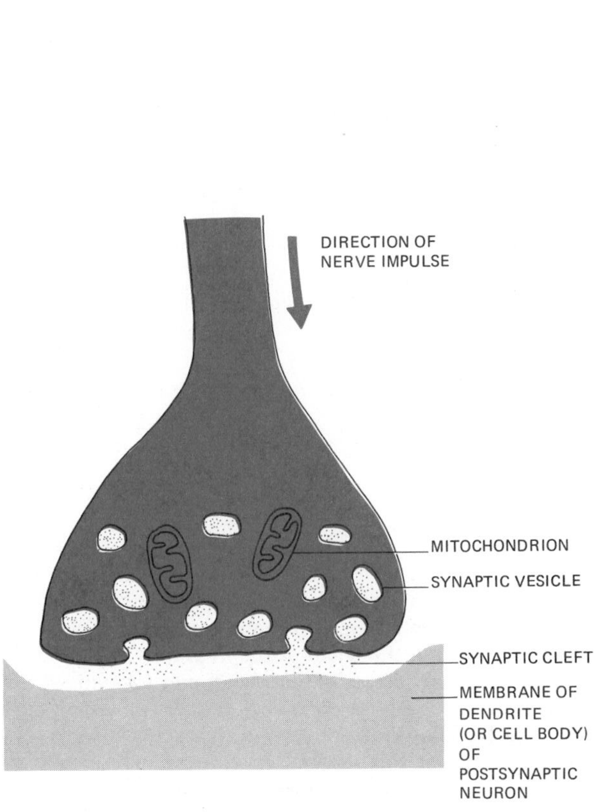

FIG. 27.3 Generalized structure of the synapse. The arrival of an impulse at the synaptic knob causes a vesicle-contained chemical (e.g. acetylcholine) to be discharged into the synaptic cleft. At excitatory synapses, this substance depolarizes the membrane of the postsynaptic neuron. At inhibitory synapses, the substance released hyperpolarizes the postsynaptic neuron. In both cases, enzymes degrade the chemical soon after its release, thus restoring the synapse to its initial state.

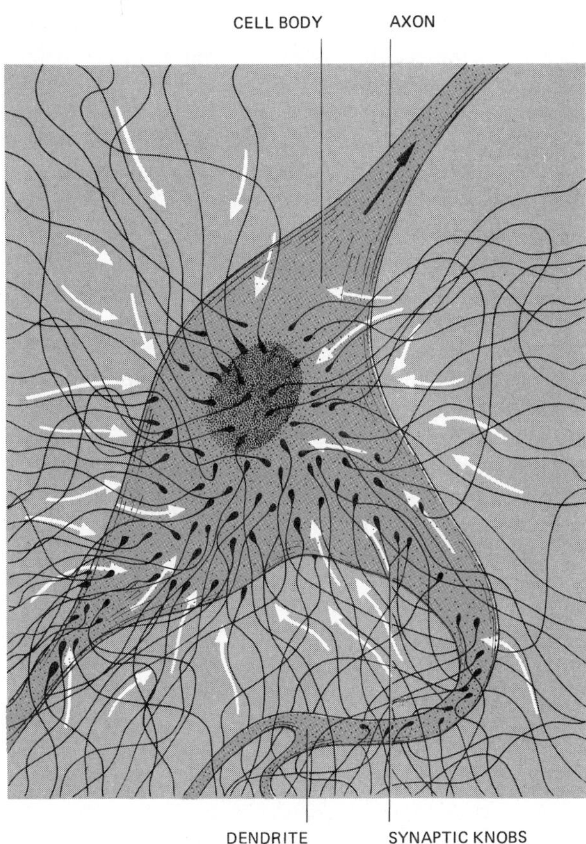

FIG. 27.4 The dendrites and cell body of a single motor neuron in the human may have as many as 5000 interneurons converging on them. Some of these release excitatory substances at the synapse. Others release inhibitory substances. Several excitatory impulses must reach the motor neuron at once in order to initiate a nerve impulse in it. The number must be increased if inhibitory impulses are reaching the motor neuron at the same time.

chemical, which, when an action potential arrives, is released into the narrow synaptic cleft or gap. At synapses outside the central nervous system, this substance is **acetylcholine (ACh)**. Accumulation of ACh in the synaptic gap reduces the size of the potential across the membrane beneath the synaptic gap by increasing the permeability of the membrane to sodium ions. This change is called an *e*xcitatory *p*ost*s*ynaptic *p*otential or **EPSP.** It behaves just like a generator potential. If depolarization of the membrane reaches threshold, an action potential is triggered in the postsynaptic neuron. Thus ACh serves as a chemical "transmitter" of nerve impulses. Because dendrites do not release a chemical transmitter, any action potential passing along a neuron from axon to dendrites would die out at the synapse. Thus the synapse acts as a valve permitting only one-way flow of nerve impulses.

Chemical transmission of the nerve impulse across the synapse also occurs in the central nervous system but positive identification of the substances involved has been very difficult to achieve. ACh, four amino acids (glutamic acid, glycine, aspartic acid, and gamma aminobutyric acid—GABA), and five amino acid derivatives (noradrenaline, adrenaline, serotonin, dopamine, and histamine) are all likely candidates. Dopamine is an intermediate in the synthesis of noradrenaline from the amino acid tyrosine. Serotonin is synthesized from the amino acid tryptophan.

There is also evidence that certain polypeptides may serve as transmitters. One of the strongest candidates is a polypeptide—containing 11 amino acids—called "Substance P." It appears to be a transmitter at the terminals of sensory neurons leading into the spinal cord and may serve a similar function in parts of the brain.

Impulses arriving at the synaptic knobs of some interneurons in the central nervous system *inhibit* the depolarization of the neurons with which they synapse. The chemical "transmitter" used in these cases (glycine? GABA?) seems to exert its effect by increasing the permeability of the neuron membrane to *potassium* ions. As additional K+ diffuses out of the cell, the interior becomes even more negative with respect to the exterior. This *hyper*polarization of the postsynaptic membrane is called an *i*nhibitory *p*ostsynaptic *p*otential (**IPSP**). It may reach a value of −80 mV. A neuron that is hyperpolarized by the influence of an inhibitory transmitter substance appears to have an increased threshold, that is, the cell is less easily stimulated. Actually, the threshold *voltage* (about −50 mV) has not changed. It is simply a matter of whether the depolarization produced by the excitatory synapses *minus* the hyperpolarizing effect of the inhibitory synapses can reach this value or not.

A single interneuron or motor neuron may have thousands of synaptic knobs terminating on its dendrites and cell body (Fig. 27.4). Some of these re-

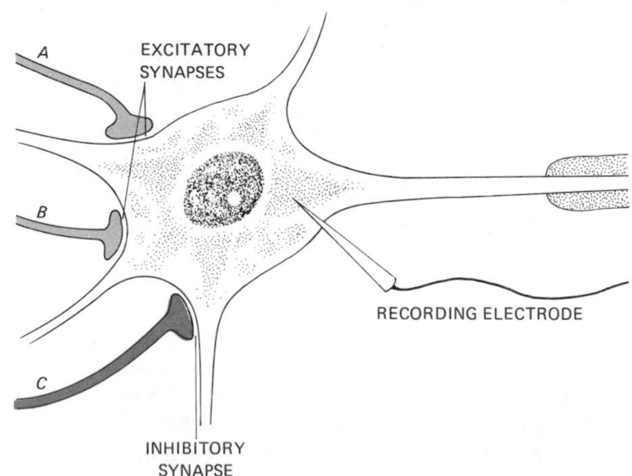

FIG. 27.5 Effects of excitatory postsynaptic potentials (EPSPs) and inhibitory postsynaptic potentials (IPSPs) on the creation of action potentials in a neuron. (1) The EPSP created by a single excitatory synapse is insufficient to reach the threshold of the neuron. (2) EPSPs created in quick succession, however, add together. If they reach threshold, an action potential is created. (3) The EPSPs created by *separate* excitatory synapses (*A* and *B*) can also be added together to reach threshold. (4) Activation of inhibitory synapses (*C*) lowers the resting potential of the neuron. The IPSP created may also prevent what would otherwise have been effective EPSPs from triggering an action potential. Normally, the number of EPSPs needed to reach threshold is greater than shown here.

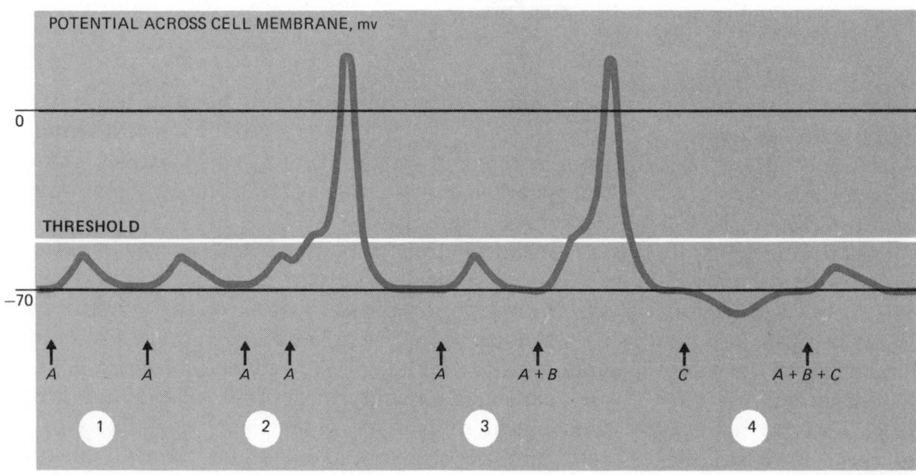

lease excitatory transmitter substances. But because it creates an EPSP of only about 0.5 mV, a single excitatory synapse cannot generate an action potential in the postsynaptic neuron. However, if a number (say 40) of excitatory synapses are active at once, their respective EPSPs are added together and may be able to reach the threshold of the postsynaptic neuron. Or if just a few excitatory synapses are activated repeatedly in a brief interval, the EPSPs again add up and may reach threshold (Fig. 27.5).

The inhibitory synapses work the same way. Rapid, repeated stimulation of many of them results in their individual IPSPs adding together. Whether the postsynaptic neuron fires or not is thus dependent upon the balance between all the EPSPs and IPSPs created over its surface.

You might well ask why a few excitatory synapses on one dendrite could not create an action potential there despite extensive inhibitory signals on another dendrite. This might occur. But many neurons have a nice device by which they are usually able to integrate *all* the excitatory and inhibitory signals reaching them. The point where the axon emerges from the cell body is called the axon hillock (Fig. 27.1). The portion of the cell membrane covering the axon hillock has a lower threshold than the membrane covering the cell body and dendrites. It is at the axon hillock membrane that the action potential is usually generated in the neuron. Having neither excitatory nor inhibitory synapses of its own, the axon hillock is in a position to evaluate the total picture of EPSPs and IPSPs created in the dendrites and cell body. If over any brief interval the sum of all the EPSPs minus the sum of all the IPSPs exceeds the threshold of the axon hillock, an action potential is generated in it.

Of what use are inhibiting neurons? Inhibition of muscles is just as important as stimulation, if coordinated movements are to be made. Imagine trying to catch a ball if all your muscles contracted at once. While *failure* to *stimulate* does result in inhibition, specific inhibitory neurons in the central nervous system provide even more precise control.

The proper operation of the synapse requires that the chemical transmitter be removed from the synaptic gap as soon as it has done its job. If it is not, it will "fire" the neuron over and over again. The removal of ACh is accomplished by the enzyme acetylcholinesterase, which hydrolyzes the molecule into inactive fragments. Substances have been discovered which interfere with the action of acetylcholinesterase. These seriously upset normal nervous activity. The "nerve" gases developed for possible use during World War II and the organophosphate insecticides which were the civilian "spinoff" from nerve gas development are powerful acetylcholinesterase inhibitors.

27.4 THE REFLEX ARC

In humans, the simplest unit of nervous response is the reflex arc. The individual neuron is the unit of structure of the nervous system, but the reflex arc is the unit of function. We can illustrate the reflex arc by examining what goes on when you touch a hot stove and quickly pull your hand away. This response is called a withdrawal reflex. In order for it to be carried out, the following actions take place:

1. The stimulus is detected by receptors in the skin.
2. These initiate nerve impulses in the *sensory* neurons leading from them to the spinal cord.
3. These impulses enter the spinal cord and initiate impulses in one or more *interneurons.*
4. Interneurons initiate impulses in the appropriate *motor* neurons.
5. When these impulses reach the junction between the motor neurons and the muscles, the *muscles* (called *flexors*) are stimulated to contract. Your hand is withdrawn.

As mentioned earlier, efficient withdrawal of the hand also requires that some of the muscles in your arm (the *extensors*) be inhibited. This is accomplished by inhibitory interneurons in the spinal cord. When they are stimulated by the sensory neurons, they inhibit the motor neurons running to the extensor muscles.

The structural basis for these actions is shown in Fig. 27.6, although it is a vastly oversimplified representation of what actually takes place. The organization of sensory, inter-, and motor neurons is not a simple 1:1:1 pattern. The axon terminals of a given sensory neuron form synapses with several interneurons. And a single interneuron may, in turn, have hundreds of different axons (both of sensory neurons and other interneurons) converging upon it. A motor neuron may have as many as 5000 different interneurons (and some sensory neurons) converging upon it (Fig. 27.4). Furthermore, as we saw in the previous section, some interneurons may *inhibit* synaptic transmission from other interneurons. It is this multitude of interconnections which provide for complex responses and which also enable us to keep informed

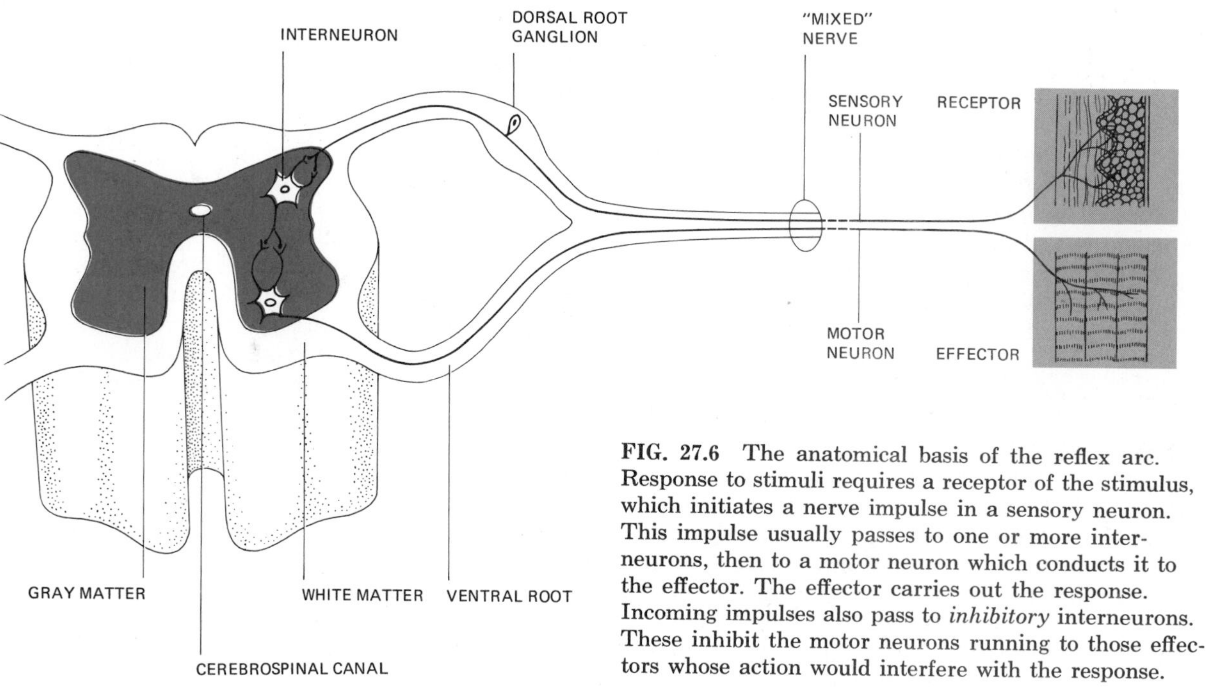

FIG. 27.6 The anatomical basis of the reflex arc. Response to stimuli requires a receptor of the stimulus, which initiates a nerve impulse in a sensory neuron. This impulse usually passes to one or more interneurons, then to a motor neuron which conducts it to the effector. The effector carries out the response. Incoming impulses also pass to *inhibitory* interneurons. These inhibit the motor neurons running to those effectors whose action would interfere with the response.

of what our body is doing. The withdrawal reflex does not require the participation of the brain at all. However, you are aware of what you have done. You can even learn to inhibit your response to the stimulus. These actions would be impossible were it not for the many interneurons running up and down the spinal cord informing the brain of what is happening and perhaps relaying modifying commands to the motor neurons.

The multitude of interconnections in the central nervous system also provides the basis for variability of response. Many of our complex activities, for example those involved in an athletic maneuver, can be markedly improved with practice. There is improvement in the speed of the response as well as in its precision. This improvement suggests that the nerve impulses are finding ever-shorter, but already existing, routes through the brain and spinal cord. These alternative routes are made possible by the large number of interconnections between sensory, inter-, and motor neurons.

In most animals one can distinguish two major divisions of the nervous system. The **central nervous system** of such different forms as planarians, the earthworm, and the grasshopper consists of clusters of cell bodies, the ganglia. Generally the ganglia are located in parts of the body where a good deal of sensory information is being received (e.g. the head) or precise control of muscles (e.g. near the mouth parts) is needed. The ganglia are connected to one another by one or more nerve cords consisting chiefly of the fibers (axons) of interneurons (Fig. 27.7). Sensory and motor axons run to and from the ganglia. They are bundled together in cables (nerves) and make up the **peripheral nervous system.** Because most nerves contain both sensory and motor axons, they are called **mixed nerves.**

The peripheral nervous system serves to inform the central nervous system of stimuli that have been detected and to cause the muscles and glands to carry out a response. The central nervous system serves as a coordinating center for the actions to be carried out. From what we have said, it should be clear that neither the central nor the peripheral nervous system can function independently of the other. However, with an organism as complex as man, we will better understand the special features of each system if we study them separately.

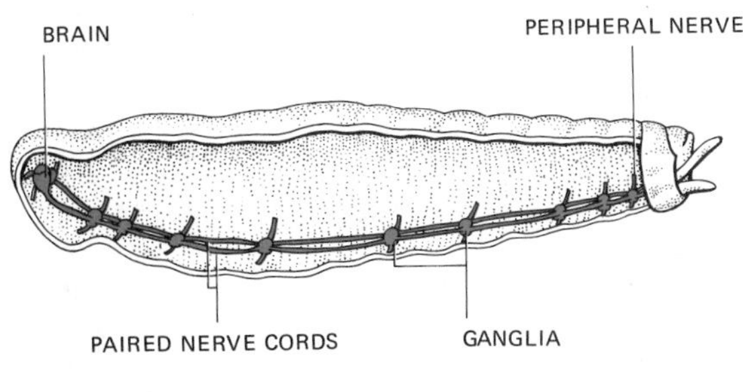

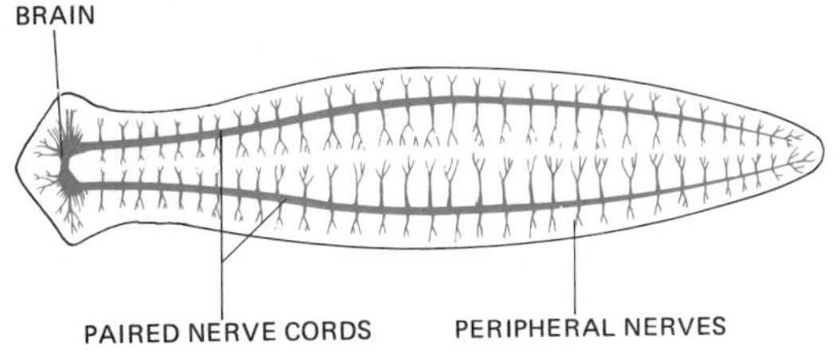

FIG. 27.7 Nervous system of the grasshopper (above) and planarian (below).

The Human Central Nervous System

Our central nervous system consists of the spinal cord and the brain.

27.5 THE SPINAL CORD

The spinal cord is a glistening white cord which runs from the base of the brain down through the backbone. A cross section of the spinal cord reveals that only the outer portion, the "white matter" of the cord, is white. The inner portion is "gray matter." Running vertically through the gray matter is a central canal filled with cerebrospinal fluid; the canal connects with cavities (ventricles) in the brain which are also filled with this fluid.

The white matter consists chiefly of long, myelinated nerve fibers running up and down the cord. The gray matter is closely packed with the cell bodies of interneurons and motor neurons.

At evenly spaced intervals along *each* side of the spinal cord are 31 pairs of projections, the nerve **roots.** These unite directly to form the mixed nerves of the peripheral nervous system. All the sensory neurons reaching the spinal cord in a mixed nerve pass into a dorsal root and then into the gray matter of the cord itself. The cell bodies of these sensory neurons are located in ganglia in the dorsal roots (Fig. 27.6). All our motor neurons originating in the spinal cord pass out through ventral roots before uniting with the sensory axons to form the mixed nerves.

The separation of sensory and motor axons in the roots is easily demonstrated when they become accidentally cut or otherwise damaged. Destruction of dorsal roots causes a loss of sensation in that part of the body which supplied sensory impulses to the damaged roots. Destruction of ventral roots, on the other hand, causes a muscular paralysis of that part of the body supplied by the motor neurons running through those roots. To achieve substantial anesthesia or paralysis, several adjacent roots must be cut because there is considerable overlapping of function between the mixed nerves they form.

The spinal cord carries out two main functions in nervous coordination. First, it connects the peripheral nervous system to the brain. Information reaching the spinal cord through sensory neurons may be transmitted up the cord by means of interneurons. In the brain all this information can be compared and then appropriate action dictated. Impulses leaving the brain travel back down the cord by way of other interneurons and then leave the cord in the motor neurons.

The many interneurons carrying impulses *from* specific receptors or *to* specific effectors are not organized randomly in the spinal cord but are, instead, grouped together in **tracts.** Impulses from the body's touch receptors, temperature receptors, proprioceptors, etc., each pass up the cord in their own special tracts. Impulses passing down the cord to the motor neurons are also localized in tracts. Curiously enough, impulses reaching the cord from the left side of the body eventually pass over to tracts running up to the right side of the brain, and vice versa. In some cases, this "crossing over" occurs as soon as the impulses enter the spinal cord. In other cases, it does not take place until the tracts actually enter the brain.

The second function of the spinal cord is to act as a minor coordinating center itself. Simple reflex responses, like the withdrawal reflex, can take place through the sole action of the spinal cord. The brain does not need to receive or initiate any nerve impulses for the action to be carried out successfully. Although only relatively simple coordination can be carried out by the spinal cord alone, its actions are a good deal more complex than we have suggested. Even for such a "simple" response as the withdrawal reflex, many motor neurons must be stimulated at the proper moment while other motor neurons are inhibited.

27.6 THE BRAIN

The activity of the brain is even less well understood than the activity of the spinal cord. Basically, the brain receives nerve impulses from the spinal cord and from cranial nerves leading directly to it from the eyes, inner ear, etc. It then organizes these impulses. This organization process is the key to brain function: conscious sensation, memory, the association of one stimulus with another or with a memory, and the coordinated body action necessary for proper responsiveness all depend upon the circuits taken by nerve impulses within the brain. Furthermore, the initiation of impulses to be sent to the motor neurons of the body does not necessarily depend upon sensory impulses reaching the brain. It is quite clear that our brain can initiate body responses simply as a result of its own self-contained activity. An example of this would be action taken as a result of something suddenly remembered. The evidence suggests that our brain is almost unique in this respect. The earthworm, grasshopper, and frog, for example, seem far more dependent on specific stimuli for initiating responses. We say that these creatures are more sense dominated.

The human brain consists of two large hemispheres (Fig. 27.8). Because of the crossing over of the spinal tracts, the left hemisphere of the brain

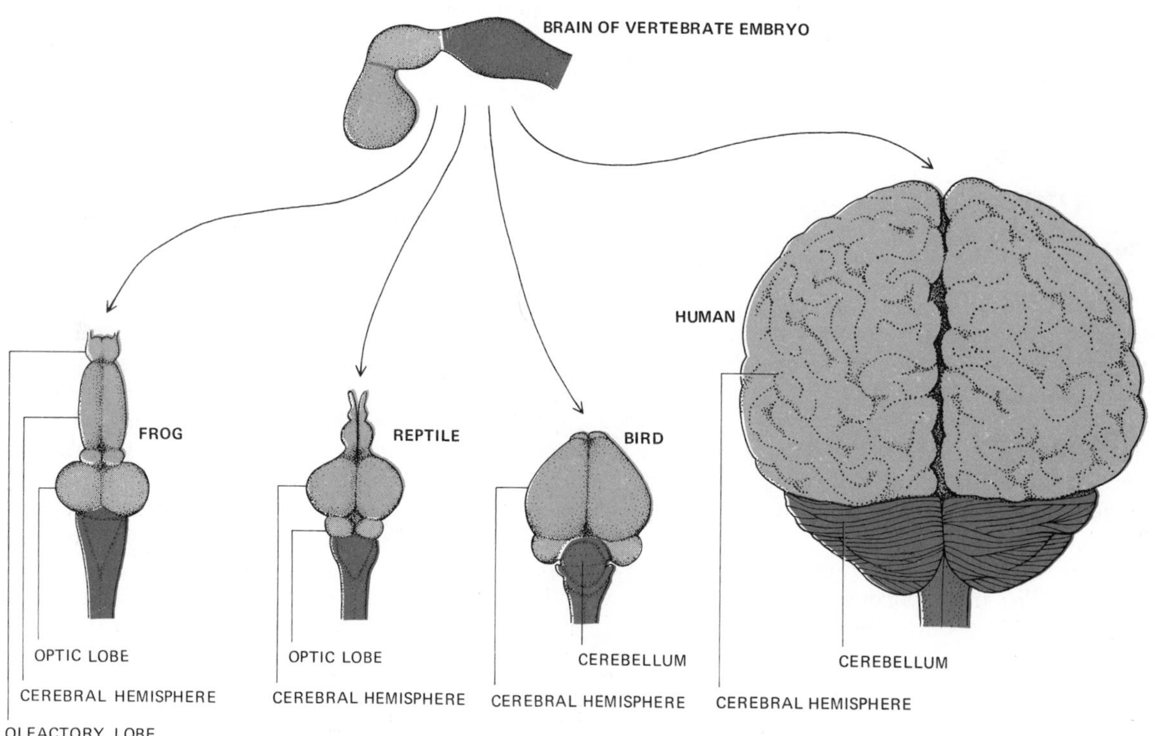

FIG. 27.8 The developing brain of the vertebrate embryo consists of three lobes. From these arise the structures of the forebrain (light color), midbrain (gray), and hindbrain (dark color). The human brain is shown from behind so that the cerebellum can be seen.

controls the right side of the body and vice versa. The brain, as well as the spinal cord, is covered with three protective membranes, the **meninges** (Fig. 27.9). Cerebrospinal fluid is found between the inner two membranes and helps to cushion the brain from blows to the skull. Within the brain are four chambers, the ventricles, also filled with cerebrospinal fluid. There are large capillary beds within two of these ventricles, permitting the exchange of materials between the blood and the cerebrospinal fluid. The cells lining the ventricles are ciliated and keep the cerebrospinal fluid circulating.

Extending from the brain are twelve pairs of cranial nerves. Not all of these are mixed nerves; the optic nerves and olfactory nerves, for example, contain sensory neurons only.

The brain is divided into three regions: forebrain, midbrain and hindbrain. These divisions are not immediately obvious in the adult human brain because each is, in itself, made up of several parts or lobes. However, the pattern is clearly visible during the development of the brain in the embryo. The brains of all vertebrates are constructed on this basic plan (Fig. 27.8).

27.7 THE HINDBRAIN

The two main portions of the hindbrain are the **medulla oblongata** and the **cerebellum.** The medulla oblongata has the appearance of simply a swollen tip of the spinal cord. Though small in size, it is absolutely essential to life. The nerve impulses which stimulate the intercostal muscles and diaphragm, and thus permit breathing, originate in the medulla. Nerves regulating heartbeat, diameter of the arterioles and other important functions also arise here. It is not surprising that destruction of the medulla brings about immediate death. We have no direct

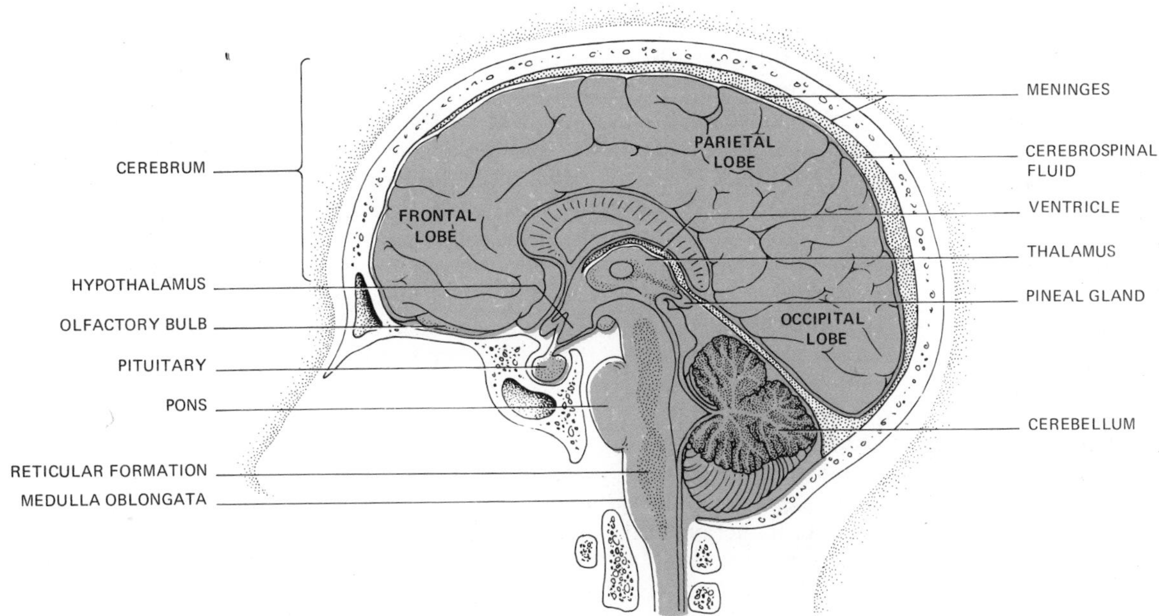

FIG. 27.9 The human brain cut lengthwise between the two cerebral hemispheres.

conscious control over the functions of the medulla although we can modify its action somewhat through the use of other brain centers.

The cerebellum consists of two deeply convoluted hemispheres. Its most important function seems to be coordinating locomotor activity in the body. Such activity is initiated by impulses arising in the motor area of the forebrain. These impulses not only travel down the spinal cord to the motor neurons but also pass into the cerebellum. As the body action is carried out, sensory impulses from the proprioceptors, the eyes, the semicircular canals, etc., are also sent to the cerebellum. The cerebellum then "compares" the information on what the body is actually doing to what the forebrain had instructed it to do. If a discrepancy exists, the cerebellum sends modifying signals to the forebrain so that the appropriate corrective signals can be sent out to the muscles. It is not surprising that birds have relatively large cerebella (Fig. 27.8) when we consider that they must be capable of moving swiftly and accurately in three dimensions of space while we and other earthbound animals spend most of our lives moving about on fairly flat surfaces.

27.8 THE MIDBRAIN

The human midbrain is quite small and inconspicuous. It relays nerve impulses between the forebrain and hindbrain and between the forebrain and the eyes. It also participates in the maintenance of balance. Some other vertebrates have relatively large midbrains. The prominent optic lobes of fishes, frogs, reptiles, and birds are part of their midbrains (Fig. 27.8).

Running up through the center of the medulla oblongata and the midbrain is a network of nerve fibers known as the **reticular formation.** It serves to activate or awaken the forebrain. Sensory tracts of the spinal cord lead both to the forebrain and to the reticular formation, but the forebrain cannot respond to impulses reaching it from the sensory tracts unless it is first awakened by the reticular formation. As you might guess from your own experience, the reticular formation is quite selective in its action. It may not arouse the forebrain when large, but familiar, stimuli (such as traffic sounds) are received. A creaking floorboard, on the other hand, may produce instant wakefulness. Destruction of the reticular formation results in a permanent coma followed eventually by death.

27.9 THE FOREBRAIN

The most prominent part of the human forebrain is the **cerebrum.** It is made up of two large, deeply convoluted hemispheres. Each of these is subdivided into four lobes: frontal, parietal, occipital, and temporal (Fig. 27.9). In most other vertebrates (e.g. the frog) there are large olfactory lobes which are present as outgrowths of the cerebrum (Fig. 27.8), but this portion of the brain is relatively small in humans. The forebrain also includes the thalamus, hypothalamus, part of the pituitary gland, and the pineal gland.

Surely no other body structure sets humans so far apart from the other vertebrates as their cerebrum. The volume of the two cerebral hemispheres averages approximately 1350 ml. While a few large mammals such as whales have still larger cerebrums, the ratio of the size of the cerebrum to that of the rest of the central nervous system is far greater in humans than in any other vertebrate.

We and the other mammals have still another important feature in the organization of the cerebrum. The exterior of the cerebrum, the cortex, is made up of gray matter, masses of cell bodies. The myelinated nerve fibers of which white matter is composed are located *within* the cerebral hemispheres. This, you will recall, is just the reverse of the arrangement in the spinal cord. It is also the reverse of the arrangement in other vertebrate brains. The surface of the frog's cerebrum is glistening and white like all the rest of the central nervous system.

The significance of this reversal of pattern probably lies in the greater surface area it provides for the cell bodies. The extraordinary properties of our brain are surely dependent upon the enormous (over ten billion) number of cell bodies present in the cerebral cortex and the unimaginably large number of possible connections which can be made between them. The many deep convolutions of the cerebral cortex provide additional surface area for these cell bodies to occupy.

Fewer than 1% of the neurons in the cerebrum send nerve fibers out of the cerebrum to other parts of the brain. What then are these vast numbers of self-contained neurons accomplishing? To date, we can do little more than speculate. We do know that a great deal of electrical activity goes on in the cerebrum. Through use of the electroencephalograph, an instrument that detects and records brain "waves," we know that this electrical activity changes during sleep, wakefulness, excitement, etc. The instrument has even been used successfully to diagnose disorders of the brain such as a tendency toward epileptic seizures. Despite the limited success of this technique, we still know practically nothing of the internal activity of the cerebrum.

Although the detailed electrical activity of the cerebrum is only dimly understood, some general functions of the cerebrum have been discovered. These discoveries have been made as a result of two kinds of studies. One is simply to destroy a portion of the cerebrum and see what happens to the victim. While this has been used with success on laboratory animals (using all the surgical precautions a human patient would receive), it is obviously risky to apply such findings to humans. However, many cases of brain damage as a result of injury or infection have been studied in humans and related to the presence of specific symptoms. The second technique is to expose the brain and then stimulate tiny portions of it with electrodes. Although this is of only limited use in laboratory animals, many humans undergoing brain surgery have volunteered to let such experiments be carried out on them while their brains were exposed. No pain is involved and when not under general anesthesia, the patients can report their sensations to the experimenter. Experiments of this sort have given us great insight into the functions of the cerebrum. For example, they have revealed the presence of a band of cortex running parallel to and just in front of the fissure of Rolando (Fig. 27.10), which controls the action of the body's skeletal muscles. Stimulation of discrete areas within this band results in the contraction of the muscles controlled by that area. The larger the area of cortex involved, the more abundant the supply of motor neurons to the part of the body controlled by it.

A similar region has been discovered in a parallel band of cortex posterior to the fissure of Rolando. This region is concerned with *sensations* from the various parts of the body. When isolated spots are

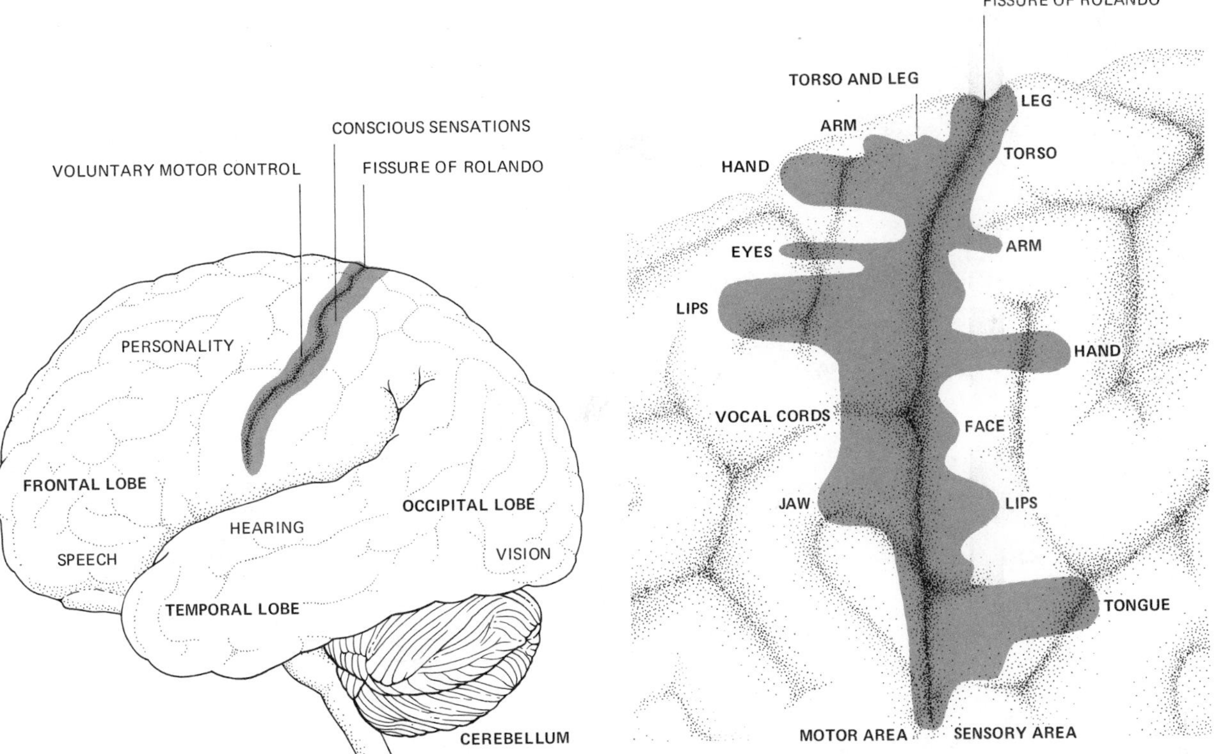

FIG. 27.10 Functions of the human cerebrum. The motor and sensory areas adjacent to the fissure of Rolando are shown in detail in the right-hand drawing. The left side of the brain (shown here) is concerned with the right side of the body, and vice versa.

stimulated electrically, the patient reports sensation in some specific area of the body. A map can be prepared in accordance with such reports (Fig. 27.10).

When portions of the occipital lobe are stimulated electrically, the patient reports the sensation of light. Not only does this region seem to be necessary for the primary act of seeing, but certain regions within it are necessary for associations to be made with what is seen. Damage to these regions may result in the person's being perfectly capable of seeing objects but incapable of associating them with previous experiences—of recognizing them. Such a defect is known as aphasia. The centers of hearing and understanding what is heard are located in the temporal lobes.

Although a few other functions have been assigned to specific areas of the cortex, the function of large areas of the cortex still remains unknown. Manipulation of these "silent" areas fails to reveal any specific activity. This is especially true of the large frontal lobes. Perhaps these large areas are responsible for carrying out some of our "higher" mental activities. Learning, memory, logical analysis, foresight, creativity, and some emotions may all depend upon the nervous activity of the frontal lobes and other silent areas. The dependency must be quite general, though, because no one has yet succeeded in relating any one of these mental activities with any specific spot in the cortex. Perhaps the vast number of neurons involved enables one portion of these silent areas to take over the function of some other portion that has become damaged.

Damage to the frontal lobes may produce changes in human behavior. Evidence of this was acquired in 1848 when Phineas P. Gage, the foreman of a crew excavating rock for the Rutland and Burlington Railroad in Vermont, accidentally exploded some blasting powder by tamping it with a metal bar. The explosion drove the bar right through the front portion of his head, severely damaging his frontal lobes. Miraculously, he survived the accident. Furthermore, he lost none of the clearly defined functions of the brain. His vision, hearing, other sensation, speech, and body coordination were unimpaired.

FIG. 27.11 Top: Skull of Phineas P. Gage, showing where a tamping bar entered (left) and exited (right) in an accident that occurred 12 years before he died of natural causes in 1861. Bottom: The tamping bar. (Courtesy of the Warren Anatomical Museum, Harvard University Medical School.)

Nevertheless, a marked change in personality was soon noted. Formerly a reasonable, sober, conscientious person, Phineas became thoughtless, irresponsible, fitful, obstinate, and profane. In short, certain special, hard-to-measure human qualities had been radically altered by his accident.

In 1935, 74 years after Phineas died (his skull now resides in the museum of Harvard Medical School—with the bar resting nearby—Fig. 27.11), it was learned that destruction of the frontal lobes alleviates certain forms of mental illness. Presumably these mental derangements were brought on by excessive worry, feelings of guilt, etc., and were thus helped by destruction of the frontal lobes. In the operation, called prefrontal lobotomy, the lobes are not actually destroyed. Instead, the fibers connecting the frontal lobes to the thalamus are simply severed.

Today, the operation is performed only infrequently. For one thing, many desirable human attributes are irrevocably lost as a result of the operation. Furthermore, the development of tranquilizing drugs has permitted doctors to achieve the same type of improvement in a much less drastic way.

Our discussion of the forebrain would not be complete without mention of the thalamus and hypothalamus (Fig. 27.9). The thalamus is the "gatekeeper" of the cerebral cortex. All sensory messages reaching the brain must pass through the thalamus in order to be sensed consciously.

We have already discussed some of the important functions carried out by the hypothalamus. In addition to monitoring and regulating the temperature and water content of the blood, the hypothalamus is the coordinating center for many of the activities of our internal organs. In other animals, and perhaps in humans, the hypothalamus is the center of such "feelings" as thirst, hunger, satiety, sex drive, and rage. The hypothalamus not only has nervous activity but also, as we have seen, produces hormones. Two of these (oxytocin and ADH) are stored in the posterior lobe of the pituitary before being released into the bloodstream. Others (the "releasing" hormones—see Section 25.10) pass to the anterior lobe of the pituitary in veins draining the hypothalamus. There they stimulate the release of the anterior lobe's own hormones (e.g. TSH, LH).

27.10 THE PROCESSING OF VISUAL INFORMATION

In most respects, the human brain remains a "black box." We know something about what goes in and what comes out, but very little of what actually goes on *in* it. We know that the brain is constructed of many interconnecting neurons. We know that neurons can either "fire" or not. These features cannot help but remind us of the organization of a computer: myriads of interconnecting circuits that are, at any moment, either "on" or "off." But what exactly does the brain do with sensory impulses reaching it? How does the brain "process" information?

The question remains to a large extent unanswered, but at least a solid start has been made in determining how the brain processes information reaching it from the eyes. In fact, the processing starts within the eyes. (This should not be surprising inasmuch as the retina is actually an extension of the brain.)

The basic strategy is to use a tiny electrode like those described earlier in the chapter to look for impulses in the neurons of interest. Work with a variety of animals has demonstrated that any single visual receptor will fire if light of sufficient intensity falls upon it. **Rods** and **cones** synapse in the retina with **bipolar cells** which, in turn, synapse with **ganglion cells** (Fig. 26.7). It is the axons of the ganglion cells that make up the optic nerve. As we learned in the last chapter, each receptor cell does not enjoy its own private circuit back to the brain. There are some 10^8 rods and cones in one human eye but only some 10^6 ganglion cell axons that make up the optic nerve. Thus a given ganglion cell must receive inputs from a number of receptor cells. Examination of the retina shows that a single bipolar cell receives input from several receptor cells and, similarly, a single ganglion cell receives inputs from several bipolar cells.

By inserting a delicate recording electrode into a single ganglion cell and then shining light on the retina, Stephen W. Kuffler demonstrated what goes on in the early steps of visual processing. Even in the dark, ganglion cells have a slow, steady rate of firing. Diffuse light directed on the retina has little effect on this rate. A tiny spot of light, however, can either increase *or decrease* the rate of firing. A given ganglion cell either (1) increases its activity when light falls on a small circular area of the retina and *decreases* it when the light falls on an area of the retina concentric with the first area (Fig. 27.12), or (2) does just the reverse. We assume that light shining upon "off" areas triggers bipolar cells that release an *inhibitory* transmitter at the ganglion cells stimulated by the "on" areas. Diffuse light which illuminates both "on" and "off" areas causes them to cancel each other. Thus the optic nerve is not tell-

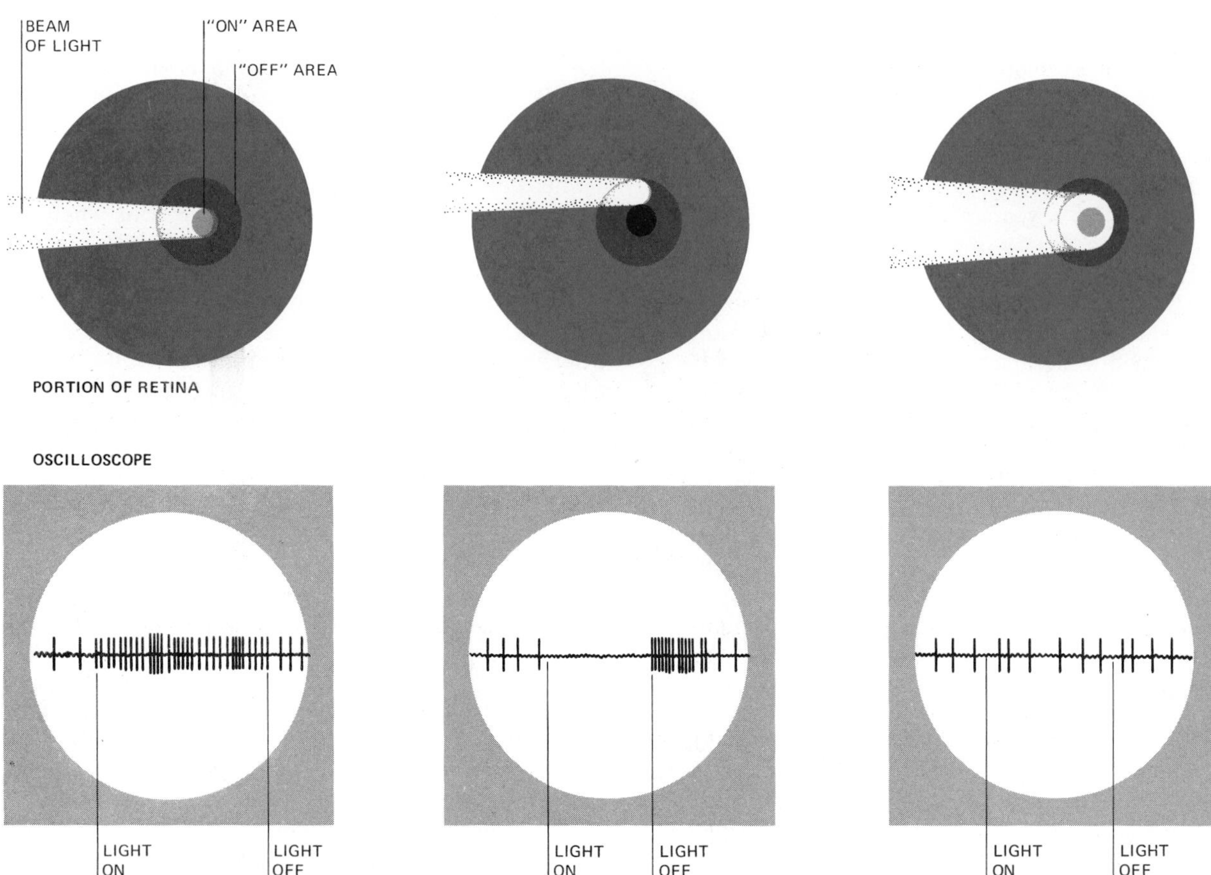

FIG. 27.12 Response of a ganglion cell to illumination of the retina. Left: Light falling on a small circular area of the retina increases the activity of the ganglion cell. Center: Light directed around the perimeter of the "on" area suppresses that ganglion cell. Right: Light shining on both areas produces no effect. Other ganglion cells have a central "off" area surrounded by an "on" area.

ing the brain that light has been detected but that contrasts between light and dark (i.e. shapes) have been detected.

The axons of the optic nerve pass back close to the center of the base of the forebrain to the so-called **lateral geniculate body** (Fig. 27.13), where they form synapses with a new set of interneurons. The axons of these cells lead up and back to the visual cortex. Two associates of Kuffler, David H. Hubel and Torsten N. Wiesel, have pursued the processing of visual information in these areas. Again they insert microelectrodes in the areas of interest, but instead of directing light into the eye, they project images on a screen placed in front of the animal (a cat or a monkey). The animal is anesthetized and the eye prevented from moving. Using this procedure, Hubel and Wiesel found that the cells of the lateral geniculate body respond about the same way that the ganglion cells do, but even more so. The story was quite different in the visual cortex. The first cells in the visual cortex, the so-called "simple cells," no longer respond to circles of light but only to bars of light (or dark) or to straight-line edges between dark and light areas (Fig. 27.13). Only when the stimulus is directed at one area of the screen, and only at a given angle, will a single simple cortical cell respond. (Of course, the ineffective positions for one cortical cell result in stimulation of *other* cortical cells.)

The preference that simple cortical cells have for lines can be explained if we assume that these cells can be activated only if they receive inputs from a

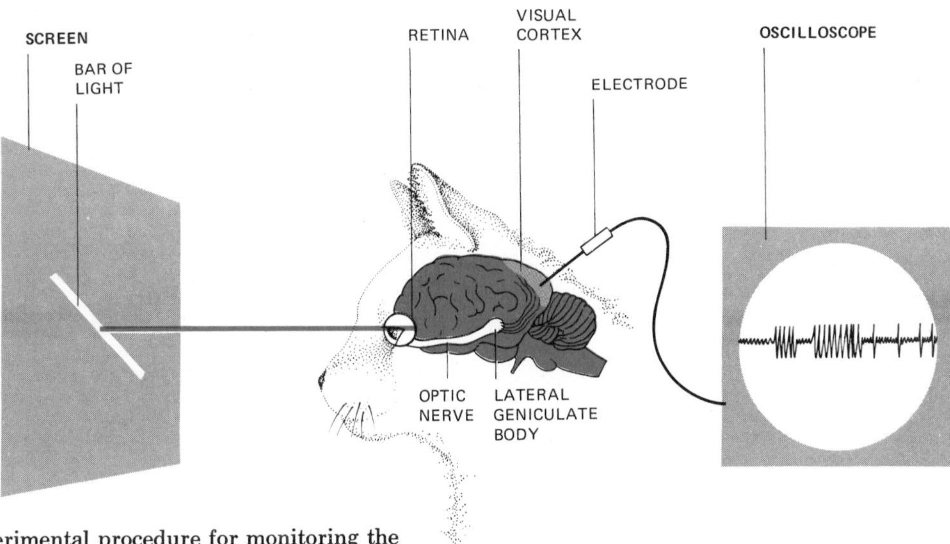

FIG. 27.13 Experimental procedure for monitoring the response of neurons in the visual cortex to various visual stimuli. In a similar way, the activity of ganglion cells (in the optic nerve) and of neurons in the lateral geniculate body can be studied by placing the electrode in those areas.

number of lateral geniculate cells whose areas (circular) of response are arranged in a line (Fig. 27.14).

Probing elsewhere in the visual cortex, Hubel and Wiesel have demonstrated the existence of still other types of interneurons ("complex" and "hypercomplex") that carry out still more processing. Complex cells still want their edges oriented in one direction, but the edges can now be moved over a substantial area of the screen. This makes sense if a series of simple cortical cells—all responding to an edge of the same slope, but each responsible for a different part of the visual field—converge on a single complex cell. Thus a complex cell continues to respond to a given stimulus even though its absolute position on the retina changes (Fig. 27.14).

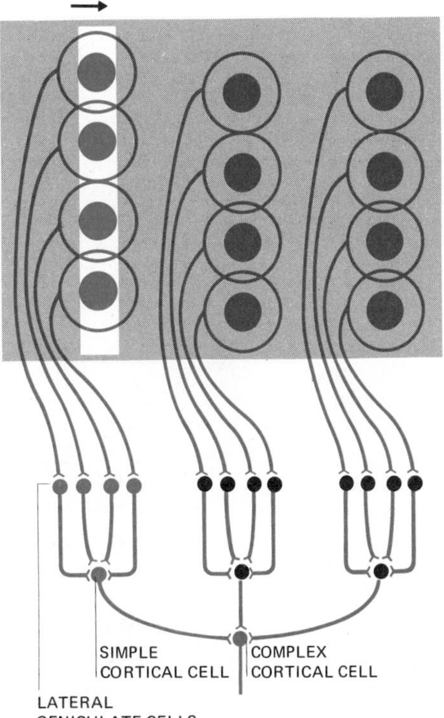

FIG. 27.14 Mechanism by which the circular response areas of ganglion and lateral geniculate body cells may be converted into the rectangular response areas characteristic of simple cortical cells. Movement of the bar of light in the direction shown would continue to activate the complex cortical cell.

The "hypercomplex" cells respond best when the straight-line stimulus, still with a definite slope, is limited in length at one or both ends. Right-angle corners often serve as powerful stimuli for these cells.

The story of how the brain processes visual information is obviously not complete. But even so, we can now get a glimpse of the essential job that the interneurons in the brain carry out. At each level of processing (bipolar cells to ganglion cells, lateral geniculate cells to simple cortical cells, etc.), the inputs of a number of interneurons are funneled into a single output. Thus at each step some of the visual information is selectively destroyed. Simple cortical cells, for example, will fire only if a number of lateral geniculate cells converging on them are simultaneously active (see Section 27.3). If not, the excitation dies out at the synapses. In this way, each level of the brain serves as a filtering device and, in so doing, provides the mechanism by which certain features of what might be a very complex stimulus can be discriminated. This pioneering work on visual processing suggests that the mammalian brain responds not to particular impulses generated in particular circuits, but to the spatial and temporal organization of many impulses passing along a multitude of converging circuits.

The Peripheral Nervous System

The peripheral nervous system is made up of the sensory and motor nerve fibers which run to and from the central nervous system and the rest of the body. It can be subdivided into the sensory-somatic system and the autonomic system.

27.11 THE SENSORY-SOMATIC SYSTEM

The sensory-somatic system consists of 12 pairs of cranial nerves, not all of which are *mixed* nerves (see Section 27.4), and 31 pairs of spinal nerves, all of which are mixed. These nerves transmit impulses from our receptors (chiefly of external stimuli) to the central nervous system. They also transmit impulses from the central nervous system to all the skeletal muscles of the body.

All our conscious awareness of the external environment and all our motor activity to cope with it operate through the sensory-somatic portion of the peripheral nervous system (Fig. 27.15). In the following chapter, we shall examine the mechanism by which the motor commands of the sensory-somatic system are executed by the muscles.

27.12 THE AUTONOMIC NERVOUS SYSTEM

The autonomic nervous system consists of sensory and motor neurons running between the central nervous system (especially the hypothalamus) and the various internal organs: the heart, viscera, and many glands, both exocrine and endocrine. It is thus responsible for detecting certain conditions in the internal environment and bringing about appropriate changes in them.

The actions of the autonomic nervous system are largely involuntary in contrast to those of the sensory-somatic system. Another difference between the two systems is that two groups of motor neurons are used to stimulate the effectors instead of just one. The first, or *preganglionic*, neurons arise in the central nervous system and run to a ganglion in the body. Here they synapse with the second, *postganglionic*, neurons which run to the effector. The autonomic nervous system has two subdivisions, the

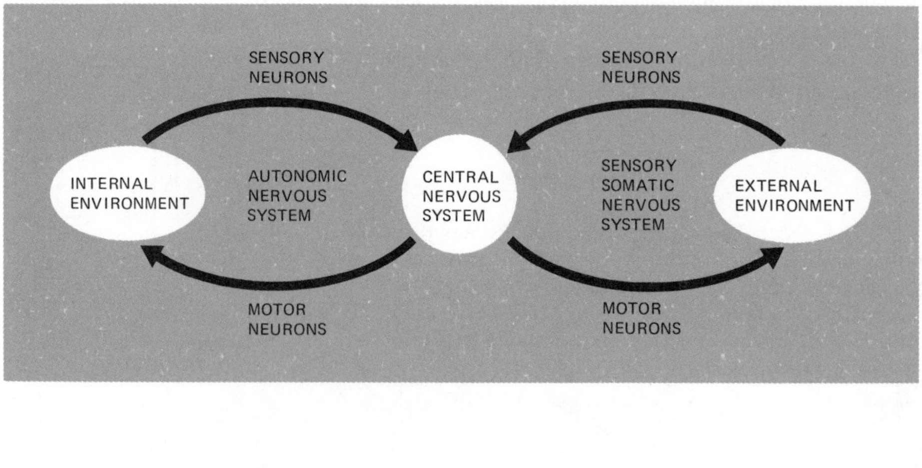

FIG. 27.15 Relationship between the three major divisions of the nervous system.

sympathetic and parasympathetic nervous systems, each with its special organization and functions.

27.13 THE SYMPATHETIC NERVOUS SYSTEM

The preganglionic motor neurons of the sympathetic nervous system arise in the spinal cord. They leave by way of the ventral root of a spinal nerve and pass into a sympathetic ganglion (Fig. 27.16). These ganglia are organized into two chains running parallel to and on either side of the spinal cord (Fig. 27.17). The preganglionic neuron may do one of three things in the sympathetic ganglion. It may (1) synapse with postganglionic neurons which then reenter the spinal nerve and ultimately pass out to the sweat glands and the walls of blood vessels near the surface of the body, (2) pass up or down the sympathetic chain and finally synapse with postganglionic neurons in a higher or lower ganglion, or (3) leave the ganglion by way of a cord leading to special ganglia (e.g. the solar plexus) in the viscera (Fig. 27.17). *Here* it may synapse with postganglionic neurons running to the muscular walls of the viscera. However, some of these preganglionic sympathetic neurons pass right on through this second ganglion and into the adrenal medulla. Here they synapse with the highly modified postganglionic cells which make up the secretory portion of the adrenal medulla.

The transmitter substance of the preganglionic sympathetic neurons is ACh. It serves to transmit impulses to the postganglionic neurons. A chemical stimulator is also released by the terminals of the postganglionic neurons. In most cases, this is noradrenaline, although in some other animals (for example the frog) the sympathetic fibers release adrenaline instead. The action of noradrenaline (or adrenaline) on a specific gland or muscle is excitatory in some cases, inhibitory in others. Its release by these terminals stimulates heartbeat, raises blood pressure, dilates the pupils, dilates the trachea and bronchi, and stimulates the conversion of liver glycogen into glucose. Sympathetic stimulation also shunts blood away from the skin and viscera to the skeletal muscles, brain, and heart. It inhibits peristalsis in the alimentary canal and contraction of the bladder and rectum. In short, stimulation of the sympathetic branch of the autonomic nervous system duplicates most, if not all, of those actions carried out by the adrenaline and noradrenaline released

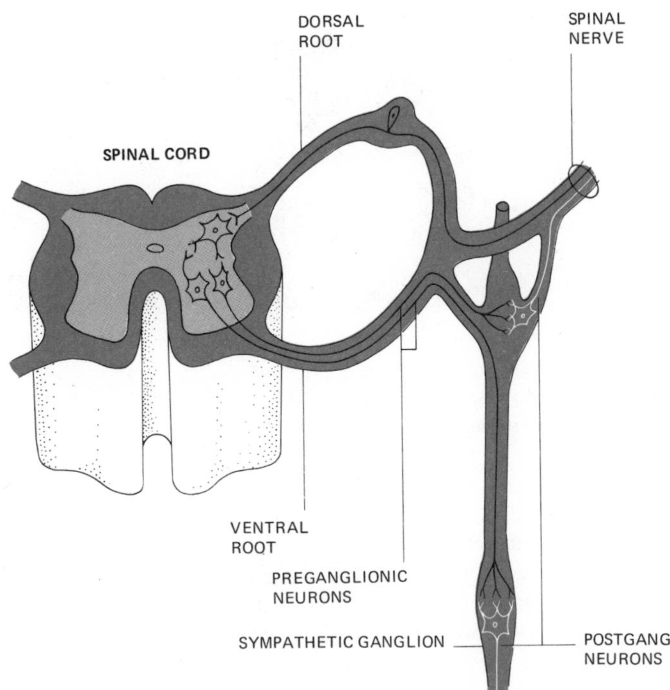

FIG. 27.16 Pathways of the sympathetic neurons. The preganglionic neurons are shown in black, the postganglionic neurons in white.

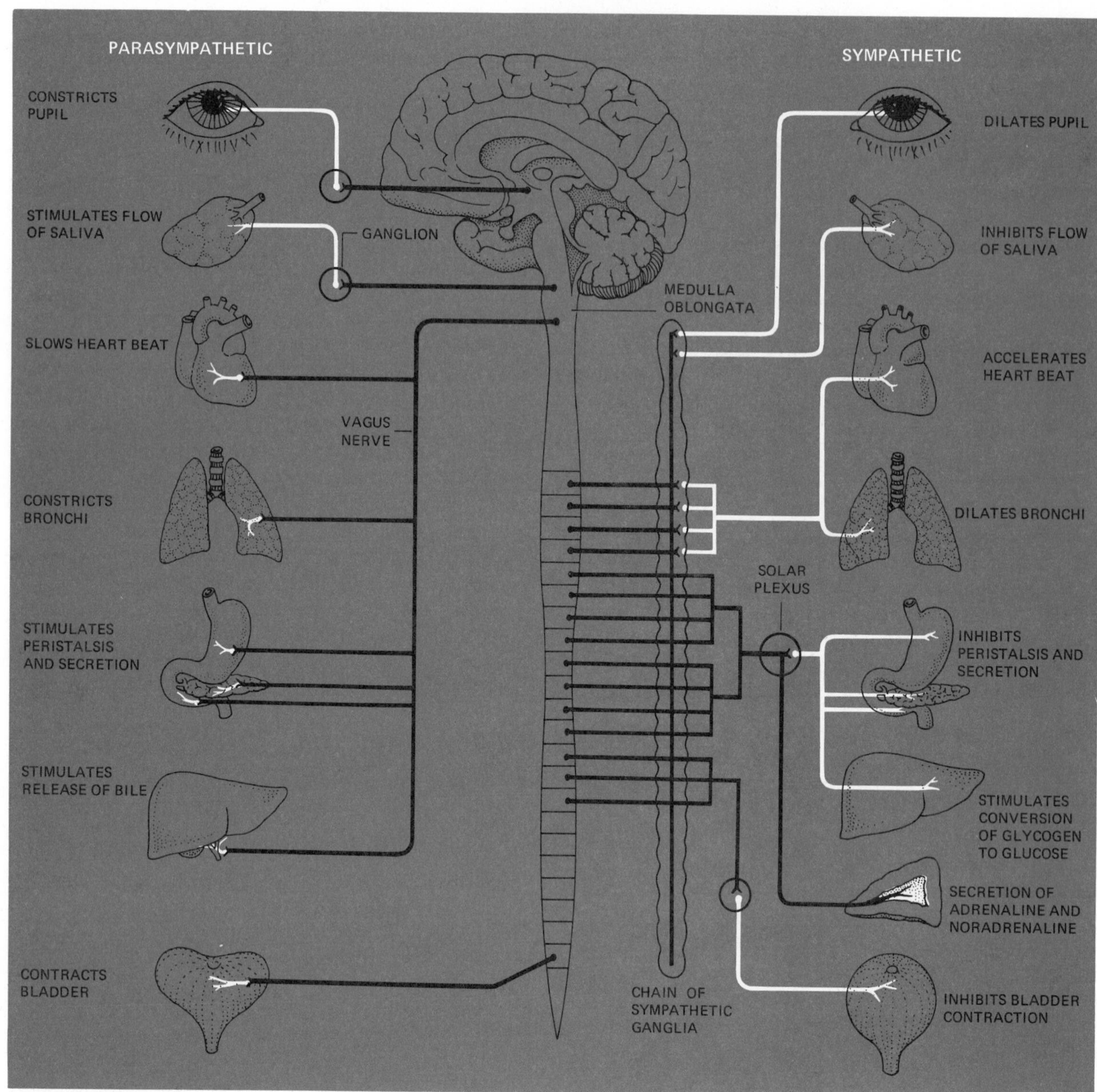

FIG. 27.17 The autonomic nervous system. The sympathetic nerves prepare the body for emergencies. The parasympathetic nerves reverse the effects of sympathetic stimulation. The preganglionic neurons are shown in black, the postganglionic neurons in white.

into the blood by the adrenal medulla. This should not be surprising when we remember that the adrenal medulla really is a part of the sympathetic nervous system, its secretory cells being modified postganglionic cells.

The actions produced by stimulation of the sympathetic nervous system are quite general. There are two reasons for this. One is that a single preganglionic neuron usually synapses with many postganglionic neurons. What starts as a single impulse becomes magnified. Second, the release of adrenaline and noradrenaline into the bloodstream ensures that every cell of the body will be exposed to these substances when necessary, even if no postganglionic neurons reach them directly.

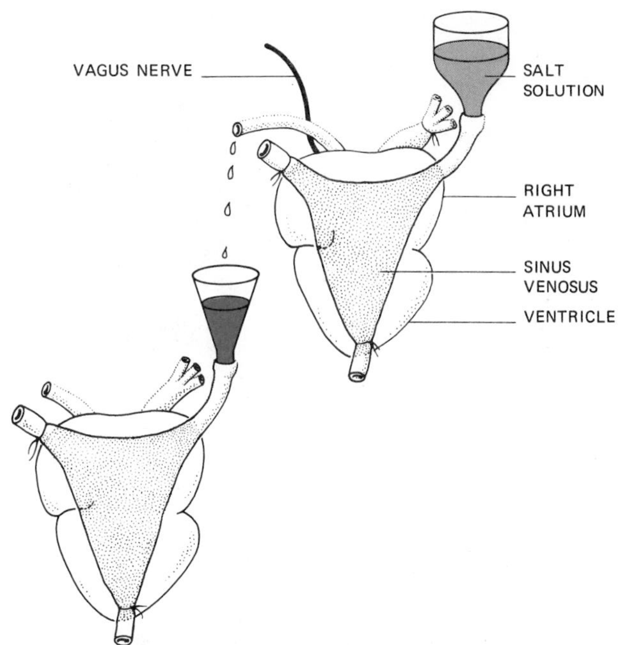

FIG. 27.18 Loewi's experiment, which showed that nerves exert their effects through the release of chemicals. Electrical stimulation of the vagus nerve leading to the first heart not only slowed its beat but, a short time later, slowed that of the second heart, too.

27.14 THE PARASYMPATHETIC NERVOUS SYSTEM

The main nerves of the parasympathetic system are the tenth cranial nerves, the **vagus** nerves, which arise in the medulla oblongata. Other preganglionic parasympathetic neurons also extend from the brain as well as the lower tip of the spinal cord (Fig. 27.17).

Each preganglionic parasympathetic neuron synapses with just a few postganglionic neurons, which are located near or in its effector organ, a muscle or gland. The synaptic knobs of the preganglionic neurons and the terminals of the postganglionic neurons both release acetylcholine. Stimulation of the parasympathetic nerves causes a slowing down of heartbeat, lowering of blood pressure, constriction of the pupils, increased blood flow to the skin and viscera, and promotes peristalsis in the alimentary canal. In brief, the parasympathetic nervous system serves to return our body functions to normal after they have been altered by sympathetic stimulation. In times of danger, the sympathetic nervous system prepares us for violent physical activity such as fighting or fleeing. These changes would be harmful if prolonged unnecessarily, and the parasympathetic system reverses them when the danger is over. We must therefore include the antagonistic activity of these two branches of the autonomic nervous system among the most important body mechanisms for maintaining homeostasis.

The discovery that specific chemical substances are released when either branch of the autonomic nervous system is stimulated was made by the Nobel Prize-winning physiologist Otto Loewi in 1920. He carefully removed the living heart of a frog with its sympathetic (*accelerator*) and parasympathetic nerve supply intact. As was expected, electrical stimulation of the first speeded up the heart while stimulation of the second slowed it down.

Loewi also found that these two responses would occur in a second frog heart if he simply bathed it with Ringer's solution taken from the stimulated heart (Fig. 27.18). This showed that chemical substances were produced by the first heart which duplicated the action of nervous stimulation. In the case of parasympathetic (vagus) stimulation, the substance was later identified as acetylcholine. During sympathetic stimulation, adrenaline (in the frog) is released.

While the autonomic nervous system is considered to be involuntary (as its name suggests), this is not entirely true. A certain amount of conscious control can be exerted over it, as has long been demonstrated by some practitioners of the Eastern religions of yoga and Zen Buddhism. During their periods of meditation, these people are clearly able to alter a number of autonomic functions including rate of heartbeat and rate of oxygen consumption. These changes are probably not simply a reflection of decreased physical activity, because they exceed the amount of change which occurs during sleep or hypnosis (Fig. 27.19). Neal Miller and his coworkers, working in the United States, have demonstrated that dogs and laboratory rats can be trained to alter, either by an increase or a decrease (as the experimenter desires), such accepted autonomic functions as blood pressure, rate of heartbeat, peristalsis, and

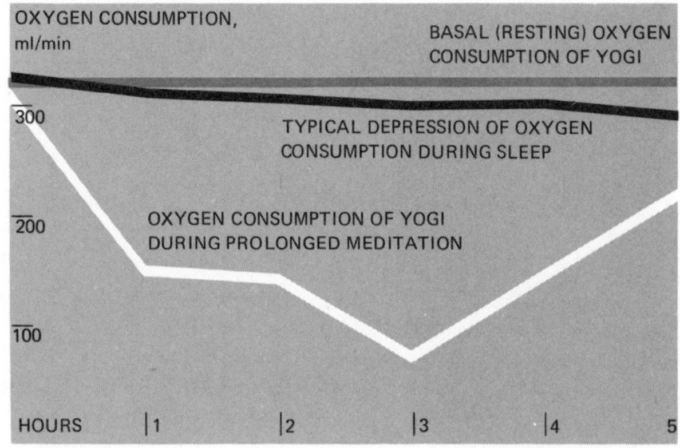

FIG. 27.19 Effect of meditation on oxygen consumption. Properly trained individuals can consciously alter a number of other functions that are controlled by the autonomic nervous system.

distribution of blood to various capillary beds. As a consequence of these discoveries, attempts are now being made to train humans with high blood pressure or heart rate to lower these to normal levels. While such learning has been successfully acquired for a time, whether it can be retained for useful periods after the training program is stopped is yet to be demonstrated.

27.15 DRUGS AND THE NERVOUS SYSTEM

The activity of the nervous system depends upon a well-orchestrated ensemble of biochemical activities. Some of these, e.g. energy release by cellular respiration, closely resemble those found elsewhere in the body. Others, such as the synthesis, release, and action of transmitter substances, are unique to the nervous system. In either case, there are many opportunities for altering nervous system activity by administering chemicals that mimic or block, or in some other way alter, one or more of its biochemical activities. In view of the manifold activities of the nervous system, it should be no surprise that alterations of its activity lead to alterations in behavior and consciousness. Perception, muscular coordination, and emotions are among the functions of the brain that chemicals can alter.

There is plenty of evidence that humans have deliberately consumed certain chemicals (e.g. ethyl alcohol, cocaine) since early in their history for the sake of achieving what they felt were beneficial or pleasurable alterations in feelings and behavior. In recent years, the success of organic chemists in synthesizing an enormous variety of psychoactive chemicals or drugs has greatly enlarged the possibilities for altering nervous system activity. Much of this research has led to advances in helping the mentally ill. But, concurrently with these advances, there has grown up an increasing interest in and consumption of these substances among people with no history of mental illness. While the use of psychoactive drugs is found among all ages and all peoples, the recent increase in the variety of drugs used has been especially dramatic among young people in the developed countries of the world.

The growth of what is often called the drug culture has provided the impetus for additional research into the mode of action of these drugs. However, the knowledge gained is as yet fragmentary. Several approaches have been taken. One is to administer the drug, either to humans or to experimental animals, and then to attempt to correlate the physiological and behavioral changes observed with known brain functions. Another approach is to administer radioactively labeled drugs to an experimental animal and see if and where they become concentrated within the brain. A good deal of effort has also gone into analyzing isolated preparations, that is, the *in vitro* effect of a drug on enzyme systems, properties of cell membranes, etc.

In considering the extraordinary diversity of psychoactive drugs, it soon becomes apparent that they fall into three major categories with respect to their physiological and behavioral effects. These are the stimulants, the depressants, and the hallucinogens.

Stimulants

The most widely used stimulants are caffeine (in coffee, tea, and cola beverages), nicotine (in cigarettes), and the amphetamines. Each of these stimulates the sympathetic nervous system, presumably through controlling centers in the hypothalamus. Each of the activities (e.g. acceleration of heart rate, dilation of the pupils, increase in blood sugar) described in our discussion of the adrenal medulla (see Section 25.11) and of the sympathetic nervous system (see Fig. 27.17) is promoted by these drugs. The sympathetic stimulation brought about by caffeine is very mild; by nicotine somewhat less so; and by the amphetamines, e.g. dexedrine, methylamphetamine ("speed"), quite powerful. Because of the role of the adrenal medulla and the rest of the sympathetic nervous system in preparing the body for stress, it is not surprising that many athletes have turned to amphetamines in an effort to improve their performance. The few studies that have been made indicate that some types of athletic performance (e.g. running) may be enhanced following use of amphetamines—perhaps mainly from a reduced sense of fatigue. Activities requiring complex interactions with teammates are not improved and, in fact, deteriorate following use of amphetamines.

The amphetamines also affect other functions associated with the hypothalamus, such as increasing thirst and decreasing hunger and sleepiness. Because of their depressant effect on appetite, amphetamines have been widely used to help people lose weight. What little long-term success they have brought about seems to be far outweighed by the psychological and physical deterioration that constant stimulation of the sympathetic nervous system produces.

Cocaine also stimulates the sympathetic nervous system. It has been used for thousands of years by certain tribes in the Andes of South America. It has few medical uses, and its nonmedical use in other countries has not (yet?) approached that of the other stimulants.

Depressants

As the name suggests, the depressants reduce nervous system activity. There are five major categories of depressants. These are:

1. Ethyl alcohol (ethanol).
2. Barbiturates. These include such drugs as Seconal, Nembutal, and Amytal.
3. Tranquilizers. The most widely used tranquilizers are meprobamate (trade names: Miltown and Equanil), chlorpromazine (Thorazine), chlordiazepoxide (Librium), and diazepam (Valium).
4. Opiates. These include opium, morphine, heroin, codeine, and methadone.
5. Anesthetics. These include ether, chloroform, and a number of other volatile hydrocarbons that are used as solvents (e.g. benzene, toluene, and carbon tetrachloride).

Ethyl alcohol is, by a wide margin, the most widely used depressant drug, not only in the United States but in most of the world. While its precise mode of action is obscure, its actions suggest a general reduction in neuron function in the brain. It appears that the sensitivity of the brain to inhibition decreases from the "top" of the system to the "bottom," i.e. from the frontal lobes of the forebrain to the medulla oblongata of the hindbrain. Thus the first effects of alcohol intoxication occur in the frontal lobes. The resulting removal of inhibitions may well give rise to the illusion that the drug is actually stimulatory. As the concentration of alcohol in the blood increases, one can watch the progressive changes resulting from the depression of lower and lower brain centers. Loss of dexterity and insensitivity to touch occur as the motor and sensory regions of the cortex become inhibited. Depression of the visual, auditory, and speech areas of the cortex leads to distorted vision, interference with hearing, and difficulty in speaking. At still higher alcohol levels, coordination and balance are lost, presumably reflecting inhibition of the cerebellum. Depression of the reticular formation produces unconsciousness and, later, coma. In rare cases, people have ingested enough alcohol to depress the medulla oblongata to the point where breathing stops and death occurs.

The barbiturates mimic some of the actions of alcohol, particularly in their ability to depress the reticular formation (thus promoting sleep) and, in high doses, the medulla oblongata (thus producing respiratory failure). As one might expect, barbiturates and alcohol act synergistically, the combination producing a depression greater than either produces alone. The combination is a frequent cause of suicide, both accidental and planned. It is estimated that barbiturate-alcohol poisoning is responsible for two-thirds of the suicides in the United Kingdom.

There are several other depressant drugs that share most of the properties of the barbiturates but are not related to them chemically. Prominent among these is methaqualone, a widely used (and abused) sedative. Like barbiturates, methaqualone acts synergistically with alcohol, and the combination may so depress the medulla oblongata that breathing stops.

Tranquilizers constitute another group of depressants that have seen greatly expanded use in recent years. These drugs act like barbiturates in reducing anxiety and tensions, but they do not share their sleep-inducing effects. Some of the more potent tranquilizers have been of enormous medical benefit in the treatment of psychotic patients.

Opium is a mixture of substances from which a number of "opiates" can be derived. These substances seem to depress the thalamus—the gateway for sensory impulses reaching the cerebral cortex—which would explain their effectiveness as pain killers. Morphine and codeine are used in large amounts as part of legitimate medical practice. Heroin is still more effective as a pain killer but is so highly addictive that its use is illegal in the United States. Methadone is a synthetic opiate that is now receiving extensive trials as a means of breaking addiction to heroin.

Although the opiates (chiefly heroin) are perhaps the most dramatically addictive drugs, they share this property with *all* the depressants described so far. Each of these drugs produces two related physiological effects. One is tolerance, that is, the necessity for a steadily increasing dose in order to achieve the same physiological and psychological effects. The second is physical dependence. Addiction implies an unwillingness to give up the drug. In this sense, virtually all psychoactive drugs are addictive. The depressants, however, produce physical as well as psychological dependence. After a period of regular use, stopping the drug precipitates a whole complex of "withdrawal" symptoms. These are always unpleasant and sometimes fatal.

A number of volatile hydrocarbons are such effective depressants of the central nervous system that they can be used as anesthetics. They would have none but a medical interest if it were not for the fact that several of them (e.g. benzene, toluene) are too dangerous to be used in medicine but make excellent industrial solvents. They find widespread use as components of cleaning fluid, glues, and other materials. In a number of countries, "sniffing" glue has become popular among very young people. It produces an effect like alcohol intoxication, and sometimes more. A number of cases have been reported where inhaling volatile hydrocarbons has caused sudden death. Laboratory studies indicate that the cause may be a blockage in the transmission of the electrical impulse that triggers contraction of the ventricles (see Section 11.18).

FIG. 27.20 Peyote cactus (*Lophophora williamsii*) seen in flower. The cactus head contains several hallucinogenic substances, of which mescaline is the most important. The dried cactus heads ("mescal buttons") have been used by Mexican Indians in their religious practices since pre-Columbian times. About a century ago, the religious use of mescal buttons spread to several Indian tribes in the United States and Canada who, in 1922, became incorporated into the Native American Church. (Courtesy of Dr. Richard Evans Schultes.)

The Hallucinogens

In moderate doses the hallucinogens have a powerful distorting effect on the subject's visual and auditory perceptions. There is also a marked enhancement of emotional responses. With higher doses, true hallucinations may occur—i.e. the subject "sees" and "hears" things that are not there at all. All the hallucinogens—mescaline (Fig. 27.20), psilocybin, lysergic acid diethylamide (LSD), and dimethoxymethylamphetamine—share a striking similarity of chemical structure (Fig. 27.21). While their precise mode of action within the brain is still in question, it is certainly significant that their structure closely resembles that of serotonin, a natural component and presumed transmitter substance in parts of the brain. Dimethoxymethylamphetamine (STP) is also of interest in that it shares (as does mescaline) the structure and the properties of *both* the hallucinogens, e.g. LSD, *and* the amphetamines (Fig. 27.21).

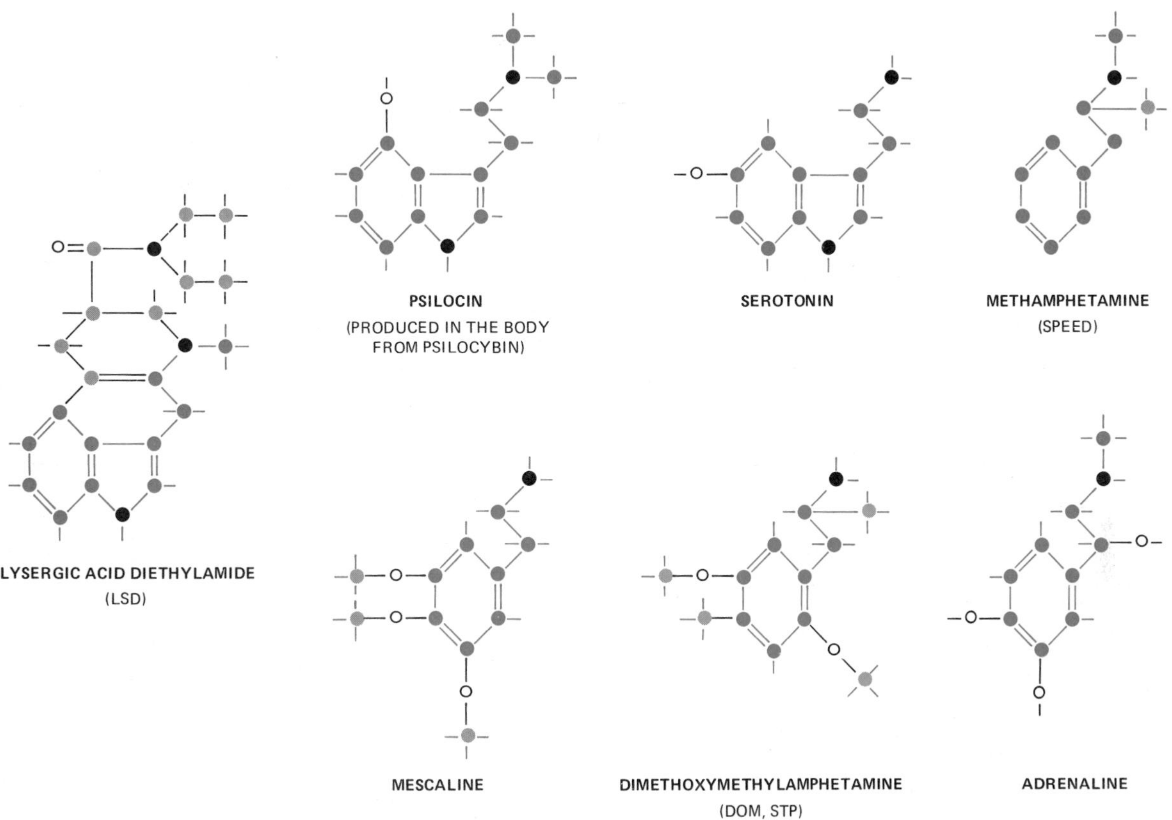

FIG. 27.21 Molecular mimicry may account for the effects of certain psychoactive drugs. The hallucinogens are similar in structure to serotonin, which is probably a transmitter substance in parts of the brain. The similarity of the amphetamines to adrenaline undoubtedly accounts for their ability to duplicate the effects of sympathetic nervous system activity. (Carbon atoms are shown in color, nitrogen atoms in black, and hydrogen atoms as short dashes.)

The place that marijuana should occupy in our long list of psychoactive drugs is uncertain. Until the isolation of its most active component, tetrahydrocannabinol (THC), little research had been carried out on the effects of marijuana. THC concentrates in certain areas of the brain where it has some of the milder effects of depressants like alcohol and, in high doses, some of the perception-distorting effects of the hallucinogens. Unlike the other depressants, tolerance and physical addiction to THC do not occur. In fact, the drug is excreted so slowly from the body that, with repeated use, a given response is achieved with a lower dose.

As yet we have acquired only a tiny glimpse of the workings of the brain and the mechanisms by which psychoactive drugs exert their effects. As desirable as more such knowledge would be, we should not wait for it before searching for ways to reduce the enormous human waste that abuse of psychoactive drugs entails.

27.16 SUMMARY

In these chapters we have distinguished between chemical coordination and nervous coordination of body functions. Now we see that all coordination is really chemical. The only thing which differs is the means used to distribute the chemical. Hormones are transported throughout the body by the bloodstream. Transmitter substances are deposited in specific, localized spots by the action of motor neurons.

By its very organization, the latter system permits more rapid, more specific, more variable body activity than the former. There is still another advantage to it. Localized release of transmitter substances by motor neurons permits the build-up of a higher concentration of the substance than could safely be tolerated by the body as a whole. It has recently been shown that the concentration of noradrenaline necessary to stimulate cells in the fatty tissue of laboratory mice and rats would be very harmful if allowed to circulate freely in the bloodstream. Localized production of noradrenaline by sympathetic neurons leading to the fatty tissue avoids this danger.

The body cannot respond to changes in its environment without the third essential feature of nervous coordination: the effectors. These are the structures that take action. Their morphology and physiology is the topic of the next chapter.

EXERCISES AND PROBLEMS

1. What parts of the nervous system participate in the maintenance of balance and coordinated body movements?
2. The drug mecholyl stimulates parasympathetic nervous activity. What is its effect on (a) the pupil, (b) the salivary glands?
3. The drug atropine inhibits parasympathetic activity. Why is it used by ophthalmologists who wish to examine the interior of a patient's eye?
4. Distinguish between cranial nerves and spinal nerves.
5. Describe the changes that occur in the body in times of emergency. Tell how each change helps prepare the body to cope with the emergency.
6. Distinguish between a nerve and a neuron.

REFERENCES

1. ADRIAN, R. H., *The Nerve Impulse,* Oxford Biology Readers, No. 67, Oxford University Press, Oxford, 1974.
2. GRAY, E. G., *The Synapse,* Oxford Biology Readers, No. 35, Oxford University Press, Oxford, 1973.
3. NICHOLLS, J. G., and D. VAN ESSEN, "The Nervous System of the Leech," *Scientific American,* Offprint No. 1287, January, 1974. The small number and large size of the neurons in the leech make this an excellent animal with which to study the physiology of nerve circuits.
4. HUBEL, D. H., "The Visual Cortex of the Brain," *Scientific American,* Offprint No. 168, November, 1963. An account of the early experiments on visual processing in the cat done in collaboration with Torsten Wiesel.
5. STENT, G. S., "Cellular Communication," *Scientific American,* Offprint No. 1257, September, 1972. Includes an up-to-date review of the work of Hubel and Wiesel and its philosophical implications.
6. PETTIGREW, J. D., "The Neurophysiology of Binocular Vision," *Scientific American,* Offprint No. 1255, August, 1972. Application of the techniques of Hubel and Wiesel to determine the way in which visual information from both eyes is processed so as to enable the animal to locate an object in three dimensions.
7. MICHAEL, C. R., "Retinal Processing of Visual Images," *Scientific American,* Offprint No. 1143, May, 1969. Demonstrates that the processing of visual information at the level of the retina is much more elaborate in frogs and ground squirrels than is the case with cats and primates.

8. GORDON, BARBARA, "The Superior Colliculus of the Brain," *Scientific American,* Offprint No. 553, December, 1972. A former student of Hubel and Wiesel extends their techniques to analyzing the nature of the visual processing that occurs in this small portion of the midbrain.

9. GLICKSTEIN, M., and A. R. GIBSON, "Visual Cells in the Pons of the Brain," *Scientific American,* Offprint No. 573, November, 1976. These cells, which receive signals from the visual cortex and pass them on to the cerebellum, play a vital role in the visual guidance of locomotion.

10. GESCHWIND, N., "Language and the Brain," *Scientific American,* Offprint No. 1246, April, 1972. Correlations between the location of brain damage and type of speech disorder provide clues to the way the language areas of the brain are organized.

11. DICARA, L. V., "Learning in the Autonomic Nervous System," *Scientific American,* Offprint No. 525, January, 1970.

12. MILLER, N. E., "Learning of Visceral and Glandular Responses," *Science,* Reprint No. 16, 31 January, 1969.

13. WALLACE, R. K., and H. BENSON, "The Physiology of Meditation," *Scientific American,* Offprint No. 1242, February, 1972. Demonstrates the effects of "transcendental meditation" on body functions generally thought to be under autonomic control.

14. AXELROD, J., "Neurotransmitters," *Scientific American,* Offprint No. 1297, June, 1974. With emphasis on noradrenaline and dopamine.

15. SCHULTES, R. E., "Hallucinogens of Plant Origin," *Science,* Reprint No. 106, 17 January, 1969.

16. GIRDANO, D. A., and DOROTHY D. GIRDANO, *Drug Education,* 2d ed., Addison-Wesley, Reading, Mass., 1976. A fine review of all aspects of drug use. Each chapter contains many references to the technical studies that have been made.

17. BRECHER, E. M., ed., *Licit and Illicit Drugs,* Little, Brown and Company, Boston, 1972. The Consumers Union report on narcotics, stimulants, depressants, inhalants, hallucinogens, and marijuana—including caffeine, nicotine, and alcohol.

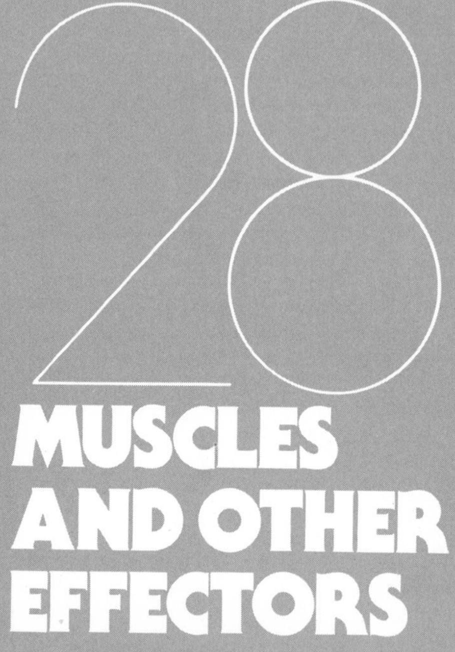

28 MUSCLES AND OTHER EFFECTORS

When we say that all living organisms respond to stimuli, we mean that they do things when the external or internal environment changes in a significant way. The structures with which animals carry out actions are called effectors. The most important effectors are those that secrete substances (the glands) and those that bring about motion. In vertebrates, the most important effectors for creating motion are the muscles.

28.1 KINDS OF MUSCLES

Three distinctly different kinds of muscles are found in vertebrates. One is **cardiac muscle,** the muscle that makes up the wall of the heart. We examined some features of cardiac muscle in the chapter on circulation and will have a little more to say about it later in this chapter.

Smooth muscle is found in the walls of all the hollow organs of the body (except the heart). Its contraction, which is generally not under voluntary control, reduces the size of these hollow structures. The blood vessels, the intestine, the bladder, and the uterus are a few examples of structures whose walls are largely made up of smooth muscle. Contraction of smooth muscle thus accomplishes such diverse tasks as moving your breakfast along through your gastrointestinal tract, expelling urine, and sending babies out into the world.

Skeletal muscle, as its name implies, is the muscle that is attached to the skeleton (Fig. 28.1). It is under voluntary control. Its contraction makes possible such voluntary acts as running, swimming, manipulating tools, and playing the violin.

But whether cardiac, smooth, or skeletal, all muscles share one feature: they are all devices that use the chemical energy of food to accomplish mechanical work.

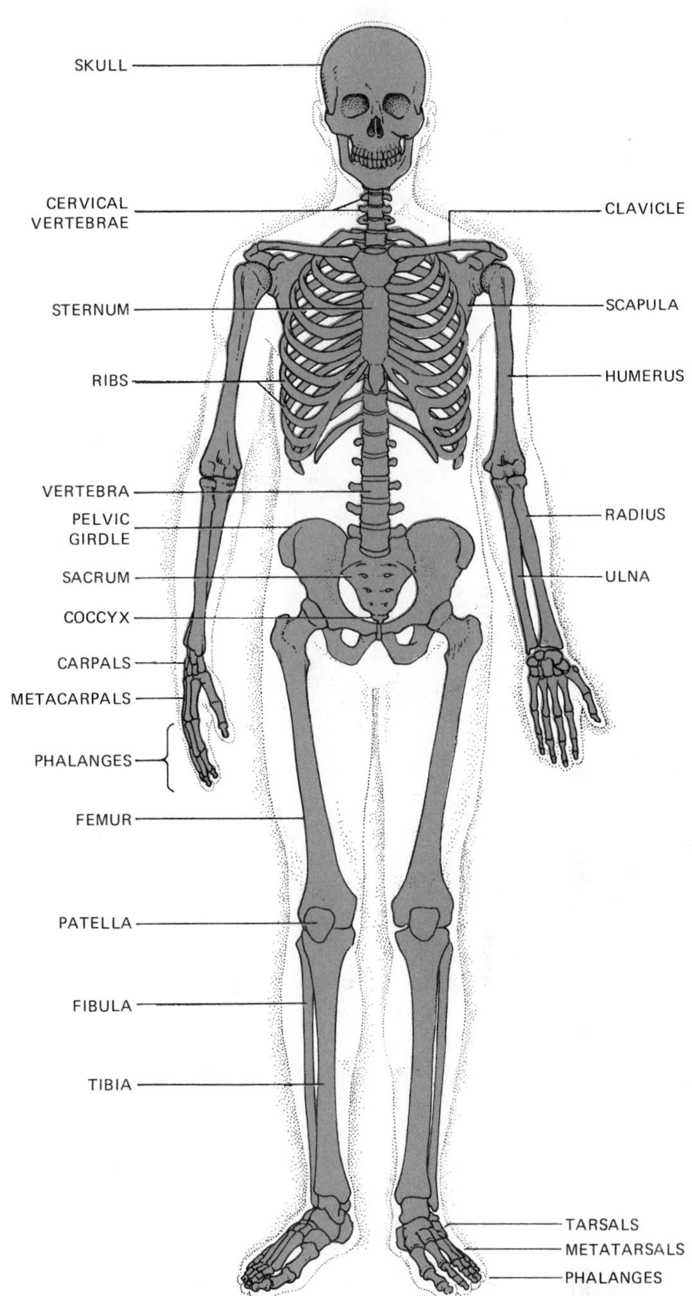

FIG. 28.1 The human skeleton.

28.2 THE STRUCTURE AND ORGANIZATION OF SKELETAL MUSCLE

A single skeletal muscle, such as the triceps muscle (Fig. 28.2), consists of a thickened muscle *belly* attached at each end to bone. At one end, called the *origin,* the muscle is attached directly to a large area of bone, in this case, the humerus. The other end, called the *insertion,* tapers into a glistening white **tendon,** which is attached to the ulna, one of the bones of the lower arm. During contraction, the origin remains stationary while the insertion does the moving. In this case, the arm is straightened or extended

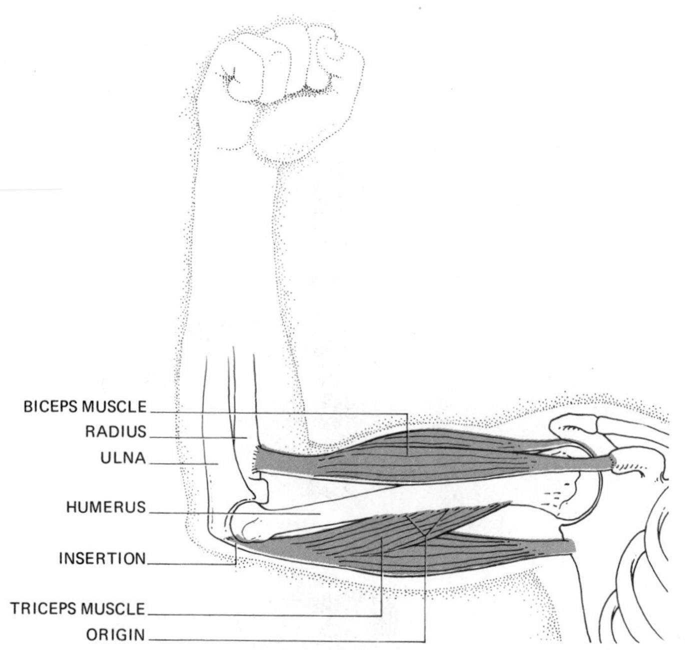

FIG. 28.2 Mechanism of movement of the human forearm. Antagonistic pairs of skeletal muscles accomplish movement at the various joints of the body.

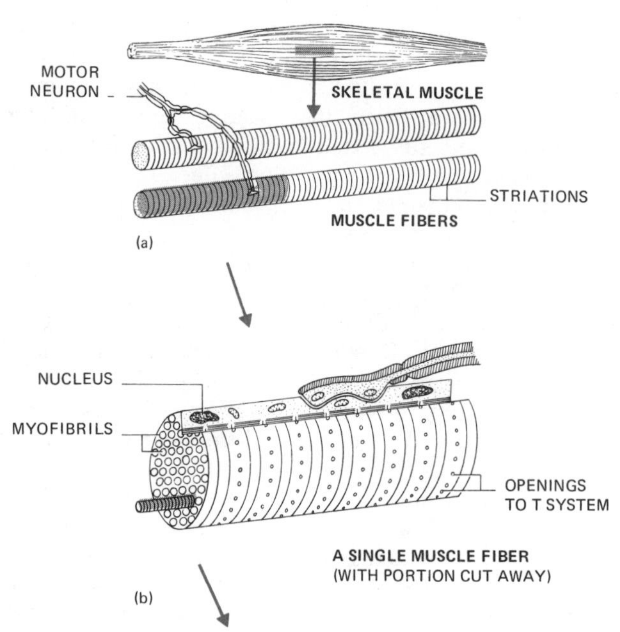

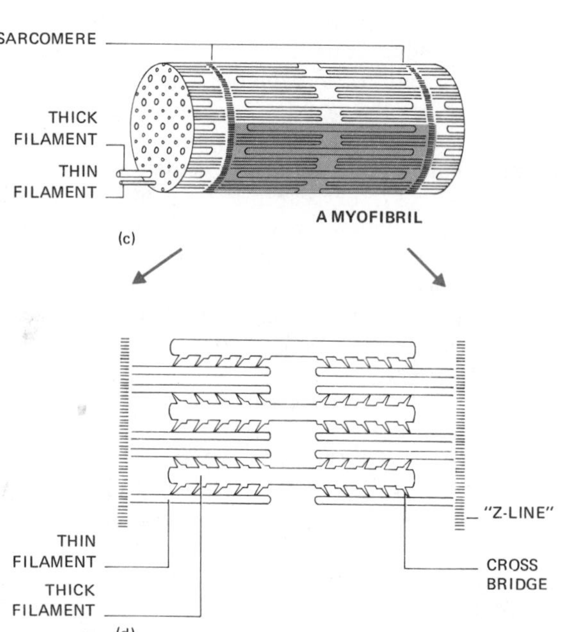

at the elbow joint. Thus the triceps is called an **extensor.** Because muscles exert force only when contracting, not when relaxing, a second muscle, a **flexor,** is needed to crook or bend the joint. The biceps muscle is the flexor for the lower arm. Together the biceps and triceps constitute an **antagonistic pair** of muscles. Similar pairs of muscles, working antagonistically across other joints, enable us to carry out almost all our skeletal movements.

A cross section through the belly of the muscle reveals thousands of muscle fibers (Fig. 28.3a). These are arranged in parallel bundles and, in some cases, extend uninterrupted from origin to insertion. They range in diameter from 10μm to 100 μm. The fibers are bound together by connective tissue through which run blood vessels and nerves.

The number of fibers in a muscle is probably fixed. Increased strength and size of the muscle is produced by an increase in the thickness of the individual fibers and an increase in the amount of other tissue, such as blood vessels and connective tissue, in the muscle.

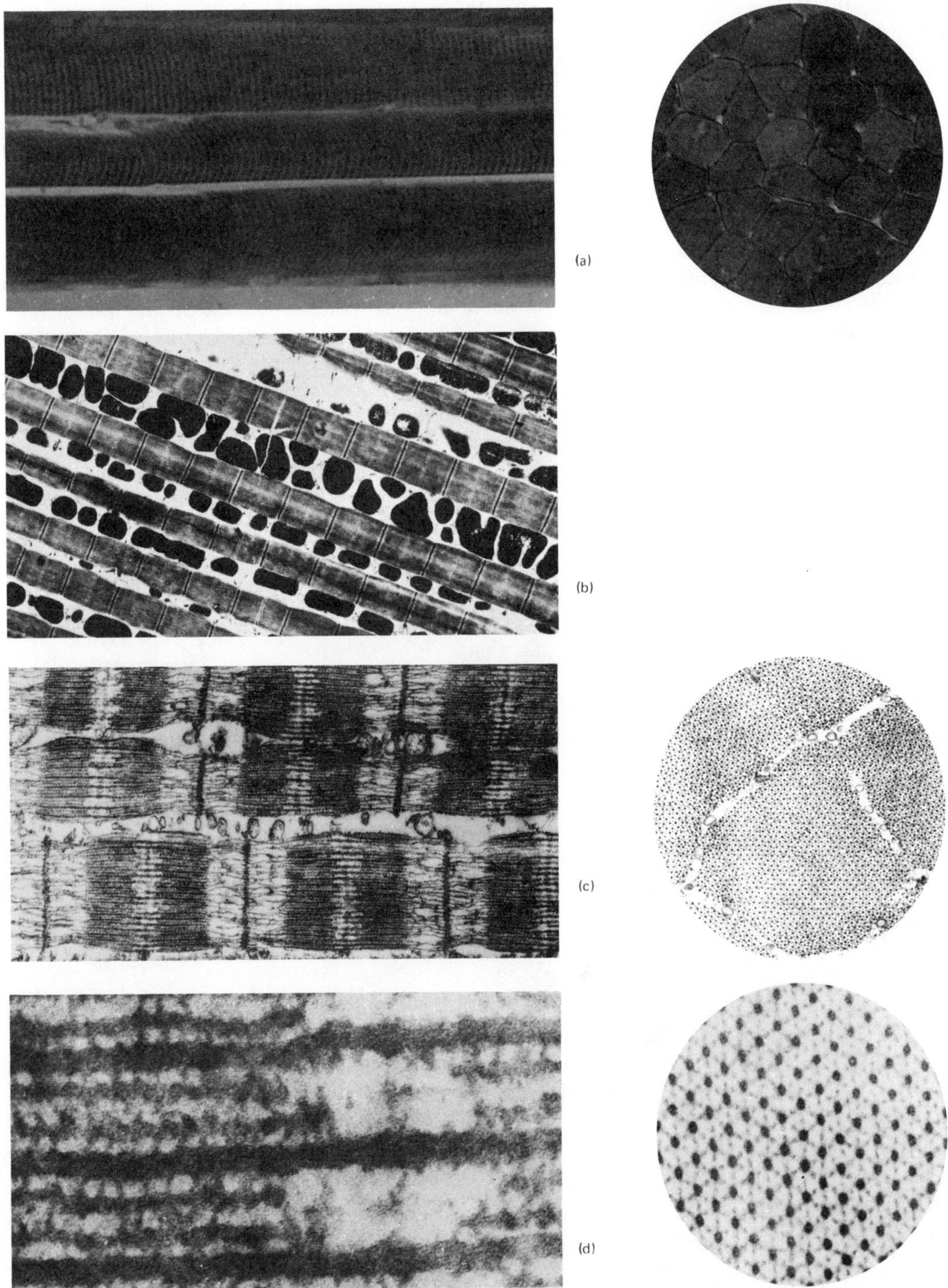

FIG. 28.3 Organization of skeletal muscle at various levels of magnification. The photos on the left above are longitudinal views corresponding roughly to the drawings opposite. The other photos are corresponding cross sections. (All electron micrographs courtesy of Dr. H. E. Huxley.)

Seen from the side, the muscle fibers show a pattern of cross-banding or striations. This appearance gives rise to another common name for skeletal muscle: striated muscle.

28.3 THE ACTIVATION OF SKELETAL MUSCLE

The contraction of skeletal muscle is controlled by the nervous system. If anything blocks the passage of nerve impulses through the motor neurons leading to a muscle, the muscle becomes paralyzed (Fig. 28.4). Skeletal muscle differs somewhat in this respect from smooth muscle and, especially, cardiac muscle. Both cardiac and smooth muscle can contract without being stimulated by the nervous system. Nerves (sympathetic and parasympathetic—see Sections 27.13 and 27.14) lead to smooth and cardiac muscle, but their effect is that of modifying the rate and/or strength of the contractions. Skeletal muscle, in contrast, is totally dependent upon nervous stimulation for its contractility.

Impulses traveling down motor neurons of the sensory-somatic system (see Section 27.11) cause the skeletal muscle fibers at which they terminate to contract. The junction between the terminal of a motor neuron and the muscle fiber is called the **neuromuscular junction.** Its properties are quite like those of a synapse. The tips of the motor axons are called motor end plates (Fig. 28.5). They contain thousands of tiny vesicles that store ACh. When a nerve impulse reaches the motor end plate, hundreds of vesicles discharge their ACh onto the surface of the muscle fiber. Muscle contraction follows. When the job is completed, acetylcholinesterase destroys the liberated ACh and leaves the field cleared for another impulse.

With the use of microelectrodes, we can analyze the electrical events at the neuromuscular junction just as we did for the synapse (see Section 27.3). Let

FIG. 28.4 The Dying Lioness, an Assyrian relief dating from about 650 B.C. Injury to the spinal cord has paralyzed the otherwise undamaged hind legs. (Courtesy of The Trustees of the British Museum.)

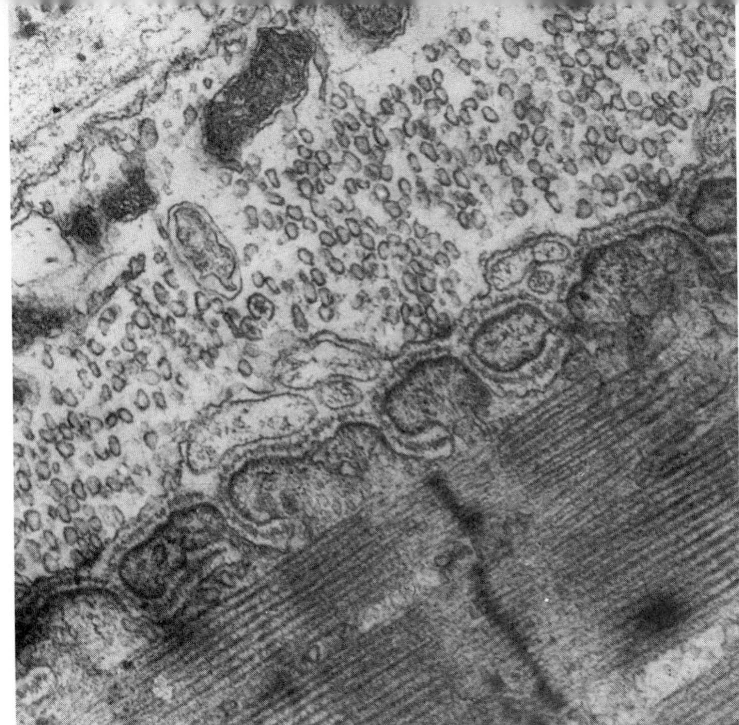

FIG. 28.5 The neuromuscular junction. Many vesicles can be seen in the portion of the motor end plate shown. These contain acetylcholine (ACh). The arrival of an action potential at the motor end plate causes ACh to be released into the gap next to the folded membrane of the muscle fiber (running diagonally upward from the lower left). Binding of ACh to receptors on this membrane causes the initiation of an action potential in the fiber, followed by the other events leading to contraction. (Courtesy of Prof. B. Katz.)

us pierce a single muscle fiber in the vicinity of a neuromuscular junction. When we do so, we discover that the muscle fiber has a resting potential of about 95 millivolts, with the interior of the fiber negatively charged just as is the interior of a neuron (Fig. 28.6). Now let us give a weak electrical shock to the motor neuron leading to that neuromuscular junction. The shock initiates a nerve impulse which, when it arrives at the motor end plate, can be detected by a recorder (Fig. 28.6). About 0.7 msec later, the charge across the membrane of the *muscle fiber* begins to decrease. This is called the *e*nd *p*late *p*otential **(EPP).** If the end plate potential reaches approximately −50 mV (and normally it always does), it triggers an **action potential** in the fiber.

The molecular and electrical events at the neuromuscular junction are very similar to those at the synapse. The arrival of an impulse at the end plate

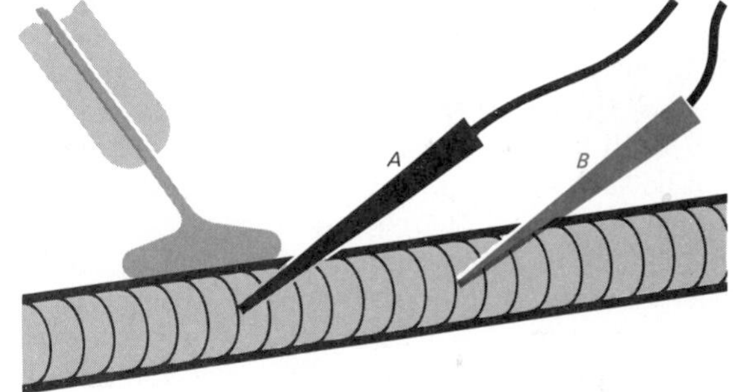

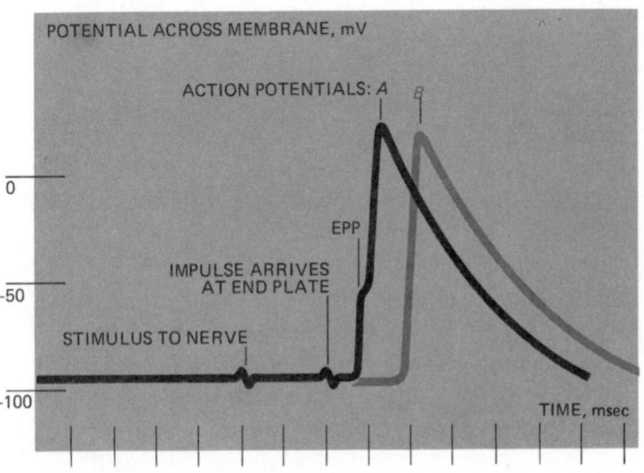

FIG. 28.6 Intracellular recording from a muscle fiber being stimulated by its motor neuron. Arrival of a nerve impulse at the motor end plate creates an end plate potential (EPP) in the membrane beneath it (*A*) but not farther away (*B*). When the EPP reaches the threshold of the fiber, an action potential is generated that sweeps along the fiber (*B*).

causes the release of a transmitter substance (ACh). The ACh increases the permeability of the membrane of the muscle fiber to the influx of sodium ions. The initial inflow of Na^+ shows up as the end plate potential. At threshold (about −50 mV), sodium ions suddenly flow in with a rush and an action potential is generated. The action potential sweeps down the length of the muscle fiber just as it does down the length of an axon. Note that an electrode placed a few millimeters away from the neuromuscular junction detects the passing of the action potential but not the creation of the end plate potential (Fig. 28.6). There is, then, really only one significant difference between the physiology of the synapse and that of the neuromuscular junction. Under normal conditions, a single impulse arriving at the motor end plate will lead to an action potential in the muscle fiber. At a synapse, in contrast, a number of excitatory postsynaptic potentials must be created in order to generate an action potential in the postsynaptic neuron (see Section 27.3).

There are a number of chemical agents that affect end plate potentials. The anaerobic bacterium *Clostridium botulinum,* an occasional inhabitant of canned food, secretes a deadly poison called botulinus toxin. This substance inhibits the release of ACh at the motor end plate. Thus the EPPs become too small to reach threshold and the muscle fiber will not be activated. Paralysis and death (from asphyxiation) is the all-too-frequent outcome.

Curare, a poison used by some South American tribes on their arrowheads, blocks the action of ACh on the fiber membrane. This, too, reduces the size of the EPPs and, when they fail to reach the threshold of the fibers, leads to paralysis.

The size of the EPP can thus be altered by drugs. It is a graded potential just like EPSPs and, like them, is dependent upon the amount of transmitter (ACh) used. But these drugs have no effect on the action potential. As long as threshold is reached, an action potential will be generated. The action potential of the muscle fiber, like that of the neuron, is all-or-none. An EPP greater than threshold produces no greater action potential than one that just reaches the threshold of the fiber.

No visible change takes place in the muscle fiber during (and immediately following) the period in which the action potential is sweeping along it. This period, which may be from 3 to 10 msec long, depending on the particular fiber studied, is thus called the **latent period.** It is followed by contraction of the fiber. Over a period of about 50 msec, the fiber exerts a tension and, if it is permitted to, shortens. Contraction, like the action potential, is all-or-none. Either the fiber contracts maximally or it doesn't contract at all.

The contraction brought on by a single stimulus is followed spontaneously by relaxation. Over a period of 50–100 msec, the fiber relaxes and, if it had shortened, resumes its original length. However, relaxation of the fiber can be prevented if it is restimulated before the cycle of contraction and relaxation is complete. It takes only 1–2 msec for the action potential to sweep down the muscle fiber. Then the resting potential is restored by the diffusion of potassium ions out of the fiber, and the fiber is ready to fire again. These electrical events are completed before the fiber has even begun to contract. Because the "refractory period" is so much shorter than the time needed for contraction and relaxation, the fiber can be *maintained* in the contracted state so long as it is stimulated frequently enough (e.g. 50 stimuli/sec—see Fig. 28.7). Such sustained contraction is called **tetanus.** As we normally use our muscles, the individual fibers go into tetanus for brief periods rather than simply undergoing single twitches.

In vertebrates, motor neurons have only a stimulating effect on muscular contraction. Inhibition of muscles occurs as a result of preventing nerve impulses in the central nervous system from reaching specific motor neurons. However, in crustaceans like the crayfish, some motor neurons release a hyperpolarizing "transmitter" at the neuromuscular junction, thus inhibiting contraction of the muscle fiber.

28.4 THE PHYSIOLOGY OF THE ENTIRE MUSCLE

Although the individual muscle fiber is the *structural* unit of skeletal muscle, it is not the *functional* unit. All motor neurons leading to skeletal muscles have branching axons, each of which terminates in a neuromuscular junction with a single muscle fiber. Nerve impulses passing down a single motor neuron will thus trigger contraction in all the muscle fibers at which the branches of that neuron terminate. This minimum unit of contraction is called the **motor unit.**

For muscles over which we have very precise control, the size of the motor unit is small. For example, a single motor neuron triggers fewer than 10 fibers in the muscles controlling eye movements. The motor unit of the muscles controlling the larynx are as small as two or three fibers per motor neuron. On

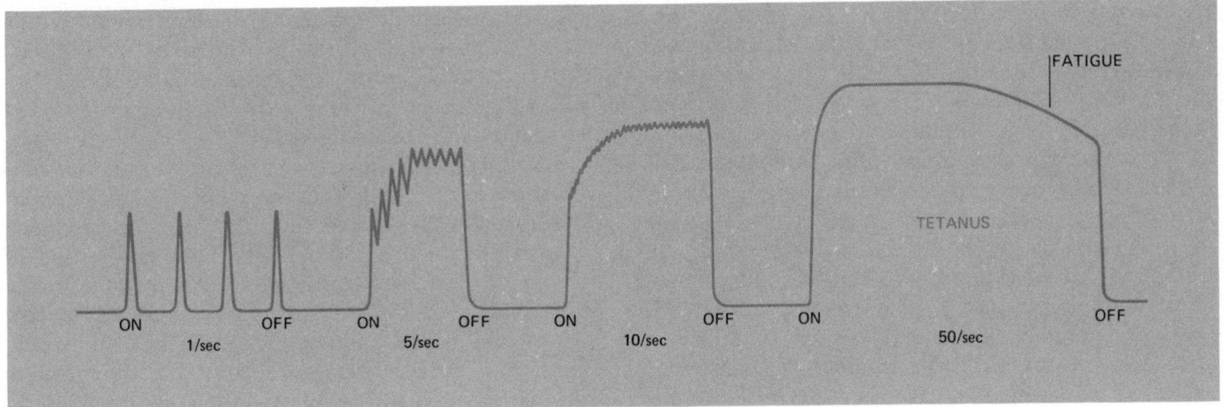

FIG. 28.7 Recordings of isotonic contractions of skeletal muscle as they might be displayed on the kymograph shown in Fig. 28.8. At each shock (1/sec) given when the electrical stimulator is turned on, the muscle gives a single twitch. When the frequency of stimulation is increased (5/sec and 10/sec), the individual twitches begin to fuse together, a phenomenon called clonus. At 50 shocks per second, the muscle goes into tetanus, a smooth sustained contraction. Clonus and tetanus are possible because the refractory period of the muscle is much briefer than the time needed for a complete cycle of contraction and relaxation. Note that the amount of contraction is greater in clonus and tetanus than in the single twitch.

the other hand, for muscles over which our control is less precise (e.g. our calf muscles) a single motor unit may include one to two thousand muscle fibers (scattered fairly uniformly through the muscle belly).

We have seen that the response of a single muscle fiber is all-or-none. Nevertheless, we know that an entire muscle does not behave in this fashion. It is possible to contract a muscle any desired degree from practically relaxed to maximally contracted. This can be demonstrated in the laboratory by stimulating the calf muscle (gastrocnemius) of a frog with an electrical stimulator and measuring, by means of a writing lever (Fig. 28.8), the amount of

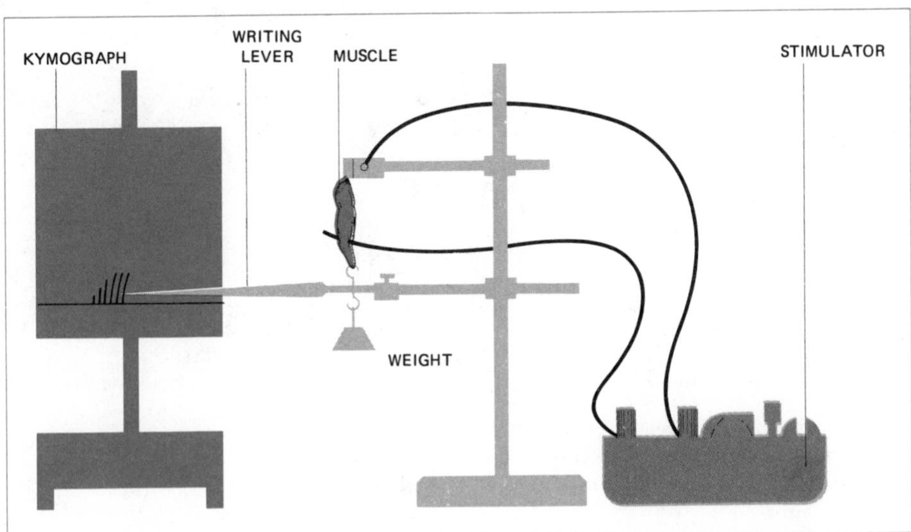

FIG. 28.8 Measuring isotonic contractions of the calf muscle of a frog in response to single shocks of increasing voltage. Contraction of the muscle lifts the writing lever, which marks the paper wrapped around the kymograph drum. Note that the muscle responds more vigorously—up to a point—as the voltage increases. This is because of the increasing *number* of muscle fibers that are activated.

contraction of the entire muscle. Too weak a shock will have no effect at all. When threshold is reached, the muscle twitches slightly. Then as the strength of the stimulus is increased, the amount of contraction increases up to a maximum. Still greater stimuli are no more effective (Fig. 28.8). How can we reconcile this graded response of the entire muscle with the all-or-none properties of the individual fibers which make it up? The answer is that the strength of contraction of an entire muscle increases as the number of individual contracting fibers increases. Thus, in the intact animal, the strength of the muscular response is controlled by the *number of motor units* activated by the central nervous system.

Even at rest, most of our skeletal muscles are in a state of partial contraction called **tonus.** If this were not so, we would have a very difficult time maintaining posture. The action of the motor units provides us with a physical basis for tonus. A few motor units are activated at all times even in the resting muscle. As one set of motor units relaxes, another set takes over. This synchronous activation and deactivation of a small percentage of the total number of motor units maintains the tonus of our skeletal muscles.

The apparatus shown in Fig. 28.8 allows the muscle to shorten as it lifts the weight. Thus the force the muscle exerts while contracting is unchanging. Such contraction is described as **isotonic** ("same tension"). But what if we added so much weight to the lever that the muscle could not lift it at all? The muscle would exert a tension when stimulated, but could not shorten. Such a contraction is described as **isometric** ("same length").

Let us place a muscle in an apparatus that permits no appreciable shortening of the muscle but does permit us to measure the tension produced when the muscle is stimulated. Thus the muscle is limited to isometric contraction. We place it in the apparatus, stretch it to a certain length, give it a series of maximal, tetanizing shocks, and measure the "active" tension, that is, the tension produced when the muscle is stimulated. (Muscles are elastic, so if we stretch the muscle in putting it in the apparatus, it will already be exerting a "resting" tension.) Now let us plot the effect of muscle length on the active tension that can be created. We find that the muscle produces the highest active tension when it is stimulated while held at the length it normally has in the intact animal. If held in the machine at shorter-than-normal or greater-than-normal lengths, the active tension produced is less (Fig. 28.9). At this point, you may simply conclude that nature knows best. (And, in fact, she does. If we surgically reattach a skeletal muscle in an animal so that its length is changed, the muscle gradually adapts to its new length and, after a few weeks, is able to exert its maximum isometric contractions at

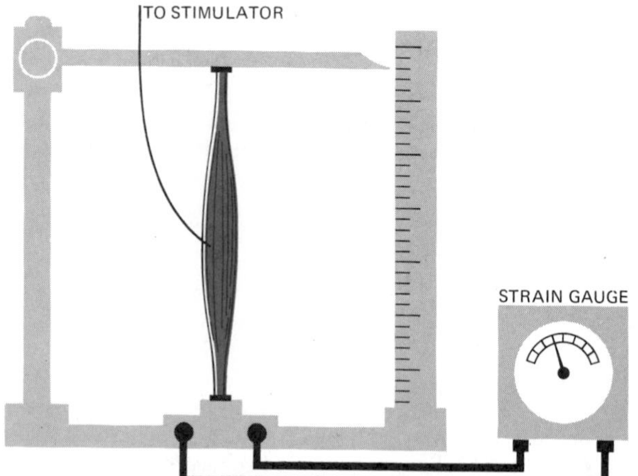

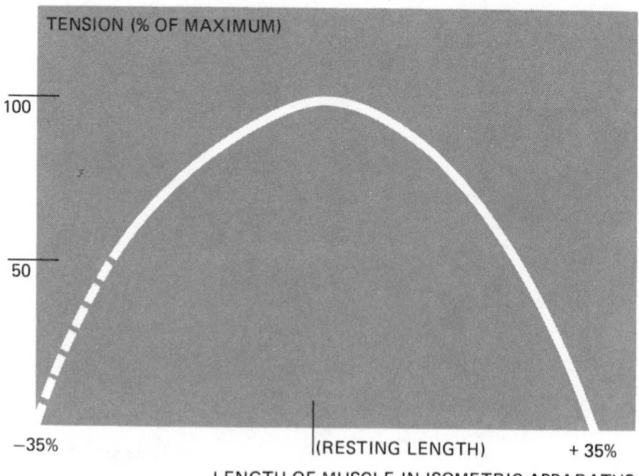

FIG. 28.9 Measuring isometric contractions. The force exerted by the muscle is maximal when the muscle is held in the apparatus at its normal ("resting") length. The active tensions produced decline at shorter lengths and when the muscle is stretched beyond its normal length.

the new length.) But there is more to be learned from this phenomenon. We are now going to examine the molecular basis of skeletal muscle contraction. When we have done so, I think that you will agree that the results we have plotted in Fig. 28.9 are just what we would have predicted.

28.5 THE MUSCLE FIBER

A magnified cross section of a single muscle fiber reveals that it is packed with **myofibrils** in much the same way the muscle belly is packed with muscle fibers. The myofibrils are all stacked lengthwise and seem to run the entire length of the fiber. The myofibrils are surrounded by cytoplasm containing many nuclei, mitochondria, and an extensive endoplasmic reticulum. The presence of many nuclei is a consequence of the fact that during embryonic development, each muscle fiber is formed by the fusion of a number of individual cells. Strictly speaking, then, the muscle fiber is not a cell but a syncytium (see Glossary). For this reason, its parts are often given special names (e.g. sarcolemma for cell membrane, sarcoplasm for cytoplasm, sarcoplasmic reticulum for endoplasmic reticulum), although this tends to obscure the essential similarity in structure and function of these structures and those found in "true" cells. The nuclei and mitochondria are located just beneath the cell membrane (Fig. 28.3b). The endoplasmic reticulum extends between the myofibrils.

The pattern of striations in a muscle fiber arises from the striations present in the individual myofibrils. Normally these are all in register with one another, although in specimens prepared for electron microscopy, the parallel registration is often disrupted.

The dark bands in the myofibrils are called A-bands. Often a somewhat lighter zone, the H-band, bisects them (Fig. 28.3c). The light bands are called I-bands. Each is bisected by a dark line, the so-called Z-line.

The electron microscope provides an explanation for these bands. A cross section in the region of the A-band (but not through the H-zone) reveals a precise geometric array of dots (Fig. 28.3d). These are of two sizes, the larger about 15 nm in diameter, the smaller about 5 nm. If the section is made through an I-band, the pattern of 5-nm dots remains, but the larger ones are gone. Conversely, a section through the H-zone shows only the 15-nm structures. The explanation for these results becomes apparent in high-quality electron micrographs of *longitudinal* sections of the myofibrils (Fig. 28.3d). The myofibril is seen to be made up of arrays of parallel filaments. The larger filaments, called the **thick filaments,** make up the A-band. Their diameter is 15 nm. The **thin filaments,** 5 nm in diameter, extend in each direction from the Z-line. Where present alone, they create the I-band. The H-zone is that portion of the A-band where the thick and thin filaments do not overlap (Fig. 28.10). The array of thick and thin filaments between the Z-lines is called a **sarcomere.** Sarcomeres vary in length from 1.5 to 3 μm. And, as we shall see, it is the ability of each of the sarcomeres in a myofibril to shorten from 3 to 1.5 μm that produces the overall shortening of the myofibril and, in turn, the muscle fiber of which it is a part.

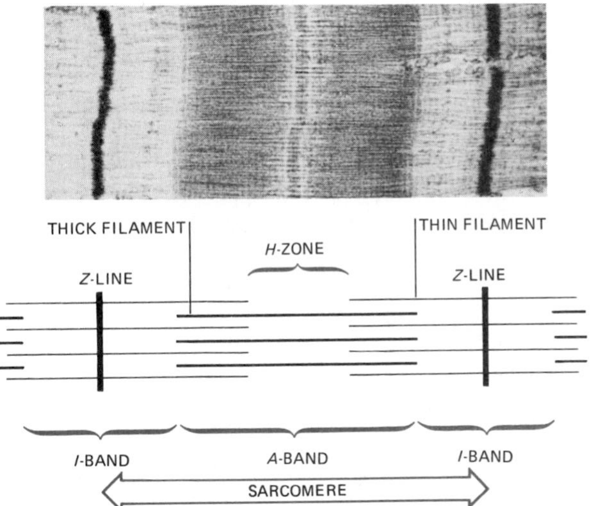

FIG. 28.10 The anatomy of the sarcomere. The thick filaments are composed of myosin. The thin filaments are composed of actin, tropomyosin, and troponin. (Courtesy of Dr. H. E. Huxley.)

28.6 THE CHEMICAL COMPOSITION OF SKELETAL MUSCLE

If we take some skeletal muscle fibers and soak them in glycerol or a dilute salt solution, a number of proteins leak out of the fibers. These include the oxygen-binding protein myoglobin (see Section 4.4) and many enzymes, including those of glycolysis (see Section 8.5). But microscopic examination of the fiber shows that no change has occurred in the appearance of the sarcomeres. And, in fact, such a preparation will still contract if we supply it with magnesium ions, ATP, and calcium ions.

If we then treat the preparation with a higher concentration of salt, large quantities of a protein called **myosin** are removed. Now the fibers can no longer contract and microscopic examination reveals that their A-bands have disappeared. Evidently, the thick filaments are composed of myosin and are essential for contraction.

More drastic treatment removes another protein called **actin,** together with two associated proteins, **tropomyosin** and **troponin.** This process destroys the structure of the *thin* filaments. On these and other grounds, it is apparent that these three proteins make up the thin filaments.

If we mix a solution of actin with a solution of myosin, the resulting mixture becomes very viscous. This is because the two proteins bind together. If we now add some ATP to the mixture, the viscosity is eliminated and the ATP is hydrolyzed into ADP and inorganic phosphate. When all the ATP has been broken down, the viscosity returns. This suggests that the complex of actin and myosin (called actomyosin) is an ATPase. Furthermore, ATP seems to alter the binding between the actin and myosin.

The latter point can also be shown with the dead muscle fibers (described earlier) that have had all their soluble proteins removed by soaking them in glycerol. Such fibers are stiff and brittle. Exposed to ATP, however, they quickly become soft and supple. (This observation explains the phenomenon of *rigor mortis,* the stiffening of skeletal muscles following death. With death, production of ATP ceases, and the proteins of the muscle fibers become irreversibly bound together.)

28.7 THE SLIDING-FILAMENT HYPOTHESIS

A portion of the myosin that makes up the thick filaments consists of globular subunits. It is these subunits that bind to actin (and also to ATP). Under the electron microscope, these subunits appear like cross bridges between the thick and thin filaments (Fig. 28.3d). We assume that in some way, perhaps by a swiveling motion, the cross bridges draw the thin filaments a tiny distance past the thick filaments. Then, the cross bridges must break (for which ATP is needed) and reform farther along the thin filament to repeat the process. As a result, the filaments are pulled past each other in a ratchet-like action. According to this mechanism, there is no need for the individual filaments to shorten, thicken, or fold during the process.

How could one test this so-called "sliding-filament" hypothesis? Let us first ask what the theory would lead us to predict about the appearance of the sarcomeres in stretched and contracted muscle. Would we expect that the width of the A-bands would change as the fiber length changed? The I-band? The H-zone? If a muscle fiber is stretched, the thin filaments should be pulled out from between the thick filaments. This would make the I-band and the H-zone wider. However, the width of the A-band would remain constant. As you can see in Fig. 28.11, the prediction is fulfilled exactly.

Now let us reexamine the active tensions produced by isometrically contracting muscle (Fig.

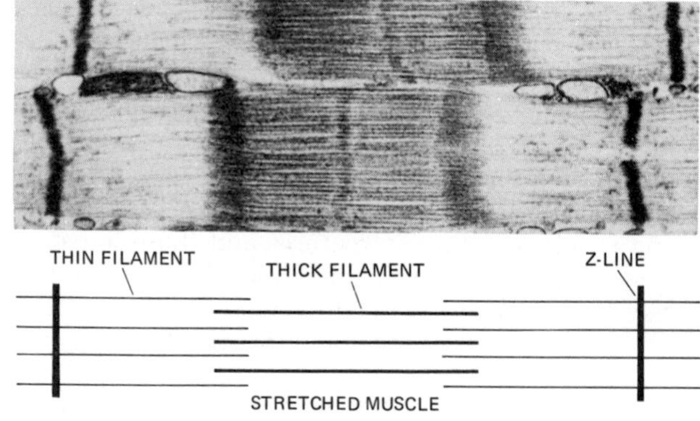

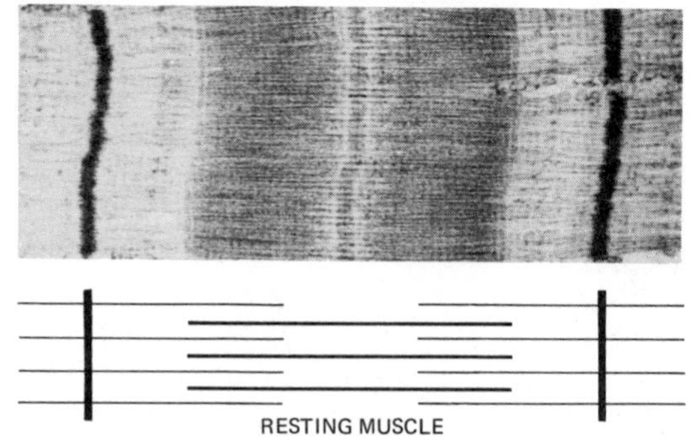

FIG. 28.11 Pattern of striations in stretched muscle and resting muscle. In stretched muscle, there is less overlap of the thick and thin filaments. Consequently the I-bands and H-zone become wider. The width of the A-band remains unchanged. (Courtesy of Dr. H. E. Huxley.)

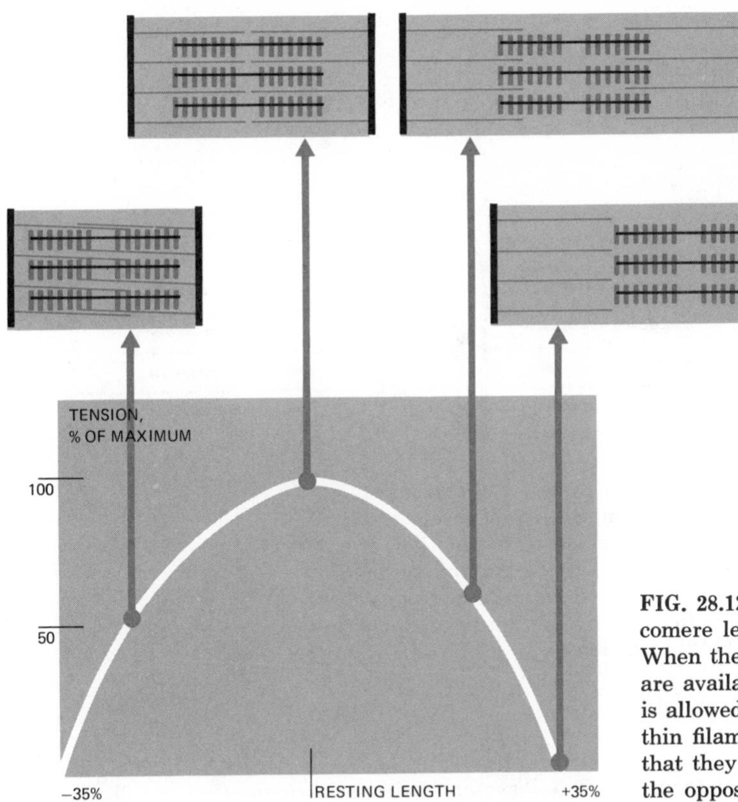

FIG. 28.12 Relationship between tension and sarcomere length in muscle contracting isometrically. When the muscle is stretched, fewer cross bridges are available to contribute to the tension. When it is allowed to shorten below its resting length, the thin filaments extend so far across the sarcomere that they interact with cross bridges exerting force the opposite way. This reduces the tension generated.

28.9). If the cross bridges exert the ratchet-like action, we must assume that the more cross bridges at work, the greater the tensions produced. We found that a muscle is capable of exerting its greatest tension when held at its normal resting length. Examination of its sarcomeres shows that the amount of overlap of thick and thin filaments is such that the maximum number of cross bridges are brought into play (Fig. 28.12). As the muscle is stretched, we find that the active tensions are reduced. This makes sense because the area of overlap of thick and thin filaments decreases as the sarcomeres are elongated (Fig. 28.11). In fact, if we stretch the muscle far enough, the thin filaments are pulled entirely away from the thick filaments. At this length, the stimulated muscle exerts no tension at all (Fig. 28.12). Thus the effect of length on the tension produced by an isometrically contracting muscle is entirely consistent with the idea that the tension is created by the cross bridges trying to pull the thin filaments into greater overlap with the thick filaments.

28.8 COUPLING EXCITATION TO CONTRACTION

A major gap remains in our story. We have examined the events leading to the creation of an action potential in a muscle fiber. We have developed a model to explain the shortening that takes place when skeletal muscle is stimulated. Now we must ask how the first event—excitation—leads to the second—contraction.

It turns out that there is another way in which we can get a muscle fiber to contract. This is to inject into the fiber (using a micropipet) a solution of calcium ions (Ca^{++}). When we do so, the fiber contracts. But there has been no action potential. Alternatively, we can treat a muscle fiber with a chemical that soaks up any calcium ions already present in the fiber. When such a muscle is stimulated, an action potential is generated but no contraction follows. Evidently, calcium ions play a role in linking the two processes.

Now, using radioactive calcium, we can look for

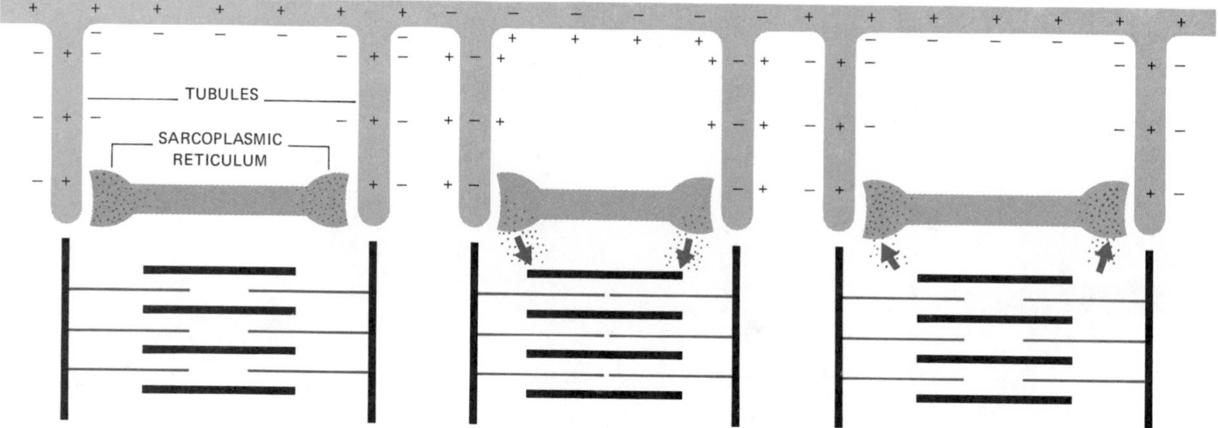

FIG. 28.13 Coupling excitation to contraction. Left: The T-system and sarcoplasmic reticulum in resting muscle. Middle: Propagation of an action potential reverses the polarity of the tubules of the T-system. This triggers the release of calcium ions from the sarcoplasmic reticulum. The binding of these calcium ions to the troponin on the thin filaments "turns on" the interaction between actin and myosin, and the sarcomere shortens. Right: Restoration of normal polarity is followed by the return of calcium ions to the sarcoplasmic reticulum and relaxation of the sarcomere.

the distribution of the ions in muscle fibers. It turns out that in *resting* fibers all the calcium is sequestered in the sacs of the endoplasmic ("sarcoplasmic") reticulum. These are particularly well developed in the regions of the Z-lines (Fig. 28.13), in other words, at the boundaries of the sarcomeres. When the fiber is activated, the calcium is released from the sarcoplasmic reticulum and diffuses among the thick and thin filaments of the sarcomere. Here it binds to troponin, one of the proteins of the thin filament. This union of calcium ions with troponin "turns on" the interaction between actin and myosin and thus movement of the filaments. When the process is over, the calcium is taken back into the sarcoplasmic reticulum.

A problem remains. A single muscle fiber is packed with myofibrils. Each myofibril is a long array of repeating sarcomeres. But if each sarcomere is "turned on" individually, how can we assure that all the sarcomeres of a myofibril *and* all the myofibrils of the fiber turn on at once? There is a neat solution. Spaced evenly along and around each fiber are inpocketings of its membrane. These are called the T-system (Fig. 28.3b). The tubes of the T-system terminate at or near the Z-lines, that is, right at the calcium-filled sacs of the sarcoplasmic reticulum. The T-system, being an extension of the fiber membrane, carries the action potential deep into the fiber. The speed at which the action potential travels assures that it will arrive at all the ends of the T-system at about the same time—triggering the release of calcium ions. Thus contraction of all the sarcomeres throughout the muscle fiber is virtually spontaneous.

28.9 THE CHEMISTRY OF MUSCULAR CONTRACTION

The immediate source of energy for muscle contraction is ATP: the source of free energy for most living processes. But a muscle fiber contains only enough ATP to power a few twitches. To what source of energy can the fiber turn in order to maintain a steady supply of ATP? The best source is, of course, the cellular respiration of nutrient molecules brought to the fiber by the blood. However, if we place an isolated muscle in an anaerobic atmosphere so that cellular respiration cannot occur, it is still able to contract vigorously for a long period. This is because the muscle fiber is able to produce energy by glycolysis (Fig. 28.14). Glycogen represents about 1% of the weight of a muscle fiber. When the fiber is

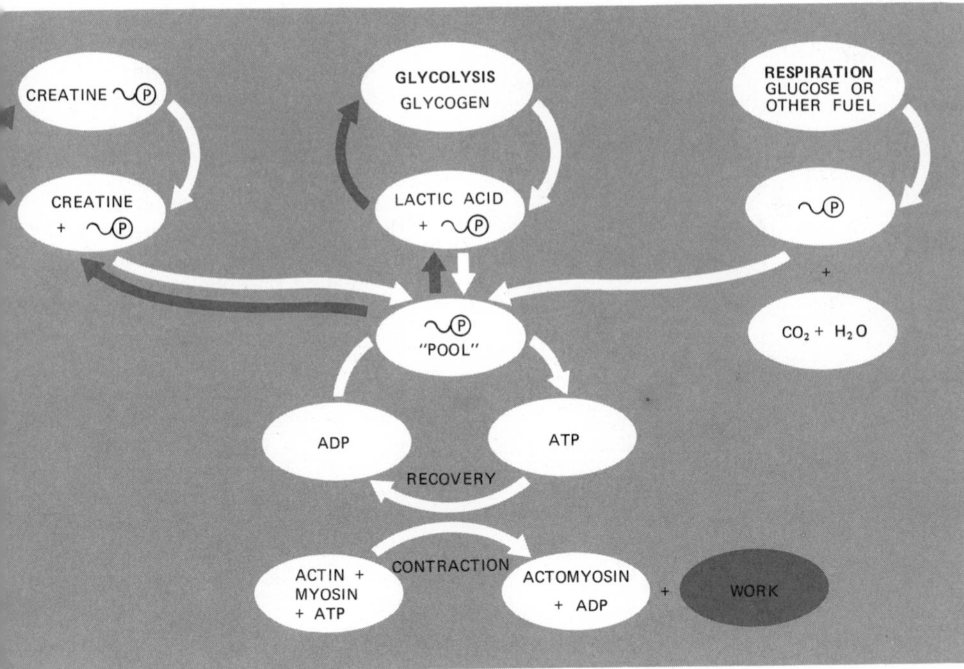

FIG. 28.14 The chemistry of muscular contraction. (Pathways in color function during payment of the oxygen debt.)

deprived of oxygen, its glycogen is broken down into glucose 1-phosphate. This is converted into its isomer glucose 6-phosphate, which enters the pathway of glycolysis that you studied in Chapter 8 (look back at Fig. 8.5). The NAD-catalyzed oxidation of 3-phosphoglyceraldehyde eventually yields pyruvic acid. With no oxygen available to which NADH can give up its electrons, it turns around and reduces the pyruvic acid to lactic acid (Fig. 8.5). While ATP is needed for the process, ATP is also generated. The net yield is two molecules of ATP for each pair of lactic acid molecules produced. Not much, perhaps, but enough to enable the muscle to keep functioning if oxygen is absent or, more commonly, if the supply of oxygen is inadequate to meet the needs of the muscle.

If we take our muscle preparation and treat it with iodoacetic acid, the muscle continues to contract. This is a significant finding because iodoacetic acid poisons the enzyme 3-phosphoglyceraldehyde dehydrogenase (Fig. 8.5) and thus puts an abrupt stop to glycolysis. Is there, then, yet another source of ATP production? The answer is yes: the compound **creatine phosphate** (CP) (Fig. 28.14). The phosphate group in this compound is attached by a "high-energy" bond like the high-energy bonds in ATP. Creatine phosphate is able to donate its high-energy phosphate to ADP, converting it into ATP.

$$CP + ADP \rightleftharpoons C + ATP$$

Thus the pool of creatine phosphate in a muscle fiber (about ten times larger than that of ATP) serves as a modest reservoir of energy for ATP synthesis.

To help us understand how these several energy sources are used, let us examine the biochemical events that occur in the skeletal muscles of an athlete running a mile. Let us assume (1) that our runner is very good, requiring only 4 minutes to complete the race, and (2) that he runs the race at an even pace (which is probably the best tactic on biochemical grounds if not always the best racing strategy). Figure 28.15 shows the rate of oxygen consumption while the runner is at rest. The theoretical amount of oxygen needed to supply energy during the race from cellular respiration *alone* is shown in black. The *actual* curve of oxygen consumption is shown in color. As you can see, when the race begins there is an increase in the rate of oxygen consumption (hence cellular respiration), but it takes a minute or more to reach a maximum rate and even that is insufficient for the job. What makes up the deficit? First, there

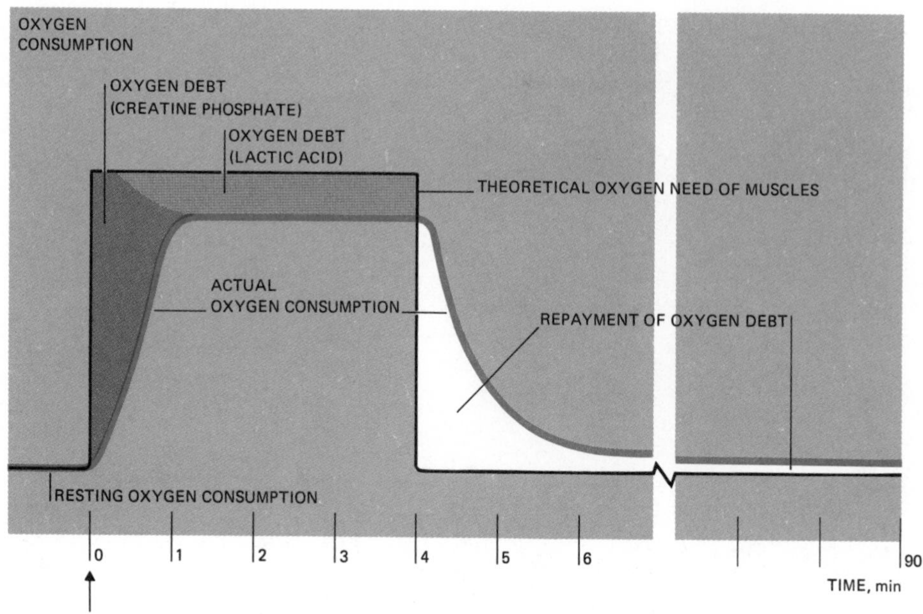

FIG. 28.15 Relationship between oxygen need and actual oxygen consumption during and following four minutes of violent physical activity such as running a mile. While the heart, lungs, and blood vessels adjust to distributing more oxygen to the muscles, the energy deficit is met primarily by the breakdown of creatine phosphate. As the race progresses, lactic acid fermentation makes up the energy deficit. After the race, cellular respiration continues at greater than "resting" levels in order to pay back the oxygen debt. Resynthesis of creatine phosphate in the muscles occurs rapidly. Resynthesis of glycogen from lactic acid, which occurs in the liver, takes place more slowly.

is the breakdown of creatine phosphate. The sudden increased demand for ATP, in the instant after the starter's gun fires, is supplied by creatine phosphate. At the same time, the rate of delivery of oxygen to the muscles begins to increase rapidly as the runner's lungs, heart, and blood vessels adapt to the new demands. Soon the supply of creatine phosphate becomes depleted. After about 15 seconds, its contribution to ATP production begins to fall off. Something must take up the slack and that is glycolysis. After a minute or more, the rate of cellular respiration reaches a steady value. If this is insufficient to maintain the pace—as is likely—continued glycolysis makes up the deficit (Fig. 28.15).

You know that once the runner crosses the finish line, his rate and depth of breathing do not immediately drop back to resting values. Instead there is a period of continued high levels of cellular respiration. It is during this period that the **oxygen debt**—the difference between the oxygen theoretically needed for the race and that actually consumed—is paid off (Fig. 28.15). Biochemically, it is a measure of the elevated rate of cellular respiration needed, first, to replenish the muscle's store of creatine phosphate and, second, to provide the energy for the resynthesis of glycogen from the lactic acid produced by glycolysis.

The rebuilding of the pool of creatine phosphate is very rapid. It takes place right within the muscle fiber. The resynthesis of lactic acid into glycogen takes much longer. One reason for this is that the process does not occur in the muscles, but in the liver. The lactic acid liberated by the exhausted muscles is carried in the blood to the liver. Here it is resynthesized into glycogen, using the energy provided by cellular respiration. Much later, *muscle* glycogen is restored by transport of glucose from the liver to the muscles. The slightly increased oxygen consumption needed to pay off the glycolysis-produced portion of the oxygen debt may last for several hours (Fig. 28.15).

The graph in Fig. 28.15 demonstrates a biochemical reason for the success of the use of "interval training" by runners. Interval training consists of alternating short (e.g. 30 sec) runs with a recovery period perhaps two to three times as long. If the length of the run is short enough, only the breakdown of creatine phosphate will contribute to the oxygen debt. Unlike the lactic acid debt, this debt can be paid off quickly during the recovery period at the conclusion of the run. Then the runner repeats the sequence. In this way a serious runner may accumulate 10 miles or more of running at very nearly top speed during the course of an afternoon. This is far more than he could have achieved by attempting the same total distance in one or only a few runs.

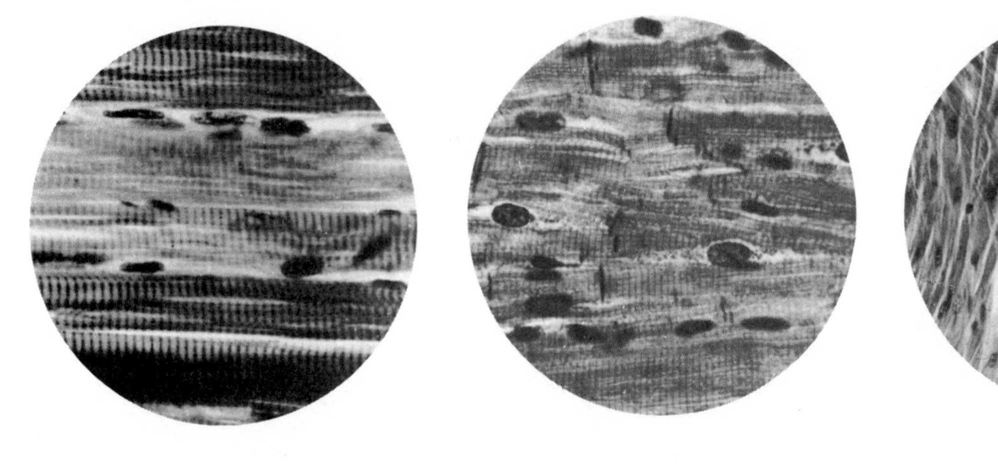
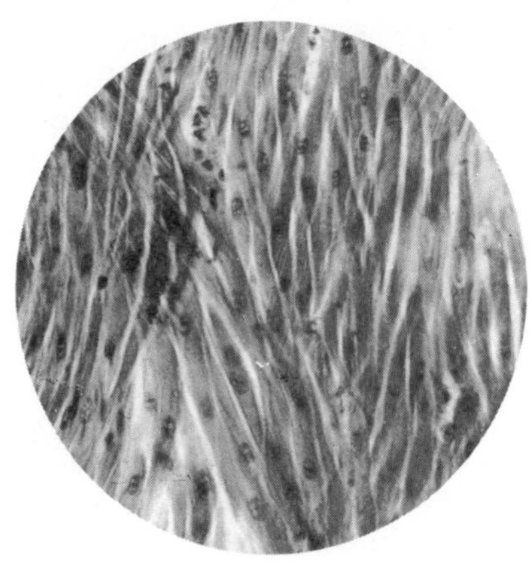

FIG. 28.16 The three kinds of muscle found in vertebrates as they appear under the light microscope. Skeletal and cardiac muscle are made up of long, multinucleate fibers. Smooth muscle is made up of single, spindle-shaped cells. (Skeletal muscle courtesy of Ward's Natural Science Establishment, Inc. Others courtesy of Turtox/Cambosco, Macmillan Science Company, Chicago.)

28.10 CARDIAC MUSCLE

Cardiac muscle is made up of interlocking, striated fibers (Fig. 28.16). If you look back at Fig. 11.17, you will see that the myofibrils run roughly parallel to one another and that the organization of the sarcomeres is like that found in skeletal muscle. However, the myofibrils in cardiac muscle are often branched and mitochondria are more abundant than in our skeletal muscle fibers.

Cardiac muscle differs from skeletal muscle in its physiology as well. The impulse which causes cardiac muscle to contract is self-generated. It is true that sympathetic and parasympathetic nerves run to the heart, but their control is purely accessory. Even when they are destroyed, the heart continues to beat as long as glucose and oxygen are available. Any deprivation of oxygen, such as that which occurs in a coronary occlusion (see Section 11.8) quickly results in death of the portion of heart muscle so deprived. Cardiac muscle, unlike skeletal muscle, has a longer refractory period than relaxation period. Hence, no tetanus is possible.

28.11 SMOOTH MUSCLE

Each smooth muscle fiber is a single spindle-shaped cell with a single nucleus (Fig. 28.16). The cells are arranged in sheets. Smooth muscle is also called nonstriated muscle because no pattern of cross striations can be seen under the light microscope. However, the cells do contain both thick and thin filaments of actin and myosin, and these are organized into contractile fibrils. Exactly how the smooth muscle fibrils produce the force of contraction is still under active investigation.

Smooth muscle may contract spontaneously, but it is primarily controlled by the motor neurons of the sympathetic and parasympathetic nervous systems. As we have seen, parasympathetic postganglionic neurons release ACh at their end plate just as the motor neurons of the sensory-somatic system do. Sympathetic postganglionic neurons usually release noradrenaline. The effect of these substances on a given piece of smooth muscle varies with the particular muscle but is always antagonistic. If ACh stimulates the muscle, noradrenaline inhibits it and vice versa. For example, ACh stimulates contraction of the muscular wall of the intestine while noradrenaline inhibits it. On the other hand, ACh inhibits the contraction of the muscular walls of the arterioles supplying blood to the viscera while noradrenaline stimulates their contraction.

The action of smooth muscle is much slower than that of skeletal muscle. It may take anywhere from three seconds to three minutes for it to contract. Furthermore, smooth muscle differs from skeletal muscle in its ability to remain contracted at varying lengths. This state is also called tonus, but unlike tonus in skeletal muscle, it does not seem to require continual stimulation and energy production.

Other Effectors

28.12 CILIA AND FLAGELLA

Cilia and flagella are whiplike appendages found on the exterior surface of many kinds of eukaryotic cells. If there are only one or a few of these structures, we call them flagella. If there are many of them, we call them cilia. Flagella tend to be longer than cilia, but otherwise are similar to them in construction.

The function of cilia and flagella is to move the extracellular fluid surrounding the cell. In the case of single cells such as *Chlamydomonas* (Fig. 14.5) or a sperm cell, this action causes the cell to be propelled through the medium. Thus cilia and flagella are locomotor structures for many single cells.

In most animals (except insects), cilia are found on the exposed surfaces of some epithelial cells. In these situations, their function is to move a liquid medium past the cell. In Chapter 10, we saw how the rhythmic beating of cilia on the gills of the clam draws in a continuous supply of fresh water. In the same chapter, we noted the important function of the ciliated epithelia of our air passages in ridding them of inhaled dust particles and the like. In this case, the rhythmic beating of the cilia sweeps a film of particle-laden mucus towards the throat.

The electron microscope reveals that cilia and flagella have the same basic structure. Each is made up of a cylindrical array of long protein filaments. Nine of these are evenly spaced around the periphery of the cylinder. Each peripheral filament consists of a microtubule (see Section 5.15) with a partial microtubule attached to it. This gives the filament a "figure 8" appearance when viewed in cross section (Fig. 28.17). The partial microtubule of each filament does not extend as far into the tip of the cilium as the complete microtubule does. Running up through the center of the bundle are two single microtubules, completing the so-called "9 + 2" arrangement. The entire assembly is sheathed in a membrane that is simply an extension of the cell membrane.

Each cilium (and flagellum) is attached to a **basal body** embedded in the cytoplasm. Basal bodies are identical to centrioles (see Section 5.16) and are, in fact, produced by them. For example, one of the centrioles in developing sperm cells becomes a basal body and produces a flagellum after it has completed its role in the distribution of chromosomes during cell division (see Section 14.4).

As we learned in Chapter 5, microtubules are synthesized from repeating units of the protein tubulin (look back at Fig. 5.19). Lesser amounts of other proteins are also present in cilia. One of these, called dynein, forms arms that extend from the complete microtubule of one filament to the partial microtubule of the adjacent filament.

(Courtesy Peter Satir)

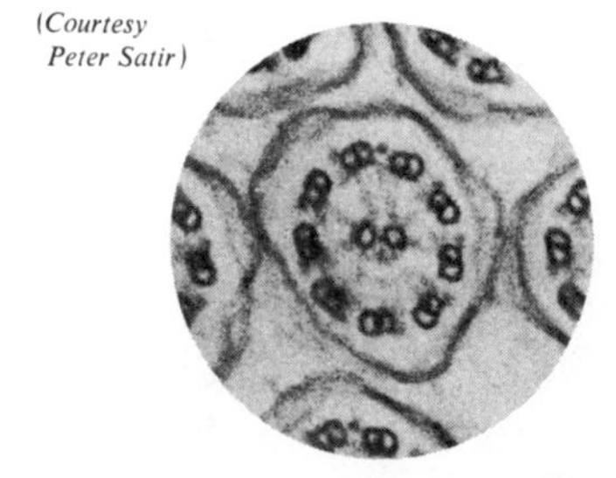

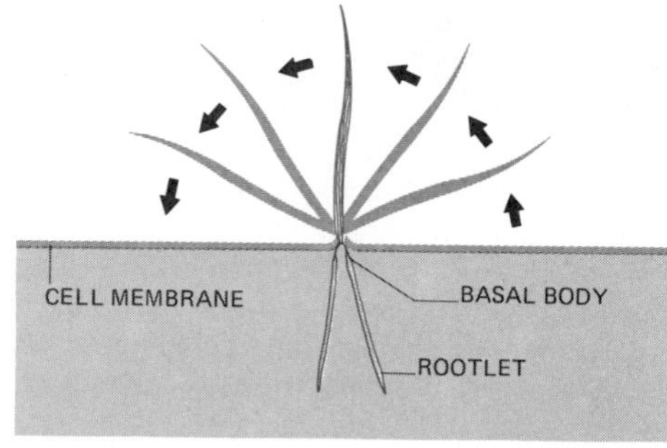

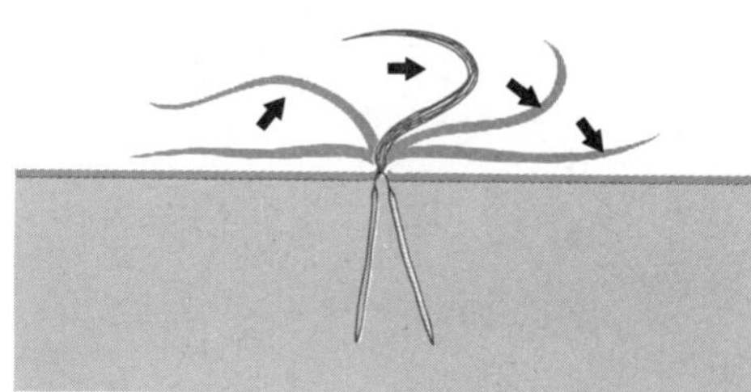

FIG. 28.17 Top: Electron micrograph of a single cilium in cross section. Note the characteristic pattern of microtubules. Bottom: Drawings of a single cilium. The power stroke is shown above, the recovery stroke below.

Several kinds of beating action are produced by cilia and flagella. For cilia, one of the commonest is a rowing motion with the cilium extended during the "power stroke" and bent during the "recovery stroke" (Fig. 28.17). The motion of flagella is usually more complex, involving rippling and lashing movements.

How do cilia work? A precisely-arranged system of filaments linked by cross bridges of dynein may well remind you of the sliding-filament hypothesis of skeletal muscle contraction. Could sliding of the microtubular filaments of cilia produce bending? The answer is yes, if we assume that there is some friction between the sliding microtubules. How could such a model be tested? You remember that the partial microtubules do not extend as far out into the tip of the cilium as do the complete microtubules to which each is attached. If we should make a slice a short distance back from the tip of a straight cilium, we should see the typical "9 + 2" pattern. But if we make a similar slice through a bent cilium (Fig. 28.18–right), approximately half the filaments (on the upperside) should be retracted. This is because of the greater length of the arc on the convex side. In such a case, we might expect that their partial microtubules would have been drawn below the plane of the slice. Thus only the complete microtubules of those filaments should be visible. When the cilium bends the other way (Fig. 28.18–left), the doublets on the opposite side disappear while they reappear on what is now the lower or concave side. Electron micrographs verify these predictions exactly.

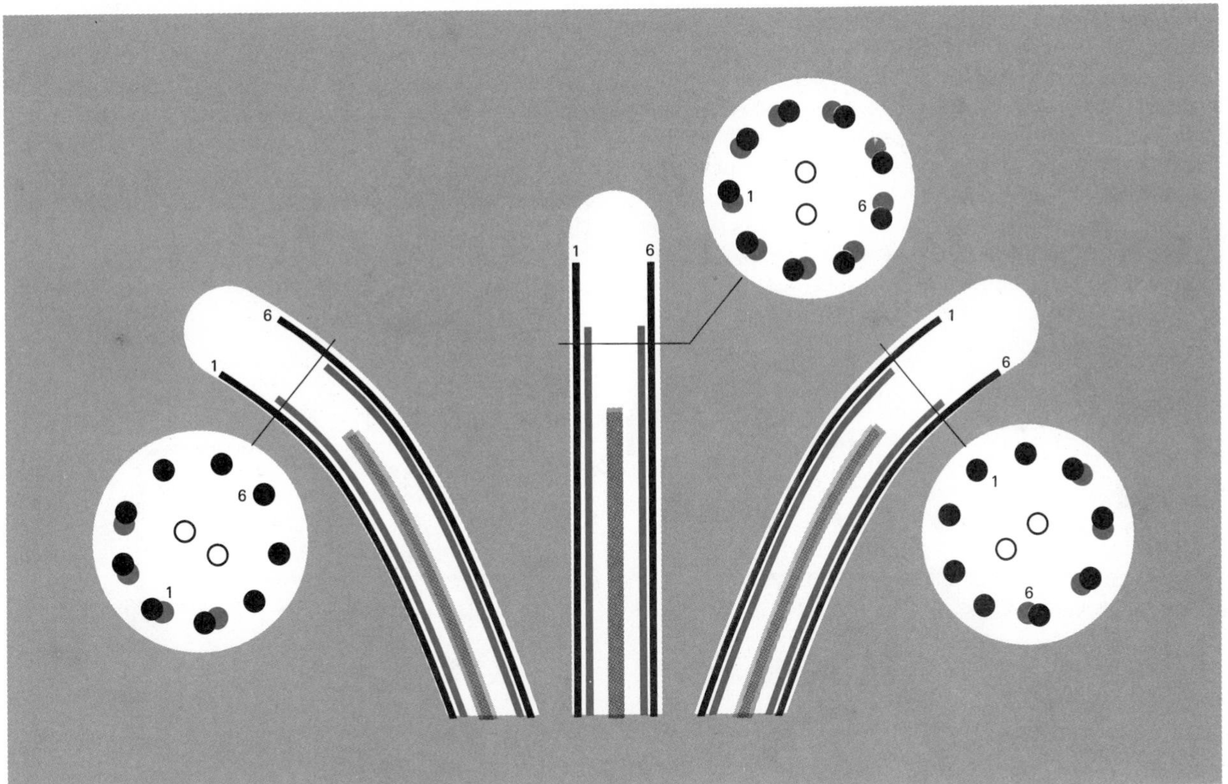

FIG. 28.18 Predictions of the sliding-microtubule hypothesis of ciliary movement. Bending of the cilium should force the filaments on the concave side toward the tip (because of the shorter arc formed). A cross section just back of the tip should reveal one set of double microtubules (Nos. 3–6) during one stroke (right) and the opposite set (Nos. 9, 1, 2, and 3) during the recovery stroke (left). Electron micrographs have verified this prediction exactly. (Based on the work of Peter Satir.)

There are other parallels between the sliding filaments of skeletal muscle and the sliding microtubules of cilia. The process is powered by ATP in each case. And the cross bridges of dynein, like the cross bridges of myosin, are the ATPase. There is also strong evidence that calcium ions play an important role in regulating ciliary movement just as they do in skeletal muscle fibers.

28.13 ELECTRIC ORGANS

Certain fishes (for example, the South American electric eel) possess electric organs. These are masses of flattened cells, called electroplates, which are stacked in neat rows along the sides of the animal. The posterior surface of each electroplate is supplied with a motor neuron; the anterior surface is not. At rest, the interior of each electroplate, like a muscle or nerve cell, is negatively charged with respect to the two exterior surfaces. The potential is about 0.08 volt but, because the charges alternate (Fig. 28.19), no current flows. When a nerve impulse reaches the posterior surface, however, its polarization reverses. The liberation of ACh by the neuron increases the permeability of the posterior membrane to the inflow of Na^+ ions so that a momentary reversal of charge occurs just as it does in the action potential of nerves and muscles. (In most fishes, electroplates are, in fact, modified muscle cells.) The anterior surfaces remain positively charged but the posterior surfaces become negatively charged. The charges now reinforce each other and a current flows just as it does through an electric battery with cells wired in "series." Voltages as high as 650 volts are produced by the several hundreds of thousands of electroplates in the South American electric eel. The flow (amperage) of the current is sufficient (¼–½ ampere) to stun a man if not kill him. The pulse of current can be repeated several hundred times each second.

Such powerful electric organs as those of the electric eel are used as both offensive and defensive weapons (to shock prey and potential predators respectively). However, the electric discharge of many electric fishes is too weak to serve such purposes. In these animals, the electric organs serve a remarkable variety of *signaling* functions. Most of these fishes emit a continuous train of electric signals in order to detect objects in the water around them. The system operates something like an underwater radar and, of course, depends for its success on the presence of sensitive electrical *receptors* (located in the skin). The presence of objects in the water distorts the electric fields created by the fish, and this alteration is detected by the receptors. Electric fishes also use their electric transmitters and receivers for such communication functions as locating mates, courtship, the defense of territories against rivals of

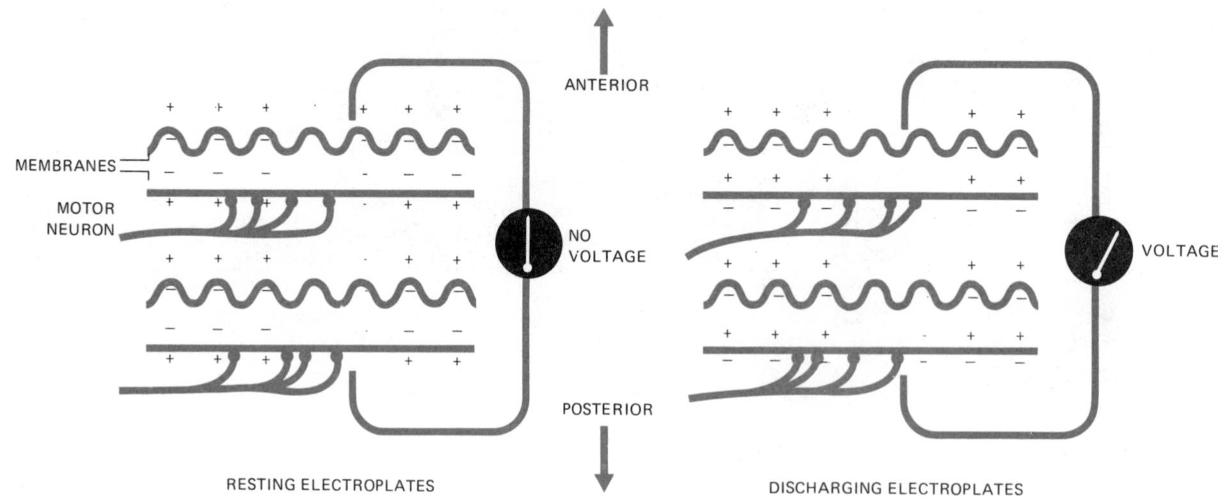

FIG. 28.19 Method of action of the electroplates in the South American electric eel. Nerve impulses reaching the posterior membranes reverse their potential and cause each electroplate to function like a cell in a battery. With several thousand of these cells arranged in "series," voltages as high as 650 volts are produced.

FIG. 28.20 This winter flounder resting on a checkerboard pattern shows how versatile is its use of chromatophores for camouflage. (Courtesy of the Field Museum of Natural History.)

the same species, and—in some cases—attracting other members of the species into schools and other aggregations.

Not only do electric fishes have receptors for detecting electric currents in the water, but at least one *nonelectric* species does as well. Presumably it uses these to detect the approach of electric fishes.

28.14 CHROMATOPHORES

Chromatophores are irregularly shaped, pigment-containing cells such as the melanophores of the frog (Fig. 25.9). By changing the distribution of pigment in these cells, the animal is able to alter its body color and blend into its surroundings. Figure 28.20 shows how remarkably effective this mechanism can be. Many experiments have shown that these protective changes in body coloration require the presence of eyes and nervous system. However, motor neurons do not activate the chromatophores directly in amphibians and crustaceans. Instead, hormones transmit the stimulus from the central nervous system to the chromatophores.

28.15 LUMINESCENT ORGANS

Many marine animals give off visible light. This is particularly common in those species that live in the dark depths of the ocean. In some cases, the light is produced by luminescent bacteria which adhere to special structures on the animal's body (Fig. 28.21).

The widespread occurrence of luminescent organs among deep-sea animals reflects the perpetual darkness in which these creatures live. At least one fish has its luminescent organ located at the tip of a protruding stalk and uses it as bait to lure prey within reach of its jaws. When disturbed, one species of squid emits from its mantle cavity a cloud of luminous water instead of the protective smokescreen of ink that its shallow-water relatives employ. Some marine animals living nearer the surface have their luminescent organs on their underside. These probably make it more difficult for predators lurking beneath to see them against the light background of the surface. Luminescent organs also serve to attract mates and thus assure that eggs and sperm will be released close to one another.

Most of us are more familiar with the luminescence of the firefly. The periodic flashes of light which we see so frequently on warm summer evenings are produced by special luminescent organs in the abdomen of the insect. The mechanism by which the light is produced is now fairly well understood. It involves a substrate (**luciferin**), an enzyme (**luciferase**), oxygen, and ATP (Fig. 28.22). The more ATP available, the brighter the light. In fact, at least one manufacturer of biochemicals sells ground

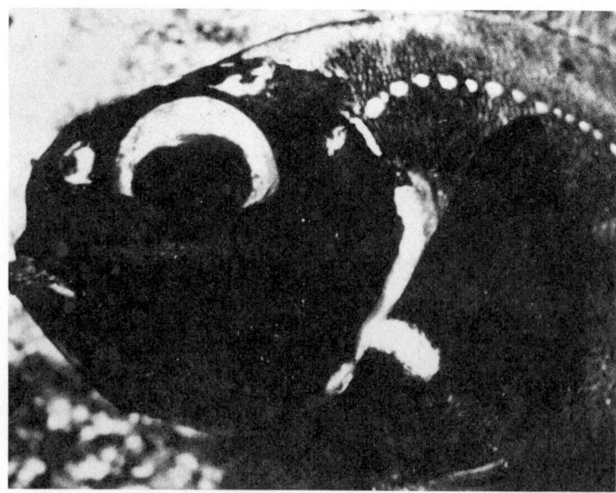

FIG. 28.21 The flashlight fish, ***Photoblepharon palpebratus,*** with the lid of its luminescent organ open (left) and closed (right). The light is produced by continuously-emitting luminescent bacteria within the organs, but its display is controlled by the fish. These animals, which were photographed along reefs in the Gulf of Elat, Israel, appear to use their luminescent organs for such varied functions as attracting prey, signalling other members of their species, and confusing potential predators. (Courtesy of Prof. J. W. Hastings from Morin *et al., Science,* **190**: 74, October 3, 1975.)

FIG. 28.22 Chemistry of luminescence in the firefly. (AMP = adenosine monophosphate. Subterminal ∽ bond is used.)

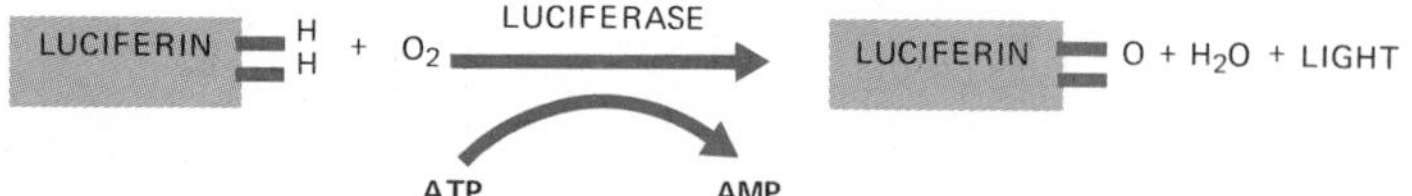

firefly "tails" to be used for determining the concentration of ATP in different biological materials. The ATP-containing material is added to a carefully prepared extract of the firefly tails, and the amount of light given off is measured with a sensitive light meter. The company itself is kept supplied with frozen fireflies by youngsters in the southern part of the United States who earn extra money by trapping the insects and selling them.

It has been well demonstrated that male and female fireflies attract each other with their flashing lights. The pattern of flashing differs from species to species. In one species, the females sometimes mimic the pattern used by the females of other species. When the males of the second species respond to these "femmes fatales," they are eaten!

Receptors, nerves, effectors: these are the structures which enable animals to detect changes in their environment, both external and internal, and to respond in appropriate, coordinated ways to these changes.

When these responses are directed externally, as in motion or locomotion, we say that the organism is displaying behavior. Some of the varied ways in which organisms behave will be considered in the next chapter.

EXERCISES AND PROBLEMS

1. Outline the chemical changes that occur in the leg muscles of a mile runner from the time the starter's gun is fired until the runner is completely recovered.
2. List the various functions carried out by the skeleton and give a specific example of bones that carry out each function.
3. How does fermentation in skeletal muscles differ from yeast fermentation?
4. In what ways does the physiology of a skeletal muscle fiber resemble that of a motor neuron? In what ways does it differ?
5. Distinguish between tetanus and tonus in skeletal muscles.
6. How can you reconcile the graded response of skeletal muscles with the all-or-none law?
7. In what ways does the physiology of the electroplates of electric organs resemble that of skeletal muscle?
8. How would you describe the response of the muscle shown in Fig. 28.8 if so much weight were added to the lever that the muscle could not lift it?
9. Oxalic acid forms an insoluble precipitate with calcium ions. What would happen to (a) the size of the action potential and (b) the strength of contraction if oxalic acid were introduced into an isolated muscle fiber?
10. What would be the effect on the size of the end plate potentials of bathing an isolated nerve-muscle preparation (from a frog) with a solution containing each of the following: (a) neostigmine (which inhibits the action of acetylcholinesterase), (b) *d*-tubocurarine (which competes with ACh for binding sites on the muscle membrane), (c) hemicholinium (which interferes with the synthesis of ACh within the nerve endings), (d) only 1/10th the normal concentration of sodium ions found in the extracellular fluid, (e) ouabain (which blocks the active transport of both sodium and potassium ions), (f) decamethonium (which mimics the action of ACh but is not hydrolyzed by acetylcholinesterase)? What would be the effect of each of the above on the size of any action potentials generated?

REFERENCES

1. BULLER, A. J., *The Contractile Behaviour of Mammalian Skeletal Muscle,* Oxford Biology Readers, No. 36, Oxford University Press, Oxford, 1975.
2. HUXLEY, H. E., "The Mechanism of Muscular Contraction," *Scientific American,* Offprint No. 1026, December, 1965.
3. MURRAY, J. M., and ANNEMARIE WEBER, "The Cooperative Action of Muscle Proteins," *Scientific American,* Offprint No. 1290, February, 1974. An excellent and well-illustrated summary of the interactions of calcium ions, troponin, tropomyosin, ATP, myosin, and actin in muscle contraction.
4. PORTER, K. R., and CLARA FRANZINI-ARMSTRONG, "The Sarcoplasmic Reticulum," *Scientific American,* Offprint No. 1007, March, 1965.
5. MARGARIA, R., "The Sources of Muscular Energy," *Scientific American,* Offprint No. 1244, March, 1972.
6. SATIR, P., "How Cilia Move," *Scientific American,* Offprint No. 1304, October, 1974.
7. LISSMANN, H. W., "Electric Location by Fishes," *Scientific American,* Offprint No. 152, March, 1963.
8. MC ELROY, W. D., and H. H. SELIGER, "Biological Luminescence," *Scientific American,* Offprint No. 141, December, 1962.
9. MC COSKER, J. E., "Flashlight Fishes," *Scientific American,* Offprint No. 693, March, 1977. Describes their remarkable luminescent organs and the uses to which they are put.

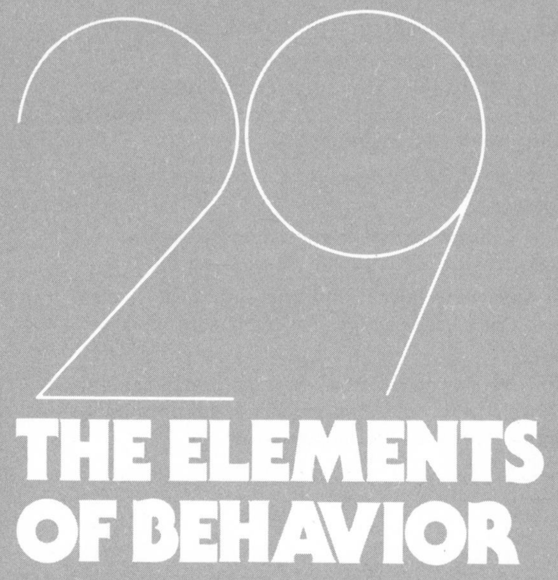

29 THE ELEMENTS OF BEHAVIOR

29.1 WHAT IS BEHAVIOR?

Behavior is action which alters the relationship between the organism and its environment. It is externally directed activity and does not include the many internal changes which are constantly taking place in living things.

Behavior may occur as a result of an external stimulus. Receptors are necessary to detect the stimulus, the nerves are needed to coordinate the response, and the effectors actually carry out the action. Behavior may also occur as a result of an internal stimulus. A hungry animal searches for food. A thirsty animal behaves in a way which will lead to satisfying its thirst. More often than not, the behavior of an organism results from a combination of both external and internal stimuli. Internal stimuli, such as hunger, provide the *motivation* for the action taken when food is actually seen or smelled.

In studying behavior, it is worthwhile to try to distinguish between innate and learned forms of behavior. The former are responses whose nature is, in large measure, determined by inherited pathways of nervous coordination. They are usually quite inflexible, a given stimulus giving rise to a given response. A salamander raised away from water until long after its cousins begin swimming successfully will swim every bit as well as they the very first time it is placed in water. Clearly this rather elaborate response is "built in" the species and not something that must be acquired by practice. Learned behavior, on the other hand, is behavior which has become more or less permanently altered as a result of the experience of the individual organism. Only by diligent practice can one learn to play baseball well. Between these extremes, there are patterns of behavior that partake of both innate and learned components. Hailman has demonstrated that the pecking behavior of newly hatched sea gulls, long thought to be innate, is nonetheless strongly influenced by the chick's early experience. (You may read of his findings in his article listed at the end of this chapter.)

Innate Behavior

29.2 BEHAVIOR IN PLANTS

Behavior in plants seems to be innate. All the members of any one species respond in the same inflexible way to a given stimulus. There is nothing voluntary about any plant response.

Plants do not possess a nervous system and, as we saw in Chapter 24, plant behavior is restricted to growth movements and turgor movements. Movements whose direction is determined by the direction from which the stimulus strikes the plant are called **tropisms.** The bending of the oat coleoptile toward a source of light is a positive phototropism (see Section 24.2). Generalized movements, that is, movements which are not oriented in a particular direction, are called nastic movements or nasties. The opening of tulip flowers on warm days is a thermonasty.

29.3 TAXES

Some organisms respond to a stimulus by automatically moving directly toward or away from or at some definite angle to it. These responses are called **taxes.** They are similar to the tropisms of plants except that actual locomotion of the entire organism is involved. Even as "simple" an organism as *E. coli* displays this behavior. When a capillary tube containing a substance such as glucose is placed in a medium containing *E. coli,* the bacteria alter their pattern of locomotion in such a way that they congregate near the source of the substance (Fig. 29.1). This re-

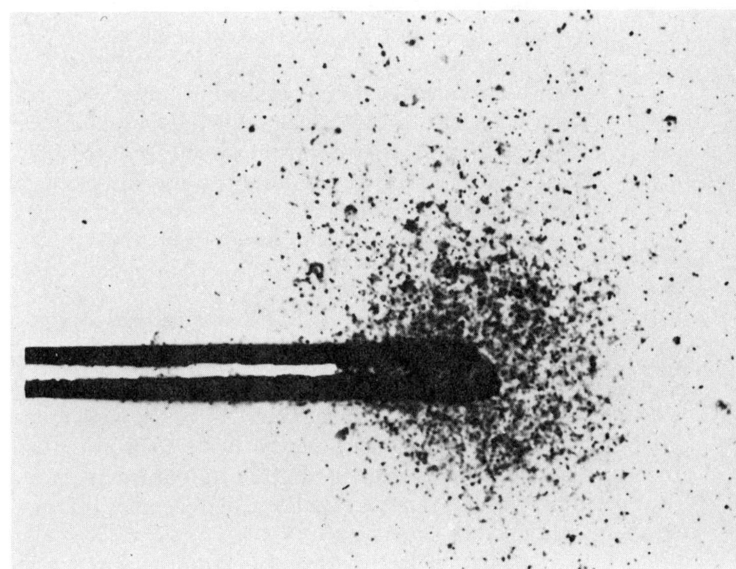

FIG. 29.1 Chemotaxis in *E. coli.* The bacteria have congregated at the opening of a capillary tube filled with a weak solution of an amino acid. (Courtesy of Dr. Julius Adler from J. Adler, *Science,* **166,** 1588–1597, December 26, 1969.)

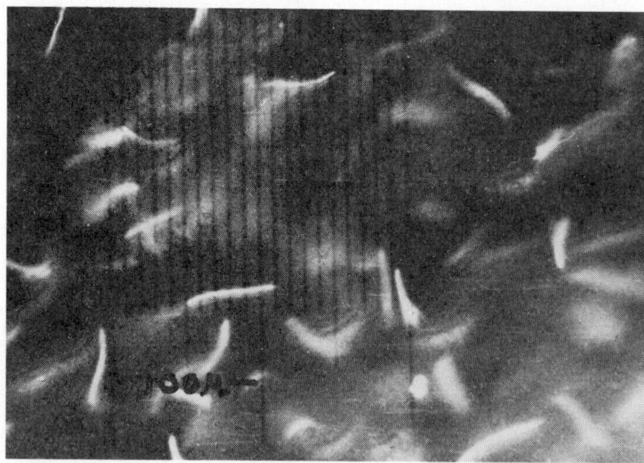

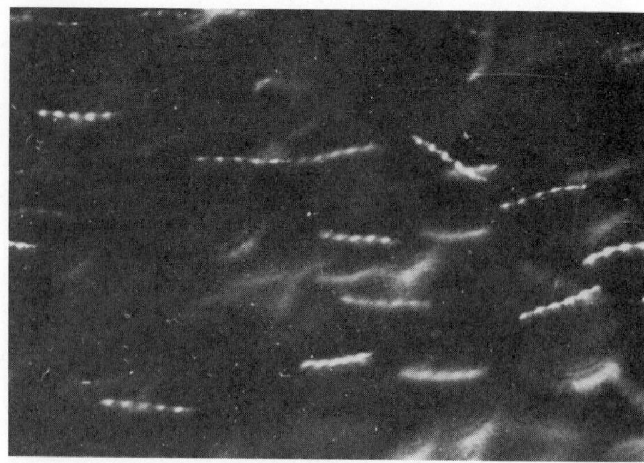

FIG. 29.2 Phototaxis in *Chlamydomonas*. Left: Randomly oriented tracks formed during 1/3 sec by these unicellular algae swimming about in red light (to which they are insensitive). Right: Upon adding a beam of blue-green light from one side, the tracks become oriented in its direction. (Courtesy of Dr. Mary Ella Feinleib, Tufts University.)

sponse, called a **chemotaxis,** does not depend on the organism's ability to metabolize the substance, although presumably that is the value of the response under normal conditions. *E. coli* responds strongly to a number of organic molecules besides glucose, including galactose and the amino acids serine and aspartic acid.

Photosynthetic microorganisms often display **phototaxis.** Both the green alga *Chlamydomonas* (Fig. 29.2) and *Euglena* swim directly toward light of moderate intensity. However, as the intensity of light increases, a point is reached where they abruptly reverse direction and swim away from the light (a negative phototaxis).

29.4 REFLEXES

Reflexes are the simplest innate responses found in animals having a nervous system. A reflex is an automatic response of part of the body to a stimulus. The response is inborn; that is, its nature is determined by an inherited pattern of receptors, nerves, and effectors.

In Chapter 27, we used the withdrawal reflex as our example of a reflex. Now that we have learned more about the properties of skeletal muscle, let us examine another example: the **stretch reflex.** The familiar knee jerk reflex is a stretch reflex. You have undoubtedly had your doctor tap you just below the knee cap with a rubber-headed hammer. Your response should have been a sudden kick with the lower leg. The response is quite automatic. It requires a properly functioning spinal cord, but the brain need play no role (although you can use your brain to try to override the reflex).

The doctor aims the hammer at a tendon that inserts an extensor located in the front of the thigh into the lower leg (Fig. 29.3). Tapping this tendon stretches the thigh muscle. This activates stretch receptors located within the muscle. These receptors consist of sensory nerve endings wrapped around special muscle fibers called **spindle fibers.** The entire structure is called a **muscle spindle.**

Stretching a spindle fiber triggers a volley of impulses in the sensory neuron (called a "I-a" neuron) attached to it. These impulses are carried into the spinal cord. The I-a axons branch within the spinal cord and form several kinds of synapses:

1. Some of the sensory (I-a) axons synapse directly with motor neurons (called "alpha" motor neurons) that carry impulses back out to the muscle that was stretched (Fig. 29.3). Contraction of this muscle, an extensor, causes the leg to straighten. The reflex is completed. Note that the response uses no interneurons (in contrast to the withdrawal reflex we examined in Chapter 27).

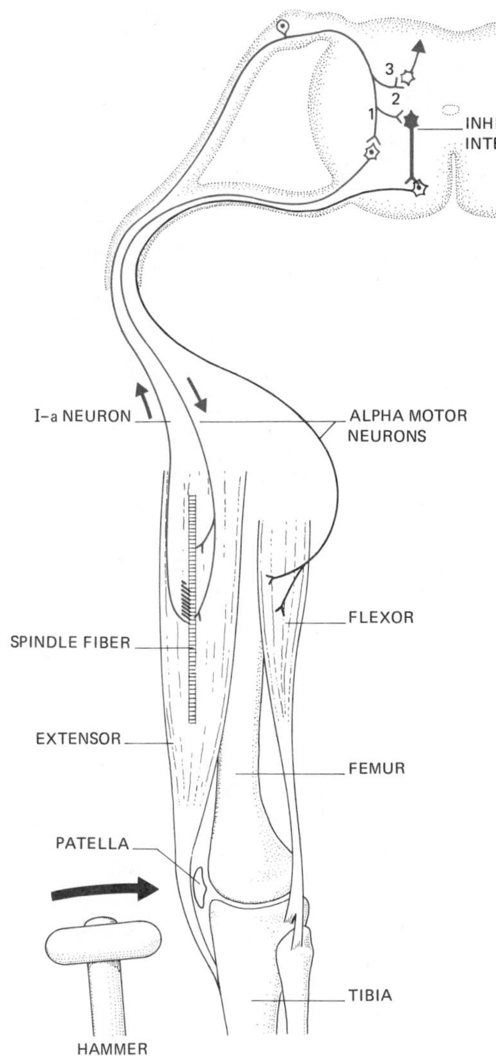

FIG. 29.3 Mechanism of the knee-jerk reflex, a stretch reflex. Stretching of a spindle fiber in the extensor sends a volley of nerve impulses along the I-a neuron into the spinal cord. The terminals of I-a neurons activate (1) alpha motor neurons leading to the same extensor, which cause it to contract; (2) inhibitory interneurons that inhibit the alpha motor neurons leading to the antagonistic flexor; and (3) interneurons that notify the brain of what has occurred.

You may wonder what is the significance of this response (aside from giving your physician some clues as to the state of your nervous system). Actually, the reflex is very important in maintaining balance. If, for example, you stub your toe, the triggering of this reflex keeps your leg from buckling under you.

2. Some of the branches of the I-a axons leading from the muscle spindles *do* synapse with interneurons in the spinal cord. These interneurons, in turn, synapse with motor neurons leading to the antagonistic muscle, a *flexor* in the back of the thigh (Fig. 29.3). However, these interneurons are inhibitory, i.e., their activation inhibits those motor neurons. Thus movement of the leg by your extensors is facilitated by simultaneous inhibition of the antagonistic flexors.
3. Still other branches of the I-a sensory axons synapse with interneurons leading to the brain. While we do not become conscious of the information arising from the muscle spindles, they do provide an additional mechanism by which the brain can monitor the performance of the muscles.

By itself, the stretch reflex plays an important but rather simple role in behavior. A muscle is stretched and responds by contracting. But the machinery of the stretch reflex opens up greater possibilities. The fibers of the muscle spindles are, like other muscle fibers, capable of contracting. However, the central portion of the spindle fiber, where the I-a neurons arise, is not contractile. The spindle fibers are activated by motor neurons—called gamma motor neurons (Fig. 29.4).

Now what if the brain, e.g. the motor cortex, wants to contract the extensors of a leg? All it has to do is to send signals down the spinal cord and out the alpha motor neurons to those muscles. They will contract. But the stretch reflex provides the machinery for a more subtle approach. The interneurons

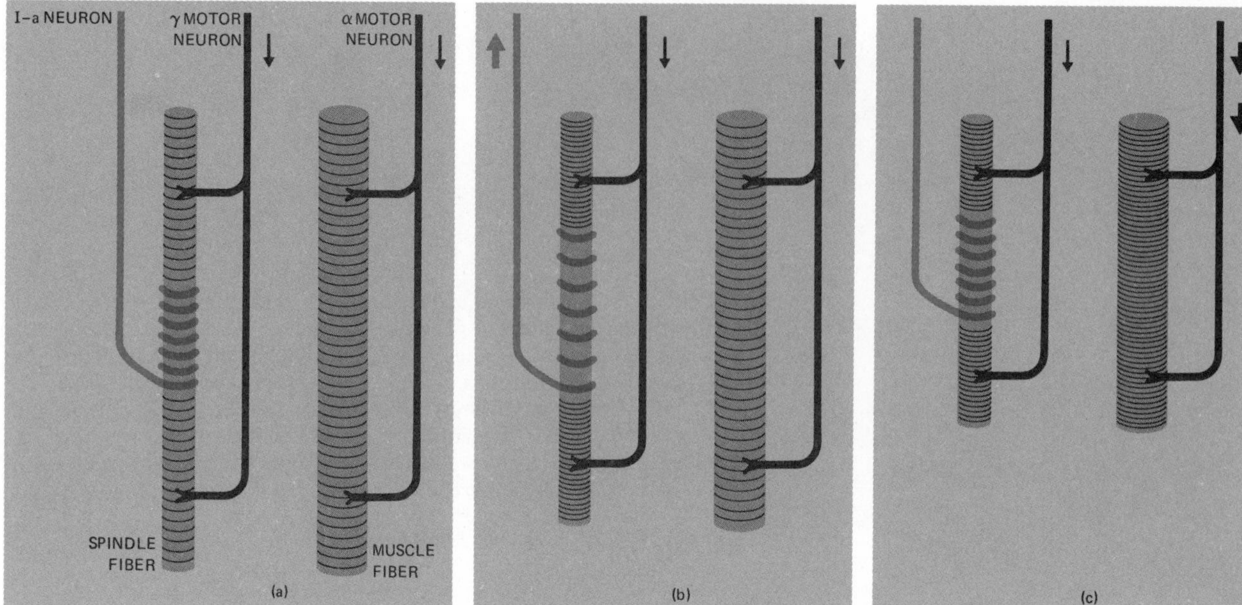

FIG. 29.4 Adjusting skeletal muscle response by means of the stretch reflex. (a) Signals from the brain activate both the major muscle fibers and the spindle fibers. (b) Because the signals to the muscle fibers are inadequate to overcome the resistance to muscle shortening, the central portion of the spindle fiber is stretched. This activates the I-a neuron. (c) The additional volley of motor impulses created by activation of the stretch reflex brings about shortening of the major muscle fibers until they relieve the stretch in the spindle fibers.

bringing signals down from the brain also synapse with the *gamma* motor neurons. Thus the spindle fibers are stimulated to contract at the same time as the other fibers of the muscle belly. Now, what if you misjudge the amount of resistance that must be overcome in order to extend your leg? Or, to put it another way, what if the signals from the brain fail to stimulate a sufficient number of alpha motor neurons to bring about the desired degree of contraction? In such a case, the spindle fibers contract more than the surrounding muscle (Fig. 29.4). This stretches the middle of the spindle fiber where the I-a neurons are attached, just as the tap of the doctor's hammer did. The stretch reflex is initiated. A volley of impulses in the I-a neurons activates additional alpha motor neurons and increases the contraction of the muscle. When the entire muscle has contracted to match the contraction of its spindle fibers, activation of the reflex ceases. In a real sense, then, the muscle spindles measure the difference in length between their own fibers and those of the remainder of the muscle. When a difference occurs, the muscle spindle creates the response that eliminates that difference.

The power-assisted steering in an automobile works on the same principle. If you are coasting down a hill with the engine off, you know how hard you must twist the steering wheel in order to turn the front wheels of the car. Then if the engine starts, the great difference between the strength of your command and the weakness of the response is detected at once, and a hydraulic system goes to work to help you get the wheels turned.

In summary, we find that the machinery of the stretch reflex provides a finely-adjusted control mechanism that:

1. directs the reflex contraction of muscle,
2. inhibits the contraction of antagonistic muscles,
3. continuously monitors the success with which commands from the brain are carried out, and swiftly and automatically makes any necessary compensatory adjustments.

29.5 INSTINCTS

Instincts are complex behavior patterns which, like reflexes, are inborn, rather inflexible, and of value in adapting the animal to its environment. They differ from reflexes in their degree of complexity. The entire body participates in instinctive behavior, and an elaborate series of actions may be involved. Instinctive behavior is probably the most important type of behavior in insects. Fishes, amphibians, reptiles, and birds also depend to a large degree on inborn, instinctive patterns of behavior.

The spider's building of a web is an example of instinctive behavior. A long series of complex actions are required to spin the web, but these, and thus the final shape of the web, are entirely dependent upon instinct. A spider spins a web which is characteristic of its species even though never before exposed to that particular pattern. Nest building in birds is also an example of instinctive behavior. Even when it has never been allowed to see another member of its species or any nest, the weaverbird builds a nest characteristic of the species. Instincts, then, are inherited just as the structure of tissues and organs is inherited (Fig. 29.5).

FIG. 29.5 The scratching behavior of a dog and a European bullfinch is part of their genetic heritage and is not changed by training. The widespread habit of scratching with a hindlimb crossed over a forelimb is common to most birds, reptiles, and mammals. (Courtesy of Rudolf Freund and *Scientific American.*)

This fact has been beautifully demonstrated by breeding experiments between closely related species with differing instinctive behavior patterns. The African peach-faced lovebird carries nesting materials to the nesting site by tucking them in its feathers. The closely related Fischer's lovebird uses its bill to transport nesting materials. When these two species are mated, the offspring are successful at transporting materials in the bill only. Nevertheless, they invariably go through the motions of trying to tuck the materials in their feathers first.

The relative inflexibility of instinctive behavior has been demonstrated with many organisms. The tropical army ant (genus *Eciton*) gets its name from the foraging marches that the entire colony makes over the jungle floor. While the motions of the colony suggest elaborate military maneuvers, in reality they arise from the simple interplay of three factors: (1) the stimulus to move, (2) the tendency of individual ants to remain close to one another following the pheromone laid down by those in front, and (3) the presence of obstacles or food in the line of march. On occasion, these factors become arranged in such a way that the behavior of the ants is truly revealed as blind and instinctive rather than the result of conscious, warlike motives. On a flat surface, such as a paved road, the lead ants begin to move away from the swarm, but their conflicting tendency to stay with the group results in their walking a circular path. The chemical trail laid down is followed by the others and soon the entire swarm is marching around and around in a circle (Fig. 29.6). Unless some obstacle interrupts the path established, the ants will march themselves to their own destruction.

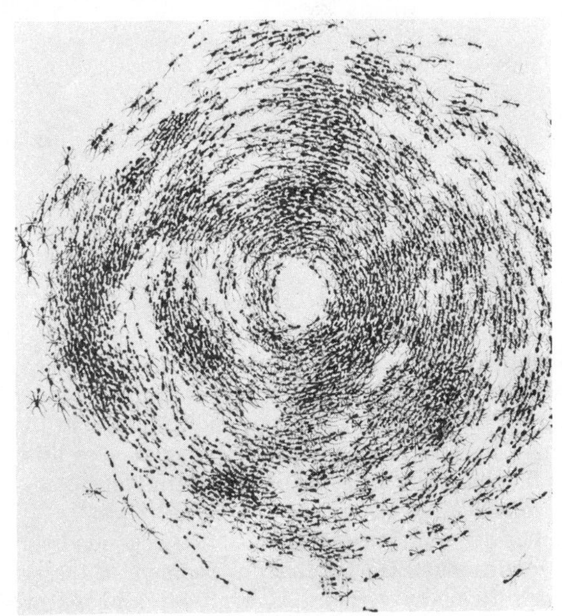

FIG. 29.6 The inflexibility of instinctive behavior. The circular column of army ants (*Eciton*) formed spontaneously and lasted for over 30 hours. (Courtesy of Dr. T. C. Schneirla, American Museum of Natural History, November, 1944.)

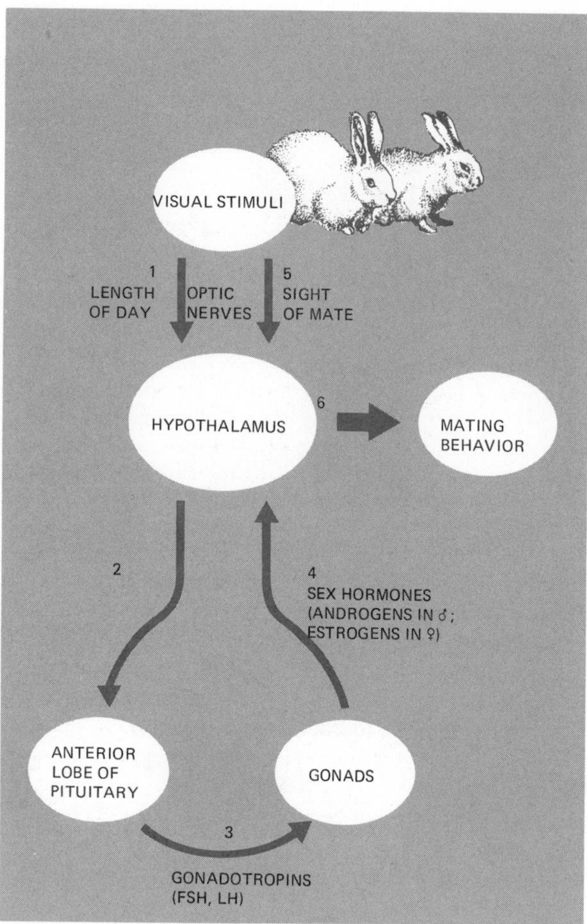

FIG. 29.7 Interaction of external and internal stimuli leading to mating behavior in the rabbit.

In most circumstances, instinctive behavior promotes the survival of the species. To the human observer, the complexity and usefulness of the behavior suggests the presence of reason and foresight in the animal. It is only when unusual circumstances arise that the true, inflexible, unthinking nature of instinctive behavior is revealed.

The carrying out of instincts often depends upon the conditions in the internal environment of the organism. In many vertebrates, courtship and mating behavior will not occur unless sex hormones are present in the bloodstream. The target organ, in at least some cases, appears to be a small region of the hypothalamus. When stimulated by the presence of sex hormones in its blood supply, the hypothalamus initiates the activities leading to mating. The level of sex hormones in the blood is, in turn, regulated by the activity of the anterior lobe of the pituitary gland (see Section 25.9). Figure 29.7 outlines the interactions that lead an animal, such as the rabbit, to seek a sexual partner and mate with it.

29.6 RELEASERS OF INSTINCTIVE BEHAVIOR

Once the body is prepared internally for certain types of instinctive behavior, an external stimulus is needed to initiate the response. Nobel Prize-winner (1973) N. Tinbergen and his students at Oxford, England, have found that this external stimulus need not necessarily be appropriate to be effective. During the breeding season, the female three-spined stickleback (Fig. 29.8) normally follows the red-bellied male to the nest that he has prepared and lays eggs in it. She will, however, follow almost any small red object presented to her. Once she is within the nest, neither a male nor a red object need any longer be present. Any object touching her near the base of the tail will cause her to liberate her eggs. It is as though the stickleback were primed internally for each item of behavior and needed only one specific signal to release the behavior pattern. For this reason, signals which trigger instinctive acts are called **releasers.** Once a given response has been released, it usually runs to completion even though the effective stimulus is immediately removed. One or two prods at the base of the tail will release the entire sequence of muscular actions involved in the liberation of the eggs.

Perhaps the most remarkable releasers of instinctive behavior yet discovered are the stars. The experiments of the German ornithologist E. G. F. Sauer indicate that European warblers, migrating at night, navigate by means of them. This is particularly remarkable in view of the fact that in the fall the young birds fly to their winter home in Africa independently of their parents. With the patient help of his wife, Sauer has even raised young warblers entirely apart from other members of their kind. (This is no small task. Warblers eat insects only and a young bird will consume dozens each day.) When fall comes, these young hand-reared birds become restless. Presented with a view of the "stars" inside a planetarium, they orient themselves toward the southeasterly course they would ordinarily follow. Although the constellations are in apparent motion throughout the night, the young birds are able

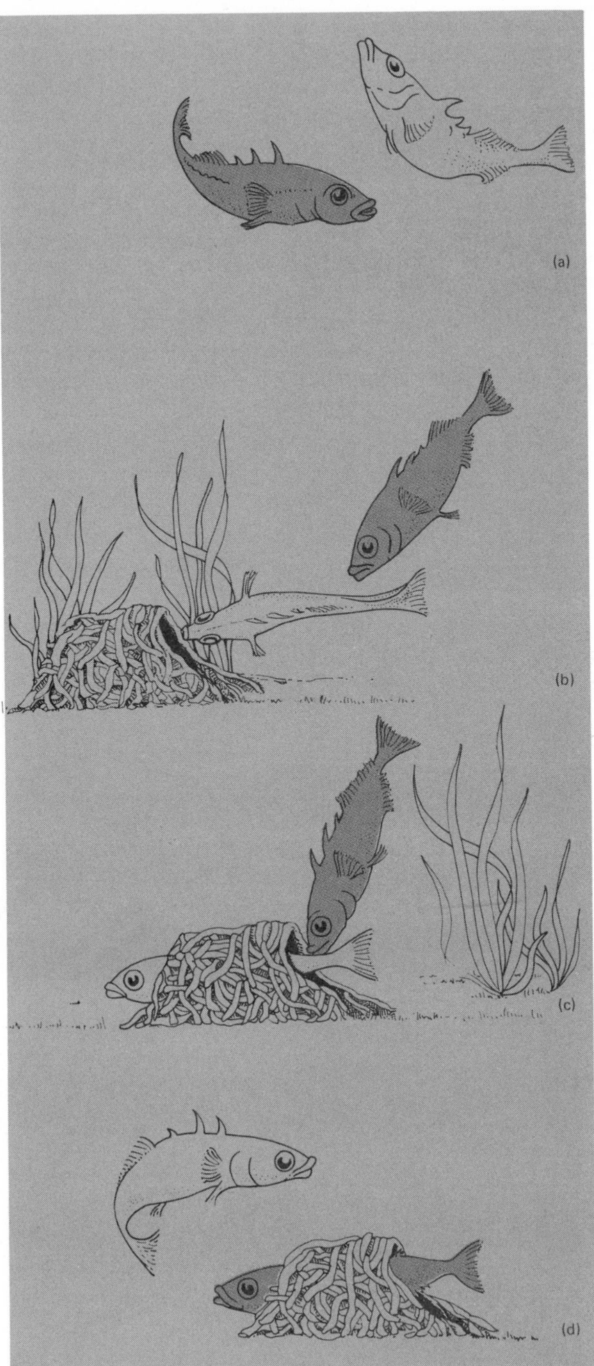

FIG. 29.8 Courtship behavior of sticklebacks. Male leads female toward nest (a), guides her into it (b), then prods the base of her tail (c). After the female lays her eggs, the male drives her from the nest, enters it himself (d), and fertilizes the eggs.

to compensate for this and continue to maintain their proper orientation.

29.7 RHYTHMIC BEHAVIOR AND BIOLOGICAL "CLOCKS"

The migratory "urge" of birds in the fall is one of many examples of behavior that is repeated at definite intervals. Such behavior is described as rhythmic or periodic. The cycles of rhythmic behavior may be as short as two hours or as long as a year. The common laboratory rat provides an example of a very short cycle. Even when food is available at all times, it generally feeds at two-hour intervals.

A great many animals alter their behavior on a daily basis. Nocturnal animals, for example, become active once every 24 hours. Fruit flies hatch in greatest numbers just at dawn. You might well claim that such periodic behavior is simply a response to the daily changes of light and darkness. This is not the complete answer, though. Even when the animals are kept under constant environmental conditions (e.g. under continuous illumination in the laboratory), many of these rhythms continue. They may, however, tend to drift one way or the other. In other words, the cycles may occur every 23 or every 25 hours instead of exactly at 24-hour intervals. For this reason, such rhythms are called *circadian,* from the Latin words *circa* (about) and *dies* (day). Under natural conditions of alternating night and day, most of these rhythms remain correctly adjusted to a 24-hour cycle.

Cycles of approximately two and four weeks are known. In Section 21.2, we discussed how the male and female California grunions (fishes) come on the beaches to spawn at the time of the full moon and new moon, that is, at intervals of about two weeks. This behavior clearly seems to be triggered by the phase of the moon and/or the height of the tides (which reach a maximum at full moon and new moon). The human menstrual cycle is a 28–30 day cycle which is now, at least, independent of the phase of the moon (and now more physiological than behavioral).

The reproductory behavior of the California grunion does not occur throughout the year but only in the spring. Thus its two-week cycle is superimposed on an annual cycle. Other animal activities also occur on a yearly basis. A few examples are the preparations for hibernation carried out by many mammals, the migration of birds, and the onset and termination of diapause in insects. As we have seen in the case of insect diapause (see Section 25.2), these responses generally depend on the only reliable indicator of the time of year: the relative length of day and night. In other words, most of these activities are regulated by changes in the photoperiod.

The ability to respond to photoperiod requires that the organism have some mechanism for measuring the hours of daylight or the hours of darkness. (Some organisms seem to measure one, some the other.) In other words, these organisms must have some sort of biological "clock." Although the exact nature of the clock mechanism is still unknown, various physiological activities have been found to fluctuate on a daily basis within the organs, tissues, and even the individual cells of organisms with circadian rhythms. Perhaps it is these internal circadian rhythms, which can maintain their periodicity independent of fluctuations in the external environment, that provide the clockwork by which photoperiod can be measured.

29.8 THE LIFE HISTORY OF THE HONEYBEE

Probably no group of animals has developed such a varied repertory of instinctive behavior as the colonial insects—the ants, termites, and honeybees. There is not space in a book of this sort to describe the incredibly diverse and elaborate kinds of behavior which all these animals exhibit, but we shall

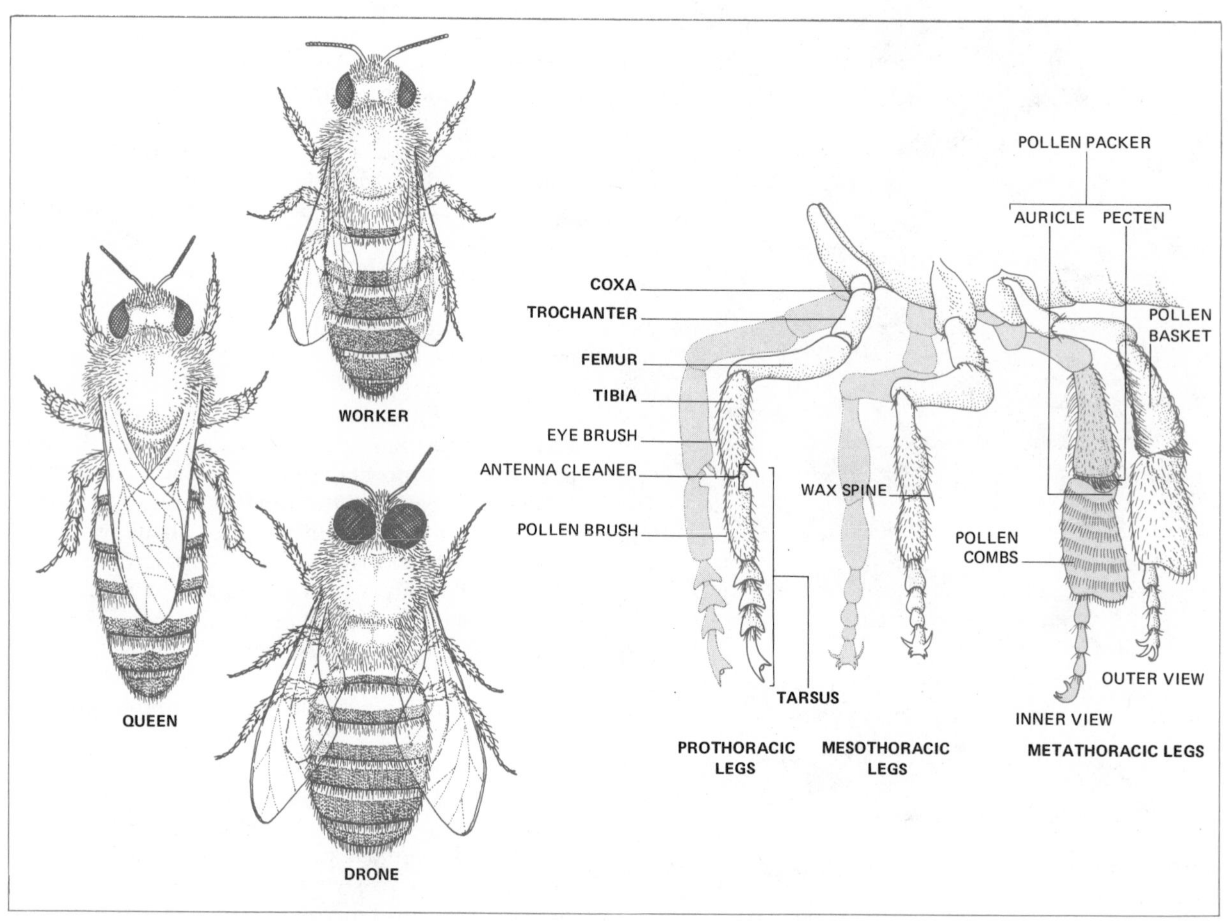

FIG. 29.9 Honeybee anatomy. The three body forms found in the colony are at the left. The legs of the worker bee are shown at the right.

examine some of the behavior displayed by the domestic honeybee, *Apis mellifera*. (If you are interested in pursuing further the behavior of other colonial insects, reading one or two of the pertinent references listed at the end of this chapter will be well worth your effort.)

The life of the honeybee colony revolves around the activities of its single **queen** (Fig. 29.9). During the spring and summer months, she spends most of her time laying eggs in the wax cells of the honeycomb. The queen fertilizes most of these eggs, just before they are deposited in the cells, by releasing sperm from storage sacs (sperm receptacles), which were filled at the time of her mating flights. These eggs hatch into larvae (the grubs) after three days. The larvae are carefully tended and fed by **worker** bees for six days. At the end of this time, the workers cap the cells with wax, and the larvae undergo metamorphosis. Three weeks after the egg is laid, a new worker bee emerges. She is a female but does not have functional sex organs.

Early in the spring the queen bee deposits a fertilized egg in each of several special cells, the queen cells (Fig. 29.10). These develop into fertile females, one of which will become the future queen of the hive. The different fate of these eggs is probably accounted for by differences in the diet supplied to the larvae. Young larvae are fed a protein-rich secretion from the salivary glands of adult bees. The composition of the secretion fed to the larvae destined to be-

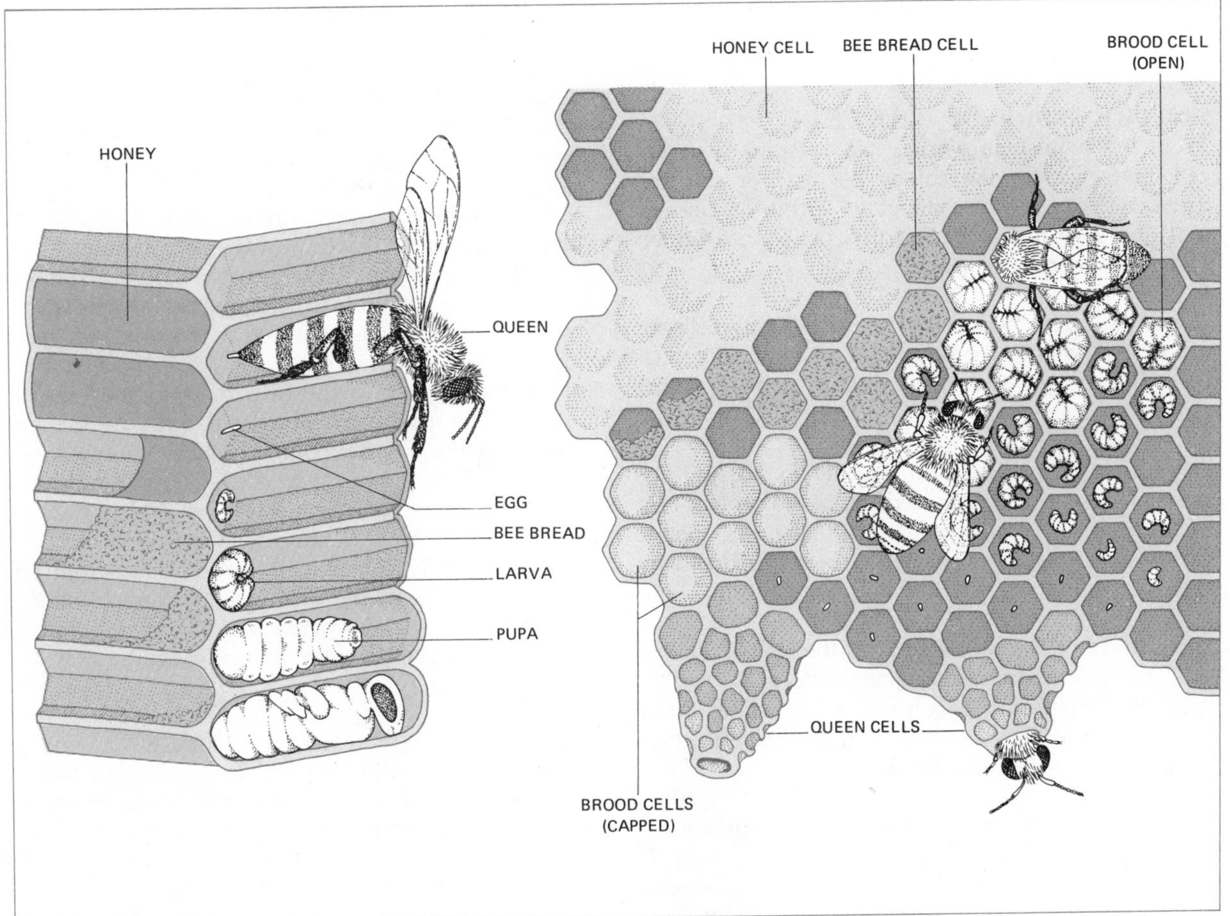

FIG. 29.10 Honeybee comb as seen face on (right) and in section (left). The cells are constructed by the workers out of wax secreted by glands in their abdomen.

come workers differs from that fed to larvae in the queen cells. The latter material is called "royal jelly."

The queen bee can also lay unfertilized eggs. These hatch and develop into male bees, the **drones.**

The active, normal beehive thus contains one queen, a few hundred drones and many thousands of workers. Before a new queen emerges, the old queen leaves the hive, taking a substantial proportion of the workers with her. This is the phenomenon of swarming. After a few days, the new queen leaves the hive, too, but just long enough to mate (while flying high in the air) with several drone bees. Then she returns to the hive to begin her egg-laying duties. Other developing queens are killed unless the hive is so crowded that additional swarms are appropriate.

The appearance of several distinct body types in a species is called **polymorphism.** In the case of the honeybees, the different forms (worker, drone, and queen) are each adapted to carry out specific functions.

29.9 THE WORK OF THE HIVE

From the moment a worker bee emerges from her cell, her work for the colony begins. For the first three weeks of her life, she stays within or close to the hive. Her duties during this period are undertaken in a definite sequence.

Days 1–3. This period is spent in cleaning cells for reuse. The worker feeds upon a pollen-honey mixture, bee bread, and her salivary glands become greatly enlarged.

Days 4–9. This period is devoted to nursing the grubs. At first the worker feeds them the protein-rich secretion of her salivary glands. After her glands shrink to normal size, she feeds them bee bread instead.

Days 10–16. The wax glands on the ventral side of the last four abdominal segments begin to secrete beeswax. The worker uses her wax spine (Fig. 29.7) to remove the plates of wax and her mandibles to mold them into new cells for the comb.

Days 17–19. During this period, the workers receive the nectar which foraging bees bring back to the hive, convert it into honey, and store it in cells. Honey production involves evaporating water from the nectar and digesting the sucrose into glucose and fructose. If the hive gets too hot, the workers receive *water* from the foragers and spread it on the surface of the combs. By fanning the water with their wings, they hasten its evaporation, thus cooling the hive. During this period, the workers also remove any debris (e.g. dead workers) that may collect in the hive.

Day 20. This is spent patrolling in front of the hive and attacking and stinging any intruders. A worker can often sting another insect and survive, but her sting usually cannot be removed from the elastic skin of vertebrates. It is pulled from her body, inflicting fatal damage to her abdominal organs.

Two additional points should be noted about the work of the hive. The sequence just outlined is somewhat flexible. Worker bees spend a good deal of time simply patrolling the hive, and they will shift their activities to remedy special needs that they discover. Although bees are proverbially busy, they actually spend only about 40% of their time in the activities described. The remainder is spent simply standing around in the hive. They work hard enough, however, so that their life span as an adult rarely exceeds six weeks in the summer. Workers hatching in the fall have virtually nothing to do, and most of them remain alive throughout the entire winter.

After three weeks in or near the hive, the summer workers take to the field as **foragers** of nectar and pollen. The nectar is collected in a special chamber (the honey stomach—Fig. 7.8) of the digestive tract and brought back to the hive to be converted into honey. The pollen is moistened with nectar and brought back in the pollen baskets of the metathoracic (hind) legs (Fig. 29.9). This material is bee bread. The collection of nectar and pollen are vitally important to the plants as well as to the bees. By accidentally transferring some pollen from one flower to the next, the bees enable cross-pollination of the plants to take place with subsequent sexual reproduction and the development of seeds.

29.10 TOOLS OF THE HONEYBEE

The intricate, well-coordinated activities of honeybees would be impossible without sense receptors, nerves, and effectors. The large, multifaceted compound eyes (see Section 26.2) provide a fair amount of pattern discrimination (Fig. 29.11) as well as four-color vision. The odor receptors, which are located on the antennae, enable the bee to discriminate among a wide variety of odors and to detect rather

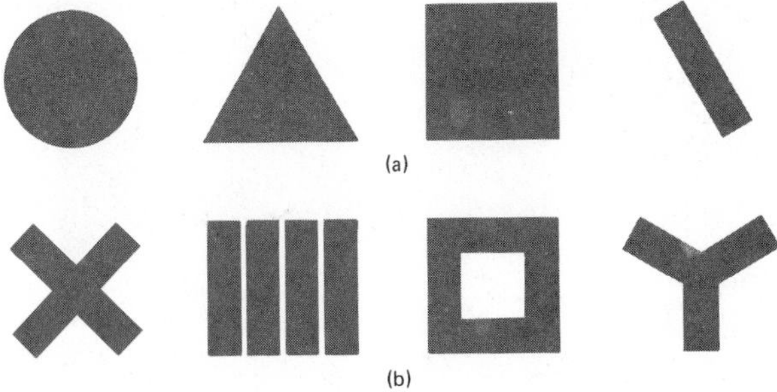

FIG. 29.11 Pattern discrimination in honeybees. Bees cannot distinguish between any of the figures in row (a) or row (b) but can distinguish any figure in row (a) from any in row (b).

FIG. 29.12 Worker honeybee foraging for pollen. Note the full pollen baskets on the metathoracic legs. (Courtesy of W. T. Davidson, National Audubon Society.)

faint ones. Despite many claims to the contrary, though, the bee's ability to detect and discriminate among odors is probably no better developed than ours. The mouth parts of the bee contain taste receptors. The bee seems to distinguish the same four taste sensations that we do: sweet, sour, salty and bitter. The bee's threshold to sour and salty is lower than ours while its threshold to bitter and sweet is higher. The relatively high threshold to sweet is of value to the bee as it prevents her from collecting nectar too dilute to be converted efficiently into honey.

The nervous system of the worker bee is well organized to receive information from the sense receptors and to coordinate the action of the muscles controlling legs, wings, and mouth parts.

The legs of the honeybee are far more than simply structures to walk on. They are highly specialized manipulative organs which enable the bee to carry out many of its tasks.

The anterior legs have soft hairs (the eye brush) on the tibia (Fig. 29.9). As the name implies, the eye brush is used for removing pollen and other debris from the eyes. The pollen brush on the first tarsal segment is made of rather stiff bristles. It is used to brush pollen off the body hairs. The first tarsal segment also contains a notch in which the bee can place its antenna. Then when the bee crooks the leg, a spur on the tibia holds the antenna in the notch while the bee cleans off pollen, etc., in a wiping motion. The entire apparatus is called the antenna cleaner.

The middle, or mesothoracic, legs also contain pollen brushes on the first tarsal segments. In addition, a wax spine sticks down from the tibia. This is used to dig plates of wax out of the wax glands.

Each hind or metathoracic leg contains pollen combs on the inner surface of the first tarsal segment, a pollen packer in the joint between the tibia and the first tarsal segment, and a pollen basket on the outer surface of the tibia. The pollen combs collect pollen from the pollen brushes of the other legs. Then with the stiff spines (the pecten) of the pollen packer of one leg, the bee removes the pollen from the pollen combs of the opposite leg. Straightening of the leg then forces the anvil-shaped auricle up against the pollen caught under the pecten. The pollen is squeezed through the joint and up into the pollen basket on the outside of the tibia.

All these manipulations are carried out in the brief interval it takes the bee to fly from one flower to another. When pollen is plentiful, the amount transported in the pollen baskets is truly impressive (Fig. 29.12).

The mouth parts of the honeybee (look back at Fig. 7.8) are also highly specialized and enable the bee to manipulate the materials essential to the life of the colony. The mandibles are used for kneading wax to make more honeycomb. The maxillae, the labium, and the tongue are modified to form a long proboscis for sucking nectar from the nectaries of flowers.

29.11 COMMUNICATION AMONG HONEYBEES

The colonial way of life places great emphasis upon the home. The hive is the center of activity and, no matter how far afield the workers roam, they almost invariably return to their own hive. Considering that bees may forage several miles from the hive, they must be good navigators to find their way back successfully. Certainly bees have sufficiently good vision to be able to navigate by means of prominent landmarks, and evidence suggests that they often do. It has also been discovered, however, that most foragers do not leave the hive until food has actually been discovered by so-called **scout bees.** As soon as the scouts find food, they return to the hive. Shortly after their return, many foragers leave and fly directly to the food supply. The remarkable thing about this behavior is that the foragers do not follow the scouts back to their discovery. Instead, they fly to the food while the scouts are still within the hive. This implies two things. First, the foraging bees must have been told in some way how to locate the food source. Second, they must have some method of navigating over unfamiliar territory as they follow these instructions.

The discovery of how scout bees communicate with the foragers and how the foragers navigate over unfamiliar territory was made by the German zoologist Karl von Frisch. Throughout his entire professional career, he has patiently studied and experimented with honeybees. Many of the facts of bee life which we have been discussing were discovered by him.

By marking scout bees with paint and watching them upon their return to an observation hive (Fig. 29.13), von Frisch discovered that the scouts perform a little dance on the vertical surface of the combs after depositing their load of nectar or pollen. This dance seems to stimulate the foragers and soon they begin to leave the hive and fly to the food source. Unless the food source is quite close (less than 75 meters) to the hive, the scout bees dance a so-called tailwagging dance (Fig. 29.14). It did not take von Frisch long to realize that the speed with which the scout bees perform this maneuver is related to the distance of the food from the hive (Fig. 29.15); the scouts even compensate for wind speed. However, the knowledge that food is present 6 kilometers from the hive is not very useful when you consider the

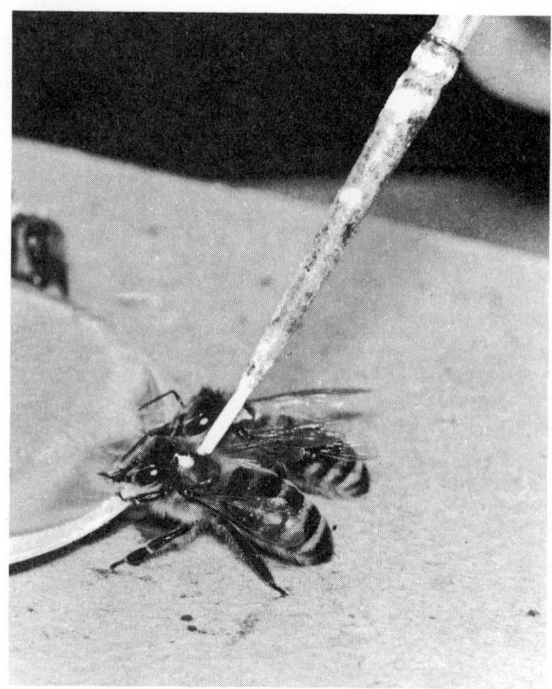

FIG. 29.13 Applying paint to foraging bees so that they can be identified on their return to the observation hive. (Courtesy of Dr. M. Renner.)

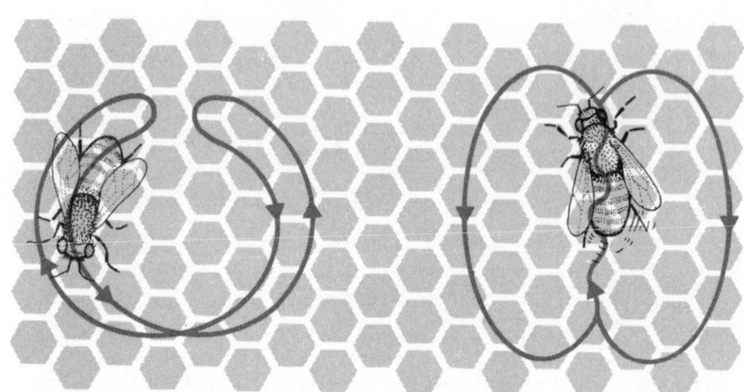

FIG. 29.14 Left: The round dance of the bee is used when food is found close to the hive. Right: The tail-wagging dance is used when food is found more than 100 m from the hive. The speed of the dance indicates just how far away the food is, and the direction of the straight portion indicates in what direction.

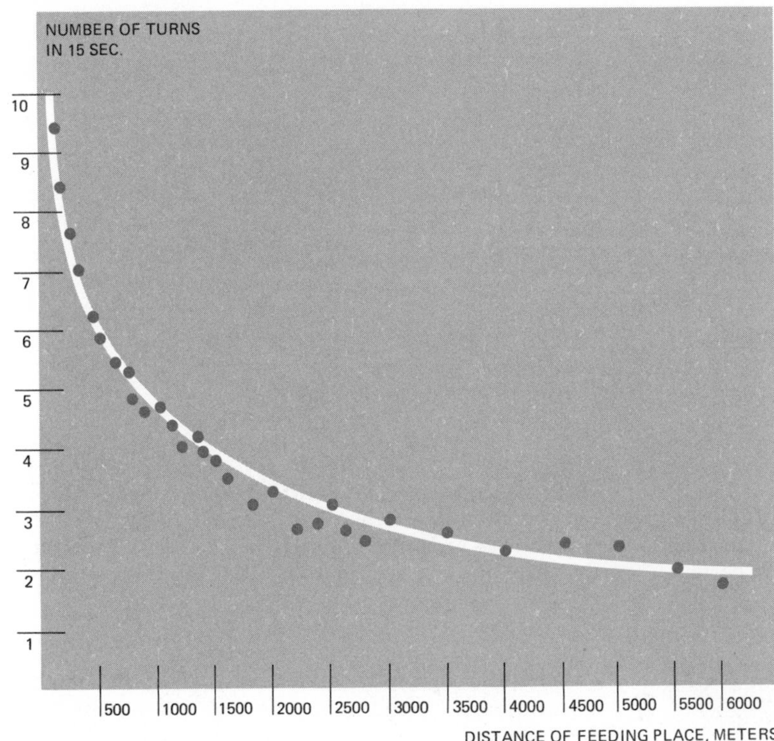

FIG. 29.15 Graph showing relationship between distance of food source and rapidity of tail-wagging dance in the honeybee (3885 observations).

long circumference included. But von Frisch also noted that the direction of the tailwagging portion of the dance varies with the direction of the food source from the hive and with the time of day. At a given time of day, the direction of the dance changes with varying locations of food. With a fixed source of food, the direction of the dance changes by the same angle as the sun during its passage through the sky. This suggests that the scout bees indicate the direction of the food source in relation to the direction of the sun. The sun is not visible in normal hives, however, and the bees dance on the vertical surface of the combs. How, then, do they translate flight angles in the darkened hive? When the food source is in the same direction as the sun, they orient the straight portion of the wagging dance up and down with their heads pointing up. It is as though they translate a positive phototaxis as a negative geotaxis. If the food source is at some angle to the right or left of the sun, the scouts dance at the same angle to the right or left of the vertical (Fig. 29.16).

The foraging bees cluster around the dancing scouts and presumably learn the direction and distance of the food in this way. If the food source is scented, the foragers also learn what odor to search for. Thus the language of the bees enables one bee to tell the others: (1) that food is available, (2) the direction to the food, (3) the distance to the food, and (4) the odor of the food.

If the dancing bees are confined to the hive for a long period of time, they shift the direction of their dance as the direction of the sun shifts. Remember, though, that they cannot observe the apparent motion of the sun while within the hive. This suggests that the bees are able to make the necessary corrections because they are "aware" of the passage of time. This fact has long been known to people who like to have tea, toast, and jam in their gardens at a fixed time every day. Within minutes of the appointed hour, foraging bees arrive in large numbers for their share of the jam. Here, then, is another example of a biological "clock." Its rate of ticking seems to be related to the bee's rate of metabolism. If a group of normally punctual bees are chilled (to lower their rate of metabolism) or exposed to anesthetizing concentrations of carbon dioxide, they will arrive at the tea table correspondingly later (Fig. 29.17).

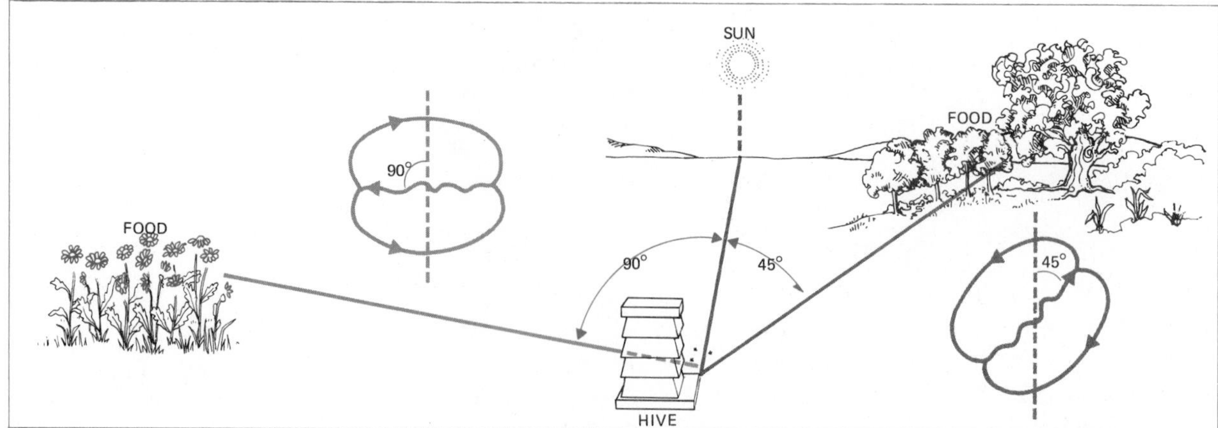

FIG. 29.16 Relationship between the angle of the dance on the vertical comb and the bearing of the sun with respect to the location of food. When the food and the sun are in the same direction, the straight portion of the dance is directed upward. When the food is at some angle to the right (black) or left (color) of the sun, the bee orients the straight portion of its dance at the same angle to the right or left of the vertical.

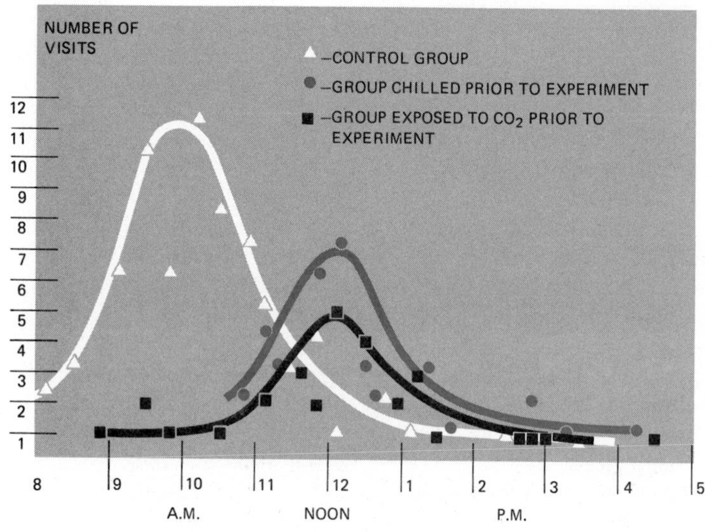

FIG. 29.17 Effect of chilling and exposure to CO_2 on the time sense of honeybees. Both chilling and CO_2 slow the metabolic rate of these animals. Each group had been fed between 9 and 11 a.m. for 4 days prior to the experiments.

These discoveries made by von Frisch are deceptively easy to describe. Like most scientific discoveries, they are the fruit of years of patient work, cleverly designed experiments, and carefully interpreted results. For a better idea of the actual process of scientific discovery, you should read von Frisch's own account (listed at the end of this chapter) of the long series of experiments which led him to his many discoveries concerning the behavior of bees.

When people first learn about the elaborate behavior of honeybees, they are apt to credit these little creatures with a high degree of intelligence, foresight, etc. Actually, this would be a mistake, since most of their behavior is instinctive and thus relatively inflexible. Within certain narrow limits, however, they can alter their behavior. When behavior is more or less permanently modified as a result of the experiences of the organism, we say that learning has occurred. The bee's association of food with a certain location, odor, or time of day is certainly an example of learning.

Learned Behavior

Learned behavior is behavior that is more or less permanently acquired or modified as a result of the experiences of the individual.

29.12 HABITUATION

Almost all animals are able to learn not to respond to repeated stimuli which have proved to be harmless. This phenomenon is known as habituation and is an example of true learning. If you make an unusual noise in the presence of the family dog, it will respond—usually by turning its head toward the source of the sound. If, however, the stimulus is given repeatedly and nothing either pleasant or unpleasant happens to the dog, it will finally cease to respond. That this is a case of true learning rather than simply the result of the adaptation of sense receptors is indicated by the fact that the response is long-lasting. When fully habituated, the animal will not respond to the stimulus even though weeks or months have elapsed since it was last presented.

29.13 IMPRINTING

One of the most narrowly specialized, clear-cut examples of learning is imprinting. If newly hatched geese (goslings) are exposed to a moving object of reasonable size emitting reasonable sounds, they will begin following it just as they would normally follow their mother. This is called imprinting. The time of exposure is quite critical. A few days after birth imprinting does not take place. Prior to this time, though, the results are quite remarkable. The gosling imprinted to a moving box or clucking man will, forsaking all others, try to follow this object for the rest of its life. In fact, when the gosling reaches sexual maturity, it will make the imprinted object, rather than a member of its own species, the goal of its sexual drive. Much of our knowledge of imprinting has been gained from the patient research of Konrad Lorenz (Fig. 29.18).

FIG. 29.18 Konrad Lorenz with imprinted goslings. In 1973, Lorenz shared a Nobel Prize with N. Tinbergen and Karl von Frisch. (Tom McAvoy—courtesy of LIFE Magazine, © 1955, Time, Inc.)

29.14 THE CONDITIONED RESPONSE

Perhaps the simplest form of learned behavior is the conditioned response. Basically, this is a response which, as a result of experience, comes to be caused by a stimulus different from that which originally triggered it. We owe our understanding of the mechanism of the conditioned response to the research of the Russian physiologist Ivan Pavlov. Pavlov found that the placing of food in a dog's mouth causes it to salivate. This is probably a simple, inborn reflex involving taste buds, sensory neurons, networks of neurons in the brain, and motor neurons running to the salivary glands. Pavlov found further that if he rang a bell every time he introduced meat into the dog's mouth, the dog eventually salivated upon hearing the bell alone. This is the *conditioned response.* The dog has learned to respond to a substitute stimulus, the *conditioned stimulus.*

We assume that the physiological basis of the conditioned response is the transfer, by appropriate neurons, of nervous activity in the auditory area of the brain to the motor neurons controlling salivation. This process involves the use of new circuits, which, we may also assume, is characteristic of all forms of learning. In fact, some psychologists feel that all learned behavior arises from the development of conditioned responses. They feel that the conditioned response is the fundamental unit of even the most complex forms of human behavior. While our knowledge of our own higher mental processes is still far too poor to accept or reject this theory outright, there is no question that humans can be conditioned to some degree.

Experiments on conditioning have taught us a good deal about the learning process in humans. Conditioning occurs most rapidly (1) when the unconditioned stimulus and conditioned stimulus are presented together frequently, (2) when there are no distractions, and (3) when a reward of some sort is given for successful performance of the conditioned response. Similarly, as every student knows, learning proceeds most successfully with repetition, lack of distraction, and good motivation.

The conditioned response has proved to be an excellent tool for determining the sensory capabilities of other animals. As we have seen from Fig. 26.4, honeybees can be trained to seek food on a piece of blue cardboard. They learn to associate the color blue with the presence of food. This is a conditioned response. By offering other colors to the blue-conditioned bee, we can discover which she confuses with blue and which she does not. In this way, Karl von Frisch (who shared a Nobel Prize with Lorenz and Tinbergen in 1973) determined that honeybees can see only four distinct colors: yellow-green, blue-green, blue-violet, and ultraviolet. The ability of animals to discriminate between similar shapes and musical tones has also been studied by the technique of conditioning.

29.15 INSTRUMENTAL CONDITIONING

Pavlov's dog was restrained in one position and the response that was being conditioned (salivation) was an innate one. But the principles of conditioning can also be used to train animals to perform tasks that are not innate. In this case, the animal is placed in a setting where it can move about and engage in a number of different behavioral activities. The experimenter chooses to reward only one, for example turning to the left. By first rewarding (e.g. with a pellet of food) even the slightest movement to the left and then only more complete turns, a skilled experimenter can, in about two minutes, train a naive pigeon to make a complete turn. A little more work and the pigeon will pace out a figure eight. Such training is known as **instrumental conditioning** or **operant conditioning** (Fig. 29.19). The latter term was coined by the psychologist B. F. Skinner (whose skill with the technique enabled him to train pigeons to play ping-pong and even a toy piano!) because it indicates that through its behavior the animal is operating on its environment, i.e. influencing its situation. It is also called trial-and-error learning because the animal is free to try various responses before finding one that will be rewarded.

Maze problems are a form of instrumental conditioning in which the animal is faced with a sequence of alternatives. All animals with bilaterally symmetrical nervous systems can learn to make some simple, consistent choices when confronted with alternatives. In a T-maze (Fig. 29.20) an earthworm or a planarian can eventually learn to take the arm of the maze which leads to a reward (e.g. food, moisture) and/or away from punishment (e.g. sandpaper, electric shock). The development of this response is a slow process, however, and never reaches the point of 100% correct choices. Interestingly enough, the earthworm learns to solve this maze even after its cerebral ganglia have been removed; but once new cerebral ganglia regenerate, the worm forgets its prior learning and must be retrained. Ants and mice are excellent solvers of mazes. It takes some ants as few as 28 trials to solve the maze shown in Fig. 29.21. As for Julia (Fig.

FIG. 29.19 Instrumental conditioning. Presented with two spots of light, the pigeon pecks at the brighter and then reaches down to pick up the grain of food that is its reward. (Courtesy of Roy DeCarava and *Scientific American.*)

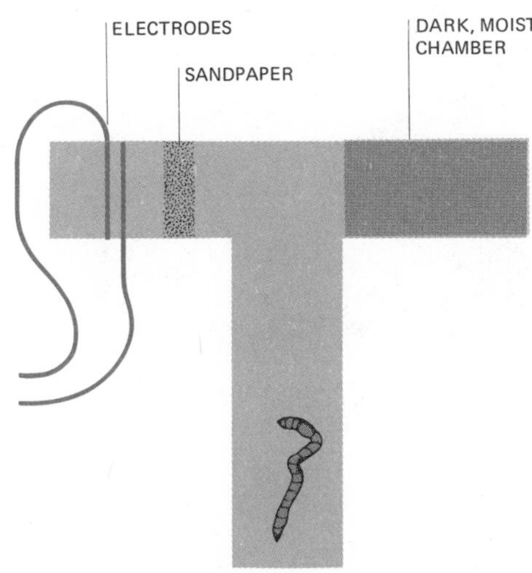

FIG. 29.20 A T-maze. Earthworms can learn to make the correct choice 90% of the time.

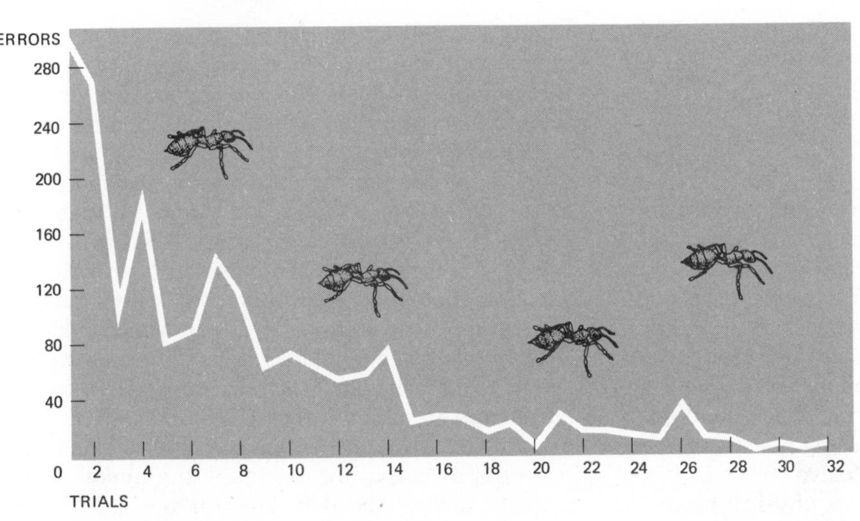

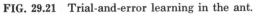

FIG. 29.21 Trial-and-error learning in the ant.

FIG. 29.22 Julia, a chimpanzee, uses a magnet to move an iron ring through the maze. Although biology students *usually* solve such a maze faster than Julia can, this is not always the case. (Courtesy of B. Rensch.)

29.22), she is able to solve mazes like the one shown on her *first* attempt most (86%) of the time and sometimes faster than biology students can!

29.16 MOTIVATION

Anyone who has tried to train an animal in a procedure, such as running a maze, that requires the animal to participate actively and to make choices knows that it can be a very frustrating procedure. The student psychologist is apt to find that when rat and newly constructed maze are brought together, the rat simply curls up in a corner. The basic problem here is one of motivation. The animal must "want" to participate in the learning process. Among most animals, motivation (or "drive" as it is sometimes called) is connected with its physical needs. A thirsty animal will search for water and a hungry animal will search for food. It is standard practice, for example, to deprive a rat of food for 23 hours prior to using it for instrumental conditioning (when, of course, the reward is food).

The satisfaction of its drives is the motivating force behind the animal's behavior. At times, the internal drive can be very precise. A rat with sugar, or salt, or even thiamine missing from an otherwise adequate diet will seek out food containing the missing substance in preference to that which does not. Perhaps we can go so far as to say that most of the spontaneous behavior of these animals results from an attempt to maintain homeostasis. Many of these drives have their origin in the hypothalamus. In some cases (thirst, for example) the hypothalamus actually detects the deficiency in the ECF. In all

FIG. 29.23 Curiosity in a monkey. The monkey repeatedly works the puzzle with no other motivation than that created by the task itself. (Courtesy of Myron Davis and *Scientific American*.)

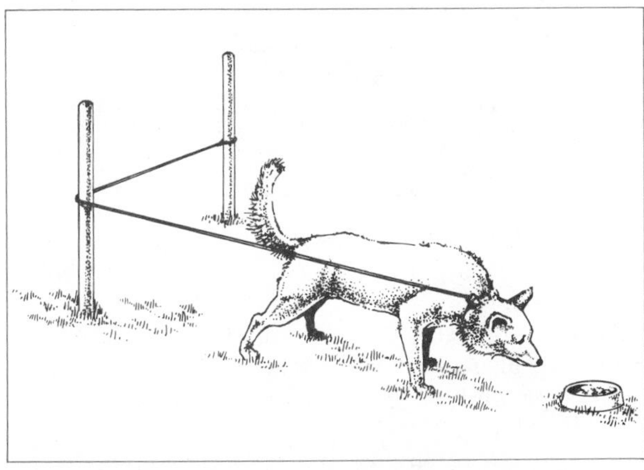

FIG. 29.24 A detour problem. A dog finally solves it by trial and error.

cases, the hypothalamus seems to initiate the responses which lead to a reduction of the drive. It may also inhibit some of these responses when the point of satiation is reached.

While we can trace a good deal of human behavior to the desire to satisfy physical needs, not all of it can be explained in such terms. Many of the things we do we seem to do for their own sake. Goats, monkeys, and chimpanzees, too, have been found to engage in problem-solving activity even when no external reward or punishment is involved (Fig. 29.23). The carrying out of the process itself seems to be its own reward.

Chimpanzees and humans are unusual also in that they will both work for distant goals. Both chimps and humans can learn to work for coins even though the coins themselves satisfy no physical need. Of course, the coins can be converted into food, and this expectation provides sufficient motivation for brief periods of work in chimpanzees and a lifetime of work in humans.

29.17 CONCEPTS

Most animals solve mazes and other problems by trial and error. As long as sufficient motivation is present, they try each alternative and gradually, through repeated successes and failures, learn to solve the problem (Fig. 29.24). Trial-and-error learning plays a negligible role in humans, however. When presented with a problem, we may make one or two random attempts at solving it and then, all of a sudden, we "get it." We call this response **insight.** (Sometimes it is referred to as the "aha!" reaction.)

Responses that occur as a result of insight are quite different from anything we have considered up to now. While they depend upon previously learned material, they are an entirely new response for the individual. Insight involves putting familiar things together in new ways. It is thus a truly creative act. Insight also depends upon the development of concepts or principles. This can be illustrated by a hypothetical experiment (Fig. 29.25). A rat is placed in front of a semicircle of doors, three of which are opened at a time. No matter which particular doors are opened, if the rat enters either the left-hand or right-hand door, it receives a shock. If it enters the middle door, it finds a reward of food. If the rat ever should learn to go immediately to the middle door (even if that door led to a shock on the previous trial), the rat would have learned a **concept.** In this

FIG. 29.25 If the rat could learn always to go to the middle one of whichever three doors happen to be open, it would have learned a concept. Although a chimpanzee can solve such a problem with ease, the rat cannot.

FIG. 29.26 A young rhesus monkey solving an oddity problem. It has learned that food will be found under whatever object is the odd one of the three presented. (Courtesy of H. F. Harlow, University of Wisconsin Primate Laboratory.)

case, the concept would be the idea of *middleness.* The rat would not be responding to any concrete, specific stimulus but rather to an idea. From its experience with particular doors it would have developed an idea about doors in general.

Actually, a rat would fail this test. Rats and most animals have little or no ability to make abstract generalizations about specific things. A human can solve this sort of problem, of course. Chimpanzees are also capable of developing abstract ideas. Monkeys and elephants can solve simple oddity problems (Fig. 29.26).

Problem solving of the type illustrated by our hypothetical rat in front of the semicircle of doors involves some form of reasoning. Chimpanzees, which can solve this problem, are probably reasoning without the aid of a language. Nevertheless, two different, but related, thought processes are involved. One, *inductive* reasoning, involves learning the general principle (middleness) from experience with the specific, concrete situation. The other, *deductive* reasoning, involves applying the general principle to some new specific situation. If the chimp had a language, it might say, "If food is always located behind the middle door, then this time it must be located behind *that* door."

29.18 LANGUAGE

All humans, even in the most primitive societies, possess a highly developed language of some sort. This involves a second level of abstraction. The very idea of a chair or door, like the idea of middleness, is an abstraction. When these ideas are then represented by symbols, a second level of abstraction occurs. There is nothing chairlike about the letters *c h a i r*.

Of course, it would be perfectly possible to create a written language with a unique symbol, such as a sketch, for each concept. This occurred to some degree in the development of language among the Chinese, the Japanese, and the Indians of North America. A more flexible arrangement is to use a relatively small number of symbols which can then be assembled in unique ways for each concept. Our written language is made up of 26 letters whose varied arrangement makes our written language possible. Contrast, if you will, the meanings associated with *r a t* and *t a r*. In a similar way, our spoken language is made up of some three to five dozen (depending on who is doing the counting) distinct sounds or **phonemes**.

For many years it was generally agreed that one feature that distinguished humans from all other animals was their use of language. Of course, we now realize that many other animals (e.g. bees) have signaling mechanisms that possess some of the features of a language. But until recently it was felt that no other creature could, as we do, take a small number of arbitrary symbols and assemble them in unique ways for particular concepts. However, two remarkable series of investigations with chimpanzees suggest that this is not a uniquely human attribute. The husband-and-wife team of Ann and David Premack chose to investigate the language potentialities of the chimpanzee by using metal-backed plastic symbols for words and a magnetized blackboard on which these could be assembled into "sentences." After some half-dozen years of training, their subject—a female chimp named Sarah—had a vocabulary of about 130 words. These included not only nouns (e.g. Sarah, apple, pail) but verbs (is, give, take, wash) and adjectives like red and yellow (the symbols for these were *not* so colored). Sarah also learned to deal with symbols representing such concepts as "same," "different," "not," "if-then," and a question. With these tools, she could clearly grasp some essentials of grammar. Furthermore, she appeared to be "thinking" in her symbolic language. When presented with her word for apple (a blue triangle) she described it as red, round, and with a stem even though no apple was present at the time.

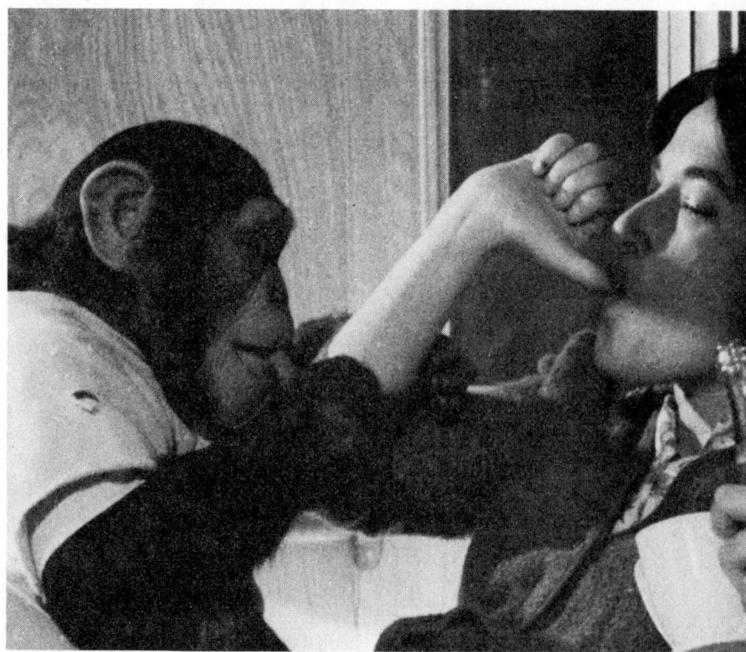

FIG. 29.27 Washoe, a young female chimpanzee, gives the sign for "drink" in the American Sign Language of the deaf. (Courtesy of R. Allen Gardner and Beatrice T. Gardner.)

The use of symbols that can be manipulated (instead of spoken words) to explore the language ability of a chimpanzee rests on the evident manual dexterity of chimps and their inability to vocalize. This was also the reason that the Gardners elected to explore chimp language ability by use of the American Sign Language, which is used by the deaf in North America. While many of the gestures in this language are somewhat representational (Fig. 29.27), many others are not. After 22 months of training, their young female chimp, named Washoe, had learned more than 30 signs. With these she was able to make a variety of requests and to answer simple questions. When she thought she was not being observed, she would "talk" to herself or name the objects in a picture book. Often, on the way to the toilet, she made the sign for "hurry."

While the skills displayed by Sarah and Washoe may appear rudimentary, they are every bit as good as those of a two-year-old child. Those who believe that human linguistic ability differs qualitatively, not just quantitatively, from that of a chimpanzee would now seem to carry the burden of proof.

29.19 MEMORY

All learning depends upon memory. If the organism is to modify its behavior from experience, it must be able to remember what its experience has been. Once something is learned, memory is necessary for the learning to be retained.

Two basic theories of memory have been proposed. One is that memory is a dynamic process. According to this theory, sensations give rise to nerve impulses, which then circulate indefinitely through the network of neurons in the central nervous system. When one considers the vast network of interneurons within the human cerebrum, such a theory seems plausible. The delay circuits used to store information in some modern computers operate on the same principle. This dynamic theory of memory is supported by the startling fact that no specific area of the human brain has ever been found essential to the retention of old memories. Whatever defects may occur as the result of damage to one region of the brain or another, loss of memory does not seem to be involved. On the other hand, a dynamic memory must always be in action. If all nerve impulses in the brain ceased, even for an instant, then this type of memory would be lost. However, when maze-trained rats have been carefully chilled to the point where no electrical activity in the brain can be detected and then rewarmed, they remember their prior training.

This fact supports the second theory of memory, which is that each remembered sensation results in some permanent, physical alteration in the brain. Perhaps a change in the resistance of certain synapses occurs, although the fact that memory does not seem to be localized anywhere in the brain argues against this view. Recently, several biologists have suggested that our memories may be stored in a chemical code within the brain. Some look to RNA, some to proteins, as the substances in which memories are coded. As you remember, these macromolecules can be built in an almost infinite number of ways.

It is a demonstrated fact that the RNA content of neurons increases with their activity so long as this activity is within normal ranges. Furthermore, it has been found that the RNA content of neurons in the human spinal cord increases from age 3 to age 40. After age 40 the RNA content of these cells remains quite constant until age 55 or 60, when it declines rapidly.

That memory might be specifically encoded in RNA has been proposed as the explanation of certain learning experiments which have been carried out on planarians. As we mentioned, these creatures can learn to solve a *T*-maze. They can also learn a conditioned response. The head of the worm dominates the tail in these responses. If the worm is cut into two pieces, both the head and the tail regenerate the missing parts. Both regenerated worms retain some memory of their prior training. However, if the pieces are allowed to regenerate in pond water to which some of the RNA-digesting enzyme, ribonuclease, has been added, only the dominant, trained head retains its memory. The head that is regenerated by the severed tail performs no better than a totally untrained worm. Another experimenter has claimed that when trained (conditioned) planarians are ground up and fed to untrained ones, the latter can be similarly trained in a much shorter time than normally fed individuals.

Such remarkable reports as these have produced a flurry of related experimentation in many different laboratories. These experiments have been done not only with planarians but also with "higher" animals such as rats and mice. Some workers claim that learning can be transferred from one animal to another with RNA extracted from the trained animals and injected into untrained ones. Other investigators claim to have accomplished the same feat with proteins and even, in one case, a polypeptide containing 15 amino acids. Some workers have found drugs which, they claim, speed up RNA synthesis and also learning. Unfortunately, in every one of these cases, workers in other laboratories have often been unable to duplicate these results. Perhaps this should not really be too surprising when you consider the many subtle factors that could influence the behavior of an animal like a rat but be overlooked by the experimenter. Furthermore, the transfer of learning that is claimed really involves simply *faster learning* on the part of the recipients than in the controls. Perhaps these extracts and chemicals that are thought to transfer or speed up learning may simply be speeding up the *general activity* of the recipients.

It is still too soon to say just what the nature of memory is. Perhaps both dynamic processes as well as physical-chemical alterations are involved. This would certainly be consistent with the discovery that the acquisition of a memory seems to occur in at least two distinct steps. In humans, for example, damage to the temporal lobes may result in the loss of ability to remember new learning for more than about one hour. Such damage has no effect on memories acquired in the years before the damage occurred. Mental patients undergoing electroshock therapy are unable to remember events that occurred just prior to the treatment, but the memory of earlier events is unimpaired. In goldfish and mice, the application of chemicals that inhibit protein synthesis has been shown to prevent the *acquisition* of long-term memories but not of short-term ones. However, inhibition of protein synthesis does not inhibit the *recall* of old memories. This has been demonstrated by training a group of mice to choose one (e.g. the left) arm of a *T*-maze. Three weeks later they are retrained to choose the other (the right). Then some are given an inhibitor of protein synthesis and others simply saline. When tested three days later, the mice whose protein synthesis had been suppressed chose the left arm; the mice that received saline went right!

It should be apparent from this chapter that we know a great deal more about the simple forms of behavior, such as the simple reflex and the conditioned response, than we do about the more elaborate processes involved in insight and reasoning. We can be reasonably certain that further study will gradually tell us more about the mechanisms of these more elaborate mental activities so characteristic of humans. While we look forward to these discoveries, we must also be warned by the memory of the ways in which past scientific discoveries, such as nuclear fission, have sometimes been used. The knowledge of how the human mind functions is knowledge that will have to be used with great wisdom.

EXERCISES AND PROBLEMS

1. Design an experiment to show *whether or not* scout bees can communicate to other foragers information about the direction of a food source 200 meters from the hive.
2. How could you prove that the odor receptors of the honeybee are located on the antennae? Describe your procedure in detail.
3. If a person is shocked through the foot, simple reflex withdrawal of the foot will occur. Describe how this reflex might be "conditioned." What changes in the functioning of the nervous system occur during this process?
4. On March 20th, the sun rises in the east at 6:00 a.m. and sets in the west approximately 12 hours later. At 9:00 a.m., a scout bee does the tail-wagging dance so that the straight portion of the dance is 135° to the right of the vertical. In which direction is the food that it is reporting?
5. How will a scout bee dance in the hive at noon on March 20th when it has discovered food 1000 m to the north of the hive?
6. How will it dance when food has been discovered 5000 m northeast of the hive?
7. Summarize the various methods of intraspecific communication found among animals.
8. Distinguish between a tropism and a taxis.
9. Design an experiment to show whether a dog can discriminate between two adjacent whole notes on the piano.
10. Design an experiment to show whether ants have any color vision.

REFERENCES

1. HAILMAN, J. P., "How an Instinct Is Learned," *Scientific American,* Offprint No. 1165, December, 1969.
2. ADLER, J., "The Sensing of Chemicals by Bacteria," *Scientific American,* Offprint No. 1337, April, 1976.
3. MERTON, P. A., "How We Control the Contraction of Our Muscles," *Scientific American,* Offprint No. 1249, May, 1972. Primarily concerned with the role of the muscle spindle in controlling the stretch reflex.
4. BENZER, S., "Genetic Dissection of Behavior," *Scientific American,* Offprint No. 1285, December, 1973. Fruit flies that are genetic mosaics of normal and mutant parts reveal clues to the structural basis of their behavior patterns.
5. EMLEN, S. T., "The Stellar-Orientation System of a Migratory Bird," *Scientific American,* Offprint No. 1327, August, 1975. Experiments with an indigo bunting in a planetarium.
6. SAUNDERS, D. S., "The Biological Clock of Insects," *Scientific American,* Offprint No. 1335, February, 1976.
7. BENTLEY, D., and R. R. HOY, "The Neurobiology of Cricket Song," *Scientific American,* Offprint No. 1302, August, 1974. The song pattern of each cricket species is stored in its genes.
8. VON FRISCH, K., *Bees: Their Vision, Chemical Senses, and Language,* Cornell University Press, New York, 1950. In this small book (available in a paperback edition) von Frisch describes the experiments which led him to so many discoveries in the fascinating behavior of bees.
9. VON FRISCH, K., "Dialects in the Language of the Bees," *Scientific American,* Offprint No. 130, August, 1962. Different kinds of bees vary in the details of their dances. These variations provide clues to the evolution of this system of communication.
10. WEHNER, R., "Polarized-Light Navigation by Insects," *Scientific American,* Offprint No. 1342, July, 1976.
11. MORSE, R. A., "Environmental Control in the Beehive," *Scientific American,* Offprint No. 1247, April, 1972.
12. LINDAUER, M., *Communication Among Social Bees,* Harvard University Press, Cambridge, Mass., 1961. A report by a former student of von Frisch of further discoveries in the role that communication plays in the lives of bees.
13. MICHENER, C. D., and MARY H. MICHENER, *American Social Insects,* Van Nostrand, Princeton, N.J., 1951. Beautifully illustrated accounts of the lives of wasps, bees, ants, and termites.
14. CHEESMAN, EVELYN, *Insects: Their Secret World,* William Sloan Associates, New York, 1953. Includes many fascinating examples of insect behavior.
15. HESS, E. H., " 'Imprinting' in a Natural Laboratory," *Scientific American,* Offprint No. 546, August, 1972.
16. BLOUGH, D. S., "Experiments in Animal Psychophysics," *Scientific American,* Offprint No. 458, July, 1961. Instrumental conditioning is used to assess the discriminatory abilities of pigeons.
17. FISHER, A. E., "Chemical Stimulation of the Brain," *Scientific American,* Offprint No. 485, June, 1964. Predictable behavior patterns in rats can be triggered by the application of certain chemicals to specific areas of the brain.
18. PREMACK, ANN J., and D. PREMACK, "Teaching Language to an Ape," *Scientific American,* Offprint No. 549, October, 1972. The story of Sarah.
19. HOCKETT, C. F., "The Origin of Speech," *Scientific American,* Offprint No. 603, September, 1960. The characteristics of human language and how it compares with other forms of animal communication.

30 THE IMMUNE RESPONSE

30.1 INTRODUCTION

For many centuries people have been aware that certain diseases never strike the same person twice. A pockmarked survivor of one smallpox epidemic could safely handle patients during a subsequent epidemic. As a result of contracting the disease, some change had occurred within the body making that person henceforth *immune* to the disease. The responsiveness of the individual had been permanently altered as a consequence of experience—a form of learning.

What is the nature of this change? In Chapter 17, you learned that pneumococci, the agents that cause bacterial pneumonia, exist in a large number of different types (I, II, III, etc.). A survivor of a bout of Type III pneumonia becomes immune to pneumonia of *that type only*. If one injects no more than two or three Type III pneumococci into the bloodstream (not their usual habitat) of a mouse, the bacteria proliferate quickly and the mouse soon dies of septicemia (blood poisoning). However, if a small amount of blood serum from an immune human is injected along with hundreds of thousands of Type III pneumococci, the mouse remains healthy. The blood of a human who has never been sick with *Type III* pneumonia has no such protective effect. This experiment tells us that at least one manifestation of immunity is to be found in the blood. We say that the immune serum contains **antibodies** against the disease.

30.2 WHAT ARE ANTIBODIES?

Prominent among the various components of the blood are proteins such as albumin, fibrinogen, and globulins (see Section 11.13). There are a number of techniques by which these proteins can be isolated from one another (see Fig. 11.13). When this is done to immune serum, the fraction containing the globulins is the only fraction that can, to pursue our example, protect a mouse from death by Type III pneumococci. These antibodies (and all others) are, then, serum globulins.

Thanks to the pioneering efforts of Sanger (see Section 4.4), it is now possible to determine the primary structure of a protein, that is, its precise sequence of amino acids. This has been done for insulin (Fig. 4.17), hemoglobin (see Section 18.2), ribonuclease, the coat protein of the virus MS2 (Fig. 18.14), and many others. Why not do the same for an antibody?

Unfortunately, it turns out that there is not just a single kind of globulin molecule that is an antibody against, for example, Type III pneumococci. There are, in fact, many of them. Insulin, hemoglobin, ribonuclease, and the MS2 coat protein all consist of just a single kind of molecule which can be isolated and analyzed. Not so for antibodies. Isolation of the antibodies against Type III pneumococci presents one with a mixture containing a number of *different* protein molecules, each with its own primary structure. Until quite recently, no techniques existed for further fractionation of this mixture so that one or more purified single components could be analyzed.

The solution to this dilemma came from the realization that a rare form of human (and mouse) cancer involves the antibody-producing system. A victim of this disease, called multiple myeloma, produces large amounts of a *single* type of antibody molecule. This is not (so far as we know) the result of an infection but, rather, another example (see Section 23.2) of a cell entering upon a period of unchecked proliferation. In this case, the cell is an antibody-secreting cell, and the **clone** (see Section 14.3) that develops yields a single protein product.

Even though one doesn't know what the antibody is an antibody *against,* the large amounts and purity of this material have enabled a number of research workers to slowly piece together a picture that we have reason to believe represents the basic structure of many kinds of antibody molecules.

Figure 30.1 shows the structure of a myeloma protein as worked out by Edelman and his colleagues. The molecule consists of (1) two long polypeptide chains (the heavy chains) containing approximately 446 amino acids and identical to each other; (2) two short (approximately 214 amino acids) chains (the light chains), which are also identical to each other; and (3) some carbohydrate attached to the heavy chains. The heavy chains are attached to each other and to the light chains by S—S bridges. There are also S—S bridges *within* both the light and heavy chains and these force the molecule into a number of compact, globular regions (Fig. 30.2).

The analysis of a number of *different* human myeloma proteins has revealed some curious uniformities. Half of each light chain and three-quarters of each heavy chain (the —COOH terminal portion in each case—Fig. 30.1) have virtually identical sequences in every human myeloma protein. These regions are thus called the constant (C) regions. The constant region of the heavy chains consists of three very similar sequences (each containing about 108 amino acids) arranged in tandem. These sequences

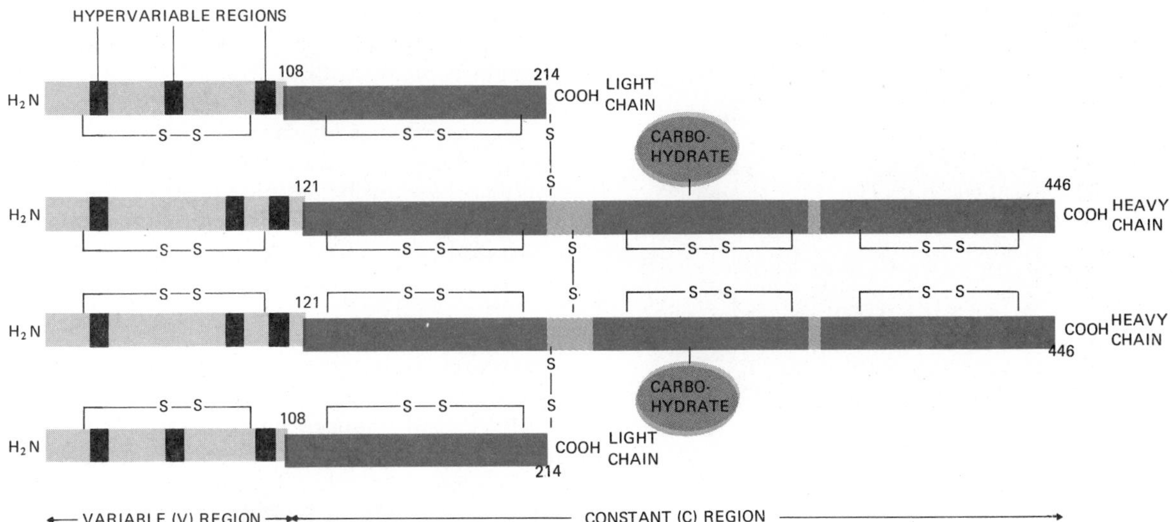

FIG. 30.1 Polypeptide chain structure of an antibody molecule. The numbers indicate the number of amino acids (counting from the amino terminal, $-NH_2$, end). The two light chains are identical, as are the two heavy chains. In the intact molecule, the chains are folded so that each cysteine is brought close to the partner with which it forms a disulfide (S—S) bond (cf. Fig. 4.20). In a given species, different antibody molecules (of the type shown here) have the same or almost the same sequence of amino acids in the constant regions, but show marked differences in the variable regions. These differences are especially pronounced in the hypervariable regions. The particular antibody molecule shown here is of the type designated as IgG. Other types (e.g. IgE—see Fig. 30.20) differ somewhat in the construction of their heavy chains.

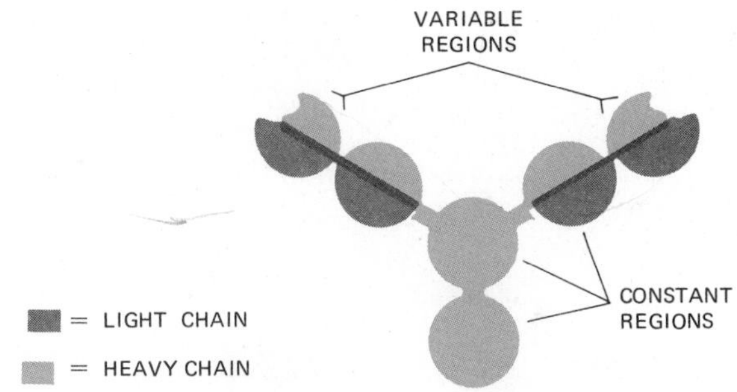

FIG. 30.2 Schematic representation of the three-dimensional structure of an antibody molecule. The specificity of an antibody, i.e. its ability to bind a particular antigenic determinant, resides in the globular domains created by the variable regions of the light and heavy chains. The domains formed by the constant regions accomplish other functions of the molecule.

are also similar to that found in the constant region of the light chain.

The situation is quite different in the amino (NH_2—) terminal (Fig. 30.1) segment of each light and heavy chain. Here each human myeloma protein examined shows different amino acids in many of the positions. Because of the variety of amino acid substitutions found, these portions of the chains are called the variable (V) regions. The sequence variability in the V region is especially pronounced in certain areas. These so-called *hypervariable* regions are located approximately at positions 24–34, 50–56, and 89–97 in light chains (Fig. 30.3) and 31–37, 51–68, 84–91, and 101–110 in heavy chains. In the next section, we shall see evidence that these hypervariable regions are crucial to antibody function.

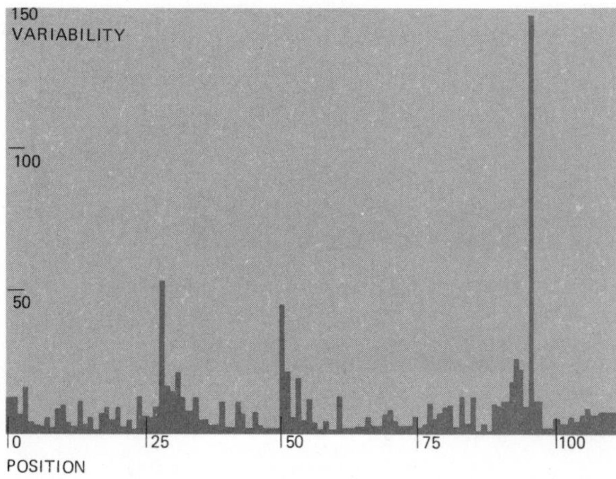

FIG. 30.3 Degree of variability found at each position in the variable (V) region of a large number of light chains derived from myeloma proteins. Where only one amino acid has ever been found at a given position, the variability is 1. If each of the 20 amino acids occurred with equal frequency at a given position, the variability would be 400. The three regions with high variability are the hypervariable regions.

You might wonder if myeloma proteins—the result of a disease process—are truly representative of the structure of antibodies. Recently, rabbit antibodies of *known* specificity (e.g. antibodies against Type III pneumococci) have been prepared that are as homogeneous as myeloma proteins. The primary structure of these antibodies fully confirms the validity of the model of antibody structure based on the myeloma proteins.

30.3 WHAT DO ANTIBODIES DO?

The type (III, IV, VIII, etc.) of a pneumococcus is determined by the molecular structure of the polysaccharide capsule that surrounds the organism (Fig. 30.4). The capsule can be removed from the bacterium and the polysaccharide purified. When a solution of pure Type III polysaccharide is added to a solution of anti-Type III pneumococcal antibodies, a dense white precipitate forms (Fig. 30.5). If the quantities used are carefully chosen, the solution remaining above the precipitate can then be shown to be completely devoid of either antibody or polysaccharide. All the polysaccharide and all the anti-

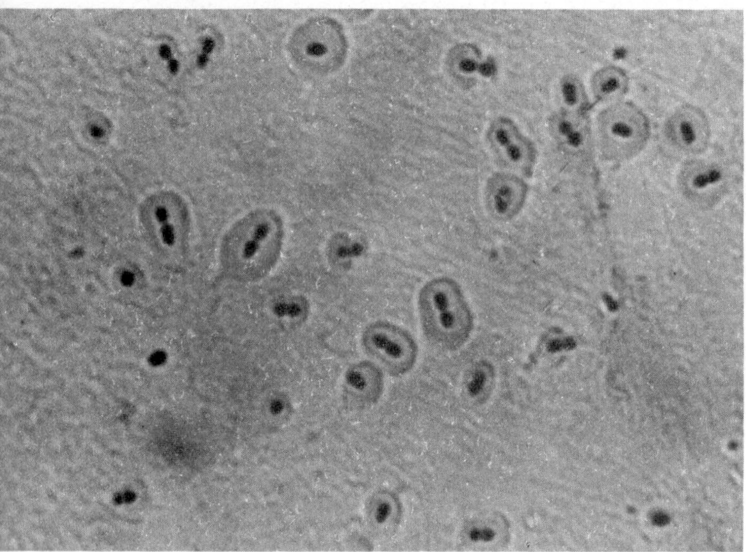

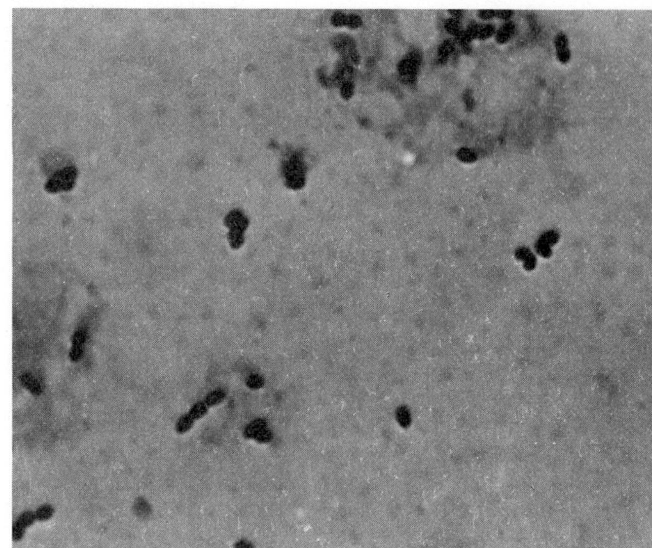

FIG. 30.4 Encapsulated (left) and nonencapsulated (right) pneumococci. The encapsulated (S) forms give rise to smooth colonies and are pathogenic because they are not easily engulfed by phagocytes (unless antibodies are present—Fig. 30.8). The nonencapsulated (R) forms produce rough colonies and are relatively harmless because they are easily engulfed by phagocytes. (Courtesy of Robert Austrian, *Journal of Experimental Medicine,* 98: 21–40, July, 1953.)

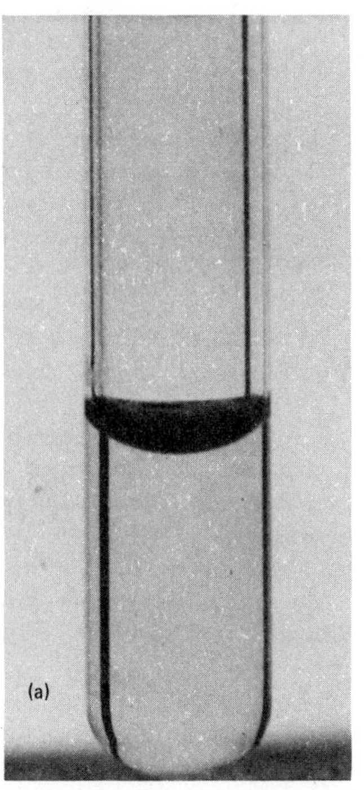

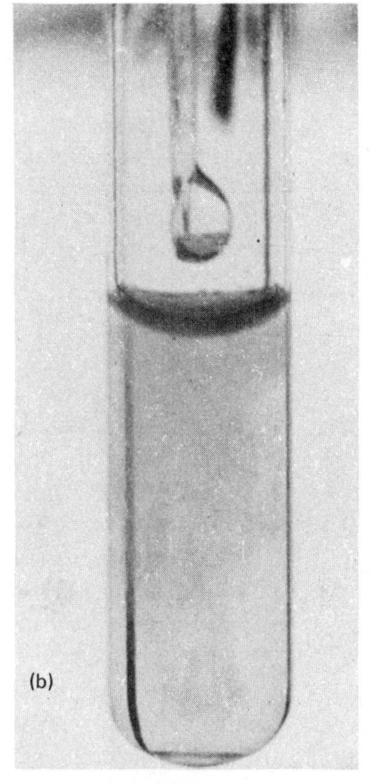

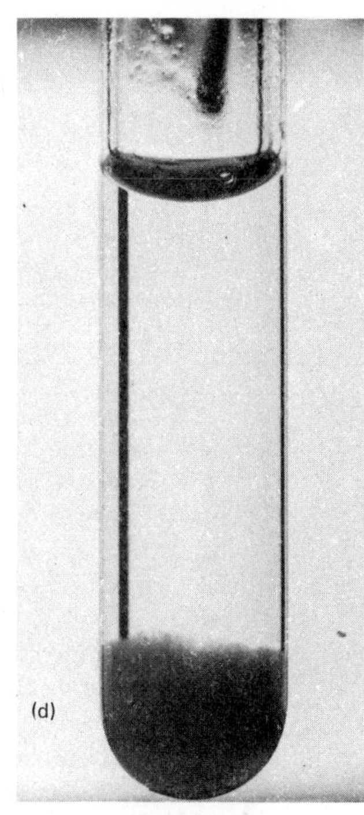

FIG. 30.5 Antigen-antibody interaction. The test tube at the left (a) contains a solution of antibodies to the Type III pneumococcal polysaccharide. A solution of the polysaccharide is added (b) and the formation of insoluble antigen-antibody complexes is revealed by the almost instantaneous appearance of turbidity (c). After an hour, the complexes settle out as a precipitate (d). If the proportion of antigen to antibody in the mixture was selected properly, the fluid above the precipitate will now be devoid of either.

bodies that were present initially are found complexed together in the precipitate. Antibodies, therefore, combine with the material that elicited them. We call this material the **antigen.** This behavior of antibodies is like that of an enzyme when it combines with its substrate (Fig. 6.17).

The Type III pneumococcal polysaccharide is an enormous polymer consisting of a linear array of thousands of glucose and glucuronic acid (a close relative of glucose) units alternating with each other (Fig. 30.6). If small lengths of this molecule (e.g. a disaccharide, trisaccharide, etc.) are added to anti-Type III antibodies, combination again occurs. However, the combining ability of the antibodies is not exhausted. They are still able to combine with the intact Type III polysaccharide. However, if we try lengths containing five to seven sugar units, the antibody is no longer capable of combining with the natural polysaccharide. This suggests that the antibody molecules combine with a portion of the polysaccharide that is only five to seven sugar units long (or approximately 24–30 A). It is interesting to note that antibodies directed against other types of mole-

COO⁻ CH₂OH COO⁻ CH₂OH

FIG. 30.6 Molecular structure of the polysaccharide capsule surrounding Type III pneumococci. The molecule is an extremely long, unbranched polymer consisting of alternating units of glucose and its relative glucuronic acid (with the —COO^- group).

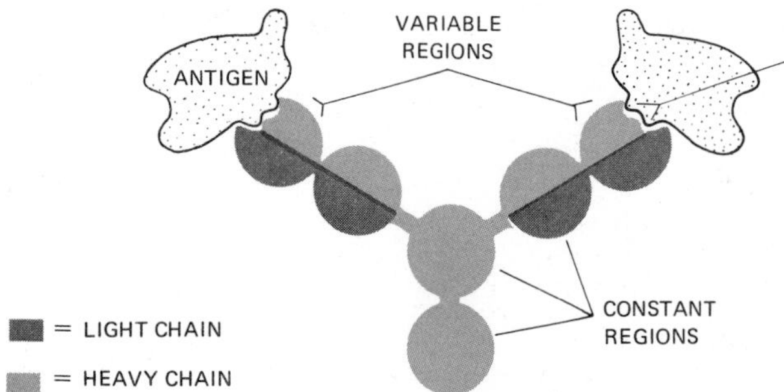

FIG. 30.7 Schematic representation of the binding of two antigen molecules to the antigen-combining sites in the variable domains of the antibody. The V regions of both the light and heavy chains participate in forming the antigen-binding site. The two sites in a molecule are identical.

cules (e.g. proteins) also unite with only a portion of the molecule, and that portion is of similar length. Thus it is not enough to say that antibodies combine with antigens, but rather they combine with a small part of the antigen, the so-called **antigenic determinant.**

We assume that the antigenic determinant fits into a groove or cleft on the antibody molecule. This cleft has a shape complementary to that of the determinant (Fig. 30.7). Thus we are dealing with the same sort of lock-and-key arrangement that we postulated for enzyme-substrate complexes (Fig. 6.14). It should be noted that the hypervariable regions of both the light and heavy chains are located so that at least 6 of the 7 participate in the making of a cleft, which is the antibody-combining site.

Most antigens are far more complex in structure than the Type III polysaccharide. The toxin produced by the diphtheria bacillus and the outer coat of a polio virus are just two examples of protein antigens that have several antigenic determinants. Exposure to these materials will elicit the formation of antibodies against each determinant. While this accounts for only a part of the story of antibody diver-

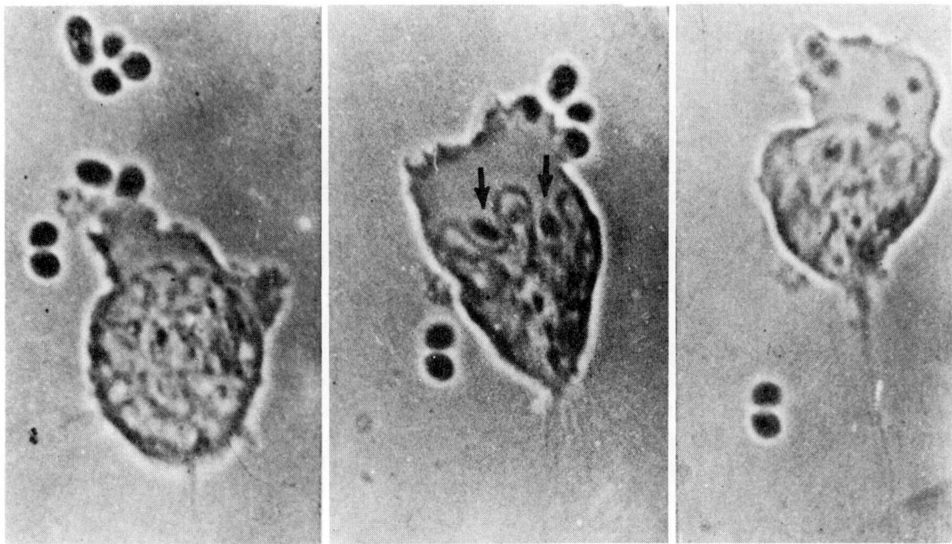

FIG. 30.8 Phagocytosis. Left: A neutrophil extends a pseudopod towards two pneumococci. Center: These bacteria have now been engulfed (arrows) and the neutrophil is beginning to engulf four more pneumococci at the upper right. Two pneumococci were spared (right). Antibodies to the polysaccharide capsule surrounding the pneumococci were added to enhance phagocytosis. (From W. B. Wood, M. R. Smith, and B. Watson, *Journal of Experimental Medicine,* **84,** 387, 1946.)

sity, it is no wonder that early workers were frustrated in their search for a single antibody molecule to analyze.

Of what biological significance is the union of antigen and antibody? Once the union occurs, the body has several mechanisms by which it can deal with the antigen-antibody complex. One of these is shown in Fig. 30.8. The capsule that surrounds pneumococci greatly interferes with their being engulfed by phagocytes. In fact, it is this property of the capsule that makes pneumococci so virulent. You remember (see Section 17.1) that capsule-less pneumococci (R forms) do not cause disease. If the appropriate anti-pneumococcal antibodies are present, they combine with the capsule and render the cells easy to engulf by endocytosis. In the days before the discovery of antibiotics, the start of antibody production by the immune system of the patient marked the turning point in the progress of the disease.

Another way of assessing the importance of antibodies is to examine the fate of those rare babies born without any immune system. For the first six weeks of life, they get along fine, thanks to their mother's antibodies, which they had received across the placenta. Soon this protection becomes inadequate, however, and the babies fall victim to raging bacterial and viral infections. The former can be aided by antibiotics but not the latter and, until quite recently, death was inevitable. Now, however, techniques have been developed so that some of these victims can be given the machinery for developing immunity and their lives saved.

30.4 ANTIGENS

Although we have up to now emphasized immunity against infectious organisms, the immune system is by no means limited to dealing with such material. The immune system is capable of making antibodies against a variety of noninfectious materials such as ragweed pollen, the active component of the poison ivy plant, and the Rh antigen, to name three troublesome examples. The Rh antigen is located on the surface of the red blood cells of "Rh positive" individuals. Before the discovery of the Rh antigen, "Rh negative" women (whose red blood cells lack the antigen) were sometimes inadvertently given transfusions of Rh positive blood. Unknown at the time, they responded by making anti-Rh antibodies. If such a woman later became pregnant with an Rh positive baby (the trait is inherited as a dominant), the fetus would often become dangerously anemic as the mother's antibodies (which cross the placenta) attacked its red cells.

Are there, then, limits to what can serve as an antigen? We have seen that polysaccharides and proteins serve as excellent antigens. Somewhat less effective but still antigenic are nucleic acids, the third main category of macromolecules. On the other hand, small molecules (e.g. glucose) do not elicit antibody formation. Size, therefore, seems to be one requirement for provoking an immune response. But what about the small molecule that causes poison ivy poisoning? What probably happens in this case is that the molecule first joins with some protein in the skin, and it is the resulting complex that becomes antigenic.

Although only macromolecules stimulate the immune mechanism, the antibodies that are formed combine with just a small portion, the determinant, of the antigen. The interaction between the two is remarkably precise. Our immune system can make antibodies that discriminate between pig insulin and human insulin (where only one amino acid out of 51 is different) and even between galactose and glucose present in a determinant (isomers that are identical except for the orientation of an —H and —OH group —see Fig. 4.9).

There is one group of macromolecules which do not act as antigens. These are macromolecules that are normal components of the body. We make antibodies in response to injections of albumin from cows but not of albumin from humans. The reason is that the albumin molecules in one human have the same structure as those in another. Thus we must add the additional criterion that a macromolecule can act as an antigen only when it is "foreign" to the body.

Although humans share many identical macromolecules, each of us has others that are *unique* to us (unless we have an identical twin). This is why grafts of skin (or organs) from one individual to another are attacked by the recipient's immune system while grafts of skin from one part of the body to another are not. The immune system thus discriminates between "self" and "nonself."

It can be fooled. Fetal calf twins share a common placenta and thus, until birth, a common blood system. Although developed from separate fertilized eggs and thus not in any sense genetically identical, such animals act like identical twins in their inability to make antibodies against each other's macromolecules. Transplants of skin from one to the other "take" just as well as they do between identical twins (see Section 13.8).

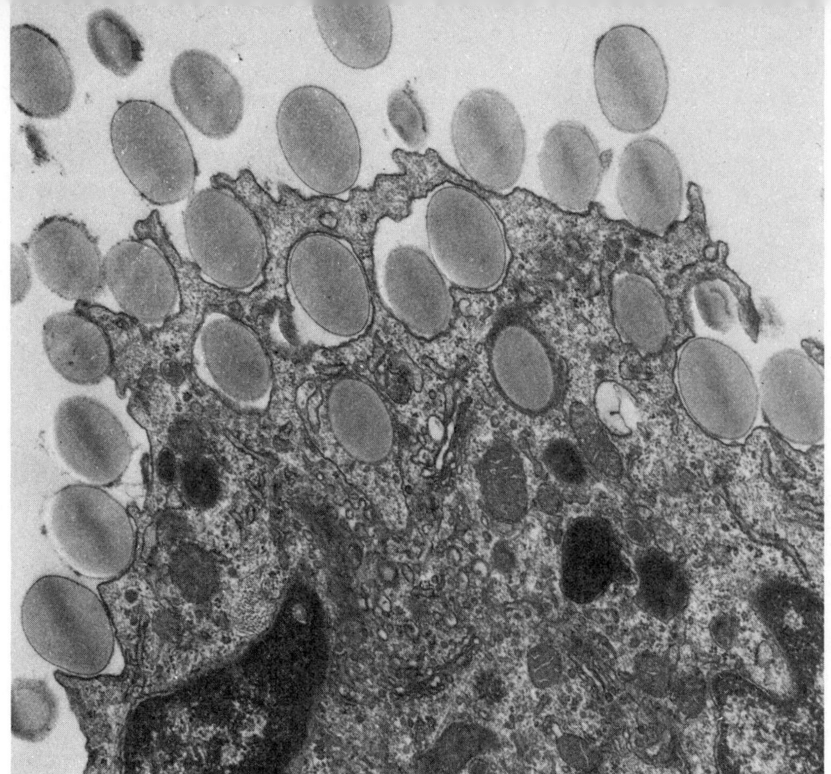

FIG. 30.9 Guinea pig phagocyte ingesting polystyrene beads. Several beads are already enclosed in vacuoles, while others are in the process of being engulfed. (22,000 X, courtesy of Dr. Robert J. North.)

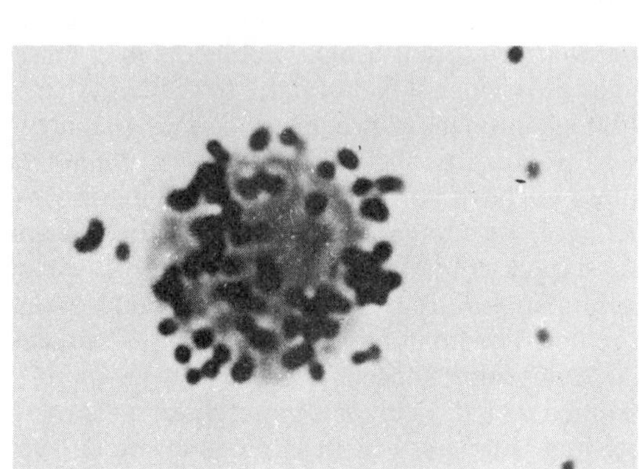

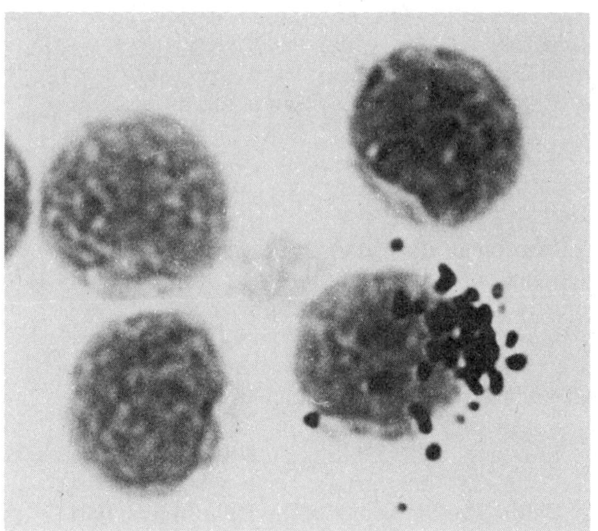

FIG. 30.10 Left: Autoradiograph of a lymph node cell to which molecules of a radioactive antigen have adhered. This cell was taken from an *unimmunized* mouse. Right: After 15 minutes of incubation at 37°C, the radioactivity is clustered in one region. This indicates that the cell's antigen-binding receptors are free to migrate over the surface of the cell. Note the other cells that have not bound any of the radioactive antigen. Presumably they have receptors for other antigens. The work pictured here represents an extension of the findings of Ada and Byrt. (Courtesy of E. Diener from E. Diener and V. H. Paetkau, *Proceedings of the National Academy of Sciences, U.S.A.*, **69**, 2364–2368, September, 1972.)

Nature's experiment with cattle twins can be duplicated with other animals. Injections of foreign molecules into a newborn mammal—before its immune machinery has been set up—will create a situation of **tolerance.** Later, when the animal becomes able to make antibodies, it will fail to do so against the material given to it at birth. It has accepted such material as "self."

30.5 HOW ARE ANTIBODIES ELICITED?

An enormous amount of research has gone into trying to piece together the steps that take place between the introduction of an antigen into the body and the start of antibody synthesis. It is now possible to carry out the entire process *in vitro* (in the "test tube"). Two kinds of cells are usually required, phagocytes that are the tissue equivalent of monocytes (Fig. 11.11), and lymphocytes. The first step in the process is the engulfment of the antigen by phagocytes (Figs. 30.8 and 30.9). What happens next is still uncertain, but in some way the phagocytes interact with lymphocytes.

You might assume, as many did at one time, that the phagocytes instruct lymphocytes how to go about the next step in antibody synthesis. But such is surely not the case. There is a good deal of evidence that *before* the introduction of an antigen, the body already has lymphocytes capable of recognizing that antigen. The lymphocytes demonstrate this capability by actually combining with the antigen, possibly after it has been processed in some way by phagocytes. The reaction between lymphocytes and antigen occurs at receptors on the cell surface that are probably identical (at least in the combining site) to the antibody molecules that will ultimately appear in the blood.

Let us examine two of the experiments that lead to this remarkable conclusion. A number of years ago, the Australian immunologists Ada and Byrt prepared a highly radioactive antigen. When this material was mixed with lymphocytes taken from mice that had never been exposed to the antigen, it stuck to a small percentage (0.02%) of the cells (Fig. 30.10). The radioactivity was so great that these cells (but *only* these) were killed. When the remaining cells of the original population were returned to a mouse whose own immune system had been destroyed (by X-rays), that mouse was *incapable* of synthesizing antibodies to the original antigen. However, it was perfectly capable of synthesizing antibodies against any other, even closely similar, antigens. Thus it appears that the unimmunized mouse already has a few lymphocytes carrying antibody-like receptors that are ready to detect and interact with the corresponding antigenic determinant when and if it appears.

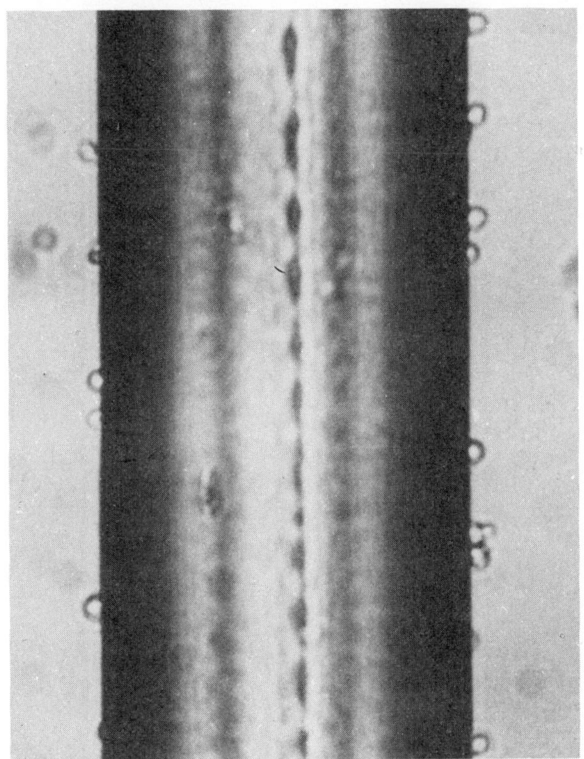

FIG. 30.11 Mouse spleen cells (lymphocytes) adhering to the surface of an antigen-coated nylon fiber. No cells would stick (1) if the nylon fiber had no antigen attached to it; (2) if the cells were first exposed to the same antigen in solution (thus "filling" all the receptors); or (3) if the cells were first exposed to antibodies against mouse antibodies (thus suggesting that the receptors are, indeed, antibody molecules). (Courtesy of G. M. Edelman from U. Rutishauser, C. F. Millette, and G. M. Edelman, *Proceedings of the National Academy of Sciences, U.S.A.*, **69**, 1596, 1972.)

In 1972, Edelman and his colleagues at the Rockefeller University demonstrated the existence of antigen-reactive cells in nonimmune mice by another technique. They attached an antigen to the surface of nylon fibers and then placed the fibers in a dish containing spleen cells (mostly lymphocytes) from an unimmunized mouse. A small percentage (about 1%) of these stuck to the fibers, indicating the presence of complementary antibody-like molecules on their surface (Fig. 30.11). In fact, once

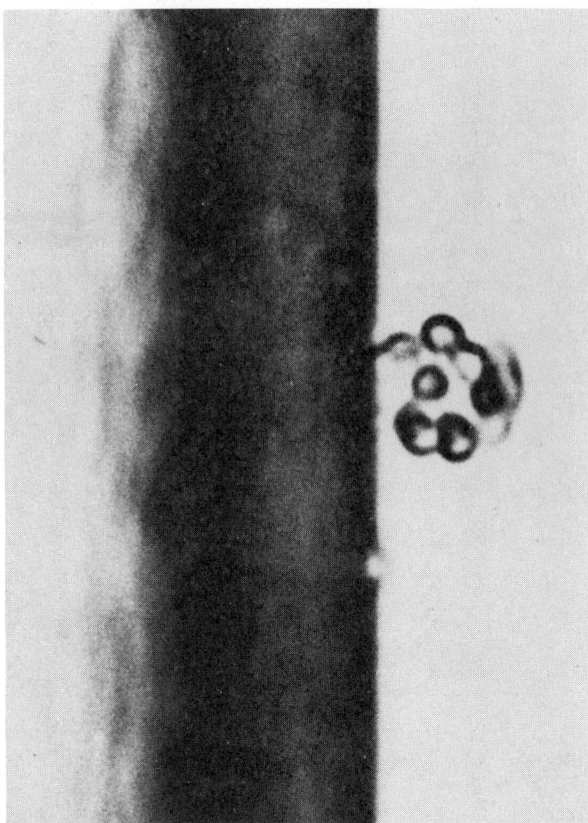

FIG. 30.12 In this photo, a mouse spleen cell stuck to an antigen-coated nylon fiber has been exposed to sheep red blood cells coated with the same antigen. The results (like those in Fig. 30.10) suggest that the antigen-binding receptors of the cell are distributed over its entire surface. Untreated sheep red blood cells would not stick to this mouse cell. (Courtesy of G. M. Edelman, from the same article as that of Fig. 30.11.)

stuck, the rest of the cell surface was still capable of interacting with additional antigen (Fig. 30.12). However, if the cells were first treated with antigen *or* with antibodies directed against mouse antibodies ("anti-antibodies") no sticking occurred.

Once an antigen has attached itself to the receptor of the appropriate lymphocyte, that cell becomes stimulated to divide (Fig. 30.13). With repeated mitotic divisions, a *clone* of cells capable of synthesizing that particular antibody develops. The idea that antibody-producing clones arise not by instruction but as a result of the selective stimulation of *preexisting* cells is referred to as the clonal selection theory.

The metabolic demands of mitosis are so great that not much antibody is synthesized during this period. But a time comes when more and more of the cells in the clone stop dividing and begin tooling up for massive antibody production. A large "rough" endoplasmic reticulum develops (see Section 5.8) and soon the cell is synthesizing one major protein component: the antibody. Such a cell is called a **plasma cell.** After assembly in the Golgi apparatus (where some of the carbohydrate is added), the antibody is secreted from the cell (by exocytosis).

Figure 30.14 shows plasma cells taken from the spleen of a rabbit immunized with Type III pneumococci. The cells were treated *first* with the Type III polysaccharide, *then* with anti-Type III antibodies that had been coupled to a fluorescent dye. Viewed under ultraviolet light, each cell synthesizing anti-Type III antibody glows brilliantly.

Antibody production occurs in the spleen, in the bone marrow, and, perhaps most important, in the lymph nodes. Lymph nodes are ideally situated to encounter antigens that enter the body through the skin, lungs, and intestine. Processing, as they do, the tissue fluids from these (and most other) organs, they are quickly exposed to foreign material. Not only do they prevent such material from reaching the bloodstream, but they have all the machinery for mounting an immune response. The secreted antibodies are transported in the lymph that drains from the node and enter the blood at the subclavian veins (look back at Fig. 11.20).

30.6 THE SECONDARY RESPONSE

After recovery from an infection, the concentration of antibodies against the infectious agent gradually declines over a period of weeks, or months, or even years. A time *may* come when no further antibody can be detected. Does this mean that the individual is once again in danger of contracting the disease? In many cases, no. Whereas the production of a significant level of antibodies upon first exposure to the antigen takes, on average, four or five days, reexposure—even after many years—calls forth a much

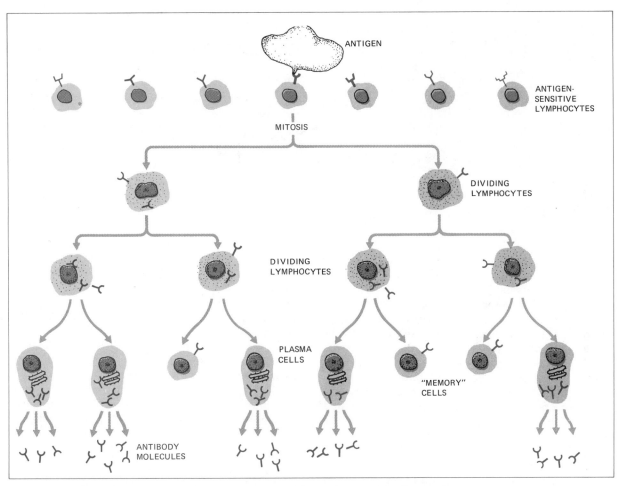

FIG. 30.13 Schematic representation of the clonal selection theory. Clonal selection leads to the production of antibody-secreting plasma cells *and* an enlarged pool of specific antigen-sensitive ("memory") cells. The exact form in which antigen is presented to the antigen-sensitive cells is not known.

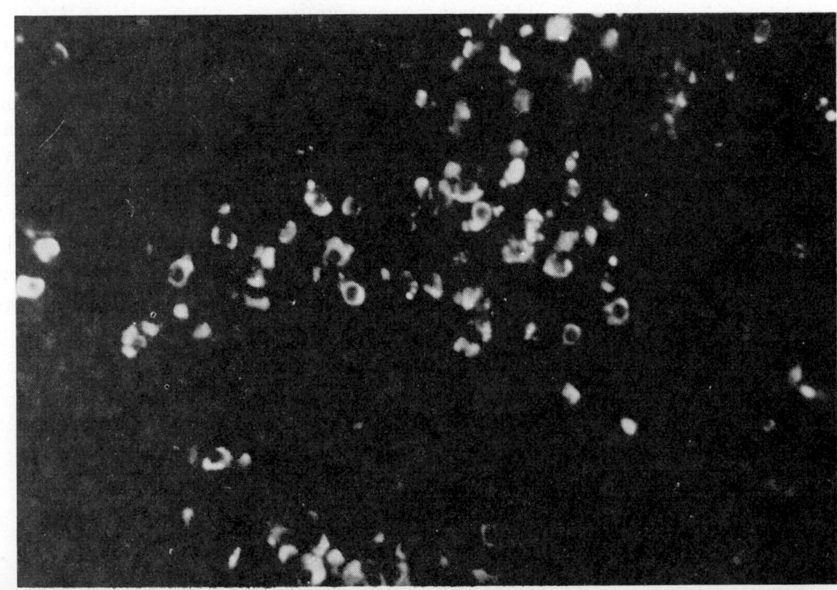

FIG. 30.14 Antibody-containing plasma cells from the spleen of a rabbit immunized with Type III pneumococci. These cells were first treated with the antigen, Type III pneumococcal polysaccharide, and then with anti-Type III antibodies attached to a fluorescent dye. Viewed under ultraviolet light, those plasma cells manufacturing anti-Type III antibodies glow brightly. (Courtesy of Dr. Albert H. Coons.)

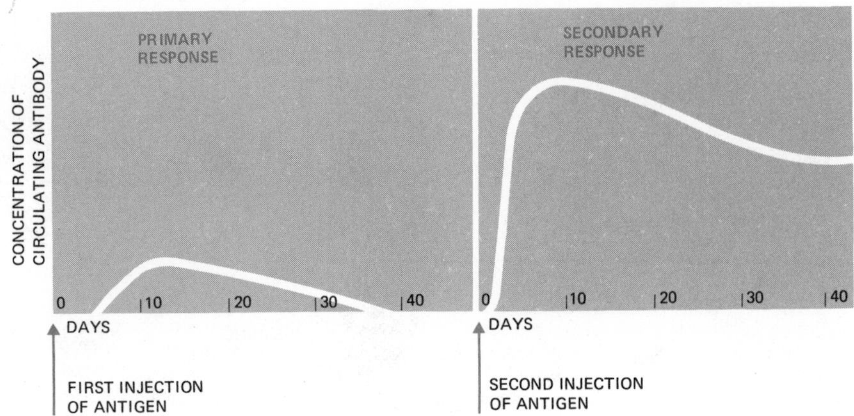

FIG. 30.15 Antibody production during the course of a typical primary (left) and secondary (right) response. The actual time course varies with such factors as the nature of the antigen and the way in which it is administered to the animal.

faster and larger response to the antigen. This is called the **secondary response** (Fig. 30.15).

What mechanism is at work here? The key seems to be the presence, during the intervening period, of a larger population of antigen-sensitive lymphocytes ("memory" cells) than existed before the primary response. Probably during the initial development of clones, some of the progeny cells neither went on dividing nor developed into plasma cells. Instead, they reverted to small lymphocytes bearing the same antibody molecules on their surface that their ancestor had (Fig. 30.13). In this way, the total number of such antigen-sensitive cells is substantially increased and the foundation laid for a bigger (and thus sooner detectable) response the next time the antigen appears. This idea is supported by the discovery that when spleen cells from *immunized* mice are mixed with the appropriate antigen-coated nylon fibers, the number of cells that stick *is* increased (and these are *not* antibody-*secreting* cells).

Not only does the secondary response require a pool of memory cells, but these must have long life spans if we assume that they will never divide again unless they encounter the appropriate antigen. A number of studies have shown that among the lymphocytes that circulate from lymph nodes to the blood and back again are some that are, indeed, long-lived. There is evidence in humans that some small lymphocytes live as long as 20 years.

Immunological memory is exceedingly precise. As you are probably painfully aware, influenza is caused by a variety of closely related viruses. From time to time, new virus variants (e.g. the "Hong Kong flu") appear and sweep through the population. In 1918, a worldwide flu epidemic killed an estimated 20 million persons. When survivors of that epidemic are exposed today to influenza virus, the antibodies they produce are not as effective against the present-day virus as they are against the virus with which they were first infected in 1918. Although no one even knew in 1918 that flu was caused by a virus, we can now identify the 1918 virus simply by examining the specificity of the antibodies that survivors produce today in their secondary response.

The secondary response thus represents a long-term alteration in the responsiveness of the organism as a result of its experience. Is, then, immunological memory so different from memory stored in the central nervous system?

Immunological memory provides the main foundation for the use of **vaccines.** These are preparations of infectious or toxic agents that have been altered so as not to cause disease. The alteration is not, however, so drastic that antibodies produced in response to the altered material will not protect against the unaltered, disease-producing agent. This usually means that the alteration must leave some of the antigenic determinants intact. Formaldehyde is often used to render bacteria and viruses noninfectious without damaging all the antigenic determinants. The Salk polio vaccine is prepared in this manner.

The first scientific vaccination was performed by the English physician Edward Jenner in 1796. He made medical history by inoculating a human subject with material squeezed from the sores of a girl afflicted with cowpox and then, a few months later, with material from the sores of people afflicted with the dread disease smallpox. His subject remained perfectly well. Jenner performed this daring experiment because he (and others) had long noted that milkmaids who had contracted cowpox from their charges were protected from the ravages of smallpox. Today, we realize that they gained this protection by developing antibodies which were effective not only against the cowpox virus but against the closely related smallpox virus as well.

You have probably been given smallpox vaccine, polio vaccine, and vaccines against tetanus, diphtheria, and whooping cough as well. In the case of tetanus and diphtheria, it is the toxins we must guard against. Their vaccines thus consist of chemically altered toxins, called **toxoids.** Vaccines against typhoid fever, typhus fever, bubonic plague, Asiatic cholera, and yellow fever are all available for people who live or travel in regions where these diseases are present.

30.7 THE BASIS OF ANTIBODY DIVERSITY

Our discussion of the immune response certainly makes the system seem an open-ended one. There seems to be no limit to the number of different antigenic determinants against which the system can operate. But is there a limit to the number of different kinds of antibody molecules that can be synthesized? Probably so, but just what it is, at present is anyone's guess. Not only does a given antigen give rise to many different antibody molecules (as we have seen), but even a single antigenic determinant will elicit the production of a diverse population of antibody molecules. But these different antibody molecules are not all equally effective in combining with the determinant. They represent a spectrum extending from those that do so exceedingly well to those that do so poorly. Presumably this reflects the degree of *precision of fit* between the shape of the antigenic determinant and the shape of the combining site of the antibody. But could not some of the antibody molecules with a poor fit to that determinant have a better fit to some other determinant? It is quite unlikely that all the 1% or so of nonimmune mouse spleen cells that stick to an antigen-coated fiber are sticking equally tightly and, in fact, probably the interaction for most of them is not strong enough to trigger a proliferative response. Most antibodies react to some degree with a variety of related determinants. So, as a first approximation, the immune response need not be responsible for synthesizing an unlimited number of different molecules. What the limits are, however, is as yet unknown.

Antibodies are proteins, and we have no reason to believe that they are synthesized any differently from any other protein. That is to say, we must assume that the cell nucleus contains DNA coding for that antibody, and that this DNA is transcribed into complementary messenger RNA molecules from which the polypeptide chains are transcribed (see Section 18.3).

But do we inherit a gene for each antibody that we can make? Let us assume that we can synthesize 10,000 different antibody molecules (maybe the figure is larger, maybe it is smaller). Among this population would be found only about a dozen different kinds of constant regions. Our problem, then, is one of accounting for the diversity of the variable (V) regions, in which, after all, the antibody specificity (the combining site) resides. (But if vertebrates do, indeed, have many genes for V regions and only a dozen genes for C regions, we will also have to look for a mechanism by which their products end up in a single polypeptide; i.e. our old rule of one gene = one polypeptide will have been modified to two genes = one polypeptide!) We might not even need 10,000 different V region genes to make 10,000 different antibodies. Heavy chains and light chains are synthesized separately and the V regions of both participate in making up the combining site. Thus only 100 different kinds of light chains combining at random with 100 different kinds of heavy chains would give us 10,000 possible combinations. There is evidence, however, that not all light chains and heavy chains combine with equal ease, and we should perhaps raise the number to account for the combinations that would not make functional antibodies of any sort. But even if we raised the figure to 1000 V_L (light chain) and 1000 V_H (heavy chain) genes, the DNA needed to code this information would represent only about 0.02% of the DNA found in each gamete.

Apparently there is no molecular configuration that the organic chemists can devise that cannot serve as an antigenic determinant. But why would we have passed on from generation to generation a gene for a V region that *might* someday meet the corresponding antigenic determinant? Our knowledge of cross-reactivity tells us, however, that a given antibody can interact to some degree with a variety of configurations. The antibody made today to a recently synthesized macromolecule may have been made repeatedly in the past to a similar antigenic determinant introduced into the body on, for example, some parasite.

The possibility that we inherit a gene for every V region we can make should not lead us to ignore other possibilities. If some mechanism existed for generating new V region sequences during the lifetime of the individual, we would be able to meet our antibody needs with a smaller number of inherited genes. A number of hypothetical mechanisms for the somatic generation of V region diversity have been proposed. These range from mechanisms in which a moderate number of V region genes undergo internal recombination events (analogous to those occurring in meiosis—see Section 14.7) to produce new sequences, to the postulation that a very small number of V genes undergo point mutations (just like those that in germplasm give rise to mutant offspring). (We have no reason to feel that germ cells are the only ones to undergo mutations simply because it is only by examining *offspring* that the presence of such mutations is ordinarily detected.) According to this second mechanism, each mutation in a lymphocyte V gene that produced a codon calling for a new amino acid could lead to the production of a light (or heavy) chain with one amino acid substitution. If present in, or near, the combining site, this change might produce an antibody molecule complementary to a new antigenic configuration. While there is no shortage of theories to explain the origin of antibody diversity, the data which will unequivocally support one mechanism over all the others are simply not yet at hand.

30.8 TRANSPLANTATION AND THE UNIQUENESS OF THE INDIVIDUAL

The first serious medical attempts at transplanting living tissue from one human to another involved the transfusion of blood into victims who had suffered serious blood loss. Sometimes the procedure was successful; sometimes it precipitated a serious, even fatal, crisis.

In 1900, Karl Landsteiner discovered the explanation for these erratic results. He found that antigenic substances may be present on the surface of red blood cells. The first ones discovered by him were the A and B antigens. People with red blood cells carrying the A antigen are said to have group A blood; people carrying the B antigen belong to blood group B. Some people have both antigens (group AB) and others (group O) have neither (Fig. 30.16).

When red blood cells carrying one or both antigens are brought in contact with the corresponding antibodies, they agglutinate, i.e. clump together (Fig. 30.17). People usually have antibodies against those red cell antigens that they lack. Figure 30.16 shows this and illustrates which blood types can and cannot be safely transfused. The basic principle to be observed is that the blood introduced into the body must not contain red blood cells which the patient's antibodies can clump. It is not too serious if the donor's blood contains antibodies against the recipient's red blood cells, because those antibodies will quickly be diluted by the plasma into which they are being introduced. Hence type O blood has been called the "universal donor," because O red cells cannot be clumped and the antibodies will be quickly diluted by the recipient's plasma. Similarly, AB blood is called the "universal recipient," because it contains no antibodies to clump red blood cells introduced into it. In practice, however, doctors prefer to match blood groups exactly when carrying out transfusions.

The Landsteiner blood groups are inherited. The group is controlled by two alleles, one inherited from each parent. However, there are three alleles, *A*, *B*, and *O*, present in the population as a whole. Figure 30.16 shows the genotypes that produce each of the blood groups. For example, persons with either genotype *AA* or *AO* will have A antigens on their red blood cells and hence group A blood. Only the genotype *AB* produces both A and B antigens and hence blood group AB.

You may well wonder why we have antibodies against those red cell antigens we lack. The answer is not entirely certain. Perhaps we are all exposed to substances (in our food, from infections, or from bacteria living in our intestine) that have antigenic determinants similar to A and B. We would proceed to synthesize antibodies against them if they did not resemble "self," but not if they did. Certainly materials carrying A-like or B-like antigenic determinants turn up from time to time, e.g. as contaminants in vaccines.

Blood transfusions are a special, limited example of tissue transplantation. After three to four weeks the transfused cells have been eliminated from the recipient. But what about transplants, e.g. kidney or heart, that we would hope would survive in their new host indefinitely? In 1956, one kidney was transplanted from a healthy young woman to her ailing *identical* twin. The recipient's life was saved and today both women continue in excellent health. However, when kidney transplants have been made between people other than identical twins, a long-term successful outcome has not been so easy (Section 13.8).

If we remove a piece of skin from a mouse of one strain and fit it into a site prepared for it on a

BLOOD GROUP	APPROX. % IN U.S. POPULATION	ANTIGENS ON RBC'S	ANTIBODIES IN PLASMA	CAN DONATE TO	CAN RECEIVE FROM	GENOTYPES
A	42	A	ANTI-B	A, AB	O, A	*AA* or *AO*
B	10	B	ANTI-A	B, AB	O, B	*BB* or *BO*
AB	3	A AND B	NEITHER	AB	O, A, B, AB	*AB*
O	45	NEITHER	ANTI-A AND ANTI-B	O, A, B, AB	O	*OO*

FIG. 30.16 The Landsteiner (ABO) blood groups and the genotypes that give rise to each.

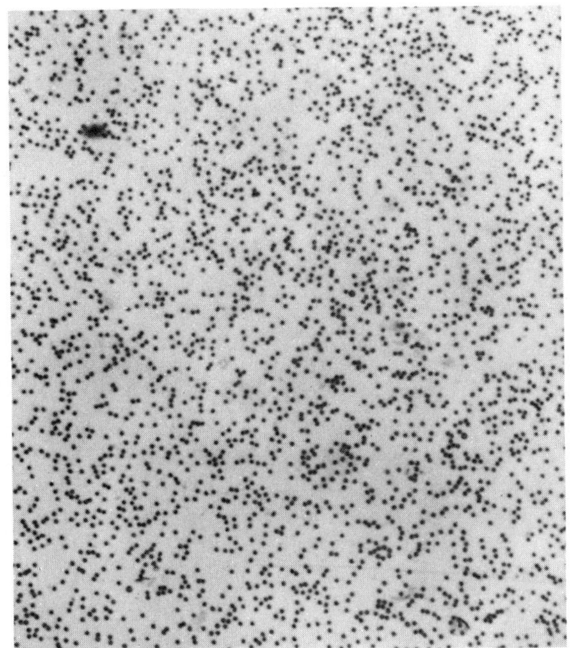

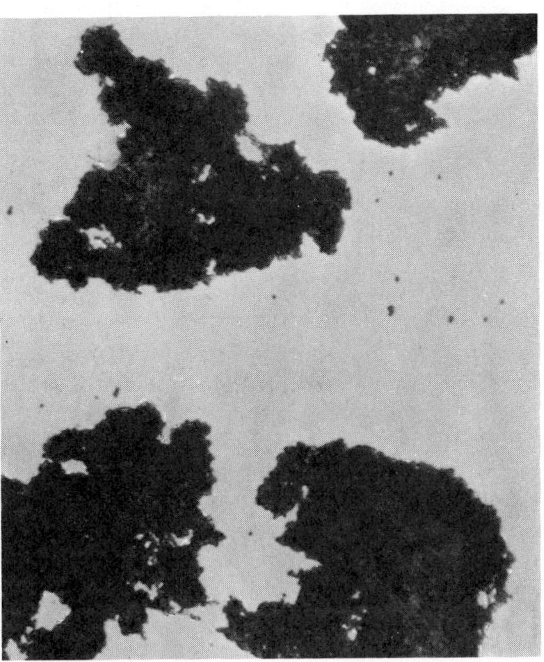

FIG. 30.17 Red blood cells before (left) and after (right) adding serum containing anti-A antibodies. The agglutination reaction indicates the presence of the A antigen on the cells.

mouse of another strain, the graft at first does very well. Blood vessels of the host grow into it, and it functions normally. Some 10–14 days after the operation, however, matters take a turn for the worse. The blood supply to the graft breaks down and quickly the graft itself degenerates. Finally, it is sloughed off like an old scab.

A clue to the nature of this rejection phenomenon can be gained by trying the experiment again. If we replace the graft with skin from a mouse of a *third* strain, the story repeats itself. But if we try again with skin from the *original* donor strain, the graft may not even last long enough to acquire a blood supply. It is rejected in a much shorter period

(5–6 days). This "second-set" phenomenon appears to be the equivalent of the secondary, "memory" response we discussed earlier.

The rejection of a graft is not usually accompanied by any striking appearance of new antibodies in the blood. Instead, it is antigen-sensitive ("killer") cells that play the major role in the rejection process. These cells (lymphocytes) recognize the foreign graft and, by mobilizing other cells in the circulatory system (chiefly monocytes), proceed to kill the cells of the graft.

The lymphocytes that are responsible for graft rejection are not the same as those that serve as the precursors of plasma cells. In order to reject a graft, a mouse must have had a functioning thymus gland at the time its immune machinery was being set up (close to the time of birth). However, the ability of the mouse to secrete *antibodies* in response to injections of Type III pneumococcal polysaccharide is *not* dependent on the thymus. Thus there appear to be two functional divisions of the immune system. One, which is responsible for cell-mediated immune reactions like graft rejection, is dependent on the thymus. The other, which is responsible for the secretion of circulating antibodies, is not.

The lymphocytes in each division are formed (like the other blood cells) in the bone marrow. Lymphocytes that will function in cell-mediated immunity, i.e. potential "killer" cells, must then migrate to the thymus before they are able to do their job. Once "processed" by the thymus, these cells are called **T cells.** The lymphocytes responsible for the secretion of antibodies are not processed by the thymus. They are called **B cells.** Although they look alike under the light microscope, there are several ways of discriminating between, and even separating, T cells and B cells.

Our understanding of the graft rejection process is increased by examining the circumstances under which it does *not* occur. First, as we have seen, grafts between genetically identical individuals are not rejected. This can be observed in human identical twins and also in strains of mice that have been inbred for so long that all individuals are virtually identical even though they arise from separate eggs. (Even among such mice, rejection occurs when male skin, *XY,* is grafted onto female, *XX,* but not the reverse. Because of its *Y* chromosome, male skin has genes not possessed by female skin. With this exception, these creatures are essentially homozygous and, thus, the females have no genes foreign to the males.)

If we breed a mouse of inbred strain A with one of inbred strain B, the F_1 offspring accept grafts from each other and *also* from either of the parental strains (Fig. 30.19). However, neither parental strain will accept an F_1 graft. These findings suggest that whenever genetic differences exist, antigenic differences do also. In this light, we can view the immune system as a device which recognizes the uniqueness of the individual and works to keep it that way.

What can we do for humans who desperately need a new kidney but have no identical twin to whom they can turn? Two approaches have proved sufficiently workable that kidney transplants now restore thousands to substantial periods of good health. One is tissue typing. Matching blood by ABO (and Rh) blood groups permits successful transfusions even though many other red cell antigens exist. While the A and B antigens can provoke serious reactions, the other antigens (e.g. MN) are so weakly antigenic that no problems arise from mismatches. In a similar way, the antigens which the immune system recognizes on the surface of the cells of transplanted organs can be characterized as strong or weak. The rejection mechanism works far more powerfully against the first than against the second. Although a perfect match of all antigens will probably only be possible between identical twins, reasonably good matches—at least with respect to the strong antigens—can often be made, especially between members of the same family.

Those rare infants born without an immune system, mentioned earlier, can sometimes be *given* one by injections of bone marrow (the source of all the white blood cells) from a healthy donor, usually a brother or a sister. In this case, you might predict that tissue typing would be unnecessary, since the recipients have no mechanism to destroy the transplanted tissue. Actually, tissue typing is just as important in this situation. If strong antigenic differences exist between host and transplant, the *transplant* detects them and mounts an immune response against the *host.* This "graft versus host" reaction is invariably fatal. Although the converse of the usual rejection problem, it illustrates once again the fundamental role that the immune mechanism plays in recognizing "nonself" and destroying it.

A second way in which the graft rejection mechanism can be minimized is to treat the recipient with drugs that inhibit the immune response. Because of the need for rapid cell proliferation during an immune response, any drug which inhibits mitosis sup-

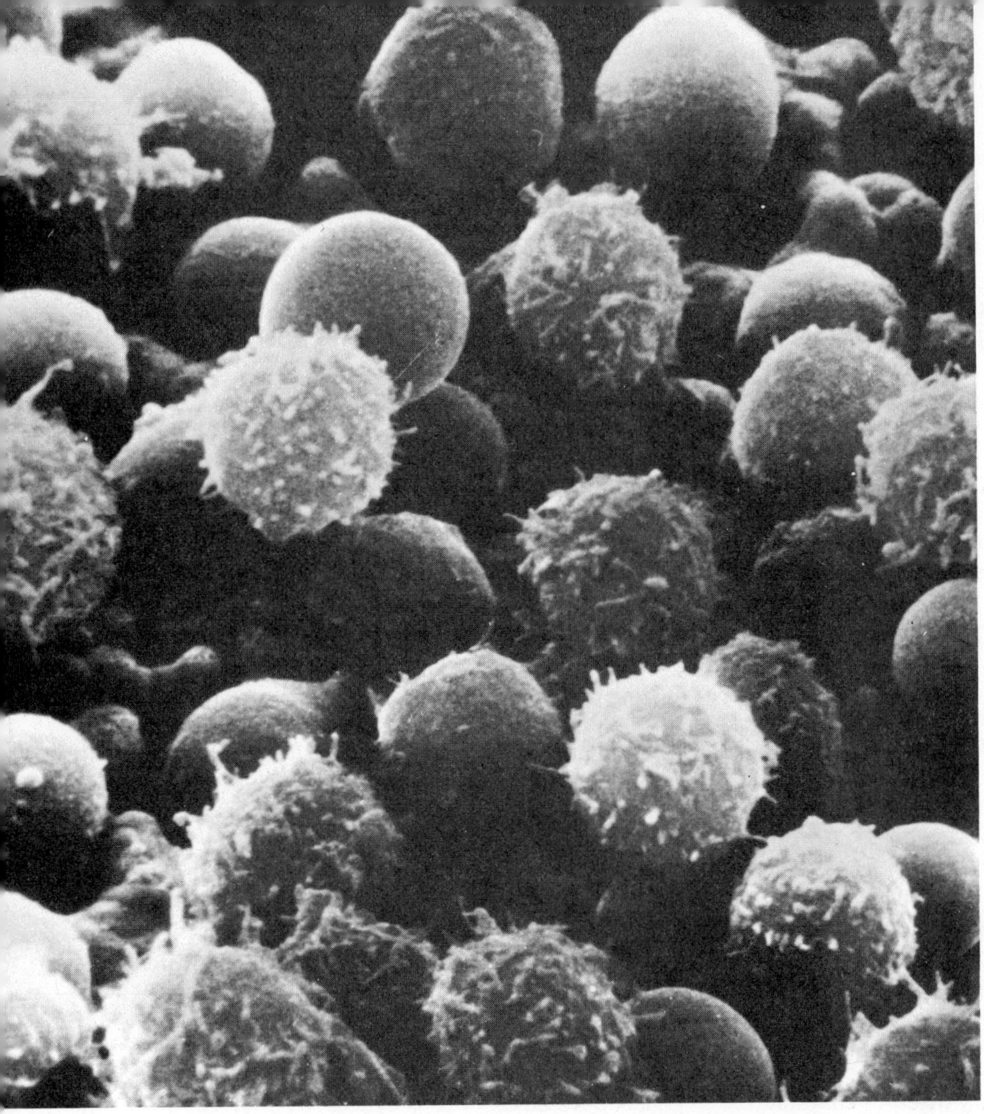

FIG. 30.18 Lymphocytes viewed with the scanning electron microscope (6000 ×). These lymphocytes are from a patient with lymphocytic leukemia in which there was an excessive production of both B and T cells (in almost equal proportions). (Courtesy of Dr. A. Polliack from Polliack et al., *Journal of Experimental Medicine* **138,** 607, 1973.)

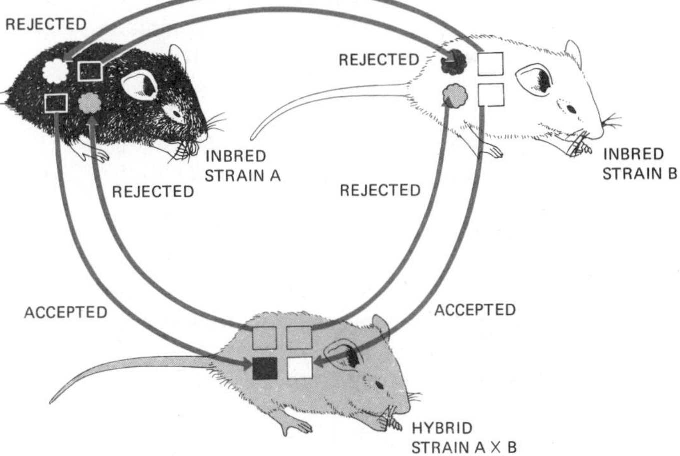

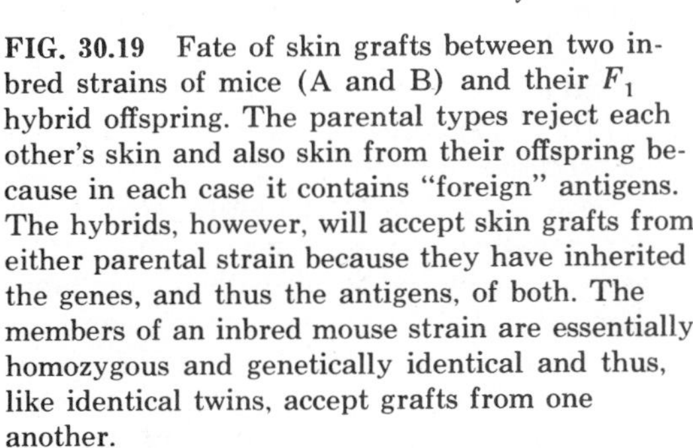
FIG. 30.19 Fate of skin grafts between two inbred strains of mice (A and B) and their F_1 hybrid offspring. The parental types reject each other's skin and also skin from their offspring because in each case it contains "foreign" antigens. The hybrids, however, will accept skin grafts from either parental strain because they have inherited the genes, and thus the antigens, of both. The members of an inbred mouse strain are essentially homozygous and genetically identical and thus, like identical twins, accept grafts from one another.

presses the response. For reasons that are not entirely clear, the glucocorticoids (see Section 25.11) are also potent immunosuppressives.

While the use of immunosuppressives has been indispensable in enabling transplant recipients to keep their new kidney or heart for substantial periods of time, use of these drugs creates new problems. The drugs now available are not in any sense selective in their action. The more effective they are at suppressing graft rejection, the more effective they are at suppressing the immune response to invading bacteria, fungi, and viruses. While antibiotics and other drugs can help with bacterial and fungal infections, viral infections do not yield to such treatment and may therefore become life-threatening.

30.9 CANCER AND IMMUNE SURVEILLANCE

Another complication that turns up in transplant recipients is a disproportionate incidence of cancer. On occasion the cancerous cells are of donor origin, having been inadvertently transplanted along with the organ. The cure in these cases is simple: stop the use of immunosuppressant drugs. The host's rejection mechanism then promptly destroys the cancerous cells. (The transplanted organ is rejected as well.)

Most of the time, however, the cancer is of host origin. Why should transplant recipients have an increased risk of developing cancer? Could it be related to their immunosuppressed status? This possibility gains plausibility when we realize that there are a number of different hereditary diseases characterized by a poor immune response which are also associated with a disproportionately high incidence of cancer. Perhaps, then, one of the functions of the immune mechanism is to destroy cancerous cells. If so, we would expect that such cells would be recognized by the immune system as "foreign" even though they had *arisen* from "self." Is there any evidence that cancer cells bear antigenic determinants not found on the normal cells of the host?

The availability of inbred mouse strains and chemical agents which induce cancers in them provides the tools with which to test this idea. We have seen that a mouse will accept as "self" a skin graft from another mouse of the same inbred strain. Will it also accept as "self" a tumor from such an animal? The answer is a qualified no. A transplanted tumor may grow so rapidly that it kills the host before the host can mount an effective immunological attack. Nevertheless, the machinery is there. If such a transplanted tumor is removed before it kills the host, or if the host is first injected with dead tumor cells, the host has a chance to get ready for a "second-set" reaction. Then, when the tumor is reimplanted, it is quickly rejected. The ability of the recipient to recognize tumor cells of its strain as foreign while it accepts normal tissues from that strain as "self" demonstrates that these cancers do bear *new* antigenic determinants.

Many tumors in experimental animals and some, perhaps, in humans are caused by viruses. In these cases, also, the cancerous cells bear antigenic determinants on their surface that are foreign to the host. Known cancer-causing agents (e.g. radiation, many chemicals, many viruses) are distributed widely through our environment. Assuming that cancers start as a single aberrant cell, we may well suspect that one of the functions of the immune mechanism is to recognize these aberrant cells and their progeny as foreign and destroy them before they have time to grow into a detectable tumor. Such a function has been described as **immune surveillance.**

If such is the case, why should we ever succumb to cancers? There are several possibilities. The incidence of cancer (after a small peak in childhood) increases with advancing age. The effectiveness of the immune mechanism (as with so much of the body's machinery—see Section 23.3) decreases with advancing years. Furthermore, many of the chemicals known to induce cancer (e.g. hydrocarbons such as benzopyrene—see Section 10.10) are also immunosuppressives. The same is true of some of the cancer-causing viruses. Perhaps part of the very basis of the carcinogenic activity of these agents is their effectiveness in suppressing immune surveillance.

The immune mechanism may itself be responsible for the breakdown of immune surveillance and the success of a tumor in gaining the upper hand. Tumor rejection, like graft rejection, is primarily a matter of specific "killer" lymphocytes (T cells) leading an attack on the foreign cells. Antibodies circulating in the bloodstream appear not to be much help in the process and may, indeed, be detrimental. There is much evidence accumulating that some type(s) of circulating antibodies may *inhibit* the rejection phenomenon by combining with determinants on the "foreign" cells. In so doing, they cover or mask these determinants so that they are not able to trigger the cell-mediated rejection process. In this way, a state of tolerance to the tumor may develop.

30.10 ALLERGIES

We cannot survive without a functioning immune system. Nevertheless, there are circumstances (e.g. when you have picnicked in a patch of poison ivy) when the immune system *seems* to be more trouble than it is worth. Sensitivity to poison ivy is an immune response. So, too, are the unpleasant reactions that some of us suffer after eating certain foods (e.g. shellfish, strawberries) or inhaling pollen from certain plants (e.g. ragweed, many grasses). Each of these immune responses is an example of an **allergy.** Each is triggered by one or more particular antigenic determinants.

There are several immune mechanisms that give rise to allergies. T cells, for example, play the more important role in some; B cells in others. Let us, however, examine one of the most common allergies: hayfever.

In our discussion of the immune system, we have emphasized up to now those antibodies that circulate in the blood plasma. However, antibodies are found elsewhere in the body. One kind of antibodies, known as IgE antibodies, bind tightly to the surface of tissue cells called mast cells (and also to basophils—see Fig. 11.11). IgE antibodies have receptors on the constant region of their heavy chains that bind to receptors on the mast cells (Fig. 30.20). These cell-bound antibodies exert no effect until and unless they meet antigenic determinants that can bind to their antigen-binding sites. When this occurs, the mast cells to which they are attached explosively discharge their cytoplasmic granules (Fig. 30.20). The major ingredient in these granules is histamine. The liberation of histamine causes swelling, redness, itching, etc. of the surrounding tissues. In effect, then, each IgE-sensitized mast cell is a loaded bomb which can be triggered to explode by a particular antigen.

Some people tend to respond to antigens in their environment (e.g. pollen grains, dust) with an unusually vigorous production of antibodies of the IgE type. As a result, subsequent exposure to those antigens triggers the rapid release of histamine. If this occurs in the tissues of the respiratory passages and around the eyes, all the unpleasant symptoms of hay fever appear.

The propensity to produce high levels of IgE antibodies appears to be inherited. Thus we find that hay fever and/or other allergies of the same type tend to run in families.

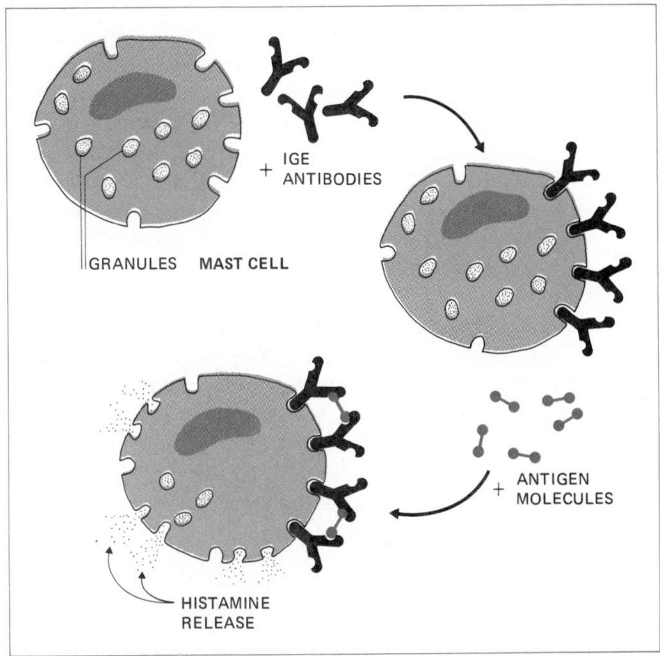

FIG. 30.20 Mechanism of sensitization and response in allergies like hay fever. Antibodies of the IgE type bind to mast cells and make them sensitive to whatever antigen (e.g. ragweed pollen) the antibodies are directed against. When exposed to the antigen to which they are sensitized, the mast cells discharge histamine. Histamine causes redness, swelling, and itching of surrounding tissues.

What can be done to help these sufferers? Avoiding the antigens is effective, of course, but may be impractical. Administration of **antihistamines** gives relief by blocking the effect of histamine. Or, one can attempt to **desensitize** the victim. This involves the deliberate, repeated introduction of gradually increasing doses of the offending antigen(s) into the body, i.e. by injection. One goal of such a procedure is to cause a build-up of other antibodies of the *same specificity* that do not bind to mast cells. If this is accomplished, the new *circulating* antibody population may be able to intercept the antigen before it reaches the tissues and triggers histamine release. Unfortunately, such desensitization is a tedious process, and the effects are usually short-lived.

30.11 AUTO-IMMUNITY

Why do we not make antibodies against our own tissue components? Or, to put the question another way, what is the mechanism of self-tolerance?

A theory of self-tolerance that has long been popular is that of "forbidden" clones. Forbidden clones are those clones of lymphocytes directed against self-components. According to this theory, they are destroyed, in some manner, by the time the immune machinery is set up shortly after birth. This theory explains nicely the fact that, as we have noted earlier, antigens introduced into the animal prior or close to birth are henceforth accepted as self. Presumably their presence had established additional forbidden clones, which were then destroyed.

There is now a wealth of evidence, both in humans and in laboratory animals, that we *can* under certain circumstances manufacture antibodies and/or sensitized ("killer") lymphocytes against self-components. The serum from patients with multiple sclerosis contains antibodies which lead to the destruction of the myelin sheath of nerve cells, even in tissue culture. Physical damage to the choroid layer of one eye may later lead to immunological damage to the choroid layer of the second eye. Some people develop both antibodies and sensitized lymphocytes directed against the proteins and cells of the thyroid gland. In these examples, as well as a number of others, the disease process appears to stem from the immunological mechanism attacking self-components. Such disorders (another form of allergy) are now called **auto-immune** diseases. Their existence does not necessarily rule out the forbidden-clone hypothesis. In each of the examples cited, the tissue under attack is not normally exposed to the system of lymph nodes where the immune response occurs. The axons of the spinal cord, for example, are bathed in cerebrospinal fluid, not blood, and there is no lymphatic drainage. One can argue, therefore, that antigenic determinants in these "privileged" sites were not exposed to the immune system at the time that the forbidden clones were being eliminated.

An all-too-common consequence of rheumatic fever is damage to the heart valves. In many ways, this too resembles an auto-immune disease. Antibodies are present which react not only with the cell walls of the streptococci that cause rheumatic fever, but also with components in the heart muscle. But here, too, it may be that active clones are not *primarily* directed against self-components but rather against the antigenically similar streptococcal determinants.

A better test of the forbidden-clone hypothesis is to examine a remarkable animal we met in an earlier chapter (see Section 21.9): the tetraparental mouse. Here is a creature constructed from two antigenically different kinds of cells. A-strain and C-strain mice reject each other's grafts. What about a mouse that is a mixture of A cells and C cells? Will the A lymphocytes mount an attack against C cells and C lymphocytes attack A cells? This does not happen; the mice—like the cattle twins discussed earlier—tolerate the mixture of cells with no signs of auto-immunity. Does this mean, then, that *two* sets of forbidden clones were eliminated during the development of the mouse? The answer is no. Spleen cells removed from a tetraparental mouse, and placed in a tissue culture that contains *no serum* from the mouse, readily attack both A-strain and C-strain cells. Adding the tetraparental serum to the cultures protects the cells from attack. The situation is thus similar to that we found with many cancerous tumors: the presence in the blood of substances (presumably antibody molecules) that, instead of attacking cells, protect them from attack.

It seems there are no forbidden clones in tetraparental mice. Are there forbidden clones in biparental vertebrates? Or is self-tolerance instead a matter of developing special antibodies that protect our own components from attack by those lymphocytes sensitized to their antigenic determinants? Only more research will tell us the answer.

30.12 SUMMARY

For many decades the study of the immune response centered upon its role in protecting vertebrates (and immunity *does* appear to be exclusively a vertebrate attribute) from infectious disease agents. But now we can see the immune response in broader perspective. The immune mechanism enables the vertebrate body to synthesize a very large (*how* large is as yet unknown) number of different kinds of protein molecules. Among this diverse collection will be found a few that are capable of physically combining with, as far as we can tell, any given three-dimensional molecular configuration (determinant) that can be devised. This machinery does not ordinarily lead to destruction of macromolecules that are normal constituents of the host. The machinery *does* act to inactivate and destroy macromolecules that enter the internal environment from the outside world. In so doing, it protects us from infectious agents such as viruses and bacteria. It may also produce unpleasant symptoms of allergy when noninfectious agents (e.g. ragweed pollen) gain entrance. It also works to destroy the surgeon's hopeful efforts to replace damaged organs. And, in all likelihood, the immune system stands ready to destroy aberrant (cancerous) cells that arise in the body. In carrying out these actions, the system learns from its experience and usually develops an enhanced capacity to react to subsequent encounters with the same configuration.

The immune mechanism thus acts to guard the body against invaders from without and the cancerous cells that threaten to invade us from within. In the words of F. M. Burnet, one of the great immunologists of the twentieth century, the immune response works to preserve "the integrity of the body."

EXERCISES AND PROBLEMS

1. A child and its mother are both blood group O. What blood groups might the father have?
2. An accident victim of blood group A should receive a transfusion from a donor of what blood group? What other blood groups could be used without serious danger?
3. What are the possible blood groups of children resulting from the marriage of a group O father with a group AB mother?

REFERENCES

1. CAPRA, J. D., and A. B. EDMUNDSON, "The Antibody Combining Site," *Scientific American,* Offprint No. 1350, January, 1977.
2. PORTER, R. R., "The Structure of Antibodies," *Scientific American,* Offprint No. 1083, October, 1967.
3. EDELMAN, G. M., "The Structure and Function of Antibodies," *Scientific American,* Offprint No. 1185, August, 1970. Edelman and Porter (author of the preceding article) shared a Nobel Prize in 1972 for their work.
4. REISFELD, R. A., and B. D. KAHAN, "Markers of Biological Individuality," *Scientific American,* Offprint No. 1251, June, 1972. On transplantation antigens.
5. NOTKINS, A. L., and H. KOPROWSKI, "How the Immune Response to a Virus Can Cause Disease," *Scientific American,* Offprint No. 1263, January, 1973.
6. BURNET, F. M., *The Integrity of the Body,* Atheneum, New York, 1966. A classic by one of the founders of modern immunology and now available in paperback.
7. ROITT, I. M., *Essential Immunology,* Blackwell Scientific Publications, 1971. Distributed in the United States by J. B. Lippincott Company, Philadelphia. A highly technical but concise and up-to-date survey of the field.
8. LERNER, R. A., and F. J. DIXON, "The Human Lymphocyte as an Experimental Animal," *Scientific American,* Offprint No. 1275, June, 1973.
9. JERNE, N. K., "The Immune System," *Scientific American,* Offprint No. 1276, July, 1973.
10. COOPER, M. D., and A. R. LAWTON III, "The Development of the Immune System," *Scientific American,* Offprint No. 1306, November, 1974.
11. PORTER, R. R., *Chemical Aspects of Immunology,* Carolina Biology Readers, No. 85, Scientific Publications Division, Carolina Biological Supply Company, Burlington, N.C., 1976.
12. RAFF, M. C., "Cell-Surface Immunology," *Scientific American,* Offprint No. 1338, May, 1976.

Fossil of *Archaeopteryx*, a "link" in the evolution of modern birds from reptilian ancestors. (Courtesy of American Museum of Natural History.)

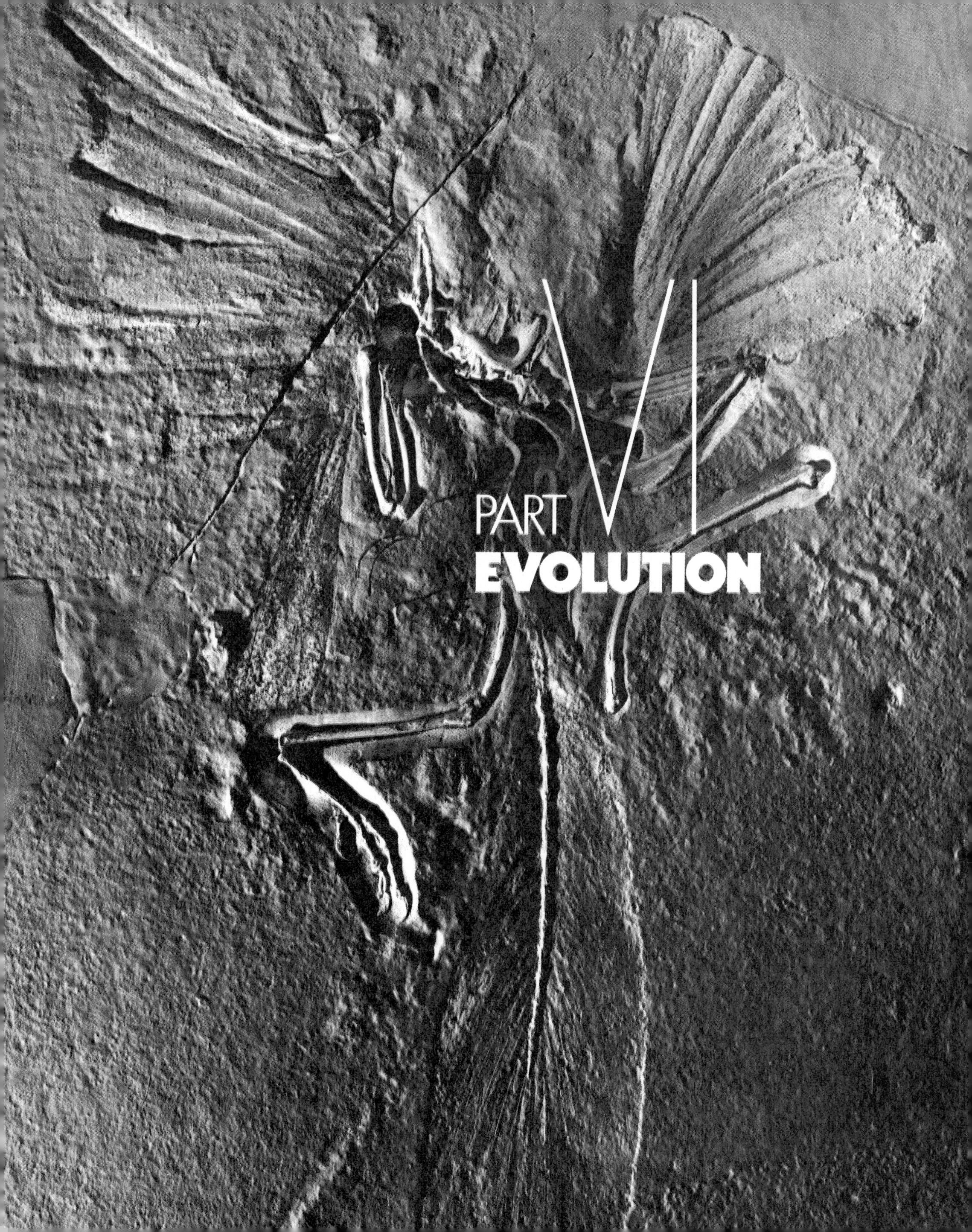

PART VI
EVOLUTION

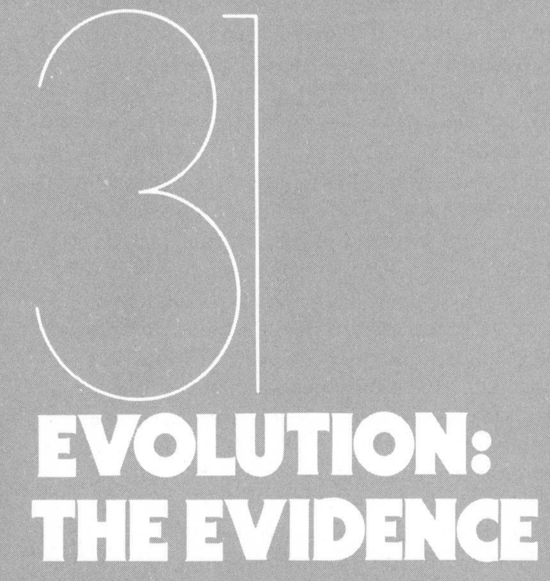

31 EVOLUTION: THE EVIDENCE

In 1859, the British naturalist Charles Darwin published *The Origin of Species.* It has been claimed that this book ranks second only to the Holy Bible in its impact on the thinking of the western world. What did it say that made it so influential?

First, *The Origin of Species* said that all living things on earth are here as a result of descent, with modification, from a common ancestor. This is the theory of **evolution.** Expressed another way, it tells us that species are not fixed, unchanging things but have, on the contrary, evolved through a process of gradual change from preexisting, different species. The theory implies, too, that all species are cousins, that is, any two species on earth have shared a common ancestor at some point in their history. This theory of evolution directly contradicts the still widely accepted idea that species are unchangeable, each species having been placed on earth in its present form.

Second, Darwin's *The Origin of Species* presented a large number of facts which Darwin felt could best be explained by a theory of evolution and could not be adequately explained by a theory of special creation. In this chapter, we shall examine some of these facts along with additional evidence that has been discovered since Darwin's time.

Finally, Darwin proposed a mechanism to explain how evolutionary change takes place. This theory, the theory of **natural selection,** is the cornerstone of *The Origin of Species.* The idea of evolutionary change is very old. Evidence to support it had been presented before Darwin's time. It was Darwin, however, who built an overwhelmingly impressive case for the existence of evolution *and* proposed a theory to explain how evolution works. This theory and its clarification and enlargement by later workers will be discussed in Chapter 32.

The idea of evolution provides a plausible explanation for a host of otherwise hard-to-explain facts. Let us now turn our attention to these.

FIG. 31.1 Dinosaur footprints in limestone near the town of Glen Rose, Texas. The tracks on the right were made by a large dinosaur walking on all four legs. The tracks on the left were made by a smaller, bipedal dinosaur, perhaps a predaceous species. The tracks were formed during the Cretaceous period some 120 million years ago. (Courtesy of Roland T. Bird.)

31.1 THE EVIDENCE FROM PALEONTOLOGY

Paleontology is the study of **fossils.** We can define a fossil as any sort of remains of a once-living organism. A variety of kinds of fossils are found. Figure 31.1 shows the footprints left by dinosaurs some 120 million years ago as they walked along the bed of a stream in the region that is now central Texas.

Under special circumstances, the **entire body** of an organism may be preserved after death. Insects, trapped in the sticky pitch of conifers growing along

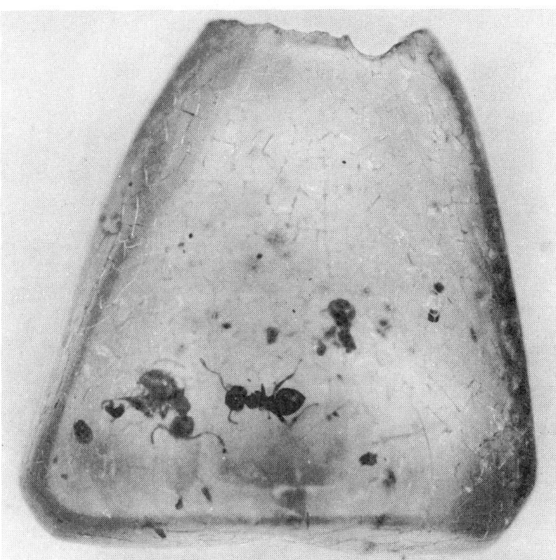

FIG. 31.2 Ants fossilized in a piece of amber.

FIG. 31.3 Partially exposed dinosaur bones in Wyoming. (Courtesy of the American Museum of Natural History.)

the Baltic Coast over 30 million years ago, can now be studied entombed in amber as easily as if they had just died (Fig. 31.2). Perhaps you have heard stories of the frozen woolly mammoths found in Siberia in the early years of this century. The meat of these animals was sufficiently well-preserved to be fed to dogs after thousands of years in nature's deep freeze.

Such total preservation of dead organisms is very rare. Usually, the soft portions of the body are quickly destroyed after death by scavengers or decomposed by decay bacteria. **Hard parts,** such as bones or shells, are more resistant to such destruction and hence more likely to be fossilized. If surrounded by sediments of clay or sand, they may yield easily recognizable fossils 500 million years later, long after the enclosing sediments have turned to rock such as shale or sandstone (Fig. 31.3). These fossils may even retain traces of organic matter for surprisingly long periods. Amino acids and small peptides have been recovered from some that are over 300 million years old.

Another common fossil is the **petrifaction.** This is a copy in stone of some plant or animal part. As the original remains disintegrated, they were replaced bit by bit with mineral deposits. This process can proceed so slowly that the original specimen is reproduced in all its detail. Figure 31.4 shows the clearly preserved annual rings in a piece of petrified wood. None of the original material is present in this specimen. It is not made of cellulose but of silica. Nevertheless, the faithfulness of the copy makes the fossil as useful as the actual specimen would be.

We know that fossils have aroused human curiosity at least since the time of the ancient Greeks. With rare exceptions, fossils are not the remains of organisms still found living on the earth. How, then, can we explain their existence? A series of special creations followed by worldwide catastrophic extinctions has sometimes been given as the explanation. The theory of evolution provides a more satisfying answer, however. The idea that all organisms alive today share a common ancestry at some period in history implies that there were fewer kinds of living things in the past and that these were less complex.

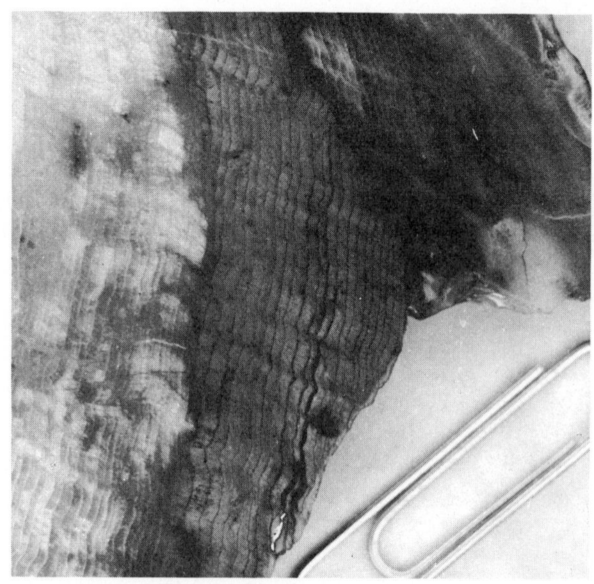

FIG. 31.4 A piece of petrified cypress wood. Although all organic matter has been replaced by silica, the pattern of annual rings stands out clearly.

FIG. 31.5 Grand Canyon as seen from the south rim. The various layers of sedimentary rock, spanning much of the Paleozoic era, are clearly visible. (Courtesy of Josef Muench.)

This describes the fossil record very well. In descending into the Grand Canyon (Fig. 31.5), one passes layer after layer of sedimentary rock, the deeper layers being the older ones. As one proceeds downward, the number of different kinds of fossils decreases. Furthermore, the complexity of the organisms represented in the deeper layers is less than in the upper layers. Fossil reptiles appear late in the geological record while fossil worms appear very early.

It must be noted that one never finds an unbroken fossil history in one location. Geological upheavals of the land (which, after all, is the only way fossils of marine organisms could ever be exposed to our view) are always followed by erosion and hence erasure of part of the fossil record.

Many of Darwin's critics cited the failure of paleontologists to find "missing links" as a serious weakness of the idea that present forms have evolved from forms now known only as fossils. The argument has steadily lost effectiveness as more and more "missing links" have been found. If a biologist should try to describe an animal intermediate between the birds of today and the reptiles from which we believe they evolved, he would almost perfectly describe *Archeopteryx,* one of the many no-longer-missing links (a creature that we shall examine in Chapter 38).

Gaps in the fossil record are still quite noticeable among the soft-bodied animals and among the prehumans. This is not surprising when you consider the remote chances of either type of organism's becoming fossilized in the first place. Soft-bodied animals are likely to decay too quickly and terrestrial ones (especially intelligent primates) are not likely to die where their remains will be quickly protected by enclosing sediments.

Then, too, we must remember that we can never expect to find more than fragments of the fossil record. Most of the fossils ever formed are still imprisoned within mountain ranges, under the earth and oceans, or have been destroyed by subsequent erosion and other geological disturbances.

Perhaps the greatest obstacle to finding missing links is that the evolution of new species of plants and animals seems, in general, to have occurred in small populations of poorly specialized forms. The fossil record is, however, filled chiefly with the remains of "climax" groups, large populations of highly specialized organisms which flourished for a period only to become extinct when conditions on earth changed.

Though we may never be able to trace the evolution of all living forms through the fossils of their ancestors, the presence and distribution of fossils already discovered provide us with some of the most direct evidence of the theory of evolution.

31.2 THE EVIDENCE FROM COMPARATIVE ANATOMY

In comparing the anatomy of one mammal with another, one cannot help being struck by the many cases in which certain parts of the body are built to the same basic plan in each specimen. This might not seem surprising if the similarly constructed parts were used in similar ways. One could argue that there was only one best way to construct the organ in question and that the Creator used it. However, these organs may actually be used in a variety of ways. Figure 31.6 illustrates the organization of bones in the forelimbs of the human, a whale, and a bat. Although these forelimbs are used for such diverse activities as lifting, swimming, and flying, the same basic structural plan is evident in all of them. Organs which have the same basic structure, the same relationship to other organs, and (as it turns out) the same type of embryonic development are said to be **homologous.**

It hardly seems reasonable that a single pattern of bones represents the best possible structure to accomplish the varied tasks to which these mammalian forelimbs are put. However, if we interpret the persistence of the basic pattern as evidence of inheritance from a common ancestor, we see that the various modifications are simply adaptations of the plan to the special needs of the organism.

The various mammalian forelimbs constitute just one example of homologous organs. Both the animal and plant kingdoms contain a large catalogue of such structures. This is hardly surprising in view of our belief that all organisms have shared a common ancestor at some time in their evolutionary history. Presumably, the more recently two species have shared an ancestor, the more homologous organs they will have in common. Homologies in more distantly related species may be harder to determine, although fossil evidence is often a great help in this.

One category of homologous organs deserves special mention for providing evidence of evolution. These are homologous organs which in some species have no apparent function. Dissection of a boa constrictor or a whale reveals bones that are thought to be homologous to the hip bones of other vertebrates.

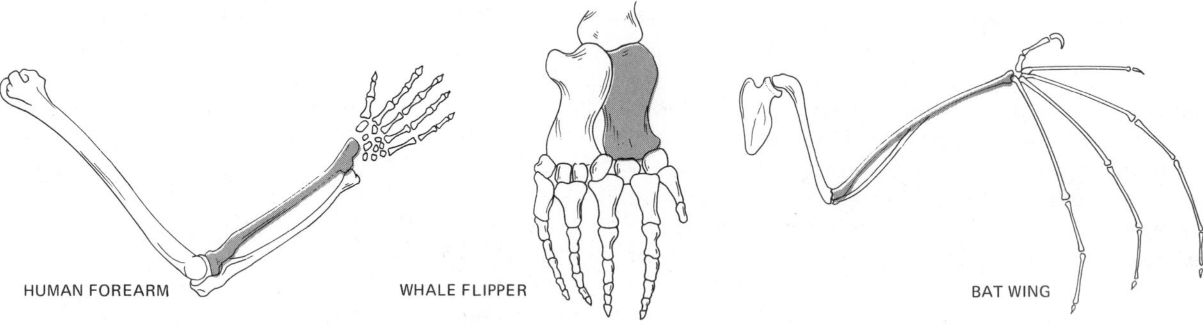

FIG. 31.6 Three vertebrate forelimbs; a study in homologous structures. In each case the bone shown in color is the radius. The forelimbs are not drawn to the same scale.

No function seems to be accomplished by these structures. If all species have been specially created, it seems like poor designing to include nonfunctional parts. If, on the other hand, we assume that both snakes and whales have evolved from four-legged ancestors, then we can understand why traces of their evolutionary heritage still remain.

Humans, too, have similar **vestigial organs.** The fused vertebrae which make up the base of the human spine are interpreted as the vestigial remnants of the tail possessed by our ancestors. In fact, human babies occasionally are born with short tails. These are, however, quickly and easily removed.

31.3 THE EVIDENCE FROM EMBRYOLOGY

The embryonic development of all vertebrates shows striking uniformities. This is particularly true during cleavage, morphogenesis, and the early stages of differentiation (Fig. 31.7). These similarities are often cited as evidence of an evolutionary relationship among the vertebrates. Of course, one might argue that there is only one effective way that a vertebrate can be constructed from a fertilized egg. There may well be considerable truth in this idea but, on the other hand, it does not adequately explain one aspect of these similarities. This is the fact that some structures appear during the development of the most advanced vertebrates just as they do in the more primitive species, only to disappear or become almost unrecognizably modified in the later stages of development. The month-old human embryo has a series of paired branchial grooves in the neck region (Fig. 31.7). These are matched on the interior by a series of paired gill pouches. This pattern appears not only in humans but in the embryonic development of all vertebrates. In the fishes, the pouches and grooves eventually meet and form gill slits, the openings which allow water to pass from the pharynx over the gills and out of the body. In the "higher" vertebrates the grooves and pouches disappear. In humans the chief trace of their existence is the Eustachian tube and auditory canal, which (interrupted only by the eardrum) connect the pharynx with the outside of the head. The temporary possession of a tail and a two-chambered heart are other examples of developmental stages through which the human embryo passes. Surely there must be more direct ways to achieve the final adult form. What explanation, then, can we give for these seemingly inefficient procedures? Again evolution provides the clue. We and the other vertebrates continue to pass through many (not all) of the embryonic stages that our ancestors passed through because we have all inherited developmental mechanisms from a common ancestor. We then go on to modify these in ways appropriate to our diverse ways of life. Therefore, it should be no surprise to find that the more distantly related two vertebrates are, the shorter the period during which they pass through similar embryonic stages (Fig. 31.7). Conversely, the more closely related two vertebrates are, the longer their embryonic development proceeds in a parallel fashion.

The idea that our embryonic development repeats that of our ancestors is called the theory of **recapitulation.** This theory is not restricted to the vertebrates. There is some evidence (which we shall examine in Chapter 37) that the ancestors of the insects had a pair of legs on each of their body segments. In this, they resembled today's millipedes.

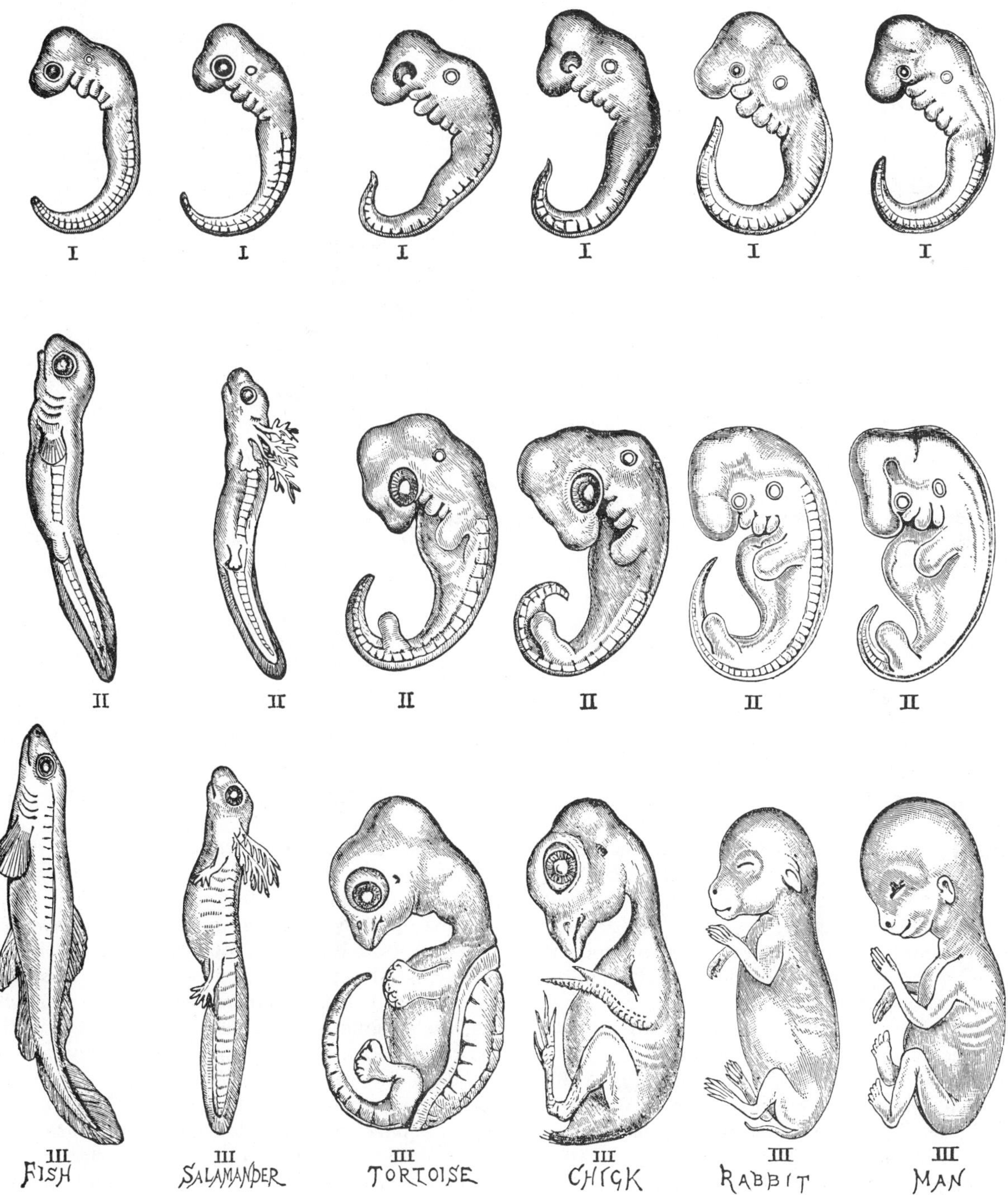

FIG. 31.7 Comparison of vertebrate embryos. (G. J. Romanes, *Darwin and After Darwin,* Open Court Publishing Co.)

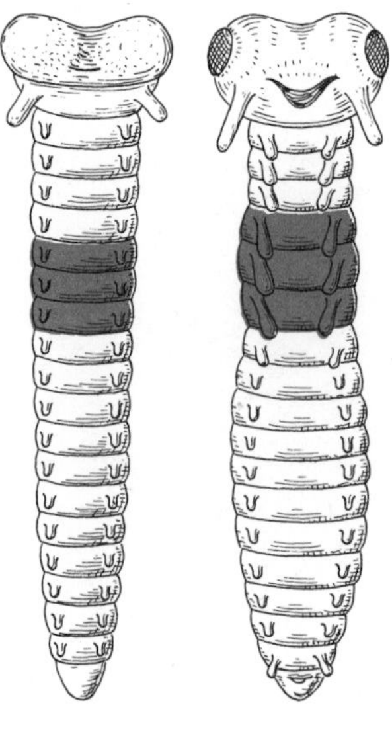

FIG. 31.8 Two stages in the embryonic development of an insect. Although limb buds appear in all segments, as they surely did in the ancestors of the insects, only those in the thorax (color) become legs. The ones in the head develop into mouth parts. The ones in the abdomen disappear.

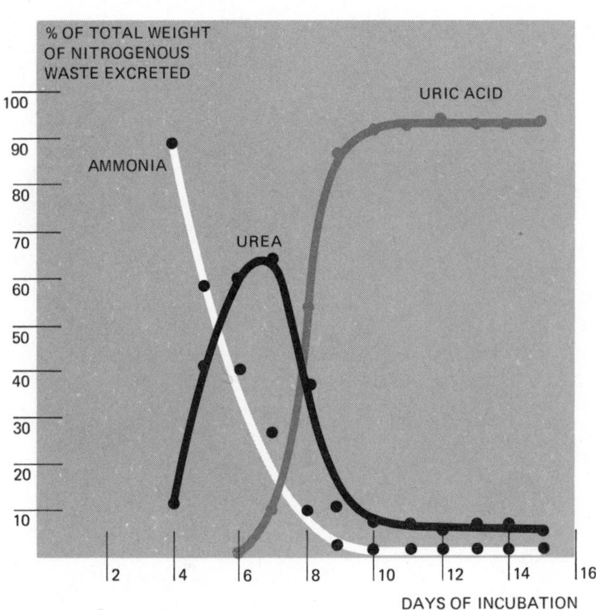

FIG. 31.9 Biochemical recapitulation in the chick embryo.

(In fact, the millipedes may well represent a second line of descent from these early forms.) In any case, during the embryonic development of insects, limb buds appear on the abdomen just as they must have in their multi-legged ancestors (Fig. 31.8). But by the time the larva hatches, only the six legs on the thorax remain.

Biochemical as well as anatomical recapitulation occurs. As you learned in Chapter 13, fish excrete a large part of their waste nitrogen as ammonia, while amphibians have the less toxic urea as their chief nitrogenous waste. Actually, the fishlike *tadpole* excretes ammonia until it undergoes metamorphosis into the adult frog. Only then does its chief nitrogenous waste become urea.

Birds go one step further and convert their waste nitrogen compounds into the almost insoluble compound uric acid. It is interesting to see (Fig. 31.9) that the developing chick embryo does not excrete uric acid from the very first but instead passes through an ammonia-excreting stage followed by a urea-excreting stage. This certainly suggests that the chick is repeating stages in the biochemical development of its ancestors.

31.4 THE EVIDENCE FROM COMPARATIVE BIOCHEMISTRY

Just as the study of comparative anatomy has shown the presence of anatomical homologies, so the study of the biochemistry of different organisms has revealed biochemical homologies. In fact, the biochemical similarity of living organisms is one of the most remarkable features of life.

Cytochrome enzymes are found in almost every living organism. One of these, cytochrome *c,* is a polypeptide chain of 104 to 112 amino acids (depending upon the organism in which it is present). In recent years the exact sequence of amino acids in these chains has been determined for the cytochrome *c*'s of such diverse organisms as humans, a rabbit, the king penguin, a rattlesnake, the tuna fish, a moth, the red bread mold *Neurospora,* and many others. Although there is considerable variation in the sequences, especially between organisms that we assume are only distantly related, there is a surprising amount of similarity as well. The sequence in humans differs from that in rhesus monkeys at only one place in the chain. Cytochrome *c* from the wheat plant differs from ours in 35 of the amino acids. However, another 35 of the amino acids in the chains have proved to be the same in every single species examined. This includes one section of 11 consecutive amino acids (Nos. 70–80) that are common to all organisms known. We have seen how the nucleotide sequence in DNA molecules codes for the amino acid sequence in proteins. The occurrence in such a wide variety of organisms of genes for cytochrome *c*—genes that contain much of the same genetic information—would be virtually inexplicable without a theory of evolution. Surely this phenomenon means that we have all inherited this gene, albeit with an accumulation of mutations, from a common ancestor.

The same argument can be applied to other biochemical similarities among organisms. Studies of the amino acid sequences in mammalian hemoglobins reveal close similarities, especially so among those species thought to be closely related (Fig. 31.10). DNA and RNA are found in every living organism and, so far as we can determine, contain the same hereditary coding mechanism. Furthermore, most vertebrates share the same, or similar, hormones. Prolactin, for example, is found in vertebrates as diverse as fishes, birds, and mammals, although its function differs in each of these classes (see Section 25.9). Thus we have a parallel on the chemical level of the homologous forelimbs we discussed earlier in the chapter: a hormone inherited from a common ancestor but with its function modified in ways appropriate to the life of each animal.

The remarkable uniformity of biochemical organization that underlies the incredible diversity of living things is difficult to interpret in any other way but an evolutionary one. Presumably these molecules appeared very early in the history of life, and almost all modern forms have inherited the ability to manufacture and use them.

If one injects human serum proteins into a rabbit (a rabbit is simply a convenient animal to use—any other mammal would do), the rabbit manufactures a wide variety of antibody molecules against all the antigenic determinants foreign to it. When rabbit-blood serum containing these antihuman antibodies is mixed in a test tube with human serum, insoluble antigen-antibody complexes form that settle out of solution as a precipitate. The amount of precipitate formed can be easily measured. What makes this reaction interesting so far as our present story is concerned is that these antihuman antibodies will also react with the blood serum of certain other mammals but to a lesser degree; that is, a smaller amount of precipitate is formed. Antihuman antibodies mixed with the serum of a human, an ape, an Old World monkey, a New World monkey, and a pig (in each of five separate test tubes) produce a precipitate in each tube. The amount produced decreases, however, from human to pig (Fig. 31.11). As we shall see in Chapter 38, this corresponds closely to the presently accepted degree of kinship between us and these other mammals.

NUMBER OF AMINO ACID DIFFERENCES FROM THE HUMAN BETA CHAIN FOUND IN THE HEMOGLOBINS OF VARIOUS SPECIES

HUMAN BETA CHAIN	0
GORILLA	1
GIBBON	2
RHESUS MONKEY	8
DOG	15
HORSE, COW	25
MOUSE	27
GRAY KANGAROO	38
CHICKEN	45
FROG	67
LAMPREY	125
SEA SLUG (A MOLLUSK)	127
SOYBEAN (LEGHEMOGLOBIN)	124

FIG. 31.10 Biochemical homology. The degree of structural similarity is roughly proportional to the closeness of kinship. All the values listed are for beta chains except for the last three, in which the distinction between beta and alpha chains does not occur. The human beta chain contains 146 amino acid residues, as do most of the others.

REACTION BETWEEN ANTIHUMAN ANTIBODIES (PREPARED IN RABBIT) AND SERUM OF VARIOUS MAMMALS, WITH HUMAN TAKEN AS 100%

HUMAN	100%
CHIMPANZEE	97%
GORILLA	92%
GIBBON	79%
BABOON	75%
SPIDER MONKEY*	58%
LEMUR	37%
HEDGEHOG (INSECTIVORA)	17%
PIG	8%

*New World species

FIGURE 31.11

Not only has this method (called comparative serology) corroborated some evolutionary relationships which had already been agreed upon, but it has been helpful in establishing relationships for which anatomical evidence had failed to provide clear-cut answers. For example, rabbits show some structural resemblances to rodents but, despite these, are placed in an order of their own, the Lagomorpha. One important reason for this is that serological tests show little affinity between rabbits and rodents; in fact, rabbits seem to be more closely related to even-toed ungulates like the pig! Whales, too, reveal by serological testing a closer relationship to the even-toed ungulates than any other mammalian order. Even plant proteins have been used as antigens and several evolutionary puzzles have been cleared up by this technique.

31.5 THE EVIDENCE FROM CHROMOSOME STRUCTURE

The differences that distinguish one species from another are, in the last analysis, genetic. Genes are located on chromosomes. If, then, we believe that all species share a common ancestor at some point in time, can we find homologies in chromosome structure just as we do in skeletal structure (Fig. 31.6) and protein structure?

The more closely related two species appear to be on the basis of such criteria as homologous organs, the more similar are their karyotypes. The karyotypes of a chimpanzee and an orangutan are practically indistinguishable and, except for the presence of 48 instead of 46 chromosomes, very similar to that of a human.

Similar examples can be found throughout the plant and animal kingdoms. Among the most interesting of these are found in insects having giant chromosomes. We saw in Chapter 16 with what precision the location of individual genes can be mapped on these chromosomes. We have also seen with what precision equivalent portions of homologous chromosomes pair with each other. Deletions, duplications, and inversions are quickly revealed by the loops they create in the paired chromosomes (Fig. 31.12).

Often (but not always) representatives of two closely related species of fruit fly can be induced, in the laboratory, to mate. Viable offspring may result although these are usually infertile (like the hybrid mule produced by mating a horse and a donkey). When one examines the giant chromosomes of these *interspecific* hybrids, one finds the explanation of their infertility. Large numbers of mismatched chromosomal regions are present, revealed as chromosome loops (Fig. 31.13). This makes successful mei-

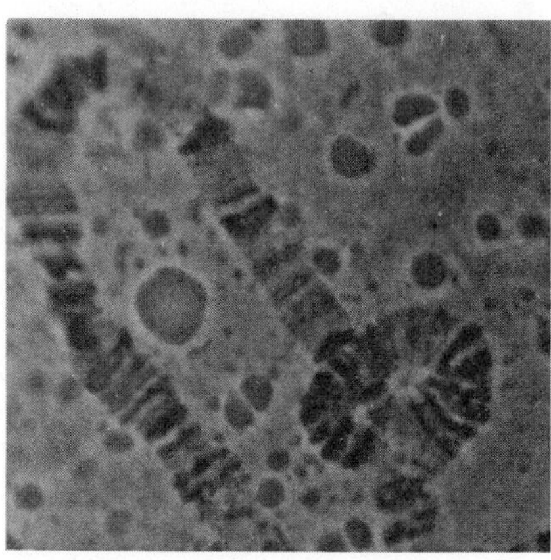

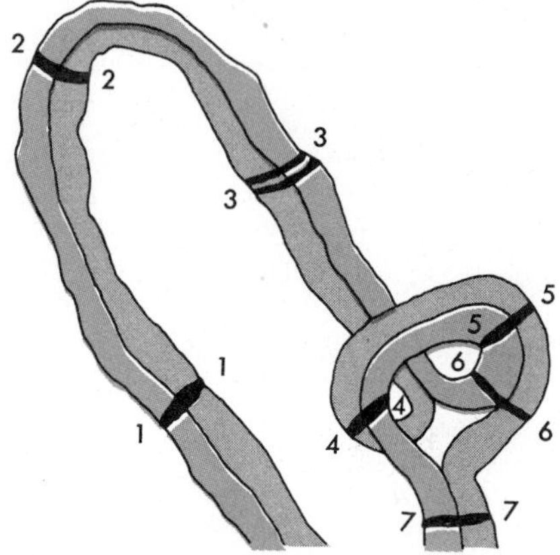

FIG. 31.12 Left: An inversion in one of the No. 3 chromosomes of *Drosophila melanogaster* has caused this loop to form. This inversion was induced by X irradiation, but similar inversions occur under natural conditions. Right: The chromosome carrying the inverted segment ("6, 5, 4") is shown in color; its normal homologue in gray. By forming the loop, the equivalent gene loci on each chromosome can pair with each other. (Photomicrograph courtesy of Edwin Vann, Ph.D., M. D.)

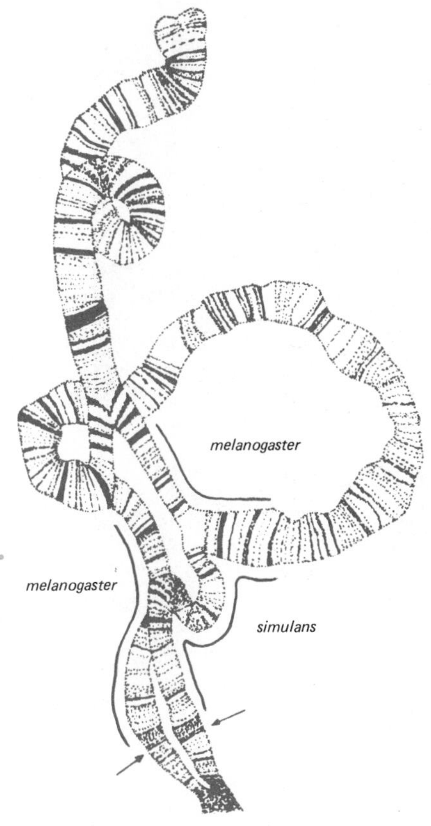

FIG. 31.13 Pairing of the No. 3 chromosomes present in the hybrid produced by mating *Drosophila melanogaster* with *Drosophila simulans.* The chromosomes pair perfectly except where an inversion is present and where there are slight differences in banding pattern (e.g. at the arrows). The extensive identity of these two giant chromosomes indicates that the two species inherited large segments of this chromosome from a common ancestor. (Based on the work of Patau, courtesy of Springer-Verlag.)

osis virtually impossible. The remarkable thing from the point of view of our story, though, is the amount of pairing that does take place. Knowing the precision with which this process occurs, we can only conclude that these *different* species share large blocks of *identical* genetic material. (Simple visual inspection of the *pattern* of bands in the chromosomes of the two species tends to confirm this.) How can we explain this except by assuming that the two species have evolved from a common ancestor from which they each inherited not only individual genes but large blocks of precisely ordered sequences of genes? In fact, detailed analysis of chromosome pairing patterns within and between various *Drosophila* species has revealed how each race and species can be related to the others through a defined sequence of chromosome break-repair events (leading to inversions and translocations). You are urged to read about these analyses in the book by Bruce Wallace cited at the end of the chapter.

31.6 THE EVIDENCE FROM PROTECTIVE RESEMBLANCE

Probably no more dramatic evidence of evolution exists than the spread of **industrial melanism.** Approximately 10% of the more than 700 species of larger moths that are found in the British Isles have been undergoing a dramatic darkening in coloration in regions of heavy industrial activity. This change is called industrial melanism. Perhaps the best-studied example is the peppered moth, *Biston betularia.* The wings and body of this moth used to be light in color with scattered dark markings (giving rise to its common name). In 1848 a coal-black mutant form of the moth was discovered near Manchester, England. Since that time, the black form (*Biston betularia* var. *carbonaria*) has gradually become more prevalent. Today over 90% of the moths in this region are black. Here certainly is an example of a change in a species taking place in nature rapidly enough to be recorded by naturalists.

FIG. 31.14 Left: The peppered moth *Biston betularia* and the dark form *carbonaria* on a lichen-encrusted tree trunk. Right: Both forms of the moth are shown resting on a blackened tree trunk typical of those near industrial areas. (Courtesy of Dr. H. B. D. Kettlewell.)

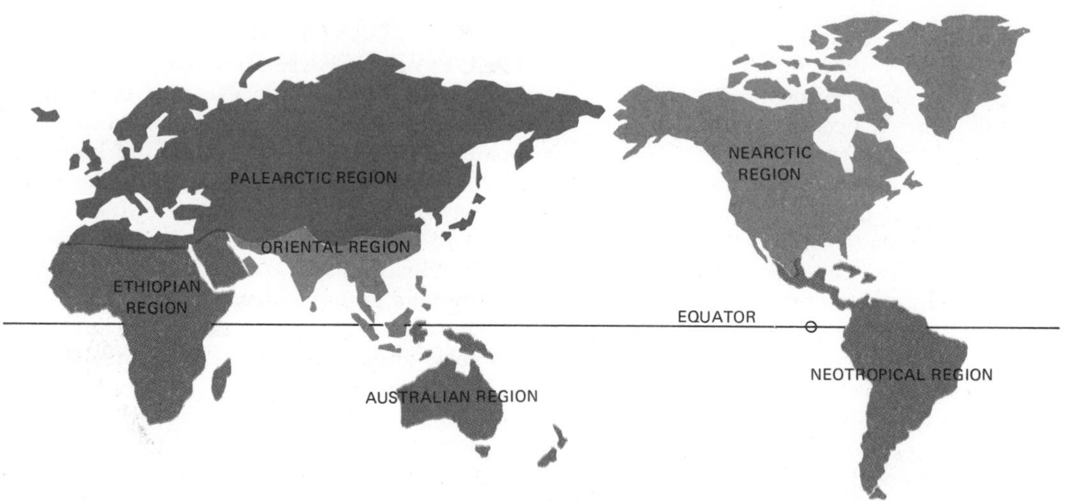

FIG. 31.15 Wallace's faunal regions. The circle on the equator shows the location of the Galapagos Islands.

What accounts for this evolutionary change? *Biston betularia* flies at night and rests by day on tree trunks in the forest. In areas far from industrial activity, the trunks of forest trees are encrusted with lichen growth. As Fig. 31.14 shows so clearly, the light-colored form of the moth is practically invisible against such a background. In areas where air pollution is severe, the combination of toxic gases and soot has killed the lichen growth and darkened branches and trunks. Against such a background, the light-colored moth stands out sharply. *Biston betularia* is preyed upon by birds which pluck it from its resting place by day. In polluted woods, it is easy to see that the dark form would have a much better chance of surviving undetected. In fact, when the English geneticist H. B. D. Kettlewell released marked moths of both kinds in these woods, he was able to recapture twice as many of the dark forms as of the light. Careful observation showed that predatory birds did, indeed, eat a much higher proportion of the light moths released than of the dark. It is no wonder that the dark form is now dominant in these regions. (Just recently, however, follow-up studies have revealed that the light form is *beginning* to make a comeback in the Manchester and London areas. This is probably the consequence of the rigid smoke abatement programs that have been placed in effect in those cities.) In unpolluted areas (e.g. Scotland and southwestern England), the woods are still lichen-encrusted and the light form of *Biston betularia* is still dominant.

31.7 THE EVIDENCE FROM GEOGRAPHICAL DISTRIBUTION

In 1876 the British naturalist Alfred Wallace (who conceived the idea of evolution and natural selection independently of Darwin), proposed that the continental areas of the world could be divided into six main regions on the basis of their animal populations (Fig. 31.15). The greatest diversity of living things is found in the two great tropical areas, the Ethiopian (tropical Africa) and the Oriental (tropical Asia and the nearby offshore islands). Fossil evidence suggests that in these regions have evolved most of our dominant plants and vertebrates. Europe and northern Asia make up the Palearctic region, while North America is the Nearctic. Spread of plants and animals into and through these regions has often been seriously limited by the harshness of the climate. The two remaining continental regions are the Neotropical (South America) and the Australian (Australia, New Zealand and New Guinea). The unusual animal life (fauna) and plant life (flora) of these regions can be explained by their intermittent isolation from the nearby land masses. The Australian region is isolated right now, of course, while the Neotropical region has had a land bridge with the Nearctic for the last few million years.

The distribution of plants and animals in the oceanic islands provides particularly strong support for the theory of evolution. Oceanic islands (e.g. the Hawaiian Islands) are those that have never been connected to one of Wallace's continental regions. Many of them have arisen from the sea in relatively recent times (geologically speaking). Nevertheless, they all support a rich and varied fauna and flora. If species were immutable, we should expect that all the creatures to have colonized such islands would be members of species represented on the mainland. As a young man of 26, Darwin visited a group of such islands, the Galapagos Islands off the coast of Ecuador (Fig. 31.15). While the marine birds were those found elsewhere in the Pacific, Darwin found scattered over the islands thirteen species of land birds which were not known anywhere else in the world. At first glance, some of these birds seemed quite diverse (see Fig. 32.11). Some had stout beaks for eating seeds. Others had beaks adapted for eating insects of one size or another. One species had a beak like a woodpecker's and actually used it for drilling holes in wood. However, it lacked the long tongue which our woodpeckers use for removing insects from the wood. Instead, it used a cactus spine, held in its beak, to dig the insect out. Underlying this superficial variety of types was a basic similarity: all these birds were finches. Although one looked far more like a warbler than a finch, its internal anatomy showed its true kinship. Here, then, were a group of birds found nowhere else on earth and all showing definite similarities. This indeed would be a curious exercise in special creation. Far more plausible is Darwin's thought that each of these birds is the result of descent with modification (evolution) from a common finch ancestor which accidentally reached the islands from its original home in South America. We have seen that Darwin marshalled several kinds of evidence to support his conviction that species are changeable and are the product of evolution. But probably no evidence was more important to his thinking than the example of his finches.

31.8 THE EVIDENCE FROM DOMESTICATION

The deliberate cultivation of plants and animals has occupied humans for thousands of years. During all of this time, but especially in the last two centuries, we have developed varieties or breeds of plants and animals which yield more and better food or which, in other ways, better serve our purposes. While, with rare exceptions, we have not created new species in this process of domestication, we have certainly created forms which differ greatly from the ancestral stock. The extraordinary diversity of the domestic dog, from Chihuahuas to Saint Bernards, provides dramatic evidence of our ability to alter a species by selective breeding procedures. In fact, the corn plant (*Zea mays*) has been altered so greatly in the course of domestication that it can no longer survive without our aid. The many breeds of horses, cows, goats, sheep, chickens, and rabbits, which can be seen at any country fair, are dramatic testimony to the variability of species and our ability to create evolutionary changes for our own benefit.

The standard by which a theory is judged is its ability to explain the *most* facts in the simplest manner and to make possible the prediction of new facts. The slowness with which most evolutionary change occurs has hampered the theory of evolution in the second respect. In its ability to provide a simple, comprehensive explanation for a vast array of facts, however, the theory of evolution justly stands as the most important theory in biology. Every aspect of the living world that humans study, from biochemistry and cytology to anthropology and history, has been nourished and broadened by it.

While the geographical distribution of animals may have provided Darwin with his most compelling evidence that evolution has occurred, it was his study of the process of domestication that gave him the clue as to how it occurs. It is his theory of the mechanism of evolutionary change which truly makes *The Origin of Species* such an epoch-making work. In the following chapter, we shall turn our attention to his theory and the elaboration it has received at the hands of more recent students of evolution.

EXERCISES AND PROBLEMS

1. The theory of recapitulation is often stated: "ontogeny recapitulates phylogeny." Does this statement accurately describe the facts? Explain.
2. Which two of the six animals whose embryonic stages are illustrated in Fig. 31.7 do you think have most recently shared a common ancestor? Explain.
3. How would you go about preparing antibodies to bovine serum albumin?
4. Frogs (*Eleutherodactylus*) were not found in Bermuda before being introduced by humans in 1880 but were prevalent in the islands of the West Indies. Can you explain why they were found in one location and not the other?
5. Would you predict on the basis of the data in Fig. 31.11 that we are more closely related to the Old World monkeys or the New World Monkeys? Explain.
6. The evolutionary history of mollusks and reptiles is known much better than that of flatworms. Why?

REFERENCES

1. BRUES, C. T., "Insects in Amber," *Scientific American,* Offprint No. 838, November, 1951.
2. ABELSON, P. H., "Paleobiochemistry," *Scientific American,* Offprint No. 101, July, 1956. Describes how amino acids have been discovered in fossils as old as 300 million years.
3. DICKERSON, R. E., "The Structure and History of an Ancient Protein," *Scientific American,* Offprint No. 1245, April, 1972. The cytochrome *c* story.
4. FRIEDEN, E., "The Chemistry of Amphibian Metamorphosis," *Scientific American,* Offprint No. 170, November, 1963. An example of biochemical recapitulation.
5. ZUCKERKANDL, E., "The Evolution of Hemoglobin," *Scientific American,* Offprint No. 1012, May, 1965. Demonstrates with mammalian hemoglobins how knowledge of the amino acid sequences of homolo-

gous proteins can be used to unravel evolutionary relationships.

6. KETTLEWELL, H. B. D., "Darwin's Missing Evidence," *Scientific American,* Offprint No. 842, March, 1959. A description of industrial melanism in *Biston betularia.*

7. WALLACE, B., *Chromosomes, Giant Molecules, and Evolution,* Norton, New York, 1966. Includes an explanation of the methodology by which inversions in giant chromosomes are used to unravel phylogenetic relationships.

8. LACK, D., "Darwin's Finches," *Scientific American,* Offprint No. 22, April, 1953.

9. FORD, E. B., *Evolution Studied by Observation and Experiment,* Oxford Biology Readers, No. 55, Oxford University Press, Oxford, 1973.

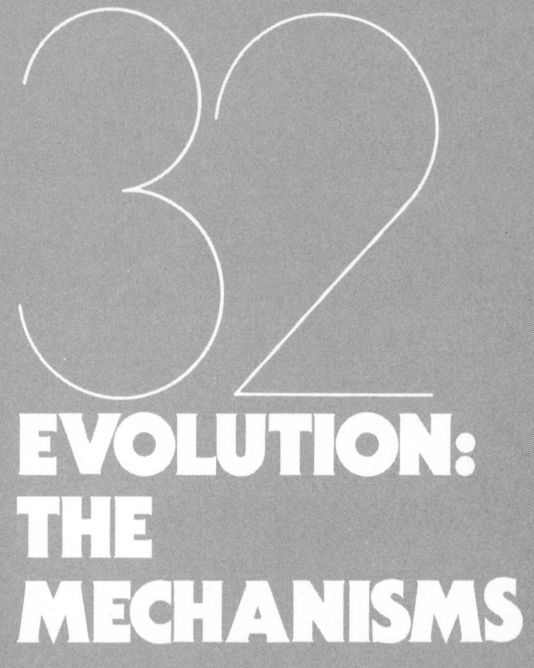

32 EVOLUTION: THE MECHANISMS

FIG. 32.1 High-yielding strain of rice growing in India. This strain is one of over two dozen developed in recent years at the International Rice Research Institute in the Philippines. A number of desirable traits have been incorporated in this strain through the use of selective breeding techniques. (USDA photo.)

32.1 INHERITABLE VARIATION: THE RAW MATERIAL OF EVOLUTION

Twenty-two years elapsed from the time that Darwin returned home convinced of the truth of evolution and the actual publication of *The Origin of Species.* During this long interval, Darwin accumulated still more data to bolster his case. Living in rural England, he had ample opportunity to watch local farmers practice the art of animal breeding. Art it was, because the science of genetics was still unborn. Nonetheless, new varieties of all sorts of domestic animals were being created. Darwin, recognizing these as evolutionary changes, sought the mechanism by which they were produced. He found that animal breeders could develop new breeds only from variants that appeared spontaneously in their livestock. That is, only if some animals were born heavier or taller or darker-coated than others could the breeders develop heavier or taller or darker-coated breeds. Darwin also realized that these variants could not arise simply from exposure to a different environment. Individuals that are larger simply as a result of better diet could not supply the raw materials for a larger breed. Only **inheritable variations** can serve this purpose.

The second vital ingredient in the formation of new varieties was **selective breeding.** Once individuals with desirable inherited traits were identified, they were mated together in the hopes that (1) a higher proportion of their offspring would have the trait than in the population as a whole and (2) the expression of the trait would be intensified in the offspring. By repeating this process of selective breeding, it eventually was possible to develop pure-breeding lines of individuals displaying the new trait (Fig. 32.1).

We can, then, bring about evolutionary change. Does a similar mechanism exist elsewhere in nature? The first thing we must look for in populations of wild organisms is inherited variations which can serve as the raw material for evolution. Those who have studied a large sample of individuals in a single species find that variation does exist in nature. Such variation takes two forms.

1. Continuous variation. Many traits found in a population of plants or animals vary in a smooth, continuous way from one extreme to the other. Body weight, body length, and coat color are just three traits in which we might expect to find a whole range of variations.

Typically, most of the individuals fall near the middle of the range with fewer individuals at the extremes (Fig. 32.2). You can demonstrate the existence of this type of variation right around you. Although your classmates may not qualify as a wild population, calculation of their heights or weights (of one sex) will reveal a wide, unbroken range of values with more individuals in the middle of the

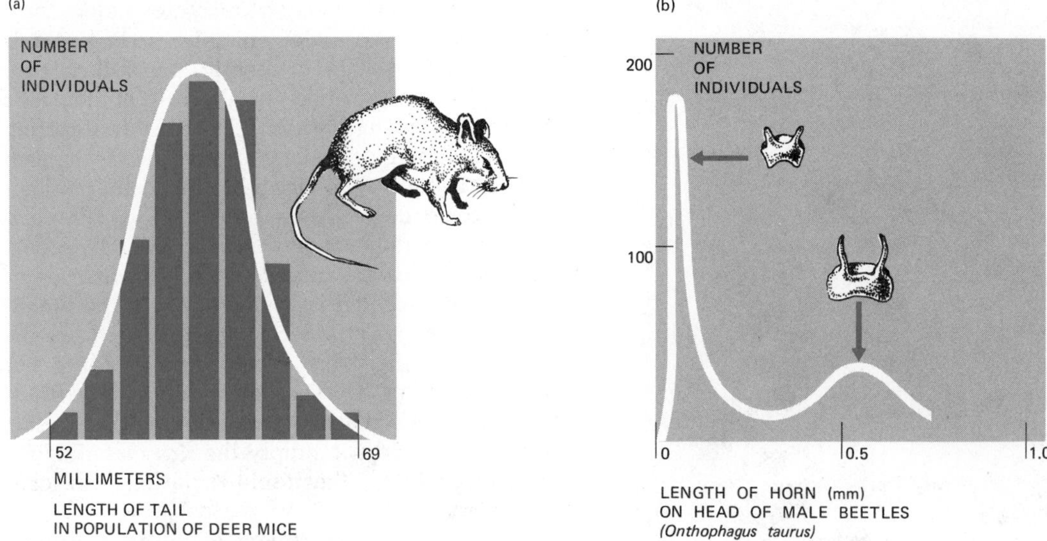

FIG. 32.2 Examples of (a) continuous variation and (b) discontinuous variation. (Redrawn by permission from D'Arcy W. Thompson, *On Growth and Form,* Vol. 1, Cambridge University Press, 1959.)

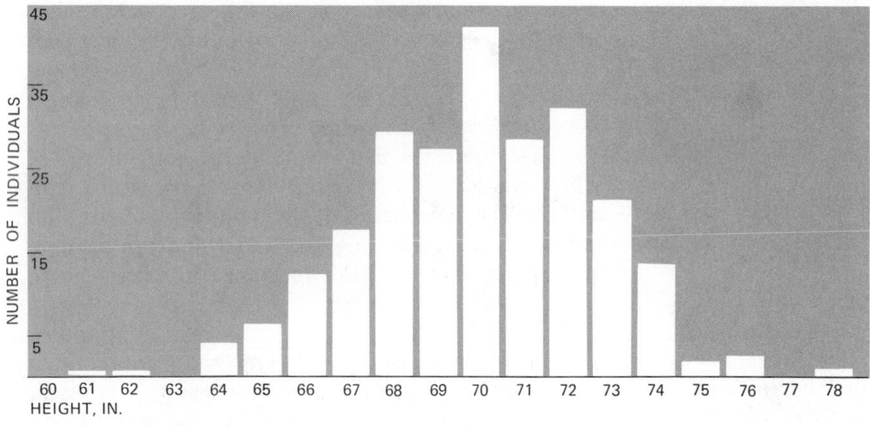

FIG. 32.3 Histogram showing distribution of heights among a group of male secondary-school seniors.

range than at the extremes. These data can be plotted on a graph (Fig. 32.3), and the curve that results is typical of those obtained by studying continuous variation in wild populations.

2. Discontinuous variation. With respect to certain traits, the individuals of some populations fall into two or more distinct categories with no intergrading between them. The fact that all humans have one of four possible ABO blood groups is an example of such discontinuous variation. Such variation is called **polymorphism.** The light and dark forms of *Biston betularia* illustrate polymorphism, too. There are no gray individuals. Plotting the dis-

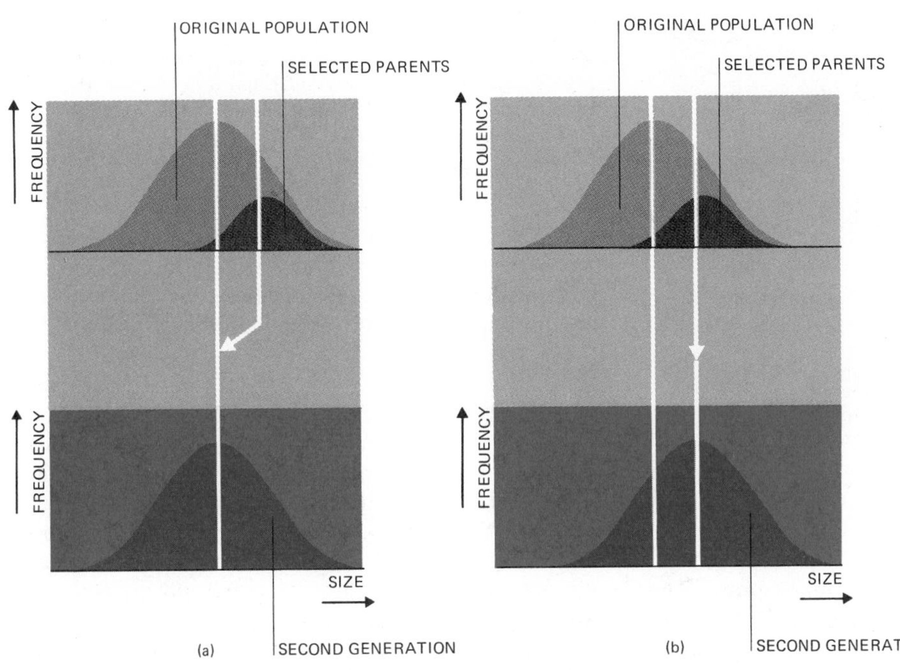

FIG. 32.4 Effect of selective breeding when variations are caused by (a) the environment alone and (b) hereditary factors acting along with environmental factors.

tribution of polymorphic traits, one gets not a single bell-shaped curve but discrete peaks for each form of the trait (Fig. 32.2).

Both continuous and discontinuous variation can provide the raw materials for evolution, but only if they arise from hereditary rather than just environmental factors. Perhaps the continuous variation in the height of your classmates is simply a reflection of variation in their diet as infants. How, then, can we determine whether variation in a wild population is to some degree inherited? As a practical matter, this is often very difficult. It involves the mating of individuals both of whom are extreme examples of the trait under study. If, for example, the offspring of two extra-large mice are substantially larger than the average for the population, hereditary factors are undoubtedly at work (Fig. 32.4). This trait is said to have a high **heritability.** On the other hand, if the offspring occupy the same range as the average for the population, environmental factors are working alone. Such a trait is said to have a zero heritability. A population of bean seeds of a pure strain will vary over a range of several millimeters in length. If extra-small beans are mated, however, the new crop will show no shift toward a smaller size. In this case, then, the heritability of size variation must be considered zero.

32.2 NATURAL SELECTION

Granted that inherited variability does occur in wild populations, is there a mechanism in nature to accomplish the selective breeding that Darwin saw was an indispensable part of the process of domestication? The answer to this question and the key to the whole theory of evolution came to Darwin as a result of reading the *Essay on Population* by the English clergyman Thomas Malthus. In this work, Malthus pointed out the remarkable reproductive potential of all organisms. There is not a species of living thing that could not completely fill its habitat with its kind if unlimited food were available to it and if the other factors in its environment were favorable.

Some would do it more rapidly than others. The ability of certain bacteria (e.g. *E. coli*) to double their weight and divide every 20 minutes would result in their blanketing the earth in a few days if nothing stopped them. Humans reproduce much more slowly but they, too, could theoretically cover the earth in a surprisingly short period of time (and are threatening to do so!). In fact, any species whose couples produce more than two surviving offspring during their lifetime will expand in numbers unless the excess die without undertaking parenthood. Under ideal conditions, then, populations grow exponentially (Fig. 32.5). This is simply a reflection of

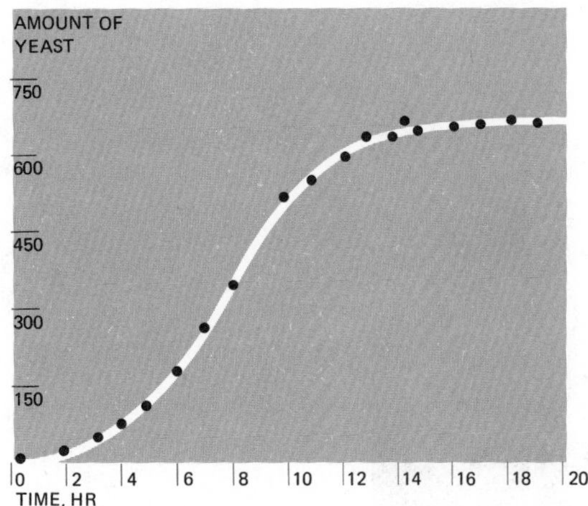

FIG. 32.5 Growth of a population of yeast cells. As long as conditions in the culture are optimum, the population grows exponentially. As conditions deteriorate, the growth rate levels off. These phases produce the S-shaped curve of growth.

the fact that the product of growth itself grows. The time scale varies from species to species, but the potentiality for exponential growth is present in them all.

What, then, keeps species from exponential population growth; what forces keep populations in check? A number of factors are at work and together they make up what Darwin called **natural selection.**

His theory of natural selection is a conclusion based upon three observable facts of nature and one preliminary conclusion. We can summarize the theory as follows:

Fact No. 1. All species have a high reproductive potential. From bacteria to elephants, they are capable of filling the earth with their kind.

Fact No. 2. Except for minor fluctuations, the population of any given species remains fairly constant from year to year.

Conclusion No. 1. Therefore, we must conclude that all creatures face a continual *struggle for existence,* a struggle that many will lose prematurely.

Fact No. 3. There is inherited variation among the individuals of any species.

Conclusion No. 2. Therefore, we may conclude that those individuals whose variations best fit them for their environment will be most likely to survive.

This idea of the *survival of the fittest* is what we call Darwin's theory of natural selection. It is the mechanism, Darwin thought, which accomplishes in nature what human selective breeding programs accomplish with domesticated plants and animals.

32.3 THE MEASURE OF "FITNESS"

Darwin's theory of natural selection has often been seriously misunderstood. To many people, it has implied that a "dog-eat-dog" existence is natural and hence justifies the most ruthless exploitation of one's fellow humans and of the other resources of our environment. Actually, fitness measured in terms of aggressiveness applies to only a portion of the world of living things and even then is only one of many criteria involved. Fitness as a biological concept can be measured only in terms of the ability to produce mature offspring. Those individuals in a population that leave the greatest number of mature offspring are the fittest. A variety of mechanisms play a role in this.

1. Survival. Perhaps the most important element of fitness is simply survival until one's reproductive years are over. Any trait that increases the organism's chances of surviving to and through this period makes that organism more "fit" than the others of its species. We call such traits **adaptations.** They may involve a change in body structure, a change in physiology, a change in behavior, or all of these. While strength, long fangs, etc. do promote survival in some cases, this kind of adaptation probably plays a minor role in evolution. For every adaptation associated with aggressive behavior, one can cite dozens of examples of adaptations which play a more inconspicuous, but every bit as important, role in promoting survival. The melanism of *B. betularia* var. *carbonaria* (Fig. 31.14) is an example of a morphological adaptation that makes its owners more fit by reducing the likelihood of their being eaten by predators.

The survival value of any trait cannot be adequately evaluated without reference to the environment in which the owner of that trait lives. The increased survival value of melanism in *B. betularia* exists only in forests darkened by industrial pollution. The presence of sickle-cell hemoglobin (see Section 18.2) in one's red blood cells reduces chances of survival in areas where malaria is not present but has a positive survival value in the areas of tropical Africa where malaria *is* prevalent. The malarial parasite spends much of its life cycle within red blood cells and presumably the presence of sickle-cell hemoglobin (Hb^S) in the red cells makes them less suitable for the parasites.

The fact that natural selection can operate only on traits that promote survival through the reproductive years has interesting consequences for those organisms, such as humans, in whom reproduction usually ceases long before death. Any traits that appear for the first time after reproduction has ceased cannot be selected for or against because no mechanism remains for their gradual increase (or decrease) in the population. Of course, traits that improve well-being late in life may improve fitness early in life, too. However, the absence of any specific selection of traits that improve physical and mental health late in life may well account for the sharp rise in degenerative changes that occurs in old age. It may also account for the fact that despite the dramatic increase in the average life span of Americans and Europeans during the last half century, our *maximum* life span remains virtually unchanged.

2. Sexual selection. Another factor affecting "fitness" is sexual selection. Any inherited trait that makes some individuals more desirable to the opposite sex than others will thus make them more successful in mating. Consequently, a larger proportion of the next generation will have the inherited trait. In 1950, Sheldon and Elizabeth Reed set up a culture of fruit flies (*Drosophila*) containing equal numbers of red-eyed males, red-eyed females, white-eyed males, and white-eyed females. Although previous experimentation had shown that white-eyed flies are just as healthy and live *just as long* as red-eyed flies (i.e. they are equal with respect to surval), after 25 generations there was not a white-eyed fly to be found in the culture. These investigators noted that not only did red-eyed females prefer red-eyed males but white-eyed females did, too. Such sexual selection was strong enough to eliminate the less "fit" flies in a remarkably short span of time.

An examination of certain traits that appear to be involved in sexual selection suggests that the evolutionary impact of sexual selection may not always work in the same direction as selection for survival. Traits that are selected sexually, and thus increase in subsequent generations, may well make their owners *less* well adapted to conditions in the environment. Brilliant plumage in male birds (which makes them more conspicuous to predators) may be an example.

3. Family size. Any trait resulting in the production of a larger number of mature offspring is also a measure of fitness. It is necessary to stress mature offspring because only mature offspring will be able to pass these traits on to still another generation. In animals, such as oysters and many fishes, which provide little or no care for their young, the way to more mature offspring is through a larger number of newborn. In these species, fitness is measured by the number of fertilized eggs they can produce. On the other hand, in species where the young are cared for by their parents, selection acts to reduce family size (to a certain point only!). Assuredly, if humans had as many offspring as oysters, they could not begin to care for their young, and fewer, rather than more, adults would be produced. With respect to family size, then, the "fittest" individuals are those with the largest families at the end, not necessarily at the beginning.

Among humans, family size appears to be a major component of "fitness." One tribe of Navaho Indians in New Mexico was established with 29 families. After eight generations, 85% of the members of the tribe were descended from just 14 of the founding families. The remaining 15 founding families thus accounted for only 15% of the descendants.

Some students of human evolution have argued that advances in medicine, sanitation, etc. have removed us from the rigors of natural selection. The facts suggest that while the forces of natural selection that act on us may be changing, they are by no means being eliminated. It has been estimated that, of all the human eggs that are fertilized, barely one-half develop into individuals who will succeed in reproducing themselves. The others are eliminated as follows: 15% by spontaneous abortion of embryos and fetuses, 5% by stillbirths and infant deaths, and 3% by childhood deaths. All of these have failed to meet the first criterion of "fitness"—survival. Another 20%, though surviving to adulthood, never marry—they are not "fit" in terms of sexual selection. Finally, of those that do marry, 10% (= 5.7% of the original number) will have no children and thus score zero on their "fitness" as measured by the criterion of family size.

One of the criticisms leveled at Darwin's theory

of natural selection was that it failed to provide any explanation of how desirable traits are passed on from generation to generation. In other words, Darwin was not able to explain the mechanism of inheritance. Such a criticism was hardly fair. As you know, the science of inheritance (genetics) had to await the rediscovery of Mendel's work at the beginning of this century. Furthermore, so long as it is granted that natural selection can work only on inheritable variation, our not knowing what the mechanism of inheritance is does not weaken the theory of evolution. In any case, today we do understand many of the principles of genetics, and they have indeed greatly increased our appreciation and understanding of the mechanism of evolution. Evolution involves a gradual change in the traits of a species. But traits are the expression of the genotype, and thus evolution must also involve a change in the genotypes of the individuals in the species. Let us now analyze the theory of natural selection in the language of genetics.

32.4 THE GENETIC SOURCE OF VARIABILITY

Genetics gives us the means to understand the origin of inherited variability in populations. The process of **sexual reproduction** creates new gene combinations, new genotypes, and hence new phenotypes or variants in the population. These new gene combinations originate in three ways: (1) At the time of gamete formation (spore formation in plants), **crossing over** of homologous chromosomes during meiosis produces new combinations of genes. (2) At metaphase of meiosis, **random assortment** of the homologous chromosomes on either side of the equatorial plane provides a further reshuffling of the genes inherited from the individual's parents. (3) The union of gametes from genetically dissimilar individuals, **outbreeding,** is also a source of variability as the maternal and paternal genes interact to produce a new phenotype, a new, unique individual.

Underlying these three processes is **mutation.** If all the individuals in a population were homozygous for the same genes at every gene locus, no amount of genetic "reshuffling" would create anything new. It is mutation that produces new alleles which can then be combined in various ways with existing genotypes.

In comparing the animals and higher plants with the lower plants (e.g. mosses) and most microorganisms, we find that the diploid generation is dominant in the former while many of the latter are haploid during the major part of their lives. If we can assume that the higher plants and animals represent a more elaborate and more successful sequence of evolutionary changes, the occurrence of diploidy may well be a factor in this. Recessive genes, as you have learned, often remain "hidden" in heterozygous individuals, the dominant allele having full expression in the phenotype. This is, however, true of diploid organisms only. In haploid organisms there is only one gene present for each gene locus. Consequently, recessive genes express themselves in the phenotype. Natural selection operates only on genes which affect the phenotype. Hence, we can expect natural selection to work more vigorously on the genes of haploid organisms than of diploid. Hidden in the recessive condition, a moderately disadvantageous gene in a diploid organism gets a chance to spread in the population and to be reshuffled in various combinations with other genes. Thus it gets a chance to prove itself advantageous in some combinations or in some environments instead of being quickly weeded out of the population by natural selection.

Our study of genetics also provides us with an understanding of the basis of hereditarily controlled variation, both continuous and discontinuous. Continuous variation that is under hereditary control arises from **multiple factors** (see Section 15.9). The distribution of multiple factors in the offspring of heterozygous parents forms a bell-shaped curve similar to the one we found that describes continuous variation as it occurs in nature. Discontinuous variation (polymorphism), on the other hand, probably is controlled by a single gene locus (e.g. human ABO blood groups) or, in the case of differences associated with sex (e.g. bright plumage in male birds), the presence or absence of a sex chromosome.

In applying the discoveries of genetics to the subject of evolution, we must consider the genotypes of entire populations rather than just individuals. This is known as **population genetics.** Students often experience some difficulty in applying the genetics of individuals to populations. For example, one common error is to assume that any phenotype (such as blood group B) which is present in only a small portion of the population will ultimately disappear. Many students seem to feel intuitively that a gene, such as *B,* which is present in only a small percentage of the population will eventually be swamped by its more prevalent allele. But intuition is insufficient in this case. Let us see why.

32.5 THE HARDY-WEINBERG LAW

If we mate two individuals that are heterozygous (e.g. *B*, *b*) for a given trait, we find that 25% of their offspring are homozygous for the dominant gene (*BB*), 50% are heterozygous like their parents (*Bb*), and 25% are homozygous for the recessive gene (*bb*) and thus, unlike their parents, express the recessive phenotype (Fig. 32.6). The reasons for these percentages are: (1) Meiosis separates the *two* alleles of each heterozygous parent. Therefore, 50% (0.5) of the gametes produced will carry one allele and 50% the other. (2) When the gametes (sperm and eggs) are brought together at random, each *B*-carrying egg will have a 1-in-2 probability of being fertilized by a *B*- (or *b*-) carrying sperm. The same holds true for the *b* eggs. The outcome of these chance encounters will thus be that shown in Fig. 32.6(a).

But in an entire *population* of breeding organisms (homozygous dominants, homozygous recessives, as well as heterozygous individuals), the total number of gametes that contribute to the formation of the next generation will in most cases not be equally divided between those carrying the dominant allele and those carrying the recessive one. As a hypothetical case, let us examine a population of hamsters in which 80% of all the gametes formed carry the dominant allele for black coat color (*B*) and 20% carry the recessive allele for gray coat (*b*). What genotypes will be produced in the next generation? Eighty percent of the sperm in the "pool" of gametes carry *B* and fertilize eggs at random. Eighty percent of *these* also carry *B*, so we predict that 64% ($.80 \times .80 = .64$) of all the zygotes will be *BB*. (The probability of an event is the *product* of the probabilities of each of the steps leading to that event. In this case, 0.64 is the product of the probability that any single sperm carries *B* (4 out of 5 or 0.80) *and* that any single egg carries *B* (also 0.80).) Twenty percent of the sperm carry *b* and fertilize the eggs at random. Eighty percent of these, you remember, carry *B* and so we would expect 16% of the zygotes formed to result from this combination and to be heterozygous (*Bb*) for the trait. Another 16% of the zygotes would also be heterozygous as they would result from the 80% of *B* sperm fertilizing the 20% of *b* eggs. Only 4% of all zygotes produced would be homozygous for gray coat (*bb*) since the 20% of the sperm carrying allele *b* had only a 1-in-5 chance of fertilizing an egg carrying *b* ($.20 \times .20 = .04$).

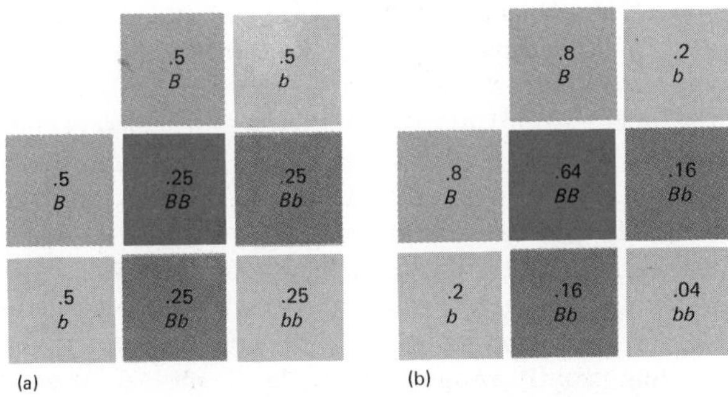

FIG. 32.6 (a) Results of random union of the gametes produced by two individuals, each heterozygous for a given trait. As a result of meiosis, half the gametes produced by each parent will carry allele *B*, the other half allele *b*. (b) Results of random union of the gametes produced by an entire *population* with a gene pool containing 80% *B* and 20% *b*.

The result, then, of random mating in such a population of hamsters would be a generation of offspring 96% of whom had black coats and only 4% gray coats. Here is a situation where intuition might suggest that gray-coated hamsters will gradually disappear. Let us see what happens, though, when this generation is mated. Sixty-four percent of all the hamsters, both male and female, will contribute only *B*-containing gametes as they are homozygous for *B*. Half of the gametes produced by the 32% heterozygous individuals, in other words 16% of the total, will also contain the *B*-allele. None of the gray hamsters will contribute *B*-containing gametes, of course. Therefore, of the entire pool of gametes available for creating the next generation, 80% will contain the *B*-allele (Fig. 32.6).

All of the gray hamsters (4% of the population) will contribute the *b*-allele to the pool of gametes. In addition, half of the gametes produced by the heterozygous individuals also contain the *b*-allele. Consequently, a total of 20% (4% + 16%) of all the gametes will contain this allele.

Bringing these gametes together at random, we find that we have duplicated our first situation exactly. Once again, a generation of 96% black hamsters and 4% gray hamsters will be formed. The proportion of allele *b* in the population has not diminished but has remained at the same level. The heterozygous individuals, although of black phenotype, help ensure that each generation will contain four gray hamsters out of a hundred.

The tables of Fig. 32.6 are a little cumbersome

for dealing with population genetics. The same results can be achieved by using the mathematical relationship $(p+q)^2 = p^2 + 2pq + q^2$. (Perhaps you remember from your study of algebra that $p^2 + 2pq + q^2$ is called the expansion of the binomial $(p+q)^2$.) The total number of genes in a population is its **gene pool.** If p represents the frequency of one gene in the gene pool and q represents the frequency of its *single* allele ($p + q$ must always equal 1—why?), the genotypes of the next generation can be quickly computed: p^2 equals the percentage of the population homozygous for the dominant gene; q^2, the percentage homozygous for its recessive allele; and $2pq$ the proportion of heterozygous individuals in the population. In our example, $p = .80$, $q = .20$, and thus

$$(.80 + .20)^2 = (.80)^2 + 2(.80)(.20) + (.20)^2$$
$$= .64 + .32 + .04.$$

This is precisely the result that we obtained by using the tables in Fig. 32.6.

Not the least advantage of the algebraic method is that we can work backward as well as forward. In fact, we were really putting the cart before the horse when we selected a population with a gene pool containing 80% *B* and 20% *b*. The only way this value can be established is by determining the percentage of the recessive phenotype and computing from it the value for q. In the case of our hamsters, $q^2 = .04$, $q = \sqrt{.04} = .20 = 20\%$, the percentage of *b*-alleles in the gene pool of the population. Since $p + q = 1$, $p = 1 - q$ and allele *B* must constitute 80% of the gene pool. Because *B* is completely dominant over *b*, we cannot distinguish the heterozygous individuals from the homozygous dominants by their phenotype. However, substituting in the middle term ($2pq$) of our expansion, we can determine the percentage of heterozygous individuals in the population:

$$2pq = (2)(.80)(.20) = .32 = 32\%.$$

In cases of *incomplete dominance* in such a polymorphic species, the three genotypes will, of course, each give a distinct phenotype.

The results of our calculations show that recessive genes do not tend to be lost from a population no matter how small their representation in the gene pool. Stated more generally, we can say that so long as certain conditions (to be discussed below) are met, gene frequencies in a randomly breeding population remain constant from generation to generation. This is known as the *Hardy-Weinberg law* in honor of the two men who first realized the significance of the binomial expansion to population genetics and hence to evolution.

What are the implications of the Hardy-Weinberg law for evolution? Certainly the mechanism described by the law is not an evolutionary one. Evolution involves changes in the gene pool, and this is precisely what does not occur with the conditions under which the Hardy-Weinberg law operates. What the law does tell us is that populations are able to maintain a **reservoir of variability** so that if future conditions require it, the genetic constitution of the population can change. If recessive alleles were continually tending to disappear from a population, the population would soon become homozygous. Such a uniform genotype would provide no variability and hence no opportunity for evolutionary change. On the other hand, under the conditions of the Hardy-Weinberg law, genes which have no present selective value will nonetheless be retained. At some future time, these genes might hold the key to survival in a changed environment, and their presence then could enable the population to avoid extinction.

32.6 WHEN THE HARDY-WEINBERG LAW FAILS TO APPLY

To see what forces lead to evolutionary change, we must examine the circumstances in which the Hardy-Weinberg law may fail to apply. There are five.

Mutation

The DNA molecule is remarkably stable, thanks in part to its redundancy and the availability of machinery to correct some of the nucleotide alterations that occur from time to time. However, *uncorrected* base shifts also occur (e.g. CG → TA), and these are mutations. In Chapter 17 we examined a number of agents that cause mutations and found that some of these, e.g. natural radiation, are unavoidable. It is difficult to estimate just how frequently a given gene mutates, but such estimates that have been made for humans usually range from 10 to 40 mutations at a given gene locus per million (10^6) gametes per generation. In other words, a mutant allele of a particular gene would be expected to appear in 10–40 of every 500,000 children born (it takes two gametes to make a child!). One reason that such estimates are difficult to make is that new mutations are apt to be recessive and thus not immediately apparent (unless they are *X*-linked, i.e. occur on the *X* chromosome and thus are expressed in males).

Why are most mutations recessive? The answer is surely historical. Every form of mutation has undoubtedly occurred repeatedly in the past. Those mutations that produced a better gene product were retained by natural selection and became part of our genetic makeup. Any mutation today will, more likely than not, produce an inferior gene product. Since we are diploid organisms, the presence of the mutant gene will usually be masked by the presence of the "normal" allele which is still making the normal gene product and thus is dominant.

Mutations are inevitable but probably exert no great effect on gene frequencies. This is because the production of harmful genes is generally counterbalanced by their removal from the gene pool by natural selection. However, such an equilibrium situation will inevitably be disturbed if the *rate* of mutation increases. Therefore, it is certainly worth while to exercise vigilance over the degree to which we expose ourselves to mutagenic agents in our environment.

Although mutation may play only a small role in evolution at any given time, it is ultimately upon mutation that evolution depends. It is only through mutation that new alleles are produced which, after being shuffled in various combinations with the remainder of the gene pool, provide the raw materials upon which natural selection can operate.

Drift

Although many species of plants and animals have large, widespread populations, the number of individuals with which any one plant or animal may mate is usually limited to those in the immediate vicinity. Insect-pollinated plants are apt to be fertilized by individuals growing relatively near to them. Despite their powers of locomotion, animals also tend to breed in a rather restricted area and hence have available a relatively small number of potential mates. Field studies show that most small mammals and insects stay close to the area in which they were born. Even birds, which may migrate over thousands of miles, usually return, season after season, to the same small area for breeding.

If geographical barriers also prevent the members of a small local population from breeding with individuals outside the locality, the local population becomes genetically isolated from the rest of its species. The gene pool of a population that combines small size and genetic isolation may not observe the Hardy-Weinberg equilibrium. Why?

The Hardy-Weinberg law is a statistical law and, in order to be valid, a sample size sufficiently large to minimize chance deviations must be taken. A hundred or fewer breeding pairs may very well not reassort their genes according to our expectations. Chance may eliminate certain individuals (e.g. homozygous recessives) *out of proportion* to their numbers in the population. In such a case, the frequencies of certain genes in the population may begin to drift toward higher or lower values. Ultimately they may represent 100% of the gene pool or, just as likely, disappear from the population entirely. This phenomenon is known as drift. Its significance to evolution was first made clear by the American geneticist Sewall Wright, who called it "random genetic drift."

Genetic drift produces alterations in the gene pool of the population, and hence it produces evolutionary change. However, unlike Darwin's idea of evolution, the new individuals in the population may very well be no more "fit" than their ancestors. Evolution by drift is aimless, not progressive.

Granted that many breeding populations of plants and animals are small enough for drift to occur, is there any evidence that it does? This is a difficult question to answer, for who can say whether an observed change in the gene pool came about because it conferred greater "fitness" on the individuals or simply as a result of drift? Although it may not invariably be true, geneticists generally feel that no combination of *A*, *B*, and *O* blood-group alleles in the human population confers any significantly greater degree of fitness than any other. Therefore, marked changes in the distribution of these alleles might indicate the operation of drift. Some years ago, a study of genetic traits, including blood group, was made of a religious sect called the Dunkers, many of whose members live in Franklin County, Pennsylvania. These people (formally known as the Old Order German Baptist Brethren) have, since the establishment of their sect in West Germany in the early eighteenth century, tended to marry exclusively within the group. In the communities studied, the number of pairs of parents in each generation has been about ninety, a breeding population small enough for drift to occur. Sixty percent of the Pennsylvania Dunkers studied had blood group A. In the United States population as a whole, the figure is 42%. In West Germany, the region from which the Dunkers emigrated, the figure is 45%. We can rule "fitness" out as a factor here because the non-Dunkers living in the same region show no such preponderance of group A blood. Some marriages outside the group have occurred, but these would naturally tend to reduce the percentage of the *A* allele in the population. Despite this tendency, the Dunkers display a distribution of blood groups found elsewhere only in such populations as Eskimos, Polynesians, and American Indians. We may confidently conclude then that drift has accounted for this unusual pattern.

The underlying cause of genetic drift is chance. Let us now examine another example of the way chance can operate to alter gene frequencies. Consider what would happen if a small group of organisms left its large parental population and served as the nucleus for starting a new population—on a previously unoccupied island, for example. The gene pool of the small founding population might well *not* be representative of the gene pool of the parent population. Thus the gene pool of the new population would be different from that of the old. This phenomenon is called the **founder effect.** There is some evidence that the first humans to reach North America (across the Bering Straits land bridge) brought with them gene frequencies not representative of the Asiatic population they left.

In Section 17.2, we examined another example of the founder effect: that of Ariaantje and Gerrit Jansz, who emigrated from Holland to South Africa in 1688, one of them bringing along a gene for the metabolic disease porphyria. Their reproductive success in their new home was sufficiently great that over 8000 descendants carry that gene today. This represents a frequency of the gene far higher than that found in the Dutch population from which it was derived.

Just how important a role drift plays in evolution is not yet known. Where drift does occur in a small population, variability in that population usually diminishes. In other words, drift is apt to lead to the complete loss of certain alleles from the gene pool. In this respect, then, drift works counter to the force of mutation. Mutation produces a new supply of alleles, a source of variability, while drift results in the elimination of all but one (*which* one cannot be predicted) at a given gene locus.

Although drift leads to loss of variability in a particular population, it may increase variability within the species as a whole. Small, isolated populations may develop traits quite different from those generally characteristic of the species. With a sudden environmental change, these new traits might enable the small group to survive while the members of the larger population succumbed.

Gene Migration

The total isolation of a local population represents an extreme that probably occurs rarely. On the other hand, there are species with large, wide-ranging populations in which distinctive, local subpopulations do not occur at all. Between these two extremes, we find many cases where local populations within a species do become established. The members of these populations generally breed only with one another, but from time to time they may breed with immigrants from adjacent populations of the same species. Although a given local population may have its own distinctive gene pool, when immigrants arrive from other populations with different gene pools, new genes are introduced. This phenomenon is called gene migration. It acts to maintain genetic variability in small populations that might otherwise lose it by drift or natural selection.

It is usually difficult to detect evidence of gene migration unless the adjacent subpopulations differ in some easily visible trait controlled by one or a few gene loci. Such a situation occurs with the dark and light forms of the peppered moth. *Biston betularia,* which differ from one another by a single gene. In general, the proportion of dark forms in an area is related to the degree of industrial pollution in that area. However, the moth population in some polluted areas has a higher proportion of light forms than would be expected. Where this situation has been analyzed, geographical barriers (such as an intervening body of water) separating that particular population from adjacent dark populations have been more effective in preventing interbreeding than the barriers separating it from adjacent light populations. Gene migration from the light population has prevented the population in question from showing the proportion of dark forms that natural selection in its habitat would normally be expected to produce. (The light population may also be affected by migration of the gene for melanism into *it.*)

Just how important gene migration has been in the evolution of animals is debatable. In plants, it seems to have played a significant role. Here we find that it occurs not only between subpopulations of the same species but, even more important, between different (but still related) species. Breeding between species is called **hybridization** and it is quite commonly observed in plants. When the hybrid formed by the mating of individuals of two different species then breeds with one of the parental types, new genes are passed into the gene pool of that parent. This phenomenon is often called introgression. It is simply a case of gene migration *between* species rather than *within* them.

Many good examples of hybridization and introgression have been studied. It occurs most often in areas where the habitat has been so disturbed (usually by gardens, roads, pastures, etc.) that hybrid individuals can survive there as well as or better than either parental type. The many "weed" species, which are so characteristic of unsettled habitats, are particularly apt to show its effects. One species of wild lettuce, a weed common along our roadsides, often shows signs of genes acquired by hybridization with the domesticated species growing in nearby gardens. As a result of repeated crossing, these genes spread through the wild-lettuce population, giving rise to increased variability as they go. Gene migration, then, in contrast to random genetic drift, brings about increased rather than decreased variability in the gene pool.

Nonrandom Mating

One of the cornerstones of the Hardy-Weinberg law is that mating in a population must be random. If certain phenotypes (and thus genotypes) are choosy in their selection of mate(s), the gene frequencies in the population may become altered. Natural selection working by means of nonrandom mating is simply another way of describing a criterion of fitness described by Darwin: sexual selection.

There is abundant evidence of nonrandom mating occurring in natural populations. The establishment of breeding territories, the use of courtship displays, and the existence of "pecking orders" may all lead to nonrandom mating. In each case, certain individuals do not get to make their proportionate genetic contribution to the next generation.

Humans often do not mate at random. Individuals of a certain phenotype may consciously or unconsciously seek mates of a similar phenotype. This is called **assortative mating** and, being nonrandom in its operation, may disturb the Hardy-Weinberg equilibrium in the gene pool. In a number of studies it has been demonstrated that humans tend to marry mates with whom they share to some degree such traits as stature, age (which can have genetic consequences such as family size or, perhaps, incidence of mutations) and even hair and eye color. The strong tendency of many ethnic groups to marry within the group is also an example of assortative mating.

A special case of assortative mating is marriage between relatives. The closer the relationship, the more genes shared and the greater the degree of inbreeding. While inbreeding by itself need not shift the gene pool, as a practical matter it often does so. This is because continued inbreeding predisposes to *homozygosity*. Potentially harmful recessive alleles—invisible in the parents—will exert their harmful effects on those offspring who inherit one from each parent. Thus inbreeding exposes such alleles to the other forces of natural selection.

Differential Reproduction

The fifth circumstance under which the Hardy-Weinberg law fails to apply occurs when differential reproduction takes place. If individuals having certain genes are better able to produce mature offspring than those without them, the frequency with which those genes appear in the population will increase. This is simply a modern way of defining natural selection: the alteration of gene frequencies in the gene pool as a result of differential reproduction.

Differential reproduction results from differential mortality and/or differential fecundity.

1. Differential mortality. Certain genotypes are less successful than others in surviving through their period of reproductive activity. The evolutionary effects of differential mortality can be felt anytime from the formation of a new zygote to the end (if there is one) of the organism's period of fertility. Natural selection working by means of differential mortality is simply another way of describing one of Darwin's criteria of fitness: survival.

Is there evidence of natural selection occurring in human populations? The distribution of the gene for sickle-cell hemoglobin (Hb^S) provides one clear example. When present in a double dose (i.e. homozygous), the gene is sufficiently lethal that its high incidence in central Africa at first seems inexplicable. But as you learned in Section 18.2, the red blood cells of heterozygous individuals, i.e. those having one sickle-hemoglobin gene and one gene for normal hemoglobin (Hb^A), contain *both* hemoglobins. This confers greater protection against one type of malaria (falciparum) common in these regions than that enjoyed by people homozygous for Hb^A. Whenever a heterozygous genotype confers a greater selective advantage than either homozygous genotype, the result is a situation of **balanced polymorphism.**

2. Differential fecundity. Another way in which certain phenotypes (and genotypes) may make a disproportionate contribution to the gene pool of the next generation is by producing a disproportionate number of young. Natural selection working by means of differential fecundity is simply another way of describing a third criterion of fitness described by Darwin: family size.

For each of these cases, the outcome is differential reproduction. In each case, certain phenotypes are better able than others to contribute their genes to the next generation. Thus, by Darwin's standards, they are more "fit." The result is a gradual change in the gene frequencies in that population. And that is evolution—in these cases, evolution by natural selection.

32.7 KIN SELECTION

So far we have emphasized the way in which natural selection works to favor the more fit and disfavor the less fit in a population. Our emphasis has been on the survival, or mating success, or fecundity of *individuals*. But what of the worker honeybee who gives up her life when danger threatens her hive? Or the mother bird who, feigning injury, flutters away from her nestful of young, thus risking death at the hands of a predator? How can evolution produce the genes for such instinctive patterns of behavior when the owners of these genes risk failing the first test of fitness: survival?

A possible solution to this dilemma lies in the effect of such altruistic behavior on the overall ("inclusive") fitness of the family of the altruistic individual. Linked together by a similar genetic endowment, the altruistic member of the family enhances the chance that many of its own genes will be passed on to future generations by sacrificing itself for the welfare of its relatives. It is interesting to note in this connection that most altruistic behavior is observed where the individuals are linked by fairly close family ties. Natural selection working at the level of the family rather than the individual is called **kin selection.**

How good is the evidence for kin selection? Does the behavior of the mother bird really increase her chances of being killed? After all, it may be advantageous to take the initiative in an encounter with a predator that wanders near. But even if she does increase her risk, is this anything more than another example of *maternal* behavior? Her children are, indeed, her kin. But isn't natural selection simply operating in one of the ways Darwin described: to produce larger mature families?

Perhaps a clearer example of altruism and kin selection is to be found in the Florida scrub jays (*Aphelocoma coerulescens coerulescens*). These birds occupy well-defined territories. When they reach maturity, many of the young birds remain for a time (one to four years) in the territory and help their parents with the raising of additional broods (Fig. 32.7). How self-sacrificing! Should not natural selection have produced a genotype that leads its owners to seek mates and start raising their own families (to receive those genes)? But the idea of kin selection suggests that the genes guiding their seemingly altruistic behavior have been selected for because they are more likely to be passed on to subsequent generations in the bodies of an increased number of younger brothers and sisters than in the bodies of their own children. To demonstrate that this is so, it is necessary to show (1) that the "helping" behavior of these unmated birds is really a help and (2) that they have truly sacrificed opportunities to be successful parents themselves.

Thanks to the patient observations of Glen Woolfenden, the first point is established. He has shown that parents with helpers raise larger broods than those without. But the second point still remains unresolved. Perhaps by waiting (1) until they have gained experience with guarding nests and feeding young and (2) until a suitable territory becomes available, these seemingly altruistic helpers are actually improving their chances of *eventually* raising larger families than they would have if they had started right at it. If so, then once again we are simply seeing natural selection working through one of Darwin's criteria of individual fitness: ability to produce larger mature families.

The honeybees and other social insects provide the clearest examples of kin selection. They are also particularly interesting examples because of the peculiar genetic relationship among the family members. You remember that male honeybees develop from the queen's unfertilized eggs and are haploid. Thus all their sperm will contain exactly the same set of genes. This means that all their daughters will share exactly the same set of *paternal* genes, although they will share—on average—only one-half

FIG. 32.7 Florida scrub jays defending their nest (which contains four young hidden under the mother). The father is crouched on top of the mother. Their helper is a two-year-old bachelor at the right. Most helpers are prior offspring of the mated couple, but this one is a brother of the breeding male. The young are thus being defended by their uncle rather than, as is more commonly the case, an older brother. (Courtesy of Glen E. Woolfenden and John W. Fitzpatrick, Archbold Biological Station.)

of their mother's genes. (Human sisters, in contrast, share one-half of their father's as well as one-half of their mother's genes.) So any behavior that favors honeybee sisters (75% of genes shared) will be more favorable to their genotypes than behavior that favors their children (50% of genes shared.) Since that is the case, why bother with children at all? Why not have most of the sisters be sterile workers, caring for their mother as she produces more and more younger sisters, a few of whom will someday become queens? As for their brothers, worker bees share only 25% of their genes with them. Is it surprising, then, that as autumn draws near, the workers lose patience with the lazy demanding ways of their brothers and finally drive them from the hive?

32.8 THE EFFECTS OF SELECTION ON POPULATIONS

The pressure of natural selection may affect the distribution of phenotypes in a population in several ways.

1. Stabilizing selection. Earlier in the chapter, in discussing continuous variation, we examined the distribution of tail lengths in a population of deer mice. We found that plotting tail length (on the *x*-axis) as a function of frequency of that length in the population (on the *y*-axis) produced a bell-shaped curve (Fig. 32.2a). The curve tells us that the majority of mice in the population have tails close to the average and only a minority have either very long or very short tails. Frequently natural selection works to weed out individuals at both extremes of the range of phenotypes while enhancing the reproductive success of those phenotypes near the mean (Fig. 32.8). In such cases, natural selection is a conservative force working to maintain the *status quo*. Just why both long tails and short tails should be disadvantageous to mice is not clear. Possibly such factors as attractiveness to mates, ease of movement, or liability to predation are involved. But the general principle is well documented. In humans, for example, the incidence of infant mortality is higher for very heavy as well as for very light babies. Thus babies of average birth weight are selected for, and babies at either extreme are selected against.

The balanced polymorphism created by the superior fitness of heterozygotes provides another example. In the malarial regions of Africa, individuals homozygous for Hb^A are selected *against* (because of their greater susceptibility to falciparum malaria)

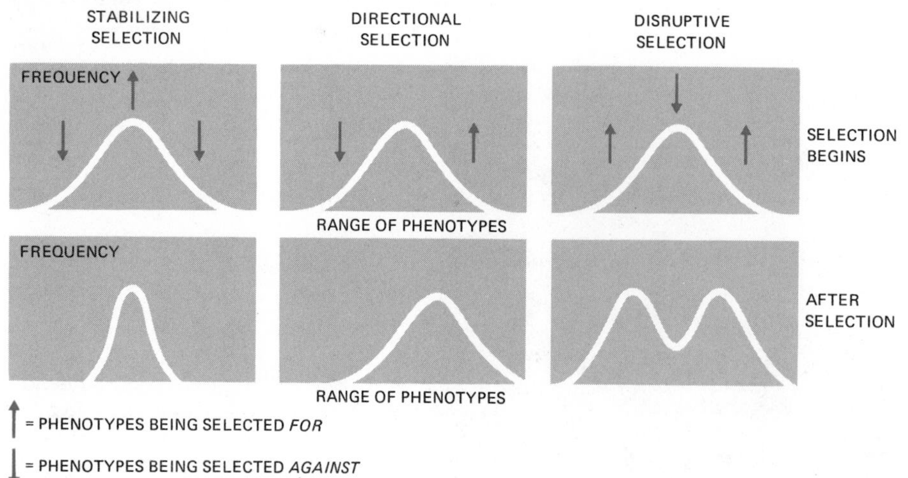

FIG. 32.8 Three ways in which natural selection can alter the distribution of phenotypes in a population. In each case, the horizontal axis represents the range of variation of the trait being considered; the vertical axis represents the number of individuals in the population at that place in the range. Left: Stabilizing selection works against individuals that are extreme examples of the trait under selection. Balanced polymorphism (see Section 32.6) is one example of stabilizing selection. Middle: Directional selection favors phenotypes at one end of the range, leading to a gradual shift in the distribution of phenotypes in the population. An example of directional selection is shown in Fig. 32.13. Right: Disruptive selection favors the extreme types over the intermediate forms. It may lead to a splitting of the population into two subpopulations.

as are individuals homozygous for Hb^S (because of the severe red blood cell destruction of sickle-cell anemia). Heterozygous individuals (Hb^A and Hb^S) are selected *for* because of their increased resistance to malaria coupled with only a mild anemia.

2. Directional selection. A population may find itself in circumstances where individuals occupying one extreme of the range of phenotypes are favored over all others. This might occur as a result of a change in the physical environment. The air pollution caused by the industrial revolution in Great Britain led to the evolution of darker populations in scores of moth species—the industrial melanism we examined in the last chapter. Later in this chapter, we shall examine the way in which competition for food between two of Darwin's finches resulted in the selection of individuals at one extreme of beak size. This shift in phenotype is called *character displacement*. It is a result of directional selection (Fig. 32.8). The principle of selective breeding illustrated by Fig. 32.4(b) is based on the deliberate use of directional selection. Directional selection, then, is a dynamic force. It leads to a progressive change in genotypes and therefore to evolutionary change.

3. Disruptive selection. There appear to be certain circumstances where individuals at both extremes of a range of phenotypes are favored over those in the middle. This is called disruptive selection (Fig. 32.8). Its evolutionary significance lies in the fact that it may lead to the splitting of a single gene pool into two distinct gene pools. This may be *one* way in which new species are formed. (We shall examine another later in the chapter).

The residues ("tailings") of mining operations often contain such high concentrations of toxic metal ions (e.g. copper, lead) that most plants are unable to grow on them. However, some hardy species, e.g. certain grasses, are able to spread from the surrounding uncontaminated soil onto such waste heaps. Examination of these plants shows that they have evolved a high degree of resistance to the toxic ions while at the same time evolving a reduced ability to grow on uncontaminated soil. Because the grasses are wind pollinated, breeding between the resistant and nonresistant populations goes on. But evidently disruptive selection is at work. A higher death rate of *less resistant* plants growing on *contaminated* soil coupled with a higher death rate of *more resistant* plants growing on *uncontaminated* soil leads to increasing divergence of the population into two subpopulations displaying the extreme manifestations of this trait (Fig. 32.8).

32.9 CAN HUMANS DIRECT THEIR OWN EVOLUTION?

If we have been so successful at manipulating the evolution of our domestic animals, what about doing the same to ourselves? Can we improve the human species by using the techniques of domestication? As mentioned in Section 32.1, one requirement for the successful production of animals and plants with new characteristics is the presence of inherited variability. And we have abundant evidence of a vast reservoir of genetic variability in humans. So the ingredients are there.

But successful animal breeding requires techniques that even though they *could* be applied to ourselves are unlikely (we should hope) to be so. In order to develop traits that breed "true," i.e. to develop homozygosity for the genes being selected for, animal breeders must practice extensive inbreeding. Most human societies have shunned close inbreeding. While the motivation behind this antipathy is debatable, it has at least had the advantage of reducing the number of offspring homozygous for, and thus affected by, deleterious alleles. Animal breeders cope with this problem by ruthlessly culling all offspring that fail to meet the standards sought. In human terms, then, application of the techniques of animal breeders would mean dictating who could mate with whom and, for most of their offspring, preventing their breeding.

Are there less stern measures that could be taken to improve the human gene pool? The application of genetic knowledge to improving the human species is called **eugenics.** Eugenic proposals can be subdivided into positive and negative measures. Positive eugenic measures involve promoting the breeding of phenotypes that are considered desirable, i.e. promoting assortative mating between those with high IQs or what have you. Some tools for such procedures are already at hand. Many women who cannot conceive by their husbands have been artificially inseminated with sperm from unknown donors. The choice of donor can, of course, be made on the basis of certain desired traits. The rapid growth of sperm banks may greatly increase the potentiality of this procedure as the donors will no longer have to be nearby or even alive at the time artificial insemination is carried out. However, sperm, fresh or frozen, carry recessive alleles as well as dominant ones. It seems unlikely that—barring a degree of coercion that we should find unthinkable—any substantial improvement in the human gene pool could be made by such measures even if we could agree as to what constitutes improvement.

Negative eugenic proposals are concerned with preventing the owners of undesirable genes from passing them on to future generations. The experience in Massachusetts and elsewhere has indicated that if babies that are born afflicted with PKU (see Section 18.2) are identified early and raised on a diet in which the intake of phenylalanine is restricted, the mental retardation so characteristic of the disease can be avoided. Should such people later have children? The answer might be "no" if we are concerned with the personal demands on the parents of trying to raise successfully their own PKU baby (the other parent would have to have been heterozygous for the trait). But should we invoke the welfare of the human gene pool in this question?

Approximately one baby in 10,000 is born with PKU. Assuming that the gene is in equilibrium in the gene pool, this tells us that one gene in every 100 in the gene pool is a defective allele:

$$q^2 = 0.0001, \quad q = 0.01.$$

Ninety-nine genes out of every hundred in the gene pool are the normal allele:

$$p = 1 - 0.01 = 0.99.$$

Thus the proportion of heterozygotes in the population is 0.0198:

$$2pq = (2)(0.01)(0.99) = 0.0198,$$

or about *one person in every 50.* This means that the gene pool of the United State population contains some four million PKU alleles. Preventing each year's crop of PKU victims from reproducing when they reach maturity would have a vanishingly small effect on the gene pool and thus on the appearance of new victims. In fact, it would probably only keep up with the appearance by mutation of new PKU genes.

But what if we prevented *carriers* of the gene from reproducing? Although PKU is a recessive trait as far as the *health* of the heterozygotes is concerned, it is now possible in many cases to identify heterozygotes by the relative inefficiency with which they remove phenylalanine from their blood (Fig.

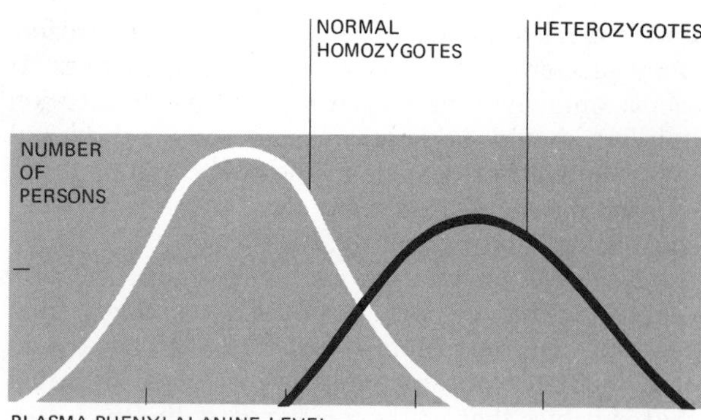

FIG. 32.9 The phenylalanine tolerance test. A short time after administering a measured amount of phenylalanine to the subject, the concentration of phenylalanine in the blood plasma is determined. The level is usually substantially higher in people who carry one PKU gene (even though they show no signs of disease) than in individuals who are homozygous for the normal gene. Both parents must be heterozygous for the trait to produce a PKU child. The chance of their doing so is 1 in 4.

32.9). Preventing the breeding of the one person in 50 that is heterozygous would, indeed, begin to purge the gene pool of the PKU genes. But what of heterozygotes for galactosemia and the numbers of other genetic ailments humans are heir to? It has been estimated that each one of us is a carrier of several deleterious alleles. The application of breeding restrictions to heterozygotes would thus eliminate us all from participating in the reproductive process.

So even if we instituted strong self-imposed selection against certain alleles, the heterozygous individuals in the population provide an enormous reservoir that would frustrate our efforts. But the matter is more complex even than that. As we have seen with the gene for sickle-cell hemoglobin (Hb^S), heterozygous individuals (i.e. those possessing one gene for Hb^S and one for Hb^A) are—in malarial regions—better fit than *either* homozygote. Individuals homozygous for Hb^S die young from an overwhelming anemia but individuals homozygous for "normal" hemoglobin (Hb^A) are far more susceptible to disability and death from falciparum malaria than are heterozygous individuals. This, then, explains the high incidence of the gene in malarial regions—a case of balanced polymorphism.

As analytical techniques have improved, it has become evident that humans (as well as many other creatures) exhibit polymorphisms at a substantial number of their gene loci—perhaps 20–30% of them. Some geneticists feel that these simply represent an accumulation of mutations that are selectively neutral, i.e. neither selected for nor against. Others, however, think that the maintenance of such high levels of heterozygosity suggests balanced polymorphism, that is, a selective advantage of heterozygotes over homozygotes. If this is the case, any positive eugenic programs run the real danger of reducing overall fitness. Inbreeding or other assortative mating necessary to produce some sought-for uniformity will, at the same time, lead to homozygosity. At the very least, this would reduce the diversity of the gene pool by exposing recessive alleles to the full rigors of selection. And, perhaps, it would also lower overall fitness by eliminating the advantages of balanced polymorphisms. In this connection, both plant and animal breeders have long realized that continued inbreeding reduces the vigor of the population. But, when two inbred strains are then allowed to breed together (outbreeding), the offspring—now heterozygous at many gene loci—are of unusual vigor. This effect of *hybrid vigor* is so dramatic that it is now routinely exploited. Most of the corn planted in the United States represents hybrid varieties produced by mating two inbred strains.

Can hybrid vigor be demonstrated in humans? Probably most human populations are sufficiently heterozygous that the phenomenon does not occur. However, there is evidence that when members of some small, highly inbred groups (like the Dunkers —Section 32.6) mate with people outside the group, their offspring do display increased stature and other signs of enhanced vigor.

Perhaps, then, the greatest danger inherent in most eugenic proposals is the possibility of reducing our genetic diversity. It may well be our most precious asset today and, at the very least, it provides the raw material for successfully coping with any changes in the pressures of natural selection that may occur in the future.

32.10 THE EFFECTS OF INCREASED SELECTION PRESSURE

The vigor with which the forces of natural selection operate varies from time to time and place to place. Other species besides humans undergo both "hard times" and "easy times." When times are hard, the forces of natural selection operate with greatest efficiency. We describe this as increased selection pressure. It arises from one or more of several causes. An increase in the number or efficiency of its predators challenges the existence of a species. More frequently, perhaps, its existence may be threatened by other species that compete with it for food, nesting areas, sunshine and water (in the case of plants), or some other essential aspect of its environment. If the population of the species is large, its members are also thrown into direct competition with each other for the necessities of life. Furthermore, a crowded population is more susceptible to parasitism by some other species, and a devastating epidemic may occur. An increase in the severity of any or all of these conditions constitutes an increase in selection pressure. The result of increased selection pressure is a reduction in the variability of the species. Predators and parasites will weed out the weak, the poorly camouflaged, or those who in any other respect are less well adapted to their environment. Only those genotypes in the population that confer the best camouflage, the most effective offensive or defensive weapons, the most efficient food-gathering structures, the greatest resistance to disease, etc., will perpetuate themselves successfully. The deviant, less efficient genotypes will be the first to suffer from increased selection pressure and will become less common.

The usual outcome of increased pressure is that the species becomes more narrowly specialized in its requirements and hence more narrowly specialized in its structural, physiological, and behavioral adaptations. The species comes to occupy an ever-narrower portion of its habitat but exploits its diminished share more effectively. The moth with the 25-cm proboscis which pollinates the tropical orchid with the 25-cm nectary has certainly followed this evolutionary trend. Most parasites also show an extreme degree of specialization in a very limited environment (Fig. 14.2).

Such an evolutionary path may provide temporary success for the species and a resulting increase in the size of the population. If the trend to increased specialization continues, however, the species commits itself to a rather uncertain future. You can well imagine what would happen to the moth with the 25-cm proboscis if its orchid should succumb to environmental changes. The study of paleontology reveals a large number of extinct animals whose fossil remains suggest that they were extremely specialized. In most cases, we can only speculate as to what environmental change led to their downfall. The principle seems clear, though. Any species that becomes very uniform and increasingly efficient at exploiting an ever-narrower aspect of its environment runs the risk of extinction when its environment changes.

32.11 THE EFFECTS OF RELAXED SELECTION PRESSURE

The evolutionary picture changes when selection pressure is relaxed. When times are easy, a species can increase its population without at the same time becoming more specialized.

Reduction in the severity of predation or parasitism or an increase in available food enable deviant genotypes to compete on more nearly equal terms with their better-adapted cousins. A sudden population explosion may occur and with it a marked rise in the variability within the species. History tells us that human societies develop their most flourishing culture in times of prosperity. So, too, relaxed selection pressure enables other populations to experiment with new, perhaps inefficient, genotypes. The genetic variability permitted by sexual reproduction now has a chance to express itself fully.

The evidence of both the past and the present suggests that the most frequent cause of relaxed selection pressure is entry into a new environment which is relatively free of (1) competing species, (2) predators, (3) parasites, or all three. When the ancestor of Darwin's finches reached the Galapagos Islands, it found at most a handful of songbird species established there. With few reptiles or birds of prey and probably no mammals present, predation must have been practically nonexistent, too. Only under such unusual circumstances were a warblerlike and a woodpeckerlike finch able to evolve. If true warblers or woodpeckers had been present in the islands, we can be sure that their efficiency would have prevented the appearance of competing forms among Darwin's finches.

In more recent times, the introduction of rabbits into Australia and destructive crop insects into this country have provided just two of many dramatic examples of the explosive success a species can achieve when introduced into a new environment where selection pressure is reduced. Unfortunately, there have not been sufficient field studies to prove that reduced selection pressure goes hand in hand with increased intraspecific variability, but there is some evidence to support this.

The Origin of Species

The idea of evolution involves two processes. First is the gradual change in genotype and phenotype of a population of living organisms. Usually these changes are adaptive; that is, the organisms become increasingly efficient at exploiting their environment. Second is the formation of new species. Assuming that life has arisen only once on the earth, the 1.2 million known species of microorganisms, plants, and animals living today (not to mention all the species that have become extinct) must have arisen from ancestors that they shared in common. So a theory of evolution must tell us not only how organisms become better adapted to their environment but also how *new* species are produced.

32.12 WHAT IS A SPECIES?

The zoologist Ernst Mayr defines a species as an actually or potentially interbreeding natural population which does not interbreed with other such populations even when there is opportunity to do so. We must qualify this somewhat by adding that if on rare occasions breeding between species does take place, the offspring produced are not so fertile and/or efficient as either of the parents. Although a horse and donkey can breed together, the mule resulting is sterile. (In plants, even this restriction sometimes fails to apply if conditions in the habitat have altered. Plant hybrids *may* be more successful than either parent in such areas.)

It seems quite clear that the process of evolutionary change in a population and the process of species formation are related. Over a period of time, the accumulation of changes in the gene pool of a population must reach a point where we may say that a new species has been formed. This involves a purely arbitrary judgment, however. Who is to say just when the transition was made from one species to the next? Even if we could resurrect some of the ancestral forms to see if they could breed successfully with their modern descendants, our question would remain unanswered. An unbroken line of forms stretches back in time from each living species, and the breaking up of this line into distinct species is an entirely arbitrary (although useful) operation.

Evolution has not, however, been just a matter of gradual change in a *single* genealogical line. The fossil record tells us that there has been a marked expansion in the number of species present on the earth. To put this another way, we believe that all our present species have diverged from common ancestors and initially from a single, first form of life. What can the theory of evolution tell us about **speciation,** that is, the formation of many species from few?

32.13 THE ROLE OF ISOLATION IN SPECIATION

The first requirement for speciation is reduced selection pressure and the intraspecific variability it permits. But this is only the start of the story. Even with increased variability in a species, there is a continuous intergrading of traits from one extreme to the other. The formation of two or more species from one requires that gaps appear in the population where no intermediate forms exist. For this to occur, some form of geographical isolation of the various subpopulations seems essential. Only then can natural selection or perhaps genetic drift produce a definite shift away from the gene frequencies of the parental type. It is no accident that the various races or subspecies of living organisms almost never occupy the same territory. The seven distinct phenotypes (and thus genotypes) associated with the seven American subspecies of the Yellowthroat *Geothlypis trichas* (Fig. 32.10) would quickly merge into one if these subspecies occupied the same territory and interbred with one another. Only in isolation can distinctive genetic differences become established.

Darwin's finches illustrate this point nicely. Not only did the absence of competitors and predators on the Galapagos Islands provide a greatly relaxed selection pressure for the ancestral bird that reached them, but the relative isolation of the various islands permitted the establishment of unique island races or subspecies. From these have arisen the thirteen species which inhabit the islands today (Fig. 32.11). Many of these species are now, in turn, made up of several distinct subspecies isolated on the various islands.

The importance of geographical isolation in the formation of species is vividly illustrated by the single finch species that is found on Cocos Island, some 500 miles to the northeast of the Galapagos. The ancestral bird arriving there must also have found a very relaxed selection pressure with few predators or competitors. How different the outcome here than in the Galapagos, though! One species gave rise to thirteen in the various Galapagos Islands, but no such divergence has occurred on Cocos Island. One finch species is still all that exists there.

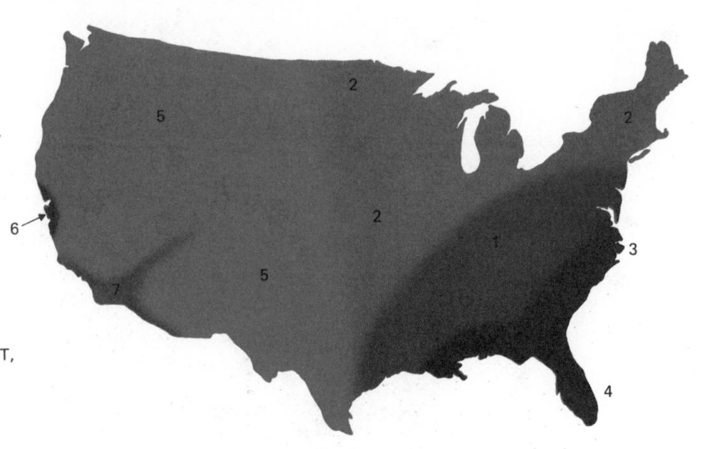

1. MARYLAND YELLOWTHROAT, *Geothlypis trichas trichas*
2. NORTHERN YELLOWTHROAT, *G. t. brachidactyla*
3. ATHENS YELLOWTHROAT, *G. t. typhicola*
4. FLORIDA YELLOWTHROAT, *G. t. ignota*
5. WESTERN YELLOWTHROAT, *G. t. occidentalis*
6. SALT MARSH YELLOWTHROAT, *G. t. sinuosa*
7. TULE YELLOWTHROAT, *G. t. scirpicola*

FIG. 32.10 Geographical distribution of the seven subspecies of *Geothlypis trichas*. As a general rule, two subspecies of a single species do not breed in the same territory.

FIG. 32.11 Darwin's finches. The finches numbered 1–7 are ground finches. They seek their food on the ground or in low shrubs. The finches numbered 8–13 are tree finches. They live primarily on insects. 1. Larger cactus ground finch. 2. Large ground finch (*Geospiza magnirostris*). 3. Medium ground finch (*G. fortis*). 4. Cactus ground finch. 5. Sharp-beaked ground finch. 6. Small ground finch (*G. fuliginosa*). 7. Woodpecker finch. 8. Vegetarian tree finch. 9. Large insectivorous tree finch (*Camarhynchus pauper*). 10. Large insectivorous tree finch (*Camarhynchus psittacula*). 11. Small insectivorous tree finch. 12. Warbler finch. 13. Mangrove finch. (From Biological Sciences Curriculum Study, *Biological Science: Molecules to Man,* Houghton Mifflin Co., 1963.)

The distribution of cottontail rabbits in this country provides still another example of the role played by geographical isolation in the formation of species. In the eastern half of our country, only eight species are found, while in the mountainous western states there are 23. In this case, high mountains have provided geographical isolation as effective as the ocean water between the various Galapagos Islands.

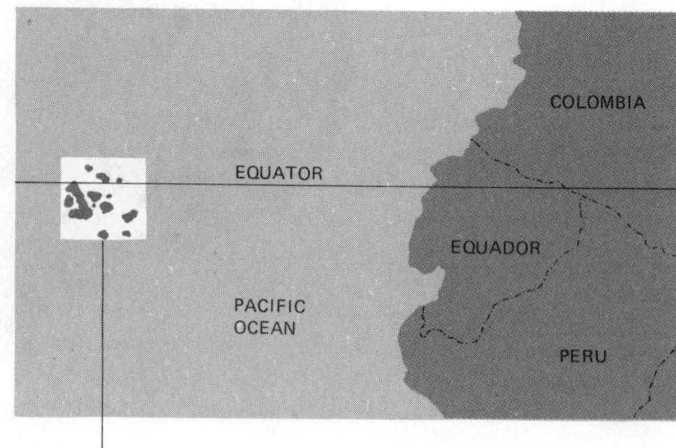

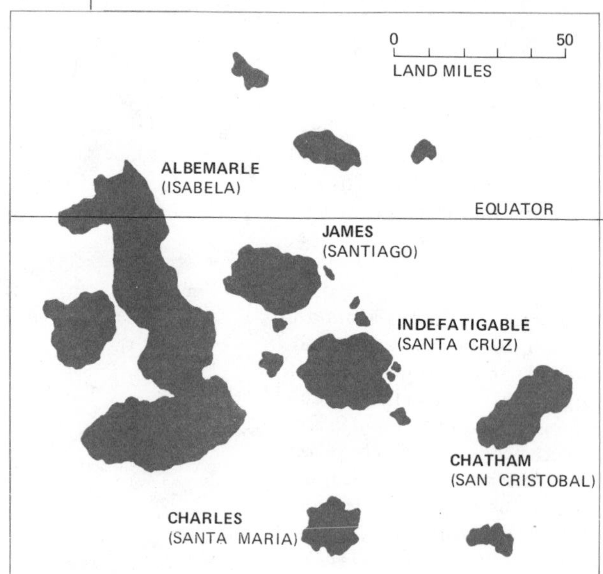

FIG. 32.12 The Galapagos Islands. Now belonging to Ecuador, each island has been renamed. The modern names are shown in parentheses.

32.14 REUNION

We have learned that consistent, distinctive differences in phenotype (and thus genotype) are found within subpopulations of a species that are somewhat isolated from one another. The island races of some of Darwin's finches and the races of *Geothlypis trichas* are cases in point. So long as these forms remain geographically separate, we prefer to consider them different subspecies of the same species. If brought together, they might not, in fact, breed together, but so long as we lack this information, we assume that they would. The question of their status, if they ever do come to occupy the same territory again, is soon answered. If successful interbreeding occurs, the differences will gradually disappear, and a single population with continuously intergrading traits will be formed again. Speciation will not have occurred. If, on the other hand, the two reunited subpopulations do not interbreed successfully, then speciation will have occurred.

This is well illustrated by the large insect-eating, tree-dwelling finches on the Galapagos, *Camarhynchus pauper* and *Camarhynchus psittacula.* Of the two, *C. pauper* appears to be more primitive in its traits and thus the earlier form. Today it is found only on Charles Island (Fig. 32.12). The closely related *C. psittacula* is found on all the central islands, including Charles. Were it not for its presence on Charles, both forms would be considered subspecies of the same species. Because they do coexist and maintain their separate identity on Charles, we know that speciation has occurred. It looks as though *C. psittacula* evolved in isolation from *C. pauper* to the point where merging of the two groups did not occur when *C. psittacula* recolonized Charles Island.

What might keep two subpopulations from interbreeding when reunited geographically as *C. psittacula* and *C. pauper* have been on Charles Island? At first it may simply be a matter of one group's lacking the releasers to evoke mating behavior in the other. On Charles, *C. psittacula* has a longer beak than *C. pauper* and the evidence indicates that Darwin's finches choose their mates primarily on the basis of beak size. In other birds, plumage may be the big factor.

In the course of time, other factors that tend to prevent interbreeding may arise. The two groups may come to occupy different habitats in the same area and thus fail to meet at breeding time. The timing of reproductive activity in the two groups may not be the same. Differences in time of flowering, for example, may prevent otherwise closely related species of angiosperms from interbreeding. Structural differences in the sex organs may also be an isolating mechanism. Finally, genetic differences between the groups may become so great that even if mating should occur, any hybrids produced are less viable or less fertile than either parental type. At this point, all would agree that speciation had occurred.

This evolutionary process may be hastened when two formerly isolated groups are reunited. Even if they no longer interbreed, they probably are still similar in many ways, including their requirements for the necessities of life. Thus the reuniting of the two groups may create an intense selection pressure, so intense in fact, that one species is eliminated entirely. Perhaps this has happened to *C. pauper* on some of the central Galapagos Islands. On the other hand, increased selection pressure may, as we have seen, lead to character displacement (see Section 32.8) and so lessen the competition between them.

The range in beak sizes of *C. pauper* on Charles and *C. psittacula* on Albemarle is roughly the same. This probably means that these two groups feed on the same type and size of insect. On Charles, however, character displacement has occurred: *C. psittacula* has a substantially larger beak than *C. pauper* and thus presumably differs enough in its food requirements for the two species to coexist there (Fig. 32.13).

It would be rash to conclude from the study of Darwin's finches that speciation in all organisms has occurred in exactly the same way. But his finches do illustrate beautifully what are probably the absolute essentials for the process. They are:

1. Reduced selection pressure (in this case achieved by moving into a new, unoccupied territory).
2. Increased variability within the species (arising from the process of sexual reproduction, working with genes no longer being selected against).
3. Isolation of a subpopulation of the species so that interbreeding with the main stock is prevented. This produces geographic races or subspecies.
4. Reuniting of the isolated group with the parental stock but without the resumption of interbreeding. At this point, true speciation has occurred.
5. Intense competition between the two reunited forms, so that further divergence in their various traits is hastened.

This process need not occur simply once. Indeed, in the case of Darwin's finches, we are sure that it has been repeated several times, giving rise to new species, which gradually have divided the available habitats between them. From the first arrival have come a variety of ground-feeding and tree-feeding finches including a warblerlike finch and the extraordinary, tool-using, woodpeckerlike species (Fig. 32.11). The production of a number of diverse species from a single ancestral one is often referred to as **adaptive radiation.** We shall study several other well-illustrated examples in later chapters.

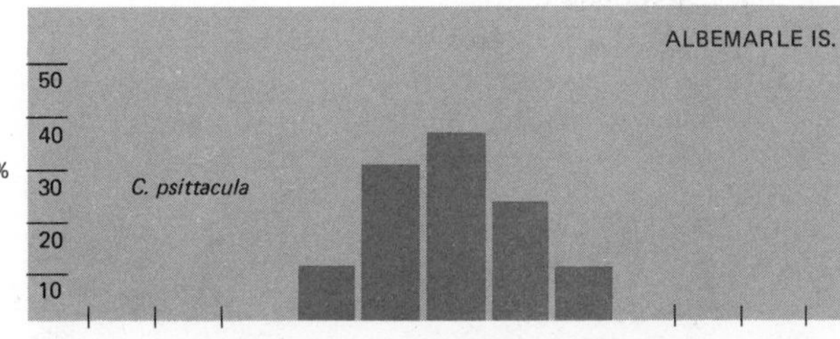

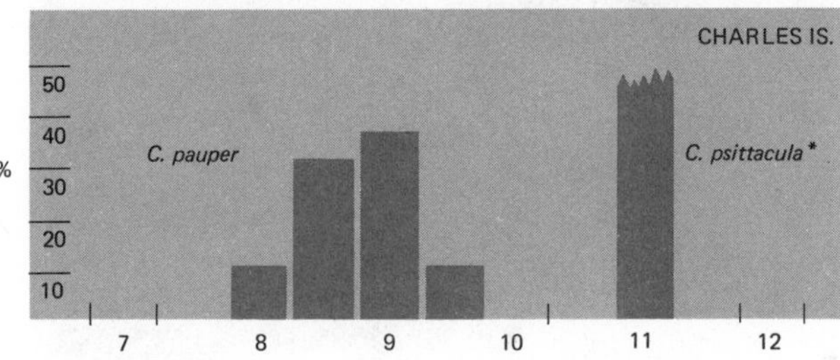

*The number of collected specimens—16—was too small to compute reliable percentages, but all fell within the size range indicated.

FIG. 32.13 Histograms of depth of beak in *C. pauper* and *C. psittacula.* The larger beak size of *C. psittacula* on Charles Island is an example of character displacement. Presumably it reduces competition with *C. pauper* (which is not found on Albemarle Island). (Redrawn by permission from David Lack, *Darwin's Finches,* Cambridge University Press, 1947.)

32.15 SPECIATION BY POLYPLOIDY

Speciation, as we have described it, is a gradual process. It begins with differences between individuals in a population, proceeds to the accumulation of consistent differences found at the subspecies level, and ends with the fixed differences associated with species formation. The entire procedure may take hundreds or thousands of years to be completed. There is one kind of species formation, however, which occurs much more rapidly. This arises from the chromosomal aberration of polyploidy (see Section 16.4).

Meiosis in polyploid (for example, $4n$) individuals produces gametes ($2n$) that are fertile with one another but often not with gametes of the parental type (n). Thus polyploid individuals are immediately isolated genetically from the parental type and a new species has been created almost overnight. In 1927, the Soviet plant geneticist Karpechenko reported the artificial production of a plant containing a full ($2n$) set of radish chromosomes and a full ($2n$) set of cabbage chromosomes produced as a result of crossing these two species (which are from different genera). Ordinarily, hybrids of such different parents would be sterile since their gametes would contain only a partial assortment of the chromosomes from each of the two parents. However, Karpechenko's hybrid being polyploid, each gamete contained one of each of the chromosomes found in both the radish and the cabbage ancestors. Fertilization produced vigorous plants which combined traits of both cabbages and radishes (unfortunately, the roots of the cabbage and the leaves of the radish). These plants were capable of breeding successfully with one another but not with either ancestral type. Karpechenko had thus produced in a very brief period a new species of plant. There is abundant evidence that speciation by polyploidy, with or without hybridization, has often occurred naturally in plants. The phenomenon is, however, very rare in animals.

32.16 CONVERGENT EVOLUTION

We have noted that species (e.g. bird and bat, whale and fish, woodchuck and wombat) are sometimes found which resemble each other superficially but

FIG. 32.14 Convergent evolution. The flying squirrel (top right) and woodchuck (bottom right) are more closely related to you than they are to the marsupials on the left. (This woodchuck, or groundhog, is a rare albino form.) (Courtesy of the New York Zoological Society.)

FLYING PHALANGER

FLYING SQUIRREL

WOMBAT

WOODCHUCK

fail to reveal the homologies that would indicate close kinship. Evolution as a result of which two species of different genealogy come to resemble one another closely is termed *convergent evolution.* It can be explained on the basis of the same forces of natural selection working in a similar way on two originally different phenotypes. There are certain structural and physiological requirements that must be met before any organism, no matter what its ancestry, can fly or swim.

Convergent evolution is in no sense the opposite of speciation. While two unrelated species may come to resemble one another closely as similar selective forces work on them, each species is, at the same time, diverging from its own ancestral stock. The many Australian marsupials which resemble placental mammals in both appearance and habits (Fig. 32.14) illustrate convergent evolution with respect to the placental mammals. With respect to the ancestral marsupials, however, they represent a most dramatic example of adaptive radiation, in other words, the multiplication of species.

EXERCISES AND PROBLEMS

1. The Kaibab squirrel has long ears and a white belly and lives on the north rim of the Grand Canyon. The Abert squirrel, which lives on the south rim, is quite similar in appearance although it has a black belly.
   a) How can you account for the presence of these two similar, but distinguishable, populations?
   b) What possible futures might await these squirrels if the canyon should be obliterated?
2. Fossil evidence indicates that evolutionary changes in honeybees have taken place unusually slowly. Can you think of a possible reason why natural selection might operate particularly slowly in this group?
3. Most of the so-called "living fossils" are found in the sea. Can you think of a reason for this?
4. How can you account for the superficial resemblance of a porpoise and a fish?
5. What term would Darwin have applied to differential reproduction?
6. List three mechanisms by which variant genotypes are produced in nature.
7. Which of these is denied to a hermaphroditic organism that practices self-fertilization exclusively?
8. What phenomenon must occur for the mechanisms in Question 6 to be effective in producing variant genotypes?
9. The proportion of homozygous recessives of a certain trait in a large population is 0.09. Assuming that the gene pool is in equilibrium and all genotypes are equally successful in reproduction, what proportion of heterozygous individuals would you expect to find in the population?
10. Why are most mutations harmful?
11. Considering the strong selection pressure against the gene for hemophilia, what reasons can you give for the continuing presence of the gene in the human population?
12. What is the genetic effect of continued inbreeding in a population?
13. Why does continued inbreeding often lower the vigor of the individuals in a population?
14. Define a species.
15. How can the Darwinian theory of evolution account for the long neck of the giraffe? How does this differ from Lamarck's explanation?
16. Approximately 36% of a certain population is unable to taste the compound phenylthiocarbamide (PTC). Everyone else finds it very bitter. Assuming that the nontasters are homozygous for the recessive gene t, what is the frequency of homozygous dominants (TT) in the population? Of heterozygotes (Tt)?
17. In a certain population, approximately one child in 2500 is homozygous for the gene z. All these children die before reaching sexual maturity. Assuming Hardy-Weinberg equilibrium, what is the proportion of "carriers" (heterozygotes) in the population?
18. The frequency of sickle-cell anemia (Hb^SHb^S) among the children in one malarial region in Africa is 144 out of 10,000. Assuming Hardy-Weinberg equilibrium, what is the approximate frequency of heterozygotes (Hb^SHb^A) in this population?

REFERENCES

1. EISELEY, L. C., "Charles Darwin," *Scientific American,* Offprint No. 108, February, 1956. A brief biography.
2. DARWIN, C., *The Origin of Species,* The New American Library, New York, 1958. One of three paperback editions of this classic.
3. VOLPE, E. P., *Understanding Evolution,* 2nd ed., Wm. C. Brown, Dubuque, 1970. A paperback providing detailed and clearly written elaborations of much of the material of this chapter.
4. HARDY, G. H., "Mendelian Proportions in a Mixed Population," *Great Experiments in Biology,* ed. by M. L. Gabriel and S. Fogel, Prentice-Hall, Englewood Cliffs, N.J., 1955. In this brief note Hardy outlines the principle that became known as the Hardy-Weinberg law.
5. CAVALLI-SFORZA, L. L., " 'Genetic Drift' in an Italian Population," *Scientific American,* Offprint No. 1154, August, 1969.
6. BISHOP, J. A., and L. M. COOK, "Moths, Melanism and Clean Air," *Scientific American,* Offprint No. 1314, January, 1975. How both selection and gene migration are reversing industrial melanism in the moth populations of Great Britain.
7. WILLS, C., "Genetic Load," *Scientific American,* Offprint No. 1172, March, 1970. The title refers to the sum of all the mutations in the gene pool of a species. The author discusses the evolutionary advantages and disadvantages of genetic load.
8. CLARKE, B., "The Causes of Biological Diversity," *Scientific American,* Offprint No. 1326, August, 1975. An analysis of two mechanisms by which natural selection maintains balanced polymorphism in populations.
9. STEBBINS, G. L., *Processes of Organic Evolution,* 2nd ed., Concepts of Modern Biology Series, Prentice-Hall, Englewood Cliffs, N.J., 1971. An excellent survey of the mechanisms at work in evolution. Paperback.
10. DAWKINS, R., *The Selfish Gene,* Oxford University Press, Oxford, 1976. This little book provides lively and thought-provoking arguments that all behavior is the product of natural selection working to enhance the success of the owner's genes. The chicken and its behavior are simply the devices of one egg for laying another egg. Weismann would have enjoyed this book and you will, too.
11. LERNER, I. M., and W. J. LIBBY, *Heredity, Evolution, and Society,* 2nd ed., Freeman, San Francisco, 1976.
12. KING, J. C., *The Biology of Race,* Harcourt Brace Jovanovich, New York, 1971.

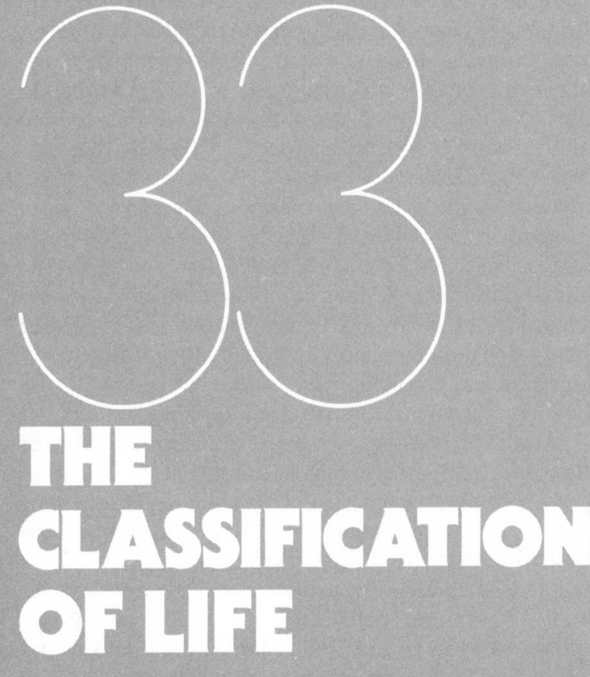

33 THE CLASSIFICATION OF LIFE

33.1 IMPORTANCE

Life occurs on our earth in a bewildering number of different forms. At least 1.2 million kinds of living organisms have already been discovered, and the list grows longer every year. In addition, fossil remains tell us of many other kinds of organisms that once inhabited the earth but are now extinct.

Before biology could be placed on any sort of scientific, comprehensible basis, it was necessary to bring some order out of the chaos of such large numbers. This was done by attempting to place the various forms of life in categories, in other words, to classify them. Organisms that were similar to one another were placed in a group together. In the earliest classification systems, for example, all green organisms which had no power of locomotion were grouped in the plant kingdom, while nongreen organisms capable of locomotion were placed in the animal kingdom. A few organisms, mushrooms for example, did not quite fit into either category, but these occasional difficulties were solved simply by assigning the perplexing forms to what seemed to be the more appropriate of the two kingdoms.

With the discovery in the seventeenth century of the world of microorganisms, the problems of classification became more troublesome. The Dutch lens grinder van Leeuwenhoek and the microscopists who came after him discovered myriads of tiny organisms that possessed characteristics typical of neither kingdom. These, the bacteria for example, were taken rather arbitrarily into the plant kingdom by the botanists. They also discovered forms such as the microscopic, green, swimming organism *Euglena* (Fig. 33.1), that had *both* plant and animal features. These became a serious bone of contention between the botanists and the zoologists. Even today, most books on botany claim *Euglena* and similar debatable forms, while the zoology books claim them also.

About one hundred years ago, the German biologist Haeckel suggested a way out of this confusing state of affairs. He proposed setting up a third kingdom, the **Protista,** to include those organisms which do not clearly fit into either the plant or the animal kingdoms. Since then, it has become apparent that even such a move fails to account for the very special properties of two groups: the prokaryotic bacteria and blue-green algae (see Section 5.19) and the fungi. The prokaryotes differ from all other creatures in a large number of ways, the most conspicuous of which is that their cells lack nuclei. This fully justifies the establishment of a fourth kingdom, the **Monera.** Fungi are eukaryotic, but they certainly are neither animals nor plants. They also differ in many ways from protists like the amoeba and *Euglena*. Thus many biologists have concluded that the fungi should be given a kingdom of their own.

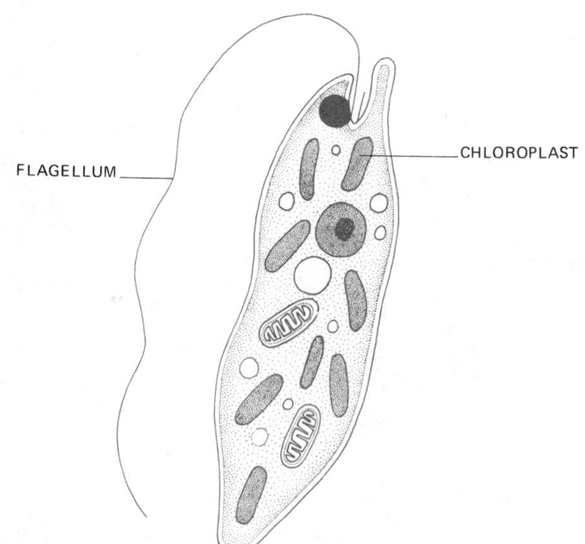

FIG. 33.1 *Euglena,* 1200 ×. This microscopic, unicellular organism carries on photosynthesis (like a plant) and moves rapidly about (like an animal). It is usually described in textbooks of both botany and zoology. In fact, it is neither a plant nor an animal but an organism only distantly related to those taxonomic groups.

At this point you may well feel that it is highhanded, to say the least, to create five kingdoms. Many students feel that to do so is also an admission of a failure to resolve our difficulties and to determine once and for all just what—plant or animal—each of these little creatures is. While such a feeling is understandable, it completely misses the point of what a classification system is and does.

33.2 THE PRINCIPLES OF CLASSIFICATION

Classification consists of placing together in categories those things that resemble each other. While this sounds simple, in actual practice it may be quite difficult. First of all, we have to decide what kind of similarities are the most important for our purpose. One of the earliest classification schemes placed in one category all those animal forms which lived in the same habitat. Thus fishes, whales, and penguins were classified as swimming creatures. This type of classification was often based on the principle that creatures possessing **analogous organs** should be classified together. Analogous organs are organs that have the same function. The fins of fishes and the flippers of whales and penguins are analogous organs because they are all used for swimming. The wings of birds, bats, and insects are analogous organs that make flying possible.

As more knowledge was gained about the anatomy of living things, it became apparent that similarities of habitat and of analogous organs were often rather superficial. The fact that bats have fur and nurse their young, birds have feathers and lay eggs, while insects are cold-blooded and have no internal skeleton suggested that these organisms differ from one another in more important ways than they resemble one another. An appreciation of the truly significant ways in which organisms resemble or differ from one another enabled the Swedish naturalist Carolus Linnaeus to found the modern system of classification. In 1753 he published a classification of the plants which was followed, in 1758, by a classification of the animals. For this work he is often called the father of **taxonomy,** the name given to the study of classification. His system of classification is fundamentally the system we use today. It is based on the principle of **homology** (see Section 31.2).

Why is a classification based upon homology so significant? As the preceding chapters should have convinced you, a classification based upon homologous organs is a classification based upon kinship. All creatures sharing homologous organs are related to one another, having inherited their homologous organs from a common ancestor. Thus humans, bats, and whales all had a single ancestor who possessed the basic forelimb structure that these creatures possess today—although obviously in a quite modified form (Fig. 31.6).

33.3 AN EXAMPLE

The essence of any modern classification scheme, then, is to group all those organisms that are related to one another. However, in the frequent absence of fossil remains of common ancestors, we have to guess the evolutionary relationships by studying the degree of similarity of the organisms we are attempting to classify. Figure 33.2 shows ten organisms. It is perfectly clear that all ten are animals and even that all ten are birds. But how should they be classified with respect to one another? Closer inspection reveals that some of the birds (*A* and *I*, for example) are quite similar in appearance. Perhaps, then, these should be placed together in one category. Another category might include *D*, *G*, and *J*, which show similarities of bill shape, size, and general body contour. *C*, *F*, and *H* seem to fall into still a third group, while *B* and *E* seem quite different in appearance. In fact, our first three groups seem to have more features in common with each other than they do with either *B* or *E*. Perhaps these three subgroups should be placed together in a larger group which excludes *B* and *E*.

This, then, is the basic process involved in classification. Of course, we have been somewhat limited in our efforts here by being restricted to a view of the external anatomy. Were a knowledge of all homologous features available to us, we could tabulate the distribution of each of these throughout the entire group (Fig. 33.2). Those birds sharing the largest number of homologous features would be grouped together. This reflects our belief that they are most closely related, in other words, have most recently descended from a common ancestor. These first groups would then be organized into larger groups on the basis of all their members sharing a smaller number of homologous features. We assume that the members of these larger groups are more distantly related than the members of the smaller groups. The common ancestor from which they are descended existed in a more remote time.

In the case of our birds, each one represents a single kind, or **species,** of bird. This is the fundamental unit of classification. The groups in which we place the various species are called genera (singular, **genus**). The two sparrows, *A* and *I*, are classified in the genus *Spizella*. Related genera are, in turn, placed in **families.** Related families comprise **orders,** and orders are grouped together in **classes** (in this case, Aves, or birds). Similar classes are placed in a given **phylum** and all related phyla make up a kingdom (Fig. 33.3).

In addition to these major groups, taxonomists often find it convenient to establish others: phyla may be divided into subphyla, and super- as well as sub-classes and -families are also formed in some cases. The currently accepted classification of our ten birds is shown in Fig. 33.4.

DISTRIBUTION OF EIGHT HOMOLOGOUS STRUCTURES

SPECIES	A	B	C	D	E	F	G	H	I	J
1. FEATHERS	+	+	+	+	+	+	+	+	+	+
2. SERRATED BILL					+					
3. PARTIAL WEBBING BETWEEN TOES		+								
4. PERCHING FEET	+		+	+		+	+	+	+	+
5. SLENDER BILL				+			+			+
6. SHORT, STOUT BILL	+		+			+		+	+	
7. YELLOW PLUMAGE ON TOP OF HEAD				+			+			
8. REDDISH-BROWN PLUMAGE ON BACK	+								+	

GROUPINGS MADE ON THE BASIS OF THESE STRUCTURES

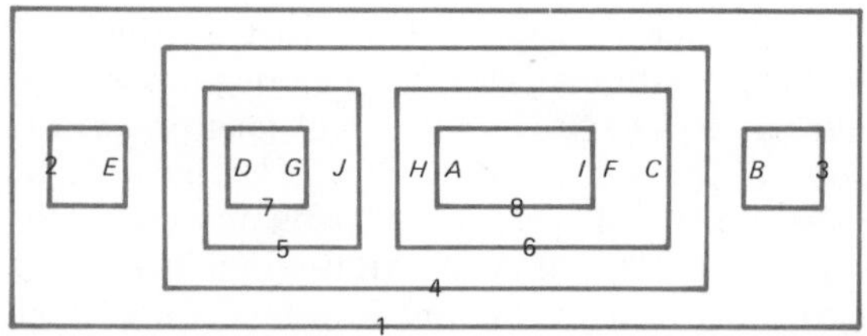

FIG. 33.2 Ten birds; a study in the techniques of classification.

THE CLASSIFICATION OF HUMANS

KINGDOM	ANIMAL
PHYLUM	CHORDATA
SUBPHYLUM	VERTEBRATA
CLASS	MAMMALIA
ORDER	PRIMATES
FAMILY	HOMINIDAE
GENUS	*Homo*
SPECIES	*Homo sapiens*

FIG. 33.3 The major taxonomic categories as they are used in the classification of humans. Most plant taxonomists use the term *division* instead of *phylum*.

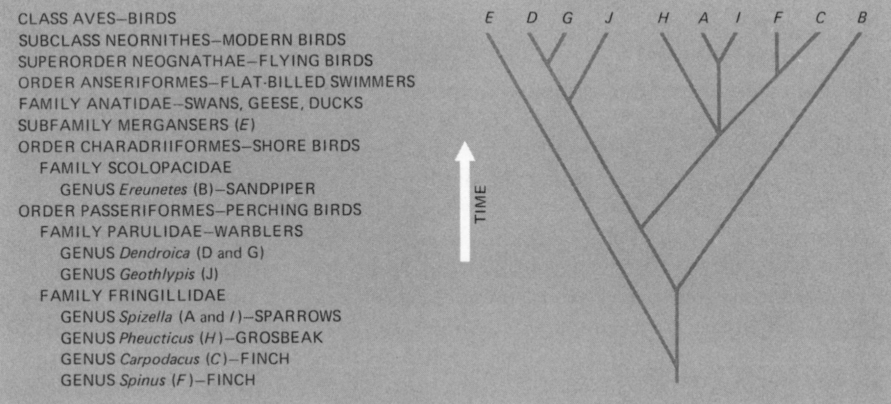

FIG. 33.4 The actual classification of the ten birds shown in Fig. 33.2 and a deduction as to their evolutionary history.

33.4 EVOLUTIONARY IMPLICATIONS OF MODERN TAXONOMY

If, as we hope, our classification scheme reflects degree of kinship, then we should be able to set up a family tree showing the evolutionary history of the group (Fig. 33.4). The species are all placed at the tips of the branches, and the common ancestors at the forks. It is important to remember two things when dealing with such a family tree. First, if the genealogy is correct, all the species in any one group, no matter what its rank, have shared a common ancestor among themselves more recently than they have with species in any other group. Second, no living organism is the ancestor of other organisms. This seems obvious, but many biology students (and sometimes their professors) slip into the error of thinking that the more primitive organisms living today are the ancestors of the more complex organisms. This is not true, although we may assume that during the course of evolutionary descent from a common ancestor, the more primitive organisms have changed less than the more complex ones. The only direct evidence of common ancestors ("missing links") that we can ever hope to find is their fossilized remains.

This second point is related to a third: the general history of life on earth seems to have been the formation of more and more different kinds of organisms. Each fork in the family tree represents speciation—two kinds of organisms *diverging* from one, the common ancestor.

Occasionally, evolution seems to work the other way. Two unrelated lines may come to resemble each other closely. This **convergent evolution** explains the superficial similarity of penguins, whales, and fishes, and of birds, bats, and insects. The need to swim or fly efficiently imposes definite limitations on body form. As unrelated forms have taken to the water or to the air, they have evolved in ways appropriate to the new medium. This has resulted in the development of many similarities of structure and, as we have seen, it has posed something of a puzzle for the taxonomist. Fortunately for him, careful inspection of the forms usually reveals their true affinities. Even the layman knows that the penguin is a bird, not a fish. The whale seems more puzzling, but its real ancestry is quite clear. Unlike true fishes, it is warm-blooded, possesses fur, bears living young, and nurses them with milk. This makes the whale more closely related to us (mammals) than to the fishes. Furthermore, dissection of the whale reveals the presence of rudimentary bones that are homologous to the hind legs of the four-legged land animals. Despite convergent evolution, the whale has not been able to erase the evidence of its true ancestry.

33.5 SCIENTIFIC NAMES

Although Linnaeus did not believe in evolution, his intuitive grasp of the importance of homologies gave us a system of classification that still works today. He also deserves the gratitude of biologists everywhere for providing a system for the naming of species. Every language has its own words for plants and animals. Our *dog* is the German's *hund* and the Frenchman's *chien*. Knowledge in biology, as in the other sciences, is discovered independently of national boundaries. It is important, therefore, that biologists in each country know just what organism their colleagues in other countries have been working with. The system of scientific names, established by Linnaeus, achieves this goal.

The scientific name of each species consists of two parts. The first is the name of the genus to which the organism belongs; the second, called the "specific epithet," identifies the particular species within the genus. Thus the family dog is *Canis familiaris*. Latin names were used by Linnaeus, but so many species have been discovered since then that now taxonomists simply coin new words and cast the genus name in the form of a Latin noun and the specific epithet as a Latin adjective. Both names are printed in italics, the genus name capitalized, but not the specific epithet. Note, too, that the characters of our Roman alphabet are always used for this purpose even by biologists in countries such as Japan, where different characters are used for ordinary purposes (Fig. 33.5).

Frequently, the specific epithet may be derived from the name of the discoverer. Thus Brewer's sparrow is *Spizella breweri*. Sometimes you will see another name or initial, not italicized, written after the scientific name. This is the name or initial of the taxonomist who coined the scientific name. Linnaeus himself proposed the name we use for the dog, so it is often written *Canis familiaris* L. You can well imagine that with the vast number of species to be named, two or more taxonomists may independently propose different names for the same organism. Definite rules, enforced by international commissions, have been set up to resolve such difficulties.

——旗口水母類 Semaeostomae——

3. **ミズクラゲ Aurelia aurita** (Linnaeus) [Aureliidae]

傘は円盤状で，直径10〜17 cm，ときには30 cmを越えるものもある。寒天質は比較的柔軟で，放射管を白く残し，やや青味をおびる程度で，大体無色透明。但し生殖腺は雌では褐色，雄では紫がかった青色を呈する。刺胞毒は認められない。三大洋に広く分布するが，本邦では北海道の西岸忍路及び関東地方以南の暖流区域に多産する。北海道以北の寒流区域には，これに代わって，もっと大形で放射管が網状に連絡しているキタミズクラゲ *A. limbata* Brandt が見られる。

FIG. 33.5 The description, as it appears in a Japanese guide to marine life, of a common jellyfish. (Courtesy Hoikusha Publishing Company, Ltd., Osaka, Japan.)

Occasionally a third italicized and Latinized word is included in the scientific name of an organism. This is the subspecies name; it serves to distinguish a particular, often local, form of the species from other forms of that same species. Despite their apparent differences, all breeds of dog belong to one species. Other species are known which contain two or more fairly distinct "breeds." These different breeds are technically called races, varieties, or subspecies. *Geothlypis trichas*, the yellowthroat, a warbler, (species *J* in our classification—see Fig. 33.2) actually occurs as seven rather distinct subspecies (Fig. 32.10). The significance of subspecies in evolution was analyzed in the preceding chapter (see Section 32.13).

33.6 HIGHER CATEGORIES

Now that you have seen how the system of classification works, it should be easier to realize just how arbitrary it is. The only taxonomic group which exists outside of our own minds is the species. All members of a given species should be capable of breeding successfully with one another, although definite information on this point is often difficult to secure. For breeding to be considered successful, the offspring should be just as fertile as the parents. Thus the fact that a horse and a donkey can mate does not prevent us from classifying them in separate species, because the mule which they produce is sterile. The species, then, is the only unit which exists in nature. The creation of genera, families, orders, etc., is a matter of human judgment.

This would be true even if all taxonomists agreed on the evolutionary relationships of the species they were studying. Refer, for example, to the family tree of our ten bird species (Fig. 33.4): should species *D, G,* and *J* each be placed in a separate genus? Or do *D* and *G* share certain features not possessed by *J,* so that they should be placed together in a single genus? Or should all three perhaps be in one genus?

A similar problem arises with the creation of families. Should all the birds but *B* and *E* be placed in one family or should *D, G,* and *J* be placed in one family and *A, C, F, H,* and *I* in another? Note that either choice would not violate our rule that all the members of any one group must have had a common ancestor more recent than any they have shared with species in other groups. (Placing *D, G, J,* and *E* in one genus would, however, violate this rule.) Thus, even though we think we know the evolutionary history of our species, there is still plenty of room for disagreement as to just how closely the different species should be grouped.

Those taxonomists who are particularly impressed by the differences between species tend to increase the number of higher categories. Taxonomists with this bias are known fondly as "splitters." On the other hand, there are taxonomists who marvel at the uniformities they see among species. These people, the "lumpers," try to keep the number of higher categories at a minimum. Thus the splitters would quickly put *D, G,* and *J* in separate genera and would put these genera in a family separate from the family containing *A, C, F, H,* and *I.* Lumpers would put *D, G,* and *J* in one genus, and all of the birds (except *B* and *E*) in a single family.

While the goal of taxonomy is always to develop a "natural" classification, that is, one based on evolutionary history, that goal is frequently not achieved. Many species, especially the microorganisms and the soft-bodied animals, have left practically no fossil remains. Consequently, taxonomists must try to reconstruct the evolutionary history of many groups on the basis of indirect evidence—chiefly, supposed homologies exhibited by living forms. And despite their best efforts, they often must admit failure.

What can one do when there simply is not enough evidence to determine evolutionary relationships? Well, a classification scheme that fails to reflect true kinship is better than no classification scheme at all. So, we create a classification scheme simply using *convenient* criteria even though our categories will probably not be "natural" ones. In the next few chapters, we shall examine several phyla that surely do not represent a "natural" grouping of organisms. As the creatures in these phyla become better known, we can hope that they will finally be classified on the basis of kinship. This will probably result in greatly increasing the number of phyla (as well as lower categories).

The general trend in classification has been to raise the status of the more primitive groups. The blue-green algae, for example, were once considered an *order* within the *class* Algae which in turn was part of the *phylum* Thallophyta of the plant *kingdom.* Today we place the blue-green algae in the Monera, a kingdom they share only with the bacteria. All this is really another way of saying that we are gradually realizing that many of the major groups of living things have had a very long and independent evolutionary history. There may be even more than five main branches to the family tree of life. However, we shall stick with five. In the remaining five chapters of this section of the text, we shall examine the major groups of organisms to be found in each kingdom.

EXERCISES AND PROBLEMS

1. Set up a classification system, as we did for the birds, that includes the horse, cat, donkey, zebra, tiger, and rattlesnake.
2. What evidence did you use in setting up this system?
3. What does your system tell us about the evolution of these animals?
4. Draw a family tree which is consistent with the judgments you make in Question 3.
5. The wildcat of the eastern United States is called *Lynx ruffus* Güld. What can you deduce about the following: (a) *Lynx ruffus floridanus* Rafinesque, (b) *Lynx ruffus texensis* Allen, (c) *Lynx ruffus californicus* Mearns?

REFERENCES

1. WHITTAKER, R. H., "New Concepts of Kingdoms of Organisms," *Science,* **163**: 150, January 10, 1969. Arguments for the five-kingdom system, presented by its creator.
2. DE BEER, SIR GAVIN, *Homology, An Unsolved Problem,* Oxford Biology Readers, No. 11, Oxford University Press, Oxford, 1971.

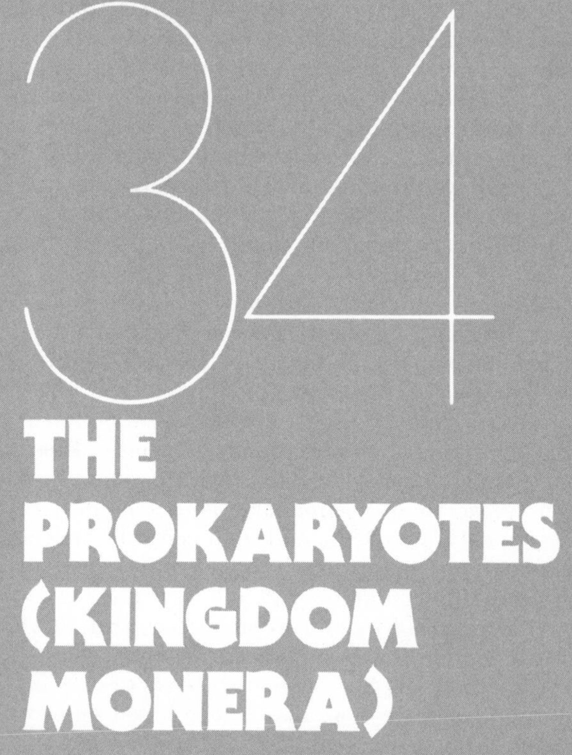

34 THE PROKARYOTES (KINGDOM MONERA)

34.1 THE NATURE OF PROKARYOTES

If one examines the anatomy of the cells of the diverse creatures that populate the earth, a basic dichotomy becomes apparent. The cells of the majority of organisms contain nuclei, mitochondria, plastids (if they are photosynthetic), and most of the other structures that we examined in Chapter 5. Such cells are called eukaryotic ("truly nuclear") and their owners are the **eukaryotes.** However, the cells of bacteria and the cells of blue-green algae do *not* have these organelles. These creatures are called **prokaryotes** ("prenuclear").

Prokaryotes differ from eukaryotes in a number of other ways. They have only a single chromosome, and it has no histones associated with it. They have no microtubules (there may be one exception) and hence no centrioles, spindle, or basal bodies. While some prokaryotes have "flagella," these are not constructed from microtubules as are the flagella and cilia of eukaryotes. The prokaryotic ribosome is different in structure from the ribosomes of eukaryotes and, as we shall see, prokaryotes share a number of biochemical peculiarities not found in eukaryotes.

Reproduction in the prokaryotes is most commonly asexual. Those species that are able to accomplish genetic recombination do so in a manner entirely different from eukaryotes. One individual, the donor, simply transfers some of its genes to a second individual, the recipient. There is no meiosis. Further details of sexual reproduction as it occurs in certain bacteria (e.g. *E. coli*) were given in Section 14.6.

The prokaryotes differ in such fundamental ways from the eukaryotes that they must be classified in a separate kingdom: Monera. Within this kingdom we shall place two phyla: one containing the bacteria (Phylum **Schizomycetes**) and one containing the blue-green algae (Phylum **Cyanophyta**).

34.2 THE BACTERIAL CELL

The cells of bacteria are encased in a cell wall. Like the cell wall of aquatic plants, their wall provides the strength needed to keep the cell from bursting when in a hypotonic medium (like a pond). The wall is also sufficiently rigid that it gives a relatively fixed shape to the cell. Bacteria that are rod-shaped are called **bacilli;** bacteria whose walls are spherical are called **cocci. Spirilla** have curved walls (Fig. 34.1).

The walls of bacteria are composed of a complex polymeric material called **peptidoglycan.** As the name may suggest to you, it contains both amino acids and sugars. The sugars are of two kinds—a nitrogen-containing hexose called N-acetylglucosamine (NAG) and its close relative, N-acetylmuramic acid (NAM). These two form a linear polymer, NAG alternating with NAM (Fig. 34.2). The links are between the number 1 and number 4 carbons and are oriented in the same way that they are in cellulose (see Fig. 4.14). It is this linkage that is attacked by lysozyme (look back at Figs. 6.16 and 6.18).

Attached to each NAM is a side chain which contains four amino acids. The first (proximal) one is alanine (Ala). The second is glutamic acid (Glu). But this glutamic acid is the D-isomer, not the L-isomer that is universally found in proteins! (You may wish to review the discussion of the optical isomers of amino acids in Section 4.4.) The third amino acid

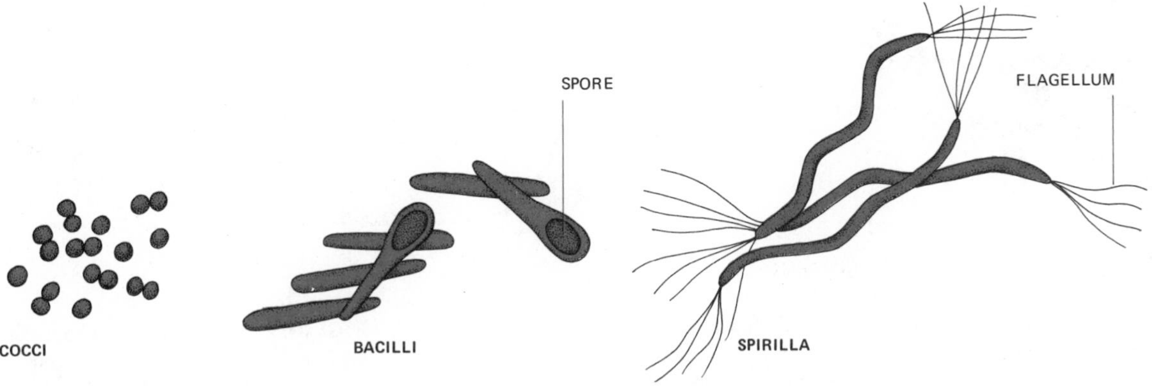

FIG. 34.1 Three common forms of bacteria. Each maintains its shape by means of its rigid cell wall.

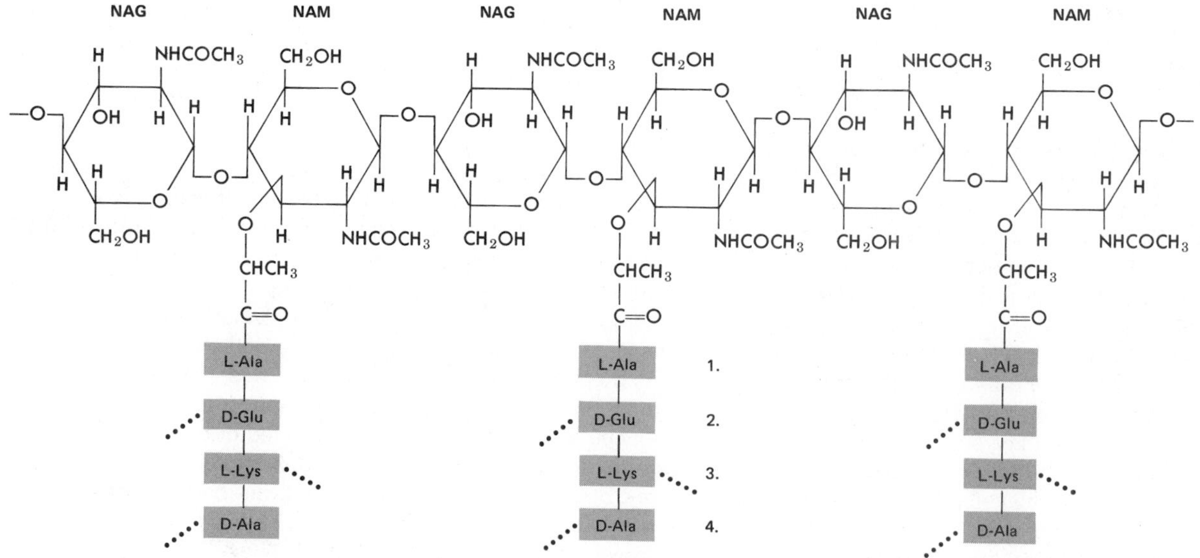

FIG. 34.2 The structure of peptidoglycan, the chief component of the bacterial cell wall. Note the similarity to chitin (Fig. 35.15). The linkage between the amino sugars in this polymer is like that in cellulose (Fig. 4.14). Some variation in the tetrapeptides attached to NAM are found from one species to another. In some species, for example, diaminopimelic acid—an amino acid found exclusively in prokaryotes—may occur at position 3. The dotted lines indicate places where a covalent linkage to an adjacent chain may be established. These bridges may be direct or may include additional short peptides, e.g. five glycine residues.

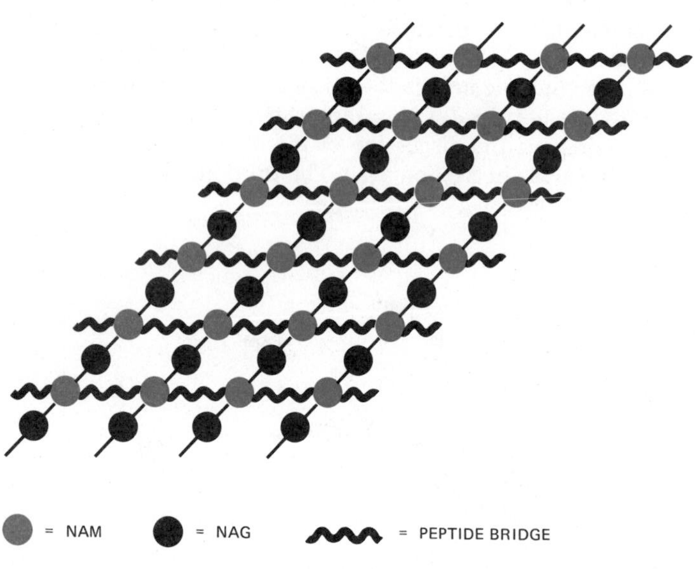

FIG. 34.3 The wall structure of bacteria. The peptide cross bridges provide covalently bonded links between adjacent chains, not only in the same plane (as shown here) but also in the plane above and below. This construction gives great strength to the cell wall.

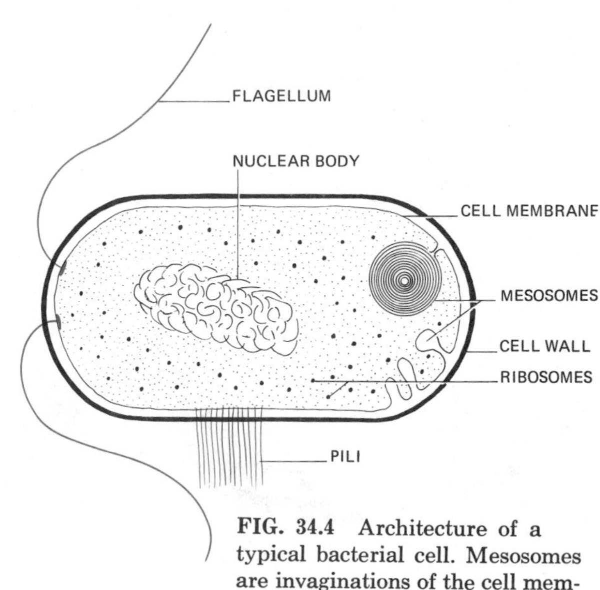

FIG. 34.4 Architecture of a typical bacterial cell. Mesosomes are invaginations of the cell membrane.

in the tetrapeptide may be the L-isomer of lysine or, instead, diaminopimelic acid, an amino acid that is never found in proteins (or anywhere in eukaryotes). The terminal amino acid is the D-isomer of alanine.

The important thing about the tetrapeptides attached to NAM is that covalent bonds can be established between them. These may be direct peptide bonds between the $-NH_2$ group of an amino acid (e.g. lysine) in one chain and a $-COOH$ in another (e.g. D-Ala). Often, however, short bridges of amino acids are inserted with peptide bonds between the tetrapeptides of adjacent chains. The cross links between NAMs can link adjacent chains in the same plane (Fig. 34.3) and also in planes above and below. It is this elaborate, covalently cross-linked structure that provides the great strength of the cell wall. It also leads to the remarkable conclusion that the bacterial cell wall meets the definition of a single molecule (albeit one of indefinite and changeable size)!

The cell membrane lies directly beneath the cell wall (Fig. 34.4). It contains a phospholipid bilayer similar to that which we found in the eukaryotic cell membrane (look back at Fig. 5.4). However, there is no cholesterol or any other steroids in the phospholipid layer of bacterial cell membranes.

Many bacteria are able to swim by means of flagella. Their flagella are anchored in the cell membrane and project out through the cell wall. The structure and composition of bacterial flagella is entirely different from that of the eukaryotic flagella we examined in Chapter 28. There is no "9 + 2" arrangement of tubulin-containing microtubules. Instead there is simply a single filament composed of a protein called flagellin. The mechanism of action of bacterial flagella is unknown.

Some bacteria (*E. coli,* for example) have a second set of protein filaments that are anchored in the cell membrane and project through the cell wall. These are called **pili** (Fig. 34.5). The function of most of the pili is quite unknown. However, certain pili are necessary for sexual reproduction (conjugation) to occur. In some manner not yet clearly understood, these pili enable a portion of the donor's chromosome to be injected into the recipient.

Portions of the cell membrane may fold into the cytoplasm to form **mesosomes.** Mesosomes serve a variety of functions. When the bacterial cell gets ready to divide, the formation of a new cross wall occurs with the aid of mesosomes. A mesosome probably assists in the orderly separation of the duplicated bacterial chromosome (which is attached to it). Cell secretions are packaged in and discharged from mesosomes. In short, mesosomes serve those functions that require a reserve supply of cell membrane.

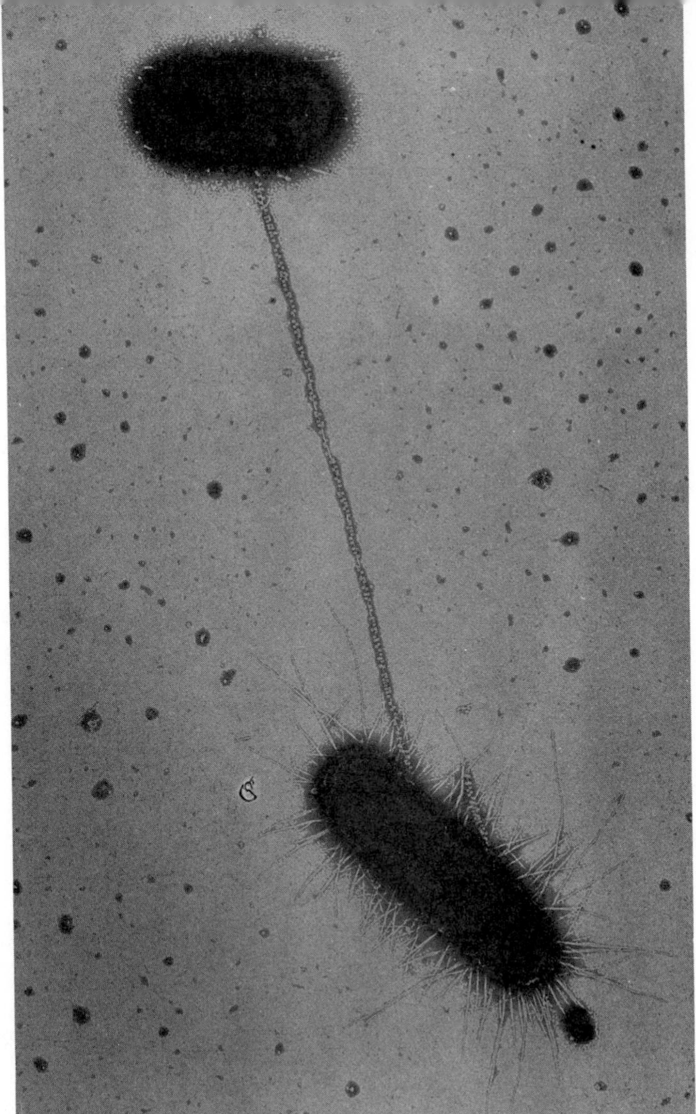

FIG. 34.5 Pili of *E. coli.* The long pilus between the cells is essential for the transfer of genetic material from the donor or "male" (bottom) to the recipient. This male cell also has many other pili that do not participate in conjugation. (Electron micrograph prepared by Judith Carnahan and Charles Brinton. Reproduced from R. Y. Stanier, M. Doudoroff, and E. A. Adelberg, *The Microbial World,* 3rd edition, 1970, by permission of Prentice-Hall, Inc., Englewood Cliffs, N.J.)

No membrane-bounded nucleus occurs in bacteria. However, the bacterial chromosome—a closed

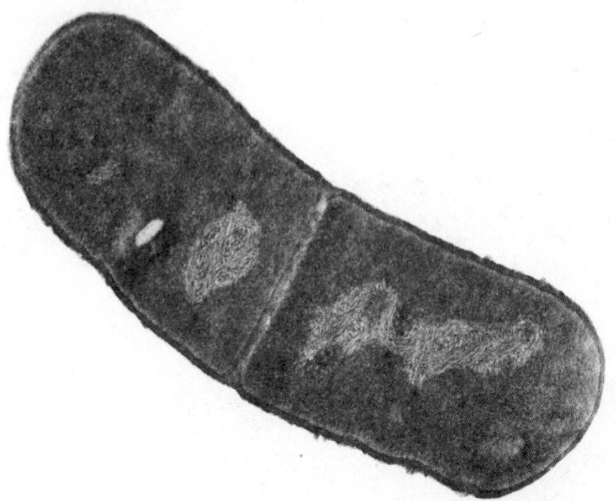

FIG. 34.6 Electron micrograph of a dividing bacterial cell, approximately 28,000 ×. The light areas in each daughter cell are the nuclear bodies. (Courtesy of C. F. Robinow and J. Marak.)

loop of double-stranded DNA some 1 mm in circumference—is folded together in a "nuclear body" within the cell (Fig. 34.6).

Many bacteria form **spores** when their food supply begins to run low. Each cell forms but a single spore (thus bacterial spores are not agents of asexual reproduction). A freshly duplicated chromosome, some ribosomes, and a wide variety of enzymes are wrapped in a double layer of cell membrane and layers of peptidoglycan. Most of the water is removed from the spore and metabolism ceases.

Spores are exceedingly resistant to adverse environmental conditions such as drying and extremes of temperature. If and when conditions improve, the spore becomes metabolically active and germinates into a vegetative cell. The spores of soil bacteria have been shown to still be viable after 50 years of dormancy.

34.3 THE CLASSIFICATION OF BACTERIA

The classification of bacteria is exceedingly difficult. It is based to a large degree on such features as shape, ability to form spores, method of energy production (glycolysis for anaerobes, cellular respiration for aerobes), and reaction to the Gram stain. The Gram stain was devised many years ago (in 1884) by the Danish bacteriologist Christian Gram. The cells are first stained with a purple dye called crystal violet. Then the preparation is treated with alcohol or acetone. This washes the crystal violet out of *gram-negative* cells. To see them at this point requires the use of a counterstain of a different color (e.g. the pink of safranin). Bacteria that are not decolorized by the alcohol/acetone wash are called *gram-positive*.

At first the Gram stain might seem to be a most arbitrary criterion to use in taxonomy. However, work in recent years has revealed that the stain does distinguish between two fundamentally different kinds of bacterial cell walls and thus probably reflects a significant and natural division among the bacteria.

In general, though, the present classification of bacteria (and it is under continual revision) can scarcely be considered a "natural" one, that is, one that reveals evolutionary affinities.

The most promising approach to making the system more natural is to investigate the ratios of deoxyribonucleotides in the organisms. You remember that in any sample of DNA, the amount of A equals the amount of T, and, similarly, the amount of C equals the amount of G. But the ratio of C + G to A + T varies widely among the bacteria. Put in another way, some bacteria favor codons containing C and G; others favor A and T. The C + G:A + T ratio in bacteria varies from a low of around 30% in some species to a high of 70–80% in others. (In humans and other vertebrates, it is 40%).

It is hard to see how two closely related bacterial species could differ much in such a fundamental property as their use of the genetic code. So, whenever we find two bacterial species that have widely divergent C + G:A + T ratios, we must assume that they are distantly related even though we may have previously classified them together on other grounds such as shape, etc. Despite such clues, the classification of bacteria, particularly categories above the level of family and order, remains unnatural. We shall touch briefly on a dozen groups.

34.4 THE PHOTOSYNTHETIC BACTERIA

Like green plants, the photosynthetic bacteria use the energy of sunlight to reduce carbon dioxide to carbohydrate. In contrast to green plant photosynthesis, however, the source of electrons is *never* water. The **purple sulfur bacteria** and the **green sulfur bacteria** use hydrogen sulfide (H_2S) to supply the electrons needed to synthesize NADPH and ATP. In

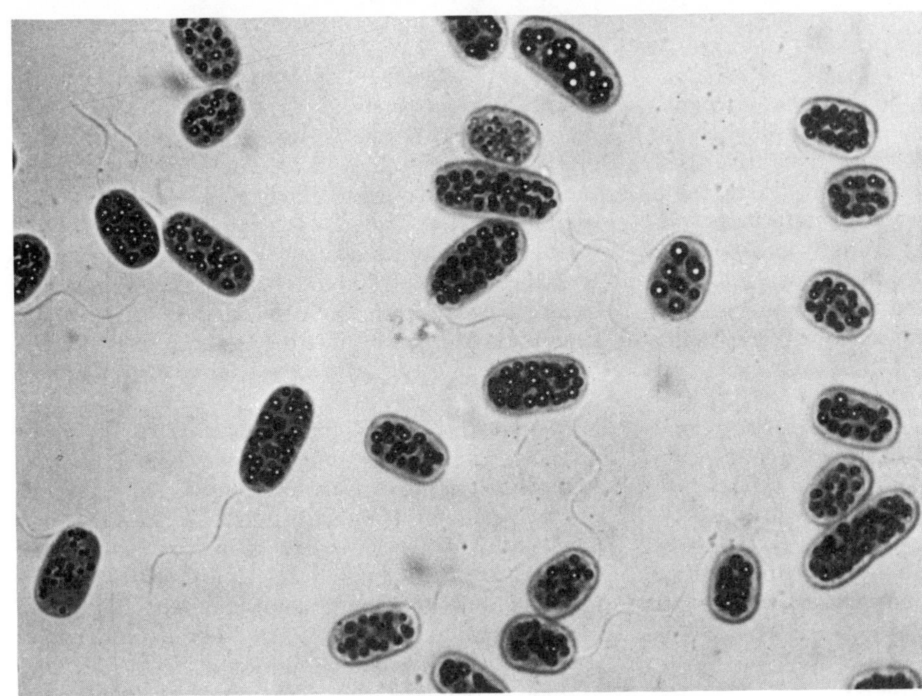

FIG. 34.7 *Chromatium,* one of the purple sulfur bacteria. The refractile bodies within the cells are granules of sulfur. Note the flagella. (From H. G. Schlegel and N. Pfennig, *Arch. Microbiol.,* 38 (1), 1961.)

the process they produce elemental sulfur (often stored as granules within the cell—Fig. 34.7).

$$2H_2S + CO_2 \xrightarrow{\text{light}} (CH_2O) + H_2O + 2S$$

It was this process that gave Van Niel the idea of the role of light in green plant photosynthesis (see Section 9.7).

The photosynthetic bacteria contain special forms of chlorophyll (called bacteriochlorophylls). These are incorporated in the membranes of their mesosomes. With this machinery, they can run Photosystem I but not Photosystem II (which explains their inability to use H_2O as a source of electrons).

Most of the photosynthetic bacteria are obligately anaerobic, i.e., they cannot tolerate free oxygen. Thus they are restricted to such habitats as the surface of sediments at the bottom of shallow ponds and estuaries. In such locations, they may well have to make do with whatever radiant energy gets through the green algae and aquatic plants growing in the water above them. Interestingly enough, the absorption spectrum of their bacteriochlorophylls lies mostly in the infrared region of the spectrum so they can trap energy that is missed by the algae above them (character displacement?).

In Chapter 2, we speculated about the nature of the earliest forms of life on earth. We worried about what would happen when the "primeval soup" of organic molecules had been consumed by the first fermenting heterotrophs. What was needed then was some organisms that could synthesize a fresh supply of organic molecules. What better candidates than the photosynthetic bacteria, with their ability to live in an anaerobic atmosphere?

But a problem remains. Living organisms cannot live by carbohydrates alone. They must have a supply of nitrogen compounds (e.g. NH_3, NO_3^-) in order to synthesize proteins and nucleic acids. Assuming that the atmosphere of the primitive earth contained nitrogen gas (N_2) as it does today, a way had to be found to "fix" this nitrogen into compound form. The photosynthesizing bacteria are able to do just this. Thus in this respect as well, they would have been able to thrive under the conditions existing early in the evolution of life on earth.

34.5 THE CHEMOAUTOTROPHIC BACTERIA

Certain colorless bacteria share the ability of chlorophyll-containing organisms to manufacture carbohydrates from inorganic raw materials, but they do not use light energy to do this. Our study of photosynthesis has shown us that in itself there is nothing unique about this. The conversion of carbon dioxide into carbohydrate can go on in animal cells as well as in plant cells. The "dark" reactions responsible are also found to go on in the cells of the chemoautotrophic bacteria. The question is: what is the source of the energy and electrons that drive these energy-consuming dark reactions? How do the colorless chemoautotrophic bacteria manufacture sufficient ATP and NADPH to produce glucose from carbon dioxide? They do so by carrying out an oxidation of some reduced substance present in their environment. The free energy made available by this oxidation is then harnessed to the manufacture of carbohydrate.

The chemoautotrophic **sulfur bacteria** oxidize H_2S in their surroundings (e.g. the water of sulfur springs) to produce energy:

$$2H_2S + O_2 \rightarrow 2S + 2H_2O; \quad \Delta F = -100 \text{ kcal.}$$

They can then use this energy to reduce carbon dioxide to carbohydrate in the same manner as the photosynthetic sulfur bacteria do:

$$2H_2S + CO_2 \rightarrow (CH_2O) + H_2O + 2S.$$

Another group of chemoautotrophic bacteria are the **iron bacteria.** (These are responsible for the brownish scale that forms inside the water tanks of flush toilets.) They complete the oxidation of partially oxidized iron compounds and are able to couple the energy produced by this oxidation to the synthesis of carbohydrate.

The **nitrifying bacteria** are also chemoautotrophic. They accomplish the oxidation of NH_3 (produced from proteins by the heterotrophic bacteria of decay) to nitrates. These oxidations provide the energy to drive the synthetic reactions of the bacteria. The nitrates produced as a byproduct supply the nitrogen needs of plants. The role of the nitrifying bacteria in the overall cycle of nitrogen on the earth will be considered again in Chapter 41.

34.6 GRAM-POSITIVE RODS

Several gram-positive, rod-shaped bacteria are of special interest. The genus *Clostridium* consists of spore-forming bacteria that are obligate anaerobes (they cannot survive exposure to oxygen). Some of these liberate potent toxins. The spores of *Clostridium tetani* are widespread in the soil and frequently gain access to the body through wounds. Puncture wounds (e.g. by splinters, nails, on the dirty needles of drug addicts) are particularly dangerous because they provide the anaerobic conditions needed for germination and growth of the organism. When this occurs, a protein toxin is liberated. It appears to block inhibitory synapses in the spinal cord and in the brain. Consequently reciprocal inhibition of antagonistic pairs of muscles ceases (see Section 29.4), and the victim suffers violent muscle spasms. Fortunately, the disease (called tetanus) is now very rare in developed countries, thanks to almost universal immunization against the toxin. Chemical alteration of the toxin produces a harmless *toxoid* that still retains the antigenic determinants of the toxin. Incorporated in a vaccine, the toxoid provides a relatively long-lasting immunity against the effects of the toxin.

Although *Clostridium botulinum* does not infect humans, it is notorious nonetheless. This is because of the toxin it produces while it grows slowly in improperly canned food. As little as 1 μg of this toxin eaten along with an uncooked bean or mushroom can cause death. The toxin blocks the release of ACh from synaptic knobs and motor end plates. Thus the victim shows evidence of sympathetic nervous activity (dilation of pupils, inhibition of urination) and skeletal muscle weakness. When the intercostal muscles become affected, breathing ceases. The toxin is a protein and is quickly (10 minutes) denatured at 100° C. Thus boiling of home-canned products makes them safe to eat.

The term bacillus not only refers to all rod-shaped bacteria but, unfortunately, is also used as the genus name for a single group of gram-positive rods. *Bacillus anthracis* causes anthrax. This is primarily a disease of domestic animals like cows, sheep, and goats. On occasions it is contracted by people working with these animals or their products (e.g. unsterilized pig bristles in shaving brushes.) Before the days of antibiotics, the mortality rate of human infection was quite high.

B. anthracis was the first bacterium to be shown (by the pioneering microbiologist Robert Koch) to cause disease. This discovery paved the way for Louis Pasteur to develop a vaccine. The success of this vaccine, the first since Jenner, opened the modern era of immunology (in 1881).

Because of the dramatic impact of the relatively few disease-causing bacteria, it is easy to overlook the many beneficial bacteria that exist. *Bacillus subtilis,* a common soil bacterium, is the source of the antibiotic bacitracin. Gram-positive rods of the genus *Lactobacillus* are essential in the converting of milk into cheese and butter (and yogurt).

34.7 GRAM-POSITIVE COCCI

Many organisms in this group grow in characteristic colonies. Staphylococci (genus *Staphylococcus*) form

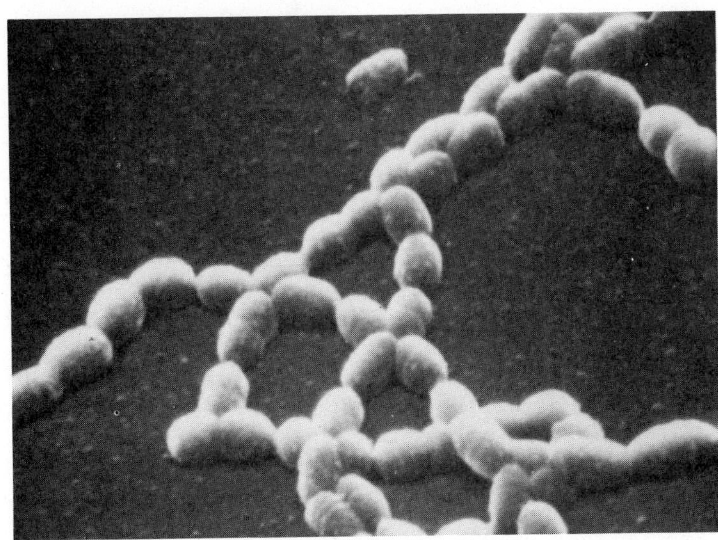

FIG. 34.8 Streptococci as seen under the scanning electron microscope (10,000 ×). This species is *Streptococcus mutans,* a common inhabitant of the mouth and the major cause of dental caries. (Courtesy Naval Dental Research Institute, Great Lakes, Illinois.)

flat packets of cells. Two species are especially common. *Staphylococcus albus* is probably growing right now on your skin. *Staphylococcus aureus* is also a frequent inhabitant of the skin, nasal passages, and gastrointestinal tract. Aside from the indignities of acne and an occasional pimple, we live in relative harmony with these organisms. However, if they are introduced *under* the skin because of injury, burns, etc. they can create serious abscesses. At one time, these were readily treated by antibiotics, but the evolution—primarily within hospitals—of antibiotic-resistant strains of staphylococci poses a real threat, especially to newborn babies and surgical patients.

Staphylococci grow luxuriantly in food, especially cream products. They secrete a toxin as they do so. Ingestion of the food, e.g. a Boston cream pie, can produce violent stomach upsets. The best way to prevent this common form of food poisoning is refrigeration of food and being sure that food handlers do not have open sores on their hands.

Streptococci grow in chains (Fig. 34.8). They are responsible for such common ailments as strep throat, impetigo, and middle ear infections. More rarely, streptococcal infections can lead to the serious complications associated with scarlet fever (a result of toxin production by the organism) and rheumatic fever. Fortunately, prompt antibiotic therapy can usually prevent the serious complications from developing.

Pneumococci, the bacteria usually responsible for bacterial pneumonia, are probably also members of the streptococcal group. However, they tend to grow in pairs, a feature that causes some workers to classify them in the genus *Diplococcus.* Bacterial pneumonia usually responds quickly to antibiotic therapy and except for the aged and infirm is seldom life-threatening any more.

The cell wall of virulent pneumococci is surrounded by a polysaccharide capsule (look back at Fig. 30.4). In Chapter 17, we examined the role this capsule played in discovering that the genetic code resides in DNA. In Chapter 30, we examined the important role that the pneumococcal capsule plays in the onset of the disease (by inhibiting phagocytosis) and in recovery from the disease (by the induction of antibodies directed against the polysaccharide).

34.8 GRAM-NEGATIVE RODS

The number of species of gram-negative bacilli is enormous. But any list would have to start with that most thoroughly studied of all creatures (with the possible exception of ourselves) *Escherichia coli.* Only rarely does this universal inhabitant of the human gut cause damage to its host. It may, in fact, help us by synthesizing vitamin K and some of the B vitamins, which we can then absorb from our intestinal contents.

Some of the gram-negative inhabitants of the human intestine are not so beneficient. *Salmonella typhi* causes typhoid fever, a common and serious epidemic disease where sanitation is poor. Infected individuals that recover (most do) may become "carriers" of the organism, that is, they continue to harbor them—usually in their gall bladder. The bacteria pass from there to the intestine (in the bile) and out in the feces. For obvious reasons, typhoid carriers should stay out of kitchens. They pose a sufficient threat, in fact, that public health departments like to keep close tabs on their whereabouts and activities.

Vibrio cholerae is the notorious agent of cholera, one of the most devastating of the intestinal diseases. The organism liberates a toxin that causes massive diarrhea (10–15 liters/day) and loss of salts. Unless the water and salts are replaced quickly, the victim may die (of shock) in a few hours. Like other intestinal diseases, cholera is contracted by ingestion of food or, more often, water that is heavily contaminated with the organism.

Perhaps no microorganism has caused more devastation to and panic in human populations than *Yersina pestis.* This is the bacillus that causes plague (also called the "black death"). The organism is ordinarily transmitted to humans by the bite of an infected rat flea. As it spreads into the lymph nodes, it causes them to become greatly swollen, hence the name "bubonic" (bubo = swelling of a lymph node) plague. Once in the lungs, however, the organism can be spread directly from human to human, causing the rapidly lethal (2–3 days) "pneumonic" plague. Albert Camus vividly describes both forms of the disease in his novel *The Plague* (*La Peste*).

Untreated, 50–75% of the cases of bubonic plague end fatally, while the mortality rate of untreated pneumonic plague is virtually 100%. No wonder that the recurrent epidemics of plague in Europe that began in the fourteenth century caused such devastation. In just three years (1348–50), at least one-quarter of the population of Europe succumbed to the disease. In some cities, the death rate was even higher. It is estimated that the "great dying" of this period reduced the population of Siena from 42,000 to 15,000. In Florence, the epidemic and the terror it created provided the setting for Boccaccio's *Decameron.*

No major epidemics of plague have occurred in this century although small outbreaks occur in Asia. The threat is not entirely gone, however. *Yersina pestis* still flourishes in some rodent populations (e.g. ground squirrels, prairie dogs) in the western United States. Each year a few cases of human plague—primarily among small-game hunters—occur. Fortunately, prompt treatment with antibiotics usually brings a swift recovery. Whether this reservoir of *Y. pestis* in the rodent population will ever serve as the source of a major human epidemic is an open question.

34.9 GRAM-NEGATIVE COCCI

Two members of this group of special concern to us are *Neisseria meningitidis* and *Neisseria gonorrhoeae.* The first causes meningococcal meningitis, an extremely serious infection of the meninges common in very young children and in military camps. The second causes one of the most widespread of human diseases: gonorrhea (a million cases are reported in the U.S. each year). The organism is spread directly from person to person by sexual contact. In the male, it invades the urethra—causing a discharge of pus—and often establishes itself in the prostate gland and epididymis. In the female, it spreads from the vagina to the cervix and Fallopian tubes. If the infection is untreated, the resulting damage to the Fallopian tubes may obstruct the passage of eggs and thus cause sterility.

For many years, penicillin has provided a quick (if often only temporary) cure for the disease. However, in 1976, a penicillin-resistant strain appeared in the U.S. (acquired in the Philippines). Since then resistant strains have cropped up in many parts of the world. In a way, it is surprising that years of exposure of the organism to penicillin had not led sooner to the evolution of resistance.

34.10 SPIRILLA

The rigid cell wall of spirilla gives them a helical shape (Fig. 34.1). They are gram-negative and motile. Most of them are found in water, both fresh and salt. One species, however, is a frequent inhabitant of the human mouth. Perhaps with a toothpick, slide, stain, and a microscope, you can find it.

34.11 ACTINOMYCETES AND THEIR RELATIVES

Most of the members of this group grow as thin filaments—like a mold—rather than as single cells. For this reason, they were long thought to be fungi. Despite their similarities of growth pattern, they are certainly not fungi. Fungi are eukaryotes and the actinomycetes are prokaryotes, with all that fundamental distinction implies with respect to cell structure and biochemistry (see Section 34.1).

While pathogenic actinomycetes exist, the group

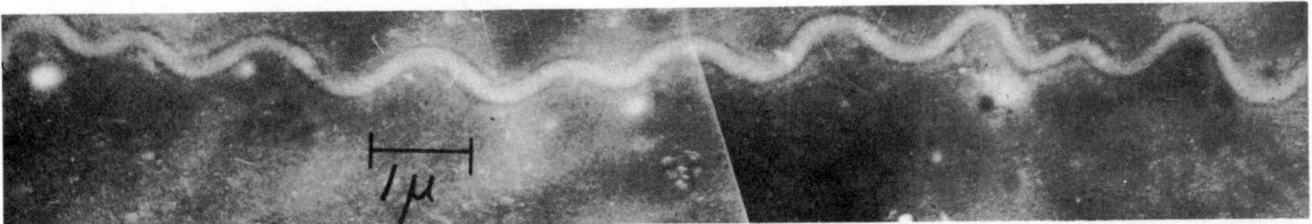

FIG. 34.9 The spirochete that causes syphilis. (Courtesy of Harry E. Morton.)

is far more significant in other ways. Actinomycetes are dominant members of the microbial population in soil. Here they play a major role in the decay of organic wastes. Many of these soil inhabitants are also important sources of antibiotics. Streptomycin, erythromycin, chloramphenicol (sold as "Chloromycetin"), and the tetracyclines ("Aureomycin" and "Terramycin") are all products of actinomycetes. So is actinomycin D. Although too toxic for general human use, the ability of actinomycin D to block DNA replication and transcription has provided molecular biologists with a tremendously powerful research tool (see, for example, Sections 19.2 and 24.6).

The **mycobacteria** and the **corynebacteria** are close relatives of the actinomycetes. Two species of mycobacteria cause serious and chronic diseases in humans: tuberculosis and leprosy. *Corynebacterium diphtheriae* is the agent that causes diphtheria. As in tetanus, the danger in diphtheria comes not from the invasion of tissues (in the throat) by the bacteria, but from the lethal toxin they produce. Diphtheria toxin exerts its poisonous effect in a most specific manner. It catalyzes the inactivation of a factor necessary for amino acids to be added to the polypeptide chain being synthesized on the ribosome. Sensibly enough, the toxin has no such effect on the ribosomes of prokaryotes (or those of chloroplasts and mitochondria!).

Diphtheria toxin is a protein, but as A. M. Pappenheimer, Jr. and his colleagues demonstrated in 1971, the structural gene coding for it is not the property of the bacterium but, instead, a bacteriophage that can infect the bacterium and become incorporated into its genome.

Diphtheria toxin can be harvested from cultures of the organism. Treatment with formaldehyde converts it into the harmless *toxoid*. Immunization with diphtheria toxoid (usually incorporated with tetanus toxoid and an inactivated preparation of the whooping cough germ in a "triple" vaccine) has caused a great reduction in the incidence of the disease, although sporadic outbreaks still occur in the U.S.

34.12 SPIROCHETES

Spirochetes are long, thin, helix-shaped bacteria that range in length from just a few to as many as 500 μm (Fig. 34.9). Their cell wall is not so rigid as that of spirilla so they are able to bend very easily. Although some spirochetes live harmlessly in fresh water, soil, or in the bodies of animals, others are serious parasites. Perhaps the most notorious of these is the spirochete that causes syphilis, a sexually transmitted disease of humans.

34.13 MYCOPLASMAS

Mycoplasmas are tiny, nonmotile bacteria without cell walls. Some are free-living while others parasitize plants, insects, and other animals. The first to be discovered causes a type of pneumonia, called pleuropneumonia, in cattle. For this reason, the mycoplasmas are also known as the pleuropneumonia-like organisms or PPLO. Another member of the group causes a disease in humans called primary atypical pneumonia.

The mycoplasmas have the distinction of including the smallest free-living organisms known. Although many are so small (0.1 μm) as to be visible only under the electron microscope, they contain everything needed to carry on all the activities of life.

34.14 RICKETTSIAS AND CHLAMYDIAE

Rickettsias, like mycoplasmas, are so small (Fig. 34.10) that they can barely be distinguished in the light microscope. They differ from the mycoplasmas in that almost all of them are obligate intracellular parasites. This means that they can grow and reproduce only so long as they are present *within* the

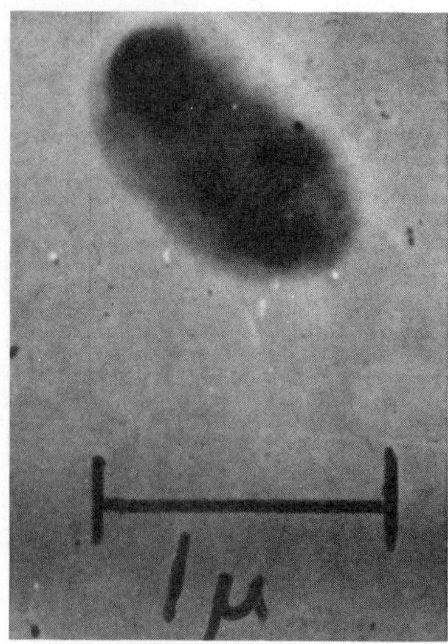

FIG. 34.10 The rickettsia that causes typhus fever. (Courtesy of Thomas F. Anderson.)

living cells of their host—certain arthropods (ticks, mites, lice, and fleas) and mammals. While rickettsia cells contain all the metabolic machinery of life, they probably depend on their host cell to keep them well supplied with coenzymes like ATP.

Typhus fever is caused by a rickettsia transmitted from human to human by the blood-sucking habits of the body louse. Rocky Mountain spotted fever is caused by a rickettsia transmitted through the bite of an infected tick. Despite its name, this disease occurs throughout the U.S. However, the severity of the disease varies from area to area (the mortality rates of untreated cases ranges from 5% in some regions to as high as 90% in the Bitterroot Valley area of Montana). Fortunately, antibiotic therapy is very effective against these organisms.

Chlamydiae resemble rickettsias in a number of ways. They, too, are tiny, obligate, intracellular parasites. The best-known example in North America is the agent that causes **psittacosis** or "parrot fever." Actually, many kinds of birds (in addition to occasional humans) serve as hosts for this organism, so the name ornithosis is now preferred.

Of far greater social and economic impact are the chlamydiae that cause **trachoma.** This is an eye infection that all too frequently results in permanent blindness. It is a serious public health problem in many parts of the world but is especially prevalent in the desert regions of Asia, Africa, and the Near East. It has been estimated that 400 million people now suffer from trachoma and 6 million have been blinded by it.

34.15 THE GLIDING BACTERIA

These prokaryotes receive their name from their mode of locomotion: gliding over their substrate. The actual mechanism by which they do this is unknown. Many species of gliding bacteria are unicellular while others form long filaments of cells. The cells in the filament share a common wall.

Most gliding bacteria are heterotrophic, but a few (Fig. 34.11) are chemoautotrophs. The latter oxidize H_2S for energy just as the chemoautotrophic sulfur bacteria described in Section 34.5 do.

The filamentous gliding bacteria are of special interest to us because they so closely resemble the other major group of prokaryotes; the blue-green algae. The resemblances are so close, in fact, that the filamentous gliding bacteria may simply represent blue-green algae that have lost the power of photosynthesis.

34.16 THE BLUE-GREEN ALGAE (PHYLUM CYANOPHYTA)

The blue-green algae are photosynthetic and prokaryotic. But they differ from the photosynthetic bacteria in several important ways. Their chlorophyll is chlorophyll *a,* the same molecule found in plants (and other algae). Furthermore, they are able to use water as a source of electrons with which to reduce carbon dioxide to carbohydrates.

$$CO_2 + 2H_2O \xrightarrow{\text{light}} (CH_2O) + H_2O + O_2$$

This is because they possess Photosystem II as well as Photosystem I.

Like gliding bacteria, the blue-green algae are encased in a wall of peptidoglycan surrounded by a gummy sheath. Some species are unicellular; some grow as filaments of connected cells (Fig. 34.12). Those that are capable of locomotion do so by gliding. Approximately 2000 species have been identified.

A number of filamentous blue-green algae can fix atmospheric nitrogen. This is done in *heterocysts* —colorless cells interspersed among the photosynthetic cells of the chain (Fig. 34.13). It is these species that "bloom" when phosphates (but not nitrates) become available in lakes and other bodies of fresh water. Nitrogen-fixing blue-green algae are also important in maintaining the fertility of rice paddies.

Blue-green algae are exceedingly hardy. Some species grow luxuriantly in the hot springs of Yellow-

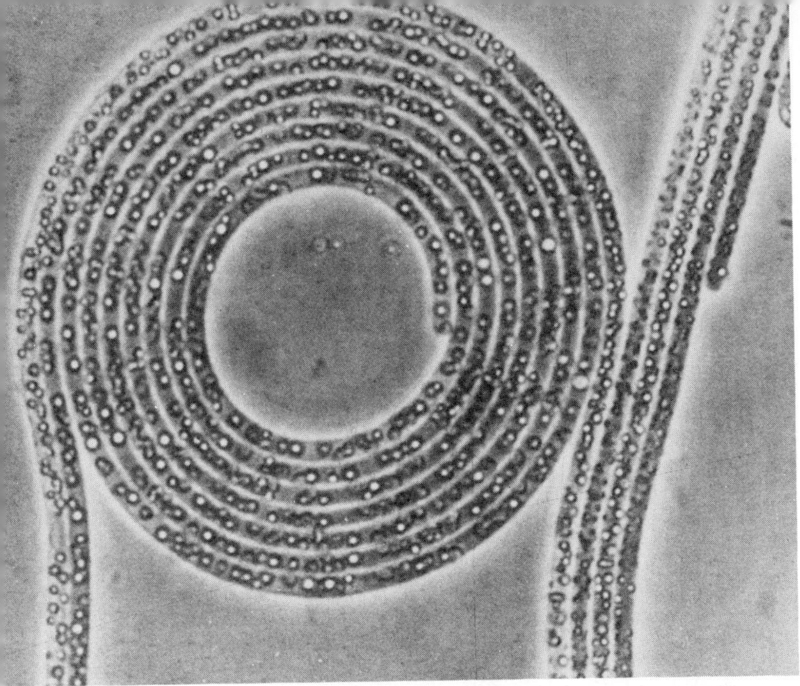

FIG. 34.11 *Beggiatoa,* one of the gliding bacteria. This chemoautotrophic organism secures its energy by oxidizing H_2S to free sulfur (visible as granules within the cells). (Courtesy John M. Larkin, Louisiana State University.)

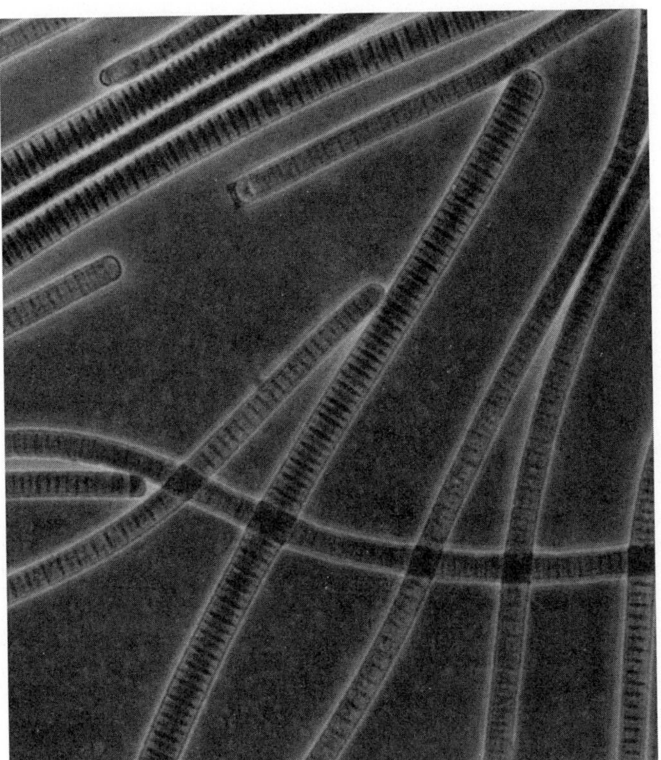

FIG. 34.12 *Oscillatoria,* a filamentous blue-green alga (1000 ×). Each of the disks in the chains is one cell.

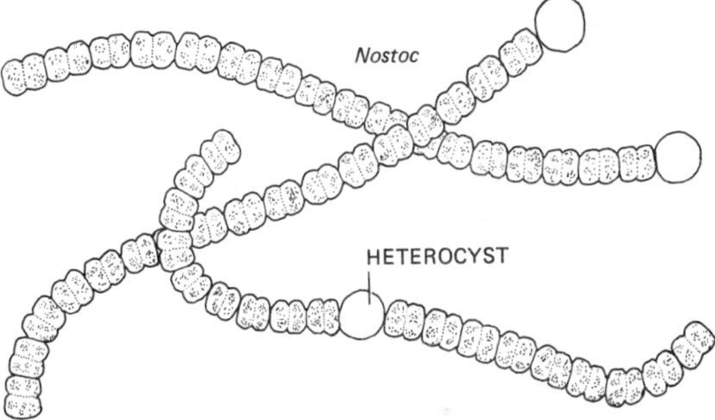

FIG. 34.13 *Nostoc,* a common blue-green alga. The colorless heterocysts fix atmospheric nitrogen. If the heterocysts carried on photosynthesis like the other cells, the liberation of oxygen would poison the nitrogenase enzyme responsible for nitrogen fixation.

FIG. 34.14 A hot spring in Yellowstone National Park. Floating at the edge of the water is a slimy mat of blue-green algae. The temperature of the water on the day this picture was taken was 72°C, hot enough to cook an egg. (The light areas at the right are caused by steam rising from the water.)

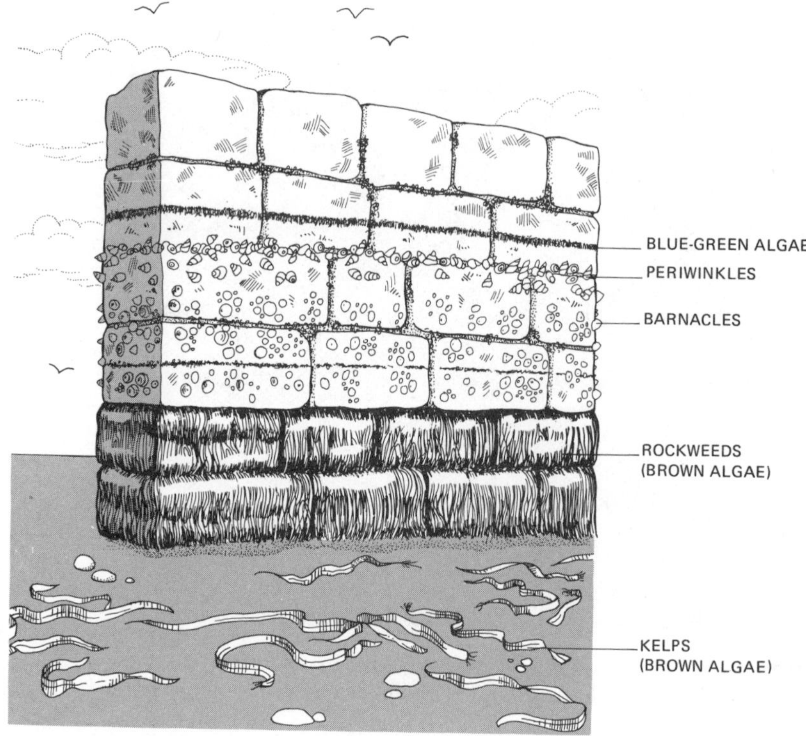

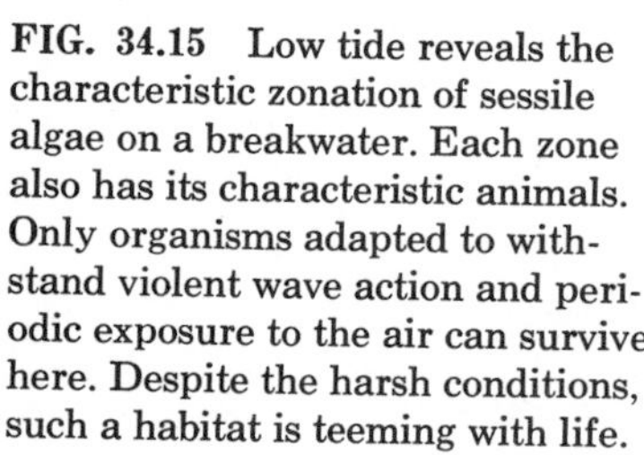

FIG. 34.15 Low tide reveals the characteristic zonation of sessile algae on a breakwater. Each zone also has its characteristic animals. Only organisms adapted to withstand violent wave action and periodic exposure to the air can survive here. Despite the harsh conditions, such a habitat is teeming with life.

stone National Park at temperatures hot enough to cook an egg (Fig. 34.14). The black band found at the high-tide mark on rocks along the sea coast is formed by blue-green algae (Fig. 34.15). When you realize that all they need to live is light, air (for N_2 and CO_2), water, and a few inorganic ions (e.g. $PO_4^{\equiv}$), their ability to live in such harsh locations is understandable. Even when deprived of water (e.g. at low tide) their gelatinous sheath helps to keep them from drying out.

In addition to chlorophyll and beta-carotene, the blue-green algae contain one or both of two additional pigments: a blue pigment called phycocyanin and a red pigment called phycoerythrin. The simple mixture of chlorophyll and phycocyanin in some species imparts the blue-green color that gives the entire group its common name. But those species that contain phycoerythrin appear red, purple, brown, or even black. The Red Sea gets its name from the periodic blooms of a red colored blue-green alga that occur in its waters.

Among the oldest fossils that have been discovered to date are filamentous structures found in limestone formations in South Africa. These formations are thought to be approximately 2.3 billion (2.3×10^9) years old. Many of the filaments resemble modern blue-green algae (e.g. *Oscillatoria*). Some even reveal what look like heterocysts.

In our discussion of the origin of life (Chapter 2), we noted the evidence that the atmosphere of the early earth contained little or no oxygen. In this chapter we have seen that photosynthetic bacteria—being anaerobic—could have survived in such circumstances. But if blue-green algae had appeared by 2.3 billion years ago, their water-based photosynthesis must have been giving off molecular oxygen (O_2) to the atmosphere. With the accumulation of oxygen in the atmosphere, the way was open for heterotrophic organisms to secure their energy by cellular respiration instead of the inefficient method of fermentation. With an inexhaustible source of organic molecules (produced by the autotrophs) and an efficient method of extracting energy from them, the full potentialities of life could be realized. In the following chapter, we shall examine the possible directions that evolution took once the prokaryotes had laid the foundations.

34.17 THE VIRUSES

Viruses are not prokaryotes. They have almost none of the features characteristic of prokaryotes—peptidoglycan wall, ribosomes, etc. They have no enzymatic machinery with which to manufacture ATP or carry out other aspects of metabolism. They are incapable of reproducing themselves. In fact, one could well argue that they fail to meet so many of the criteria of life (see Chapter 1) that they cannot even be considered to be living organisms.

So why study them here? Why study them at all? To answer the second question first, they are undeniably related to living things. If you have ever suffered a bout of influenza or measles, or know of victims of polio, you should have a keen interest in learning about the agents—viruses—that cause these diseases. As for the first question, the answer is largely historical and methodological. Many of the scientists who study bacteria also study viruses. And many of the techniques employed in studying bacteria are also used in studying viruses—especially those viruses, the bacteriophages, that infect bacteria.

If viruses are not living organisms, what are they? Let us examine their properties and, in so doing, we may be able to arrive at an understanding of what they are. Viruses have two phases to their existence, one inside of living cells, one outside. Outside of their host, viruses consist of discrete particles. Generally, these are very small; some have diameters of only 9 nm—half the size of a ribosome. However, a few viruses are relatively large. The vaccinia virus, for example, has a diameter of about 230 nm (0.23 μm), which makes it larger than some bacteria.

It was the tiny size of viruses that led to their discovery. In 1892, the Russian scientist Iwanowsky prepared an extract of tobacco leaves that were afflicted with *tobacco mosaic disease*. He then filtered the extract through a filter with pores so fine that any bacteria present would be trapped. Surprisingly, the bacteria-free filtrate was fully infective, transmitting the disease when rubbed on healthy tobacco leaves. Thus was discovered tobacco mosaic virus **(TMV)**.

In the years that followed, other "filterable viruses" were discovered that were responsible for infections in other plants, in animals, and in bacteria (the bacteriophages). In 1935, Wendell Stanley electrified the scientific world by crystallizing TMV and showing that the crystals retained their infectivity indefinitely, stored on a shelf in a bottle. This was hardly the thing one would expect to be able to do with a living organism!

Virus particles (called virions) consist of:

1. An interior core of nucleic acid. For some, the nucleic acid is DNA; for some it is RNA. None has

both. In most cases, the nucleic acid is present as a single molecule.

2. A surrounding coat of protein, called a **capsid.** The capsid protects the nucleic acid core, determines what kind of cell the virus particle will be able to attach to, and assists in one way or another with the insertion of the virion (or at least its core) into the host cell. Some viruses have other ingredients, e.g. lipids, in their capsids, and these substances may be derived from cell constituents of their host.

The DNA viruses. Most of the viruses containing DNA cores have their DNA in the form of the Watson-Crick double helix. Some important DNA viruses are the smallpox (and vaccinia) viruses, the herpes simplex virus (HSV—the cause of "cold sores"), the SV40 virus that infects primate cells and causes tumors in rodent cells, and a number of bacteriophages.

The essential elements in the infective cycle of the DNA bacteriophages was examined in Chapter 17 (look back at Fig. 17.4). The process consists of five stages.

1. The virions attach to the surface of their host cell. The proteins of the capsid serve to inject the DNA core into the cell. It was this separation of core and capsid that was the basis of the demonstration by Hershey and Chase that DNA, not protein, was the genetic material (see Section 17.1).

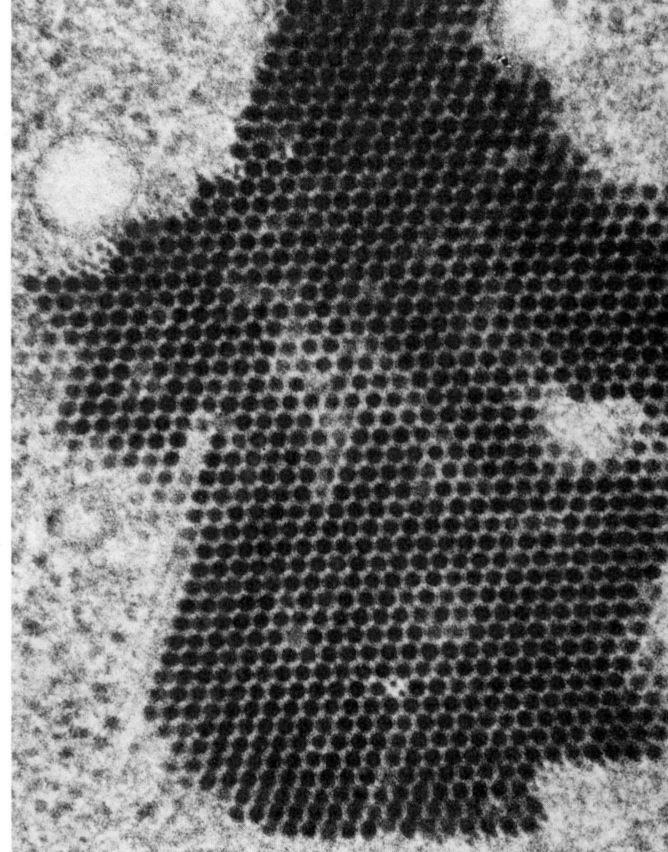

FIG. 34.17 Virions of polio virus packed in a crystalline array within the cytoplasm of a mammalian cell. (102,000 ×, courtesy of Dr. Samuel Dales, University of Western Ontario.)

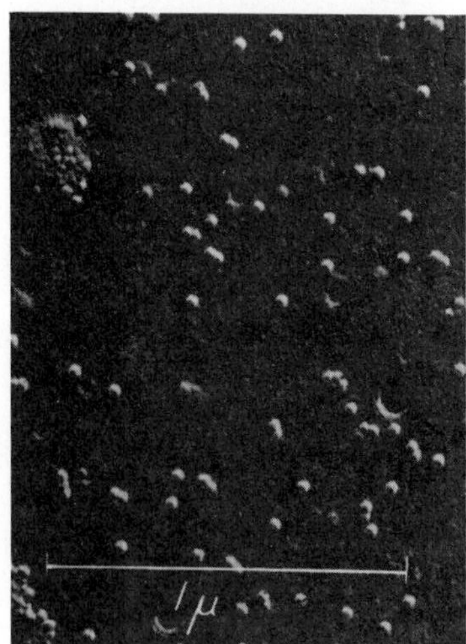

FIG. 34.16 Electron micrograph of one of the polio viruses. (Courtesy A. R. Taylor.)

2. Once within the cell, some of the bacteriophage genes (the "early" genes) are *transcribed* (by the host's RNA polymerase) and *translated* (by the host's ribosomes, *t*RNA, etc.) to produce enzymes that will make many copies of the phage DNA and will turn off (even destroy) the host's DNA.

3. As fresh copies of phage DNA accumulate, other genes (the "late" genes) are transcribed and translated to form the proteins of the capsid.

4. The stockpile of DNA cores and capsid proteins are assembled into complete virons.

5. Another "late" gene is transcribed and translated into molecules of lysozyme. The lysozyme attacks the peptidoglycan wall (from the inside, of course). Eventually the cell ruptures and releases its

content of virions. The cycle is now complete and ready to be repeated.

The RNA viruses. In most RNA-containing viruses, the RNA is in single-stranded form. Some important RNA viruses are those that cause polio (Fig. 34.16), yellow fever, rabies, equine encephalitis, influenza, mumps, and measles. TMV is an RNA virus, as is the bacteriophage MS2 which we studied so thoroughly in Chapter 18 (see Section 18.5).

The infection cycle of RNA viruses, like MS2, is similar in many respects to that of DNA viruses. After the RNA strand gets into its host (*E. coli,* for MS2), it serves as a messenger RNA and a portion is *translated* into an RNA polymerase which proceeds to manufacture additional copies of the original strand. To do so, however, requires that the original strand first serve as the template for the synthesis of its complementary "mirror image," the so-called "replicative intermediate." From *this* is synthesized RNA strands identical to the original. As the newly manufactured RNA molecules accumulate, those portions coding for the capsid protein (or proteins) are translated. Assembly of the complete virions (Fig. 34.17) and their release from the cell follow.

Disappearing viruses. The infective cycles described in the previous paragraphs end in the death of the host cell. Animal cells, in most instances, disintegrate, releasing their content of newly assembled virions. Bacterial cells literally burst, a process called lysis. Because of these fates, these infective cycles are called **lytic cycles.**

Sometimes, however, after a bacterial cell is infected by a DNA bacteriophage, the intracellular events of the lytic cycle are not completed. Soon the bacterium resumes its normal existence, including reproducing itself. Where has the virus gone? It is still there, and, in fact, is present in every one of the descendants (with occasional exceptions) of the originally infected cells. These cells constitute an infected clone. That they still harbor the virus can be demonstrated by irradiating the cells with ultraviolet rays or treating them with certain chemicals. Such treatment restores the normal lytic cycle (the phage is said to have been "rescued"—hardly the case for its host!).

The stable relationship between a virus and its bacterial host is called **lysogeny.** The viral DNA actually becomes incorporated into the genome of its host and becomes replicated when the host's DNA is replicated prior to each cell division. During lysogeny, the phage is called a **prophage.** In some cases, the prophage becomes inserted directly into the chromosome of its host. This occurs at a fixed location on the chromosome, which can be mapped by standard genetic techniques. In fact, when the phage is "rescued," the released virions may contain some host genes as well as their own. When these virions infect new hosts, they insert these *bacterial* genes into them. This process of genetic transfer—a virus-mediated "transformation"—is given the special name **transduction.**

What does the prophage do while it is a part of its host genome? It may well express certain of its genes. For example, the structural gene for the synthesis of diphtheria toxin is the property of the prophage (see Section 34.11), not of its host.

A similar disappearing act can occur with animal viruses (Fig. 34.18). Simian virus 40 (SV40) is a DNA virus that produces a lytic infection in kidney cells of the African green monkey. It can also infect other types of cells (e.g. human, mouse, rat, hamster), but in that case, the lytic cycle is seldom completed. Instead, the virus disappears in much the same way that lysogenic bacteriophages do. As in lysogeny, however, it leaves some clues to its presence. New antigens appear on the surface of the host cell (Fig. 34.18). And if the host cell is fused with an African green monkey kidney cell (see Section 16.18), the latter becomes infected and lyzed.

These nonlytic infections resemble lysogeny in another way as well. The DNA of the virus becomes incorporated in the genome of its host (e.g. SV40 becomes incorporated in chromosome No. 7 in human cells) and is replicated whenever the host cell goes through an S phase.

These hidden virus infections often have another result. The infected cells become *transformed.* The metabolism of the cell changes in a number of ways. Most important, though, is that the cell no longer divides in a controlled fashion. Consequently, these cells grow into malignant tumors = cancers. Therefore, such viruses are *oncogenic* or cancer-producing when introduced into the appropriate animals (e.g. rodents, birds). Viral transformation can also be carried out *in vitro,* that is, in cell culture. If the transformed cells are then introduced into an appropriate host, they cause cancers to develop. (Note that the use here of the term "transformation" is quite different from the meaning we gave it in discussing bacterial transformation in Chapter 17.)

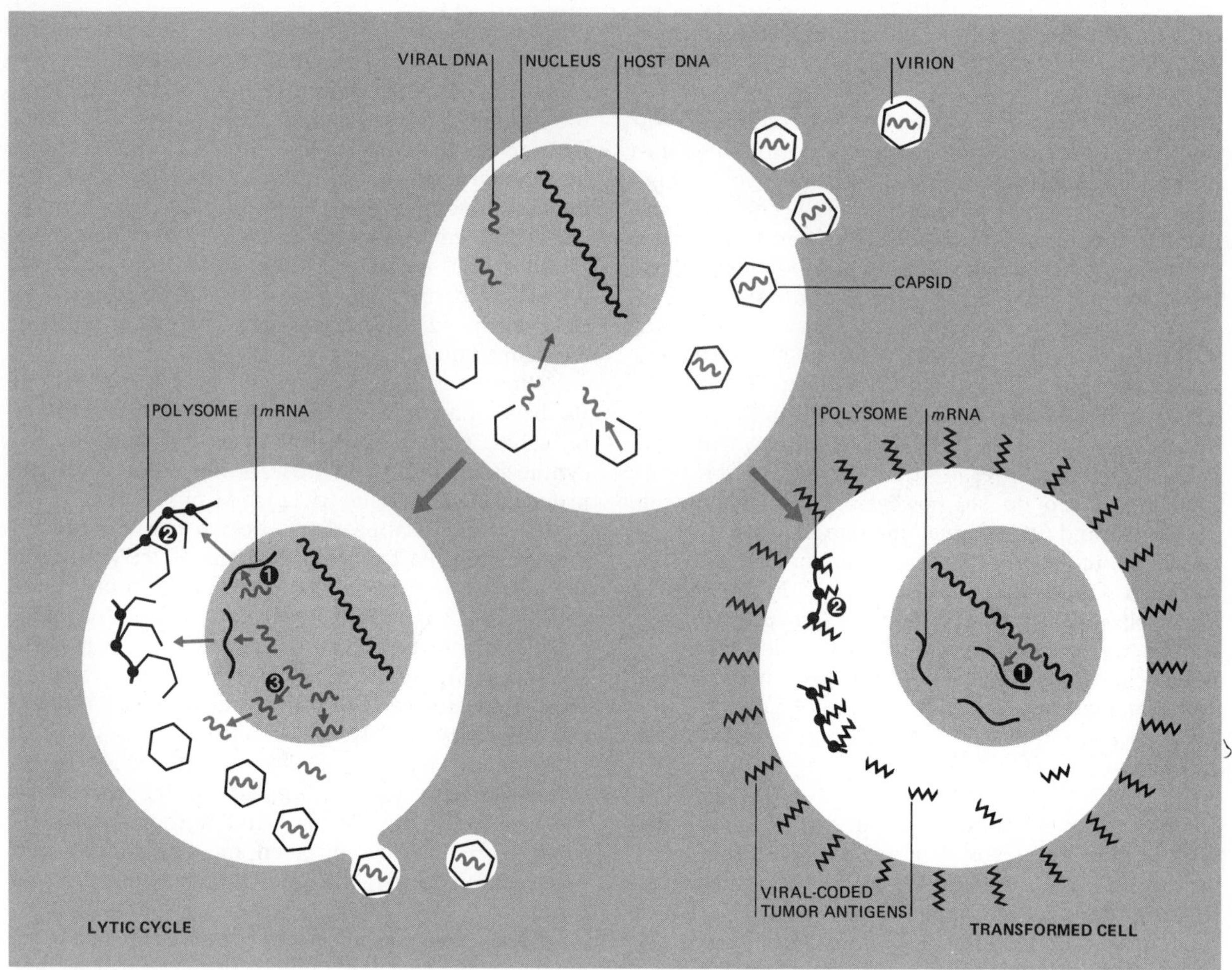

FIG. 34.18 Alternative fates of certain DNA viruses that infect mammalian cells. In some cells or under some conditions, the virus establishes a lytic cycle. Taking over the metabolic machinery of its host, the viral DNA is transcribed (1), translated (2), and replicated (3) to provide the materials for the assembly of new virons (lower left). In other cells or under other conditions (lower right), the viral DNA becomes incorporated into the DNA of its host. Transcription (1) and translation (2) of one or more of its genes produces antigens that are displayed on the surface of the cell. Such a transformed cell may begin to divide uncontrollably, that is, it may become cancerous.

A few RNA viruses are oncogenic; they, too, can transform animal cells. However, before their genetic information can become incorporated into the genome of their host, it must first be transcribed into DNA. It was this realization that led Temin and Baltimore to search for an RNA-dependent DNA polymerase in transformed cells. Their search revealed the existence of **reverse transcriptase** (see Section 18.7)—"reverse" because it causes genetic information to flow from RNA to DNA instead of in the usual direction of DNA to RNA.

Summary. So what is a virus? It is an infectious nucleic acid, packaged in an assembly of macromolecules (the capsid) that largely determines what cells it can infect. Once inside its host cell, this infectious nucleic acid may do one of two things. It may subvert virtually all the metabolic machinery of its host (e.g. enzymes of ATP production, ribosomes, *t*RNA) to one purpose: making additional copies of itself (in its lytic cycle). Or, it may go into hiding—masquerading for a time as part of its host's own genome. The consequences to the host *may* be minor, e.g. the occasional outbreak of cold sores as stress or exposure to light convert a latent herpes simplex virus (HSV) infection into a lytic one. But the consequences may be grave—as an infected host cell escapes from the normal regulatory mechanism of the body and grows uncontrollably into a cancer.

I leave it to you to decide if viruses are alive or not.

EXERCISES AND PROBLEMS

1. Distinguish between (a) true bacteria, (b) mycoplasmas, (c) rickettsias, (d) viruses.
2. For each of the following, indicate whether it is characteristic of prokaryotes or eukaryotes or both: (a) endoplasmic reticulum, (b) ribosomes, (c) spindle, (d) DNA, (e) D-amino acids, (f) histones, (g) mitosis, (h) poisoned by diphtheria toxin.
3. Distinguish between transformation, transduction, and conjugation in bacteria. What does each accomplish?
4. In what ways are bacteria and blue-green algae similar?
5. Some biologists have maintained that virus-like particles were the first form of life on earth. What evidence supports this hypothesis? What evidence weakens it?
6. The cells of *Oscillatoria erythraea* are bright red in color. Why, then, is this species classified as a blue-green alga?
7. In what ways are RNA and DNA viruses similar? In what ways are they different?

REFERENCES

1. STANIER, R. Y., E. A. ADELBERG, and J. INGRAHAM, *The Microbial World,* 4th ed., Prentice-Hall, Englewood Cliffs, N.J., 1976. A comprehensive treatise with emphasis on the bacteria.
2. BROCK, T. D., ed., *Milestones in Microbiology,* Prentice-Hall, Englewood Cliffs, N.J., 1961. Includes original reports by Pasteur (e.g. on lactic acid fermentation), by Koch (e.g. on anthrax and tuberculosis), by Stanley (on the crystallization of TMV), by Gram (on his stain), and many, many others.
3. ECHLIN, P., "The Blue-Green Algae," *Scientific American,* Offprint No. 1044, June, 1966.
4. SANDERS, F. K., *The Growth of Viruses,* Oxford Biology Readers, No. 64, Oxford University Press, Oxford, 1975.
5. SPECTOR, DEBORAH H., and D. BALTIMORE, "The Molecular Biology of Poliovirus," *Scientific American,* May, 1975. A detailed account of the lytic cycle of this RNA virus.
6. RAFFERTY, K. A., JR., "Herpes Viruses and Cancer," *Scientific American,* October, 1973. Describes how herpes viruses, on occasion, fail to complete their normal lytic cycle and, instead, induce malignancy in their host cell.
7. CAMPBELL, A. M., "How Viruses Insert Their DNA into the DNA of the Host Cell," *Scientific American,* Offprint No. 1347, December, 1976.

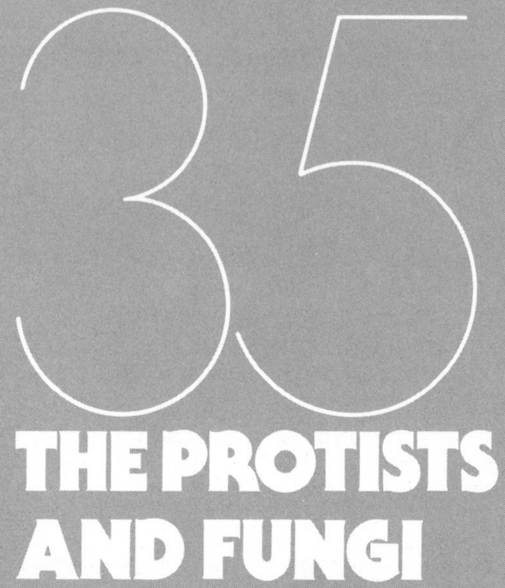

35 THE PROTISTS AND FUNGI

By any criterion, the eukaryotes dominate the earth. Of the 1.2 million or more kinds of living organisms found on the earth today, all but a few thousand are eukaryotes. How should this enormous assemblage be classified? Some biologists would prefer two kingdoms: the plants and the animals. However, such a scheme means that some forms (e.g. kelps, yeast) must be included with the plants even though they are probably no more closely related to a grass plant than you are. We shall, then, divide the eukaryotes into four groups or kingdoms: the **protists,** the **fungi,** the **plants,** and the **animals.** Of these, the protists were the first to evolve.

The Kingdom Protista

35.1 CHARACTERISTICS

Every phylum that we place in the Protista contains some unicellular members (with one exception). Several phyla also contain species whose members are made up of many cells, but in none of these is there the development of specialized tissues, organs, etc. that we find in plants and animals.

The name *Protista* means literally "the very first." Although their evolutionary relationships are quite obscure, it is reasonably certain that most of the phyla that we include in the Protista (11 of them) appeared on earth before either the plants or the animals. One or two may have evolved later, but independently of the plants and animals and without ever achieving the latter's structural complexity and diversity.

35.2 THE EVOLUTION OF EUKARYOTES

How did the protists evolve? Perhaps the most intriguing hypothesis that attempts to answer this question is that these first eukaryotes evolved by the *symbiotic association* of two or more kinds of prokaryotes. According to this theory, the mitochondria of eukaryotic cells were once aerobic bacteria living symbiotically within their host ("endosymbiosis"). Chloroplasts are thought to be derived from prokaryotic algae (e.g. the blue-green algae). It has even been suggested that endosymbiotic spirochetes provided the first flagella and cilia, and, with these, the basal bodies from which the spindle could be constructed and mitosis made possible. Let us examine some of the evidence which supports this theory.

Both mitochondria and chloroplasts (also basal bodies) are *semi*autonomous in the sense that they can duplicate themselves independently of the duplication of the cell in which they reside. This property becomes more remarkable when we realize that both mitochondria and chloroplasts have rudimentary genetic systems quite separate from the genetic system in the nucleus. Each has DNA and its own protein-synthesizing machinery. The extraordinary thing from the point of view of our hypothesis is that this genetic machinery is so similar to that of prokaryotes. The DNA is present as a single molecule, just as it is in bacteria and blue-green algae. It is not complexed with histones like the DNA of eukaryotic chromosomes. The relative proportion of $G + C$ to $A + T$ in mitochondrial DNA is like that of bacterial DNA, not eukaryotic DNA. Chloroplast DNA from *Euglena,* a eukaryotic protist, shows extensive sequence homologies with the genes for ribosomal RNA in blue-green algae.

The parallels extend to the protein-synthesizing machinery. The size and properties of the ribosomes found within mitochondria and chloroplasts are like those of bacteria and blue-green algae, not like those around them in the eukaryotic cytoplasm. A number of antibiotics, e.g. streptomycin, exert their effect by interfering with protein synthesis in bacteria but not in their eukaryotic host. These antibiotics do, however, inhibit protein synthesis within mitochondria and chloroplasts. Conversely, inhibitors of protein synthesis in eukaryotic cytoplasm such as diphtheria toxin have (sensibly enough) no such effect *either* on bacterial protein synthesis *or* on protein synthesis within chloroplasts and mitochondria. When ribosomes from *Euglena* chloroplasts are dissociated into their components and then allowed to reassociate with complementary subunits from *E. coli* ribosomes, the hybrid complexes that result work perfectly. The antibiotic rifampicin, which inhibits the RNA polymerase of bacteria, has no such effect on the RNA polymerase within the eukaryotic nucleus, but does inhibit the RNA polymerase within mitochondria.

All of this tells us that chloroplasts and mitochondria share many properties with prokaryotic cells. But what of evidence that prokaryotes can indeed form endosymbiotic relationships? A number of heterotrophic organisms exploit photosynthetic endosymbionts. In many cases, these endosymbionts are prokaryotes, i.e. blue-green algae. Figure 35.1 is an electron micrograph of a unicellular green alga that has lost its own *chloroplasts* but which is still able to carry on photosynthesis, thanks to the acqui-

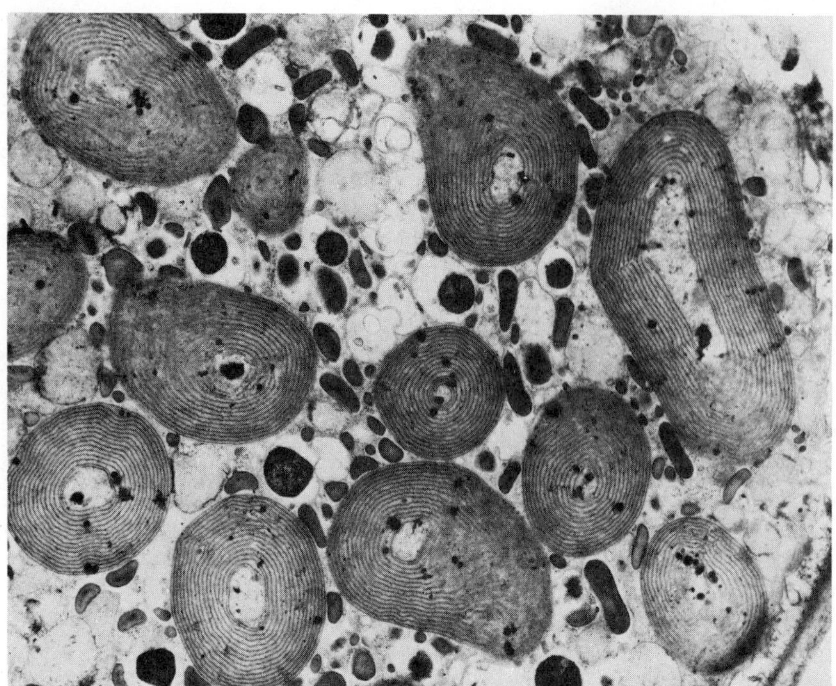

FIG. 35.1 *Glaucocystis nostochinearum,* a unicellular green alga (Chlorophyta). It has no chloroplasts of the type normally found in green algae, but carries on photosynthesis by means of organelles that resemble *blue-green* algae (large ovals). Note portion of the cell wall at the lower right. (10,000 ×, courtesy of Patrick Echlin.)

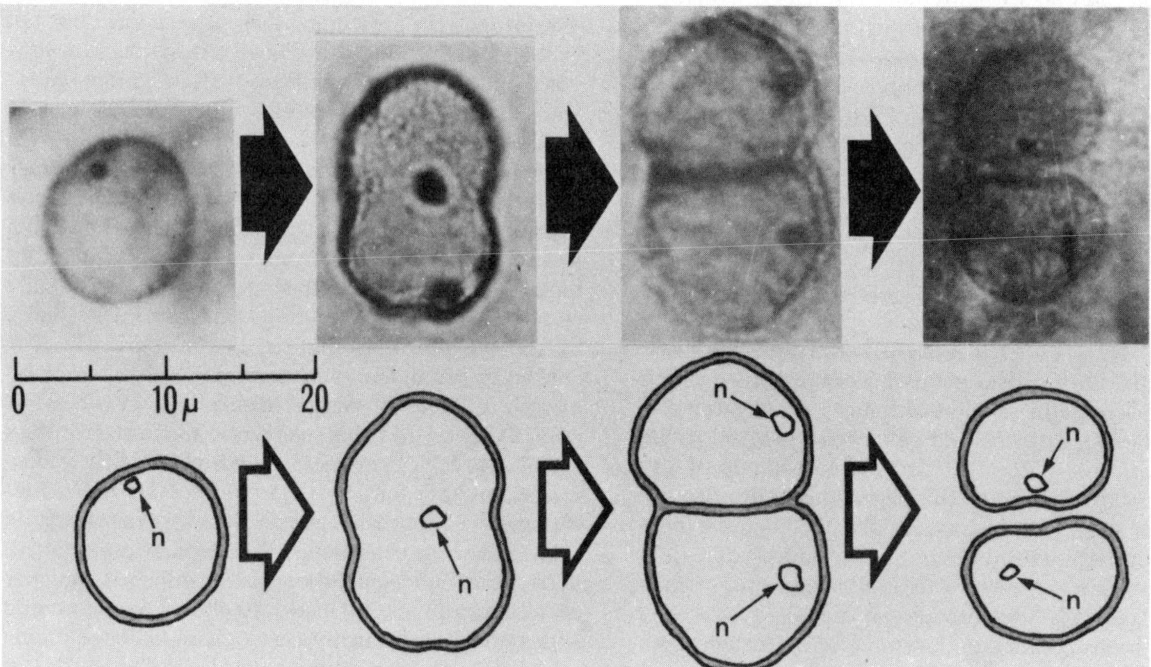

FIG. 35.2 Fossil algae from Australia. These fossils are 850 ± 100 million years old. If, as is suggested by the drawings, the dark bodies within the cells are nuclei, then these fossils (named *Glenobotrydion aenigmatis*!) are the earliest eukaryotes known. The sequence shown suggests cell division by mitosis. (Courtesy of J. William Schopf.)

sition of structures that resemble—both in structure and in biochemistry—unicellular blue-green algae. The relationship between the cell and its unusual "organelles" is now, at least, obligatory. When separated from one another, neither can grow alone. But their intimate relationship may simply reflect an intermediate stage in the process that has culminated in modern chloroplasts, which—although retaining their own genes—also depend upon nuclear genes of their "host" for essential functions.

Let us look at an example. A major step in the dark reactions of photosynthesis is the uptake of carbon dioxide by ribulose diphosphate (look back at Fig. 9.18). This step is catalyzed by the enzyme ribulose diphosphate carboxylase. This protein is made up of multiple copies of two kinds of subunits, one large, one small. The structural gene encoding the polypeptide of the large subunit is part of the DNA of the chloroplast. This subunit is synthesized on the chloroplast (prokaryotic) ribosomes. The small subunit, on the other hand, is encoded by a nuclear gene and is synthesized on eukaryotic ribosomes in the cytoplasm of the cell.

A similar situation occurs in mitochondria. Mitochondrial genes are responsible for certain mitochondrial components (e.g. the RNA for its own ribosomes, and the polypeptides for the synthesis of cytochrome *b*). Nuclear genes, on the other hand, are responsible for other essential mitochondrial components such as its cytochrome *c* and its RNA and DNA polymerases. If chloroplasts and mitochondria are truly the evolutionary outcome of an early endosymbiosis, it is clear why these organelles are no longer capable of an independent existence.

When did the eukaryotic protists first appear on the earth? Because almost all modern eukaryotes are strictly aerobic, we may conclude that they probably did not appear until prokaryotic algae (the blue-greens) had produced an oxygen-rich atmosphere. On geological grounds, this probably occurred some two billion (2×10^9) years ago. But just when life took advantage of the opportunity for aerobic existence and became eukaryotic is uncertain. Fossils approximately one billion years old show striking resemblances to modern eukaryotic algae (Fig. 35.2). However, deteriorating cultures of some modern blue-green algae develop inclusions that resemble the "nuclei" seen in Fig. 35.2. But even if *these* fossils are not eukaryotic, the chances are that eukaryotes evolved long before the appearance of unquestioned eukaryotic fossils—worms and the like—some 600 million (0.6×10^9) years ago.

What was the first eukaryotic protist? We simply do not know. But if we put into a cell a nucleus, mitochondria, some microtubules, and perhaps a flagellum, we would have made a creature that today we would call a protozoan.

The word *protozoa* is no longer a formal taxonomic term but rather a common name for some 30,000 tiny, single-celled, nongreen organisms. Beyond these simple traits they share, the group is extraordinarily diverse. We shall assign its members to one of four phyla on the basis of the method of locomotion they use. This is probably an inadequate criterion for claiming kinship. Thus we should remember that our classification is, at least in part, "unnatural." If a natural classification could be achieved, it probably would require the creation of several more phyla of protozoa.

35.3 THE RHIZOPODS (PHYLUM SARCODINA)

The members of this phylum move by the flowing of their cell contents into temporary projections called **pseudopodia.** The amoeba (Fig. 35.3) is the classic example of the group and has caused this type of movement to be described as *amoeboid*. The amoeba is about the size of one of the periods printed on this page. It lives in fresh water (including the fresh water in home humidifiers, where it has been implicated as the cause of an allergic reaction—"humidifier fever"—in some susceptible individuals). Far more serious are the parasitic amoebas found in the tropics. These intestinal parasites cause amoebic dysentery. Amoebas feed by phagocytosis. With rare exceptions, they reproduce only asexually.

Two large groups of marine protozoa are frequently included in this phylum. The first, the **foraminifera,** are protected by a many-chambered external skeleton made of calcium carbonate. The chalk cliffs of Dover, England, were formed from deep sediments of foraminifera shells. The second, the **radiolaria,** are especially abundant in the Indian and Pacific Oceans. These organisms possess an internal skeleton made of silica which is often of remarkable intricacy and beauty (Fig. 35.3). Although pseudopodia are present in both these groups, these organisms are probably not really closely related to the amoeba or even to each other. They should probably be assigned phyla of their own. The same holds true for the **heliozoans.** Most of these beautiful creatures (Fig. 35.4) are found in fresh water. The long axopods radiating out from the cell are supported by a precisely oriented array of microtubules (Fig. 35.5).

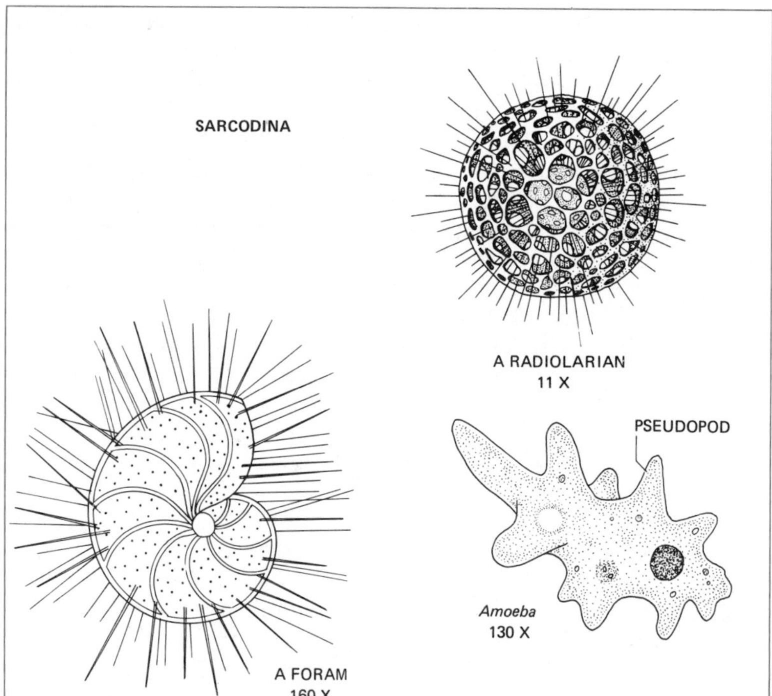
SARCODINA
A RADIOLARIAN
11 X
PSEUDOPOD
Amoeba
130 X
A FORAM
160 X

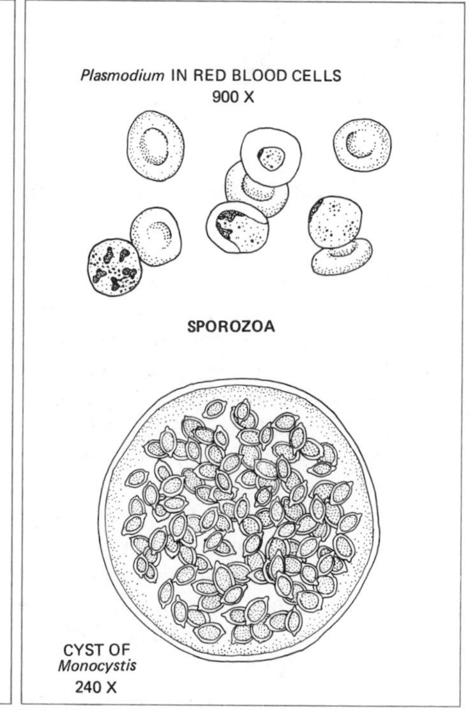
Plasmodium IN RED BLOOD CELLS
900 X
SPOROZOA
CYST OF
Monocystis
240 X

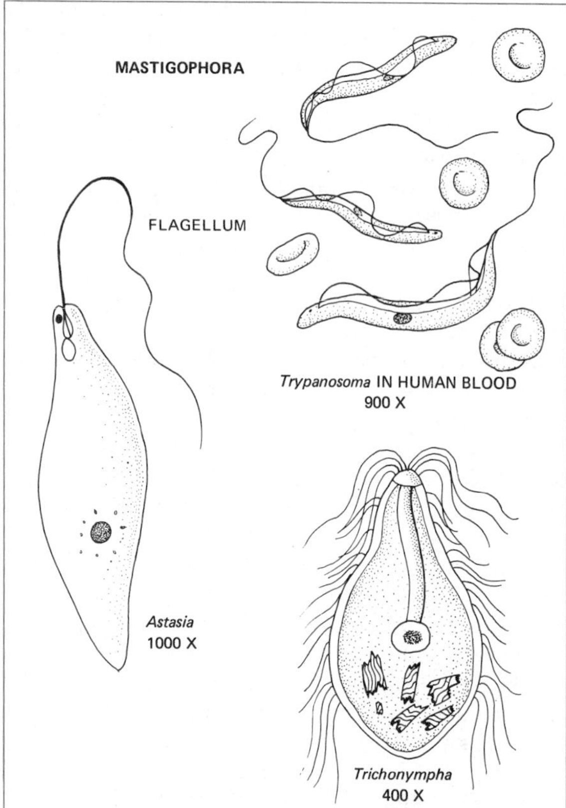
MASTIGOPHORA
FLAGELLUM
Trypanosoma IN HUMAN BLOOD
900 X
Astasia
1000 X
Trichonympha
400 X

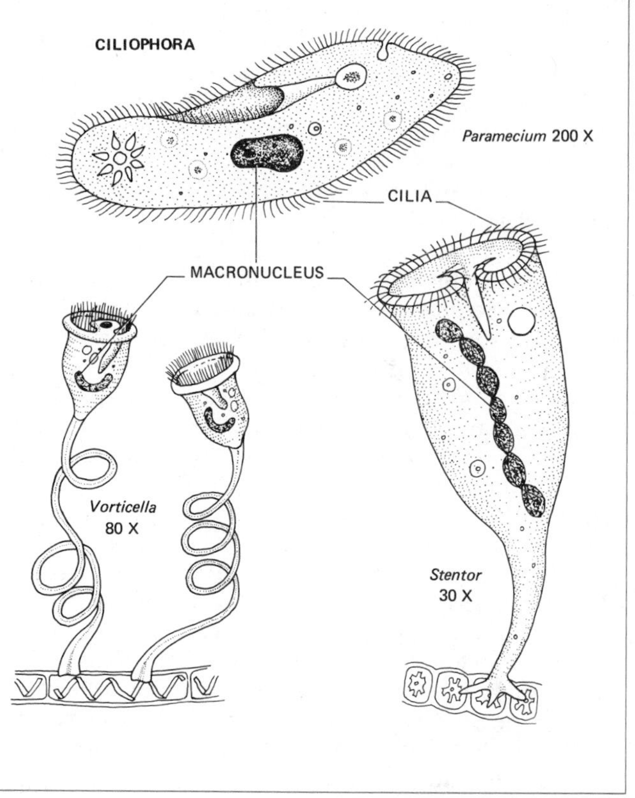
CILIOPHORA
Paramecium 200 X
CILIA
MACRONUCLEUS
Vorticella
80 X
Stentor
30 X

◄ **FIG. 35.3** Representative protozoans from each of four phyla.

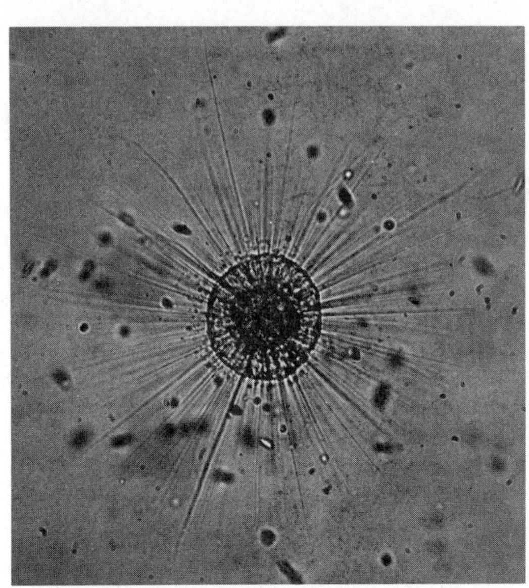

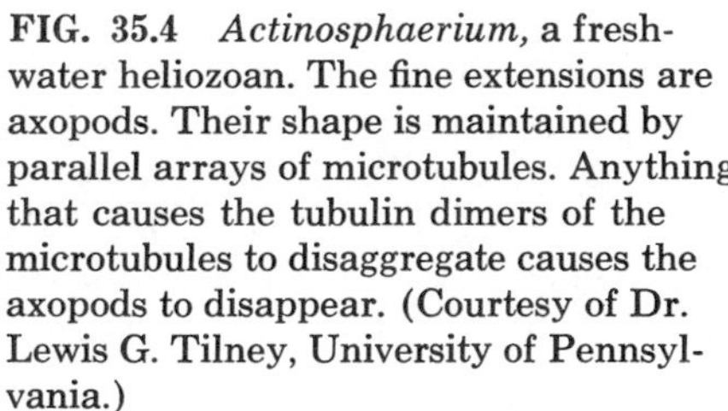

FIG. 35.4 *Actinosphaerium,* a fresh-water heliozoan. The fine extensions are axopods. Their shape is maintained by parallel arrays of microtubules. Anything that causes the tubulin dimers of the microtubules to disaggregate causes the axopods to disappear. (Courtesy of Dr. Lewis G. Tilney, University of Pennsylvania.)

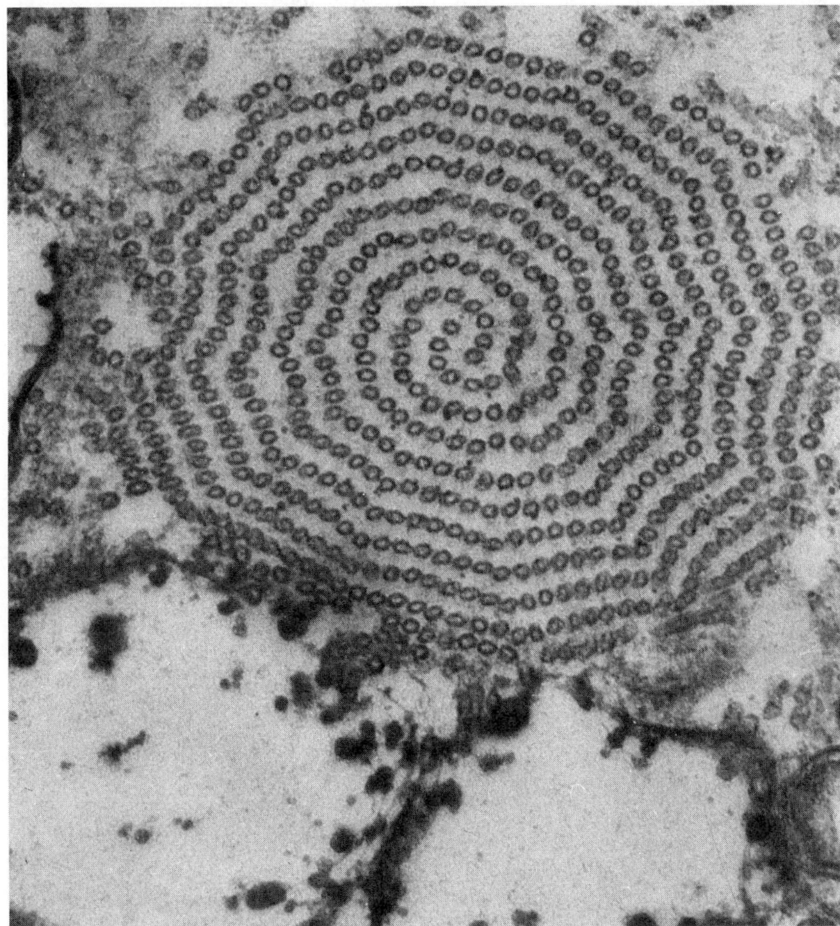

FIG. 35.5 Cross section through a heliozoan axopod as seen under the electron microscope. The orderly array of microtubules is essential to maintain the structure of these protoplasmic extensions. (Courtesy of Dr. Lewis G. Tilney.)

35.4 THE FLAGELLATES (PHYLUM MASTIGOPHORA)

As their name implies, the flagellates move by means of one or more flagella. These are constructed from microtubules with the "9 + 2" pattern that we examined in Section 28.12. Except for this common trait, it is an extraordinarily heterogeneous group. It includes forms such as *Astasia* (Fig. 35.3), which would be called *Euglena* and classified as an alga if only it had chloroplasts.

A number of species of flagellates (e.g. *Trichonympha*—Fig. 35.3—and *Myxotricha*) live within the gut of termites and digest the wood particles that the termites eat but cannot digest for themselves. *Myxotricha* appears to be covered with flagella but actually owns only four of them. The other "flagella" are actually spirochetes attached to its surface. Their lashing movements enable the creature to swim about; it uses its own flagella simply for steering. This remarkable relationship is particularly interesting because it reflects a way in which flagella and microtubules might have evolved. In fact, there is evidence that these spirochetes contain tubulin, the raw material from which microtubules are constructed.

In large parts of Africa, humans and cattle are parasitized by another group of flagellates of the genus *Trypanosoma*. Trypanosomes cause African sleeping sickness in humans and nagana in cattle. In each case, they are injected into the bloodstream of their host by the bite of a tsetse fly.

The trypanosomes live in the blood (Fig. 35.3). The disease they cause is protracted as well as serious. Why doesn't the immune system respond to these invaders and destroy them? In fact, the body does respond to the presence of the trypanosomes by making antibodies against them. But just when the response is ready to become effective, the trypanosomes change the antigenic determinants on their surface and escape destruction! The immune system tries again, and again the trypanosomes hide by altering their antigenic determinants.

Some flagellates also move by means of pseudopodia. They may have both at one time (Fig. 35.6) or develop one or the other as conditions dictate. The occurrence of such forms suggests a close relationship between these phyla. In fact, the name Rhizoflagellata has been proposed to encompass both the rhizopods and the flagellates. Except for the absence of chloroplasts, between them they have all the apparatus of eukaryotes and, for this reason, some consider that early rhizoflagellates were the stock from which all other eukaryotes evolved.

35.5 THE CILIATES (PHYLUM CILIOPHORA)

Ciliates move by the rhythmic beating of cilia. Cilia, like flagella, are constructed from the "9 + 2" pattern of microtubules. Although all ciliates are single-celled, some are sufficiently large to be seen by the unaided eye. The slipper-shaped *Paramecium* is the classic example of this group. It is commonly found in fresh water along with other ciliates such as *Stentor* and *Vorticella* (Fig. 35.3). In contrast to the other protozoans, the ciliates are probably all closely related to each other; in other words, the group is a "natural" one.

Ciliates feed by sweeping a stream of particle-laden water into a "mouth" and "gullet." Food vacuoles fill at the base of the gullet and move into the cytoplasm, where their contents are digested. Indigestible materials in the vacuoles (e.g. diatom shells) are discharged to the outside by exocytosis (through a permanent pore). Those ciliates that live in fresh water cope with the continuous influx of water by

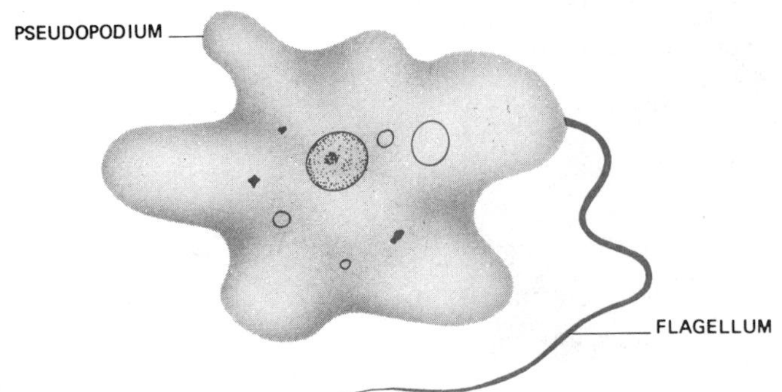

FIG. 35.6 *Mastigamoeba* moves by means of pseudopodia and a flagellum.

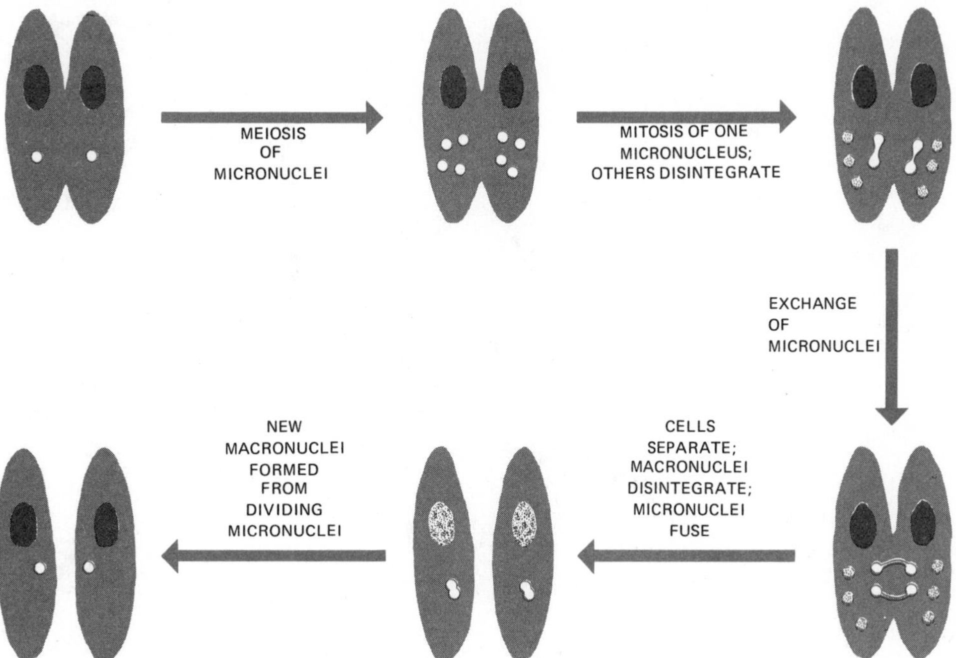

FIG. 35.7 Conjugation in *Paramecium caudatum*. This process is like other kinds of sexual reproduction in that it allows genetic recombination to occur. In this case, the new "offspring" are identical twins. Shortly after the cells separate, the old macronuclei are replaced by new ones carrying the new set of genes.

pumping it out with one or more contractile vacuoles. Parasitic ciliates, which live in isotonic surroundings, have no contractile vacuoles.

Ciliates have one or more *micronuclei* and a large, polyploid *macronucleus*. The macronucleus is responsible for running the general metabolic activities of the cell. Its large size is probably a reflection of the large volume of cytoplasm in these creatures. The micronuclei are responsible for sexual reproduction and, from time to time, the formation of a new macronucleus.

Asexual reproduction in the ciliates is by fission. The chromosomes of the micronuclei are separated on a spindle. The macronucleus may simply be pinched in two.

Ciliates also reproduce sexually. Let us examine the process of sexual reproduction as it occurs in a representative ciliate, *Paramecium caudatum*. The process is called **conjugation.** Two paramecia come together side by side (Fig. 35.7), and a cytoplasmic bridge forms between them. The micronucleus of each goes through the two consecutive divisions of meiosis (see Section 14.7). Three of the resulting haploid nuclei disintegrate. The fourth divides by mitosis. One daughter nucleus in each cell moves across the cytoplasmic bridge and fuses with the remaining daughter nucleus in the other cell (Fig. 35.7). Then the cells separate. While the old macronucleus disintegrates, the micronuclei go through several mitotic divisions. Four of these daughter nuclei form a new macronucleus.

Two parents came together and two parents separated. What kind of reproduction is that, you may well ask? But the process they have been through is the very essence of sexual reproduction—genetic recombination. The "offspring" are not the same as the parents; they are new individuals and their macronuclei will soon reflect this fact. Curiously enough, they have also become identical twins. Each parent formed two identical haploid nuclei—gave one away and kept the other. Thus when the cells separate, their new diploid micronuclei are identical. As the twins begin asexual reproduction, they contribute jointly to the formation of a single **clone.**

There is nothing primitive or simple about the ciliated protozoans. Forms such as *Paramecium* are not only large for single cells, but have many organelles which parallel in function the organs found in multicellular creatures. Indeed, the complexity of the ciliates has led some biologists to consider them acellular (not cellular) rather than unicellular organisms. What they wish to stress is that the *Paramecium* "body" is far more elaborate in its organization than any cell out of which multicellular organisms

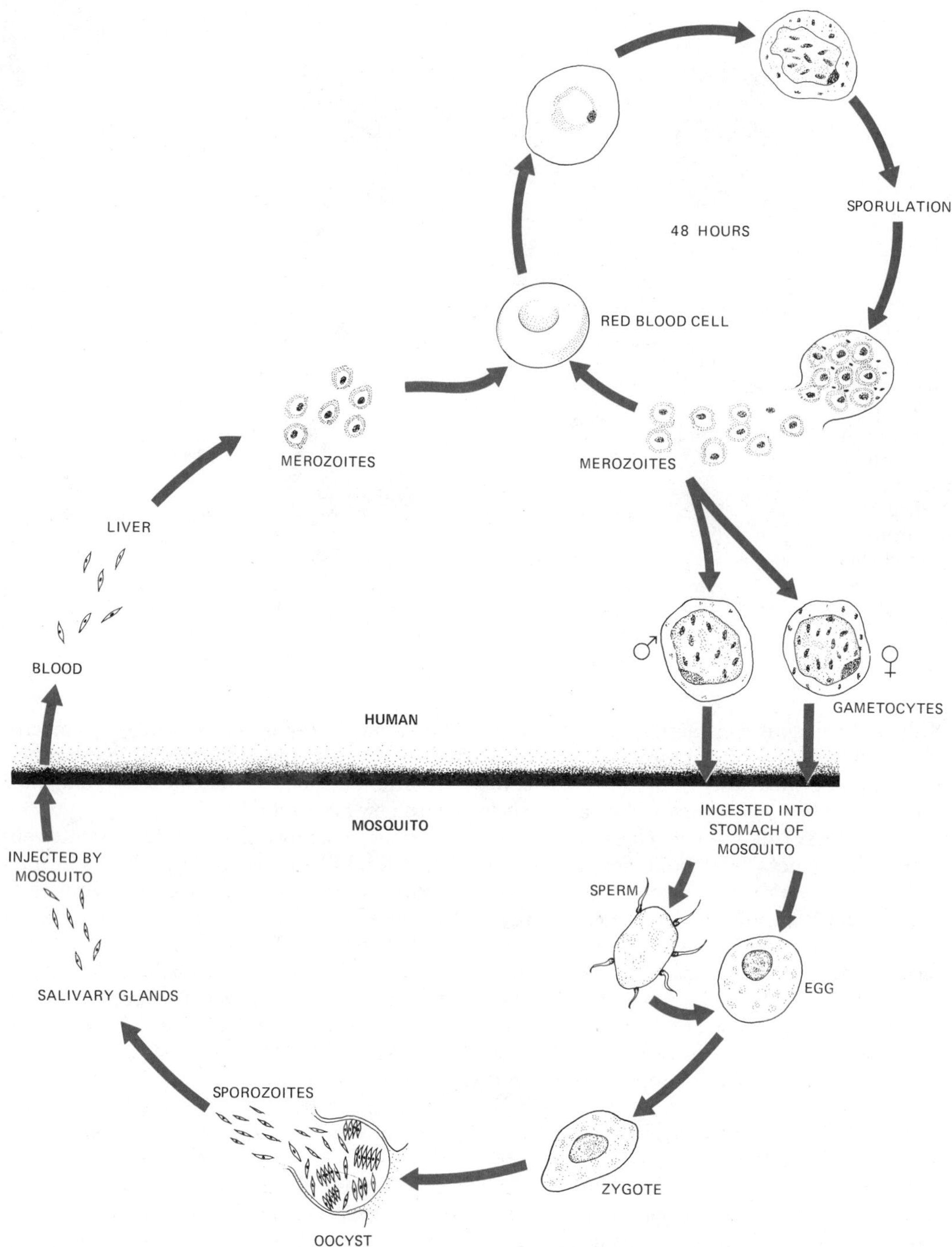

FIG. 35.8 The life cycle of *Plasmodium vivax,* a sporozoan that causes one of the commonest forms of malaria. The synchronous release of merozoites from the parasitized red blood cells produces the chills and fever characteristic of the disease. In *P. vivax,* this occurs approximately every 48 hours. The organism can remain within its human host for many years but cannot complete its life cycle unless it is able to alternate between humans and mosquitoes of the genus *Anopheles*.

are made. Whether this is a sound approach or not, the fact remains that the ciliates are the most elaborate of all the protozoans.

The high degree of specialization of modern ciliates has suggested to some biologists that they represent an evolutionary dead-end. However, it is possible to make a case for the evolution of the metazoans from early ciliates. If this is true, the ciliates will have to be moved from their present obscure position to the very center of the stage of evolution. In Chapter 37 we shall examine the grounds upon which this theory is based.

35.6 THE SPOROZOANS (PHYLUM SPOROZOA)

All sporozoans are parasitic. They nourish themselves by absorbing nutrients from their host. They lack the power of locomotion during most (in some cases, all) of their life cycle.

The most notorious members of this phylum belong to the genus *Plasmodium.* These sporozoans invade red blood cells (Fig. 35.3), causing malaria. Malaria has the dubious distinction of having caused more human deaths than any other infectious disease. It is transmitted from human to human through the bite of mosquitoes of the genus *Anopheles.*

The life cycle of *Plasmodium* is quite complex (Fig. 35.8). The symptoms of the disease are caused by the *merozoites* which reproduce asexually within the red blood cells of their human host. Periodically they all break out of the cells together and when they do so, they bring on the chills and fever characteristic of the disease. This occurs approximately every 48 hours with *Plasmodium vivax* (Fig. 35.8) and also with *P. falciparum,* the most dangerous member of the group. Eventually, some merozoites develop into either male or female gametocytes. These will soon die unless they are sucked up by the bite of an anopheline mosquito. Once within the stomach of their new host, the gametocytes develop into gametes. If both sperm and egg are present, they fuse with each other to form a zygote. The zygote invades the stomach wall of the mosquito and soon produces thousands of *sporozoites* (Fig. 35.8). These migrate to the salivary gland, ready to be injected into a new human host. If this occurs, they are carried to the liver and invade its cells. Here a first crop of merozoites is produced by asexual reproduction. The merozoites then move on to the red blood cells and the cycle of chills and fever begins.

Most forms of malaria are chronic. The organism may coexist with its host for years. While tucked inside a red cell, the parasites are protected from attack by antibodies. But when they leave, why doesn't the immune system destroy them? Like the trypanosomes discussed earlier in the chapter, the sporozoites keep changing their antigenic determinants and in this way evade effective attack by antibodies.

35.7 THE EUKARYOTIC ALGAE

"Algae" (like "protozoa") is no longer a formal taxonomic term. It is now just a common name for a large number of chlorophyll-containing, but rather simple, organisms. Because they are photosynthetic, most botanists claim them as members of the plant kingdom. Indeed, some are quite plantlike, but others bear only the most superficial resemblances to the organisms we generally know as plants.

Most algae live in the ocean, but freshwater forms are abundant, too. Some are unicellular; some grow as simple colonies of cells. Some are truly multicellular, but there is very little differentiation of cells. We shall examine six phyla of eukaryotic algae.

35.8 THE RED ALGAE (PHYLUM RHODOPHYTA)

The red algae are almost exclusively a marine group. Some are unicellular, but most are multicellular forms that grow anchored to rocks, wharves, etc. below the level of the average low tide (Fig. 35.9). Approximately 1500 species have been identified.

Red algae carry on photosynthesis using chlorophyll *a.* (Some species have a second chlorophyll, called chlorophyll *d,* but there is no chlorophyll *b.*) Being good eukaryotes, they incorporate their chlorophyll in one or more chloroplasts. However, the system of membranes in these chloroplasts looks very much like that in the cells of blue-green algae (Fig. 35.10). The resemblance between the two extends to their pigments. Red algae, like blue-green algae, have phycocyanin and phycoerythrin in their photosynthetic membranes. These serve as antenna pigments, passing absorbed energy on to chlorophyll *a* (see Section 9.7). Their similarities to the prokaryotic blue-green algae suggest that the red algae are a very primitive group. It may thus be significant that the red algae seem never to have acquired the "9 + 2" flagella so common to other eukaryotes.

Some red algae are used as food in coastal regions, particularly in the Orient. Agar-agar, which is widely used as a base for culturing bacteria and other organisms, is extracted from a red alga.

FIG. 35.9 Seaweeds exposed at low tide. Two species of brown algae and one red alga (the mossy material at the water line) grow here, tightly attached to the rocks.

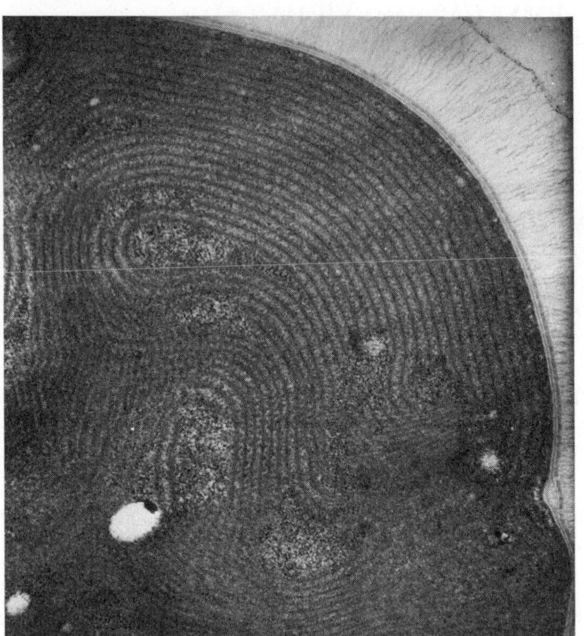

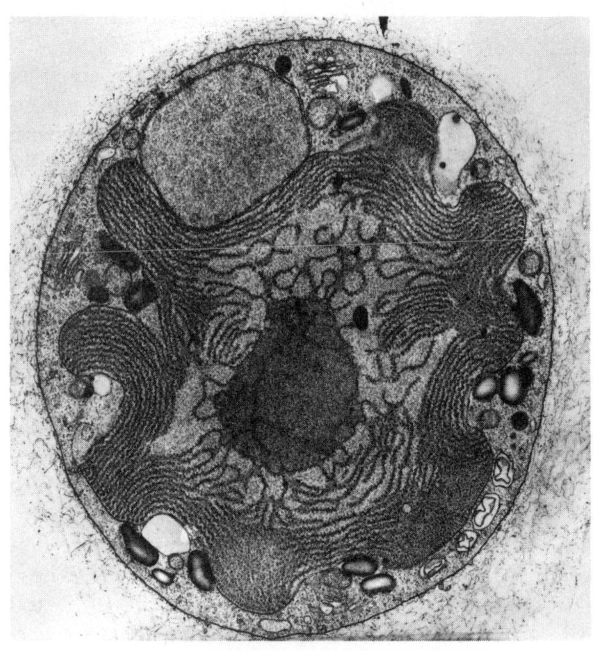

FIG. 35.10 Left: Electron micrograph (25,000 ×) of a blue-green alga. (Courtesy of Dr. G. Cohen-Bazire.) Right: *Porphyridium,* a unicellular red alga. (Electron micrograph 10,000 ×, courtesy of Dr. E. Gantt.) The organization of the membranes in the single chloroplast of the red alga and the pigments they contain are quite unlike the chloroplasts of other photosynthetic eukaryotes, but quite similar to prokaryotic blue-green algae.

35.9 THE DINOFLAGELLATES (PHYLUM PYRROPHYTA)

Almost all of the dinoflagellates (there are some 900 of them) are unicellular. Like the red algae, they exhibit a number of traits that appear intermediate in nature between those of prokaryotes and those of the more advanced eukaryotes. For example, they have no histones on their chromosomes, and their mitosis is much less complex than that of higher eukaryotes. Dinoflagellates do have the eukaryotic type ("9 + 2") of flagellum (two of them, in fact—see Fig. 35.11), another major acquisition of life.

Aside from their scientific interest, dinoflagellates become of concern to us when from time to time

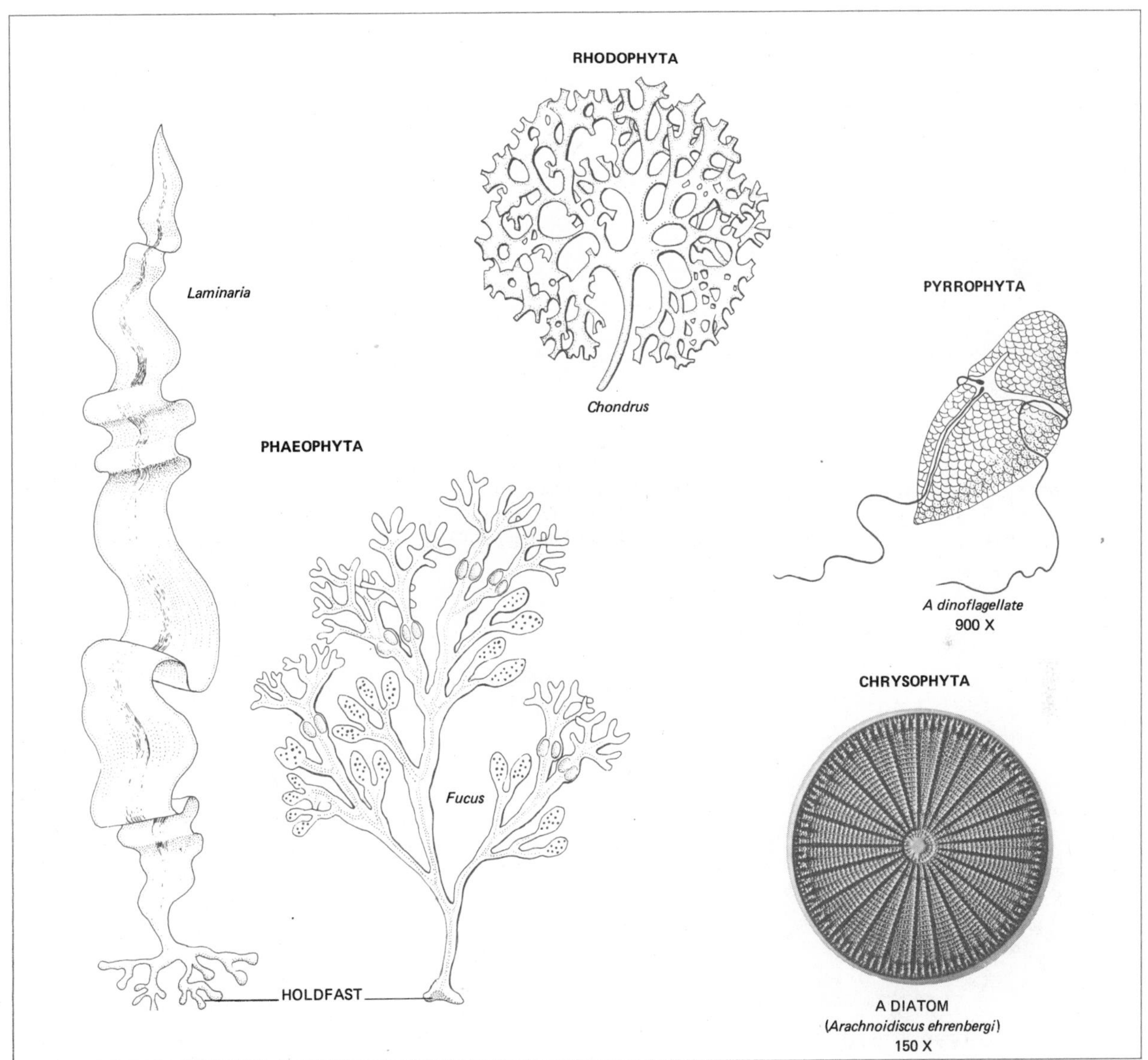

FIG. 35.11 Representative algae from each of four phyla. (Photo courtesy Turtox/Cambosco, Macmillan Science Company, Chicago.)

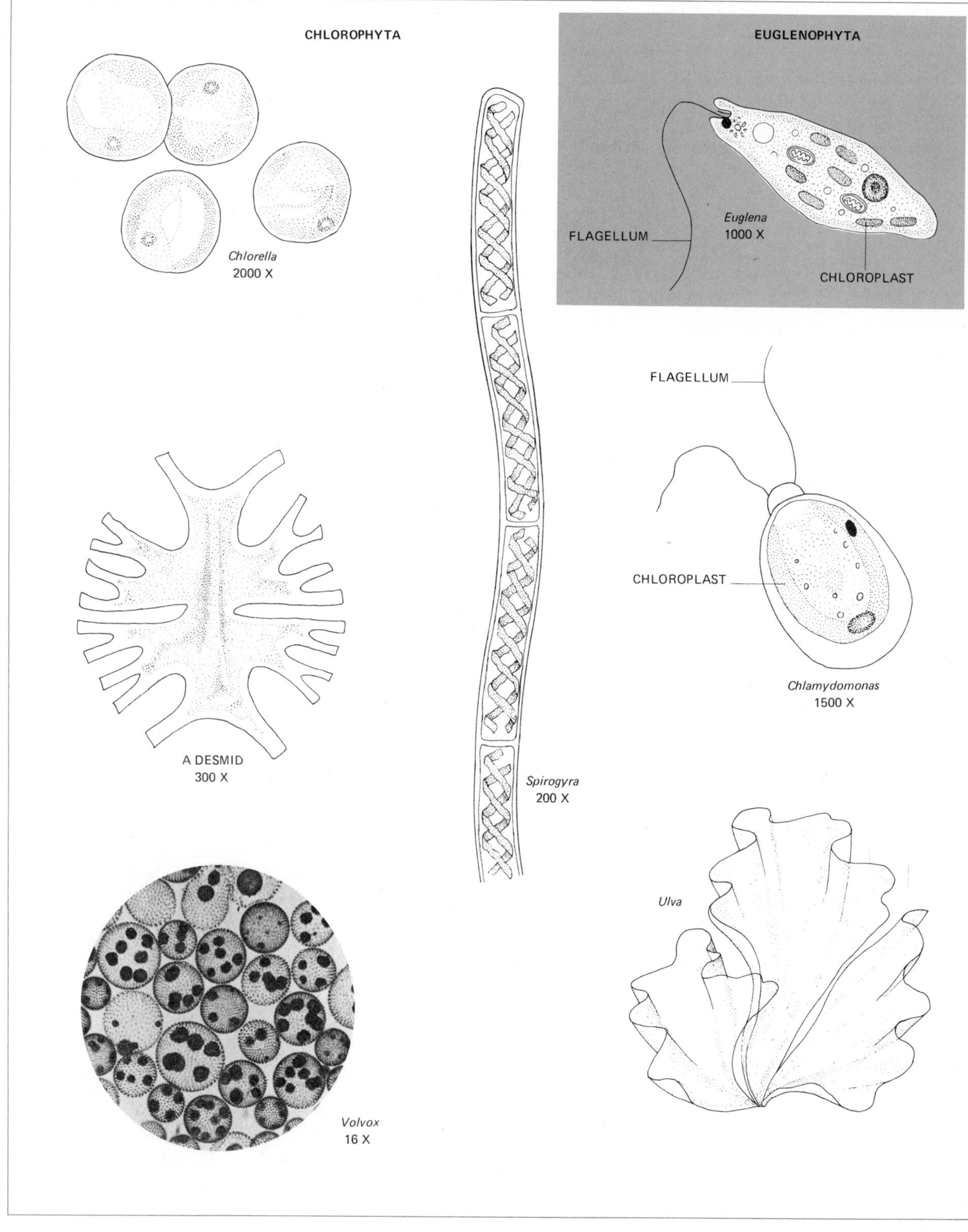
CHLOROPHYTA
EUGLENOPHYTA
Chlorella
2000 X
FLAGELLUM
Euglena
1000 X
CHLOROPLAST
FLAGELLUM
CHLOROPLAST
Chlamydomonas
1500 X
A DESMID
300 X
Spirogyra
200 X
Ulva
Volvox
16 X

◀ FIG. 35.12 Representative green algae (Chlorophyta) and, at upper right, *Euglena* (Euglenophyta). (Photo courtesy Turtox/Cambosco, Macmillan Science Company, Chicago.)

they reproduce explosively and create the poisonous red tides that may cause extensive kills of marine fish and make filter-feeding marine animals like clams unfit for human consumption.

35.10 THE EUGLENOPHYTES (PHYLUM EUGLENOPHYTA)

These are photosynthetic flagellates that are not encased in a rigid cell wall. *Euglena* is a typical member of the group (which numbers some 400 species). Lacking a cell wall, *Euglena* can change shape easily. It moves rapidly by means of a long flagellum located at its anterior end (Fig. 35.12). The euglenophytes have both chlorophyll *a* and chlorophyll *b*.

Were it not for its plastids, *Euglena* would be practically indistinguishable from *Astasia* (see Section 35.4). In fact, if you treat *Euglena* with streptomycin or ultraviolet light, you can destroy its chloroplasts and produce a colorless heterotroph that you would promptly assign to the Mastigophora. Perhaps many of the protozoa are the evolutionary outcome of a spontaneous loss of the chloroplasts possessed by their ancestors. In any case, we surely violate our principles of taxonomy by placing such forms in separate phyla when it is only the presence or absence of chlorophyll that distinguishes them.

35.11 THE GREEN ALGAE (PHYLUM CHLOROPHYTA)

These algae resemble the euglenophytes in their photosynthetic pigments (chlorophyll *a* and *b*). However, their cells are encased in a rigid wall of cellulose. Some of them are flagellated (e.g. *Chlamydomonas*) and even those that are not (e.g. *Ulva*) produce flagellated gametes and/or zoospores. You may wish to review the life cycle of *Chlamydomonas*, a typical unicellular green alga (see Section 14.2).

Some 6500 species of green algae have been identified. In addition to many unicellular members like *Chlamydomonas*, the phylum contains colonial and multicellular forms. *Spirogyra* grows as a green filament of cells, each of which leads an independent life. *Volvox* and *Ulva* (sea lettuce) probably should be considered multicellular forms; that is, the hollow sphere of *Volvox* makes up one individual as does the flat double layer of cells of *Ulva* (Fig. 35.12). The line between colonies of individual cells and a single multicellular organism is not always easy to draw. Even in those members of the phylum that seem to be truly multicellular, the constituent cells are not specialized to form specific tissues or organs.

The manner in which colonial and multicellular algae may have evolved from unicellular ancestors is suggested by the series of flagellated forms (*Chlamydomonas, Gonium, Pandorina, Eudorina, Pleodorina,* and *Volvox*) that were examined in Chapter 5 (look back at Fig. 5.26). The story they seem to tell is that colonial forms arose when the daughter cells of unicellular forms remained attached to one another following mitosis. As colonies became larger and more elaborate, some specialization among their constituent cells produced truly multicellular organisms, eventually even forms as large as *Ulva*.

Many of the nonmotile members of this phylum are quite plantlike in appearance. Their cellulose cell walls and the presence of chlorophyll *a* and *b* are also characteristics of plants and suggest that the green algae are the closest protistan relatives of the plants. Most biologists believe that the plant kingdom evolved from early members of this phylum.

Green algae are important as a source of food for many protozoans and aquatic animals. The unicellular *Chlorella* has received a great deal of attention from biologists, both as the organism with which many of the details of photosynthesis were worked out and as a possible source of food in areas unsuited for conventional agriculture (Fig. 35.13).

FIG. 35.13 Pilot plant for growing single-celled algae in mass culture. Such techniques hold special promise for desert areas where conventional agriculture is impractical. (Courtesy of Arthur D. Little, Inc.)

When water supplies are "fertilized" with phosphates and nitrates (e.g. from sewage), freshwater green alga often form extensive algal blooms.

35.12 THE GOLDEN ALGAE (PHYLUM CHRYSOPHYTA)

The golden algae get their color from a yellow-brown carotenoid called fucoxanthin. Their chlorophylls are *a* and *c*. Most of the members of the group are unicellular and many are flagellated. The group contains some 5300 species, 5000 of which are **diatoms.**

Diatoms have a cell wall or shell that is made up of two overlapping halves. These shells are impregnated with silica and are often beautifully ornamented. In fact, the delicate sculpturing of the shells of some species provides a good test of the quality of microscope lenses (Fig. 35.11).

Diatoms play an important role in the economy of nature. Both in fresh water and in the oceans, they accomplish a major part of all the photosynthesis that occurs. Thus they serve as a vital source of food materials for many colorless protists and small animals. These, in turn, serve as food for still larger animals. As the main food producers in aquatic environments, the diatoms thus sustain a multitude of other, nonphotosynthetic organisms.

The glassy shells of the diatoms do not decay after death. Consequently they may accumulate in deep layers at the bottom of the ocean in regions where diatoms are especially abundant. In some cases, these deposits have, millions of years later, been lifted above sea level by geological forces. This has occurred in California and the deposits are now mined and sold as "diatomaceous earth." This material is used as a filtering agent for clarifying liquids and is also incorporated in soundproofing materials. The countless numbers of tiny glassy shells make a fine abrasive, and scouring powders, silver polish, etc. often contain diatomaceous earth.

The diatoms form such a homogeneous grouping that the splitters are probably fully justified in removing them from the Chrysophyta and placing them in a phylum of their own (which has been named Bacillariophyta).

35.13 THE BROWN ALGAE (PHYLUM PHAEOPHYTA)

The members of this phylum (of 1500 species) contain fucoxanthin, which masks the green of their chlorophylls (*a* and *c*). They are all multicellular, somewhat plantlike forms found almost exclusively in salt water. The rockweeds, which form a dense mat on rocks exposed to the changing tides (Fig. 35.9) and the kelps (Fig. 35.11) are large and widespread members of the phylum. There is some spe-

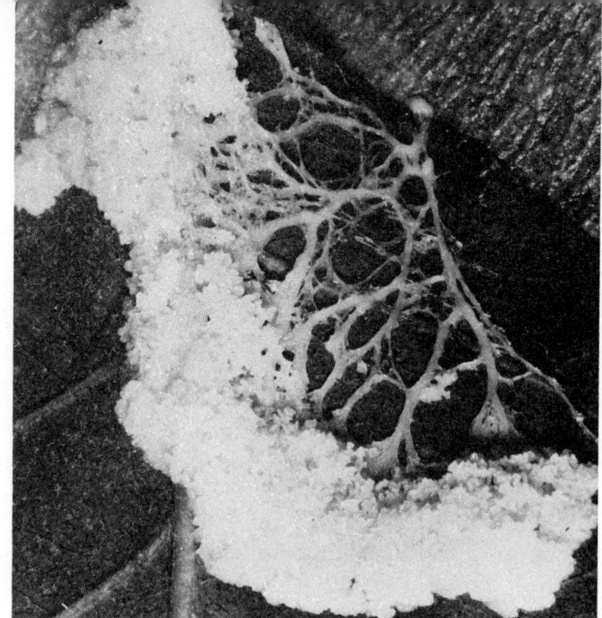

FIG. 35.14 *Stemonitis,* a common plasmodial slime mold. Top: The plasmodium stage just prior to the formation of sporangia. (Courtesy of Prof. I. K. Ross.) Bottom: Fully developed sporangia. (Courtesy of Turtox/Cambosco, Macmillan Science Company, Chicago.)

cialization of parts in these organisms, and many have rather complex life cycles. Giant kelps along the Pacific coast grow as long as 30 meters. Despite their great size, the organization of these forms is still quite simple compared with that of true plants.

Brown algae are used for food in some coastal areas of the world and, in the U.S., as a source of fertilizer and iodine.

35.14 THE SLIME MOLDS (PHYLUM MYXOMYCETES)

The members of this group are popularly known as slime molds because at one stage in their life cycle they consist of a spreading slimy mass.

In the *plasmodial* slime molds (e.g., *Stemonitis,* Fig. 35.14) the slimy mass, called a plasmodium, contains thousands of nuclei. The plasmodium moves slowly over its substrate (e.g., a rotten log) engulfing food and growing as it does so. Eventually, the plasmodium develops elaborate stalks that produce and release spores (Fig. 35.14). If the spores land in a suitable location, they germinate to form single cells which move by both flagella and pseudopodia. These fuse in pairs and start the formation of a new plasmodium.

In the *cellular* slime molds, thousands of individual amoebalike cells aggregate into a slimy mass. However, there is no fusion of the individual cells. The aggregating cells are attracted to each other by the cyclic AMP (Fig. 25.12) that they release. In the course of their life cycle, these curious organisms are unicellular, multicellular, funguslike (spores), and protozoalike (amoeboid). Such a collection of traits makes them of great scientific interest. With the exception of one species that causes powdery scab in potatoes, however, these organisms are of little economic importance.

While the life cycles of the slime molds share many similarities, the splitters are probably wise to assign the plasmodial slime molds to one phylum (Myxomycetes) and the cellular slime molds to another (Acrasiomycetes).

The Kingdom Fungi

35.15 CHARACTERISTICS

Most of these eukaryotes grow as tubular filaments called **hyphae.** An interwoven mass of hyphae is referred to as a **mycelium.** The hyphae are coenocytic, that is, they are not compartmentalized into separate cells. Although cross-walls are found in some hyphae, these are perforated so that the cytoplasm and many nuclei within the hyphae are free to flow throughout the mycelium. The walls of the hyphae are strengthened by chitin, a polymer of N-acetylglucosamine (NAG—see Section 34.2). The linkage between the sugars (Fig. 35.15) is like that in cellulose (look back at Fig. 4.14) and peptidoglycan (Fig. 34.2) and provides the same sort of structural rigidity that those polymers do.

Fungi have no chlorophyll and hence are heterotrophic. They secure their nourishment by absorbing food molecules from their surroundings (often having first digested them by secreting extracellular hydrolytic enzymes). Their food may be derived from such sources as rich soil, manufactured food products, and the bodies of plants and animals (both dead and living). Those that live within a living host may contribute something to their host in return for their meal, for example helping a plant secure minerals from the soil. More often though, they damage their

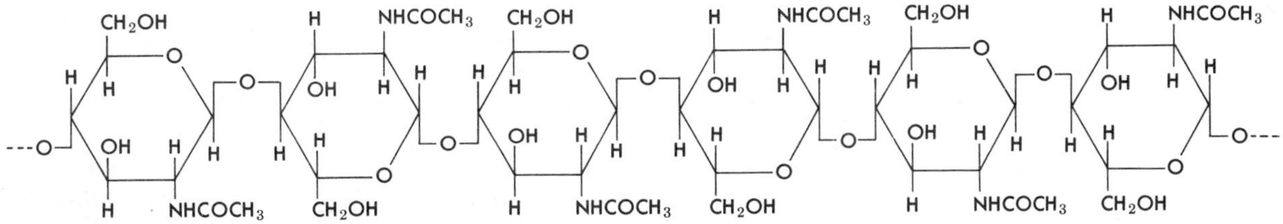

FIG. 35.15 Molecular structure of chitin. The monomers are N-acetylglucosamine. The linkage is like that in cellulose (see Fig. 4.14). Chitin is also the chief structural component of the exoskeleton of arthropods.

host. The disappearance of the American chestnut (perhaps to be followed by the American elm), the ravages caused by wheat rusts, and athlete's foot are just a few examples of the damage caused by parasitic fungi. But our justifiable concern with the fungi that cause disease should not blind us to the absolutely essential—but undramatic—role that fungi play in decomposing dead organisms and releasing their nutrients for reuse by the living.

Fungi disperse themselves to new locations by releasing spores. In some aquatic species, these swim by means of flagella. However, the spores of the many terrestrial fungi are wind-blown. They are so light and are produced in such numbers that they are present almost everywhere. The spores of the wheat rust fungus have been found 4,000 meters in the air and more than 1450 km (900 miles) from the place where they were released.

Approximately 30,000 species of fungi have been identified. Traditionally, these have been assigned to one or another of four taxonomic groups, primarily on the basis of the kinds of spores they produce. These groups are the phycomycetes, the ascomycetes, the basidiomycetes, and the "fungi imperfecti."

36.16 THE PHYCOMYCETES

These are the "algal" fungi (Gk. *phyco* = seaweed). In fact, some are aquatic and some are not. The aquatic members are commonly called water molds. They produce flagellated spores (zoospores) and/or gametes.

Two common water molds are *Saprolegnia* and *Achlya*. A reliable way to secure these organisms for study is to drop some boiled rice grains in a container of pond water. In a few days, a mycelium will be visible around the grains and will grow out into the water.

A number of water molds are economically important. One species of *Saprolegnia* is a parasite of fish and can be a serious problem in fish hatcheries. The downy mildews of grapes and other valuable crops are caused by water molds. But surely no water mold has created economic dislocation to compare with that caused by *Phytophthora infestans,* the cause of "late blight" in potatoes. In 1845 and again in 1846 it was responsible for the almost total destruction of the potato crop in Ireland. These repeated failures of their main source of calories led to the great Irish famine of 1845–1860. During this period approximately a million people starved to death and many more emigrated to the New World. By the end of the period, death and emigration had reduced the population of Ireland from 9 million to 4 million.

FIG. 35.16 Representative fungi. ►

The terrestrial phycomycetes have neither motile spores nor gametes. Their spores are dispersed on air currents. The classic member of the group is *Rhizopus* (Fig. 35.16). We have used *Rhizopus* as an example of saprophytic nutrition (see Section 7.3) and of airborne spore dispersal (see Section 14.2). Although *Rhizopus stolonifer* occasionally produces a slice of moldy bread, other members of the genus more than make up for its depradations. *Rhizopus oryzae* is used in the fermentation of sake, the rice wine of the Orient. *Rhizopus* is also used in the commercial production of glucocorticoids (see Section 25.11).

Fungi are assigned to the phylum Phycomycetes on the basis of two criteria: (1) forming their spores *within* a *sporangium* (Fig. 35.16) and (2) having no septa (cross-walls) in their hyphae. These are probably an inadequate basis for claiming kinship. Most mycologists now assign the terrestrial forms like *Rhizopus* to the phylum Zygomycetes. The water molds like *Achlya* and *Saprolegnia* are assigned to the phylum Oomycetes. The splitters have even established additional phyla in which to place still other water molds.

35.17 THE PHYLUM ASCOMYCETES

Ascomycetes produce two kinds of spores. Those that are formed asexually, called **conidia,** develop in chains at the tips of the hyphae (Fig. 35.16). These spores are equivalent to the spores produced by the sporangium of the phycomycetes. A second kind of spore is produced as a result of sexual reproduction. Four or eight of these spores, the **ascospores,** develop inside a saclike **ascus.** We studied the life history of *Neurospora,* a typical ascomycete, in Chapter 18.

The ascomycetes play many important roles in our lives. On the deficit side of the ledger, they attack many valuable plants. The powdery mildews, which belong to this class, parasitize a variety of crop and ornamental plants. The chestnut blight is caused by an ascomycete. This organism has literally eliminated from our country what just a few decades ago was one of our major forest trees. The Dutch elm

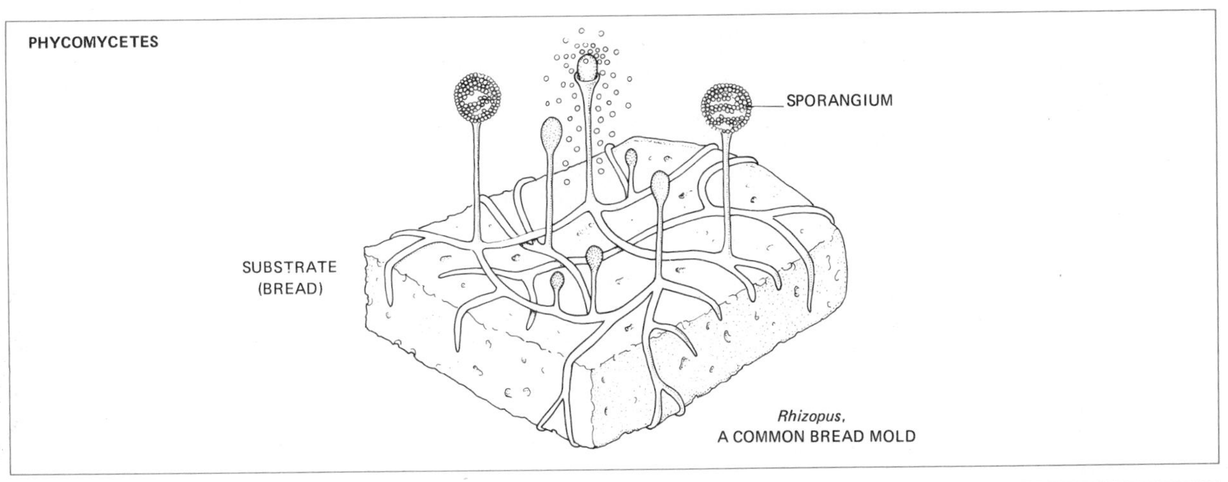

Rhizopus,
A COMMON BREAD MOLD

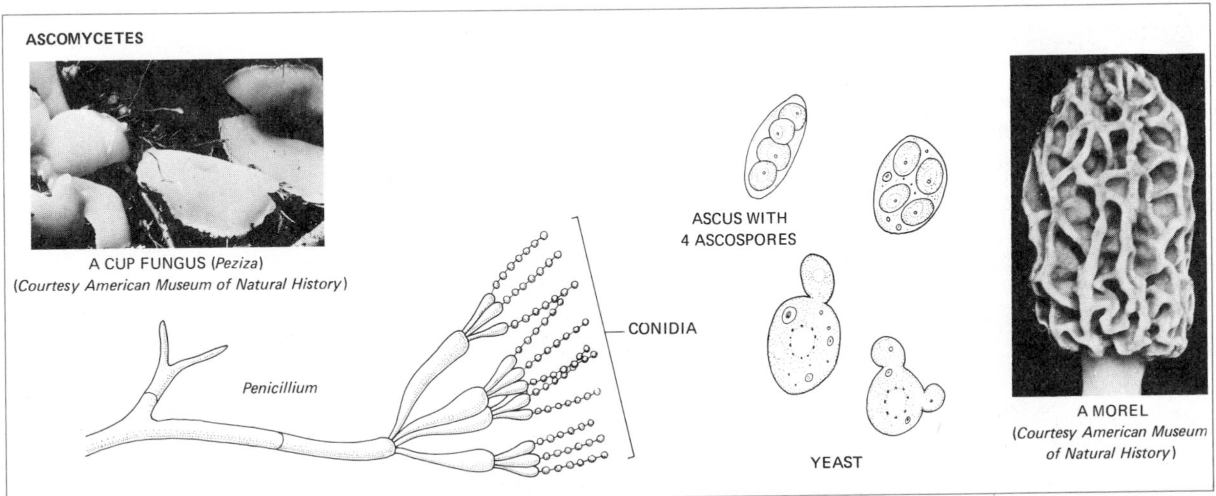

A CUP FUNGUS (*Peziza*)
(*Courtesy American Museum of Natural History*)

A MOREL
(*Courtesy American Museum of Natural History*)

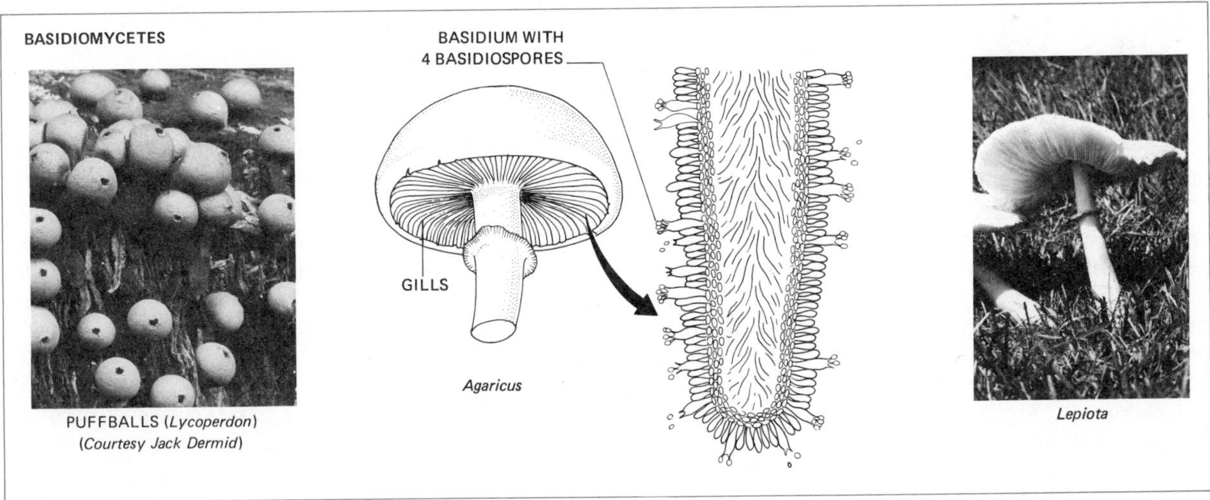

PUFFBALLS (*Lycoperdon*)
(*Courtesy Jack Dermid*)

Agaricus

Lepiota

FIG. 35.17 A farmer from the Périgord (in southwestern France) admires a truffle. Truffles are ascomycetes that establish a symbiotic relationship with the roots of such trees as oaks. (Courtesy French Embassy Press & Information Division, New York.)

disease, also caused by an ascomycete, shows promise of doing the same to our stately American elms.

On the asset side, the ascomycete *Penicillium* produces the life-saving antibiotic penicillin. Other species in the same genus are used in the production of Camembert and Roquefort cheeses. The tasty morel, which is often thought to be a mushroom, is actually an ascomycete (Fig. 35.16). So, too, is the truffle, considered by some to be one of the world's great delicacies (Fig. 35.17).

Yeasts are unicellular members of this class. They are of enormous commercial value not only as a source of alcoholic beverages (beer, wine, etc.) but as a source of alcohol for industrial purposes. The same reaction is exploited in the baking industry, but here the desired product is the gas, not the alcohol. The carbon dioxide makes bread and cakes "rise" and gives them a spongy, palatable texture. The alcohol produced by the yeast evaporates during the baking process. Yeast is also used in the commercial production of vitamins.

35.18 THE PHYLUM BASIDIOMYCETES

The fungi of this phylum are dispersed by spores borne at the tips of club-shaped structures called **basidia** (Fig. 35.16). This phylum includes the mushrooms, shelf fungi, puffballs, rusts, and smuts.

The familiar mushroom or toadstool is only a portion of the fungus body. The main part of the mycelium grows beneath the surface of the ground. Only when conditions are suitable does the mycelium send mushrooms up above the surface. These are simply masses of interwoven hyphae. The basidia develop on the undersides and release their spores (four from each basidium) into the air.

The basidiomycetes are of great economic importance to man. Mushrooms are widely used for food and their cultivation is an important business. Many wild mushrooms are edible, too, but unless one takes the trouble to learn to recognize the individual species, it is best to leave them alone. A few species such as *Amanita muscaria* are extremely poisonous.

Although some basidiomycetes are good to eat, they can never compensate for the food losses caused by their cousins, the smuts and rusts. The latter, especially, are responsible for serious losses to such important crops as wheat, oats, and rye.

Some of the rusts have rather complex life histories. Wheat rust parasitizes not just wheat but also barberries. Where the winters are cold, the rust cannot complete its life cycle unless it invades wheat plants and barberry plants alternately. Four different kinds of spores are produced during the process. The eradication of barberry plants has been attempted in wheat-growing areas in an effort to control this rust.

White pine trees are attacked fatally by the white-pine blister rust. This rust, too, requires an alternate host, gooseberry or wild currant bushes, in order to complete its life cycle. In areas where white pines are an important lumber or ornamental tree, it is often illegal to raise gooseberries or wild currants.

35.19 THE FUNGI IMPERFECTI (PHYLUM DEUTEROMYCETES)

The ascomycetes and basidiomycetes are distinguished by the type of sexual spore they produce, ascospores in the first case, basidiospores in the second. However, many of these fungi cannot produce sexual spores unless two different strains are present to serve as the two parents. There are several thousand fungi of which only one strain is known or which for some other reason fail to reproduce sexually. In this situation, there is no way of telling whether the fungus is an ascomycete or a basidiomycete. Consequently, it is placed in a special "dustbin" category, the **fungi imperfecti**. The parasites which cause ringworm and athlete's foot in humans are so classified. It is significant, however, that from time to time the sexual stage of one of these "imperfect" fungi is discovered and then it is immediately reclassified, usually to the ascomycetes.

35.20 THE LICHENS

Lichens are not single organisms at all, but a composite consisting of a fungus mycelium within which algal cells are imbedded (Fig. 35.18). In some lichens, the fungus is a basidiomycete or an imperfect fungus, but in the majority it is an ascomycete. The alga is either green or blue-green and is unicellular. Although lichens are given scientific names as though they were a single organism, it probably makes more sense to think of the name as applying to the fungus and then to specify the algal partner separately if desired. Some 18,000 "species" of lichens have been identified.

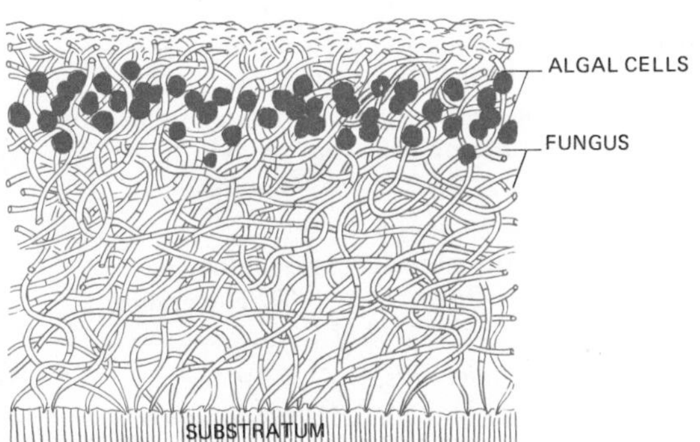

FIG. 35.18 Left: A common rock lichen. Right: Its structural organization.

Some of the algae found in lichens (e.g. *Nostoc*) also grow independently in nature. Almost all of the fungi, however, are found only in lichens, although they generally can be cultured in the laboratory.

Most fungi are found in areas where there is plenty of organic matter to serve as food. The fungi found in lichens are, in contrast, capable of living in the most forbidding environments. Some lichens grow profusely right on the surfaces of rocks (Fig. 35.18). Lichens are a dominant feature of the vegetation in the Arctic and Antarctic. What enables fungi to thrive in such locations? It is clearly their symbiotic partner, the alga. Using $NaH^{14}CO_3$ as a tracer, it can be demonstrated that the bulk of the sugars photosynthesized by the alga are translocated to the fungus. If the algal partner is blue-green, fixed nitrogen is donated to the fungus as well. And what does the alga receive in return? While there has been much speculation on the matter, no benefit has as yet been demonstrated.

The dispersal of lichens is not yet well understood either. The fungus member produces wind-blown spores but these are not accompanied by the alga. Perhaps the spores meet the appropriate alga by chance when they land in a new location, but this seems rather unlikely. Probably dispersal in the lichens is accomplished when fragments of the lichen, containing both fungus and alga, become detached from the parent body and are transported to new locations.

Lichens are of interest in the economy of nature because they are among the first organisms to colonize harsh, newly created environments. Rocks exposed by retreating glaciers, landslides, etc., soon develop lichen growth. As portions of the lichen body die and decay, organic matter, or humus, is produced. In time, enough soil may form in the crevices of the rock so that plants such as mosses can become established. Eventually what was once a desolate area may support a luxuriant growth of vegetation.

EXERCISES AND PROBLEMS

1. Summarize the evidence that supports the endosymbiotic theory of the evolution of eukaryotes. By what alternative mechanisms might eukaryotes have evolved?
2. Do you think that all the organisms on the earth today have evolved from a *single* first form of life? What evidence supports your case? What evidence weakens it?
3. Summarize the animallike and plantlike characteristic of *Euglena*.
4. What materials important to humans are derived from ascomycetes?

REFERENCES

1. MARGULIS, LYNN, "Symbiosis and Evolution," *Scientific American,* Offprint No. 1230, August, 1971. One of its chief proponents examines the hypothesis that mitochondria, chloroplasts, and flagella have all evolved from prokaryotic endosymbionts.
2. GOODENOUGH, URSULA W., and R. P. LEVINE, "The Genetic Activity of Mitochondria and Chloroplasts," *Scientific American,* Offprint No. 1203, November, 1970. And how it resembles that of the prokaryotes.
3. LEEDALE, G. F., *The Euglenoids,* Oxford Biology Readers, No. 5, Oxford University Press, Oxford, 1971.
4. CURTIS, H., *The Marvelous Animals: An Introduction to the Protozoa,* The Natural History Press, Garden City, N.Y., 1968.
5. ALEXOPOULOS, C. J., and H. C. BOLD, *Algae and Fungi,* Macmillan, New York, 1967. A compact survey.
6. SMITH, D. C., *The Lichen Symbiosis*, Oxford Biology Readers, No. 42, Oxford University Press, Oxford, 1973.

36 THE PLANT KINGDOM

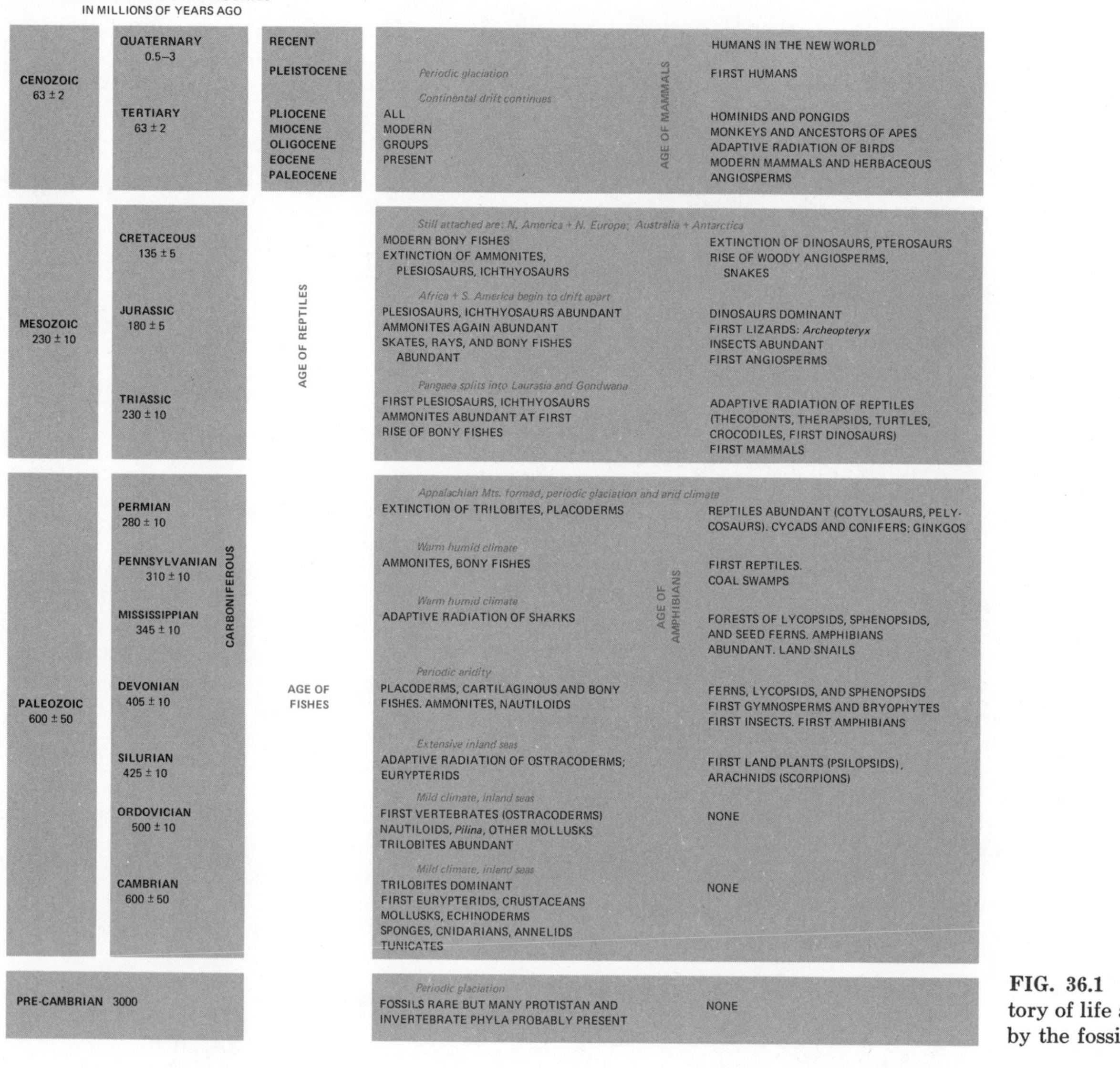

FIG. 36.1 The history of life as revealed by the fossil record.

36.1 THE GEOLOGICAL ERAS

Fossils of ancient prokaryotes such as *Eobacterium* (Fig. 2.6), which is approximately three billion years old, and possible eukaryotes such as *Glenobotrydion* (Fig. 35.2), which is some one billion years old, are exceedingly rare. It is not until we examine sedimentary rocks formed about 600 million years ago, at the start of the Paleozoic era, that we find a wide variety of fossil organisms. This means that through more than four-fifths of the unimaginable span of time that life has existed on this earth, it left little or no record of its presence. Since that time, the record of evolving life has been quite well preserved.

The geological and biological history of the earth since the first appearance of abundant fossils is divided into three major eras (Fig. 36.1). Each of these is further subdivided into periods. At first glance it might seem surprising that marked geological changes in the earth should coincide with marked changes in the species present. Consider, though, that changes in geology (e.g. mountain formation, lowering of the sea level) bring changes in climate, and both of these alter the habitats available for living organisms. The forces of natural selection must surely have changed when the geology of the earth changed. As you can see in Fig. 36.1, the dates of the various periods are not precisely known. Only a few of the rocks (e.g. those of the early Permian) contain enough uranium to be dated accurately. The others must be dated on the basis of their relative thickness, and we have already noted the uncertainties of this method.

As we discuss the history of evolutionary change in this chapter and those that follow, you will want to refer frequently to Fig. 36.1 to see how the details fit into the whole panorama of life.

36.2 THE EVOLUTION OF PLANTS

Although direct fossil evidence is lacking for some, we believe that almost all of the groups of organisms discussed up to this point appeared in the waters of the earth in the Pre-Cambrian period, before the start of the Paleozoic era (Fig. 36.1). During the Cambrian and Ordovician periods, the oceans and bodies of fresh water must have continued to support an elaborate array of protists and (as we shall see in the next chapter) aquatic invertebrates. But the land must have presented a desolate scene with few, if any, autotrophs adapted to life there and hence no source of food to support heterotrophs. Where sea met land, however, green algae may have been evolving features that would enable them to survive intermittent periods of dryness. By the end of the Silurian, descendants capable of life on dry land had appeared and begun to colonize this new environment. These were the plants.

We shall include in the plant kingdom all those organisms that (1) contain chlorophyll *a* and *b*, (2) lack the power of motion or locomotion by means of contracting fibers, (3) have bodies made up of many cells differentiated to form tissues and organs, (4) have sex organs composed of many accessory cells, and (5) produce offspring that, as partially developed embryos, are protected and nourished for a time within the body of the parent plant. We shall not be dismayed if, as a result of a secondary loss, an occasional plant lacks chlorophyll. The Indian pipe (see Fig. 43.12) contains none, but in all other respects is plantlike and is certainly descended from green ancestors.

Many, if not most, botanists would also include the algae in the plant kingdom. By including them, however, they have to modify requirement (1)—only the Chlorophyta and Euglenophyta have chlorophyll *b*—and eliminate requirements (3) through (5). None of the algae meets requirements (4) and (5), and although some are multicellular, there is little differentiation of cells. For reasons already mentioned, we shall cast our lot with Haeckel and place the algae (except the blue-greens) in the kingdom Protista. Having done so, we find that the plant kingdom is conveniently divided into two phyla (botanists usually call them *divisions* rather than phyla): the Bryophyta and the Tracheophyta.

36.3 THE MOSSES AND LIVERWORTS (PHYLUM BRYOPHYTA)

Some 23,000 species of living mosses and liverworts have been identified. These are small, fairly simple plants that are usually found growing in moist places. Most liverworts have a thin, leathery body that grows flat upon the supporting medium—still water or moist soil. Figure 36.2 shows a common liverwort growing on the surface of pond water.

The plant body of mosses is a little more elaborate than that of liverworts. It consists of an erect shoot bearing tiny leaflets arranged in spirals. Neither mosses nor liverworts have any woody tissues for support and thus they never grow very large. They have no specialized vascular system for the transport of water and food throughout the plant.

In Chapter 20 we learned that sexual reproduction in mosses can occur only if the sperm cells are able to swim or be splashed from the plant that produces them to the plant where the egg is. The lack of a special water transport system and the need for free water for sexual reproduction are two reasons why the bryophytes are restricted to habitats that, periodically at least, have abundant moisture.

Approximately 14,000 species of mosses are known. The haircap moss, *Polytrichum commune* (Fig. 36.2), is a widely distributed and commonly studied form. *Sphagnum* moss grows luxuriantly in bogs and, when partially decomposed, is sold as peat moss to gardeners wishing to improve their soil.

Mosses play an important role in the economy of nature. With the lichens, they are among the first plants to grow in barren areas such as rock ledges exposed by a retreating glacier. They grow rapidly and form a great deal of decomposed plant material, humus, which rapidly forms soil suitable for the growth of more complex plants. In established forests, the spongy texture of moss beds absorbs water from rain and melting snow. This reduces the likelihood of floods in the springtime and of dried-up streams in the summer. It also reduces the loss of soil by water erosion.

FIG. 36.2 Representative bryophytes. Left: The common haircap moss *Polytrichum commune.* Right: A common liverwort, *Ricciocarpus natans.* (Courtesy of William C. Steere and *AIBS Bulletin.*)

The bryophytes have sometimes been considered the ancestors of the vascular plants. Their simplicity of structure, lack of vascular tissue, and restriction to damp locations do suggest that they are intermediate forms between the algae and the vascular plants. The fossil record indicates, however, that this interpretation is false. No bryophyte fossils are found in rocks formed before the Devonian period, but as we shall see, vascular plants were already present in the Silurian. We may conclude, then, that the mosses and liverworts either represent a second, rather unsuccessful, colonization of the land by aquatic ancestors or possibly (as some have suggested) are descendants of land plants that have lost many of the adaptations of their ancestors. In either case, the lack of a vascular system and woody tissue and the necessity for surface water in which the sperm can move from antheridia to archegonia have limited the evolutionary potentialities of these organisms.

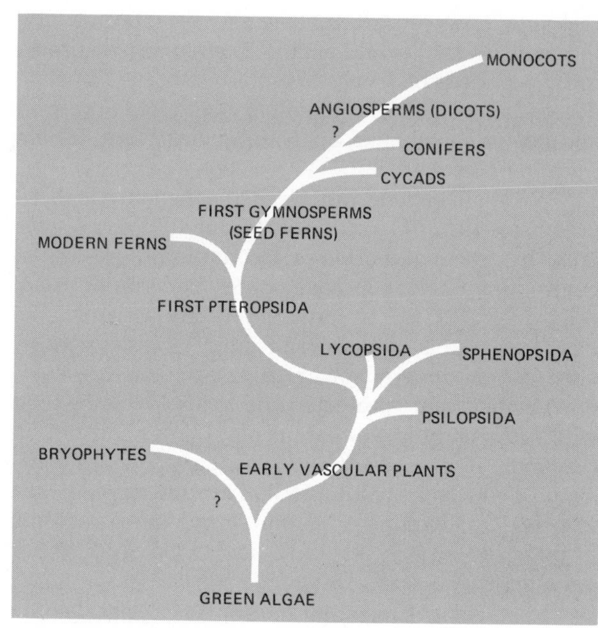

FIG. 36.3 The suspected evolutionary relationships of the plants.

36.4 THE VASCULAR PLANTS (PHYLUM TRACHEOPHYTA)

Although the anatomy of the first plants is not well known, the earliest fossils indicate that these organ-

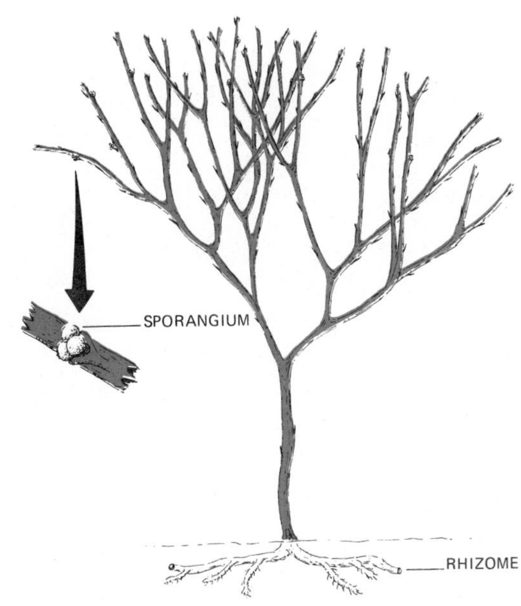

FIG. 36.4 Left: A reconstruction of the Devonian landscape, showing three species of psilopsids in the foreground. (Courtesy of the Brooklyn Botanic Garden.) Right: *Psilotum nudum,* a psilopsid that still grows in Florida and elsewhere.

isms had already evolved a vascular system for the transport of water and food throughout the plant. Therefore, they would be bona fide members of the phylum Tracheophyta, the phylum of the vascular plants. By the end of the Devonian period, four distinct groups had appeared, each of which has left some descendants right up to the present time. We assign these groups to the subphyla Psilopsida, Lycopsida, Sphenopsida, and Pteropsida (Fig. 36.3). Their living descendants total some 260,000 species.

A. Subphylum Psilopsida

The psilopsids had no roots or leaves. They did, however, have both an underground stem (rhizome) and an erect, aerial stem, each of which contained xylem and phloem. Photosynthesis was carried out in the erect stem, and this also produced the sporangia.

Four species alive today closely resemble the fossil psilopsids and, although not all botanists agree, we shall include them in the subphylum. If *Psilotum nudum* (Fig. 36.4) is a genuine psilopsid, we can deduce that the early psilopsids produced only one kind of spore. These developed into tiny gametophytes which produced both antheridia and archegonia. Fertilization was by swimming sperm and hence these plants must have been restricted to habitats that were quite wet, at least part of the time.

B. Subphylum Lycopsida

The members of this subphylum are commonly called clubmosses. This name is derived from their superficial resemblance to mosses (they commonly grow close to the ground and have tiny leaves) and from the fact that they produce spores in clublike structures (Fig. 36.5) called **strobili.** They are not mosses at all but vascular plants with xylem and phloem in the roots and leaves connecting with that in the stem. The leaves are quite simple and small, with the vascular tissue occurring in a single, unbranched vein.

About 1000 species of lycopsids exist today. The genera *Selaginella* and *Lycopodium* are fairly common in North America. Some members of *Lycopodium* are commonly called ground pines and are used for Christmas decorations (Fig. 36.5).

Although all the members of this subphylum living today are quite small, fossils have been found of species that reached heights of 100 feet. These trees were especially abundant during the Mississippian

FIG. 36.5 *Lycopodium obscurum* growing close to the forest floor. These plants stand about 8 in. (20 cm) tall. Although commonly called "ground pines," they are actually members of the subphylum Lycopsida, an entirely different group.

FIG. 36.6 Carboniferous forest. An artist's reconstruction of what the swamp forests of the Mississippian and Pennsylvanian periods looked like. The large trunks on the left and lying flat in the foreground are of tree-sized lycopsids. The trees on the right are sphenopsids. Ferns and seed ferns are also present. (Courtesy Field Museum of Natural History.)

and Pennsylvanian periods (Fig. 36.6), and their remains contributed to the formation of extensive coal deposits. Much of the world's present coal was formed at that time, so the two periods are often collectively referred to as the Carboniferous period. When we burn this coal today, we are releasing energy that was stored by photosynthesis some 300 million years ago.

Some of the lycopsids produced (and still do) not one but two kinds of spores: microspores (male) and megaspores (female). These developed into separate male and female gametophytes, respectively. In some cases, the megaspore was retained within the tissues of the parent sporophyte and, as in the gymnosperms and angiosperms of today, this provided a protected environment for fertilization to take place.

FIG. 36.7 *Equisetum palustre,* a common horsetail. It is a member of the subphylum Sphenopsida.

C. Subphylum Sphenopsida

The plants in this subphylum are commonly called horsetails or scouring rushes. The first name is derived from their characteristic method of branching: whorls or rings of branchlets arising from an aboveground shoot (Fig. 36.7). This shoot develops each season from an underground stem. The second name is derived from the fact that these plants were once used for cleaning pots and pans. Horsetails often grow in sandy locations and incorporate substantial quantities of *silica* in their stems. The silica makes the stems quite abrasive and hence an effective scouring material. The leaves are very small and are arranged in whorls or rings around the stem. Only one genus of sphenopsids, *Equisetum,* containing about 25 species, survives today. Many other, much larger species were, however, once dominant features in the landscape (Fig. 36.6) and, like the early lycopsids, contributed to the formation of coal.

D. Subphylum Pteropsida

The plants in this subphylum differ from those in the other three in possessing relatively large leaves with many, often branching, veins. The subphylum is further subdivided into three classes: the ferns, the gymnosperms (many of which later reduced their leaves to "needles"), and the angiosperms.

1. Class Filicinae (ferns). The first pteropsids, the ferns, also contributed a large array of species to the flora of the Devonian landscape. Like most of our temperate-climate ferns of today, they were **homosporous;** that is, only one kind of spore was produced. Each fern spore, you remember, develops into a prothallus that bears both male and female sex organs. Fertilization requires moisture in which the ciliated sperm can swim to the egg. For this reason, the ferns are still restricted to habitats where abundant water is present during part of the growing season. Approximately 9500 species of ferns live on the earth today. Most of these are found in the tropics, where some may grow to heights of 40 feet (thirteen meters) or more (Fig. 36.8). A smaller number are found in temperate regions. These generally grow in damp, shady locations. Their stem (the **rhizome**), as well as their roots, is located beneath the ground. The leaves grow up from the rhizome each spring and manufacture food by photosynthesis. In our largest temperate species they may grow as tall as three or four feet. Although the leaves are generally killed by the first frost, the rhizome and the root system remain alive through the winter.

Ferns are dispersed to new locations by means of tiny, wind-blown spores. These are formed in sporangia which develop on the leaves. (You may wish to review the life cycle of ferns in Chapter 20.)

The delicate, elaborate branching patterns of

FIG. 36.8 Ferns. Left: Tree ferns in Puerto Rico. Right: Ferns of temperate climates have underground stems.

fern leaves make these plants highly prized for ornamental use. Earlier in the earth's history, ferns grew in great profusion; their remains also contributed to the formation of coal.

2. Class Gymnospermae. Fossil remains from the Devonian indicate that some early fernlike plants were **heterosporous,** that is, they produced both microspores and megaspores. The megaspore was retained within the moist tissue of the parent sporophyte. Fertilization took place here, freed from dependence on a supply of surface water. On the other hand, the necessity for the microspores to be carried from one plant to another in order to reach the female gametophyte robbed them of their value as agents of dispersal. This function was probably taken over by seeds—dormant, protected, embryo sporophytes.

The seed ferns, as these plants are called, were among the earliest of the **gymnosperms.** Although seed ferns are now extinct, some of their descendants, the cycads (which resemble them closely—Fig. 36.9) and the ginkgo, survive to this day. These modern seed-bearing plants reveal their ancient lineage by

FIG. 36.9 A cycad (*Cycas revoluta*) growing in the U.S. Virgin Islands.

FIG. 36.10 The ginkgo (courtesy of USDA, Forest Service) and its leaves.

the fact that after the microspore reaches the ovule, it liberates a ciliated sperm which, by swimming in moisture supplied by the parent sporophyte, reaches the egg.

Cycads are tropical plants that bear a superficial resemblance to palms. Only one genus grows wild in the United States, and it is found only in Florida. The ginkgo (Fig. 36.10) is the last surviving species of a once prominent group of plants. Although the ginkgo may still grow wild in the interior of China, it grows only in cultivation elsewhere. It is widely cultivated in temperate regions, however, because it grows rapidly and seems especially tolerant of smoke and other conditions of city life. The sexes are separate in the ginkgo, and male trees are generally preferred for shade-tree planting. This is because the seeds, which are produced by the female trees only, give off an unpleasant odor when they are crushed underfoot.

The early gymnosperms flourished during the Mississippian and Pennsylvanian periods. Along with the ferns, lycopsids, and sphenopsids, they contributed to the formation of coal. Toward the end of these periods, the **conifers** appeared.

Today, the conifers (e.g. pines, spruces, and firs) are the most abundant of the gymnosperms, but they generally dominate the landscape only in regions where the winters are very cold. The group includes the largest and the oldest of all living organisms. One redwood (genus *Sequoia*) growing in California is 368 feet high. Bristlecone pines growing high in the mountains of eastern California have been shown to be over 4000 years old.

Most conifers are evergreen, with their leaves modified to form "needles." The microspores and megaspores of gymnosperms are formed in **cones** (Fig. 36.11). When they reach the female cones, the microspores develop into a male gametophyte which produces (1) a pollen tube and (2) the sperm nucleus which accomplishes fertilization. No motile sperm are produced.

The coniferous forests of North America supply a large part of the lumber used for building. The paper industry also depends to a large degree on conifers as a source of paper pulp.

3. Class Angiospermae. Although some evidence of angiosperms appears in the fossil record in Juras-

FIG. 36.11 Branch of the loblolly pine, a conifer. Note the female cones produced during the current season (upper right), last season (middle), and two seasons before (lower left). The two-year-old cones are about to release their content of seeds. (Courtesy USDA, Forest Service.)

FIG. 36.12 A common "duckweed," one of the smallest angiosperms.

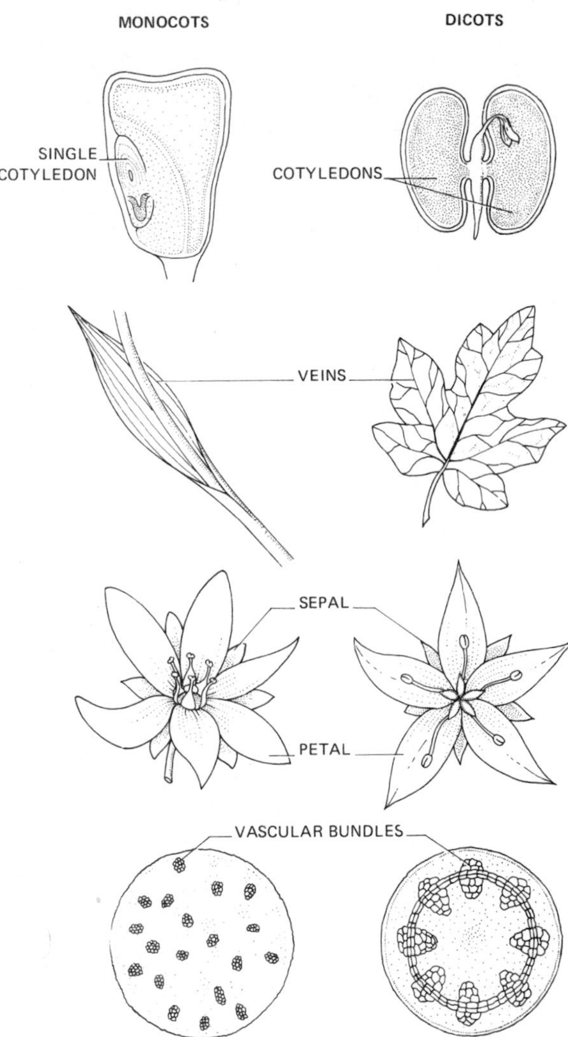

FIG. 36.13 A comparison of the structural patterns found in dicots and monocots.

sic deposits, it was not until the end of the Mesozoic era that the angiosperms became the dominant plants of the landscape. That they are the dominant plants on earth today is unquestionable. There are some 250,000 species of living angiosperms. The rest of the plant kingdom includes only about 34,000 living species. The angiosperms are found in practically every habitat. Although they incorporate a variety of features that enable them to live even in arid locations, some have returned to an aquatic existence (Fig. 36.12).

The angiosperms are divided into two subclasses, the **dicots** and the **monocots.** These names are derived from the number of cotyledons present in the seeds: two in those of the first group, one in those of the second. Dicots and monocots also differ in several other respects (Fig. 36.13). The veins in dicot leaves are arranged in a network pattern while in monocots they lie parallel to one another. The vascular bundles in the dicot stem are arranged in a radial pattern like the spokes of a wheel, while in monocots they are scattered randomly throughout the stem. The various parts which make up the dicot flower (petals, for example) usually occur in fours, fives, or some multiple thereof. Flower parts in monocots occur in threes or some multiple of three.

The dicots are the larger (and older) of the two groups, with approximately 200,000 species known. The buttercup, snapdragon, carnation, magnolia, poppy, cabbage, rose, pea, poinsettia, cotton plant, cactus, carrot, blueberry, mint, tomato, elm, oak, and maple represent 18 of the 250 families of dicots.

About 50,000 species of monocots are known. These include the lilies, palms, orchids, irises, tulips, sedges (a grasslike plant of marshy locations), onions, asparagus, and the grasses. The grasses include corn, wheat, rice, and all the other cereal grains upon which we depend so heavily for food.

36.5 ADAPTATIONS OF ANGIOSPERMS

The evolutionary success of angiosperms and their importance to human existence justify our using them to illustrate the main features of plant structure. As in most of the tracheophytes, the plant body is made up of three main organs: **roots, stem,** and **leaves.**

Roots anchor the plant to the soil and absorb water and minerals from it. These materials are then transported to stem and leaves by the vascular system. Some angiosperms, the carrot, for example, also use their roots for the storage of food (Fig. 24.19).

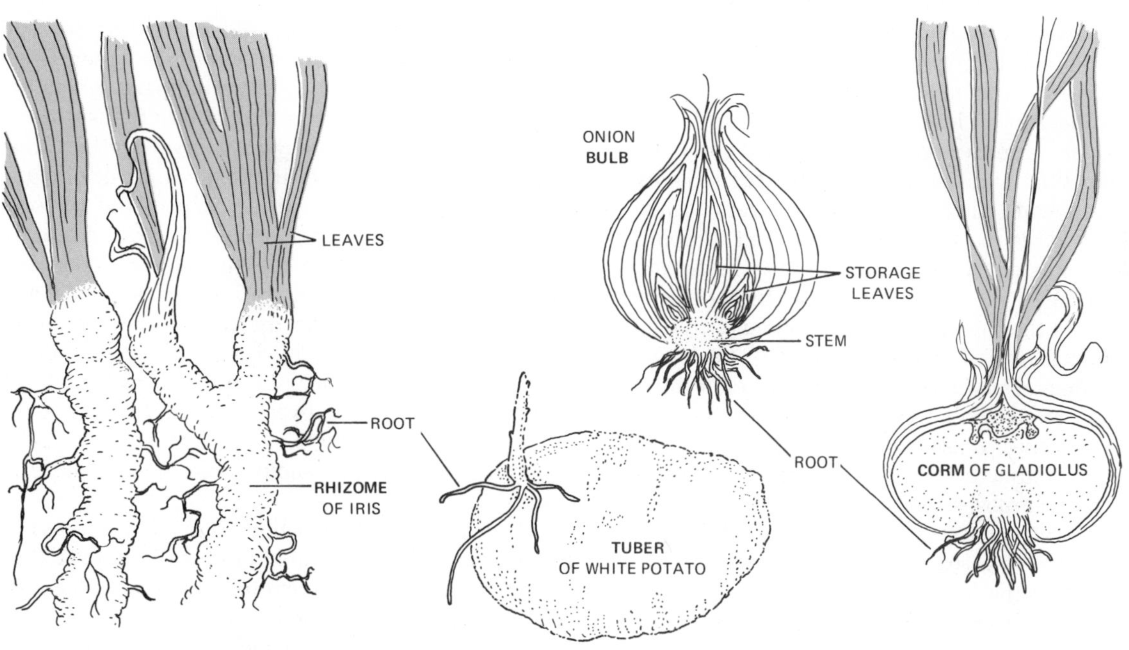

FIG. 36.14 Four kinds of underground stems.

The **stem** produces the leaves and supports them so that they are properly exposed to the sun. Under special circumstances, stems also produce flower buds. These develop into flowers and enable the plant to carry on sexual reproduction.

Stems are also important simply as a connecting link between the roots and the leaves. Leaves manufacture food without which the roots would starve. Roots absorb soil water and minerals without which the leaves could not produce food. The vascular system of the stem insures that these essential materials travel rapidly between the roots and the leaves.

Stems are sometimes used for the storage of food. The most striking examples of this are found in the swollen underground stems of such plants as the iris, gladiolus, and white potato (Fig. 36.14).

The main function of **leaves** is the manufacture of food by photosynthesis. This occurs in the thin flat blade of the leaf. In most dicots, the blade is attached to the stem by a petiole. The vascular system of the stem extends into the petiole and, as veins, into the blade itself. The veins not only transport materials to and from the leaf but also provide a skeleton to strengthen the leaf.

Angiosperm leaves may be **simple** or **compound.** In simple leaves the blade consists of only a single part attached to the petiole. Compound leaves are subdivided into leaflets. Palmately compound leaves, as in the strawberry, are made up of leaflets radiating from a single point. In pinnately compound leaves, the leaflets are arranged along either side of the midrib of the leaf (Fig. 36.15).

Many angiosperm leaves are modified for functions other than photosynthesis. The **tendrils** by which the pea vine clings to its support are modified leaves. The large petioles of celery and rhubarb leaves serve to store food. So, too, do the fleshy underground leaves of the onion bulb. Insect-eating plants such as the pitcher plant and Venus's flytrap use modified leaves to catch their prey. The leaves of the ornamental plant *Bryophyllum* develop miniature plants along their margins and thus serve to reproduce the species asexually (Fig. 36.16).

Patterns of growth vary widely among the angiosperms. Trees, shrubs, and some vines are **perennial.** Their aerial, woody stems grow longer and thicker during each growing season. Many of these are deciduous, i.e., all the leaves are shed in the autumn. Other perennials produce only soft, herbaceous stems. In temperate climates, the portions above ground die in the autumn. The underground portions (roots and often a rhizome) remain alive, however, and send up new shoots (stems and leaves) the following spring. Many of our grasses and wildflowers are herbaceous perennials.

Some angiosperms are **biennials,** completing their life cycle in two years. The carrot, for example, sends leaves above the ground during the first season (Fig. 24.19). These die in the fall, but the roots and short stem remain alive. The following year, the stem sends up a stalk that bears flowers in which the seeds are produced. After the seeds are liberated, the entire plant dies.

The **annual** angiosperms are herbaceous plants that develop from seeds, grow, flower, produce seeds, and die all in one season. Their seeds serve not only to disperse the plant to new locations but also to enable the species to appear again the next season. Many cultivated flowers, weeds, and some grasses, including the cereal grains, are annuals.

It is difficult to overestimate the importance of angiosperms to our existence. They provide us with all but a tiny fraction of our food, either directly or as the chief source of food for our livestock. Angiosperms also provide us with such important products as cotton, natural rubber, paper, wood, tobacco, and linen. A number of valuable industrial chemicals and drugs are extracted from angiosperms.

As the dominant plants on the earth today, angiosperms play a crucial role in providing a suitable physical habitat for humans and other terrestrial animals. Their root systems aid in holding the soil against the eroding forces of wind and water (look back at Fig. 10.25 to see what happens when the vegetation is removed from the land). Transpiration and the absorption of the sun's rays by the forest canopy moderate the severity of the summer climate. A thick carpet of angiosperms reduces the force of falling rain and acts as a sponge to hold water in the soil. This diminishes the chances of floods and water erosion while helping to keep wells, ponds, and streams filled throughout the dry summer months.

And we do not live by bread and water alone. So we must not overlook the esthetic role that angiosperms play in our lives. Woods, gardens, even house plants enrich our lives in ways that have little to do with the harsh economic and caloric needs of survival.

To what can we attribute the great success of the angiosperms? Probably the answer is that they incorporate the most efficient adaptations to dry-land living with the most varied and efficient methods of sexual reproduction and dispersal. Let us review the features which have permitted angiosperms (and, in varying degrees, other plants) to invade the dry land.

The possession of roots permits the extraction

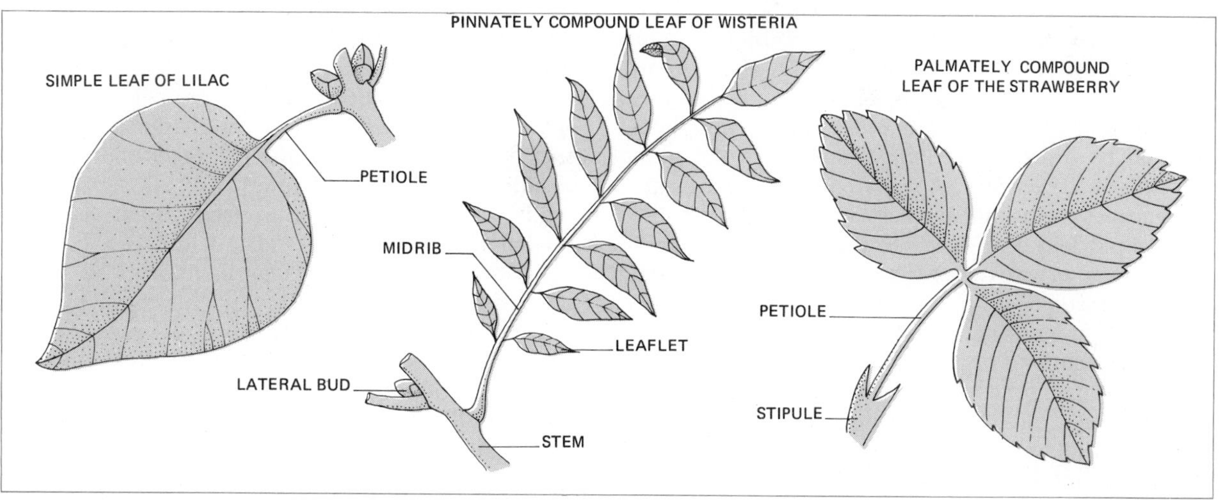

FIG. 36.15 Leaf forms.

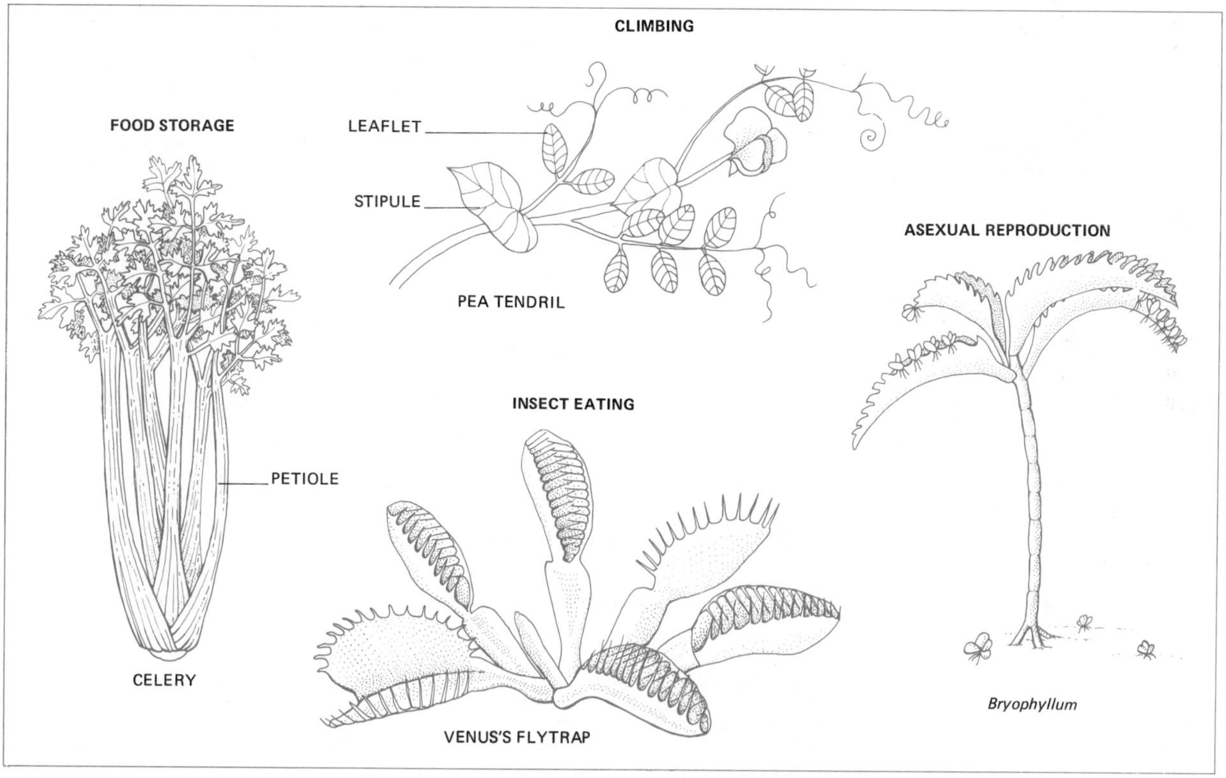

FIG. 36.16 Various leaf modifications.

of moisture and minerals from beneath the surface of the land. Roots also serve to anchor the plant against the wind. The presence of a cambium capable of producing woody tissue provides for the support, high in the air if need be, of leaves and flowers. Xylem and phloem enable water, foods, hormones, etc., to be translocated long distances in the plant quickly and efficiently. A waxy cuticle on leaves and herbaceous stems and the cork on woody stems prevent rapid loss of water from the plant by evaporation. Such a waterproof covering is also gasproof, but the needs of gas exchange are taken care of by stomata and lenticels. The shedding of leaves in most temperate-climate angiosperms further reduces loss of water during the winter (when soil water may be frozen) and also helps reduce the likelihood of mechanical damage to the plant from the accumulation of ice and snow.

The angiosperms share with the gymnosperms the characteristic of retaining the female gametophyte within the megasporangium. As we have seen, this device puts an end to the need for surface water to effect the transfer of sperm from one plant to another. Instead, pollen grains can be transported by air to the megasporangium, thus bringing the gametes together. In the gymnosperms and some angiosperms, the pollen grains are simply blown from plant to plant by the wind. Many other angiosperms attract insects or other animals by their flowers and thus exploit the animal's power of locomotion to aid cross-pollination. The quantity of pollen produced by animal-pollinated angiosperms is usually far less than that produced by wind-pollinated species.

The embryo sporophyte plant, the seed, is also an effective adaptation to dry-land living. Protected by seed coats and supplied with food reserves, the seed can withstand harsh, dry conditions for long periods while still remaining ready to germinate when conditions finally become suitable. Seed production in the angiosperms is more efficient than in the gymnosperms in that food is translocated into the seed only if fertilization has taken place. In the gymnosperms, the food reserves of all the potential seeds are deposited before fertilization and thus wasted if pollination should fail to occur.

Most of the adaptations mentioned so far are not unique to the angiosperms. They have played a role in allowing the other plants to colonize the land, too. It is only in the production of **flowers** and **fruits** that the angiosperms stand alone. The efficient methods of pollination and seed dispersal which flowers and fruits make possible, added to all the other features which reach their fullest development in the angiosperms, have enabled these plants to invade nearly every conceivable habitat on this earth. In the angiosperms, we see the most adaptable group of plants yet produced by evolutionary change in organisms that were once restricted to an aquatic world.

The colonization of the land by vascular plants in the Silurian had far-reaching consequences. Plants are autotrophic and their presence on land provided there for the first time an abundance of food for heterotrophic organisms. The fungi and many animals were quick to exploit this and soon followed the plants in adapting to the special conditions of dry-land living. In the next chapters we shall examine how animals accomplished this transition. It will be an important part of our larger story of the evolution of the animals as a whole.

EXERCISES AND PROBLEMS

1. Check the plant classification schemes in a number of different textbooks. What are the major similarities? The major differences? Which do you prefer? Why?
2. In what ways are the members of the plant kingdom different from the green algae? From the blue-green algae? From the red algae?

REFERENCES

1. BELL, P. R., and WOODCOCK, C. L. F., *The Diversity of Green Plants,* Addison-Wesley, Reading, Mass., 1968. The algae are also discussed.
2. DOYLE, W. T., *The Biology of Higher Cryptogams,* Macmillan, New York, 1970. A succinct treatment of the bryophytes, psilopsids, lycopsids, sphenopsids, and ferns.
3. WATSON, E. V., *Mosses,* Oxford Biology Readers, No. 29, Oxford University Press, Oxford, 1972.
4. SPORNE, K. R., *The Mysterious Origin of Flowering Plants,* Oxford Biology Readers, No. 3, Oxford University Press, Oxford, 1971.

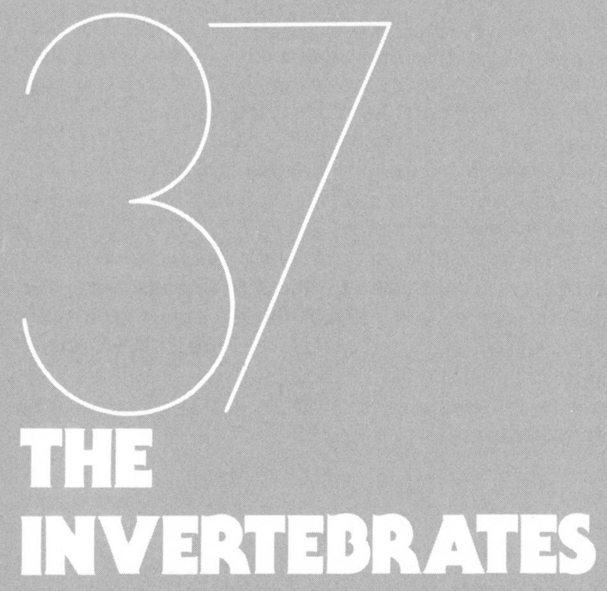

37 THE INVERTEBRATES

37.1 INTRODUCTION

Animals are organisms that (1) do not have chlorophyll, (2) are capable of locomotion or, at least, body movement by means of contracting fibers and (3) are multicellular. A few organisms fail to meet the second criterion but show such close similarities to creatures that do, that we cannot fail to recognize them as animals.

The animal kingdom is generally divided into some 20 to 30 different phyla. The animals that make up all but one of these have no backbone and hence are commonly described as invertebrates. One phylum (our own) contains a few invertebrates, and these will be discussed in this chapter. The remaining members of the phylum are the extraordinarily successful vertebrates. Their evolution will be studied in the following chapter.

Three of the animal phyla include creatures built on rather simple patterns of organization. For this reason, we suspect that these animals are primitive, or to put it another way, are little-changed descendants of the earliest forms of animal life. These animals are the sponges, the cnidarians, and the flatworms.

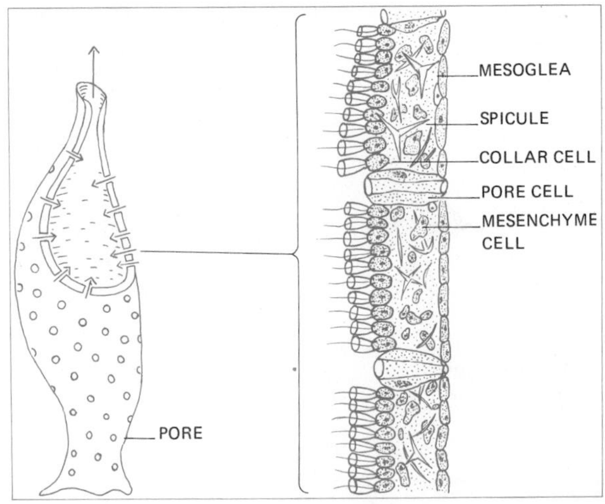

FIG. 37.1 Structure of a simple sponge. Sponges secure food and oxygen from the water which is continuously drawn in through their pores.

37.2 THE SPONGES (PHYLUM PORIFERA)

These are simple animals that spend their lives anchored to a rock or other solid surface underneath the water. About 10,000 species are known, a few of which live in fresh water, but most live in the ocean. The phylum gets its name from the many small openings or pores that perforate the sponge body. The animals feed by drawing water in through these pores and filtering out tiny food particles that may be present (Fig. 37.1).

The sponge body consists of two layers of cells with a layer of jellylike material, the **mesoglea,** between them. The cells of the inner layer have flagella which set up the water currents. These cells also consume the filtered food particles.

The shape of the sponge is maintained by a skeleton of spicules formed by scattered cells in the mesoglea. The spicules are quite hard, being composed of either silica or limestone (calcium carbonate). Some sponges have no spicules but instead are supported by a network of flexible, tough fibers. These sponges, which occur in shallow tropical waters, are harvested by divers and after processing are sold for cleaning purposes.

Organisms that are anchored to one spot must have some means of dispersing their offspring to new locations. The sponges accomplish this by producing small, free-swimming larvae. These swim away from the parent sponge and, after finding a suitable new surface, attach themselves to it and develop into adults.

Fossil remains indicate that sponges were one of the earliest forms of animal life to appear on earth. There is no evidence, however, that any other animal forms have evolved from sponges since then. The sponges seem to occupy a rather unique place in the animal kingdom and have, in fact, been placed in their own subkingdom, Parazoa, by some taxonomists.

37.3 THE CNIDARIANS (PHYLUM CNIDARIA)

All the members of this phylum possess specialized stinging cells called **cnidoblasts** that have given rise to the phylum name. Each cnidoblast contains a poison-filled, barbed thread, the nematocyst (Fig. 37.2). When the trigger of the cnidoblast is touched, the nematocyst is discharged. It is used for trapping and paralyzing prey as well as for defense against enemies.

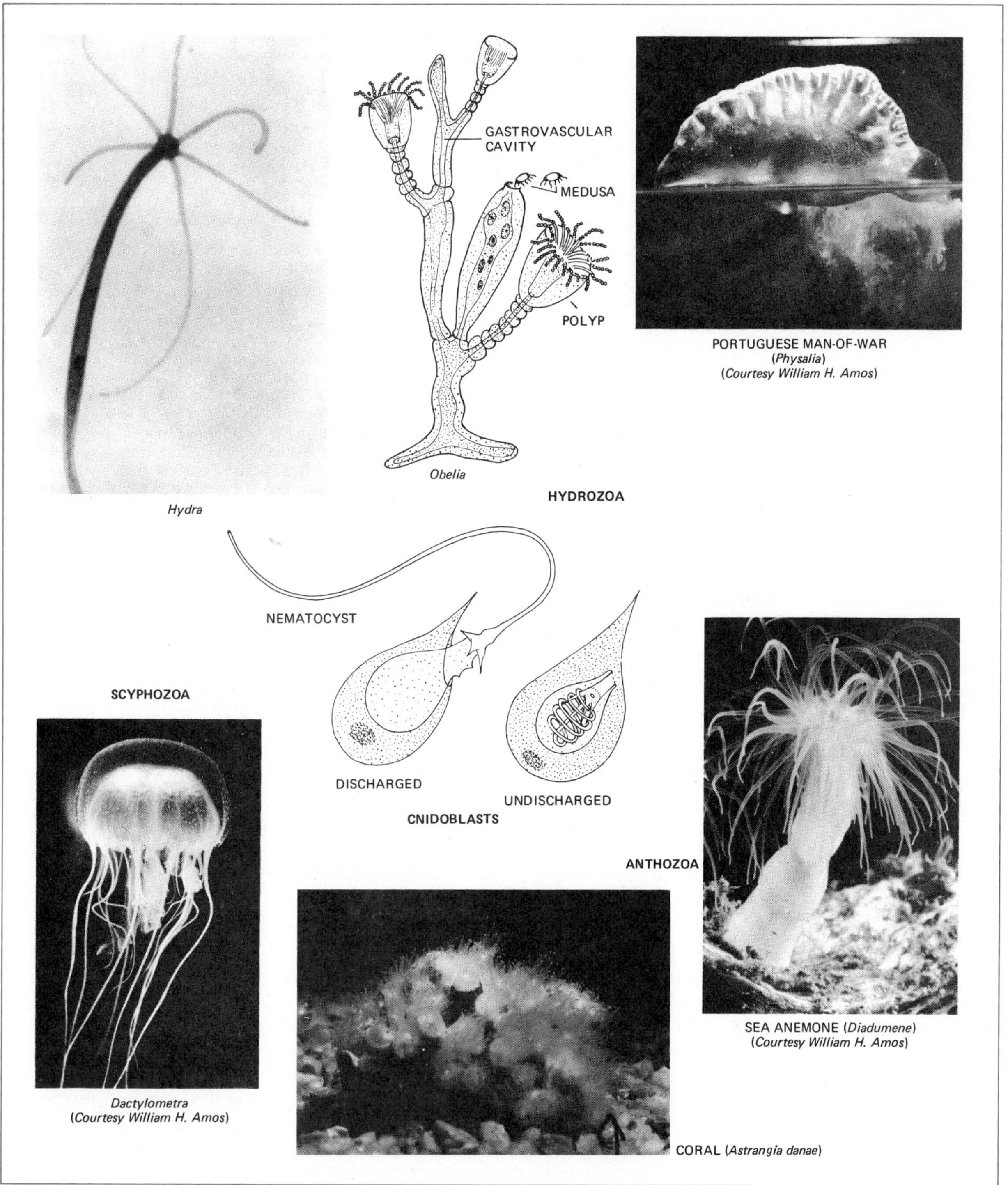

FIG. 37.2 Representative cnidarians. Most of the members of this phylum live in salt water.

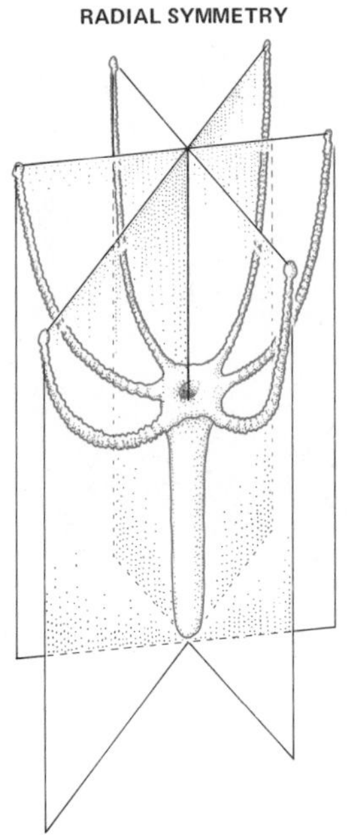

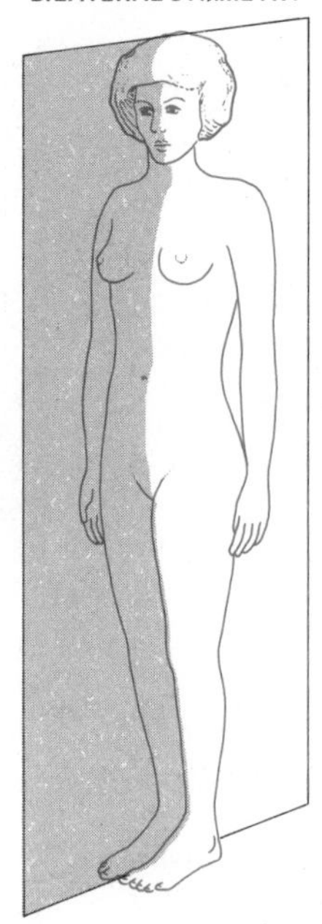

FIG. 37.3 Bilateral symmetry is characteristic of animals that move about actively. They have right and left sides, a front and a back. Animals with radial symmetry spend their lives anchored to one spot or move about quite sluggishly.

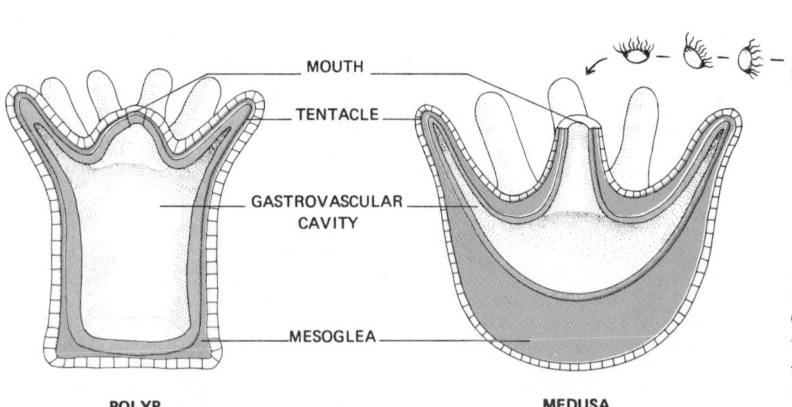

FIG. 37.4 The two body forms found in cnidarians. The medusa has been turned upside down to show its basic similarity to the polyp. The jelly of a jellyfish (medusa) is a greatly enlarged mesoglea.

The body of all members of this phylum consists of two layers of cells with a jellylike mesoglea between them. The mesoglea has cells scattered through it, however, and is considered by some biologists to be a third cell layer. The body is organized as a hollow cylinder with a single opening at one end. Food is introduced through this opening, the mouth, and into the inner cavity, called the **gastrovascular cavity.** This cavity is also called a coelenteron and for many years the name of this phylum was **Coelenterata.** Another group of animals (called comb jellies) were included then as they also have a coelenteron. They have no cnidoblasts, however, and are no longer thought to be closely related to the forms we are discussing.

All the parts, such as tentacles, in the Cnidaria are arranged in a circle around the cylindrical body. This pattern of organization is known as radial symmetry. If we should cut a hydra from the head (anterior) to the base (posterior) in any plane passing through the midline, the organism would be divided into two equal halves. Contrast this with the bilateral symmetry of humans. There is only one plane passing through the midline that will divide the human body into two equal halves. This is the plane running from the back (dorsal) surface to the front (ventral) surface (Fig. 37.3). It divides the body into right and left halves. Radially symmetrical animals like the cnidarians have neither dorsal and ventral surfaces nor right and left sides.

About 9000 species of Cnidaria are known. Most of these are found in the oceans, although a few species, such as *Hydra* (Fig. 37.2), occur in fresh water. The phylum is divided into three classes.

1. Class Hydrozoa. The ready availability of the common freshwater hydra makes it the most common cnidarian studied by biology students. Although it exhibits the major structural features of the phylum, it is really not typical of either its phylum or its class. First of all, unlike most of the other members of the phylum it is found in fresh water. Second, it exists as a single individual. Most members of its class are colonial, that is, many individuals live attached to one another. Third, *Hydra* exhibits only one body form, the **polyp** (Fig. 37.4). Most members of the class produce a second body type, the **medusa,** that floats or swims freely in the water and aids in the dispersal of the species. Despite its superficial difference in appearance, the medusa is basically an upside-down polyp. Figure 37.2 shows the two body forms as they occur in *Obelia,* a more nearly typical member of the class. The Portuguese man-of-war (*Physalia*), whose nematocysts may cause serious, even fatal, poisoning in humans, is also a member of this class. It consists of a floating, gas-filled bag from which dangle long chains of polyps.

2. Class Scyphozoa. The main body form of jellyfishes, which make up this class, is the medusa. The jelly is simply a greatly enlarged mesoglea. The tentacles of the medusa bear the cnidoblasts, which in some species can cause considerable pain to unwary swimmers.

3. Class Anthozoa. This class contains the sea anemones and the corals. These creatures consist of a polyp stage only. Corals secrete limestone shelters that are responsible for the development of great reefs and atolls in tropical waters. Figure 37.2 shows *Astrangia danae,* one of the few corals found in northern waters.

37.4 THE FLATWORMS (PHYLUM PLATYHELMINTHES)

This phylum contains some 9000 species. These organisms are aptly named: many are almost ribbon-like in shape. They are bilaterally symmetrical with right and left sides and a dorsal and ventral surface as well as an anterior and posterior. This type of symmetry seems to be associated with active locomotion. Many of the common freshwater flatworms, called **planarians,** can move quite rapidly. When attached to an underwater surface, they secrete a slimy mucus layer underneath them and then propel themselves forward on the mucus by the beating of myriads of cilia on their ventral surface. When suspended freely in the water, planarians swim by undulating body movements. The efficient locomotion of the planarians enables them to seek their food actively, in contrast to the radially symmetrical cnidarians.

Bilateral symmetry is also associated with a concentration of sense organs at the anterior end of the animal. Planarians have light, touch, and vibration receptors at their anterior end, the end that meets changes in the environment first. Such a concentration of sensory equipment in the head is called **cephalization.**

The food of the planarian is taken in through a mouth on the ventral surface and into a gastrovascular cavity. Although this cavity is much more elaborate in shape than that of the hydra, it is built upon the same saclike plan. Undigested material must still leave by the mouth as it does in the hydra.

Many of the free-living flatworms (class **Turbellaria**) are not nearly as complex as the common freshwater planarians that we have been discussing. In fact, some of them are constructed so simply that biologists feel they are the most primitive bilaterally symmetrical animals and perhaps all the other phyla we shall discuss evolved from early members of this group. Today these worms can be found in moist soil and salt water as well as in fresh water.

Two classes of flatworms are composed exclusively of parasitic forms. Adult **flukes** (class **Trematoda**) remain attached to their host by means of suckers on the ventral surface. Many produce larvae that are also parasitic but in a different host, usually some species of snail. Lung flukes and liver flukes (Fig. 37.5) are serious parasites of humans and other animals. But at the present time blood flukes represent the greatest hazard to humans. Several species in the genus *Schistosoma* infect humans,

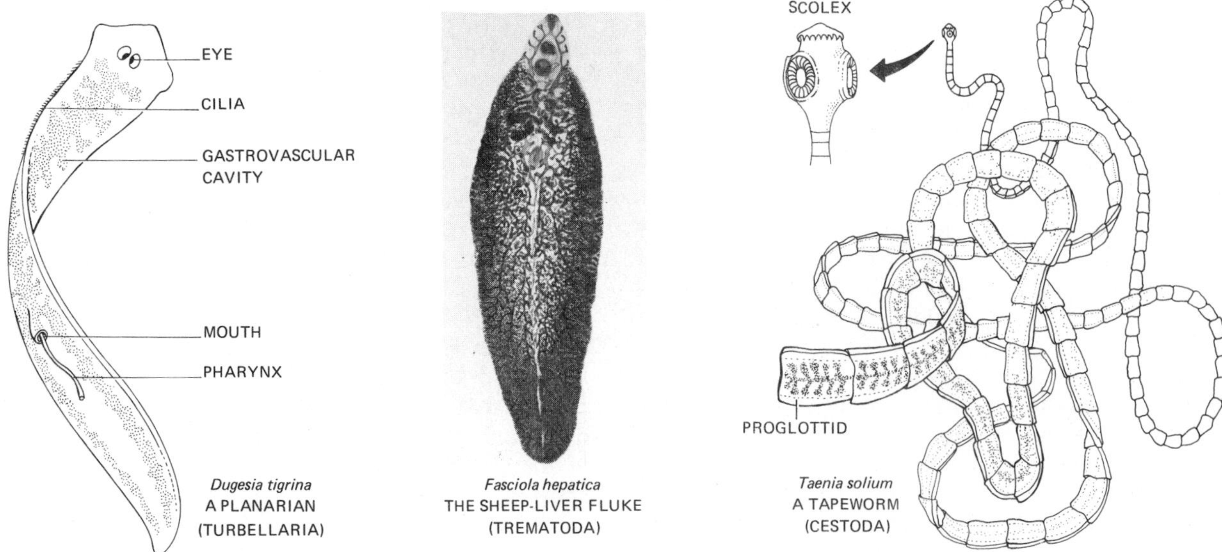

FIG. 37.5 Representative flatworms. The tapeworm has no digestive organs. (Photo courtesy Turtox/Cambosco, Macmillan Science Company, Chicago.)

causing schistosomiasis. This disease has probably been a serious health problem in tropical regions since before the dawn of history. But the recent building of irrigation systems in formerly desert regions (in Egypt, for example, supplied by the waters impounded by the Aswan Dam) has greatly increased the habitat needed by blood flukes to complete their life cycle (Fig. 37.6). As a result, schistosomiasis has become one of the greatest public health problems of our times.

Tapeworms (class **Cestoda**), like flukes, are exclusively parasitic. The adults live in the intestine of their host and absorb nourishment from their surroundings. They form ribbonlike colonies of relatively independent proglottids. In some species, these grow to lengths of 60 feet or more. Most tapeworms need two or more different hosts in order to complete their life cycle. Humans can acquire tapeworms from eating undercooked fish, beef, and pork. The life cycle of the pork tapeworm, *Taenia solium,* is shown in Fig. 14.2.

37.5 THE ORIGIN OF ANIMALS

From what did the animals evolve? We simply do not know. The event occurred in Pre-Cambrian times and no fossils that might enlighten us have been found. Thus we are forced to examine evidence such as homologous organs, homologies in the pattern of embryonic development, even biochemical homologies—all in living forms—in order to provide a basis for our speculations.

While a number of theories of animal origins have been proposed, they fall into two rather distinct categories. On the one hand, there are those who believe that the multicellularity characteristic of animals arose as a consequence of the formation of colonies of cells. With the development of specialized cells within the colony, interdependence of parts follows, and the boundary between a colony and a multicellular individual has been crossed. In the sponges the presence of collar cells, which closely resemble certain unicellular flagellates (Fig. 37.7), suggests that sponges did evolve this way.

What about the other animals? Perhaps they, too, arose this way, but independently of the sponges. Some biologists think that colonial green algae were not only the ancestors of the plants but also the ancestors of the animals. This theory is supported by the existence of larval cnidarians whose structure—a ciliated, hollow ball of cells—is suggestive of the structure of *Volvox*. Inward migration of some of the surface cells of this larva produces two tissue layers, and later the gastrovascular cavity of the mature cnidarian develops within. The flatworms, which also have a gastrovascular cavity with a single opening,

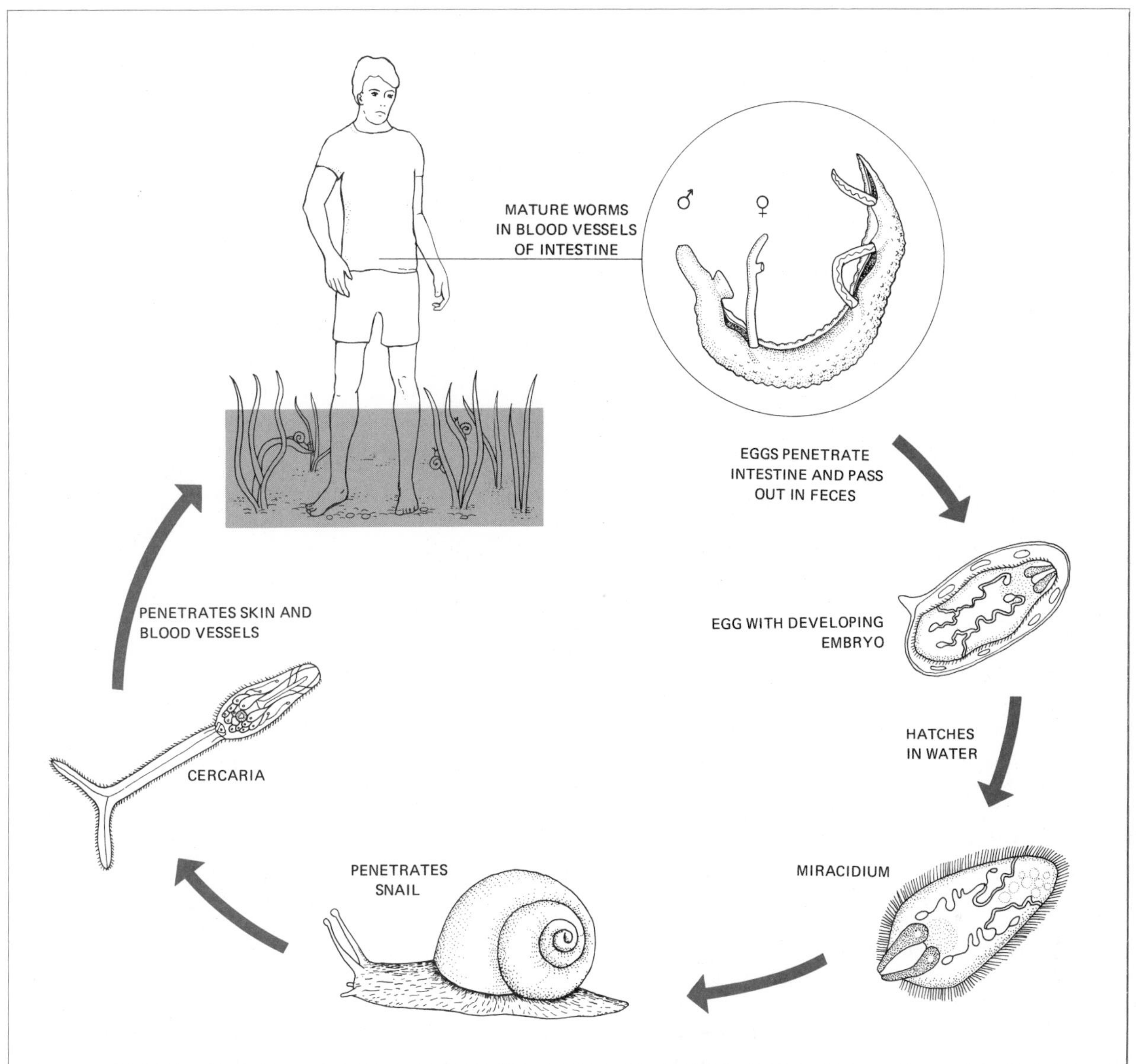

FIG. 37.6 Life cycle of the blood fluke, *Schistosoma mansoni.* Once within the alternate host, a snail, a single miracidium may produce as many as 200,000 infectious cercariae. Both sexes must infect the human if the cycle is to continue. With the increasing use of irrigation in tropical regions, the incidence of human infection is increasing alarmingly.

are considered descendants of the cnidarians. (One wonders, though, what happened to that most distinctive feature of the cnidarians, the cnidoblast. It is true that some flatworms have cnidoblasts, which they use as weapons, but they do not grow them. They simply incorporate them into their own epidermis after first securing them by eating hydras!) With the penetration of the gastrovascular cavity

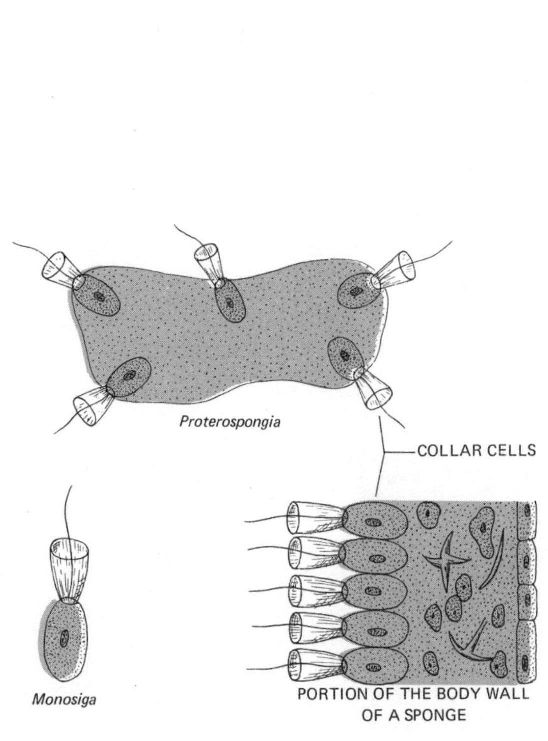

FIG. 37.7 The presence of collar cells in sponges suggests that these multicellular animals have evolved from colonial forms like *Proterospongia.* They, in turn, probably evolved from unicellular forms.

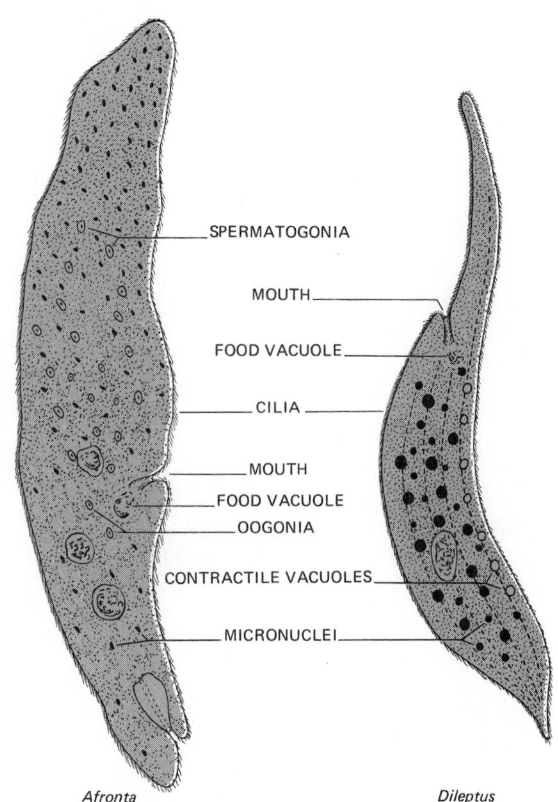

FIG. 37.8 A syncytial flatworm (left) compared with a ciliated protozoan (right). The only true cells in *Afronta* are the oogonia and spermatogonia.

through the other end of the organism, the tube-within-a-tube body plan of the other animals could have been formed.

An alternative explanation for the origin of animals is that they arose by the cellularization of ciliated protozoans. As we noted in Chapter 35, many ciliates are quite large (for single cells) and exceedingly complex in their anatomy. Often these forms have a large number of nuclei scattered through the cytoplasm. Developing an internal system of membranes to partition these nuclei into separate cells would transform these creatures into bona fide animals.

In fact, some tiny flatworms (relatives of the planarians) resemble ciliated protozoans in a number of ways, including size (Fig. 37.8). These worms are syncytial; that is, most of the animal body is a continuous mass of cytoplasm in which are suspended many nuclei. In this feature, and in their methods of locomotion and nutrition, these animals are very like the ciliates and provide a plausible link with the protists. Then, from a syncytial flatworm, it is not much of a jump to more specialized forms such as the planarians.

37.6 THE ROUNDWORMS (PHYLUM NEMATODA)

Roundworms, or nematodes, are elongated, cylindrical animals. They show two evolutionary advances over the flatworms (from which they may have evolved). They have a one-way digestive tract running from a mouth at the front to an anus at the rear. This one-way digestive system is advantageous as it eliminates the mixing of incoming food with outgoing wastes. After food enters the mouth, it can be processed step-by-step as it passes from one section of the digestive tract to another. Finally, undigested remains are eliminated at the anus. Nematodes also have a cavity between the digestive tract and the body wall. It develops from the blastocoel cavity during embryonic development and is not therefore related to the coelom which, as you remember, develops totally surrounded by mesoderm. For this reason it is often called a *pseudocoel.* In it are found various internal organs including those of reproduction. The presence of a definite body cavity rather than a solid mass of mesoderm allows for greater freedom of movement although the whiplike motion so characteristic of nematodes does not seem to exploit this.

The majority of nematodes are quite small. A few, such as *Ascaris* (Fig. 37.9), may be almost a foot in length and one, a veritable giant that parasitizes whales, reaches a length of 30 feet! Most nematodes, however, are not much bigger than small bits of thread. They can nevertheless be easily identified by the thrashing, whiplike movements they make. Rich soil literally teems with these little organisms and examination with a lens of a little soil and water is sure to reveal their presence.

About 10,000 species of nematodes have been identified so far, but the list is surely far from complete. Nematodes live practically everywhere. They are found in fresh water and salt water, as well as soil. As parasites, they live in the bodies of plants and other animals. It has been said that if all the matter of our earth disappeared except for roundworms, we would still be able to recognize all the formerly existing features—living as well as nonliving—by the nematodes they had contained.

Most nematodes are free-living. It is the parasitic forms, however, that are of greatest interest to us. One of the most serious of these in warm climates is the hookworm. This organism lives attached to the wall of the intestine and sucks blood and tissue fluids from it. A heavy infestation causes great weakness and lethargy. The disease is usually contracted by walking barefoot on soil contaminated by human excrement. At one time the American hookworm infected over two million inhabitants of the southeastern part of the U.S.A. Now proper sanitation and the wearing of shoes have done much to reduce the incidence of infection.

The Southeast has no monopoly on nematode infections. According to recent estimates, over four million people throughout the United States have been more or less heavily infected with the nematode *Trichinella spiralis.* Infection comes from eating raw or undercooked pork that contains the organisms (Fig. 37.9). One is wise to assume that all pork contains them, because federal meat inspectors make no attempt to check for the presence of this parasite. Infection of humans by *T. spiralis* is something of a biological mistake. Inasmuch as we do not practice cannibalism or leave the bodies of our dead around carelessly, there is no way for the worms to escape to another host animal. They simply form a resting stage in our muscles and ultimately die. In a heavy infection, however, the worms may cause such serious body disturbances that death of the host from **trichinosis** results.

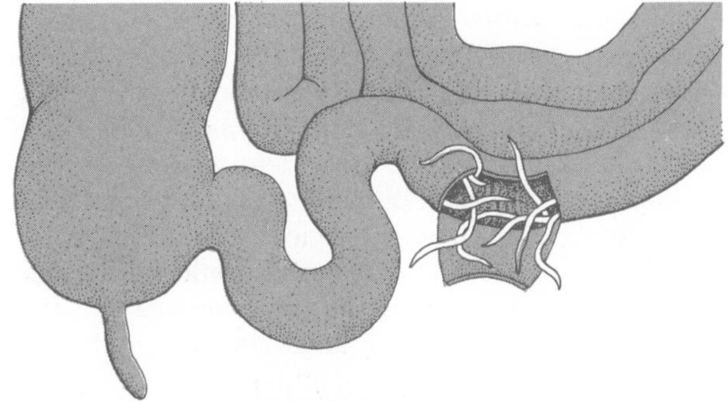

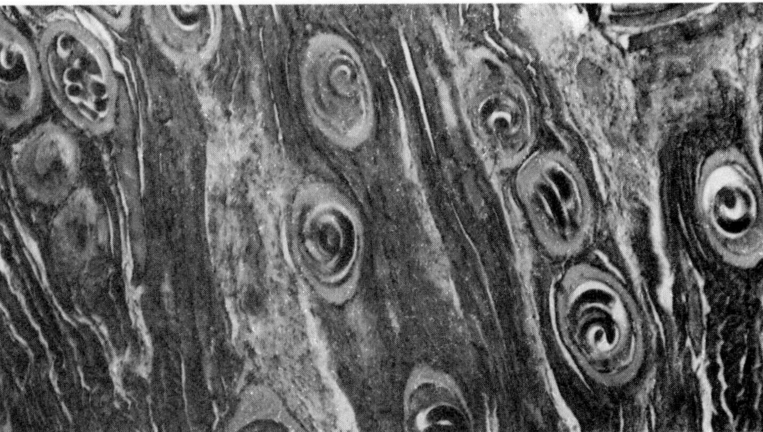

FIG. 37.9 Two parasitic nematodes. Top: *Ascaris* in the human intestine. Bottom: Larvae of *Trichinella spiralis* encysted in pig muscle (20 ×).

Humans, especially children, are also frequently infected with other kinds of nematodes, such as *Ascaris,* whipworms, and pinworms. These all live in the intestine and are contracted by careless sanitary habits. Infestation is usually not serious, however, and can be easily cured. The "worms" for which we have the family dog treated periodically are usually also nematodes of the genus *Ascaris*. The filarial worm, which causes the gross deformity of elephantiasis, is one of several parasitic nematodes that afflict humans living in tropical regions (Fig. 11.21).

Our survey of the nematodes would not be complete without mentioning the great damage that they do to crop plants such as oranges, tobacco, and strawberries. These parasites seldom kill their host outright, but they may weaken it so that it succumbs to some other invader. Only in recent years have agricultural scientists come to realize that nematode attack may be the basic cause of a large percentage of our annual crop losses.

37.7 THE ANNELID WORMS (PHYLUM ANNELIDA)

The worms in this phylum are segmented; that is, their bodies are made up of repeating units. Although some structures, such as the digestive tract, extend the entire length of the worm, others such as the excretory organs are repeated in segment after segment. The segmentation is revealed externally as a series of rings (Fig. 37.10).

Other characteristics of the annelids are bilateral symmetry, an efficient circulatory system with blood pumped through a closed system of blood vessels, and a fairly elaborate nervous system. The major nerve trunk runs along the ventral side.

Another feature of the annelid worms not found in more primitive animals is the presence of a large fluid-containing body cavity. This permits the internal organs to slide easily against one another and this, in turn, makes extensive movements of the body easy to accomplish. The cavity, called a **coelom,** is lined throughout with mesoderm. However, its embryonic development is quite different from the way in which the coelom develops in vertebrates (see Section 22.3). Early in cleavage, special mesodermal cells are formed within the embryo. Mitotic division of these cells produces a mass of mesoderm tissue.

FIG. 37.10 Representative annelids. Left: An earthworm (Oligochaeta). Center: A clam worm (Polychaeta). Right: A leech (Hirudinea).

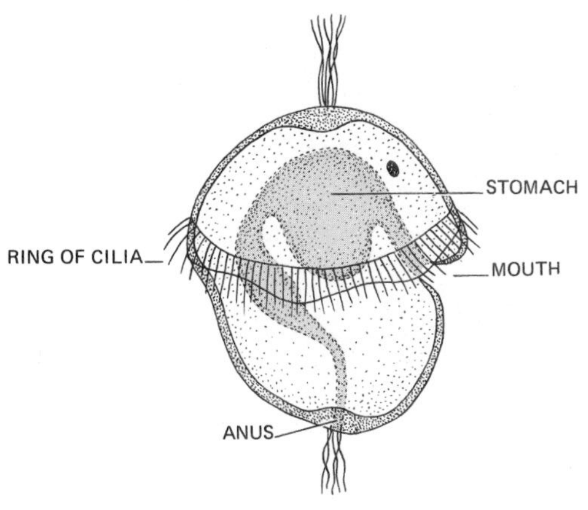

FIG. 37.11 A trochophore larva. These are produced by members of several animal phyla, including annelids and mollusks.

Eventually a space develops within this mass, and this gradually enlarges to form the coelom.

Some 8900 species of annelids have been described. These are distributed among three classes. The largest class, the **Polychaeta,** consists of marine worms like the clam worm (Fig. 37.10). These animals produce free-swimming larvae called *trochophores* (Fig. 37.11). The common earthworm and a number of freshwater forms are members of the class **Oligochaeta.** The third class, **Hirudinea,** is composed of the leeches (Fig. 37.10).

37.8 THE MOLLUSKS (PHYLUM MOLLUSCA)

With about 100,000 living species, this phylum must be included among the most important animal phyla on earth today. It consists of soft-bodied, unsegmented (with one exception) animals many of whom are protected by one or more shells made of limestone (calcium carbonate). These shells (or valves) are formed by a special fold of the body wall called the **mantle.** The great majority of the mollusks live in salt water, but many are found in fresh water and some occur on land. The phylum is divided into three major and several minor classes.

1. Class Bivalvia. The clams, oysters, mussels, scallops, etc., are commonly known as **bivalves** because two shells encase the body. The bivalves have bilateral symmetry, but in their case it is not associated with rapid locomotion. Those that move at all do so by extending a thick, muscular foot from between the valves. All members of the class feed by filtering food particles from water which they draw in under the mantle.

Many bivalves are prized as food. In addition, pearls are produced by certain species of oysters. On the negative side of the ledger, the "shipworm" does great damage to wooden wharves and boats. This organism is not a worm at all, but a bivalve which uses its two valves to bore tunnels in wood that is submerged in salt water.

2. Class Gastropoda. A second large class of mollusks includes all the **snails** and their shell-less relatives, the **slugs.** Snails are often referred to as **univalves** because of their single shell. This shell is coiled, as are the internal organs of the animal. There is no plane of symmetry in adult snails although they develop from bilaterally symmetrical larvae. Snails feed by scraping food with a rasping, tonguelike *radula.* They have a distinct head with two eyes which are often located on stalks. The great majority of snail species live in salt water but some are also found in fresh water and even on land. The latter breathe air by means of a lunglike adaptation of the mantle. Slugs are found in salt water and on land. Terrestrial ("garden") slugs are rather dark in appearance but marine slugs are often brilliantly colored and decorated.

Although some species are prized as food, snails cause far more food loss for humans than their consumption compensates for. Many marine snails, such as the oyster drill, feed upon commercially valuable bivalves. But far more serious is the destruction brought to crops in some parts of the world by the terrestrial slugs and snails. Some snails also serve as an intermediate host for flukes. Despite the damage caused by some species, the group as a whole should be prized if only for the extraordinary array of beautiful snail shells that are cast up on our beaches, bringing delight to casual and serious shell collectors alike.

3. Class Cephalopoda. The various species of octopus and squid (Fig. 37.12) as well as the chambered nautilus belong to the class Cephalopoda. All of these organisms have a large, well-developed head bearing prominent eyes and surrounded by a ring of arms (eight in the octopus, ten in the squid) which aid in locomotion and the grasping of prey. They are found exclusively in salt water. Although the cephalopod eye looks and functions much like the vertebrate eye (such as ours), these are analo-

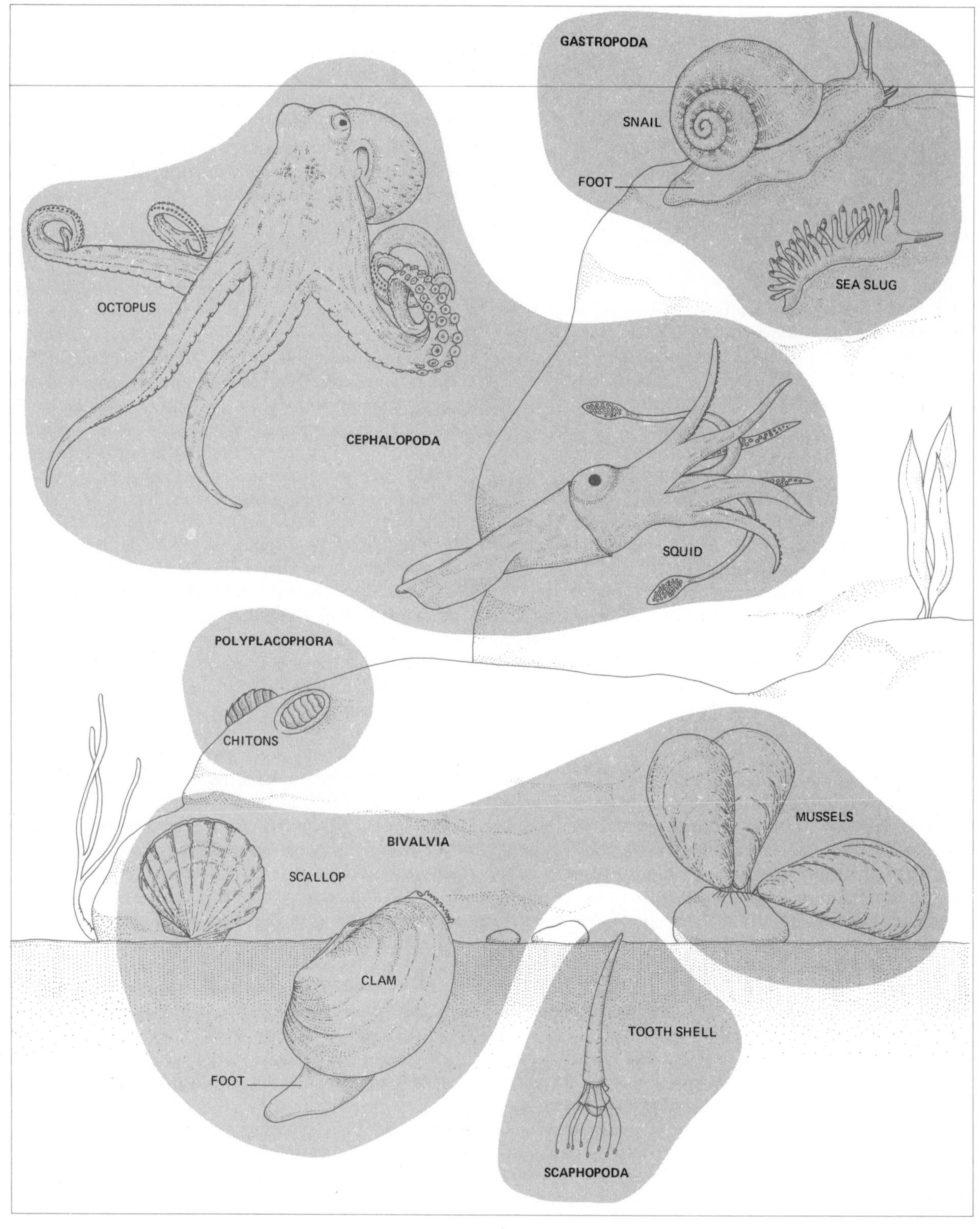
GASTROPODA
SNAIL
FOOT
SEA SLUG
OCTOPUS
CEPHALOPODA
SQUID
POLYPLACOPHORA
CHITONS
BIVALVIA
SCALLOP
MUSSELS
CLAM
FOOT
TOOTH SHELL
SCAPHOPODA

◄ FIG. 37.12 Representative mollusks.

gous, not homologous, organs and have arisen as a result of convergent evolution. Only the chambered nautilus has a large shell. The shell of squids is reduced to a thin plate imbedded in the mantle, while octopuses have no shell at all.

The cephalopod mollusks are the most complex members of the phylum. They also include the largest of all the invertebrates. Records exist of a 28-foot octopus and a 50-foot squid. Squids have curious methods of locomotion and defense. They can move very rapidly by forcibly ejecting a stream of water from under the mantle. When in danger, the squid supplements its jet-propelled escape by squirting a black, inky fluid into the water and thus distracting its enemy. Both octopuses and squid are important foods for humans in some parts of the world.

4. Class Scaphopoda. The "tooth shells" constitute a small class of marine mollusks that spend their adult lives almost buried in the sand. They feed by filtering tiny organisms out of the water that is drawn in through the opening in the protruding end of their shell (Fig. 37.12).

5. Class Polyplacophora. Chitons are sluggish organisms that live inconspicuously at the seashore. Their shell consists of several (usually eight) separate, overlapping plates. Although this gives them the appearance of being segmented, their internal organs are not.

6. Class Monoplacophora. Until *Neopilina* (Fig. 1.7) was discovered in 1952, the class to which it belongs was thought to have been extinct for millions of years. This mollusk is of great interest because in addition to possessing all the typical molluscan features, it is segmented internally. In this latter respect, it is like the annelids and thus supports the widely held view that the mollusks and the annelids are closely related. The fact that mollusks also produce trochophore larvae provides additional evidence of their kinship with the annelids. However, the coelom—so characteristic of annelids—is greatly reduced in size in the mollusks. The main body cavity is the **hemocoel,** derived from the embryonic blastocoel.

Although the earliest mollusks were aquatic, the colonization of the land by plants opened up niches into which the snails and slugs were able to move. The protection of a shell, internal fertilization (sometimes embryonic development as well), and the modification of the mantle into a "lung" were all adaptations enabling them to shift from the aquatic to the terrestrial environment.

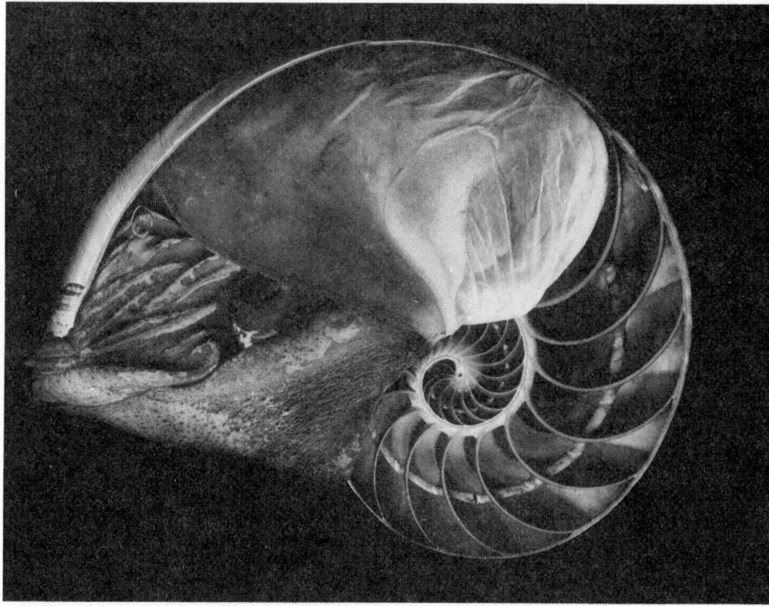

FIG. 37.13 Shell of the chambered nautilus cut to show its construction. As it grew, the cephalopod inhabitant kept extending the shell at the open end and walling off those portions at the rear that had become too cramped. (Courtesy of the American Museum of Natural History.)

It is the evolutionary success of the snails and bivalves that makes the phylum one of the three that are dominant on the earth today. Of some 100,000 species of mollusks, all but a few hundred belong to these two classes. It was not always so. The Ordovician and Silurian seas were well populated by the nautiloids, a group of shelled cephalopods whose only living descendants are three species of chambered nautilus (Fig. 37.13). Some of these forms possessed shells 15 ft long. In the Mesozoic era, the ammonites, another group of cephalopods, were the dominant invertebrates in the sea. Although now sharply reduced in numbers, the surviving cephalopods are still impressive. The 50-ft (15 m) squid ranks as the largest invertebrate that ever lived.

37.9 THE ARTHROPODS (PHYLUM ARTHROPODA)

If number of species is any criterion, this is the major phylum on earth today. Over 765,000 different spe-

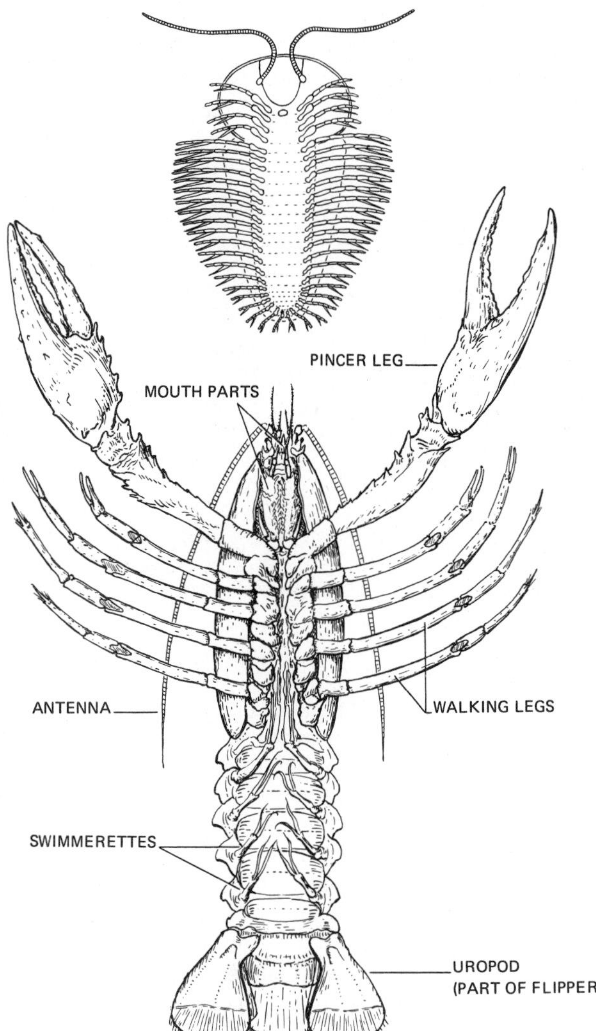

FIG. 37.14 A trilobite (above) and crayfish compared. The paired, jointed appendages of the crayfish show far more structural and functional specialization than do those of the trilobite.

FIG. 37.15 An artist's reconstruction of aquatic life of the Silurian period. The segmented animals are eurypterids, a now-extinct group of aquatic merostomes. (Courtesy of the Field Museum of Natural History.)

FIG. 37.16 Remains of a eurypterid. (Courtesy of the American Museum of Natural History.)

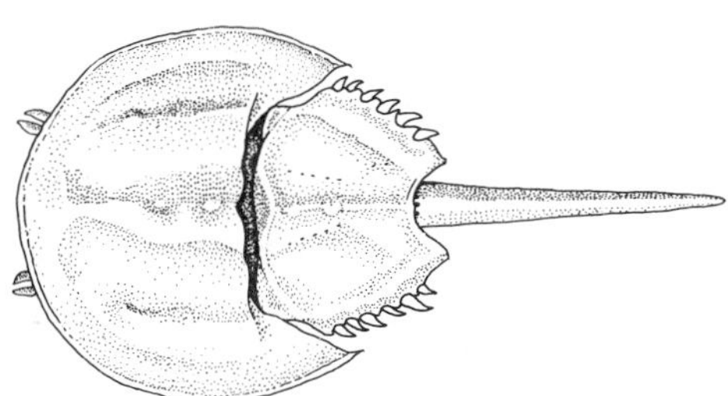

Limulus, THE HORSESHOE "CRAB"

FIG. 37.17 This "living fossil" has existed in the sea virtually unchanged for 200 million years.

cies of arthropods have been identified. That is more than all the other species of living things put together. Each year still more arthropod species are found, living in every conceivable habitat. Fresh water, salt water, soil, and practically the entire surface of the earth abound with them. They are almost the only animals to be found in the interior of Antarctica and on the snow- and rock-strewn slopes of our highest mountains.

All members of this phylum have a segmented body enclosed in a tough, jointed external skeleton made principally of **chitin,** a polymer of N-acetylglucosamine (NAG—see Fig. 35.15). The symmetry is bilateral and is clearly marked by the pairs of jointed appendages arranged along either side of the midventral axis. In all living arthropods, the various appendages in a given species show considerable variety of structure and function. In addition to locomotion, they may aid in food-getting, in sensation, and as offensive and defensive weapons.

Unlike those of annelids, the segments of arthropods show considerable variation in structure from front to back. They are usually combined into three main regions: the **head, thorax,** and **abdomen.** Arthropods have a circulatory system which is considered "open" inasmuch as the blood (in contrast to that of the annelids) is not confined within blood vessels at all times. The main part of the nervous system of arthropods, like that of the annelids, runs along the ventral side of the organism.

A. Subphylum Trilobitomorpha

Among the most abundant fossils of the Cambrian period are the **trilobites.** These early arthropods were segmented and had paired, jointed appendages and an exoskeleton just like those of their modern descendants. Although we suspect that the anterior appendages aided in food-getting while the posterior ones were used for locomotion, they were quite uniform in structure (Fig. 37.14).

B. Subphylum Chelicerata

In the members of this group, the head and thorax are fused into a **cephalothorax.** The first pair of appendages are adapted for feeding. These structures are called chelicerae and are responsible for the subphylum name. The members of this group have no antennae.

1. Class Merostomata. Most of the members of this class are now extinct. They were aquatic animals called **eurypterids** that first appeared in the Cambrian (Figs. 37.15 and 37.16). Eurypterids were the largest of all arthropods, one Silurian species reaching a length of over 9 feet. The eurypterids probably evolved from early trilobites. In any case, both groups became extinct by the end of the Paleozoic era.

The only member of the Merostomata to survive into modern times is *Limulus,* the horseshoe crab (Fig. 37.17). It has done so with practically no evolutionary change since the genus first appeared in Triassic seas over 200 million years ago.

2. Class Arachnida. Although themselves extinct, the eurypterids have left descendants. These are the arachnids. The earliest arachnids were aquatic, but by Silurian times, terrestrial forms had evolved. The scorpions, mites, ticks, spiders, and daddy longlegs of today are their descendants (Fig. 37.18). They are air-breathing animals. Locomotion is by four pairs of legs.

The arachnids are not a well-loved group. This is perhaps unfair although some do cause problems. Mites and ticks are annoying parasites of humans and other animals. Ticks transmit a number of diseases such as Rocky Mountain spotted fever. Scorpions and some spiders can inflict painful stings and bites, but with the possible exception of that of the

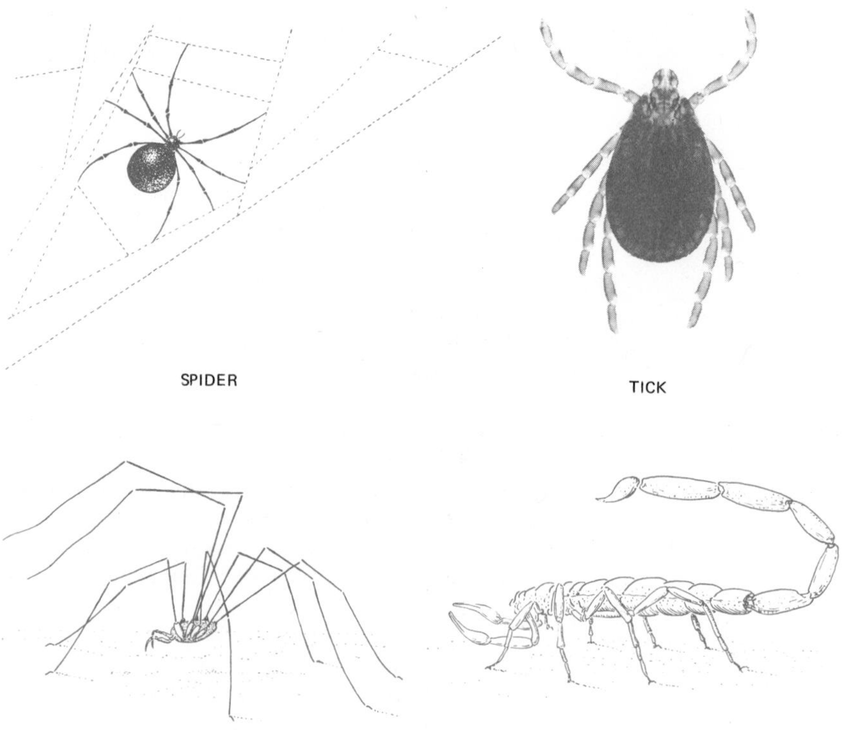

FIG. 37.18 Representative arachnids. These arthropods have four pairs of legs and no antennae.

black widow spider, the bites of species found in North America are not usually dangerous to adults.

C. Subphylum Mandibulata

The arthropods in this group possess mandibles—paired mouth parts used in feeding. They also have antennae. There are four major classes in this subphylum.

1. Class Crustacea. The earliest crustaceans appeared in the Cambrian. Their nearest relatives may have been the trilobites. However, they differed from the trilobites in the variety of modifications of their appendages. The modern crayfish with its flipper, swimmerettes, walking legs, pincer legs, specialized mouth parts (including mandibles), and antennae (two pairs) stands in sharp contrast to the trilobite (Fig. 37.14). The structural diversity of these numerous appendages reflects the variety of functions carried out by them.

Crustaceans, like chelicerates, have their head and thorax fused into a cephalothorax. The class includes crayfish, lobsters, barnacles, crabs, and an enormous variety of tiny shrimplike creatures. With the exception of the "pill bug," which lives under logs and stones, the crustaceans are aquatic, being found in both fresh water and salt water. All members of the class breathe by gills. They range in size from microscopic forms to lobsters and crabs weighing several kilograms (Fig. 37.19).

In certain areas of the world, lobsters, crabs, and shrimp are important items in the human diet. Crustaceans are also the chief source of food for many fish and mammals. The great blue whale, whose 100-ft (30 m) length makes it the largest animal that ever lived, feeds exclusively on tiny crustaceans.

2. Class Chilopoda. These are the **centipedes** (Fig. 37.20). They are elongated and flattened in appearance. Each segment behind the head bears one pair of legs but, despite their name, the total number is considerably less than one hundred. The centipedes are carnivorous, feeding upon other animals with the aid of strong jaws and a poisonous bite.

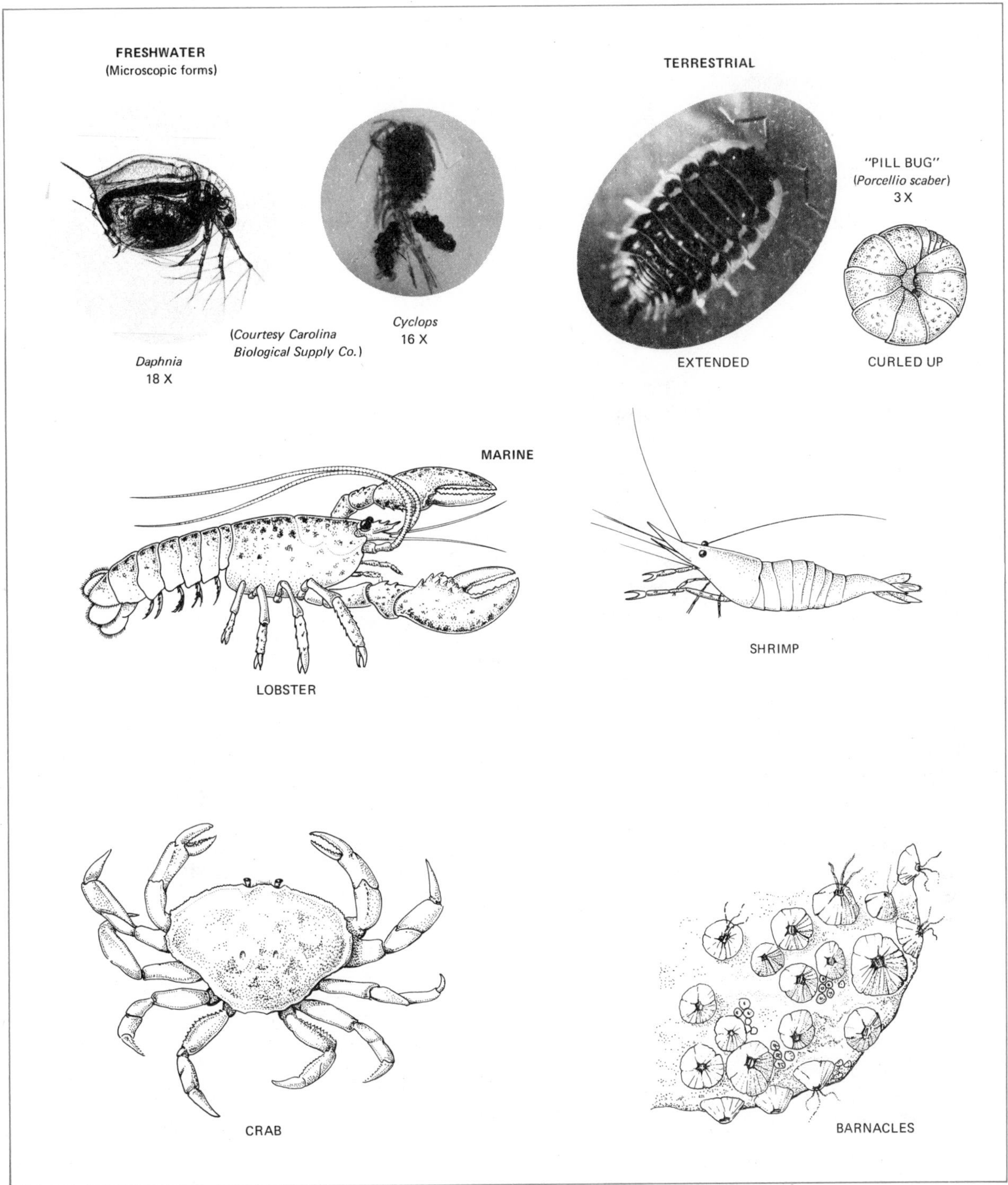

FIG. 37.19 Representative crustaceans. This class of arthropods has been most successful in the sea.

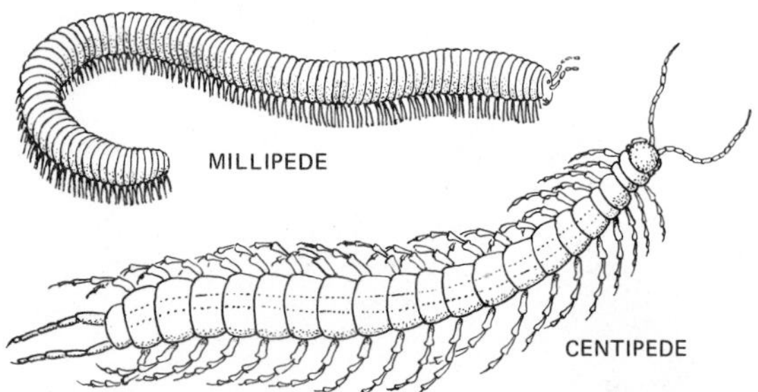

FIG. 37.20 Millipedes have two pairs of legs on each segment; centipedes have only one pair per segment.

DIPTERA
HOUSEFLY

HOMOPTERA
CICADA

ODONATA
DRAGONFLY

HYMENOPTERA
BEE

LEPIDOPTERA
BUTTERFLY

ORTHOPTERA
GRASSHOPPER

COLEOPTERA
POTATO BEETLE

ANOPLURA
LOUSE

SIPHONAPTERA
FLEA

ISOPTERA
TERMITE

THYSANURA
SILVERFISH

HEMIPTERA
SHIELDBUG

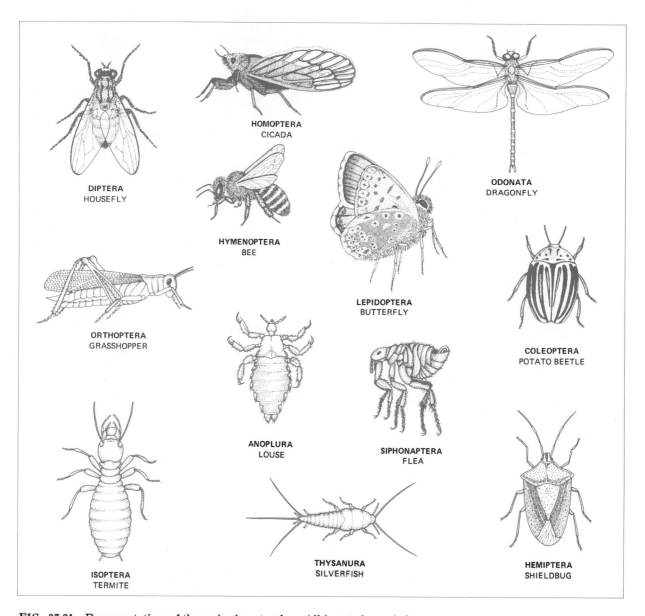

FIG. 37.21 Representatives of the major insect orders. All insects have six legs.

3. Class Diplopoda. The members of this class, the **millipedes** (Fig. 37.20), differ from the centipedes in having two pairs of legs in each segment. In this case, too, the total number is far less than the thousand which their name implies. Millipedes are more cylindrical in shape than centipedes and have an herbivorous (vegetarian) diet. Both millipedes and centipedes probably arose in the Devonian period, although clear fossil remains are found only as far back as Carboniferous times.

4. Class Insecta. Insects were definitely established by the Devonian. This versatile, fast-reproducing, fast-evolving group has become the dominant class of arthropods in every habitat but salt water (where the crustaceans reign supreme). Over 700,000 currently living species of insects have been identified, which is over one-half of all the kinds of living things found upon the earth.

The insect body is divided into three regions: head, thorax, and abdomen. The thorax itself consists of three distinct segments, each of which bears a pair of legs. Thus the insects are the six-legged creatures. Most of them, when adult, also have one or two pairs of wings on the thorax. They have a single pair of antennae mounted on the head.

With 700,000 species in one class, you can well imagine that a great deal of attention has been paid to the lesser taxonomic categories. About two dozen orders have been established on the basis of differences in metamorphosis, wing structure, and mouth parts. Figure 37.21 illustrates some of the major insect orders.

Of the several classes of arthropods, no other solved the problems of dry-land living so successfully. The exoskeleton is quite impervious to water and thus prevents fatal dehydration of the body when the surrounding air is dry. Gas exchange is accomplished by a system of tracheal tubes (see Section 10.5), which penetrate to every portion of the body. The paired, jointed, clawed appendages provide not only for locomotion but also for ingestion of food. Mandibles, maxillae, and labia are fashioned in remarkably diverse ways to produce sucking, biting, chewing, and rasping mouth parts.

Other adaptations to a terrestrial life included the development of wings for flying. (The insects were the first animals to use this medium for locomotion.) Internal fertilization and embryonic development within a waterproof egg solved the problem of bringing the gametes together and protecting the embryo without the aid of water. Finally, the ability to excrete their nitrogen-containing wastes as uric acid must be counted as one of the important evolutionary adaptations to a dry habitat. Not only does uric acid require very little water for its excretion, but the fact that what little is used is then reclaimed in the rectum enables these creatures to deposit wastes which are practically dry.

The efficiency of insect structure and function cannot be disputed. Both in terms of number of species and number of individuals, the insects earn a position as one of the most successful groups on earth. They are our chief competitors for food. Except for the oceans, insects have adapted to practically every habitat available on the earth.

37.10 THE PHYLUM ONYCHOPHORA

The fossil record tells us little about the evolutionary relationships of crustaceans, millipedes, centipedes, and insects. However, a living animal, *Peripatus,* supports the view that they evolved from a primitive segmented worm. *Peripatus* (Fig. 37.22) has internal segmentation, pairs of nephridialike excretory organs in each segment, a cuticle, and muscular body wall, all quite suggestive of the annelids. Furthermore, the excretory and sex organs are ciliated as they are in annelids, and the structure of the digestive tract,

FIG. 37.22 *Peripatus.* This little animal has features of both annelid worms and insects. It is classified in its own phylum, Onychophora. Once a widespread group, it is now found only in a few isolated tropical areas. (Courtesy of Ward's Natural Science Establishment, Inc.)

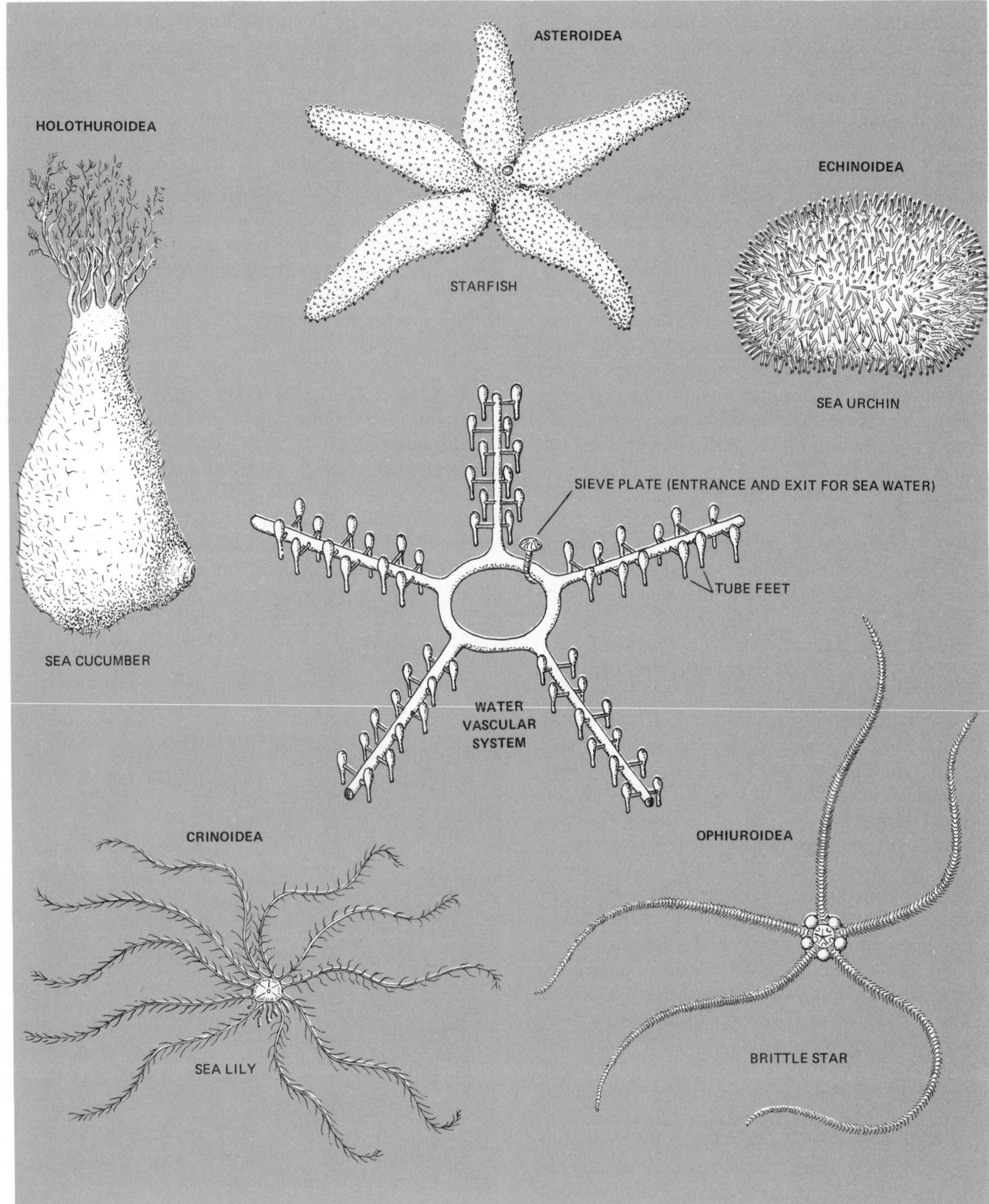

FIG. 37.23 Representative echinoderms. All members of the phylum live in salt water, have a spiny skin and radial symmetry (as adults), and move slowly with the aid of their tube feet. Extension and retraction of the tube feet is accomplished by the movement of sea water in and out of the water vascular system.

central nervous system, and legs is similar to that of annelids. On the other hand, the legs of *Peripatus* bear insectlike claws, and the circulatory system is open with the blood passing through an extensive hemocoel. Furthermore, these little creatures breathe air by means of tracheae, and the embryonic development of their eggs resembles that of arthropods.

Peripatus is certainly not the ancestor of the millipedes, centipedes, and insects. However, its curious anatomy suggests that it may be a relatively little-changed descendant of forms which also led to these arthropods. Fossil members of its phylum have been found side-by-side with fossil trilobites, crustaceans, and annelids. The fact that *Peripatus* occurs in small, isolated parts of the jungles of Central and South America, Africa, Asia, and Australia indicates that it may represent the last surviving remnants of a once-flourishing group.

37.11 THE ECHINODERMS (PHYLUM ECHINODERMATA)

This phylum consists of about 6000 species, all of which live in salt water. Their most conspicuous characteristics are their spiny skin and their radial symmetry. Perhaps their most interesting feature, though, is their water vascular system (Fig. 37.23). Sea water is taken into a system of canals and is used to extend the many **tube feet.** These latter structures have suckers on their tips and aid the animal in attaching itself to solid surfaces. The phylum is subdivided into five classes.

As their name suggests, sea lilies (class **Crinoidea**) are rather plantlike in appearance. Many of them are sessile, that is, they live firmly attached by a stalk to some underwater object. It is this way of life that probably accounts for their radial symmetry, and not any evolutionary relationship to the other radially symmetrical animals, the Cnidaria. Bilateral symmetry is associated with rapid locomotion, as we have seen; radial symmetry is more appropriate for a sessile creature that must concern itself with all directions. However, all the echinoderms do produce free-swimming larvae that are bilaterally symmetrical. This is interpreted to mean that the echinoderms have evolved from bilaterally symmetrical ancestors.

The **starfish** body consists of a central disk containing the mouth and surrounded by five arms. Starfishes (class **Asteroidea**) are able to move about with the aid of their tube feet but only very slowly. Of all the echinoderms, they are the only ones of much practical importance to man. This stems from their habit of preying on commercially valuable bivalves such as oysters. Brittle stars (class **Ophiuroidea**) differ from starfishes in having long, thin arms which are distinct from the central disk and in being able to move about quite rapidly (for an echinoderm!). Photographs taken of the ocean bottom sometimes reveal fantastic numbers of these organisms living side by side.

Sea urchins and sand dollars (class **Echinoidea**) have a hollow, rigid, boxlike skeleton. Hinged to it are spines, which in some sea urchins are quite long. Rows of openings in the skeleton permit the long, slender tube feet to extend outward. Together, the spines and tube feet enable the sea urchin to move about slowly. Sea urchins are eaten occasionally in some coastal regions of the world.

Sea cucumbers (class **Holothuroidea**) have a leathery (rather than spiny) skin, no arms, and virtually no skeleton. Although generally rather unremarkable in appearance and behavior, their response to potential predators is quite startling. When sufficiently disturbed, they contract their muscular body wall until the internal pressure becomes so great that the body wall splits open. With this, their internal organs, along with a good deal of sticky, gelatinous material, spill out into the water. Often, while the potential predator is coping with this, the remains of the sea cucumber crawl away and begin the process of rebuilding its missing organs.

37.12 THE CHORDATES (PHYLUM CHORDATA)

The chordates include more than 44,000 species. All of them have bilateral symmetry, some degree of segmentation, and an internal skeleton. What sets them apart from all the other animals, however, is three unique features (Fig. 37.24): (1) All chordates possess a flexible, rodlike structure, the **notochord,** at some stage of their development. This is located dorsal to the digestive tract and provides internal support for the body. In most chordates, it is replaced by a vertebral column or backbone long before maturity is reached. (2) At some stage in development, all chordates possess pairs of **gill pouches.** These are lateral outpocketings of the throat. They are matched on the exterior by paired grooves. In aquatic chordates, the gill pouches break through to the exterior grooves, forming **gill slits.** These provide an exit for water that is taken in through the mouth and passed over the gills. In the land chordates, the gill pouches do not break through but become greatly modified during the later course of development. (3) All chordates possess a hollow

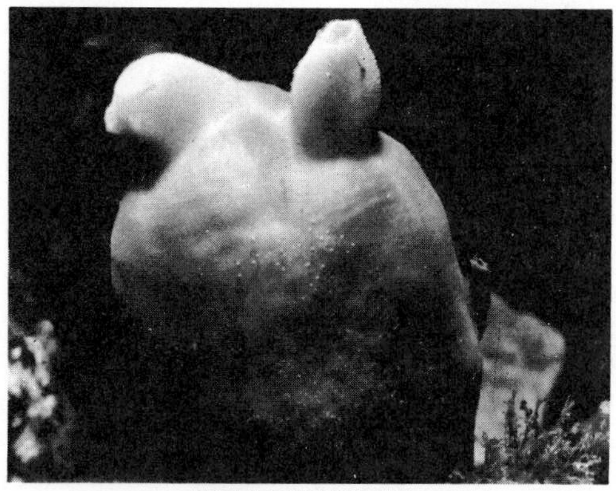

ADULT TUNICATE (*Halocynthia*)
(*Courtesy Ralph Buchsbaum*)

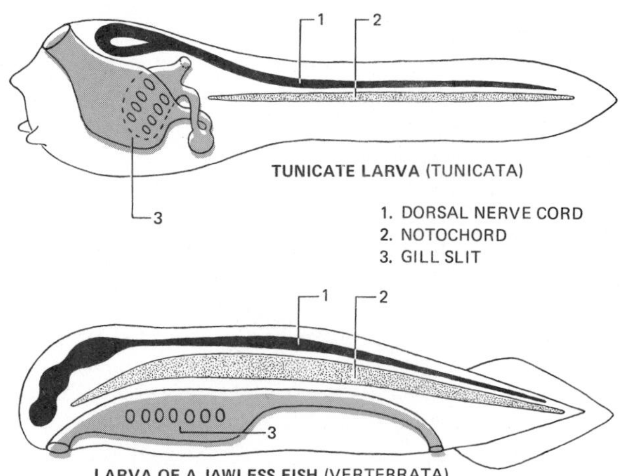

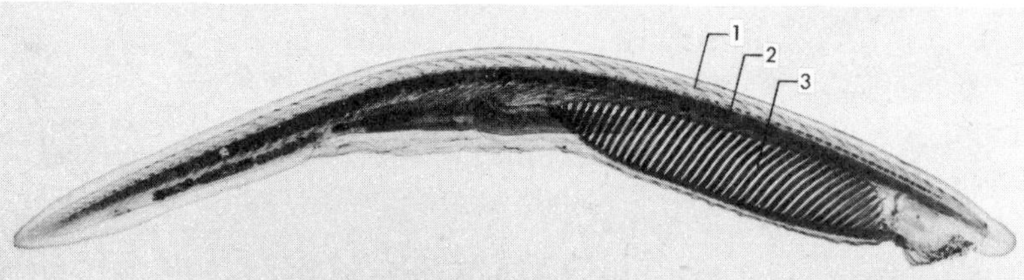

AMPHIOXUS (Cephalochordata)

FIG. 37.24 Three chordates. Although the adult forms of these three animals differ strikingly, at some time each shows the three basic chordate characteristics listed. (Amphioxus courtesy of Turtox/Cambosco, Macmillan Science Company, Chicago.)

nerve cord that develops on the dorsal side of the body above the notochord. At its anterior end, it becomes enlarged to form the brain.

The phylum is composed of three subphyla. The largest of these is the subphylum **Vertebrata,** the group in which all animals with backbones are placed. It is the subphylum to which we belong (along with fishes, frogs, snakes, birds, and other mammals). The vertebrates are such dominant animals on the earth that we often describe the animal kingdom as being composed of vertebrates and "invertebrates," the latter including all the phyla we studied in this chapter. We shall study the vertebrates in the next chapter. But now let us examine the two subphyla of "invertebrate chordates," not only for their own sake, but in the hope that they may provide clues to the evolutionary origins of the vertebrates.

1. Subphylum Cephalochordata

The representative member of this tiny subphylum is a small (five-cm), fishlike organism, the **amphioxus.** Throughout its entire life, the amphioxus possesses a notochord, a dorsal tubular nerve cord, and functional gill slits (Fig. 37.24). Although capable of swimming, it spends most of its time partially buried in the sand while it filters microscopic food particles from the water. It is a marine form living close to the shore.

2. Subphylum Tunicata (Urochordata)

The members of this subphylum are also found exclusively in salt water. They are sessile animals that live by straining sea water through their gill slits and ingesting any food particles that are trapped in the process. The phylum gets its name from the tough covering, the tunic, which surrounds the body.

Curiously enough, the tunic contains substantial quantities of cellulose, a substance generally confined to the plant kingdom and certain protists. Many tunicates are commonly known as sea squirts because of the streams of water they expel when the body is contracted suddenly.

Aside from the presence of gill slits, it is hard to see what makes these animals chordates. There is neither notochord nor dorsal tubular nervous system. Like all sessile animals, however, the tunicates produce free-swimming larvae that can disperse the species to new locations, and the structure of these larvae is distinctly chordate. They have both a notochord and a dorsal nervous system (Fig. 37.24). Only when the larva settles on some underwater object and develops into the adult does it lose these features that reveal its true chordate affinities.

For many years biologists included still a third group of primitive marine animals as a subphylum of the Chordata. The acorn worms (Fig. 37.25) achieved this status by virtue of possessing gill slits, a dorsal nerve cord (as well as a ventral one) and what was thought to be a rudimentary notochord. There is now, however, so much doubt as to whether the structure is homologous to the notochord of the other chordates that the acorn worms have been placed in a phylum (Hemichordata) of their own.

FIG. 37.25 An acorn worm (Phylum Hemichordata). This marine animal has both echinoderm and chordate features. (Courtesy of William H. Amos.)

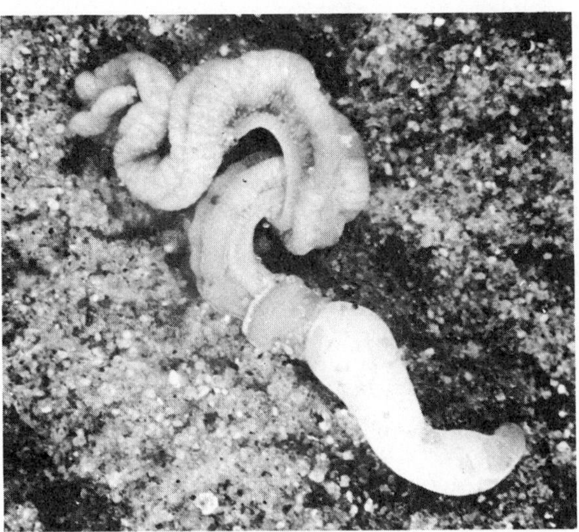

37.13 THE EVOLUTION OF THE ECHINODERM SUPERPHYLUM

Echinoderms and chordates share a number of features which do not occur in the other animal phyla. This is particularly apparent in their embryonic development.

As we saw in Fig. 22.1, the cleavage furrows in chordate eggs occur in regular planes, either perpendicular or parallel to one another. This is also true of echinoderm eggs. However, in all the other animals we have discussed (with the possible exception of the arthropods, whose method of cleavage is unusual), the planes of cleavage are progressively tipped with respect to one another so that a *spiral cleavage* results.

The cells produced by the first cleavage or two in echinoderm and chordate eggs are fully capable of developing into a complete embryo if isolated from each other. For this reason identical twins can and do occur in both groups. On the other hand, cleavage in the invertebrates which we have studied up to this point results in cells that have lost some of their potentiality. Identical twins do not occur.

In the echinoderms and chordates, gastrulation begins at that portion of the embryo which will become the anus of the adult (see Section 22.3). However, in the other phyla with a complete digestive tract, it begins at the end which ultimately becomes the mouth. Both groups possess a coelom, but the mechanism of its formation is quite different. In annelids, arthropods, and mollusks, the coelom arises by a splitting of a mass of mesodermal cells which have their origin within the embryo. In echinoderms and chordates, the coelom develops within a sheet of cells which pushes in from the exterior during gastrulation. One other distinction between these two groups which deserves mention is a chemical one. As you learned in Chapter 28, the compound creatine phosphate serves as a reservoir of high-energy phosphate in human muscles. It also performs the same function in all other vetebrates and in many of the invertebrate chordates (e.g. *Amphioxus*—Fig. 37.24) and in some echinoderms (e.g. brittle stars). Most of the other invertebrates that have been examined have a related substance, arginine phosphate, which we assume performs the same function for them. The sea urchins, acorn worms (Fig. 37.25), and tunicates (Fig. 37.24) have one or the other or both of these substances.

The occurrence of so many sharp contrasts between the echinoderms and chordates on the one hand and most of the remaining animals on the other has led some taxonomists to establish two superphyla: the echinoderm superphylum (to which we belong) and the annelid superphylum. It has also made it quite clear that we cannot expect to find clues to our own evolutionary origins by examining such advanced and distantly related forms as mollusks and arthropods. How, then, can we account for the origin of vertebrates and, thus, of ourselves? Fossil echinoderms were present in the Cambrian so we know that the echinoderm superphylum was established before that time. We do not know from which more primitive group it evolved. One that has been suggested as possibly ancestral to the echinoderms and chordates is the **Nemertina.** This phylum, which has not been discussed before, contains a small group of flattened, ciliated worms that are generally thought to be closely related to the flatworms. But one important way in which they differ from the flatworms is that gastrulation produces a one-way digestive tract with both mouth and anus. In some nemertines, however, the site of gastrulation (the blastopore) becomes neither the mouth nor the anus. Instead it becomes sealed over in later embryonic development, and mouth and anus formation occur on opposite sides of it. Perhaps these worms are descendants of a group in which the site of gastrulation was in transition from the mouth end of the archenteron (annelid superphylum) to the anus end (echinoderm superphylum). The hypothesis is supported by the discovery of some two dozen structures in the nemertines which seem to be homologous to structures found in one of the most primitive vertebrates, the hagfish.

Assuming that the echinoderm superphylum is descended from nemertine ancestors, we must now attempt to unravel the evolutionary relationships within the superphylum. All of the chordates are bilaterally symmetrical. The echinoderms, on the other hand, are radially symmetrical. Their radial symmetry should not fool us, though, for the larvae of the echinoderms are bilaterally symmetrical. Adhering to our conviction that embryonic features are the more primitive ones, we conclude that the radial symmetry of adult echinoderms is a secondary adaptation to a sessile or, at most, sluggish way of life. (Remember the cnidarians.) The fossil record tells us that the earliest echinoderms were sessile animals attached by means of a stalk. Some of the modern descendants of these early echinoderms (the sea lilies) are still sessile, and all of them at least pass through a stalked stage.

The echinoderms have been moderately successful in their rather limited habitat of the oceans. From the earliest stalked members of the phylum arose the other classes: the sea urchins, brittle stars, starfish, and sea cucumbers (Fig. 37.23). The abundance of these organisms, their internal skeleton, and their marine habitat have resulted in a remarkably complete fossil record of their evolution.

The position of the acorn worms (Fig. 37.25) in this story is puzzling. A notochordlike structure, gill slits, and what has been claimed to be a dorsal central nervous system have led to their being classed as chordates by some biologists. Their affinity with the echinoderms is undisputed, however. The larva of the acorn worm is practically indistinguishable from the larva of the sea cucumber. Acorn worms have both creatine phosphate and arginine phosphate as do the sea urchins. Finally, comparative serology clinches the argument. Antibodies developed in response to acorn worm proteins also react strongly with sea cucumber proteins.

Despite the superficial resemblance of *Amphioxus* to fishes (Fig. 37.24), it is to the tunicates (subphylum Tunicata) that we look for the origin of the vertebrates. Actually, examination of an adult tunicate such as the sea peach, *Halocynthia,* could not reveal a more unlikely-looking candidate for vertebrate ancestry. This sessile, filter-feeding organism with its "tunic" containing cellulose has no notochord and only a rudimentary nervous system. However, the larvae produced by these organisms are strikingly different in form. They are free-swimming, bilaterally symmetrical "tadpoles" with gill slits, a dorsal tubular central nervous system, and a tail through which runs a notochord. In fact, if we try to visualize what a primitive, idealized chordate should have looked like, we arrive at an organism not much different from a tunicate larva.

One difficulty remains. The fossil evidence suggests that vertebrates arose in fresh water while the tunicates are exclusively marine. The tunicates do, however, live close to shore and their larvae could have entered the mouths of rivers. Furthermore, the steady one-way current of river water may well have altered selection pressure in favor of an active, bilaterally symmetrical form. Thus, the ancestral tunicate larva may have abandoned metamorphosis to the adult and provided the start of vertebrate evolution. The abandonment of metamorphosis is not unknown elsewhere. Remember the axolotl (Fig. 25.6), which reaches sexual maturity as a gill-breathing larva. Here, too, the environment (lack of iodine) has selected for the retention of juvenile traits.

EXERCISES AND PROBLEMS

1. If someone brought a radially symmetrical animal to you, what conclusions might you draw about (a) its habitat, (b) its power of locomotion, (c) its feeding habits?
2. For each of the following, tell whether it belongs to the annelid or echinoderm superphylum: (a) trochophore larva, (b) tunicate, (c) identical twins, (d) squid, (e) sea cucumber.
3. The largest vertebrate and the largest invertebrate live in the ocean. Can you think of an explanation for this?
4. Hemoglobins consist of a protein to which is attached the prosthetic group, heme. Hemoglobins are found in vertebrates, *certain* mollusks, nemertines, annelid worms, insects, crustaceans, nematodes, echinoderms, and even in leguminous plants. Does their presence in these forms indicate close kinship? If not, can you account for the repeated independent evolution of hemoglobins?

REFERENCES

1. GLAESSNER, M. F., "Pre-Cambrian Animals," *Scientific American,* Offprint No. 837, March, 1961.
2. RUSSELL-HUNTER, W. D., *A Biology of Lower Invertebrates,* Macmillan, New York, 1968. Includes the sponges, cnidarians, flatworms, roundworms, nemertines, annelids, and mollusks.
3. BUCHSBAUM, R., *Animals Without Backbones,* University of Chicago Press, Chicago, 1948. Chapters 27 and 28 examine the fossil record of the invertebrates and their possible evolutionary relationships.
4. NICHOLS, D., *The Uniqueness of the Echinoderms,* Oxford Biology Readers, No. 53, Oxford University Press, Oxford, 1975. A brief, well-illustrated survey of echinoderm features.
5. BONE, Q., *The Origin of Chordates,* Oxford Biology Readers, No. 18, Oxford University Press, Oxford, 1972. Examines several theories of chordate evolution.
6. BARNES, R. D., *Invertebrate Zoology,* 3rd. ed., Saunders, Philadelphia, 1974.

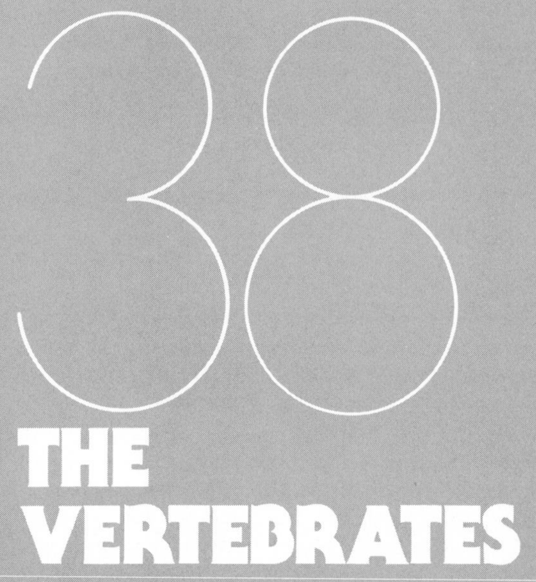
38
THE
VERTEBRATES

38.1 THE JAWLESS FISHES (CLASS AGNATHA)

The first vertebrates to appear in the fossil record are jawless fishes, the **ostracoderms.** Some appear in Ordovician rocks, although they become far more abundant in the Silurian. These creatures were relatively small (6 to 12 inches) flattened fishes (Fig. 38.1), which probably lived by sucking up organic debris from the bottom of the streams they inhabited. Gas exchange occurred at the pairs of internal gills, each gill supported by a bony arch. Water entered the mouth, passed over the gills and left through a series of gill pouches which opened out at the surface. There were no fins, the fish swimming by means of undulating movements.

The body was encased in an armor of bony plates. This may have provided some protection against the large eurypterids which shared the same habitat. It may also have reduced the inflow of water in the hypotonic environment. The gills had to be exposed to the water, however, so some continual inflow was unavoidable. The solution to the problem was to use the pressure created by the contraction of the heart to pump water back out of the body. The ostracoderm probably had inherited nephridialike excretory tubules from its invertebrate ancestors. The development of a glomerulus near the opening (the nephrostome) of each permitted the transfer of fluid (nephric filtrate) from the blood to the tubule and thence to the exterior. Of course, valuable materials (e.g. salts) would also be lost by such a mechanism so we must assume that reabsorption of such materials occurred in the tubules. The presence of a capillary network draining the glomeruli provided the increased blood supply necessary for increased tubular reabsorption. Thus evolved the first kidneys.

They were not primarily organs for excretion but for maintaining water balance in a hypotonic environment. Nitrogenous wastes (chiefly ammonia) were probably excreted at the gills. This is certainly true of the modern freshwater fishes. This first kidney did, however, establish a pattern which was to be adapted to the excretory and water-balance needs of each of the later vertebrate classes.

The only jawless fishes left today are the lampreys (Fig. 38.2) and the hagfishes. These creatures are still the most primitive of the vertebrates. Besides lacking jaws, they have no paired fins. The notochord persists throughout the life of the organism. It is never completely replaced by the skeleton, which is made of cartilage. There are no scales on the body.

Despite its isolated position in the world of the vertebrates, the sea lamprey has proved to be a great nuisance to humans. It feeds by attaching itself with its sucking mouth to the bodies of bony fishes and sucking their tissue fluids. With the construction of the Welland Canal around Niagara Falls, the sea lamprey has been able to move into the Great Lakes from its former marine habitat. In a remarkably short time, it has virtually wiped out the large lake trout fishing industry that once existed. The recent discovery of a chemical that selectively kills young lampreys while sparing other organisms gives promise that this pest may be brought under control and the fish population of the Great Lakes restored.

38.2 THE PLACODERMS

A second group of armored fishes appeared early in the Devonian. These placoderms (Fig. 38.1) differed from their agnath ancestors in two fundamental respects. They had jaws and they had paired fins. The former enabled them to prey actively on smaller animals; the latter aided locomotion by stabilizing the fish in the water. The resulting shift away from bottom feeding was also reflected in the development of a cylindrical body in contrast to the flattened body of the agnaths.

The fossil record reveals an extensive adaptive radiation of these fishes during the Devonian. Most of them later became extinct, but some produced lines of descent which led to the two major classes of modern fishes, the cartilaginous and bony fishes, the Chondrichthyes and Osteichthyes respectively. (Actually *bony* fish is not a uniquely descriptive name for the latter because their placoderm and agnath ancestors also had skeletons of bone.)

The Devonian was marked by periods when many lakes and streams dried up or became much smaller and warmer. This environmental change must have imposed a rigorous selection pressure on the freshwater Devonian fishes. Two solutions to the problem lay open to these animals, and both were taken. On the one hand, they could return to the ocean. On the other, the development of lungs for air-breathing would enable them to survive temporary periods of diminished water supply.

38.3 THE CARTILAGINOUS FISHES (CLASS CHONDRICHTHYES)

Retreat to the oceans was the method of escape taken by the first cartilaginous fishes. These fishes

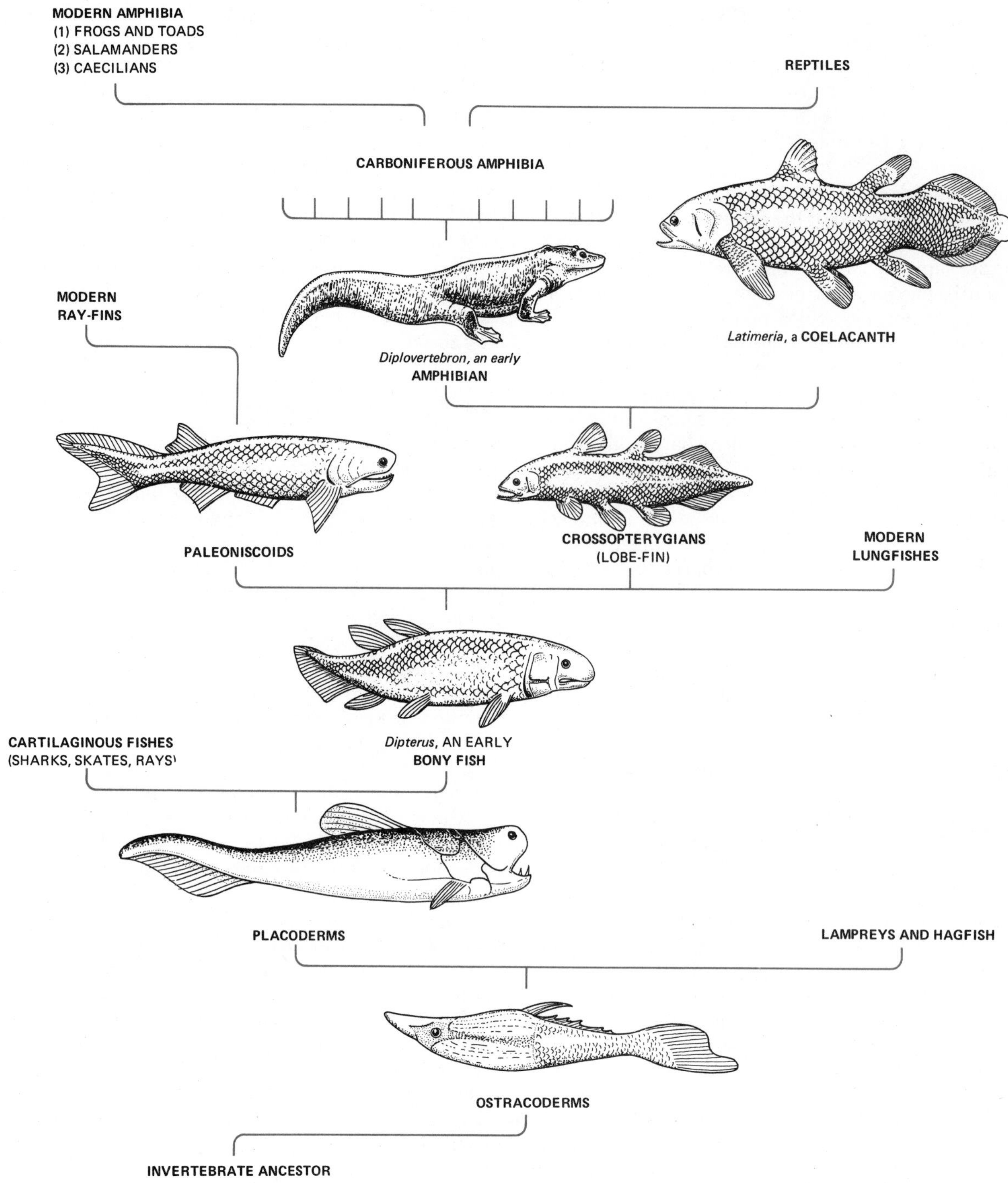

FIG. 38.1 The evolution of the fishes and amphibians.

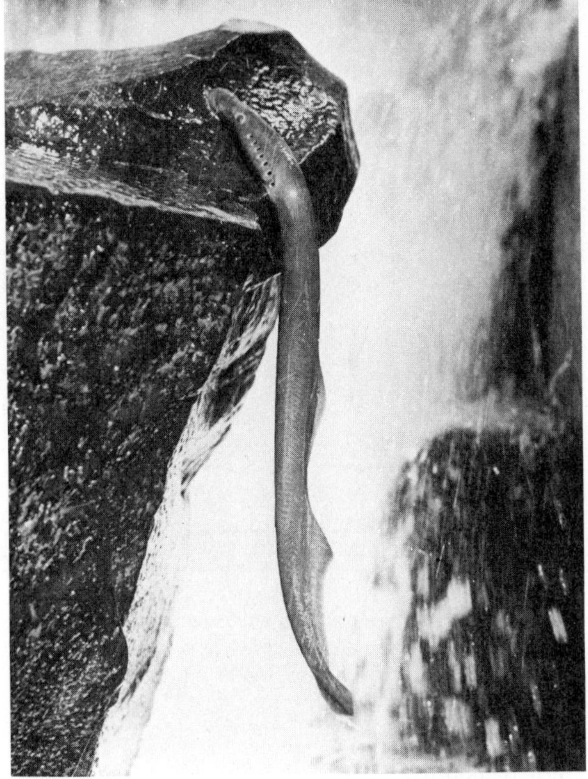

FIG. 38.2 The sea lamprey. Note the presence of gill slits and the absence of paired fins. (Courtesy of the Carolina Biological Supply Co.)

(the earliest were sharks little different from our modern forms) get their name from the fact that their skeleton is made of cartilage, not bone. Like the placoderms, the sharks have jaws. The jaw "bones" developed from the first two pairs of gill arches (Fig. 38.3). It is worth noting that in so doing, one pair of gill slits was no longer required. However, the opening still persists in some modern fishes. It is called the spiracle. In addition to sharks, the skates and rays are members of this class.

When the first cartilaginous fishes returned to the sea, they were exchanging a hypotonic environment (fresh water) for a hypertonic one. Instead of being faced with the problem of getting rid of excess water, they had to develop a method of conserving body water against the dehydrating effects of the sea.

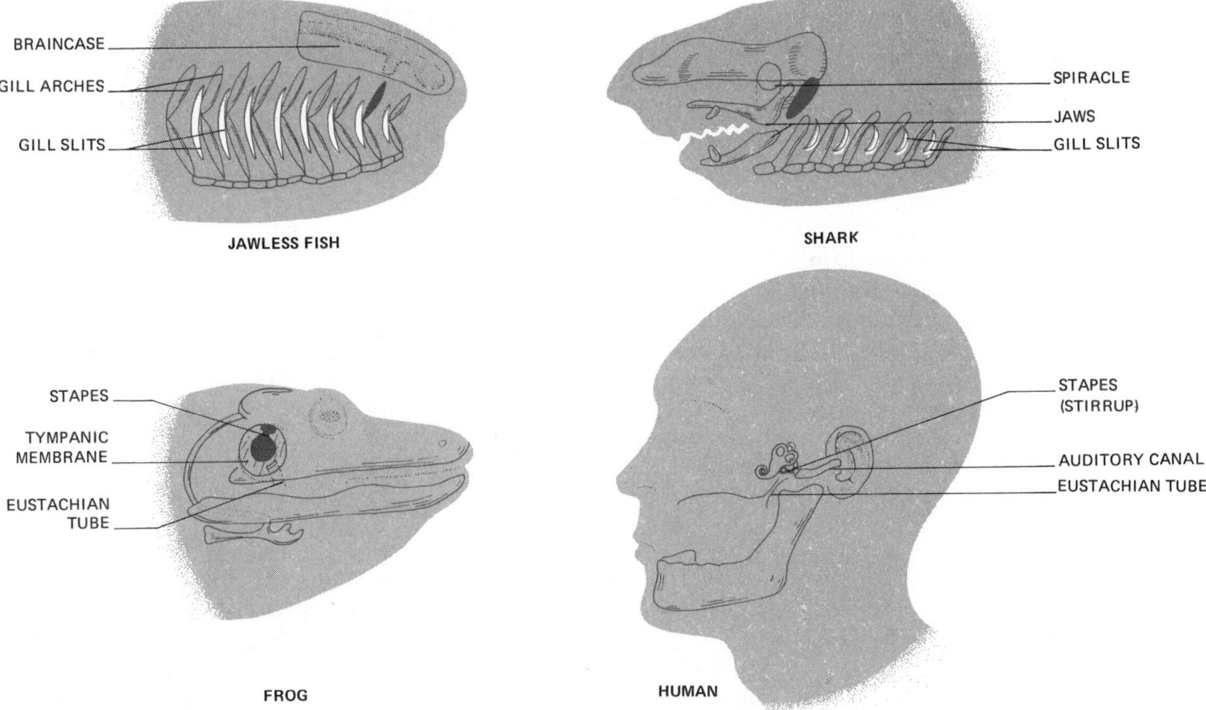

FIG. 38.3 Evolutionary transformation of the first gill slit and second (hyoid) gill arch of the jawless fishes. The bones homologous to the upper portion of the hyoid are shown in color.

As you know, they accomplished this by converting their nitrogenous wastes into urea and allowing its concentration to build up in the blood until the blood was isotonic to the sea water. Today this requires a concentration of 2.5%, which far exceeds that found in other vertebrates (0.02%). Because the ability to reach osmotic balance with the sea by this method develops late in the embryology of these animals, the eggs cannot simply be deposited unprotected in the water.

One solution to this problem is to deposit the egg, surrounded by a suitable fluid medium, in an impervious case in which embryonic development can occur. A second solution is simply to retain the eggs, and the embryos into which they develop, within the mother's body until they are capable of coping with the sea-water environment. Both these solutions require that internal fertilization take place, and the cartilaginous fishes were the first vertebrates to develop this. The pelvic fins of the male are specially modified for depositing sperm in the reproductive tract of the female.

FIG. 38.4 The African lungfish surfaces from time to time to fill its lungs with air. Its gills are vestigial. (Courtesy of the American Museum of Natural History.)

38.4 THE BONY FISHES (CLASS OSTEICHTHYES)

The so-called bony fishes employed the second solution to the problem of periodic droughts. These developed a pair of pouched outgrowths from the pharynx which served as primitive lungs. They were inflated by air taken in through the mouth. The bodies of these fishes were covered with scales, the only vestige of their ancestors' armor being the bones of the cranium.

These fishes quickly (still in the Devonian) diverged into three distinct groups, the paleoniscoids, the lungfishes, and the crossopterygians.

The **paleoniscoids** were distinguished by the presence of ray-fins (fins in which neither muscle nor bone was present) and the fact that ventilation of the lungs continued through the mouth. Many of them migrated to the sea during the late Paleozoic and the Mesozoic eras. In a stable aquatic environment, lungs were not necessary, and they were transformed into a swim bladder by means of which the fish could alter its buoyancy in the water. All of our modern commercial fish, both marine and freshwater (e.g. salmon, tuna, mackerel, trout, and bass) are descendants of this group.

The **lungfishes** developed a significant innovation not possessed by their ancestors. Their nostrils, which in the first Osteichthyes had opened only to the outside and were used for smelling (as in all modern paleoniscoid descendants), developed internal openings to the mouth cavity. This made it possible to breathe air with the mouth closed. Once a rather successful group, the lungfishes of today are found in a few restricted localities of Africa, Australia, and South America, where their lungs still enable them to survive periodic droughts (Fig. 38.4). The widely scattered distribution of these fishes indicates that they are isolated remnants of a once-widespread group.

Judging from presentday lungfishes, two other significant adaptations evolved in this group. One was the development of two atria and a partial septum in the ventricle of the heart. This permitted at least a partial separation of oxygenated blood returning from the lung(s) and deoxygenated blood returning from the rest of the body and thus provided a marked improvement in circulatory efficiency (see Section 11.6). The second adaptation was the development of the enzyme system necessary to convert ammonia into the less toxic urea. This is particularly well developed in the African and South American species. While in the water, these fishes excrete their waste nitrogen as ammonia just as all the ray-fins do. In times of drought, however, these animals burrow into the mud and switch over to urea production.

The **crossopterygians** also had internal nostrils through which the lungs could be inflated. In addition, their pectoral and pelvic fins were lobed; that is, they were fleshy and were supported by bones. A glance at Fig. 38.5 shows that the pattern of bones (a bony girdle articulating with a single bone which,

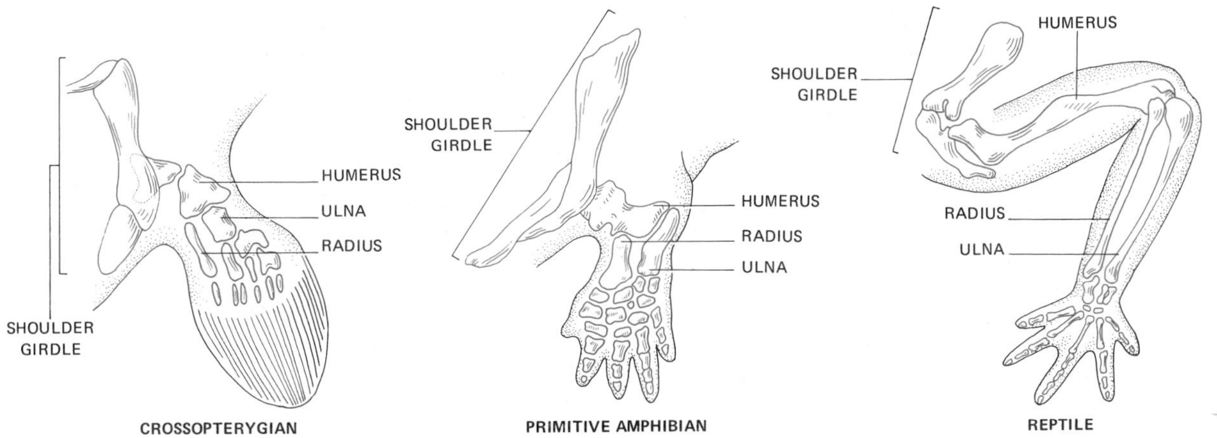

FIG. 38.5 Evolutionary development of the tetrapod forelimb from the pectoral fin of a crossopterygian.

in turn, articulates with two others, etc.) is the one we find in all four-legged vertebrates. Each of our arm and leg bones is homologous to a bone in the crossopterygian pectoral and pelvic fins, respectively.

Fossil remains tell us that the ancient lungfishes, like some of their modern descendants, burrowed in the mud and became dormant during periods of drought. The crossopterygians with their lobe-fins were able to attack the problem more aggressively. Their lobe-fins enabled them to waddle from one shrinking pool to another in search of better conditions. The combination of lungs and lobe-fins thus provided the necessary equipment to enter a new, uncrowded habitat, the land. One group of crossopterygian descendants did just that, in a rather tentative manner. These were the amphibians. Another group turned instead, as so many other Devonian fishes had done, to the sea. These, the coelacanths, were quite successful for a time in their new habitat but became extinct by the end of the Mesozoic era, some 70 million years ago. Or so everyone believed until 1938. In that year a living coelacanth (*Latimeria*) was pulled up from the ocean depths off the East Coast of Africa. Since then more than two dozen specimens have been caught. These "living fossils" still have paired lungs like those of their ancestors, but they are no longer functional.

The Devonian has been called the "Age of Fishes." This is surely appropriate in view of the extraordinary adaptive radiation which took place in the group during that period. Both fresh water and salt water became well populated by them. Toward the end of the Devonian, however, a new group of vertebrates appeared. These were the amphibians, the first of the four-legged vertebrates or tetrapods.

38.5 THE AMPHIBIANS (CLASS AMPHIBIA)

The amphibians were the vertebrate pioneers of the land. Their lungs and bony limbs, inherited from their crossopterygian ancestors, provided a means of locomotion and a means of breathing air. A second atrium in the heart permitted oxygenated blood to be returned directly to it for pumping to the body under full pressure. While some mixing of oxygenated and deoxygenated blood occurred in the single ventricle, the three-chambered heart must have provided a substantial increase in circulatory efficiency and thus ability to cope with the more changeable, demanding land environment.

On land, the ability to detect sound is of great importance, and the amphibians developed simple ears out of structures inherited from their ancestors. The spiracle (Fig. 38.3) was covered with a membrane that served as an eardrum and a no-longer-used jawbone (itself originally derived from an agnath gill arch) served to transmit vibrations from this membrane to the inner ear. The innermost ossicle (the stirrup) of our middle ear is homologous to this bone.

As their name implies, the amphibians were only semiterrestrial. They had to return to water in order

FIG. 38.6 Representative amphibians. Upper left: The spotted salamander is a common inhabitant of moist woods. Right: The leopard frog is a favorite laboratory animal. (Courtesy of Ward's Natural Science Establishment, Inc.) Lower left: The wormlike caecilians are found in the tropics. (Courtesy of the New York Zoological Society.)

to lay eggs, and their modern descendants, at least, cannot withstand long exposure to dry air. The periodic shift from water to land and back again imposes additional problems in maintaining water balance and excreting nitrogenous wastes. In the water, the continual inflow must be pumped out of the glomeruli, as in the freshwater fishes. On land, water must be conserved, and the amphibian accomplishes this by reducing the blood supply to the glomeruli and, hence, the filtration rate. Of course, this also diminishes the blood flow from the glomeruli to the tubules. Tubular function must be maintained, however, and increased activity of the auxiliary renal portal system (Fig. 13.10) permits this.

The earliest amphibians were large enough by modern standards (*Diplovertebron,* Fig. 38.1, reached a length of about 2 ft), but some of the later forms were truly of impressive size. Some fossil specimens are 8 ft long. These amphibians flourished during Carboniferous times. Vast swamps covered the earth, plant life was abundant, and there were plenty of insects on which the amphibians could feed. This period is often referred to as the Age of Amphibia.

It was followed by a period (the Permian) when the earth became colder and dryer. A decline in the fortunes of the amphibians set in that has continued to this day. Only three orders remain today. They are: (1) frogs and toads (order **Anura**), (2) salamanders and newts (order **Urodela**), and (3) caecilians (order **Apoda**), which are limbless, wormlike creatures (Fig. 38.6). Lacking a waterproof skin and waterproof eggs, none of the amphibians has ever fully adapted to the conditions of land life.

38.6 THE REPTILES (CLASS REPTILIA)

The first truly terrestrial animals were the reptiles. These evolved from amphibians during Carboniferous times. With the arrival of the Permian, they were able to cope with the new conditions better than the amphibians. The chief advance the earliest reptiles (the "stem" reptiles or **cotylosaurs**) showed over the amphibians was the development of a

FIG. 38.7 American chameleon emerging from egg. (Courtesy of the Carolina Biological Supply Co.)

shelled, yolk-filled egg (Fig 38.7) which could be deposited on land without danger of its drying out. A shell impervious to water is just as impervious to sperm and thus the development of the shelled egg coincided with the development of internal fertilization.

The embryo developing within the egg produced four extraembryonic membranes. Protected by the fluid contained within the **amnion,** it secured food through the **yolk sac,** breathed through the **chorion** and **allantois,** and stored its metabolic wastes in the sac formed by the allantois.

Probably these earliest reptiles, whose short legs extended sideways from the body, still spent most of their time in the water and simply laid their eggs on land where they could be better hidden from predators. With the increasing dryness of the Permian, however, other modifications for dry-land living evolved. Development of a dry skin enabled them to leave the water safely. But dry skin could not be used for gas exchange. Improved lungs, inflated by expansion of the rib cage, solved this problem. A partial septum in the ventricle reduced the mixing of oxygenated with deoxygenated blood and thus improved the efficiency of circulation. The success of these evolutionary developments was immediate. The stem reptiles underwent an adaptive radiation, producing five major lines of descent (Fig. 38.8).

Pelycosaurs

Although the pelycosaurs, too, probably spent most of their time in the water, their legs were positioned under, rather than at the sides of, the body. This position allowed for more rapid and effortless running on land. From the pelycosaurs evolved a group of small, active, land-dwelling reptiles, the **therapsids.** At the start of the Mesozoic, these were the most abundant of the reptiles, but they soon were overshadowed in importance by other groups. As it turned out, however, this was only a temporary (about 100 million years) eclipse, since the descendants of the therapsids, the mammals, ultimately inherited the earth.

Turtles (Order Chelonia)

Although the evolution of the reptiles produced animals capable of living on dry land, many of them did not do so. From their origin early in the Mesozoic era right down to the present time, most turtles have lived in either fresh water or the ocean. Despite their habitat, they have not abandoned their heritage of land adaptations. They breathe air (although the ribs now form the bony box covering the animal and hence cannot help to ventilate the lungs) and lay shelled eggs on land. It is interesting to realize that while our terrestrial toads are returning to fresh water each spring to lay their eggs, the freshwater turtles are crawling up on land to scoop out a hole in sand or soil in which to lay theirs. Although never a prevalent group, the turtles deserve recognition for still being with us after over 200 million years on the earth, a span of time which has seen most of their early reptile contemporaries vanish.

Plesiosaurs and Ichthyosaurs

The members of these two lines of descent were marine reptiles which flourished during the Jurassic but became extinct by the end of the Mesozoic era. They were undoubtedly fisheaters and quite specialized for life in the sea. In fact, their finlike appendages (Fig. 38.8) were so unsuited for locomotion on land that the ichthyosaurs, at least, took to retaining their eggs within the mother's body rather than laying them on land. The young were born alive and active, as many shark young are born today.

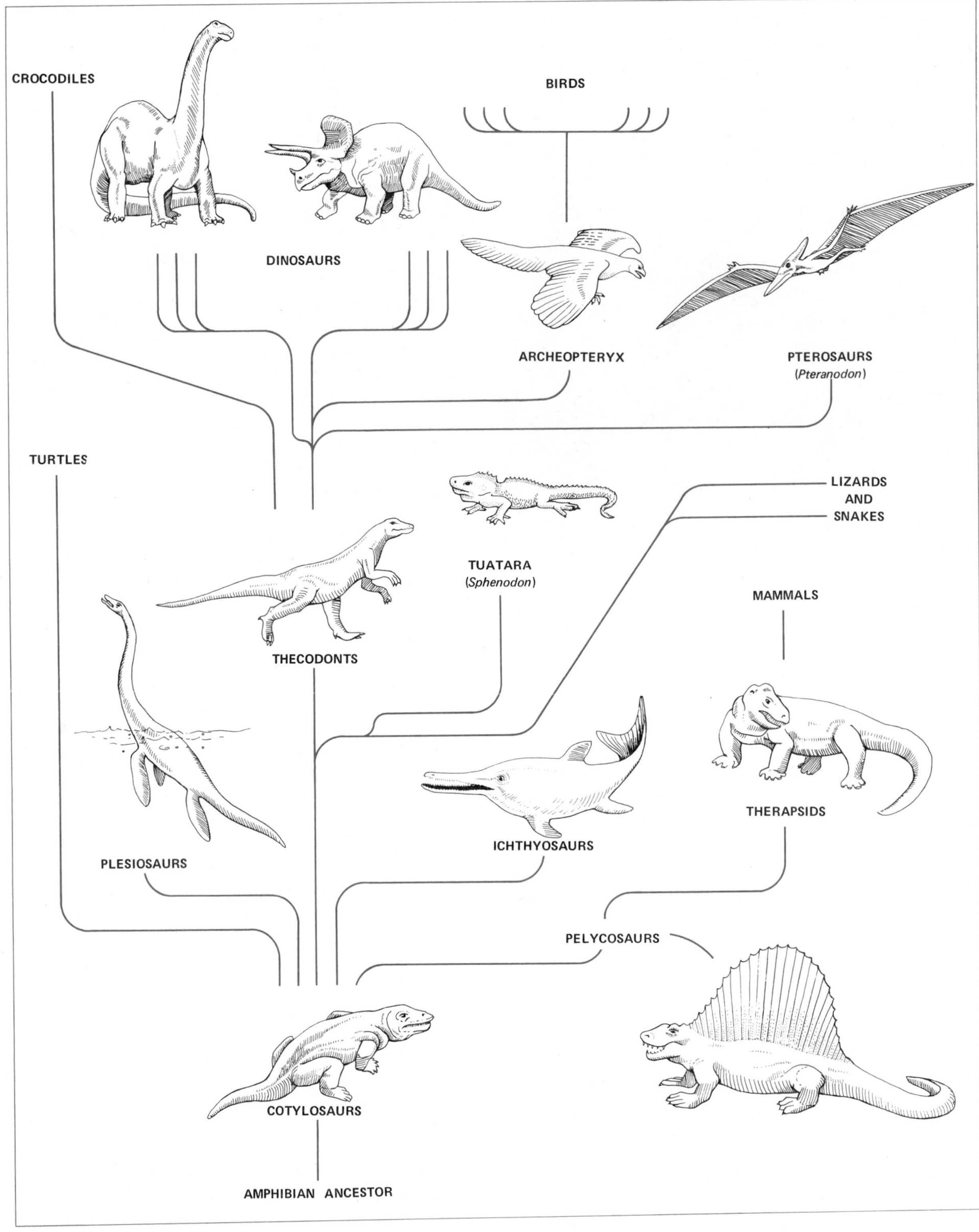
CROCODILES
BIRDS
DINOSAURS
ARCHEOPTERYX
PTEROSAURS
(*Pteranodon*)
TURTLES
LIZARDS
AND
SNAKES
TUATARA
(*Sphenodon*)
MAMMALS
THECODONTS
THERAPSIDS
ICHTHYOSAURS
PLESIOSAURS
PELYCOSAURS
COTYLOSAURS
AMPHIBIAN ANCESTOR

◀ **FIG. 38.8** The evolution of the reptiles.

Diapsids

The animals found in the fifth line of descent from the stem reptiles are called diapsids because they possessed a characteristic double-arched bony structure in the temporal regions of their skull (di = two; apsid = arch). We assume (based on their present-day descendants) that these reptiles also possessed an important physiological adaptation for land living that was not present in the other groups. This was the ability to convert nitrogenous waste into the almost-insoluble compound, uric acid. As you learned in Chapter 13, this physiological modification allows potentially toxic nitrogenous wastes to be excreted safely and with almost no loss of water. Some uric acid is filtered at the glomeruli, but most of it is excreted directly into the tubules. In fact, in the interests of water conservation, the glomeruli are reduced to the point where they produce little more nephric filtrate than is needed to flush the uric acid into the cloaca. Here even this small amount of water is largely reabsorbed. The residue of uric acid is a whitish paste that passes out with the feces. The ability of this group of reptiles and their descendants to excrete their nitrogenous wastes in this manner has almost totally freed them from dependence on drinking water. The water in their food plus that produced by cellular respiration is usually sufficient for their needs.

Evolution in this group of reptiles soon diverged along two separate branches (Fig. 38.8).

1. One branch, by further branching, produced two of the four surviving orders of reptiles on the earth today: the order **Rhynchocephalia** and the order **Squamata.** The rhynchocephalians are hardly a flourishing group; the rare tuatara of New Zealand (Fig. 38.9) is the only surviving member of the order. The squamata, on the other hand, are doing reasonably well. They are the **lizards** and **snakes.**

Our modern lizards (Fig. 38.10) first appeared in the Jurassic. They are still important colonizers of the deserts and forests of warm parts of the world. One group of Cretaceous lizards became burrowing animals. Their legs eventually disappeared and thus the **snakes** were established. (Vestigial remnants of hind legs can still be detected in the boa and python.) Although they are able to survive in temperate regions by hibernating during the winter, the snakes also have been more successful in the subtropical and tropical regions of the world.

FIG. 38.9 The tuatara (*Sphenodon*). This "living fossil" is found on a few small islands near the coast of New Zealand. It is the only surviving member of its order (Rhynchocephalia). (Courtesy of the New York Zoological Society.)

FIG. 38.10 The Komodo monitor lizard, a native of Malaya, is the largest of the lizards. This specimen weighed over 200 pounds. (Courtesy of the New York Zoological Society.)

FIG. 38.11 The American alligator is *able* to move rapidly on land although it usually does not. (Courtesy of the American Museum of Natural History.)

2. The **thecodonts** made up the second branch of the uric-acid-excreting, fully terrestrial reptiles. These animals were able to run quickly over the land by rising up on their hind legs and using their long tail for balance. The hind legs became larger and more powerful than the front. Fossils of some of the more highly evolved thecodonts reveal the presence of an insulating covering on the body and a bone histology that suggest that these animals were able to maintain a relatively high and well-regulated body temperature. This, combined with their agility and tolerance of arid conditions, provided an enormously successful evolutionary plan.

Five orders of reptiles evolved from the thecodonts. The members of this remarkable adaptive radiation are often referred to as the **ruling reptiles** because they completely dominated both land and air during the rest of the Mesozoic era.

a) The **crocodiles** and **alligators** (order **Crocodilia**). These animals abandoned the bipedal locomotion of their thecodont ancestors but retained (and still have) the larger hind limbs. Although specimens in zoos do not often do so, these animals are capable of moving quickly with the entire torso lifted off the ground (Fig. 38.11). They are the only reptile descendants of the thecodonts that have not become extinct.

FIG. 38.12 Fossil skeleton of *Tyrannosaurus rex,* one of the largest bipedal dinosaurs. Note the small forelimbs of this carnivore. (Courtesy of the American Museum of Natural History.)

b) and **c)** The **dinosaurs.** Late in the Triassic, two orders of dinosaurs appeared, each of which underwent an extraordinary adaptive radiation. Throughout the remainder of the Mesozoic era, the earth was populated by dinosaurs of every size, shape, and description. Many of the larger forms were so huge that they probably were restricted to wallowing about in marshes, letting the buoyancy of the water support some of their body weight. The abundance and habitat of these forms have given us a remarkably complete fossil record of them. The discovery and mounting of fossil dinosaurs was the most active branch of paleontology for many years and captured the imagination of people everywhere. When we look at the reconstructed skeletons of such forms as *Tyrannosaurus* (Fig. 38.12), which was 47 ft (14.3 m) long and stood 19 ft (5.8 m) high (note the greatly reduced forelimbs), and *Brachiosaurus,* which prob-

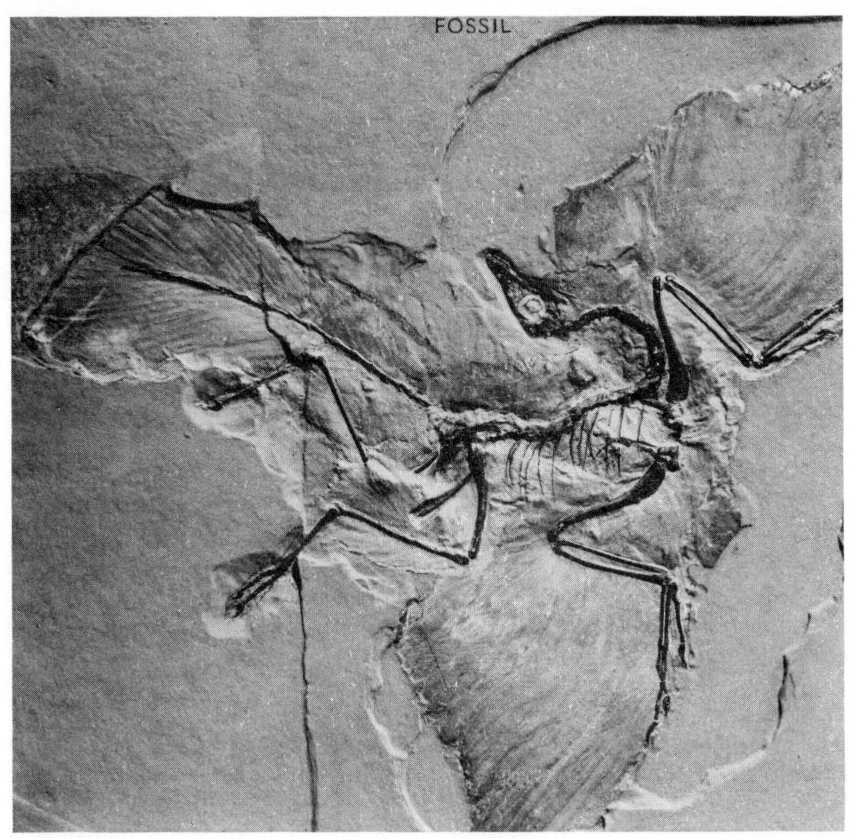

FIG. 38.13 Cast of fossil of *Archeopteryx*. The long tail, teeth, and claws on the wings are reptilian features. The feathers, clearly visible, cause it to be classified as a bird. (Courtesy of the American Museum of Natural History.)

ably weighed close to 50 tons, we can well understand why. Although representing only two out of some 15 orders of reptiles in existence then, the dinosaurs alone fully justify our calling the Mesozoic era the "Age of Reptiles."

d) and **e)** The **flying reptiles.** Two groups of Mesozoic reptiles took to the air. The two-legged gait of the thecodonts had freed the forelimbs for use as wings. At first these probably were used only for gliding, but later, sustained flight became possible. One of these groups, the **pterosaurs,** ruled the air throughout most of the Mesozoic era. *Pteranodon* (Fig. 38.8), with its wingspread of 8.2 m, was until recently thought to be the largest member of the order. Then, in the early nineteen seventies, fossils of a pterosaur that had a wingspread of 15.5 meters (almost 51 feet!) were discovered in Big Bend National Park in Texas. Picture, if you will, this creature gliding in over your backyard! The second group of flying reptiles were the ancestors of our modern birds.

38.7 THE BIRDS (CLASS AVES)

The second group of reptiles that took to the air developed a modification not found in the pterosaurs: **feathers.** These scaly outgrowths provided a broad, light, but strong, surface for the wings. They also provided heat insulation for the body, making it possible to be small and yet maintain a relatively high—and constant—body temperature even in cold climates. These were the first birds.

The discovery of *Archeopteryx* in Jurassic rock (Fig. 38.13) has provided us with one of the best examples of a "missing link." This creature possessed feathers, and thus we may arbitrarily call it a bird. But its relationship to the reptiles is obvious. The rather rudimentary wings had claws. There were teeth in the mouth, and it had a long tail. These are reptilian features no longer found in living birds.

Although they were well established by the end of the Mesozoic era, it was in the Cenozoic that the birds underwent such an extensive adaptive radiation. The large number of species and their wide distribution attest to their success.

The structure and physiology of birds is adapted in a variety of ways to efficient flight. Chief among these, of course, are the wings. Although wings now enable birds to travel long distances in search of suitable and abundant food, they probably first arose as an adaptation which helped them escape from predators. The presence of wingless birds in the Antarctic, New Zealand, and other regions where predators are rare gives indirect evidence of this.

An efficient bird, like an efficient airplane, needs to be light and powerful. Lightness has been achieved by feathers, hollow bones, and a single gonad (in the females) which enlarges and is active only during the breeding season. Loss of teeth has reduced the weight of the head. Their function has been taken over by the gizzard, which is located near the center of gravity.

Power is achieved by large breast muscles attached to a greatly enlarged sternum (Fig. 38.14). Birds also have a four-chambered heart, and its efficiency has permitted the development of a fixed body temperature (homeothermy). Homeothermy, in turn, permits a high metabolic rate under all environmental temperatures. Birds can remain active in cold weather, unlike their reptile cousins, which become sluggish as the temperature falls. The high metabolic rate reflects a rapid release of energy for flying. Birds secure their energy from concentrated foods, particularly those rich in fats such as seeds, insects, and other animals.

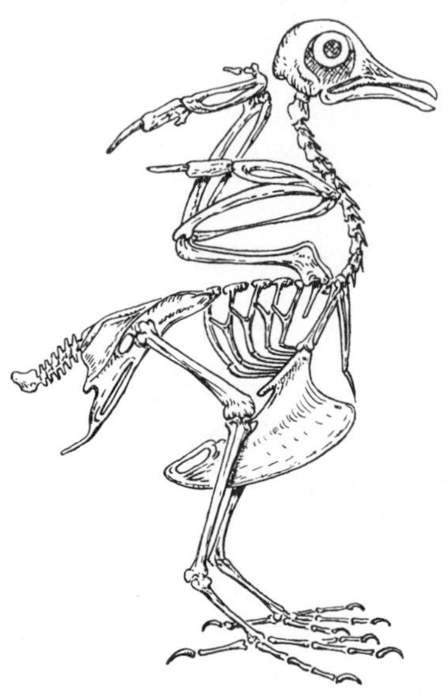

FIG. 38.14 Skeleton of a pigeon. Note the large sternum to which the flight muscles are attached. Compare the tail with that of *Archeopteryx* (Fig. 38.13).

The end of the Mesozoic era was marked by great geological and biological changes on the earth. We are sure that the two were interrelated although as yet we can only speculate as to how. Geologically, this period was marked by an uplifting of mountain ranges in several parts of the world. The Rockies, the Andes, and the Himalayas achieved their present eminence at this time. Biologically, it was marked by the extinction of the reptilian orders that had flourished during the Mesozoic. By the dawn of the Cenozoic, the plesiosaurs, the ichthyosaurs, the pterosaurs, and every single dinosaur had vanished from the face of the earth. While we do not yet understand just why these reptiles were unable to survive the geological changes of those times, we do know that the widespread extinction of the reptiles opened up habitats on land, in the air, and in the waters of the earth. Ready to move into these habitats were the birds and the mammals.

38.8 CONTINENTAL DRIFT

One geological change that began early in the Mesozoic era and has continued to the present is the drifting of the continents over the surface of the earth. The existence of such a phenomenon, though long suspected, is only now beginning to be widely appreciated. It provides at last a satisfying explanation for a number of puzzling facts, both geological and biological. Each of these observations lends support to the idea that at the start of the Mesozoic era, all the continents were attached to one another in a single land mass, which has been given the name Pangaea (Fig. 38.15). During the Triassic, Pangaea began to break up, first into two major land masses: Laurasia in the northern hemisphere and Gondwana in the southern hemisphere. Although the exact timetable is still uncertain, the present continents separated at intervals throughout the rest of the Mesozoic and through the Cenozoic era, ultimately reaching the positions that they occupy today. Although a careful study of the physical evidence for continental drift is beyond the scope of this book, let us at least glance at some of it.

The shapes of the present continents have long suggested to astute observers that they might at one time have been in direct contact. The east coast of South America and the west coast of Africa are strikingly complementary. This complementarity is even more dramatic when one attempts to fit the continents together using the edges of the continental

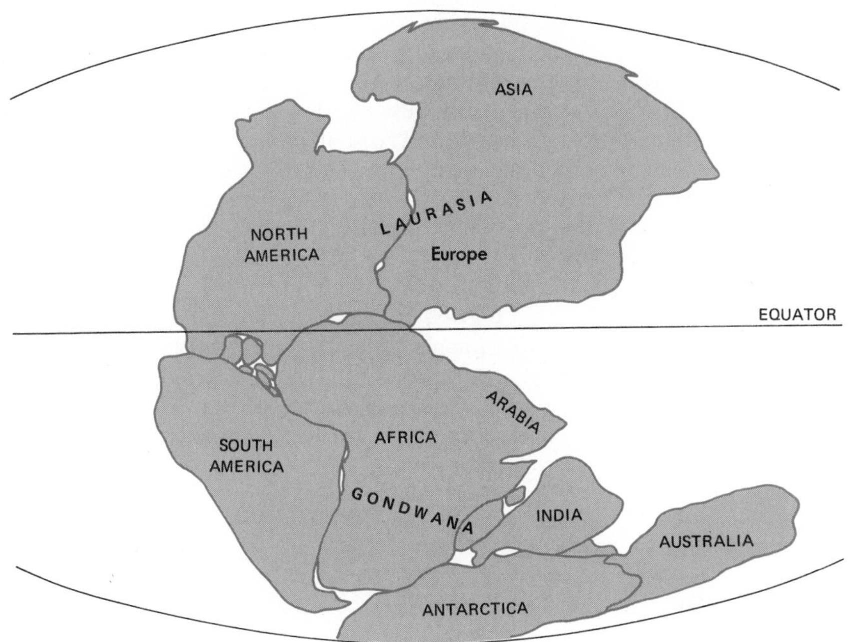

FIG. 38.15 Pangaea. Reconstruction of the single land mass thought to have existed 200 million years ago during the Triassic period and from which our present continents were formed. The reconstruction is based on a computer-generated fit of the continents as they would look if the sea level were lowered 6,000 feet. By the start of the Jurassic period, 180 million years ago, Pangaea had split into Laurasia to the north and Gondwana to the south. There is considerable evidence that continental drift has continued to the present time and, indeed, is still continuing. (Adapted from the data of R. S. Dietz and J. C. Holden.)

shelf (at 500 or 100 fathoms) as the boundaries rather than the present shorelines (Fig. 38.15).

If the continents were once a single land mass, we would expect to find similar geological features in areas now separated but once presumed to have been in contact. This is indeed the case. In both mineral content and age (as determined by isotope dating), the rocks in a small area on the east coast of Brazil match precisely those found in Ghana on the west coast of Africa. The low mountain ranges and rock types in New England and eastern Canada appear to be continued in parts of Great Britain, France, and Scandinavia. India and the southern part of Africa both show evidence of periodic glaciation during Paleozoic times. This is a remarkable finding, indeed, for land that now lies close to the equator. Not only do the *patterns* of glacial deposits match in each of these areas, but they also match those found in South America, Australia, and Antarctica.

When rocks containing iron ores are first formed, they become permanently magnetized in the direction of the earth's magnetic field at that time. The study of such rocks formed at different times in the earth's history in different areas of the earth has revealed them to be magnetized in directions that corroborate the story of continental wandering we have been discussing. Faced with such an impressive body of evidence for the existence of continental drift, is there any mechanism which would provide the forces necessary to move the continents about? One mechanism that has been postulated involves spreading of the floors of the oceans; it has been given the name **plate tectonics.** A large body of geophysical evidence has been gathered in recent years—ranging from the analysis of earthquakes to measurements of the magnetism and topology of the ocean floors—that supports the concept of plate tectonics and thus provides a plausible mechanism for continental drift. You may examine some of this evidence in the references cited at the end of the chapter.

What does continental drift tell us about the history of animal and plant evolution and vice versa? The hypothesis that a single land mass has broken up into the present continents explains many paleontological puzzles. It explains, for example, why certain fossil reptiles that are found in South Africa are also found in Brazil and Argentina. It makes sense of the discoveries in 1969 and 1970 of fossil amphibians, cotylosaurs, and therapsids in Antarctica. In many cases the species that were identified there are also present as fossils in South Africa, India, and China. It even explains why the humble earthworm of eastern North America has such close relatives in western Europe but not in western North

America. These are only a few of many examples. The distribution of plants and animals from Mesozoic times to the present, on the one hand, can best be understood in terms of a gradual fragmentation of Pangaea into the continents of today; on the other, it reinforces the evidence acquired by geophysicists that our continents have indeed been on the move.

38.9 THE MAMMALS (CLASS MAMMALIA)

We think that the first mammals arose late in the Triassic period from therapsid (Fig. 38.8) ancestors. They were small but very active animals whose diet probably consisted chiefly of insects. Associated with their active life was their ability to maintain a fixed body temperature (homeothermy). As in the birds (which did not appear until the Jurassic) this went hand in hand with the development of a four-chambered heart and completely separate oxygenating and systemic circuits. Conservation of body heat was made possible by the development of **hair.** While the earliest mammals laid eggs as did their reptile ancestors, the young, after hatching, were nourished with **milk** secreted by glands in the mother's skin.

In contrast to their reptile ancestors, the teeth of the mammals became specialized for cutting (incisors), tearing (canines), and grinding (molars) their food. The gray matter of the cerebrum, which is covered by white matter in the reptiles, grew out

FIG. 38.16 Representative marsupials. The euro kangaroo (top right) and the koala bear (below) are two of many marsupial species found in Australia. Bottom right: The common opossum is the only marsupial found in North America. These young are sufficiently well developed to leave their mother's pouch and climb on her back. (Koala bear, courtesy American Museum of Natural History. The others, courtesy New York Zoological Society.)

and over the surface of the brain. This modification was to have far-reaching consequences.

Evolution of the earliest mammals took several different routes. Of the groups that evolved only three have survived to the present. These are: (1) **monotremes,** the egg-laying mammals (Subclass **Prototheria**), (2) **marsupials,** the pouched mammals (Subclass **Metatheria**), and (3) **placental mammals** (Subclass **Eutheria**).

Each of these groups is distinguished by the manner in which the young are cared for during embryonic development. The monotremes continue to lay eggs as their therapsid ancestors did. The duckbill platypus and the spiny anteaters (echidnas —Fig. 21.9) are the only monotremes left on earth.

In the marsupials (Fig. 31.16), the young are **retained for a short period within the reproductive tract of the mother. During this brief time, some** nourishment is secured by means of the yolk sac which grows into the wall of the uterus. At a very early stage of development, though, the young are born. They then crawl into a pouch on the mother's abdomen and attach themselves to a milk-dispensing nipple. Development is completed here.

The earliest marsupials probably resembled our opossum. They *may* have evolved in North America but, while Gondwana still existed, spread into the areas that were to become South America, Australia, and New Zealand. Their presence in South America and Australia is difficult to explain without assuming a unified land mass during at least the early part of the Mesozoic. We may with some confidence predict that fossil marsupials will some day be found in Antarctica.

In the relative isolation of Australia and South America, the marsupials underwent elaborate adaptive radiations. But the reuniting of North and South America through the isthmus of Panama some two million years ago brought the third and most advanced group of mammals, the **placental mammals,** into direct competition with the South American marsupials, all but 69 species of which have since become extinct. Only in Australia and New Zealand, areas where placental mammals have only just recently gained entrance, does a rich diversity of marsupials survive today. The monotremes also survive only in these areas.

The placental mammals retain their young within the uterus of the mother until development is well advanced. There is little yolk in the egg, but the extraembryonic membranes form an umbilical cord and placenta through which the developing young are able to secure nourishment directly from the mother.

1. INSECTIVORA: SHREWS, MOLES
2. SCALY ANTEATERS
3. EDENTATA: ARMADILLOS
4. CHIROPTERA: BATS
5. RODENTIA: SQUIRRELS, RATS, MICE, BEAVER
6. LAGOMORPHA: RABBITS AND HARES
7. CARNIVORA: LION, CATS, DOGS, SKUNK, WALRUS, SEA LION
8. ODD-TOED UNGULATES: ZEBRA, HORSE, RHINOCEROS
9. EVEN-TOED UNGULATES: CAMELS, ELK, GIRAFFE, BISON, SHEEP, HIPPOPOTAMUS
10. CETACEA: WHALES, DOLPHINS, PORPOISES
11. SIRENIA: MANATEE, DUGONG
12. PROBOSCIDEA: ELEPHANTS
13. AARDVARKS
14. PRIMATES: LEMURS, NEW WORLD MONKEYS, OLD WORLD MONKEYS, GREAT APES, HUMANS

FIG. 38.17 Orders of placental mammals (Eutheria) and representatives of each.

For some 70 million years in the Mesozoic era, the placental mammals were represented by just a single order. However, by the end of the second epoch, the Eocene, of the Cenozoic era, the placentals had radiated into at least fourteen different orders (Fig. 38.17. What explains their long period of insignificance followed by such an explosive burst into prominence? The first requirement for speciation is relaxed selection pressure, and it was not until the extinction of the ruling reptiles that the mammals secured this. With most of the reptiles gone, a variety of habitats was available for occupancy by the mammals. Ungulates replaced the herbivorous dinosaurs. Carnivores preyed upon them. The cetaceans (whales, etc.) and some carnivores replaced the plesiosaurs in the oceans. The ancestors of the bats even found a new habitat. With the aid of echo location, they were able to prey upon night-flying insects. This ability and the lack of competition in this area have resulted in their order's becoming second only to the rodents in number of species.

38.10 THE EVOLUTION OF *HOMO SAPIENS*

We are members of the order of placental mammals known as the **primates.** This group had its origin some 60 million years ago. The early members were probably arboreal (tree-living) animals of which today's lemurs and tarsier are the little-changed de-

FIG. 38.18 A tarsier. Its large, forward-looking eyes and grasping hands are typical of the primates. (Courtesy New York Zoological Society.)

scendants. Grasping hands (with nails) and good vision are both of great value for life in the trees, and as Fig. 38.18 makes clear, these animals are equipped with both. Not only is travel through the tree tops made easier and safer, but these features pay an extra dividend in permitting the animal to feed while keeping a constant watch for enemies.

By roughly 30 million years ago, these early primates had, through repeated speciations, split into three separate stocks: the "New World" monkeys, the "Old World" monkeys, and the ancestors of the apes. Their most significant evolutionary advance was the development of stereoscopic ("3D") vision as well as color vision. The New World monkeys seem always to have been confined to tropical America. The other two groups spread over Africa, parts of Europe, and Asia.

The ancestors of the apes differed from the Old World monkeys in their longer arms, in a more erect posture, and in losing their tails. These modifications reflected their method of locomotion: swinging from branch to branch. By 14 million years ago, they had subdivided again into the **hominids,** who came down out of the trees and took to walking over the plains in search of food, and the true apes (pongids), who remained in the trees. From these latter forms have evolved today's "great apes," the gorilla, chimpanzee, and orangutan (Fig. 38.19).

Although we are certainly descendants of the early hominids, the scarcity of fossils makes it exceedingly difficult to trace our genealogy exactly. Perhaps as early as four million years ago, at least two independent "man-ape" populations coexisted. One was a stocky form that is called *Australopithecus robustus* by some workers and *Paranthropus* by others. The second was a smaller, slightly built form called *Australopithecus africanus* or, by some, *Homo africanus.* (These alternative names provide a hint of the controversy that has surrounded the naming of and deducing of affinities between various pre-*Homo sapiens* fossil discoveries.) *Australopithecus robustus* may have used crude tools, but *Australopithecus africanus* appears to have been more inventive in this respect.

Now only one further adaptation was necessary for the emergence of humans: the great enlargement of the cerebral cortex and with this the full exploitation of their tool-making and tool-using capabilities. By five hundred thousand years ago (and perhaps much earlier), *Homo erectus* existed throughout much of Africa and Asia (e.g. "Java man" and "Peking man"). The cranial capacity of these early humans averaged some 950 cm^3, as compared with the 450 cm^3 of *Australopithecus africanus* and the 800–2000 cm^3 range found in modern humans.

The spread of *Homo erectus* through Africa, Asia, and the Near East created a number of relatively isolated gene pools. Thus may have begun the process of subspeciation that gave rise to the racial groups of today. We have no reason to think that the process operated any differently from the process that has given rise to the seven subspecies of *Geothlypis trichas* in North America (Fig. 32.10). We thus assume that the modern human, *Homo sapiens,* represents the outcome of evolutionary trends that occurred in all *Homo erectus* groups wherever they were. If this is indeed the case, it explains the continuum that archeologists find between the features of *Homo erectus* and those of *Homo sapiens* in Africa, in Asia, and in Europe.

How many races of humans did the isolation of gene pools create? There is no simple answer to

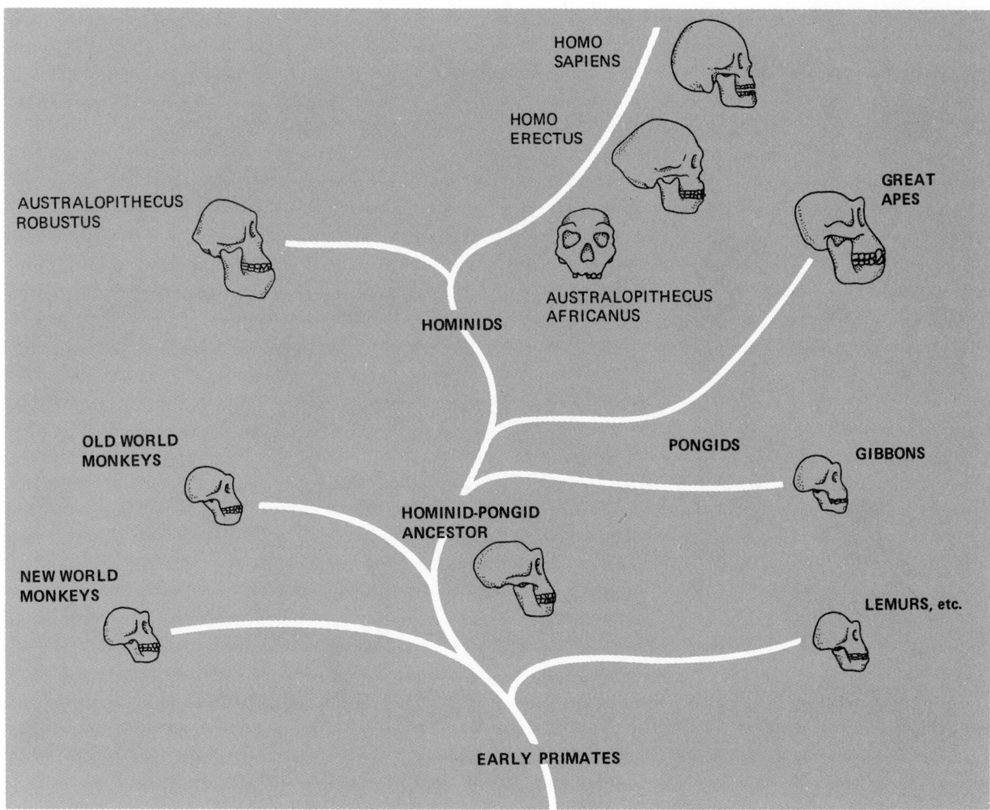

FIG. 38.19 The evolution of the primates.

this question. The number picked is simply a reflection of how impressed the investigator is by measurable differences between the gene pools of various populations. Some workers list five races. These are (1) Negroid (the subpopulations that arose in Africa), (2) Mongoloid (those in Asia), (3) Caucasoid (those in the near East and, later, Europe, (4) American Indian (those *Homo sapiens* that arrived in the Western hemisphere via the land "bridge" which has periodically connected Asia and North America in the Bering Strait region), (5) Australoid (a subpopulation isolated in Australia).

But even if one simply studies gene pools as reflected in the distribution of such blood groups as ABO, it is apparent that there is considerable genetic diversity *within* the groups listed above. The European group can be subdivided by a number of criteria. Figure 38.20 shows the cline in the prevalence of the allele for blood group B as one proceeds across Europe from the northeast to the southwest. Distinct human subpopulations are also found in Africa. The inhabitants of India are genetically different from those in Asia. Distinct gene frequency differences occur between the Indians of South America, of North America, and the Eskimos. What we are seeing, then, is the outcome of a time when the human population existed as a substantial number of relatively isolated subpopulations. Natural selection, and perhaps drift, working on these isolated gene pools gave rise to the racial differences we observe today.

The formation of geographically isolated subspecies is a first step toward speciation. Did *Homo erectus* and, in turn, *Homo sapiens* continue along the path? Clearly, the answer is no. Humans have always been among the most mobile of animals. Various human populations have migrated repeatedly throughout virtually the entire world. Whenever one such population has met another, some interbreeding has occurred and thus some fusing of their respective

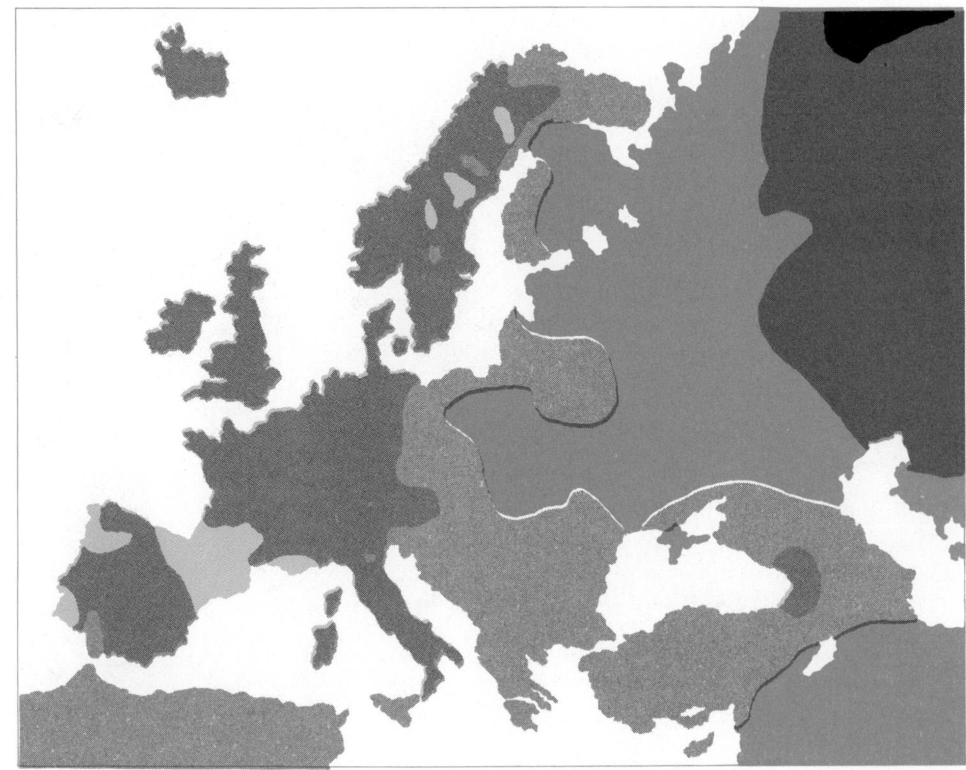

BLOOD-GROUP B GENE
PERCENTAGE

25–30
20–25
15–20
10–15
5–10
0–5

FIG. 38.20 Gene frequencies for the blood-group B gene in the populations of Europe and parts of Asia. Note the gradual reduction in the frequency of this gene as one proceeds from northeast to southwest. Such a continuous gradation in gene frequency is called a *cline*. The frequency of the blood-group A gene exhibits a cline in the opposite direction. [Redrawn from Mourant *et al.* (1958).]

gene pools. To cite one example, in the Americas, the migration of genes from the European-derived gene pool into the African-derived gene pool has given rise to a gene pool substantially different from either.

While cultural barriers to breeding between gene pools have often existed, there is no evidence that the early steps of human subspeciation ever proceeded to a point where any genetic barriers to interbreeding had become established. *Homo sapiens* is today a single species, albeit of considerable genetic diversity, of worldwide distribution. The partial fusing of gene pools that has occurred repeatedly in the past appears to be accelerating as geographical barriers to migration break down and, some time later, cultural barriers to gene migration fall as well. Barring a worldwide holocaust that might create small, isolated populations once again, the outlook for the future is for an accelerating process of gene migration through the various human gene pools leading ultimately to a worldwide randomly breeding population. If and as this prediction is fulfilled, we may hope that many of the social barriers that have set one human in conflict with another will disappear as well.

38.11 SUMMARY

In tracing the evolutionary history of the animals, it is tempting to see in the story a guiding hand working out the destiny that ultimately produced humans from the first flatworms. According to such a view, lungs, limbs, and the shelled egg evolved so that animals could colonize the land, and grasping hands arose so humans could become tool users. Such a view of evolution is called teleological. It interprets the process of evolution as one of working toward a predetermined goal. There is no evidence in the fossil record to support such a view. On the contrary, we see that the evolution of lungs and limbs was the result of selection pressure acting to enable the crossopterygians to remain in the water. The shelled egg probably evolved because the aquatic, early reptiles that produced it could hide it on land out of sight of hungry aquatic predators. The grasping hand evolved in response to the needs of life in the trees. But note that each of these adaptations, which arose in response to certain selection pressures, then provided a plan that enabled a new environment to be colonized. The amphibians found the earth filled with plants and insects. So did the more efficient reptiles, and the amphibians began a decline which has continued to this day. Two-thirds of the history of the mammals was spent in obscurity. With the widespread extinction of reptiles, however, the land was left open for the mammals to begin their period of dominance which still continues. Could the reptiles ever take over again? Certainly not until or unless the now-dominant mammals disappear in large numbers from the face of the earth. The fossil record is filled with examples, both major and minor, of animal groups which arise, flourish for a time, finally become extinct and, in so doing, pave the way for an adaptive radiation of some other animal group. It is a story of adaptations, evolving to meet some immediate need, which by chance alone turn out to open new opportunities, new, less crowded ways of life to their owners.

EXERCISES AND PROBLEMS

1. In what ways does the evolutionary history of the amphibians parallel that of the mosses? In what ways does it differ from it?
2. Summarize the specific structural and physiological modifications that make birds so successful in their way of life.
3. Summarize the adaptations which have made reptiles more successful at colonizing the land than the amphibians have been.
4. "Java man" was once classified as *Pithecanthropus erectus*. Now most anthropologists classify this creature as *Homo erectus*. What change in their thinking does this reflect?
5. Would you consider humans the most specialized of the mammals? Defend your view.
6. Should caecilians be considered primitive amphibians? Defend your view.
7. What features are unique to the mammals?
8. What trait is unique to birds?
9. How does a trout differ from a shark? In what ways are they similar?

REFERENCES

1. ROMER, A. S., *The Vertebrate Story,* University of Chicago Press, Chicago, 1959. Traces the evolution of the vertebrates from their suspected origins to the modern forms, including humans, that inhabit the earth today. A wealth of information on the fossil links between the various groups is also in-included.
2. SMITH, H. W., *From Fish to Philosopher,* Doubleday, New York, 1961. Tells the story of vertebrate evolution with special emphasis on the structural and physiological changes in the kidneys.
3. COLBERT, E. H., *Evolution of the Vertebrates,* Science Editions, New York, 1961. A history of vertebrate evolution in a paperback edition.
4. BAKKER, R. T., "Dinosaur Renaissance," *Scientific American,* Offprint No. 916, April, 1975. Presents evidence that some thecodonts and all their dinosaur descendants were homeothermic. The author believes that birds are the surviving descendants of one order of these "hot-blooded" dinosaurs.
5. DE BEER, SIR GAVIN, *The Evolution of Flying and Flightless Birds,* Oxford Biology Readers, No. 68, Oxford University Press, Oxford, 1975. Includes a detailed examination of the place of *Archeopteryx* in the evolution of birds.
6. NAPIER, J. R., *Primates and Their Adaptations,* Oxford Biology Readers, No. 28, Oxford University Press, Oxford, 1972.
7. HOWELLS, W., *Evolution of the Genus Homo,* Addison-Wesley, Reading, Mass., 1973. A succinct paperback account of evolution from the early hominids to modern man.
8. HALLAM, A., "Continental Drift and the Fossil Record," *Scientific American,* Offprint No. 903, November, 1972.
9. DIETZ, R. S., and J. C. HOLDEN, "The Breakup of Pangaea," *Scientific American,* Offprint No. 892, October, 1970.

The biosphere consists of populations of living organisms interacting with each other as well as with the nonliving features of their environment. (Courtesy of American Airlines.)

PART VII

ECOLOGY: THE BIOLOGY OF POPULATIONS AND THEIR ENVIRONMENT

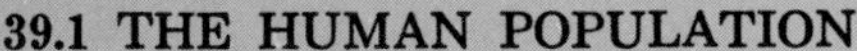

39 THE GROWTH OF POPULATIONS

39.1 THE HUMAN POPULATION

Figure 39.1 is a graph on which has been plotted estimates of the size of the world's human population over the last two millennia. Unlike many of the illustrations used in this text, this one has had to be redrawn for each new edition. Each time the change has consisted of extending the solid part of the line into the dotted portion. This is because the projected population estimates in one edition have become realized by the time of the next edition.

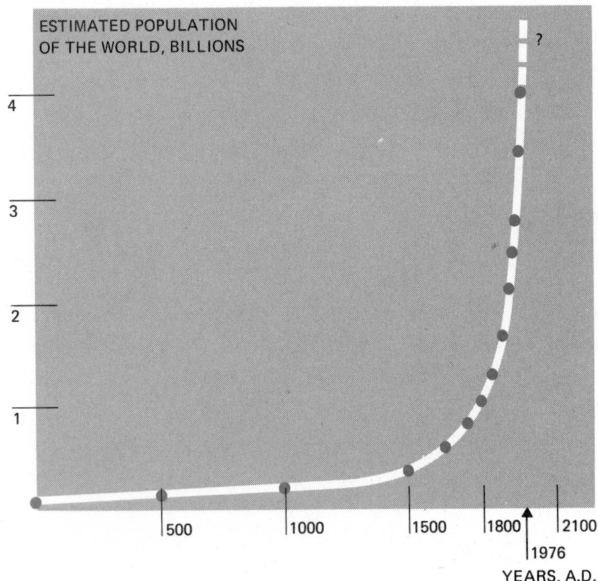

FIG. 39.1 Growth of the human population. The estimates since 1800 are based on more accurate data than those before.

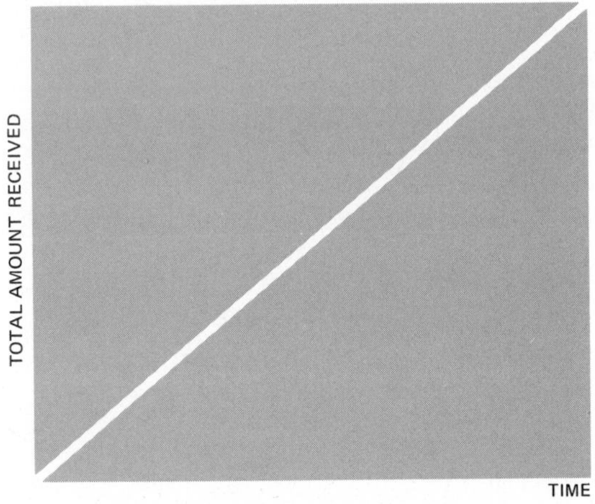

FIG. 39.2 Linear growth. If you lent someone \$100 at 6% interest per year, a graph of your income would look like this. Populations seldom grow in this fashion.

As you can see, the size of the world's human population is increasing very rapidly at the present time. But the shape of the graph has more to tell us than that. If over a given interval of time, the world always gained a given number of new bodies, the curve would have been a straight line with a rising slope (Fig. 39.2). The steepness of the slope would depend on how large the increments in population were. The shape of the curve in Fig. 39.1 tells us that the *incremental gains* in population have been getting larger with each passing interval of time. To examine some actual figures, let us look at the recent population history of Mexico.

In 1973, the population of Mexico was estimated to be 56.2 million. In 1974 the number had risen to 57.3 million, an increase of 1.1 million during the year. In 1974, the value was 59.2 million, a larger increase (1.9 million) than in the previous year. The 1976 estimate was 62.3 million, an increase of 3.1 million over the previous year. Thus the size of the population has not simply been increasing from year to year, but the size of the annual increase has itself been increasing.

The rapid and accelerating growth of the world's population has become of enormous concern to every thoughtful person. How long will the present trend continue? Indefinitely? Surely not. But even if it continues only for 38 more years, the world's population will have doubled from 4 to 8 billion people. How large a population can the resources of the earth support? How large a population can live *decently* on this planet? We simply do not know. Some demographers (students of population) claim that we have already exceeded the number. Others say that the earth can hold many billions more. In fact, there are too many unknown and unpredictable factors to be able to give a reliable estimate. What we can do, however, is to examine some of the principles of population dynamics and, in so doing, perhaps gain a better appreciation of what happened to bring about this situation and what the future may hold.

39.2 PRINCIPLES OF POPULATION GROWTH

To help you grasp what is involved in population growth, think about what is happening in your town. Imagine that you have chosen to keep track of all the birth announcements in your local paper for one year. At the end of the year, you divide the total number of births by the number of thousands of people in your town. The result is the *birth rate* in number of births per 1000 per year. If your town is representative of the United States average, the figure will be 15. Now you do the same with a year's accumulation of death notices. The number of deaths per 1000 population at the end of the year is the annual *death rate* for that year. The present United States average is 9. So, what happened to your town's population during the year? It grew by the difference between the birth rate and the death rate. For the country as a whole, this difference is 6 per thousand. Expressed as a decimal, $6/1000 = 0.06$ per year. This value is called the rate of natural increase (r).

$$\text{birth rate } (b) - \text{death rate } (d) = \text{rate of natural increase } (r)$$

But surely during the year some people moved into town and others moved away. If there was a *net* migration, that has to be figured in as well. For the United States as a whole, net immigration produced an actual rate of increase in 1976 of 0.08, two percentage points higher than r, the rate of natural increase.

In order to simplify our analysis, let us ignore the complications of migration. Certainly as far as the whole planet is concerned, migration is irrelevant. (This, despite the optimistic illusions of some who think that we might start colonizing other planets; if we do, I certainly hope it won't be Mars!)

Now, if each year your town simply added a constant number of people, the population would grow, but a plot of its growth would resemble Fig. 39.2. That would be growth by "simple interest." But populations do not grow that way. As I have emphasized earlier, the product of growth grows itself. So the growth of populations is a problem in *compound* interest.

$$\text{next year's population} = \text{current population} + r \times \text{current population}$$

At the end of each year (or whatever period you wish to use), the base against which the rate is applied has *grown*. Try it yourself. Whatever figures you pick, a plot of population as time elapses will produce a curve like that in Fig. 39.3.

Thus we can predict population growth in much the same way that we can predict the ever-more-rapidly growing balance in a savings account when

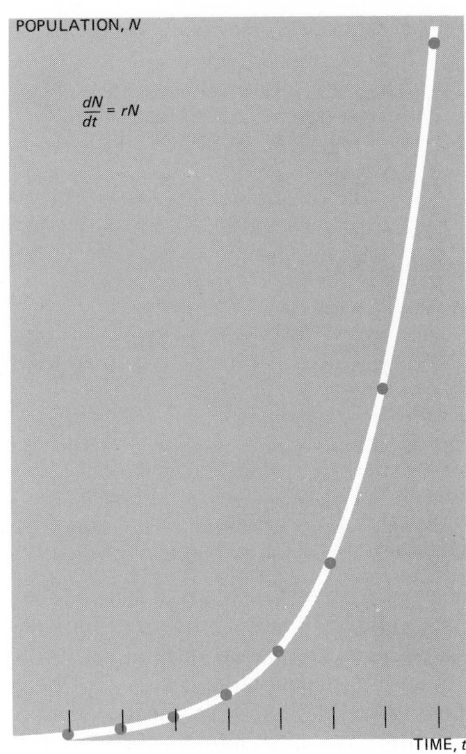

FIG. 39.3 Exponential growth. With a positive value—no matter how small—for r, the growth of N will accelerate with the passage of time.

the interest is allowed to accumulate. There are, however, improvements we can make in the procedure. Savings banks *declare* interest only at certain stated intervals, e.g. once a month. If your bank is offering an annual rate of 6%, the interest in the balance in your account is calculated at the end of each month by multiplying the balance by 0.005 ($0.06 \div 12$). This procedure gives you a slightly better year-end balance than one annual calculation would (Fig. 39.4). Perhaps your savings bank has even begun offering "daily compounding" and proudly notes that thus they give an "effective rate" of interest that is slightly higher than the annual rate. But the repetitive calculations (today's amount = yesterday's amount + $0.06/365 \times$ yesterday's amount) are tedious to say the least. Furthermore, populations do not grow in annual or monthly or even daily spurts, but all the time. If you have studied calculus, you know that it provides a tool for solving such problems. We express our compound interest problem or our population growth

GROWTH OF $100 COMPOUNDED AT 6%

	COMPOUNDED ANNUALLY	COMPOUNDED MONTHLY	COMPOUNDED CONTINUOUSLY*
END OF FIRST YEAR	106.00	106.17	106.18
END OF SECOND YEAR	112.36	112.72	112.75
END OF THIRD YEAR	119.10	119.67	119.72
END OF FOURTH YEAR	126.25	127.05	127.12
END OF FIFTH YEAR	133.82	134.89	134.99

*Using equation $N = N_0 e^{rt}$

FIGURE 39.4

problem by the equation

$$\text{rate of growth in the population at any instant} \quad \left(\frac{dN}{dt}\right) = rN,$$

where r is the rate of natural increase in some stated interval of time (t) and N is the number of individuals in the population at a given instant. A useful form of this differential equation is $N = N_0 e^{rt}$ where N_0 is the starting population, N is the population after a certain time (t) has elapsed, and e is the constant 2.71828 . . . (the base of natural logarithms). In this form, we have an equation that quickly (especially if you have access to a hand calculator) enables us to *predict* population growth (and the growth of your savings account—Fig. 39.4). Plotting the results gives us a curve like that in Fig. 39.3. This curve is known as an **exponential growth curve** because it reflects the growth of a number raised to an exponent (rt).

Now you may feel that we have not made your savings account grow very rapidly. But instead of 100 dollars, let us take 100 people and allow them to grow at the same rate of compound interest ($r = 0.06$/year) for 400 years. (This is a value for r that is substantially higher than the present value for the world, $r = 0.018$, but not impossible to achieve. In 1971, Kuwait came within a hair's breadth of doing so with an r of 0.059.) At the end of the first year, there would only be 106 people. At the end of 10 years there would still be only 182. But after 100 years, the population would have risen to over 40,000. At the end of the next century, the population would exceed 16 million. A century later it would have grown to over six billion. Finally, with 400 years elapsed, the population would be over two trillion (2.6×10^{12}). That is what exponential growth can do for you.

In our example, it took a few hundred years to achieve really impressive numbers of humans, but that is because we are relatively slow breeders. *E. coli,* doubling its weight and dividing every 20 minutes, could blanket the earth in a few days if nothing stopped it. So whether fast or slow, all living things have the theoretical capacity to increase their population to the absolute limits of their environment.

But do populations really keep on growing exponentially like that? The answer is in two parts: (a) yes, sometimes, and right now is one of those times for the human population, as you can see in Fig. 39.1, but (b) only for a period, never indefinitely. What happens, then, to put a stop to exponential population growth?

39.3 RISING DEATH RATES

One common situation encountered in studying populations is shown in Fig. 39.5. This graph shows the rise and fall of the deer population from 1907 to 1939 in the Grand Canyon National Game Preserve in Arizona. Prior to 1907, the deer population had been relatively stable. But then it began to grow exponentially. There is still some debate about the precipitating factors, but it is probably significant that during this period a program of competitor and predator control was in effect. When the preserve was created, thousands of sheep and cattle—both of which compete with deer for food—were removed from it. From 1907 to 1939, 30 wolves, 816 cougars, 863 bobcats, and 7388 coyotes were killed. By 1918, the deer population had seriously overgrazed the range. By 1924, there was not enough food left to support the population through the winter, and during that winter and the next, thousands of deer died of starvation. Although killing of

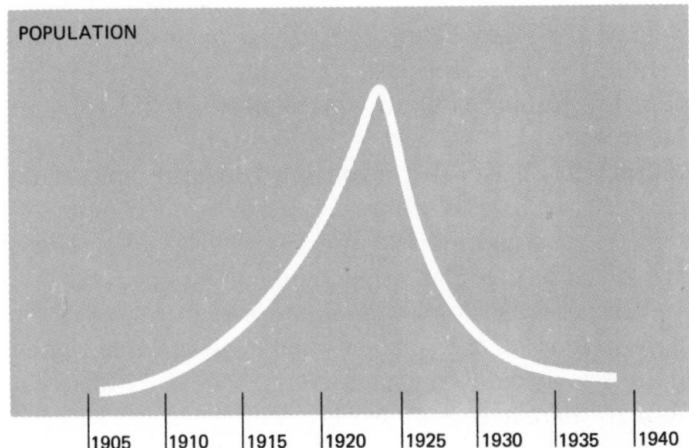

FIG. 39.5 Changes in the deer population of the Grand Canyon National Game Preserve. Although estimates of the absolute number of deer in the game preserve at the peak of the population explosion vary from 30 to 100 thousand, most observers agree that the population went through the type of boom and bust cycle illustrated by this curve.

predators continued until 1939, the deer population did not rebound from its "crash." Perhaps overgrazing had caused such deterioration of the range that shortage of food became the chief limiting factor on population growth.

Many other populations, particularly rodent populations, regularly go through such periods of boom and bust. The periodic lemming migrations, which mark the end of a period of exponential growth, are a famous example.

A second possible outcome of exponential growth is shown in Fig. 39.6. This curve shows the growth of a population of yeast cells living in a flask in the laboratory. After a period of exponential growth, the size of the population begins to level off and soon reaches a stable value.

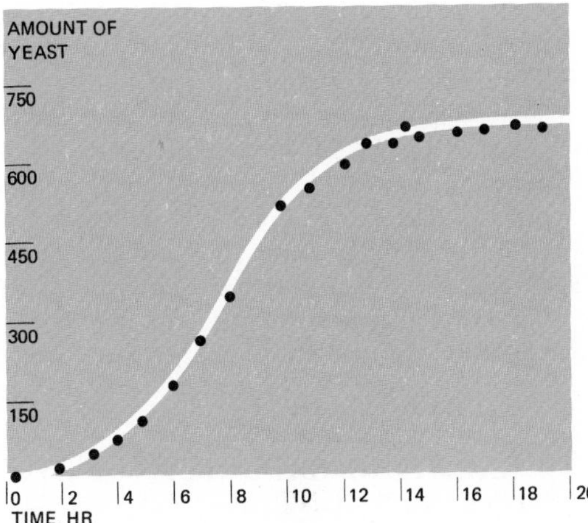

FIG. 39.6 The growth of a yeast population maintained in culture. This type of growth curve is often described as sigmoid or S-shaped.

What has happened? We can gain a clue by adding fresh culture medium to the flask. When we do so, exponential growth resumes until a new, higher, plateau is reached. Evidently, the growth rate of this population declines as the density of the population approaches a certain critical value. Perhaps the death rate increases; perhaps the birth rate decreases; perhaps both. In either or both cases, r declines. When $r = 0$, $dN/dt = 0$ and the population ceases to grow. The yeasts have achieved *zero population growth* ("ZPG").

Why might the death rate rise in a crowded yeast population? One clear possibility is starvation as the sugar in the medium begins to be used up. Another is that the yeast cells begin to poison one another by their metabolic wastes. This is, in fact, the case here. Their chief metabolic waste is ethanol and when its concentration reaches 12–14%, the yeasts succumb (which explains the maximum alcohol content of natural alcoholic beverages like wine). So here are two effects—competition for food and fouling of their environment—that lead to an increase in death rate. Such effects on populations are known as "density-dependent" because they are the result of a change (in this case, an increase) in population density.

The efforts of predators and parasites to satisfy their own needs constitute important density-dependent controls on many populations. As the prey population increases, predators are able to harvest it more easily. Parasites are able to pass from individual to individual more easily as the population density of their host increases. It is no accident that throughout most of human history, the population of the world's cities has been maintained only through continual immigration from the countryside. Not until the institution of community sanitation, immunization, and other public health measures did cities become able to avoid repeated sharp drops in population as a result of epidemics. (Remember what happened to the population of Siena during the "great dying" of the fourteenth century—see Section 34.8). Epidemics, whether in human or other populations, are a dramatic demonstration of a density-dependent check on population size achieved through increased death rates.

39.4 DECLINING BIRTH RATES

The size of r depends upon birth rate as well as death rate. The production of *new* yeast cells probably declines as the culture becomes crowded and the food supply runs low. Thus declining birth rate joins with increasing death rate to produce zero population growth. But is the only way to check the reproductive potential of a species to limit food and other essential resources? Probably not.

We know that humans make deliberate family planning choices, but does analogous behavior occur in other animal species? Perhaps so. Under laboratory conditions, one species of flatworm has been shown to produce fewer young as the population of adults in its container increases. This phenomenon, which may be caused by an inhibitory chemical released in the water, occurs even when ample food, oxygen, and water are available. Fruit flies living under crowded conditions lay fewer eggs. Laboratory rats living in a confined area soon reach a stable population size even though abundant food is available. This comes about through a sharp rise in infant mortality. Reduced maternal care and even cannibalism take a heavy toll of the newborn.

In Chapter 21 we reviewed a number of the methods by which humans, through the application of technological advances, are able to limit family size. In affluent societies any other deliberate attempts at keeping the population in check are generally avoided. It has not always been so. Reduced maternal care—sometimes inadvertent, sometimes deliberate—and even infanticide have at certain times and in certain places been important factors in influencing population growth. (See Langer's article, "Checks on Population Growth: 1750–1850," cited at the end of this chapter.)

An alternative to limiting the number of offspring per parent is to limit the number of parents. Some mammals and birds achieve this by establishing breeding territories. Each mating pair occupies an area of a size sufficient to supply all its needs including those of its offspring (see the discussion of scrub jays in Section 32.7). One or both members defend this area against intrusion from other members of the *same* species. Thus they not only ensure that the resources upon which they depend will not be exceeded but they may also keep the general population in check by preventing breeding among its surplus members. Celibacy, an analogous phenomenon, is more common in human societies when times are difficult (as Langer documents in his article).

In general, closely regulated societies keep their populations in check by reducing birth rates rather than by succumbing to increased death rates. Although there are many worker females in the beehive to accomplish the work of the hive, only the queen has functional reproductive organs. Her rate of egg-laying is adjusted to the overall needs of the hive. Whenever bad weather or poor flowering reduce the amount of food being brought into the hive, she lays fewer eggs. In the late summer, she ceases her egg-laying entirely and thus avoids the necessity of using the winter stores of food for raising young.

Social conventions among humans also have a marked influence on birth rates. These include attitudes concerning the proper age of marriage and the most popular family size. Unfortunately, social conventions—and the technological birth control techniques which may supplement them—are most effective in population control among just those people least in need of it. In the poorer countries of the world, early marriage, a desire for large families, and inability to employ birth control devices successfully are the rule.

39.5 MIGRATION

In our initial discussion of population dynamics, we agreed to ignore the effects of migration on population size. But migration is often a major density-dependent factor in reducing population size. As population levels rise, many of its members emigrate. The migration of the lemmings has already been mentioned. Many insects respond similarly. The swarms of desert locusts for example (see Fig. 39.7), are a response to high population densities.

39.6 THE CARRYING CAPACITY OF THE ENVIRONMENT

Our exponential growth equation and the curve it yields have been only partially successful at describing population growth. Let us try to improve on them.

As our culture of yeast became densely popu-

FIG. 39.7 A locust (grasshopper) plague in Algeria. (Photo by Jean Manuel.)

lated, the rate of population growth declined and finally reached zero. We attributed this to inadequate food and accumulation of poisonous wastes. Put another way, the particular conditions in our culture represented an environment that could not support a population greater than a limiting value. We shall call this value K, the **carrying capacity** of the environment.

When the population is far below K, its growth is exponential, but as the population approaches K, it begins to bump up against ever-stronger "environmental resistance." Let us use the expression

$$\frac{K-N}{K}$$

as a "growth realization factor," that is, a factor representing the degree to which the population can actually realize its maximum possible rate of increase. Introducing this factor into our original (exponential) growth equation, we get:

$$\frac{dN}{dt} = rN\left(\frac{K-N}{K}\right).$$

What does the equation tell us? If the size of a given population (N) is far below the carrying capacity of the environment (K), the growth realization factor will be close to 1 and the population will show exponential growth. However, as N begins to approach K, our growth realization factor approaches zero and the rate of population growth drops to zero.

$$\frac{dN}{dt} = 0 = \text{“ZPG”}$$

Plotting the growth of a population from an initial growth realization factor of 1 to a final growth realization factor of 0 yields a curve shaped like those in Figs. 39.6 and 39.8. This curve is called the **logistic growth curve,** or, often, the S-shaped curve of growth.

The growth curve of the deer population in the Grand Canyon National Game Preserve was clearly not logistic (see Fig. 39.5). The population rose well above the carrying capacity of the environment. But what does our equation of logistic growth tell us would happen if N should—as it did here—exceed K? The growth realization factor would become *negative* and so the population growth rate would become negative as well. And it certainly did.

We have used K to represent the ability of the environment to support a population. Many factors might contribute to K. Availability of food is certainly one of the most important, but other essential ingredients, e.g. availability of nesting sites, might be involved as well. The value of K is also determined by interspecific competition, that is, the presence of other species competing for the same resources.

As the population density of a species increases, the K of the predators and the parasites of that

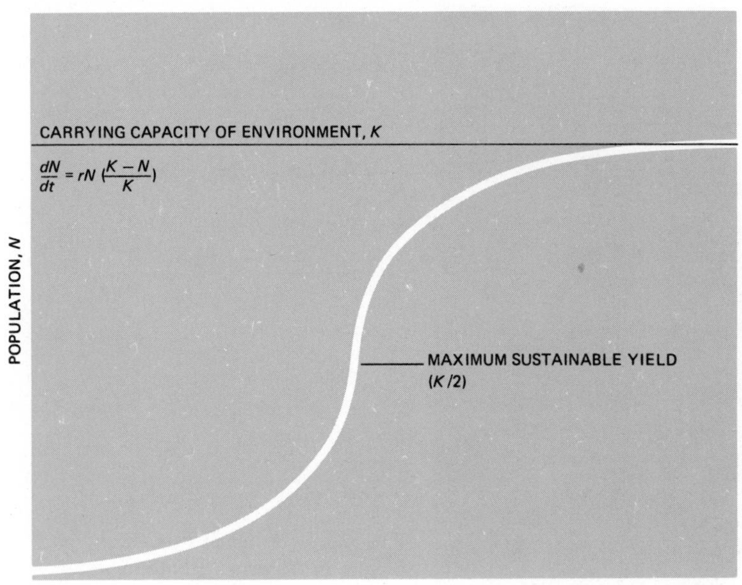

FIG. 39.8 The logistic curve of growth. Compare this curve with the growth curve of the yeast population shown in Fig. 39.6. A population can be harvested most vigorously when it is growing most rapidly. This occurs with a population size equal to $K/2$.

FIG. 39.9 Cyclical fluctuations of the varying hare and lynx populations in the Hudson Bay region from 1850 to 1910. The size of the lynx population was closely dependent upon the size of the population of its principal prey, the hare. The factors causing the hare population to go through these boom and bust cycles are still poorly understood, but variation in the degree of predation by lynxes was probably *not* a major factor.

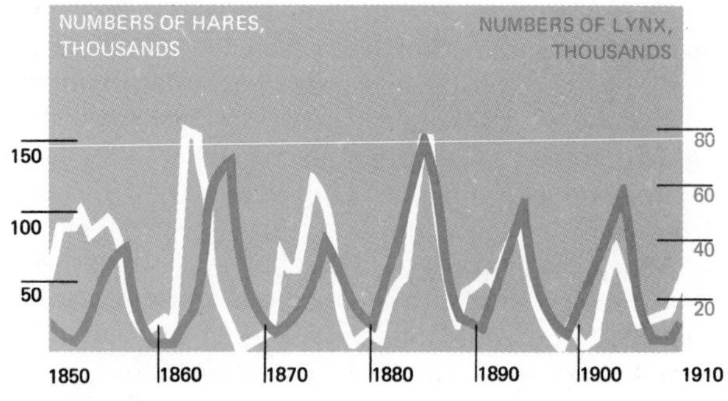

population increases as well. The stage is thus set for greatly increased predation and/or parasitism of the prey species. The outcome of this may simply be a cropping of surplus individuals in the prey population. More likely, though, increased efficiency of predation (and/or parasitism) will drive the prey population back below its K. But with a now-reduced *prey* population, the K of the *predator* population has been reduced, and their population must decline accordingly. The result is a cyclical fluctuation in *both* populations. There is apt to be some lag in these shifts so that the populations of predators and prey not only fluctuate, but the periods of maximum population of the two species do not always coincide. Figure 39.9 illustrates the cyclical relationship between the populations (measured by counting the hides offered for sale at the Hudson Bay trading posts) of the varying hare ("snowshoe rabbit") and its chief predator, the lynx, in the region around Hudson Bay in Canada.

While actual populations often deviate from our theoretical logistic growth curve, the curve can provide us with a number of valuable insights as to the best ways of managing other species.

Say, for example, you are overrun by some pest such as house rats. The logistic curve tells us that you will never solve your problem by setting rat traps. No matter how many you put out, the r for rats is so high (perhaps 0.0147 per *day*) that they will be reproducing much faster than you can catch them. No, the thing to do is to prevent them from gaining access to food in and around your house. With a sharply reduced K, their population must decline.

The converse of the pest problem is the problem of how to keep endangered species from becoming extinct. The solution that usually first comes to mind is to outlaw hunting of the species. But barring hunting will have no appreciable impact if the habitat upon which that species depends for its K—saltmarsh, or woods, or whatever—disappears under the parking lot of a shopping plaza.

In recent years, the introduction of more efficient fishing methods in the world's ocean fisheries has been followed by ominous declines in the catch of many species. It appears that harvesting has been so intense that the populations have been unable to maintain themselves. The logistic growth curve provides a goal to strive for in the managing of fisheries: to harvest at only such a rate that the fish population is maintained at $K/2$. Why? Because it is at this population size, that the population is growing most rapidly (Fig. 39.8). The value $K/2$ is known as the **maximum sustainable yield.**

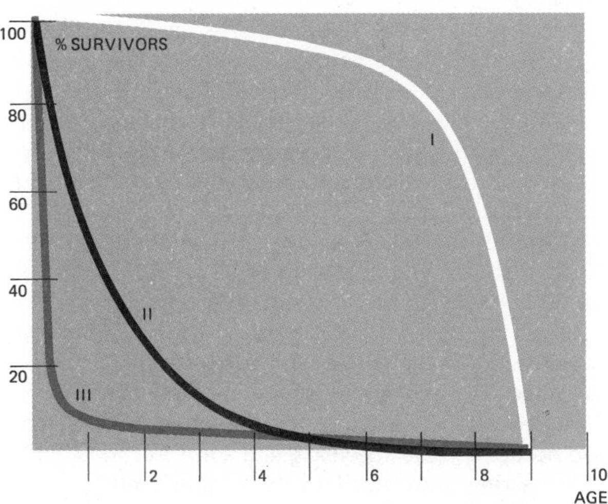

FIG. 39.10 Theoretical survivorship curves. The vertical axis represents the fraction of survivors at each age (horizontal axis). The Type I curve is characteristic of organisms that age but show little random mortality before then. The chance of death is the same at all ages for organisms with a Type II curve. K strategists usually have survivorship curves somewhere between Type I and Type II. The Type III curve is typical of organisms that produce huge numbers of offspring accompanied by high rates of infant mortality. Many r strategists have such a survivorship curve.

39.7 r STRATEGISTS AND K STRATEGISTS

A number of years ago, I plowed up a section of an old field and allowed it to lie fallow. In the first season it grew a luxuriant crop of ragweed. Ragweed is a plant that is beautifully adapted to exploiting its environment in a hurry, before competitors can become established. It grows rapidly and produces an extraordinary abundance of seeds (after producing its enormous output of wind-blown pollen, the bane of many hayfever sufferers!). Because ragweed's approach to evolutionary success is through rapid reproduction, i.e. through a high value of r, it is called an r strategist. Other weeds, many insects, and many rodents are also r strategists. In fact, if we consider an organism a pest, it is probably an r strategist.

In general, r strategists share a number of qualities:

1. They are usually to be found in disturbed and/or transitory habitats. In the second season of my field, perennial grasses and wildflowers had produced a dense carpet of mixed vegetation and there was not a ragweed plant to be found.
2. They have short life spans. The house mouse, with a maximum life span of three years, is an r strategist.
3. They usually have short generation times, that is, they have short gestation periods and are soon ready to produce another crop of young. The housefly can produce seven generations each year (each of about 120 young).
4. They produce large numbers of offspring. The American oyster, releasing a million eggs in one season, is an r strategist. Most of its offspring will die, but the sheer size of its output increases the probability that some offspring will disperse successfully to new suitable habitats.
5. They take very little care of their offspring. Thus infant mortality is huge. If we plot a survivorship curve for an r strategist, it takes the form of the curve labeled Type III in Fig. 39.10. The

high reproductive rates in countries like India may well be a response to the high rates of infant mortality (look back at Fig. 23.8).

When a habitat becomes filled with a diverse population of creatures competing with one another for the necessities of life, the r strategist is at a disadvantage. Now the premium is placed on the efficiency with which a species can exploit the resources of its environment. Now conditions favor the K strategist.

Species that are K strategists have stable populations that are very close to K. Therefore there is nothing to be gained by employing a high r. The goal of the species is to be as closely adapted as possible to the habitat in which it lives. This habitat is called the **ecological niche** of the organism. It includes all the features of the environment that are needed by the species, e.g. a source of food, suitable nesting sites, availability of water, etc.

Typically, K strategists share the following qualities.

1. They are usually found in stable habitats. Most of the species in a mature forest will be K strategists.
2. They have long life spans. The oak, the elephant, and the tortoise are all K strategists.
3. They usually have long generation times. It takes nine months to produce a human baby.
4. They produce small numbers of offspring. The Florida scrub jay (see Section 32.7), with its handful of offspring and rigid rule of one breeding pair to a territory, is a K strategist.
5. They take good care of their offspring. Infant mortality tends to be low. If we plot survivorship curves for K strategists they are usually somewhere between a Type I curve (where most of the population dies from old age) and a Type II curve (where all ages are equally at risk of being struck down by random hazards).
6. Evolution in K strategists has produced a close adaptation of the organism to its niche. K strategists typically evolve in such a way that they exploit with ever-increasing efficiency an ever-narrower slice of the environment. The moth with the 25-cm proboscis that can feed only from the orchid with the 25-cm nectary is a K strategist.

39.8 IN CONCLUSION

The next six chapters will be devoted to examining some of the density-dependent factors that act to keep populations in check.

In Chapter 40, we shall examine the flow of energy through populations and see how for each species the availability of energy is a major determinant of the carrying capacity of the environment (K) for that species.

In Chapters 41 and 42, we shall examine the mechanisms by which several of the essential elements of life (e.g. nitrogen, phosphorus) cycle through populations of living things. The availability in its environment of these essential ingredients is another determinant of the K of a species.

In Chapter 43, we shall examine some of the ways in which organisms interact with each other. We shall find that to the extent that organisms *compete* for the same resources in their environment, they act to reduce the carrying capacity, K, of the environment for each other. If one species *preys* upon or *parasitizes* another, that species may lower the population of the prey species below its K. In so doing, it shifts its own K downward. The population size the lynx can hope to achieve is as much limited (perhaps more so) by the varying hare population as the varying hare population is limited by the lynx (Fig. 39.9).

In the final two chapters, we shall return to the human population. We shall examine the roles that parasitism (in Chapter 44) and competition for food (in Chapter 45) have played in placing the human population on the steepest part of what appears today to be an exponential growth curve. Since exponential growth rates cannot continue indefinitely, that steep slope must in time turn out to have been merely a segment in either the logistic curve or the curve that describes a boom-and-bust cycle (Fig. 39.5) of population growth. Let us devoutly hope that it proves to be the former rather than the latter.

EXERCISES AND PROBLEMS

1. The population of Nicaragua in 1977 was estimated to be 2.3 million. The birth rate was 48/1000; the death rate 14/1000. What was r (per year)? What was the approximate annual increase in population? At these rates, what will be the approximate population of Nicaragua in 15 years?
2. If your bank account is earning interest at an annual rate of 3% compounded continuously, how many years will it take to double your money? How many years at 7%?
3. The current birth rate in Sweden is 13/1000; the death rate 11/1000. At these rates, how many years will it take for the population to double?
4. The current birth rate in Libya is estimated to be 48/1000; the death rate 9/1000. If these rates continue, how many years will it take for the population to double?
5. The population of India in 1977 was estimated to be 622.7 million, and r was estimated at 0.021 (per year). If this rate continues, what will be the population of India in the year 2000?
6. What factors regulate the rate of growth of a population of animals started by a single breeding pair introduced into a new territory? What factors limit the ultimate size of this population?

REFERENCES

1. WYNNE-EDWARDS, V. C., "Population Control in Animals," *Scientific American*, Offprint No. 192, August, 1964. The author cites several examples of animals whose populations seem to be controlled by forms of group behavior (such as territoriality) rather than by food supply, predation, or parasitism.
2. LANGER, W. L., "Checks on Population Growth: 1750–1850," *Scientific American*, Offprint No. 674, February, 1972. The author presents evidence that the most important of these were celibacy and infanticide.
3. FREJKA, T., "The Prospects for Stationary World Population," *Scientific American*, Offprint No. 683, March, 1973. On the basis of current trends in population growth and a world population in 1973 of 3.6 billion, the author predicts that the world population will stabilize at some 8.4 billion by the year 2100.
4. EHRLICH, P. R., and ANNE H. EHRLICH, *Population Resources Environment*, 2d ed., Freeman, San Francisco, 1972. A well-written and thoroughly documented study of these crucial issues in human ecology.
5. The American Assembly, *The Population Dilemma*, 2d ed., Prentice-Hall, Englewood Cliffs, N.J., 1969.
6. WILSON, E. O., and W. H. BOSSERT, *A Primer of Population Biology*, Sinauer Associates, Inc., Stamford, Conn., 1971. Chapter 3 includes mathematical models of the growth of populations.
7. COALE, A. J., "The History of the Human Population," *Scientific American*, September, 1974. Reprinted in *The Human Population: A Scientific American Book*, Freeman, San Francisco, 1974. Available in paperback.
8. MYERS, JUDITH H., and C. J. KREBS, "Population Cycles in Rodents," *Scientific American*, Offprint No. 1296, June, 1974. Presents evidence that the boom and bust cycle characteristic of many rodent populations is a consequence of changes in the gene pool caused by emigration.
9. KORMONDY, E. J., *Concepts of Ecology*, 2nd ed., Concepts of Modern Biology Series, Prentice-Hall, Englewood Cliffs, N.J., 1976. Chapter 6 deals with the ecology of populations.

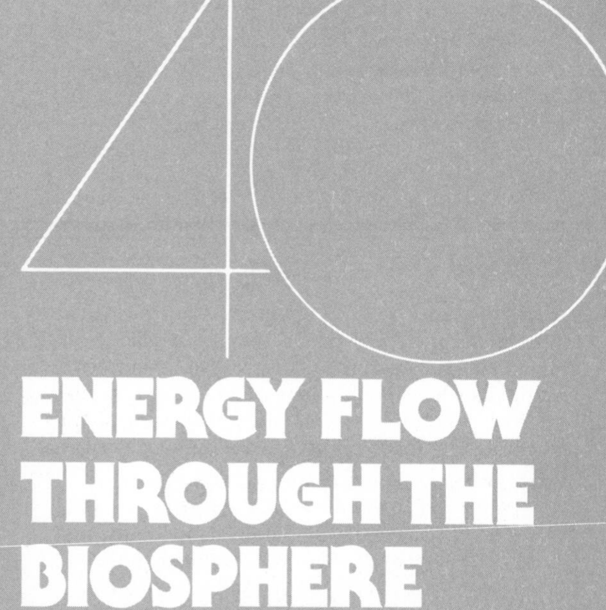

40 ENERGY FLOW THROUGH THE BIOSPHERE

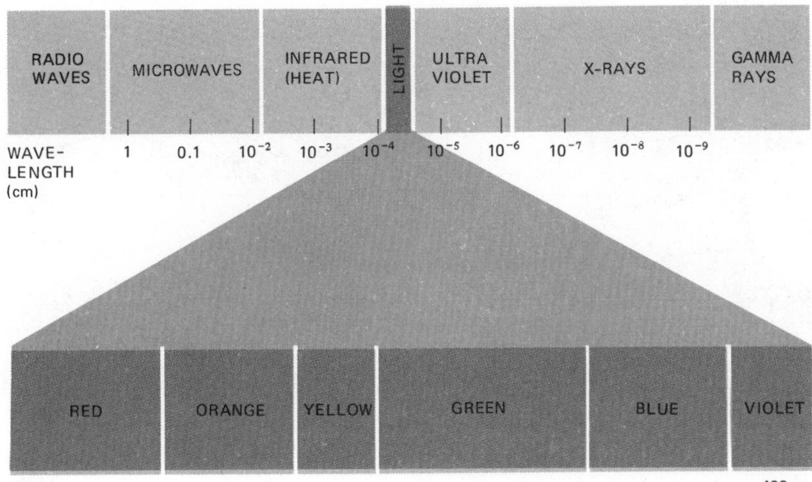

FIG. 40.1 The spectrum of electromagnetic radiation. Approximately one-half of the energy reaching the earth from the sun is light, that is, it is confined to the visible portion of the spectrum. The remainder arrives as heat and a small amount of ultraviolet light.

40.1 THE INPUT OF ENERGY

Earlier in the book, we examined how green plants store the energy of the sun in photosynthesis (in Chapter 9) and how *all* organisms use this stored energy to live (in Chapter 8). We examined these processes as they occur in an individual plant or animal. Now we shall turn our attention to the broader picture of how the sun's energy is trapped and used within entire communities of plants, animals, and microorganisms. Such a community, together with the nonliving features of its environment (e.g. air, water) with which it interacts, we shall call an **ecosystem.** The total of all the world's ecosystems (minus any overlaps between them) constitutes the **biosphere.** Our shift in emphasis from individuals to ecosystems will be like shifting one's attention from the way a single person earns and spends his money to the economics of a whole region or country.

The energy that reaches the earth from the sun constitutes a small region in the spectrum of electromagnetic radiation (Fig. 40.1). It includes radiation with wavelengths between 400 and 700 nm (1 nm = 10^{-9} meter) to which the human eye is sensitive and which we thus call visible light. It also includes a small amount of shorter wavelength ultraviolet light (responsible for tanning our skin) and longer wavelength infrared rays or heat. The intensity of radiation striking the earth varies with latitude and with the season of the year. The axis of the earth is tipped 23½° with respect to the plane in which the earth moves about the sun. For this reason, the northern hemisphere receives more than 12 hours of sunlight during the six months (approximately March 21 to September 23) when the axis of the earth is tipped toward the sun and less than 12 hours during the remaining months when the axis is tipped away (Fig. 40.2). The reverse situation occurs in the southern hemisphere. This phenomenon results in a net gain of solar radiation during one half of the year and a net loss during the other half, and hence is responsible for the seasons. The farther north (or south) from the equator one travels, the more pronounced the cooling of the earth in winter and the longer it takes for warm temperatures to return in the spring.

At extreme latitudes, not only is there great seasonal *variation* in the amount of sunlight received, but the total amount received during the course of the year is less than that in the tropics. This is true even though the total number of hours that the sun is above the horizon is the same. The long days of summer do not compensate for the short days of winter. There are several reasons for this. The more extreme the latitude, the closer the sun is to the horizon, even at midday. Its rays, entering the atmosphere at an angle, must traverse a longer path before they reach the surface of the earth, and increased losses by absorption thus occur. Furthermore, the intensity of the light is reduced because a given amount of light shines on a larger area at extreme latitudes than it does at the equator (Fig. 40.2). The low angle at which the rays strike the earth also results in greater losses by reflection, thus further reducing the amount of energy available for photosynthesis.

Other things being equal (which they often are *not*), we would expect, then, that tropical regions

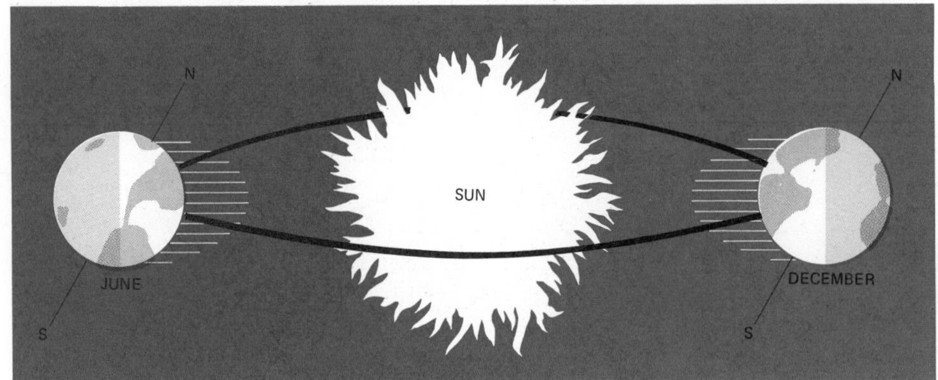

FIG. 40.2 The origin of seasons. Not only are the days shorter during winter but the intensity of the sunlight at the surface of the earth is diminished. The *total* amount of energy received during the year declines with increasing latitude, that is, the long days of summer do not compensate for the short days of winter.

would be more productive than temperate areas. They receive some 8000–10,000 kilocalories (kcal) of energy a day on each square meter of surface throughout the year. A kilocalorie, you remember, is the amount of heat needed to warm 1 kilogram of water 1 degree Celsius (°C). Because all of the light trapped in photosynthesis is ultimately released as heat, it makes sense to follow the flow of energy through the biosphere in units of heat.

Thanks to the long days of summer, temperate regions also receive some 8000–10,000 kcal/m²/day during the growing season. Thus we would expect their daily productivity during the summer months to equal that of the tropics. It may, in fact, be somewhat higher, although the few studies that have been made on the productivity of the tropics have yielded contradictory data.

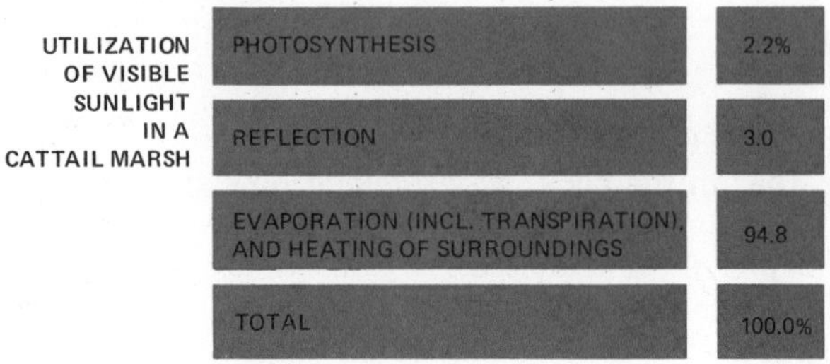

UTILIZATION OF VISIBLE SUNLIGHT IN A CATTAIL MARSH	
PHOTOSYNTHESIS	2.2%
REFLECTION	3.0
EVAPORATION (INCL. TRANSPIRATION), AND HEATING OF SURROUNDINGS	94.8
TOTAL	100.0%

FIG. 40.3 At least one-half of the energy trapped by photosynthesis is later released by the cellular respiration of the plants.

40.2 ECOSYSTEM PRODUCTIVITY

How efficient is photosynthesis in converting the energy of the sun into the chemical energy stored in food molecules? One way to answer this question is to collect and weigh the amount of plant material produced by 1 m² of land over a given interval of time. One gram of dried plant material (stems, leaves, etc.), which is largely carbohydrate, yields about 4.25 kcal of energy when burned (or respired).

A number of studies have been carried out using this approach and they demonstrate that an average temperate-climate forest stores approximately 5000 kcal/m² of energy during the course of a year. This is roughly equivalent to the production of 1 kg of dry plant material for every square meter of forest. Considering the *total* amount of radiation received and correcting for the fraction of that usable in photosynthesis, the efficiency of conversion is approximately 1%. This value would be somewhat higher (Fig. 40.3) if one included the fraction of trapped energy that has to be used by the plant to run its own metabolic activities, that is, the amount consumed by the plant in cellular respiration. In order to arrive at this figure, it is necessary to measure the rate of respiration of the plants being studied. This can be done by enclosing them in a transparent container and measuring the rate at which the CO_2 content of the surrounding air increases during the night because of respiration. Measuring the rate of CO_2 *decrease* during the day (because of a rate of photosynthesis greater than the rate of respiration) enables one to calculate the increase in organic matter, at least in terms of carbohydrate. The amount of trapped energy that plants ultimately lose to respiration varies markedly from type to type, but 50% is about average for forests. Whatever the figure, though, the quantity of greatest concern to us is **net productivity,** that is, the amount of energy *stored* by the plant community over a given period of time.

Desert plant communities store some 500 kcal/

ESTIMATED NET PRODUCTIVITY OF CERTAIN ECOSYSTEMS (IN KILOCALORIES/m²/YEAR)

TEMPERATE DECIDUOUS FOREST	5,000
TROPICAL RAIN FOREST	15,000
TALL-GRASS PRAIRIE	2,000
DESERT	500
COASTAL MARSH	12,000
OCEAN CLOSE TO SHORE	2,500
OPEN OCEAN	800
CLEAR (OLIGOTROPHIC) LAKE	800
LAKE IN ADVANCED STATE OF EUTROPHICATION	2,400
SILVER SPRINGS, FLORIDA	8,800
FIELD OF ALFALFA (A LEGUME)	15,000
CORN FIELD, U.S.	4,500
RICE PADDIES, JAPAN	5,500
SUGAR CANE, HAWAII	25,000

FIG. 40.4 Estimated net productivity of certain natural and artificially created ecosystems. The values are only approximations and are subject to marked fluctuations because of variations in temperature, fertility, and availability of water.

m^2/yr and thus average only 10% of the productivity of temperate forests. Estimates of the productivity of tropical forests vary widely. Some measurements have indicated productivities four to five times that of temperate forests, while others have yielded much lower values.

A great deal has been said in recent years about the potential of the oceans to assist in the feeding of the growing human population. Such measurements of productivity as have been made suggest that we may already be harvesting the oceans at a rate close to their maximum sustainable yield. The oceans are most productive where nutrients are readily available close to the surface. This is true of shallow, coastal waters and of certain areas, for example, off the coast of Peru, where upwelling ocean currents bring nutrients up to a level where sunlight penetrates sufficiently to make photosynthesis possible. Although an estimate of average productivity for the oceans as a whole is very uncertain, it is probably in the range of 500–1000 kcal/m^2/yr, or not much better than the productivity of deserts. Even granting that the oceans cover three-fourths of the earth's surface, they probably account for no more than one-third of its productivity.

Most human food comes directly or indirectly from agriculture. Under optimum conditions, the productivity of certain types of agriculture may be extraordinarily high. Fields of sugarcane may store 25,000 or more kcal/m^2/yr. Planting three crops a year of the new strains of rice that are part of the "green revolution" (Fig. 32.1) can produce similar rates of productivity. However, the edible portions of the plant, the rice grains, store only one-fourth of the net productivity of the plants. As for agriculture in temperate climates, the productivities range from 5000 to 12,000 kcal/m^2/yr, about the same as for mature forests in the same areas (Fig. 40.4). Note again, however, that these values represent the net productivity achieved by the *entire* plant, not just the portion we find edible (which would reduce the value by roughly two-thirds).

Although intensively managed agricultural land is capable of productivities that exceed those of most natural plant communities, the differences are somewhat illusory. The input of energy in natural communities comes almost entirely from the sunlight that shines on them. This is not so for high-yield agriculture. In order to achieve the yields that hybrid strains of corn, rice, and wheat have made possible, there must be not only an input of sunlight but also inputs of fertilizer, irrigation water, pesticides, and the mechanical energy needed to plow and cultivate the soil as well as to harvest the crop. The manufacture of fertilizer not only requires *direct* inputs of energy, but much of our fertilizer is made from petroleum and natural gas. (These raw materials serve as the source of the hydrogen atoms with which atmospheric nitrogen is "fixed" to form such fertilizers as ammonia, urea, and ammonium nitrate.) The manufacture, transportation, and application of pesticides as well as fertilizer takes energy. Soo, too, does the pumping of irrigation water and the running of machinery for plowing, cultivating, and harvesting the crop. The major source of energy for all these purposes is fossil fuel: coal, petroleum, and natural gas. All of these represent the stored productivity of ecosystems existing earlier in the earth's history. Therefore, a realistic figure for the net productivity of agricultural land should take into account the number of kcal/m^2/yr of energy from fuel (fossil sunlight) that were expended in the process.

Such estimates have been made. They reveal that in the United States, for example, there has

FIG. 40.5 A member of the Tsembaga prepares a garden site in the jungle of New Guinea. Burning the brush cleared from the plot releases minerals that will be used by the crops to be planted. Many tree stumps are left alive. They will first be used to support certain crop plants. Later they will restore the garden site to forest. (Courtesy of Dr. Roy A. Rappaport.)

been a steady increase through the years of this century in the number of calories of fuel needed to produce one calorie in food ready to eat. Today, the ratio is about 9:1. This means that for every calorie set on the table before you, nine calories were used to grow, process, transport, market, and cook the food containing that calorie. In a real sense, modern agriculture is a process of converting fossil fuel into food. Quantitatively, sunlight plays only a minor part.

High agricultural yields are possible without large inputs of fossil energy. A small tribe in New Guinea, the Tsembaga, practices intensive agriculture in small, temporary clearings which they cut in the jungle where they live. The only input of energy (aside from sunlight) is that provided by their muscles. No additional energy is needed for fertilizer. This is because the soil is fertilized with the ashes produced by burning the vegetation they cut to make the clearing (Fig. 40.5). As the crops mature, natural vegetation is allowed slowly to grow back until, after one to two years, the field is abandoned to the regrowing forest. Thus, cropping does not continue long enough to seriously deplete the fertility of the soil.

The calorie input needed to clear each plot, plant it, weed it, and harvest the crop can be estimated. The calorie content of the harvest is also easily determined. The ratio of the two turns out to be better than 1:16. In other words, while United States agriculture expends nine calories to gain one calorie of food, Tsembaga agriculture produces sixteen calories of food for every calorie of human input. Nor is Tsembaga agriculture inefficient in terms of yields per acre. The average yield of edible plant parts is 2300 kcal/m^2, which is as good as or better than the yields of edible portions of wheat and corn as grown in the United States. Then why do not developed countries return to growing crops the way the Tsembaga do? Because we are unwilling or unable to make the investment in human labor that they make. Less than 5% of the United States population grows the food for all the rest. Every Tsembagan is a farmer.

40.3 FOOD CHAINS

What happens to the net productivity of a plant community? Some is harvested by plant-eating animals, the herbivores. These include not only animals like deer and cattle but many small herbivores such as insects. Some of the net production of plants is consumed by organisms of decay, chiefly fungi and bacteria. In certain plant communities, a fraction of the net production is stored. In bogs, for example, much of the plant debris escapes decay and accumulates as peat. Such deposits gave rise in the past to coal.

In an immature forest, the total amount of organic matter increases yearly as the woody perennials increase in size. This, too, represents storage. When the forest becomes mature, however, the loss of organic matter by death and decay, when added to the loss by grazing, equals the net productivity. At this point, there is no further increase in biomass from year to year. The term **biomass** (or **standing crop**) is used to describe the total organic matter present in an ecosystem.

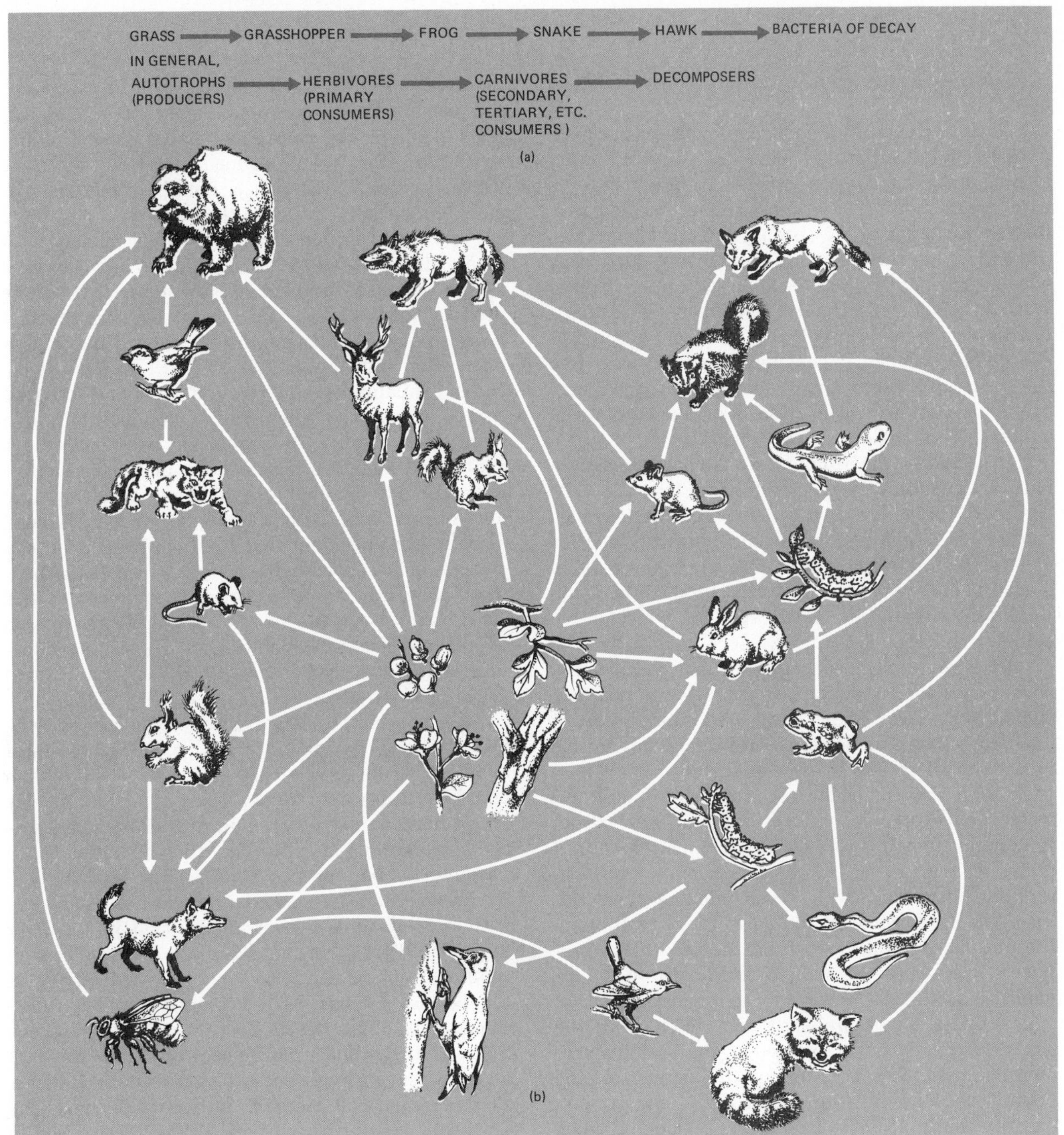

FIG. 40.6 (a) Food chains. (b) A food web in a temperate deciduous forest. (Courtesy of Dr. V. E. Shelford.)

When some of the standing crop of a plant community is eaten, the energy passes to a heterotroph, which depends on it for its very existence. Grasshoppers, for example, grow and carry on all their activities thanks to the energy stored in the plant food they consume. Herbivores, in turn, provide food for carnivorous animals. Grasshoppers, to take our example, may be consumed by toads. The process of transferring energy from creature to creature may continue. The toad may be eaten by a black snake, which may, in turn, be consumed by a hawk.

Such a pathway of food consumption is called a **food chain.** All food chains start with an autotrophic organism, a photosynthesizing organism such as a green plant. These organisms are called **producers** because only they can manufacture food from inorganic raw materials. Any organism, such as a cow or grasshopper, that feeds directly upon a plant is an herbivore or **primary consumer** (Fig. 40.6). Carnivores, like the toad, that feed on herbivores are called **secondary consumers.** Carnivores, like the snake, that eat secondary consumers are **tertiary consumers,** and so on. Each level of consumption in a food chain is called a **trophic level.**

In actually determining what eats what in a natural community, one soon realizes that the various food chains are interconnected. Most animals consume a varied diet and, in turn, provide food for a variety of other creatures that prey upon them. Thus the energy present in the net production of the producers passes into **food webs** of considerable complexity (Fig. 40.6).

At each level of consumption in a food chain, some of the net production of that level is not consumed by the next higher level, but, after the death of the organisms, is decomposed by microorganisms of decay. These are chiefly fungi and certain bacteria, which are found in large numbers in soil and wherever else organic matter is present. They extract the energy remaining in the organic matter and, in doing so, release the inorganic products of its degradation (e.g. CO_2, NH_3) back into the surroundings. We have seen that the flow of energy through the biosphere is one-way: from the sun to the producers, then to the consumers, and finally to the organisms of decay. However, the materials out of which living things are built and which store the sun's energy must be recycled if the system is to continue operating. It is through the activity of the organisms of decay—working at every link in food chains—that much of this recycling is made possible. We shall examine in Chapter 41 some of the special roles that organisms of decay play in the recycling of matter.

40.4 ENERGY FLOW THROUGH FOOD CHAINS

How efficient is the conversion of the net production of one trophic level to the net production of the next? It is certainly far from 100%, although in some organisms, such as carefully bred varieties of poultry, the percentage of conversion may be remarkably high. Broilers can gain about half a pound of live weight for every pound of food ingested. (Since the water content of the two is not the same, the efficiency is actually less than the apparent 50%.)

It is unlikely that efficiencies of conversion much greater than this will ever be possible, because much of the energy consumed by an organism must be used to keep it alive and cannot be stored as net productivity. This energy is transferred by cellular respiration into the energy of ATP. As such it becomes available to (1) drive the metabolic activities that convert some of the ingested food molecules into more chicken, and (2) enable the chicken to move about, and maintain homeostasis, including keeping warm. In fact, the final form of energy resulting from all the metabolic activities of the chicken is heat. Thus it is no accident that the high conversion efficiencies achieved with these birds are possible only when they are reared in relative confinement and warm temperatures. The more physical activity an organism engages in, the smaller will be the percentage of its food used in growth, that is, contributed to net productivity.

A few attempts have been made to measure the flow of energy through food chains. One of the most elaborate of these was carried out by H. T. Odum on a river ecosystem, Silver Springs, in Florida. He found that the net production of the producers was 8833 kcal/m^2/yr (Fig. 40.7). Much of this material (5465 kcal) became plant debris that was decayed by decomposers or carried downstream out of the system. Herbivores consumed 3368 kcal/m^2/yr. Over half (1890 kcal) of this was lost, chiefly through cellular respiration. Thus the net productivity of the herbivores was 1478 kcal/m^2/yr. This represented 17% of the net productivity of the producers. Some of the primary consumers died and their remains decayed there or were transported downstream. Only 383 kcal/m^2/yr were consumed by secondary consumers. Of this, 316 kcal were used in their respiration, leaving only 67 kcal/m^2/yr of net productivity at this trophic level. This is only 4½% of the net productivity of the prior level. Such a low conversion efficiency is characteristic of carnivores, which must expend a high percentage of their energy in seeking prey from which to secure more energy.

Of the net productivity of 67 kcal/m^2/yr at the level of secondary consumers (first carnivores), 46 kcal were subsequently lost by decay and downstream transport. Only 21 kcal/m^2/yr passed to ter-

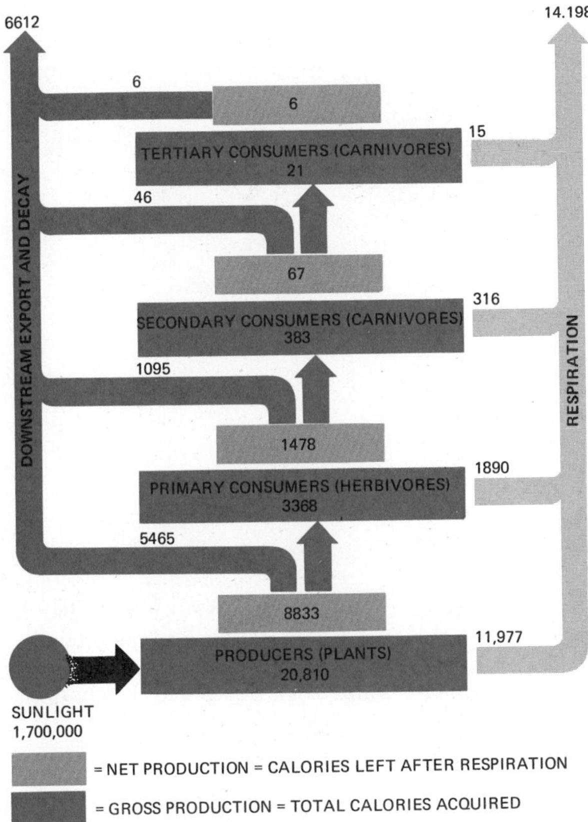

FIG. 40.7 The flow of energy through Silver Springs, Florida. The figures are given in kilocalories per square meter per year (kcal/m²/yr). Note the substantial losses in net production as energy passes from one trophic level to the next. (Based on data obtained by Howard T. Odum.)

tiary consumers. Of this, they used 15 kcal in respiration, and had a net productivity of only 6 kcal/m²/yr (Fig. 40.7). Since there was no higher level of carnivores and no storage of energy (i.e. the biomass of the tertiary consumers did not increase from year to year), the net production was ultimately passed to decay organisms, either in the Silver Springs ecosystem or downstream. In contrast to the bog or young forests mentioned earlier, the Silver Springs ecosystem had no energy storage at any trophic level. To put it another way, neither the total biomass nor the biomass of any trophic level increased from year to year.

In the Silver Springs system, the efficiency of transfer of energy from one trophic level to the next varies from 17 to 4.5%. In fact, from similar studies in other ecosystems, we may use as a rule of thumb an average efficiency of transfer (net production at one level to net production at the next) of about 10%. Whatever the small variations that may occur from case to case, this rough approximation is of enormous significance. It explains, for example, why it costs more to buy a pound of beefsteak than a pound of corn. It took many pounds of the latter to produce one pound of the former. Most of us include both plants and animals in our diet. Whenever we eat meat, however, we are securing our energy third-hand. Its production has entailed substantial losses of energy. Of course, the situation is even worse for carnivores that prey upon other carnivores (e.g. when we eat a trout). Thus, at each link in a food chain a substantial portion of the sun's energy, originally trapped by the photosynthesizing autotroph, is dissipated back to the environment (ultimately as heat). We can conclude, then, that the total amount of energy stored in the bodies of a given population is dependent upon its trophic level. The total amount of energy present in a population of toads must necessarily be far less than that in the insects upon which they prey. The insects, in turn, possess only a fraction of the energy stored in the plants upon which they feed. This decrease in the total available energy at each higher trophic level is sometimes described as a **pyramid of energy.** In the food chain we discussed earlier, the autotrophic grasses provide the base of the pyramid. The carnivorous hawk occupies the narrow apex (Fig. 40.8). Using Odum's data on net productivity at the various trophic levels in Silver Springs, we get the pyramid of energy shown in Fig. 40.9.

Another consequence of the ecological principles we have been discussing is that smaller animals are more numerous than larger ones. Figure 40.10 shows the **pyramid of numbers** resulting when a census of the populations of autotrophs, herbivores, and two trophic levels of carnivores is taken in an acre of grassland. The pyramid arises principally from the fact that each species is limited in its total biomass by its trophic level. If the size of the individuals is small, their numbers can be greater, and vice versa. Predators are *usually* larger than their prey, however, but being at a higher trophic level, their biomass must be smaller. Consequently, the number of individuals in the predator population is distinctly smaller than that in the prey population.

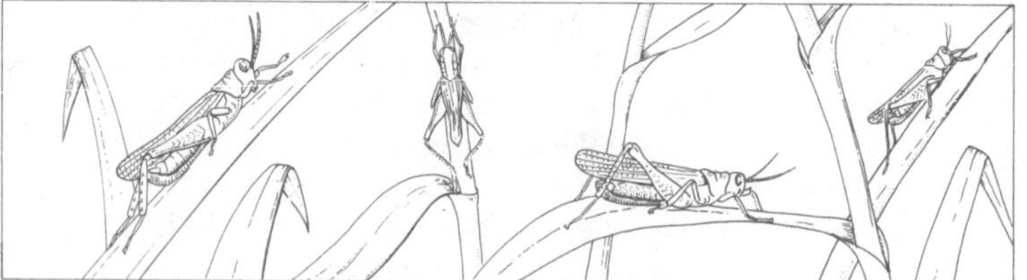

FIG. 40.8 A pyramid of energy. At each link in the food chain, energy that was originally stored by the autotrophic grass plants is dissipated. What other relationships exist in such a food chain?

FIG. 40.9 Pyramid of energy in Silver Springs, Florida. The figures represent net production at each trophic level expressed in kcal/m^2/yr. (Based on data obtained by Howard T. Odum.)

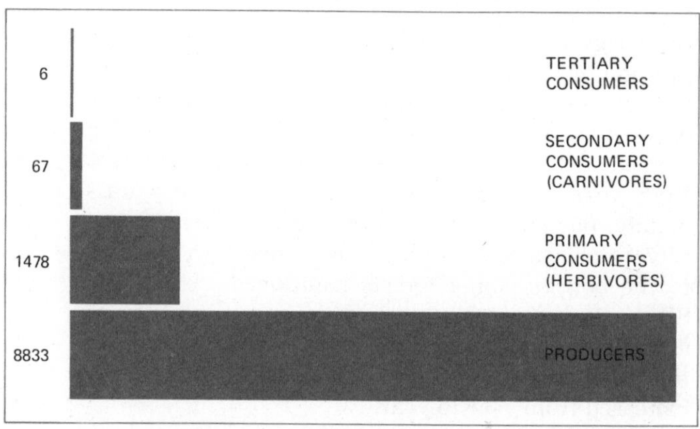

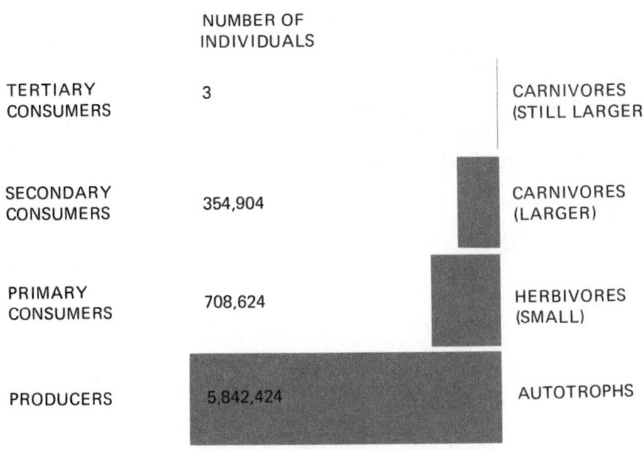

FIG. 40.10 Pyramid of numbers in an acre of bluegrass. At each successive trophic level the number of individuals in the population must be sharply reduced *if* each individual is larger than those creatures on which it feeds. This is because of the losses in biomass as one proceeds from one trophic level to the next. (Redrawn by permission from E. P. Odum, *Fundamentals of Ecology,* 2nd ed., W. B. Saunders Co., Philadelphia, Pa., 1959. Based on data acquired by Evans, Cain, and Walcott.)

40.5 THE BIOMES

If you were a resident of Pennsylvania and traveled to northern France, you might be impressed by the cultural differences you found there but would probably feel that the general landscape is very much like home. A botanist would recognize that the species of trees and shrubs in northern France are in many cases not the same as those back in Pennsylvania but that the general types of plants and their growth habits are similar. Traveling south to the Mediterranean coast of France would be a different story. There the plants would present a landscape totally different from that of Pennsylvania (and one that would make a resident of southern California feel right at home). Not only would the plants be different, but a keen observer would find that the wild animals in the two regions differed as well. Such distinctive plant and animal communities are called **biomes.**

The exact number of different kinds of biomes on the earth cannot be given. This is simply because no area is completely homogeneous in its plant and animal life. If you like to emphasize the differences, you increase the number of biomes. If you are the type who overlooks differences in favor of uniformities, you choose a smaller number of biomes. For our purposes, however, we can consider that North America contains seven biomes.

A number of climatic features interact in the creation and maintenance of a biome. Where precipitation is moderately abundant (40 inches or more per year) and distributed fairly evenly throughout the year, the major determinant is temperature. Earlier in the chapter we examined the mechanism which creates lower average temperatures at increasing latitudes. Actually, it is not average temperature that is so important, but such limiting factors as (1) whether it ever freezes or (2) length of growing season.

Assuming adequate rainfall, we find four characteristic biomes as we proceed from a region of high average temperatures (the tropics) to a region of low (the Arctic).

1. The tropical rain forest. In the western hemisphere, the tropical rain forest reaches its fullest development in the jungles of Central and South America. The trees are very tall and of a great variety of species. One rarely finds two trees of the same species growing close to one another. The vegetation is so dense that little light reaches the forest floor. Most of the plants are evergreen, not deciduous. The branches of the trees are festooned with vines and epiphytes. Epiphytes are plants that live perched on sturdier plants. Unlike vines, their roots do not reach the ground. Nor do they take nourishment from their host. Many orchids and many bromeliads (members of the pineapple family like Spanish "moss") are epiphytes.

The diversity of animal life as well as plant life in the tropical rain forest is very great. A large proportion of the animal species, mammals and reptiles as well as birds, live in the forest trees.

The closest thing to a tropical rain forest biome in the continental United States is the little wooded "islands" found scattered through the Everglades in the southern tip of Florida (Fig. 40.11). Their existence depends on the fact that it never freezes and on protection from the fires that periodically sweep the Everglades. These fragments of subtropical rain forest are protected in the Everglades National Park.

2. The temperate deciduous forest. The temperate deciduous forest biome of North America occupies the eastern half of the United States. It is characterized by hardwood trees (beech, maple, oak, hickory, etc.) which shed their leaves in the autumn. The number of different tree species here is far more

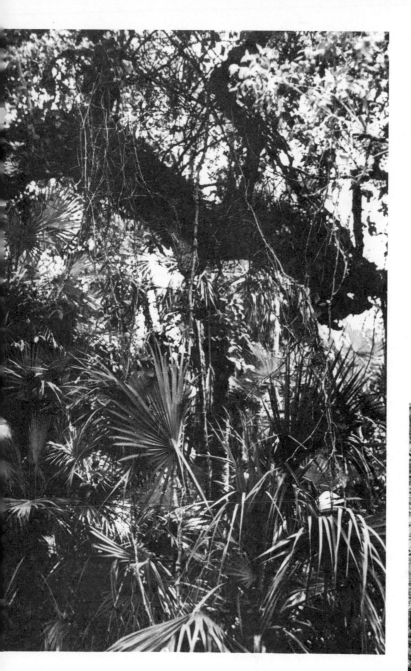

FIG. 40.11 Vegetation in the Everglades of Florida. The abundance of vines and epiphytes is characteristic of the tropical rain forest biome.

FIG. 40.12 Beeches and maples growing in the temperate deciduous forest of New Hampshire. (Courtesy of Eliot Porter.)

limited than in the tropical rain forest biome, and large stands of a single species may often be found. As for animal life, deer, raccoons, and salamanders are especially characteristic of this biome (Fig. 40.12).

3. The taiga. As one moves north into Canada, a new picture presents itself: a landscape dominated by conifers, especially spruces and firs (Fig. 40.13). This is the taiga, named after the similar biome found in the U.S.S.R. It is a land dotted with lakes and populated by bears, rodents (e.g. squirrels), birds, and moose. The latter is such a typical member that it has led some to call this the "spruce-moose" biome. During the cold winters, many of the mammals hibernate, while many of the birds migrate south.

4. The tundra. Still farther north, the trees of the taiga become stunted by the harshness of the subarctic climate. Finally they disappear to reveal a land of bogs and lakes, a land so cold in winter that

FIG. 40.13 The taiga, popularly known as the "spruce-moose" biome. (Courtesy of Lee E. Yeager.)

FIG. 40.14 Caribou grazing in the tundra of northern Canada. (Courtesy of Fritz Goro, *Life*.)

even the long days of the brief arctic summer are unable to thaw the permafrost beneath the surface layers of soil. Sphagnum moss, a wide variety of lichens, and some grasses and fast-growing annuals dominate the landscape during the short growing season (Fig. 40.14). Caribou feed upon this growth. So do vast numbers of insects. Swarms of migrating birds, especially waterfowl, invade the tundra in the summer to raise their young, feeding them with insects and a large variety of aquatic invertebrates and vertebrates. As the arctic summer draws to a close, the birds fly south, and all but a few of the permanent residents, in one way or another, prepare themselves to spend the winter in a dormant condition.

There is another way to visit the different biomes which does not require traveling hundreds of miles. Temperature is the major influence in establishing the four biomes already discussed and, as we have seen, temperature varies with altitude as well as latitude. A trip up New Hampshire's Mt. Washington on the famous cog railway will, in a short time, enable one to pass from a temperate deciduous hardwood forest in the valley, through a region dominated by spruces, up into an alpine, lichen- and moss-dominated region near the summit. While the physical and biotic features of these zones are not identical with the biomes established by latitude, there are many close parallels (Fig. 40.15). As a rule, a climb of 1000 ft is roughly equivalent, in changed flora and fauna, to a trip northward of some 600 miles.

Three other major biomes are found in North America: grassland (prairie or plains), desert, and the chaparral. The controlling factor in these three biomes is not so much temperature but the amount and distribution of rainfall. The prevailing winds in the western half of our country blow in from the Pacific laden with moisture. Each time this air rises up the western slopes of, successively, the Coast Ranges, the Sierras and Cascades, and finally the Rockies, it expands and is cooled. Its moisture condenses to rain, which drenches the mountain slopes beneath. When the air reaches the eastern slopes, it is relatively dry, and much less rain falls. How much falls and when determine whether the biome in these areas will be grasslands, desert, or chaparral.

5. Grasslands. The annual precipitation in the grasslands averages 20 inches per year. A large proportion of this falls in the growing season. This promotes a vigorous growth of grasses (Fig. 40.16) but, except along river valleys, is barely adequate for the growth of forests. The factor that probably tips the balance from forests to grass is fire. There is considerable evidence that fires—set both by lightning and by humans—regularly swept the plains in

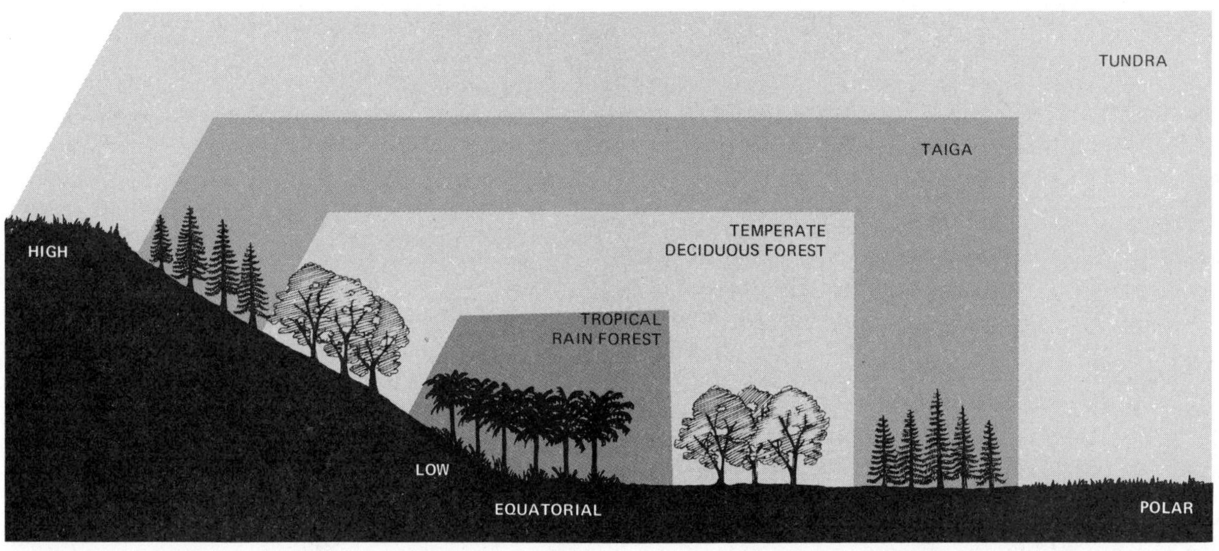

FIG. 40.15 The biomes change not only with latitude (right) but also with altitude (left).

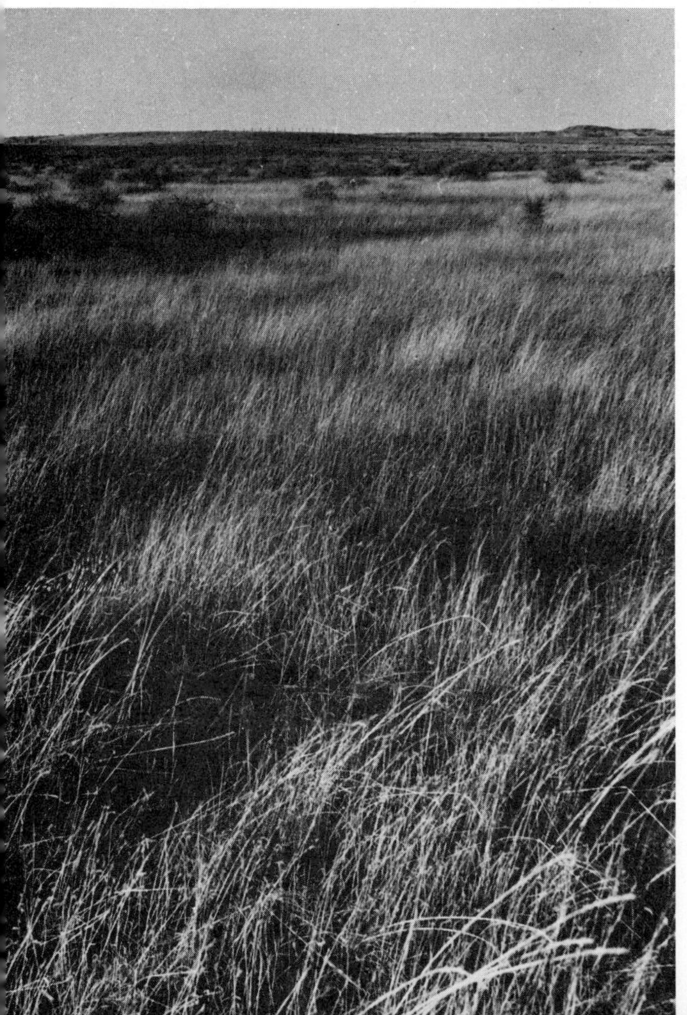

earlier times. Thanks to their underground stems and buds, grasses are not harmed by the fires that destroy most shrubs and trees.

The abundance of grass for fodder, coupled with the lack of shelter from predators, produces similar animal populations in grasslands throughout the world. The dominant vertebrates are swiftly moving, herbivorous ungulates. In this country, bison and antelope were conspicuous occupants of the grasslands before the coming of the white settlers. Now the level grasslands supply our nation with an abundance of corn, wheat, and other grains, while the hillier areas support our domesticated ungulates, cattle and sheep.

6. The desert. Annual rainfall in the desert is less than 10 inches and, in some years, may be zero. Because of the extreme dryness of the desert, its colonization is limited to (a) plants such as cacti, sagebrush, and mesquite that have a number of adaptations by which they conserve water over long periods (Fig. 40.17) and (b) fast-growing annual plants whose seeds can germinate, develop to maturity,

FIG. 40.16 The grasslands biome. This plains area in Weld County, Colorado, has been only lightly grazed by cattle. (Courtesy of the U.S. Forest Service.)

FIG. 40.17 The desert biome in Monument Valley on the Utah-Arizona border. (Photo by Hiroji Kubota, Magnum Photos.)

flower, and produce a new crop of seeds all within a few weeks following a rare, soaking rain. Most of the animals of the desert (mammals, lizards and snakes, insects, and even some birds) are adapted for burrowing to escape the scorching heat of the desert sun. Many of them limit their forays for food to the night hours.

7. The chaparral. The annual rainfall in the chaparral biome may reach 20 to 30 inches but almost all of this falls in the winter. Summers are very dry and all the plants—trees, shrubs, and grasses—are more or less dormant then. This biome is found in California. Similar biomes (with other names, e.g. scrub biome) are found around much of the Mediterranean, and along the southern coast of Australia. The trees of the chaparral are mostly oaks, both deciduous and evergreen. Scrub oaks and shrubs like manzanita and the California lilac (not a relative of the eastern lilac) form dense, evergreen thickets (Fig. 40.18). All of these plants are adapted to drought by such mechanisms as waxy, waterproof coatings on their leaves.

The chaparral has many plants brought from similar biomes elsewhere. Vineyards, olives, and figs flourish just as they do in their native Mediterranean biome. So, too, do eucalyptus trees transplanted from the equivalent biome in Australia.

40.6 FIRE

We usually think of fire as having only a harmful effect on biomes and work very hard to suppress it. But there is growing evidence that fire plays an important role in maintaining some biomes (or parts of biomes). We mentioned earlier that the grasslands biome probably requires periodic fires for its maintenance unless intensive grazing or mowing substi-

FIG. 40.18 Chaparral-clad foothills of the Sierra Nevada in California.

tutes for them. Some of the subdivisions of the taiga and temperate deciduous forest biomes exist because of fire. The pine woods of our southeastern coastal region are an example of the latter. These pines are not damaged by ground fires and indeed their very existence seems to depend on them. When fires are excluded, deciduous hardwoods grow up under them, eventually replacing them. The transition may be hastened by fungi which thrive in the litter of the unburned forest floor and then move on to infect the pines. The ponderosa pine and redwood forests of our western mountains also appear to be dependent on fire to keep competing hardwoods at bay.

The Everglades in Florida and the entire chaparral biome owe their very existence to fire. All the chaparral plants are adapted to fire and, after being burned, quickly sprout fresh growth (Fig. 40.19) which can support a larger population of herbivores (e.g. deer) than before. Tree-ring analyses indicate that fires swept through the chaparral in prehistoric times every few years.

The main reason for our aversion to forest fires is their great destructiveness, but this destructiveness occurs only with "crown" fires, that is, when the fire races through the tops of the trees. The vegetation destroyed by ground fires is quickly replaced, as we have seen. The irony of the matter is that by excluding fire, so much plant material accumulates close to the forest floor that when a ground fire does get started, it is so hot that it may quickly become a crown fire, destroying everything in its path. On the other hand, periodic ground fires keep the fuel level low. Dry chaparral plants are filled with resins and waxes and burn with great intensity. Almost every year enormously destructive fires occur in California. Perhaps the reintroduction of controlled, frequent fires will turn out to save more homes, watersheds, and ecosystem productivity than the present attempts to rigidly exclude fire from the chaparral.

When fire is excluded from an ecosystem where it has been common, changes in the plant community begin to take place. The changes in the southeast coastal pine forests, mentioned earlier, are an example. Any such long-term changes in a plant community are called plant successions.

FIG. 40.19 Chaparral in Southern California in the spring, after a fire had swept the area the previous autumn. Many of the shrubs have already sent up fresh shoots from the ground line.

40.7 PLANT SUCCESSION

Many plant communities are not self-sustaining. A field in the temperate deciduous forest biome will remain a field only as long as it is grazed or mowed regularly. If these activities cease, the grass plants begin to be replaced by other species. The elimination of an environmental factor such as mowing or fire tips the balance in favor of other species, and the newcomers may, in turn, by their own growth sow the seeds of their own destruction. For example, the gray birch grows well only in sunny locations. The shade it casts prevents other gray birches from growing up beneath it but permits white pine seedlings to become established. When mature, white pines then cast so deep a shade that the gray birch is no longer able to survive in that area. By its growth, one population of plants alters the environment in a way unfavorable to it and favorable to some other plant population.

The process of plant succession begins just as soon as a land area capable of supporting plant life is formed. The exposure of rocks by a retreating glacier, the formation of beaches at the edge of oceans and lakes, and the gradual filling in of ponds all provide fresh terrain for plant colonization. Let us examine the plant succession which occurs on bare rock in the colder parts of the temperate deciduous forest biome.

The first colonizers are the lichens and certain mosses. Acids secreted by the lichens attack the rock substrate and provide bits of soil. Additional soil particles may be formed by weathering or be blown to the rock from elsewhere and lodge in its crevices. Damage and decay to the lichen growth supplies some humus. Soon sufficient soil is present in the cracks for other mosses to become established. They produce fresh growth each season, the old growth decaying and quickly providing additional amounts of humus. Soon there is enough soil for grasses and, later, low shrubby growth such as blueberries. These, in turn, provide excellent cultural conditions for the seeds of such sun-loving, fast-growing *r* strategists (see Section 39.7) as gray birch trees and poplars (quaking aspens). As we noted above, white pine soon replaces these short-lived species. In the dense shade of white pines, only shade-tolerant maple and beech seedlings thrive. When these large trees finally take over, the succession comes to an end. Maple and beech seedlings are able to develop under the conditions imposed by their parents, and the population (now mostly *K* strategists) becomes self-sustaining. It is known as a **climax** forest.

A similar process occurs as shallow ponds gradually fill in with soil washed in from the surrounding terrain and organic matter produced by underwater plants. As we walk from the edge of a poorly drained, boggy pond back into the forest, we pass through a series of zones that recreate in space the plant succession which has been occurring in time (Fig. 40.20). From the swamp loosestrife and arrowhead at the water's edge, past sphagnum moss and pitcher plants, then shrub-sized blueberries and poison sumac followed by black spruce and American larch and, finally, swamp maples and white pines, one passes concentric zones, each representing a later stage of plant succession as the soil has become firmer, dryer, and the shade denser.

Lumbering, grazing, farming, fires, and hurricanes interrupt the process of succession by removing the dominant plants in the community. Elimination of these factors sets the stage for a new succession to begin. The many abandoned farms in New England bear eloquent testimony to this. People often wonder why our pioneers built stone walls through the deep woods. The answer is they did not. The walls one sees in the woods today once marked the boundaries of fields and pastures, but when cultivation and grazing ceased, a **secondary succession** began. The grass of abandoned fields and pastures gave way to rank weeds and low shrubby growth. Soon gray birch, poplars, or cedars flourished. White pines or, in sandy well-drained locations, oaks followed. If they are left alone, we will someday see once again a climax forest of maples and beeches.

The colonization of bare rock, the filling in of a pond, and the secondary succession following the abandonment of a field each involve different species in the early stages of succession. In any given region, though, the species in the final, self-sustaining climax forest are the same. The tendency for all plant successions to end in the same climax community is called **convergence.**

There are few parallels to plant succession in animal populations. An animal population may drastically alter the conditions of its environment so that it is no longer suitable for that population, but this is not inevitable. Nevertheless, succession in animal populations does occur simply as a consequence of succession in plant populations. As fields revert to woods, the kinds of birds, mammals, and invertebrates present change, too.

In general, the process of plant succession is a reflection of the increasing efficiency of the community at intercepting the energy of the sun and con-

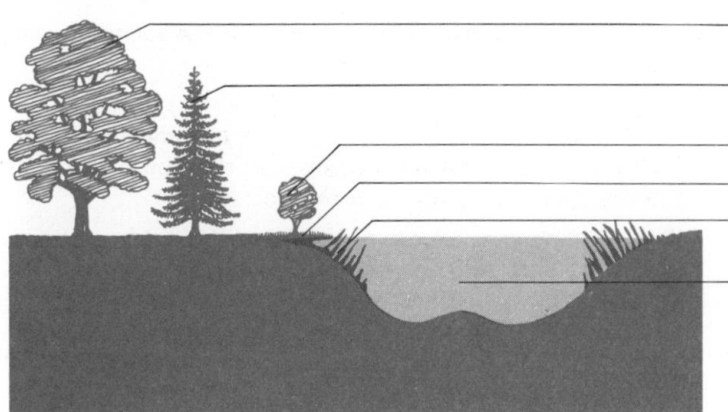

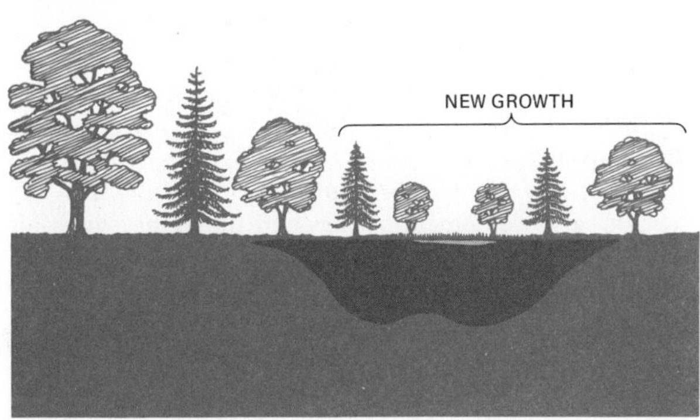

FIG. 40.20 Plant succession in a bog. Starting with the stems of the swamp loosestrife, which grow right out of the water, an increasingly thick mat of vegetation is formed as the years go by and the area of open water diminishes. Where water is still underneath, the bog "quakes" when walked on. The rate of decay in the bog is low so that there is a net storage of energy in this ecosystem from one season to the next. This occurs through an increase in biomass and also by the accumulation of peat. Peat represents a first stage in the formation of coal. In climax communities, there is no net gain in energy from one season to the next.

verting it into chemical energy. As one stage of succession follows another, the biomass of the community increases. This is simply the outcome of the increasing amount of annual net productivity, i.e. calories, stored by the plant community. This, in turn, may provide the caloric support for an increasing population of consumers. When the climax forest is finally fully established, no further increases in net productivity occur. All of the annual net productivity of the plants is consumed by the heterotrophs, animals and microorganisms, of the community. The system comes into equilibrium and reaches peak efficiency at channeling the energy of the sun into all the food webs of the community.

EXERCISES AND PROBLEMS

1. Which of the following sources of energy depend ultimately on the energy of sunlight: (a) wood, (b) coal, (c) oil, (d) hydroelectric power, (e) atomic power?
2. Assuming a growing season of 100 days, what fraction of the radiant energy falling on a temperate deciduous forest (10,000 kcal/m^2/day) is trapped in net productivity? (See Fig. 40.4.)
3. It has been estimated that in 1963 it took a subsidy of 8 million kcal to feed each person in the United States. Assuming a daily requirement of 3000 kcal, what was the ratio of energy subsidy to energy acquired? In what ways was the energy subsidy spent?
4. In the food chain shown in Fig. 40.6(a), how many square meters of grass are needed to support one hawk? Assume that (a) the hawk has a daily requirement of 548 kcal, (b) the net productivity of the grass is 2000 kcal/m^2/year, (c) the efficiency of energy transfer from one trophic level to the next is 10% in all cases.
5. In one temperate deciduous forest, the gross production was determined to be 11,000 kcal/m^2/yr. The total losses to respiration and decay of all trophic levels was 8875 kcal/m^2/yr. By what amount (grams/m^2) was this forest growing each year? Was this a climax forest? Explain.

REFERENCES

1. KORMONDY, E. J., *Concepts of Ecology,* 2nd ed., Prentice-Hall, Englewood Cliffs, N.J., 1976. A small text with excellent discussions of many of the topics surveyed in this chapter.
2. WOODWELL, G. M., "The Energy Cycle of the Biosphere," *Scientific American,* Offprint No. 1190, September, 1970.
3. GATES, D. M., "The Flow of Energy in the Biosphere," *Scientific American,* Offprint No. 664, September, 1971.

References 2 and 3 describe the flow of energy through, and the net productivity of, a variety of ecosystems.

4. KEMP, W. B., "The Flow of Energy in a Hunting Society," *Scientific American,* Offprint No. 665, September, 1971. Energy-flow analysis applied to two Eskimo households.
5. RAPPAPORT, R. A., "The Flow of Energy in an Agricultural Society," *Scientific American,* Offprint No. 666, September, 1971. The Silver Springs-type analysis applied to the Tsembaga of New Guinea.

References 3–5 also appear in *Energy and Power,* Freeman, San Francisco, 1971. Available in paperback.

6. HORN, H. S., "Forest Succession," *Scientific American,* Offprint No. 1321, May, 1975.
7. WOODWELL, G. M., "Toxic Substances and Ecological Cycles," *Scientific American,* Offprint No. 1066, March, 1967. Shows how toxic materials such as radioactive fallout and pesticides are concentrated—sometimes to harmful levels—as they pass from one link in a food chain to another.
8. WATTS, MAY T., *Reading the Landscape,* Macmillan, New York, 1957. A popular account of how the landscape reveals to the alert eye the forces in the physical and biotic environment that have affected it.
9. WENT, F. W., "The Ecology of Desert Plants," *Scientific American,* Offprint No. 114, April, 1955.
10. COOPER, C. F., "The Ecology of Fire," *Scientific American,* Offprint No. 1099, April, 1961.
11. WOODWELL, G. M., "Effects of Pollution on the Structure and Physiology of Ecosystems," *Science,* Reprint No. 53, 24 April, 1970.
12. DEEVEY, E. S., JR., "Bogs," *Scientific American,* Offprint No. 840, October, 1958. Describes the historical development of bogs and the plant succession that occurs in them.
13. ODUM, E. P., "The Strategy of Ecosystem Development," *Science,* Reprint No. 42, 18 April, 1969. The interrelations between energy flow, productivity, and stage of succession.
14. RICHARDS, P. W., "The Tropical Rain Forest," *Scientific American,* Offprint No. 1286, December, 1973.

41 THE CYCLES OF MATTER

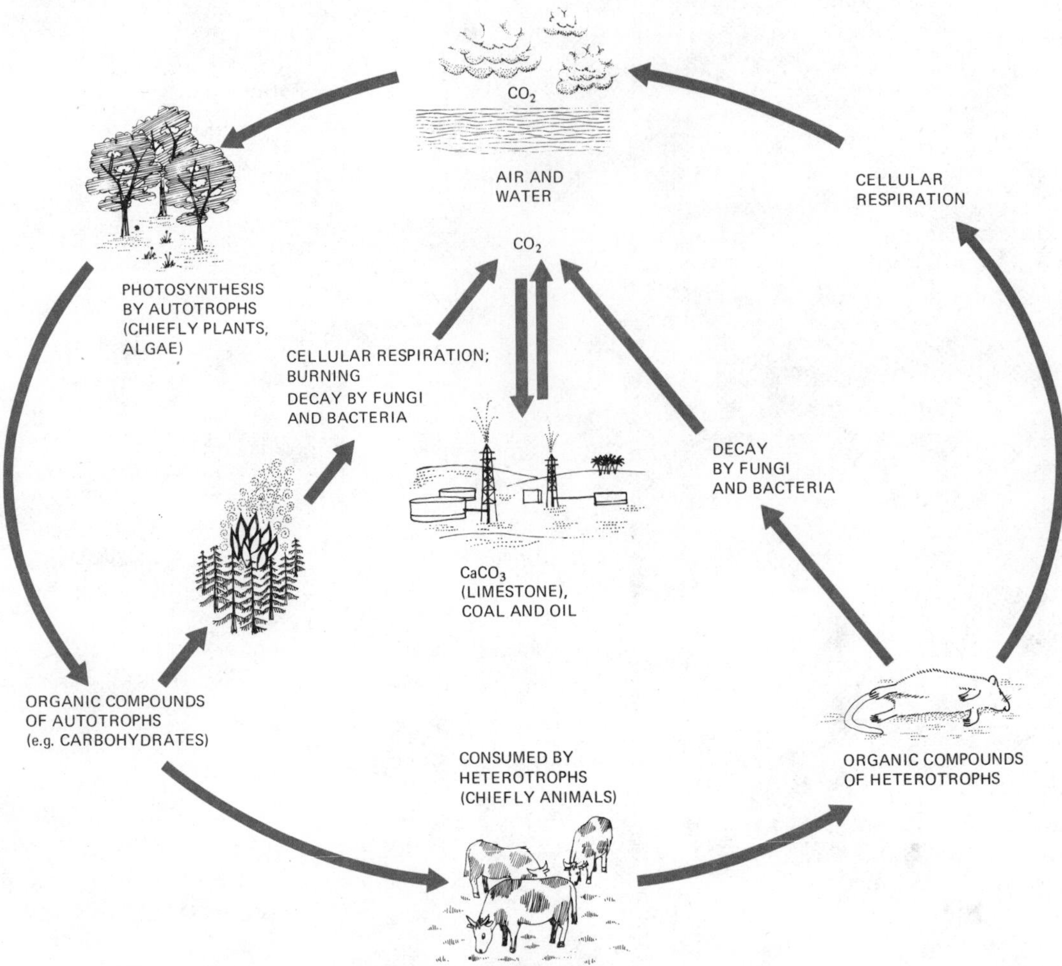

FIG. 41.1 The carbon cycle. The heterotrophic microorganisms of decay produce CO_2 by the respiration of organic molecules secured from the bodies of plants and the bodies and excreta of animals.

We have seen that the flow of energy through the biosphere is one-way. Radiant energy from the sun is intercepted by the biosphere. After passing through the transformations which keep all living things alive, it returns to outer space as heat. Thus there is no "energy cycle."

The interception and use of energy by living organisms depends upon storing energy in chemical bonds and exploiting this energy when the bonds are broken. Living matter is built from a limited number—perhaps 25—of kinds of atoms. Some of these are abundant in the nonliving world. Some are very scarce (look back at Fig. 3.7). In either case, however, life has persisted on this planet for a period exceeding three billion years because mechanisms exist by which these atoms can be used over and over again. Thus the atoms of life do cycle, and it is the nature of some of these cycles to which we now turn our attention.

41.1 THE CARBON CYCLE

Although carbon is a very rare element in the nonliving sector of the earth, it represents some 18% of living matter. The ability of carbon atoms to bond to each other provides the basis for the molecular diversity and molecular size without which life as we know it could not exist.

Outside of organic matter, carbon is found as the gas carbon dioxide and as carbonate rocks (limestone, coral). It is the autotrophs—chiefly green plants—that take carbon dioxide and reduce it to organic compounds: carbohydrates, proteins, lipids, and others. Terrestrial producers get their carbon dioxide from the atmosphere and aquatic producers use that dissolved (as bicarbonate, HCO_3^-) in water. In Chapter 40, we traced the path of fixed carbon through food webs—all of which depend on it not only for their structure but also for their energy.

At each trophic level in a food web, carbon returns to the atmosphere or water as a result of respiration. Plants, herbivores, and carnivores all respire and in so doing liberate carbon dioxide (Fig. 41.1). Much of the organic matter at each trophic level is not consumed by a higher trophic level but passes instead to the "final" trophic level, the decay organisms. This occurs as plants and animals or their parts (e.g. leaves in the autumn) die. Bacteria and fungi serve the absolutely vital function of unlocking carbon from corpses and debris that will no longer serve as food for other trophic levels. Through their metabolism, carbon dioxide is released and the cycle of carbon can turn again.

Are the processes of carbon dioxide uptake and release in balance? The answer is clearly, no. The carbon dioxide content of the atmosphere is gradually increasing (Fig. 41.2). Presumably this increase began with the start of the industrial revolution. By burning ever-increasing amounts of coal, oil, and natural gas, we are returning to the air carbon that had been locked within the earth for millions of years. However, the increase in atmospheric carbon dioxide is only one-third or so of that which would have been expected from the well-established data on fossil fuel use. Where has the rest disappeared to? Perhaps some of it has stimulated and been consumed by a greater worldwide rate of photosynthesis. Investigators growing crop plants in controlled environments have demonstrated that moderate increases in carbon dioxide availability increase photosynthesis. Therefore, some of the carbon dioxide released by our use of fossil fuel may have increased worldwide primary productivity. Another probable depository for our carbon dioxide production is the sea. Carbon dioxide in the air exchanges readily with carbon dioxide dissolved in the sea. The dissolved carbon dioxide is, in turn, in equilibrium with carbonate deposits in the sea. If more carbon dioxide is added to sea water, the excess is precipitated as carbonate sediments like coral and limestone—$CaCO_3$. The converse would also be true, with the decomposition of these sediments restoring any reduction in dissolved carbon dioxide. Thus these oceanic deposits provide a huge reservoir of carbon and a buffer which helps to minimize changes in the atmospheric concentration of carbon dioxide.

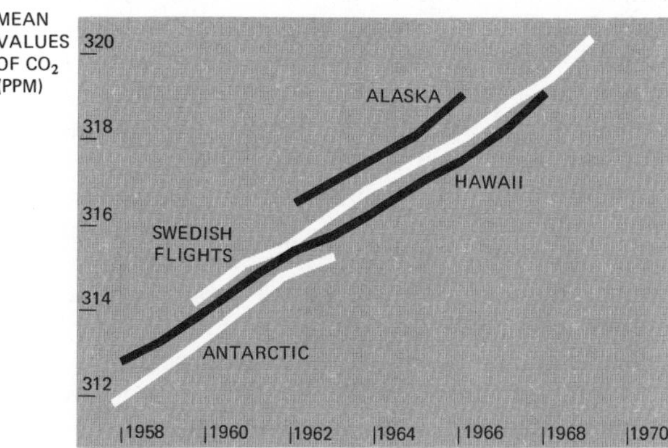

FIG. 41.2 Annual increase in the carbon dioxide concentration of the atmosphere in four locations. The values represent yearly *averages* (in parts per million, ppm) and do not reveal the shifts (e.g. 5–6 ppm in Hawaii) that occur during the course of a year.

Despite these "sinks" for our greatly increased carbon dioxide production, the concentration of atmospheric carbon dioxide has increased significantly in recent decades. Does this increase pose a threat to the earth and life on it? At this point, we simply do not know. Carbon dioxide is transparent to light but rather opaque to heat rays. Therefore the carbon dioxide in the air retards the radiation of heat from the earth back into outer space. Some have predicted that the increase in atmospheric carbon dioxide will so increase the "thickness" of this thermal blanket that the earth will become warmer. This would lead to the melting of the great ice caps at the poles. It has been estimated that if all the ice on earth were melted, the sea level would rise somewhere between 75 and 150 meters, enough to inundate most of the coastal cities of the world. Before taking to higher ground, though, we must bear in mind two other things. One is that human activities may have also increased the *reflectance* of the earth. The very activities that deposit carbon dioxide in the air also deposit dust particles that impede the passage of light through the atmosphere. Air pollution thus appears to reduce the arrival of energy as well as its departure. Whether these opposing effects cancel each other out is not known. However, weather data from the last quarter-century show that average temperatures on earth are declining. There is abundant geological evidence of climatic fluctuation throughout the history of life on the earth. Whether the present weather change is a result of human activities or of long-term changes over which we have no control is as yet an unanswered question.

41.2 THE OXYGEN CYCLE

Molecular oxygen (O_2) represents 20% of the earth's atmosphere. This pool serves the needs of all terrestrial respiring organisms and, as it dissolves in water, the needs of aquatic organisms as well. In the process of respiration, oxygen serves as the final acceptor for electrons removed from the carbon atoms of food. The product is water. The cycle is completed in photosynthesis as the energy of light is harnessed to the stripping of electrons away from the oxygen atoms of water molecules. The electrons reduce the carbon atoms (of carbon dioxide) to carbohydrate. Molecular oxygen is left over and the cycle is complete (Fig. 41.3).

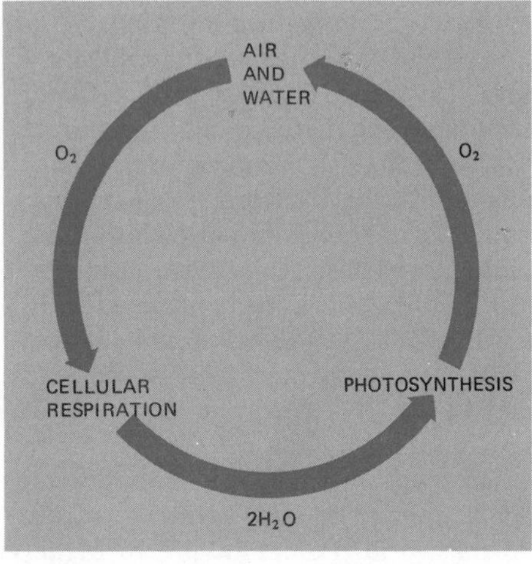

FIG. 41.3 The oxygen cycle. Virtually all of the oxygen present in the atmosphere today has been placed there by photosynthesizing plants and algae. Oxygen serves as the final acceptor of electrons in cellular respiration.

For every molecule of oxygen used in cellular respiration, a molecule of carbon dioxide is liberated. Conversely, for every molecule of carbon dioxide taken up in photosynthesis, a molecule of oxygen is given off. The study of minerals formed very early in the earth's history indicates that at one time there was no oxygen in the earth's atmosphere. With the evolution of photosynthesis using water, oxygen first appeared. Assuming the early development of a mature biosphere with no net productivity, that is, with respiration and photosynthesis in balance, we may well ask, what accounts for the pool of oxygen present today? Each molecule of oxygen accumulated in the atmosphere must represent a carbon atom that was once reduced in photosynthesis but has since escaped oxidation. These are the carbon atoms sequestered away in coal and oil and other organic deposits. These are also the carbon atoms that make up the bodies of the worldwide living biomass and those dead portions of plants and animals that have so far escaped decay.

In burning fossil fuels, we use that amount of oxygen that had been placed in the atmosphere when the carbon atoms of the fuel were first reduced. This realization raises the spectre that as we burn ever-increasing amounts of coal, oil, and natural gas we may be seriously depleting the oxygen concentration of the air. Estimates are made from time to time of the earth's reserves of these fuels. Even if we accept the most generous estimates, the total combustion of this material would deplete the atmospheric pool of oxygen by no more than 2–8%. Why no more? One answer is that most of the reduced carbon of the earth is too thinly distributed and/or too deeply buried to serve as fuel. However, its presence provides a large buffer against drastic drops in oxygen levels. Even if the oxygen concentration of the atmosphere did drop 8%, the effect on humans would be less than that of moving from New York City to Denver. (At an altitude of 5000 ft, the concentration of oxygen in the air is 18% less than that at sea level.) The combustion of all our fossil fuels would surely create problems (e.g. air pollution) far more serious than the effect on the oxygen pool.

While human activities show no signs of having a significant effect on the oxygen content of the *air*, this is not true for aquatic environments. Many of the streams and lakes of crowded, industrial countries suffer periodic shortages of dissolved oxygen. These are often so severe that certain aquatic organisms are no longer able to survive. The precipitating factor is the dumping of organic and other wastes into the water. These are degraded by decay organisms, which use dissolved oxygen in the process. In fact, the most widely used indicator of water pollution is the *b*iochemical *o*xygen *d*emand, BOD. This is the measure of the oxygen needed to completely oxidize the substances present in the water. The higher the BOD of a stream or lake, the less oxygen available for the organisms that normally live there. For those whose oxygen requirements are high (e.g. most fishes), a rising BOD threatens their ability to survive.

41.3 THE NITROGEN CYCLE

All living things require nitrogen atoms for the building of proteins and a variety of other essential organic molecules. Air, which is 79% nitrogen, serves as the reservoir of this substance. Despite the great size of the nitrogen pool, it is often a limiting ingredient for living things. This is because most organisms cannot use nitrogen in its elemental form, that is, as the gas N_2. In order for plants to manufacture protein, they must secure nitrogen in "fixed" form, i.e., incorporated in compounds. The most commonly used form is as nitrate ions, NO_3^-. Nevertheless, other substances such as ammonia, NH_3 and urea, $(NH_2)_2CO$, are used successfully, both in natural systems and as fertilizers in agriculture.

Nitrogen fixation. The nitrogen molecule, N_2, is quite inert. To break it apart so that its atoms can combine with other atoms requires the input of substantial amounts of energy. Three processes play an important part in the fixation of nitrogen in the biosphere. One of these is lightning. The enormous energy of lightning breaks nitrogen molecules and enables them to combine with the oxygen in the air. (The process is analogous to that occurring in the internal combustion engine, see Section 10.10.) Nitrogen oxides are formed which dissolve in rain, forming nitrates. In this form they are carried to the earth (Fig. 41.4). Atmospheric fixation of nitrogen probably represents some 5–8% of the total.

The need for nitrates in the manufacture of conventional explosives led to the development in Germany, on the eve of World War I, of an industrial process of nitrogen fixation. In this process hydrogen (usually derived from natural gas or petroleum) and nitrogen are reacted to form ammonia, NH_3. In

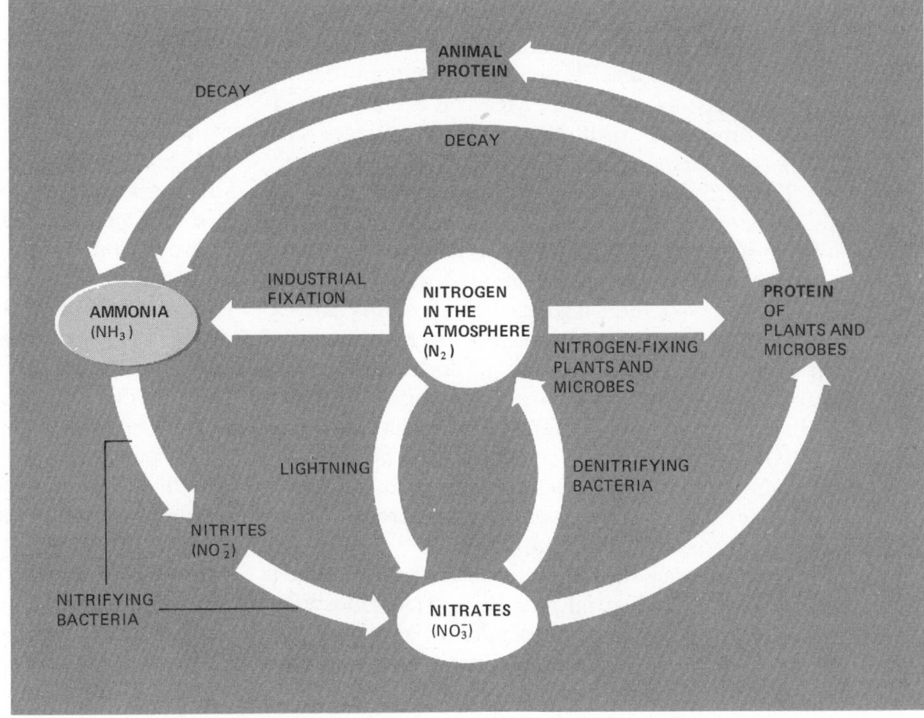

FIG. 41.4 The nitrogen cycle. Microorganisms play several essential roles in the cycling of nitrogen through the biosphere. It is estimated that half of the nitrogen fixed on the earth today is the direct result of two human activities: industrial fixation and the planting of legumes.

FIG. 41.5 Soybean plants growing in Missouri. The ability of soybeans to incorporate atmospheric nitrogen into their proteins makes them a valuable source of this essential food. (Courtesy of American Soybean Association.)

order for the reaction to proceed efficiently, it must be carried out at a high temperature (600°C), under enormous pressure, and in the presence of a catalyst. Today, most of the industrially fixed nitrogen is used as fertilizer. The initial product, ammonia, can be used directly as fertilizer. Most of it, however, is further processed to other common fertilizers such as urea and ammonium nitrate, NH_4NO_3.

The growing needs of agriculture have led to an ever-increasing production of industrially fixed nitrogen. Perhaps as much as one-third of all the nitrogen fixation occurring today in the biosphere is accomplished industrially. This is really an extraordinary human intrusion into the functioning of the biosphere. Whether such a perturbation of a natural cycle will work to our ultimate advantage is difficult to predict. Certainly our agricultural productivity depends upon the current high rates of nitrogen fixation. However, damaging side-effects in our water supply are becoming apparent, as we shall see in the following chapter.

Our influence on the rate of nitrogen fixation is not limited to industrial activities. The widespread cultivation of legumes, especially alfalfa and soybeans (Fig. 41.5), has greatly increased the worldwide rate of nitrogen fixation. Legumes are a family of plants (including peas, alfalfa, and clover), the roots of which harbor gram-negative bacilli of the genus *Rhizobium*. The bacteria are able to fix atmospheric nitrogen, for their host as well as for themselves. Further details of this remarkable symbiotic association will be presented in Chapter 43.

Certain other microorganisms can fix atmospheric nitrogen. In fact, the ability to fix nitrogen appears to be exclusively a prokaryotic skill. Some of the actinomycetes (see Section 34.11) live in association with plants other than legumes. Other nitrogen-fixing bacteria (e.g. *Azotobacter, Clostridium*) live free in the soil. Some blue-green algae are also capable of fixing nitrogen and play an important role in maintaining the fertility of semiaquatic environments like rice paddies.

Despite extensive research, it is still not clear just how nitrogen fixers are able to overcome the high energy barriers inherent in the process. They require an enzyme, called nitrogenase, and a huge expenditure of ATP. Although the first stable product of the process is ammonia, this is quickly incorporated into protein and other nitrogen-containing organic compounds. For our purposes, then, nitrogen fixation leads to the incorporation of nitrogen into plant (and microbial) protein. Plants lacking the benefits of association with nitrogen-fixers manufacture their proteins from nitrogen taken in from the soil—generally as nitrates.

Decay. The proteins manufactured by plants enter and pass through food webs just as carbohydrates do. At each trophic level, there are losses back to the environment, chiefly in excretions. The final beneficiaries of organic nitrogen compounds are the microorganisms of decay. Through their activities, the organic nitrogen molecules in excretions and corpses are broken down to ammonia.

Nitrification. Ammonia can be taken up directly by plants—through their roots and, as demonstrated in some species, through their leaves. (These latter, when exposed to isotopically labeled ammonia gas, incorporate the label in their proteins.) However, most of the ammonia produced by decay is converted into nitrates. This is accomplished in two steps. Bacteria of the genus *Nitrosomonas* oxidize NH_3 to nitrites (NO_2^-). Nitrites are then oxidized to nitrates (NO_3^-) by bacteria of the genus *Nitrobacter*. These two groups of chemoautotrophic bacteria are called nitrifying bacteria. Through their activities (which supply them with all their energy needs), nitrogen is made easily available to the roots of plants.

Denitrification. Were the processes described above the complete story of the nitrogen cycle, we would be faced with a steady reduction in the pool of free, atmospheric nitrogen as it became fixed and began cycling through various ecosystems (Fig. 41.4). Another process, denitrification, reduces nitrates to nitrogen, thus replenishing the atmosphere. Once again, bacteria are the agents involved. They live deep in the soil and in aquatic sediments where oxygen is in short supply. They use nitrates as an alternative to oxygen for the final electron acceptor in their respiration. In so doing they close the nitrogen cycle. Whether their activities are keeping up with our ever-increasing efficiency at promoting nitrogen fixation remains to be seen.

41.4 THE SULFUR CYCLE

Sulfur is incorporated in virtually all proteins and thus is an absolutely essential element for all living things. It moves through the biosphere in two cycles, an inner one and an outer one (Fig. 41.6). The inner cycling involves passing from soil (or water, in aquatic environments) to plants, to animals, and back to soil or water. There are, however, leaks in this inner cycle. Some of the sulfur compounds present on land (e.g. in soil) are carried away to the sea in river water. This sulfur would be lost to the

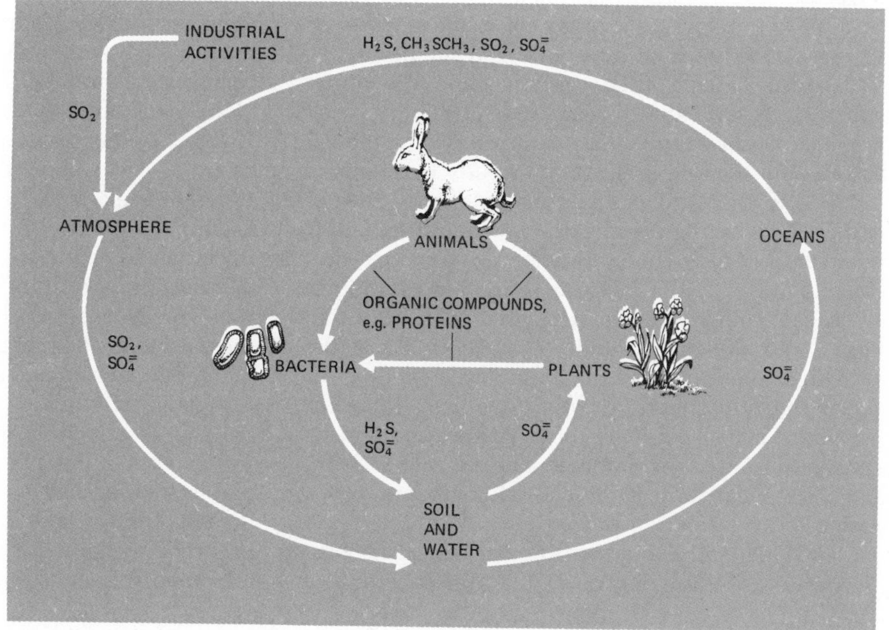

FIG. 41.6 The sulfur cycle. The burning of fossil fuels and the smelting of sulfur-containing ores have added substantially to the natural input of gaseous sulfur compounds into the atmosphere.

FIG. 41.7 Sandstone figure over the portal of a castle in Westphalia, Germany, photographed in 1908 (left) and again in 1969 (right). Acid rain produced by the air pollution generated in the heavily industrialized Ruhr region of Germany probably accounts for the severe damage. The castle was built in 1702. (Photo courtesy of Herr Schmidt-Thomsen.)

terrestrial cycle were it not for a mechanism to return it to land. This consists of converting it to gaseous compounds such as hydrogen sulfide (H_2S) and sulfur dioxide (SO_2). These enter the atmosphere and are carried over land. Generally, they are then washed out of the air in rainfall, although some sulfur dioxide is absorbed directly by plants.

Bacteria play a crucial role in the cycling of sulfur. When air is present, the breakdown of sulfur compounds (including the decay of protein) produces sulfate ($SO_4^{=}$). Under anaerobic conditions, hydrogen sulfide (a gas with the odor of rotten eggs) and dimethyl sulfide (CH_3SCH_3) are the principal products. When these latter two gases reach the atmosphere, they are oxidized to sulfur dioxide. Further oxidation of sulfur dioxide and its solution in rainwater produces sulfuric acid and sulfates, the chief form in which sulfur returns to terrestrial ecosystems.

Coal and oil contain sulfur and their combustion releases sulfur dioxide to the atmosphere. The smelting of sulfur-containing ores, such as copper ore, also adds enormous amounts of sulfur dioxide to the air. In fact, sulfur dioxide pollution around copper smelters may destroy all vegetation for miles (look back at Fig. 10.25). While estimates vary, 15–25% of the gaseous sulfur in the atmosphere has been placed there through industrial activities. Whether this poses a serious threat to the global cycling of sulfur is not known. We have already reviewed some of the problems caused by sulfur dioxide air pollution. In industrial regions of Europe and North America, and in areas downwind of these regions, the increased sulfur dioxide concentration produces rain and snow of decreased pH. Normal rain is slightly acid (pH 5.7) thanks to the carbon dioxide dissolved in it. Where sulfur dioxide pollution exists, the pH of rain is regularly about 4 and values as low as 2.1 have been recorded. We do not know whether this "acid rain" is a health hazard or whether it will accelerate the release of minerals from soils and rocks. What is clear is that it is accelerating the decomposition of statuary and buildings made of limestone and marble (Fig. 41.7).

41.5 THE PHOSPHORUS CYCLE

Although the proportion of phosphorus in living matter is relatively small, the part it plays is absolutely indispensable. Nucleic acids, the substances which store and translate the genetic code, are rich in phosphorus. Many of the intermediates in photosynthesis and cellular respiration are combined with phosphorus, and phosphorus atoms provide the basis for the high-energy bonds of ATP, which is the energy currency of both photosynthesis and cellular respiration.

Phosphorus is a rather rare component in the nonliving world. The productivity of most terrestrial ecosystems can be increased if the amount of phosphorus available in the soil is increased. Because agricultural yields are also limited by the availability of nitrogen and potassium, fertilization programs often include all three of these nutrients. In fact, the composition of most fertilizers is expressed by three numbers. The first gives the percent of nitrogen in the fertilizer. The second gives the phosphorus content (as though it were present as P_2O_5), and the third the content of potassium (expressed as though it were the oxide, K_2O). Thus 100 lb of a 5-10-10 fertilizer contains 5 lb of nitrogen and an amount of phosphorus and potassium equivalent to that in 10 lb of P_2O_5 and K_2O respectively.

Phosphorus, like nitrogen and sulfur, participates in an inner cycle as well as in a global, geological cycle (Fig. 41.8). In the smaller cycle, organic matter containing phosphorus (e.g. plant debris, animal excrement) is decayed and the phosphorus becomes available for uptake by plant roots and reincorporation into organic matter. After passing through food chains, it once again passes to the decomposers and the cycle is closed. There are leaks from the inner cycle to the outer one. Water leaches phosphorus not only from phosphate-containing rocks but also from soil. Some of this is intercepted by aquatic life, but ultimately it finds its way to the sea.

The global cycling of phosphorus differs from that of carbon, nitrogen, and sulfur in one major respect. Phosphorus forms no volatile compounds by which it can pass from the oceans to the atmosphere, and thence back to land. Once in the sea only two mechanisms exist for recycling it to terrestrial ecosystems. One is through sea birds that harvest the

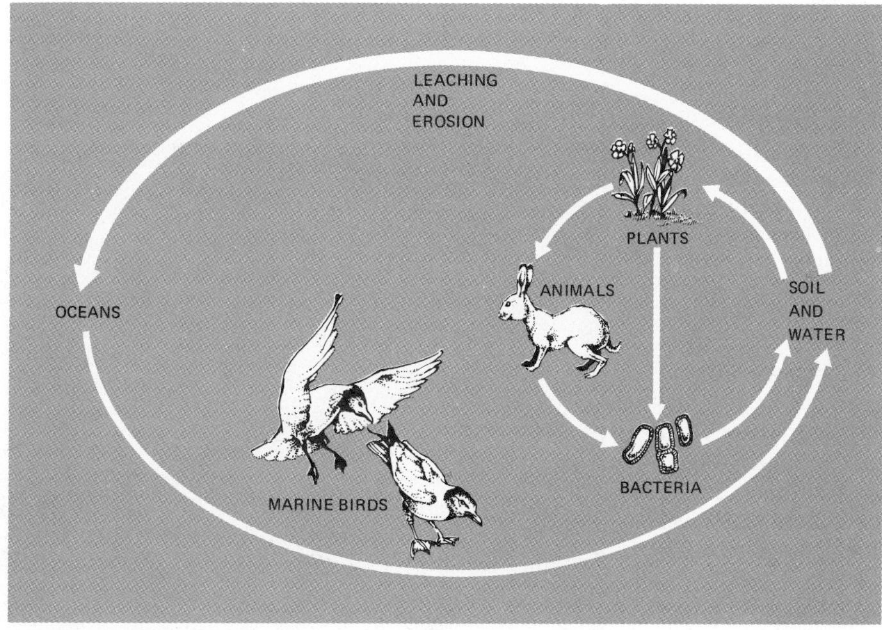

FIG. 41.8 The phosphorus cycle. Phosphorus is usually the limiting nutrient for organisms living in fresh water. Much of the phosphorus reaching the oceans is lost for long periods to terrestrial organisms.

FIG. 41.9 North Guanape Island, off the coast of Peru, is covered with guano—thick deposits of the excrement of sea birds such as cormorants. Because of its rich content of phosphates and nitrates, guano is a valuable fertilizer and is mined for that purpose. The dark patches in the photograph are flocks of nesting cormorants. (Courtesy of the American Museum of Natural History.)

FIG. 41.10 An algal bloom in Lake Minnetonka, Minnesota. Drainage into the lake of nutrients from cesspools, sewage treatment plants, and fertilized agricultural land has supported this explosive growth of algae. (Courtesy of the Minneapolis Star.)

phosphorus passing through marine food chains and may return it to land in their excrement. In fact, the great guano deposits on islands off the coast of Peru represent an important source of phosphorus-containing fertilizer (Fig. 41.9). Aside from the activity of these animals, we must wait for the slow geological uplifting of ocean sediments to form land—a process measured in the millions of years.

We affect the cycling of phosphorus by mining phosphate-containing rock from which we manufacture fertilizer and a great variety of industrial products. Among the latter are water-softening agents that make up 35–50% of many household detergents. After their work is done and the plug is pulled, these soluble phosphates join the general flow of sewage into our waterways. In many lakes and rivers, phosphorus is the limiting nutrient. Put another way, the growth of producers (e.g. algae) is regulated by the amount of phosphorus available. The popularity of phosphate detergents is now so great that their use may well account for a doubling of the phosphate content of waste water in the United States. Much of the nation's waste water is discharged into waterways without any treatment. Even where sewage treatment is used, most methods do not remove any appreciable amount of phosphate. The result has been a substantial increase in the phosphate content of many water supplies, of which Like Erie is a notorious example. With rising phosphate levels have come rising rates of primary productivity, chiefly in the form of explosive growths ("blooms") of algae (Fig. 41.10). Productivity in many contexts is a good thing, but in fresh water it means floating mats of algae, depressed oxygen levels at night, followed by fish kills and, finally, the decay—with its accompanying odor—of much of the increased algal biomass. We use the term **eutrophication** to describe the rise in productivity of water that follows its enrichment with nutrients. Eutrophication reduces the quality of the water supply for both drinking and recreation.

Compounding the problem is the fact that some of the phosphorus in fertilizer applied to cropland fails to be incorporated by the plants and is instead washed by rainfall and irrigation water into the local waterways. While estimates of the total amount involved are terribly uncertain, it is probably safe to say that in areas where population density is great or agriculture is intense, our activities have tripled the movement of phosphates into water supplies. Possible ways to control this movement, which harms the system it leaves as well as the system it enters, will be examined in Chapter 42.

41.6 OTHER MINERAL REQUIREMENTS

In addition to the major components of living matter, the cycling of which we have been examining, there are a variety of other elements that are also essential to life. Some, such as calcium, sodium, chloride, and magnesium, are needed in modest amounts. For others, the amounts needed are very small, so small in fact that we speak of these as trace elements. Because most of them are abundant in the nonliving world and are needed in only small amounts in the biosphere, their cycling poses relatively few problems of adequate supply. Perhaps only iodine is likely to be so scarce in the soil and water of some areas as to pose a hazard to human health. One or more of several other trace elements, such as iron, copper, zinc, cobalt, and molybdenum, may be sufficiently scarce in certain geographical locations that they limit plant productivity or prevent the successful raising of livestock. Molybdenum, for example, is absolutely essential for nitrogen fixation. But the amounts required are remarkably small. One ounce of molybdenum broadcast over an acre of cropland in Australia was found to be sufficient to restore fertility for over ten years (Fig. 41.11).

FIG. 41.11 Demonstration that clover grows normally only where the supply of molybdenum is adequate. The soil shown here (in eastern Australia) is naturally deficient in molybdenum. Although the entire fenced-in plot was seeded to clover, the plant was able to flourish and fix nitrogen only where molybdenum fertilizer had been added. (Courtesy of A. J. Anderson.)

FIG. 41.12 Measuring the cycling of materials in a forest in New Hampshire. The input of minerals into the valley is determined by measuring, with rain gauges, the amount of precipitation and its mineral content. The output from the valley is determined at this installation by measuring the discharge of water from the single stream that drains the valley and determining its mineral content. In recent years, the forest has been accumulating $SO_4^{=}$, NO_3^-, and K^+, while having a net loss of Na^+ and silica. (Courtesy of USDA, Forest Service.)

41.7 ANALYTICAL TECHNIQUES

Throughout this chapter and Chapter 40, we have examined estimates of the amounts and rates of energy and materials passing through the biosphere or some of its compartments. Given the size and complexity of these ecosystems, you may well wonder how such estimates are arrived at. A number of kinds of measurements must be made. Let us examine one particular example. In the White Mountain National Forest of New Hampshire a series of adjacent valleys have been set aside for a long-term study of the input and export of a large number of materials including ammonia (as NH_4^+), nitrate (NO_3^-), and sulfate ($SO_4^{=}$). This ecosystem was chosen because the only way in which significant amounts of material could enter was by way of rain or snow. Because the forest was underlain with a bed of water-impervious rock, the only way in which materials could leave was suspended or dissolved in the stream tributary that drained each valley (Fig. 41.12). It was assumed that if any animals brought materials into the valley, this would be balanced by animals taking materials away.

The analytical procedure had two major components. Rain gauges were distributed throughout each valley in order to compute the annual rainfall. Periodically the collected rain was analyzed for the presence of the various materials under study (e.g. sulfate). As for export from the system, the tributary draining each valley was led through a settling basin, equipped so that the rate of water flow could be measured (Fig. 41.12). Periodic analysis of this water provides data from which the quantities of materials leaving the ecosystem can be computed.

It turned out that the single greatest input into the ecosystem was sulfate. Three grams were added each year for each square meter of the forest, whereas only 2.9 g/m^2 left each year in the stream water. Thus this ecosystem appears to be accumulating sulfate. (The same is true of ammonium, nitrate, and potassium ions.) For other materials, such as sodium (Na^+) and silica (SiO_2), the rate of export was greater than the rate of import. Presumably, the first group was accumulating in the ecosystem as the forest grew, while the second group was following the inevitable progression of minerals from the land to the sea (see Chapter 42).

The limited scope of this ecosystem has provided

data in which one can place a high degree of confidence. It should be easy to see, however, that to extrapolate these data to a global scale introduces enormous uncertainties. Therefore, the figures presented in our studies of energy flow and materials cycling must be considered more or less crude estimates. As better data become available, many will have to be revised (just as many of them have been revised repeatedly in the past).

The conclusions we have drawn are therefore based on the assumption that the data—imperfect though they are—give reasonable approximations of the functioning of the biosphere. These conclusions must be regarded as tentative. The growing concern about human impact on the functioning of the biosphere has, on occasions, led to the use of certain data as the basis for gloomy predictions of imminent catastrophe. Our examination of these questions certainly leaves little ground for complacency. On the other hand, once you see how limited are the data from which many extrapolations and predictions are made, you can appreciate how uncertain our knowledge of the biosphere is and approach dramatic claims with a dose of healthy skepticism.

EXERCISES AND PROBLEMS

1. List four different kinds of bacteria that participate in the nitrogen cycle and tell specifically what each does.
2. What is the percentage of nitrogen by weight in a sample of 8-6-4 fertilizer?
3. Why do plants need nitrates and phosphates?
4. Approximately how many milligrams of carbon dioxide are present in each kilogram of air?

REFERENCES

1. BOLIN, B., "The Carbon Cycle," *Scientific American,* Offprint No. 1193, September, 1970.
2. CLOUD, P., and A. GIBOR, "The Oxygen Cycle," *Scientific American,* Offprint No. 1192, September, 1970.
3. DELWICHE, C. C., "The Nitrogen Cycle," *Scientific American,* Offprint No. 1194, September, 1970.
4. DEEVEY, E. S., JR., "Mineral Cycles," *Scientific American,* Offprint No. 1195, September, 1970.

All of the above articles can also be found in *The Biosphere,* Freeman, San Francisco, 1970. Available in paperback.

5. BORMANN, F. H., and G. E. LIKENS, "The Nutrient Cycles of an Ecosystem," *Scientific American,* Offprint No. 1202, October, 1970. How the input and output of minerals were measured in a forest in New Hampshire.
6. JANICK, J., C. H. NOLLER, and C. L. RHYKERD, "The Cycles of Plant and Animal Nutrition," *Scientific American,* September, 1976. Also available in *Food and Agriculture: A Scientific American Book,* Freeman, San Francisco, 1976.

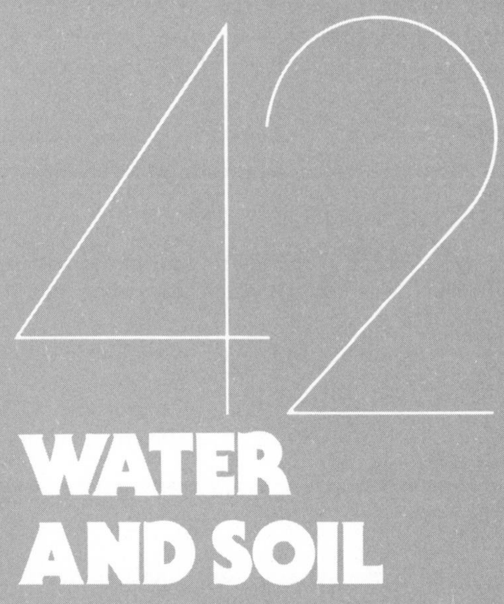

WATER AND SOIL

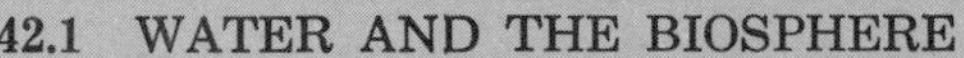

42.1 WATER AND THE BIOSPHERE

The importance of water to life is so pervasive that one ecologist has described the biosphere as a region (1) in which liquid water can exist in substantial quantities and (2) that receives an ample supply of energy from an external source. Water is the single most abundant molecule in living things. It represents some 60% of the weight of the adult human body and as much as 95% of the weight of such delicate forms as jellyfish and embryos. Virtually all of biochemistry is aqueous chemistry; that is, water provides the medium in which most biochemical interactions occur. Water also participates directly in many biochemical reactions, including those involved in cellular respiration, digestion, and photosynthesis.

Water is the habitat for many species of microorganisms, plants, and animals. Being terrestrial creatures we are apt to overlook the size, the complexity, and the special problems of freshwater and marine ecosystems.

On both a local and a global basis, water participates in the cycling of all other materials needed by living things. Every substance discussed in Chapter 41 cycles through the biosphere with some assistance from water.

42.2 THE PROPERTIES OF WATER

Water has a number of unique or unusual properties that make possible the many special roles it plays in the biosphere. It is unsurpassed as a solvent, able to dissolve a wide range of inorganic and organic substances. It is water's versatility as a solvent that enables it to serve as the medium of transport of so many materials. This is true within individual organisms as well as for the biosphere as a whole. Blood, lymph, and urine are three water-based liquids with essential transport functions, as we have seen.

Water exists in liquid form over a range of temperatures that are found over much of the earth and outside of which life processes are slowed down or halted entirely. Even when temperatures drop below its freezing point (0°C), the ice that forms floats on top of the water, thus providing a thermal blanket for the liquid water—and its denizens—beneath. The solid form of all other substances is denser than the liquid form. If this were also true of ice, bodies of water would freeze from the bottom up and in many cases would never thaw completely during summer.

Water has the largest heat capacity of any common substance. What this means is that it takes more heat (energy) to raise the temperature of water a given number of degrees than for any other material. Conversely, in cooling through a given number of degrees, water gives up more heat than any other common material. This property is important both on an individual and on a global basis. For the individual organism, the high heat capacity of water buffers it from sudden extreme shifts in temperature. On a larger scale, oceans and other large bodies of water moderate seasonal temperature fluctuations. It is for this reason that coastal areas are cooler in summer and warmer in winter than inland regions of the same latitude.

Water is also remarkable in its extraordinarily high heat of vaporization. It takes, on average, some 600 calories to convert a gram of water into water vapor. This value is higher than for any other common substance. It is of enormous significance to many animals in regulating body temperature. Sweating and panting are two common behavioral activities by which mammals (and some reptiles) lower their body temperature. The effectiveness of these activities stems from the large amount of body heat that must be used to evaporate water.

42.3 THE WATER CYCLE

Like all the other ingredients of the biosphere, water cycles (Fig. 42.1). Its major pool is the oceans, which contain an estimated 97% of the earth's water. The remainder is fresh water in liquid, solid, and vapor forms. About 75% of the fresh water is immobilized as the ice of glaciers and the pack ice of the polar regions. Recall the estimate given earlier, that if all glacial ice melted, the sea level would rise some 75–150 meters.

Something less than 1% of the earth's water exists as fresh, liquid water. Much of this is present as ground water in rock and soil layers beneath the surface of the earth. The remainder is the surface water of lakes and rivers. Most river water ultimately reaches the sea, bearing a load of salt, and mingles with the salt water of the sea. This continuing loss of fresh water is made up by the evaporation of sea water, using the energy of the sun. Evaporation is simply low-temperature distillation and thus the salts are left behind. As a global average, the quantity of sea water evaporated each year is equivalent to a layer 120 cm deep over the entire surface of the oceans. When the water vapor is cooled, it condenses and returns to the earth as fresh water in the

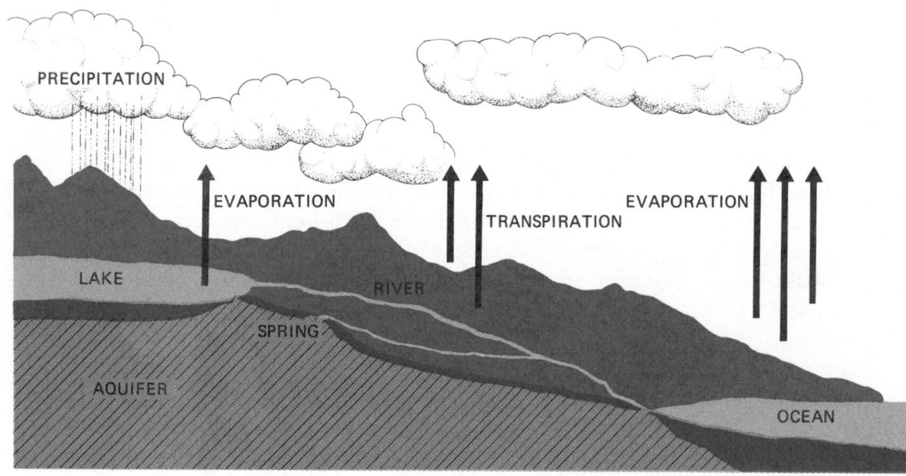

FIG. 42.1 The distribution of water. Evaporated sea water as it falls on land replaces the fresh water that flows constantly into the sea.

form of rain or snow. Of course it rains over the oceans as well as over land, but the topography of land areas promotes precipitation and, as a consequence, they receive more precipitation than the oceans. The difference is made up by river flow back to the oceans.

Evaporation also occurs on land. This takes place from the surface of standing water, from exposed moist soil, and especially, through the stomata of plants. Over a growing season, crops and forests transpire an amount of water equivalent to a layer 50–80 cm deep and, for many land areas, this represents well over half of the total water lost to evaporation.

The water falling on land first enters the upper layers of soil, from which it may pass to such bodies of surface water as ponds and streams. Where the necessary conditions are met, water may also pass down to the **water table.** The water table is simply the upper surface of a subterranean zone of water-saturated soil and/or rock. Such a zone is known as an **aquifer** (Fig. 42.2). At certain places and times, the water table reaches the surface of the earth, forming springs. Many of these occur in water-filled depressions, adding to the flow of surface water in such "spring-fed" lakes and streams. Conversely, water in such bodies can move downward and thus replenish or enlarge the aquifer.

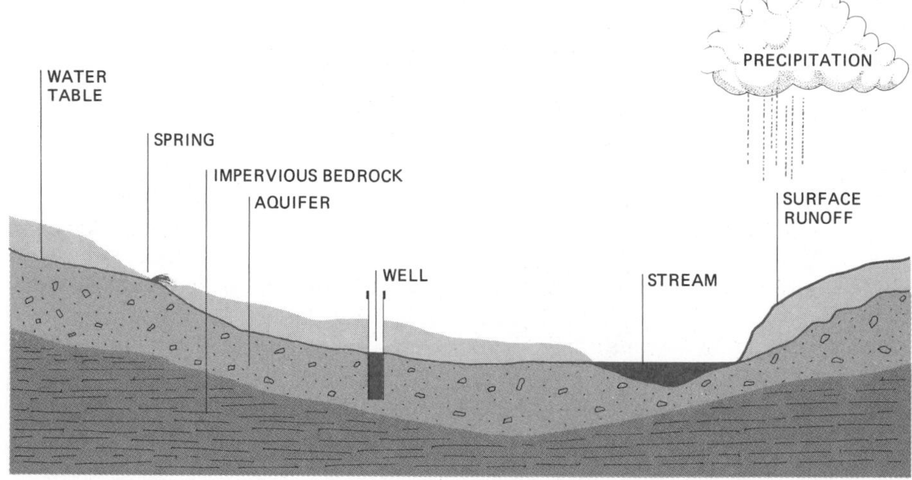

FIG. 42.2 Relationship between ground water and surface water. Pumping from streams and wells withdraws water from the aquifer. Water added to streams (and wells) replenishes the aquifer.

42.4 OUR USE OF WATER

A fast-increasing proportion of the earth's fresh water is being used to meet the needs of a single species: humans. The largest single use of water is in the irrigation of crops. In the United States alone, we use some 75–100 × 10^9 gal of irrigation water each day. This is approximately one-half of our total water consumption. Most animals must consume about ten times their weight in water each year. The demands of plants are much greater. Losses through transpiration are usually so great that crop plants may consume 100 times their weight in water during the growing season. In order to supply this need through irrigation, allowance must also be made for water loss through direct evaporation from irrigation ditches and cultivated soil. Enough irrigation water must also be used to keep salts moving downward toward the water table. As a practical matter, then, the ratio of weight of irrigation water needed to weight of crop produced is more apt to range from 1000:1 for such crops as sugarcane and corn to 10,000:1 for cotton.

Manufacturing and the generation of electrical power together require some 70 × 10^9 gal/day of water, an amount not far short of that needed for irrigation. In contrast to irrigation, however, much of this water is returned to the source from which it was taken. Power plants use water for cooling and then discharge the water, now somewhat warmer, back into the river from whence it came. Industry uses water in a multitude of ways, but sheer economics has forced manufacturers to find ways to recycle water to some degree through their processes. The paper industry is one of the largest users of water; it takes 250–600 lb of water to manufacture 1 lb of paper. The lower values are characteristic of mills with the most advanced recycling techniques.

The consumption of water for household use in the United States runs to some 27 × 10^9 gal/day or about 135 gal/day for each person. As in manufacturing, most of this water could theoretically be purified and returned quickly to the supply once it had served its function of washing or flushing away wastes. The actual process of recycling is made more difficult, however, because domestic waste water involves such an enormous number of widely dispersed point sources of pollution.

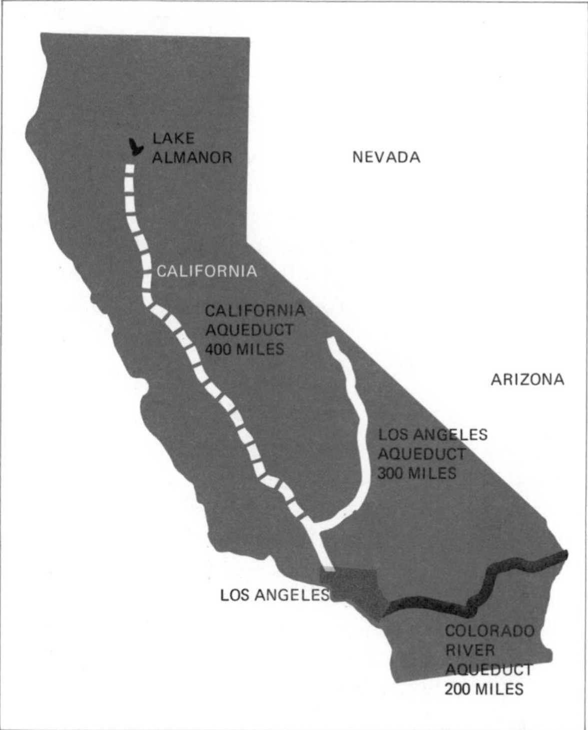

FIG. 42.3 The water supply of the Los Angeles basin. The low rainfall and large population of this area has required the building of aqueducts to bring in water from other regions. The Los Angeles Aqueduct, which brings water from the Owens Valley east of the Sierras, was completed in the 1920's. The Colorado River Aqueduct was built in the 1930's. Because neither of these will meet the future needs of the region, the California Aqueduct has been built to bring water from less densely populated northern California.

42.5 SOURCES OF WATER

The need for water is so great that every possible source has been exploited. In regions where rainfall is abundant, the surface water of lakes and rivers is the major source. Even where these supplies are inadequate, surface water may be brought by aqueduct from more ample sources several hundred miles away (Fig. 42.3).

Underground water is tapped extensively in arid locations. A substantial part of the water needs of Texas, Arizona, and New Mexico is met by withdrawing water from aquifers by means of wells. The discovery of extensive aquifers under the Sahara has raised the promise of bringing fertility and prosperity once again to that region. However, the experience already gained in tapping underground water supplies suggests the need for caution. Much of this water is truly fossil water, some of it 2000 or more years old. This indicates that it was a long time accumulating there. Before extensive withdrawals are made, it will be wise to determine just how fast the aquifers can be replenished. Certainly the experience in the southwestern United States has been ominous. The pumping of ground water in many regions is proceeding at a faster rate than the recharging of the aquifers. Thus the water table is declining. Not only does this mean that wells must be periodically driven deeper—and costs of pumping increase—but it raises the threat of ultimate bankruptcy of the system.

42.6 DESALINATION OF WATER

With 97% of the earth's water in the sea, it is no wonder that we have long searched for ways of economically converting salt water to fresh. A number of the world's arid regions are located close to the sea, e.g. along the west coast of South America and Australia and around the eastern Mediterranean, the Red Sea, and the Persian Gulf. Since the early 1950's the United States has led a multi-pronged attack on the problems of developing efficient, large-scale desalination methods. A number of techniques have been tried. The most obvious approach was to mimic nature and distill the sea water. A number of plants using various distillation techniques have been in operation. The most efficient of these are able to produce fresh water at a price sufficiently low to make it practical for industrial and domestic uses in locations that would otherwise go without or would have to import their water from other areas.

As the crystal structure of ice forms during freezing, impurities are excluded. In some parts of the world sea ice is actually harvested and used as a source of acceptable fresh water. This principle has now been used to desalinate water on an industrial scale. The energy requirements and ultimate cost appear to be similar to those for distillation processes.

Another approach to desalination is to mix salt water with ion-exchange resins. These materials remove positive and negative ions (e.g. Na^+, Mg^{++}, Cl^-), exchanging H^+ and OH^- (hence, H_2O) ions for them. When the capacity of the resins is exhausted, their activity can be restored by chemical treatment. Such resins can produce water of high quality (many laboratories produce their "distilled"

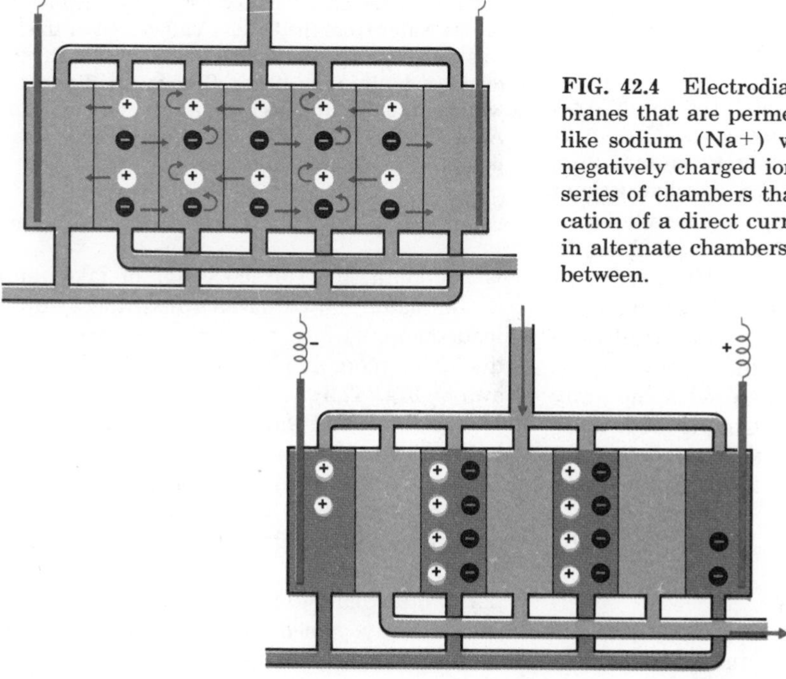

FIG. 42.4 Electrodialysis. The alternation of membranes that are permeable to positively charged ions like sodium (Na^+) with membranes permeable to negatively charged ions like chloride (Cl^-) forms a series of chambers that are filled with salt water. Application of a direct current causes the ions to accumulate in alternate chambers, leaving fresh water in the ones between.

water this way) but the expenses involved indicate that the process will be practical only for the conversion of brackish water for small-scale, local needs.

Desalting by electrodialysis requires two kinds of semipermeable membranes, one permeable only to positive ions, the other only to negative ions. By alternating these and applying a direct current, the ions accumulate in alternate chambers, leaving fresh water in the chambers between (Fig. 42.4). Again, this technique appears to be too expensive to do more than treat brackish water in critical areas. Such economic limitations do not, however, preclude its use as a final step in the purification of sewage, as we shall see later.

Semipermeable membranes can also be manufactured that allow the passage of water molecules but prevent the passage of salt ions. When such a membrane separates sea water from fresh water, the fresh water diffuses through the membrane (see Section 6.3). The diffusion pressure is about 350 lb/in^2. If an external pressure greater than 350 lb/in^2 is applied to the side containing sea water, the process reverses. Water molecules are forced from the side containing sea water to the opposite side, thus converting sea water to fresh. This filtration process is called reverse osmosis. Like electrodialysis, the process may hold more promise for sewage treatment than for desalting sea water.

Our appetite for fresh water is so great that it has even led to serious proposals that Antarctic icebergs (glacial ice) be towed to arid, coastal regions (e.g. Australia, southern California) as a source of fresh water. It has also intensified our concern with developing ways to intercept water that is not dissipated by our activities (e.g. in flushing toilets) and purifying this water sufficiently so that it may be used repeatedly as it passes on its way to the sea.

42.7 WATER POLLUTION

The great solvent power of water makes the creation of absolutely pure water a theoretical rather than a practical goal. Even the highest-quality distilled water contains dissolved gases and, to a slight degree, solids. The problem, therefore, is one of determining what quality of water is needed to meet a given purpose and then finding practical means of achieving that quality.

The problem is compounded because every use to which water is put—washing, irrigation, flushing away wastes, cooling, making paper, etc.—adds something to the water. Unless the effluent from these uses evaporates, this "used" water eventually finds its way back into surface (and ground) water supplies and thus is subject to reuse. Those downstream have to use the effluent of those upstream. To what degree, then, can various water users add materials to the water they use without (1) making its reuse downstream prohibitively costly because of needed purification and (2) destroying it as a habitat for desirable organisms? Not only in the United States, but in many other areas of the world, the growing needs for water to support growing populations and a growing technology have produced an overwhelming need for the implementation of water purification practices. Many of the world's rivers are today little more than open sewers.

What are the contaminants that degrade water supplies? They include suspended solids such as the silt from eroding farmland, certain kinds of industrial wastes, and organic waste ranging from plant debris to human fecal matter. They include microorganisms such as bacteria and viruses that may render the water a hazard to health. They include a vast array of dissolved materials, both inorganic and organic. Among the many inorganic solutes are nitrates, phosphates, and sodium, calcium, and chloride ions. The use of sodium chloride and calcium chloride to melt ice on winter roads has, in certain localities, raised the salt content of domestic water supplies above a level considered safe for heart patients restricted to a low-salt diet. Water containing over 100 ppm of calcium and magnesium ions is considered "hard" water. These ions form precipitates which interfere with the proper operation of boilers, washing machines, and other water-using equipment.

The dissolved organic wastes in water are exceedingly varied and, except for certain industries, poorly identified. Human sewage, food-processing plants, and many industrial operations all add large amounts of dissolved organic matter to water supplies. Aside from the noxious aspects of certain organic pollutants, these materials present a serious problem as a group. This is because they serve as nutrients for bacteria and other microorganisms in the water. Actually, up to a point, we should be grateful for this, because the activity of these microorganisms oxidizes the pollutants to harmless products (e.g. carbon dioxide, sulfate). In this way, a lake or stream can literally purify itself. The problem comes when the system is, as is so often the case, overwhelmed by such wastes. The decay of these materials requires oxygen, and with heavy loads the oxygen content of the water may diminish below the point where most fish can survive. As conditions in the water become anaerobic, the breakdown products become reduced rather than oxidized molecules, many of which (e.g. hydrogen sulfide) produce offensive odors and tastes.

Because of this relationship between organic pollutants and dissolved oxygen, the most widely used measure of water pollution is the biochemical oxygen demand (BOD), referred to in Chapter 41. This is a test of the amount of oxygen needed for bacteria to break down the organic matter in a water sample over a period of five days. Typical BOD values for raw sewage run from 200 to 400 mg of oxygen per liter of water (therefore, 200–400 ppm). Water for drinking should have a BOD less than 1.

For convenience, we can divide the major sources of water pollution into three groups: industrial, agricultural, and domestic. Industrial pollution is as varied as industry itself. Although much remains to be done, the high cost of water has prompted many industries to treat their waste water at the plant site so that some of it (about two-thirds, on average) can be reused. About 8% of industrial wastes are processed in community sewage treatment plants. But, unfortunately, many older industries that are located along rivers continue to dump untreated waste directly into the river.

One of the most intractable sources of industrial water pollution is the acid water that drains from abandoned coal mines into the rivers of much of Appalachia. These abandoned mines contain ferrous sulfide, FeS. Exposed to water and oxygen, the ferrous sulfide is oxidized to ferric ions (Fe^{+++}) and sulfuric acid—both of which seriously degrade the quality of the water for most purposes. pH values as low as 1.0 occur in acid mine water and extensive fish kills are caused by it. A number of processes for treating acid mine water have been developed and are used in active mines. For abandoned mines, however, the greatest hope is to prevent formation of the acid water in the first place. This requires that steps be taken to prevent water from reaching the mine or, failing that, to flood the mine completely so that oxygen cannot reach the ferrous sulfide. The inability to assess damages on the former owners of abandoned mines and the lack of public funds have made progress in this direction exceedingly slow.

The raising of livestock often creates serious water pollution. With respect to waste output, the animal population of the United States is estimated to outproduce the human population 20:1. The problem is particularly severe when the animals are raised in relative confinement. Over half the cattle

FIG. 42.5 Fish killed by careless disposal of pesticides. Some unused insecticides (a mixture of chlordane—a chlorinated hydrocarbon—and malathion—an organophosphate, see Chapter 45) were dumped one hundred yards away from this creek in Missouri. Washing of the material into the creek killed an estimated 349,000 fish, as well as uncounted numbers of aquatic insects, worms, crayfish, etc. (Courtesy of Steven B. Fuller.)

slaughtered in the United States spend their last weeks being fattened in feed lots. The water runoff from a feedlot carrying thousands of head of cattle can impose an enormous BOD on local water sources.

Agriculture degrades the quality of water supplies in two other important ways. Pesticides washing from crops and soil into streams and rivers have repeatedly caused enormous fish kills (Fig. 42.5). The use of large amounts of quick-acting (i.e. highly soluble) fertilizer also causes water pollution problems, although these are not as dramatic nor as well documented. The problems appear to be greatest with nitrate fertilizers applied in the spring. Rainwater leaching through fertilized fields dissolves nitrates and carries them into streams and ground water supplies. In some localized areas, the concentration of nitrates in ground water (hence in well water) has become so high as to pose a threat to health. Ingested nitrate is reduced in infants' stomachs to nitrite, which unites with hemoglobin, preventing it from carrying oxygen. Illness and sometimes death have occurred in both children and domestic animals after drinking well water containing high levels of nitrate.

Phosphate pollution of water supplies is an exceedingly serious matter, as phosphorus is usually the limiting nutrient in freshwater habitats. However, phosphate, unlike nitrate, binds tightly to soil particles. Crop fertilization is thus probably not a major source of phosphate pollution *except* as poor cultivation permits soil erosion to take place and the eroded, fertilized soil is washed into water supplies. Most phosphate pollution in water supplies probably stems from domestic sewage (see Section 41.5).

42.8 THE TREATMENT OF DOMESTIC SEWAGE

The wastes generated by approximately 60% of the United States population are collected in sewer systems and carried along by some 14 billion gallons of water a day. Of this enormous volume, some 10% is allowed to pass untreated into rivers, streams, and the ocean. The rest receives some form of treatment to improve the quality of the water (which constitutes 99.9% of sewage) before it is released for reuse. It should be noted, though, that many communities combine their sanitary and storm sewers. As a result, heavy rains usually exceed the capacity of the treatment system and, at such times, sewage escapes untreated into surface waters.

The simplest, and least effective, method of treatment is simply to allow the undissolved solids in raw sewage to settle out of suspension. Such **primary treatment,** as it is called (Fig. 42.6), removes only one-third of the BOD and virtually none of the dissolved substances. However, a slim majority

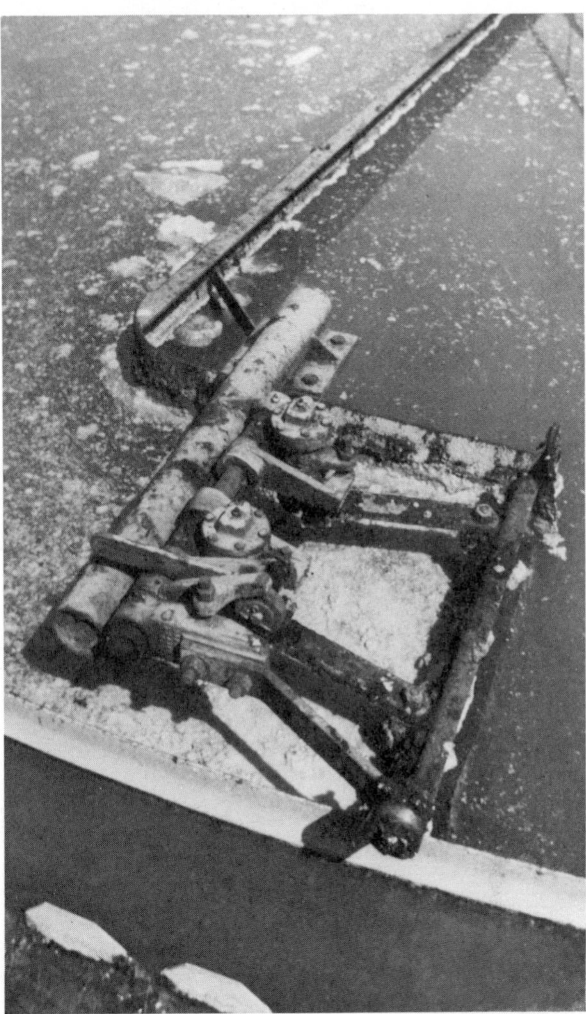

FIG. 42.6 Primary treatment of sewage. Suspended solids settle to the bottom of this basin while fats and other floating materials are skimmed off the top by the revolving boom. In this installation, there is no further treatment of the sewage, and the effluent is returned to a local stream. Primary treatment reduces the BOD of sewage by only one-third. Dissolved minerals like nitrates and phosphates are not removed at all. (Courtesy of Dr. Edward S. Hodgson, Tufts University.)

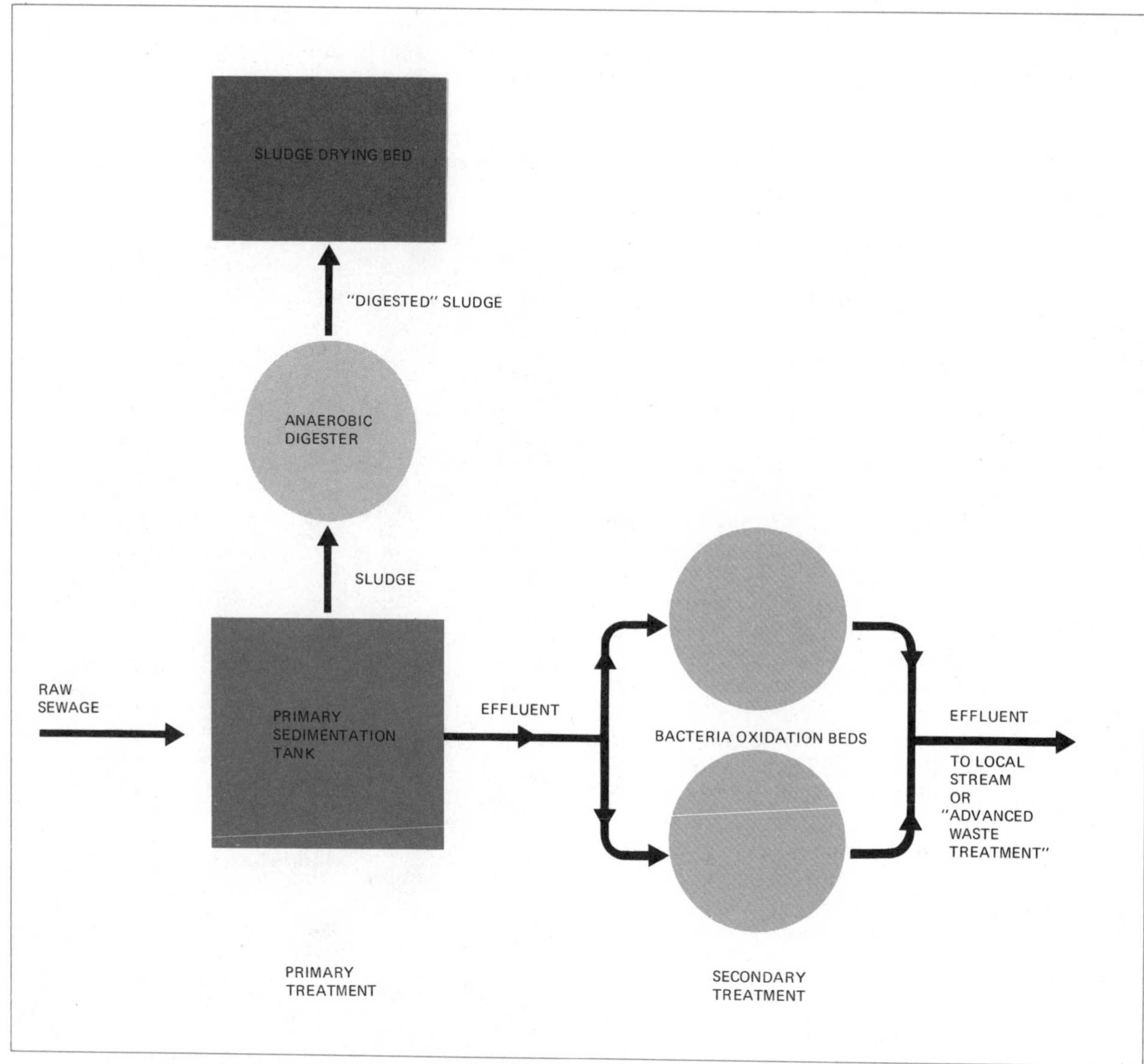

FIG. 42.7 Sewage treatment. In secondary treatment, *dissolved* organic matter is oxidized to such products as carbon dioxide and water by bacteria and other microorganisms. This reduces the BOD of the effluent. At a few plants in the United States, this effluent is then treated to remove dissolved nitrates and phosphates which would otherwise hasten the eutrophication of the body of water receiving the effluent. Attempts to use digested sludge as a low-grade fertilizer have usually not succeeded because of its content of toxic chemicals derived from industrial wastes.

of sewage treatment plants then pass the effluent from the primary treatment to a **secondary treatment** process (Fig. 42.7). Here the effluent is brought in contact with oxygen and aerobic microorganisms. Their metabolic activities break down much of the organic matter to harmless substances such as carbon dioxide and water. Primary and secondary treatment together can remove up to 90% of the BOD. With chlorination to reduce its content of bacteria, the effluent from secondary treatment is returned to the surface water supply.

42.9 ADVANCED WASTE TREATMENT

In recent years, it has become apparent that even the secondary treatment of sewage is inadequate to protect our surface (and underground) water supplies. Reservoirs and lakes in many parts of the world have been undergoing a frightening and accelerating deterioration in quality. Symptoms include a loss of clarity and periodic "blooms" of algae (look back at Fig. 41.10). Not only do algal blooms reduce the aesthetic values of the water, but they have other effects as well. After a period of explosive growth, the algae die and decay. The decay uses up so much oxygen in the water that most of the fish population is killed. A shift from aerobic to anaerobic conditions also creates end products that impart offensive odors and tastes to the water.

The growth of algae is limited by the availability of sunlight and nutrients. Phosphate is especially apt to be the limiting factor in fresh water although nitrates may be in short supply as well. Algal blooms occur when the water becomes enriched with phosphates and/or nitrates. The productivity of the water ecosystem is greatly, if erratically, increased—the process known as eutrophication. The eutrophication of surface water is probably inevitable. What has become ominous is the greatly accelerated rate of eutrophication of reservoirs and lakes.

While runoff from agricultural activities contributes to eutrophication, the major culprit appears to be the discharge of inadequately treated sewage. Most of the nitrogen and phosphorus in raw sewage is still present in the effluent from secondary treatment. How to remove these substances has become of prime concern. A number of advanced (or "tertiary") waste treatment procedures have been tried. As the problem is chiefly one of removing dissolved salts, it should be no surprise that the various techniques being explored for desalting sea water have also been studied for use in advanced waste treatment. These include electrodialysis, ion-exchange resins, and reverse osmosis (see Section 42.6). In addition, chemical methods which absorb dissolved organic molecules and precipitate phosphorus compounds show great promise.

The problem of nitrogen in the effluent might be solved if the activities of denitrifying bacteria could be harnessed to the task (see Section 41.3). However, most of the nitrogen is in the form of ammonia (NH_3). Once released into streams it is converted into the nitrates and nitrites upon which denitrifiers act, but by then the damage is done. However, by increasing the availability of oxygen at the secondary treatment, *nitrifying* bacteria growing in the system can convert most of the ammonia to nitrate. If the effluent is then brought in contact with denitrifiers (and a carbon source such as methanol to feed them), they convert most of this nitrate to nitrogen gas, N_2. This simply escapes to the atmosphere. Whether this manipulation of the nitrogen cycle will turn out to be practical on a large scale remains to be seen.

42.10 THERMAL POLLUTION

Eighty percent of the water used by industry is used simply for cooling. And the lion's share of this is used in the generation of electricity. To operate efficiently, a steam turbine must receive its steam at a high temperature and release it to a condenser kept at a low temperature. A water-cooling system is the most practical way to maintain this temperature differential. One kilowatt-hour of electricity produced by coal- or oil-fired generators dissipates some 1.5 million calories of waste heat. Nuclear power, which shows great promise of reducing air pollution, is 50–60% more wasteful of heat. One nuclear generating plant located on the Connecticut River, and typical of those now in operation, uses 372,000 gal of cooling water each minute and discharges it back to the river as much as 12°C warmer (Fig. 42.8). The ecological effects of this discharge have been under intensive study since 1965. Although effects have been seen, they do not appear to be harmful. This is partly because the river is sufficiently deep where the plant is located so that the warmed water never reaches the bottom. Thus heat-intolerant fish and other species can survive and migrate in the deeper water. What is known about the effects of heat on aquatic life suggests the need for caution, however. Metabolic rates and hence oxygen demand go up as temperature goes up. On the other hand, oxygen is less soluble in warm water than in cold. In the late summer, when river temperatures are naturally high and water flow low, the increased temperature burden created by larger and more numerous plants may have serious consequences.

The rapid growth of the power industry and the increasing part that nuclear power will play in

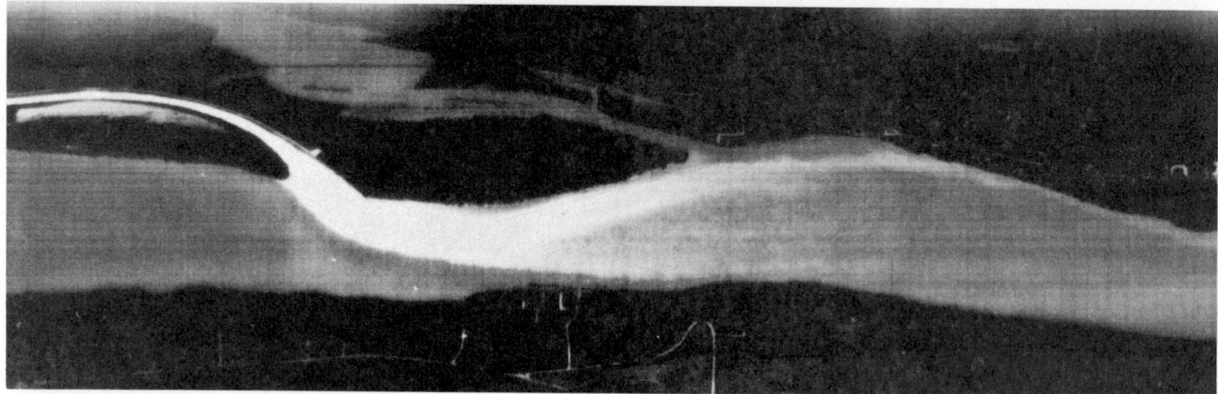

FIG. 42.8 Discharge of heated water by the Connecticut Yankee Nuclear Power Plant. River water is taken in to cool steam condensers and then discharged into a canal that empties back into the Connecticut River. In this area, the flow of the river is affected by the ocean tides. The top photo was taken at high tide and the bottom photo three hours later at ebb tide. Both photos were taken from an altitude of 2000 feet with equipment sensitive to infrared (heat) radiation. The heated water and other warm objects appear as light areas in these photographs. The plant was producing 479 megawatts of electricity at the time. (Photos courtesy of Environmental Analysis Department, HRB-Singer, Inc.)

FIG. 42.9 Cooling towers at the Paradise Steam Generating Plant of the Tennessee Valley Authority. The towers supplement the use of cooling water from the Green River that flows behind the plant. Each tower is 437 feet high and could enclose a football field in its base. This plant, in western Kentucky, has a generating capacity of 2558 megawatts of electricity, the largest of any fuel-fired plant in the world. (Courtesy of Tennessee Valley Authority.)

it indicate that the possible ecological consequences of each new station should be evaluated before construction is authorized and that alternative heat dissipating systems should be used whenever there appears to be a real risk of damage to an aquatic ecosystem. Cooling towers and holding lagoons are two alternatives that have been used (Fig. 42.9). Both approaches are expensive, the towers because of construction and operating costs, the artificial lagoons in terms of the land needed. A 1000-megawatt nuclear power plant, and plants of this size are now being planned, would require a lake extending over one to three square miles in order to dissipate its waste heat output.

42.11 SOIL

For terrestrial ecosystems, soil is the point of entry of most materials into living matter. Through their roots plants absorb water, nitrates, phosphates, sulfates, potassium, copper, zinc, and other essential minerals. With these, they convert carbon dioxide (taken in through the leaves) into proteins, carbohydrates, fats, nucleic acids, and the vitamins upon which they and all heterotrophs depend. Along with temperature and water, soil is a major determinant in the earth's productivity.

A vertical section through the soil usually reveals several layers or horizons (Fig. 42.10). The very top layer consists of partially decayed organic debris like leaves. Beneath this is the topsoil. This horizon is usually dark in color because of the decayed organic matter, **humus**, which has been incorporated in it from above. Productive agriculture depends upon a thick layer of topsoil. This is because humus gives the soil a loose texture that holds water and allows air to diffuse through it. Air is essential in order to meet the respiratory needs of plant roots, decay microorganisms, and all the other inhabitants of the soil.

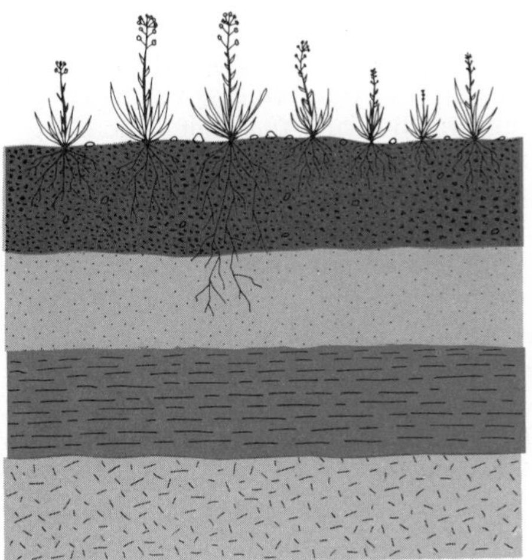

FIG. 42.10 A soil profile characteristic of the plains states. The rainfall is so light that soil minerals are seldom carried below the subsoil. As a result the fertility of this soil is easy to maintain.

Beneath the topsoil is often found a distinct horizon called the **subsoil.** This is usually lighter in color than topsoil and often contains a rich accumulation of inorganic nutrients.

Beneath the subsoil is a layer of weathered parent material. This represents the first steps in the chemical breakdown of rock into soil. Often the weathered parent material is underlain by the parent material itself, although sometimes it has been carried to its present site from some other location by wind, water, or glaciers. The chemical nature of the parent material, whether granite, limestone, or sandstone for example, has a great influence on the properties of the soil derived from it.

The formation of soil is a dynamic process. It is the outcome of: (1) the disintegration, through chemical action, of the parent material, (2) the formation (by microorganisms) and incorporation (by burrowing animals) of humus from above, and (3) the movement of minerals dissolved in the water percolating through it.

The amount of water falling on soil influences its properties and productivity tremendously. In humid areas (75–100 cm or more of precipitation per year), enough water falls on the soil so that much of it passes on down to the water table. As it does, it carries minerals with it. Such soils tend to be acid, and of low and (if unattended) diminishing fertility. Only by regular fertilization and liming (to restore calcium and raise pH) can productive agriculture be carried out in them. The soils in the eastern United States are of this sort.

In the plains states, the annual rainfall is sufficiently low (about 50 cm) that little or no rainwater ever percolates down to the water table. Calcium and other minerals are not carried below the reach of plant roots and consequently remain available for use. This keeps the pH and general fertility high. Except to the extent that minerals are lost when crops are removed, the minerals simply cycle

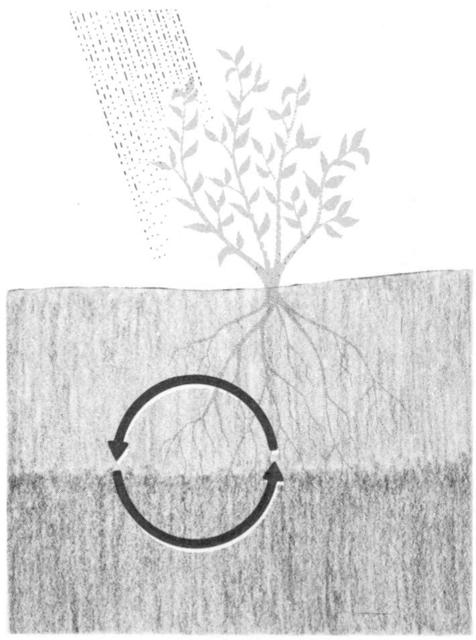

FIG. 42.11 Cycling of minerals in the soil of the plains states. The only loss of minerals occurs with the removal of crops.

from subsoil to plant parts to topsoil and back to subsoil (Fig. 42.11). The self-restoring fertility of the soils of the plains states accounts for this region's nickname of "breadbasket" of the nation.

In deserts, the rainfall is so low (25 cm per year or less) that water remains near the surface and is largely lost in evaporation. The salts it carries are left near the top of the soil. Their accumulation may make the soil so alkaline and so salty that most crops cannot be grown. The situation is especially bad in the Great Basin states (Utah and Nevada) because surface water flowing down from the mountains—bearing its load of dissolved salts—cannot flow on to the sea but simply flows out onto the valley floors and evaporates.

42.12 PROSPECTS FOR INCREASING THE WORLD'S CROPLAND

In attempting to meet the food needs of the earth's growing population, one of the first places to look for increased productivity is to increase the amount of land under cultivation. Three possibilities have been explored. One of these is to bring desert soils into production by irrigation. The southwestern United States and Israel are two regions that have converted large areas of formerly nonproductive land into fertile fields. (Fig. 42.12). However, even assuming an abundant source of pure water, irrigation is no panacea. Even the best irrigation water contains dissolved salts. If just enough water is applied to satisfy the needs of the crop, the salts are never carried deep into the soil. The high rates of evaporation characteristic of such areas hastens the accumulation of salts in the upper layers of the soil. If uncorrected, the condition may become so severe that only salt-tolerant crops, like sugar beets, can be grown.

The situation can be corrected by using enough additional irrigation water to flush the salts deep into the soil. Unfortunately, many desert soils are shallow and are underlain by layers that are relatively impervious to water. Irrigation water that does not evaporate accumulates in the soil and slowly the water table rises to the surface (if not at the point where the irrigation is applied, then further down the valley). Soon fields become waterlogged with salty water and, unless steps are taken to drain away this water, productivity quickly declines.

The Aswan Dam on the upper Nile River in Egypt was built in the hope that the impounded water could open up vast new areas to irrigated farming. To date, the project has not fulfilled the hopes of its planners. The extraordinarily high rate of evaporation in the region makes farming by irrigation a risky venture for reasons we have mentioned. The evaporation rates are, in fact, so high (up to 250 cm or 100 in. of water per year) that the lake behind the dam is not even filling as expected. Prior to the building of the dam, the annual flooding of the Nile brought silt and fresh nutrients to the land along its lower reaches, thus maintaining fertility. Now flooding does not occur and the continued productivity of these lands—which have been under cultivation for over 6000 years—has suddenly become dependent on intensive use of fertilizers. In the long run, the total agricultural productivity of Egypt may be no greater than before.

A great deal of the earth's surface is covered with hills and mountains. Most steep land is used to grow trees or for grazing animals. Attempts to enlarge the world's food production by converting such land to the growing of conventional crops are risky at best. Cultivated soil is highly susceptible to erosion by wind and water and its susceptibility rises exponentially with the steepness of the slope. While terracing is a solution that has been used since the dawn of agriculture, it hinders the large-scale use of machinery. The productivity of steep

FIG. 42.12 Aerial view of the Imperial Valley in southern California, taken by the crew of Apollo 9. The Salton Sea is visible at top center. Below it are the farms of the Imperial Valley (the many small squares). Were it not for the irrigation water brought by canal from the Colorado River, the entire area would be a desert. (NASA photo.)

lands is best maintained by keeping them covered at all times by grass or trees.

Perhaps the fondest hope of those who are looking for new land to cultivate is the vast tracts of jungle in Africa and South America. In one sense, these regions are quite productive. But the soil itself is not. Because of the high rainfall, nutrients are quickly washed out of the topsoil unless they have been incorporated in the forest plants. As plant and animal debris falls upon the ground, it is quickly decomposed because of the warmth and moisture. What is happening then, is that the minerals are held mainly by the forest cover, not by the soil itself.

FIG. 42.13 Intricate carvings decorate a pediment of the temple of Angkor Wat in Cambodia. The temple was constructed approximately one thousand years ago by the now-vanished Khmer civilization. The construction materials are sandstone and laterite. Laterite is a brick-like material produced when the local soil is exposed to sunlight. It has been suggested that the Khmer civilization failed because of the inability of the lateritic soil of the region to support successful agriculture. (Courtesy of Chhut Chhoeur, Permanent Mission of the Khmer Republic to the United Nations.)

When the jungle cover is removed, and cultivation attempted, the soils quickly lose fertility.

The situation is rendered even worse by the lack of humus (the topsoil may be no thicker than two inches) and the high iron and aluminum content of most of these soils. Once exposed to the sun, these so-called **lateritic** soils soon bake into a brick-like material which simply cannot be cultivated (Fig. 42.13). Virtually all attempts at using lateritic soils for what we would call "conventional" agriculture have failed. But, as we learned from the Tsembaga (see Section 40.2), jungle soils *can* be productive. The most ancient (some might say primitive!) way of working with these soils is still the best: clearing a small area of jungle, growing crops for only a year or two, and then abandoning the area to the jungle once again. In this way, laterization of the soil is avoided and its fragile fertility is maintained.

Attempts to export the agricultural technology of one region and apply it to another have failed more often than they have succeeded. The factors affecting successful agriculture—nutrients, water, pests, length of growing season, temperature, adaptability of the local farmers, and soil—vary from region to region. Only from careful on-site research will come improvements in productivity and most of these will be incremental gains rather than dramatic breakthroughs.

EXERCISES AND PROBLEMS

1. A sample of pond water containing 7 ppm of dissolved oxygen was stoppered and placed in the dark. After 5 days, the concentration of dissolved oxygen had dropped to 2 ppm and remained at that level. What was the BOD of the initial sample?
2. Over the course of one growing season (100 days), a corn field transpired an amount of water equivalent to a layer 60 cm deep. How much energy from the sun (in $kcal/m^2$) was needed to do the job? Approximately what fraction of the total radiant energy falling on the field was used in this process?
3. Why is the layer of topsoil much thicker in grasslands than in the tropical rain forest?
4. The time required for organic debris (e.g. newly fallen leaves) to decay and become incorporated in the topsoil was determined in four different ecosystems: tropical rain forest, temperate deciduous forest, taiga, and grassland. The values, *not* in the same order, were: 1 year, 3 years, 7 years, and 6 weeks. Match the values with the appropriate ecosystems.

REFERENCES

1. REVELLE, R., "Water," *Scientific American,* Offprint No. 878, September, 1963.
2. PENMAN, H. L., "The Water Cycle," *Scientific American,* Offprint No. 1191, September, 1970.
3. CLARK, J. R., "Thermal Pollution and Aquatic Life," *Scientific American,* Offprint No. 1135, March, 1969.
4. MERRIMAN, D., "The Calefaction of a River," *Scientific American,* Offprint No. 1177, May, 1970. A study of the effects (minimal) of thermal pollution in the Connecticut River.
5. KELLOGG, C. E., "Soil," *Scientific American,* Offprint No. 821, July, 1950.
6. PEIXOTO, J. P., and M. A. KETTANI, "The Control of the Water Cycle," *Scientific American,* Offprint No. 907, April, 1973.
7. MCNEIL, MARY, "Lateritic Soils," *Scientific American,* Offprint No. 870, November, 1964.
8. REVELLE, R., "The Resources Available for Agriculture," *Scientific American,* September, 1976. Also available in *Food and Agriculture: A Scientific American Book,* Freeman, San Francisco, 1976.
9. AMBROGGI, R. P., "Underground Reservoirs to Control the Water Cycle," *Scientific American,* Offprint No. 924, May, 1977. How aquifers might be mined in times of water shortage and recharged when water is abundant.

43 INTERACTIONS BETWEEN SPECIES

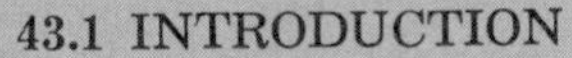

43.1 INTRODUCTION

The life of every organism is affected by the lives of other organisms. Just as each organism must cope with the conditions in its physical environment, so it must cope with the problems and opportunities presented by other organisms, in other words, the conditions in its biotic environment. The influence of some parts of the biotic environment is quite direct. Without suitable plant species on which to feed, the cow dies. Without the flagellate *Trichonympha* within its digestive tract, the termite cannot secure nourishment from the cellulose it consumes (Section 35.4).

Other influences of the biotic environment are less direct. An English naturalist of Darwin's time dramatized this point by saying that England's greatness was dependent upon her old maids. His reasoning went like this. England's power was dependent upon her navy, which in turn depended upon the physical well-being of her sailors. Good English roast beef took care of that and was supplied abundantly by cattle feeding on lush fields of clover (a legume and hence rich in protein). Reproduction of clover follows pollination by bumblebees. Bumblebee nests are preyed upon by field mice. Field mice are kept in check by cats and, of course, old maids keep cats! While this argument is somewhat fanciful, it does illustrate the point that any change in the biotic environment has not only direct effects (more cats = fewer field mice) but also indirect effects, which spread throughout the community of living organisms.

Most of the interactions between species involve food. Competing for food, eating, and avoiding being eaten are the most common ways in which organisms affect one another. Often these interactions are of brief duration. Two species may simply compete with each other for a certain type of food. Or, the relationship may be one of predator and prey, with one species attempting to feed on the other. There are many cases, however, where two species live in close association for long periods of time. Such associations are called **symbiotic** ("living together"). In all cases, at least one member of the symbiotic pair benefits from the relationship. The other member may be injured by the presence of the first member **(parasitism)**, may be relatively unaffected by it **(commensalism)**, or may also benefit from it **(mutualism)**. These various types of interactions between populations constitute the topic of this chapter.

43.2 INTERSPECIFIC COMPETITION

When two species each depend upon a particular resource in their environment, they are thrown into competition with each other for that resource. Most often, the resource involved is food, but other things such as a place in the sun (for plants) may be involved as well. Taken together, all the ecological requirements of a given species constitute its **ecological niche.**

Some organisms have relatively broad niches. The hawk alters its diet according to the relative abundance of the several kinds of animals that can serve as prey. The niche of the cotton boll weevil, in contrast, is a narrow one. It feeds exclusively on cotton plants. Where there is no cotton, there are no boll weevils. The occurrence of similar niches in different parts of the world helps explain the phenomenon of convergent evolution. The flying phalanger and wombat occupy in Australia the niches that are filled in North America by the flying squirrel and woodchuck (Fig. 32.14).

While the niches of many animal species in a community overlap, it may well be that no two species ever occupy *exactly* the same niche in the same location. If they did, we would expect that one species would be more efficient than the other at exploiting the niche and would eventually replace the less well-adapted species completely. This principle of **competitive exclusion** may not always hold. For example, two species of insects might occupy the same niche but other factors (weather, parasitism, predation) could be so severe that neither population could become large enough to diminish the food supply seriously. Generally, however, a close study of the feeding habits of two species which seem to occupy the same niche reveals some differences. While three of Darwin's finches, *Geospiza magnirostris, G. fortis,* and *G. fuliginosa* (see Fig. 32.11), eat a certain amount of the same food, each includes in its diet seeds of sizes not usually consumed by the others. We can thus conclude that the number of niches occupied by heterotrophs in any region is approximately equal to the number of species there.

The limited capacity of the environment to furnish energy establishes an absolute limit to population size. However, this limit is not apt to be reached under natural conditions. Every source of food is exploited by more than one species. Grasshoppers, rabbits, and cows all compete for the available grass. In other words, the presence of rabbits reduces the carrying capacity of the environment, K, for both grasshoppers and cows. They, in turn, reduce the K for rabbits as well as for each other. Thus the population of each is kept in check by competition with the others. Under these conditions, we can appreciate that any inherited traits which diminish the severity of interspecific competition will, by natural selection, tend to become established in the population. The usual outcome is the evolution of adaptations which increase the feeding efficiency of the species. The intense competition between two of Darwin's finches, *Camarhynchus pauper* and *C. psittacula,* for a particular size of seed on which to feed has resulted in powerful directional selection. The outcome has been **character displacement** and thus a reduction in the degree of overlap of their two niches so that the species can now coexist (see Fig. 32.13).

Increased efficiency of feeding goes hand in hand with increased specialization, however, and the result is a progressive narrowing of the niche of each species. The cotton boll weevil and the moth with the 25-cm proboscis each represent an extreme of feeding specialization. Their niche is narrow, but they exploit it more efficiently than their competitors can.

Plants, too, are always in competition with other plant species for sunlight, soil, water, and minerals. There are many specialized adaptations by which plants reduce interspecific competition for these necessities. Species with shallow root systems are able to coexist with deep-rooted species because each is tapping a different region in the soil. The seeds of shade-tolerant species may fail to germinate in sunny locations where they would be at a competitive disadvantage and, similarly, the seeds of shade-intolerant species will not develop successfully in shady locations. In the desert, the leaves shed by the brittlebush leave a poison in the soil that keeps competing annuals at their distance. The evolution of epiphytism, vines, and (in the case of many woodland shrubs) growth before the forest trees reach full leaf, represent adaptations to compete more effectively for sunlight.

43.3 PREDATION

Another major check on population size is predation. The majority of heterotrophic species secure their food by preying on other organisms. Although exceptions exist, most predators are larger than the prey they consume. Their relationship with their prey is usually a temporary one, just long enough to consume it or at least part of it. Animals which prey upon other animals almost always kill their prey. On the other hand, herbivorous predators (e.g. deer, rabbits, insects) usually just browse on a portion of the prey's body. The regenerative ability of most green plants then ensures that additional food will be available at a later time.

When we consider the central role that food plays in the lives of all animals, we should not be surprised at the many adaptations which (a) increase the effectiveness of predation and (b) minimize the risk of being preyed upon! Let us examine some examples of devices that help their owners to avoid being eaten.

1. Camouflage (Cryptic Coloration)

Many animals are patterned in ways that help them blend in with their surroundings. Some, like the winter flounder, are able to quickly alter their patterns as they move from one background to another (Fig. 28.20). The ptarmigan, with its white winter plumage and mottled summer plumage, keeps itself inconspicuous despite the changing seasons (Fig. 43.1). Many insects have evolved remarkably precise

FIG. 43.1 Ptarmigan in summer plumage, incubating her eggs. In the winter these birds are white. (Courtesy of the American Museum of Natural History.)

FIG. 43.2 Note the dark line on *Kallima's* wing that matches the midvein of the leaves.

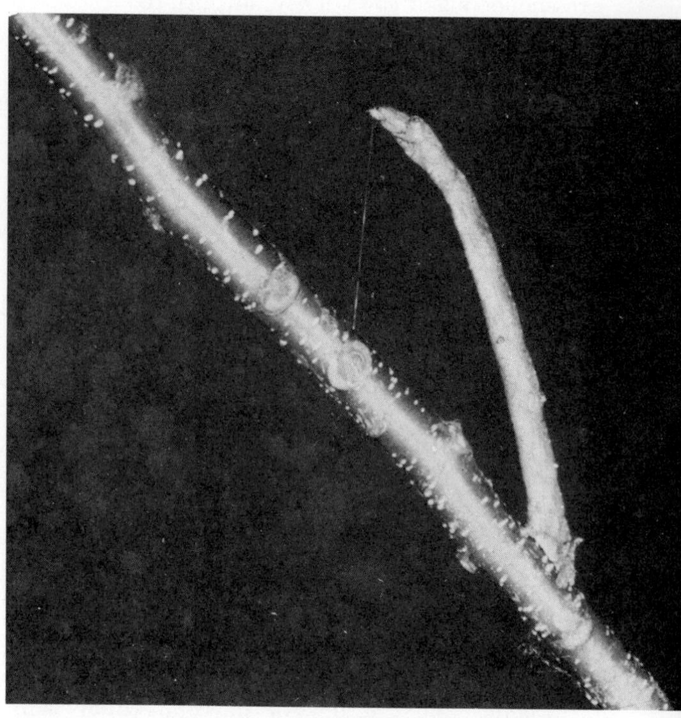

FIG. 43.3 The twig caterpillar. (Courtesy of Muriel V. Williams.)

techniques of camouflage. In Chapter 31, we examined the evolutionary response of the peppered moth, *Biston betularia,* to the darkening by air pollution of the tree trunks on which it rests (look back at Fig. 31.14). Although the true veins of a butterfly's wings radiate from the point of attachment, *Kallima*'s resemblance to a leaf extends even to having a dark line crossing its veins that matches the midvein of a leaf (Fig. 43.2). The motionless twig caterpillar complete with "buds" and "lenticels" (Fig. 43.3) is beautifully adapted to escape detection by birds (but pays for its cleverness by occasionally having some other insect lay eggs on it by mistake).

2. Defense

Many plants and animals have adaptations by which they protect themselves against attack by predators. The quills of the porcupine and the scent glands of the skunk are familiar examples. Millipedes secrete poisonous hydrocyanic acid when disturbed. Many insects have special glands and aiming devices which enable them to squirt potential predators with such noxious substances as 85% acetic acid or 40% formic acid. Toxic chemicals (including hydrocyanic acid) and thorns or spines are defensive mechanisms frequently found in plants.

What if you have a powerful defensive weapon but no potential predator notices until it has already launched an attack on you? One obvious answer is for the species to evolve **warning coloration.** Figure 43.4 shows the conspicuous larva of the monarch butterfly. There is no question of camouflage here. Rather, this creature is advertising its presence. It can afford to do so because it stores within its body toxic (to vertebrates) chemicals that it acquires from the milkweed plants upon which it commonly

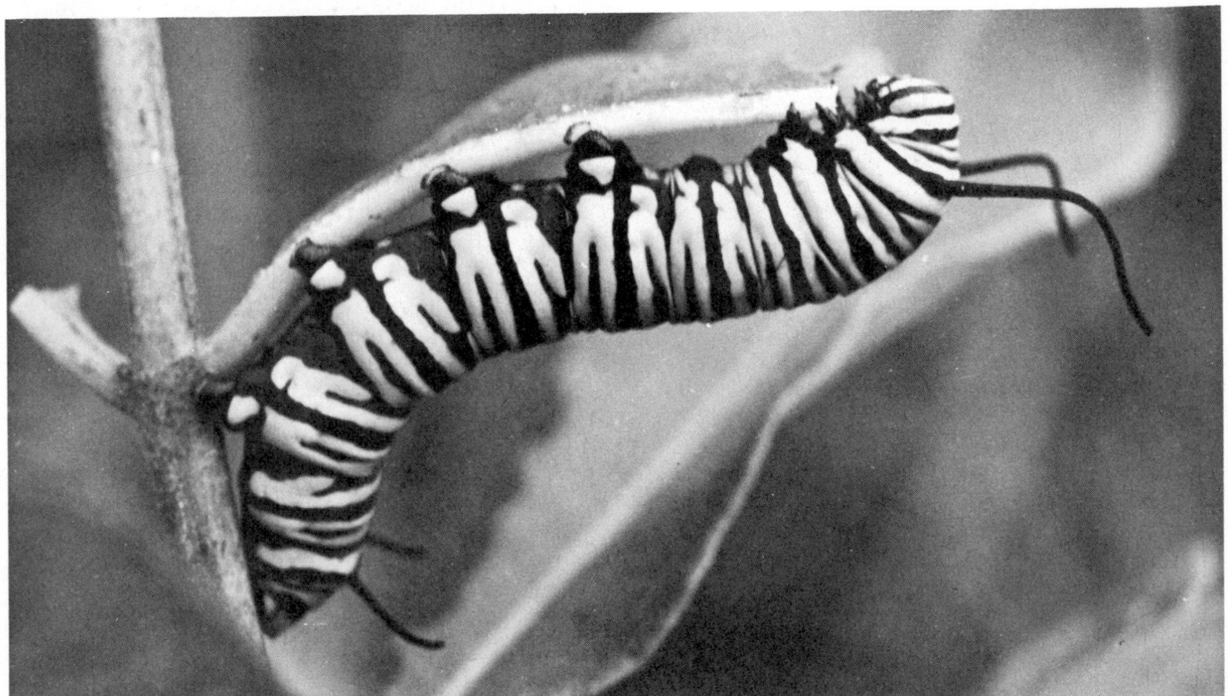

FIG. 43.4 Larva of the monarch butterfly: an example of warning coloration. The milkweed leaves upon which this animal is feeding contain several substances that are toxic to vertebrates. The larva stores these within its body and thus becomes unpalatable to vertebrate predators. Warning coloration is also called aposematic coloration.

feeds. Furthermore, the larvae retain these chemicals during metamorphosis so that the adults are unpalatable as well. Lincoln P. Brower has demonstrated that after just one encounter with an unpalatable monarch (Fig. 43.5), a blue jay will refuse to sample another. Clearly it is a valuable adaptation for the monarch to be easily recognizable by the blue jay.

3. Mimicry

If an animal is not lucky enough to be noxious to potential predators, why not look like an animal that is? This phenomenon is called mimicry. The viceroy butterfly contains no toxic substances in its body and presumably is quite palatable (one entomologist has tried it and declared that it tasted like dried toast!). However, the viceroy's striking resemblance to the monarch (Fig. 43.6) enables it to capitalize on the monarch's unpalatability. The robber fly mimics the well-armed bumblebee (Fig. 43.7) and thereby gains protection from predators. It has been demonstrated that after a toad has suffered the consequences of snapping up a bumblebee, it not only avoids bumblebees in the future but the robber fly as well. It is interesting to note that the robber fly, despite its resemblance to the bumblebee, has not been able to dispense with its evolutionary heritage. Like all flies, it has two wings instead of the four possessed by the bumblebees and other hymenopterans.

A number of harmless snakes mimic closely the bright warning coloration of the coral snake—the most poisonous snake in the United States. All of these cases in which a harmless species resembles a harmful species, are examples of **Batesian mimicry** (named after Henry W. Bates, a nineteenth century naturalist who studied many such cases).

FIG. 43.6 The monarch (top) and viceroy (bottom), an example of Batesian mimicry. The palatable viceroy gains protection from its close resemblance to the frequently unpalatable monarch. Batesian mimicry is successful only so long as the population of the mimic (the viceroy) remains smaller than the population of the model (the monarch). (Courtesy of Dr. Thomas Eisner, Cornell University.) ►

FIG. 43.5 A blue jay eats a portion of a monarch butterfly that had fed (in its larval stage) on poisonous milkweed (above). A short time later, the blue jay vomits (right). Following this episode, the blue jay refused to eat any other monarch offered to it. (Courtesy of Prof. Lincoln P. Brower, Amherst College.)

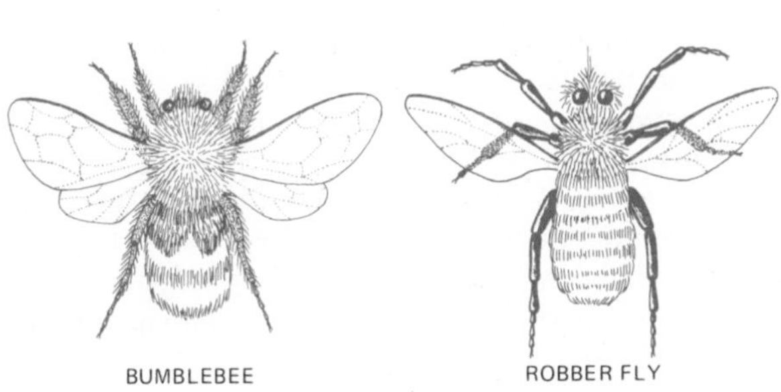

FIG. 43.7 The stingless robber fly is more closely related to the mosquito than to the bumblebee.

Some unpalatable animals closely resemble other equally unpalatable species. Such mimicry is called **Müllerian mimicry** (in honor of the German zoologist Fritz Müller who studied it). Presumably each species gains a measure of protection from the occasional, but educational, losses of the other species to predators.

4. Group Behavior

Cooperation between members of the same species often reduces the severity of predation. Grazing ungulates are usually organized so that the strong are on the outside of the herd, the weak within. The presence of many sets of eyes and ears in the herd makes the predator more easily detected and its approach more difficult. Smelt, a kind of fish, have been shown to release a pheromone into the water when alarmed. This serves to warn the other members of the school. When a honeybee stings an enemy, she releases isoamyl acetate, which excites other bees to join the attack.

5. Escape Responses

The balance between prey and predator is a delicate one and somewhat self-regulating. An increase in the population of the victim permits the predator population to increase. Increased predation then lowers the population density of the prey species and is followed by a reduction in the concentration of predators. Of course, that opens the way for another build-up in the prey population, and so it goes.

It is obviously important to the balance of nature for the rise and fall of predator and prey populations not to be so extreme that the prey are completely exterminated. If this should occur, the predator population would soon follow suit unless an alternate source of food was available. This might be impossible for a predator occupying a narrow niche. In a restricted habitat such as a small freshwater pond, mutual extinction could thus be a real threat.

We know practically nothing as yet about how this danger is generally avoided in such situations, but at least one fascinating control mechanism has recently been discovered. In the eastern United States, freshwater ponds are often populated by tiny mosquito fish (*Gambusia*), which feed on mosquito larvae (thus serving a useful function for us). *Gambusia* is, in turn, preyed upon by the pickerel, a fast-swimming, voracious predator. Mosquito fish do not hide in underwater vegetation, and, after watching a hungry pickerel at work, it seems miraculous that any mosquito fish could escape being eaten by it. Escape they do, thanks to a nice self-protective mechanism. Whenever mosquito fish detect the presence of a pickerel, they move up to the surface of the water and begin splashing about. So long as this goes on, the pickerel makes little effort to attack them and, when it does attack, often misses. Let one member drift beneath the surface, though, and it is devoured in an instant. The signal to the mosquito fish is a chemical one. When water where pickerel have been swimming is added to an aquarium containing mosquito fish, they immediately carry out their self-protective response.

It is not yet fully understood just why the pickerel is so often unsuccessful while its prey is at the surface, but the value of this mechanism is perfectly clear. By failing to exterminate its prey, the pickerel avoids exterminating itself. The pickerel population simply crops the surplus *Gambusia* population, and thus both populations are maintained at a stable level. It is also easy to see how natural selection maintains this response in *Gambusia*. Any individuals in the population whose genotype leads to a diminished response will be the first ones to disappear, genotype and all, down the throat of the pickerel.

43.4 PARASITISM

A parasite is an organism that lives on or in the body of another organism (the host) from whose tissues it derives its nourishment and to whom it does some degree of damage. The distinction between parasites and predators is not always clear. The tick or leech attaches itself to the body of its host for a short period while it sucks blood. In a sense, this is simply a form of predation in which the smaller and weaker organism is browsing on the larger and stronger one. In the case of blood-sucking hookworms, the relationship is of long duration and clearly one of parasitism.

There is probably not an organism on this earth that is not parasitized at some time during its life. Animals are parasitized by bacteria, fungi, viruses, protozoans, flatworms (tapeworms and flukes), nematodes, insects (fleas, lice), and arachnids (mites). Plants are parasitized by nematodes, fungi, bacteria, viruses, and a few other plants. Even microorganisms have parasites—amoebas, for example, may harbor bacteria and bacteria harbor viruses.

Parasites damage their host in two major ways. One is by actually consuming its tissues. Hookworms, parasitic amoebas, and the malarial parasites all cause this sort of damage. Some parasites do not consume enough of the material of their host to be dangerous, but in the process of their metabolism they release toxins which poison the host. The bacteria that cause tetanus, diphtheria, and scarlet fever are dangerous for this reason. Tetanus toxin interferes with synaptic transmission in the central nervous system. Diphtheria toxin exerts its poisonous effect on the protein-synthesizing machinery of the cell.

Of course, most intestinal parasites are competing with their host for its ingested food, and when the host's diet is inadequate to start with, the loss of nutrients to a parasite may in itself have serious consequences. Probably competition for particular vitamins or amino acids is more serious than competition for calories.

Although there are obvious exceptions, a parasite usually does not kill its host. To do so is to deprive itself of a meal ticket. Instead, the well-adapted parasite consumes just enough host tissue to supply its own needs without destroying the host. It has been said that a parasite lives on the income of the host while a predator lives off the host's capital.

So parasitism, like predation, is a two-way relationship. It is the product of evolutionary adaptations on the part of both host and parasite. The fact that each species of parasite is restricted to one or, at most, a few species of hosts suggests this. The extreme specialization of some parasites to conditions in their host and the host's ability to tolerate their presence is further evidence of a long history of mutual adaptations.

The significance of mutual adaptations between host and parasite is often nicely demonstrated when a parasite gains accidental entrance into an abnormal host. Larval tapeworms, hookworms, and flukes, whose normal host is some vertebrate other than humans, nonetheless may occasionally invade one of us by mistake. When they do so, they migrate widely through the body, causing extensive damage as they go. It is almost as though they were hunting for normal conditions. The rickettsia that causes Rocky Mountain spotted fever and the virus that causes yellow fever are two examples of parasites which exist in animal "reservoirs" (rodents and monkeys, respectively) and do little damage to these hosts. The outcome of their entry into humans is often quite a different story.

Parasites are often referred to as degenerate. In one sense, this is true. During the course of adapting to conditions in their particular niche, parasites have lost structures essential to the welfare of their free-living relatives. Looked at from the parasite's point of view, though, the loss of organs that are no longer appropriate represents a gain in efficiency and, thus, improved specialization. The tapeworm has no eyes, no digestive tract, and only vestiges of nervous, excretory, and muscular systems. But what good would these be anyway in the human intestine? On the other hand, the tapeworm has a scolex, a cuticle resistant to digestive enzymes, a highly efficient reproductive system, and a shape which practically ensures that it will not accidentally obstruct the intestine and endanger its host.

Loss of nonuseful structures and functions is characteristic of all parasites. *Rafflesia* is a parasitic angiosperm found in Malaya, which does not have roots, stems or leaves. It does, however, have tubes which penetrate the tissues of its host and it has one of the largest flowers (3 to 5 feet in diameter) known. This extreme emphasis on reproduction is also found in *Sacculina,* a barnacle which parasitizes crabs. The adult consists of no more than a sac (hence the name) containing reproductive organs, and not until the larvae were identified could it even be determined that the organism was a crustacean.

Viruses *may* represent the extreme of degeneration. They lack most of the enzyme systems necessary for a free life. Most consist simply of (1) sufficient genes to instruct the host cell how to manufacture more of themselves and (2) a protein capsid which helps the virus invade its host.

The mutual evolutionary adaptations of parasite and host may lead to a situation in which the parasite becomes less damaging to its host at the same time that the host becomes more resistant to the destructive effects of its parasite. A striking example of this has been taking place in Australia.

In 1859, the European rabbit was introduced into Australia for sport. It multiplied explosively in a land where it had no important predators (Fig. 43.8). The raising of sheep (another imported species) suffered severely as the rabbits competed with the sheep for forage. The situation became steadily worse until 1950, when rabbits infected with the myxoma virus were brought from Brazil and released. The epidemic which followed was fantastic. Millions of rabbits (perhaps as many as 99.5% of the population) died. Green grass returned. Sheep raising became easy once again. But the rabbits were not eliminated. In fact, although small epidemics still occur, the rabbit population has recovered to about 20% of its pre-1950 value. What has happened? Thanks to the careful plans that were made when the virus was introduced, it is possible to find out. Today's rabbits are markedly more resistant to the virus than were the rabbits in 1950. Each time

FIG. 43.8 Rabbits in Australia. Having removed all the forage plants, which ordinarily supply them with water as well as food, the rabbits must drink water from a pool. (Courtesy of Dunston, from *Black Star*.)

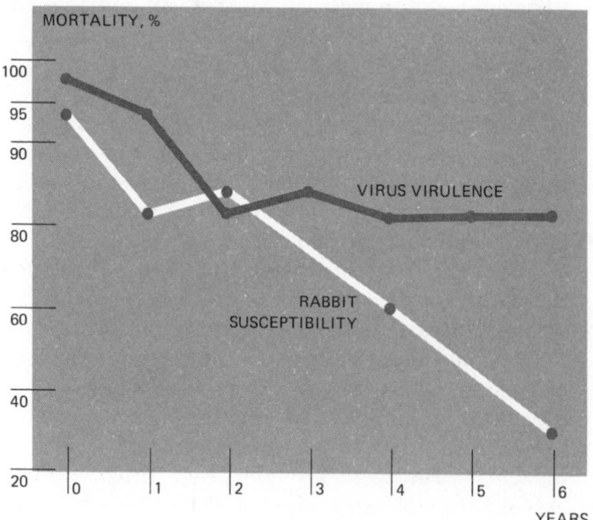

FIG. 43.9 Evolutionary changes in the virulence of the myxoma virus and the susceptibility of the European rabbit to the virus in the years following the introduction (in 1950) of the virus into Australia. The virulence of the virus was measured by determining the percent mortality of laboratory rabbits when they were infected with virus isolated each year from the wild rabbit population. The susceptibility of the wild rabbit population was measured by determining the percent mortality of young wild rabbits when deliberately infected with a strain of the virus isolated a short time after the first epidemic and maintained after that in the laboratory. (Redrawn by permission from Sir Macfarlane Burnet and D. O. White, *Natural History of Infectious Disease*, 4th ed., Cambridge University Press, 1972.)

young wild rabbits are trapped and deliberately infected with a laboratory strain of the virus (isolated in 1951), their mortality rate is, in general, lower than before (Fig. 43.9). Any student of evolution would have expected this. An epidemic with an initial mortality of 99.5% is certainly a powerful force for the directional selection of rabbits with a measure of innate resistance. Perhaps more surprising is the evolutionary change that has occurred to the virus. When laboratory-raised rabbits are infected with strains presently infecting the wild rabbit population, their mortality is substantially lower than it was when the original virus was used (Fig. 43.9).

While the relationship between rabbit and virus is still clearly one of parasitism, will it always be so? Probably. But there is evidence that the outcome of a long history of co-evolution of host and parasite may lead to a situation in which the parasite ceases to cause any significant damage to its host. Many bacteria that live in our throat and intestines do so without causing—usually—any harm. Such a relationship, in which they are benefitted and we are not harmed, is an example of commensalism. But the balance may be precarious. Something may happen, e.g. a suppression of the immune response, that enables these organisms to gain the upper hand and a commensalistic relationship reverts to a parasitic one.

What began as parasitism may even evolve to a stage where both partners benefit from the relationship. Such a mutually beneficial symbiosis is called mutualism. In 1966, K. W. Jeon discovered a culture of amoebas that had become infected with bacteria (60,000 to 150,000 per cell). The infection affected the amoebas adversely, slowing their rate of growth and making them much more fragile. Five years later, the amoebas still had their bacteria but now no ill effects could be seen. Most interesting of all, the amoebas—or at least their nuclei—had become dependent upon the bacteria. With micromanipulators it is a relatively easy task to remove the nucleus from an amoeba and replace it with another nucleus. When the nucleus was removed from an infected amoeba and replaced with one from a noninfected strain, the new combination worked very well. However, when the nucleus from an infected cell was placed in noninfected cytoplasm, the combination generally failed to survive. Evidently, proper functioning of these nuclei now depended upon some (unknown) contribution from the bacteria. What started as parasitism had become mutualism (the bacteria cannot be grown outside their host).

43.5 COMMENSALISM

Commensalism means "at table together." It is used to describe the symbiotic relationship in which one organism consumes the unused food of another. The relationship between the remora and the shark is a classic example. The dorsal fin of the remora is modified into a sucker by means of which the remora can form a temporary attachment to the shark (Fig. 43.10). The shark does not seem to be inconvenienced by this and makes no attempt to prey upon

FIG. 43.10 The sand tiger shark with two remoras attached. The presence of the remoras does not seem to inconvenience the shark, and they benefit from scraps of food left uneaten by their host. (Courtesy of New York Zoological Society.)

the remora. When the shark does feed, however, the remora is in a position to pick up scraps the shark fails to consume. There are certain species of barnacles that are found only as commensals on the jaws of whales. There are even species of barnacles found only as commensals on the barnacles that grow on whales!

Many of the bacteria living in our large intestine should be classed as commensals. They feed on undigested food materials and generally do not harm us. In fact, the experience gained from rearing germ-free laboratory animals suggests that some, at least, of the bacteria that live in the intestine benefit their host. Germ-free animals (e.g. mice) are abnormal in a number of ways, and it is now standard practice to deliberately infect such animals with several species of microorganisms so that the animals will develop normally. Evidently, the relationship between these microorganisms and their host is to some degree mutualistic.

43.6 MUTUALISM

Symbiotic relationships in which *each* species benefits are called mutualistic. (Some biologists prefer to restrict the term symbiosis to such relationships.) Algae are often found living within the bodies of heterotrophic organisms. The ciliate *Paramecium bursaria* harbors unicellular green algae within its cells (Fig. 43.11). The alga supplies food to its host. The alga presumably benefits from the availability of carbon dioxide and the fact that it can be trans-

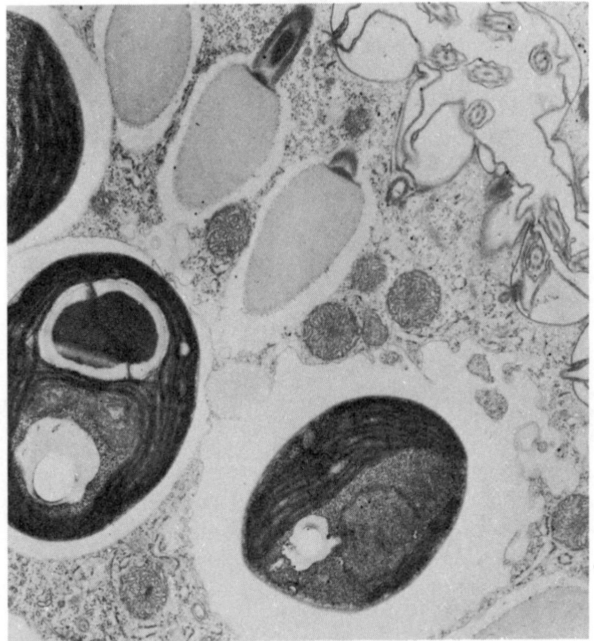

FIG. 43.11 Endosymbiosis. A protozoan, *Paramecium bursaria,* which harbors unicellular green algae (the dark ovals) within its cell. The algae supply food to their host. Both organisms can be cultured apart from each other, but then the paramecium must be given extra food. When brought together again, the paramecium ingests the algae and incorporates them in vacuoles. Many other aquatic heterotrophs, such as certain sponges, planarians, and clams, also harbor algae within their cells. (11,000 ×, courtesy of Stephen Karakashian.)

FIG. 43.12 Indian pipe, a flowering plant that lacks chlorophyll. It gets its nourishment through mycorrhizal fungi that connect its roots to those of a chlorophyll-containing species such as a spruce or pine tree. (Courtesy of J. W. Thompson.)

ported by its host to a spot where ample light is present. Both organisms can be cultured apart from each other but then the paramecium must be given extra food. When brought together again, the paramecium ingests the algae and incorporates them in vacuoles. Many other aquatic heterotrophs, such as certain sponges, sea anemones, planarians, and clams, also harbor algae within their cells.

Mutualistic relations between plants and fungi are very common. The fungus invades and lives in or among the cortex cells of the secondary roots. The association of fungus and root is called a **mycorrhiza.** A number of experiments have demonstrated clearly that the presence of mycorrhizal fungi greatly increases the efficiency with which the host plant absorbs minerals from the soil. Some mycorrhizal fungi also secrete antibiotics which may help protect their host from invasion by parasitic fungi and bacteria.

The advantage of the mycorrhizal relationship to the fungus has also been shown. The fungus gets its nourishment from the sugars that the plant translocates to its roots. In fact, mycorrhizal fungi may establish conduits for nutrients *between* plant species. The colorless, and hence heterotrophic, Indian pipe (Fig. 43.12) secures its nourishment from mycorrhizal fungi that are attached simultaneously to the roots of some autotroph such as a pine or spruce. Radioactive carbon administered to the spruce tree eventually turns up in the Indian pipes, even though they may be growing some distance away on the forest floor.

Many of the mushrooms that we find growing on the forest floor are the spore-forming bodies of mycorrhizal fungi. The truffle, prized by gourmets, is often found in oak forests because the fungus that produces it establishes mycorrhiza on oak roots.

Mutualistic relationships often involve intricate adaptations of structure, function, and even behavior on the part of the two species. Some species of ants secure their food by growing gardens of a particular fungus in their underground nests (Fig. 43.13). The fungus is nourished by leaves the ants collect, bring underground, and chew into a nutritious pulp. The fact that the fungus is never found except in ant

FIG. 43.13 An ant (*Mycetosoritis hartmanii*) feeding in its fungus garden. The white filaments surrounding the ant are the fungus which is growing on leaf fragments harvested by the ant and placed in its garden. (Courtesy of John C. Moser, United States Forest Service.)

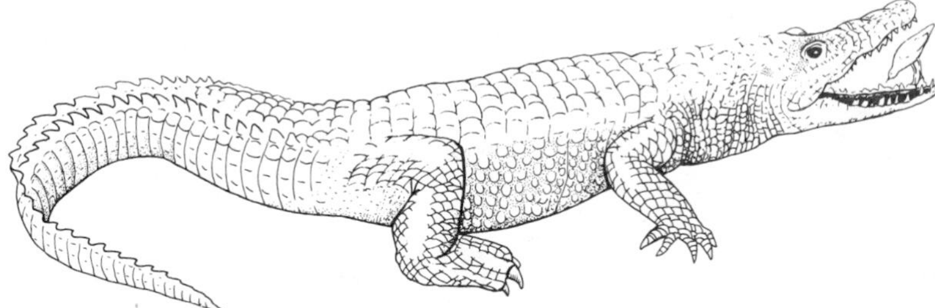

FIG. 43.14 Cleaning symbiosis. The Nile crocodile opens its mouth and permits the Egyptian plover to feed on any leeches attached to its gums. Cleaning symbiosis is more common in fishes.

gardens indicates a close physiological adaptation to the conditions of its culture. The behavior of the ants is certainly adapted to the needs of the mutualistic relationship. Even a special structural adaptation exists: a pouch in the head of the queen into which she stuffs some of the fungus before going off to start a new colony! Figure 43.14 illustrates still another kind of mutualistic relationship that involves elaborate behavioral adaptations.

Symbiotic Nitrogen Fixation

In terms of the overall economy of the biosphere, one of the most important examples of mutualism is that established between soil bacteria of the genus *Rhizobium* and their legume hosts. Although each organism is able to survive independently (soil nitrates must then be available to the legume), growth together is clearly beneficial to them both. Only together, can nitrogen fixation occur. A substantial fraction of the nitrogen fixation occurring on the earth today is accomplished by the symbiotic union of rhizobia and legumes. Because of the importance of symbiotic nitrogen fixation, it has been studied intensively. Let us examine some of the principal features of the process.

Rhizobia are gram-negative bacilli that live freely in soil (especially in soil where legumes have been grown). However, as long as they are living in the soil, they are incapable of fixing nitrogen. Only when they have invaded the roots of the appropriate legume can they begin to fix nitrogen.

The invasion of the legume's roots by the rhizobia could be considered an infection. But an infection with a difference. The legume roots secrete into the soil a substance that attracts the rhizobia to them! Once the bacteria enter an epithelial cell of the root, they migrate into the cortex. Their migration occurs within an intracellular channel that grows through one cortex cell after another (Fig. 43.15). This channel, called the "infection thread," is actually constructed by the root cells, not the bac-

FIG. 43.15 Rhizobia-filled infection thread growing into a cell (from the upper left to lower right). Note how the wall of the infection thread is continuous with the wall of the cell. Once the thread ruptures, rhizobia are released directly into the cytoplasm of the cell (dark ovals). (Electron micrograph courtesy of Dr. D. C. Jordan.)

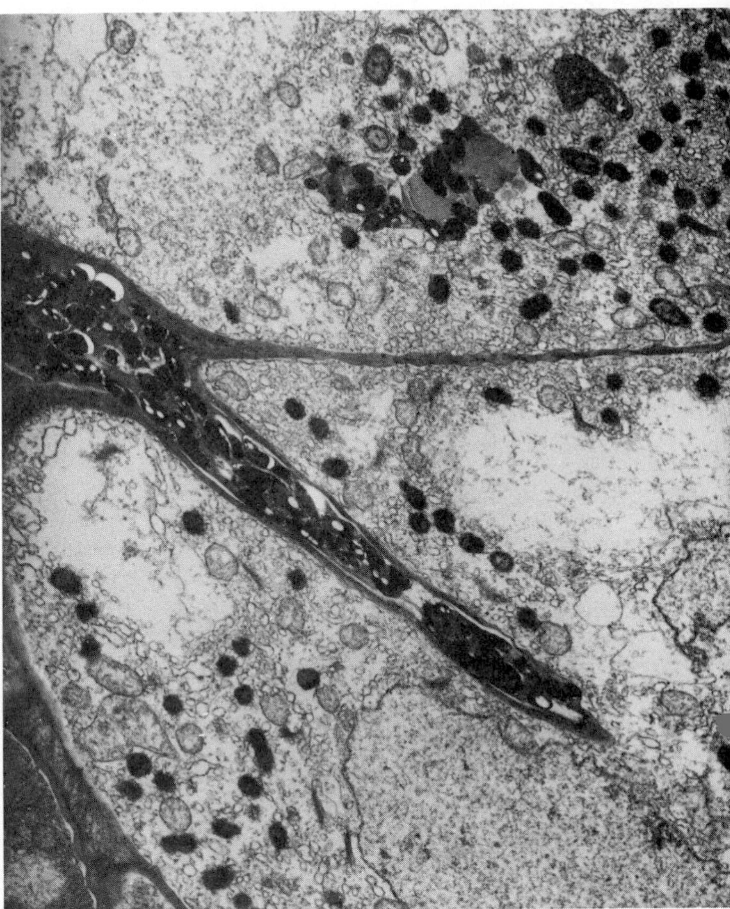

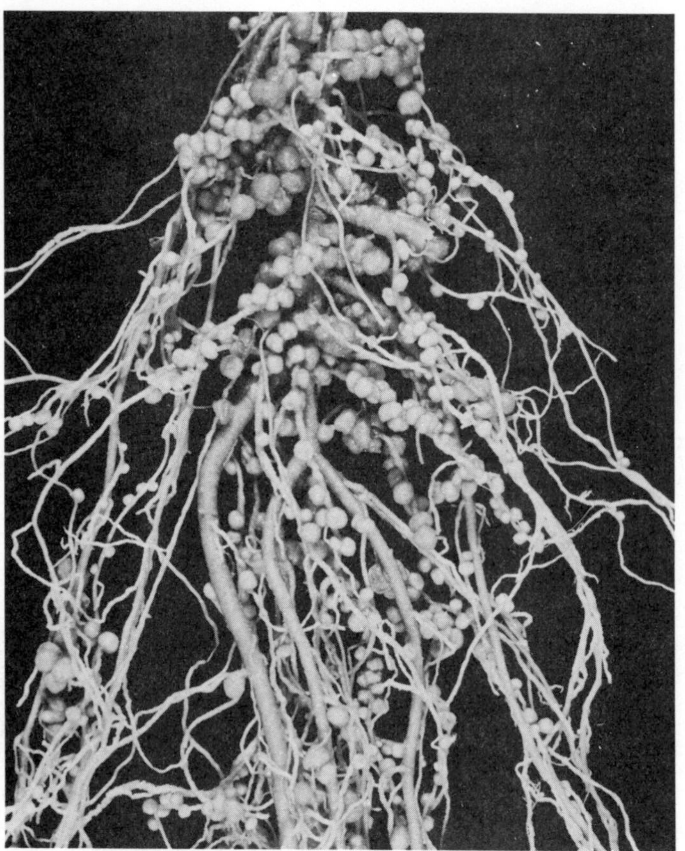

FIG. 43.16 Nodules on the roots of the birdsfoot trefoil, a legume. (Courtesy of The Nitragin Co., Milwaukee, Wisconsin.)

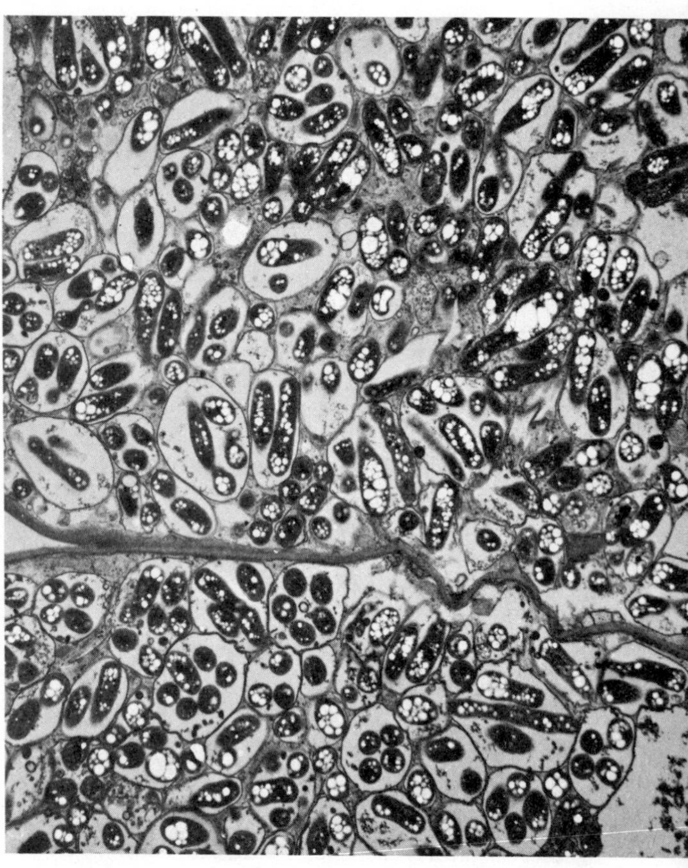

FIG. 43.17 Bacteroid-filled cells from an active soybean nodule. The horizontal line marks the walls between two adjacent nodule cells. (10,400 ×, courtesy of R. R. Hebert.)

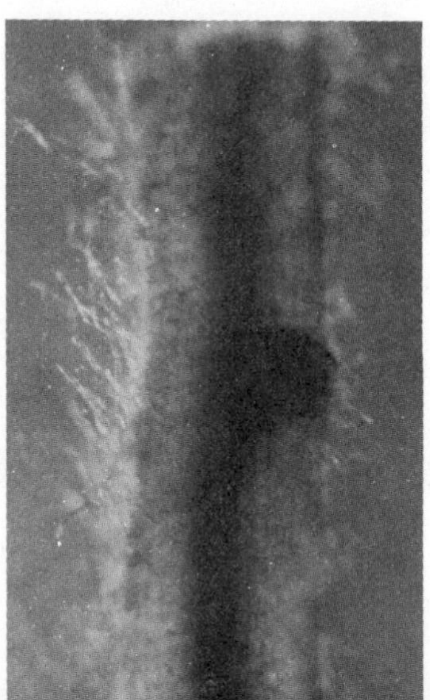

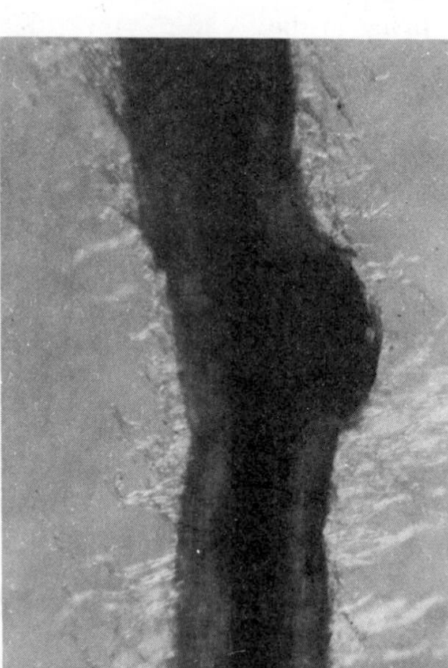

FIG. 43.18 Left: Segment of a pea root showing a developing lateral root. Right: Segment of a pea root showing a developing nodule (12 days after the root was infected by rhizobia). Both structures are connected to the nutrient transport system of the plant (dark area extending through the center of the root). (Photomicrographs 12 ×, courtesy of John G. Torrey.)

teria, and is formed only in response to the infection. When the infection thread reaches a cell deep in the cortex, it bursts and the bacteria swarm through the cell. Their presence causes the cell to go through several rounds of mitosis—without cell division—so the cell becomes polyploid. Then the cell begins to divide rapidly, ultimately forming a nodule (Fig. 43.16). The stimulus for these changes is probably the secretion of cytokinins (see Section 24.8) by the rhizobia.

The rhizobia also go through a period of rapid multiplication within the nodule cells. Then they begin to change shape and lose their motility. The *bacteroids*, as they are now called, may almost fill the cell (Fig. 43.17). Only at this time does nitrogen fixation begin.

Root nodules are not simply structureless masses of cells. Each becomes connected by xylem and phloem to the vascular system of the rest of the plant. Thus the development of nodules, while dependent on the presence of the rhizobia, appears to be a well-coordinated developmental process of the plant (Fig. 43.18).

Without rhizobia, legumes cannot fix nitrogen. But without legumes, rhizobia cannot do it either. This is so, even though a number of other kinds of soil bacteria (e.g. *Azotobacter*) are able by themselves to fix nitrogen, having no need to establish any sort of symbiotic relationship with another organism. Clearly, legumes and rhizobia are mutually interdependent with respect to their ability to fix nitrogen. What does each organism contribute to the process?

Biochemical studies of isolated bacteroids have demonstrated that they contain all the metabolic machinery—including nitrogenase—needed to fix nitrogen. Then why is the legume necessary? The legume is certainly *helpful* in that it supplies nutrients to the bacteroids with which they synthesize the large amounts of ATP that are needed to convert the inert molecules of nitrogen (N_2) into ammonia (NH_3).

In order to secure adequate amounts of ATP, the bacteroids need oxygen to carry on cellular respiration. However, the activity of nitrogenase is strongly inhibited by oxygen. Thus the bacteroids must walk a metabolic tightrope between too much and too little oxygen. Their job is made easier by a second contribution from their host: hemoglobin. Nodules are filled with hemoglobin. So much is present, in fact, that a freshly cut nodule is bright red in color! The hemoglobin of the legume (called leghemoglobin), like the hemoglobin of vertebrates, probably supplies just the right concentration of oxygen to the bacteroids to satisfy their conflicting requirements.

Beyond these contributions, though, the plant must also, in a way as yet unknown, make possible the conversion of rhizobia that cannot fix nitrogen into bacteroids that can.

How did two such organisms ever work out such an intimate and complex living relationship? Assuming that the ancestors of the rhizobia could carry out the entire process themselves—as many other soil bacteria still do—they must have gained some real advantage from evolving to share the duties with the legume. Perhaps the environment provided by their host, e.g. lots of food and just the right amount of oxygen, enabled the rhizobia to do the job more efficiently than before and *also* to stop carrying out inefficiently certain functions that the legume can carry out more efficiently.

Not all rhizobia can successfully form nodules on the roots of all legumes. Some strains will infect only peas, some only clover, some only alfalfa, etc. The deliberate exposure of legume seeds to the proper strain of rhizobia is now a routine agricultural practice. The photograph in Fig. 43.16 was taken by research workers in a company that specializes in producing rhizobial strains appropriate to each leguminous crop.

The implications of genetic engineering, which we discussed in Section 18.8, have not escaped research workers interested in nitrogen fixation. In 1972, two scientists in England, R. A. Dixon and J. R. Postgate, succeeded in transferring the genes required for nitrogen fixation (the *nif* genes) into *E. coli*. While this maneuver probably will not enable *us* to give up our dependence on plants and animals for our nitrogen needs, it does raise the possibility of inserting these genes into other kinds of cells. The major sources of energy for the human species are the cereal grains: rice, wheat, and corn. While these are superb at meeting our calorie requirements, the protein they produce (and for which they need large amounts of high-nitrogen fertilizer) is inadequate to meet human dietary needs. If functioning *nif* genes could be introduced into these plants, it would represent an extraordinary advance in the ability of the human species to feed itself.

EXERCISES AND PROBLEMS

1. List one human parasite from each of seven phyla of eukaryotes.
2. How would you attempt to demonstrate that *Paramecium bursaria* actually receives nourishment from the unicellular green algae it harbors?
3. The fixation of nitrogen to ammonia requires approximately 147 kcal/mole of N_2. What is the minimum number of ATP molecules that you would expect to be needed for the process? (Recent measurements indicate that the actual value is close to this.)

REFERENCES

1. FEDER, H. M., "Escape Responses in Marine Invertebrates," *Scientific American,* Offprint No. 1254, July, 1972.
2. HOWARD, W. E., *The Biology of Predator Control,* Addison-Wesley Modules in Biology, No. 11, Addison-Wesley, Reading, Mass., 1974.
3. EHRLICH, P. R., and P. H. RAVEN, "Butterflies and Plants,"*Scientific American,* Offprint No. 1076, June, 1967. The unpalatability of certain butterflies (which makes them good models to be mimicked) arises from chemicals in the plants they have eaten during their larval stage.
4. BROWER, L. P., *"Ecological Chemistry," Scientific American,* Offprint No. 1133, February, 1969. Demonstrates that the unpalatability of the monarch butterfly comes from toxic (to vertebrates like the blue jay) substances in the milkweed plants upon which monarch larvae feed.
5. SMITH, D. C., *Symbiosis of Algae with Invertebrates,* Oxford Biology Readers, No. 43, Oxford University Press, Oxford, 1973.
6. HARLEY, J. L., *Mycorrhiza,* Oxford Biology Readers, No. 12, Oxford University Press, Oxford, 1971.
7. BATRA, SUZANNE W. T., and L. R. BATRA, "The Fungus Gardens of Insects," *Scientific American,* Offprint No. 1086, November, 1967.
8. LIMBAUGH, C., "Cleaning Symbiosis," *Scientific American,* Offprint No. 135, August, 1961. Gives several examples of mutualism in which animals of one species secure food by cleaning animals of another species.
9. BRILL, W. J., "Biological Nitrogen Fixation," *Scientific American,* Offprint No. 922, March, 1977.

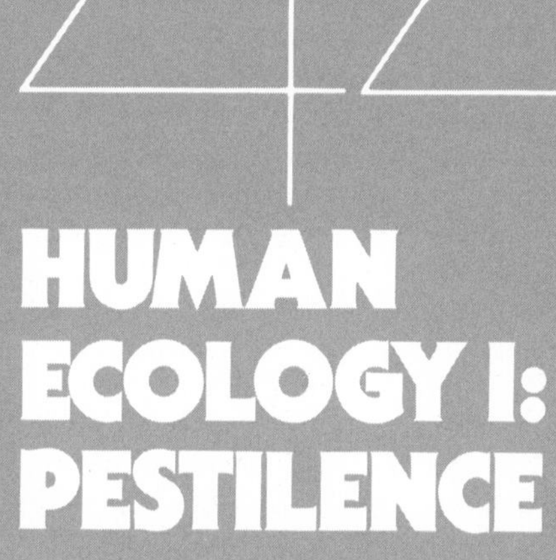

44 HUMAN ECOLOGY I: PESTILENCE

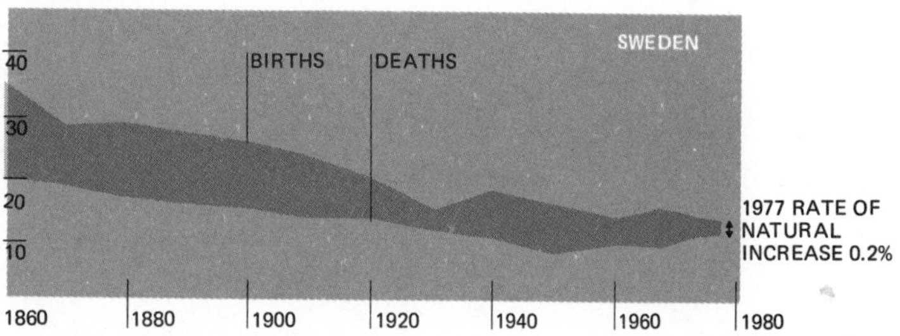

FIG. 44.1 Birth rates and death rates in Sweden since 1860. The difference between the two at any time gives the rate of increase (or if deaths should ever exceed births, decrease) of the population. In 1977, the birth rate was 2 per thousand greater than the death rate, representing a rate of natural increase (r) of 0.2% per year. Declining birth rates that roughly parallel declining death rates are characteristic of developed countries. (Data from the Population Reference Bureau.)

44.1 MORTALITY: FAMINE, WAR, AND PESTILENCE

The growth of a population is dependent on both birth rate and death rate. The human species is right now in the grip of a worldwide population explosion. However, an increase in birth rates cannot account for this unprecedented, and accelerating, rate of population growth. Birth rates appear to have remained about the same in the less developed countries and, excepting for minor fluctuations, to have dropped in developed parts of the world (Fig. 44.1). What, then, accounts for the population explosion? The answer is a dramatic decline in death rates.

The human machine is certainly capable of running for 70 to 100 years before it wears out. However, primitive humans probably had a life expectancy at birth of little more than 20 years. What, then, were the forces that so often led to death before the deterioration of age set in? By and large, these were the same forces that keep death rates high in any population of living organisms. They are: (1) the **limited capacity of the environment** to supply the needs of the species (for heterotrophs, this usually means the capacity of the environment to supply food); (2) the efficiency of **predators** at cropping the species to meet *their* food needs; (3) the inefficiency of the **parasites** of that species which, by precipitating the death of their host, deprive themselves of a meal ticket. Translated into human terms, the causes of premature death and the most efficient checks on population growth are famine, war (only humans prey on humans), and pestilence.

The dawn of agriculture and animal husbandry ushered in a period of human history marked by increasing efficiency at securing food and thus an increasing capacity of the environment to support the human species. The gains have been particularly marked during the last three centuries. Improved farm machinery, more land under cultivation, greater crop yields through an expanded use of fertilizers, pesticides, and irrigation, the merging of small farm units into large ones, and the means to transport food from regions of surplus to regions of scarcity have all played their roles. Even as recently as the eighteenth century, it took 80 people on United States farms to feed 100 individuals—themselves and 20 city cousins. Today, it takes fewer than 5, i.e. 5 people on the farm support themselves and 95 nonfarm people.

It is difficult to isolate the role that war has played in holding human population in check. Certainly war has been an almost continual aspect of the human condition, and death rates attributable directly to warfare have at times been staggering. Perhaps, though, war has had its greatest impact on the human population by its indirect effects: starvation and epidemics brought on by the breakdown of normal sanitary practices. Even in such a devastating war as the United States Civil War, more soldiers lost their lives to disease than to the immediate trauma of battlefield injuries.

But it is to the effectiveness of our struggle against parasites that we can lay most of the praise (or blame) for today's skyrocketing human population. Improved sanitation, immunization programs, the use of antibiotics and other chemotherapeutic measures, and the use of pesticides to kill the animals that transmit parasites to us have all played substantial roles in reducing deaths from infectious

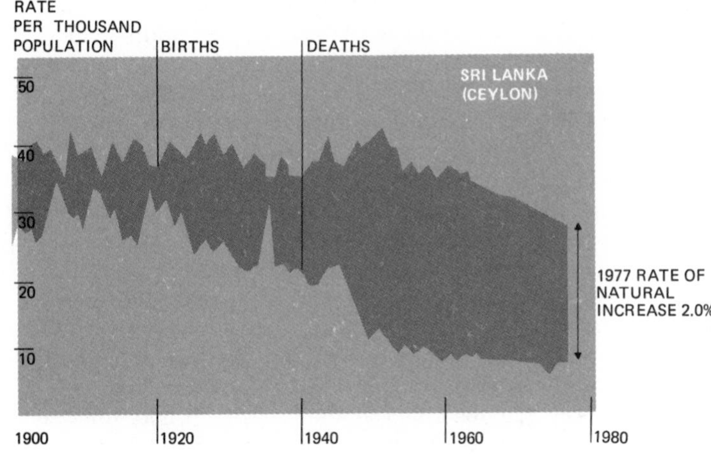

FIG. 44.2 Birth rates and death rates in Ceylon (now called Sri Lanka) since 1900. The sharp drop in death rates after 1945 reflects the widespread introduction of malaria control programs. The failure of the birth rate to decline significantly has led to a period of rapid population growth. (From the Population Reference Bureau.)

diseases. Prior to World War II, these advances were pretty much confined to the affluent, industrialized countries of the world, which have responded with a concurrent drop in birth rate (Fig. 44.1). But since the 1940's this technology has been exported to the impoverished regions of the world, with dramatic results. The annual death rate in Ceylon (now called Sri Lanka) in 1945 stood at 22 per thousand people. In 1946, a large-scale program of mosquito control—through the use of DDT—was begun. By virtually eliminating the mosquitoes that transmit malaria from person to person, the cycle of the disease was broken. In nine years, the death rate had dropped to 10 per thousand (Fig. 44.2) and in 1977 it stood at 8. But no *compensating* decline in the birth rate has taken place (1977 rate = 28 per thousand) so that in 1977 the population was increasing at an annual rate of 2.0% (20/1000/yr). Remembering the principle of exponential growth we realize that this is *compound* not simple interest and, as in a savings bank, will if continued cause the population to double in size in 35 years. Sharply declining death rates accompanied by birth rates that are holding relatively steady are today characteristic of most of the less developed regions of the world (Fig. 44.3).

Let us now examine in greater detail the role of parasitism in population control and how this has been altered by our technological advances.

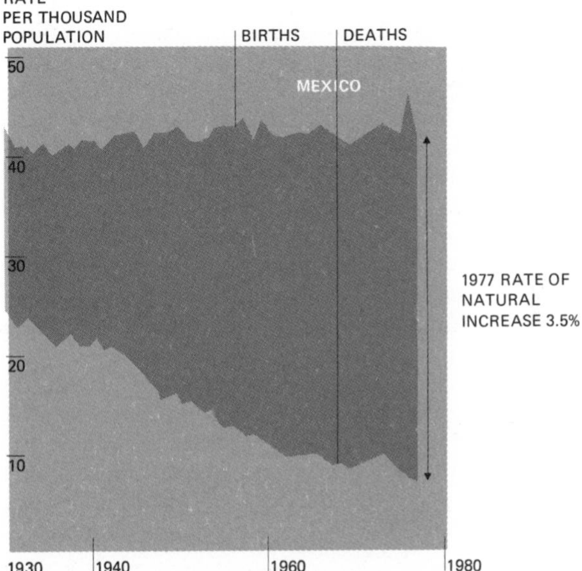

FIG. 44.3 Birth and death rates in Mexico since 1930. Introduction of a variety of public health measures has produced a rapid decline in death rates. There has been no corresponding decline in birth rates. This situation, with its attendant rapid rate of population growth, is characteristic of many of the less developed regions of the world. (From the Population Reference Bureau.)

44.2 PARASITISM: INVADING THE HOST

The list of human parasites is long. We may harbor a number of kinds of worms (hookworms, tapeworms, flukes, etc.) and protozoans (e.g. parasitic amoebas, the malarial organisms, and the trypanosome that causes African sleeping sickness—Fig. 35.3). Bacteria, fungi, and many viruses are also parasitic in humans.

Almost all locations on the host's body may become occupied by parasites. Viruses, some bacteria, and the malarial parasites invade and multiply right within the very cells of the host. Most parasitic bacteria and protozoans, as well as many roundworms and flukes, invade the tissues of their host. One finds tapeworms, roundworms, protozoans, and bacteria within the alimentary canal. Most insect parasites (e.g. the louse) are ectoparasites, that is, they live on the surface of the body.

One problem faced by all parasites is how to invade their host. The vertebrate body presents a number of formidable barriers against invasion by parasites. The outer layers of the skin consist of dead, keratinized cells that normally prevent penetration by parasites. The corneas of the eyes are about the only place where living cells are directly exposed to the external environment. But they are kept continuously moistened with tears, one of the chief ingredients of which is lysozyme—the enzyme that digests the peptidoglycan of bacterial cell walls (see Section 6.9).

Even though the exterior of the body is well-armored, the requirements of metabolism demand that a number of exchange organs (e.g. lungs, intestine, kidney tubules) have direct *access* to the external world. They *are* covered with living cells and thus risk invasion by microorganisms. The lungs achieve substantial protection from bacterial invasion by (a) the cilia that drive inhaled particles—trapped in mucus—back up to the throat, (b) phagocytic cells, e.g. macrophages, that engulf inhaled particles that escape ejection, and (c) the secretion of IgA, a type of antibody that is particularly abundant in secretions. In this case the IgA provides a protective barrier against airborne antigens that have previously elicited an immune response.

The low pH of the gastric juice is lethal for many potential pathogens that are swallowed with food or water. Of course, an enormous population of bacteria does live in the intestine, but these ordinarily do not invade the tissues because of (a) the activity of phagocytic cells and (b) the presence of an immune barrier of IgA. When the immune response is suppressed—following massive doses of radiation or the use of immunosuppressive drugs—one of the first effects is ulceration of the wall of the gastrointestinal tract as the resident bacteria escape from the normal checks and begin to invade living tissues.

The kidney tubules and bladder are also at risk of invasion by pathogens. However, urine is normally sterile and its continual production and flow acts to flush microorganisms back down the plumbing, thus minimizing the danger of bladder and kidney infections.

Of course these barriers are often breached. Parasites ingested in food or water often are able to establish themselves in the gastrointestinal tract and cause disease. Bacteria, fungi, or viruses dispersed in the air may be inhaled and overwhelm the normal defense mechanisms of the lungs. The bites of insects and arachnids (e.g. ticks) can act as hypodermic needles, introducing protozoans, bacteria, and viruses into the internal environment.

Once the parasite has gained entrance into the body of the host, there are three possible outcomes. The host may be killed by the parasite, it may tolerate it for an indefinite period of time, or it may destroy the parasite completely. Which of these three possible outcomes actually happens is dependent upon four factors: (1) the number of invading parasites, (2) their nature, (3) the defenses of the host and, where humans are involved, (4) the nature of any treatment given.

Ingestion of undercooked pork containing a few *Trichinella* cysts (Fig. 37.9) usually causes such a mild case of trichinosis that it is not even diagnosed as such. Eating heavily infected meat frequently results in death. Often the body copes successfully with small numbers of invading parasites but succumbs to heavier infestations. Of course the absolute number of parasites which will overwhelm the host varies with the severity of the damage caused by the particular parasite.

Parasites do vary in their pathogenicity, that is, in the severity of the damage they do to the host. We may harbor a tapeworm for years with little harm done. However, 90% of the infections by a certain strain of rickettsias causing Rocky Mountain spotted fever end in death if no treatment is given.

44.3 HOST RESISTANCE TO PARASITES

What happens when a parasite enters its host also depends on the defensive mechanisms that the host puts into operation to combat the invasion. As one example, let us examine the lines of defense that are brought into play when bacteria, carried on a splinter, are introduced beneath the skin.

Mechanical damage by the splinter will kill some cells and as these disintegrate, they provide nourishment for the bacteria which were introduced into the wound. These begin to multiply. However, damaged cells also liberate substances which signal the body to begin taking action against the invader. One of these, histamine, increases the blood supply to the capillaries in the area. Furthermore, the walls of the capillaries become "leakier." The result of these two actions is an accumulation of lymph in the tissue spaces. This increase is not drained away by the lymph capillaries as they become plugged with a mesh of fibrin. Histamine also stimulates the conversion of a serum globulin into polypeptide *kinins* which pass into the tissue spaces and accelerate these changes. Some substance, perhaps a kinin, is produced which attracts **phagocytes** from the blood stream. These, mostly neutrophils, migrate through the capillary walls into the tissue spaces and begin ingesting the bacteria and damaged cells. After consuming a number of bacteria, they die, and their disintegrated bodies are one component of pus. Other phagocytes establish a living barrier between the region of tissue destruction and intact healthy tissue. In this way, they isolate the site of infection.

All of these localized responses constitute **acute inflammation.** The external symptoms are redness (caused by the increased blood supply), tenderness (caused by the kinins), and if the wound is open, the oozing of pus. If all goes well, the process of acute inflammation will prevent the infection from spreading, and after all the bacteria have been destroyed, healing can begin.

Localized inflammation sometimes fails to contain the invading parasites. They may escape into the lymph and/or blood capillaries. A second line of defense comes into play at this time. The sinuses of the **lymph nodes** are lined with "fixed" phagocytes, which engulf bacteria passing through and thus prevent them from reaching the bloodstream. If the invasion is especially heavy, the lymph nodes may themselves become infected and swollen. Hidden infections are sometimes first detected by the appearance of these "swollen glands" in such areas as the neck, armpits, and groin.

Bacteria that reach the bloodstream are engulfed by fixed phagocytes which line the sinuses of the **spleen.** The blood sinuses in the **liver** are similarly equipped to screen out the steady stream of bacteria which penetrate the lining of the intestine and reach the liver by way of the hepatic portal veins. In this way, these organisms are kept from entering the general circulation.

For many infections, the thing that finally tips the balance in favor of the host is the production of **antibodies.** In the days before the discovery of drugs with which to combat bacterial pneumonia, the outcome of the disease often depended upon which occurred first: suffocation from the accumulation of fluid in the lungs, or the appearance of antipneumococcal antibodies. Physicians (and parents) anxiously awaited the period of "crisis"—usually on the fifth or sixth day. If the patient survived that long, the appearance of antibodies quickly turned the tide of battle. Phagocytes now were able to engulf antibody-coated pneumococci (look back at Fig. 30.8) and the patient made a dramatic and speedy recovery.

As for infections by organisms that the immune system has encountered before, the ability to mount a **secondary response** (see Fig. 30.15) may knock down the reinvasion so quickly that no symptoms are even noticed. The individual is **immune** to that parasite.

Some of our parasites reside intracellularly and thus are protected from exposure to antibodies. Several protozoan parasites, including the malarial parasites, use this trick. However, the malarial parasites break out of their (our) red cells on a regular basis (this produces the recurrent fever) and are then exposed to antibodies. Nevertheless, they may survive for years within the host. One reason for this is that the parasite is able to alter the nature of its surface antigenic determinants from time to time. In this way, it escapes attack by antibodies until a new primary response develops. Trypanosomes employ the same trick. Blood flukes (schistosomes) circumvent the host's immune response in another way. They actually incorporate *host* antigens into their surface. Thus coated, the immune system sees them as "self" rather than "nonself" and fails to mount an attack against them.

It has been known for some time that people sick with one virus possess a temporary immunity against infection by other viruses. This protection occurs because cells attacked by the first virus produce an antivirus agent called *interferon.* Interferon is a protein molecule produced shortly after infection and, in every case, faster than antibodies can be produced. Unlike antibodies, interferon is not specific, that is, it is a single substance which seems to be effective against all viruses. Once synthesized, it in turn induces the synthesis of a second protein which interferes with cellular RNA synthesis or protein synthesis (or both). Therefore, viral replication is blocked. If safe ways can be found to enhance the body's production of interferon, it should provide an excellent weapon against virus diseases (just the ones against which other drugs are so ineffective).

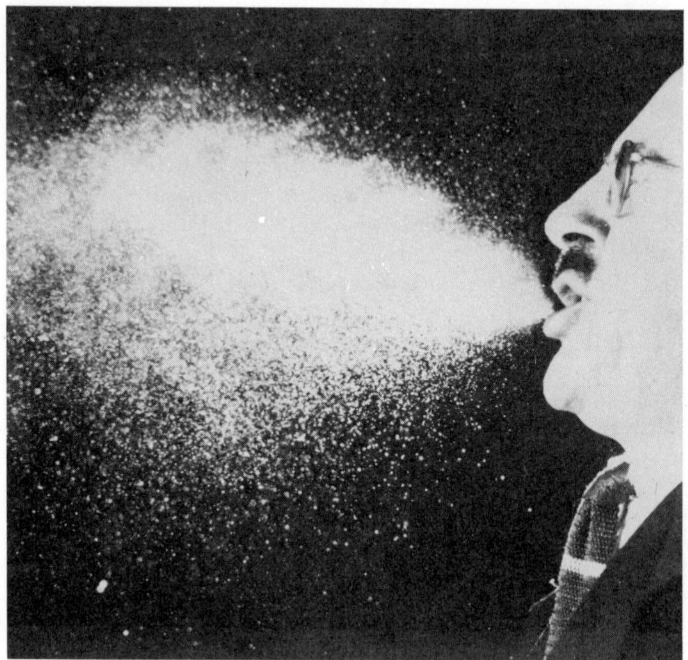

FIG. 44.4 A violent, unstifled sneeze. High-speed flash illumination reveals the cloud of droplets produced. These are a major factor in the spread of bacteria and viruses that enter the body by way of the respiratory tract. (Courtesy of Dr. Marshall W. Jennison, *Scientific Monthly*, Vol. 52, pp. 24–33, Jan., 1941.)

44.4 INTERFERING WITH THE TRANSMISSION OF PARASITES

A serious problem facing all parasites is how to get from one host to another. Exit from the host body can be made by way of the lungs (during coughing and sneezing—Fig. 44.4), by way of the anus (in the feces) and, in a few cases (e.g. the spirochete that causes syphilis), directly through the skin and mucus membranes. Bacteria and viruses that cause infection of the air passages and lungs are examples of the first case. They leave by the same route they entered. Intestinal parasites such as bacteria, viruses, and protozoans, leave in the feces or, in the case of tapeworms, hookworms, etc., deposit eggs which are expelled in the feces. Where personal and public sanitary conditions are poor, there is a good possibility that some of these organisms will be ingested by other hosts.

Some of the worms which inhabit our alimentary canal produce larval stages which must develop in a second, intermediate host before they are capable of infecting humans again. The **intermediate host** of the pig tapeworm (*Taenia solium*) is the pig. It seems pretty clear that the intermediate host in this case is a device that improves the chances of the parasite's reentering the **primary host,** a human. We are far more likely to ingest undercooked pork than human feces. This is not true for pigs raised under unsanitary conditions. The largest (up to 18 m = 60 ft!) tapeworm of humans, the fish tapeworm, requires three hosts in order to complete its life cycle: a freshwater crustacean (*Cyclops*—Fig. 44.5), a fish, and a human. When you consider that human feces often reach bodies of water, that freshwater fish eat *Cyclops* and humans may eat improperly cooked fish, it is easy to see how this complicated life cycle actually improves the chances of this parasite's getting from one human to another.

The problem of transport is even more acute for those parasites that live in our blood or other tissues. The malarial parasite, the yellow fever virus, and the roundworm which causes elephantiasis can pass from one human to another only with the help of an intermediate host, the mosquito, capable of withdrawing the parasite from the blood of one host and inoculating it into the blood of another. We use the term **vector** for any animal (most of them are insects) that transmits pathogenic microorganisms in this way.

Even with elaborate life cycles that make reentry into the primary host easier, the chances of any one individual parasite's doing so are extremely small. It is therefore not surprising that many of our parasites are amazingly fecund, that is, they produce enormous numbers of offspring. The fish tapeworm, mentioned above, discharges up to one million fertilized eggs into its host's feces each day. A female hookworm may lay 25 to 35 thousand eggs each day throughout her five-year life span. Once we understand the problem a parasite faces in getting from host to host, we can appreciate why so many parasites (e.g. tapeworms and hookworms) are little more than efficient machines for sexual reproduction.

The earliest successes at coping with human diseases revolved around interrupting the means by which parasites pass from one host to the next. Careful disposal of human sewage and the development of municipal water supplies carrying purified (e.g. chlorinated) water have greatly reduced the incidence of such intestinal diseases as typhoid fever, amoebic dysentery, and cholera. Thorough cooking of pork, beef, and fish prevents transmission of para-

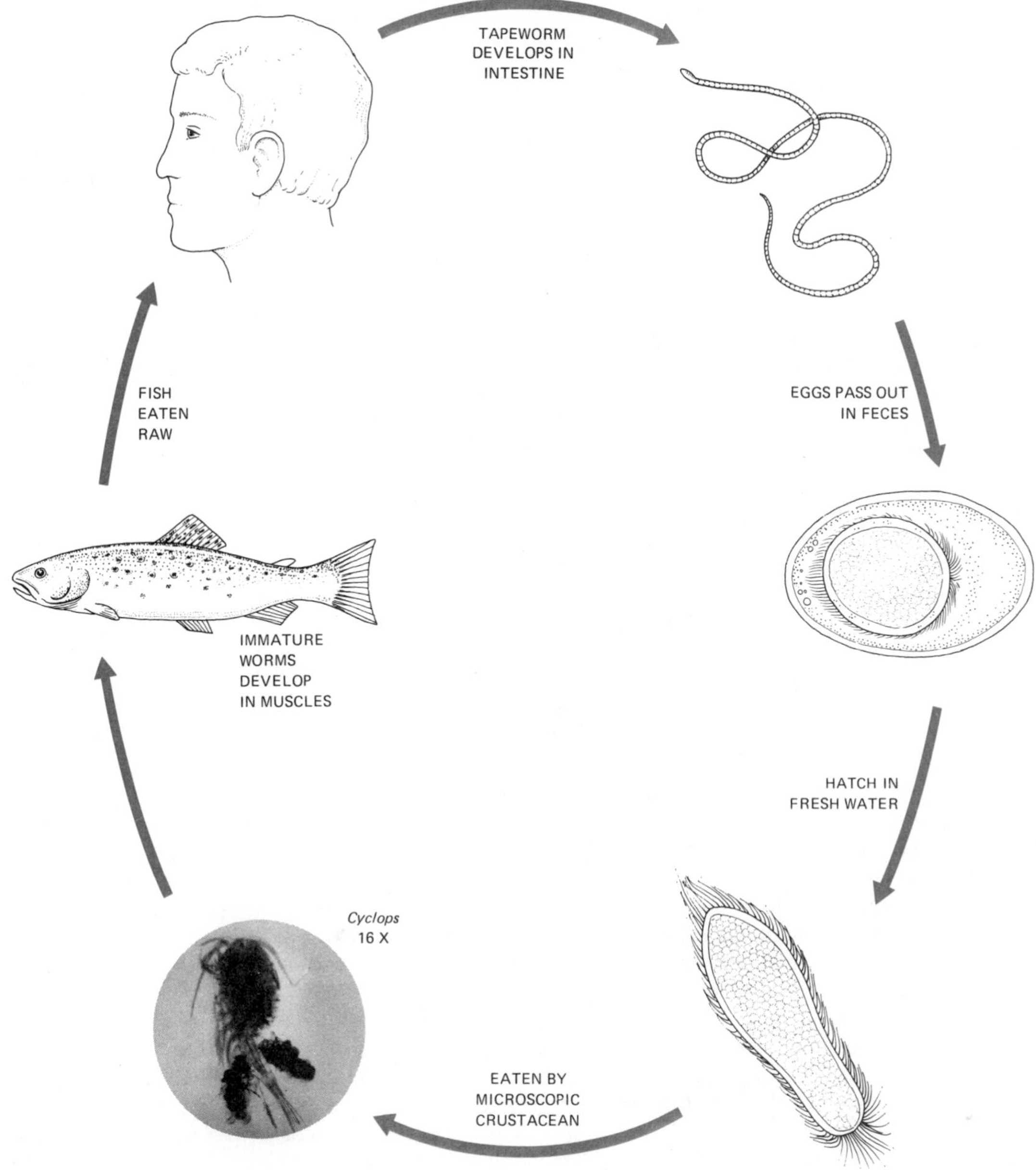

FIG. 44.5 Life cycle of *Diphyllobothrium latum,* the broad fish tapeworm. Adults as long as 18 meters (60 ft) have been removed from the human intestine. The incidence of infection is particularly high in countries along the Baltic Coast and in the Great Lakes region.

sites that gain entry to the body through ingestion of intermediate hosts.

The killing of vectors is another public health measure that has been instrumental in reducing the incidence of a number of infectious diseases. In 1943 and 1944, widespread application of DDT dusts to the civilian population of Italy nipped in the bud what threatened to be a devastating epidemic of typhus fever. The vector of the microorganism that causes typhus is an insect, the body louse. We have already mentioned the role that DDT played in virtually eliminating the mosquitoes that serve as the vector of malaria in Sri Lanka. This story was repeated throughout many of the malarial regions of the world in the years following World War II.

44.5 DENSITY-DEPENDENCE OF HUMAN PARASITES

With the widespread acceptance of vaccination against smallpox in the nineteenth century, another public health victory was achieved (see Section 30.6). This was the first use of deliberate immunization to reduce the severity of epidemics. It has been followed by successful immunization programs for diphtheria, whooping cough, tetanus, polio (Fig. 44.6), yellow fever, and others. These have been so successful that once-major killers like diphtheria have been virtually eliminated from the globe. In 1976, in fact, the World Health Organization reported that smallpox had been eliminated from the earth except for one tiny pocket of cases remaining in Africa. By the time you read these words, this disease—historically one of our most dreaded scourges, and the first disease to be combatted by vaccination—may be gone forever.

Your vaccination with tetanus toxoid helps only you. Your vaccination with polio vaccine or diphtheria toxoid helps your neighbors as well. This is because the spread of polio virus or the diphtheria bacillus from person to person requires a *high density* of *susceptible individuals*. Other common examples of human diseases where transmission is highly density-dependent are measles, smallpox, rubella ("German measles"), and mumps. All of these diseases share a number of features.

They all cause an acute illness of relatively brief duration. The period of time during which the victim is contagious to others is also short. The disease ends either in death or in the complete elimination of the parasite and, often, a lifelong immunity. For this reason, the parasite cannot survive unless it can reach other *susceptible* individuals during the brief period when its host is contagious. Thus the success of the parasite is enhanced by a high population density if that population is not already immune. But an earlier epidemic of the same disease or a moderately successful vaccination program may have produced so many immune people that transmission of the parasite fails. The introduction in 1941 of diphtheria immunization in England and Wales caused a precipitous decline in the number of cases (Fig. 44.7). This situation has remained stable since, even though not all children are protected by any means. However, once over 50% of the children are immune, the chances that an epidemic will become established are very slight. This phenomenon is called herd immunity. It tells us that your immunization against certain density-dependent parasites represents a public as well as a private health achievement.

These strongly density-dependent diseases survive best in cities. And even then, they are dependent upon the periodic arrival of new crops of susceptible individuals. This may occur through immigration. Until modern times, in fact, immigration from the countryside was the only way in which cities could even maintain their populations in the face of repeated epidemics. The other sources of new susceptible individuals is childbirth. Figure 44.8 shows the incidence of measles in South Australia from 1916 through 1920. Note the relative absence of measles after the epidemic of 1916—an absence that was terminated in 1919 when a fresh crop of susceptible children had achieved the required density.

The requirement for a high population density and the periodic arrival of a fresh crop of susceptible individuals makes these density-dependent parasites tailor-made for military camps. Measles, bacterial meningitis, bacterial pneumonia, and influenza are some of the epidemic diseases that regularly sweep through large military installations with transient populations. The worldwide epidemic of "Spanish" influenza that is thought to have killed over 20 million people in 1918–1919 was spawned in the military populations waging World War I. Not only spawned but quickly transmitted around the globe following the movement of troops.

Many of these density-dependent diseases have caused the most devastating of human epidemics. We have already mentioned the "great dying" in the 15th century as the plague bacillus swept through all of Europe. Even diseases that we consider little more than a nuisance can cause enormous mortality if they gain access to a totally unprotected population. When the Pilgrims settled in New England in 1620 they found an American Indian population that was perhaps only 10% of that of just a few years earlier. The Indian population had been decimated by some introduced parasite—probably the measles virus. In the nineteenth century, the introduction of measles into Polynesia caused a series of devastating epidemics as a result of which approximately one-fifth of the population died. Presumably these people had never before encountered the measles virus.

One could argue that a parasite that either kills or confers lifelong immunity on its host is a poorly

FIG. 44.6 Number of cases of polio (per 100,000 population) in the United States from 1940 to 1965. Note the dramatic decline brought about by the introduction of the Salk vaccine. (Based on J. R. Paul, 1971, and redrawn from Sir Macfarlane Burnet and D. O. White, *Natural History of Infectious Disease,* 4th ed., Cambridge University Press, 1972.) ▼

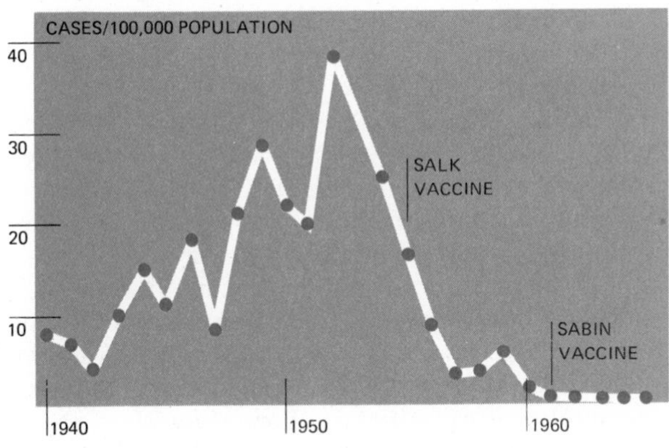

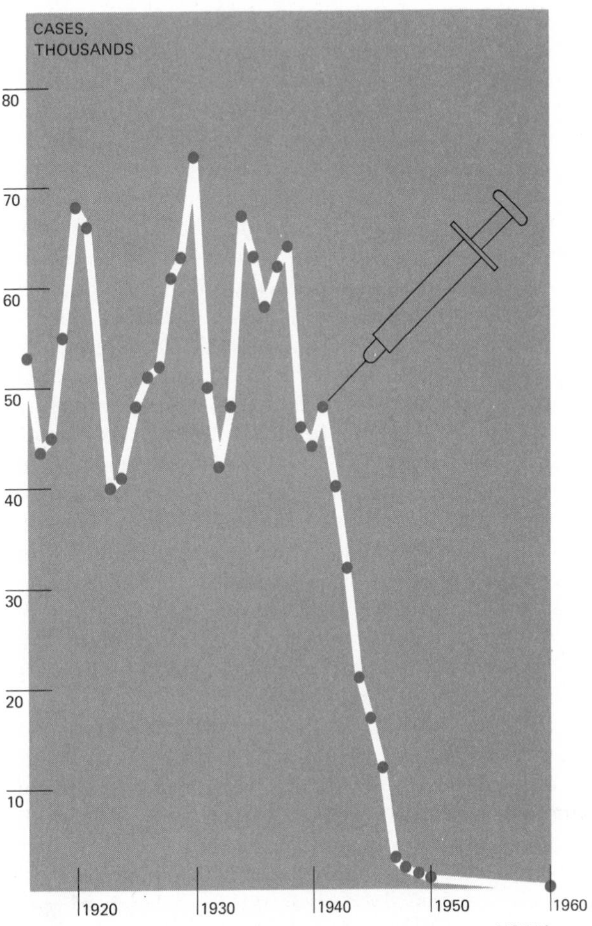

▲
FIG. 44.7 Incidence of diphtheria in England and Wales from 1916 to 1960. Note the dramatic decline in cases following the introduction of immunization in 1941. Although not all children are immunized, the effect of herd immunity has prevented further epidemics. (From F. T. Perkins, *J. Royal Soc. Arts,* **63:** 82, 1965.)

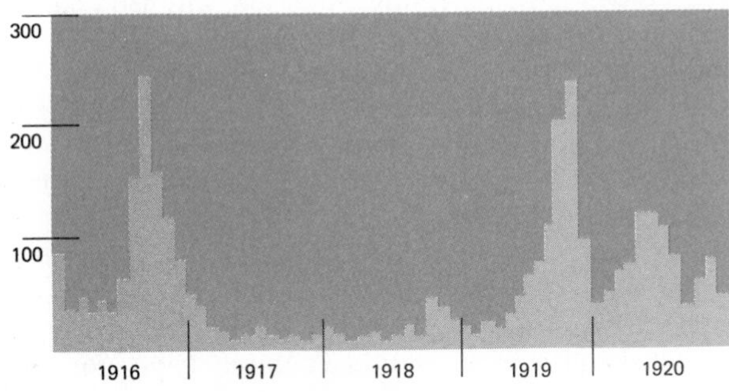

FIG. 44.8 Monthly notifications of measles in South Australia from 1916 to 1920. After a major epidemic (1916), most of the population is immune. A new epidemic must await a fresh crop of susceptible children. (Redrawn by permission from Sir Macfarlane Burnet and D. O. White, *Natural History of Infectious Disease,* 4th ed., Cambridge University Press, 1972.)

adapted parasite indeed. The examples we have been examining may well be examples of diseases that did not even exist until humans began living in cities. Thus only a few thousands of years have been available for humans and these organisms to coadapt.

But surely humans have always been attacked by parasites. Yes, and probably by parasites that are quite different in their behavior from the ones we have been discussing. Let us now examine a second group of human diseases, a group that includes such diseases as tuberculosis, leprosy, malaria, amoebic dysentery, and schistosomiasis. These diseases are generally chronic. The parasite remains for very long periods in its host, and the host remains contagious for much or all of this time. These diseases produce only a weak immunity. For all of these reasons, the parasites that cause these diseases do not exhibit the extreme density-dependence of those we discussed earlier. These parasites can exist for indefinite periods in tiny, isolated human populations. They may well have been parasitizing humans since the time humans first evolved. Thus much more time has been available for the mutual evolutionary adaptation of host and parasite.

We have seen how the protozoa that cause malaria and African sleeping sickness and the blood flukes that cause schistosomiasis evade the feeble immune response mounted by their host. While the resident parasites keep ducking their host's immune response, the antibodies that are produced *may* be effective against the arrival of a fresh batch of parasites. It is as though the first arrivals are working to ensure that their welfare is not endangered by "superinfection," the crowding in of competitors. The distribution of the gene for sickle hemoglobin (Hb^S) must also represent the outcome of a long history of coevolution of the parasite (*P. falciparum*) and host. So these parasites are old friends. While they do not produce such dramatic epidemics as some of our more recently acquired parasites, their control and/or elimination pose very difficult problems.

44.6 CHEMOTHERAPY

The period of World War II marked the first successful use of chemical agents that, when introduced into the body of a victim of an infectious disease, destroyed the parasite without destroying the host. These were the sulfa drugs and the antibiotics. It is somewhat startling to realize with what a small number of weapons we are able to combat parasites directly. But a moment's thought should make it clear why this is so. Many of the biochemical activities of living things are shared by all organisms, from microbes to humans. Many ways exist to interfere with these, for example, blocking electron transport in the cytochrome enzyme by the use of cyanide. A dose of cyanide, however, will not only kill the parasites but will kill the patient as well. The problem, then, is to find some essential metabolic activity of the parasite that is not shared by its host and to use this to get at the parasite selectively.

Histologists (biologists who study tissues) have known for years that some biological stains are quite selective in their action. They will stain certain types of cells and not others. Following this clue, the German bacteriologist Paul Ehrlich attempted to find a substance which would combine selectively with the spirochete that causes syphilis and, in so doing, kill it without killing the host. After a long patient search, he found such a substance. Thus was born (in 1910) Salvarsan, the first synthetic chemotherapeutic agent.

Another German scientist, Gerhard Domagk, followed his lead and, in 1932, found that a red dye called prontosil was effective against the streptococci ravaging his daughter's body. Later, a French chemist discovered that all the activity of prontosil resides in just a portion of the molecule. This portion (which is not a dye at all) was **sulfanilamide,** the first of a large family of sulfa drugs, which were developed just in time to play a major role in reducing the number of casualties in World War II.

The action of the sulfa drugs is dependent upon molecular mimicry. All organisms require folic acid (one of our B-vitamins) in order to grow. Most bacteria are able to synthesize folic acid from simpler building blocks, one of which is para-aminobenzoic acid (PABA). A glance at Fig. 44.9 shows the similarity between the PABA and sulfanilamide molecules. When an excess of sulfanilamide is present in the medium, the bacteria use it instead of PABA but, in so doing, fail to produce the folic acid they must have. If we need folic acid, too, why isn't sulfanilamide lethal to us as it is to the bacteria? The answer is that the folic acid molecule is a vitamin for us; that is, we cannot synthesize it. Here, then, is a biochemical difference between host and parasite which can be exploited in killing one without killing the other.

Although the discovery and exploitation of such differences has been slow, some of our worm and protozoan parasites can now be eliminated chemically. Chloroquine and primaquine have, for ex-

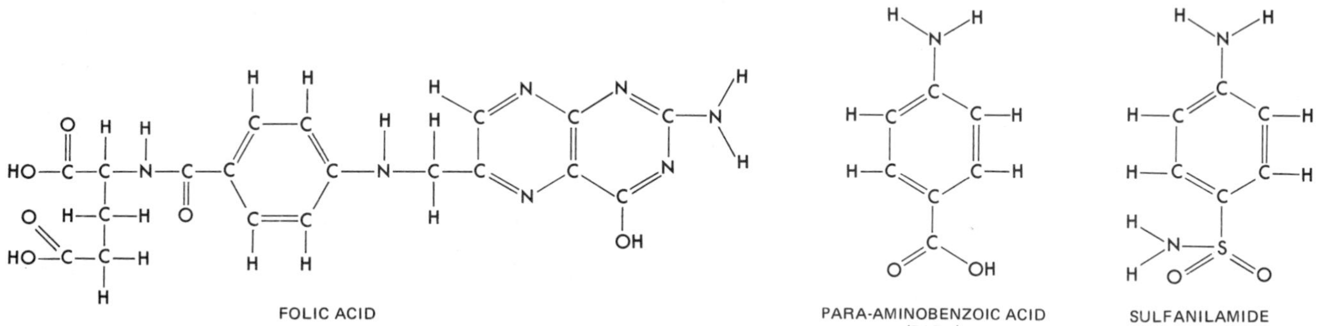

FOLIC ACID

PARA-AMINOBENZOIC ACID (PABA)

SULFANILAMIDE

FIG. 44.9 Molecular mimicry accounts for the effectiveness of sulfa drugs. Presented with sulfanilamide, bacteria are tricked into using it instead of PABA to synthesize their folic acid. We are not affected similarly because we secure our folic acid ready-made in our diet, that is, for us it is a vitamin.

ample, proved of great value in curing malarial infections. One by one, effective chemical weapons are being developed against other parasitic protozoans.

The development of antiviral drugs has lagged far behind that of antibacterial drugs. However, a few drugs that combine efficacy and safety have been discovered and are beginning to be used in therapy. Idoxuridine is effective against infections of the cornea by herpes simplex virus (HSV). Methisazone showed promise against the smallpox virus, but it now appears that the heroic immunization campaign mounted by the World Health Organization will make methisazone a drug without a disease. Amantadine hydrochloride has shown some promise as an agent against influenza viruses of the A type.

44.7 ANTIBIOTICS

There is no fundamental difference between the antibiotics and the chemotherapeutic drugs discussed above. In both cases, specific molecules are involved, and laboratory synthesis of antibiotics is often possible, too. However, each antibiotic was discovered originally as the secretion of a fungus or a bacterium, which inhibited the growth of other bacteria. The first, and in many ways still the best, antibiotic was penicillin. Its discovery stemmed from the observation by Alexander Fleming that bacterial growth was inhibited in a culture dish which had accidentally become contaminated with a mold of the genus *Penicillium* (Fig. 44.10). The significance of this discovery was appreciated in time to develop techniques of mass production of penicillin for use in World

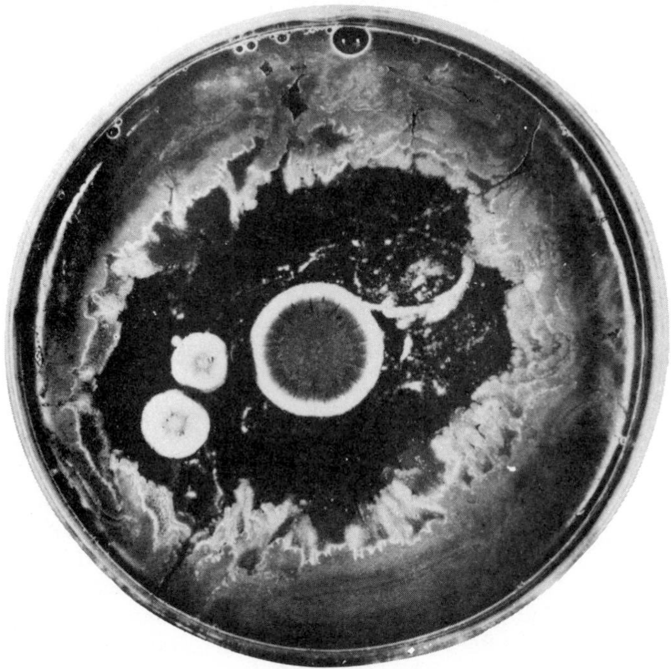

FIG. 44.10 The growth of bacteria on the agar in this culture dish has been inhibited near the three circular colonies of the fungus *Penicillium notatum*. The antibiotic penicillin, diffusing outward from the colonies, is responsible for this effect. (Courtesy of Merck & Co., Inc.)

War II. Since then, laboratories throughout the world have tested hundreds of thousands of specimens of fungi and bacteria for antibiotic activity. When such has been found, the substance has been tested for safety. Very few antibiotics have passed all the hurdles, but chloramphenicol (sold as "chloromycetin"), streptomycin, the tetracyclines ("aureomycin" and "terramycin"), and a few others have now joined penicillin as weapons in the fight against bacterial infections.

The discovery of each antibiotic has been an empirical process, that is, one of simply trying to find a safe and effective agent without concern for its method of action. However, later research has revealed that the selectivity of each antibiotic—like that of a chemotherapeutic agent—rests on interference with a metabolic process in the bacterium that is not found in the host. Penicillin, for example, exerts its bacteriocidal effect by blocking synthesis of the peptidoglycan wall of bacteria—a structure that is unique to prokaryotes (look back at Fig. 34.3). A number of other antibiotics interfere with those steps in protein synthesis that are characteristic of bacteria but not of "higher" organisms. Even then, the selectivity is not absolute. Chloroplasts and (more significant for human biology) mitochondria have bacterialike protein-synthesizing machinery and thus are susceptible to antibiotic inhibition of their activities.

44.8 PASSIVE IMMUNITY

Antibodies represent an additional weapon against rampaging parasites. In Chapter 30, we saw how selective antibodies are in their action. Following infection, the development of a primary immune response (see Section 30.5) *may* occur rapidly enough to turn the tide of battle. In certain cases, however, there may be sound reasons for not waiting for the host to mount a primary immune response against the pathogen. If chemotherapeutic agents are ineffective (e.g. against virus-caused diseases like polio), it is possible to introduce into the patient antibodies that were produced by another animal such as a sheep, horse, or another human. This use of **antiserum** of another animal confers an immediate, temporary, immunity on the patient. This immunity is described as *passive* because it is not produced by the patient's own immunological mechanism.

To give an example, if sublethal quantities of tetanus toxin are injected into a horse or sheep, the animal develops **antitoxin.** By bleeding the animal periodically and extracting its serum, the antitoxin can be harvested and used to treat human patients who may have been infected with tetanus bacilli from a splinter or other wound. Unfortunately, there is some danger to the use of antisera prepared in the bodies of other animals, because we can become allergic to other foreign proteins present in the mixture. Therefore, it is far wiser to use tetanus toxoid to develop one's own active immunity, although this must be done before infection actually occurs.

The gamma-globulin fraction of adult human blood is apt to contain antibodies against a number of common disease organisms. For this reason, it is often separated from donated blood and used as an antiserum. Temporary protection for children or adults exposed to hepatitis or measles is commonly achieved by injections of gamma-globulin pooled from many blood donors.

44.9 PUBLIC HEALTH MEASURES: THE OUTLOOK

The public health and therapeutic measures we have been discussing make up a success story of which we can be proud. But they have had undesirable effects as well. For example, the widespread use of antibiotics has altered the pressure of natural selection so as to *select for* those spontaneous mutants capable of withstanding the effects of the drugs. Staphylococci that have developed resistance to penicillin are now a serious problem in hospitals. In fact, many pathogenic bacteria have become resistant to a number of antibiotics and these pose a serious threat.

The introduction of effective sanitation has repeatedly lowered the incidence of such diseases as typhoid, but *raised* the incidence of polio. This is because polio virus, which is excreted in the feces, generally produces a mild disease—and subsequent immunity—in infants. With improved sanitation, however, infants often escape exposure to the virus. Then, when they meet the virus for the first time as children or young adults, the course of the disease is far more severe. Fortunately, worldwide immunization against polio now raises the hope of driving the virus to extinction.

Finally, our remarkable victories over infectious disease have greatly increased life expectancies and thus the size of the human population. Whether we have the environmental capacity to nourish this exploding population has now become the crucial issue for humankind.

EXERCISES AND PROBLEMS

1. What problems must be solved by all internal parasites? How have these been solved by the pig tapeworm? by *Plasmodium vivax?* by *Schistosoma mansoni?*
2. With respect to the pig tapeworm (*Taenia solium*), why do we say that we are the final host and the pig is the intermediate host rather than the other way around?
3. What was the rate of natural increase (r) in Ceylon (Sri Lanka) in 1920? 1950? Explain the change.
4. Contrast tetanus toxoid and tetanus antitoxin with respect to (a) method of manufacture, (b) mechanism of action in the body, (c) appropriate medical use.
5. Evaluate measles and malaria with respect to (a) duration of illness, (b) ability to persist in isolated populations, (c) duration of period of contagiousness, (d) transmission by a vector, (e) duration of acquired immunity, (f) duration of residence of parasite within its host.

REFERENCES

1. BROCK, T. D., ed., *Milestones in Microbiology,* Prentice-Hall, Englewood Cliffs, N.J., 1961. A paperback that includes:
   a) Edward Jenner's paper on producing immunity against smallpox by inoculation with cowpox virus.
   b) Alexander Fleming's report on the antibacterial action of material from cultures of *Penicillium.*
   c) Gerhard Domagk's report on the antibacterial action of prontosil, a dye that is converted in the body into sulfanilamide.
   d) Donald D. Wood's analysis of the molecular mimicry by which sulfanilamide exerts its antibacterial effect.
2. HILLEMAN, M. R., and A. A. TYTELL, "The Induction of Interferon," *Scientific American,* Offprint No. 1226, July, 1971.
3. BURNET, SIR MACFARLANE, and D. O. WHITE, *Natural History of Infectious Disease,* 4th ed., Cambridge University Press, Cambridge, 1972. This little classic should be read by every serious student of biology. Available in paperback.
4. LANGER, W. L., "Immunization against Smallpox before Jenner," *Scientific American,* January, 1976. Before Jenner, many doctors deliberately inoculated the smallpox virus into the skin in order to produce a mild but immunizing case of the disease. The procedure was called variolation.
5. WINSLOW, OLA ELIZABETH, *A Destroying Angel: The Conquest of Smallpox in Colonial Boston,* Houghton Mifflin Company, Boston, 1974. The ravages of smallpox and the controversies surrounding variolation are vividly portrayed in this brief history.
6. HENDERSON, D. A., "The Eradication of Smallpox," *Scientific American,* October, 1976. At the time this article was written, the last known cases of smallpox were confined to a few tiny villages in Ethiopia. In 1977, however, hundreds of new cases were discovered in Somalia.
7. BURKE, D. C., "The Status of Interferon," *Scientific American,* Offprint No. 1356, April, 1977. Its mechanism of action and therapeutic promise.

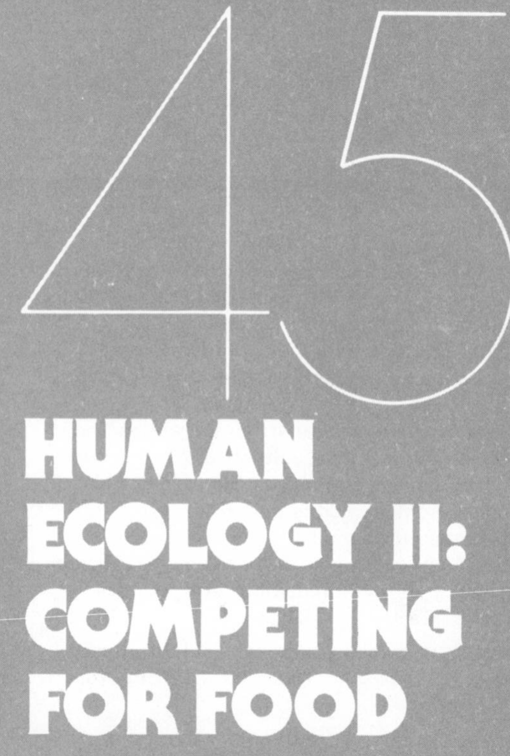

45 HUMAN ECOLOGY II: COMPETING FOR FOOD

45.1 THE CARRYING CAPACITY OF THE ENVIRONMENT

The capacity of the environment of a given species depends ultimately on the amount of energy that species can trap. In Chapter 40, we examined the flow of energy through populations, beginning with autotrophs and ending with heterotrophs. We learned of the sharp losses of energy that occur as it flows from one link to another in a food chain. These losses set strict limits to the biomass of populations at each trophic level. The problem for the growing human species is, from a selfish point of view at least, how to deflect more energy into our own biomass. Such agri-technological developments as the breeding of higher-yielding crops, the development of larger and more efficient farm equipment, and the increased use of fertilizers and irrigation have all played major roles in increasing the capacity of our environment. Unfortunately, these advances have often represented short-term benefits derived at the expense of long-term damage to the ecosystem from eutrophication, depletion of fossil fuel reserves, soil erosion, etc. Inasmuch as the happy and not-so-happy facets of deflecting energy into the human population have been examined in earlier chapters, we shall say no more here. Instead, let us devote this chapter to examining another aspect of the problem: how to minimize the impact of those creatures that compete with us for energy.

The list of our competitors for food is very long, ranging from the raccoon that strips off an ear of corn in the family corn patch to the nematodes (roundworms) that feed off the roots of the corn plant. In terms of total damage, however, our major competitors are the insects (Fig. 45.1), the fungi, and the weeds, and it is these three groups to which we will direct our attention.

FIG. 45.1 A cornfield before and after invasion by grasshoppers. (Courtesy of U.S.D.A.)

45.2 THE HAZARDS OF MONOCULTURE

As our efficiency at growing crops has increased, so has the effort that we must expend to keep competitors for those crops in check. The reason for this is really quite simple. Efficient cropping demands monoculture, that is, pure stands of a single plant. Monoculture maximizes the efficiency with which cultural practices and farm machinery can be employed, at every step from preparing the land for planting to harvesting. But monoculture also greatly increases the capacity of the environment to support our competitors for that crop. Unending acres of cotton plants provide a vast amount of food for the

FIG. 45.2 A member of the Tsembaga weeds the family garden. Note the polyculture that is characteristic of the Tsembaga garden. A variety of crop plants grow intermixed. Their roots and foliage exploit different levels of soil and light exposure respectively. The absence of large stands of a single plant species reduces the opportunity for pests to develop large, destructive populations. (Courtesy of Dr. Roy A. Rappaport.)

cotton boll worm and thus permit it to enter upon a period of rapid, exponential population growth. Because in monoculture the plants of a single species grow close together, an expanding pest population spreads easily throughout the entire stand.

The agricultural practices of the Tsembaga of New Guinea, which we examined in Chapter 40 (see Section 40.2), provide a nice contrast to monoculture. Each one of their scattered small gardens is apt to contain close to two dozen different kinds of food plants (such as sweet potatoes, taro, yams, cassava, bananas, hibiscus, beans, sugarcane, etc.). These are not set out in rows or pure stands but are intermixed with one another. Two advantages follow from this. First, the root systems and foliage systems of the different species exploit different layers of soil and different levels of exposure to light respectively (Fig. 45.2). Thus competition between the plants is reduced, and the net productivity of the garden maximized. Second, the absence of pure stands of a single plant reduces pest build-up and spread. Thus the Tsembaga have no need to use pesticides to secure adequate crop yields. But, as we have seen, such gardening does require a degree of intensive hand labor that is simply impractical and unavailable in modern industrialized nations.

45.3 EARLY PEST CONTROL TECHNIQUES

In earlier times, when farm labor was abundant and moderate crop damage could be tolerated, the fight against insect pests was often as simple as removing the pests from individual plants. S. W. Cole, writing in his *American Fruit Book* of 1849, suggested ridding plum trees of weevils (plum curculio) by spreading a cloth beneath the tree and shaking the insects down onto it with "a sudden jar of the tree or branches, using a mallet covered by a pad." The use of chemicals was limited to such innocuous materials as solutions of whale oil soap or water in which tobacco leaves had steeped (even today nicotine is used as an insecticide).

However, by the end of the nineteenth century the widespread use of pesticides had begun. Insect pests were attacked with inorganic substances such as lead arsenate and with the organic plant extracts pyrethrum and rotenone. Fungi-caused plant diseases were attacked with such inorganic materials as copper sulfate, mercuric chloride, and sulfur. Weeds were usually controlled mechanically (e.g. by hoeing) but sodium arsenite was sometimes used.

The inorganic pesticides have the disadvantages of high toxicity to organisms other than the target organism and great persistence. Even though lead arsenate has been largely abandoned by apple growers for a decade or more, remarkably high concentrations of lead are still found in orchard soils.

Organic insecticides, like pyrethrum and rotenone, that are extracted from plants are still highly prized. They are exceedingly toxic to insects and relatively harmless to other creatures (except fish!). They decompose readily so that residues do not accumulate on crops or in the soil. Unfortunately, they are also very expensive and their molecular structures are so complex that commercial synthesis has not yet provided inexpensive substitutes for the natural materials. However, the advantages of pyrethrins (natural or synthetic components of pyrethrum) can be had economically for certain purposes by mixing the material with a synergist such as piperonyl butoxide. Piperonyl butoxide is not an effective insecticide by itself, but when mixed with pyrethrins, it markedly enhances their effectiveness and thus reduces the amount needed.

45.4 DDT

With the onset of World War II, pyrethrum, much of which was imported from Kenya, became very scarce. The importance of insect control in wartime dictated a hurried search for an alternative. The alternative was DDT, a compound (Fig. 45.3) that had been sitting on laboratory shelves for many years but whose insecticidal properties were first appreciated in the late thirties.

DDT can be synthesized in huge quantities at very low cost. Its rapid introduction by the armed forces certainly has to be ranked along with the development of penicillin and sulfa drugs as a major contributor to the fact that World War II was the first war in history where trauma killed more people —combatants and noncombatants alike—than infectious disease. We have already mentioned the role it played in battling a devastating typhus epidemic in Italy in 1943 and 1944.

With peace (such as it was) restored, DDT was pressed into worldwide use against crop predators and, especially, the vectors of human diseases such as malaria, yellow fever (mosquitoes), and plague

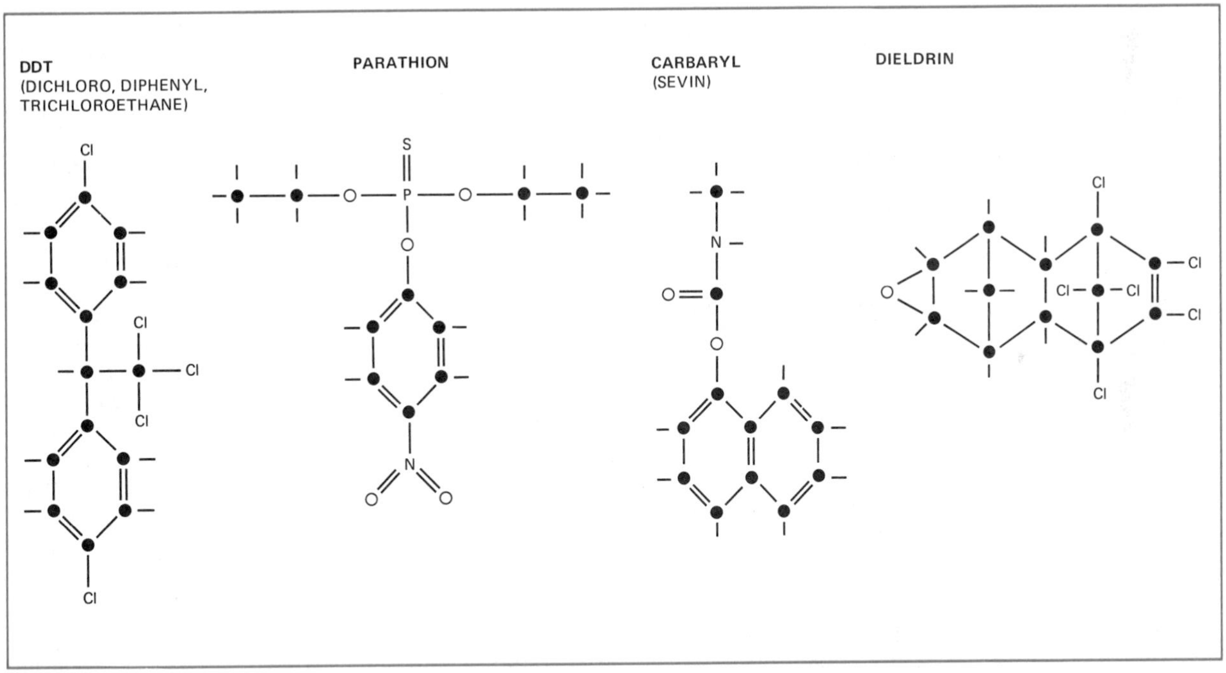

DDT (DICHLORO, DIPHENYL, TRICHLOROETHANE) — PARATHION — CARBARYL (SEVIN) — DIELDRIN

FIG. 45.3 Molecular structure of several important insecticides. DDT and dieldrin are representative chlorinated hydrocarbons. Parathion is a widely used and highly toxic organophosphate. Carbaryl is a carbamate. Solid circles represent carbon atoms; dashes represent hydrogen atoms.

(fleas). DDT was particularly effective against malarial mosquitoes because of its high resistance to degradation, i.e. its persistence. One or two sprays a year on the walls of native homes kept them clear of mosquitoes. Prior to the introduction of DDT, the number of cases of malaria in Ceylon ran in excess of a million a year. By 1963, the disease had been practically eliminated from the island. However, growing concern about the hazards of DDT use led to its abandonment in the mid-1960's and soon thereafter, a resurgence of the incidence of malaria occurred.

Despite the early and undisputable successes achieved with DDT, it quickly became apparent that it had serious drawbacks. As early as 1946, Swedish workers reported the appearance of populations of houseflies resistant to the effects of DDT. Similar reports soon appeared all over the world. Approximately 100 insect species, including mosquitoes and many crop pests, have populations resistant to DDT. This unhappy situation should have been no surprise. The more intense the use of DDT, the higher the pressure of natural selection for those occasional mutant flies or mosquitoes that were resistant. It did not take them long to replace their susceptible cousins.

The response to the development of DDT resistance was to accelerate the search for substitute insecticides. DDT is one of a whole family of **chlorinated hydrocarbons** having insecticidal properties, of which methoxychlor is another widely used example. However, insects resistant to DDT are apt to be resistant to its close relatives as well.

However, there are other chlorinated hydrocarbons, such as aldrin, dieldrin, and endrin, that have molecular structures quite different from DDT and some of these have been useful against DDT-resistant species. But these materials are more toxic—both to humans and wildlife—than DDT, while sharing with it prolonged persistence on crops and in the soil.

On December 31, 1972, the use of DDT in the United States was forbidden for any but public health purposes (for which it has not been needed). This action was taken as a result of a growing awareness that DDT was influencing the biosphere in ways that went beyond the reduction in the populations of some pests and the enhancement of the populations of other (DDT-resistant) pests. Improvements in analytical techniques, which permitted the detection of DDT in very low concentrations, revealed that DDT was more widely distributed through the biosphere than had been suspected. Examination of the tissues of people directly exposed to DDT, through working in plants manufacturing it or on farms applying it, showed the presence of DDT in their blood and also in other tissues, especially fatty tissue. However, the concentration of DDT in the fatty tissue was far higher than in the blood, often as much as 1000 times higher. DDT is very soluble in fat solvents and poorly soluble in water so it is not surprising that it is preferentially stored in fatty tissue.

One should have anticipated finding DDT in the bodies of those occupationally exposed to it; however, every study that has been made of the general population has revealed that we, too, harbor DDT residues in our fat deposits. While the average concentration of DDT in the general United States population is now some 6 ppm, this average masks some striking differences in the distribution of these DDT burdens. Southerners have twice the concentrations that northerners do (more household pests in hot climates?). American blacks have twice the concentrations that whites of the same age, sex, and location do. Where did these DDT deposits come from? One likely possibility is that we ingested the DDT in the form of residues on food. But how, then, can one explain the presence of 3 ppm DDT in the fatty tissue of Eskimoes, whose diet contains virtually no DDT residues? Another possible means of entry into the body is through inhalation of DDT-containing dust. Such dust might arise from the local use of DDT, e.g. against household pests, or, perhaps, represent DDT that had been carried—much like radioactive fallout—many miles from its point of application.

Are such body burdens of DDT harmful to humans? To date, there is no direct evidence that they are. People accidentally or deliberately exposed for extended periods to amounts of DDT far greater than these values represent seem to suffer no long-term effects. In the early stages of exposure, the body levels of DDT (and its metabolite DDE) rise rapidly at first and then reach a steady level, the magnitude of which is dependent on the amount of exposure. From that point on, the body excretes the material as fast as it acquires it.

Body burdens as high as those characteristic of people occupationally exposed to DDT do have one well-documented effect. This is the induction of an enhanced level of enzyme synthesis by the liver. Whether this represents a potential hazard is not known. However, the effect has led to the successful

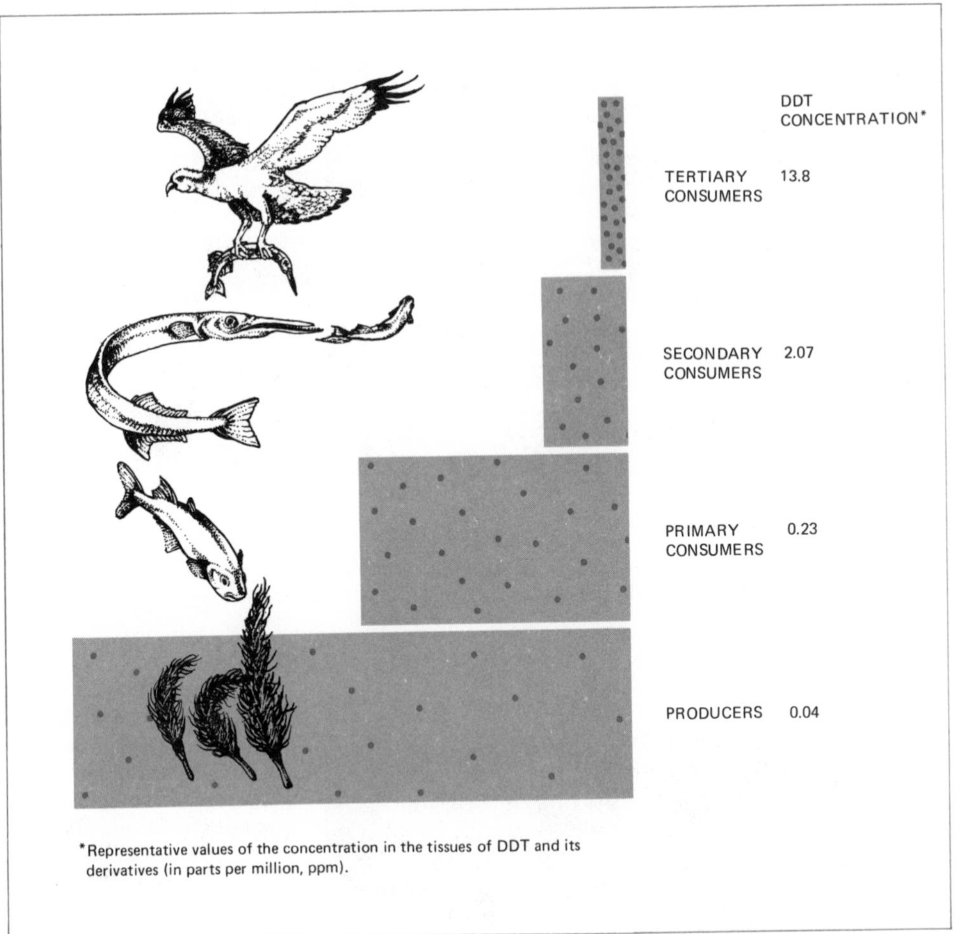

FIG. 45.4 Concentration of DDT in the tissues of organisms representing four successive trophic levels in a food chain. The concentration effect occurs because DDT is metabolized and excreted much more slowly than the nutrients passed from one trophic level to the next. Thus most of the DDT ingested as a part of *gross* production is still present in the *net* production that remains at that trophic level (see Fig. 40.7).

therapeutic use of DDT as a drug to aid victims of impaired liver function.

While harmful effects from average exposures to DDT have not been demonstrated in humans, DDT and the other chlorinated hydrocarbons have been repeatedly shown to be damaging to other species, especially certain innocent, i.e. unintended, victims of DDT sprays such as fish, earthworms, robins, etc. The hazard of DDT to innocent organisms is particularly acute for those species that live at the end of food chains. Because DDT is metabolized and excreted so slowly, it accumulates in the bodies (especially in fat) of other creatures just as it does in humans. For example, spraying a marsh to control mosquitoes will result in the accumulation of traces of DDT in the cells of microscopic aquatic organisms (the plankton) in the marsh. In feeding on these, filter-feeders like clams harvest DDT as well as food. Concentrations of DDT ten times greater than those in the plankton have been measured in clams. This process of concentration goes right up the food chain from one trophic level to the next (Fig. 45.4). Gulls, which feed upon clams, may accumulate DDT to 40 or more times the concentration in their prey. This represents a 400-fold increase in concentration along the length of this short food chain. While gulls seem to be surviving such pesticide burdens, there is abundant evidence that carnivores at the ends of longer food chains (e.g. ospreys, pelicans, falcons, and eagles) are suffering serious declines in fecundity and hence population because of this phenomenon. High body levels

SPECIES	LOCATION	AVERAGE CONCENTRATION OF DDE[1] IN EGGS (PPM)	REDUCTION IN SHELL THICKNESS[2]
PEREGRINE FALCON	ALASKAN TUNDRA (NORTH SLOPE)	889	-21.7%
PEREGRINE FALCON	CENTRAL ALASKA	673	-16.8%
PEREGRINE FALCON	ALEUTIAN ISLANDS	167	-7.5%
ROUGH-LEGGED HAWK	ALASKAN TUNDRA (NORTH SLOPE)	22.5	-3.3%
GYRFALCON	SEWARD PENINSULA, ALASKA	3.88	0

[1] Synthesized in the body from DDT
[2] Thickness compared with shells collected prior to 1947

FIG. 45.5 Correlation between DDE concentrations in the eggs of Alaskan falcons and hawks and reduction in the thickness of the egg shells. DDE is a metabolite of DDT. (Data from T. J. Cade *et al., Science* **172,** 955, 1971.)

FIG. 45.6 Thin-shelled eggs in the nest of a brown pelican in a colony off the coast of California. DDE concentrations as high as 2500 parts per million were found in the eggs of this colony. The shells were so thin that most eggs were crushed by the parent's body as it attempted to incubate them. No young birds hatched in the entire colony the year this photo was taken. (Courtesy of Joseph R. Jehl, Jr.)

of chlorinated hydrocarbons interfere with production of egg shells of normal thickness (Fig. 45.5) with resulting high mortality (Fig. 45.6). DDT may interfere at other points in the reproductive process as well.

Another group of non-target victims of DDT and other pesticides are those insects that prey upon harmful or destructive insects, i.e. their natural enemies. Killing these has serious ecological—and economic!—effects. Once apple growers learned to control codling moth larvae and apple maggots with DDT, they quickly found their orchards under devastating attack by scale insects and mites. The reason was simply that DDT killed not only the intended victims, but also killed the natural insect enemies of the scale insects and mites. Without this check on their population, their numbers increased alarmingly until new sprays, especially designed to control them, were applied.

45.5 THE ORGANOPHOSPHATES AND CARBAMATES

With the increasing incidence of DDT-resistant insects, the search was on for alternative insecticides. Among the most successful of these were the organophosphates. These substances, e.g. parathion (Fig. 45.3) and malathion, are related to the "nerve" gases developed during World War II. They react irreversibly with the enzyme acetylcholinesterase which, you may recall, is responsible for inactivating acetylcholine (ACh) at the neuromuscular junctions (see Section 28.3) and at certain synapses in the central and autonomic nervous systems (see Section 27.3). Thus, the organophosphate insecticides are, in general, exceedingly toxic to humans. Parathion, to cite a notorious example, is gram for gram 30 times more toxic than DDT. It is not surprising, then, that serious illness and a substantial number of fatalities have been caused by organophosphates. Children playing with discarded containers of parathion or the accidental contamination of food are often at fault. In 1968, 300 cases of parathion poisoning in Tijuana, Mexico, with 17 deaths, were traced to accidental contamination of sugar used in baking. A decade earlier, contaminated wheat in India led to 360 cases of parathion poisoning with 102 deaths.

Unlike chlorinated hydrocarbons, the organophosphates break down in the environment quite quickly and thus residues on crops have not been a problem. Nor are they stored in animal tissue, so accumulation in food chains has not been a problem either. But the development of resistance among the target pest populations is a problem with organophosphates just as it is with the chlorinated hydrocarbons. In an attempt to keep ahead of this problem, the carbamates entered in the fray.

Carbamate insecticides, e.g. carbaryl (Sevin) are also acetylcholinesterase inhibitors, but, in contrast to the organophosphates, their inhibition is reversible. Furthermore, these compounds are rapidly detoxified and excreted. Thus their hazard to warm-blooded animals appears to be far less than the other agents we have discussed. They also become degraded rapidly in the environment, so persistence does not appear to be a problem. There is, however, the continuing problem of danger to useful insects, especially honeybees, and we should certainly anticipate the appearance of carbamate-resistant pests.

The search for substances with which to exert chemical control over insect pests goes on. Recently, compounds which interfere with the development of the insect exoskeleton (e.g. a substance designated PH-60-38) have shown promise as insecticides. It is often said that the evolution of insecticide-resistant species requires the introduction of ever-more-powerful insecticides. But this obscures the essential ingredient in the process, which is the development of a new substance that attacks some other chink in the insect's armor. Whether gram-for-gram the new substance is more or less toxic to the insect (and to us) is quite a separate issue.

45.6 "THIRD-GENERATION" PESTICIDES

In Chapter 25 (see Section 25.2), we examined the hormones that control growth and metamorphosis in insects. We took note of the potentialities of analogues of **juvenile hormone (JH)** as possible insecticides. Their usefulness should stem from the expectation that insects would not be apt to develop resistance to a substance that for them is a normal (almost) body constituent. This new approach to insect control led Carroll Williams, one of the pioneers in the study of the endocrinology of insects, to call these substances "third-generation pesticides" (because they came after the inorganic chemicals like lead arsenate—"first generation"—and the organic chemicals like DDT and parathion—"second generation"). One JH mimic is now available commercially and has been registered by the U.S. Environmental Protection Agency for use against mosquitoes and flies. Others will probably follow.

In 1976, William S. Bowers and his coworkers reported the discovery of substances with *anti*-JH activity. Because one effect of these substances was to induce precocious metamorphosis in certain insects, they were named *precocenes*. The potentialities of precocenes for reducing insect damage were examined in Section 25.2. Because of these possibilities, Bowers anticipates that they will usher in a "fourth generation" of pesticides.

45.7 BIOLOGICAL CONTROLS

As early as the nineteenth century, some observant naturalists were aware of the important role played by the predators and parasites of insect pests in keeping them in check. Could we enlist the aid of these natural enemies in our battle for increased food production?

In case after case, the pests that pose the greatest threat to agriculture in a given region are those that are foreigners, i.e. pests that have been imported, usually accidentally, from elsewhere. Often these same pests are not a serious problem in their native land. Asa Fitch, a prominent American entomologist, guessed the reasons for this. In 1856, he wrote: "Why is it so severe and unremitting a pest in our country (referring to the wheat midge, a European import) when it is so slight and transitory in its native land? There must be a cause for this remarkable difference. What can that cause be? I can impute it to only one thing. We here are destitute of nature's appointed means for repressing and subduing this insect. Those other insects which have been created for the purpose of quelling this species and keeping it restrained within its appropriate sphere have never yet reached our shores. We have received the evil without the remedy."

In 1887, the cottony cushion scale insect, an Australian import, was devastating the citrus groves of California. Following Fitch's clue, an entomologist went to Australia to find a natural enemy of this pest. He returned with a species of lady beetle. Released in California the beetle quickly brought the cottony cushion scale under control. At least until 1946. In that year, the pest made a dramatic resurgence. This coincided with the first use of DDT in the groves and the quick disappearance of the lady beetles. Only by altering spray procedures and reintroducing the beetle was the cottony cushion scale once more brought in check.

This story of successful biological control has not been repeated as often as one would like. One problem is that for those predators or parasites that specialize in feeding on one or a very few prey species, their own population is dependent on the population of their prey. If they wipe out their prey, they then eliminate themselves. On the other hand, simply achieving moderate reductions in the pest species upon which they prey may leave a population still great enough to cause levels of crop damage unacceptable to the farmer. Furthermore, imported predators may do their job for a time but be unable to survive on an indefinite basis (e.g. get through harsh winters). Consequently, many successful biological control measures require the *periodic* release of predators. This, in turn, often means that the predators must be raised in captivity. Despite these drawbacks, biological control victories are being won. Apple growers in Maryland, Virginia, and West Virginia are finding that release of lady beetles will control mites in their orchards and make it possible to eliminate some sprays. A mite that attacks apple leaves is being controlled in Australia by means of a second, *predatory* mite. As a special bonus, the predatory mite happens to have developed resistance to the organophosphate insecticide used in the orchards to control other pests.

Biological control measures have also been useful in controlling some weed pests. The introduction into Australia of a cactus-feeding moth brought under control an epidemic spread of prickly pear cactus that was ruining millions of acres of rangeland. In 1944, two species of beetles were introduced into California to control the Klamath weed that was ruining some five million acres of rangeland in California and the Pacific Northwest (Fig. 45.7). Before its release, exhaustive tests were made to be certain that the beetles would not turn to valuable plants once they had consumed all the Klamath weed they could find. The beetles succeeded beautifully in California, restoring about 99% of the endangered rangeland to usefulness (Fig. 45.8).

Parasites, as well as predators, have been exploited in attempts to achieve biological control over destructive species. The bacterium *Bacillus popilliae* is grown commercially to help control the Japanese beetle by infecting it with "milky disease." Another bacterium, *Bacillus thuringiensis,* is also marketed by several firms in the United States to aid in controlling the larvae of certain harmful moths and butterflies.

In some cases, this agent infects the pests and eventually kills them. In other cases, it is the toxin manufactured by the bacterium while it is growing

FIG. 45.7 Rangeland in Blocksburg, California in 1948. The plant in bloom in the foreground is Klamath weed. The weed has just been killed in the far portion of the field following the introduction of its natural predator, the *Chrysolina* beetle. By 1950, the weed had been entirely eradicated and replaced by a lush cover of grass. (Photo by the late J. K. Holloway, courtesy of Prof. Carl B. Huffaker.)

FIG. 45.8 Plaque commemorating the successful biological control achieved over Klamath weed through the release of its natural predator, the *Chrysolina* beetle. The plaque is located at the Agricultural Center Building, Eureka, California. (Courtesy of John V. Lenz.)

in culture that does the job. In the latter case, then, one is really dealing with another form of chemical control, albeit one with high selectivity for certain insects and no known danger to other kinds of animals. Approximately 60% of the lettuce crop in California is now treated with *B. thuringiensis*. Field studies indicate that it may provide an effective control mechanism for the gypsy moth as well.

Viruses also show promise as pest control devices. They are even more selective in their action than bacteria, acting on only one or a few species and not harming other organisms in the environment. Unfortunately, they must be grown in living host cells, which means having to cultivate the pest species. Once released in the field, however, the infection may spread naturally, or the early victims can be harvested, ground up, and used to disseminate the virus to new locations. Viruses have been successfully used to control such insects as the cotton boll worm, tobacco budworm, European pine sawfly, cabbage looper, and alfalfa caterpillar.

45.8 BREEDING RESISTANT SPECIES

A promising approach to increasing crop yields is to breed pest resistance into the crop in the same way that genes for other qualities such as yield, color, protein content, etc. are introduced into domestic species. A wild tomato plant found growing in a sugarcane field in Peru provided the genetic material with which resistance to several fungus diseases was introduced into more than 30 commercial tomato varieties. Some two dozen commercial wheat varieties have been bred for resistance to the Hessian fly (a serious pest of wheat thought to have reached the United States in the straw used as bedding for the Hessian soldiers fighting for Britain during the Revolutionary War). Alfalfa varieties (e.g. Lahontan) resistant to aphids, and corn varieties resistant to attack by corn earworm (Fig. 45.9), rootworm, and borer have been produced by careful breeding programs.

The breeding of fungus-resistant species of cereal crops like wheat, oats, and barley has been singularly successful. But the development of resistant varieties is no permanent cure. Just as the use of antibiotics has been countered by the evolution of antibiotic-resistant bacteria, so the development of fungus-resistant grains by selective breeding is

FIG. 45.9 Top: Sweet corn (variety: Golden Cross) damaged by the corn earworm. Bottom: Earworm damage to a variety of sweet corn (La2W × 14S) that was bred for resistance to attack by this pest. Note the marked reduction in the severity of the damage. (U.S.D.A. photograph.)

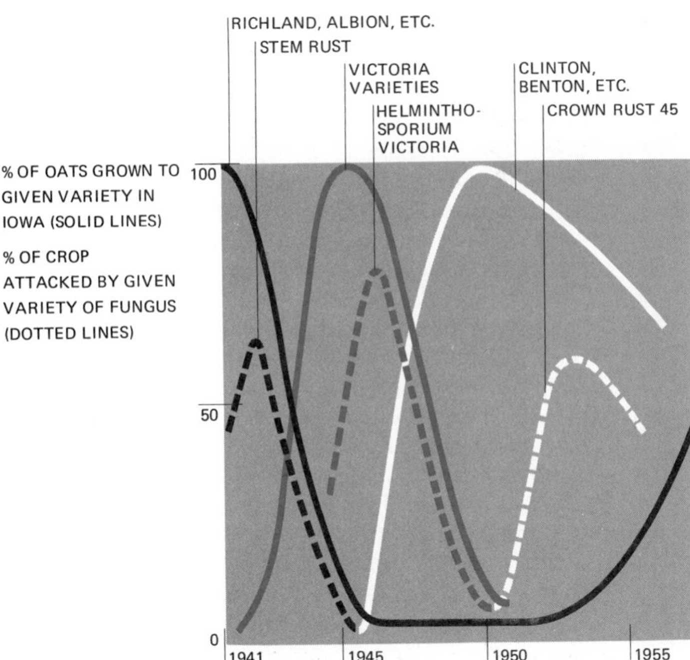

FIG. 45.10 Host-parasite relations between the domestic oat (*Avena sativa*) and some of its parasites. Each introduction of a fungus-resistant variety was followed by the appearance of a new form of a fungus able to parasitize it successfully. Each time this occurred, the oat farmers in Iowa had to turn to still another resistant variety. By 1955, so many years had elapsed since the Richland and Albion varieties had last been planted, that their most serious parasite had disappeared and they could be safely reintroduced (for a time!).

inevitably followed by the appearance of mutant fungi capable of parasitizing them. One of the major and never-ending tasks of plant breeders in the United States Department of Agriculture is to keep one step ahead of the fungi and to ensure that grain farmers will always be able to switch to a still-newer resistant variety as the older varieties begin to succumb. Figure 45.10 traces the changing tides of battle between our domestic oat and the fungi that parasitize it.

In 1970 the United States witnessed a vivid demonstration of the ability of parasites to coevolve with their host. Prior to 1970, the southern corn leaf blight, caused by the fungus *Helminthosporium maydis*, had been a relatively minor problem. The hybrid corn varieties in use were relatively resistant to the organism, although they were known to be susceptible to a *Helminthosporium* race found in the Philippines.

In 1969 corn losses due to *Helminthosporium* began to rise in Florida, presumably because of the fresh appearance of a mutant race having infective properties similar to the Philippines race. During the 1970 season, this new race moved out of Florida, swept through the Gulf states, and then moved on into the Midwest and eventually into Canada. In some areas, 50% or more of the crop was lost. Overall, the United States corn harvest was reduced 15%—a loss of approximately one billion dollars.

Fortunately, plant breeders had already developed corn strains resistant to another fungus and these turned out to be resistant to the new race of *Helminthosporium* also. However, the lessons are clear: the development of host resistance is always temporary and the more widespread the cultivation of a genetically uniform crop, the greater the risk of disastrous epidemics.

45.9 OTHER APPROACHES TO PEST CONTROL

Crop rotation is a simple but often effective means for reducing pest attack. By planting a field to one crop one season and to a different crop the next, any pests that survive the winter in the soil or plant stubble find no prey to support renewed population growth the following season. The removal and destruction of parasitized plant debris, even simple tillage techniques, can often keep pests to acceptable levels without harm to the environment.

The use of insect repellents to protect crops is another technique that shows promise. In some cases, such repellents can be absorbed by the plant, making it unpalatable to its predators.

Conversely, insect *attractants* can be harnessed to the cause of pest control. Some 200 chemicals have been discovered by which one sex (usually the female) of an insect species attracts its mates. A

FIG. 45.11 A research technician with the U.S. Forest Service's Pacific Southwest Forest and Range Experiment Station puts up a sticky trap in the woods near Bass Lake in the Sierra National Forest. Trap contains a vial of the sex attractant for the western pine beetle, a major pest of pine forests in the west. The insects are caught in the trap as they fly upwind toward the attractant. (U.S. Forest Service photo.)

number of these sex attractants, or their close chemical relatives, have been synthesized and are available commercially. These include the sex attractants of the cabbage looper, the cotton boll weevil, the pink bollworm, and the gypsy moth. Extensive field experimentation is being conducted with them. In general, two approaches have been tried. In some field studies, distributing the attractant throughout the area appears to mask the insect's own attractant so that the sexes fail to get together. This has been called the "male confusion" or "communication disruption" technique. Alternatively, traps can be "baited" with the pest's attractant and the males (or, in some cases, females) lured to their death (Fig. 45.11). Each of these approaches looks promising, especially when infestation by the pest is not too severe.

In the fall, many insects enter a dormant period, called **diapause,** in which they spend the winter. In the spring, they end their diapause and resume their life cycle. Both entering and leaving diapause are regulated by photoperiod, that is, the relative length of day and night. In recent field studies, it was found that illuminating test plots at night prevented larvae from entering diapause. Because nondiapausing insects cannot survive the harsh conditions of winter, this raises the possibility of controlling pests by artificially manipulating the photoperiod. Further research will be needed to determine whether the levels of illumination required are economically feasible.

45.10 THE STERILE MALE TECHNIQUE

One of the cleverest and most successful means of attacking insect pests is the sterile male technique. This technique was first applied against the screwworm fly, a serious pest of livestock. The female flies lay their eggs in sores or other open wounds on the animals. After hatching, the larvae eat the tissues of the host. As they do so, they expose a still larger area to egg-laying, and death of the host is the all-too-frequent outcome. It has been estimated that prior to its eradication from the southeastern states, the screwworm was causing annual livestock losses of $20 million.

The eradication of the pest was achieved by releasing laboratory-reared and sterilized flies into the natural population. Sterilization was accomplished by exposing the laboratory flies to just enough gamma radiation to make them sterile but not enough to reduce their general vigor as potential mates.

Starting early in 1958, up to 50 million sterilized flies were released each week from aircraft flying over Florida and parts of the adjoining states. The success of the undertaking depended only on the sterile males (but it was impractical to separate the sexes prior to release). As their number was increased in an area, the chances of a fertile female in the natural population mating with a sterile male increased also. Each time this occurred, the female would lay sterile eggs. Since the female mates only once, her reproductive career was thus ended without any offspring being produced. By early 1959, the pest was totally eliminated east of the Mississippi River.

Attention then turned to the task of eradicating the screwworm in the southwestern states. This was a formidable undertaking because the pest winters in Mexico and with each new season could move across the border and through the cattle-raising states of the southwest. Nevertheless, by repeated air drops of sterilized flies along the Mexico-Texas border, totaling some ten billion flies each year, the pest was virtually eliminated from the southwestern states as well. Until 1972, that is. In that year, the pest scored a marked resurgence—over 90,000 cases of screwworm attack occurring in American livestock. What happened? Perhaps once again we have an example of nature's adaptability. If, for example, any screwworm flies in the natural population had an *inherited* disinclination to mate with released flies, natural selection would favor them overwhelmingly.

Despite the recent setback, the overall success of the screwworm program has encouraged the application of the sterile male technique to other pests. It has proven of value in controlling natural populations of the pink bollworm (a pest of cotton plants) in California, Nevada, and Arizona, and of a harmful fruit fly (not *Drosophila*) in California. Used as part of an integrated program of controls that included heavy applications of pesticides and use of the boll weevil sex attractant, it virtually eliminated the cotton boll weevil in the center of a treated area encompassing some 5000 square miles in southern Mississippi and adjacent parts of Alabama and Louisiana.

For some insect pests, the doses of radiation needed to sterilize the males harm them in other ways as well. To circumvent this problem, attempts are being made to introduce sterility or partial sterility into pest populations in more subtle ways. It is possible, for example, to develop populations of laboratory-reared insects that carry abnormal chromosomes (produced by induced breaks and rearrangements—see Section 16.4). Having a full complement of genes, these insects are perfectly healthy. Mated to wild members of the species, however, they would introduce these abnormal chromosomes into the natural population.

Introducing abnormal chromosomes would result in a substantial number (the frequency would depend on the number of chromosomes involved) of offspring that would be nonviable because of an unbalanced gene content. Those offspring that, as a result of random assortment (see Section 14.7) were lucky enough to receive a *balanced* set of abnormal chromosomes would then be in a position to repeat the process and thus spread a measure of infertility through still another generation. Vigorous efforts are now being made to adapt these techniques for use in the control of houseflies and mosquitoes.

45.11 WHAT DOES THE FUTURE HOLD?

The human population on the earth today could not be maintained if we did not deliberately upset the natural balance of other populations in our favor. Agriculture, industry, commerce, and speedy transportation have brought changes in our biotic environment which have been beneficial. They have also encouraged secondary changes in plant and animal populations (weeds, rabbits, insect pests, etc.) which threaten us continually with a sudden reduction in the capacity of the environment to support us. Furthermore, as our influence has spread to every corner of the earth, we have sometimes threatened the very existence of species (e.g. the blue whale) whose continued presence on earth, while perhaps not indispensable to our welfare, provides interest and variety in our lives.

With a fast-growing population in a fast-shrinking world, the margin for error in our handling of our environment has been sharply reduced. If we are going to avoid future catastrophe, we must bring the *long-term* capacity of our environment into balance with our population. We must learn to conserve resources that cannot be renewed (e.g. metal ores, oil, and coal) at least until satisfactory substitutes can be found for them. We must direct our efforts, as much as possible, toward the use of renewable resources for this purpose, that is, materials which are the products of photosynthesis.

Our ever-increasing dependence on renewable resources requires that we do everything we can to avoid reducing the efficiency with which the energy of the sun can be converted into these materials. Soil must be conserved and its fertility carefully maintained. Water must be conserved and its purity guarded. We should also continue to breed plants and animals capable of transforming matter and energy with minimum losses, that is, capable of growing with maximum efficiency. We must exercise the greatest care in transporting species from one region of the earth to another and in applying chemical and biological controls to other populations. All these things will require biologists, in laboratories and in the field, who can unravel the complex ecological interactions which follow any change we impose on the biosphere. It will also require men and women from all walks of life with the wisdom and foresight to put into practice the knowledge gained from ecology. We must, and will, use our environment. We should not use it up. An ecological budget which is unbalanced for too long endangers the physical and spiritual well-being, in fact the very existence, of our future generations. It also endangers the future of the many other living things which share the planet with us and which have taught us so much about ourselves and about the nature of life itself.

EXERCISES AND PROBLEMS

1. How exactly do populations of DDT-resistant flies arise?
2. Which of the following would you expect to be a symptom of parathion poisoning: (a) dilated pupils, (b) excess salivation, (c) rapid heartbeat, (d) muscular tremors, (e) diarrhea, (f) dilated bronchi?

REFERENCES

1. CARSON, RACHEL, *Silent Spring,* Houghton Mifflin, Boston, 1962. It was this landmark book that made the general public aware for the first time of the problems connected with the widespread and indiscriminate use of pesticides.
2. U.S. DEPARTMENT OF HEALTH, EDUCATION, AND WELFARE, *Report of the Secretary's Commission on Pesticides and Their Relationship to Environmental Health,* U.S. Government Printing Office, Washington, D.C., 20402. An encyclopedia of data on pesticides.
3. PEAKALL, D. B., "Pesticides and the Reproduction of Birds," *Scientific American,* Offprint No. 1174, April, 1970.
4. JACOBSON, M., and M. BEROZA, "Insect Attractants," *Scientific American,* Offprint No. 189, August, 1964. How sex attractants are being used as a weapon against certain insects.
5. WILLIAMS, C. M., "Third-Generation Pesticides," *Scientific American,* Offprint No. 1078, July, 1967. Explores the potentialities of juvenile hormone as an insecticide.
6. SCHNEIDER, D., "The Sex-Attractant Receptor of Moths," *Scientific American,* Offprint No. 1299, July, 1974. Concludes that a nerve impulse in a receptor cell can be triggered by one molecule of the attractant.

GLOSSARY

GUIDE TO PRONUNCIATION*

ə —banana, collect, abut
ər —operation, further
ä —bother, cot
ŋ —'sing(siŋ) singer('siŋ-ər)
ȯ —saw, all, gnaw
yü—youth, union, cu, few, mute
' —primary accent (before main accented syllable)
ˌ —secondary accent

ABSORPTION SPECTRUM Electromagnetic spectrum whose intensity at each wavelength is a measure of the amount of energy at the wavelength that has passed through a selectively absorbing substance.

ACETYLCHOLINE (ə-ˌsēt-əl-'kō-ˌlēn) Organic compound secreted at the ends of many neurons. A neurohumor.

ACID [L. *acidus*, sour] Molecule or ion that liberates protons, usually in water.

ACTINOMYCIN D (ˌak-ti-ˌnō-'mīs-ən) An antibiotic isolated from soil bacteria that blocks DNA-dependent RNA synthesis.

ACTION SPECTRUM Rate of physiological activity plotted against wavelength of light.

ACTIVE TRANSPORT Transfer of a substance across a cell membrane from a region of low to one of high concentration. Requires energy.

ADAPTATION [L. *adaptare*, to fit] Any characteristic of an organism that contributes to its survival in its environment.

ADAPTIVE RADIATION Evolution, from a single ancestral species, of a variety of descendant species adapted to different ways of life.

ADENOSINE TRIPHOSPHATE (ATP) (ə-ˌden-ə-ˌsēn-ˌtrī-'fäs-ˌfāt) Organic compound that is the immediate source of energy for the activities of the cell.

ADHESION Force of attraction between unlike molecules.

ADVENTITIOUS ROOT Root that arises from a stem or leaf.

AEROBIC (ˌa-(ə-)'ro-bik) [Gk. *aeros*, air; *bios*, life] Requiring the presence of free oxygen (O_2).

ALGA ('al-gə) A photosynthetic, often plantlike, organism generally found growing in aquatic or damp locations.

* *By permission. From Webster's Seventh New Collegiate Dictionary, Copyright 1963, by G. & C. Merriam Co., Publishers of the Merriam-Webster Dictionaries.*

ALLANTOIS (ə-'lant-ə-was) [Gk. *allas*, sausage] Extraembryonic membrane of reptiles, birds, and mammals that forms a pouch growing out of the posterior part of the alimentary canal.

ALLELE (ə-'lē(ə)l) Alternative form of a gene that may occur at a given gene locus.

ALLERGY ('al-ər-jē) An exaggerated immune response, i.e. hypersensitivity to an antigen.

ALLOSTERIC (ˌal-ə-'stir-ik) Refers to a change in the properties (probably including shape) of a protein following the binding of a small molecule to a site on the protein other than its active site.

AMINO TERMINAL The end of a polypeptide chain that has a free amino group ($-NH_2$) attached to the backbone of the chain. Synthesis of polypeptide chains proceeds from the amino terminal to the carboxyl terminal.

AMNIOCENTESIS ('am-nē-ō-ˌsen-'tē-ses) Removal of amniotic fluid from a pregnant woman in order to study its composition and/or culture the cells contained in it.

AMNION ('am-nē-ˌän) Extraembryonic membrane of reptiles, birds, and mammals that encloses the embryo in a fluid-filled sac.

AMOEBA (ə-'mē-bə) A unicellular protozoan that moves by means of pseudopodia.

AMPHETAMINE (am-'fet-ə-ˌmēn) A drug that is similar in molecular structure to adrenaline and noradrenaline and shares their stimulatory properties.

AMPHIPATHIC (ˌam(p)-fē-'path-ik) [Gk. both feelings] Used to describe molecules that contain both polar (hydrophilic) and apolar (hydrophobic) groups. Sodium stearate (a soap) is an amphipathic molecule.

AMYLASE ('am-ə-ˌlās) [L. *amylum*, starch] An enzyme that digests (hydrolyzes) starch.

ANABOLISM (ə-'nab-ə-ˌliz-əm) Constructive metabolism in which complex substances are synthesized from simpler ones.

ANAEROBIC (ˌan-ə-'rō-bik) [Gk. *an*, not + aerobic] Not requiring the presence of free oxygen.

ANALOGOUS (ə-'nal-a-gəs) [Gk. *analogos*, proportionate] (Of organs in different species), having a similar function but a different structure and embryonic development.

ANDROGEN ('an-drə-jən) One of a group of sex hormones of male vertebrates which promotes the development of the secondary sex characteristics.

ANEMIA [Gk. *anaimia*, bloodlessness] Deficiency of red blood cells or hemoglobin in the blood.

ANESTHETIC (ˌan-əs-'thet-ik) A substance that produces loss of sensation.

ANTIBODY A protein, produced by a vertebrate, that can unite with a specific antigen.

ANTICODON The group of three adjacent bases on a transfer RNA molecule that pairs with a complementary codon on a messenger RNA molecule.

ANTIGEN ('ant-i-jən) Macromolecule (usually a protein or polysaccharide) which, when introduced into the body of an animal it is foreign to, stimulates the formation of antibodies.

ANTIGENIC DETERMINANT Any portion of an antigen that combines with the active site of an antibody.

ANTITOXIN Mixture of antibodies formed in response to a toxin.

AQUIFER ('ak-wə-fər) A layer in the earth that is saturated with water.

ARCHENTERON (är-'kent-ə-ˌrän) Central cavity of the gastrula of an embryo, ultimately becoming the alimentary canal.

ARTHROPOD A member of the phylum of invertebrate animals characterized by jointed legs and exoskeleton. Includes insects, arachnids, and crustaceans.

ASCUS ('as-kəs) [Gk. *askos*, wineskin, bladder] Tubular spore sac of ascomycetes in which (usually) 8 ascospores are produced.

ASEXUAL ((')ā-'seksh-(ə-)wel) REPRODUCTION Reproduction without the union of gametes (or any nuclear material).

ASSORTATIVE MATING Mating between similar individuals in a population; therefore, a form of nonrandom mating.

ATOM ('at-əm) [Gk. *atomos*, indivisible] Smallest particle of an element that can enter into combination with other elements.

ATOMIC WEIGHT UNIT One-twelfth the weight of an atom of carbon-12. Also called the dalton.

AUTO-IMMUNE DISEASE Disease characterized by the mounting of an immune response against constituents of an individual's own tissues.

AUTOSOME ('ot-ə-ˌsōm) Any chromosome that is not a sex chromosome.

AUTOTROPHIC (ˌȯt-ə-'träf-ik) Capable of synthesizing organic compounds from inorganic raw materials.

AUXIN ('ȯk-sən) [Gk. *auxein*, to increase] Plant hormone that, among other effects, promotes cell elongation.

AXON ('ak-ˌsän) Single extension of a neuron (usually long and often branched), which conducts nerve impulses away from the dendrites.

BACTERIOPHAGE (bak-'tir-ē-ə-ˌfāj) [bacterium + Gk. *phagein*, to eat] Virus that infects bacteria.

BALANCED POLYMORPHISM The maintenance of two or more distinct phenotypes in a population by natural selection. Balanced polymorphism can occur as a result of disruptive selection or if (as in sickle-cell anemia) the heterozygotes are more fit than either homozygote.

BARBITURATE (bär-'bich-ə-rət) Any of several hundred derivatives of barbituric acid used as sedatives (e.g. phenobarbital).

BASE Molecule or ion that can take a proton from an acid.

BASIDIUM (bə-'sid-ē-əm) Club-shaped spore-producing structure of basidiomycetes on which four basidiospores are formed.

B CELL An immunologically active cell (lymphocyte) that is not dependent on the thymus gland for its activity. Antibody-*secreting* cells (plasma cells) are derived from B cells.

BETA-GALACTOSIDASE ('bāt-ə ˌgə-ˌlak-tə-'sī-ˌdās) An enzyme ("lactase") that hydrolyzes the disaccharide lactose.

BIOASSAY (ˌbī-(ˌ)ō-'as-ˌā) Quantitative determination of the strength of a biologically active substance from its effect on a living organism.

BIOCHEMICAL OXYGEN DEMAND (BOD) The oxygen needed (in mg/l or ppm) by bacteria and other microorganisms to oxidize the organic matter present in a water sample such as water polluted by sewage. Also called biological oxygen demand.

BIOMASS The total amount of living matter in a given population of organisms.

BIOME ('bī-ōm) A distinctive plant and animal community produced and maintained by the climate. The coniferous forest region of North America constitutes a single biome (the taiga).

BIOSPHERE The part of our planet in which life exists and with which it exchanges materials. Includes a small part of the lithosphere (the solid earth) and large parts of the hydrosphere (water) and atmosphere.

BLASTOCYST The blastula that is formed by placental mammals. The blastocyst is the embryonic stage that implants in the wall of the uterus.

BLASTULA ('blas-chə-lə) Early stage of animal development in which a single (usually) layer of cells surrounds a fluid-filled cavity (the blastocoel), thus forming a hollow ball.

BOTANY ('bät-ən-ē) [Gk. *botanē*, pasture, herb] The study of plants.

BRANCHIAL GROOVES ('braŋ-kē-əl) Series of external, paired grooves in the neck region of vertebrate embryos that correspond in position to the outpocketings of the pharynx (the gill pouches).

BUDDING Asexual reproduction in which a new organism develops from an outgrowth of the parent.

CALORIE [L. *calor*, heat] The amount of heat required to raise the temperature of 1 g of water 1°C. When capitalized, a unit of heat 1000 times larger than the above.

CAMBIUM ('kam-bē-əm) Layer of meristematic cells in the roots and stems of many tracheophytes that produces secondary xylem and phloem.

CANCER ('kan-sər) Any of a group of diseases characterized by an uncontrolled proliferation of cells.

CARBOHYDRATE (ˌkär-bō-'hī-ˌdrāt) Organic compound of carbon, hydrogen, and oxygen, generally with a 2:1 ratio of hydrogen atoms to oxygen atoms. Sugars, starches, and cellulose are carbohydrates.

CARBOXYL TERMINAL The end of a polypeptide chain with a free carboxyl (−COOH) group attached to the backbone of the chain. Synthesis of a polypeptide chain proceeds from the amino terminal to the carboxyl terminal.

CARCINOGEN (kär-'sin-ə-jən) A cancer-inducing substance.

CATABOLISM (kə-'tab-ə-ˌliz-əm) Destructive metabolism in which complex molecules are broken down into simpler ones, with the liberation of energy.

CATALYST ('kat-əl-əst) Substance that accelerates the rate of chemical reaction without being used up in the process.

CELL-MEDIATED IMMUNITY An immune response (e.g. graft rejection) that is dependent on the presence of specifically sensitized (i.e. able to recognize a particular antigen) lymphocytes.

CEPHALIZATION (ˌsef-ə-lə-'zā-shən) Evolutionary tendency toward the concentration of sense receptors and central nervous system at the anterior end of an animal.

CEPHALOTHORAX (ˌsef-ə-lō-'thō(ə)r-ˌaks) Fused head and thorax found in arachnids and many crustaceans.

CHALONE ('kal-ˌōn) A substance secreted by a tissue that inhibits mitosis in that tissue.

CHAPARRAL (ˌshap-ə-'ral) The scrub forest biome found in southern California.

CHARACTER DISPLACEMENT Evolutionary divergence of two species that reduces the amount of niche overlap between them.

CHEMOAUTOTROPHIC ('kem-ō-ˌȯt-ə-'träf-ik) Autotrophic, using energy secured by oxidizing some inorganic substance. Characteristic of certain bacteria.

CHIASMA (kī-'az-mə) [Gk. cross] The attachment of two nonsister chromatids in a bivalent that is first seen in diplotene of prophase I of meiosis. Each chiasma results in the exchange of genetic material between the nonsister chromatids, i.e. in crossing over. The plural is chiasmata.

CHITIN ('kīt-ᵊn) Nitrogen-containing polysaccharide that forms the exoskeleton of arthropods and the cell walls of many fungi.

CHLOROPHYLL [Gk. *chloros*, green] Green pigment used in photosynthesis.

CHLOROPLAST ('klōr-ə-ˌplast) Plastid containing chlorophyll.

CHOLESTEROL (kə-'les-tə-ˌrȯl) The most abundant steroid in the human body. It probably serves as the starting material for the synthesis of the other steroids found in the body.

CHOLINESTERASE (ˌkō-lə-'nes-tə-ˌrās) Enzyme that hydrolyzes and thus inactivates acetylcholine.

CHORION ('kor-ē-ˌän) Extraembryonic membrane of reptiles, birds, and mammals that covers the embryo and, in mammals, contributes to the formation of the placenta.

CHROMATID ('krō-mə-təd) Each of the two strands ("sister chromatids") of a duplicated chromosome while they remain attached by their shared centromere.

CHROMATIN ('krō-mə-tən) The diffuse chromosomes present in eukaryotic nuclei during interphase.

CHROMATOGRAPHY (ˌkrō-mə-'täg-rə-fē) The process of separating the components of a mixture by their differential adsorption to an insoluble matrix (e.g. paper) as the mixture is passed through the matrix.

CHROMOSOME ('krō-mə-ˌsōm) Elongated structures in the cell nucleus, containing DNA and protein and bearing the genes. The number in the nucleus is usually constant for each species.

CIRCADIAN (ˌsər-'kā-ˌdē-ən) [L. *circum*, round about + *dies*, day] Occurring approximately once a day.

CLEAVAGE Repeated mitotic division of the zygote which forms the many-celled blastula.

CLINE ('klīn) Continuous gradation of structural or physiological differences exhibited by the members of a species along a line extending from one part of their range to another.

CLOACA (klō-'ā-kə) Posterior part of the alimentary canal into which the urinary and reproductive ducts empty in birds, reptiles, amphibians, and many fishes.

CLONE The descendants produced asexually from a single cell or a single organism. Characterized by an identical genetic constitution.

CNIDARIA (nī-'da-rēə) [Gk. *knidē*, nettle] The phylum of animals that includes the hydra, jellyfishes, sea anemones, and corals. Formerly called Coelenterata.

CODOMINANCE The independent expression of each of two alleles in a heterozygote.

CODON The three adjacent bases in a molecule of DNA or messenger RNA that code for a particular amino acid.

COELOM ('sē-ləm) [Gk. *koilōma*, cavity] Main body cavity of many animals. It is lined with an epithelium derived from mesoderm.

COENOCYTE ('sē-nə-ˌsīt) Mass of cytoplasm, containing many nuclei formed by the repeated division of the nucleus of a single cell with no division of its cytoplasm.

COENZYME ((')kō-'en-ˌzīm) Organic compound that by combining temporarily with an enzyme makes its active.

COHESION Force of attraction between like molecules.

COLEOPTILE (ˌkō-lē-'äp-tᵊl) [Gk. *koleos*, sheath] Protective sheath surrounding the plumule of grass seedlings.

COLLENCHYMA (kə-'leŋ-kə-mə) Supporting plant tissue consisting of living cells whose walls are thickened at the corners. Often found in young, growing stems and petioles.

COLLOID ('käl-ˌȯid) [Gk. *kolla*, glue] Substance whose particles (macromolecules or aggregates of smaller molecules) range from 1 nm to 100 nm in size.

COMMENSALISM (kə-'men(t)-sə-ˌliz-əm) [L. *com*, together + *mensa*, table] A close living relationship between two species, in which one benefits from the other without harming or benefiting it.

COMPOUND Substance that can be decomposed into simpler substances. The elements of a compound are present in definite proportions by weight.

CONIFER A cone-bearing gymnosperm. Includes pines, spruces, firs, etc.

CONJUGATION Form of sexual reproduction in which genetic material is exchanged during the temporary union of two cells. Occurs in many ciliates (e.g. *Paramecium*) and some bacteria.

CONVERGENCE [L. *convergere*, to turn together] Evolution of superficially similar traits in unrelated organisms that live in a similar environment.

COPULATION (ˌkäp-yə-'lā-shən) Physical union of two animals during which sperm cells are transferred from one to the other.

COREPRESSOR A small molecule that joins with a repressor molecule to block gene action.

CORTEX ('kȯr-ˌteks) [L. *cortex*, bark] The outer part of an organ.

COVALENT BOND ((')kō-'vā-lənt) Chemical bond formed by one or more shared pairs of electrons.

CUTIN ('kyüt-ᵊn) Waxy material secreted by the exposed epidermal cells of plants.

CYTOCHROME ('sīt-ə-ˌkrōm) One of several iron-containing proteins, found in mitochondria and chloroplasts, that transfer electrons in the process of cellular respiration and photosynthesis respectively.

CYTOKINESIS The division of the cytoplasm—as contrasted with the nucleus—during mitosis or meiosis.

CYTOKININ ('sī-tō-ˌkī-nin) One of a group of adenine-containing compounds that stimulate mitosis in plants.

CYTOSOL The fluid in which the organelles of the cytoplasm are suspended. Also called hyaloplasm and ground substance.

CYTOPLASM ('sīt-ə-ˌplaz-əm) General term for all the contents of a cell outside of the nucleus and within the cell membrane.

DALTON Unit of weight equal to 1/12th the weight of an atom of ^{12}C.

DEAMINATION ((ˌ)dē-ˌam-ə-'nā-shən) Removal of an amino ($-NH_2$) group from a compound.

DECARBOXYLATION The removal of carbon dioxide from the carboxyl group of an organic acid.

DEDIFFERENTIATION ((')dē-ˌdif-ə-ˌren-chē-'ā-shən) Reversion of a specialized cell to a more generalized, embryonic type.

DENATURATION ((ˌ)dē-ˌnā-chə-'rā-shən) Alteration of the physical properties and three-dimensional structure of a protein by agents too mild to break the peptide bonds.

DENDRITE ('den-ˌdrīt) Branching, usually short, extension of a neuron in which the generator potential is created.

DENITRIFICATION Reduction of nitrates to free nitrogen (N_2).

DEOXYRIBONUCLEIC ACID (DNA) (dē-ˌäk-sē-'rī-bō-n(y)u̇-ˌklē-ik) Nucleic acid found in chromosomes that stores the hereditary information of the organism.

DIALYSIS (dī-'al-ə-səs) The separation of solute molecules by virtue of their differing rates of diffusion through a semipermeable membrane.

DIAPAUSE A period of dormancy, commonly occurring in insects.

DIASTOLE (dī-'as-tə-lē) Phase of relaxation of the heart.

DIFFERENTIATION (ˌdif-ə-ˌren-chē-'ā-shən) Structural and functional modification of an unspecialized cell into a specialized one.

DIFFUSION Migration of molecules or ions, as a result of their own random movements, from a region of higher to a region of lower concentration.

DIGESTION (dī-'jes(h)-chən) Breakdown of macromolecules of food by hydrolysis.

DIHYBRID ((')dī-'hī-brəd) Being heterozygous at two different gene loci.

DIOECIOUS ((')dī-'ē-shəs) Having male sex organs on one plant, female on another of the same species. The holly is dioecious.

DIPLOID ('dip-ˌlȯid) Having two of each kind of chromosome (except for the sex chromosomes); $2n$.

DISACCHARIDE ((')dī-'sak-ə-ˌrīd) A sugar (e.g. sucrose) that can be hydrolyzed into two monosaccharides.

DISSOCIATION ((ˌ)dis-ˌō-s(h)ē-'ā-shən) Separation of ions from a molecule or crystal lattice.

DISTAL ('dist-əl) Situated away from the place of origin or attachment.

DNA POLYMERASE An enzyme that catalyzes the linking together of deoxyribonucleotides to form DNA complementary to a template (either DNA or, in the case of reverse transcriptase, RNA).

ECOLOGY (ˌi-'käl-ə-jē) Study of the interrelationships of organisms and their environment.

ECOSYSTEM ('ekō-(or ēkō)ˌsis-təm) A community of organisms interacting with each other and with their nonliving surroundings.

ECTODERM ('ek-tə-ˌdərm) [Gk. *ektos*, outside + *derma*, skin] Outermost layer of cells of an animal embryo.

EDEMA (i-'dē-mə) [Gk. *oidema*, swelling] Abnormal accumulation of lymph in the tissue spaces.

EFFECTOR (i-'fek-tər) Body structure by which an organism acts. In humans the chief effectors are the muscles and glands.

EGESTION (i-'jes(h)-chən) [L. *egerere* to carry outside, discharge] Elimination of undigested materials from the alimentary canal.

ELECTRON (i-'lek-trän) Negatively charged particle present outside the nucleus of an atom.

ELECTRONEGATIVE Having an affinity for electrons.

ELEMENT Any of about 100 substances that consist of only one kind of atom and cannot be decomposed into simpler substances.

EMBRYO An animal or plant in an early stage of development from a zygote.

EMPHYSEMA (ˌem(p)-fə-'sē-mə) [Gk. *emphysēma*, bodily inflation] A condition of the lungs characterized by a reduction in the surface available for gas exchange.

EMULSION (i-'məl-shən) [L. *emulgere*, to milk out] Mixture consisting of droplets of one liquid suspended in a second.

ENANTIOMORPH One of two mirror-image optical isomers. Also called enantiomer.

END PLATE POTENTIAL (EPP) The partial depolarization created in a muscle fiber in the region of an activated neuromuscular junction.

ENDOCYTOSIS The engulfment by a cell of extracellular material accompanied by the invagination and pinching off of a portion of the cell membrane. The engulfed material is thus enclosed within a vacuole.

ENDODERM ('en-də-ˌdərm) [Gk. *endon*, within + *derma*, skin] Innermost layer of cells of an animal embryo.

ENDOSPERM ('en-də-ˌsperm) [Gk. *sperma*, seed] Nutritive tissue that surrounds and nourishes the developing embryo of seed plants.

ENDOSYMBIONT An organism living within the body of its symbiotic partner.

ENERGY [Gk. *energos*, active] Capacity for doing work.

ENZYME ('en-zīm) [Gk. *enzymos*, leavened] Protein catalyst produced by a living organism.

EPICOTYL ('ep-ə-ˌkät-əl) That portion of the shoot of a plant embryo or seedling above the node at which the cotyledons are attached.

EPIPHYTE ('ep-ə-ˌfīt) [Gk. *epi*, on + *phyton*, plant] Plant that grows entirely on another plant but for position and support only.

EQUILIBRIUM State of balance between opposing actions.

ESTROGEN ('es-trə-jən) One of a group of female sex hormones which, among other effects, promotes the development of the secondary sex characteristics.

ETIOLATION ('ēt-ē-ə-ˌlā-shən) A phenomenon exhibited by plants grown in darkness, characterized by pale color, long internodes, and small leaves.

EUGENICS (yü-'jen-iks) [Gk. *eugenēs*, wellborn] The application of genetics in an attempt to "improve" the hereditary qualities of humans.

EUKARYOTE (yü-'ka-rē-ˌōt) An organism whose cells contain a membrane-bounded nucleus. Often spelled eucaryote.

EUTROPHICATION (ˌyü-ˌträf-ə-'kā-shən) The process in which a body of water becomes enriched in dissolved nutrients.

EXCITATORY POSTSYNAPTIC POTENTIAL (EPSP) The partial depolarization created in a neuron by the arrival of an action potential at the terminals of a neuron synapsing with it.

EXCRETION [L. *excretus*, sifted out] Elimination of metabolic wastes by an organism.

EXOCYTOSIS The discharge of vacuole-enclosed materials from a cell by the fusion of the vacuole membrane with the cell membrane.

EXTENSOR (ek-'sten-sər) A muscle that extends a limb.

EXTRACELLULAR FLUID (ECF) The fluid which bathes the cells.

FAUNA ('fon-ə) Animal life in a certain environment.

FEEDBACK INHIBITION The inhibition of the first enzyme in a metabolic pathway by the final product of that pathway.

FERMENTATION Anaerobic decomposition of an organic compound (e.g. glucose) by a living organism.

FETUS ('fēt-əs) [L. offspring] Unborn mammal after it has largely completed its embryonic morphogenesis and differentiation (in humans, after three months of development).

FISSION ('fish-ən) [L. *fissus*, split] Asexual reproduction by division of the body into two or more equal parts.

FLATWORM Any member of the phylum Platyhelminthes. Includes turbellarians, tapeworms, and flukes.

FLAVIN ('flā-vən) [L. *flavus*, yellow] Yellow pigment that, when combined with a protein, transports electrons to the cytochromes.

FLEXOR ('flek-sər) A muscle that bends a limb.

FLORA Plant life in a certain environment.

FLUORESCENCE ((ˌ)flù(-ə)r-'es-ən(t)s) Emission of light by a substance following the absorption of radiation of a different wavelength.

FOOD CHAIN Sequence of organisms in which each uses the next lower member of the sequence as a food source and is eaten by the one above.

FOSSIL [L. *fossilis*, dug up] Any remains of an organism or evidence of its presence that has been preserved in the earth.

FOVEA ('fō-vē-ə) [L. pit] Shallow depression in the retina, containing no rods or blood vessels but richly supplied with cones and providing the most acute vision.

FRUIT [L. *fructus*, fruit] Ripened ovary (and sometimes accessory parts) of a flower.

FUNGUS ('fən-gəs) A simple, nonmotile, nonphotosynthetic eukaryotic organism. Lives as a saprophyte or parasite. Examples are molds, yeasts, mushrooms, etc.

GAMETE (gə-'mēt) [Gk. *gametes*, husband] Haploid reproductive cell which, after fusion with another gamete, initiates the development of a new individual.

GAMETOPHYTE (gə-'mēt-ə-ˌfīt) Haploid, gamete-producing stage in the life cycle of a plant.

GANGLION ('gaŋ-glē-ən) Small mass of nerve tissue containing the cell bodies of neurons.

GASTRULA ('gas-trə-lə) Stage in the embryonic development of animals during which endoderm, mesoderm, and the archenteron are formed.

GENE LOCUS Location of a particular gene (or one of its alleles) on a chromosome.

GENE POOL All the genes in a given population of a species.

GENERATOR POTENTIAL Tiny current created across the membrane of a stimulated receptor cell. Its strength increases with the strength of the stimulus, and at a certain level (the threshold) it initiates one or more nerve impulses in an adjacent neuron.

GENETIC MOSAIC (mō-'zā-ik) An individual containing cells of more than one genotype.

GENOME ('jē-ˌnōm) A complete (haploid) set of genes.

GENOTYPE ('jē-nə-ˌtīp) Genetic constitution of an individual.

GENUS A taxonomic category that includes (usually) several closely related species. Similar genera are grouped in a family.

GERMINATION [L. *germinatus*, sprouted] Resumption of growth of the embryo within a seed, or of a spore.

GILL SLITS Paired openings from the pharynx to the exterior that occur in many aquatic chordates when the gill pouches open out at the branchial grooves.

GLYCOLYSIS (glī-'käl-ə-səs) The anaerobic catabolism of glucose.

GLYCOPROTEIN A protein with covalently attached sugars and/or polysaccharides.

GONAD ('gō-ˌnad) Gamete-producing organ.

GROWTH Increase in the size of an organism, resulting from an increase in its number of cells, their size, the amount of intercellular matrix, or all of these.

GYMNOSPERM [Gk. *gymnos*, naked + *sperma*, seed] Any seed-producing tracheophyte whose seeds are not enclosed in an ovary. Includes conifers, cycads, and ginkgos.

HABITUATION (hə-ˌbich-ə-'wā-shən) [L. *habitus*, habit] The process of becoming accustomed to anything.

HALLUCINOGEN (hə-'lüs-ən-ə-ˌjen) A substance that induces hallucinations.

HAPLOID ('hap-ˌlȯid) [Gk. *haploeidēs*, single] Having only a single set of chromosomes (n) as is present in gametes. Also called monoploid.

HEMOGLOBIN ('hē-mə-ˌglō-bən) A red, iron-containing protein that transports O_2 and CO_2 in the blood of vertebrates and some invertebrates.

HEPARIN ('hep-ə-rən) A polysaccharide that inhibits clotting.

HERBACEOUS (ˌ(h)ər-'bā-shəs) [L. *herbaceus*, grassy] Nonwoody.

HERBIVORE ('ər-bə-ˌvōr) An animal that eats plants.

HERTZ Cycles per second.

HETEROGAMY (ˌhet-ə-'räg-ə-mē) Condition in which the two gametes are unlike in structure, e.g. sperm and eggs.

HETEROTROPHIC (ˌhet-ə-rə-'träf-ik) Requiring a supply of organic compounds (food) from the environment.

HETEROZYGOUS (ˌhet-ə-rō-'zī-gəs) Having two different alleles (e.g. *A*, *a*) at the corresponding gene loci on homologous chromosomes.

HEXOSE A sugar containing six carbon atoms. Glucose is a hexose.

HISTONE ('his-ˌtōn) A basic protein associated with the nuclear DNA of eukaryotes.

HOMEOSTASIS (ˌhō-mē-ō-'stā-səs) Maintenance of constancy of the internal environment (ECF).

HOMEOTHERMIC ('hōmēō-ˌthərmik) Having a constant body temperature above that of the usual surroundings; therefore, "warm-blooded."

HOMINID ('häm-ə-ˌnid) [L. *homo*, man] A humanlike—as opposed to apelike—creature.

HOMOLOGOUS (hō-'mäl-ə-gəs) [Gk. *homologos*, agreeing] (Of organs in different species), showing a fundamental similarity of structure, embryonic development, and relationship.

HOMOZYGOUS (hō-mō-'zī-gəs) Having identical alleles (e.g. *AA* or *aa*) at the corresponding gene loci on homologous chromosomes.

HORMONE ('hȯr-ˌmōn) [Gk. *hormōn*, stirring up] Organic compound produced by cells in one part of a body, which, after being transported by body fluids, exerts an effect on the activities of cells elsewhere in the body.

HUMUS ('hyü-məs) Organic matter in the soil.

HYBRID ('hī-brəd) Organism produced by genetically dissimilar parents. It is heterozygous for one or (more often) many pairs of genes.

HYDROCARBON A compound containing only carbon and hydrogen. Characteristic of petroleum and coal.

HYDROLYSIS (hī-'dräl-ə-səs) [Gk. *hydōr*, water + *lysis*, loosening] Decomposition of a substance by the insertion of water molecules between certain of its bonds. Extracellular digestion is accomplished by hydrolysis.

HYDROPHILIC [L. water-loving] Used to describe molecules or molecular groups that are attracted to water and other polar solvents.

HYDROPHOBIC [Gk. fear of water] Used to describe molecules or molecular groups that mix poorly with water. Hydrocarbons and fats are hydrophobic.

HYPERTONIC ('hī-pər-'tän-ik) Having a lower water concentration than the solution under comparison.

HYPOCOTYL ('hī-pə-ˌkät-əl) That portion of the shoot of a plant embryo or seedling below the node at which the cotyledons are attached.

HYPOTONIC ('hī-pō-'tän-ik) Having a greater water concentration than the solution under comparison.

IgA A class of antibody molecules abundant in tears, colostrum, and other secretions.

IgE A class of antibody molecules that bind to basophils and mast cells and are responsible for many allergic reactions.

IgG The class of antibody molecules that is most abundant in the blood.

IMMUNITY A state of enhanced responsiveness to a particular molecular configuration (e.g. present on the surface of an invading bacterium) induced by prior exposure to that configuration. Probably occurs only in vertebrates.

IMMUNOGLOBULIN A protein molecule that acts as an antibody.

IMMUNOLOGICAL TOLERANCE The inability to produce antibodies and/or a cell-mediated immune response to a specific antigen.

IMMUNOSUPPRESSION The use of a drug or other agent (e.g. X-rays) to inhibit the immune response.

INDUCER A molecule that activates genes, perhaps by blocking the action of a repressor.

INDUCTION Process in the embryo whereby one tissue directs the differentiation of another.

INFLAMMATION [L. *in*, into + *flamma*, flame] Response of a tissue to injury, characterized by increased blood flow, increased temperature, redness, accumulation of leucocytes, and pain.

INGESTION [L. *ingerere*, to carry in] Taking of food or water into the body.

INHIBITORY POSTSYNAPTIC POTENTIAL (IPSP) The hyperpolarization created in a neuron by the arrival of an action potential at the terminals of a neuron synapsing with it.

INORGANIC Term describing all compounds that do not contain carbon as well as a few simple carbon-containing substances such as carbon dioxide and the carbonates.

INTERMEDIATE HOST Host normally used by a parasite just during an immature, or larval, stage of the parasite's life cycle.

INTERNEURON Any neuron that is activated by and, in turn, activates other neurons. In vertebrates, most interneurons are confined to the central nervous system. Also called association neuron.

INTERSTITIAL FLUID (ˌint-ər-'stish-əl) Derived from the blood, the fluid lying between, and thus bathing, the cells of animals. Lymph is derived from it.

INTROGRESSION (ˌin-trə-'gresh-ən) Introduction of the genes of one species into the gene pool of another species.

INVERTEBRATE (in-'vərt-ə-brət) An animal that has no backbone.

IN VITRO ('vē-trō) [L. in glass] Done in the "test tube."

IN VIVO ('vē-vō) [L. in life] Refers to experiments performed in a living organism.

ION ('ī-ən) Atom or group of atoms that has an electrical charge arising from the gain or loss of electrons.

IONIC BOND (ī-'än-ik) Chemical bond formed between ions of opposite charge.

ISOGAMY (ī-'säg-ə-mē) [Gk. *isos*, equal + *gamos*, marriage] Condition in which the two gametes are alike in structure, as in *Chlamydomonas*.

ISOMER ('ī-sə-mər) [Gk. *isomeres*, equally divided] Molecule with the same molecular formula as another but with a different structural formula, e.g. glucose and fructose.

ISOMETRIC [Gk. same measure] The contraction, without shortening, of a muscle.

ISOTONIC (ˌī-sə-'tän-ik) [Gk. same tension] (1) Adjective to describe the contraction of a muscle that is allowed to shorten as it exerts a steady force. (2) Having the same water concentration as the solution under comparison.

ISOTOPE ('ī-sə-ˌtōp) Atom that differs in weight from other atoms of the same element because of a different number of neutrons in its nucleus.

KARYOTYPE ('kar-ē-ə-ˌtīp) The total chromosome complement of a cell.

KININ ('kī-nin) One of a group of polypeptides produced in the blood or tissues which dilate blood vessels and produce the pain associated with inflammation.

LACTOSE ('lak-ˌtōs) A disaccharide (milk sugar) that hydrolyzes to give one molecule of glucose and one of galactose.

LARVA ('lär-və) [L. specter] Immature stage of many animals that must undergo metamorphosis to become an adult.

LATENT PERIOD ('lāt-ənt) Interval between the application of a stimulus and the first detectable response.

LATERAL GENICULATE BODY One of a pair of brain centers where the axons of the optic nerve synapse with interneurons leading to the visual cortex.

LEGUME ('leg-ˌyüm) A member of a family of pod-bearing plants that includes peas, beans, clovers, alfalfa, etc.

LEUKEMIA (lü-'kē-mē-ə) A cancer characterized by an uncontrolled increase in the number of leucocytes in the blood.

LICHEN ('lī-kən) Mutualistic association of a fungus and an alga.

LIGNIN ('lig-nən) [L. *lignum*, wood] Complex substance found in the cell walls of sclerenchyma and xylem tissue, which are strengthened by it.

LINKAGE Tendency of two genes to be inherited together because they are located on the same chromosome.

LIPASE ('lī-ˌpās) Enzyme that digests fats.

LITHOSPHERE [Gk. *lithos*, stone] The rocks and other solid materials that comprise the crust of the earth.

LUMEN ('lü-mən) [L. opening] The cavity inside a tubular structure such as a blood vessel or kidney tubule.

LYMPH [L. *lympha*, water goddess] Fluid present in the vessels of the lymphatic system. Lymph is derived from interstitial fluid and contains many lymphocytes.

LYSIS ('lī-səs) [Gk. dissolution] The disintegration of a cell following destruction of its cell membrane.

LYSOGENY (lī-'säj-ə-nē) The stable incorporation of a prophage into the genome of a bacterium.

MACROMOLECULE A molecule with a molecular weight of several thousand or more. Proteins, nucleic acids, cellulose, and starch are macromolecules.

MARSUPIAL (mär-'sü-pē-əl) Any of the order of pouched mammals such as opossums, wombats, and kangaroos.

MATRIX ('mā-triks) [L. womb] Intercellular material in which animal cells are imbedded, especially those of connective tissue.

MEDULLA (mə-'dəl-ə) Inner part of an organ.

MEDUSA (mi-'d(y)ü-sə) Jellyfish form occurring in the life cycle of some cnidarians.

MEIOSIS (mī-'ō-səs) [Gk. *meiōsis*, diminution] The two successive cell divisions, with only one duplication of the chromosomes, which produce four cells, each containing one-half the number of chromosomes in the original cell.

MERISTEM ('mer-ə-ˌstem) [Gk. *meristos*, divided] Embryonic plant tissue which produces new cells by repeated mitosis.

MESODERM ('mez-ə-ˌdərm) [Gk. *mesos*, middle + *derma*, skin] Layer of cells in an animal embryo, located between the ectoderm and endoderm.

MESOGLEA (ˌmez-ə-'glē-ə) [Gk. *glia*, glue] Gelatinous layer located between the two cell layers of sponges and cnidarians.

METABOLISM (mə-'tab-ə-ˌliz-əm) [Gk. *metabolē*, change] Exchange of matter and energy between an organism and its environment and the transformation of this matter and energy within the organism.

METABOLITE A substance used in or produced by the metabolism of an organism.

METAMORPHOSIS (ˌmet-ə-'mȯr-fə-səs) [Gk. *metamorphoun*, to transform] Process of change (usually abrupt) from larval to adult form.

MICROORGANISM An organism of microscopic size such as bacteria, protozoans, and many algae. Also called a microbe.

MITOSIS (mī-'tō-səs) [Gk. *mitos*, thread] Cell (or simply nuclear) division following duplication of the chromosomes, whereby each daughter cell (or nucleus) has exactly the same chromosome content as the parent.

MIXTURE Material containing two or more substances, each of which retains its characteristic properties. The composition of a mixture is variable. A solution is a mixture.

MOLE Quantity of a substance whose weight in grams is numerically equal to the molecular weight of the substance, e.g. 18 g of water is 1 mole.

MOLECULAR WEIGHT (mə-'lek-yə-lər) Sum of the atomic weights of the atoms in a molecule.

MOLECULE ('mäl-i-ˌkyü(ə)l) [L. *moles*, mass] Smallest particle of a covalently bonded element or compound that retains the properties of that substance, e.g. O_2, H_2O.

MOLT To shed the outer covering.

MONOCULTURE The growing in large stands of a single kind of crop plant.

MONOECIOUS (mə-'nē-shəs) Having both male and female cones or flowers on the same plant.

MONOGLYCERIDE Glycerol carrying a single fatty acid.

MONOMER ('män-ə-mər) Simple molecular unit which can be linked with others to form a polymer. The glucose molecule is the monomer of starch.

MONOSACCHARIDE (ˌmän-ə-ˈsak-ə-ˌrīd) [Gk. *monos*, single + *sakcharon*, sugar] A simple sugar, e.g. glucose ($C_6H_{12}O_6$).

MORPHOGENESIS (ˌmȯr-fə-ˈjen-ə-səs) Development of body form.

MORPHOLOGY (mȯr-ˈfäl-ə-jē) [Gk. *morphē*, form + *logos*, study] Study of the structure of organisms.

MOTOR UNIT All the skeletal muscle fibers stimulated by a single motor neuron.

MULTIPLE ALLELES More than two alleles found at a given locus in a population.

MULTIPLE FACTORS Nonallelic genes that affect the same trait in an additive fashion.

MUTATION (myü-ˈtā-shən) [L. *mutare*, to change] Stable, inheritable change in a gene.

MUTUALISM Close, mutually beneficial association between two organisms of different species.

MYCELIUM (mī-ˈsē-lē-əm) Mass of interwoven hyphae of a fungus.

MYCORRHIZA (ˌmī-kə-ˈrī-zə) The symbiotic association of a fungus with the roots of a plant.

MYELIN SHEATH (ˈmī-ə-lən ˈshēth) Fatty covering found around many axons.

MYELOMA PROTEIN (ˌmī-ə-ˈlō-mə) A pure immunoglobulin (antibody) produced by a cancerous clone of plasma cells.

MYONEURAL JUNCTION (ˌmī-ōˈn(y)u̇r-əl) Junction between a motor neuron and a muscle fiber. Also called a neuromuscular junction.

NECTAR (ˈnek-tər) Sugar solution, secreted by plants, from which bees make honey.

NEPHRON (ˈnef-rän) [Gk. *nephros*, kidney] Function unit of vertebrate kidneys.

NERVE Bundle of axons.

NET PRODUCTIVITY The amount of energy trapped in organic matter during a specified interval at a given trophic level less that lost by the respiration of the organisms at that level.

NEUROHUMOR (ˌn(y)u̇r-ō-ˈhyü-mər) Substance released at the synaptic knob or end plate of a neuron that either stimulates or inhibits the next neuron or muscle fiber. Acetylcholine and noradrenaline are important neurohumors.

NEUROMUSCULAR JUNCTION (ˌn(y)u̇r-ō-ˈməs-kyə-lər) Junction between a motor neuron and a muscle fiber. Also called a myoneural junction.

NEURON (ˈn(y)ü-ˌrän) Nerve cell.

NEUTRON (ˈn(y)ü-ˌträn) Electrically neutral particle found in the nuclei of all atoms except hydrogen-1.

NICHE (ˈnich) [L. *nidus*, nest] The place in a biological community occupied by a particular species in its relation to other species.

NICOTINAMIDE ADENINE DINUCLEOTIDE (NAD) (ˌnik-ə-ˈtē-nə-ˌmīd ˈad-ən-ˌēn ˈdī-ˈn(y)üklēəˌtīd) Coenzyme that transfers electrons within the cell. Formerly called diphosphopyridine nucleotide (DPN) and Coenzyme I.

NITROGEN FIXATION The conversion of atmospheric nitrogen (N_2) into nitrogen-containing compounds (e.g. NH_3).

NODE [L. *nodus*, knot] In plants, the point on a stem at which one or more leaves develop.

NONDISJUNCTION Failure of two homologous chromosomes to separate during meiosis.

NOTOCHORD (ˈnōt-ə-ˌkȯ(ə)rd) [Gk. *nōton*, back + *chorda*, cord] Longitudinal, flexible rod located between the central nervous system and the alimentary canal at some stage in the development of all chordates. In vertebrates it is generally replaced by a column of vertebrae.

NUCLEIC ACID (n(y)ü-ˈklē-ik) A polymer of nucleotides; DNA and RNA.

NUCLEOTIDE (ˈn(y)ü-klē-ə-ˌtīd) Molecule consisting of (1) a purine or pyrimidine, (2) a 5-carbon sugar, and (3) a phosphate group—all linked together.

ONCOGENIC Cancer-causing.

ONTOGENY (än-ˈtäj-ə-nē) Process of development of an individual organism.

OPERATOR GENE A gene that turns the structural genes adjacent to it "on" and "off."

OPERON A set of adjacent structural genes and the operator gene that controls them.

OPIATE (ˈō-pē-ˌāt) A narcotic material prepared or derived from opium.

OPSIN (ˈäp-sən) The protein part of the visual pigments of the eye.

ORGAN Group of tissues that perform a specific function for an animal or plant, e.g. stomach, leaf.

ORGANELLE (ˌȯr-gə-ˈnel) Specialized part of a cell, e.g. contractile vacuole, analogous to an organ.

ORGANIC (ȯr-ˈgan-ik) Term describing all compounds whose molecules contain carbon, with a few exceptions such as carbon dioxide and the carbonates.

ORGANISM Individual living being.

OSMOSIS (ä-ˈsmō-səs) Diffusion of a solvent (usually water) through a semipermeable membrane.

OSSICLE (ˈäs-i-kəl) [L. *ossiculum*, a little bone] Small bone such as those that transmit vibrations through the middle ear.

OUTBREEDING Mating of genetically dissimilar, relatively unrelated, individuals.

OVOVIVIPAROUS (ˌō-(ˌ)vō-ˌvī-ˈvip-(ə-)res) Having embryos that develop to adult form within the mother's body while securing nourishment from the egg rather than directly from the mother's tissues. Many insects, snails, fishes, lizards, and snakes are ovoviviparous.

OVULATION (ˌō-vyə-ˈlā-shən) Release of one or more eggs from the ovary.

OVULE (ˈō-(ˌ)vyü(ə)l) [L. *ovum*, egg] Megasporangium found within the ovary of a seed plant. After fertilization of the egg inside, it develops into a seed.

OXIDATION Process of removing electrons from a substance.

PARASITE Organism living on or in another organism from which it derives its food and which it harms to some extent.

PARENCHYMA (pə-ˈreŋ-kə-mə) Plant tissue consisting of thin-walled cells, often loosely packed, that function in photosynthesis and/or food storage.

PARTHENOGENESIS (ˌpär-thə-nō-ˈjen-ə-səs) [Gk. *parthenos*, virgin + *genēs*, born] Development of an unfertilized egg into a new individual. Often occurs naturally in certain plants and animals, e.g. aphids.

PATHOGEN (ˈpath-ə-jən) Disease-causing organism or virus.

PENTOSE A sugar containing 5 carbon atoms. Ribulose is a pentose.

PERISTALSIS (ˌper-ə-ˈstòl-səs) Successive waves of contraction passing along the walls of tubular organs, such as the intestine, thus forcing their contents along.

pH Negative logarithm of the hydrogen-ion concentration (in moles/liter) of a solution, which thus provides a measure of acidity and alkalinity.

PHAGOCYTOSIS (ˌfag-ə-(ˌ)sī-ˈtō-səs) [Gk. *phagein*, to eat] Engulfing of solid particles by a cell.

PHENOTYPE (ˈfē-nə-ˌtīp) [Gk. *phainein*, to show + *typos* type] Appearance of an organism, resulting from the interaction of its genotype and its environment.

PHEROMONE (ˈfer-ə-ˌmōn) [Gk. *pherein*, to carry] Compound secreted externally by an animal which influences other members of the same species.

PHLOEM (ˈflō-ˌem) [Gk. *phloios*, bark] Complex vascular tissue of plants that translocates food throughout the plant.

PHONEME (ˈfō-ˌnēm) [Gk. *phōnēma*, utterance] One of the basic sounds of which speech is made up.

PHOSPHOLIPID (ˌfäs-fō-ˈlip-ˌid) A fat derivative in which one fatty acid has been replaced by a phosphate group and one of several nitrogen-containing molecules.

PHOTOPERIODISM (ˌfōt-ə-ˈpir-ē-ə-ˌdiz-əm) Developmental or behavioral response of an organism to the duration of daylight or darkness.

PHOTORESPIRATION The oxidation of 2-carbon compounds to CO_2 that occurs in many plants concurrently with photosynthesis. Accomplished by peroxisomes, not mitochondria. Its occurrence reduces the efficiency of photosynthesis.

PHYLOGENY (fī-ˈläj-ə-nē) Evolutionary history of a species.

PHYLUM (fī-ləm) [Gk. *phylon*, tribe] Major taxonomic category comprising one or more classes. In plant classification, the term *division* is often used instead.

PHYSIOLOGY (ˌfiz-ē-ˈäl-ə-jē) Study of the processes occurring in living organisms.

PIGMENT [L. *pingere*, to paint] Substance that absorbs light, often selectively.

PLANARIAN A free-living, usually aquatic, flatworm comprising one subdivision of the class Turbellaria.

PLANKTON (ˈplaŋ(k)-tən) [Gk. *planktos*, drifting] Floating, generally microscopic, protistan and animal life in a body of water.

PLASMA Fluid matrix of the blood.

PLASMODIUM (plaz-ˈmōd-ē-əm) Multinucleate, motile mass of protoplasm.

PLASMOLYSIS (plaz-ˈmäl-ə-səs) Shrinkage of the cytoplasm away from the wall of a plant cell, placed in a hypertonic medium, because of the loss of water by osmosis.

PLEIOTROPY (plī-ˈä-trə-pē) [Gk. *pleiōn*, more] The production by a single gene of more than one effect on the phenotype.

PLUMULE (ˈplü-(ˌ)myü(ə)l) Terminal bud of a plant embryo, usually consisting of embryonic leaves and the epicotyl.

POIKILOTHERMIC (ˈpòikəlōˌthərmik) Having a body temperature that fluctuates with that of the surroundings. Commonly, "cold-blooded."

POLARITY (pō-ˈlar-ət-ē) Intrinsic anterior-posterior orientation in an organism that seems to account, for example, for the regeneration of missing body parts (as in a planarian) in proper relationship to the rest of the body.

POLYMER (ˈpäl-ə-mər) [Gk. *polymerēs*, having many parts] Compound whose molecule consists of many repeating units linked together.

POLYMORPHISM (ˌpäl-i-ˈmor-ˌfiz-əm) Occurrence of several distinct phenotypes in a population, e.g. queen, drone, and worker bees.

POLYP (ˈpäl-əp) Anchored, tubular body form characteristic of most cnidarians, at least during one stage of their lives. The hydra and corals are polyps.

POLYPEPTIDE (ˌpäl-i-ˈpep-ˌtīd) Molecule consisting of fewer than 100 amino acids linked together in a single chain.

POLYPLOIDY (ˈpäl-i-ˌplȯid-ē) Having 3 or more complete (= haploid number) sets of chromosomes.

POLYSACCHARIDE (ˌpäl-i-ˈsak-ə-ˌrīd) Carbohydrate, e.g. starch, cellulose, that is made up of 3 or more monosaccharides linked together.

POLYTENE Used to describe the multistranded ("giant") interphase chromosome found in certain metabolically active cells of insects.

PRECURSOR (pri-ˈkər-sər) [L. *praecurrere*, to run before] Substance from which another substance is formed.

PRECURSOR ACTIVATION The activation of the last enzyme in a metabolic pathway by the substrate of the first enzyme in that pathway.

PREDATION (pri-ˈdā-shən) Living by devouring other organisms.

PRIMARY HOST Host normally used by a parasite during the adult stage of its life cycle.

PRIMITIVE Similar to that occurring in the early evolutionary history of the structure or organism in question.

PROBOSCIS (prə-ˈbäs-əs) [Gk. *pro*, before + *boskein*, to feed] Tubular extension at the anterior end of an animal, usually used in feeding.

PRODUCER An organism that can synthesize organic molecules from inorganic ones, i.e. an autotroph. Producers start food chains.

PROGLOTTID ((ˈ)prō-ˈglät-əd) One of the segments of a tapeworm.

PROKARYOTE (prō-ˈka-rē-ˌōt) An organism whose cell contains neither a membrane-bounded nucleus nor other membrane-bounded organelles such as mitochondria and plastids. The bacteria and blue-green algae. Often spelled procaryote.

PROSTHETIC GROUP [Gk. *prosthetos*, added] The nonprotein portion of a conjugated protein. Metal ions and a variety of organic molecules (e.g. vitamins, sugars, lipids) can serve as prosthetic groups. Prosthetic groups are usually bound covalently to their protein.

PROSTAGLANDIN Any of a number of 20-carbon organic acids that are synthesized in the body from unsaturated fatty acids and are responsible for a variety of metabolic activities.

PROTEASE (ˈprōt-ē-ˌās) Enzyme that hydrolyzes peptide bonds whether in proteins or peptides.

PROTEINASE (ˈprō-ˌtē-ˌnās) Enzyme that hydrolyzes the peptide bonds of proteins.

PROTON (ˈprō-ˌtän) [Gk. *prōtos*, first] Positively charged particle found in the nuclei of all atoms. The hydrogen ion (H^+) is a proton.

PROTOPLAST A cell (plant or bacterial) from which the wall has been removed.

PROTOZOAN A unicellular, heterotrophic, often motile eukaryote. The amoeba is a common example.

PROXIMAL (ˈpräk-sə-məl) Situated near the place of origin or attachment.

PUPA (ˈpyü-pə) [L. *pupa*, doll] Stage (usually dormant) between the larva and the adult of insects having complete metamorphosis.

PURINE (ˈpyu̇(ə)r-ˌēn) Double-ring, nitrogen-containing base that is a component of nucleic acids and several other biologically active substances.

PSEUDOCOEL (ˈsüdō-ˌsēl) Body cavity found in some animals, e.g. roundworms, between the body wall (mesoderm) and the alimentary canal (endoderm). It is not lined with a sheet of mesodermal cells as is a true coelom.

PYRIMIDINE (pī-ˈrim-ə-ˌdēn) Single-ring, nitrogen-containing base that is a component of nucleic acids.

RADICLE (ˈrad-i-kəl) Root portion of the embryo of seed plants.

REACTANT (rē-ˈak-tənt) Substance that enters into a chemical reaction.

RECAPITULATION Occurrence, in the embryonic development of an individual, of stages thought to have occurred in the embryonic development of its ancestors.

REDOX POTENTIAL A measure (in volts) of the affinity of a substance for electrons compared with hydrogen (which is set at 0). Substances more strongly electronegative than (i.e. capable of oxidizing) hydrogen have positive redox potentials. Substances less electronegative than (i.e. capable of reducing) hydrogen have negative redox potentials.

REDOX REACTION A chemical reaction in which electrons are transferred from one atom (which is thereby oxidized) to another (which is thereby reduced).

REDUCTION Process of adding electrons to a substance.

REFRACTORY PERIOD (ri-ˈfrak-t(ə-)rē' pir-ē-əd) Brief interval following the response of a neuron or muscle fiber during which it is incapable of a second response.

REGENERATION Regrowth of lost or injured parts of an organism.

REGULATOR GENE A gene that produces a repressor.

RELEASER Stimulus that initiates instinctive behavior.

REM Roentgen equivalent man. The amount of absorbed radiation that will cause as much damage in human tissue as one roentgen of X-rays.

REPRESSOR A protein that blocks gene action by combining with an operator gene.

REVERSE TRANSCRIPTASE An enzyme that catalyzes the synthesis of DNA that is complementary to an RNA template, i.e. an RNA-dependent DNA polymerase.

RHIZOBIA (rī-ˈzō-bē-ə) Soil bacteria of the genus *Rhizobium* which are able to fix nitrogen *after* establishing a symbiotic relationship with the roots of a legume.

RHIZOID (ˈrī-ˌzȯid) Hairlike structure that serves as a root for bryophytes, fern prothallia, and certain fungi and lichens.

RHIZOME (ˈrī-ˌzōm) Underground stem.

RIBONUCLEIC ACID (RNA) (ˈrī-bō-n(y)ů-ˌklē-ik) Nucleic acid found in both the nucleus and the cytoplasm that functions in the synthesis of protein.

RIBONUCLEOPROTEIN (ˈrī-bō-ˌn(y)ü-klē-ō-ˈprō-ˌtēn) A complex of RNA and protein.

RNA POLYMERASE An enzyme that catalyzes the linking together of ribonucleotides to form RNA complementary to a template (either DNA or RNA).

SAPROPHYTE (ˈsap-rə-ˌfīt) [Gk. *sapros*, rotten + *phyton*, plant] Heterotrophic plant (or fungus) that secures its food by the extracellular digestion of nonliving organic matter.

SARCOMERE (ˈsär-kə-ˌmi(ə)r) The repeating, contractile unit of the myofibril. It is bounded at each end by a "Z-line."

SCAVENGER Organism that feeds on dead organisms or the wastes of organisms.

SCION (ˈsī-ən) Detached part of a plant, e.g. a piece of stem, that is grafted on to another plant.

SCLERENCHYMA (sklə-ˈreŋ-kə-mə) Supporting plant tissue consisting of cells whose walls are uniformly thickened and often lignified.

SEED Embryo plant, supplied with food and protected by seed coats, that serves as the agent of dispersal of gymnosperms and angiosperms. It develops from the fertilized ovule.

SEQUENCE The linear order of amino acids in a polypeptide chain or nucleotides in a nucleic acid.

SERUM The clear fluid that can be squeezed out from a blood clot; hence blood plasma from which the fibrinogen and other clotting factors have been removed.

SESSILE (ˈses-əl) [L. *sessilis*, fit for sitting] (1) In plants: lacking a stalk, e.g. a petiole-less leaf. (2) In animals: attached to the substratum; anchored.

SEXUAL REPRODUCTION The production of new individuals by the union of the genetic material (DNA) of two different cells, usually gametes and usually from different parents.

SOLUTE (ˈsäl-ˌyüt) Dissolved substance in a solution.

SOLUTION [L. *solutus*, loosened] Mixture consisting of molecules or ions less than 1 nm in diameter, suspended in a fluid medium (water in most biological systems).

SOLVENT Dissolving medium in a solution.

SOMITE (ˈsō-ˌmīt) [Gk. *sōma*, body] One of the blocks of mesoderm that develop in a longitudinal series on either side of the notochord in vertebrate embryos.

SPECIATION (spē-s(h)ē-ˈā-shən) Formation of species.

SPECIES (ˈspē(ˌ)shēz) [L. kind] Taxonomic category consisting of a group of actually or potentially interbreeding natural populations which ordinarily do not interbreed with other such groups even when there is opportunity to do so. (The singular and plural are spelled alike.)

SPIRACLE (ˈspī-ri-kəl) [L. *spirare*, to breathe] (1) In insects, the external opening of a trachea. (2) In many fishes, the vestigial remnant of the first gill slit of their agnath ancestors.

SPORANGIUM (spə-ˈran-jē-əm) Structure within which asexual spores are produced.

SPORE [Gk. *spora*, seed] Asexual reproductive structure, usually unicellular, which serves to disperse the species and/or enable it to survive unfavorable conditions and which can develop into a new individual.

SPOROPHYTE (ˈspōr-ə-ˌfīt) Diploid, spore-producing stage in the life cycle of a plant.

STATOCYST (ˈstat-ə-ˌsist) Organ of balance, found in some aquatic invertebrates.

STEROID (ˈsti(ə)r-ˌȯid) One of many fat-soluble, biologically active compounds whose molecules contain a system of four rings made up of 17 carbon atoms. (See Fig. 4–8.)

STIMULUS [L. *stimulus*, goad] Change in the environment of an organism that initiates a response.

STOCK Part of a plant (usually including roots) to which a scion is grafted.

STOLON ('stō-lən) [L. *stolon*, branch, sucker] Horizontal stem that produces new plants at its nodes.

STROBILUS ('strä-bə-ləs) Aggregation of modified leaves bearing sporangia; a cone.

STRUCTURAL GENE The sequence of nucleotides coding for a single gene product, i.e. that is transcribed into an RNA molecule.

SUBERIN ('sü-bə-rən) Waxy material present in the walls of cork cells that makes them waterproof.

SUBSTRATE ('səb-ˌstrāt) [L. *substratus*, spread under] (1) Substance that is acted upon by an enzyme. (2) Base (e.g. soil, rock) upon which an organism lives. Also called the substratum.

SUCCESSION Progressive change in the nature of the plant population of an area.

SUSPENSION Mixture containing solid particles larger than 100 μm distributed throughout a fluid. The particles will ultimately settle out under the force of gravity.

SYMBIOSIS (ˌsim-ˌbī-'ō-səs) [Gk. *symbioun*, to live together] The living together in close association of organisms of different species. Mutualism, commensalism, and parasitism are forms of symbiosis.

SYNAPSE ('sin-ˌaps) [Gk. *synapsis*, juncture] Gap between two neurons across which the nerve impulse is transmitted.

SYNAPSIS (sə-'nap-səs) Union, side by side, of homologous chromosomes early in meiosis.

SYNCYTIUM (sin-'sish-(ē)-əm) Mass of cytoplasm containing many nuclei, formed by the fusion of cells.

SYNDROME ('sin-ˌdrōm) [Gk. combination] A collection of symptoms and signs characteristic of a particular disease.

SYNGAMY ('siŋ-gə-mē) Union of gametes in sexual reproduction.

SYNTHESIS ('sin(t)-thə-səs) [Gk. *syntithenai*, to put together] Formation of a compound from other, usually simpler, substances.

SYSTEM Group of organs that perform one or more functions as a unit, e.g. the organs of the digestive system.

SYSTOLE ('sis-tə-lē) Phase of contraction of the heart.

TAIGA (tī-'gä) [Russian] The northern coniferous forest.

TAXIS ('tak-səs) Automatic locomotion of a motile organism in a direction determined by the direction from which the stimulus strikes it.

TAXONOMY (tak-'sän-ə-mē) [Gk. *taxis*, arrangement + *nomos*, law] Classification of living organisms.

T CELL A cell (lymphocyte) that requires the presence of the thymus gland in order to acquire immunological activity. The cells active in cell-mediated immunity are T cells.

TETANUS ('tet-ən-əs) [Gk. *tetanus*, stretched, rigid] Sustained, maximal contraction of a muscle.

THRESHOLD ('thresh-ˌ(h)ōld) Minimum intensity of a stimulus to which a given structure responds.

THROMBIN ('thräm-bən) [Gk. *thrombos*, clot] Enzyme that converts fibrinogen to fibrin.

THYLAKOID ('thī-lə-ˌkoid) A pair of chlorophyll-containing membranes which form disklike structures within chloroplasts. Stacks of thylakoids are called grana.

TISSUE [L. *texere*, to weave] Association of cells bound together by cell walls (plants) or intercellular matrix (animals) that performs a particular function.

TONUS ('tō-nəs) [L. *tonus*, tension] Sustained, partial contraction of a muscle.

TORR A unit of pressure equal to that exerted by a column of mercury one mm high; hence equal to the unit mmHg.

TOXIN Metabolic product (usually a protein) of an organism that is poisonous to another organism.

TOXOID Toxin treated to destroy its poisonous quality but leave it capable of stimulating the production of antibodies.

TRACHEOPHYTE ('trā-kē-ə-ˌfīt) A plant with a vascular system of xylem and phloem. Includes all plants except mosses and their relatives.

TRANQUILIZER A drug used to reduce anxiety or other emotional disturbance. "Major" tranquilizers (e.g. chlorpromazine) are used to quiet mental patients; "minor" tranquilizers (e.g. meprobamate) are widely used to reduce anxiety and tension.

TRANSCRIPTION The synthesis of a sequence of ribonucleotides complementary to the sequence of deoxyribonucleotides in a molecule of DNA.

TRANSFORMATION The modification of the genotype of a cell by the introduction of DNA from another source. Also the conversion of a normal cell into a cancerous one.

TRANSLATION The synthesis of a polypeptide in accordance with the genetic information encoded in a molecule of messenger RNA.

TRANSLOCATION (1) Transport of materials from one part of a plant to another. (2) Transfer of a piece of one chromosome to another, nonhomologous, chromosome.

TRANSPIRATION [L. *trans*, across + *spirare*, to breathe] Evaporation of water from plants.

TROPHIC LEVEL Position in a food chain such as primary consumer, secondary consumer, etc.

TROPISM ('trō-ˌpiz-əm) Automatic response of growth or orientation in a direction determined by the direction from which the stimulus strikes the organism.

TUNDRA ('tən-drə) [Russ.] Relatively flat, treeless plain north of the taiga and south of the polar region.

TURGOR ('tər-gər) [L. *turgere*, to swell] Distention of the wall of a plant cell by the accumulation of water within the cell.

VACCINE (vak-'sēn) [L. *vaccinus*, of cows] Preparation of dead or weakened pathogens which, when introduced into the body, stimulates the production of antibodies without causing the symptoms of the disease.

VALENCE ('vā-lən(t)s) [L. *valens*, being strong] Number of electrons gained, lost, or shared by an atom in bonding with one or more other atoms.

VASCULAR ('vas-kya-lər) [L. *vasculum*, small vessel] Containing vessels that conduct fluid.

VASECTOMY (va-'sek-tə-ˌmē) Surgical removal of a portion of each vas deferens, thus blocking the addition of sperm to the seminal fluid.

VECTOR [L. *vectus*, carried] An animal, e.g. an insect, that transmits parasites.

VERTEBRATE ('vərt-ə-brət) Any backboned animal. Includes fishes, amphibians, reptiles, birds, and mammals.

VESTIGIAL (ve-'stij-(ē)əl) Term applied to a degenerate or incompletely developed structure which was more fully developed at an earlier stage of the organism and/or in its ancestors.

VIRION A complete virus particle. It consists of genetic material (DNA or RNA) surrounded by a coat of protein (and sometimes other materials as well).

VISCERA ('vis-(ə)rə) Organs in the body cavity.

VITAMIN [L. *vita*, life] Organic compound which is needed in small quantities by an organism in its metabolism and which it cannot synthesize from the carbohydrates, fats, and proteins in its diet.

VIVIPAROUS (vī-'vip-(ə)res) Having embryos that developed to adult form within the mother's body while securing most of their nourishment from the mother's tissues rather than from the yolk of the egg.

XYLEM ('zī-ləm) Vascular tissue of plants that conducts water and dissolved minerals from the roots upward and often provides mechanical support to the plant. Wood is composed of xylem.

YEAST ('yēst) A unicellular fungus. One species is used in brewing and baking because of its efficiency at fermenting carbohydrates to ethyl alcohol and CO_2.

ZOOLOGY (zō-'äl-ə-jē) [Gk. *zōion*, animal] Study of animals.

ZOOSPORE ('zō-ə-ˌspō(ə)r) Flagellated, swimming spore produced asexually.

ZYGOTE ('zī-ˌgōt) [Gk. *zygotes*, yoked] Cell formed by the union of two gametes.

THE INTERNATIONAL SYSTEM OF UNITS

LENGTH

Basic unit is the meter (m), which equals 39.37 in.

Common Multiples and Subdivisions

kilometer (km) = 10^3m
decimeter (dm) = 10^{-1}m
centimeter (cm) = 10^{-2}m
millimeter (mm) = 10^{-3}m
micrometer (μm) = 10^{-6}m*
nanometer (nm) = 10^{-9}m
angstrom (A) = 10^{-10}m

* also called the micron (μ)

VOLUME

Basic unit is the liter (l), which equals 1 cubic decimeter, dm^3 (1.06 quart). A liter of water at its maximum density weighs *almost* 1 kg. Therefore, 1 ml (1 ml = 10^{-3} l) of water weighs, for all practical purposes, 1 g. One cubic centimeter (cm^3 or cc) is 10^{-3} dm^3. Therefore, it is equal to 1 ml and the units ml and cc are used interchangeably.

MASS

Basic unit is the gram (g).

Common Multiples and Subdivisions

kilogram (kg) = 10^3g = 2.2 lb
centigram (cg) = 10^{-2}g
milligram (mg) = 10^{-3}g
microgram (μg) = 10^{-6}g

TEMPERATURE

Basic unit is the Celsius (formerly known as centigrade) degree, °C. 0°C is the freezing point of water; 100°C is the boiling point of water. To convert from °C to °F (Fahrenheit) or vice versa: °F − 32 = 9/5 °C.

USEFUL EQUIVALENTS

1 in. = 2.54 cm
1 oz = 28.35 g
1 lb = 453.6 g
1 U.S. fluid oz = 29.57 ml
1 U.S. liquid qt = 0.946 l

INDEX

Note: Numbers in boldface indicate that the reference appears in a figure.